13TH EDITION

Modern Carpentry

R. Jack Jones • Willis H. Wagner • Howard Bud Smith • Mark W. Huth

Publisher

The Goodheart-Willcox Company, Inc.

Tinley Park, IL

www.g-w.com

Preface

Congratulations—you are about to gain knowledge and skills that will serve you well throughout your life. You are preparing to learn the basics of residential and light construction. You will learn how a house is constructed from foundation to roof, and you will begin to develop the basic skills needed to build and maintain a structure. Additionally, you will learn subject matter that will help you understand green construction, electrical and plumbing systems, as well as heating, ventilation, and air conditioning (HVAC) principles.

You have several choices of how you will make use of this knowledge and skill in your future. The construction industry is a broad field, and career options are vast. You may choose to pursue a career in the construction trades, perhaps as a carpenter, plumber, electrician, or HVAC technician. Some tradesmen and women transition into fields that are related to construction but are not actively building. Many companies eagerly seek out people with knowledge in construction. Insurance adjusters, inspectors, safety professionals, trade journalists, and salespeople are just a few examples of careers that you can qualify for with experience and knowledge of the construction field. You may continue your education at a four-year college and pursue a career in construction management or engineering. Even if you choose a career outside of construction, your study of carpentry will benefit you as a future homeowner.

Modern Carpentry is a well-illustrated, easy-to-understand source of authoritative and up-to-date information on building materials, tools, and construction methods. It provides detailed coverage of all aspects of light construction. Included are site development, print reading, site layout, excavating, foundation work, framing, sheathing, insulating, roofing, window and door installation, exterior finishing, interior finishing, and mechanical systems. Because images are key to understanding, there are many updated and new images in this edition.

Chapters are arranged in a logical sequence similar to the order builders use to complete various phases of home construction. Special emphasis is given to the construction careers and workplace skills, including attitudes and relationships with coworkers. This textbook strongly focuses on safe work habits in handling tools, materials, and prefabricated components.

Green construction practices are emphasized and have been updated in this thirteenth edition. Thinking Green tips throughout the textbook highlight items related to sustainable construction within the context of the chapter topic. In addition, a new chapter *Building Envelope and Control Layers*, provides a foundational understanding of the basic building science needed by today's carpentry professionals.

This edition also includes increased coverage of safe work practices and building code requirements. Safety is critical when working with construction tools and materials. *Modern Carpentry* provides an overview of safe work practices in a dedicated *Safety* chapter and then supplements this general knowledge with Safety Notes throughout the textbook. Code Notes highlight construction requirements found in model building codes.

Modern Carpentry also provides chapters introducing mechanical systems: electrical wiring, plumbing, and heating and cooling (HVAC). More than cursory discussions of principles, these chapters explain how to install such systems.

Enjoy your study of construction, the accomplishment of developing new skills, and the satisfaction of building something with your own hands. Most importantly, always work safely.

R. Jack Jones

Willis H. Wagner

Howard Bud Smith

Mark W. Huth

About the Authors

R. Jack Jones is an Associate Professor of Building Construction at Alfred State College of Technology, a State University of New York institution, where he teaches courses in carpentry, blueprint reading, and construction safety. Mr. Jones serves as the Chair of the Building Trades Department, overseeing one of the finest teams of construction faculty in higher education. He has over 25 years of experience working in the construction industry as a carpenter, general contractor, and educator. He has been a member of the United Brotherhood of Carpenters and Joiners of America and has experience in both commercial and residential construction. Mr. Jones earned a Bachelor of Science degree in Business Administration through Mansfield University of Pennsylvania and serves as an Outreach Trainer for the Occupational Safety and Health Administration. He is an honorary member of The National Society of Leadership and Success, where he received the Excellence in Teaching Award. Additionally, Mr. Jones is a member of the Rochester Builder's Exchange and has been designated as a Certified Installer by the Vinyl Siding Institute.

Willis H. Wagner was Professor Emeritus for the Department of Industrial Technology at Iowa State University. Mr. Wagner earned a Bachelor of Science degree from Central Missouri State University and a Master of Science degree from the University of Missouri. He worked in industry as a drafter and woodworker and taught at the high school and collegiate levels. Mr. Wagner was a member of IAITA and ITEEA, and an honorary member of the Waterloo Technical Society.

Howard Bud Smith spent most of his career as a magazine and textbook editor. He also worked as a carpenter for several years while earning a PhD in journalism from Marquette University. Mr. Smith has published articles in a variety of magazines over his career. He is currently retired.

Mark W. Huth brought to *Modern Carpentry* a great deal of construction education experience from multiple perspectives. Besides working as a carpenter and contractor, he worked as a carpentry instructor and as a consultant to both secondary and postsecondary educators. Mr. Huth also wrote textbooks in the areas of construction print reading, construction principles, and construction math.

Reviewers

The authors and publisher wish to thank the following industry and teaching professionals for their valuable input in the development of *Modern Carpentry*.

Robert Anderson
Montana State University–Northern
Havre, Montana

Dean M. Bortz
Columbus State Community College
Albany, Oregon

Ron Care
Linn Benton Community College
Hugo, Oklahoma

Homer Chapman
Kiamichi Technology Center
Hugo, Oklahoma

Mike DeMattei
John A. Logan College
Carterville, Illinois

Jodie Eiland
Gordon Cooper Technology Center
Shawnee, Oklahoma

Mark D. Enger
Meridian School District #2
Meridian, Idaho

Clinton J. Gray
Alfred State College
Alfred, New York

Bob Gresko
Pennsylvania College of Technology
Williamsport, Pennsylvania

Gary J. Levesque
Dighton-Rehoboth Regional High School
North Dighton, Maine

Don Lucas
Fulton-Montgomery Community College
Johnstown, New York

Joe Musheno
Johnson College
Scranton, Pennsylvania

Mark A. Parrott
Portage High School
Portage, Wisconsin

Joseph L. Percevecz
Claudia Taylor Johnson High School
San Antonio, Texas

Bobby Richardson
Robert E. Lee High School
Tyler, Texas

Randolph Stevens
Wayne County Community College
Detroit, Michigan

Jim Wiater
Middlesex County Vocational
Piscataway, New Jersey

Acknowledgments

The authors and publisher would like to thank the following companies, organizations, and individuals for their contribution of resource material, images, or other support in the development of *Modern Carpentry*.

A & C Paslode

ABTco, Inc.

Accufooting

Agricultural Extension Service, University of Minnesota

Alfred State College

Alum-A-Pole Corp.

American Building Components

American Institute of Timber Construction

American Olean Tile Co.

American Standard

American Tool

Amerock Corp.

Amoco Foam Products Co.

Andersen Corp.

APA-The Engineered Wood Association

Arizona Tile

Ark Seal, Inc.

Armstrong World Industries, Inc.

Arxx Building Products

Asphalt Roofing Manufacturer's Association

A. Weyerhaeuser

Baldwin Hardware Corp.

Bayfield Lumber Company

Benjamin Obdyke, Inc.

Bird Division, CertainTeed Corp.

Black & Decker

Boen Hardwood Flooring, Inc.

Boise-Cascade Corp.

Brady Adams

Brammer Mfg. Co.

BuildSMART, LLC

Bullard-Haven Technical School

Cambria USA

Carol Electric

Carrier Corporation, subsidiary of United Technologies Group

Cedar Shake and Shingle Bureau

CEMCO and Dale Industries Inc.

C-E Morgan

CertainTeed Corporation

Chief Architect

Christa Construction LLC

Clopay Building Products

Council of Forest Industries of British Columbia

CPG Building Products— TimberTech ReliaBoard

Crown Aluminum Industries

Dauphin Island

David White

Deckmaster

DeckWise

Delta International Machinery Corp.

Des Moines, Iowa, Public Schools

DeVilbiss Co.

DeWalt

Dörken Systems Inc.

Dutcher Glass and Paint

E.I. duPont de Nemours & Co.

Empire Level, Inc.

Enercept, Inc.

Estwing Manufacturing

Express Products, Inc.

Forest Products Laboratory

Formica Group North America

Forney Industries, Inc.

Frank Paxton Lumber Co.

Gang-Nail Systems, Inc.

General Products, Inc.

Georgia-Pacific Corp.

Gold Bond Building Products

Graco Inc.

Greco Painting

Haas Cabinet Co., Inc.

Hadley-Hobley Construction

Handyman Construction

Harrington

Hilti Tools

H.L. Stud Corp.

Hohmann and Barnard, Inc.

Huber Engineered Woods

Ideal Co.

Independent Nail and Packing Co.

Infinity Cutting Tools

Innovative Building Systems

Insulspan

International Code Council

Iowa Energy Policy Council

I-XL Furniture Co.

Jack Klasey

James Hardie Building Products

Johns Manville

Johnson-Manley Lumber Co.

Kasten-Weiler Construction

KCPL

Keuffel & Esser

Klein Tools

K.N. Crowder

KraftMaid Cabinetry, Inc.

Kunkle Valve Co. Inc.

Leviton

Lifetile

Lift-All Company, Inc.

Lite-Form International

Loren LaPointe Drywalling

Macklanburg-Duncan

Majestic Co.

Manville Building Products, Inc.

Marvin Windows and Doors

Masonite Corp.

McDaniels Construction Co., Inc.

Merillat Industries, Inc.

Mike Hess

Modular Genius

Moen Incorporated

Mohawk Finishing Products

Monier Group

National Association of Home
Builders

National Concrete Masonry
Association

National Decorating Products
Assn.

National Fenestration Rating
Council

National Forest Products Assn.

National Gypsum Co.

National Woodwork
Manufacturers Association

N.C. Brown Center for
Ultrastructure Studies

North Bennett Street School,
Boston

Oak Flooring Institute

OSHA

Owens Corning

Patent Scaffolding Co.

Pease Industries, Inc.

Pella Corp.

Pennsylvania College of
Technology

Pergo, Inc.

PHIUS

Pierce Construction, Ltd.

Porter-Cable

Portland Cement Association

Preway, Inc.

Reed Manufacturing Co.

RenewAire

Rev-A-Shelf, Inc.

Rheem Manufacturing Company

Ridge Tool Company

Rik Vandermeulen, Unalam

Robert Bosch Tool Corporation

Rock Island Millwork

Rollex Corp.

RotoZip

Schlage

Shakertown Corp.

Simpson Strong-Tie Company, Inc.

SkillsUSA

Slant/Fin Corp.

Solatube North America, Ltd.

Southern Forest Products
Association

Southern Pine Council

Spectra-Physics Laserplane, Inc.

SR Sloan, Inc.

Stanley Door Systems

Stephen P. Kielar

Steve Olewinski

Stihl USA

Sto Finish Systems

StructureCraft

Symons by Dayton Superior

TECO

The Energy Conservancy

The Garlinghouse Company

The Panel Clip Co.

The Stanley Co.

The US Environmental Protection
Agency's Energy Star Program

Tibbias Flooring Inc.

Timber Engineering Co.

Timberpeg

Top Tread Stairways

Trex Company, Inc.

Trimble

Trudeau Construction Co.

TrusJoist MacMillan

Truss Plate Institute

Trussworks, Inc.

TruStile Doors

TrusWal Systems Corp.

Tru Value Hardware, Ashland, WI

T.W. Lewis Construction
Company

Typar

United States Federal Trade
Commission

United States Steel

United Technologies Carrier

US Department of Energy

US Green Building Council

US Gypsum Co.

Vaughan and Bushnell
Manufacturing Co.

Vermont Frames

Village of Flossmoor, Illinois

Wausau Homes, Inc.

Weiser Lock

Wermers Design and Architecture

Western Forms

Western Wood Products Assn. and
Timber Engineering Co.

Weyerhaeuser

William Powell Co.

Wilsonart LLC

Wolmanized® Wood

Yankee Barn Homes, Inc.

TOOLS FOR STUDENT AND INSTRUCTOR SUCCESS

Student Tools

Student Text

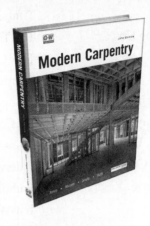

Modern Carpentry provides detailed coverage of all aspects of light construction, including site preparation and layout, foundations, framing and sheathing, roofing, windows and doors, exterior finish, stairs, cabinetry, and finishing. New author R. Jack Jones revised this edition for the modern classroom with new features, hundreds of new images, and expanded coverage of green building practices that highlight industry trends toward greener processes. Special topics, such as chimneys, fireplaces, and decks, are also covered. The text familiarizes students with other aspects of the building trades with chapters on plumbing, electrical, and HVAC.

Workbook

The workbook that accompanies *Modern Carpentry* includes instructor-created activities to help students recall, review, and apply concepts introduced in the book.

Certification

Precision Exams by YouScience

Goodheart-Wilcox is pleased to partner with YouScience to correlate *Modern Carpentry* with their Carpentry certification standards. Students who pass the exam and performance portion of the exam can earn a Career Skills certification. Precision Exams by YouScience and Career Skills Exams were created in partnership with industry and subject matter experts to align real-world job skills with marketplace demands. Students can showcase their skills and knowledge with industry-recognized certifications—and build outstanding resumes to stand out from the crowd!

And for teachers, Precision Exams by YouScience provides:

- Access to a library of Career Skills Exams, including pre- and post-assessments for all 16 National Career Clusters
- Suite of on-demand reporting to measure program and student academic growth
- Easy-to-use, 100% online administration

To see how *Modern Carpentry* correlates to Precision Exams by YouScience standards, visit the Correlations tab at www.g-w.com/modern-carpentry-2022. For more information about Precision Exams by YouScience, visit www.youscience.com/precision-exams.

Instructor Tools

LMS Integration

Integrate Goodheart-Willcox content within your Learning Management System for a seamless user experience for both you and your students. LMS-ready content in Common Cartridge® format facilitates single sign-on integration and gives you control of student enrollment and data. With a Common Cartridge integration, you can access the LMS features and tools you are accustomed to using and G-W course resources in one convenient location—your LMS.

G-W Common Cartridge provides a complete learning package for you and your students. The included digital resources help your students remain engaged and learn effectively:

- **eBook Content.** G-W Common Cartridge includes the textbook content in an online, reflowable format. The eBook is interactive, with highlighting, magnification, note-taking, and text-to-speech features.
- **Workbook Content.** Students can have access to a digital version of the Workbook.
- **Videos.** Students can access dozens of professional videos that demonstrate important procedures in the text. These videos clarify steps and aid students in visualizing important skills.
- **Animations.** These digital resources engage students and aid in comprehension and understanding.
- **Drill and Practice.** Learning new vocabulary is critical to student success. These vocabulary activities, which are provided for all key terms in each chapter, provide an active, engaging, and effective way for students to learn the required terminology.

When you incorporate G-W content into your courses via Common Cartridge, you have the flexibility to customize and structure the content to meet the educational needs of your students. You may also choose to add your own content to the course.

For instructors, the Common Cartridge includes the Online Instructor Resources. QTI® question banks are available within the Online Instructor Resources for import into your LMS. These prebuilt assessments help you measure student knowledge and track results in your LMS gradebook. Questions and tests can be customized to meet your assessment needs.

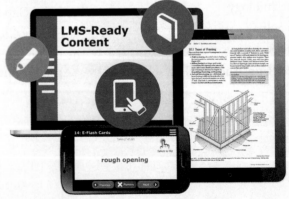

Online Instructor Resources (OIR)

Online Instructor Resources provide all the support needed to make preparation and classroom instruction easier than ever. Available in one accessible location, the OIR includes Instructor Resources, Instructor's Presentations for PowerPoint®, and Assessment Software with Question Banks. The OIR is available as a subscription and can be accessed at school, at home, or on the go.

Instructor Resources One resource provides instructors with time-saving preparation tools such as answer keys, editable lesson plans, and other teaching aids.

Instructor's Presentations for PowerPoint® These fully customizable, richly illustrated slides help you teach and visually reinforce the key concepts from each chapter.

Assessment Software with Question Banks Administer and manage assessments to meet your classroom needs. The question banks that accompany this textbook include hundreds of matching, completion, multiple choice, and short answer questions to assess student knowledge of the content in each chapter.

Using the assessment software simplifies the process of creating, managing, administering, and grading tests. You can have the software generate a test for you with randomly selected questions. You may also choose specific questions from the question banks and, if you wish, add your own questions to create customized tests to meet your classroom needs.

G-W Integrated Learning Solution

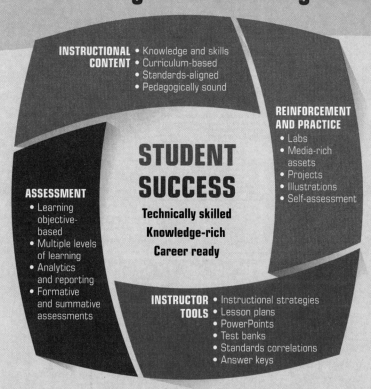

INSTRUCTIONAL CONTENT
- Knowledge and skills
- Curriculum-based
- Standards-aligned
- Pedagogically sound

REINFORCEMENT AND PRACTICE
- Labs
- Media-rich assets
- Projects
- Illustrations
- Self-assessment

STUDENT SUCCESS

Technically skilled
Knowledge-rich
Career ready

ASSESSMENT
- Learning objective-based
- Multiple levels of learning
- Analytics and reporting
- Formative and summative assessments

INSTRUCTOR TOOLS
- Instructional strategies
- Lesson plans
- PowerPoints
- Test banks
- Standards correlations
- Answer keys

The G-W Integrated Learning Solution offers easy-to-use resources that help students and instructors achieve success.

▶ EXPERT AUTHORS
▶ TRUSTED REVIEWERS
▶ 100 YEARS OF EXPERIENCE

EMPLOYABILITY SKILLS · TECHNICAL SKILLS · ACADEMIC KNOWLEDGE · INDUSTRY RECOGNIZED STANDARDS

Features of the Textbook

The instructional design of this textbook includes student-focused learning tools to help you succeed. This visual guide highlights these features.

Chapter Opening Materials

Each chapter opener contains a list of learning objectives and a list of technical terms. The **Objectives** clearly identify the knowledge and skills to be gained when the chapter is completed. **Technical Terms** list the key words to be learned in the chapter.

Additional Features

Additional features are used throughout the body of each chapter to further learning and knowledge. **Safety Notes** alert you to potentially dangerous materials and practices. **Code Notes** point out specific items from typical building codes. **Procedures** are highlighted throughout the textbook to provide clear instructions for hands-on service activities. **Pro Tips** provide advice and guidance that is especially applicable for on-the-job situations. **Green Notes** highlight key items related to sustainability, energy efficiency, and environmental issues.

CHAPTER **7**

Plans, Specifications, and Codes

OBJECTIVES

After studying this chapter, you will be able to:
- Identify the elements commonly included in a set of house plans.
- Demonstrate the use of scale in architectural drawings.
- Identify architectural symbols.
- Explain the use of building specifications.
- Describe the application of building codes, standards, and permits.
- List the items required by building officials to obtain a building permit.

TECHNICAL TERMS

architectural drawing
authority having jurisdiction (AHJ)
building code
building permit
computer-aided drafting and design (CADD)
CADD-CAM
computer-aided manufacturing (CAM)
detail drawing

dimensions
drawn to scale
electrical plan
elevation
floor plan
footprint
foundation plan
framing plan
list of materials
model code
pictorial sketch

plot plan
print
schedule
section drawing
set of plans
setback
specifications
stock plan
symbol
truss plan

A good plan and well-defined contract are important in building construction. Expectations should be specified in the contract in no uncertain terms. For the project to go as seamlessly as possible and to avoid possible future litigation, both sides should know *exactly* what is expected of the other.

Every carpenter must know how to read and understand **architectural drawings** (plans) and correctly in-

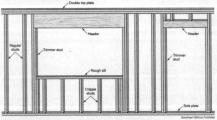

286   Section 2 Foundations and Framing

Safety Note
When framing a roof, use extra care to prevent a fall. Fall protection devices must be used. Erect solid scaffolding wherever it is helpful. Avoid working directly above another person.

Code Note
The connection between the top plate and the rafter provides resistance to uplift. Uplift can occur when strong winds push against the side of a building and roof. In some cases, additional tie-connections may be needed based on the rafter spacing, roof span, and design wind speed. In the United States, these additional tie-down connections are most commonly needed along the gulf coast and Atlantic coast regions. Be sure to check the requirements in your local building code.

PROCEDURE
Laying Out a Gable End Frame
1. Using a level and straightedge, mark the top plate to show the location of a stud directly below the ridge, if possible. Often, a ventilator is located in the center of the end frame. If this is the case, mark a distance 1/2 of the ventilator width to either side of the plumb line. This mark is the location of the first stud.
2. Position the second stud on the top plate and use a level to recheck for plumb. Mark the location of the rafter on the side of the stud.
3. Lay out the next stud, plumb it, and mark the rafter location on the stud's side. The difference between the two stud lengths is the *common difference.* **Figure 12-17.**
4. Using the common difference, continue marking, cutting, and placing studs until the edge of the frame is reached. If the rafters on both sides have the same rise and run, the end frame should be symmetrical (both sides identical).

PROCEDURE
Finding the Common Difference with the Framing Square
The common difference can also be obtained using the framing square. See **Figure 12-18.**
1. Set the square on a stud. Align the unit run on the blade with the edge of the stud. Then align the unit rise on the tongue to the same edge. (This is the same as the step method used in determining the length of a rafter.)
2. Mark a line on the stud along the outside edge of the blade.
3. Slide the square away from you. Keep the outer edge of the blade on this line. Watch the inch marks on the blade. Stop when the inch number for the stud spacing (16" or 24") aligns with the edge of the stud.
4. Read the inch mark on the tongue of the square that aligns with the edge of the stud. This is the common difference.

12.10 Gable End Frame
The end frames of a gable should be assembled after the rafters and ridge have been installed. The end frame consists of vertical studs running from the top plate of the bearing walls to the end rafters. The framing of the gable end is most easily done while it lies flat over the ceiling joists. Overhang, brick racks, frieze trim, louvers, housewrap, and siding can all be installed at this time. After it is erected, the end gable should be well braced before installing the ridge board and rafters. The following procedure can be used to determine the length and location of studs.

Roof designs often include an *extended rake* (gable overhang). Typical framing is illustrated in **Figure 12-19.** It requires the gable end frame be constructed before the roof frame is completed.

Figure 12-17. Lay out studs for a

Chapter 11 Wall and Ceiling Framing   249

In conventional framing, extra studs are included around rough openings, as shown in the assembly in **Figure 11-10.** The studs and trimmers support the header and provide a nailing surface for window and door casing.

11.1.4 Alternate Header Construction
In large window openings, the size of the header may reduce the length of the upper cripple studs to a point where they cannot be easily assembled. In this case, the cripple studs can be replaced with flat blocking. Another solution is to increase the header size to completely fill the space to the plate. Some builders follow this practice and extend it to include all openings, regardless of the span. The cost of labor required to cut and fit the cripple studs is usually greater than the cost of the larger headers. A disadvantage of such construction is extra shrinkage. Shrinkage may cause cracks above doors and windows unless special precautions are taken when applying the interior wall finish.

Green Note
As described in the previous paragraph, headers are often sized to fill the space above the rough opening to the wall plate. Without any added insulation, this creates a tremendous thermal bridge. To cut heat loss and build greener headers, size headers properly and add a layer of rigid foam insulation between a built-up header's layers or to the exterior face.

11.2 Plate Layout
Sole plates and top plates are the same size as the studs, typically 2×4 or 2×6. Use only straight stock for plates. Select two pieces of equal length and lay them side by side along the location of the outside wall. The length is determined by what can be easily lifted off of the floor and into a vertical position after it is assembled. Remember that the weight may include all of the framing for rough openings, bracing, and sheathing. If wall jacks or a forklift are available for lifting, sections can be made larger. Where they must be lifted by hand, attach sheathing after the wall is up. Always locate joints over a stud. The centerlines of rough openings are marked first.

Pro Tip
Carefully check over your rough opening layouts for errors. Do the math before cutting and framing.

11.2.1 Laying Out the Second Exterior Stud Wall
Laying out the second exterior wall follows the same procedure as the first outside wall, with one exception. If sheet material is used for rough siding, then the location of the first stud from the corner post must allow for the edge of the panel to be flush with the outside edge of the siding. If the siding is 3/4″ thick and the studs are 16″ OC, lay out the first stud 15 1/4″ from the end of the plate.

Figure 11-10. Framing door and window openings. Notice how the wide header above the window eliminates cripple studs between the header and the top plate.

Copyright Goodheart-Willcox Co., Inc.

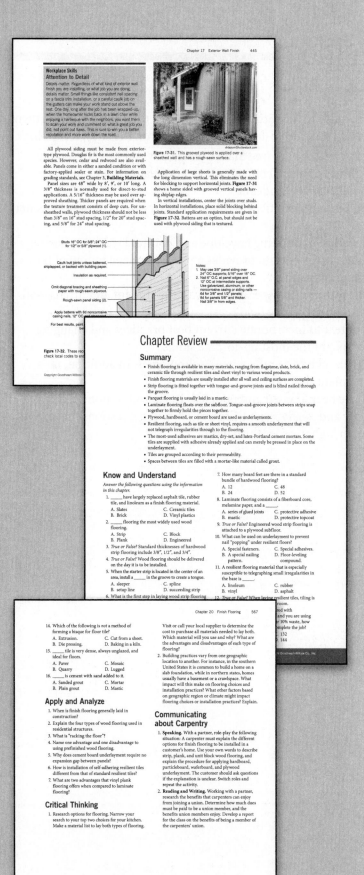

Illustrations

Illustrations have been designed to clearly and simply communicate the specific topic. Illustrations have been updated for this edition. Photographic images have been updated to show the latest equipment.

Expanding Your Learning

Workplace Skills highlight the professional behaviors and traits that employers look for and that will help you succeed in the workplace. **Construction Careers** features and profiles can provide a path for career success.

End-of-Chapter Content

End-of-chapter material provides an opportunity for review and application of concepts. A concise **Summary** provides an additional review tool and reinforces key learning objectives. This helps you focus on important concepts presented in the text. **Know and Understand** questions enable you to demonstrate knowledge, identification, and comprehension of chapter material. **Apply and Analyze** questions extend learning and help you analyze and apply knowledge. **Critical Thinking** questions develop higher-order thinking and problem solving, personal, and workplace skills. **Communicating about Carpentry** activities are designed to improve and develop language skills such as reading, speaking, and listening.

New to This Edition

The following changes have been made to the 13th edition of *Modern Carpentry* to strengthen the integrated learning solution and provide up-to-date information on the latest technology.

- New author R. Jack Jones, a professor at Alfred State College, revised this edition for the modern classroom. Mr. Jones updated content, wrote new features, and supplied hundreds of new images to help engage students.
- Expanded coverage of green building practices highlight industry trends toward greener processes.
- A new chapter providing coverage of building envelope and control layers emphasizes today's focus on green building practices and provides a proper understanding of how environmental conditions impact a home.
- Addition of Workplace Skills features that speak directly to the student and discuss soft skills required for working in the carpentry field.
- Content updated throughout text to incorporate current best practices and changing codes to ensure students are taught the most up-to-date skills.
- Chapter illustrations have been revised to reflect new technologies. In addition, new photos were added to complement the content.
- End-of-chapter questions have been revised and updated. Review Questions include Know and Understand, Apply and Analyze, and Critical Thinking questions that extend students' learning and develop higher-order thinking.

Brief Contents

Contents

Section 4
Finishing

Feature Contents

Workplace Skills

Construction Careers

Procedures

David Papazian/Shutterstock.com

CHAPTER 1

The Carpenter's Workplace

OBJECTIVES

After studying this chapter, you will be able to:

- Cite the projected demand for carpenters in coming years.
- List job possibilities for the trained carpenter.
- Describe the sequence of carpentry training and apprenticeship.
- Discuss abilities and characteristics needed by those in the carpentry field.
- Describe the qualities employers value in an entry-level carpenter.
- Describe the advantages and disadvantages of being an entrepreneur.
- List other occupations that are related to carpentry.

TECHNICAL TERMS

apprenticeship
discrimination
entrepreneur
estimator

journeyman
liability
manual dexterity
self-employment

sexual harassment
SkillsUSA

A laborer is a worker who does manual labor, working with their hands and back. A tradesman or tradeswoman is someone who uses specialized training, while still doing some manual labor. A carpenter is a tradesman who is able to incorporate creativity and heart into their work.

Carpentry is a trade that dates back thousands of years. It has helped to advance our society from living in tents and caves to the astonishing structures and homes that we enjoy today. Tradesmen who practice carpentry enjoy a rewarding career. Builders often feel a connection to the structures and environments they helped create. These structures are tangible evidence of their passions, skills, and effort. Unlike many careers today, carpenters can see and touch the fruit of their labor. Their work creates utility, something that is useful, and more than that, carpentry often results in something beautiful. It is ideal for a person who has an aptitude for creating and building with tools and materials, **Figure 1-1**.

A successful career in the trade requires the development of physical skill, mental aptitude, and soft skills. Soft skills are personal and career skills that allow you

to work effectively with others toward a common goal. Carpentry also requires a thorough knowledge of the materials and methods used in construction work.

Monkey Business Images/Shutterstock.com

Figure 1-1. Along with their tools, carpenters must use interpersonal skills and training to be successful on the jobsite.

Skilled workers are important to the maintenance and development of the economy and infrastructure. Carpenters can expect to work at a variety of construction activities that range from the construction of traditional buildings, skyscrapers, highways, bridges, industrial plants, and marine structures to more creative niche markets, such as construction of movie sets, amusement parks, timber frame structures, or custom furniture and cabinetry. They cut, fit, and assemble wood, steel, composite, and other materials. If carpenters work in commercial building construction, they will likely work with metal framing materials. Increasingly, metal framing is found in residential construction.

Carpenters working for general building contractors become involved in many different tasks. These often include laying out building lines, excavating, installing footings and foundations, framing, roofing, building stairs, insulating, drywalling, laying flooring, and installing trim work and cabinetry.

Code Note
A well-trained carpenter is expected to know the requirements of local building codes. These codes dictate where materials may be used and how a structure is to be built. Failure to follow the codes can prove costly, since inspectors may require work that does not meet code requirements to be redone. This means more time and money spent on materials and labor.

1.1 Economic Outlook

The population of the United States continues to grow. It surpassed 300 million in the first decade of the 21st century. This continued growth increases the demand for housing and the construction workers who build it. Building construction, especially residential structures, continues to increase as the population grows. Much of this growth in the demand for carpenters will come from the need to improve the nation's infrastructure—roads, bridges, and tunnels. See **Figure 1-2**.

A   *Zhengzaishuru/Shutterstock.com*   B   *Alice Lowe/Shutterstock.com*

Figure 1-2. A—Infrastructure work, such as bridge and road repairs, will be a major source of construction employment in the next ten years. Next to services, construction occupations are among the fastest-growing in the United States. B—Miami's landmark Scorpion Tower uses revolutionary design and building techniques, rather than using a traditional skeletal frame network. Construction workers from many trades had to learn new methods and work with new materials to accomplish this design.

1.2 Employment Outlook

Carpenters are the largest group of trades workers. According to the United States Bureau of Labor statistics, there are more than 1 million carpenters in the United States today. This number is expected to increase by more than 80,100 by 2028. That is a faster growth rate than is projected for the average of all occupations. Thirty percent of carpenters work in residential construction and nearly as many work in commercial building construction. The remaining carpenters work in a variety of industries, including roofing, foundation work, building exteriors, government agencies, and manufacturing companies.

The median pay for carpenters in the United States in 2018 was $46,590 per year, nearly $8,000 more than the median annual wage for all workers in the nation. Some of the highest-paying carpentry jobs are in motion picture and video production, promoters of performing arts and sports, and performing arts companies. Jobs in carpentry are expected to be plentiful in the coming years because of increased building demand and the need to replace those who transfer into other jobs or retire. Construction growth also means more opportunities for building trades workers such as surveyors, electricians, bricklayers, plumbers, cabinetmakers, landscapers, and operating engineers (operators of heavy construction equipment). Carpenters commonly develop the skills needed to make a switch into any of these career paths. **Figure 1-3** shows a variety of construction-related careers available to anyone who has the interest and aptitude to pursue them.

On the negative side is the prospect of cyclic economic downturns that affect the construction industry. When the economy is weak, demand for construction workers is reduced because there are fewer construction projects. At such times, many people skilled in construction find it hard to get or hold a job. Those who have been in such occupations for a long time have come to expect both growth and decline in the industry. A carpenter should be prepared for this type of volatility.

Carpenters can expect that their jobs will become increasingly technical and automated as time goes on. Thus, apprentices and journeymen alike should expect to be learning new tasks throughout their working life.

With experience, new opportunities may come along. Often, because of their holistic understanding of the job, carpenters find opportunities for advancement into supervisory and management positions. In addition to progression into supervisory positions, carpenters can become educators, consultants, business owners, technical representatives for construction materials manufacturers, and union leaders.

1.3 Working Conditions

Careers in the building trades can be physically demanding. However, technology, equipment, and safety laws have improved over time. These improvements help to mitigate the physical strain and hazards associated with working in this field.

A — *Bannafarsai_Stock/Shutterstock.com* B — *Phovoir/Shutterstock.com* C — *sirtravelalot/Shutterstock.com*
D — *Vlad Teodor/Shutterstock.com* E — *Dmitry Kalinovsky/Shutterstock.com* F — *Vladim Ratnikov/Shutterstock.com*

Figure 1-3. Carpentry is related to a wide variety of skilled occupations. A—Surveying. B—Electrical trades. C—Construction superintendent. D—Cabinetmaking. E—Mason. F—Operating engineer.

Carpenters are expected to handle heavy and cumbersome materials, sometimes while working at great heights. Climbing, stooping, kneeling, and prolonged standing are all part of the job. On many sites, work continues in the heat of summer and through the cold of winter. Carpenters continually use tools that are dangerous when operated improperly. These tools can be loud, sharp, and many can produce flying debris. Jobsites can be hazardous places filled with threats to safety. The Occupational Safety and Health Administration (OSHA) is a government agency that was created to help protect workers from incurring injury on the job. When properly followed, OSHA standards reduce safety risks tremendously. All employers and employees are required to follow these standards.

1.4 Job Opportunities

Job opportunities as a carpenter depend somewhat on your choice of work. There are many areas where you can look for a job:

- Building construction—The majority of carpenters work in this area.
- Remodeling, maintenance, and renovations—This category contains many self-employed people as well as those who work for manufacturers and building management firms.
- Prefabrication of buildings and building components—While this area employs many workers with fewer skills than carpenters, the latter may hold higher-level jobs in this type of fabrication.
- Specialization and related occupations—It is relatively easy for a carpenter to move into specialties within the carpentry field.

Workplace Skills
Work Ethic

What is your idea of good work ethic? Your answer to this question may determine the course of your career. Regardless of your occupation, if you see work as drudgery, it will be. Come to work each day with a positive attitude and put your best effort into any task. Work can be rewarding, engaging, and even fun. Not every day on the job will be awesome. Use the hard days as motivation, you can overcome the challenges and gain confidence in yourself. With the right attitude toward work, you will develop a strong work ethic.

A general building contractor may subcontract various phases of construction. As a result, carpenters may specialize in foundation work, framing, drywalling, insulating, siding, or roofing. **Figure 1-4** shows two of these specialties. A carpenter may also choose to move into related fields such as terrazzo work, heavy equipment operation, bricklaying, or electrical work.

Four of every ten carpenters are self-employed. *Self-employment* has a number of advantages. Often, profits are higher due to lower overhead. That is, all money earned goes directly to the worker as opposed to being shared with an employer. Self-employment also allows for job mobility. In fact, some carpenters change employers at the conclusion of every construction job. Others alternate between working for a contractor and being the contractor themselves on small jobs. Self-employment also has some disadvantages. A self-employed individual is responsible for keeping track of all tax liabilities and maintaining appropriate insurances. Additionally, they receive no retirement or medical benefits from an employer.

A   *Jwwaddles/Shutterstock.com*   B   *Orange Line Media/Shutterstock.com*

Figure 1-4. Often, carpenters specialize in one particular phase of building. A—This crew specializes in putting in foundations. This involves excavating, building forms, and laying down vapor barrier and reinforcing mat. B—This crew does only framing.

After gaining experience, a carpenter may want to undertake a small construction contract. This may be a first step toward a general contracting business. An experienced carpenter is usually able to handle the work of an estimator. *Estimators* calculate labor, material costs, overhead, and profit before a contractor bids on a job.

A carpenter who prefers the sales and service aspects of construction work can often find a position with a lumber yard or building supply center. A carpenter can also join the customer service department of a company producing structures in a factory. Customers may need advice on installation or choice of prefabricated units.

Green Note

Career opportunities in construction's green building sector continue to expand. Technology and materials that support super-insulated structures, photovoltaic systems, and commercial wind generation all require qualified personnel to perform installations. These are some of the fastest growing occupations in the United States.

1.5 Training

A complete high school education is highly desirable for a successful career in carpentry or other areas of building construction. Take as many woodworking and building construction courses as possible. Other courses, such as print reading, career education, and computer-aided drafting (CAD), are also valuable.

Students interested in the building trades sometimes avoid science and math. This is unfortunate, because these studies are essential if you are to understand the technical aspects of modern methods and materials used in construction work. Include social studies and English in your studies because being competent in reading, writing, and speaking will allow you to communicate well with colleagues and customers alike. This will come in handy if you need to write a contract or advertise your business.

After graduating from high school, you will have the skills needed to enter into an apprenticeship training program, **Figure 1-5**. If circumstances permit, you may decide to enroll in a technical school in your area. A technical college education is helpful, particularly if one has no avenue to gain on-the-job training. A formal education can give a broad knowledge base and training without the fear of being dismissed for making mistakes. You can take advanced courses in carpentry and related areas. If possible, take classes in concrete work, bricklaying, plumbing, sheet metal, and electrical wiring.

A carpenter often works closely with people in these areas. A basic understanding of their methods and procedures is very valuable. This is especially true if you want to become a supervisor.

1.5.1 Apprenticeship

Many carpenters learn their trade on the job, taught by their employer or the lead carpenter. However, many learn the trade through apprenticeship or through formal training in technical training centers or high school programs. *Apprenticeship* is a program that allows a student to both work and attend school to learn a trade—in this instance, carpentry.

Apprenticeship training has its beginning in early times. It started as an arrangement between children and their parents. This system passed the knowledge and skill of a trade on to each succeeding generation. As society and the economic structure became more complex, the trainee often left home and was placed under the guidance and direction of another master of the trade in the community. The student was called an apprentice and learned the trade of this master. During the period of apprenticeship (sometimes as long as seven years) the apprentice often lived in the master's house. There were no wages, but board, room, and clothing were provided.

SkillsUSA

Figure 1-5. An apprenticeship program provides training on the job and in the classroom. Here an apprentice is competing in a skill competition.

When the training was complete, the apprentice earned the status of journeyman and could work for wages. The word *journeyman* arose from the fact that an apprentice who had completed the training period was then free to "journey" to other places in search of employment. The term journeyman still means a tradesperson who is fully qualified to perform all tasks of the trade. To gain more experience, the journeyman worked with other masters as an equal. Then this seasoned worker would often set up his or her own business. Today, apprenticeship programs are much more organized and standardized. The United Brotherhood of Carpenters and Joiners of America, a carpenters union formed in 1881, has a very structured apprentice program. Local committees consisting of labor and management provide direct control. There is help from schools and state and federal organizations and agencies, **Figure 1-6**. There are also many strong nonunion apprentice programs, such as those offered by Associated Builders and Contractors.

1.5.2 Apprenticeship Stages

An apprentice works under a signed agreement with an employer. The agreement includes the approval of local and state committees on apprenticeship training.

Applicants must be at least 17 years of age and satisfy the local committee that they have the ability or aptitude to master the trade. Then, they are placed on a waiting list. The waiting period can last from one to five years, depending on local demand.

The term of apprenticeship for the field of carpentry is normally four or five years. This may be adjusted for applicants with significant experience or those who have completed certain advanced courses in technical schools.

In addition to the instruction and skills learned through regular work on the job, an apprentice attends construction classes in subjects related to the trade. These classes are usually held in the evening and total at least 144 hours per year. They cover technical information about tools, machines, methods, and processes. The classes provide practice in mathematical calculations, print reading, sketching, layout work, estimating, and similar activities. A great deal of study is required to master the technical knowledge needed in carpentry work.

An apprentice works and is paid while learning. The wage scale is determined by the local committee. The apprentice pay rate while in the field usually starts at about 50% of the amount received by a journeyman carpenter. This scale is advanced regularly and may approach 90% of a journeyman's pay during the last year.

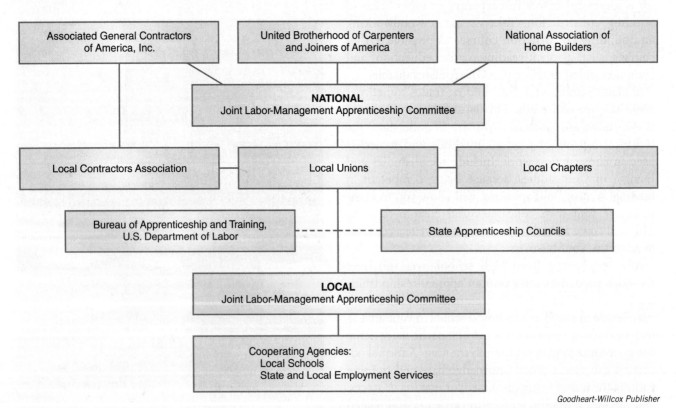

Goodheart-Willcox Publisher

Figure 1-6. The apprenticeship and training system for the carpentry trade is well organized. Additional information may be obtained from the US Department of Labor.

Apprentices are usually paid a reduced rate while they are in the classroom.

When the training period is complete and the apprentice has passed a final examination, they become a journeyman carpenter. A certificate that affirms this status is issued by the program that conducts the apprenticeship. This document is recognized throughout the country. While many carpenters take further classes to help them stay up-to-date in the industry, no further training is required to maintain their journeyman status.

1.6 Skill Development and Job Competency

Reading and comprehension skills are important for carpenters. They are crucial for following construction drawings, building codes, and specifications. Not so obvious is the need to carefully read instructions. As modern engineered materials become more common, following manufacturers' printed instructions becomes more important. This principle holds true even if you are switching brands on a product that you have installed before. Although any construction worker faces time constraints on a daily basis, paying close attention to manufacturers' instructions on application of their products actually saves time. A manufacturer will only honor their warranty if the product is installed to their specific instructions. The old adage applies: If you don't have time to do it right the first time, how will you find the time to do it over?

sirtravelalot/Shutterstock.com

Figure 1-7 Doing it right the first time demands accuracy in taking measurements. Carpenters "measure twice and cut once."

Workplace Skills
Reputation

A job well done has more benefits than a happy customer or employer. Word travels fast in the construction world. It is exceedingly helpful to have a good reputation as a carpenter. A reputation for quality work will win you more jobs and favor with employers as your career develops.

Customers will often write reviews of your work online or tell others about their experience with you. This can greatly help or hinder you as you try to grow your business. Take pride in your work. At the end of the job, you can rest satisfied knowing that you did it well, because you did it right.

Doing it right the first time has other benefits for a carpenter and his or her employer, **Figure 1-7**. When the job is done right, it reduces the amount of warranty work the contractor must do and promotes greater customer satisfaction.

1.6.1 What Employers Expect of Beginning Carpenters

Results of a survey of union and nonunion indicate that what construction contractors most value in entry-level carpenters is not always a high level of knowledge, or even well-honed carpentry skills. **Figure 1-8** shows the qualities contractors most value in beginning carpenters. The most valued qualities were separated into three categories or behaviors:

- Affective behavior—Attitude and interpersonal relationships.
- Cognitive behavior—Knowledge of any sort; in this case, carpentry work.
- Psychomotor skills—Eye and hand coordination, meaning how well the worker uses tools and the quality of his or her work.

Skills/Behaviors Contractors Value Most in Beginner Carpenters

Skill or Behavior	Value*	Type
Has respect for customers and owners	4.77	Affective
Able to measure material accurately	4.73	Cognitive
Carries basic tools (pencil, tape, knife)	4.68	Affective
Notifies supervisor when unable to report for work due to personal emergency	4.62	Affective
Cooperates with coworkers	4.59	Affective
Observes safety rules	4.58	Affective
Uses safe lifting/carrying techniques	4.49	Affective
Takes proper care of power tools	4.49	Affective
Follows company rules/procedures	4.49	Affective
Operates circular saw	4.49	Psychomotor
Reports for work properly rested/ready to work	4.49	Affective
Uses leveling instruments	4.25	Cognitive
Keeps work area clean	4.24	Affective
Attends to details	4.23	Affective
Operates pneumatic nailer	4.07	Psychomotor
Follows fall protection procedures	4.04	Cognitive
Uses the right nail for the application	3.97	Cognitive
Reports for work 10 or more minutes early	3.91	Affective
Wears suitable clothing	3.88	Affective
Recognizes fire hazards	3.87	Cognitive
Performs first aid	3.86	Psychomotor
Volunteers for undesirable jobs	3.81	Affective
Records number of hours worked on the job	3.80	Cognitive
Absent from work no more than four days per year	3.80	Affective
Cares for electrical extension cords	3.75	Affective
Toe-nail with minimum splitting of lumber	3.70	Psychomotor
Reads blueprints	3.65	Cognitive
Operates screw gun	3.62	Psychomotor
Uses telephone etiquette	3.54	Affective
Applies adhesives	3.51	Psychomotor
Scales dimensions from a blueprint	3.48	Cognitive
Operates powder-actuated tool	3.44	Psychomotor
Uses jobsite sanitation facilities	3.41	Affective
Applies caulk	3.41	Psychomotor
Installs door hardware	3.35	Psychomotor
Installs doors	3.25	Psychomotor
Installs moulding	3.17	Psychomotor
Operates radial arm saw	3.17	Psychomotor
Establishes building lines	3.15	Cognitive
Installs cabinets	3.13	Psychomotor
Lays out stairways	3.07	Cognitive
Installs metal studs	3.01	Psychomotor

* 5=crucial, 1=unimportant

Goodheart-Willcox Publisher

Figure 1-8. Responding contractors were asked to rate 54 qualities on a scale of 1 to 5, with 1 being of little or no value for success and 5 being crucial for success. Contractors listed these 42 items of attitude, knowledge, or skill as most valued in a beginning carpenter. Of the 42, 18 (almost half) had to do with attitude and personal relationships, 14 with skills in carpentry, and 10 with knowledge of carpentry. Other qualities were rated, but these were most frequently mentioned.

Critical thinking skills are also important to employers. Critical thinking skills allow you to make decisions and solve problems. These higher-level skills enable you to think beyond the obvious. You learn to interpret information and make judgments. Supervisors appreciate employees who can analyze problems and suggest workable solutions.

Computer and technology skills are also important. Many types of computer applications are used in the construction industry, including the following:

- Drafters use CAD software to create building plans.
- Estimators use spreadsheet software and other programs to prepare estimates.
- Construction managers use scheduling software to keep track of progress on projects.
- Architects use word processing software to create specification documents.
- Carpenters use the internet to find product installation information.
- Carpenters use mobile apps for calculation tasks, such as measurement conversions, rafter layout, and stair layout.

More digital tools are constantly being developed and used in the field. Throughout your working career, you will need to continue to learn about and apply the latest technology.

1.7 Conduct on the Job

As mentioned earlier, maintaining a proper relationship with others on the building project is important for a construction worker. It avoids potential problems in meeting production schedules and needless friction between workers.

Federal law protects workers from various forms of discrimination and abuse in the workplace. *Discrimination* exists when a person or group of people is treated differently because of race, age, religion, sexual orientation, or gender. Sexual harassment is considered a type of discrimination. *Sexual harassment* is any unwelcome sexual advances or requests for sexual favors. It also includes any verbal or physical conduct of a sexual nature under specific conditions.

Appropriate jobsite behavior includes avoiding practical jokes or horseplay that could cause injury. Avoid distracting other workers with casual conversation—except during breaks and lunch periods, communications should be brief and related to the job at hand.

Stress can distract you from your task and cause a hazardous situation. If you are not concentrating on your work, you are more likely to create an error that could require rework or, more importantly, could create a dangerous situation for yourself or a coworker.

Never work while under the influence of alcohol or drugs. Doing so makes you a danger to yourself and others and will likely result in your immediate termination.

On the jobsite, if you witness unacceptable behavior—such as sexual harassment or distraction due to stress or substance abuse—by a coworker, you must report it to your supervisor. In these cases, your coworker is putting everyone on the jobsite at risk. If you do not report your coworker and an accident results from the unacceptable behavior, you will regret your inaction.

1.8 Personal Qualifications

As in other occupations, carpentry requires a level of knowledge and skill that allows the worker to produce quality work in a timely manner.

To become a successful carpenter, you must have certain characteristics:

- Be physically able to do the work.
- Display *manual dexterity*. This means being skilled with your hands and having a talent for working with the tools and materials of the trade.
- Have a sincere interest and enthusiasm that will intensify your efforts as you study and practice the skills and know-how required.
- Possess certain character traits. Honesty in all of your dealings is very important, especially in the quality and quantity of work performed. You must show courtesy, respect, and loyalty to those with whom you work. Punctuality and reliability reflect your general attitude and are important not only during your training program, but later when you enter regular employment.

Appropriate personal conduct on the job site is important. The ability to cooperate with others—students, coworkers, supervisors, employers, vendors, and customers—is essential to success. Most people who fail in the carpentry trade do so because of a weakness in personal characteristics, not because of their skill level.

Safety Note
The hazards associated with carpentry require that you develop a good attitude toward safety. This means that you must be willing to spend time learning the safest way to do your work. You must be willing to follow safety rules and regulations at all times.

Elena Elisseeva/Shutterstock.com

Figure 1-9. Carpenters must keep up to date with the latest tools, materials, and techniques. This worker is installing a solar panel on a residence.

Even after you complete your training program, you must continue to improve your skills and adjust to new methods and techniques, **Figure 1-9**. Each day brings new materials and improved procedures to the construction industry. This presents a special challenge to those in the carpentry trade. You should read and study new books and manufacturer literature, along with trade journals and magazines in the building construction field. Much information can be obtained at association meetings and conventions where new products are exhibited. To be a successful carpenter, you also need to keep informed on code changes, new zoning ordinances, safety regulations, and other aspects of construction work that apply to the local community.

For those who have ability and are willing to work, the field of carpentry is a satisfying, fulfilling, and well-paying occupation. Advancement is determined by your willingness to develop new skills and to seize opportunities as they present themselves.

1.9 Entrepreneurship

Earlier in this chapter, you read that four out of every ten carpenters are self-employed. In other words, they are *entrepreneurs*. They operate a business of their own.

There are advantages and disadvantages to operating your own construction company. First, you may find that you gain more satisfaction in certain carpentry tasks. As your own boss, you can choose to specialize in these areas and thus offer your services to the community as a subcontractor. You are free to concentrate in the area of your specialty and develop superior skills and competence that allow you to compete with other firms.

Of course, there are disadvantages and risks to owning a company. Failure is possible and you can spend years repaying loans and debts incurred. Your income could be uncertain, depending on how well the business is doing or how healthy the economy is. You will have to work longer hours and many of the tasks—paperwork, maintenance, customer relations, setting prices, and organizing your day—are difficult.

There is also the matter of *liability*. An employer is legally responsible, or liable, for the safety of all the people in the company. The employer pays for insurance in case something goes wrong in the workplace involving an employee or anyone else present. As an employee of a company, you are covered under the company's insurance plan. When you work for yourself, the entire burden of liability—for you and everyone on your crew— falls on you.

An entrepreneur must possess certain characteristics to be successful. This list summarizes the major characteristics:

- Healthy—Long hours and physical labor make heavy demands on a carpenter or construction contractor.
- Knowledgeable—To make a profit, he or she must know all aspects of the trade or trade specialty. In addition, the person must understand the industry and the products being used in the trade.
- Good planner—Running a successful business means that nothing is left to chance. He or she must be able to foresee difficulties as well as plan how to take advantage of opportunities.
- Willing to take calculated risks—Once a plan has been conceived that takes into account those events likely to occur, the person must have the courage to risk money, time, and resources on making the plan work.
- Organized—Profits in the future often depend on the records kept in the past. Organization of receipts, invoices, tax records, and more is critical to maintaining a successful business.
- Innovative—Successful entrepreneurs are always finding ways to improve and produce better work, thus gaining the confidence of customers.

- Responsible—Responsibility means being willing to accept the consequences of a decision, whether good or bad. This includes paying debts, keeping promises, and accepting the responsibility for mistakes of his or her employees.
- Goal-oriented—In order for a business to reach its goal, he or she must set goals and work hard to achieve them.
- Forward-thinking—In recent years, there has been a large population increase of people over the age of 60. This population group demands many services. As a result, accessible, low-maintenance housing is one of the areas predicted to be in great demand. Thus, a business designed to offer services such as home construction or rehabilitation, or rehabbing, of homes may do well.

1.10 Teaching as a Construction Career

It is not unusual for successful and skilled carpenters to become teachers. For some, this is a part-time occupation. Others may leave the construction field to teach their skills to others. This may involve returning to school to take courses leading to certification or getting a college degree in education. There are also some who, as students, take construction courses to become proficient as preparation for teaching career and technical courses.

Carpentry instructors not only work in high schools, technical schools, and universities, but may be employed by tool and equipment manufacturers. In such cases, they may visit educational institutions to instruct other teachers and trainees on how to use the products manufactured by their employers, or engage in demonstrations at trade shows such as the International Builders Show, or JLC Live. See **Figure 1-10**.

Additionally, some carpenters teach their trade through writing. Magazines, trade journals, and technical writing are areas where tradesmen who have the ability to write can combine their skills and experience to fill this need in the industry.

1.11 Career Advancement

Working with your hands as a carpenter is a rewarding occupation that many choose for a life-long career. However, if a change in pay or occupation is desired, potential for advancement is broad. A carpenter's outlook for career advancement can depend on many factors and can lead to a lucrative career.

A successful entrepreneur can build their enterprise into a thriving business. If managed well, it can evolve into a large corporation. In this case, you may not only make a career and a good living for yourself, but you can also provide a rewarding career to many others.

On the other hand, a carpenter working for a large construction firm can prove their worth and management skills to make their way up the chain of command and become a foreman or superintendent, running multimillion dollar jobs.

A   *Pennsylvania College of Technology*   B   *Simpson Strong-tie*

Figure 1-10. Teaching as a profession. A—Teaching construction to beginners is a rewarding profession. B—A representative of a company conducts training on the proper use of hangers and connectors.

Many companies have office personnel who specialize in specific areas, such as estimators, customer service representatives, project managers, schedulers, and health and safety officers. Carpenters are a good fit for these positions because they have experience and understanding that spans across other trades. Many of these careers offer rewarding work and good pay.

For those who choose to continue their education, carpentry is a trade that pairs well with construction-related college degrees, such as architecture, construction management, or civil engineering. Understanding challenges that occur in the field as a builder is an enormous advantage in these careers. All of these options and many more are potential career paths for an aspiring carpenter.

1.12 Carpentry-Related Occupations

Construction offers a number of other occupations with skills closely allied to carpentry. Opportunities and training are also similar. The following sections discuss several occupations that are related to carpentry.

1.12.1 Bricklayers and Stonemasons

Bricklayers, or brickmasons, work with manufactured masonry units. They construct walls, floors, partitions, chimneys, and patios. A beginner may work as a hod carrier (helper) while learning the trade, **Figure 1-11**.

In addition to bringing bricks and other materials to the bricklayer, a hod carrier is expected to mix mortar and set up a scaffold.

Stonemasons construct walls out of stone, set stone for veneer walls on buildings, and lay stone floors. They work with natural stone, such as marble, granite, and limestone, as well as with artificial stone made with concrete, marble chips, and other masonry materials. The mason often works with a set of drawings on which each stone is numbered. Helpers may be employed to locate and carry these stones to the mason.

Employment in these occupations is expected to be excellent. Job openings are growing faster than the number of persons in training.

1.12.2 Electricians

Electricians normally work from prints as they install, connect, test, and maintain electrical wiring, **Figure 1-12**. Much of their work is in residential construction. However, they also work in factories, office buildings, or exterior wiring installations.

Electric power companies employ some electricians who work mostly as linemen. Their work is always outdoors, dealing with high-voltage electricity. The most common task of a lineman is installing and maintaining electrical service to customers.

Safety Note
Being an electrician is often strenuous, sometimes involving work on scaffolds and ladders. They are also at risk from electrical shock. To avoid injury, they must carefully work, deliberately move about, and follow strict safety procedures.

Goodheart-Willcox Publisher

Figure 1-11. Beginners in the masonry trade work with a journeyman as a helper (called a hod carrier) while learning bricklaying or stonemason skills.

Lisa F. Young/Shutterstock.com

Figure 1-12. Electricians must be highly skilled in electrical theory. They must also keep up-to-date with the National Electrical Code.

1.12.3 Plumbers

Plumbers are among the most highly paid construction trades workers. Employment opportunities are good for the foreseeable future. The work requires some heavy lifting, much standing, and working in close quarters. Clearing clogged pipes is often disagreeable and can be messy.

In general, residential plumbers install and repair water, waste disposal, drainage, and gas systems. They also install fixtures—sinks, bathtubs, showers, and toilets—and appliances such as dishwashers and water heaters. Some specialize in installing sewers, septic tanks, and drain fields.

1.12.4 HVACR Technicians

Heating, ventilating, air conditioning, and refrigeration systems involve many mechanical, electrical, and electronic components. See **Figure 1-13**. Among these are electric motors, pumps, compressors, fans, ductwork, pipes, thermostats, and switches.

Though trained to do both installation and maintenance/repair, HVACR technicians usually specialize in one or the other. Some may narrow their specialization further to one type of equipment.

Furnace installers, or heating equipment technicians, install gas, oil, electric, solid-fuel, or multiple-fuel systems. They follow prints or manufacturers' specifications. After placing the units, they install fuel and water supply pipes, ductwork, vents, pumps, and other parts of the system.

Air conditioning technicians, like furnace installers, install and service central air conditioning systems. They must be able to follow manufacturers' instruction manuals, as well as read prints and specifications. After installing all lines and other parts, they charge the system with refrigerant, program the controls, and test for proper operation.

There are many job opportunities in this field. The heating and air conditioning service industry is rapidly expanding. Demand for well-trained technicians should be better than average for all occupations for a number of years. Specialists in service will be in the best position for secure employment.

1.13 Organizations Promoting Construction Training

Various organizations promote development of excellence in construction. One of them, *SkillsUSA*, holds state and national competitions to encourage students in a variety of occupations, including building trades, **Figure 1-14**.

In the carpentry trade, associations actively promote and support training in construction fields. Two of the

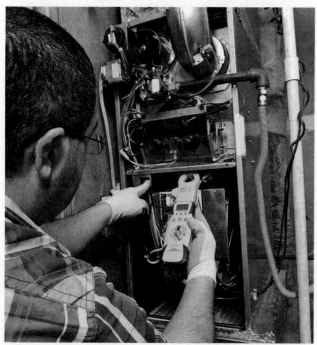

Geo Martinez/Shutterstock.com

Figure 1-13. HVAC systems require periodic service—far more than other building systems require. HVAC technicians often specialize in either system installation or service.

Goodheart-Willcox Publisher

Figure 1-14. Carpentry trainees from all states compete at the SkillsUSA national contest.

most influential are the Associated General Contractors (AGC) and the National Association of Home Builders (NAHB). The major emphasis of the AGC is to represent contractors who do heavy construction work. The NAHB concentrates its efforts on residential and light commercial construction.

The United Brotherhood of Carpenters and Joiners of America looks after the interests of construction workers. While most members are carpenters, a wide spectrum of construction workers also belong. Membership includes cabinetmakers, pile drivers, millwrights, floor covering workers, and a host of industrial workers employed by factories turning out plywood, lumber, and other construction products. Information about related occupations is available from several organizations or unions.

Masonry trade

- International Union of Bricklayers and Allied Craftsmen
- Portland Cement Association

Electrical trade

- Independent Electrical Contractors, Inc.
- National Electrical Contractors Association (NECA)
- International Brotherhood of Electrical Workers (IBEW)

Plumbing trade

- Plumbing-Heating-Cooling Contractors National Association
- Mechanical Contractors Association of America

HVACR

- The Sheet Metal and Air Conditioning Contractors National Association
- Air Conditioning Contractors of America
- American Society of Heating, Refrigerating, and Air Conditioning Engineers (ASHRAE)
- Plumbing-Heating-Cooling Contractors National Association

Construction Careers
Construction Manager

Sometimes called "constructors," construction managers are in overall charge of planning and coordinating a construction project. The construction manager may work for the general contractor or may be the representative of a separate management firm hired by the developer or owner to oversee the entire project. There are also some self-employed construction managers.

Minerva Studio/Shutterstock.com

Construction managers plan, coordinate, and oversee all aspects of a construction project.

A construction manager's duties usually span the entire life of the project, from conceptual development to the building's completion. Working from the project design documents, they oversee all aspects of organization, scheduling, and implementation. Very large and complex projects are beyond the managerial abilities of a single person. In such situations, an overall management firm places individual construction managers in charge of specific aspects of the job, such as site preparation, steel erection, concrete work, mechanical systems, plumbing, HVAC, and so on.

Organizational ability, math and computer skills, and a thorough knowledge of construction processes are vital to success as a construction manager. Managers are usually responsible for creating and adhering to schedules, estimating and projecting costs, and scheduling the work of the various tradespeople. Communication skills are also important. Construction managers must be able to effectively communicate with everyone involved in the project, from the building owner to subcontractors and their workers. They must also communicate with materials suppliers and government representatives, such as building inspectors.

The traditional route to construction management positions led through the building trades, since the work requires solid experience in various aspects of construction. Increasingly, management firms are seeking candidates who combine such experience with a four-year college degree in construction management or civil engineering. Computer software experience in such areas as CAD, cost estimating, and project management is highly desirable. Certification in this field is becoming more common. Several organizations offer this professional recognition based on education, experience, and a written exam.

Advancement opportunities are best with large management firms, since a construction manager will gain experience and eventually be able to assume responsibility for larger and more complex projects. Upper-level managers in such firms may have broad responsibility for a very large project, supervising the work of a number of construction managers.

Chapter Review

Summary

- There are many employment opportunities for carpenters in residential, commercial, and heavy construction. With continued population growth and demand for new housing, the number of carpentry jobs should continue to grow.

- In addition to new construction, carpenters are employed in remodeling and renovating work, prefabrication of buildings, and related construction fields.

- Entry into the carpentry field is often through an apprenticeship program, typically taking four to five years. When the apprenticeship period is completed, the carpenter has journeyman status.

- Personal qualifications for carpentry work include manual dexterity, the ability to physically do the work, sound work habits, and good interpersonal skills.

- Many carpenters are entrepreneurs, operating their own businesses. Numerous additional career paths and advancements are available to carpenters. These opportunities include teaching or moving to an administrative, supervisory, or management role. Carpenters may also shift specialties and move into other building trades areas, such as electrical or plumbing work.

- Organizations such as SkillsUSA, contractors' associations, and labor unions actively promote the carpentry trade.

Know and Understand

Answer the following questions using the information in this chapter.

1. A trained carpenter would *not* be expected to _____.
   A. lay out building lines
   B. install a septic system
   C. insulate a home
   D. install cabinetry

2. More than _____ people are employed as carpenters in the United States.
   A. 50,000
   B. 650,000
   C. 1,000,000
   D. 3,500,000

3. Approximately _____ of all carpenters work in residential construction.
   A. 15%
   B. 30%
   C. 50%
   D. 75%

4. *True or False?* Entrepreneurs are guaranteed to succeed because they can set their own prices.

5. *True or False?* For high school students considering a career in carpentry, English is not an important class.

6. An apprenticeship in carpentry normally covers _____ years.
   A. 4–5
   B. 5–7
   C. 1–3
   D. 2–3

7. Today, a journeyman is considered to be _____.
   A. a tradesman who is in an official training program
   B. a carpenter who is working from home
   C. someone who has been trained to work at heights
   D. a tradesman who is capable of performing all tasks without supervision

8. Which of the following is *not* a quality that construction contractors most value in beginning carpenters?
   A. Affective behavior.
   B. Psychosocial skills.
   C. Psychomotor skills.
   D. Cognitive behavior.

9. In the coming years, demand for carpenters is expected to _____.
   A. increase
   B. decrease
   C. remain the same
   D. phase out completely

10. A trained carpenter may be qualified for which of the following career paths?

    A. Self-employment
    B. Teaching
    C. Estimator
    D. All of the above.

11. *True or False?* Personal conduct on the job is not important if you have the physical skill to get the job done quickly.

12. A(n) _____ is someone who operates their own business.

    A. journeyman
    B. apprentice
    C. entrepreneur
    D. All of the above.

13. SkillsUSA, Associated General Contractors, and the National Association of Homebuilders are _____.

    A. organizations that support and promote construction training
    B. organizations that loan money to construction entrepreneurs
    C. trade schools
    D. All of the above.

Apply and Analyze

1. Define *sexual harassment*.
2. List three uses for computers in the construction industry.
3. Why is it important to maintain a proper relationship with your coworkers?
4. List some of the roles, activities, or tasks performed by a carpenter in the construction process.

Critical Thinking

1. Perform an internet search for carpentry jobs. Sort through the job listings until you find one with 10–15 qualifications. These may also be listed as job duties, experience, or job qualifications. Categorize these qualifications as *affective* (executive function, soft skills), *cognitive* (knowledge, mental ability), or *psychomotor* (physical skills). Looking over your list, what skills do you believe are the most important for an aspiring carpenter to develop? Why?

2. Find out how to enter a carpentry apprenticeship program in your locality. Contact a carpentry contractor, union local, and state apprenticeship agency. What skills do you have now? What skills must you learn? Report your findings to the class.

3. Secure literature from the main associations in construction. Report what they say about opportunities in construction.

4. Find job postings or help-wanted ads for five jobs related to the construction field.

Communicating about Carpentry

1. **Speaking and Listening.** Interview a carpenter to learn their attitudes about what they do for a living. Prepare a list of possible questions to ask before you conduct the interview.

2. **Speaking and Listening.** Interview a carpenter. Ask the person to describe a typical day at work. Here are some of the questions you might ask:

    A. What is the work environment like?
    B. What are the job duties?
    C. What is your favorite part about being a carpenter? Least favorite?
    D. What kinds of injuries are you at risk for?
    E. What other types of professionals do you work with?

3. **Reading and Writing.** Create an informational pamphlet on the different opportunities available in the carpentry field. Describe how much time is required to become employed full-time in these positions. Include images in your pamphlet. Present your pamphlet to the class.

4. **Speaking and Listening.** Working with a partner, role-play the following scenario: you are a carpenter who has just observed a coworker under the influence of drugs or alcohol. One student acts as the carpenter and reports the unacceptable behavior to the supervisor. The other student acts as the carpenter's supervisor. Role-play in front of the class and ask the class for feedback on the communication.

CHAPTER 2

Safety

OBJECTIVES

After studying this chapter, you will be able to:
- Explain what OSHA is and its purpose.
- Explain housekeeping measures that promote safe working conditions.
- List and describe clothing safety as it applies to carpenters.
- List other personal protective equipment recommended for carpenters to use.
- List safety measures relating to shoring and scaffolding.
- Cite safety measures relating to hand and power tools.
- Explain how electrical power is used safely on a jobsite.
- Describe proper methods of lifting and carrying to avoid personal injury.
- Describe the classes of fires.

TECHNICAL TERMS

angle of repose	ground-fault circuit interrupter (GFCI)	pneumatic tool
Class A fire	hard hat	pressure-treated lumber
Class B fire	hearing protection	respirator
Class C fire	hot conductor	safety boots
Class D fire	neutral conductor	Safety Data Sheet (SDS)
Class K fire	Occupational Safety and Health Administration (OSHA)	safety factor
competent person		safety glasses
dust mask	OSHA standards	shoring
electric shock		trench
ground		trench box

Carpenters work with tools and materials that can cause serious injury if not maintained properly and used correctly. The carpenter's workplace—a construction site—can be a dangerous place if safety precautions are not observed. According to recent Occupational Safety and Health Administration (OSHA) statistics, 20.7% of annual fatalities in private industry were construction related. The leading causes of construction worker deaths were falls, electrocution, struck by object, and being caught between two solid, fixed objects. These "Fatal Four" as OSHA calls them, were responsible for 59.9% of construction worker deaths. Eliminating these factors would save hundreds of workers' lives every year.

2.1 Occupational Safety and Health Administration

Every worker in any industry has a right to a safe and healthful workplace. That is why, in 1970, Congress created the ***Occupational Safety and Health Administration (OSHA)*** to help protect workers from incurring injury on the job. The organization enforces something called the OSHA General Duty Clause, requiring employers to provide a place of employment that is free from recognized hazards that could cause physical harm. The clause also requires employees to comply with all safety standards, rules, and regulations.

OSHA protects workers by the following actions:

- Setting standards for workplace safety.
- Providing information and training about safety and health.
- Providing training to employers and workers for best safety practices.
- Inspecting workplaces.

If you work for a private company, you are covered by an OSHA regional office. The Occupational Safety and Health Act, which created OSHA, authorizes states to establish their own safety and health programs, subject to OSHA approval, **Figure 2-1**. Some states write their own standards. Others adopt the federal OSHA standards. *OSHA standards* are rules that describe the methods that employers must use to protect their employees from hazards. OSHA 1926 CFR Subpart C contains the basic requirements of a safety and health program, and defines terms used in construction standards.

The Occupational Safety and Health Act grants workers important rights. Workers play a vital role in identifying and correcting problems at their workplace. Most employers are quick to correct problems when they are called to their attention. If a worker feels that his or her employer is not correcting a hazardous condition, he has the right to file a complaint with OSHA.

2.1.1 Employers' Responsibilities under OSHA

Employers have the responsibility to provide a safe workplace following certain guidelines.

- Provide their workers with a workplace that does not have serious hazards.
- Follow all OSHA safety and health standards.
- Find and correct safety and health problems.

OSHA further requires that employers must first try to eliminate or reduce hazards by making feasible changes in working conditions rather than merely relying on personal protective equipment such as masks, gloves, or earplugs. Switching to safer chemicals, enclosing processes to trap harmful fumes, or using ventilation systems to clean the air are examples of effective ways to eliminate or reduce risks.

Many OSHA standards require an employer to provide a competent person to ensure safety standards are being met. This *competent person* should be capable of identifying existing and predictable hazards in the work area surroundings and identifying working conditions that are unsanitary, hazardous, or dangerous to employees. This competent person also must have the authority to correct or eliminate safety hazards.

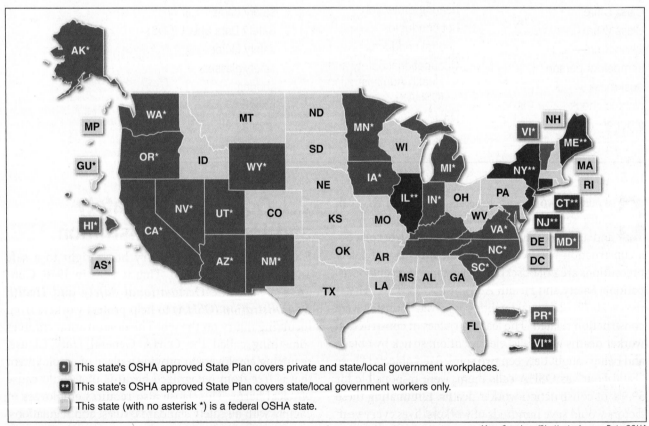

This state's OSHA approved State Plan covers private and state/local government workplaces.

This state's OSHA approved State Plan covers state/local government workers only.

This state (with no asterisk *) is a federal OSHA state.

Map: CarryLove/Shuttestock.com; Data:OSHA

Figure 2-1. Twenty-eight states plus Puerto Rico and the Virgin Islands have OSHA approved programs.

2.1.2 Workers' Responsibilities under OSHA

Workers are expected to comply with all safety and health standards that apply to them on the job. The following lists workers' responsibilities:

- Become familiar with the OSHA poster, **Figure 2-2**, that every employer is required to display

- Follow employer's rules regarding safety and health and use all required safety gear
- Follow safe work practices as required by their employer
- Report all safety hazards to their supervisor or their employer's safety committee

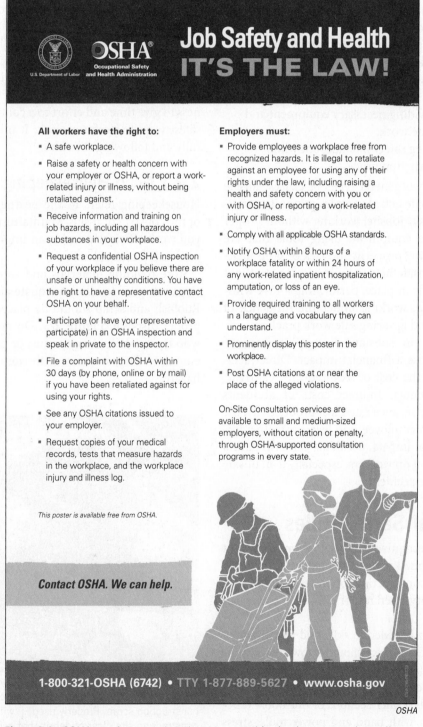

Figure 2-2. OSHA requires every employer to post this sign in an area where all employees can read it.

2.1.3 Safety Culture

Together, employers and employees create a safety culture. A safety culture is the attitude and behaviors on the jobsite with respect to safety issues. A good safety culture is critical to preventing and avoiding injuries in the workplace. Some characteristics of a jobsite with a good safety culture include the following:

- Workers constantly watching for potential hazards (hazard recognition), assessing potential hazards (risk assessment), and taking action to eliminate potential hazards.
- Workers using proper safety equipment and performing tasks in a safe manner.
- Employers providing safety training throughout the duration of the project.
- Employers providing necessary equipment and materials for safe work.
- Employers taking safety into account when planning construction processes.

Accidents have two general types of causes: unsafe conditions and unsafe acts. Examples of unsafe conditions include a messy jobsite, working without proper personal protective equipment, using tools that are worn or not operating properly, and lack of signs identifying hazards. Unsafe acts include working without proper safety guards in place, performing a task without proper training, working under the influence of drugs or alcohol, and ignoring safe work practices.

Accidents result in substantial negative impacts. First, accidents cause a financial impact. Direct accident costs include the cost of required medical treatments and settlements. Indirect costs of accidents include increased insurance rates, lost productivity, the cost of replacing an employee, and legal costs. In addition to the financial impact, accidents can also have a demoralizing impact on workers, especially if an unsafe condition led to the accident.

2.2 General Safety Rules

Good carpenters recognize that safety is an important part of the job. They know that accidents can occur easily in building construction, often resulting in partial or total disability. Even minor cuts and bruises can be painful.

Safety is based on knowledge, skill, and an attitude of care and concern. Carpenters should know correct procedures for performing the work. They should also be familiar with the potential hazards and how to minimize or eliminate hazards. Other sections of this book, especially those dealing with hand and power tools, stress proper safety rules. Read and follow these safety rules.

A good attitude toward safety is important. This includes belief in the importance of safety and willingness to give time and effort to a continuous study of the safest ways to perform work. It means working carefully and following the rules.

2.2.1 Good Housekeeping

Housekeeping refers to the neatness and good order of the construction site. Maintaining a clean site helps you work efficiently and is an important factor in the prevention of accidents.

Store building materials and supplies in neat piles. Organize piles to allow adequate aisles and walkways. Rubbish and scrap should be placed in containers for proper disposal, **Figure 2-3**. Do not permit blocks of wood, nails, bolts, empty cans, or pieces of wire to accumulate. They interfere with work and are a tripping hazard.

Gena Melendrez/Shutterstock.com

Figure 2-3. An on-site dumpster provides safe disposal for construction scrap. Properly used, it will help to keep the construction site clear of debris that could cause accidents and injuries.

Keep tools and equipment in panels or chests when they are not being used. This provides protection for the tools, as well as for workers on the jobsite. In addition to improving efficiency and safety, good housekeeping helps maintain a better appearance at the construction project. This will, in turn, contribute to the good morale of all workers.

2.3 Clothing

Wear clothing appropriate for the work and weather conditions. Wear rain gear in wet weather, winter weight clothing in cold weather, and cool clothing in hot weather. Working when you are uncomfortable or numb from the cold detracts from your alertness and is an invitation to accidents. It is equally unsafe to work in open-toed shoes or clothing that does not protect you from the work environment. For example, shorts and short-sleeve shirts do not provide adequate protection when welding.

Trousers or overalls should fit properly and have legs without cuffs. Avoid loose-fitting or ragged clothing. They can catch on nails or pull hands or other parts of the body into cutting tools or moving machinery. Keep shirts and jackets buttoned. Sleeves should also be buttoned or rolled up. All clothing should be maintained in a good state of repair and washed when dirty.

Shoes should be sturdy and have thick soles to protect feet from protruding nails. Never wear shoes with leather soles such as tennis or lightweight canvas shoes. They will not provide satisfactory traction on smooth wood surfaces or on sloping roofs. When you work at a site where heavy objects can be dropped on your feet, you should wear safety shoes.

OSHA requires protective headgear to be worn, especially if there is work being performed overhead.

Headgear should provide the necessary protection, be comfortable, permit good visibility, and shade your eyes.

2.4 Personal Protective Equipment

Safety glasses are to be worn whenever work involves even the slightest potential hazard to your eyes, **Figure 2-4.** American National Standards Institute (ANSI) sets many of the standards that have been adopted by industry. Safety glasses that satisfy OSHA requirements are stamped with ANSI standard number Z87.1-2003. Clean your safety glasses as needed to remove dirt, dust, and other debris that may reduce your vision. Store safety glasses in a bag, box, or case when not in use. If lenses become scratched or pitted or if hinges become loose, replace the safety glasses.

Safety boots and shoes are required on heavy construction jobs. OSHA standards require protective footwear in work areas where foot injuries could occur from falling or rolling objects, sharp objects, or electrical hazards. This footwear consists of special reinforced toes that will withstand a load of 2500 lb.

A *hard hat* is to be worn whenever you are exposed to any possibility of falling objects. Many construction sites require all persons on the site to wear a hard hat at all times.

gpointstudio/iStock/Thinkstock

Figure 2-4. Safety glasses must meet the specifications of ANSI Z87.1-2003.

Jon Rehg/Shutterstock.com

Figure 2-5. Hard hats are required wherever there is danger from falling objects. In cold weather, a winter liner can be worn under the hard hat.

Since carpenters must often work outside during cold weather, warmer protective headgear such as winter liners should be available. See **Figure 2-5**. Standard specifications require that hard hats withstand a certain degree of denting. They must be able to resist breaking when struck with an 80 lb ball dropped from a 5′ height.

A *dust mask*, or particulate mask, is a disposable filter-type mask worn over the nose and mouth to protect the worker from inhaling dust and other fine particles in the air, **Figure 2-6**. A particulate mask is not a respirator and is not adequate protection from contaminants that are immediately hazardous to your health.

Prior to 1978, paint often contained lead. The dust created when sanding lead paint is hazardous. Asbestos, once common for insulating pipes and as exterior siding, is also hazardous. A *respirator* is required when removing paint (including non-lead paint), working with asbestos-containing materials, or spraying finishing materials, **Figure 2-7**. Many other instances require a respirator. Employers are responsible for developing and implementing a respiratory protection program if respirator use is required.

Hearing protection should be used whenever working in the vicinity of loud equipment. Approximately one out of ten people have suffered some hearing loss, and exposure to loud noise is the main cause. According to the Environmental Protection Agency (EPA), continued exposure to 70 decibels of noise (the noise of a freeway 50′ from the pavement edge) can cause hearing loss. The OSHA standards for hearing protection require hearing protection when the noise level is at or above 85 decibels (approximately the noise level of a food blender) for 8 hours. The best way to protect your hearing is to wear hearing protection whenever you are exposed to loud noise. Hearing protection can be foam earplugs or, for greater protection, earmuffs.

Finally, hand protection is also important. Hand injuries directly affect a carpenter's ability to work, so proper protection is vital. Cuts, nicks, and scrapes are not only painful, but could also lead to infection or reduce the ability of a worker's hands to use tools. Wear gloves of an appropriate type when handling rough materials.

Photographee.eu/Shutterstock.com

Figure 2-6. A particulate mask filters out dust.

il21/Shutterstock.com

Figure 2-7. An approved respirator is required for certain jobs.

2.5 Fall Protection

In the construction industry in the United States, falls are the leading cause of worker fatalities. In recent years, 300–400 construction workers died annually as a result of a fall. OSHA requires all contractors to provide fall protection for anyone who is working within 6′ of the edge of a surface that is more than 6′ from the ground or a stable surface. This protection may be in the form of a fall-arrest harness, **Figure 2-8**, or guardrail, **Figure 2-9**.

M2020/Shutterstock.com

Figure 2-8. This worker wears a harness that allows full range of movement. A clip on the harness attaches to an anchorage point, which must be capable of supporting the impact of a fall.

An acceptable guardrail must have a top rail strong enough to support a 200 lb load and be 42″ high, plus or minus 3″. A midrail should be located at half that distance. In addition, floor openings larger than 2′ square must be covered with material that can safely support the working load. Roof openings, including skylights, must be covered securely or protected by guardrails.

2.6 Scaffolds and Ladders

The variety of scaffolds used on construction sites is vast. OSHA standards address different classifications of scaffolds individually; however, many of the standards are universal. Scaffolds must have a minimum *safety factor* of four. This means that the scaffold is capable of carrying a load four times greater than its load rating. Inspections must be made daily before the scaffold is used.

Ladders must be inspected for damage at frequent and regular intervals. Their use should be limited to climbing from one level to another. The ground where the ladder is set must be firm and even to prevent the ladder from wobbling. Working while being supported on a ladder is hazardous and should be kept to a minimum.

Take special care to protect against possible electrocution that can result from contact with overhead power lines while erecting, moving, or working from metal or conductive ladders and scaffolds. Avoid tasks that require excessive force while on a ladder.

Joe Gough/Shutterstock.com

Figure 2-9. This guardrail system protects workers at this level from falling over the edge.

When a worker is applying great force to a task, he is likely to be off balance and a small slip can cause a fall. There are many more safety rules that must be observed in the use of scaffolds and ladders. Refer to Chapter 6, **Scaffolds, Ladders, and Rigging**, for more specific rules.

2.7 Hand Tools

Always select the correct type and size of tool for your work. Be sure it is sharp and properly adjusted. Guard against using any tool if the handle is loose or in poor condition. Dull tools are hazardous to use because additional force must be applied to make them cut. Oil or dirt on a tool may cause it to slip, resulting in injury.

Here are a few rules to follow regarding hand tools:

- Hold hand tools correctly. Most edge tools should be held in both hands with the cutting action away from your body.
- Be careful when using your hand or fingers as a guide to start a cut with a handsaw.
- Keep edged and pointed tools turned downward.
- Carry only a few tools at one time unless they are mounted in a special holder, **Figure 2-10**.
- Do not carry sharp tools in pockets of your clothing.
- When not in use, tools should be kept in special boxes, chests, or cabinets.

TFoxFoto/Shutterstock.com

Figure 2-10. A tool belt is a safe, handy method of keeping hand tools close at hand.

2.8 Power Tools

Before operating any power tool or machine, you must be thoroughly familiar with the way it works and the correct procedures to follow. Read the directions in the owner's manual, if there is one. Make sure you understand them. Be alert and, above all, use common sense as you work. In general, when you learn to use equipment the correct way, you also learn to use it the safe way, **Figure 2-11**. If the tool you are using has a power cord, check it to make sure there are no breaks in the insulation. Power cords must *not* be taped or otherwise repaired. If the cord is damaged, it should be replaced.

Safety Note

There are a number of general safety rules that apply to power equipment. In addition, special safety rules must be observed in the operation of each individual tool or machine. Those that apply directly to the power tools commonly used for carpentry are listed in Chapter 5, **Power Tools**. Study the rules and follow them carefully.

Goodheart-Willcox Publisher

Figure 2-11. Using a power tool the safe way means wearing safety glasses and making sure all guards are in place and working properly. Do not use a corded tool with a damaged cord.

2.9 Electrical Power

Electrical power is almost always essential on a construction site. Its source may be an electrical generator or a temporary service. Safe use of this power is important—even a small amount of electrical current is capable of causing serious injury or death.

On the jobsite, electrical power travels through power cords. From its source, the current travels through one conductor, known as the *hot conductor*, to the tool. It then returns to the source through a second conductor, known as the *neutral conductor*. Insulation on hot conductors is always either red or black. Insulation on neutral conductors is always white. A third conductor in the power cord is known as the ground conductor, or *ground*, and usually has green insulation or is a bare copper wire. Its purpose is to safely carry away the current from accidental grounding. Chapter 5, **Power Tools**, explains this in more detail.

Moisture can turn many materials, including soil, into conductors of electrical current. A worker can become a part of the current-conducting loop by coming in contact with both current-carrying conductors or with the hot conductor and the ground. When this happens the worker receives an *electric shock*, a discharge of electrical current passing through the body. A person can experience different levels of electric shock depending on the amount of current flowing throw the discharge, **Figure 2-12**.

Power tools made of metal are insulated against accidental grounding. This reduces the danger, but does not eliminate it. Make sure that the tool you are using is grounded. The electrical system should also be checked for proper ground. Any break in the grounding wire makes the grounding system inoperative. When the ground conductor is broken or missing, the worker may become the easiest path for electrical current to take during a tool malfunction, causing electric shock or electrocution. The integrity of a grounding wire of a power cord can be checked with an ohmmeter or a continuity tester.

Circuit breakers and fuses provide some protection against shock. They are designed to open the circuit if a short should occur. A much more efficient protective device, shown in **Figure 2-13**, is a *ground-fault circuit interrupter (GFCI)*. It detects tiny amounts of current and opens the circuit before shock can occur, more rapidly than a fuse or circuit breaker. A GFCI device can be

A
Goodheart-Willcox Publisher

B
Leviton

Figure 2-13. A GFCI provides protection against electrical shock. A—Temporary power on the construction site is required to be GFCI-protected. B—If the outlet is not GFCI-protected, a portable power distribution center can be installed to provide a number of GFCI outlets.

Effects of Electric Shock

Level of Current	Effects on Human Body
1 mA	Threshold of feeling. Slight tingling.
5 mA	Shock felt, but not painful yet. Involuntary muscle movements.
10–20 mA	Painful shock. Sustained muscle contraction. Inability to release grip.
100–300 mA	Paralysis of respiratory muscles. Can be fatal. Severe internal and external burns.
2 A	Cardiac arrest (heartbeat stops). Internal organ damage. Death is probable.

Goodheart-Willcox Publisher; Leviton

Figure 2-12. The effects of an electrical shock can range from very slight tingling to death.

attached to a power cord supplying electricity to a power tool, or can be integrated into a receptacle or circuit breaker. All temporary electrical circuits on a construction site are required by OSHA to be GFCI protected.

2.10 Compressed Air

Some of the tools carpenters use are powered by compressed air. These are called *pneumatic tools*. The air pressure used to power these tools is usually 80–100 pounds per square inch (psi). That is enough pressure to cause particles to puncture human skin and do serious damage if directed at the eyes or ears. Working with pneumatic tools can also cause serious safety hazards from flying objects. Careful attention must always be exercised when working with or around compressed air.

Safety Note
Always wear safety glasses when working with pneumatic tools or any compressed gas. Never direct an air stream at yourself or another person. Do not use compressed air to blow dust off your skin or clothes. Pneumatic tools should be equipped with quick-disconnect fittings, **Figure 2-14**. These fittings allow for fast and easy disconnection of the air supply.

2.11 Decks and Floors

To perform an operation safely, either with hand or power tools, the carpenter should stand on a firm, solid base. The surface should be smooth, but not slippery. Do not attempt to work over rough piles of earth or on stacks of material that are unstable. Whenever possible, stay well away from floor openings, floor edges, and excavations. Where this cannot be done, install adequate guardrails or barricades. In cold weather, remove ice from work surfaces.

2.12 Excavations

A few days before beginning any excavation, the contractor must call 811 and tell the operator where they plan to dig. The 811 operator will notify the utility companies in the area and within a couple of days, utility representatives will come to the site and mark any buried utilities. 811 is operational just about everywhere, but if you work in an area where it is not available, check with the local utility companies before you dig. Excavated soil and rock must be kept at least 2' away from the edge of an excavation. Use ladders or steps to enter trenches that are more than 4' deep.

OSHA defines a *trench* as an excavation that is deeper than it is wide and is less than 15' wide at the bottom. In the United States, an average of two workers per month are killed as a result of trenches collapsing on them. Trenches 5' deep or greater require a protective system unless the excavation is made entirely in stable rock. A protective system may consist of shoring or a device called a trench box. The competent person on the jobsite will be trained to determine what excavations require protection from cave-ins. Shoring and adequate bracing can be placed across the face of an excavation when required. *Shoring* is a method of using braces and blocks to support a heavy load (the face of the excavation in this case). Inspect the excavation and shoring daily, especially after rain. Follow state and local regulations. Alternatively, a trench box can be used to protect workers in a trench, **Figure 2-15**.

Goodheart-Willcox Publisher

Figure 2-14. Pneumatic tools should always be equipped with a quick-disconnect coupling.

Alfred State College

Figure 2–15. A trench box is lowered into place to provide these workers safe access to a fire hydrant tap connection.

A *trench box* is a reinforced steel box that is lowered into an excavation. Workers can enter the trench box with a ladder to perform their work. When used properly the box will keep a trench collapse from crushing or burying the worker.

Never climb into an open trench until approved to do so by the competent person. Always be sure that proper reinforcement against cave-in has been installed or the sides have been sloped to the angle of repose of the material being excavated. *Angle of repose* is the angle the soil naturally assumes when it is unsupported.

2.13 Falling Objects

When working on upper levels of a structure, be cautious handling tools and materials so that there is no chance of them falling on workers below. Do not place tools on the edge of scaffolds, stepladders, windowsills, or on any other surface where they might be knocked off.

If long pieces of lumber must be leaned against the side of the structure, position them on enough of an angle so that they will not fall sideways. When moving through a building under construction, be aware of overhead work, and, wherever possible, avoid passing directly beneath other workers. Stay clear of materials being hoisted. Wear an approved hard hat whenever there is a possibility of falling objects.

Falling objects are one type of "struck-by" hazard. Other examples of struck-by hazards include vehicles being driven on a jobsite, loads being lifted and moved by cranes, walls under construction that may fail, and stored materials that are not properly secured.

2.14 Handling Hazardous Materials

Many of the materials used in building construction can pose a hazard. Some damage your body when inhaled, some enter your body through contact with your skin, and some are flammable or explosive. There are too many such hazardous materials to memorize them all. To know if a material is hazardous, always read the manufacturer's label.

OSHA's Hazard Communication (HazCom) Standard requires employers to notify employees about hazardous materials and provide information. Anyone who will be working with a potentially hazardous material must have access to the *Safety Data Sheet (SDS)* for that material. The SDS (formerly known as Material Safety Data Sheet, or MSDS) describes the hazard that may exist, how to avoid it, and what to do in case of exposure to it. The recommended topics and their order on an SDS are intended to put the information workers are most likely to need first. The following 16 sections are the topics on an SDS.

- Identification
- Hazard(s) identification
- Composition/information on ingredients
- First-aid measures
- Fire-fighting measures
- Accidental release measures
- Handling and storage
- Exposure controls/personal protection
- Physical and chemical properties
- Stability and reactivity
- Toxicological information
- Ecological information
- Disposal considerations
- Transport information
- Regulatory information
- Other information

Pressure-treated lumber, especially wood treated with chromated copper arsenic (CCA) preservative, requires special care in handling for safety of the construction worker. ***Pressure-treated lumber*** has been treated with chemical preservatives to protect it from rot and insects. Although wood treated with CCA has been phased out in favor of less dangerous preservatives, you may still encounter this type of lumber. When it is used, precautions are necessary to protect your health.

- Always be sure to read and follow the manufacturer's safety instructions.
- Avoid prolonged inhalation of sawdust particles.
- Sawing should be done outdoors while wearing a dust mask.
- Wear safety goggles when power sawing or machining.
- To minimize skin contact, wear gloves when handling any pressure-treated materials.
- After cutting or handling pressure-treated lumber, wash any skin that was exposed.
- Wash clothing exposed to sawdust separate from other clothes.

Homeowners should never burn CCA-treated wood or use it as compost or mulch. CCA-treated wood can be disposed of with regular municipal trash (not yard waste). Homeowners should contact the appropriate state and local agencies for further guidance on the disposal of CCA-treated wood. Never burn scraps of treated wood. Burning the wood releases hazardous chemicals into the air. The ash that remains after burning poses a health hazard.

Use great care when spraying paints, stains, or similar materials. Use an approved respirator and protect exposed skin by covering it with clothing. Many paints contain harmful chemicals that can cause health problems if inhaled or absorbed through the skin. Harmful substances such as asbestos, lead, and silica may exist in materials you are working with. Employers are required to provide proper training when working with these materials. See Chapter 23, **Painting, Finishing, and Decorating**, for more information before doing any painting.

2.15 Lifting and Carrying

Improper lifting and carrying of heavy objects may cause injuries. When lifting, stand close to the load, bend your knees, and grasp the object firmly. Then, lift by straightening your legs and keeping your body as close to vertical as possible, **Figure 2-16**. To lower the object, reverse the procedure.

When carrying a heavy load, do not turn or twist your body, but make adjustments in position by shifting your feet. If the load is heavy or bulky, have others help. Never underestimate the weight to be moved or overestimate your own ability. Always seek assistance before attempting to carry long pieces of lumber.

Pro Tip
Stretching exercises are a precaution against muscle strain. Many company safety programs include stretch training.

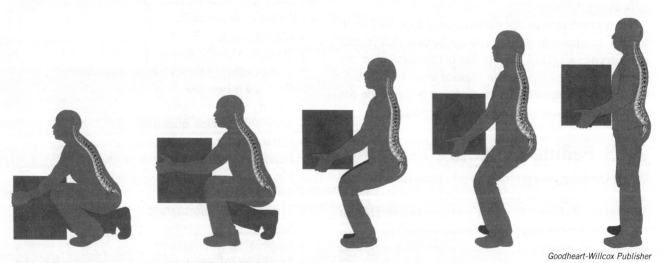

Goodheart-Willcox Publisher

Figure 2-16. Proper lifting and carrying techniques are important to avoid injury.

2.16 Fire Protection

Carpenters should have a good understanding of fire hazards. They must know the causes of fires and methods of controlling them. There are five different classes of fires.

- *Class A fires* result from burning wood and debris.
- *Class B fires* involve highly volatile materials such as gasoline, oil, paints, and oil-soaked rags.
- *Class C fires* result from faulty electrical wiring and equipment.
- *Class D fires* involve combustible metals, such as magnesium and sodium.
- *Class K fires* involve fires with vegetable or cooking oils.

Any of the first three classes of fires can occur on a typical construction site. Combustible metals and cooking oils are less often found on construction sites, but could be if the construction site is at an operational industrial center or restaurant.

Approved fire prevention practices should be followed throughout the construction project. Good housekeeping is an important aspect. Special precautions should be taken during the final stages of construction when heating equipment and wiring are being installed and when highly flammable surface finishes are being applied. Always keep containers of flammable materials closed when not in use. Dispose of oily rags and combustible materials promptly.

Fire extinguishers should be available on the construction site. Be sure to use the proper kind for each type of fire, **Figure 2-17**. All fire extinguishers have a safety pin in the handle. To use the extinguisher, remember the PASS technique: **P**ull the pin out, **A**im the discharge nozzle at the base of the fire, **S**queeze the handle, **S**weep the extinguishing agent from side to side at the fire. Start several feet away from the fire and move in as it is extinguished. Study and follow local regulations.

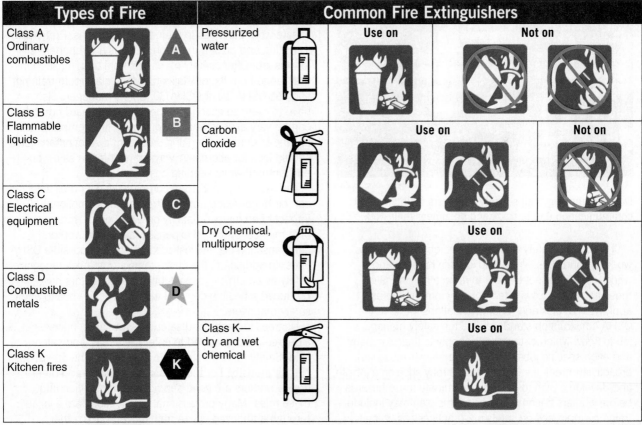

Goodheart-Willcox Publisher

Figure 2-17. Not every fire extinguisher will put out every kind of fire. Check the label on the extinguisher. Using the wrong extinguisher could strengthen the fire, electrocute you, or produce toxic fumes.

2.17 First Aid

Knowledge of first aid is important. You should understand approved procedures and be able to exercise good judgment in applying them. Remember that an accident victim may receive additional injury from unskilled treatment. Information of this nature can be secured from your local Red Cross.

As a preventive measure against infection, keep an approved first aid kit on the jobsite. Because of the nature of the material being handled and the dirty conditions of the work area, even superficial wounds should be treated promptly. Clean, sterilize, and bandage all cuts and nicks. As a precaution, it is important to maintain a current tetanus shot as a guard against infection.

Construction Careers
Health and Safety Manager

Maintaining a safe jobsite is in everyone's best interest. Employees benefit from knowing that their health and safety is valued by their employer. Safety regulations can be complex and can easily be overlooked when workers are focused on completing jobs and producing results. For this reason, many companies hire a dedicated health and safety manager to oversee their safety program and to ensure compliance with OSHA standards.

Cineberg/Shutterstock.com

Digital technology helps safety managers maintain good communication and to track and document their work.

Construction safety managers visit jobsites to assess working conditions. They help identify hazards and work with the project leaders to reduce or eliminate those hazards. This is often done by interpreting and applying the rules and recommendations found in the OSHA construction standard. It is the safety manager's job to know which of these rules apply to their company and each specific jobsite. A safety manager must work proactively, predicting what hazards may arise on a jobsite, and develop a plan to reduce or eliminate those hazards before workers become exposed. This plan may include training employees on safe work practices or putting in place administrative controls, such as a requirement to rotate job assignments to avoid repetitive motion injuries or determining appropriate warning signs. They are also responsible for writing company health and safety policies and performing jobsite inspections to observe and correct unsafe work practices. The safety manager maintains all company forms, records, and reports that pertain to health and safety.

Because a safety manager interacts with workers in a role of authority it is important that they are good leaders and good communicators. They must also be relatable as they may have to win the respect of the workers on the job and perhaps even persuade them of the importance of safe work practices in their daily operations. Proficiency in writing and presenting information is an important skill for this job. Often, safety managers hold training sessions that range from toolbox talks to formal classroom training, such as OSHA 10 or OSHA 30. Safety managers also have to exercise good judgement. Some hazard controls can be very expensive and would cost their employer a fortune to enforce. Thinking creatively, a safety manager should look for alternative ways to maintain a safe work environment while keeping costs in check.

The monetary costs for a company who ignores safety can be huge. Aside from the tremendous emotional impact of an injury or fatality on a job, employers face medical and legal expenses, as well as workers compensation claims, lost production, and possible OSHA fines. In addition to the costs, many companies recognize safety as an ethical responsibility. These factors have produced a healthy demand for construction health and safety managers.

Degrees in health and safety are available; however, a degree is not required to enter the health and safety field. Safety managers need to understand the operations of their industry. For this reason, carpenters and other trades workers are good candidates to transition into these roles. Many professional certifications are available through authorized OSHA training centers or other educational institutions

Chapter Review

Summary

- Construction sites can pose significant safety and health hazards. Safety on the site is the responsibility of both the employer and the employee.
- Good housekeeping on the site is important to preventing accidents.
- Safety rules must be observed in dress, moving about the jobsite, and proper handling of potentially dangerous tools and equipment.
- When working at heights over 6′, especially on roofs, fall protection devices must be used.
- Where needed, scaffolds and ladder should be used. Scaffolds must be inspected daily and must be able to safely and securely support loads.
- Excavations 5′ deep or greater must be shored and braced to prevent collapse.
- To prevent injury, correct lifting and carrying procedures must be observed.
- Carpenters should have a good understanding of the causes of fires and methods of controlling them.
- A first aid kit should always be available on the jobsite.

Know and Understand

Answer the following questions using the information in this chapter.

1. The purpose of OSHA is to _____.
   A. fine employers who are unsafe
   B. empower employees who are being underpaid
   C. ensure safe and healthful working conditions
   D. All of the above.

2. Which of the following is *not* a measure that promotes good housekeeping on the jobsite?
   A. Kick scrap out of the way after making a cut.
   B. Put tools away when they are not being used.
   C. Store building materials and supplies in neat piles.
   D. Do not allow scrap and waste to accumulate.

3. Which of the following is considered safe clothing for a carpenter on the jobsite?
   A. Unbuttoned sleeves.
   B. Trousers or overalls without cuffs.
   C. Sandals.
   D. All of the above.

4. When should a hard hat be worn?
   A. After a bad haircut.
   B. When the sun is bright.
   C. Wherever there is a danger of falling objects.
   D. All of the above.

5. Safety boots will withstand a load of up to _____ lb.
   A. 50   C. 2500
   B. 1500   D. 5000

6. *True or False?* A tool with a dull edge is more dangerous than one with a sharp edge.

7. When not being used, tools should be _____.
   A. kept near at hand for the next use
   B. stored on a shelf, out of reach
   C. thrown in the back of a truck
   D. stored in panels or chests

8. When excavating a trench, a protective system such as shoring or a trench box must be installed at a depth of _____.
   A. 3′   C. 5′
   B. 4′   D. 6′

9. Scaffolds must have a minimum safety factor of _____.
   A. four   C. eight
   B. six   D. ten

10. *True or False?* When a power tool has a cord that has been nicked or frayed, it should be taped with electrical tape.

11. A(n) _____ detects small amounts of electrical current and opens the circuit before electrocution can occur.
    A. circuit breaker
    B. ground-fault circuit interrupter (GFCI)
    C. fuse
    D. ohmmeter

12. When a grounding conductor is broken or missing, _____.
    A. the tool should be thrown away
    B. the circuit breaker will open
    C. the tool will be inoperative
    D. the user may become the path for electrical current

13. Where can you find information about the hazards posed by a construction material and how to combat them?
    A. Safety Data Sheet (SDS).
    B. Manufacturers database of hazardous materials.
    C. Occupational Safety and Health Administration (OSHA).
    D. Hazardous Materials Index (HMI).
14. *True or False?* When lifting heavy objects, bend over at the hips and grasp the object firmly.
15. An electrical fire is classified as a Class _____ fire.
    A. A   C. C
    B. B   D. K

Apply and Analyze

1. List the four leading causes of death for construction workers on the jobsite.
2. Explain the role of OSHA in jobsite safety.
3. What is a safety culture?
4. In what situations should safety glasses be worn?
5. List two methods of providing fall protection.
6. List three safe practices related to hand tool use.

Critical Thinking

1. Sometimes encouraging a safety culture is easier said than done. Imagine you are a crew leader for a roofing company. An experienced roofer, Ron, has just been hired and assigned to your crew. Ron is fast at laying down shingles and does a fantastic job, but says this roof is not very steep and he doesn't need a personal fall-arrest harness. Ron assures you that he has worked on hundreds of roofs and has never fallen, further stating that a harness will only slow him down and cost the company money. What do you think of Ron's attitude toward wearing a personal fall-arrest harness? As crew leader, how would you handle this situation? What consequences might there be for allowing Ron to continue to work without a harness?
2. Following safety rules can save a company a lot of money. If an accident or injury is avoided, a job is completed faster and cheaper. Ignoring safety requirements can have a large impact and high cost. What are the potential consequences for overlooking safety requirements? What are the associated costs (direct and indirect) that can occur. For help, check out the OSHA website.
3. On the jobsite, what do you think is the biggest challenge to improving the safety culture?

4. Use the internet or a catalog from a safety equipment company. Find five different types of personal protective equipment, such as safety glasses and hard hats, and identify the number of choices offered in each category. Determine why they are different.
5. Research safety measures appropriate to working with air-powered (pneumatic) tools.

Communicating about Carpentry

1. **Speaking and Listening.** Working in a group, create a poster of a carpenter practicing proper safety while working. Identify the safety measures the carpenter is taking while working by using arrow pointers. As you work with the group, discuss the meaning of each term. Afterward, display your posters in the classroom as a convenient reference aid for discussions and assignments.
2. **Speaking and Reading.** You are a construction foreman in charge of your work crew. It is your responsibility to ensure the safety of your workers in the event of a fire. You must post signs in the work area to educate your workers on the different types of fires and how to extinguish them. Research the types of fires and categorize them as Class A, Class B, or Class C. Create your signs in the form of a presentation. Share the presentation with the class, as though the class were your crew. Ask for and answer any questions your crew may have.
3. **Speaking and Listening.** Review the examples of unsafe conditions and unsafe acts listed in the chapter. Working in small groups, brainstorm to identify additional examples of unsafe conditions and unsafe acts. Share your list with the rest of class.
4. **Speaking and Listening.** Review the impacts of accident costs described in this chapter. Working in small groups, brainstorm to identify other indirect costs of workplace accidents. Share your list with the rest of class.
5. **Writing.** Write a detailed description of a situation in which you have seen a construction worker on the job—something you observed while riding by in a car would be fine. Research the OSHA standards that would apply to that person.
6. **Reading and Writing.** Research how the user can get a shock from an electrically powered tool that is improperly grounded and the effects of electrical shock on the human body. Develop a list of safety measures appropriate to working with electrically powered tools.

CHAPTER 3

Building Materials

OBJECTIVES

After studying this chapter, you will be able to:

- Describe the hardwood and softwood classifications.
- Define moisture content (MC and EMC).
- Identify common defects in lumber.
- Define lumber grading terms.
- Calculate lumber sizes according to established industry standards.
- Explain plywood, OSB, hardboard, and particleboard grades and uses.
- List precautions to observe while working with treated lumber.
- Identify types of engineered lumber and list their uses and advantages.
- Discuss the uses of light gage steel framing.
- Identify a variety of metal framing connectors and indicate where each is used.
- Identify nail types and sizing units.

TECHNICAL TERMS

annular ring	glue-laminated beam (glulam)	pitch pocket
blue stain	hardboard	plain-sawed
board foot	heartwood	polyurethane adhesive
bound water	hole	polyvinyl resin emulsion adhesive
cambium	honeycombing	premium grade
casein glue	I-joist	pressure-treated lumber
composite board	kiln dried	quarter-sawed
conifer	knot	sapwood
contact cement	laminated-strand lumber (LSL)	seasoning
core	laminated-veneer lumber (LVL)	select
cross-band	lignin	shake
decay	lumber	span rating
deciduous	lumber core	splits and checks
defect	mastic	sticker
edge-grained	moisture content (MC)	tension bridging
engineered lumber products (ELP)	No. 1 common	thermoplastic
epoxy	open time	thermoset adhesive
equilibrium moisture content (EMC)	open-grain wood	tracheid
faces	oriented strand board (OSB)	urea-formaldehyde resin glue
factory and shop lumber	parallel lamination	veneer core
FAS (firsts and seconds)	parallel-strand lumber (PSL)	wane
fiber saturation point	particleboard	warp
flat-grained	phloem	xylem
free water	photosynthesis	

Christian Delbert/Shutterstock.com

Figure 3-1. The frame of a house is usually constructed of wood, though other materials are sometimes used. As construction is completed, many different materials will be used. This chapter describes these materials.

Many different types of materials go into the construction of a modern residence, **Figure 3-1**. A carpenter should be familiar with all of these materials. Each has special properties that are suited to certain building applications.

Before being used in construction, some of these materials are treated or engineered to improve their qualities. Some are composites of different materials, which are designed to do a certain job better than a natural material. Construction materials include the following:

- Sawed lumber and logs
- Plywood
- Composition panels (composite board, particleboard, hardboard, oriented strand board, and fiberboard)
- Engineered lumber
- Wood and nonwood materials for shingles and flooring
- Steel and aluminum
- Metal fasteners
- Concrete and masonry units
- Mortar
- Adhesives and sealants
- Vapor barriers
- Gypsum board and other interior coverings

3.1 Lumber

Wood is one of our greatest natural resources. When cut into pieces uniform in thickness, width, and length, it becomes lumber. This material has always been widely used for residential construction. *Lumber* is the name given to natural or engineered sawmill products. Lumber includes the following:

- Boards used for flooring, sheathing, paneling, and trim
- Dimension lumber used for sills, plates, studs, joists, rafters, and other framing members
- Timbers used for posts, beams, and heavy stringers
- Numerous specialty items

Carpenters must have a good working knowledge of lumber. They must be familiar with type, grade, size, and other details to properly select and use lumber. To understand how to handle and treat wood, carpenters should also know something about the growth, structure, and characteristics of wood.

3.2 Wood Structure and Growth

Wood is made of long narrow tubes or cells called fibers or *tracheids*. See **Figure 3-2**. The cells are no larger around than human hair. Their length varies from about 1/25″ for hardwoods to approximately 1/8″ for softwoods. Tiny strands of cellulose make up the cell walls. The cells are held together with a natural cement called *lignin*. This cellular structure makes it possible to drive nails and screws into the wood. It also accounts for the light weight, low heat transmission, and sound absorption qualities of wood.

A tree has three growing parts:

- The tips of the roots
- The leaves
- A layer of cells just inside the bark called the *cambium*

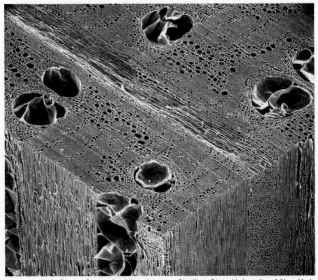

N.C. Brown Center for Ultrastructure Studies, State University of New York, College of Environmental Science and Forestry, Syracuse, NY

Figure 3-2. Photomicrograph (photo using a microscope) of wood showing the long cells that make up wood.

Water absorbed by the roots travels through the sapwood to the leaves, where it is combined with carbon dioxide from the air. Through the chemical process of *photosynthesis*, sunlight changes these elements to a food called carbohydrates. The sap carries this food back to the various parts of the tree.

Figure 3-3 shows the location of the cambium layer, where new cells are formed. The inside area of the cambium layer is called *xylem*. It develops new wood cells. The outside area, known as *phloem*, develops cells that form the bark.

3.2.1 Annular Rings

Growth in the cambium layer takes place in the spring and summer. Separate layers form each season. These layers are called *annular rings*. Each ring is composed of two layers: springwood and summerwood.

In the spring, trees grow rapidly and the cells produced are large and thin-walled. As growth slows down during summer months, the cells produced are smaller, thick-walled, and darker in color. These annual growth rings are largely responsible for the grain patterns that are seen on the surface of boards.

Sapwood is located inside the cambium layer. It contains living cells and may be several inches thick. Sapwood carries sap to the leaves. The *heartwood* of the tree is formed as the sapwood becomes inactive. Usually, it turns darker in color because of the presence of gums and resins. In some woods such as hemlock, spruce, and basswood, there is little or no difference in the appearance of sapwood and heartwood. Sapwood is as strong and heavy as heartwood, but is not as durable when exposed to weather.

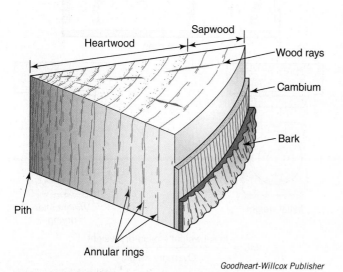

Goodheart-Willcox Publisher

Figure 3-3. A tree is made up of many different layers. Growth takes place in the cambium layer.

3.3 Kinds of Wood

Lumber is either softwood or hardwood. Softwoods come from evergreen, or needle-bearing, trees. These are called *conifers* because many of them bear cones. Hardwoods come from broadleaf (*deciduous*) trees that shed their leaves at the end of the growing season. See **Figure 3-4**. This classification is somewhat confusing because many of the hardwood trees produce a softer wood than some of the so-called softwood trees. A number of the more common kinds of commercial softwoods and hardwoods are listed in **Figure 3-5**. Hardwoods are generally, but not exclusively used in applications where durability and beauty are important. Cabinetry, flooring, furniture, and unpainted trims are often made from hardwoods. Softwoods, on the other hand, get used where economy is important. Framing lumber, siding, paneling, painted millwork, and engineered lumber are generally made from softwoods.

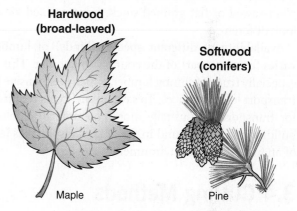

Goodheart-Willcox Publisher

Figure 3-4. Woods are classified as either hardwoods or softwoods. Hardwoods have broad leaves; softwoods have needles.

| Wood for Residential Use ||
Softwoods	Hardwoods
Douglas fir	Basswood
Southern pine	Willow
Western larch	American elm
Hemlock	*Mahogany
White fir	Sweet gum
Spruce	*White ash
Ponderosa pine	Beech
Western red cedar	Birch
Redwood	Cherry
Cypress	Maple
White pine	*Oak
Sugar pine	*Walnut

*Open-grain wood

Goodheart-Willcox Publisher

Figure 3-5. A list of woods popular for residential use. Usually, softwoods are used as framing lumber.

A number of hardwoods have large pores in the cellular structure and are called *open-grain woods*. They require additional operations during finishing. Different kinds of wood will also vary in weight, strength, workability, color, texture, grain pattern, and odor.

Study the full-color wood samples at the end of this chapter as a first step in wood identification. To further develop your ability to identify woods, study actual specimens. Several of the softwoods used in construction work are similar in appearance. Considerable experience is required to make accurate identification.

Most of the samples shown in this text were cut from plain-sawed or flat-grained stock. Edge-grained views look different.

Availability of different species (kinds) of lumber varies from one part of the country to another. This is especially true of framing lumber, which is expensive to transport long distances. To save on fuel costs required for transport, it is usually more economical to select building materials found in the area. It also makes less of an impact on the environment.

3.4 Cutting Methods

Most lumber is cut so that the annular rings form an angle of less than 45° with the surface of the board. This produces lumber called *flat-grained* if it is softwood, or *plain-sawed* if it is hardwood. This method produces the least waste and more desirable grain patterns.

Lumber can also be cut so the annular rings form an angle of more than 45° with the surface of the board, **Figure 3-6**.

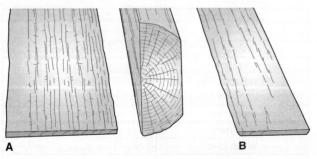

A   B

Goodheart-Willcox Publisher

Figure 3-6. Lumber may be sawed in different ways. A—Board is edge-grained or quarter-sawed. Saw cut was made roughly parallel to a line running through center of log. B—Flat-grained or plain-sawed board.

This method produces lumber called *edge-grained* if it is softwood, and *quarter-sawed* if it is hardwood. It is more difficult and expensive to use this method. However, it produces lumber that swells and shrinks less across its width and is not as likely to warp.

3.5 Moisture Content and Shrinkage

Before wood can be commercially used, a large part of the moisture (sap) must be removed. When a living tree is cut, more than half its weight may be moisture. Lumber used for framing and outside finish should be dried to a moisture content of less than 19%. That is the moisture content below which mold does not grow. Cabinet and furniture woods are dried to a moisture content of 6%–8% because the atmospheric humidity level is lower indoors.

The amount of moisture, or *moisture content (MC)*, in wood is given as a percent of the oven-dry weight. To determine the moisture content, a sample is first weighed. It is then placed in an oven and dried at a temperature of about 212°F (100°C). The drying is continued until the wood no longer loses weight. The sample is weighed again and this oven-dry weight is subtracted from the initial weight. The difference is then divided by the oven-dry weight, **Figure 3-7**.

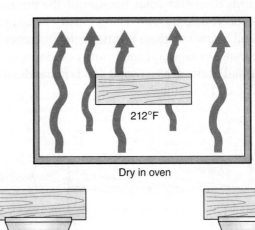

Dry in oven

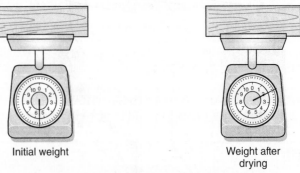

Initial weight   Weight after drying

$$MC = \frac{\text{Initial weight} - \text{oven-dry weight}}{\text{Oven-dry weight}}$$

Goodheart-Willcox Publisher

Figure 3-7. The moisture content of wood can be determined by weighing it before and after it is dried.

Moisture contained in the cell cavities is called *free water*. Moisture contained in the cell walls is called *bound water*. As the wood dries, moisture first leaves the cell cavities. When the cells are empty, but the cell walls are still full of moisture, the wood has reached a condition called the *fiber saturation point*. For most woods, this is about 30%, **Figure 3-8**. The fiber saturation point is important because wood does not start to shrink until this point is reached. As the MC drops below 30%, moisture is removed from the cell walls and they shrink. **Figure 3-9** shows the actual shrinkage in a 2×10 joist.

Wood shrinks the most along the direction of the annual rings (tangentially) and about half as much across these rings. There is little shrinkage in the length. How this shrinkage affects lumber cut from different parts of a log is shown in **Figure 3-10**. As wood takes on moisture, it swells in the same proportion as the shrinkage that took place. These changes in dimensions during drying and taking on moisture are the causes of warping and splitting.

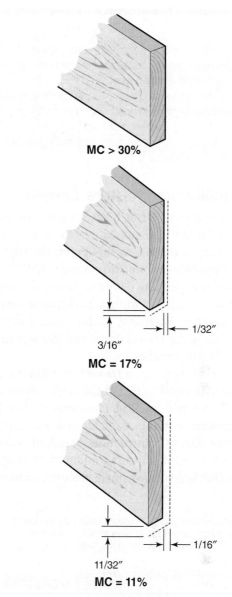

Goodheart-Willcox Publisher

Figure 3-9. A 2 × 10 may shrink 1/16″ across its shortest dimension as it dries.

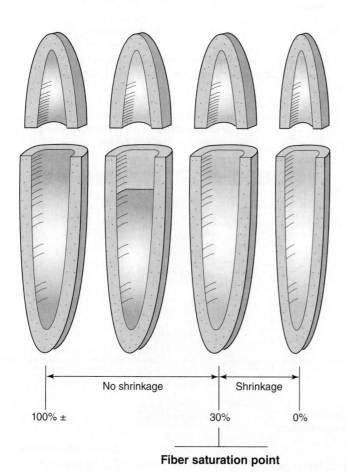

Goodheart-Willcox Publisher

Figure 3-8. How a wood cell dries. First, the free water in the cell cavity is removed. Then, the cell wall dries and shrinks.

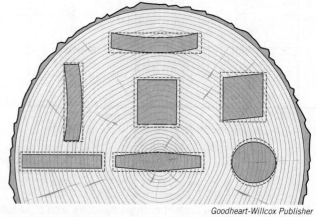

Goodheart-Willcox Publisher

Figure 3-10. Shrinkage and distortion of flat, square, and round pieces is affected by the direction of the annual rings in relation to the lumber piece.

3.5.1 Equilibrium Moisture Content

A piece of wood gives off or takes on moisture from the air around it until the moisture in the wood is balanced with that in the air. At this point, the wood is said to be at *equilibrium moisture content (EMC)*.

Since wood is exposed to daily and seasonal changes in the relative humidity of the air, its moisture content is always changing. Therefore, its dimensions are also changing. This is the reason doors and drawers often stick during humid weather.

Ideally, a wood structure is framed with lumber at a moisture content equal to that of the environment it is located in. This is not practical. Lumber with such low moisture content is seldom available and would likely gain moisture during construction. Standard practice is to use lumber with a moisture content in the range of 15%–19%. That is the approximate moisture content of lumber as it is sold. In heated structures, it will eventually reach a level of about 8%. However, this will vary in different geographical areas, **Figure 3-12**.

Carpenters understand that shrinkage is inevitable. They make allowances where needed to reduce the effects of shrinkage. The first, and by far the greatest, change in moisture content occurs during the first year after construction, particularly during the first heating season. When green lumber (more than 20% MC) is used, shrinkage is excessive. Warping, plaster cracks, nail pops, squeaky floors, and other difficulties are almost impossible to prevent.

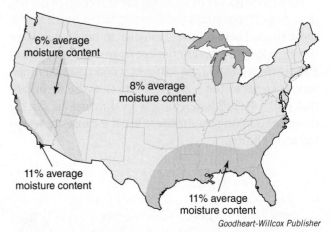

Goodheart-Willcox Publisher

Figure 3-12. This map shows average moisture content of interior woodwork for various regions of the United States.

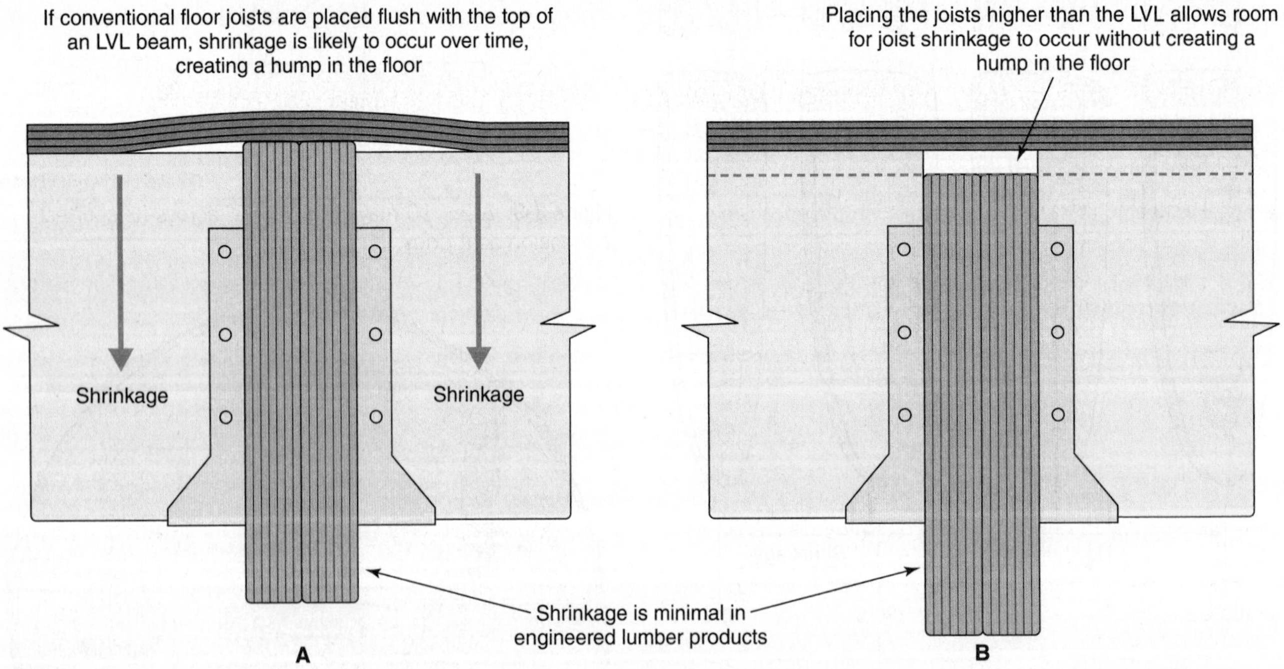

If conventional floor joists are placed flush with the top of an LVL beam, shrinkage is likely to occur over time, creating a hump in the floor

Shrinkage   Shrinkage

Placing the joists higher than the LVL allows room for joist shrinkage to occur without creating a hump in the floor

Shrinkage is minimal in engineered lumber products

A   B

Goodheart-Willcox Publisher

Figure 3-11. Moisture content changes the dimension of lumber. Different lumbers will be affected differently. A—When abutting conventional lumber to an engineered beam a hump in the floor can form if shrinkage is not accounted for. B—Conventional lumber should be held above the LVL during construction. The moisture content should be measured to determine how much shrinkage to expect.

3.5.2 Seasoning Lumber

Seasoning is reducing the moisture content of lumber to the required level specified for its grade and use. In air drying, the lumber is simply exposed to the outside air. It is carefully stacked with **stickers** (wood strips) between layers so air can circulate through the pile. Boards are also spaced within the layers so air can reach edges. Air drying is a slower process than kiln drying. It often creates additional defects in the wood.

Lumber is **kiln dried** by placing it in huge ovens where the temperature and humidity can be carefully controlled. When the green lumber is first placed in the kiln, steam is used to keep the humidity high. The temperature, meanwhile, is kept at a low level. Gradually the temperature is raised while the humidity is reduced. Fans keep the air in constant circulation. Bundles of lumber may carry a stamp to indicate that they have been kiln dried, KD for "kiln dried" and PKD for "partly kiln dried."

3.5.3 Moisture Meters

The moisture content of wood can be determined by oven drying a sample as previously described, or using an electronic moisture meter. Although the oven drying method is more accurate, meters are often used because readings can be secured rapidly and conveniently. The meters are usually calibrated to cover a range from 7% to 25%. Accuracy is generally plus or minus 1% of the moisture content.

There are two basic types of moisture meters. One determines the moisture content by measuring the electrical resistance between two pin-type electrodes that are driven into the wood. See **Figure 3-13**. The other type is pinless and uses an electromagnetic signal to measure the moisture content in the wood.

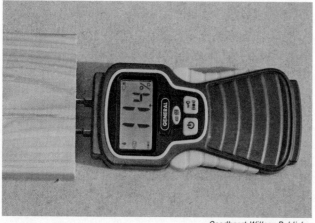

Goodheart-Willcox Publisher

Figure 3-13. An electronic moisture meter uses a digital display to indicate moisture content. This model has probes that are inserted into the cut end.

3.6 Lumber Defects

A *defect* is an irregularity occurring in or on wood that reduces its strength, durability, or usefulness. It may also detract from or improve the wood's appearance. For example, knots are commonly considered a defect, but may add to the appearance of pine paneling. Lumber with certain defects, particularly knots, should not be used where strength is important. An imperfection that impairs only the appearance of wood is called a blemish. Some common defects include the following:

- *Knots*—Caused by an embedded branch or limb. Knots reduce the strength of the wood. The extent of damage depends on the type of knot, its size, and its location. See **Figure 3-14** and **Figure 3-15**.

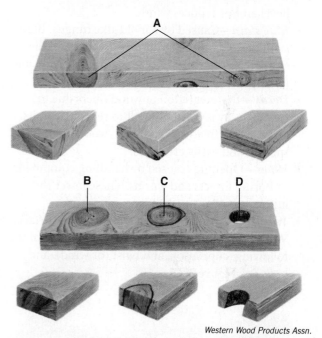

Western Wood Products Assn.

Figure 3-14. Types of knots. A—Knots may appear on one or both faces and on edges of sawn lumber. B—A sound (cannot be knocked loose), intergrown knot. C—A sound, encased knot. D—Unsound knots can fall out or be knocked out easily.

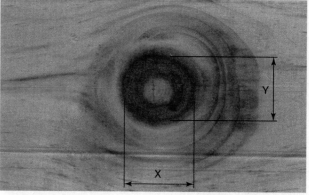

Goodheart-Willcox Publisher

Figure 3-15. Knot size is determined by adding the width and length of the knot and dividing by two.

- *Splits and checks*—Separations of the wood fibers that run along the grain and across the annular growth rings. Splits usually occur at the ends of lumber that has been unevenly seasoned.
- *Shakes*—Separations along the grain and between the annular growth rings. Shakes are likely to occur only in species with abrupt change from spring to summer growth.
- *Pitch pockets*—Cavities that contain or have contained pitch in solid or liquid form.
- *Honeycombing*—Separation of the wood fibers inside the tree. Honeycombing may not be visible on the board's surface.
- *Wane*—The presence of bark or the absence of wood along the edge of a board; wane forms a bevel and/or reduces width.
- *Blue stain*—Caused by a mold-like fungus. Blue stain discoloration has little or no effect on the strength of the wood. However, it is objectionable for appearance in some grades of lumber.
- *Decay*—Disintegration of wood fibers due to fungi. Decay is often difficult to recognize in its early stages. In advanced stages, wood is soft, spongy, and easily crumbles.
- *Holes*—Openings caused by handling equipment or boring insects and worms. These lower the lumber grade.
- *Warp*—Any variation from true or plane surface. Warp may include any one or combinations of the following: cup, crook, and twist (or wind), and bow. See **Figure 3-16**.

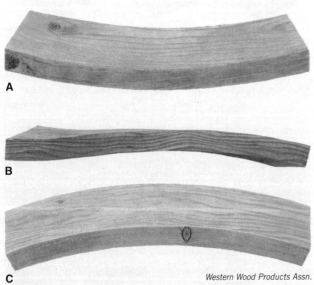

Western Wood Products Assn.

Figure 3-16. Warping affects quality, strength, and appearance. A—Crook. B—Twist or wind. C—Bow.

3.7 Softwood Grades

Various associations of lumber producers—Western Wood Products Association, Southern Forest Products Association, California Redwood Association, and similar groups—develop and apply detailed rules for lumber grading. These agencies publish grading rules for the species of lumber produced in their regions. They also have qualified personnel who supervise grading standards at sawmills.

Basic classifications of softwood grading include boards, dimensions, and timbers. The grades within these classifications are shown in **Figure 3-17**.

Another classification, called *factory and shop lumber*, is graded primarily for remanufacturing purposes. Millwork plants use it to fabricate windows, doors, moldings, and other trim items.

The carpenter must understand that quality construction does not require that all lumber be of the highest grade. Today, lumber is graded for specific uses. In a given structure, several grades may be appropriate. The key to good economical construction is matching the proper grade with its function.

3.8 Hardwood Grades

Hardwood used in building construction is primarily used for flooring, cabinets, and some interior trim. The National Hardwood Lumber Association establishes grades for hardwood lumber. *FAS (firsts and seconds)* is the highest grade. It specifies that pieces be no less than 6″ wide by 8′ long and yield at least 83.33% clear cuttings. The next lower grade, *selects*, permits pieces 4″ wide by 6′ long. A still lower grade is *No. 1 common*. Lumber in this group is expected to yield 66.66% clear cuttings.

3.9 Lumber Sizes

When listing and calculating the size and amount of lumber, the nominal dimension is always used. **Figure 3-18** lists the nominal and dressed sizes for various classifications of lumber. Note that nominal sizes are sometimes listed in quarters. For example, 1 1/4″ material is given as 5/4. This nominal dimension is its rough unfinished measurement (width and thickness), **Figure 3-19**. The dressed size is less than the nominal size as a result of seasoning and surfacing. Dressed sizes of lumber, established by the American Lumber Standards, are consistently applied throughout the industry.

Boards

Appearance grades	Selects	B & Better C select D selec	(IWP – supreme) (IWP – choice) (IWP – quality)	**Specification check list**	
	Finish	Superior Prime E		☐ Grades listed in order of quality. ☐ Include all species suited to project. ☐ For economy, specify lowest grade that will satisfy job requirement. ☐ Specify surface texture desired. ☐ Specify moisture content suited to project. ☐ Specify (WP) grade stamp. For finish and exposed pieces, specify stamp on back or ends.	
	Paneling	Clear (any select or finish grade) No. 2 common selected for knotty paneling No. 3 common selected for knotty paneling			
	Siding (bevel, bungalow)	Superior Prime		**Western red cedar**	
Boards sheathing			**Alternate board grades**	Finish Paneling and ceiling	**Clear heart** **A** **B**
		No. 1 common (IWP – colonial) No. 2 common (IWP – sterling) No. 3 common (IWP – standard) No. 4 common (IWP – utility)	Select merchantable Construction Standard Utility	Bevel siding	Clear – V.G. heart A – Bevel siding B – Bevel siding C – Bevel siding

Dimensions / Grade Markings

Dimensions		Grade Markings
Light framing 2″ to 4″ thick 2″ to 4″ wide	Construction Standard Utility Economy	
Studs 2″ to 4″ thick 2″ to 4″ wide 10′ or shorter	Stud Economy stud	
Structural light framing 2″ to 4″ thick 2″ to 4″ wide	Select structural No. 1 No. 2 No. 3 Economy	
Structural joists and planks 2″ to 4″ thick 6″ and wider	Select structural No. 1 No. 2 No. 3 Economy	

Grade Markings diagram:
- Mill identification
- Grade designation
- Certification mark
- Condition of seasoning at time of surfacing. MC-15: 15% max. M.C. S-DRY: 19% max. M.C. S-GRN: over 19% M.C.
- Species of wood

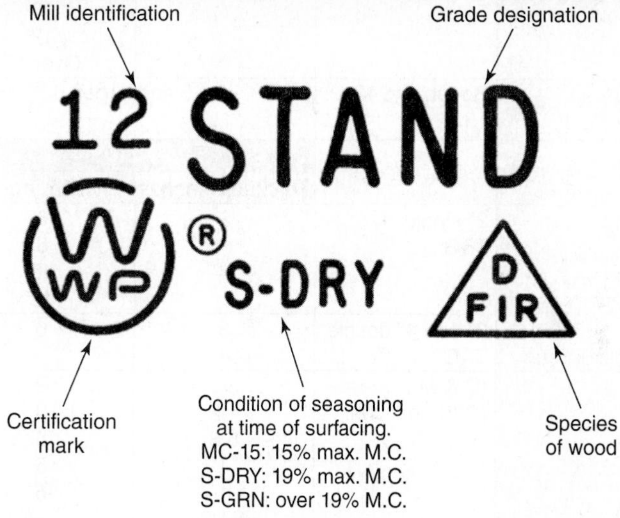

Timbers

Beams and stringers	Select structural No. 1 No. 2 (No. 1 mining) No. 3 (No. 2 mining)	Posts and timbers	Select structural No. 1 No. 2 (No. 1 mining) No. 3 (No. 2 mining)

Goodheart-Willcox Publisher

Figure 3-17. Softwood lumber classifications and grades. Names of grades and their specifications will vary among lumber manufacturers' associations and among regions producing lumber. The grade listed on top is better than the grades below it for each classification.

Board Measure	Product classification					
The term "board measure" indicates that a board foot is the unit for measuring lumber. A board foot is one inch thick and 12 inches square.		**Thickness**	**Width**		**Thickness**	**Width**
	Board lumber	1"	2" or more	Beams and stringers	5" and thicker	more than 2" greater than thickness
The number of board feet in a piece is obtained by multiplying the normal thickness in inches by the nominal width in inches by the length in feet and dividing by 12:	Light framing	2" to 4"	2" to 4"	Posts and timbers	5" x 5" and larger	not more than 2" greater than thickness
	Studs	2" to 4"	2" to 4" 10' and shorter	Decking		4" to 12" wide
$\dfrac{(T \times W \times L)}{12}$	Structural light framing	2" to 4"	2" to 4"	Siding	thickness expressed by dimension of butt edge	
Lumber less than one inch in thickness is figured as one-inch.	Joist and planks	2" to 4"	6" and wider	Moldings	size at thickest and widest points	
	Lengths of lumber generally are 6 feet and longer in multiples of 2'					

Standard Lumber Sizes/Nominal, Dressed, Based on WWPA Rules

Product	Description	Nominal Size		Dressed Dimensions		Lengths feet
				Thickness and Widths inches		
		Thickness inches	Width inches	Surfaced Dry	Surfaced Unseasoned	
Framing	S4S..................	2 3 4	2 3 4 6 8 10 12 Over 12	1-1/2 2-1/2 3-1/2 5-1/2 7-1/4 9-1/4 11-1/4 Off 3/4	1-9/16 2-9/16 3-9/16 5-5/8 7-1/2 9-1/2 11-1/2 Off 1/2	6' and longer in multiples of 1'
Timbers	Rough or S4S.....	5 and larger		**Thickness inches**	**Width inches**	Same
				1/2 off nominal		

		Nominal Size		Dressed Dimensions Surfaced Dry		
		Thickness inches	Width inches	Thickness inches	Width inches	Lengths feet
Decking Decking is usually surfaced to single T&G in 2" thickness and double T&G in 3" and 4" thicknesses	2" single T&G......	2	6 8 10 12	1 1/2	5 6 3/4 8 3/4 10 3/4	6' and longer in multiples of 1'
	3" and 4" double T&G..................	3 4	6	2 1/2 3 1/2	5 1/4	
Flooring	(D & M), (S2S and CM).....	3/8 1/2 5/8 1 1 1/4 1 1/2	2 3 4 5 6	5/16 7/16 9/16 3/4 1 1 1/4	1 1/8 2 1/8 3 1/8 4 1/8 5 1/8	4' and longer in multiples of 1'
Ceiling and partition	(S2S and CM).....	3/8 1/2 5/8 3/4	3 4 5 6	5/16 7/16 9/16 11/16	2 1/8 3 1/8 4 1/8 5 1/8	4' and longer in multiples of 1'
Factory and shop lumber	S2S..................	1 (4/4) 1 1/4 (5/4) 1 1/2 (6/4) 1 3/4 (7/4) 2 (8/4) 2 1/2 10/4 3 (12/4) 4 (16/4)	5 and wider (4" and wider in 4/4 No. 1 Shop and 4/4 No. 2 Shop)	25/22 (4/4) 1 5/22 (5/4) 1 13/22 (6/4) 1 19/22 (7/4) 1 13/16 (8/4) 2 3/8 10/4 2 3/4 (12/4) 3 3/4 (16/4)	Usually sold random width	4' and longer in multiples of 1'

Abbreviations

Abbreviated descriptions appearing in the size table are explained below.	S4S – Surfaced four sides.	D & M – Dressed and matched.
S1S – Surface one side.	S1S1E – Surfaced one side, one edge.	T & G – Tongue and grooved.
S2S – Surfaced two sides.	S1S2E – Surfaced one side, two edges.	EV1S – Edge vee on one side.
	CM – Center matched.	S1E – Surfaced one edge.

Western Wood Products Assn.

Figure 3-18. Standard lumber sizes are set by government agencies and lumber associations.

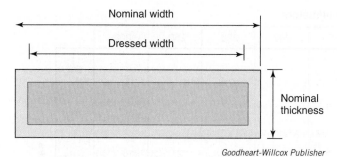

Nominal width

Dressed width

Nominal thickness

Goodheart-Willcox Publisher

Figure 3-19. Nominal size is greater than dressed size.

3.10 Calculating Board Footage

The unit of measure for lumber is the **board foot**. This is a piece 1″ thick and 12″ square or its equivalent (144 cu in). Standard-size pieces can be quickly calculated by visualizing the board feet included. For example: a board 1″ × 12″ and 10′ long contains 10 bd ft. If the piece is only 6″ wide, it contains 5 bd ft. If the original board is 2″ thick, it contains 20 bd ft.

The following formula can be applied to calculate board feet. T represents the thickness of the board (in), W is the width (in), and L is the length (ft).

$$\text{bd ft} = \frac{\text{No. pcs.} \times T \times W \times L}{12}$$

An example of the application of the formula follows. Find the number of board feet in six pieces of lumber that measure 1″ × 8″ × 14′.

$$\text{bd ft} = \frac{6 \times 1 \times 8 \times 14}{12}$$

$$= \frac{\overset{1}{\cancel{6}} \times 1 \times 8 \times 14}{\underset{2}{\cancel{12}}}$$

$$= \frac{1 \times 1 \times 8 \times 14}{\underset{1}{\cancel{2}}}$$

$$= \frac{1 \times 1 \times 4 \times 14}{1}$$

$$= \quad 56 \text{ bd ft}$$

Stock less than 1″ thick is figured as though it is 1″. When the stock is thicker than 1″, the nominal size is used. When this size contains a mixed fraction, such as 1 1/4, change it to an improper fraction (5/4) and place the numerator above the formula line and the denominator below. Find the board feet in two pieces of lumber that measure 1 1/4″ × 10″ × 8′.

$$\text{bd ft} = \frac{2 \times 5 \times 10 \times 8}{4 \times 12}$$

$$= \frac{\overset{5}{\cancel{2}} \times 5 \times \overset{2}{\cancel{10}} \times 8}{\underset{1}{\cancel{4}} \times \underset{6}{\cancel{12}}}$$

$$= \frac{\overset{1}{\cancel{2}} \times 5 \times 5 \times 2}{1 \times \underset{3}{\cancel{6}}}$$

$$= \frac{1 \times 5 \times 5 \times 2}{1 \times 3}$$

$$= \frac{50}{3}$$

$$= 16 \ 2/3 \text{ bd ft}$$

Use the nominal size of the material when figuring the footage. Items such as moldings, furring strips, and rounds are priced and sold by the lineal foot. Thickness and width are disregarded.

3.11 Metric Lumber Measure

Metric-size lumber gives thickness and width in millimeters (mm) and length in meters (m). There is little difference between metric and conventional dimensions for common sizes of lumber. For example, the common 1×4 board is about 25 mm × 100 mm. Visually, they appear to be about the same size. Metric lumber lengths start at 1.8 m (about 6′) and increase in steps of 300 mm (about a foot) to 6.3 m. This is a little more than 20′. See **Figure 3-20** for a chart of standard sizes.

3.12 Panel Materials

Wood panels for construction are manufactured in several different ways:

- Plywood—Thin sheets are laminated to various thicknesses.
- Composite plywood—Veneer faces are bonded to different kinds of wood cores.
- Nonveneered panels—Includes particleboard, fiberboard, and oriented strand board.

3.12.1 Plywood

Plywood is constructed by gluing together a number of thin layers (plies) of wood with the grain direction turned at right angles in each successive layer. An odd number (3, 5, 7) of plies is used so they are balanced on either side of a center core and so the grain of the outside layers run in the same direction. See **Figure 3-21**.

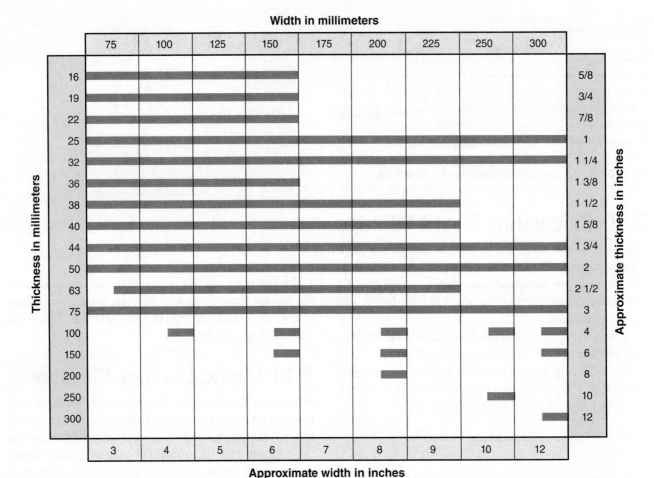

Figure 3-20. This chart compares metric and inch lumber sizes. Dimensions of metric lumber are given in millimeters. Lengths are always given in meters and range from 1.8 m to 6.3 m in increments of 0.3 m (about 1′).

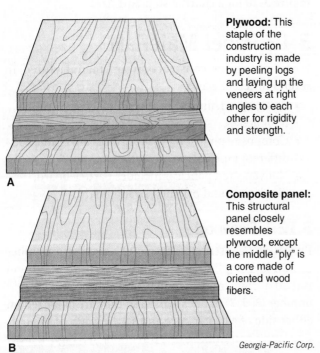

Plywood: This staple of the construction industry is made by peeling logs and laying up the veneers at right angles to each other for rigidity and strength.

Composite panel: This structural panel closely resembles plywood, except the middle "ply" is a core made of oriented wood fibers.

Georgia-Pacific Corp.

Figure 3-21. Panel construction. A—Standard plywood construction. B—Composite panel plywood construction.

The outer plies are called *faces* or face and back. The next layers under these are called *cross-bands*; the other inside layer or layers are called the *core*. A thin plywood panel made of three layers consists of two faces and a core. There are two basic types of plywood.

- Exterior plywood is bonded with waterproof glues. It can be used for siding, concrete forms, and other construction where it will be exposed to the weather or excessive moisture.
- Interior plywood may be manufactured with either interior or exterior glue. Exterior glue is generally used. The grade of the panel is mainly determined by the quality of the layers.

Plywood is made in thicknesses from 1/8″ to more than 1″. Common sizes are 1/4″, 3/8″, 1/2″, and 3/4″. Thicknesses of 15/32″, 19/32″, and 23/32″ are also available. A standard panel size is 4′ wide by 8′ long.

The metric equivalent of a 4′ × 8′ panel is a 1220 mm × 2440 mm panel. This size is commonly used for metric projects. A second metric panel size—1200 mm × 2400 mm—is also used on some metric projects.

Panel nominal thickness	
in	mm
1/4	6.4
5/16	7.9
11/32	8.7
3/8	9.5
7/16	11.1
15/32	11.9
1/2	12.7
19/32	15.1
5/8	15.9
23/32	18.3
3/4	19.1
7/8	22.2
1	25.4
1 3/32	27.8
1 1/8	28.6

A

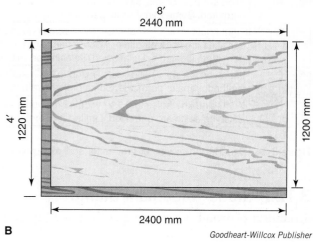

B

Goodheart-Willcox Publisher

Figure 3-22. Comparisons of conventional and metric thicknesses/sizes of plywood panels. A—Panel nominal thicknesses. B—Sizes for a standard plywood panel.

Panels at this alternative metric size are slightly smaller than the standard 4′ × 8′ panels. **Figure 3-22** compares metric and conventional panel sizes.

Softwood plywood for general construction is manufactured in accordance with *U.S. Product Standard PS 1-95 for Construction and Industrial Plywood.* This provides for designating species, strength, type of glue, and appearance.

Many species of softwood are used in making plywood. There are five separate plywood groups based on stiffness and strength. Group 1 includes the stiffest and strongest woods. See **Figure 3-23**.

Group 1	Group 2		Group 3	Group 4	Group 5
Apitong	Cedar, Port	Tangile	Alder, Red	Aspen	Basswood
Beech,	Orford	White	Birch, Paper	Bigtooth	Poplar,
American	Cypress	Lauan	Cedar, Alaska	Quaking	Balsam
Birch	Douglas	Maple, Black	Fir,	Cativo	
Sweet	Fir 2[a]	Mengkulang	Subalpine	Cedar	
Yellow	Fir	Meranti,	Hemlock,	Incense	
Douglas	Balsam	Red[b]	Eastern	Western	
Fir 1[a]	California	Mersawa	Maple,	Red	
Kapur	Red	Pine	Bigleaf	Cottonwood	
Keruing	Grand	Pond	Pine	Eastern	
Larch,	Noble	Red	Jack	Black	
Western	Pacific	Virginia	Lodgepole	(Western	
Maple, Sugar	Silver	Western	Ponderosa	Poplar)	
Pine	White	White	Spruce	Pine	
Caribbean	Hemlock,	Spruce	Redwood	Eastern	
Ocote	Western	Red	Spruce	White	
Pine, Southern	Lauan	Sitka	Englemann	Sugar	
Loblolly	Almon	Sweetgum	White		
Longleaf	Bagtikan	Tamarack			
Shortleaf	Mayapis	Yellow			
Slash	Red Lauan	poplar			
Tanoak					

(a) Refers to Douglas fir from the states of Washington, Oregon, California, Idaho, Montana, and Wyoming, as well as from Alberta and British Columbia shall be classified as Douglas Fir No. 1. Those grown in Nevada, Utah, Colorado, Arizona, and New Mexico shall be classified as Douglas Fir No. 2. (b) Red Meranti shall be limited to species with a specific gravity of 0.41 or more, based on green volume and oven-dry weight.

APA-The Engineered Wood Association

Figure 3-23. Classification of softwood plywood, rating species for strength and stiffness. Group 1 represents the strongest woods.

	Veneer Grades for Softwood Plywood
A	Smooth paintable. Not more than 18 neatly made repairs, boat, sled, or router type, and parallel to grain, permitted. May be used for natural finish in less demanding application. Synthetic or wood repairs permitted.
B	Solid surface. Shims, sled, or router repairs, and tight knots up to 1″ across grain permitted. Wood or synthetic repairs permitted. Some minor splits permitted.
C	Tight knots to 1 ½″. Knot holes to 1″ across grain and some to 1 ½″ if total width of knots and knotholes is within specified limits. Synthetic or wood repairs permitted. Discoloration and sanding defects that do not impair strength permitted. Limited splits allowed. Stitching permitted.
C (plugged)	Improved C veneer with splits limited to 1/8″ width and knotholes or other open defects limited to ¼″ by ½″. Wood or synthetic repairs permitted. Some broken grain permitted.
D	Knots and knotholes to 2 1/2" width across grain and 1/2" larger within specified limits. Limited splits are permitted. Stitching splits are permitted. Limited to Exposure 1 or interior panels.

APA-The Engineered Wood Association

Figure 3-24. Description of veneer grades for softwood plywood.

3.12.2 Grade-Trademark Stamp

Structural plywood panels are quality graded in two ways:

- A lettering system indicates the veneer quality for the face and back of the panel, **Figure 3-24**.
- A name indicates the intended use or "performance rating."

APA-The Engineered Wood Association has a rigid testing program based upon standard PS 1-95. Mills that are members of the association may use the official grade-trademark. It is stamped on each piece of plywood. **Figure 3-25** explains the parts of the APA trademark and grade stamp.

3.12.3 Exposure Ratings

The grade-trademark stamp gives an "exposure durability" classification to plywood. There are two basic categories:

- Exterior type, which has 100% waterproof glueline.
- Interior type, with moisture-resistant glueline.

However, panels can be manufactured in three exposure durability classifications:

- Exterior—Fully waterproof adhesives, for use where subject to permanent exposure to weather or moisture.
- Exposure 1—Waterproof bond designed for use where long construction delays are expected before providing protection.
- Exposure 2—Identified as interior type with intermediate glue for protected applications in which high humidity or water leaks may exist.

Most plywood is manufactured with waterproof exterior glue. However, interior panels may be manufactured with intermediate or interior glue.

3.12.4 Span Ratings of Plywood

Structural panels with thicknesses of 15/32″, 19/32″, and 23/32″ are rated according to the allowable span between supports. The trademark is placed on sheathing, floor underlayment, and siding panels with numbers called *span ratings*. This rating is the maximum recommended center-to-center distance (in inches) between supports when the long dimension of the panel is at a right angle to the supports.

APA-rated sheathing is typically manufactured with span ratings of 24/0, 24/16, 40/20, or 48/24. As shown in **Figure 3-25**, the first number is the allowable span in roofing applications and the second number is the allowable span in flooring applications. Typical Sturd-I-Floor ratings are 16″, 20″, 24″, 32″, and 48″. This single number is the recommended center-to-center spacing of floor joists when panels are installed with the long dimension at a right angle to the supports.

Workplace Skills
Communication

Do you consider yourself a good listener? Good listeners make an effort to fully understand what others are trying to communicate. All employers want employees who are good listeners. Imagine being the lead carpenter on a framing job. You have sorted your lumber and set aside the best quality 2×6s. You have told your crew that the good lumber is only to be used for corners, door, and window frames. You thought they were all listening, but after making a run to the lumber yard, you return to find all the good lumber cut up into blocking. Your crew may have heard you, but they clearly were not listening. Poor communication and listening skills can end up costing money to fix mistakes and possible injury if workers do not listen to warnings.

APA TRADEMARKS
& GRADE STAMP ANATOMY

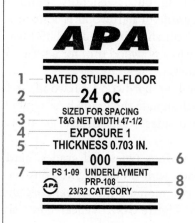

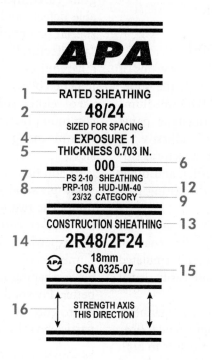

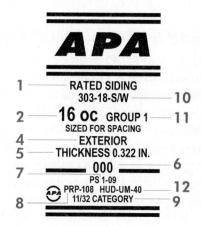

1. **Panel grade** — the term "grade" may refer to panel grade or veneer grade. Panel grades are generally identified in terms of the veneer grade used on the face and back of the panel (e.g., A-B, B-C, etc.) or by a name suggesting the panel's intended end use (e.g., APA Rated Sheathing, Underlayment, etc.).

2. **Span Rating** — Two numbers separated by a slash. The left-hand number is the maximum recommended center-to-center spacing for supports in inches when the panel is used for roof sheathing with long dimensions across supports. The right-hand number is the maximum center-to-center spacing of supports in inches when the panel is used for subflooring with the long dimension across supports. The Span Rating on APA RATED STURD-I-FLOOR and APA RATED SIDING panels appears as a single number. The Span Ratings for APA RATED STURD-I-FLOOR panels, like those for APA RATED SHEATHING, are based on application of the panel with the long dimension or strength axis across three or more supports.

3. **Tongue-and-groove** — A system of jointing in which the rib or tongue of one member fits into the groove of another. In Sturd-I-Floor panels, tongue-and-groove edges eliminate the need for blocking beneath panel edges for support.

4. **Bond Classification** — Exposure ratings for APA wood structural panels may be Exterior or Exposure 1. The classifications are based on the severity and duration of weather and moisture exposure the panels are designed to withstand, and the wood materials and adhesives used in manufacture.

5. **Mill thickness declaration** — Designated thickness subject to tolerances specified in standard.

6. **Mill number** — Manufacturing mill identification number.

7. **Product Standard** — An industry product manufacturing or performance specification. Voluntary Product Standard PS 1-09 for Construction and Industrial Plywood was developed cooperatively by the plywood industry and the U.S.

Department of Commerce. Voluntary Product Standard PS 2-10, Performance Standard for Wood-Based Structural Use Panels, establishes performance criteria for specific designated construction applications.

8. **APA's Performance Rated Panel Standard** — A standard applying to panels such as APA RATED SHEATHING, APA RATED STURD-I-FLOOR, and APA RATED SIDING. Panels manufactured to meet APA performance standards must satisfy rigorous, exacting performance criteria for the intended applications.

9. **Performance Category** — A panel designation related to the panel thickness range that is linked to the nominal panel thickness designations used in the International Building Code (IBC) and the International Residential Code (IRC).

10. **Siding face grade** — Grade identification of siding face, based on number of repairs and appearance characteristics.

11. **Species group number** — Classified according to strength and stiffness under manufacturing standard PS 1 into Groups 1 through 5. Group 1 woods are the strongest, Group 2 seconded strongest, and so on.

12. **HUD recognition.**

13. **Panel grade, Canadian Standard** — Some APA panels are manufactured to both U.S. and Canadian standards, and carry a dual mark, as shown in the Rated Sheathing mark above. Other panels may carry a single mark for either the U.S. or Canadian standard.

14. **Panel mark** — Rating and end-use designation, Canadian standard. In the Canadian span marking, the "R" signifies roofs and the "F" represents subfloors.

15. **Canadian performance-rated panel standard.**

16. **Panel face orientation indicator.**

APA-The Engineered Wood Association

Figure 3-25. Explanation of the elements of the APA trademark and grade stamp found on all rated panels.

Siding is manufactured with ratings of 16″ and 24″. Panel and lap siding rated at 16″ may be applied directly to studs spaced at 16″ OC (on-center). Panel and lap siding rated at 24″ can be applied to studs spaced 24″ OC. When applied over nailable structural sheathing, the span rating indicates the maximum recommended spacing of vertical rows of nails.

3.12.5 HDO and MDO Plywood

HDO and MDO plywood combine the properties of exterior-type plywood and plywood with the superior wearability of overlaid surfaces. MDO (medium-density overlay) plywood is an exterior-type panel manufactured with 100% waterproof adhesive. It readily accepts paint and is suited for structural siding, exterior color accented panels, soffits, and other applications where long-lasting paint or coating performance is required. See **Figure 3-26**.

MDO panels that are to be used outside should be edge-sealed as soon as possible. One or two coats of high-quality exterior house paint primer formulated for wood should be used. Edges are more easily sealed while the panels are in a stack.

HDO (high-density overlay) plywood is manufactured with a thermosetting resin-impregnated fiber surface that is bonded under heat and pressure to both sides of the panel. It is more rugged than MDO and suited for such punishing applications as concrete forming. The resin overlay requires no additional finish and resists abrasion, moisture penetration, and damage from common chemicals and solvents.

3.12.6 Hardwood Plywood Grades

The Hardwood Plywood Institute uses a numbering system for grading the faces and backs of panels. A grading specification of 1-2 indicates a good face with grain carefully matched and a good back, but without grain carefully matched. A No. 3 back permits noticeable defects and patching, but is generally sound. A special or **premium grade** of hardwood is known as "architectural" or "sequence-matched." This usually requires an order to a plywood mill for a series of matched plywood panels.

For either softwood or hardwood plywood, it is common practice to designate the grade by a symbol. G2S means "good two sides." G1S means "good one side."

In addition to the various types and grades, hardwood plywood is made with different core constructions. The two most common are the **veneer core** and the **lumber core**. See **Figure 3-27**. Veneer cores are less expensive, are fairly stable, and are warp-resistant. Lumber cores are easier to cut, have edges that are better for shaping and finishing, and hold nails and screws better. Plywood is also manufactured with a particleboard core. It is made by gluing veneers directly to a particleboard surface.

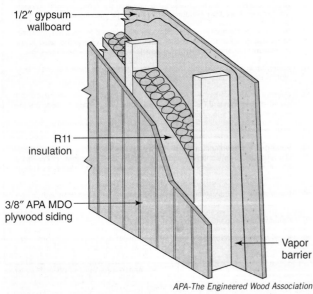

APA-The Engineered Wood Association

Figure 3-26. This sketch shows the makeup of APA Sturd-I-Wall, which uses MDO plywood as an exterior finish.

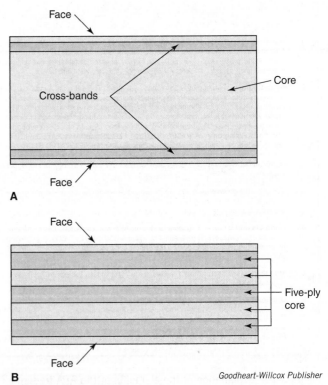

Goodheart-Willcox Publisher

Figure 3-27. Hardwood plywood is of two types. A—Veneer core. B—Lumber core.

3.12.7 Oriented Strand Board

Oriented strand board (OSB) is made of individual flakes of wood that are laid down like a blanket and adhered to each other with suitable resins and glues. The fibers are put down in successive layers arranged at right angles to one another. It is generally manufactured from aspen poplar in the northern part of North America and from southern yellow pine in the south. The panels are bonded under heat and pressure with thermosetting phenol formaldehyde or isocyanate binders. Panels range in thickness from 1/4″ to 1 1/8″. OSB manufactured in accordance with the rules of the APA bear a similar trademark and grade stamp to that found on plywood. The International Residential Code (IRC) and the National Building Code of Canada approve OSB for a variety of structural applications, provided it is used in accordance with the grade stamp, **Figure 3-28**. The applications of OSB are many:

- A combination subfloor and underlayment.
- Sheathing for roofs and exterior walls. Oriented strand board provides an excellent nailing base for siding. The large size of the panels also reduces air infiltration through the building's walls.
- APA-rated siding. When special surface treatments are used, oriented strand board is suitable as an exterior finish.
- Soffits.
- I-joists.

Goodheart-Willcox Publisher

Figure 3-28. Oriented strand board is approved by all building code organizations for a variety of construction applications.

3.12.8 Composite Board

Panels made up of a core of reconstituted wood with a thin veneer on either side are called *composite board* or composite panels. These materials are widely used in modern construction. They are good as sheathing, subflooring, siding, and interior wall surfaces.

3.12.9 Particleboard

In cabinetwork, hardboard and particleboard serve as appropriate materials for drawer bottoms and concealed panels in cases, cabinets, and chests. They are manufactured by many different companies and sold under various trade names.

Particleboard is made of wood flakes, chips, and shavings bonded together with resins or adhesives. It is not as heavy as hardboard (about 40 lb per cu ft compared to 50–80 lb per cu ft) and is available in thicker panels. Particleboard may be constructed of layers made of different-size wood particles. Large ones in the center provide strength. Fine ones at the surface provide smoothness. Particleboard is available in thicknesses ranging from 1/4″ to 17/16″. The most common panel size is 4′ × 8′.

Particleboard is used as a base for veneers and laminates. It is important in the construction of countertops, cabinets, drawers and shelving, many types of folding and sliding doors, room dividers, and a variety of built-ins.

This material is popular because it has a smooth, grain-free surface and is stable. Its surface qualities make it a popular choice as a base for laminates. Particleboard doors do not warp and require little adjustment after installation.

3.12.10 Hardboard

Hardboard is made of refined wood fibers pressed together to form a hard, dense material (50–80 lb per cu ft). There are two types: standard and tempered.

Tempered hardboard is impregnated (filled) with oils and resins. These materials make it harder, slightly heavier, more water resistant, and darker in appearance. Hardboard is manufactured with one side smooth (S1S) or both sides smooth (S2S). It is available in thicknesses from 1/12″ to 5/8″. The most common thicknesses are 1/8″, 3/16″, and 1/4″. Panels are 4′ wide and come in standard lengths of 8′, 10′, 12′, and 16′.

3.13 Wood Treatments

Wood and wood products that will be exposed to high levels of moisture should be protected from attack by fungi, insects, and borers. Millwork plants employ extensive treatment processes in the manufacture of items such as door frames and window units.

Structures that are continually exposed to weather—outside stairs, fences, decks, and furniture—should be constructed from *pressure-treated lumber* that has long-lasting resistance to termites and fungal decay. In addition, the lumber should be pressure-treated with a liquid repellent that slows the rate at which the moisture is absorbed and released. Pressure-treated lumber is used as landscape timbers and sill plates, and also in the construction of permanent wood foundations, decks, porches, and docks.

When wood is treated under pressure and with controlled conditions, the treatment deeply penetrates into the wood's cellular structure. Thus treated, the wood resists rot, decay, and termites. Even when exposed to severe conditions, the wood provides excellent service.

For many years, chromated copper arsenate (CCA) was the preservative material used in most treated wood products. The EPA ordered the phasing-out of CCA-treated lumber for residential applications by the end of 2003 because of the potential health hazards it presents.

Safety Note

The Environmental Protection Agency (EPA) warns that treated wood should *never* be used where the waterborne arsenical preservatives in the treatment may become a component of food, animal feed, or drinking water.

Replacements for CCA-treated material eliminate the arsenic hazard, but are somewhat more expensive. The most widely available replacement is lumber treated with an ammonia/copper preservative known as ACQ. Also available are copper azole-treated wood and borate-treated lumber. Borate preservatives are not toxic to animals or humans, but cannot be used in applications where moisture is constantly present.

The EPA has produced a consumer information sheet that outlines uses and handling precautions for treated wood products. Copies of the sheet are available at home improvement centers and lumberyards.

There are three major types of liquid preservatives:

- Waterborne—These preservatives are used for residential, commercial, recreational, marine, agricultural, and industrial applications.

Waterborne treatments leave the wood clear, odorless, and paintable. Waterborne preservatives (in addition to the phased-out CCA), include ACQ, borate, copper azole, acid copper chromate (ACC), and chromated zinc chloride (CZC).

- Oilborne—The best-known oilborne preservative is pentachlorophenol, which can be mixed with various solvents. It is highly toxic to insects and fungi. However, treated wood may become discolored. It is used on farms, around utility poles, and in industry.

- Creosote—This is a mixture of creosote and coal tar in heavy oil. It is suitable for treatment of pilings, utility poles, and railroad ties. Creosoted wood cannot be painted.

Often, treated lumber carries an identifying end tag or stamp, **Figure 3-29**. This tag or stamp indicates the type of preservative, proper exposure conditions, and other information. More information on treated lumber can be obtained by contacting the following organizations:

- American Wood Preservers' Association
- Wood Preservation Canada
- Southern Forest Products Association
- Southern Pine Inspection Bureau

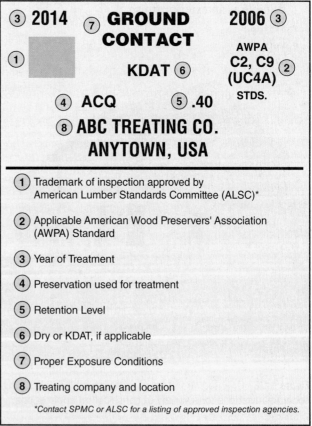

Southern Pine Council

Figure 3-29. An end tag identifies treated lumber for ground-contact use. Other tags denote lumber suitable for above ground use.

3.14 Handling and Storing Lumber

Building materials are expensive and should be maintained in the best condition that is practical. After they are delivered to the construction site, this becomes the responsibility of the carpenter. Piles of framing lumber and sheathing should be laid on level skids raised at least 6″ above the ground. Be sure all pieces are supported and lying straight. Cover the material with canvas or waterproof paper. Polyethylene film also provides a watertight covering.

If moisture absorption is likely, cut the steel banding on panel materials to prevent edge damage when the fibers expand. Keep coverings open and away from the sides and bottom of lumber stacks to promote good ventilation. Tight coverings promote mold growth.

Exterior finish materials, door frames, and window units should not be delivered until the structure is partially enclosed and the roof surfaced. In cold weather, the entire structure should be enclosed and heated before interior finish and cabinetwork are delivered and stored.

APA structural wood panels delivered on site should be properly stored and handled. Protect ends and edges from damage. Whenever possible, store panels under a cover. Keep sanded panels and other appearance grades away from high traffic areas. Weigh down the top panel to prevent warpage.

Panels stored outdoors should be stacked on a level surface atop 4 × 4 stringers or other blocking. Use at least three stringers. Never allow panels to touch the ground. Keep panels well ventilated by separating them with wood strips to prevent mildew growth.

When finish lumber is received at a moisture content different from what it will attain in the structure, it should be open-stacked with wood strips so air can circulate around each piece. Plywood, especially the fine hardwoods, must be handled with care. Sanded faces can become soiled and scarred if not protected. While stored, the panels should lie flat.

3.15 Engineered Lumber

Builders have long relied on traditional wood products for residential and light commercial framing. Modern tree-farming technology has ensured that this material continues to be available, keeping up with the increased demand for wood structures. It once took more than 80 years to grow a tree to sawmill size; tree farming has shortened the span to 20 to 30 years. In the process, unfortunately, the tree's annular rings, which determine its tensile and comprehensive strength, are much farther apart, resulting in weaker dimension lumber.

The industry has found a way to strengthen lumber and use more of the tree through the development of *engineered lumber products (ELP)*.

Engineered lumber products include those wood structural units that have been altered through manufacturing processes to make them stronger, straighter, and more dimensionally stable than sawn lumber. These processes produce more lumber products from less timber and can utilize smaller trees that are unsuitable as sawn lumber. Where sawn lumber typically uses only 40% of a log, engineered materials use 75% of the log for structural lumber. At the same time, the manufacturing processes require less energy than those used for solid lumber.

Components are glued together in different configurations—some are solid, some are shaped like I-beams, and some are in truss form. Engineered lumber products include the following:

- Laminated-veneer lumber (LVL)
- Glue-laminated beams (glulams)
- I-joists
- Open-web trusses
- Parallel-strand lumber (PSL)
- Laminated-strand lumber (LSL)

3.15.1 Laminated-Veneer Lumber

Laminated-veneer lumber (LVL), **Figure 3-30**, is produced under various trade names, such as Microlam™, Parallam™, and Gang-Lam™. It is produced much like plywood. Various species of wood may be used, but Douglas fir and southern pine are preferred for their superior strength.

Goodheart-Willcox Publisher

Figure 3-30. LVL is available in thicknesses of 1 1/4″, 1 3/4″, and 3 1/2″, widths from 7 1/4″ to 24″, and any length needed.

In the manufacturing process, a log is first debarked, then cut into thin (1/10″–3/16″ thick) sheets that are 27″–54″ wide. Once the sheets are dried, the next step removes defects and coats the veneer sheets with phenolic glue. Next, sheets are stacked with their ends randomly staggered so they overlap. A heat press compresses and cures the assembly, which then is ready to be cut to specified lengths and widths.

Another widely used process cuts the veneer into 1/2″ wide strips. These strips are coated with waterproof phenolic glue and fed into a series of automated machines. The resulting billet comes out measuring 12″ × 17″ × 66′. The billet can be cut into sizes commonly used by the construction industry. The laminate material is used as headers, beams, and columns. LVL is manufactured in grades from 1.3E to 2.0E, with 2.0E being the stiffest and the strongest. The E in LVL grades stands for modulus of elasticity or bending strength.

LVL products are dimensionally stable, uniform in size throughout, predictable in performance, and can be nailed, drilled, and cut with ordinary construction tools. LVL products easily take stain and can be left exposed, where they become a natural part of the interior decor.

Although a cross section of LVL looks somewhat like plywood, there is an important difference—grain orientation. In conventional plywood, each veneer layer is placed at a 90° angle to the previous layer. In LVL, all of the veneer panels have their grain running in the same direction. Called *parallel lamination*, this provides greater strength, allowing the member to carry the heavy loads required of beams, headers, and truss components. Standard laminated beams are 1 3/4″ thick with depths of 5 1/2″, 7 1/4″, 9 1/2″, 11 7/8″, 14″, 16″, and 18″.

3.15.2 Glue-Laminated Beams

Glue-laminated beams (glulams) are made by gluing and then applying heavy pressure to a stack of four or more layers of 1 1/2″ thick stock. The result is a beam that can be up to 30″ deep, **Figure 3-31**. The layers, called lams, are arranged to produce maximum strength.

Types

Manufacturers produce two types of glulams: balanced and unbalanced. Balanced glulams are manufactured with similar lumber grades at both the top and bottom portions of the beam. Balanced glulams are typically used where both the top and the bottom of the beam may be under tension, such as in multiple spans.

An unbalanced glulam uses higher grade lumber on one side of the beam, so it resists tension best on that side. Unbalanced glulams are most economical for single-span beams.

Rik Vandermeulen, Unalam

Figure 3-31. Glulams can be manufactured with curves and great depths. They are often an architectural highlight of a building.

Glulams always come with one edge marked "top." This edge must be installed upward.

Glulams are normally used (either exposed or hidden) for rafters, floor-support beams, and stair stringers. Manufacturers also produce glulams in curved shapes for supporting arched roofs such as those found in churches or other public buildings. The appearance grade of a glulam has no bearing on its strength. Three grades of appearance are available.

- Industrial appearance—Designed for use where appearance is not important.
- Architectural appearance—Where appearance is important.
- Premium appearance—When appearance is critical.

Fire resistance is a significant advantage of glue-laminated beams. They are slow to ignite, and they burn slowly.

Sizes

Glulams are sold in both custom and stock sizes. Manufacturers produce to customer specifications for long spans, unusually heavy loads, or other conditions. Stock beams are manufactured in commonly used dimensions. Distributors and dealers cut them to order for customers. Typical stock widths are 3 1/8″, 3 1/2″, 5 1/8″, 5 1/2″, and 4 3/4″. These meet the requirements of most residential applications.

3.15.3 I-Joists

I-joists have, as the name suggests, an I shape. They consist of flanges (also called chords) of structural composite lumber, such as laminated-veneer lumber or sawn lumber. I-joists also consist of a vertical web of oriented strand board that is 3/8″ or 7/16″ thick. The web is fastened into grooves in the flanges using waterproof glue. See **Figure 3-32**.

Goodheart-Willcox Publisher

Figure 3-32. I-joist with top and bottom flanges consisting of either laminated-veneer lumber or solid finger-jointed lumber and OSB web. To avoid damage, always store I-joists with web upright.

I-joists are light, straight, and strong. They are designed to span long distances. Though mostly used as joists and rafters, I-joists may be used as headers, too. Most mills deliver them in lengths up to 60′. Distributors and dealers cut them to frequently used lengths (16′–36′) specified by builders. Most common depths are 9 1/2″, 11 1/2″, 14″, and 16″.

An added feature of the I-joist is that it is often manufactured with knockouts scored into the web to allow access for plumbing and electrical services. To preserve the strength engineered into I-joists, the APA specifies the size and location of holes cut into the web. These are found in Appendix B, **Technical Information**. A table of dimensions and span tables is also shown.

I-joists manufactured to APA specifications carry a trademark indicating conformance to their PRI Standard 400. **Figure 3-33** is a sample of this trademark with an explanation of its markings.

3.15.4 Open-Web Trusses

Open-web trusses are sometimes used in place of floor joists, **Figure 3-34**. This is especially true when dealing with long spans. These trusses are fabricated in factories from solid 2×4 lumber. Sometimes, steel webs are used instead of wood. Depths of 14″ and 16″ are most common. The open webs of the trusses allow installation of pipes, wiring, drains, and other mechanical systems without having to cut openings, as in other types of joists.

Goodheart-Willcox Publisher

Figure 3-34. Open-web trusses as shipped from the factory. Webs and chords are fastened with metal plate connectors. The arrow points to offsets that are designed to receive a ribbon. This ribbon ties joists together and holds them upright.

SAMPLE TRADEMARK – Position of trademark on I-joist may vary by manufacturer

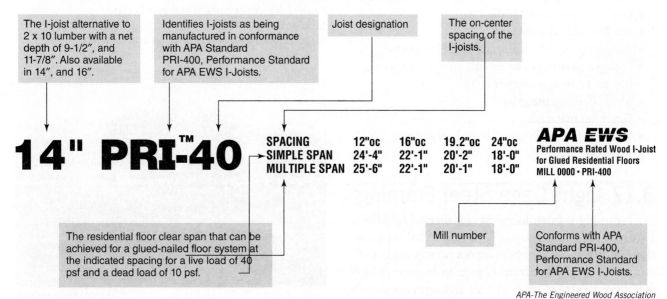

The I-joist alternative to 2 x 10 lumber with a net depth of 9-1/2″, and 11-7/8″. Also available in 14″, and 16″.

Identifies I-joists as being manufactured in conformance with APA Standard PRI-400, Performance Standard for APA EWS I-Joists.

Joist designation

The on-center spacing of the I-joists.

14" PRI™-40

	12"oc	16"oc	19.2"oc	24"oc
SPACING				
SIMPLE SPAN	24'-4"	22'-1"	20'-2"	18'-0"
MULTIPLE SPAN	25'-6"	22'-1"	20'-1"	18'-0"

APA EWS
Performance Rated Wood I-Joist
for Glued Residential Floors
MILL 0000 · PRI-400

The residential floor clear span that can be achieved for a glued-nailed floor system at the indicated spacing for a live load of 40 psf and a dead load of 10 psf.

Mill number

Conforms with APA Standard PRI-400, Performance Standard for APA EWS I-Joists.

APA-The Engineered Wood Association

Figure 3-33. This trademark appears on wood I-beams manufactured to APA specifications. Spacing and span information save builders' time.

3.15.5 Parallel-Strand Lumber

To make *parallel-strand lumber (PSL)*, the manufacturer lays up 1″ wide by 8′ long strands of veneer that have been coated with adhesive. Bonded under pressure, the veneer strips form billets up to 11″ thick by 19″ deep. Billets can be extruded to any length, but 66′ is usual. Lumber yards cut the billets to specified lengths for customers. There are five common thicknesses: 1 3/4″, 2 11/16″, 3 1/2″, 5 1/4″, and 7″. PSL products are used as exposed posts and beams. When pressure-treated, they are suitable for outdoor use in porches, decks, and gazebos.

3.15.6 Laminated-Strand Lumber

Laminated-strand lumber (LSL) is laid up from 1/32″ × 1″ × 12″ strands of wood bonded with polyurethane adhesive. It is available in two thicknesses: 1 1/4″ and 3 1/2″. Depths vary up to a maximum 16″ and lengths vary up to 35′. Uses include door and window headers, rim joists in floors, and core stock for flush doors with veneer overlays.

Green Note

Precut, custom, and fabricated components can be cut to specification at the lumber mill and shipped to the jobsite. This reduces jobsite waste due to material misuse, poor storage, carpenter error, and other mishaps.

3.16 Nonwood Materials

The carpenter works with a number of materials other than lumber and wood-based products. Some of the more common items include the following:

- Metal framing members, especially joists and studs
- Gypsum and metal lath
- Wallboard and sheathing
- Insulating boards, batts, and loose insulation
- Siding and shingles of asphalt, metal, fiberglass, tile, and concrete, **Figure 3-35**
- Metal flashing material
- Caulking materials
- Resilient flooring materials and carpeting
- Bonding materials

3.17 Light Gage Steel Framing

Steel framing is often found in commercial buildings, **Figure 3-36**, and occasionally in residential construction. Steel framing members are manufactured in various widths, and range from 12 gage to 26 gage. In light frame construction, studs over 10′ in length are usually made of 20 gage steel. Tracks and studs less than 10′ long are usually 26 gage. Joists and rafters are typically 16 or 18 gage. The lower the gage number, the thicker the metal.

Steel studs are attached to base and ceiling channels with screws. A typical stud consists of a metal channel with openings through which electrical and plumbing lines can be installed.

Wall surface material, such as drywall or paneling, is attached to the metal stud with self-tapping drywall screws. Metal stud systems are usually used for nonloadbearing walls and partitions. Metal framing materials are covered in greater detail in Chapter 13, **Framing with Steel**.

CertainTeed Corporation

Figure 3-35. Fiber cement siding is becoming increasingly popular because it is durable, resists fading, and can be manufactured in a variety of styles and patterns.

Goodheart-Willcox Publisher

Figure 3-36. Steel framing is sometimes used in modern residential and light commercial buildings.

3.18 Metal Framing Connectors

At one time, all wood-to-wood connections in a wood frame were made with nails alone. While this practice still continues and produces adequate strength for the structure, metal connectors are faster to install and improve uniformity in strength. They may be required by code in areas that experience high winds.

Basically, metal framing connectors are stamped brackets or strapping designed to make wood-to-wood, wood-to-masonry, or wood-to-concrete connections. Unless the connector is designed to be exposed and decorative, it is made from galvanized metal in various gages (thicknesses).

Strapping or ties are designed to hold parts of a frame together, **Figure 3-37**. They are often used to "quake-proof" structures by tying frames to foundations and roofs to walls.

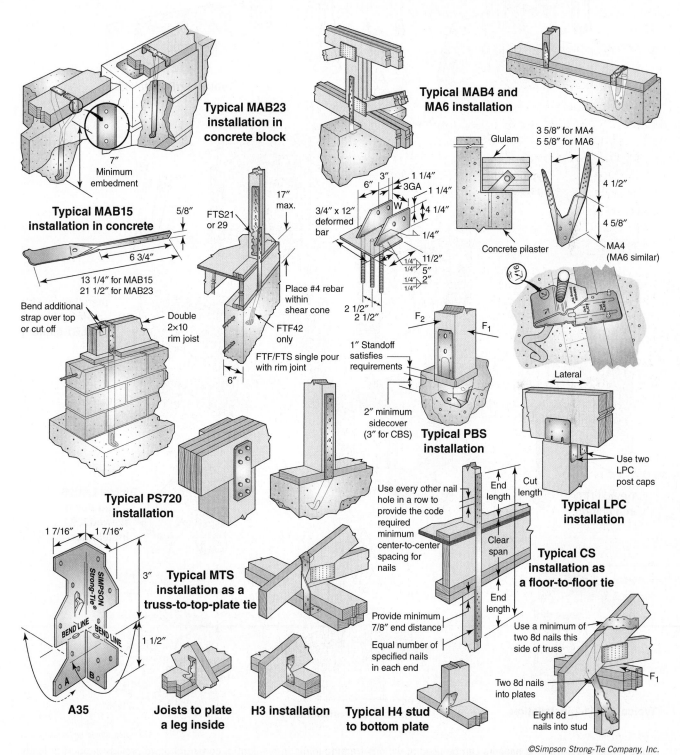

©Simpson Strong-Tie Company, Inc.

Figure 3-37. Framing connectors and ties are made in many different configurations to secure walls to foundations and roofs to walls.

Straps are perforated so that they can be fastened with nails without first drilling holes.

Hangers connect the end of one framing member to another. Hangers are used where floor or ceiling joists intersect another framing member, such as a beam,

Figure 3-38. Designs are available for solid lumber, laminated lumber, or I-joists.

Other connectors are designed for special purposes. ***Tension bridging*** can be used instead of solid wood or wood braces to transfer loads from joist to joist.

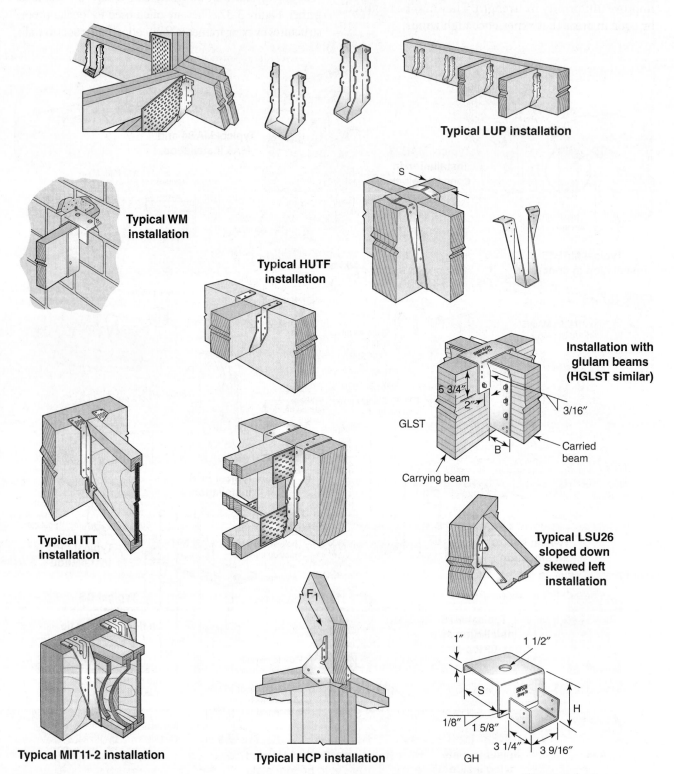

Typical LUP installation

Typical WM installation

Typical HUTF installation

S

Installation with glulam beams (HGLST similar)

5 3/4″

2″

3/16″

GLST

B

Carried beam

Carrying beam

Typical ITT installation

Typical MIT11-2 installation

F₁

Typical HCP installation

Typical LSU26 sloped down skewed left installation

1″

1 1/2″

S

H

1/8″   1 5/8″

3 1/4″   3 9/16″

GH

©Simpson Strong-Tie Company, Inc.

Figure 3-38. Wood-framing hangers can be used at various framing joints in residential construction to provide support of loads and stresses placed on joists and beams.

They can be used on either solid lumber joists or on wood I-beams. Metal corner braces are another type of tension connector. See **Figure 3-39**.

Fasteners used for metal connectors should be able to withstand shear stresses placed on them by the connectors.

Bolts may sometimes be used, but nails or screws are most common. Some builders secure the fasteners with drywall screws, which penetrate wood quickly and cannot be easily withdrawn. However, drywall screws have little shear strength and are not recommended.

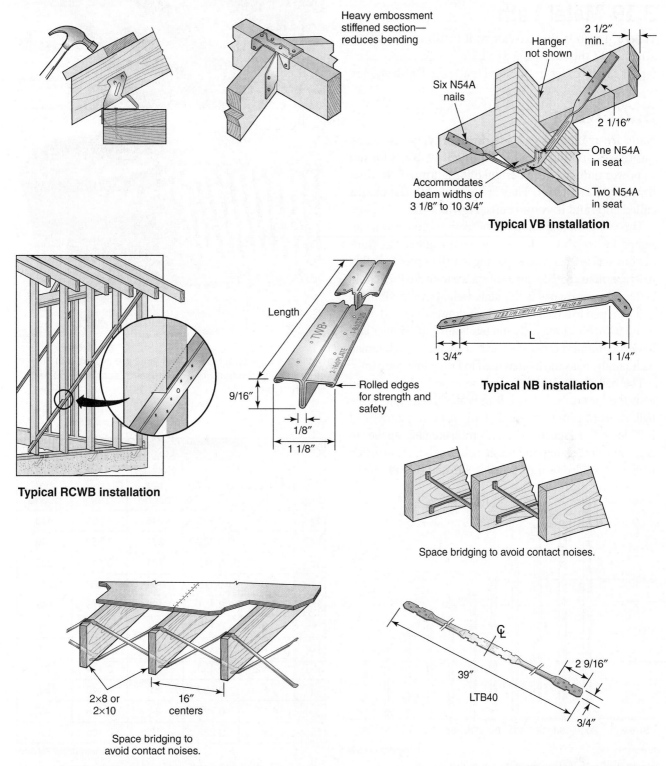

Heavy embossment stiffened section—reduces bending

Hanger not shown

2 1/2″ min.

Six N54A nails

2 1/16″

One N54A in seat

Accommodates beam widths of 3 1/8″ to 10 3/4″

Two N54A in seat

Typical VB installation

Length

TVB

9/16″

1/8″

1 1/8″

Rolled edges for strength and safety

1 3/4″   L   1 1/4″

Typical NB installation

Typical RCWB installation

Space bridging to avoid contact noises.

2×8 or 2×10   16″ centers

Space bridging to avoid contact noises.

39″

2 9/16″

LTB40

3/4″

©Simpson Strong-Tie Company, Inc.

Figure 3-39. Some types of strapping and ties are designed to work under tension to prevent joists from tipping or building frames from racking.

Generally, only nails should be used. Nail length varies with the type of connector. Manufacturer's literature should be consulted, as many manufacturers have specific requirements for the nails to be used with their connectors or hangers.

3.19 Metal Lath

Where stucco is the exterior wall finish, metal lath is attached to the sheathing as a base for the stucco plaster. The lath comes in rolls and is attached with staples.

3.20 Nails

Nails are available in a wide range of types and sizes. Basic sizes are illustrated in **Figure 3-40**. Nails for use in power nail guns are designed for automatic feeding through the nail gun. They are provided in flat blanks called clips and in round coils, **Figure 3-41**.

The common nail has a heavy cross section and is designed for rough framing and general nailing. The thinner box nail is used where splitting may be a problem. Box nails are often coated with a substance to make them hold better. The casing nail is the same weight as the common nail and used to attach door and window casings and other wood trim. Finishing nails and brads are quite similar and have the thinnest cross section and smallest head. Finish nails are the nails most often used to fasten interior trim.

The nail size unit is called a penny and is abbreviated with the lowercase letter d. It indicates the length of the nail. A 2d (2 penny) nail is 1″ long. A 6d (6 penny) nail is 2″ long. See **Figure 3-42**. This measurement applies to common, box, casing, and finish nails. Brads and small box nails are specified by their actual length and gage number.

Irra/shutterstock.com

Figure 3-41. Clips for power nail guns are available in angled or straight, and round head or clipped. Each gun requires a specific type of clip.

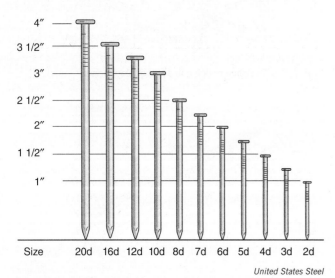

United States Steel

Figure 3-40. Nail sizes are given in a unit called the "penny." It is written as d.

Size	Length (in)	Common		Box	
		Diameter (in)	Number per pound	Diameter (in)	Number per pound
4d	1 1/2	0.102	316	0.083	473
5d	1 3/4	0.102	271	0.083	406
6d	2	0.115	181	0.102	236
7d	2 1/4	0.115	161	0.102	210
8d	2 1/2	0.131	106	0.115	145
10d	3	0.148	69	0.127	94
12d	3 1/4	0.148	63	0.127	88
16d	3 1/2	0.165	49	0.134	71
20d	4	0.203	31	0.148	52
30d	4 1/2	0.220	24	0.148	46
40d	5	0.238	18	0.165	35

Goodheart-Willcox Publisher

Figure 3-42. Nail sizes and approximate number in a pound are shown on this chart.

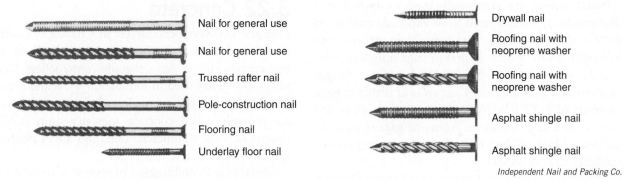

Nail for general use

Nail for general use

Trussed rafter nail

Pole-construction nail

Flooring nail

Underlay floor nail

Drywall nail

Roofing nail with neoprene washer

Roofing nail with neoprene washer

Asphalt shingle nail

Asphalt shingle nail

Independent Nail and Packing Co.

Figure 3-43. Annular- and spiral-threaded nails are designed for special purposes.

A few of the many nails designed for special purposes are shown in **Figure 3-43**. Annular or spiral threads greatly increase holding power. Some nails have special coatings of zinc, cement, or resin. Coating or threading increases nail-holding power. Nails are made from materials such as iron, steel, copper, bronze, aluminum, and stainless steel.

3.21 Screws

Screws have greater holding power than nails and are often used for interior construction and cabinetwork. The length and diameter (gage number) determine their size. Screws are classified according to the shape of head, surface finish, and the material from which they are made. There are nearly as many types of screws available for use in construction as there are uses for them. Only the most common are discussed here.

The most common construction screws have a bugle head. Deck screws are coarse thread screws with an extra sharp point for use in deck building. They may be made of steel with a corrosion-resistant coating or stainless steel. Both Philips drive and square drive are available. Coarse-thread drywall screws, also known as particleboard screws, have wide-spaced threads and are considered standard for attaching drywall to studs. Fine thread drywall screws have twinfast threads and are ideal for attaching drywall to thin (20–25 gage) metal studs. Drywall and deck screws are generally driven with an electric or battery screw gun. Their aggressive thread pattern and thin shank allow them to be installed without drilling a pilot hole.

Wood screws, like those shown in **Figure 3-44,** are generally installed by hand with a screw driver. Unlike drywall and deck screws, they do require a pilot hole and are generally countersink or counterbore, **Figure 3-45**. A countersink pilot hole is angled at the top. A counterbore pilot hole has side perpendicular to the top. This extra space below the surface creates room for the screw head to be level with or below the surface.

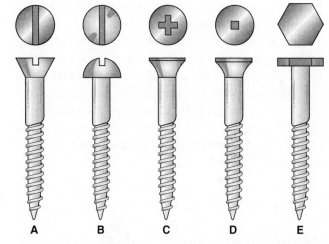

A   B   C   D   E

Goodheart-Willcox Publisher

Figure 3-44. Types of heads found on common screws. A—Flat head, straight slot. B—Round head, straight slot. C—Bugle head, Phillips drive. D—Bugle head, square drive. E—Hex head.

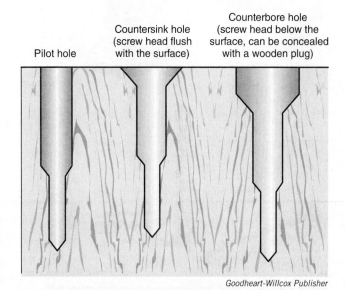

Pilot hole

Countersink hole (screw head flush with the surface)

Counterbore hole (screw head below the surface, can be concealed with a wooden plug)

Goodheart-Willcox Publisher

Figure 3-45. Countersink and counterbore pilot holes are created by using a countersink drill bit, which drills both the pilot hole and the recess for the screw head.

Wood screws are commonly used in applications that require the screw head to be exposed, such as on door hinges or wooden furniture. They are available in lengths from 1/4″ to 6″ and in gage numbers from 0 to 24. The gage number can vary for a given length of screw. For example, a 3/4″ screw is available in gage numbers of 4–12. The No. 4 is a thin screw, while the No. 12 has a large diameter. From one gage number to the next, the size of the wood screw changes by 13 thousandths (.013) of an inch.

Most wood screws used today are made of mild steel with a zinc chromate finish. Nickel- and chromium-plated screws, as well as screws made of brass or stainless steel, are available for special work. Wood screws are usually priced and sold by the box.

Sheet metal screws are often used by carpenters to assemble light-gauge metal framing. They are available in self-piercing and self-drilling varieties, **Figure 3-46**.

Additional useful fasteners include lag screws, drywall anchors, hanger bolts, carriage bolts (specially designed for woodwork), corrugated fasteners, and metal splines. Specialized fasteners are described in other sections of this book.

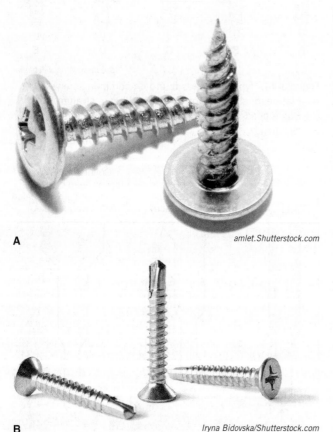

A   amlet.Shutterstock.com

B   Iryna Bidovska/Shutterstock.com

Figure 3-46. A—Self-piercing screws have a sharp point capable of penetrating thin steel. B—Self-drilling screws are equipped with a drill point to cut through steel.

3.22 Concrete

Concrete is covered extensively in Chapter 9, **Footings and Foundations**. It is the material of choice in most foundations and all slab-on-grade construction. Blocks, which are widely used in foundation walls and even for entire building walls, are made entirely of concrete. They are properly called concrete masonry units (CMUs). Frames of some commercial buildings are of poured or prefabricated concrete.

Concrete is a combination of several materials: Portland cement, sand, crushed stone, and water. The sand and crushed stone are called aggregates. Additives are included in the mix to provide desired characteristics. Portland cement is a manufactured product made from limestone mixed with shale, clay, or marl. These materials are combined in carefully measured proportions. When water is added, the mixture becomes plastic and is easily shaped in forms. The water also sets in motion a chemical reaction called hydration. It is this action, not drying, that cures concrete and imparts the desired strength. Too much water can cause the cement to separate from the aggregate and produce a weaker final concrete. Too little water will prevent full curing of the concrete.

3.23 Adhesive Bonding Agents

Bonding materials that carpenters use include glues, adhesives, cements, and mastics. Research and development have created many new products in this area. Glues come from natural materials; adhesives are developed from synthetic materials; cements and mastics are rubber-based. Some are highly specialized, designed for use with a specific material or application. Brief descriptions of several of the commonly used bonding materials follow.

Polyvinyl resin emulsion adhesive (generally called yellow glue or white glue) is excellent for interior construction. It comes ready to use in plastic squeeze bottles and is easily applied. This adhesive sets up rapidly, does not dull tools or stain the wood, and securely holds wood parts. The original polyvinyl glue was white glue. Yellow glue has additives that make it less runny and causes it to set more quickly, making it easier to use. Yellow glue and white glue have the same holding power and are stronger than the wood.

Polyvinyls harden when their moisture is absorbed by the wood or evaporates. They are not waterproof, so they should not be used for assemblies that will be subjected to high humidity or moisture. The vinyl-acetate resins used in the adhesive are ***thermoplastic***.

Under heat they will soften. They should not be used in situations where the temperature may rise above 165°F (75°C).

Yellow glues have some benefits over white glues. Since they are more heat-resistant, they do not clog sandpaper. Some newer forms of yellow glue are made with cross-linked polymers. This makes them even stronger with much better water resistance. They still can be cleaned up with water.

Thermoset adhesives are more water- and heat-resistant than the polyvinyl adhesives. In this category are urea formaldehydes and resorcinol formaldehydes. These glues have longer curing times, but much better water and heat resistance than polyvinyl glues. *Urea-formaldehyde resin glue* (usually called urea resin) is available in a dry powder form that contains the hardening agent, or catalyst. It is mixed with water to a thick, creamy consistency before use. Joints should be clamped securely during curing time. Urea resin is moisture-resistant, dries to a light brown color, and securely holds wood. It hardens through chemical action when water is added and sets in four to eight hours, depending on room temperature.

Resorcinol resin adhesive is a two-part mixture of resin and curing agent or catalyst. It is a waterproof adhesive suitable for exterior use. Resorcinol leaves a pronounced dark glue line, so it is not used where a clear finish will be applied.

Polyurethane adhesives are known as single-component, moisture-catalyzed adhesives. They use moisture from the air or wood to set. As they set up, they foam and spread themselves. This action also fills any gaps. Polyurethanes have few or no solvents, are completely waterproof, and will take stain. They can be applied at temperatures ranging from 40°F (5°C) to 90°F (30°C) and will bond with wood having up to 25% moisture content.

Surfaces must fit snugly and be free of dust, oil, wax, old glue, or paint. On wood, it is good practice to wipe each surface with a damp cloth about a minute before gluing.

To wipe up excess polyurethane adhesive, use mineral spirits or acetone. If this is not practical (since the adhesive may foam for up to an hour) trim off excess with a chisel or scraper after the material has set for several hours.

Epoxy is a polymer (synthetic material) that is often used as an adhesive. It consists of two parts that are mixed at the time of use. One part is the epoxy resin that forms the main body of the adhesive. The second part is a hardener or catalyst that causes the chemical reaction to cure the epoxy. Epoxy is exceptionally strong, forms a very strong bond with many base materials, and is waterproof. It is expensive and does not penetrate the surface of wood, so it is not used to bond wood.

Contact cement is made with a neoprene or latex rubber base. It is an excellent adhesive for applying plastic laminates or joining parts that cannot be easily clamped together. It works well for applying thin veneer strips to plywood edges and can also be used to join combinations of wood, cloth, leather, rubber, sheet foams, and plastics.

To use contact cement, apply it to each surface with a brush, short nap roller, or finishing trowel. Allow to dry until a piece of paper will not stick to the film. Surfaces must never be allowed to touch before they are in perfect alignment. Carefully align cemented surfaces and firmly press them together. Bonding takes place immediately. The pieces must be carefully aligned for the initial contact, because they cannot be moved after they touch. The bonding time is not critical and can usually be performed anytime within one hour.

Some contact cement contains volatile, flammable solvents. Thoroughly ventilate the work area where it is applied. Carefully read directions on the container for additional handling instructions.

Safety Note
Never use flammable contact cement in a building with open flame, such as a space heater, kitchen range pilot light, or furnace. The fumes are very flammable and may result in explosion. If contact cement must be used in such a space, use water-based contact cement.

Casein glue is made from milk curd, hydrated lime, and sodium hydroxide. It is supplied in powder form and is mixed with cold water for use. After mixing, it should sit for about 15 minutes before it is applied. It is classified as a water-resistant glue.

Casein glue is used for structural laminating and works well with wood that has high moisture content. It has good joint filling qualities and, therefore, is often used on materials that have not been carefully surfaced. Casein glue is used for gluing oily woods such as teak, padouk, and lemon wood. Its main disadvantages are that it stains the wood (especially such species as oak, maple, and redwood) and has an abrasive effect on tool edges.

3.23.1 Mastics

There is a wide array of *mastic* products, including waterproofing materials and adhesives. Some mastics are liquids applied with a brush or roller, such as those used for foundation dampproofing. Some are heavy, paste-like adhesives applied with a notched trowel or a caulking gun. They vary in their characteristics and application methods, and are usually designed for a specific type of material. Some are waterproof; others must be used where there is no excessive moisture.

One application method consists of placing several gobs on the surface of the material and then pressing the unit firmly in place. This causes the mastic to spread over a wider area. Some mastics are spread over the surface with a notched trowel. Still others are designed for caulking gun application.

Mastics are usually packaged in metal or plastic containers or in gun cartridges and are ready for application. Always follow the directions of the manufacturer.

3.23.2 Bonding Basics for Adhesives

In the United States, the construction industry classifies glue's resistance to water using three ratings:

- Type I includes those adhesives that can be submerged in water indefinitely with no loss of bond strength.
- Type II designates adhesives that can be soaked in water for four hours, and then dried for 19 hours with no loss of bond strength at 120°F (50°C).
- Type III adhesives have little or no resistance to water.

3.23.3 Open Time and Other Variables

Also called setting time, *open time* is the amount of time between spreading the material and when the parts must be clamped. This is not important on small parts like casing or baseboard miters, but is critical when large surfaces must be glued. Setting time varies with temperature and humidity.

Temperature is closely related to open time. Standard open time ratings are typically based on application with a temperature between 65°F (20°C) and 75°F (25°C). At these temperatures, adhesives easily spread and have open times as indicated by the manufacturer. Above or below this temperature range, viscosity and open time change.

Squeeze-out glue can create extra work for finishers if it is not promptly cleaned off after clamping. Some glues can be cleaned with a damp rag, while others require mineral spirits or thinner. Some finishers prefer that glues be allowed to dry before the excess is removed with a chisel. It is always best to ask which is preferred.

3.24 Staples

Staples are sometimes used as a fastener for cabinetry, trim, insulation, underlayment, ceiling tile, and roofing applications. Pneumatic staplers, electric staplers, and hand staplers are used to drive the staple into the materials being fastened.

Staples are available in various lengths, widths, and thicknesses (gages). Some staples are coated to provide electrical insulation or corrosion protection. Staple point types include spear, chisel, inside chisel (legs move outward as staple penetrates), outside chisel (legs move inward), crosscut chisel, divergent (legs move to either side), and outside chisel divergent (legs move inward and to the sides).

3.25 Specialty Fasteners

A variety of additional fasteners are used in carpentry and construction. Various types of bolts, wallboard anchors, and concrete anchoring systems exist. Bolts are used to fasten large wood and steel members to one another. Wallboard anchors are used to attach items to gypsum wallboard, a product that does not hold a screw or nail on its own. Concrete anchoring systems are used to make attachments to concrete. They can range from light-duty anchors, such as plastic inserts that accept a screw, to heavy-duty anchors such as expansion bolts, which are hammered into a hole drilled in the concrete. As a nut is tightened on the bolt, its tapered end is wedged tight in the concrete, giving it exceptional holding strength.

Specialty fasteners must be installed and used exactly as directed by the manufacturer. Always read product literature and follow instructions carefully. Failure to do so can lead to costly damage.

Chapter Review

Summary

- A carpenter should be familiar with the different types of materials used in construction.
- In construction, wood is one of our greatest natural resources.
- Lumber is the name given to wood when it is cut into uniform pieces for building material. Lumber can be classified as either hardwood or softwood.
- Wood is used in construction in the form of boards, dimension lumber, or panels.
- Panels may be made as laminated layers (plywood) or composite material of wood particles or chips and a resin binder.
- Dimensional lumber is solid wood sawn to size.
- Engineered lumber consists of layers of material laminated together with adhesives or I-joists combining solid and OSB components. Engineered lumber is typically stronger and straighter than traditional dimension lumber.
- Metal framing materials, fabricated from thin steel, are widely used in commercial construction, but are sometimes used in residential construction.
- Steel joist hangers and other metal framing connectors improve joint strength in lumber construction. They are fastened together with screws or welded joints.
- Various types of construction adhesives are used in construction. Applications and specific instructions vary. Always follow manufacturers recommendations.

Know and Understand

Answer the following questions using the information in this chapter.

1. Construction materials include _____.
   A. mortar, concrete, and masonry units
   B. composition panels, engineered lumber, and sawed lumber
   C. steel and aluminum, vapor barriers, and gypsum board
   D. All of the above.

2. *True or False?* Dimension lumber is used primarily for posts, beams, and heavy stringers.

3. *True or False?* All hardwood trees produce wood that is harder than that of softwood trees.

4. Which of the following kinds of wood is classified as a hardwood?
   A. Hemlock   C. Willow
   B. Redwood   D. All of the above.

5. Wood reaches its equilibrium moisture content when the _____.
   A. moisture content in the wood is the same as the moisture content in the air
   B. moisture content in the wood has been reduced to 0%
   C. cells of the wood are empty, but the cell walls are still full of moisture
   D. wood has been kiln dried

6. Wood reaches its fiber saturation point when the _____.
   A. moisture content in the wood is the same as the moisture content in the air
   B. moisture content in the wood has been reduced to 0%
   C. cells of the wood are empty, but the cell walls are still full of moisture
   D. wood has been kiln dried

7. *True or False?* The screw size unit is called a penny.

8. What are the dimensions of a 2×4 that has been dressed and dried?
   A. $2'' \times 4''$ C. $1\ 1/2'' \times 3\ 1/4''$
   B. $1\ 3/4'' \times 3\ 1/2''$ D. $1\ 1/2'' \times 3\ 1/2''$

9. What are the dimensions of a 2×8 that has been dressed and dried?
   A. $2'' \times 8''$ C. $1\ 1/2'' \times 7\ 1/4''$
   B. $1\ 3/4'' \times 7\ 1/2''$ D. $1\ 1/2'' \times 7\ 1/2''$

10. An engineered lumber product that is made up of veneer layers and is designed to carry heavy loads is called _____.
    A. glue-laminated beam
    B. laminated–veneer lumber
    C. I-joist
    D. open-web truss

11. Wood treated with _____ has been phased out of use.
    A. ammonia/copper
    B. chromated copper arsenate
    C. borate
    D. copper azole

12. *True or False?* Light gage steel framing is used only in commercial construction.

13. *True or False?* Metal framing connectors are stamped brackets or strapping designed to make wood-to-wood, wood-to-masonry, or wood-to-concrete connections.

14. *True or False?* Hangers are designed to hold parts of a frame together.

15. *True or False?* The common nail has a heavy cross section and is used where splitting may be a problem.

16. *True or False?* Tension bridging may be used to transfer loads from joist to joist.

For questions 17–26, match the lumber defect with its description:

A. Knots	F. Wane
B. Splits and checks	G. Blue stain
C. Shakes	H. Decay
D. Pitch pockets	I. Holes
E. Honeycombing	J. Warp

17. Cavities that contain or have contained pitch in solid or liquid form.

18. Separations along the grain and between the annular growth rings.

19. Disintegration of wood fibers due to fungi.

20. Caused by an embedded branch or limb.

21. Any variation from true or plane surface.

22. Separations of the wood fibers that run along the grain and across the annular growth rings.

23. Separation of the wood fibers inside the tree.

24. Caused by a mold-like fungus.

25. Openings caused by handling equipment or boring insects and worms.

26. The presence of bark or the absence of wood along the edge of a board

Apply and Analyze

1. What is the natural cement that holds wood cells together?

2. When a softwood log is cut so the annular rings form an angle greater than 45° with the surface of the boards, what is the lumber called?

3. What is the moisture content of a board if a test sample that originally weighed 11.5 oz was found to weigh 10 oz after oven drying?

4. For nearly all kinds of wood, the fiber saturation point is about what percent of the moisture content.

5. What is the best available grade of hardwood lumber?

6. Briefly state the use for each of the following types of engineered lumber: laminated-veneer lumber (LVL), glue-laminated beams (glulams), parallel-strand lumber (PSL), and laminated-strand lumber (LSL).

7. How many board feet of lumber are contained in a stack of 24 pieces of $2'' \times 4'' \times 8'$?

8. Where should a plywood panel marked "Exposure 1" be used?

9. What is a span rating for plywood?

10. Briefly describe the proper method of storing and handling the following building materials: framing lumber and sheathing, exterior finish materials, door frames, window units, and plywood.

11. How do polyvinyl glues cure and how does temperature affect them?

12. Explain how moisture-catalyzed adhesives are cured (hardened).

Critical Thinking

1. Imagine you are visiting your uncle in his new home in Pennsylvania. The home was finished about three months ago and was built during a very wet spring. Your uncle remarks that he was amazed at how fast the contractor was able to build the home, even through the wet weather. The summer has been humid, but Pennsylvania winters are known for dry air. Heating the house will start to bring the framing lumber to its equilibrium moisture content. By your appraisal, what issues might your uncle come to find this winter?

2. I-joists have some advantages over conventional framing lumber. However, in the event of a fire, I-joists are more susceptible to rapid loss of structural integrity. Design a method of protection that could extend the structural life of I-joists in the event of a fire. Justify your design as if you had to convince a code enforcement officer.

Communicating about Carpentry

1. **Reading and Speaking.** Create a collage that identifies different types of lumber and metal framing. Show examples by using pictures from magazines of things used to build with lumber and metal. Show and discuss your collage in a group of four to five classmates. Are the other members of your group able to determine the types of lumber and framing that you tried to represent?

2. **Writing.** Write an essay describing the home you lived in while growing up. Be as specific as possible while trying to identify the different materials used in the construction of the home. Compare the house you grew up in with your dream house. Point out the differences, other than size, and include in your description the materials used and the design of both houses. Be detailed in your descriptions.

3. **Reading.** Before you read the chapter, read the table and photo captions. What do you know about the material covered in this chapter just from reading the captions? After reading the chapter, determine if the knowledge you retained just from reading the captions is accurate.

Wood Identification

A key element in woodworking and in carpentry is the proper identification of the wood species used by carpenters. These samples are intended as a guide and aid to the student carpenter in learning to identify various woods. Shown are typical color and grain characteristics for 10 different species.

Ash, white. The heartwood of the white ash is brown and the sapwood is light-colored or nearly white. Wood from second-growth trees is heavy, strong, hard, and stiff and it has high resistance to shock. White ash is used principally for nonstriking tool handles, oars, baseball bats, and other sporting and athletic goods. Additional uses are decorative veneer, cabinets, furniture, flooring, millwork, and crates.

Birch. A hard, strong wood (47 lb per cu ft). Easily machine-worked with excellent finishing characteristics. Heartwood is reddish-brown with white sapwood. Fine grain and texture. Used extensively for quality furniture, cabinetwork, doors, interior trim, and plywood. Also used for dowels, spools, toothpicks, and clothespins.

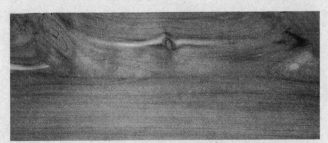

Cedar, red, eastern. The heartwood of eastern red cedar is bright or dull red and the narrow sapwood is nearly white. The wood is moderately heavy and is dimensionally stable after drying. The texture is fine and uniform and the wood has a distinctive aroma that is reputed to repel moths, but this claim has not been supported by research. Lumber is manufactured into chests, wardrobes, and closet lining. Other uses include flooring, novelties, pencils, scientific instruments, and small boats. The greatest quantity of eastern red cedar is used for fence posts.

Fir. There are several pieces of fir, with similar properties. Widely used for framing lumber. Grown in the western half of North America. Pronounced grain pattern in plain sawn lumber. Usually has a few small knots. Light brown with yellow or reddish overtones (38 lb per cu ft). Easy to work with ordinary carpentry tools, but has coarse grain, causing it to split if not worked properly. Frequently mixed with hemlock and sold as hem-fir. Trees grow quite large, so there is a good supply, but does not replenish as quickly as some other softwoods.

Hemlock. Grows throughout northern North America. There are two primary varieties of hemlock, eastern hemlock, and western hemlock. They have similar properties and appearance. It is light reddish brown in color with a distinct grain pattern (28 lb per cu ft). Splinters easily when worked. Used primarily for construction lumber. Often mixed with fir and sold as hem-fir.

Maple, sugar. Also called hard maple, it is hard, strong, and heavy (44 lb per cu ft) with a fine texture and grain pattern. Light tan color, with occasional dark streaks, the wood is hard to work with hand tools. However, it machines easily and is an excellent turning wood. Used for floors, bowling alleys, woodenware, handles, and quality furniture.

Oak, red. Heavy (45 lb per cu ft) and hard with the same general characteristics as white oak. Heartwood is reddish-brown in color. No tyloses in wood pores. Difficult to work with hand tools. Used for flooring, millwork, and inside trim.

Oak, white. Heavy (47 lb per cu ft), very hard, durable, and strong. Works best with power tools. Heartwood is grayish-brown with open pores that are distinct and plugged with a hairlike growth called tyloses. Sawing or slicing oak in a radial direction results in a striking pattern as shown. The "flakes" are formed by large wood rays that reflect light. Used where dramatic wood grain effects are desired. Used for high quality millwork, interior finish, furniture, carvings, boat structures, barrels, and kegs.

Pine, white. Soft, light (28 lb per cu ft), and even textured. Cream colored with some resin canals but not as prevalent as sugar pine. Works easily with hand or machine tools. Used for interior and exterior trim and millwork item. Knotty grades often used for wall paneling.

Poplar. Poplar is also known as yellow-poplar, tulip-poplar, and tulipwood. The greatest commercial production of poplar lumber is in the South and Southeast. The wood is generally straight grained and comparatively uniform in texture. The lumber is used primarily for furniture, interior moulding, siding, cabinets, musical instruments, and structural components. Boxes, pallets, and crates are made from lower grade stock. Poplar is also made into plywood for paneling, furniture, piano cases, and various other special products.

CHAPTER **4**

Hand Tools

OBJECTIVES

After studying this chapter, you will be able to:
- Identify the most common hand tools.
- Select the proper hand tool for a given job.
- Identify the main parts and purpose of each major hand tool.
- Explain proper methods of tool maintenance and storage.

TECHNICAL TERMS

all-hard blades	flexible-back blade	pry bar
backsaw	framing square	pull saw
bar clamp	hacksaw	ripsaw
block plane	hand screw	scratch awl
butt gauge	jointing	sledgehammer
cat's paw	kerf	speed square
C-clamp	level	tape measure
chalk line	marking gauge	T-bevel
claw hammer	measuring tape	tin snips
combination square	nail puller	try square
coping saw	nail set	utility knife
crosscut saw	oilstone	wing divider
drywall saw	pencil compass	wood chisel
flat bar	plumb bob	wrecking bar

Hand tools are essential to every aspect of carpentry work. A great variety of tools is required because carpentry covers a broad range of activities.

As a skilled worker, a carpenter carefully selects the types and sizes of tools that best suit the requirements of the job at hand. Tools become an important part of a carpenter's life, helping them perform the various tasks of the trade with speed and accuracy.

Although the basic design of common woodworking tools has changed little over many years, modern technology and industrial know-how have brought numerous improvements. Special tools have been developed to do specific jobs. Experienced carpenters appreciate the importance of having good tools and select those that are accurately made from quality materials. They know that such tools last longer and will enable them to do better work.

A detailed study of the selection, care, and use of all the hand tools available is not practical for this text. Instead, a general description of the tools most commonly used by carpenters is provided.

4.1 Measuring and Layout Tools

Measuring tools must be handled with considerable care and kept clean. Only then can a high level of accuracy be assured.

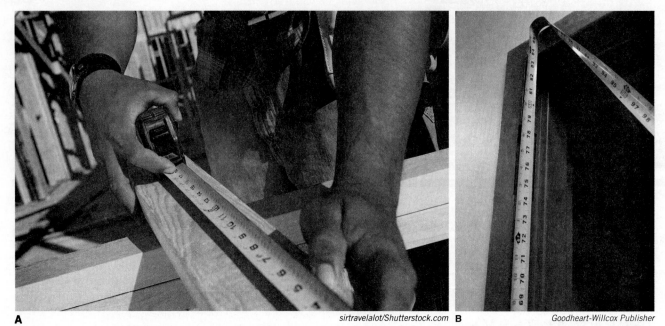

A   *sirtravelalot/Shutterstock.com*   B   *Goodheart-Willcox Publisher*

Figure 4-1. A—Pocket tape rules are available in lengths from 6′ to 30′. B—Because of its flexibility, the tape easily doubles over to make measurements that are out of reach.

4.1.1 Tapes and Rules

A *tape measure*, sometimes called a tape rule, is one of the tools carpenters use most often, **Figure 4-1**. Most carpenters frequently make use of two sizes of tape measures. A 25′–35′ tape is used for most measuring jobs. These tape measures have a blade that is 3/4″ or 1″ wide, a hook on the end, and a locking mechanism that allows the spring-loaded tape to remain extended. These tapes are typically self-retracting.

A longer tape, called a *measuring tape*, is used for building layouts. These tapes are available in lengths of 50′–300′. Long tapes are intended to be more flexible than shorter tape measures. Measuring tapes are bulkier and have a crank to wind the tape back into the case, **Figure 4-2**.

Goodheart-Willcox Publisher

Figure 4-2. Long tapes of 50′ or 100′ have a flexible tape and a larger end hook, which can slip over a small nail head.

Pro Tip

It is easy for dirt and moisture to enter the tape case when retracting the tape. To prevent damage to the tape measure, wipe the tape with a cloth as it is retracted, especially if it is wet.

4.1.2 Squares

The *framing square*, also called a rafter square, is specially designed for a carpenter. The blade (or body) is 2″ wide and 24″ long. The tongue is at a right angle to the blade. It is 1 1/2″ wide and 16″ long. The corner of the square is called the heel. One side of the square, with the tongue to the right and toward you, is called the face. See **Figure 4-3**.

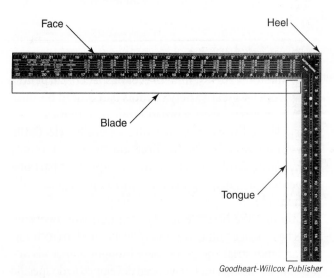

Goodheart-Willcox Publisher

Figure 4-3. The parts of a framing square are the blade, tongue, heel, and face.

Framing squares have a number of tables printed on them. The face has rafter tables printed on the body. These can be used in determining the length of rafters. The tongue carries the octagon scale. This scale is used to lay out the angles of an octagon. The opposite side of the square may have the Essex Board Measure Table and the Brace Measure Table. These tables are seldom used compared to the rafter table.

The outside edge of the face is divided into 1/16″ graduations, while the inside edge has 1/8″ graduations. The outside edge of the opposite side of the square has each inch graduated in twelfths of an inch. This is useful in making a scale drawing. Each inch represents a foot and each 1/12″ represents an inch. The inside edge of the tongue is graduated in tenths of an inch.

The framing square is a versatile tool with many uses. The following are just a few tasks a carpenter can perform using a framing square:

- Check or square lines across boards
- Check the squareness of corners
- Make a 45° angle across wide boards
- Find the length and angles of rafters
- Lay out stairs

An explanation of how the square is used in framing rafters is included in Chapter 12, **Roof Framing**. The framing square is available in aluminum or in steel with a copper-clad or blued finish.

Carpenters also use a framing tool called a ***speed square***, a variation of the framing square that performs most of the same functions. This tool can be used to mark any angle for rafter cuts by aligning the desired degree mark on the tool with the edge of the rafter being cut. See **Figure 4-4**.

Carpenters generally use a speed square more than any other square. Speed squares are available in two sizes. The smaller size, about 7″ on each of the right-angle sides, is convenient to carry in a tool belt. The larger one, usually 12″ on each of the right-angle sides, is frequently called a rafter square because it is large enough to lay out large rafters. Speed squares are graduated in 1/8″ increments on one side. The other side has a guide lip with no markings, making it easy to align with the edge of a board. The hypotenuse (long side) is marked with degrees. Set the pivot point at the edge of a board and pivot the square until the desired angle is shown on the board, **Figure 4-4**. The speed square can also be used to guide a portable circular saw in making square cuts, **Figure 4-5**.

Try squares are another variation of square tools available with blades 6″–12″ long. Handles are made of wood or metal. These are used to check the squareness of surfaces and edges. They are also used to lay out lines perpendicular to an edge, **Figure 4-6**.

Goodheart-Willcox Publisher

Figure 4-5. The speed square can be used as a guide for making square cuts. Hold the square against the edge of the wood and let the base of the saw ride against the other edge of the square.

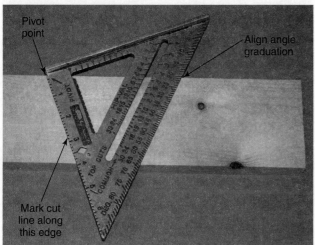

Goodheart-Willcox Publisher

Figure 4-4. To measure an angle with a speed square, (1) set the pivot point where the angle begins, (2) rotate the square until the desired angle graduation aligns with the edge of the board, and (3) mark the angle along the side of the square.

Anna Berdnik/Shuttetstock.com

Figure 4-6. The try square is handy for laying out lines perpendicular to an edge. This student is able to lay out a precise mark to square off this board.

The *combination square* serves a similar purpose as a try square. It is also used to lay out miter joints. Its adjustable, sliding blade allows it to be conveniently used as a gauging tool. It can be used as a marking gauge to make parallel lines and as a square for 90° and 45° angles. Refer to **Figure 4-7**. Although the combination square has multiple uses, it is difficult to carry in a tool belt and it is not as durable as a speed square.

The *T-bevel* has an adjustable blade making it possible to transfer an angle from one place to another. It is useful in laying out cuts for hip and valley rafters. **Figure 4-8** illustrates how the framing square is used to set the T-bevel at a 45° angle.

The *pencil compass* is available in several sizes and serves a number of purposes. Dimensions can be "stepped off" along a layout or transferred from one position to another. Circles and arcs can also be drawn on surfaces, **Figure 4-9**. A pencil compass can also be used as a scribing tool to transfer an irregular profile, **Figure 4-10**. Similar to pencil compasses are *wing dividers*. They differ from the pencil compass in that both legs have a point. Legs can be locked to prevent movement that would change the measurement being transferred.

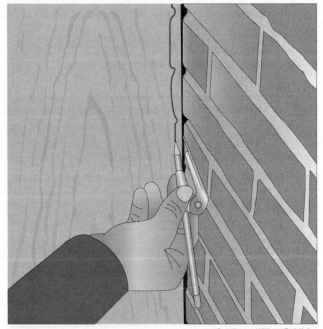

Everyonephoto Studio/Shutterstock.com

Figure 4-9. A pencil compass may be used to draw arcs and circles or to step off short distances.

pirke/Shutterstock.com

Figure 4-7. A combination square is a versatile marking and layout tool.

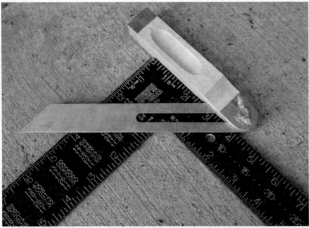

Goodheart-Willcox Publisher

Figure 4-8. The framing square can be used to set a T-bevel at a 45° angle.

Goodheart-Willcox Publisher

Figure 4–10. A pencil compass can act as an adjustable scribe to transfer a profile.

A *Christopher Williams/Shutterstock.com* **B** *Rob kemp/Shutterstock.com*

Figure 4-11. A— A marking tool scribes a line parallel to an edge. B—A carpenter lays out a mortise for a mortise and tenon joint.

4.1.3 Marking and Orientation Tools

A *marking gauge* is used to lay out parallel lines along the edges of material, **Figure 4-11.** It may also be used to perform these tasks:

- Transfer a dimension from one place to another
- Check sizes of material

The *butt gauge* can be used for the same purposes. However, it is normally used to lay out the gain (recess) for hinges.

In layout work, a *scratch awl* is used to scribe lines on the surface of the material. It is also used to mark points or form starter holes for small screws or nails.

Using a *chalk line* is an easy way to mark long, straight lines. The line is a thin strong cord that is covered with powdered chalk. It is held tight and close to the surface. Then, it is snapped, **Figure 4-12.** This action drives the chalk onto the surface forming a distinct mark. A special reel within the case rechalks the line each time it is wound.

A *level* is an instrument designed with a vial filled with a liquid to create a bubble, **Figure 4-13.** When the bubble is perfectly centered between the lines on the vial, this indicates the vial is level. Levels come in assorted sizes. Common lengths of longer levels are 2′, 4′, and 6′. The 2′ level is useful for many leveling jobs, but a 4′or 6′ level is used to install doors and where long pieces must be made level or plumb (perpendicular to the earth).

Empire Level, Inc.

Figure 4-13. Levels have one or more vials filled with mineral spirits, causing a bubble to rise to the highest point in the vial.

Goodheart-Willcox Publisher

Figure 4-12. Snapping a chalk line is a fast, simple way to lay down straight lines over long distances.

Some levels have a magnetic edge, so they can be attached to steel framing, **Figure 4-14**.

The *plumb bob* establishes a vertical line when attached to and suspended from a line. Its weight pulls the line in a true vertical position for layout and checking. The point of the plumb bob is always directly below the point from which it hangs.

4.2 Saws

The principal types of handsaws used by carpenters are illustrated in **Figure 4-15**. These are available in several different lengths as well as various tooth sizes (given as teeth per inch or points per inch). The number of teeth per inch determines how fine the blade cuts. Saws with a high number of teeth per inch generally have smaller teeth, removing less material and creating a smoother cut. Each type of saw has a specific purpose.

4.2.1 Saw Types

The most common type of handsaw may be either a crosscut saw or a ripsaw. *Crosscut saws*, as the name implies, are designed to cut across the wood grain.

Empire Level, Inc.

Figure 4-14. A long level with a magnetic edge is useful for working with steel framing.

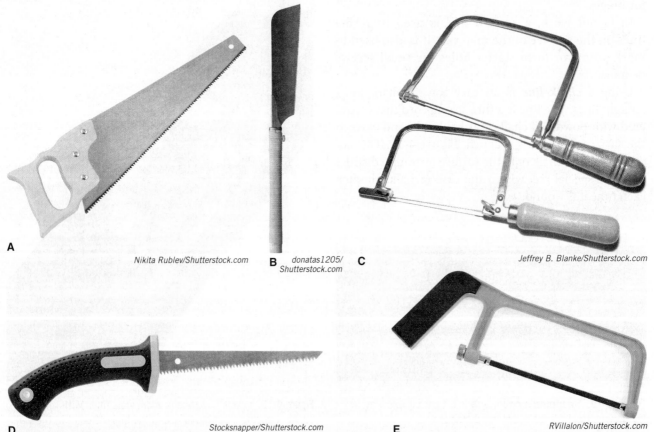

A   *Nikita Rublev/Shutterstock.com*   B   *donatas1205/ Shutterstock.com*   C   *Jeffrey B. Blanke/Shutterstock.com*

D   *Stocksnapper/Shutterstock.com*   E   *RVillalon/Shutterstock.com*

Figure 4-15. Carpenters may use several different kinds of handsaws in the course of their work. A—Handsaw. B—Pull saw. C—Coping saw. D—Drywall saw. E—Hacksaw.

Their teeth are shaped so that they come to a point on alternating sides, **Figure 4-16**. Every carpenter needs a good crosscut saw with a tooth size ranging from 8 to 11 points. A crosscut saw for general use has 8 teeth per inch.

About 65°

1″

8 Points per inch, 7 teeth

Set

15°   45°   60°

Kerf

A

90°

1″

5 1/2 Points per inch, 4 1/2 teeth

Set

60°

90°

Kerf

B   Goodheart-Willcox Publisher

Figure 4-16. The teeth of crosscut saws and ripsaws are quite different. A—Crosscut teeth cut like a knife. The top illustration shows the shape and angle of teeth. The bottom illustration shows how teeth cut. B—Ripsaw teeth cut like a chisel. The top illustration shows shape and angle of teeth. The angle is often increased from 90° to give a negative rake. The bottom illustration shows how ripsaw teeth cut.

A finishing saw used for fine cutting has 10 or 11 teeth per inch. Using the thumb as a guide will ensure getting a crosscut started on the line, as shown in **Figure 4-17**.

Ripsaws have chisel-shaped teeth, which cut best along the grain. Note that the teeth are set (bent) alternately from side to side. This is so the *kerf* (cut in the wood) will be large enough for the blade to freely run. Some saws have a taper ground from the toothed edge to the back edge. It eliminates the need for a large set on the teeth. Most ripping operations are now performed with power saws, making the ripsaw an optional tool in a carpenter's arsenal.

A saw occasionally used for cutting curves is the *coping saw*, **Figure 4-18**. This saw has a thin, flexible blade that is pulled tight by the saw frame. With a blade that has 15 teeth per inch, the coping saw can make very fine cuts.

The *backsaw* has a thin blade reinforced with a heavier steel strip along the back edge. The teeth are small (14–16 teeth per inch). Thus, the cuts produced are fine. It is used mostly for interior finish work.

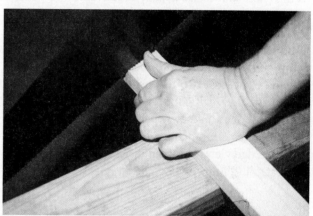

Goodheart-Willcox Publisher

Figure 4-17. Use your thumb as a guide against the saw blade when starting a cut with a handsaw.

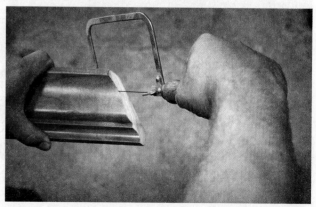

Goodheart-Willcox Publisher

Figure 4-18. A coping saw is designed to cut along curves. Its thin, flexible blade is supported under tension by the saw's frame. The blade has about 15 teeth per inch.

Pull saws come in many shapes and sizes, but generally have a handle that is in a straight line with the blade. The blade is thin and has fine teeth that come to a sharp point. The saw cuts as it is pulled *toward* the user.

The *drywall saw*, sometimes called a jab saw, has large teeth specially designed for cutting through paper facings, backings, and the gypsum core. Gullets (spaces between the saw teeth) are rounded to prevent clogging with the gypsum material.

Due to the wide range of work that a carpenter must be prepared to handle, a *hacksaw* should be included in the hand tool assortment. A hacksaw can be used to cut nails, bolts, other metal fasteners, and metal trim. Most hacksaws have an adjustable frame, permitting the use of several lengths of blades.

Hacksaw blades are made of high-speed steel, tungsten alloy steel, molybdenum steel, and other special alloys. *All-hard blades* are heat-treated, making them very brittle and easily broken if misused. *Flexible-back blades* are hardened only around the teeth. These blades are usually preferred by carpenters because they do not break as easily as all-hard blades.

The blade's cutting edge may have anywhere from 14 to 32 teeth per inch. Generally, the thinner the metal being cut, the finer (more teeth per inch) the blade should be. At least two teeth should rest on the metal. Otherwise, the sawing motion may shear off teeth.

Like woodcutting saws, hacksaw blades have a set. This provides clearance for the blade to slide through the cut. Edge views showing different types of sets are illustrated in **Figure 4-19**.

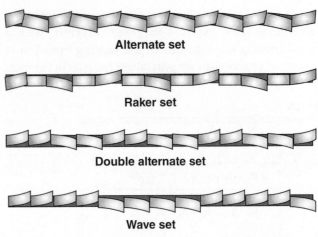

Alternate set

Raker set

Double alternate set

Wave set

Goodheart-Willcox Publisher

Figure 4-19. This enlarged view of hacksaw blades shows different types of sets.

Blades should be installed with teeth pointing forward, away from the handle. Use both hands to operate the saw. Apply enough pressure on the forward stroke to allow each tooth to remove a small amount of metal. Release pressure on the return stroke to reduce wear on the blade. Saw with long steady strokes paced at 40–50 per minute.

4.3 Planing, Smoothing, and Shaping Tools

The most important edge tools are the planes and chisels. Several other tools that depend on a cutting edge to perform work are also shown in **Figure 4-20**.

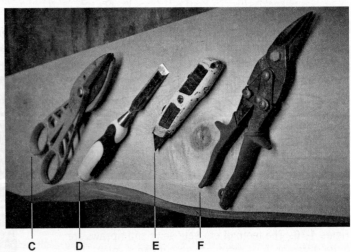

Goodheart-Willcox Publisher

Figure 4-20. Edge tools such as planes, chisels, and cutting tools are used as carpentry shaping materials. A—Jack plane. B—Block plane. C—Tin snips. D—Wood chisel. E—Utility knife. F—Aviation snips

Standard surfacing planes include the smooth plane (8″–9″ long), jack plane (14″ long), and fore and jointer plane (18″–24″ long). **Figure 4-21** illustrates the parts of a plane and how they fit together. The jack plane is usually selected for general-purpose work.

Another surfacing plane, the *block plane*, is especially useful. The block plane is small surfacing plane (6″–7″ long) that produces a fine, smooth cut, making it suitable for fitting and trimming work and can be used with one hand. The blade is mounted at a low angle and the bevel of the cutter is turned up. This plane produces a fine, smooth cut, making it suitable for fitting and trimming work.

Wood chisels are used to trim and cut away wood or composition materials to form joints or recesses, **Figure 4-22**. They are also helpful in paring and smoothing small, interior surfaces that are inaccessible for other edge tools. Chisel widths range from 1/8″ to 2″. For general work, a carpenter usually selects 3/8″, 1/2″, 3/4″, and 1 1/4″ sizes. A soft-face hammer or mallet should be used to drive the chisel when making deep cuts.

Tin snips are used to cut composition shingles and light sheet metals, such as flashing. *Utility knives* have very sharp, replaceable blades and are useful for trimming wood and cutting veneer, hardboard, gypsum wallboard, vapor barrier, and housewrap. They are also used to cut batt insulation and for accurate layout.

4.4 Fastening Tools

Much of carpentry work consists of fastening parts together. Nails, screws, bolts, and other types of connectors are used. The *claw hammer* is the most often used fastening tool. Two common shapes of hammer heads are the curved claw and ripping (straight) claw.

The curved claw is best suited for pulling nails. The ripping claw, which may be driven between fastened pieces, is used somewhat like a chisel to pry them apart. Straight claw hammers are also used for framing.

Gang Liu/Shutterstock.com

Figure 4-22. Cutting a recess with a wood chisel. The chisel should be kept sharp enough to cut easily with hand pressure or light mallet blows. It should never be struck with a hammer.

Parts of a hammer are shown in **Figure 4-23**. The face can be either flat or slightly rounded (bell face). The bell face is used most often. It will drive nails flush with the surface without leaving hammer marks on the wood. The hammer head is forged from high-quality steel and is heat treated to give the poll and face extra hardness.

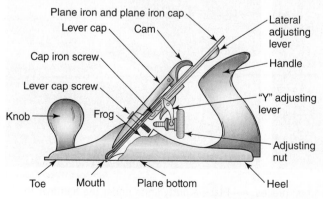
Goodheart-Willcox Publisher

Figure 4-21. A standard plane is made up of many parts.

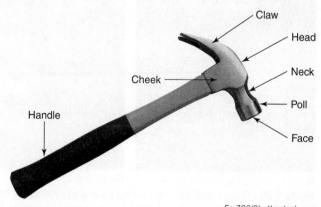
Far700/Shutterstock.com

Figure 4-23. Parts of a hammer.

Estwing Mfg. Co.

Figure 4-24. Framing hammers have longer handles and are heavier than most other hammers.

The size of a hammer is determined by the weight of its head. They are available in sizes from 7 oz to 22 oz. Carpenters generally use either the 16 oz, 20 oz, or 22 oz, depending on the work to be done. Framing hammers are generally 20 oz or 22 oz with longer handles and straight claws. The face of a framing hammer may have a checkered surface to help prevent the hammer from glancing off the nail, **Figure 4-24**.

Sledgehammers are available in weights from 2 lb to 16 lb. The most common are 2 lb and 8 lb. Carpenters use sledgehammers to align concrete forms, drive stakes, and perform other heavy work.

Pro Tip
A hammer should be given good care. It is especially important to keep the handle tight and the face clean. If a wooden handle becomes loose, it can be tightened by driving the wedges deeper or by installing a new handle. A hammer with a loose head can be a dangerous hazard on the jobsite.

Figure 4-25 illustrates the procedure for pulling nails to relieve strain on the hammer handle and protect the work.

Workplace Skills
Tool Care
Your tools are an extension of your skills and abilities. Properly maintained and organized, they will pay for themselves as they give you the ability to perform a task that you otherwise could not do. Imagine showing up to a jobsite and not being able to work because you do not have the proper tools in working condition. Always keep your tools organized and properly maintained so they are ready for use. Fumbling through your toolbox or truck wastes valuable work time and looks foolish.

Hatchets are used for rough work on such jobs as making stakes and building concrete forms. Some are designed for wood shingle work, especially wood shakes, **Figure 4-26**. See also Chapter 14, **Roofing Materials and Methods**.

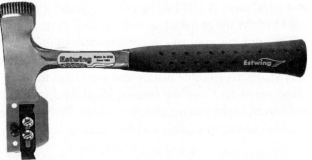

Estwing Mfg. Co.

Figure 4-26. A shingling hatchet is designed for installing wood shingles and shakes. The adjustable stops on the blade are used for measuring the exposure on shingles.

A   **B**   **C**   *Goodheart-Willcox Publisher*

Figure 4-25. Proper methods for pulling nails and separating two wood pieces. A—Place a putty knife under the hammer to protect the wood surface. B—Use a block of wood to increase leverage and protect the surface. C—To separate two nailed surfaces, a ripping claw hammer works best. The ripping claw can also be used to split short boards.

The *nail set* is designed to drive the heads of casing and finishing nails below the surface of the wood. Tips range in diameter from 1/32″ to 5/32″ by increments of 1/32″. Overall length is usually 4″.

Screwdrivers are available in a number of sizes and styles, **Figure 4-27**. Sizes are specified by giving the length of the blade, measuring from the base of the handle to the tip. The most common sizes for a carpenter are 3″, 4″, 6″, and 8″. The size of a Phillips screwdriver is given as a point number ranging from a No. 0 (the smallest), to a No. 4 (the largest). Size numbers 1, 2, and 3 will fit most of the screws used in carpentry work.

Tips of screwdrivers must be carefully picked for the job. For a slotted screw, tips must be square, have the correct width, and snugly fit into the screw slot.

The width of the tip should be equal to the length of the bottom of the screw slot. The sides of the screwdriver tip should be carefully ground to an included angle of not more than 8° and to a thickness that will fit the screw slot.

Today, mechanical tackers, staplers, and nailers perform a variety of operations formerly accomplished by hand nailing. They provide an efficient method of attaching insulation, roofing material, underlayment, ceiling tile, and many other products. An advantage of the regular stapler is that it leaves one hand free to hold the material.

4.5 Prying Tools

Sometimes, particularly when remodeling, it is necessary to take apart nailed structural members. The same is true when forms or temporary braces are removed from a building under construction. *Pry bars* are shown in **Figure 4-28**. Lengths range from 7″ for small flat bars to 30″ for pry bars.

Wrecking bars vary in length from 12″ to 36″. They are used to strip concrete forms, disassemble scaffolding, and for other rough work involving prying, scraping, and nail pulling.

Flat bars are designed for removing larger nails or spikes from lumber. Most flat bars are 13″–15″ long. The smaller ones, around 7″, are handy because they fit in a carpenter's tool belt. A *cat's paw* is used only to pull a nail above the surface of the wood. A hammer is used to drive the sharpened claw under the nail head, raising it sufficiently to be grasped by the hammer or a pry bar.

A *nail puller* is a specialty tool with a fixed jaw and a movable jaw and lever. A sliding driver on the end of the handle drives the jaws under the nail head. Leverage tightens the jaw's grip on the nail, raising it above the wood's surface, **Figure 4-29**.

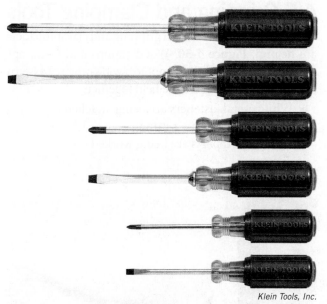

Klein Tools, Inc.

Figure 4-27. There are many types of screwdrivers and screw slots. A screwdriver set typically includes screwdrivers of different sizes and head tips, such as slotted and Phillips.

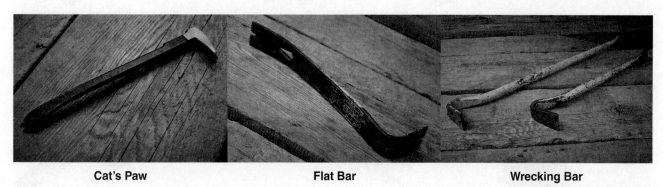

Cat's Paw   Flat Bar   Wrecking Bar

Goodheart-Willcox Publisher

Figure 4-28. Pry bars are used for separating, disassembling, and aligning heavy building components. The most common pry bars are shown here.

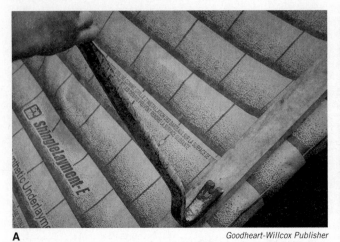

A   *Goodheart-Willcox Publisher*   B   *Goodheart-Willcox Publisher*

Figure 4-29. Using prying tools. A—The pry bar is used to remove larger fasteners. Its long handle provides considerable leverage. B—Nail puller claws are first driven under the nail with a weight that slides along the handle. A small fulcrum attached to one claw tightens the puller's grip on the nail.

Safety Note

There are certain safety precautions that should be observed while using pry bars. Be sure to have good balance before applying force to the bar. Grip the bar in such a way that fingers are not bruised by coming into contact with any part of the building structure as the fastener or piece comes free. Always wear safety glasses when using a pry bar.

4.6 Gripping and Clamping Tools

Carpenters find that certain gripping and clamping tools are helpful, **Figure 4-30**. Wood clamps and C-clamps are especially useful and adapt to a wide range of assemblies where parts need to be held together:

- While metal fasteners are being attached
- While adhesives are setting
- While wood parts are being worked on

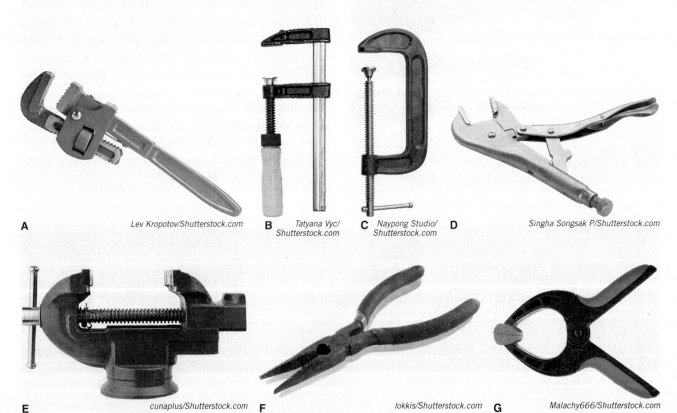

A   *Lev Kropotov/Shutterstock.com*   B   *Tatyana Vyc/ Shutterstock.com*   C   *Naypong Studio/ Shutterstock.com*   D   *Singha Songsak P/Shutterstock.com*

E   *cunaplus/Shutterstock.com*   F   *lokkis/Shutterstock.com*   G   *Malachy666/Shutterstock.com*

Figure 4-30. Clamping and gripping tools are helpful in carpentry work. A—Straight pipe wrench. B—Bar clamp. C—C-clamp. D—Clamping Pliers. E—Table vise clamp. F—Needle-nosed pliers. G—Spring clamp.

Sometimes clamps are used to hold jigs or fixtures to a machine for some special setup. **C-clamps** have a C-shaped frame and are available in sizes from 1″ to 12″. The clamp size represents the tool's largest opening.

A **hand screw** is a clamp made of two wooden jaws controlled by two screws. This clamp type is ideal for woodworking because the jaws are broad and distribute the pressure over a wide area. Because the jaws are wooden, the hand screw is less likely to damage wooden pieces being clamped. Sizes are designated by the width of the jaw opening and range from 4″ to 24″.

Bar clamps, or pipe clamps, are useful for holding larger pieces while glue is drying or receiving mechanical fasteners. Some clamps are adjustable at both jaws. Newer types, like the one shown in **Figure 4-31**, feature one-handed clamping mechanisms.

Clamping pliers are an adjustable, all-purpose tool with a lever action that locks the jaws when the handles are fully closed. The pliers can be substituted for a vise or can be used in place of a pipe wrench, open-end wrench, or adjustable wrench. Carpenters use a wide variety of wrenches to service and adjust equipment. Tool maintenance is an important responsibility. Well-maintained tools are safe tools.

4.7 Tool Belts and Storage

Many carpenters use tool belts and tool holders that hang from the waist. These items keep tools close and ready for use, while freeing hands for other tasks when the tools are not in use. See **Figure 4-32A**.

American Tool

Figure 4-31. Many bar clamps have a pistol-grip mechanism that allow quick size adjustment and one-handed clamping.

Some type of container is needed to store and transport tools to and from a jobsite. In addition to being portable, the container should protect the tools from weather damage or loss. At one time, folding tool panels met this need. Today's carpenter has a growing number of options. Some prefer to carry tools in a reinforced fabric bag or in a 5-gallon bucket fitted with a fabric holder full of pockets. The one shown in **Figure 4-32B** is typical.

Green Note
Best practice is green practice when selecting and using hand and power tools. Select tools with low impact on the environment. Maintain and handle all tools properly. Energy-efficient tools used carelessly create waste by damaging materials and the tools themselves.

A   *kurhan/Shutterstock.com*   B   *KellyLT/Shutterstock.com*

Figure 4-32. A—The tool belt can increase a carpenter's efficiency by keeping needed items in easy reach. B—A fabric bucket liner with many pockets is handy for organizing and transporting tools. The liner fits any 5-gallon plastic bucket.

4.8 Care and Maintenance of Tools

Experienced carpenters take pride in their tools and keep them in good working condition. They know that even high-quality tools will not perform satisfactorily if they are dull or out of adjustment. They are also aware that tools that are not in good condition can present a safety hazard.

Tools should be kept clean by wiping them with a clean cloth. To prevent rust, tools that get wet during use should be wiped with a cloth that has been treated with a few drops of light oil. Once the tool has been cleaned, all oil residues should be wiped away because tools left oily will attract dirt and make the tool difficult to handle.

Keep handles on all tools tight. If handles or fittings are broken, they should be replaced.

It is a simple matter to hone edge tools on an *oilstone*, **Figure 4-33**. For tools with single-bevel edges, like planes and chisels, place the tool on the stone with the bevel flat on the surface. Raise the back edge of the tool a few degrees so only the cutting edge is in contact. Then move the tool back and forth until a fine wire edge can be detected by pulling the finger over the edge. Now place the back of the tool flat on the oilstone and stroke lightly several times. Turn the tool over and again stroke the beveled side lightly. Repeat the total operation several times until the wire edge has disappeared from the cutting edge.

When the bevel becomes blunt, reshape it by grinding. An edge can be ground many times. If a cutting edge is not damaged, it can be honed several times before grinding is required. The grinding angle will vary somewhat depending on the work for which the tool is used. **Figure 4-34** shows grinding and honing angles recommended for a jack plane iron. The bevel on a properly ground jack plane iron will be about two times as wide, or even slightly wider, as the thickness of the iron. Since block planes are ground at a slightly lower angle, they can be used to plane end grain.

Some tools may be sharpened with a file. Saws require filing as well as setting. Before filing, they often require *jointing*. In this operation, the height of the teeth is evenly struck off. Filing a saw is a tedious operation. Most carpenters prefer to send their saws to a shop where they can be machine-sharpened by an expert.

When a saw is only slightly dull, the teeth can often be sharpened with a few strokes from a triangular saw file. Be sure to match the original angle of the teeth. File the back of one tooth and the front of an adjacent tooth in a single stroke. **Figure 4-35** shows how correctly filed teeth should appear.

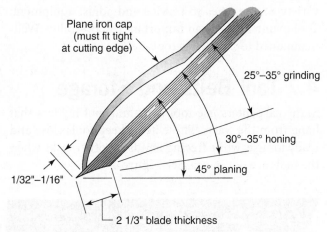

Plane iron cap
(must fit tight at cutting edge)

25°–35° grinding

30°–35° honing

45° planing

1/32"–1/16"

2 1/3" blade thickness

Goodheart-Willcox Publisher

Figure 4-34. Follow these grinding and honing angles for a plane iron

Nikodash/Shutterstock.com

Figure 4-33. Honing a chisel on an oilstone. Place a few drops of oil on the stone first.

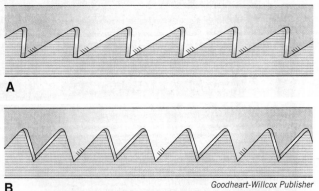

A

B

Goodheart-Willcox Publisher

Figure 4-35. Correctly filed saw teeth. A—Rip teeth. B—Crosscut teeth.

Chapter Review

Summary

- Hand tools are essential to a carpenter's work.
- As a skilled worker, a carpenter must carefully select the kind, type, and size of tools that best suit the requirements of the job.
- A carpenter works with measuring and layout tools, such as rules and squares.
- The most common carpentry tool is the claw hammer, which is used to nail wooden building components.
- A carpenter also uses saws, smoothing tools (such as planes), prying tools, and clamping tools.
- Experienced carpenters take pride in their tools and keep them in good working condition. This may involve cleaning, adjusting, sharpening, and repairing.

Know and Understand

Answer the following questions using the information in this chapter.

1. Most measuring jobs can be completed using a tape measure that is _____ long.
   - A. 6′–8′
   - B. 10′–14′
   - C. 15′–18′
   - D. 25′–35′
2. The outside edge of the face of a framing square is graduated in what fractions of an inch?
   - A. 1/10″
   - B. 1/12″
   - C. 1/16″
   - D. 1/24″
3. The speed square is used to _____.
   - A. mark an angle for rafter cuts
   - B. guide a portable circular saw in making square cuts
   - C. check the squareness of corners
   - D. All of the above.
4. The _____ establishes a vertical line when attached to and suspended from a line.
   - A. plumb bob
   - B. chalk line
   - C. marking gauge
   - D. level
5. A backsaw is used for fine work. It usually has _____ points per inch.
   - A. 8
   - B. 8–10
   - C. 12–14
   - D. 14–16
6. *True or False?* The type of saw that cuts as it is pulled *toward* the user is called a pull saw.
7. *True or False?* Carpenters prefer all-hard bladed hack saws because they are more flexible and do not break.
8. *True or False?* The bevel of a block plane blade is turned down.
9. The size of a claw hammer is determined by the _____.
   - A. length of the handle
   - B. length of the head
   - C. weight of the head
   - D. All of the above.
10. *True or False?* You should never strike a wood chisel with a hammer.
11. Tin snips are used to cut _____.
    - A. composition shingles
    - B. light sheet metals
    - C. flashing
    - D. All of the above.
12. *True or False?* The most common sledgehammer weights are 2 lb and 8 lb.
13. The operation of filing off the points of saw teeth until all are level is called _____.
    - A. planing
    - B. jointing
    - C. clamping
    - D. prying

Apply and Analyze

1. Where would you find the rafter table on a framing square? What would you use it for?
2. Name the different types of pry bars and explain the purpose of each.
3. How are sizes of slotted screwdrivers and Phillips screwdrivers determined?
4. Explain how to prevent hand tools from rusting.
5. Explain why it is important to remove oil residue from hand tools once the tool has been cleaned.

Critical Thinking

1. Depending on the climate where you live, the air bubble trapped inside of a leveling vial becomes longer in the summer and shorter in the winter. Explain why this happens.

2. Based on your answer for question #1, would you say a level's reading is consistently accurate enough for the most precise jobs, so long as the bubble is within the two lines on the vial, even if it is not perfectly centered between the two marks?

3. What are the advantages of using a speed square over a framing square to lay out a rafter? What are the disadvantages?

4. Claw hammers are available in different weights and constructions (one piece, wooden handle, fiberglass handle). Research the uses and the advantages/disadvantages of each type. Based on your research, choose a hammer with a weight and construction that would be suitable for you to use in framing a house. Explain why that is your hammer of choice.

5. Research the proper method for cutting a mortise in a door jamb, using a wood chisel and mallet. Make a poster showing the step-by-step procedure.

Communicating about Carpentry

1. **Speaking and Listening.** Work in small groups. Bring in some hand tools from your home. Describe the tool and its use. Answer questions from the group at the end of your presentation.

2. **Speaking.** Make a series of safety posters for the use of hand tools. Explain your posters to the class.

3. **Reading.** Before you read the chapter, read the table and photo captions. What do you know about the material covered in this chapter just from reading the captions? After reading the chapter, determine if the knowledge you retained just from reading the captions is accurate.

CHAPTER 5

Power Tools

OBJECTIVES

After studying this chapter, you will be able to:

- Identify common power tools.
- Explain the function and operation of a carpenter's principal power tools.
- Identify the major parts of common power tools.
- Apply power tool safety rules.

TECHNICAL TERMS

belt sander	jointer	power stapler
chain saw	magnetic starter	radial arm saw
chop saw	miter saw	reciprocating saw
current	nail gun	rotary hammer drill
drywall screw shooter	oscillating multi-tool	router
electric drill	powder-actuated tool (PAT)	saber saw
finishing sander	power block plane	table saw
impact driver	power plane	voltage

Power tools greatly reduce the time required to perform many of the operations in carpentry work. A carpenter can plane, route, bore, and saw in less time with less exertion. In addition, when proper tools are used in the correct way, a high level of accuracy can easily be achieved. There are two general types of power tools, portable and stationary:

- Portable tools are lightweight and intended to be carried around by a carpenter. They may be used anywhere on the construction site or on any part of the structure. Portable power tools may be powered through an electrical cord plugged into an outlet or extension cord, a rechargeable lithium-ion battery, or compressed air through an air hose. Battery-operated cordless power tools are becoming more powerful with batteries that require recharging less frequently. 18- to 24-volt tools deliver nearly as much power as corded tools.
- Stationary tools, usually called machines, are heavy equipment mounted on benches or stands. The workpiece is brought to them. Benches and stands rest firmly on the floor or ground. The machines must be level when used on the jobsite.

This chapter provides only brief descriptions of the kinds of power tools most commonly associated with carpentry work. It should be supplemented with woodworking textbooks and reference books devoted to power tool operation. Manufacturer bulletins and operator's manuals are also good sources of information.

5.1 Power Tool Safety

Safety must be practiced constantly. Before operating any power tool, there are three things you must thoroughly understand:

- The way it works
- The correct way to use it
- The consequences of a mistake when using it

You must be wide awake and alert. Never operate a power tool when you are tired or ill. Think through the operation before performing it. Know what you are going to do and what the tool will do. Make all adjustments before turning on the power. Be sure blades and cutters are sharp and are of the correct type for the work.

While operating a power tool, do not allow yourself to be distracted. Do not distract the attention of others while they are operating power tools. Keep all safety guards in place and wear safety glasses. See **Figure 5-1**. Never wear loose fitting clothing that could get caught in moving parts; dangling jewelry and earbuds should be removed when working with power tools. Tie back long hair before getting started.

Feed the work carefully and only as fast as the tool will easily cut. Overloading is hazardous to the operator and likely to damage the tool or work. When the operation is complete, turn the power off.

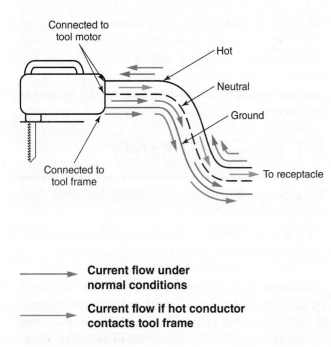

Monkey Business Images/Shutterstock.com

Figure 5-1. When operating a power tool, give your full attention to the work. A moment's distraction can have severe consequences.

Wait until the moving parts have stopped before leaving the machine.

Anytime the tool requires servicing, such as a blade change or maintenance, disconnect the power to prevent accidental start-up.

5.1.1 Electrical Safety

To understand electrical safety, one must first know how electric current flows in a circuit. Electrical energy consists of *current* (the movement of electrons in a conductor) and *voltage* (the pressure that pushes current through a conductor). A generator, battery, or some other source makes voltage. The voltage applies force to the electrons in one of two conductors (wires). If there is a path through a tool lamp or other load, and then to a return conductor, the electrons will move. The conductor from the source to the load is the hot conductor. The conductor leading back to the source is the neutral conductor. It is common to use the earth (or ground) as part of the path back to the source. If a person comes in contact with the hot conductor and the earth or ground, he or she may provide an easier path for the current to get back to the source. When this happens, the person receives an electric shock. To prevent a shock in the event of a faulty appliance or tool, most have a third conductor, called the ground conductor that connects the frame of the tool with the earth, so current is less likely to travel through the user. See **Figure 5-2**.

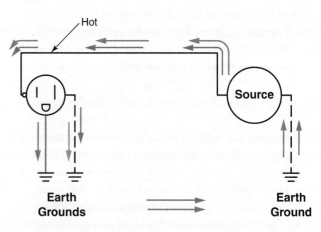

Goodheart-Willcox Publisher

Figure 5-2. When the load is in good condition, the electric current flows only in the hot (black) and neutral (white) conductors. A ground (green) conductor is added so current does not flow through the user if the frame of the appliance or tool becomes energized (hot).

Always make sure that the electric power source is of the correct voltage and that the tool switch is in the *off* position before plugging the tool into an electrical outlet. Corded power tools operate on 120 V or 240 V electric power. This information is usually shown on a plate attached to the power tool's case. If in doubt about the power source voltage, examine the receptacle, **Figure 5-3**.

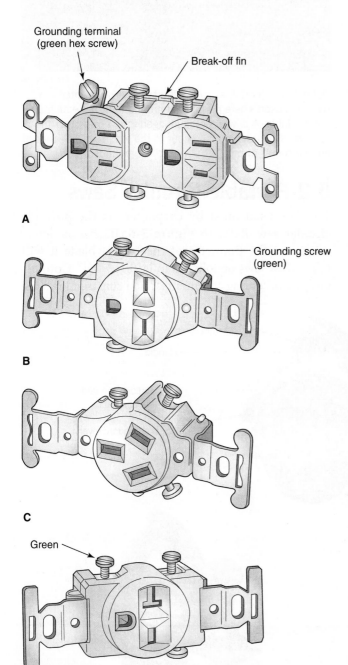

Figure 5-3. Different types of receptacles. A—Receptacles for 120 V power source always have this configuration. B—A 240 V receptacle with tandem blades and U-shaped ground prong. C—A 240 V receptacle for three-bladed plug. D—A 240 V receptacle for horizontal and vertical blades and U-shaped ground prong.

Goodheart-Willcox Publisher

Safety Note

Never carry a portable power tool by its cord or disconnect it from the receptacle by pulling on the cord. This can damage the insulation and electrical connections inside of the plug or the tool, creating a potential shock hazard.

Stationary power tools are factory equipped with **magnetic starters**. These are safety devices that automatically turn the switch to the *off* position in case of power failure. This feature is important because personal injury or damage to the equipment could result if power is re-established with the switch in the *on* position. *Always* make sure these switches are in good working order.

The electrical cord and plug must be in good condition and must provide a ground for the tool. This means that extension cords should be the three-wire type. Make sure the conducting wire is large enough to prevent excessive voltage drop, **Figure 5-4**.

Be careful in stringing electrical extension cords around the work site. Place them where they will not be damaged or interfere with other workers. If run through a window or door opening, cords must be protected from getting pinched. If on the ground, they should be secured so as not to present a tripping hazard. Make certain that extension cords maintain ground continuity. This means that the third conductor (wire) in a cord is providing an unbroken path for current back to a grounded terminal. Thus, in case of a short in the tool case, current harmlessly bleeds off to ground.

Maximum Current for Extension Cords		
Extension Cord Length (Feet)	Maximum Current (Amps)	Wire Gauge (AWG)
25	13	16
25	15	14
50	5	18
50	10	16
50	15	14
75	5	18
75	10	16
75	15	14
100	5	16
100	15	12

Goodheart-Willcox Publisher

Figure 5-4. Extension cords must be large enough to carry the required amperage (amps) without overheating.

5.1.2 Shock Protection

Electric shock is one of the potential hazards of working with power tools. Always be sure that proper grounding is provided. Receptacles should have a third terminal for grounding. Plugs and cords should be of an approved type.

Portable power tools should be double insulated or otherwise grounded to protect the worker from dangerous electrical shock. Even though a circuit may be grounded, the operator of a portable power tool could be electrocuted should a bare conductor come in contact with the metal tool case. A ground-fault circuit interrupter (GFCI) should be used on all construction jobsites. These units can be installed in a circuit or plugged into an outlet that is grounded. These units sense when there is even a slight difference between the current flowing in the hot leg and that in the neutral leg of a circuit. When a ground fault (difference in current flow) occurs, the GFCI turns off power to the tool. See **Figure 5-5**.

Safety Note

Under certain circumstances, even a double-insulated tool can provide a serious or even fatal shock. If moisture is allowed to get inside of the tool's housing, it can provide an electrical path from energized parts to the outside.

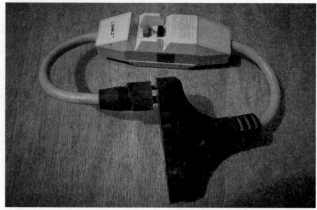

Carol Electric

Figure 5-5. A portable ground-fault circuit interrupter (GFCI) is used where permanent ground fault protection is not available. A power tool is plugged into it to protect the operator. Should a ground occur in the tool, the GFCI trips at around 5 mA, turning off power to the grounded tool.

5.2 Portable Circular Saws

The saw used most by carpenters is the portable circular saw. Refer to **Figure 5-6**. Its size is determined by the diameter of the largest blade it will take. Most saws on jobsites today are 7 1/4″. Cordless saws are becoming commonplace on construction sites.

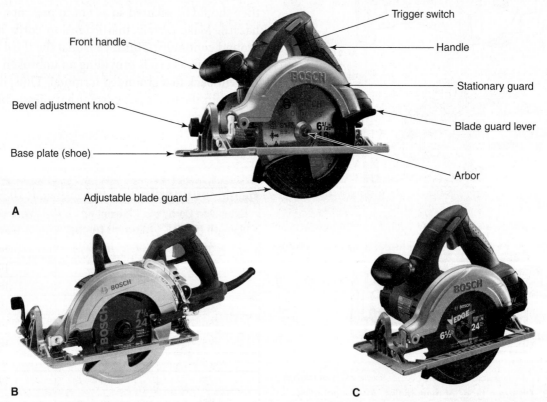

Photos courtesy of Robert Bosch Tool Corporation®

Figure 5-6. Portable circular saws. A—Parts of a portable circular saw. During use, the adjustable blade guard is pushed back by the stock (material being cut). A spring returns the guard when the cut is completed. B—Worm drive and hypoid saws use a gearbox which makes them heavier, but more powerful than direct-drive saws. C—A cordless circular saw is practical where electric service is not available.

Planning Ahead

Want to earn respect on the job? Want to be the next in line for a promotion? Then you need to learn to think like a boss. The boss is always thinking about what comes next, preparing a mental and physical list of what steps need to be taken, what tools will be needed, what materials will need to be on-site, and what information needs to be communicated to make the day go as planned today, tomorrow, next week, and beyond. You do not need to go to those great lengths right away. But, you should be thinking about what comes next. When you finish a task, give some thought to what the next job will be, and then show the initiative to take action. Whatever you do, do not stand around waiting for the boss to tell you what to do. At the very least, organize the tool chest, pick up scrap, or start pushing the broom until you can get some direction. Use company time productively. Think like the boss.

Though they do not develop as much torque as corded saws, they are convenient in situations where extension cords are difficult to use. The depth of cut is adjusted by raising or lowering the base, or shoe. Bevel cuts are made by tilting the shoe. The portable saw may be used to make cuts in assembled work. For example, flooring and roofing boards are often nailed into place before their ends are trimmed.

Various blades are available for different applications, **Figure 5-7**. The most commonly used is the framing or combination blade, which has relatively few teeth and an aggressive cut. These blades are ideal for rough framing and can be used for both rip cuts, along the grain, and cross cuts, across the grain. Specialty blades are available for cutting plywood, steel, masonry, and fiber cement.

5.2.1 Making a Cut

Before making any cut, the stock to be cut should be positioned on stable supports so that the blade will not touch anything on the opposite side of the stock. The piece should be positioned so that the waste, or piece being cut off, can fall freely without pinching the saw blade.

Safety Note

A common mistake for beginners is to place a piece of lumber or sheathing on a pair of sawhorses, then try to make a cut between the sawhorses. As the material begins to fall, it pinches the saw blade. The result can be a violent kickback of the saw and a very poor cut. Always allow the waste to fall away freely.

To use a portable saw, firmly grasp the handle in one hand with the forefinger ready to operate the trigger switch. Most saws have two handles. Holding the saw with both hands makes following your layout line easier and protects your hands from getting too close to the blade.

Rest the base of the saw on the work and align the guide mark with the layout line. Turn on the switch,

Figure 5-7. Circular saw blades. A—Coarse combination. B—Finish. C—Fiber cement. D—Masonry. E—Abrasive for metal.

allow the motor to reach full speed, then smoothly feed the blade into the stock. Release the switch as soon as the cut is finished. Hold the saw until the blade stops.

Safety Note

Follow these safety rules when working with portable circular saws.

- Wear safety glasses and hearing protection.
- Unplug the saw to replace a saw blade. Some saws have a control that locks the arbor during this operation, **Figure 5-8**.
- If the saw has two handles, keep both hands on them as much as possible during the cut. See **Figure 5-9**. Start the cut using one hand to raise the guard (when needed, usually at the beginning of a compound cut). When the cut is completed, make sure the guard returns to the closed position. Never block the guard so that it stays open—doing so will severely compromise your safety.
- Support stock being cut so that the kerf will not close up and bind the blade during or at the end of the cut. Do not prevent the waste from falling away from the cut.
- Support thin materials near the cut. Small pieces should be clamped to a benchtop or sawhorse.
- Be careful not to cut into the sawhorse or other supporting device.
- Before starting the saw, adjust the depth of cut to the thickness of the stock, plus about 1/8″.
- Check the base and angle adjustments to be sure they are tight.
- Plug the cord into a grounded outlet and be sure it will not become tangled in the work.
- Always place the saw base on the stock with the blade clear before turning on the switch.
- During the cut, stand to one side of the cutting line. Never reach under the workpiece during a cut.
- Never pull back on a running saw once the cut is started—to do so may cause a kickback.
- If the saw blade strays off the cutting line, stop the saw, pull back, and start the cut over. Otherwise, the blade may bind and cause an injury.
- Keep hands clear of the cutting line.
- Have the saw at full power and up to speed before beginning a cut.
- Always use a sharp blade of the proper type for the material and cut.
- Never force the saw through the material.
- Since they are light and portable, hand power saws can be carried anywhere on the jobsite. Use care to avoid damage or injury.
- Dropping the saw or allowing it to fall can damage or destroy it. Never use a damaged saw.
- Keep the saw guard clean and lubricated, so that it functions as designed.
- Immediately remove from service any saw that does not properly function.

Blade lock control

Black & Decker

Figure 5-8. Always disconnect the saw from the power source while adjusting or changing blades. Pressing a control locks the arbor during blade change.

Everyonephoto Studio/Shutterstock.com

Figure 5- 9. The circular saw is one of a carpenter's most often used tools. Hold the saw firmly and guide it through the cut slowly. Use both handles when cutting with a circular saw. Secure your work with a clamp or other device if necessary.

5.3 Saber Saw

The *saber saw*, also called a portable jig saw, is a portable saw useful for a wide range of light work. See **Figure 5-10**. Carpenters, cabinetmakers, electricians, and home craftspeople use it. The stroke of the blade is about 1/2″. The saw operates at a speed of approximately 2500 strokes per minute, but some have a variable speed control.

There is a wide variety of blade types available for saber saws, **Figure 5-11**. In general, narrower blades can cut tighter curves in wood or plastic. Blades with more teeth per inch make smoother cuts, but cut more slowly. Blades for wood cutting have 6–12 teeth per inch.

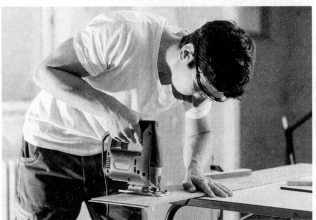

Kzenon/Shutterstock.com

Figure 5-10. A saber saw cuts with an action much like a handsaw.

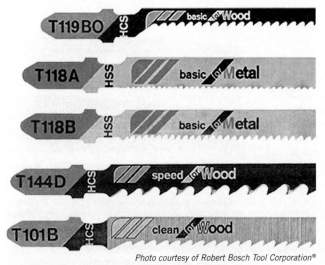

Photo courtesy of Robert Bosch Tool Corporation®

Figure 5-11. Saber saw blades come in a variety of configurations to suit different tasks. Narrow blades are used for curved cuts in wood or plastic. Some blades have small teeth for fine cuts in wood and plywood or metal. Blades with larger teeth are used for faster wood cutting.

For general purpose work, a blade with 10 teeth per inch is satisfactory. Always select a blade that will have at least two teeth in contact with the material being cut.

Saber saws vary in the way the blade is mounted in the chuck. Follow the directions and lubrication schedule in the manufacturer's manual.

The saber saw can be used to make straight or bevel cuts. Curves are usually cut by guiding the saw along a layout line. However, circular cuts may be more accurately made with a special guide or attachment.

Since the blade cuts on the upstroke, splintering takes place on the top side of the work. This must be considered when making finished cuts, especially in fine hardwood plywood. Always firmly hold the base of the saw against the surface of the material being cut.

When cutting internal openings, a starting hole can be drilled in the waste stock or the saw can be held on end so the blade cuts its own opening. See **Figure 5-12**. This is called plunge cutting and must be done with considerable care. Firmly rest the toe of the base on the work and turn on the motor. Then, slowly lower the blade into the stock.

5.4 Oscillating Multi-Tool

The *oscillating multi-tool* has a blade that is offset, making it useful for cutting, scraping, sanding, and grinding in difficult to reach places. See **Figure 5-13**. These tools are available in both corded and cordless versions and can be equipped with a wide variety of attachments. The multi-tool is designed for accuracy rather than speed. As its name suggests, the tool's blade oscillates at a rapid pace, allowing it to make fine cuts or remove material, if using an abrasive attachment.

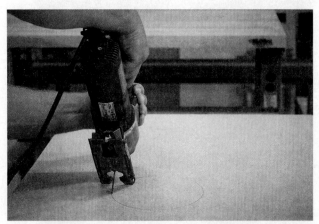

Goodheart-Willcox Publisher

Figure 5-12. On internal cuts, a saber saw can make its own opening. Be sure to rest the base on the workpiece before starting the cut. Plunge the blade in slowly while holding the trigger.

Figure 5-13. A multi-tool is useful for shortening door jams when installing a new floor. The offset blade allows for a flush cut.

Cutting blades are available for wood, plastic, metal, and carpet. The blades come in a variety of shapes to accommodate different tasks. Scraping blades are used to remove old paint or caulk. They have a sharpened straightedge, rather than cutting teeth. Sanding pads are available in a variety of shapes that can be used for profile sanding. Carbide rasps and grout removal attachments are used for tile restoration work, **Figure 5-14**.

Figure 5-14. Manufacturers produce a variety of useful multi-tool attachments.

5.5 Reciprocating Saw

Operation of the *reciprocating saw*, **Figure 5-15**, is similar to that of the saber saw. Like the saber saw, its cutting action is linear (back-and-forth), not circular. A reciprocating saw is larger and more powerful than a saber saw. It is useful under conditions where a circular saw is not safe or practical. Carpenters use it in remodeling work where sections of framing, sheathing, or inside walls must be removed. Blades are available to cut through wood, plaster, fiberglass, and all kinds of metals.

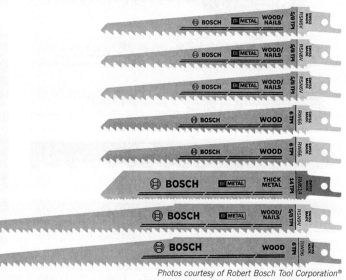

A   **B**

Figure 5-15. A reciprocating saw. A—This type of saw has a blade that moves forward and back (reciprocates) through a stroke distance of about 1 1/8″. It can be fitted with a variety of blades to suit all purposes. B—This set of all-purpose blades shows the variety of reciprocating blades available.

5.6 Chain Saws

Occasionally, a *chain saw* may be needed to cut heavy timbers, posts, or pilings. See **Figure 5-16**. Both gas and electric models are available, but gasoline-powered types are more powerful. Blade lengths vary from about 14″–24″. Chain saws are also useful for demolition during remodeling jobs. An 18″ blade is recommended for a carpenter.

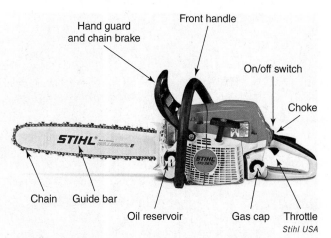

Figure 5-16. A chain saw is suitable for cutting heavy timbers.

5.7 Portable Electric Drills

Portable *electric drills* come in a wide range of types and sizes. The size is determined by the chuck capacity; 3/8″ and 1/2″ chucks are the sizes normally used by a carpenter. Cordless portable drills are handy for many jobs, **Figure 5-17**. Power is supplied by a rechargeable lithium-ion battery. Most carpenters prefer cordless drills because of the freedom of movement they provide. The speed of a variable speed drill is adjusted by the trigger. Most drills have a reversing switch that allows the drill to be run in either direction. Bits designed for use in a portable electric drill are shown in **Figure 5-18**.

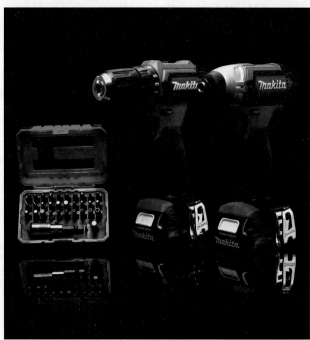

Fahroni/Shutterstock.com

Figure 5-17. Cordless drills and impact drivers have become essential tools for carpenters on the job.

Goodheart-Willcox Publisher   **A**   *Goodheart-Willcox Publisher*   **B**   *Goodheart-Willcox Publisher*   **C**

D   *Goodheart-Willcox Publisher*   **E**   *Goodheart-Willcox Publisher*

Figure 5-18. Drill bits a carpenter might use. A—Spade bits for boring holes in wood, available in sizes from 3/8″ to 1 1/2″. B—Twist drill bits for drilling holes in metal, plastic, and wood, available in sizes from tiny to 1/2″. C—Auger bits for boring holes in wood, available in many sizes and types. D—Forstner bit for flat-bottom holes in wood, available in sizes from 1/4″ to 2 1/2″. E—Hole saws are preferred for cutting larger diameter holes up to 6″.

Safety Note

Follow these safety rules when working with portable drills.

- Wear safety glasses.
- Select the correct drill or bit for your work and securely mount it into the chuck.
- Be sure stock is firmly held so it does not move during the operation.
- When using corded drills, be sure the switch is in the *off* position before connecting the drill to a properly grounded outlet.
- Turn on the switch for a moment to see if the bit is properly centered and running true.
- With the switch off, place the point of the bit in the punched layout hole.

- Firmly hold the drill at the correct drilling angle, using one or both hands as needed.
- Turn on the switch and feed the drill into the work. The pressure required will vary with the size of the drill and the kind of material being drilled.
- During operation, keep the drill aligned with the desired direction of the hole.
- When drilling deep holes, especially with a twist drill, withdraw the drill bit from the hole several times during the process to clear the cuttings.
- Always turn off the power and remove the bit from the drill as soon as you have completed your work.

5.8 Impact Driver

While drills have long been used for drilling and driving fasteners such as screws and lags, *impact drivers* have taken over the task of driving fasteners. Impact drivers are more compact and lightweight than drills. They also have an internal hammering mechanism between the motor and chuck, which ensures better contact and less slip on the fastener. Impact drivers generally have a higher torque rating than comparable drills, which allows for better battery longevity and performance when driving screws. While the chuck on a drill accepts multiple sizes of drill bits, impact drivers have a chuck that will only accept a 1/4″ hexagonal shank, **Figure 5-19**.

Green Note

While contributing significantly to increased production, the Lithium Ion batteries used in cordless tools are toxic to the environment when disposed of improperly. Contact your local recycling facility to find out the best way to dispose of or recycle your faulty batteries.

Pro Tip

If you will be using your cordless tool for more than a few minutes, throw a spare battery in the charger, even if you think it is already charged. Today's batteries and chargers regulate so they will not over-charge. It does them no harm to be put on the charger twice. You do not want to be caught off guard with no battery power to work with.

5.9 Rotary Hammer Drills

Rotary hammer drills are heavy-duty tools used to drill holes in concrete and other masonry materials. See **Figure 5-20**. Anchor devices are then placed in the holes to receive bolts or screws that fasten materials to floors, walls, or ceilings. Various bit types are available.

Photo courtesy of Robert Bosch Tool Corporation®

Figure 5-20. This cordless rotary hammer drill is fitted with a dust extraction system, which reduces exposure to harmful silica dust.

A *NEFLO PHOTO/Shutterstock.com*   **B**   *Goodheart-Willcox Publisher*

Figure 5-19. A—Cordless impact drivers are used frequently by the carpenter on the jobsite. B—While a drill chuck will accept round or hexagonal shanks, an impact driver can only accept accessories with a 1/4″ hexagonal shank.

Concrete bits include a carbide-tip bit. Carbide-tip bits are also available for faster cutting and greater durability in drilling concrete and masonry. Twist drills are also offered for use in drilling steel. Depending on the type of bit used, a rotary hammer can drill holes from 3/16″ to 6″ in diameter.

Safety Note
Follow these safety rules when working with rotary hammer drills.
- Wear safety glasses and hearing protection.
- Maintain good balance and a tight grip on the handles. Bits can bind in the hole, tearing the tool from your grasp and possibly knocking you off balance.
- Steadily push into the workpiece, pulling back frequently to clear away chips.
- Keep bits sharp.
- Disconnect the drill from power while installing or removing bits.
- Never lock the trigger in the *on* position.
- Check that the switch is in the *off* position before plugging it in.

5.10 Power Planes

The *power plane* produces finished wood surfaces with speed and accuracy, **Figure 5-21**. The motor, which operates at a speed of about 20,000 rpm, drives a spiral cutter. The depth of cut is adjusted by raising or lowering the front shoe. The rear shoe (main bed) must be kept level with the cutting edge of the cutter head. The power plane is equipped with an adjustable fence for planing bevels and chamfers. For surfacing operations, the fence is removed.

Photo courtesy of Robert Bosch Tool Corporation®

Figure 5-21. Portable power planes are sometimes used instead of hand planes because they are faster and more accurate.

Hold and operate the power plane in about the same manner as you would a hand plane. Support the work so that it cannot move, while still permitting easy operation. Start the cut with the front shoe firmly resting on the work and the cutter head slightly behind the surface. Be sure the electric cord is kept clear to prevent damage. Start the motor before the cutter head engages the stock. Move the plane forward with smooth, even pressure on the work. When finishing the cut, apply heavier pressure on the rear shoe.

The *power block plane* can be used on small surfaces. It has features and adjustments similar to the regular power plane. Since it is small, this plane is designed to be operated with one hand. When using the tool, the work should be securely held or clamped in place. In planing small stock, kickbacks may occur. Be sure your free hand is kept well out of the way.

Safety Note
Follow these safety rules when working with power planes.
- Study the manufacturer's instructions for proper adjustment and operation.
- Be sure the machine is properly grounded.
- Hold the standard power plane in both hands before you pull the trigger switch. Continue to hold it in both hands until the motor stops after releasing the switch.
- Always securely clamp the work in the best position for performing the operation.
- Do not attempt to operate a power plane with one hand if it was designed for two-handed operation.
- Be mindful of the fact that there is a high-speed, very sharp cutting head at the bottom of the plane.
- Disconnect the electric cord before making adjustments or changing cutters.

5.11 Routers

Routers are used to cut irregular shapes and to form various contours on edges, **Figure 5-22**. When equipped with special guides, routers can be used to cut dados, grooves, mortises, and dovetail joints. Important uses in carpentry include the cutting of gains for hinges when hanging passage doors and routing housed stringers for stair construction.

A collet-type chuck is used to hold a bit. When changing bits on the router, it may be necessary to remove the base. However, it is usually possible to change bits through openings in the base. See **Figure 5-23**.

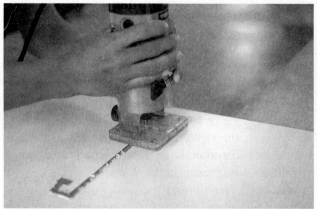

Azami Adiputera/Shutterstock.com

Figure 5-22. A portable router has a motor mounted in an adjustable base that rests on the workpiece. The motor revolves in a clockwise direction. Always use two hands when routing.

PROCEDURE
Changing a Router Bit
The following describes how to change a router bit.
1. Disconnect the router cord from power.
2. Lock the shaft or hold it with a wrench, depending on the router.
3. Loosen the chuck with a wrench.
4. Remove the bit.
5. Select the new bit. Some routers have a sleeve that is part of the bit assembly.
6. Install the bit on the arbor.
7. Tighten the chuck with a wrench.

A   Bullard-Haven Technical School

B   Bullard-Haven Technical School

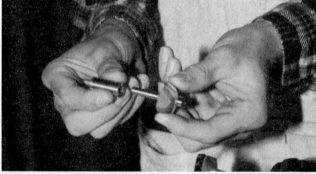

C   Bullard-Haven Technical School

D   Bullard-Haven Technical School E   Bullard-Haven Technical School

Figure 5-23. Changing a router bit. First, unplug the router from the power source. A—Use a wrench to loosen the chuck. B—Remove the bit being replaced. C—Select a new bit. D—Slide the new bit onto the arbor. E—Tighten the chuck and adjust the cutting depth.

OlekStock/Shutterstock.com

Figure 5-24. There are many router bit shapes available for everything from square dados and rabbets to decorative edges and special joinery tasks. Bits with carbide cutting edges give better results and last longer than steel cutting edges.

Figure 5-24 shows a few of the hundreds of router bits available. Straight bits are used when cutting dados and grooves. Some bits for shaping and forming edges have a pilot tip that guides the router. The router motor revolves in a clockwise direction (when viewed from above) and should be fed from left to right when making a cut along an edge. When cutting around the outside of rectangular or circular pieces, always move in a counterclockwise direction.

Safety Note

Follow these safety rules when working with routers.

- Wear safety glasses and hearing protection.
- Securely mount the bit in the chuck and make sure the base is tight.
- Be sure the motor is properly grounded.
- Be certain the work is securely clamped so it will remain stationary during the routing operation.
- Place the router base on the work, template, or guide—with the bit clear of the wood—before turning on the power. Firmly hold the router when turning on the motor. Starting torque could wrench the tool from your grasp.
- Hold the router with both hands and smoothly feed it through the cut in the correct direction.
- When the cut is complete, turn off the motor. Do not lift the machine from the work until the motor has stopped.
- Always unplug the router when mounting bits or making adjustments.

5.12 Portable Sanders

There are two basic types of portable sanders, belt and finish. The types vary widely in size and design. Carefully follow the manufacturer's instructions when mounting abrasive belts, discs, and sheets. Be sure to follow the manufacturer's lubrication schedule, as well.

5.12.1 Belt Sander

A *belt sander* is a portable tool that consists of a belt of sandpaper that is rotated by a motor. The size of a belt sander is determined by the width and length of the belt. Using the sander takes some skill. Firmly support the stock. The switch must always be in the *off* position before plugging in the electric cord. Like all portable power tools, the sander should be properly grounded. Check the belt and make sure it is properly tracking.

Hold the sander over the work. Start the motor. Then, carefully and evenly lower the sander onto the surface. When using belt and finish sanders, be sure to travel with the grain. Move the sander forward and backward over the surface in even strokes. At the end of each stroke, shift it sideways about one-half the width of the belt.

Continue over the entire surface, holding the sander level. Sand each area the same amount. Do not press down on the sander; its weight is sufficient to provide the proper pressure for the cutting action. When work is completed, raise the sander from the surface and turn off the motor.

5.12.2 Finishing Sanders

Finishing sanders are used for final sanding, where only a small amount of material needs to be removed. They are also used for cutting down and rubbing finishing coats. There are two general types of finish sanders, orbital and oscillating. An orbital sander is shown in **Figure 5-25**.

Also called a pad sander, the finishing sander has a pad backing up the sanding paper. Unlike belt sanders, where the sanding belt moves, the paper is held stationary on the pad by mechanical grippers or is attached to the pad with either a strong adhesive or hook-and-loop surfaces. The pad moves, carrying the paper.

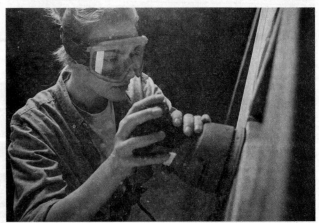

ivan_kislitsin/Shutterstock.com

Figure 5-25. Using an orbital sander to perform finish sanding on cabinetwork. Some orbital finishing sanders are equipped with a dust bag.

5.13 Power Staplers and Nail Guns

A wide variety of *power staplers* and *nail guns* are available. Most of them are pneumatic (air powered), but battery-operated and gas-operated models are also available, **Figure 5-26**.

Using power staplers and nail guns speeds up work and also allows a carpenter to work in tight places where swinging a hammer might be difficult or impossible. Staplers offer an alternative to nails, especially for securing sheathing and shingles. The following are two examples of nail gun types.

- Strip-fed—These hold a strip of nails in a long magazine located under the handle. Most framing nail guns and finish nail guns are strip-fed. Some use strips at a 20°–22° angle, others use straight strips. Some use clipped head nails, others use round head nails. Be sure you use the right type for your nail gun.
- Coil-fed—This type carries nails in a round canister, usually located ahead of a trigger handle. Most roofing nail guns are coil-fed.

All nail guns rely on two basic controls: a finger trigger and a contact safety tip located on the nose of the gun. Trigger mechanisms can vary based on different factors.

- The order in which the controls are activated.
- Whether the trigger can be held in the squeezed position to discharge multiple nails as the nose is bumped against a surface, called bump firing, or if it must be released and then squeezed again for each individual nail.

Combining these variations gives four kinds of triggers. Some nail guns have a selective trigger switch, which allows the user to choose among two or more trigger systems. The sequential fire is the safest and most accurate type of nail gun trigger. This trigger will only fire a nail when the controls are activated in a certain order. First, the safety contact tip must be pushed into the workpiece, then the user squeezes the trigger to discharge a nail. Both the safety contact tip and the trigger must be released and activated again to fire a second nail. Nails cannot be bump fired.

To operate a nail gun, first press the nose against the work to disarm the safety device that keeps it from driving the fastener. Then, press the trigger to drive the nail, as shown in **Figure 5-27**.

A   *Hadley-Hobley Construction*

B   *Trussworks, Inc.*

Figure 5-27. Pneumatic nail guns speed work and can be used in close quarters where using a hammer may be difficult. A—This finish nail gun requires no lubrication and drives 1″–2″ brads. B—Framing nail gun drives 2″–3 1/2″ smooth-shank nails, 2″–3″ screw-shank nails, and 2″ and 2 3/8″ ring-shank nails.

A   *A & C Paslode; DeWalt*   B   *A & C Paslode; DeWalt*   C   *A & C Paslode; DeWalt*

Figure 5-26. There are several types of nail guns. A—Pneumatic. B—Battery-operated. C—Gas-operated.

Most air compressors used on the jobsite are portable, **Figure 5-28**. They may be driven by either electric motors or gas engines. Manufacturers of pneumatic tools can assist carpenters in selecting a compressor large enough to handle air needs on the job.

A   *Goodheart-Willcox Publisher*

B   *Johnson-Manley Lumber Co.*

Figure 5-28. An air compressor suitable for operating pneumatic tools must deliver up to 110 pounds per square inch (psi) of pressure. A—A typical portable compressor. B—This generator and compressor on a jobsite power electrical as well as pneumatic tools.

5.14 Miter Saw

Other than portable circular saws, power *miter saws* are used most often by carpenters for sawing operations. A compound sliding miter saw, **Figure 5-29**, is a particularly versatile tool. It has a base with a fence in the back to hold the work in position.

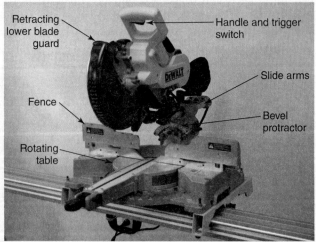

Goodheart-Willcox Publisher

Figure 5-29. Power miter saws are used extensively on the jobsite.

A portion of the base rotates left and right along with the sliding arm, motor, blade, and control handle. A 10″ or 12″ blade is common. The blade is fitted with a guard that retracts as the saw is lowered into the cut. The sliding arms allow the saw motor and blade to move from front to back while cutting wider work pieces.

Most of the sawing jobs that can be done on a radial arm saw can be done with a sliding compound miter saw, **Figure 5-30**. Unlike a radial arm saw, a power miter saw is always practical and easy to transport to a jobsite. It can be set up on saw horses and planks or on a miter saw stand.

Green Note
Evaluate waste for other uses on or off the jobsite before disposal. "Wasted" lumber can often be used elsewhere in the project, on another jobsite, or in a craft project offsite.

The *chop saw* is a variation of the power miter saw, **Figure 5-31**. Chop saws do not tilt for bevels and they do not have the sliding mechanism for moving through wide boards. Some chop saws have a vise to hold metal in position to be cut. Chop saws can be set for miters or square cut offs only. Chop saws are usually used for cutting metal.

Goodheart-Willcox Publisher

Figure 5-31. A metal cutting chop saw is almost a necessity when working with light gauge steel framing.

Abrasive blades for cutting metal and special carbide-tipped metal cutting blades are available for cutting all types of light-gauge metal. The most common blade size for chop saws is 14″.

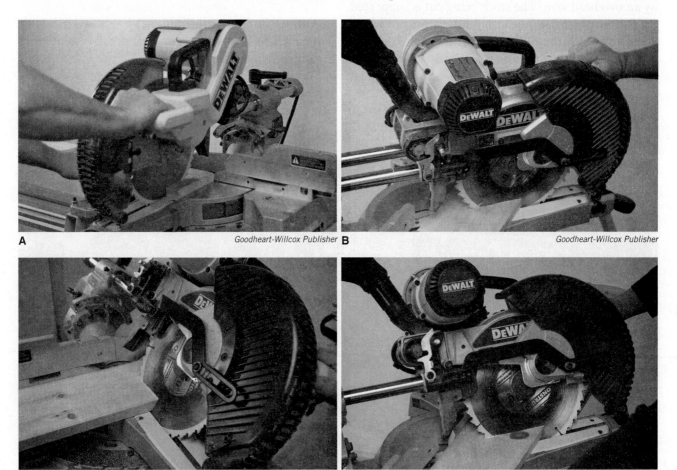

A — Goodheart-Willcox Publisher   B — Goodheart-Willcox Publisher
C — Goodheart-Willcox Publisher   D — Goodheart-Willcox Publisher

Figure 5-30. A compound sliding miter saw will perform a variety of sawing operations. A—Square cut off. B—Miter. C—Bevel. D—Compound miter (bevel plus miter).

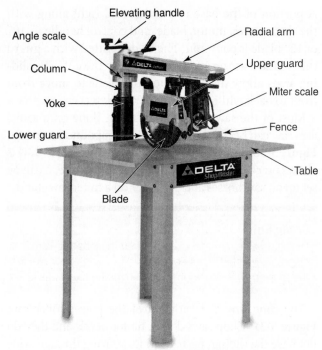

Delta International Machinery Corp.

Figure 5-32. Radial arm saws make most cuts with the material held stationary.

5.15 Radial Arm Saws

The motor and blade of the *radial arm saw* are carried by an overhead arm. The stock being cut is supported on a stationary table. The arm is attached to and supported by a column at the back of the table. By raising or lowering the overhead arm, the depth of cut can be controlled. **Figure 5-32** shows the parts of a typical radial arm saw.

The motor is mounted in a yoke and may be tilted for angle cuts. The yoke is suspended from the arm on a pivot, which permits the motor to be rotated in a horizontal plane. Adjustments make it possible to perform many sawing operations.

While crosscutting, mitering, beveling, or dadoing, the work is firmly held on the table. The saw is pulled through the workpiece, **Figure 5-33**. For ripping and grooving, the blade is turned parallel to the table and

mavo/Shutterstock.com

Figure 5-33. Using a radial arm saw to cut lumber.

locked in this position. Stock is then fed into the blade in a manner similar to feeding a table saw, except that the blade is above the stock.

For regular crosscuts and miters, first be sure the saw is resting against the column. Then, place your work on the table and align the cut. Firmly hold the stock against the table fence with your hand at least 6″ away from the path of the saw blade. Turn on the motor. Grasp the saw handle and firmly and slowly pull the saw through the cut. See **Figure 5-34**. The radial arm saw may tend to "feed itself." You must control the rate of feed. When the cut is completed, return the saw to the rear of the table and shut off the motor.

Radial arm saws are seldom used in construction. They are heavy and easy to knock out of adjustment when transporting. They do not produce as fine a cut as a power miter saw. Most importantly, they are dangerous to use. The blade is pulled through the workpiece, *toward* the operator, causing a tendency for the saw to "climb" into the workpiece. Because there are now safer and more efficient alternatives to radial arm saws, their use in the construction field has been greatly reduced.

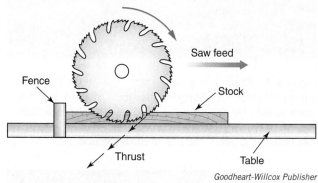

Goodheart-Willcox Publisher

Figure 5-34. On a radial arm saw, blade rotation is in the direction of the saw feed.

Safety Note

Follow these safety rules when working with radial arm saws.

- Wear safety glasses and hearing protection.
- Stock must be firmly held on the table and against the fence for all crosscutting operations. The ends of long boards must be supported level with the table.
- Before turning on the motor, be sure clamps and locking devices are tight. Check the depth of cut.
- Keep the guard and anti-kickback device in position.
- Always return the saw to the rear of the table after completing a crosscut or a miter cut. Never remove stock from the table until the saw has been returned and the blade has come to a complete stop.
- Maintain a 6″ margin of safety. Keep your hands this distance away from the path of the saw blade.
- Shut off the motor and wait for the blade to stop before making any adjustments.
- Do not leave the machine before the blade has stopped.
- Keep the table clean and free of scrap pieces and excessive amounts of sawdust. Do not attempt to clean off the table while the saw is running.
- In crosscutting, always pull the blade *toward* you while maintaining a firm grip on the handle.
- Stock to be ripped must be flat with one straight edge to guide it along the fence.
- When ripping, always feed stock into the blade so that the bottom teeth are turning toward you. This is the side opposite of the anti-kickback fingers.

5.16 Table Saws

Table saws are basic machines used in finish work, such as interior trim. There are two categories of table saws, contractor saws and cabinet saws. See **Figure 5-35**.

Contractor saw

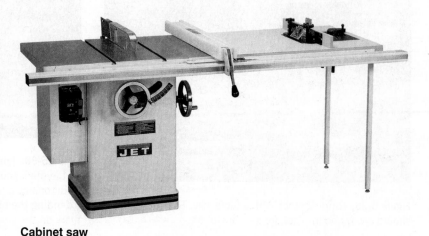

Cabinet saw

Goodheart-Willcox Publisher

Figure 5-35. Table saws are versatile, but are especially used by carpenters for ripping boards to width.

Contractor saws are lighter and more portable. As with all power tools, be sure to understand and follow the safety rules for operation. Table saws can be dangerous machines. They are powerful and operate at high rpms. These characteristics make them particularly unforgiving when a mistake is made.

The table saw is mostly used for ripping stock to width and cutting it to length. It also cuts bevels, chamfers, and tapers. Properly set up, the table saw can be used to produce grooves, dados, rabbets, and other forms that are basic to a wide variety of joints. The size of the saw is determined by the largest blade it will take. Stock to be ripped must have at least one flat face to rest on the table and one straight edge to run along the fence, thus serving as a guide for the cut. **Figure 5-36** shows the correct procedure for making a rip cut. When crosscutting on a table saw, the fence is not used. Doing so can cause violent kickback and result in serious injury. **Figure 5-37** shows a standard crosscutting operation. A line is squared across the stock to show where the cut is to be located. Make a check mark on the side of the line where the saw kerf will be located. Since the guard tends to hide the saw blade, it is helpful to use a pencil line on the table surface when aligning the cut. Since most of the work will be located to the left, it should extend back from the left side of the blade, as shown in **Figure 5-38**. A miter gauge is used to stabilize the material and set the angle of the cut. It is used to feed the stock into the blade.

A   *Goodheart-Willcox Publisher*

B   *Goodheart-Willcox Publisher*

Figure 5-37. Crosscutting on the table saw allows you to square off stock.. A—Squaring the stock. Securely hold the stock against the miter gauge. B—Guide the stock through the cut by pushing the miter gauge. One side of the guard has been lifted out of the way to show operation.

Goodheart-Willcox Publisher

Figure 5-36. Ripping stock with a table saw. The guard should always be in place for safe operation.

Goodheart-Willcox Publisher

Figure 5-38. For this cut, the right side of the stock is scrap so the mark should align with the left side of the saw kerf. Additionally, a line can be drawn on the table of the saw, extending the plane of the blade. This can help align the mark on the stock with the blade before the cut is started.

Follow these safety rules when working with table saws.
- Wear safety glasses.
- Be certain the blade is sharp and right for the job at hand.
- Make sure the saw is equipped with a guard and use it.
- Set the blade so it extends about 1/4″ above the stock to be cut.
- Stand to one side of the operating blade and do not reach across it.
- Maintain a 4″ margin of safety. Do not let your hands come closer than 4″ to the operating blade, even though the guard is in position.
- Stock should be surfaced and at least one edge jointed before the piece is cut on the saw.
- Use the fence or miter gage to control the stock. Do not cut stock freehand.
- Always use push sticks when ripping short, narrow pieces.
- Stop the saw before making adjustments.
- Do not let small, scrap cuttings accumulate around the saw blade. Use a push stick to move them away.
- Setups must be carefully made and checked before the power is turned on.
- Remove the dado head or any special blades after use.
- Workers helping to "tail-off" the saw should not push or pull on the stock, only support it. The operator must control the feed and direction of the cut.
- Once work is complete, turn off the machine and remain until the blade has stopped. Clear the saw table and place waste in a scrap box.

5.17 Jointers

Jointers are commonly used to dress the edges of boards. They may also be used for planing a face or for cutting a rabbet, bevel, chamfer, or taper. The cutter head is cylindrical and usually has three or four knives. As the cylinder rotates at high speed, the knives cut away small chips. This cutting action produces a smooth surface on the wooden workpiece.

Principal parts of a jointer are shown in **Figure 5-39**. The cutter head revolves at a speed of about 4500 rpm. The size of the jointer is determined by the length of the knives. There are three main adjustable parts of a jointer.

- Infeed table
- Outfeed table
- Fence

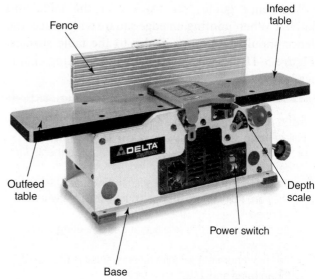

Delta International Machinery Corp.

Figure 5-39. Principal parts of a jointer. Small 4″–6″ jointers like this one can be easily moved from one jobsite to another.

The outfeed table must be exactly the same height as the knife edges at their highest point of rotation. This is a critical adjustment. See **Figure 5-40**. If the outfeed table is too high, the stock will gradually rise out of the cut and a slight taper will be formed. If it is too low, the tail end of the stock will drop as it leaves the infeed table and cause a "bite" in the surface or edge.

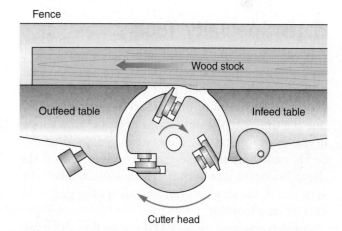

Goodheart-Willcox Publisher

Figure 5-40. Operation of a jointer. Note the direction of the wood grain. Avoid working against the grain. The outfeed table must be set at the exact height of the cutter head blades.

The fence guides the stock over the table and knives. When jointing an edge square with a face, the fence should be perpendicular to the table surface, **Figure 5-41**. The fence is tilted when cutting chamfers or bevels.

Safety Note

Follow these safety rules when working with jointers.

- Before turning on the machine, make adjustments for depth of cut and position of fence.
- Do not touch the knives (blades in the cutter head). They are razor sharp and can cut flesh, even when stopped.
- Be sure the guard is in place and is operating properly.
- The maximum cut for jointing on a small jointer is 1/8″ for an edge and 1/16″ for a face.
- Stock must be at least 12″ long. Stock to be surfaced must be at least 3/8″ thick.
- Feed the work so the knives will cut with the grain. Use stock that is free from knots, splits, and checks.
- Keep your hands away from the cutter head, even though the guard is in position. Maintain at least a 4″ margin of safety.
- Use a push block when planing a flat surface. Do not apply pressure directly over the knives with your hand.
- Do not plane end grain.
- The jointer knives must be sharp. Dull knives vibrate the stock and may cause a kickback.
- When work is complete, turn off the machine. Stand by until the cutter head has stopped.

5.18 Specialty Tools

Drywall screw shooters (also called screw guns) speed up drywall installation and provide better connections, all but eliminating "nail popping." See **Figure 5-42**. Most shooters require drywall screws to be inserted one at a time into the nose. Some drywall screw shooters have a magazine that allows strips of drywall screws to be loaded for multiple shots. Locators control the depth below the face of the drywall. Usually a belt clip attached to the shooter's housing allows the gun to be carried on a tool belt.

Powder-actuated tools (PATs) are used to drive fasteners into concrete and structural steel or concrete.

Goodheart-Willcox Publisher

Figure 5-41. Hold the board perpendicular to the table surface when jointing an edge. Keep hands at least 4″ away from the cutter head. This board is wide enough so a carpenter's hands never come near the cutter head.

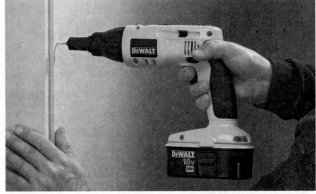

Goodheart-Willcox Publisher

Figure 5-42. This drywall screw gun has variable speeds from 0 to 4500 rpm.

See **Figure 5-43**.The tool may be either direct acting or indirect acting. Both types depend on the expanding gases from an exploded cartridge to drive the fastener. Both types have a triggering mechanism similar to that of a firearm. To use the tool, it must be firmly pressed against the material being fastened. Then, pull the trigger to activate the tool. Expanding gases from the exploding powder act on either the fastener or a piston. The fastener is driven into the concrete or metal.

A   *Goodheart-Willcox Publisher*   B   *Goodheart-Willcox Publisher*   C   *Goodheart-Willcox Publisher*

Figure 5-43. The powder-actuated tool can drive fasteners into concrete, steel, and other hard surfaces. These tools come in a range of configurations. A—Single-shot tools fire one cartridge and one pin at a time. B—Semi-automatic tools use strip load cartridges that advance into the chamber when the tool is cycled. C—Fully automatic tools use a magazine to hold strip pins and strip load cartridges for faster cycle times.

5.19 Power Tool Care and Maintenance

Care of power tools is especially important if they are to properly function and give long service. Sharp blades and cutters ensure accurate work and make the tool much safer to operate. Good carpenters take pride in the condition and appearance of their tools.

Most power tools are equipped with sealed bearings that seldom need attention. Follow the manufacturer's recommendations for lubrication schedules. Gear mechanisms for portable power tools usually require a special lubricant.

Clean and polish bare metal surfaces with 600 grit wet-or-dry abrasive paper when required. Apply a coat of paste wax to protect the surface and reduce friction. Do not apply oil to woodworking machines unless it is recommended by the manufacturer. Oil attracts dust and can become gummy over time.

Some power tools, especially those with a number of accessories, can be purchased with a case. In addition to making transport easier, such cases keep the accessories organized and protected.

Cutters and blades require periodic sharpening. Most carpenters are too busy and usually do not have the equipment to accurately grind cutters or completely fit saw blades. They usually send these items to a shop to be sharpened by an expert. Still, carpenters may lightly hone cutters before grinding operations are needed. Also, they may prefer to file saw blades several times before sending them in for a complete fitting. A fitting includes jointing, gumming, setting, and filing. Carbide-tipped tools stay sharp about ten times longer than those with regular steel teeth or edges. A special diamond grinding wheel is used to sharpen carbide edges.

Chapter Review

Summary

- Power tools greatly reduce the time needed to do many carpentry tasks.
- There are two categories of power tools used by carpenters: portable and stationary.
- Portable tools are lightweight and intended to be carried around by the worker. They include such items as drills, saws, power planes, routers, sanders, screw shooters, and nail guns.
- Stationary power tools are larger machines mounted on benches or stands. They are set up on the jobsite and work is brought to them. Stationary power tools include various types of saws, jointers, and planers.
- Observing safety rules and wearing proper protective devices are vital when working with power tools.
- Like hand tools, power tools must be properly cleaned and maintained for safe use.

Know and Understand

Answer the following questions using the information in this chapter.

1. All of the following supply power to portable power tools, *except* a(n)_____.
   A. electrical cord
   B. rechargeable battery
   C. generator
   D. air hose

2. *True or False?* Always disconnect the power before servicing a power tool.

3. The size of a portable circular saw is determined by the _____.
   A. diameter of the largest blade
   B. circumference of the saw
   C. type of blade
   D. All of the above.

4. The most commonly used saw blade is the _____ blade.
   A. curved
   B. framing
   C. scraping
   D. offset

5. For general-purpose sawing in wood, a saber saw blade should have about _____ teeth per inch.
   A. 2
   B. 6
   C. 10
   D. 12

6. The blade of a _____ saw cuts on the upstroke.
   A. portable circular
   B. saber
   C. reciprocating
   D. All of the above.

7. A(n) _____ is commonly used to remove sections of framing.
   A. portable circular saw
   B. oscillating multi-tool
   C. reciprocating saw
   D. chain saw

8. *True or False?* Electric-powered chain saws are more powerful than gasoline-powered types.

9. *True or False?* When drilling deep holes, do not withdraw a twist drill until the hole is completed.

10. _____ are currently the main tool used for driving fasteners.
    A. Portable electric drills
    B. Rotary hammer drills
    C. Power planes
    D. Cordless impact drivers

11. The depth of cut of a power plane is adjusted by raising or lowering the _____.
    A. rear shoe
    B. front shoe
    C. fence
    D. spiral cutter

12. Typically, a standard router bit is held in a _____-type chuck.
    A. collet
    B. drill
    C. hexagonal
    D. combination

13. The size of a belt sander is determined by the _____ of the belt.
    A. pressure
    B. weight
    C. width and length
    D. torque

14. *True or False?* It is safe to use bottled oxygen gas to power a pneumatic stapler.

15. When crosscutting with the radial arm saw, the blade is moved _____ the operator.
    A. toward
    B. away from
    C. above
    D. below

16. For regular work, the _____ of the jointer should be perfectly aligned with the knife edges at their highest point of rotation.
    A. infeed table
    B. outfeed table
    C. fence
    D. cutter head

Apply and Analyze

1. How can you determine the voltage required by a portable power tool?

2. Give the meaning of the term *ground continuity* and explain why it is important.

3. What is a ground-fault circuit interrupter and why should it be used in carpentry?

4. Explain why a carpenter would choose to use a reciprocating saw over a saber saw.

5. Name two advantages of a nail gun.

6. Describe how a power tool should be cleaned and maintained.

Critical Thinking

1. When operating a table saw, using the fence while crosscutting a board is very dangerous. Doing so is likely to cause a kickback that can lead to a serious injury. Why does this happen?

2. Explain why you should not use a circular saw to make an unsupported cut between two sawhorses.

3. Imagine you find a bargain on a good pneumatic finish nail gun at a yard sale. The person selling it says it works fine but it was his late father's and he does not know anything more about it. You cannot pass up the deal, so you buy it. How would you go about determining what gauge, type, and angle of nails to buy for it?

Communicating about Carpentry

1. **Reading.** With a partner, make flash cards for the key terms in Chapters 4 and 5. On the front of the card, write a term. On the back of the card, write a brief definition. Take turns quizzing one another on the definitions of key terms. Practice reading aloud the terms, clarifying pronunciations where needed.

2. **Listening.** In small groups, discuss with your classmates—in basic, everyday language—when power tools should be used instead of hand tools. Discuss the various precautions—both safety and electrical—that should be taken before using power tools. Conduct this discussion as though you had never read this chapter. Take notes on the observations expressed. Then, review the points discussed, factoring in your new knowledge of tools. Develop a summary of what you have learned about tools and present it to the class, using the terms that you have learned in the chapter.

3. **Speaking.** Make an oral presentation to the class on the hazards involved in using nail guns and how to avoid them. The OSHA publication entitled "Nail Gun Safety: A Guide for Construction Contractors" might provide useful information for your presentation. It is available on the internet.

CHAPTER 6

Scaffolds, Ladders, and Rigging

OBJECTIVES

After studying this chapter, you will be able to:

- Explain typical designs and construction of manufactured and site-built scaffolding.
- Discuss the types and uses of brackets and jacks.
- List ladder types and maintenance techniques.
- Apply ladder and scaffolding safety rules.
- Recognize and describe common safety defects in slings.
- Use standard hand signals to direct a crane operator.

TECHNICAL TERMS

bearer	rigging	scaffolding	toeboard
ladder jack	roofing bracket	sling	trestle jack
mudsill	safety shoes	tag line	

Carpenters use **scaffolding**, also called staging, to work in areas that are out of reach while standing on the ground or floor deck. Scaffolds are temporary platforms that support workers, tools, and materials with a high degree of safety, **Figure 6-1**. Their design must help a worker avoid stooping and reaching. The type of scaffold needed for a job depends on how many workers will be on it at one time, the distance it must extend above the ground, and whether it must support building materials as well as people.

A   *Alum-A-Pole Corp.*   B   *Bannafarsai_Stock/Shutterstock.com*

Figure 6-1. Scaffold is used extensively indoors and outdoors on construction sites. A—This scaffold system is used for siding work. It has pump-jack units attached to aluminum poles. A foot pedal, as shown, is used to raise the scaffold. B—This frame scaffold is equipped with wheels for mobility.

According to the US Department of Labor, an estimated 2.3 million construction workers, or 65% of the construction industry, work on scaffolds. Protecting these workers from scaffold-related accidents may prevent some of the 4,500 injuries and over 60 deaths every year. Of the workers injured in scaffold accidents, 72% said the accident was caused by slipping, the planking or support giving way, or being struck by a falling object. OSHA has very specific regulations covering the safe construction and use of scaffolding.

Safety Note

The OSHA website provides a large amount of information about scaffold and ladder safety, including standards and training materials. Explore the available resources on www.OSHA.gov.

6.1 Types of Scaffolding

Scaffolding includes a great variety of designs. In addition to the types manufactured with metal components, there are wood scaffolds that can be built onsite by a carpenter crew. Some manufactured types are meant to be assembled and disassembled so they can be reused at other building sites. Another type stays assembled, is completely mobile, and can be moved from site to site. Scaffolds of various types are shown throughout this chapter.

6.1.1 Manufactured Scaffolding

Many builders use sectional steel or aluminum scaffolding. It can be quickly and easily assembled from prefabricated frames. The pump-jack scaffold system shown in **Figure 6-2** has light aluminum uprights that

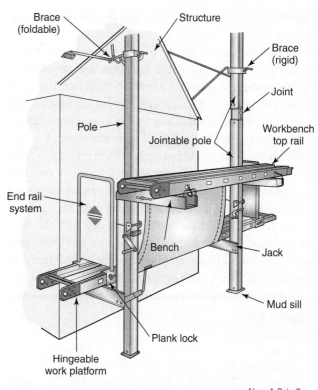

Alum-A-Pole Corp.

Figure 6-2. Details of an aluminum pole system for scaffolding.

are attached to the building with metal braces. The platform is adjusted to various heights by a jack that raises and lowers the brackets supporting the platform. The platform and bench units are also manufactured of aluminum.

Figure 6-3 shows another manufactured type of scaffold. It consists of truss frames and diagonal braces that can be horizontally and vertically assembled to build a scaffold of nearly any safe height or length.

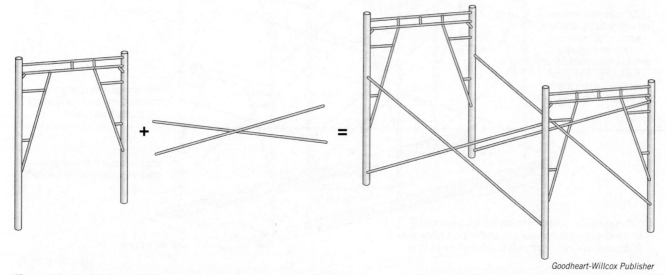

Goodheart-Willcox Publisher

Figure 6-3. Prefabricated truss frames and diagonal braces can be used to rapidly build scaffold sections.

Figure 6-4 illustrates the erection details of this scaffolding system.

There are many styles of sectional steel scaffolding. Some types have adjustable legs that are attached to the ground-level sections, **Figure 6-5**. When the scaffold is erected on soil that may allow for settling, planks commonly called *mudsills* are used to prevent settling. Frame sizes range from 2′ to 5′ wide and from 3′ to 10′ high. Various lengths of bracing provide frame spacings of 5′–10′. The basic units are set up and then vertically and horizontally joined. The frames can be equipped with casters when a rolling scaffold is desired.

Another scaffold type is the suspended scaffold. Suspended scaffolds are hung from the roof or other overhead structures. These are used mainly by painters and should be used only with light equipment and materials.

Goodheart-Willcox Publisher

Figure 6-5. Adjustable extensions on the legs of scaffolding frames allow leveling for stability and safety. Planks or other support material should be used to prevent legs from sinking into the ground.

Safety Note

In order to meet OSHA standards, all scaffolding must be capable of supporting four times the intended load. The intended load includes all persons, equipment, tools, materials, and other loads reasonably anticipated.

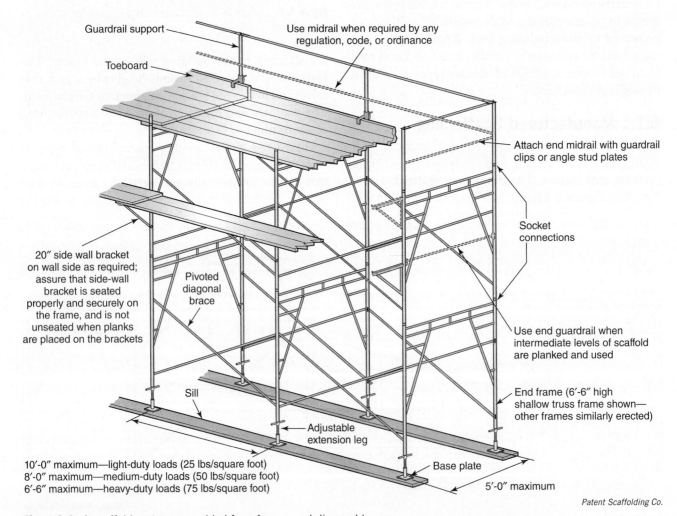

Guardrail support

Use midrail when required by any regulation, code, or ordinance

Toeboard

Attach end midrail with guardrail clips or angle stud plates

20″ side wall bracket on wall side as required; assure that side-wall bracket is seated properly and securely on the frame, and is not unseated when planks are placed on the brackets

Pivoted diagonal brace

Socket connections

Use end guardrail when intermediate levels of scaffold are planked and used

Sill

End frame (6′-6″ high shallow truss frame shown— other frames similarly erected)

Adjustable extension leg

Base plate

5′-0″ maximum

10′-0″ maximum—light-duty loads (25 lbs/square foot)
8′-0″ maximum—medium-duty loads (50 lbs/square foot)
6′-6″ maximum—heavy-duty loads (75 lbs/square foot)

Patent Scaffolding Co.

Figure 6-4. A scaffold system assembled from frames and diagonal braces.

6.1.2 Mobile Scaffolding

Movable or mobile scaffolding comes assembled and uses either a mechanical or hydraulic system to adjust its height. Some units are self-propelled and some are mounted on a trailer arrangement that is easily moved from one site to another. This is the major advantage of mobile scaffolding. **Figure 6-6** shows a mobile scaffold raised by a scissor jack that is activated by a hydraulic cylinder. **Figure 6-7** shows another type of mobile scaffold, usually called a boom lift, or basket lift. Some of these aerial lifts are set up to be towed behind a truck others are self-propelled by a gas, propane, or electric motor.

Safety Note

Many mobile scaffolds incorporate anchor points into their railing system. These anchor points are intended to be used with a positioning device as a restraint system and do not meet the impact requirements for a personal fall arrest system. Falling from the scaffold could produce enough force to overturn the unit. The railing is intended to act as the fall arrest system. Never work outside the railing unless you have proper fall protection.

6.1.3 Site-Constructed Wood Scaffolding

Where it is not possible or practical to use manufactured scaffolding, carpenters may build scaffolds from available lumber. In constructing wooden scaffolds, the uprights should be made of clear (no knots or holes), straight-grained 2×4s. The lower ends should be placed on planks to prevent settling into the ground. See **Figure 6-8**.

Imfoto/Shutterstock.com

Figure 6-6. Mobile scaffolding can be quickly positioned, raised, and lowered. This scissor lift is self-propelled.

A *Dmitry Kalinovsky/Shutterstock.com*   **B** *Bannafarsai_Stock/Shutterstock.com*

Figure 6-7. Aerial lifts increase productivity and give safe access to difficult to reach places. A—Telescoping boom lifts have an extendable boom that can reach great distances. The largest of these machines can reach up to 185′. B—Articulating boom lifts have a knuckle that allows for versatility in tight spaces.

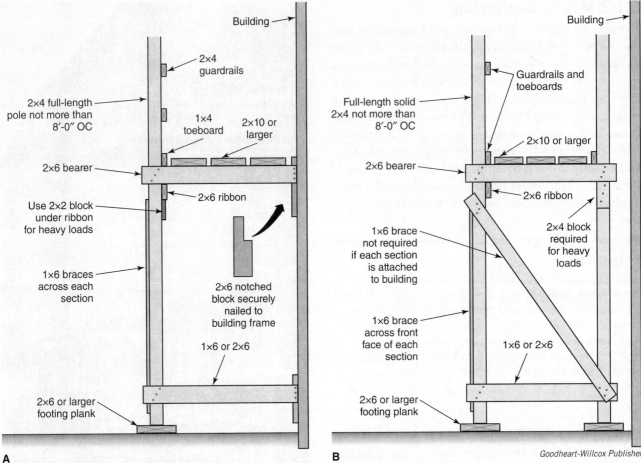

Figure 6-8. Designs for site-built scaffolding. A—Single-pole scaffold. The horizontal distance between bearer sections should never exceed 8′. B—Double-pole scaffold. The structure can be extended upward to form several platforms, but the maximum height should be limited to 18′.

Bearers, sometimes called cross ledgers, consist of 2×6s about 4′ long. Use at least three 16d nails at each end of the bearers to fasten them to the uprights. For the single-pole scaffold, one end of the bearer should be fastened to a 2×6 block securely nailed to the wall. Braces may be made of 1×6 lumber fastened to uprights with 10d nails and with 8d nails where they cross. For the platform, 2×10 planks without large knots should be used. It is good practice to spike the planks to bearers to prevent slipping. Whether lumber is used for planks or uprights, select lengths with parallel grain, **Figure 6-9**.

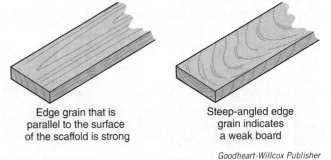

Edge grain that is parallel to the surface of the scaffold is strong

Steep-angled edge grain indicates a weak board

Goodheart-Willcox Publisher

Figure 6-9. For a safe scaffold, use straight-grain lumber for both uprights and platforms.

6.1.4 Guardrails and Toeboards

All planked areas of scaffolds greater than 10′ above a lower level are required to have guardrails and toeboards on all open sides and ends. The top rail of the guardrail must be between 36″ and 45″ above the planked surface. The top rail or must be capable of supporting a force of at least 200 pounds in any direction. There must also be a midrail halfway between the top rail and the planked surface. The midrail must be capable of supporting a force of at least 150 pounds in any direction. No guardrails are required on the side of a scaffold toward the work surface, so long as it is less than 14″ from a building wall.

A *toeboard* must be installed along the edge of the planking to prevent tools and material from slipping off the plank and possibly injuring someone below. The toeboard must be within 1″ of the planking. Additionally, all planks should be within 1″ of one another. Alternatively, a screen between the toeboard and midrail can be installed.

6.2 Brackets, Jacks, and Trestles

Metal wall brackets are sometimes used for scaffolding in residential construction, **Figure 6-10**. They can be quickly attached to a wall, easily moved from one construction site to another, and installed above overhangs or ledges.

Great care must be used when fastening the brackets to a wall. For light work at low levels, connections with nails may provide enough safety. Use at least four 16d nails. Be sure the nails penetrate sound framing lumber. Most carpenters prefer the greater safety provided by brackets that hook around wall studs. Holes in the sheathing required to accommodate such brackets must be repaired before the exterior finish is applied.

Some metal wall brackets have posts for holding guardrails and toeboards, as discussed earlier.

A wire mesh screen is sometimes added on open sides of the scaffold. The scaffold shown in **Figure 6-11** is fastened to the wall with a bolt.

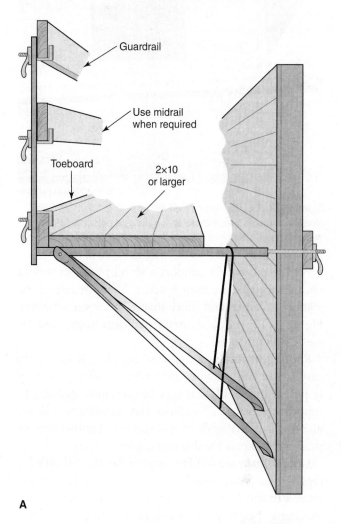

A

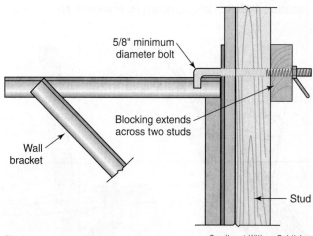

B

Goodheart-Willcox Publisher

Figure 6-11. A—This metal wall bracket is securely bolted through the wall. High scaffolds must be equipped with guardrails and toeboards. B—Detail of the anchoring system. The bolt passes through the wall alongside the stud and is anchored to blocking that must span at least two studs.

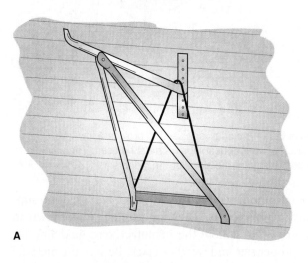

A

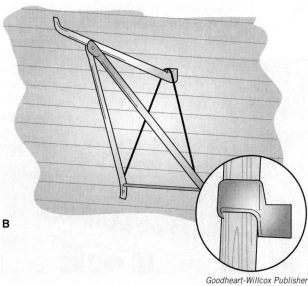

B

Goodheart-Willcox Publisher

Figure 6-10. Types of metal wall brackets. A—Attached with nails securely set in the building frame. B—Directly hooked to wall studs.

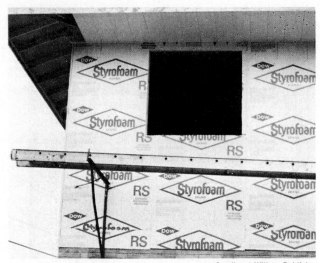

Goodheart-Willcox Publisher

Figure 6-12. Scaffolding formed with metal brackets and an aluminum plank. Since there is no railing, a personal fall arrest system should be worn when working on this type of scaffolding.

Wall brackets that support a wood platform should not be spaced more than 8′ apart. Metal or reinforced platforms can extend more than 8′ between supports. **Figure 6-12** shows a metal platform supported by metal wall brackets.

Roofing brackets, sometimes called roof jacks, are easy to use and provide safety when working on steep slopes, **Figure 6-13**. One type has an adjustable arm to support the plank. This allows a level platform to be set up on nearly any angle of roof surface. Another type of roofing bracket is fixed at one angle.

Ladder jacks are used to support simple scaffolds for repair jobs. Setup requires two sturdy ladders of the same size and a strong plank. **Figure 6-14** shows a jack that hangs below the ladder and hooks to the side rails. Another type of jack is shown in **Figure 6-15**.

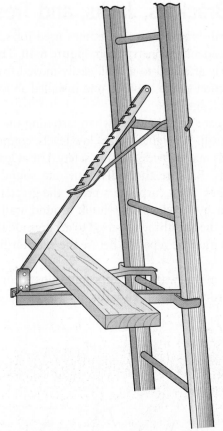

Goodheart-Willcox Publisher

Figure 6-14. A ladder jack used to provide a low-level scaffold. Two identical ladders must be used. The plank is located inside of the slope of the ladders.

For interior work, *trestle jacks* are used to support low platforms. They are assembled as shown in **Figure 6-16**. Follow the manufacturer's guideline for the proper size and weight of jack. Be sure the material used for the ledger (horizontal board held by the jacks) is sound and large enough to support the load.

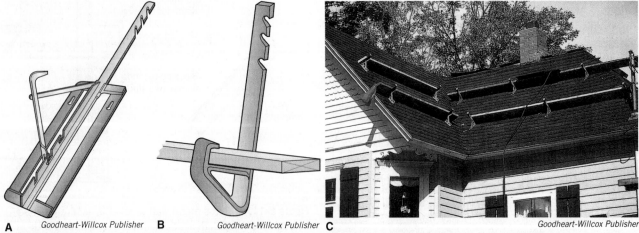

A   Goodheart-Willcox Publisher   B   Goodheart-Willcox Publisher   C   Goodheart-Willcox Publisher

Figure 6-13. Roofing brackets. A—Bracket with slots to slip over nails driven into the roof frame. The adjustable arm is used to level planking on roofs of different pitches. B—Bracket with fixed angle for steep roofs. C—Brackets and planks installed for a re-roofing job.

A   *Trudeau Construction Co.* B   *Trudeau Construction Co.*

Figure 6-15. Ladder jacks projecting from the front side of ladders. A—A personal fall arrest system should be worn at all times when using a scaffold without a guardrail. B—Ladder jacks used with a metal platform.

Safety Note

When working on or around scaffolding, observe these rules:

- All scaffolds should be built under the direction of a competent worker.
- Follow the design specifications listed in local and state codes.
- Inspect scaffolds each day before use.
- Provide adequate pads or sills under scaffold posts.
- Plumb and level scaffold members as each is set.
- Equip all planked areas with proper guardrails, toeboards, and screens when required.
- Power lines near scaffolds are dangerous. Consult the local power company for advice and recommendations.
- Never use ladders or makeshift devices on top of scaffold platforms to increase the height.
- Be certain the planking is heavy enough to carry the load with a safe span length.
- Planking should be lapped at least 12″ and extend at least 6″ but less than 12″ beyond all supports.
- Before moving a rolling scaffold, remove all persons, materials, and equipment from the platform.
- The height of the rolling scaffold platform should not exceed four times the smallest base dimension.

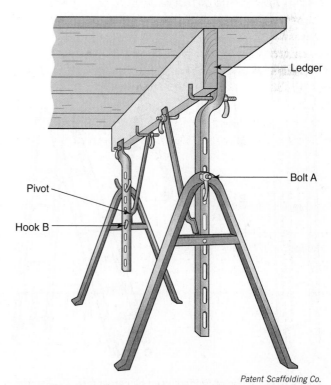

Patent Scaffolding Co.

Figure 6-16. Trestle jacks used for low platforms. To raise the jacks, loosen bolt A and move hook B to another slot. Note that the ledger is firmly clamped to the jacks.

6.3 Ladders

Types of ladders commonly used by a carpenter are illustrated in **Figure 6-17**. Stepladders range in size from 4′ to 20′. A one-piece (single straight) ladder is available in sizes of 8′–26′. Extension ladders provide lengths up to 60′.

Ladders are made of aluminum, fiberglass, and wood. Aluminum ladders are light in weight, but should not be used near electrical wires, as their high electrical conductivity can cause electrocution. Fiberglass ladders are heavier but do not conduct electricity and are not affected by moisture. Quality wood ladders are made from clear, straight-grained stock. Wood ladders should be given a clear finish. When reconditioning a wood ladder, always apply a clear finish. Never use paint, as it may conceal dangerous defects.

Ladders are classified by their duty rating, **Figure 6-18**. This rating gives the maximum weight that should be applied to the ladder, including loads that are imposed by personnel, tools, and equipment such as ladder jacks and planks. The duty rating appears on a permanent label.

Type	Duty Rating	Use	Load
1AA	Special Duty	Rugged	375 lb
1A	Extra Heavy Duty	Industrial	300 lb
1	Heavy Duty	Industrial	250 lb
II	Medium Duty	Commercial	225 lb
III	Light Duty	Household	200 lb

Goodheart-Willcox Publisher

Figure 6-18. Ladder duty ratings can be found on a sticker adhered to the ladder. Always be sure the ladder is designed to support the intended load.

Some fiberglass ladders also use a color coding system, which makes their duty rating easily identifiable.

Basic care and handling of ladders is illustrated in **Figure 6-19**. Always keep ladders clean. Do not let grease, oil, mud, or paint collect on the rails or rungs. The rails are the sides, while the rungs are the steps. Keep all fittings tight. On extension ladders, lubricate the locks and pulleys. Replace any frayed or worn rope. Inspect ladders before use. If you observe any physical damage, remove the ladder from service.

Patent Scaffolding Co.

Figure 6-17. Types of ladders used by carpenters. A—Extension ladder B—Stepladder. C—Stepladder with platform and tool holder. D—Tripod stepladder for working in tight spaces. E—Extension trestle ladder for accessing difficult-to-reach places, such as above acoustical ceilings.

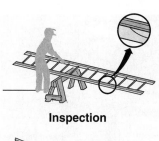

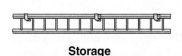

Inspection

Ladders should be inspected frequently. Those with defects should be either repaired or destroyed.

Carrying

Always carry a ladder over your shoulder with the front end elevated. Be sure not to drop it or let it fall. Such an impact weakens a ladder.

Storage

Store horizontally on supports to prevent sagging. Do not store near heat or out in the weather.

Goodheart-Willcox Publisher

Figure 6-19. Care and handling of ladders.

To erect a ladder, lay the ladder flat and perpendicular to the structure, placing the lower end (closer to the structure) against a solid base so it cannot slide. Raise the top end. Walk toward the bottom end, grasping and raising the ladder rung by rung as you proceed. When the ladder is vertical, lean it against the structure at the proper angle. Make sure the bottom end of both rails rests on a firm base.

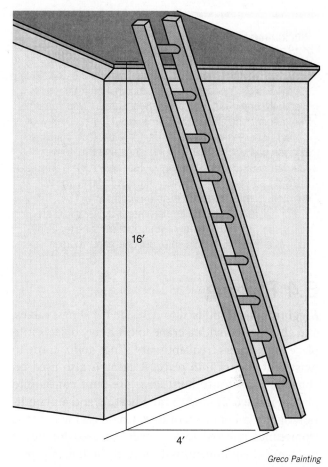

Greco Painting

Figure 6-20. Proper ladder placement. The horizontal distance from the wall to the base of the ladder should be 1/4 of the length of the ladder.

Safety Note

Follow these safety rules when working with ladders.

- Always inspect a ladder before using it.
- Place the ladder so the horizontal distance from its lower end to the vertical wall is 1/4 of the length of the ladder, **Figure 6-20**.
- If the ladder is placed against a roof or other horizontal surface, the top of the ladder should extend 3′ above the roof edge.
- Before climbing the ladder, be sure both rails rest on solid footing.
- When the ladder is used on surfaces where the bottom might slip, equip the rails with *safety shoes*.
- Never place a ladder in front of a doorway if the door can be opened toward the ladder.
- Always face the ladder when climbing up or down.
- Maintain three points of contact (two feet and one hand or one foot and two hands) while on the ladder.
- Never place a ladder on boxes or any unstable base to get more height.

- Never splice together two short ladders to make a longer ladder.
- Place ladders so work can be done without leaning beyond either side rail.
- Be sure extension ladders have sufficient lap between sections. A 36′ length should lap at least 3′; a 48′, at least 4′.
- Keep both hands free to grasp when climbing a ladder. Use a hand line to raise or lower tools and materials.
- Before mounting a stepladder, be sure it is fully open and locked and all four legs are firmly supported.
- Do not stand on either of the two top steps of a stepladder.
- Do not leave tools on the top of a stepladder unless the ladder has a special holder.
- Never use wood or metal ladders where contact with electric current is possible.

Goodheart-Willcox Publisher

Figure 6-21. A SIP is being hoisted into place using a crane and a web sling.

6.4 Rigging

Rigging is the term used to describe the slings, cables, and chains used with a crane to lift heavy objects into place. Carpenters frequently use slings and a crane to hoist roof trusses into place. Rigging is also used on construction sites to hoist large pre-built panels into place, **Figure 6-21**, and to raise heating and air conditioning equipment to roof tops. Chain and wire ropes (sometimes referred to as cable) are used for heavy mechanical equipment, but most of the hoisting performed by carpenters is with synthetic slings.

6.4.1 Synthetic Slings

Synthetic *slings* are made of nylon or polyester. They are either flat straps or round, like rope, **Figure 6-22**. They are used by carpenters because they are soft, flexible, and lightweight, making them easy to handle. They are also less damaging to wood than are chain and wire-rope slings. Synthetic slings are not affected by water, but they are damaged by exposure to excessive heat or sunlight for long periods of time. They are also easily cut or damaged by abrasion. Nylon slings are damaged by acids, but they are not affected by caustic chemicals.

Polyester slings, on the other hand, are not damaged by acids, but they are damaged by caustic chemicals. Slings should be inspected for damage by a competent person before each use. **Figure 6-23** shows some of the defects that might appear. If any of these are seen, the sling should be taken out of service.

6.4.2 Hoisting with a Crane

Safety is a very important concern when using a crane to hoist heavy objects. Even the empty hook on the end of the crane rigging can be deadly if it swings and strikes an unaware worker. All persons not directly involved in the hoisting operation should be clear of the path of the load. One person should be designated as the director. That person is responsible for the safety of all workers in the area.

Do not place your hand inside the sling as it is being tightened around the load. Also, do not attempt to guide the load by pushing or pulling with your hands. A tag line should be attached to the load. A *tag line* is a rope attached to the load and held by a person on the ground to prevent the load from twisting and swinging. An object being hoisted by a crane is usually heavy. If you try to handle it, there is a risk of being injured.

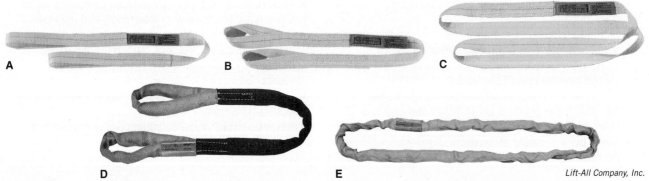

Lift-All Company, Inc.

Figure 6-22. Types of synthetic slings. A—Standard eye. B—Twisted eye. C—Endless. D—Eye-and-eye round sling. E—Endless round sling.

INSPECTION CRITERIA FOR WEB SLINGS

The following photos illustrate some of the common damage that occurs to web slings, indicating that the sling should be taken out of service.

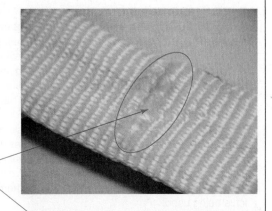

THE DAMAGE: **Surface and Edge Cuts**—It is important to realize that all the fibers in web slings contribute to the strength of that sling. When there have been a significant number of fibers broken in a web sling, as shown here, that sling should be taken out of service.

WHAT TO LOOK FOR: **Broken fibers** of equal length indicate that the sling has been cut by an edge. **Red core warning yarns may or may not be visible with cuts and are not required to show before removing slings from service.**

TO PREVENT: Always protect synthetic slings from becoming cut by corners and edges by using wear pads and other devices.

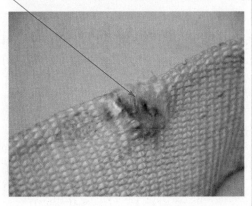

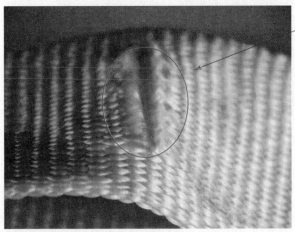

THE DAMAGE: **Holes/Snags/Pulls**

WHAT TO LOOK FOR: **Punctures or areas** where fibers stand out from the rest of the sling surface.

TO PREVENT: Avoid sling contact with protrusions, both during lifts and while transporting or storing.

THE DAMAGE: **Abrasion**

WHAT TO LOOK FOR: Areas of the sling that look and feel **fuzzy** indicate that the fibers have been broken by being subject to contact and movement against a rough surface. Affected areas are usually stained.

TO PREVENT: Never drag slings along the ground. Never pull slings from under loads that are resting on the sling. Use wear pads between slings and rough surface loads.

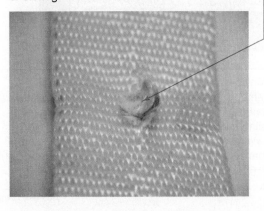

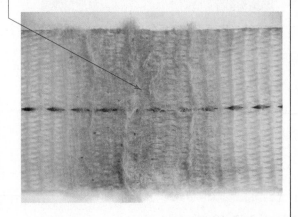

Lift-All-Company, Inc.

Figure 6-23. Common defects in synthetic slings. *(Continued.)*

INSPECTION CRITERIA FOR WEB SLINGS

THE DAMAGE: **Heat/Chemical**
WHAT TO LOOK FOR: **Melted or charred fibers** anywhere along the sling. Heat and chemical damage look similar and they both have the effect of damaging sling fibers and compromising the sling's strength. Look for discoloration and/or fibers that have been fused together and often feel hard or crunchy.
TO PREVENT: Never use nylon or polyester slings where they can be exposed to temperatures in excess of 200°F. Never use nylon or polyester slings in or around chemicals without confirming that the sling material is compatible with the chemicals being used.

THE DAMAGE: **Knots** compromise the strength of all slings by not allowing all fibers to contribute to the lift as designed. Knots may reduced sling strength by up to 50%.
WHAT TO LOOK FOR: **Knots** are rather obvious problems as shown below.
TO PREVENT: Never tie knots in slings and never use slings that are knotted.

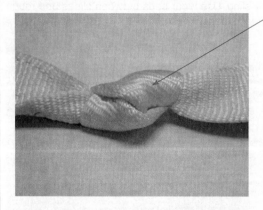

THE DAMAGE: **Broken/Worn Stitching** in the main stitch patterns of web slings has a direct adverse effect on the strength of a sling. The stitch patterns in web slings have been engineered to produce the most strength out of the webbing. If the stitiching in not fully intact, the strength of the sling may be affected.
WHAT TO LOOK FOR: **Loose or broken threads** in the main stitch patterns.
TO PREVENT: Never pull slings from beneath loads where stitch patterns can be hung up or snagged. Never overload the slings or allow the load edge to directly contact the stitch pattern while lifting. Never place a sling eye over a hook or other attachment whose width/diameter exceeds 1/3 the eye length.

THE DAMAGE: **Illegible or Missing Tags**—The information provided by the sling tag is important for knowing what sling to use and how it will function.
WHAT TO LOOK FOR: If you cannot find or read all of the information on a sling tag, OSHA requires that the sling be taken out of service.
TO PREVENT: Never set loads down on top of slings or pull sling from beneath loads if there is any resistance. Load edges should never contact sling tags during the lift. Avoid paint or chemical contact with tags.

> **Red Core Yarns**—are an **additional** aid to warn of dangerous sling damage. All standard *Lift-All* Web Sling have this warning feature. The red core yarns become exposed when the sling surface is cut or worn through the woven face yarns. When red yarns are visible, the sling should be removed from service immediately.

Figure 6-23. (*Continued.*)

Only the person designated to give instructions to the crane operator should do so. Instructions may be given by radio or by approved hand signals. The standard hand signals given in **Figure 6-24** are approved by the American National Standards Institute (ANSI).

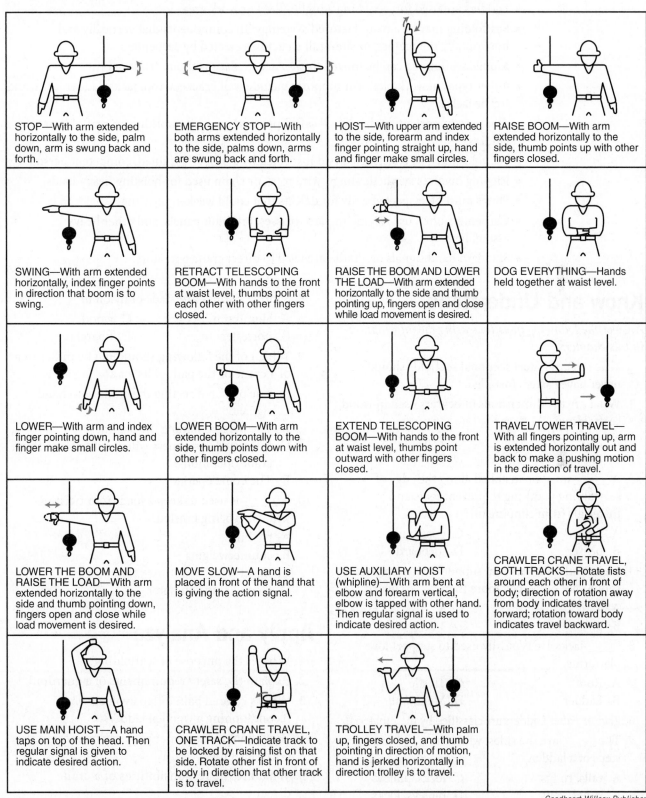

Goodheart-Willcox Publisher

Figure 6-24. OSHA-approved hand signals to direct a crane operator.

Chapter Review

Summary

- Scaffolding is used to work in areas too high to reach from the ground or floor deck.
- Ladders are used for accessing a higher level or work area.
- Scaffolding may be a manufactured system, with components that vertically and horizontally fit together, or site-built structures erected by carpenters.
- Mobile scaffolding can be moved from place to place without disassembly.
- Other types of work platform supports include wall brackets, roof jacks, ladder jacks, and trestle jacks.
- It is very important to observe all safety precautions while working on scaffolds or ladders.
- Scaffolds must be constructed and inspected by a properly trained, competent person.
- Rigging involves synthetic slings, wire rope, or chain used for hoisting heavy loads.
- Slings must be inspected daily for defects that could weaken the sling.
- Carpenters often use rigging to place trusses, pre-built panels, and other heavy assemblies.
- Standard hand signals or a radio are used to direct crane operators.

Know and Understand

Answer the following questions using the information in this chapter.

1. *True or False?* Steel sectional scaffolds should never have casters installed.

2. What are the dimensions of bearers used in wood scaffolding?
   - A. 2×4
   - B. 2×6
   - C. 1×6
   - D. 2×10

3. A _____ must be installed along the edge of scaffolding planking to prevent tools and materials from slipping off?
   - A. brace
   - B. bearer
   - C. guardrail
   - D. toeboard

4. Metal wall brackets that support a wood platform should not be spaced more than _____′ apart.
   - A. 4
   - B. 6
   - C. 8
   - D. 14

5. _____ jacks are typically used to support low platforms.
   - A. Roof
   - B. Ladder
   - C. Trestle
   - D. Metal wall

6. *True or False?* Ladders are classified by their material.

7. The _____ are the sides, while the _____ are the steps of a ladder.
   - A. rails, rungs
   - B. rungs, rails
   - C. locks, pulleys
   - D. pulleys, locks

8. Synthetic slings are made of nylon or _____.
   - A. aluminum
   - B. polyester
   - C. wood
   - D. fiberglass

9. Which of the following should *not* be in the area of a crane or the path of its load?
   - A. The person directing the crane with hand signals.
   - B. The person holding the tag line.
   - C. Interested bystanders who want to learn about crane operations.
   - D. The crane operator.

10. A _____ is used to keep a load from twisting while it is being hoisted.
    - A. tag line
    - B. counterweights
    - C. rigger
    - D. second crane operator

Apply and Analyze

1. What is the purpose of scaffolding?
2. Explain the safety requirements of a guardrail.
3. Why is colored paint often avoided when reconditioning a wooden ladder?
4. How should a carpenter properly maintain a polyester sling?
5. What are the responsibilities of a crane director?

Critical Thinking

1. Why might some contractors choose to use a boom lift on a specific job rather than a ladder or scaffolding?
2. Who do you think should identify and address safety concerns on a scaffold that is in use on the jobsite?
3. Why is it important to ensure the base of an extension ladder rests on a level surface? What are some safe options for set-up when the surface is not level?

Communicating about Carpentry

1. **Speaking.** Working with a partner, create a model of a scaffold. Use the information in this lesson and do any research necessary. Create the model so that the parts can be removed.

Use your model to demonstrate to the class how the pieces fit together, what purposes the different features serve, and how the scaffold is broken down.

2. **Reading and Speaking.** Pick a figure in this chapter. Working with a partner, tell and retell the important information being conveyed by that figure. Through your collaboration, develop what you and your partner believe is the most interesting verbal description of the importance of the chosen figure. Present your narration to the class.

3. **Reading and speaking.** The Federal Occupational Safety and Health Act requires the use of fall protection systems on many types of construction projects where people will be working above the ground. Contact the nearest OSHA office or go to the OSHA website and find information on the circumstances where fall protection systems must be used, what those systems consist of, and how they work. Present an oral report to the class.

CHAPTER **7**

Plans, Specifications, and Codes

OBJECTIVES

After studying this chapter, you will be able to:
- Identify the elements commonly included in a set of house plans.
- Demonstrate the use of scale in architectural drawings.
- Identify architectural symbols.
- Explain the use of building specifications.
- Describe the application of building codes, standards, and permits.
- List the items required by building officials to obtain a building permit.

TECHNICAL TERMS

architectural drawing
authority having jurisdiction (AHJ)
building code
building permit
computer-aided drafting and design (CADD)
CADD-CAM
computer-aided manufacturing (CAM)
detail drawing

dimensions
drawn to scale
electrical plan
elevation
floor plan
footprint
foundation plan
framing plan
list of materials
model code
pictorial sketch

plot plan
print
schedule
section drawing
set of plans
setback
specifications
stock plan
symbol
truss plan

A good plan and well-defined contract are important in building construction. Expectations should be specified in the contract in no uncertain terms. For the project to go as seamlessly as possible and to avoid possible future litigation, both sides should know *exactly* what is expected of the other.

Every carpenter must know how to read and understand **architectural drawings** (plans) and correctly interpret the information found in written specifications. Simply put, the plans tell you *how* to build and the specifications tell you *what* materials must be used. See **Figure 7-1**. It has been said that "blueprints are the language of the construction industry." Carpenters must be fluent in this language.

Figure 7-1. Carpenters must frequently study plans as they construct a building.

Copies of the architect's original drawings are usually called *prints*, or blueprints. At one time, all drawing copies were chemically made on treated paper that displayed white lines on a blue background. Today, prints are usually reproduced with dark lines on a white background. Although the term blueprint is still used, print is the preferred term.

Code Note

Architectural firms generally create an in-house set of drawing standards (guidelines). These standards may incorporate parts of drafting standards developed by organizations such as the American Institute of Architects (AIA), the American Design Drafting Association (ADDA), and the US National CAD Standard.

7.1 Set of Plans

Vast amounts of information are needed to build a house or light commercial structure. Since a single drawing could not possibly hold all that information, many drawings are needed. When printed and bound together, these sheets are known as a *set of plans*.

Usually, construction drawings, or plans, are drawn to several different scales and show different things. A person accustomed to reading plans is familiar with the kinds of information present in each. This narrows the search as the contractor determines what the architect and owner want. For example, the dimensions of a room can be found only on a floor plan. The size of a fastener can be shown only in a detail. Many things can be shown on a section drawing, such as materials used, dimensions, and elevations.

The carpenter and building contractor are not the only persons needing a set of plans. The owner receives a set as assurance that the plan reflects his or her wishes. Tradespeople, such as the electrician, plumber, and heating contractor, need sets so they can install their systems. Lumber dealers and other suppliers use drawings to determine the quantity of materials needed. A carpenter uses only parts of the set.

7.1.1 Drawings in a Set of Plans

A set of house plans usually includes the following drawings:

- Plans—Bird's-eye views that include plot plan, foundation or basement plan, and floor plans.
- Building elevations—Floor plans that show the front, rear, and sides of the building.
- Section drawings—Drawings that show interior parts and structures as though a cut had been made through the assembly.
- Detail drawings—Enlarged drawings that show the details of how a particular aspect of the building is to be constructed.
- Drawings of the mechanical systems—Drawings that describe and give the location of electrical, plumbing, heating, and air conditioning components. These are often not included on drawing sets for relatively simple houses.

Builders usually look at the elevation drawings first. These offer the best representation of what is to be built. Carpenters who specialize in the installation of acoustical ceilings look at a plan view. Since plan views are always drawn to show a view looking down from above, they look down at the floor as though it were a photograph. A contractor uses a set of plans to bid on the construction of a building, **Figure 7-2**. An estimator studies them in preparing a bid.

7.1.2 Stock Plans

When a set of plans is mass-produced to be offered for sale to many clients, it is called a *stock plan*. Today, a wide range of such ready-drawn plans is available for people who want to build. Stock plans were once sold through catalogues, but are now widely available to purchase online. A client who purchases a set of these plans will not be building a one-of-a-kind home. These same plans have likely been used around the country to build other nearly identical homes from the same set of prints.

Rido/Shutterstock.com

Figure 7-2. A contractor may discuss a building plan with the architect before submitting a bid.

The Garlinghouse Company. The drawings in this chapter cannot be reproduced or built. If copies of the plans or more information is required, contact The Garlinghouse Company.

Figure 7-3. Full-color renderings help the buyer of a stock plan visualize how a residence will look. Plans for this single-story house are shown throughout this chapter.

This chapter shows a number of drawings taken from a set of stock plans (Plan #20161) from The Garlinghouse Company, Chantilly, VA. See **Figure 7-3**. The drawings referenced as plans are discussed in more detail.

7.1.3 Scale

Drawings must be reduced so they fit on the drawing sheet. This must always be done in such a way that the reduced drawing is in *exact* proportion to the actual size. Such drawings are said to be **drawn to scale**.

Residential plan views are generally drawn to 1/4″ scale (1/4″ = 1′-0″). This means that for each 1/4″ on the plan, the building dimension is 1′. When certain parts of the structure need to be shown in greater detail, they are drawn to a larger scale, such as 1″ = 1′-0″. Others, such as framing plans, are often drawn to a smaller scale (1/8″ = 1′-0″). **Figure 7-4** shows an architect's scale and a partial list of conventional inch-foot scales.

The units of measure surveyors usually work with are feet and decimal parts of a foot. Plot plans usually use these units for property boundaries and other land measurements. The engineer's scale shows feet and 10 smaller divisions within the foot.

While carpenters are not responsible for making drawings, it is helpful to be able to make a simple sketch for a building or remodeling job. The sketch is often good enough to be used as a guide during construction. A **pictorial sketch** is a drawing showing three dimensions, much like a photograph. It is often needed so the customer can visualize the completed job, **Figure 7-5**.

7.1.4 Alphabet of Lines

Architects prepare drawings using a standard convention of line types known as the "alphabet of lines." Each type of line represents something different. In order to read drawing plans effectively, you must be familiar with the meaning of each line type. See **Figure 7-6**.

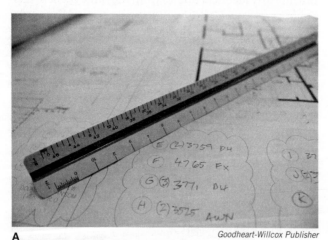

A   Goodheart-Willcox Publisher

Common-Inch Foot Scales			
Scale	Where used	Ratio	Relationship to actual size
1/32″	Site plans	1:384	1/32″ = 1′-0″
1/16″	Plot plan	1:192	1/16″ = 1′-0″
1/8″	Plot plan	1:96	1/8″ = 1′-0″
1/4″	Elevation plans	1:48	1/4″ = 1′-0″
3/8″	Construction details	1:32	3/8″ = 1′-0″
1/2″	Construction details	1:24	1/2″ = 1′-0″
3/4″	Construction details	1:16	3/4″ = 1′-0″
1″	Construction details	1:12	1″ = 1′-0″
1 1/2″	Construction details	1:8	1 1/2″ = 1′-0″
3″	Construction details	1:4	3″ = 1′-0″

B   Goodheart-Willcox Publisher

Figure 7-4. A—Architect's scale. In this view, scales of 1/8″=1′ and 1/2″=1′ can be seen starting at the left side of the scale; 1/4″=1′and 1″=1′ share this side of the tool, but originate from the right side of the scale. Gradations indicating parts of a foot can be seen to the left of the 0 mark. Each line represents 3″ on the 1/8″=1′ scale and 1/2″ on the 1/2″=1′ scale. B—Common inch-foot scales used in residential drawings.

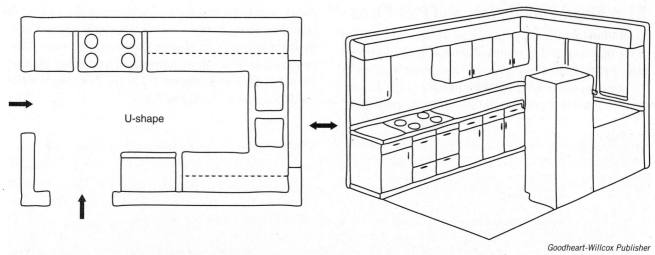

Goodheart-Willcox Publisher

Figure 7-5. It is easier to visualize how the kitchen will look if shown in a picture-like sketch, such as the one at right. This can easily be done using isometric sketch paper.

Line Types Commonly Used on House Plans

Line Name and Appearance	Description
Property line	Identifies the boundaries of a building lot.
Border line	Thick, continuous line used around the boarder of the drawing.
Object line	Thick, continuous line identifies objects such as walls, windows, doors, foundations, and other physical components of the building.
Hidden line	Thin, dashed line identifies the edge of an object located behind another object in the view shown in the drawing.
Center line	Thin line of alternating long and short dashes used to mark the locations of centers of circular features, column lines, and lines of symmetry.
Dimension line Extension line 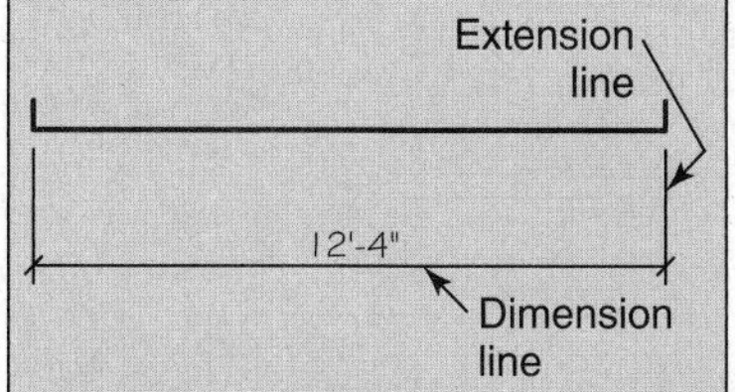 [Extension line, 12'-4", Dimension line]	Thin lines used with dimension values to show the "real life" distance between two points.
Section cutting line  [B / A4]	Identifies the place in the drawing from which a section drawing is viewed, and identifies the location of the section view in the set of plans.
Break lines 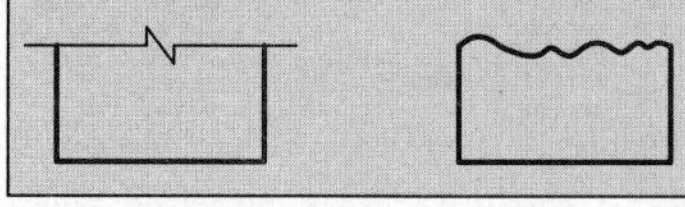	Thick or thin lines identifying a break in an object.

Goodheart-Willcox Publisher

Figure 7-6. Construction drawings contain several types of line. Each line type has a specific meaning and use.

7.1.5 Floor, Foundation, and Plot Plans

Floor plans are drawings that show the size and outline of the building and its rooms. They also give additional information useful to a carpenter and other workers in the construction trades. Floor plans have many dimensions to show the location and size of inside partitions, doors, windows, and stairs. Dimensions are explained later. Plumbing fixtures, as well as appliance and utility installations, can also be shown in this view. Carpenters use floor plans more than any other plan in the set of drawings. **Figure 7-7** is the floor plan of the residence shown in **Figure 7-3**.

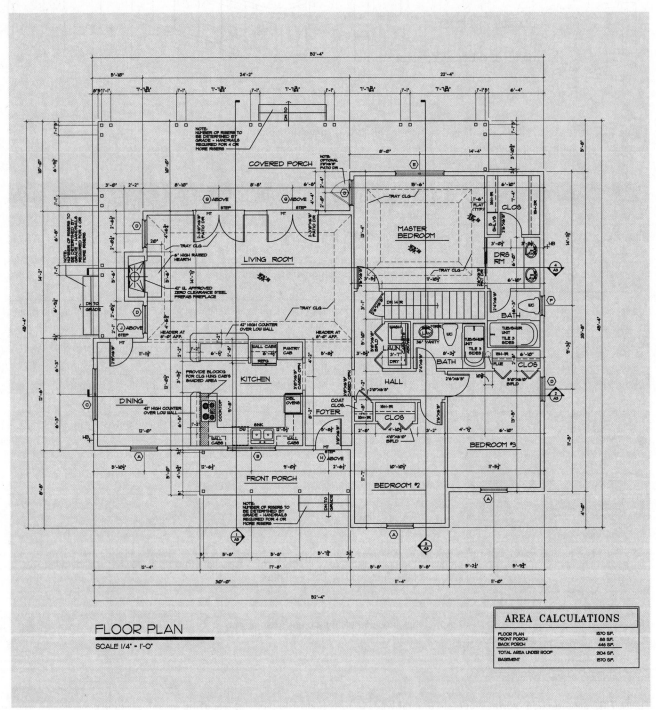

Figure 7-7. The floor plan for one-story house shown in Figure 7-3. Floor plans do not give details on how assemblies are built; however, they do give overall dimensions, shapes, and sizes of rooms. They also give placement of mechanical systems cabinetry, and other built-in features of the building.

Foundation plans are similar to floor plans, but instead show the foundation on which the building will be built. When shown, the footings are represented as a dashed line. It is assumed that the basement floor is in place and that the grade (ground) covers the footings on the outside.

Generally, hidden lines represent something that is in place but cannot be seen when the job is complete. See **Figure 7-8**.

Most plans include a *plot plan*. This drawing shows the *footprint* (size and shape) of the building and its location on the site (either a lot or acreage).

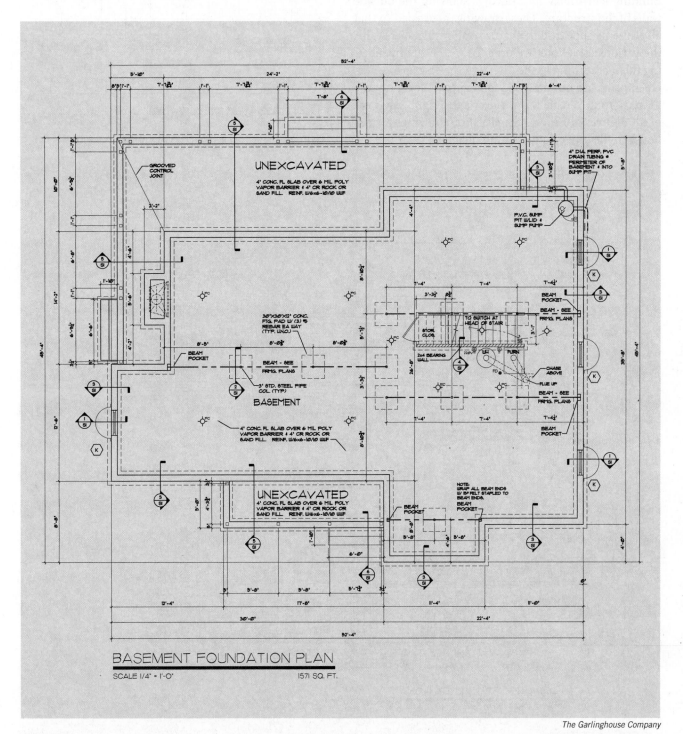

The Garlinghouse Company

Figure 7-8. A foundation plan shows the builder exactly how to construct the foundation. Note the many details included in such a plan: location of all features, including supporting posts and beams, pockets for supporting beams, window openings, sump, furnace, and electrical boxes. Often, details of special features are also shown.

The plot plan shows *setbacks*, the distances from the property lines to the building lines. See **Figure 7-9**. Allowable setbacks vary according to local zoning restrictions.

7.1.6 Elevations

Building *elevations* are drawings showing the outside walls of the structure. These drawings are scaled so that all elements appear in their true relationship. Generally, the various elevations are oriented to the site by labeling them according to the direction they face. However, when plans are not designed for a specific location, the terms "front," "rear," "left side," and "right side" are used. **Figure 7-10** shows the elevation drawings for the house in the other figures in this chapter.

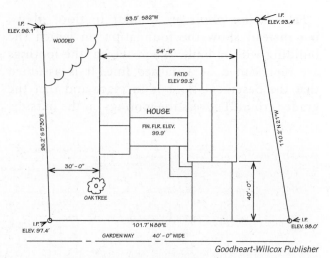

Goodheart-Willcox Publisher

Figure 7-9. Example of a site or plot plan.

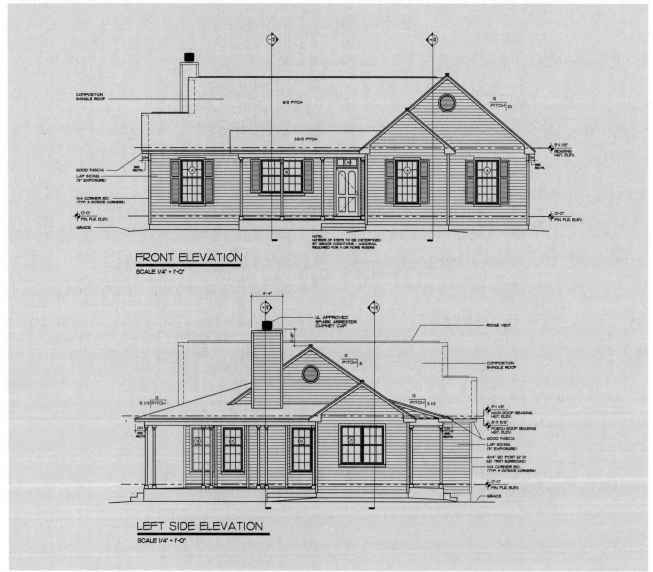

The Garlinghouse Company

Figure 7-10. In stock plans, elevations are usually marked "front," "rear," "left side," and "right side" because the architect does not know which direction the house will face. Elevations may carry notes indicating special features, code requirements, or reminders. (Drawings are not to scale.)

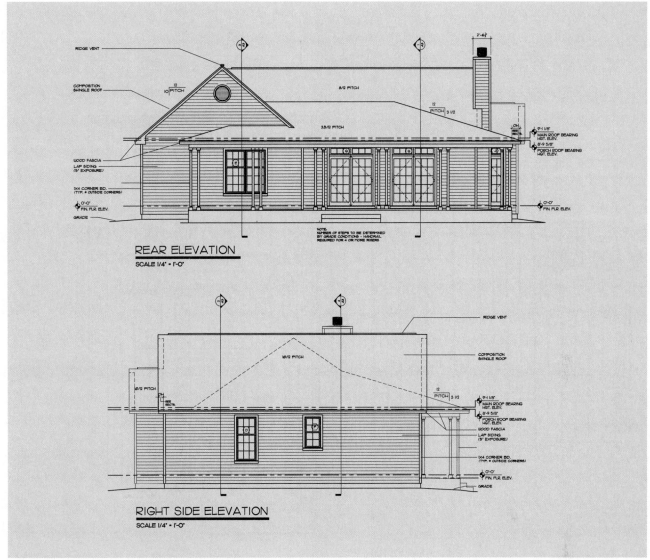

Figure 7-10. *(Continued.)*

By studying the elevation views, a carpenter can determine the following information:

- Floor levels
- Grade lines
- Window and door heights
- Roof slopes
- Types of materials used on wall and roof surfaces

7.1.7 Framing Plans

Sometimes a house plan also has drawings showing the size, number, and location of the structural members of the building's frame. These are known as *framing plans*. See **Figure 7-11**. Separate plans may be drawn for the floors, ceilings, walls, and roof. These plans specify the size and spacing of the framing members. When the roof is to be framed with trusses, a *truss plan* is included, **Figure 7-12**. The trusses and other framing members are drawn to show where they will be positioned in the building. Openings needed for chimneys, windows, and doors are shown. Dimensions are not used, but the drawings are made to scale, as with other plan drawings.

7.1.8 Section and Detail Drawings

A floor plan or elevation drawing does not show small parts of the structure or how the parts fit into the total structure. For this, a carpenter needs to consult drawings called sections and details.

A *section drawing*, or section view, gives important information about size, materials, fastening, and support systems, as well as concealed features. Parts of the structure likely to have a section drawing include walls, window and door frames, footings, and foundations.

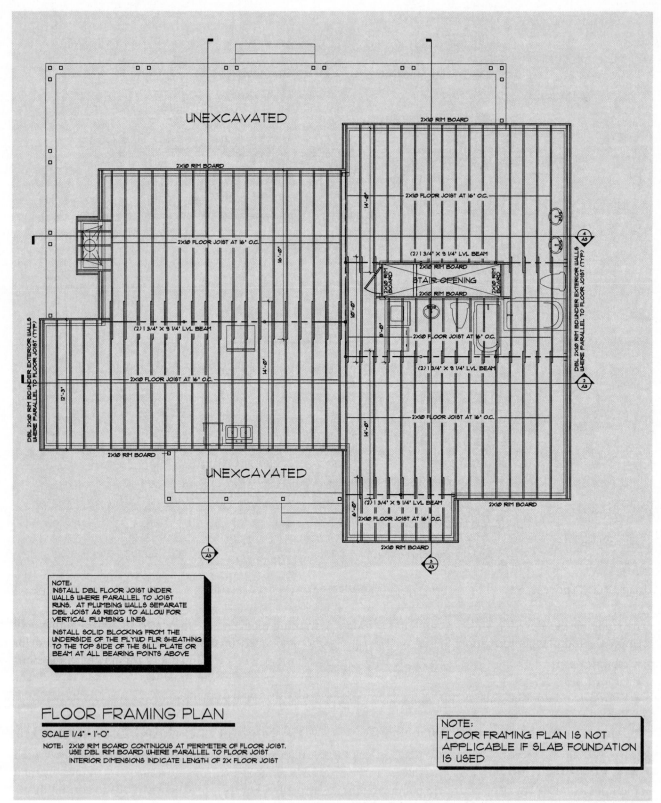

UNEXCAVATED

2X10 RIM BOARD

2X10 RIM BOARD

2X10 FLOOR JOIST AT 16" O.C.

2X10 FLOOR JOIST AT 16" O.C.

(2) 1 3/4" X 9 1/4" LVL BEAM

STAIR OPENING

2X10 RIM BOARD

2X10 RIM BOARD

2X10 FLOOR JOIST AT 16" O.C.

(2) 1 3/4" X 9 1/4" LVL BEAM

2X10 FLOOR JOIST AT 16" O.C.

(2) 1 3/4" X 9 1/4" LVL BEAM

2X10 FLOOR JOIST AT 16" O.C.

2X10 RIM BOARD

(2) 1 3/4" X 9 1/4" LVL BEAM

2X10 FLOOR JOIST AT 16" O.C.

2X10 RIM BOARD

DBL 2X10 RIM BD UNDER EXTERIOR WALLS WHERE PARALLEL TO FLOOR JOIST (TYP)

DBL 2X10 RIM BD UNDER EXTERIOR WALLS WHERE PARALLEL TO FLOOR JOIST (TYP)

2X10 RIM BOARD

UNEXCAVATED

NOTE:
INSTALL DBL FLOOR JOIST UNDER
WALLS WHERE PARALLEL TO JOIST
RUNS. AT PLUMBING WALLS SEPARATE
DBL JOIST AS REQ'D TO ALLOW FOR
VERTICAL PLUMBING LINES

INSTALL SOLID BLOCKING FROM THE
UNDERSIDE OF THE PLYWD FLR SHEATHING
TO THE TOP SIDE OF THE SILL PLATE OR
BEAM AT ALL BEARING POINTS ABOVE

FLOOR FRAMING PLAN

SCALE 1/4" = 1'-0"

NOTE: 2X10 RIM BOARD CONTINUOUS AT PERIMETER OF FLOOR JOIST.
USE DBL RIM BOARD WHERE PARALLEL TO FLOOR JOIST
INTERIOR DIMENSIONS INDICATE LENGTH OF 2X FLOOR JOIST

NOTE:
FLOOR FRAMING PLAN IS NOT
APPLICABLE IF SLAB FOUNDATION
IS USED

The Garlinghouse Company

Figure 7-11. House plans often include framing drawings. The floor framing plan specifies sizes and spacing of joists, girders, and support columns.

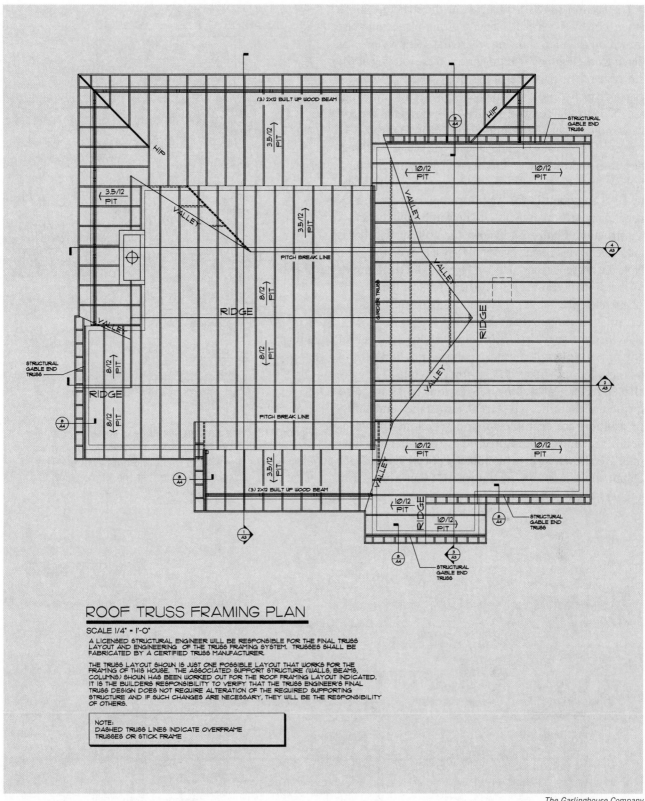

ROOF TRUSS FRAMING PLAN

SCALE 1/4" = 1'-0"

A LICENSED STRUCTURAL ENGINEER WILL BE RESPONSIBLE FOR THE FINAL TRUSS LAYOUT AND ENGINEERING OF THE TRUSS FRAMING SYSTEM. TRUSSES SHALL BE FABRICATED BY A CERTIFIED TRUSS MANUFACTURER.

THE TRUSS LAYOUT SHOWN IS JUST ONE POSSIBLE LAYOUT THAT WORKS FOR THE FRAMING OF THIS HOUSE. THE ASSOCIATED SUPPORT STRUCTURE (WALLS, BEAMS, COLUMNS) SHOWN HAS BEEN WORKED OUT FOR THE ROOF FRAMING LAYOUT INDICATED. IT IS THE BUILDERS RESPONSIBILITY TO VERIFY THAT THE TRUSS ENGINEER'S FINAL TRUSS DESIGN DOES NOT REQUIRE ALTERATION OF THE REQUIRED SUPPORTING STRUCTURE AND IF SUCH CHANGES ARE NECESSARY, THEY WILL BE THE RESPONSIBILITY OF OTHERS.

NOTE:
DASHED TRUSS LINES INDICATE OVERFRAME TRUSSES OR STICK FRAME

The Garlinghouse Company

Figure 7-12. A truss plan shows a carpenter where each truss is to be located and how special situations, like openings in the roof, are framed.

The section shows how a part of the structure looks when cut by a vertical plane. Cutting-plane lines are drawn on the plan showing where the section view was taken from, **Figure 7-13**. Imagine you are looking at the cut edge of the part after it has been sawed in two. Because of the need to show many details, section drawings are made to a large scale. **Figure 7-14** shows the section view indicated by the cutting-plane line in **Figure 7-13**.

Most sets of plans include at least one typical wall section if they do not have specific wall sections. A typical wall section shows what would be seen if the section was taken at any point in the wall. A complex structure may need many section drawings to show important details of construction. **Figure 7-15** shows the wall section that is indicated by the dashed oval at the left side of **Figure 7-14**. Note the reference mark 1/A5 in **Figure 7-14**, corresponds to the drawing number 1 on sheet A5 in **Figure 7-15**. These reference marks help the builder navigate the prints and orient them to one another.

As in sections, *detail drawings* show a carpenter any important and complex construction that cannot be included in plan and elevation drawings. They are also large scale and show how various parts are to be located and connected. Fireplaces, stairs, and built-in cabinets are examples of constructions shown as detail drawings. See **Figure 7-16**. Some detail drawings are shown at or near full size. See **Figure 7-17**. Detail drawings may be shown in plan view (from above) or elevation view (from the side).

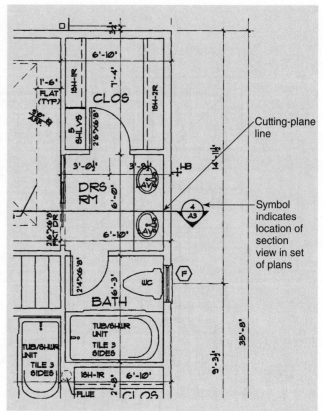

Figure 7-13. This cutting-plane line is shown on the floor plan in Figure 7-7. The arrowhead portion of the symbol indicates the direction in which the cut is viewed. The number 3 and the A3 indicate that the section view is drawing number 3 on sheet A3.

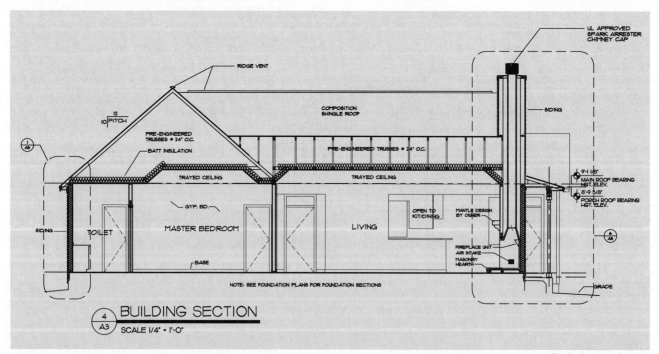

Figure 7-14. A section view of an entire house helps a carpenter understand the construction.

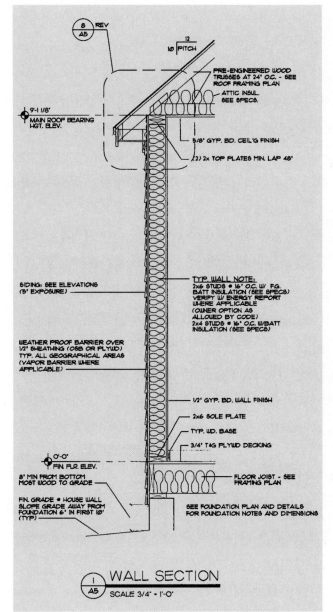

WALL SECTION

SCALE 3/4" = 1'-0"

The Garlinghouse Company

Figure 7-15. This wall section is indicated by dashed lines and a circle with a notation that this is drawing 1 on sheet A5.

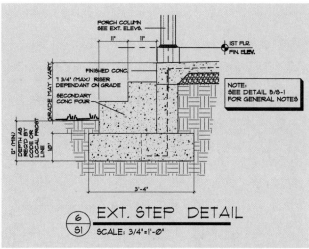

EXT. STEP DETAIL

SCALE: 3/4"=1'-0"

The Garlinghouse Company

Figure 7-16. This detail drawing shows how the stairs at the back of the house are to be built.

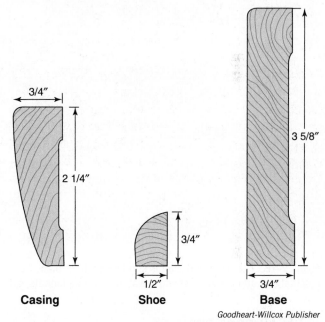

Casing   **Shoe**   **Base**

Goodheart-Willcox Publisher

Figure 7-17. Some detail drawings are made at or near full size. Shown are cross sections of special millwork needed for finishing a house.

Plans for commercial buildings and some single-family homes include drawings for electrical wiring. **Figure 7-18** is an *electrical plan*. It shows the location of every part of an electrical system and diagrams each branch circuit. By following this plan, an electrician knows where to place electrical receptacles, fixture boxes, and switches. Lines connecting the fixtures do not necessarily represent routing of the conductors, they simply let the electrician know that those fixtures share a circuit.

Most single family home plans do not include plumbing and heating drawings. Basic components of these systems are shown on the floor plans. Floor plans usually include the following mechanical and electrical items:

- Plumbing fixtures
- Washing machine location
- Furnace or boiler
- Air conditioning equipment
- Electrical circuit breaker/distribution panels
- Electrical outlets
- Switches
- Lighting fixtures
- Exhaust fans
- Smoke alarms and carbon monoxide alarms

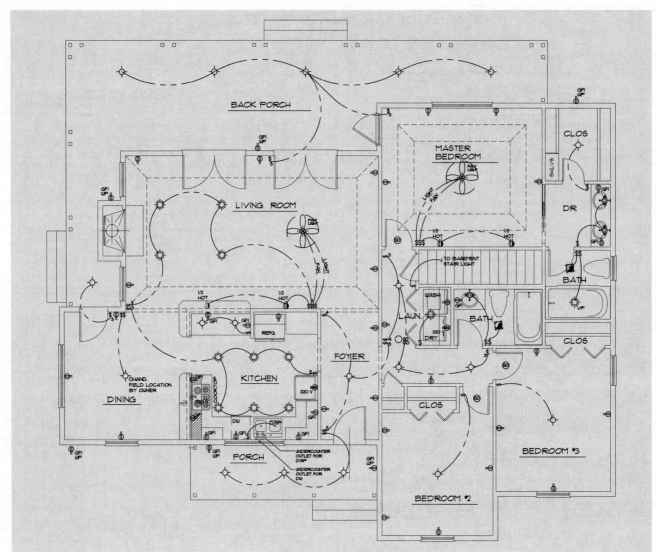

The Garlinghouse Company

Figure 7-18. The electrical plan is similar to a floor plan, but is intended specifically as a guide for the electrician in making calculations about size of service, switching arrangements, and number of circuits needed.

7.1.9 Dimensions

Dimensions show distance and size. In general, all dimensions greater than 1′ are written in feet and inches, as opposed to inches only. For example, standard ceiling height is given as 8′-0″ rather than 96″.

Carpenters using conventional measure prefer to work with feet and inches since measurement in this form is easier to visualize. When laying out various distances, they often need to add or subtract dimensions. The following are steps for making calculations.

Addition

$$\begin{array}{r} 6'\text{-}8'' \\ 4'\text{-}6'' \\ 2'\text{-}4'' \\ + 1'\text{-}2'' \\ \hline 13'\text{-}20'' \text{ or } 14'\text{-}8'' \end{array}$$

Subtraction

$$\begin{array}{r} 8'\text{-}4'' \\ - 6'\text{-}10'' \\ \hline \end{array}$$

Since 10 cannot be subtracted from 4, borrow 12″ from 8′.

$$\begin{array}{r} 7'\text{-}16'' \\ + 6'\text{-}10'' \\ \hline 1'\text{-}6'' \end{array}$$

7.1.10 Schedules and Lists of Materials

Some sets of plans also include a *list of materials*, **Figure 7-19**. It is known by other names as well: bill of materials, lumber list, or mill list. Whatever the list is called, it includes all of the materials and assemblies needed to build the structure.

MATERIALS LIST PLAN NO. 10372

4900	Face bricks for interior walls
2950	Common bricks for interior walls (bedrooms) & center of trombe wall
7 1/4	Cu. yds. concrete for basement floor slab
20	Cu. yds. concrete for 1st. floor slab (6″ thick)
7 1/4	Cu. yds. concrete for garage floor slab and apron
1	Cu. yd. concrete for entrance platforms and steps
14	Cu. yds. concrete for footings
45 3/4	Cu. yds. concrete for foundation walls
224	Lin. ft. 4″ plastic drainage tubing
45	Tons 3/4″ to 1″ crushed stone – under slabs & around drainage tubing

STRUCTURAL STEEL

1	8WF17 steel beam 22′-01/2″ long
1	4″ steel pipe columns 7′-4″ long with plates
1	4″ steel pipe columns 8′-3″ long with plates (verify)
8	3/8″×11″×24″ steel plate lintels over trombe wall openings
1	Complete gas vent
3150	Lin. ft. No. 4 reinforcing bars
2500	Sq. ft. 6″×6″×10/10 gauge reinforcing mesh
2	Galvanized steel areaways 36″ diameter 24″ high
64	1/2″ anchor bolts 10″ long
8	16×8 screened foundation vents w/dampers

CARPENTER'S LUMBER

**Note – Recommended framing lumber shall have a minimum allowable extreme fiber bending stress of 1450 PSI and a minimum modulus of elasticity of 1,400,000 PSI, except rafter joists over living room have F_b of 1500 and an E of 1,400,000.

Design Live Loads –
Floors – 40 PSF
Ceilings – 20 PSF
Roof – 40 PSF

First Floor Joists & Headers

5	2×10×8
19	2×10×12
21	2×10×14
2	2×10×16

Sub Sills

1	2×6×10
3	2×6×12
1	2×6×14
2	2×6×16
2	2×6×20
1	2×3×20

Goodheart-Willcox Publisher

Figure 7-19. One page of a typical list of materials. Such a list includes all the materials and assemblies (for example, doors, windows, and cabinets) needed for construction of the building. (This list is for a structure other than the one shown in previous drawings.)

A materials list usually includes the number of the item and its name, description, size, and the material of which it is made. Built-in items, such as cabinets, are included.

The drawing set includes *schedules*, which provide basic information about related components. The most common are a window schedule and a door schedule. Schedules give the quantity needed, size of the rough openings, and descriptions. Sometimes the manufacturer is also specified. **Figure 7-20** shows door and window schedules with marks corresponding to the plans and elevations in **Figure 7-7** and **Figure 7-8**.

7.1.11 Symbols

Since architectural plans are drawn to a small scale, materials and construction particulars can seldom be shown as they actually appear. Also, it would require too much time to produce drawings of this nature. The architect, therefore, uses *symbols* to represent materials and other items and certain approved shortcuts (called conventional representations). These simplify the illustration of assemblies and other elements of the structure. Generally accepted symbols are illustrated in **Figures 7-21** through **7-24**.

MARK	MANUF.	MODEL #	UNIT DIMENSION	REMARKS
A	ANDERSEN	3056	3'-1 5/8" X 5'-9 1/4"	EGRESS
B		2-2842	5'-7 1/4" X 4'-5 1/4"	
C		2-2846	5'-7 1/4" X 4'-9 1/4"	
D		2856	2'-9 5/8" X 5'-9 1/4"	
E		2-2856	5'-7 1/4" X 5'-9 1/4"	
F		20310	2'-1 5/8" X 4'-1 1/4"	
G		6010	5'-11 1/4" X 1'-0"	DOOR TRANSOM
H		3010	3'-0" X 1'-0"	DOOR TRANSOM
J		2810	2'-8" X 1'-0"	DOOR TRANSOM
K		2817	2'-7 5/8" X 1'-7 1/4"	BASEMENT/UTILITY WINDOW

WINDOW SCHEDULE — WINDOWS SHALL BE ANDERSEN "NARROWLINE" #200 SERIES DOUBLE HUNG WINDOWS

NOTE: ADD 1/2" TO WINDOW HEIGHT AND WIDTH TO DETERMINE ROUGH OPENING DIMENSIONS

The Garlinghouse Company

Figure 7-20. The window schedule gives complete information about all of the windows, with marks corresponding to those on the elevations.

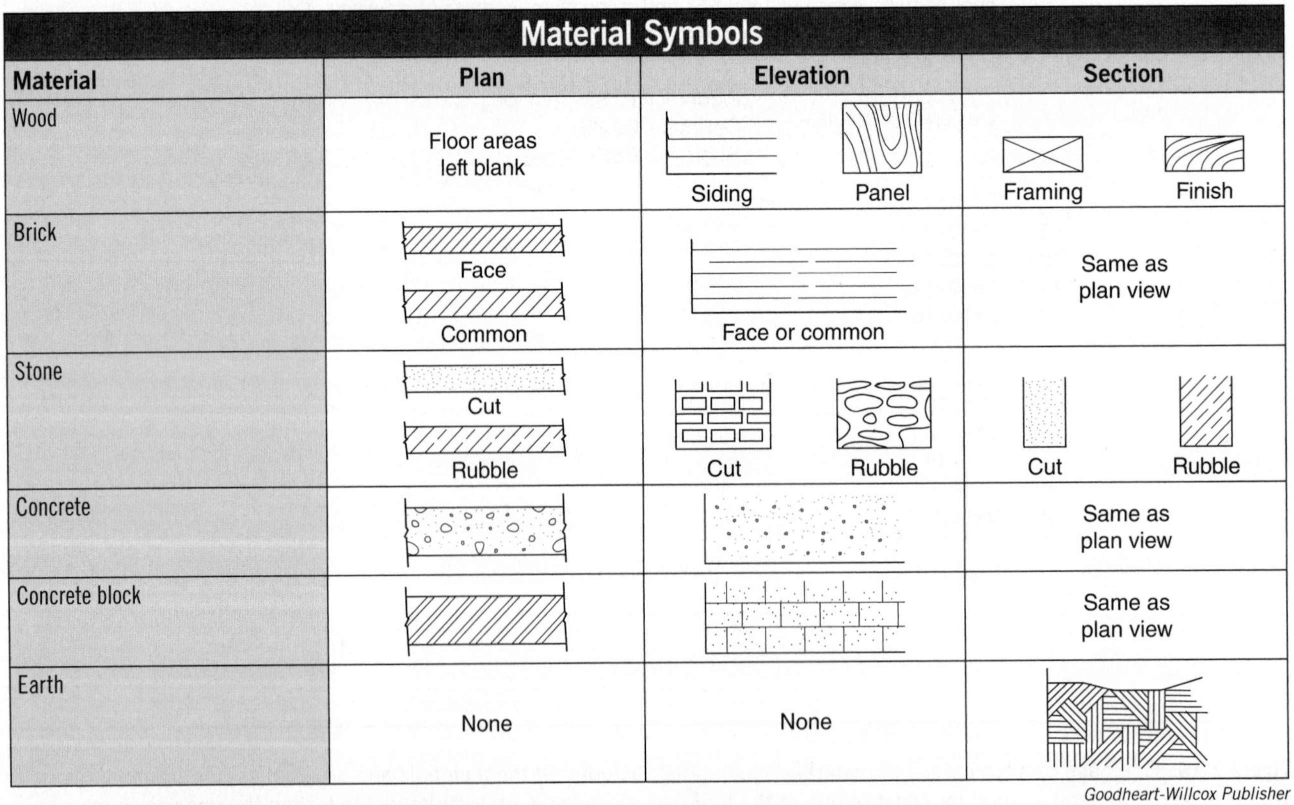

Goodheart-Willcox Publisher

Figure 7-21. Symbols are used to represent things that are impractical to draw.

Material Symbols

Material	Plan	Elevation	Section
Glass			Large scale / Small scale
Insulation	Same as section	Insulation	Loose fill or batt / Board
Plaster	Same as section	Plaster	Stud / Lath and Plaster
Structural steel		Indicated by note	
Sheet metal flashing	Indicated by note		Show contour
Tile	Floor	Wall	

Figure 7-21. (*Continued.*)

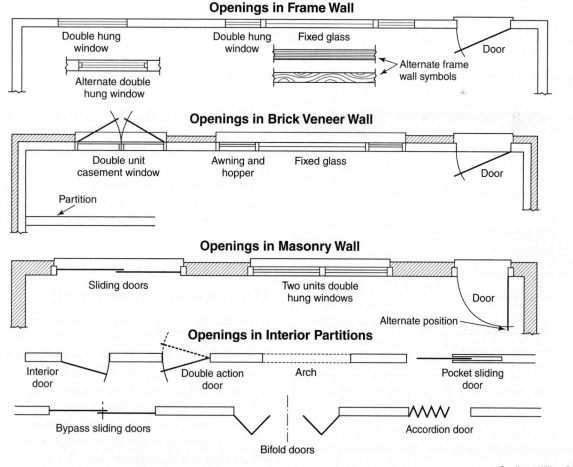

Openings in Frame Wall
Double hung window
Alternate double hung window
Double hung window   Fixed glass
Alternate frame wall symbols
Door

Openings in Brick Veneer Wall
Double unit casement window
Awning and hopper   Fixed glass
Door
Partition

Openings in Masonry Wall
Sliding doors
Two units double hung windows
Door
Alternate position

Openings in Interior Partitions
Interior door
Double action door   Arch   Pocket sliding door
Bypass sliding doors
Bifold doors
Accordion door

Figure 7-22. How windows and doors are presented in plan views. The swing of doors (the direction they swing) is always shown.

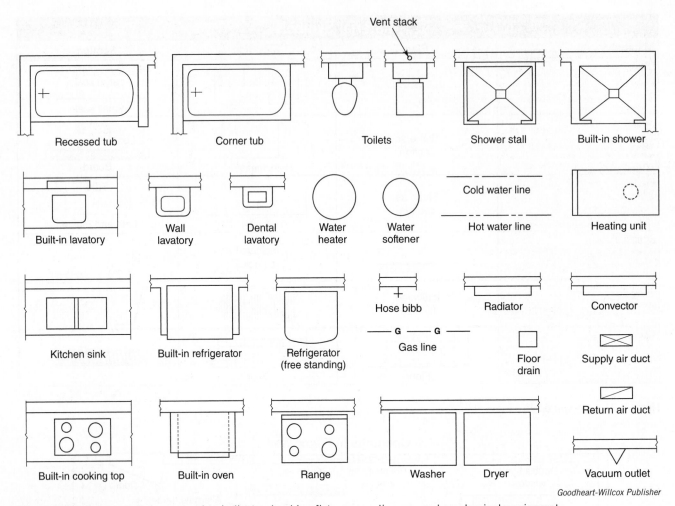

Figure 7-23. These symbols are used to indicate plumbing fixtures, appliances, and mechanical equipment.

Goodheart-Willcox Publisher

Electrical Symbols

Ceiling outlets for fixtures		Lighting panel	
		Power panel	
Wall fixture outlet		S	Single-pole switch
Ceiling outlet with pull switch ($_{PS}$)		S_2	Double-pole switch
Wall outlet with pull switch ($_{PS}$)		S_3	Three-way switch
Duplex convenience outlet		S_4	Four-way switch
Waterproof convenience outlet ($_{WP}$)		S_P	Switch with pilot light
Convenience outlet $_{1, 3}$, 1 = single 3 = triple			Push button
Range outlet ($_R$)			Bell

Goodheart-Willcox Publisher

Figure 7-24. Some electrical symbols. Although carpenters are not responsible for wiring, they should recognize the symbols.

While the symbols shown in the figures are generally accepted as standard, some drafters use variations of these symbols. Most drawing sets include a key to all of the symbols used.

Abbreviations are commonly used on plans to save space. See **Figure 7-25**. For a listing of common abbreviations, refer to Appendix B, **Technical Information**.

7.2 How to Find Unmarked Dimensions

Architectural plans include dimensions that show many distances and sizes. Still, carpenters may require a dimension that is not shown. It is always best to compute unmarked dimensions from marked or known dimensions. The plans almost always contain enough information so that any dimension can be computed. In some cases, it may be necessary to refer to more than one drawing in the set to compute the unknown dimension. For example, to find the dimension from

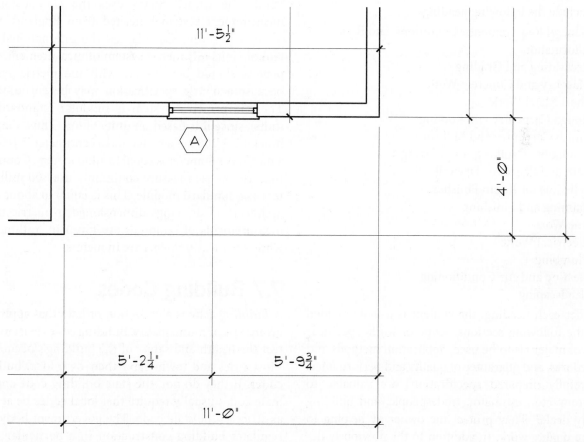

Common Abbreviations

Abbreviation	Meaning	Abbreviation	Meaning	Abbreviation	Meaning
ALT	Alternate	DR	Door	NOM	Nominal
BSMT	Basement	EL, ELEV	Elevation	REF	Reference
BM	Beam	EXT	Exterior	REV	Revision
BLDG	Building	GR	Grade	STD	Standard
CKT	Circuit	HDR	Header	THK	Thick
COL	Column	LAV	Lavatory	WC	Water closet
DIA, ∅	Diameter	LVL	Level	WWF	Welded wire fabric

Goodheart-Willcox Publisher

Figure 7-25. Many abbreviations are used on construction drawings. See Appendix B, Technical Information for a more comprehensive list of abbreviations.

the outside of the right wall to the right side of the rough opening for window A in **Figure 7-26**, follow these steps:

1. See the marked dimension to the center of the window (5'-9 3/4").
2. Check the window schedule (**Figure 7-20**) to find the size of this window (3'-1 5/8").

3. A note beneath the window schedule says to add 1/2", so the rough opening is 3'-1 5/8"+1/2"=3'-2 1/8" wide.
4. Subtract half the rough opening size from the dimension to the centerline (5'-9 3/4" minus 1'-7 1/16" = 4'-2 11/16". That is the dimension from the outside of the right wall to the inside of the rough opening.

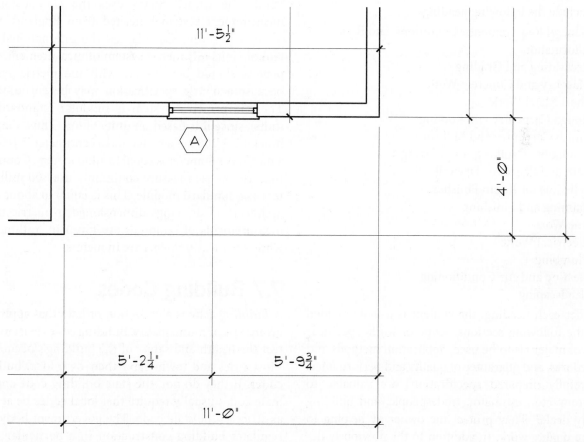

The Garlinghouse Company

Figure 7-26. It is sometimes necessary to refer to more than one drawing in the set to determine a dimension. Here, part of the floor plan from Figure 7-7 is shown.

7.3 Changing Plans

Sometimes an owner wants changes in plans as the job progresses. If the changes are minor—such as changing the size or location of a window, or making a revision in the design of a built-in cabinet—a carpenter can usually handle them. Sketches or notations should be recorded on each set of plans so there are no misunderstandings. Major changes, such as the relocation of a load-bearing wall or stairs, may generate a chain reaction of problems, not only architecturally but also added cost and delayed completion date. These changes should be undertaken only after an architect has made the necessary plan changes and the owner has approved them.

7.4 Specifications

Although the working drawings show many of the requirements for a structure, certain supplementary information is best presented in written form as *specifications*, commonly called specs. See **Figure 7-27**. A carpenter should carefully check and follow the specs. Specifications for a residential structure generally include the following headings:

- General Requirements, Conditions, and Basic Information
- Excavating and Grading
- Masonry and Concrete Work
- Sheet Metal Work
- Rough Carpentry and Roofing
- Finish Carpentry and Millwork
- Insulation, Caulking, and Glazing
- Lath and Plaster or Drywall
- Schedule for Room Finishes
- Painting and Finishing
- Tile Work
- Electrical Work
- Plumbing
- Heating and Air Conditioning
- Landscaping

Under each heading, the content is usually divided into the following sections: scope of work, specifications of materials to be used, application methods and procedures, and guarantee of quality and performance.

Carefully prepared specifications are valuable to the contractor, estimator, tradespeople, and building supply dealer. They protect the owner by helping to ensure quality work. In addition to the previously described items, specifications may include information and requirements regarding building permits, contract payment provisions, insurance and bonding, and provisions for making changes in the original plans.

7.5 Computer-Aided Drafting and Design (CADD)

Computers are usually used to create construction drawings and even perspective drawings for houses. This is known as *computer-aided drafting and design (CADD)*. Working with a variety of architectural software, the drafter creates drawings on the computer. Then, the drawings are sent to a printer or plotter to be duplicated on paper. A plotter is capable of producing drawings in larger sizes than most printers. Some software can automatically produce materials lists.

Today, the manufacture of construction components, such as truss joists and rafters or even whole houses, can be computer controlled. The combination of design and manufacturing controlled by computer is called *computer-aided manufacturing (CADD-CAM)*.

7.6 Metric Measurement

While the United States uses the US Customary measurement system inherited from England, most nations, including Canada, use the International System of Units (SI) metric system of measurement. Carpenters should be familiar with the metric system because a metric specification may be encountered. The entire metric system is modular (standardized units), since it is based on units of ten. Thus, the millimeter is 1/1000 of a meter and a centimeter is 1/100 of a meter. A kilometer is equal to 1000 meters. Countries that use metric measure commonly use 300 millimeters as a standard module. This is equal to about 1′. In architectural drawings dimensioned in metric units, measurements of buildings are given in millimeters, while site measurements are in meters.

7.7 Building Codes

A *building code* is a collection of laws that apply to a given state or municipality. Building codes exist to protect the health and safety of the building's occupants. Most cities and towns have their own local building codes. If they do not, the state building code applies. State codes usually require that local codes be at least as strict as the state code. The government body that regulates building construction in a particular area, such as a city's building department, is known as the *authority having jurisdiction (AHJ)*.

DIVISION 5   **PLAN NO. 12275**

CARPENTRY, MILLWORK, & HARDWARE

General Conditions:

This contractor shall read the General Conditions and Supplementary General Conditions which are a part of these Specifications.

Scope of Work:

Furnish and install all rough lumber, millwork, rough and finished hardware, including all grounds, furring, and blocking, frames and doors, etc., as shown on the drawings or hereinafter specified.

Rough Lumber:

All lumber used for grounds, furring, blocking, etc., shall be #1 Common Douglas Fir or Pine.

Grounds and Blocking:

Furnish and install suitable grounds wherever required around all openings for nailing metal or wood trim such as casings, wainscoting, cap, shelving, etc.; also building into masonry all blocking as required. All wood grounds must be perfectly true and level and securely fastened in place.

Doors:

Interior doors to be plain sliced red oak for stain finish. Solid core doors to be laminated 1 3/4" thick, or as called for in the door schedule. See plan for Label doors.

Temporary Doors and Enclosures:

Provide temporary board enclosures and batten doors for all entrance openings. Provide suitable hardware and locks to prevent access to work by unauthorized persons.

At any openings used for ingress and egress of material, provide and maintain protection at jambs and sills as long as openings are so used.

Rough Hardware:

All hardware such as nails, spikes, anchors, bolts, rods, etc., in connection with rough car- by this contractor. Use aluminum non-staining nails for all exterior wood. around opening edges of trim shall be lightly smoothed and machine sanded at the carpenters at the building in a first class manner before applied.

Finishing Hardware:

An allowance of $1,000.00 shall be included for finishing hardware, which will be selected by the architect and installed by this contractor. Locksets shall match existing design and finish. Adjustment of cost above or below this allowance shall be made according to authentic invoice for hardware for this building from hardware supplier. Hardware supplier shall furnish schedule and templates to frame manufacturer.

Caulking:

Caulk all exterior windows, doors, etc., and openings throughout building with polysulfide based sealant, see specifications under cut stone.

Wood Paneling:

All wood paneling shall be Weyerhauser Forestglo, Orleans Oak 1/4" prefinished. Apply over 1/2" drywall on steel studs and furring.

Figure 7-27. A portion of the specifications that are included with a set of house plans. Other topics included in specifications are plumbing, heating, and wiring.

The code is usually presented in booklet form. It covers all important aspects of the erection of a new building, as well as the alteration, repair, and demolition of existing buildings. The basic purpose is to provide for the health, safety, and general welfare of building occupants and other people in the community.

A code sets the minimum standards that are acceptable in a community for design, quality of materials, and quality of construction. It also sets requirements concerning certain design factors:

- Size, height, and bulk of buildings
- Size of escape doors and windows
- Ceiling heights
- Lighting and ventilation
- Insulation requirements

There are codes that contain detailed directions and standards for installation of a building's systems. These govern the methods and materials used when installing plumbing, wiring, and heating systems.

A carpenter should be aware that building codes might vary from community to community. What is common practice in one area may not be allowed in another. For example, some communities' codes will require one or more sumps in basement floors and perimeter drainage around the footings. Another community may not mention them. These differences should be carefully noted.

Some items in a code must be adjusted to local conditions. In northern climates, footings need to be deeper than in southern states. Structures in high-wind areas require extra bracing and strapping.

Pro Tip
In doing any kind of carpentry work, the importance of closely following all building codes cannot be overemphasized. If a building inspector finds work that does not conform to code, the work must be redone correctly. This can add considerable expense to construction.

7.7.1 Model Codes

There are continual improvements in construction methods and materials. This makes the updating of local building codes cumbersome—communities must convene officials to update the codes—and costly. Because of this, many communities have adopted *model codes*. One of the first model codes was introduced by The American Insurance Association (successor to the National Board of Fire Underwriters). This publication was known as the *National Building Code*.

The International Code Council introduced the *International Building Code (IBC)* and the *International Residential Code for One- and Two-Family Dwellings (IRC)* in 2000. These model codes were developed from the codes published by the Building Officials and Code Administrators International, Inc. (BOCA), the Uniform Codes published by the International Conference of Building Officials (ICBO), and the Standard Codes published by the Southern Building Code Congress International, Inc. (SBCCI). The IBC and the IRC are model codes, meaning that they are available for authorities having jurisdiction to use as the basis for their local codes. A community may adopt the model code, verbatim, as their code, or they may adopt it with some changes to make it suit their local needs. The IRC is the portion of the IBC that applies to one- and two-family homes. See **Figure 7-28**.

In addition to observing building codes adopted by local communities (cities, towns, counties), a carpenter must be aware of certain laws at the state level that govern buildings. Several states have adopted other codes, including electrical codes, fire codes, and energy codes.

A carefully prepared and up-to-date code is not sufficient in itself to ensure safe and adequate buildings. All codes must be properly administered by officials who are experts in the field. Under these conditions, the owner can be assured of a well-constructed building.

SECTION R305
CEILING HEIGHT

R305.1 Minimum height.

Habitable space, hallways and portions of *basements* containing these spaces shall have a ceiling height of not less than 7 feet (2134 mm).

Bathrooms, toilet rooms and laundry rooms shall have a ceiling height of not less than 6 feet 8 inches (2032 mm).

Exceptions:

1. For rooms with sloped ceilings, the required floor area of the room shall have a ceiling height of not less than 5 feet (1524 mm) and not less than 50 percent of the required floor area shall have a ceiling height of not less than 7 feet (2134 mm).

2. The ceiling height above bathroom and toilet room fixtures shall be such that the fixture is capable of being used for its intended purpose. A shower or tub equipped with a showerhead shall have a ceiling height of not less than 6 feet 8 inches (2032 mm) above an area of not less than 30 inches (762 mm) by 30 inches (762 mm) at the showerhead.

3. Beams, girders, ducts or other obstructions in *basements* containing *habitable space* shall be permitted to project to within 6 feet 4 inches (1931 mm) of the finished floor.

R305.1.1 Basements.

Portions of *basements* that do not contain *habitable space* or hallways shall have a ceiling height of not less than 6 feet 8 inches (2032 mm).

Exception: At beams, girders, ducts or other obstructions, the ceiling height shall be no less than 6 feet 4 inches (1931 mm) from the finish floor.

Portions of this publication reproduce excerpts from the 2018 International Residential Code for One- and Two-Family Dwellings, International Code Council, Inc., Washington, D.C. Reproduced with permission. All rights reserved.

Figure 7-28. This is the section of the IRC that regulates ceiling heights. Most sections of the code are much longer and many have several pages of tables and illustrations.

Also, a carpenter is protected against the unfair competition of those who are willing to sacrifice quality for an excessive margin of profit.

Communities have inspectors who enforce the building code. They make periodic inspections during construction or remodeling. The inspectors are persons who have worked in the construction trades or who are otherwise knowledgeable about construction.

Pro Tip
Building inspectors usually require the builder to make appointments for inspections at specified stages of construction. For example, most building permits require an inspection of the footing forms and reinforcement before any concrete is placed. However, inspectors may do unscheduled inspections as well, showing up without notice.

Green Note
Practices that started as 'green' are becoming 'code' across the country. Structures built with green materials are healthier and safer for their occupants. Reducing excess materials and energy loss cut costs throughout a building's lifetime. See Chapter 28, **Green Building and Certification Programs** for more on how green building practices evolve into required building practices.

7.7.2 Standards

Building codes are based on standards developed by manufacturers, trade associations, government agencies, professionals, and tradespeople. All are looking for a desirable level of quality through efficient means. A particular material, method, or procedure is technically described through specifications.

Town of Jonesville
Code Enforcement Office
123 North Main Street
Jonesville, New York 14655
(585) 555-1780 ext 202 FAX (585) 555-0046

BUILDING PERMIT

TO: CONSTRUCT_____ REMODEL_____ DEMOLISH_____ OTHER_____

AT: 3867 Foundation Drive

A: 1751 Sq ft Residential Home

ISSUED TO: John Smith

ISSUED BY: R. Fullen   **PERMIT NUMBER:** 17-06R

DATE ISSUED: 04/25/2021   **DATE EXPIRES:** 04/25/2022

INSPECTIONS REQUIRED

___X___   SEWER LATERALS, STORM AND SANITATION, **BEFORE** COVERING
___X___   TRENCH FOOTINGS, **BEFORE** CONCRETE POUR, MIN. 48" BELOW GRADE
___X___   FOUNDATION WALL, **BEFORE** BACKFILLING
___X___   PLUMBING, UNDERGROUND, **BEFORE** COVERING
___X___   ROUGH CARPENTRY AND PLUMBING, **BEFORE** COVERING
___X___   ELECTRICAL (CEO Approved Electrical Inspector)
___X___   INSULATION, **BEFORE** COVERING WALLS, CEILINGS AND FLOORS
___X___   FINAL, **BEFORE** CERTIFICATE OF OCCUPANCY OR COMPLIANCE

THE BUREAU OF CODE ENFORCEMENT **MUST** BE NOTIFIED FOR REQUIRED INSPECTIONS
CALL 585-555-1780 Ext. 202
ALLOW TWO (2) BUSINESS DAYS NOTICE FOR INSPECTIONS (excluding weekends or holidays)

- WORK TO BE DONE IN ACCORDANCE WITH TOWN/VILLAGE OF JONESVILLE & STATE BUILDING AND FIRE CODES.

- PERMIT VOID IF CONSTRUCTION IS NOT STARTED WITHIN SIX (6) MONTHS OF DATE ISSUED

POST ON PROPERTY   POST ON PROPERTY   POST ON PROPERTY

Figure 7-29. Construction cannot begin until a building permit is obtained from the community's building officials. The permit and inspection card must be displayed at the building site. This permit is combined with the card.

Specifications become standards when their use is formally adopted by broad groups of manufacturers, builders, agencies, and associations. The following are organizations devoted to the establishment of standards, many of which are directly related to the field of construction:

- The American Society for Testing and Materials (ASTM)
- American National Standards Institute (ANSI)
- Underwriters Laboratories (UL)

The Commodity Standards Division of the US Department of Commerce develops commercial standards. The chief purpose of the agency is to establish quality requirements and approved methods of testing, rating, and labeling. These standards are designated by the initials CS, followed by a code number and the year of the latest revision.

7.7.3 Building Permits and Inspections

A ***building permit*** is issued by the community's building officials stating that construction can begin, **Figure 7-29**. Steps for securing a building permit vary from one community to another. Usually, the contractor or building owner files a formal application with the appropriate local agency. This may be a village clerk, city clerk, or a county building department. The application is given to the agency with one or two sets of plans, **Figure 7-30**. Usually, the submitted drawings must include these items:

- Floor plans
- Specifications
- Site plan
- Elevation drawings

VILLAGE OF FLOSSMOOR
APPLICATION FOR BUILDING PERMIT

Type of Work:
Erection ____ Remodel ____ Addition ____ Repair ____ Demolish ____
Construction: Brick ____ Frame ____ Other ____
Color: Roof ____ Brick ____ Trim ____ Siding ____
Proposed use ____________
Livable floor area ____________

Required Service Entrance Conductors – Full Rated
100 Amp. Service and 100 Amp. Service Switch - .... Wire Size ....
150 Amp. Service and 150 Amp. Service Switch - .... Wire Size ....
200 Amp. Service and 200 Amp. Service Switch - .... Wire Size ....
400 Amp. or Larger Service and 400 Amp. Service Switch - ....
Wire Size ....
Underground Service ☐

MAJOR APPLIANCE CIRCUITS REQUIRED BY CODE

☐ Range .... Amps.	☐ 2 Laundry Circuits-20 .... Amps.
☐ Built-in Oven .... Amps.	☐ Heating Plant .... Amps.
☐ Water Heater .... Amps.	☐ Lighting Circuits .... Amps.
☐ Dishwasher .... Amps.	ELECTRICAL HEATING CIRCUITS
☐ Garbage Disposal .... Amps.	☐ Cable .... Amps.
☐ Sump Pump .... Amps.	☐ Baseboard Units .... Amps.
☐ Clothes Dryer .... Amps.	☐ Electrical Furnace .... Amps.
☐ Bathroom Heater .... Amps.	COMMERCIAL WIRING
☐ Fixed Air Conditioner .... Amps.	☐ Lighting .... Amps.
☐ Window Air Conditioner .... Amps.	☐ Motors .... Amps.
☐ Electric Door Opener .... Amps.	☐ Appliances .... Amps.
☐ 2 Kitchen Circuits–20 .... Amps.	☐ Other .... Amps.

*NOTE – No Lighting or Other Current Consuming Device Shall Be Connected to the 2 Kitchen Circuits, the 2 Laundry Circuits, Heating Plant, Sump Pump or Air Conditioner Circuits.

Street Address: ____________
Date ____________ Township ____________
Real Estate Index No. ___ — — —
Lot: ____________ Block ____________
Subdivision: ____________

Property Owner: ____________
Present Address ____________
Telephone ____________
Architect: ____________
Address ____________
Telephone ____________
General Contractor: ____________
Address ____________
Telephone ____________ Bond Expir. Date ____________
Air Conditioning & Heating Contractor: ____________
Address ____________
Telephone ____________
Electrical Contractor: ____________
Address ____________
Telephone ____________ Licensed ____________
Plumbing Contractor: ____________
Address ____________
Telephone ____________ Bond Expir. Date ____________
State License No. ____________

As owner of the property, for which this permit is issued and/or as the applicant for this permit, I expressly agree to conform to all applicable ordinances, rules and regulations of the Village of Flossmoor.

Contractor or Owners Signature: ____________

Estimated cost of building complete, including all materials and labor $ ____________

Building Permit No. ____________
Electrical Permit No. ____________
Plumbing Permit No. ____________

Village of Flossmoor, Illinois

Figure 7-30. An application for a building permit must be accompanied by information about the structure that is to be built. In addition to the information given on the form, plans must be submitted.

A filing fee and plan review fee are usually required. These are in addition to the fee for the building permit itself.

The plans are examined by building officials to determine if they meet the requirements of the local code. Some communities have an architectural committee that determines if the plans are satisfactory.

It is sometimes necessary for the builder or owner to submit supporting data to show how the building design meets the code. When the plan meets all of the requirements of the building code, a building permit is issued. Permit fees are based on cost of construction, and can amount to hundreds of dollars for large structures.

When construction begins, the building permit and an inspection card are posted on the building site in a location where they can be seen by all who visit the site. Sometimes the two are combined, as in **Figure 7-29**.

As work progresses, the building inspector makes inspections and fills out the inspection card for approval of work completed. It is important that the permit and card always be attached to the building or somewhere on the construction site. This location allow the permit to be clearly seen from the road, so passersby can see that a building permit has been issued and the project has approval from the AHJ.

Work on the structure should not proceed beyond the point indicated in each successive inspection. Carpenters on the job must pay close attention to this record. Mechanical work (heating, plumbing, and electrical wiring) may never be enclosed before the building inspector has approved the installations. When construction is complete, a final inspection must be made and an occupancy permit issued. Until then, the building may not be occupied.

Construction Careers
Inspector

There are many different kinds of inspectors whose work touches on the construction industry. Many of these can be grouped under the general title of building inspector since they deal with some aspect of the building process, such as electrical, plumbing, or mechanical systems.

KomootP/Shutterstock.com

Building inspectors usually have a background in the building trades. They determine if buildings are safe for occupation. Building inspectors must have a thorough knowledge of all local building codes and regulations.

Approximately half of all inspectors are employed by local units of government. These inspectors perform the important task of making sure that a building and its systems conform to codes, standards, and local regulations. The inspection process usually begins even before construction starts, when plans are submitted to the building inspector's department

for review. The review may result in approval or may require changes to bring the plan into compliance with codes and regulations. Once construction gets underway, inspections are made at a number of stages—foundation, framing, electrical, plumbing, and mechanical system (such as HVAC) installations. Once all work is completed to the satisfaction of the inspector, a final approval, often called an occupancy permit, can be issued.

People employed as building inspectors and in the various specialties, such as electrical or plumbing inspectors, should have a thorough knowledge of all codes and local regulations that the builder must meet. A good background in the building trades is very desirable for those seeking to become inspectors. Many tradespeople "retire" to work as building inspectors. Some inspectors have academic preparation in such areas as construction technology or construction management. A degree in engineering or architecture is helpful in advancing to supervisory duties.

Working conditions as an inspector are generally the same in most building trades jobs, although less physically demanding. Inspectors must be able to climb ladders, work in confined spaces, and carefully navigate through areas cluttered with materials and tools.

In addition to the inspectors employed by government, a significant number work for engineering or architectural service firms. A growing number are self-employed, primarily in the field of home inspection where they are hired by a buyer or seller to examine a property being offered for sale.

Chapter Review

Summary

- Carpenters must be able to read and understand building plans.
- Included in a typical set of plans are site, foundation, and floor plans; framing, electrical, plumbing, and HVAC plans; elevations; section drawings; detail drawings; and lists of materials.
- In addition to plans, there are written specifications, which must be carefully followed.
- Dimensioning may be in US Customary units (as used in the United States) or metric units.
- Local building codes often govern how a structure may be built.
- Permits for building must be obtained from local government units and periodic inspections must be done to approve the work.

Know and Understand

Answer the following questions using the information in this chapter.

1. _____ are known as the language of the construction industry.
   A. Sets of plans
   B. Architectural drawings
   C. Blueprints
   D. Specifications

2. *True or False?* The building owner generally does not receive a set of plans before construction begins.

3. _____ drawings in a set of house plans shows the interior parts and structures when cut by a vertical plane?
   A. Building elevation
   B. Section
   C. Detail
   D. Mechanical systems

4. *True or False?* When a client purchases a stock plan, they are guaranteed to get one-of-a-kind design for their building.

5. Residential plan views are usually drawn to a scale of _____″ = 1′-0″.
   A. 1
   B. 1/4
   C. 1/8
   D. 1/16

6. A _____ is a drawing showing three dimensions, much like a photograph.
   A. scale
   B. floor plan
   C. pictorial sketch
   D. building elevation

7. Carpenters use _____ more than any other in the set of plans.
   A. building elevations
   B. plot plans
   C. floor plans
   D. framing plans

8. The plot plan shows the _____ (size and shape) of the building and its location on the site.
   A. setbacks
   B. scale
   C. structure
   D. footprint

9. Elevation drawings show the _____ walls of the structure.
   A. inside
   B. outside
   C. supporting
   D. All of the above.

10. A truss plan shows all of the following, *except* _____.
    A. positioning of trusses in the building
    B. openings needed for windows
    C. drawn to scale drawings
    D. dimensions

11. Lines connecting electrical fixtures in an electrical plan inform the electrician what information about the fixtures?
    A. The flow of electrical current.
    B. The fixtures share the same circuit.
    C. The elevation of the fixtures.
    D. The type of wiring for each fixture.

12. What part of an architectural plan generally includes requirements regarding insurance and bonding?
    A. Specifications.
    B. Schedules.
    C. Conventional representations.
    D. Building codes.

13. *True or False?* Supplementary information provided by specifications includes general requirements, conditions, and basic information.

14. When taking building measurements in the metric system, the standard unit is the _____.
    A. yard
    B. millimeter
    C. inch
    D. foot

15. Which of the following organizations is directly involved in the establishment of construction standards?
    A. The American Insurance Association.
    B. The Southern Building Code Congress International, Inc. (SBCCI).
    C. The American Society for Testing and Materials (ASTM).
    D. The Occupational Safety and Health Administration (OHSA).

16. Which of these drawings is *not* generally submitted with a building permit application?
    A. Specifications.
    B. Site plan.
    C. Elevation drawings.
    D. Detail drawings.

17. When construction begins, the building permit and _____ are posted on the building site where everyone can see it.
    A. architectural plans
    B. inspection card
    C. building code
    D. construction standards

Apply and Analyze

1. Explain what drawings a builder generally looks at first on a set of plans and why it is important to start with that type.

2. Explain why a carpenter needs to use detail drawings in conjunction with plan and elevation drawings.

3. Why are materials and construction particulars not shown to scale on architectural plans?

4. Describe the conditions that must be met to make a major change in the architectural plans.

5. Describe an environmental factor that may alter a community's building codes when compared to a community in a different climate.

6. Explain the process of acquiring a building permit for construction of a new building.

Critical Thinking

1. Louis produces drawings for Zver Architectural Corporation. Why would Louis chose to draw floorplans in 1/4″ = 1′ scale rather than 1/2″ = 1′?

2. How would you solve a discrepancy on a set of plans?

3. Do you agree with the requirement to have multiple inspections as work progresses on a project? Why or why not?

4. Obtain a set of plans for an average-size residence. If you have blueprint reading textbooks, these often include sets of plans. Carefully study the views shown. Make a list of symbols, notes, and abbreviations that you do not understand. Then, go to reference books and architectural standards books to find the information. Ask your instructor for help if you have difficulty with some of the views.

Communicating about Carpentry

1. **Reading and Speaking.** Go to your town or county's website, or visit your town or county building office to view the requirements for acquiring building permits. Collect material showing the permits needed for various installations in a building. Analyze the data in these materials based on the knowledge gained from this lesson. Recommend to the class the items that are needed for various buildings.

2. **Writing and Speaking.** You are a building contractor. You have been hired to inspect a building that was recently bought by a big corporation to make sure it is "up to code." The company wants to house 1500 employees and it will be open twenty-four hours, seven days a week. Create a list in the form of a presentation on what you would need to consider. Share the presentation with the class. See if they have any recommendations on what your list may be missing.

CHAPTER **8**

Building Layout

OBJECTIVES

After studying this chapter, you will be able to:

- Explain the operation of the builder's level and level-transit.
- Explain the basic operation of a laser level system.
- Demonstrate proper setup, sighting, and leveling procedures.
- Measure and lay out angles using leveling equipment.
- Read the vernier scale.
- Explain the procedure of laying out building lines and finding the grade level.
- Use a plumb line.

TECHNICAL TERMS

auto level	fill	plumb
benchmark	grade	property line
builder's level	grade leveling	setback
building line	laser level	total station
building site	leveling rod	transit
cut	line of sight	vernier scale
elevation	lot	zoning ordinance

Before construction of a foundation or slab for a building can begin, a carpenter must know where the structure will be located on the property. This requires a series of steps involving the use of measuring tapes and building layout instruments. The place where a building is to be constructed is called a ***building site***. Often, these sites are in communities where there are streets and where small tracts of land are broken up into smaller parcels called ***lots***.

If the building site is a city lot, placing the building requires more steps. Most communities have strict requirements called zoning ordinances. ***Zoning ordinances*** are laws that define and dictate how property can be used. These laws are specific to geographic areas within a municipality, and may mandate whether a property can be used for residential, commercial, or industrial use. Zoning ordinances may control the size or height of new construction, or in some cases, preserve a historic district by limiting changes to existing buildings

or restricting new construction. Ordinances often regulate the minimum property ***setback***—a certain distance buildings must be set back from the street and the minimum clearances from adjoining properties. The local ordinances must be carefully checked for these requirements before layout begins.

Green Note

Organize a jobsite recycling program to manage efforts to minimize waste onsite. Designate a place to store materials ready for reuse on the jobsite. Include bins for separating wood, metal, and other scrap material for recycling. Many cities have stores that champion the reuse, restore, and repurpose mentality. These stores will accept salvageable materials as donations and offer them back to the community at a discounted rate. Taking advantage of these programs can reduce the cost of disposal and keep salvageable materials from filling up our landfills.

8.1 Plot Plan

Most, if not all, communities require the builder or owner to furnish a plot plan before they will issue a building permit. A surveyor provides a survey of the site and locates stakes to indicate the boundaries, also called *property lines*. The plot plan indicates the location of the structure and distances to property lines on all sides. Surveyors should always locate the property lines. They should also draw the plot plan. The work of an engineer or surveyor protects the owner and builder from costly errors in measurement. Because surveyors usually measure land distances in feet and decimal parts of a foot (Chapter 7, **Plans, Specifications, and Codes**), most plot plans show property boundaries in these units, **Figure 8-1**.

8.2 Measuring Tapes

For measurements and layouts involving long distances, steel tapes, usually called measuring tapes, may be used, **Figure 8-2**. Tapes are available in lengths of 50′–300′. There are various types with differing graduations. A carpenter will usually select one that is

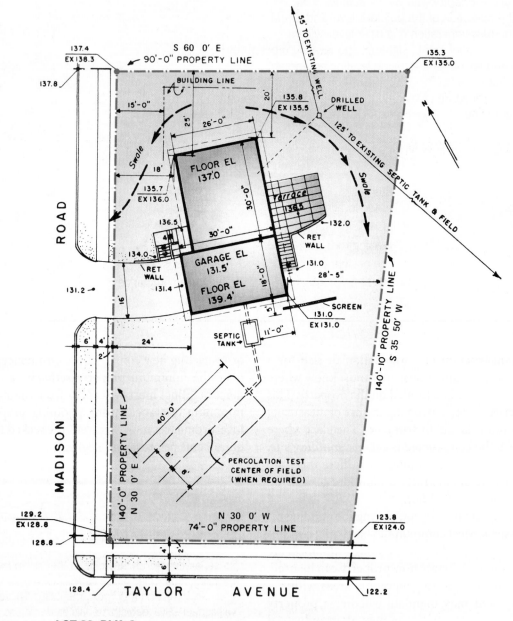

LOT 29. BLK. 2
HOMESTEAD ADDITION, LINCOLN COUNTY, COLUMBIA

Figure 8-1. Providing a plot plan is the responsibility of the architect or owner. The plan shows the property boundaries, along with the building lines for the proposed building. Many communities require a plot plan before issuing a building permit.

marked in feet, inches, and eighths of an inch. Surveyors, on the other hand, require a tape graduated in feet and decimal parts of a foot. See **Figure 8-2B**.

While they can perform the same task of accurately measuring short distances, measuring tapes differ from tape measures in that they are used to measure long distances. They are far bulkier and more cumbersome than tape measures. Measuring tapes require the user to crank a handle in order to wind the tape back in.

8.3 Establishing Building Lines

Laying out a building means locating the outside corners of its foundation, then marking them with wooden stakes. A nail or screw driven into the top of the stake will more precisely mark the corner of the building. *Building lines* are the lines marking where the walls of the structure will be. These lines must conform to setback requirements on distance of the structure from boundary lines of the property.

Once the property lines are known and marked by a surveyor, the building lines can be found by measuring distances with a tape. See **Figure 8-3**.

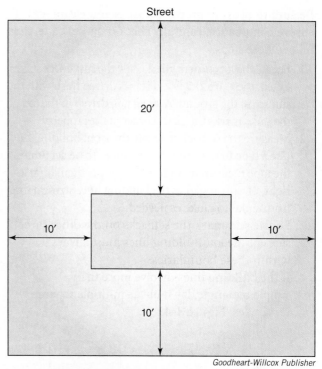
Goodheart-Willcox Publisher

Figure 8-3. A simple rectangular structure can be laid out by taking measurements, using lot lines as reference points.

The Stanley Co.

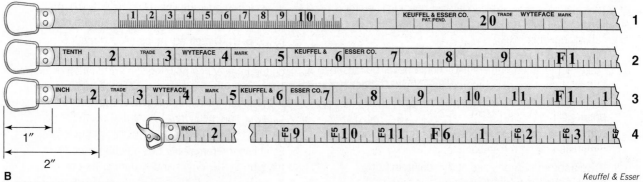
Keuffel & Esser

Figure 8-2. Measuring tape. A—Long tapes for layout use are made in lengths of up to 300′. B—Measuring tapes are made with different systems and graduations. 1. Metric. 2. Feet and decimal graduations. 3. Feet, inch, and eighth-inch graduations. 4. Feet and inches with feet repeated at each inch mark.

Be sure to observe proper setbacks and clearances. To ensure compliance, carefully check the local code, then follow these steps:

1. Locate the lot corners, marked by the surveyor's stakes. Often, these are 2×2 wood stakes driven flush with the surface of the ground. A small nail driven in the top of the stake marks the exact corner of the property.

2. Measure the setback to locate the front building line. Use a tape to measure the distance. To be accurate, the tape measurement must be perpendicular (at a right, or 90°, angle) to the building line. String a line from stake-to-stake as needed.

3. Locate and mark the setbacks on the other three sides of the lot. Building lines must always be within these boundaries.

4. If the building lines involve more than a simple rectangle, lay it out as multiple separate rectangles, **Figure 8-4**.

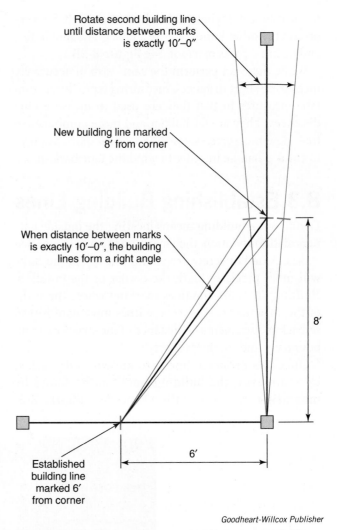

Goodheart-Willcox Publisher

Figure 8-5. To check whether a corner is square, mark 6′ along one building line and 8′ on the intersecting building line. If the distance between the marks is 10′, the corner is square.

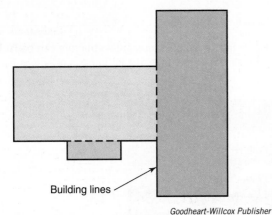

Goodheart-Willcox Publisher

Figure 8-4. Irregularly shaped buildings can usually be treated as separate rectangles added together.

PROCEDURE
Using Measuring Tapes to Establish Building Lines

In the absence of optical instruments (which are described later), it is possible to lay out building lines for small structures with measuring tapes. Follow these steps to establish building lines:

1. Drive a 2×2 stake marking one corner of the building. The exact location of the stake should never be closer to the lot lines than the intersection of two setback lines. Drive a nail in the top of the stake to mark the building line.

2. With a tape, measure off one side of the building dimension along the setback.

3. Drive another corner stake at this point and drive a nail in the top to mark the dimension. This establishes the beginning point of the second side of the building.

4. From the second stake, take a measurement for the intersecting building line. Use the 6–8–10 method shown in **Figure 8-5** to establish a square corner.

5. In the 6–8–10 method, measure and mark 6′ along one building line already established. Next, measure 8′ on the intersecting building line and mark that location. Then, measure the diagonal distance from one mark to the other. Move the building line as needed until the diagonal line measures 10′. At that point, the corner formed by the intersecting lines is square.

6. Establish the remaining two building lines by measuring the length and squaring the corners.

7. Test the accuracy of the layout by diagonally measuring from corner to corner. If these measurements are equal, the building lines are laid out square.

8.4 Laying Out with Leveling Instruments

In residential construction, it is important that building lines be accurately established in relation to lot lines. It is also important that footings and foundation walls be level, square, and the correct size.

If the building is small, such as a 24′ garage, a carpenter's level, line level, framing square, and rule are accurate enough for laying out and checking the building lines. As size increases, however, special leveling instruments are needed for greater accuracy and efficiency.

8.4.1 Leveling Instruments

The builder's level, auto level, and transit are more accurate instruments for building layout. A laser level is a particularly useful tool for leveling, but is not as useful for laying out building lines. These leveling instruments are frequently used in construction work. Builder's levels, auto levels, and transits are basically telescopes with accurate bubble levels that must be mounted and leveled on a base that can be rotated. When the job is too large for the chalk line, straightedge, level, and square, then leveling instruments should be used.

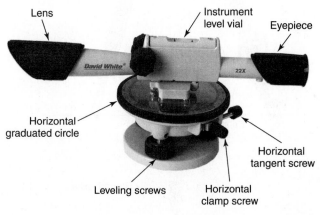
David White

Figure 8-6. The builder's level is used to sight level lines and lay out or measure horizontal lines.

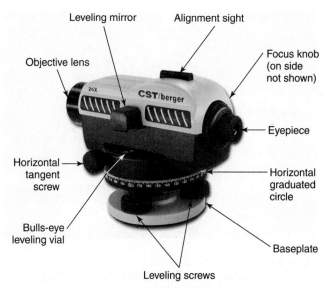
Photo courtesy of Robert Bosch Tool Corporation®

Figure 8-7. The auto level performs the same functions as a builder's level, but is faster and easier to set up.

The optical device of these instruments operates on the basic principle that a *line of sight* is a straight line that does not dip, sag, or curve. Any point along a level line of sight will be at the same height as any other point. Through the use of these instruments, the line of sight replaces the chalk line, line level, and straightedge.

The *builder's level*, also called a dumpy level or optical level, is shown in **Figure 8-6**. It consists of an accurate spirit level and a telescope assembly attached to a circular base that swivels 360°. Leveling screws are used to adjust the base after the instrument has been mounted on a tripod. The telescope is fixed so that it does not move up or down, but it rotates on the base. This permits any angle in a horizontal plane to be laid out or measured.

The *auto level*, **Figure 8-7**, is similar to the builder's level except that it contains an internal compensator that eliminates any minor variant from level.

A   *Goodheart-Willcox Publisher* B   *Goodheart-Willcox Publisher*

Figure 8-8. A—Auto levels have a circular bulls-eye leveling vial, rather than a cylindrical leveling vial. During setup the user centers the bubble in the ring by adjusting the leveling screws. B—Most auto levels are equipped with a mirror to easily check the instrument for level from the normal user position.

Most auto levels use a circular bulls-eye level, **Figure 8-8**, rather than a cylindrical leveling vial. The bulls-eye level is not as precise as the cylindrical level; however, the internal compensator ensures accuracy. This feature makes auto levels easier to set up and less likely to get jarred out of level throughout the operation.

The *transit* is like the builder's level in most respects, **Figure 8-9**. However, the telescope can be pivoted up and down 45° in each direction. The instrument has a horizontal scale and a vertical scale that can be used to accurately measure angles in both directions. Using this instrument, it is possible to accurately measure vertical angles. The transit can also be used to determine if

a wall is perfectly *plumb* (vertical) by sighting the vertical crosshair while pivoting the scope up and down. Its vertical movement also simplifies the operation of aligning a row of stakes, especially when they vary in height.

In use, the builder's level, transit, and laser level are mounted on tripods, **Figure 8-10**. The lengths of the tripod legs can be adjusted independently. This feature makes the tripod easier to level on sloping ground. It also permits the legs to be shortened for easier handling and storing.

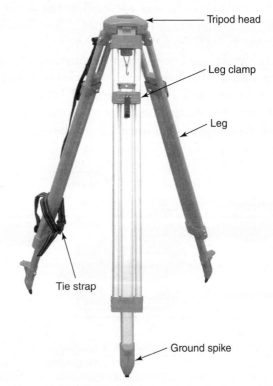

David White

Figure 8-9. A transit can be used to lay out or check both level and plumb lines. It can also be used to measure angles in either the horizontal or vertical planes.

Figure 8-10. Tripod legs hinge at top and are adjustable for use on uneven terrain.

A *leveling rod*, **Figure 8-11**, is used to measure vertical distances from a horizontal sight line. It is a long rod marked off with numbered graduations. It allows differences in elevation between the position of the level and various positions where the rod is held to be easily read. The person operating the level can make the readings, or the target can be adjusted up and down to the line of sight and then the person holding the rod (rod holder) can make the reading.

The rod shown in **Figure 8-11** has graduations in feet and decimal parts of a foot. As mentioned earlier, this is the type used for regular surveying work. Rods are also available with graduations in feet and inches.

A *laser level* is a popular choice for leveling jobs, because one worker acting alone can perform most leveling and plumbing operations with this instrument. **Figure 8-12** shows a laser transmitter and its receiver. The laser level emits a beam of light in a 360° circle. It can be adjusted so the circle is perfectly level or perfectly plumb. A separate laser detector or target signals when the laser beam strikes it. Rotary laser levels function much like a builder's level for establishing a level line. A laser level can be mounted on its side, so it projects a vertical circle of light that can be used for determining when a wall is plumb. During use, the transmitter may be left unattended. The receiver can be tended by a rod holder, **Figure 8-13**.

Receiver

Transmitter

TFoxFoto/Shutterstock.com

Figure 8-12. Laser level transmitters project a beam 360°. The beam can be set to level, plumb, or slope on many lasers. The laser receiver detects the beam and indicates whether it is higher or lower than the mark.

Photo courtesy of Robert Bosch Tool Corporation®

Figure 8-13. A rod holder moves the receiver up or down until it is on line of sight with the laser level.

The laser level is also adapted to other construction tasks to establish either level or plumb lines. Horizontal operations include leveling suspended ceiling grids and leveling raised-access computer floors. Vertical operations include plumbing partitions, curtain walls, posts, columns, elevator shafts, or any operation requiring a vertical reference point.

Safety Note

For safety, there are regulations limiting power output of a laser. Federal regulations limit Class II construction lasers to 1 milliwatt and Class IIIa units to less than 5 milliwatts. These levels are safe for human eyes as long as the user does not stare into the beam. Manufacturers' cautions should be carefully observed.

Pro Tip

Leveling instruments and equipment will vary somewhat, depending on the manufacturer. Always carefully read and study the instructions for a given brand.

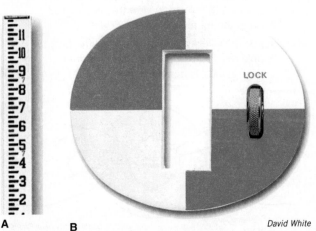

LOCK

A   **B**

David White

Figure 8-11. Leveling rod with target. The target can be moved up or down to match the line of sight of the transit.

8.4.2 Care of Leveling Instruments

Leveling instruments are more delicate than most other carpentry tools and equipment. Special precautions must be followed in their use so they will continue to provide accurate readings over a long period of time. Here are some suggestions:

- Keep the instrument clean and dry. Store it in its carrying case when not in use.
- When the instrument is set up, have a plastic bag or cover handy to use in case of rain. If the instrument becomes wet, dry it before storing.
- When moving the instrument from its case to the tripod, grip it by the base.
- Never leave the instrument unattended when it is set up near moving equipment.
- When moving a tripod-mounted instrument, handle it with care. Hold the instrument upright; never carry it in a horizontal position.
- Never overtighten leveling screws or any of the other adjusting screws or clamps.
- Always set the tripod on firm ground with the legs spread well apart. When it is set up on a floor or pavement, take extra precautions to ensure that the legs will not slip.
- For precision work, permit the instrument to reach ambient (air) temperature before making readings.
- When the lenses collect dust and dirt, clean them with a camel hair brush or special lens paper.
- Never use force on any of the adjustments. They should easily turn by hand.
- Have the instrument cleaned, lubricated, and checked yearly by a qualified repair station or by the manufacturer.

PROCEDURE
Setting Up the Instrument

Use the following procedure to set up a tripod-mounted instrument.

1. Place the tripod so it will be a firm and stable base for the instrument. The base of the legs should be about 3'-6" apart. Make sure the points are well into the ground and the head is fairly level.
2. Check the leg clamps on the adjustable legs. They should be tight enough to carry the weight of the instrument without collapsing or sinking.
3. Carefully lift the instrument from its case by the base plate. Before mounting the instrument, loosen the clamp screws. On some instruments, the leveling screws must be turned up so the tripod cup assembly can be hand-tightened to the instrument mounting stud. Set the telescope lock lever of the transit in the closed position.

Continued

PROCEDURE *(continued)*

4. Attach the instrument to the tripod. If it is to be located over an exact point, such as a corner stake, attach the plumb bob and move the instrument over the spot. Do this before the final leveling.

PROCEDURE
Leveling the Instrument

Leveling the instrument is the most important operation in preparing it for use. None of the readings taken or levels sighted will be accurate unless the instrument is level throughout the work. The tripod must be firmly set on the ground or it will not remain level during use. Follow these steps to level the instrument.

Builder's Level:

1. Release the horizontal clamp screw and line up the telescope so it is directly over a pair of the leveling screws.
2. Grasp the two screws between the thumb and forefinger, as shown in **Figure 8-14**. Uniformly turn both screws with your thumbs moving toward each other or away from each other.
3. Keep turning until the bubble of the level vial is centered between the graduations. You will find on most instruments that the bubble travels in the direction your left thumb moves. See **Figure 8-15**. Leveling screws should bear firmly on the base plate. Never tighten the screws so much that they bind.
4. When the bubble is centered, rotate the telescope 90° (so it is over the other pair of leveling screws) and repeat the leveling operation.
5. Recheck the instrument over each pair of screws. When the instrument is level at both positions, the telescope can be turned in a complete circle without any change in the bubble.

Auto Level:

1. Most auto levels have a mirror placed at eye level that is angled to give the user a horizontal path of sight to the leveling vial. See **Figure 8-8**. Position yourself so that you can view the leveling vial through the mirror.
2. Adjust the leveling screws one at a time until the bubble rests within the center ring of the bullseye level. The level is ready for use.

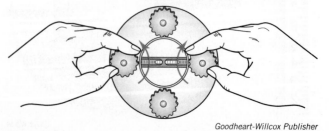

Goodheart-Willcox Publisher

Figure 8-14. Adjust leveling screws to center the bubble in the level vial.

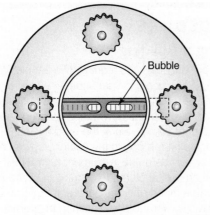

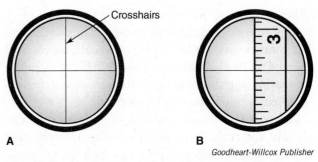

Figure 8-17. View through the telescope. A—Crosshairs vertically and horizontally split the image area in half. B—The object in the view should be centered on the crosshairs.

Figure 8-15. The bubble of the level vial will generally move in the same direction as the left thumb. This bubble needs to move left.

8.5 Using Leveling Instruments

A carpenter can use a leveling instrument to prepare the building site for excavation and grade leveling. The following are jobs that can be done with leveling instruments:

- Locating the building lines and laying out horizontal angles (square corners)
- Finding grade levels and elevations
- Determining plumb (vertical) lines

PROCEDURE

Sighting

Most builder's levels have a telescope with a power of about 20×. This means that the object being sighted appears to be 20 times closer than it actually is. The procedure for sighting is easy to learn:

1. Line up the telescope by sighting along the barrel and then look into the eyepiece, **Figure 8-16**.
2. Adjust the focusing knob until the image is clear and sharp.
3. When the crosshairs are in approximate position on the object, **Figure 8-17**, tighten the horizontal-motion clamp.
4. Make the final alignment by turning the tangent screw.

8.5.1 The Horizontal Graduated Circle

Laying out corners with the transit requires an understanding of how the horizontal graduated circle is marked. It is divided into spaces of 1°, **Figure 8-18**.

Photo courtesy of Robert Bosch Tool Corporation®

Figure 8-16. Sighting a level line with an auto level. Both eyes are kept open during sighting. This reduces eyestrain and provides the best view. Hand signals are used when the rod holder is out of earshot.

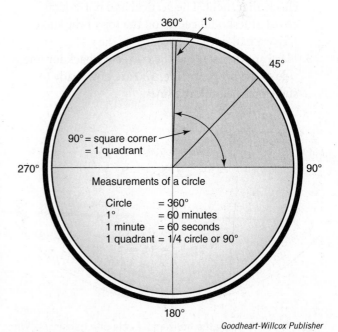

Figure 8-18. The graduated circle of a transit corresponds to the 360° of a full circle; 90° represents a quadrant, which would give you a square corner for a building.

When you swing (rotate) the telescope of the builder's level or transit, the graduated circle remains stationary, but another scale, called the *vernier scale*, moves. It is marked off in 15-minute intervals. When laying out or measuring angles where there are fractions of degrees involved, you will use this vernier scale.

Figure 8-19A shows a section of the graduated circle and the scale. It reads 75°. Notice that the zero mark on the vernier exactly lines up with the 75° mark. Now look at **Figure 8-19B**. The zero mark has moved past the mark for 75°, but is not on 76°. You need to read along the vernier scale until you find a mark that is closest to being directly over a degree mark on the circle. That number is 45. The reading is 75°, plus the number on the vernier, 45 minutes.

Vernier scales will not be the same on all instruments. Study the operator's manual for instructions about the particular model you are using.

8.5.2 Laying Out and Staking a Building

Staking out is done before establishing the grade level. It begins with locating the lot lines. Corners of the lot should normally be marked with stakes. Then, take the following actions:

1. Center and level the instrument (builder's level or transit) over the lot corner stake. Measure the setback called for by local codes. Sight across to the opposite corner stake.
2. Drive a 2×2 stake at the setback in line with the lot stakes. Use the transit or builder's level to check alignment. The vertical and horizontal crosshairs should center on the top of the stake. Drive a nail in the top-center of the stake.
3. Place another stake at the correct setback for one side of the property line. You are now ready to stake out the building lines.

PROCEDURE
Staking a Building

Staking out building lines requires two people. When a builder's level is used, the second person will use a rod that must be plumbed along the line of sight. Since a transit can pivot up and down, the second person uses a stake to locate corners along the building line.

1. Attach a plumb bob to the center screw or hook on the underside of the instrument. Some instruments have an optical plumb for zeroing in over a point. Shift the tripod until the point of the plumb bob is directly over the point marking the corner of the building lines. This is at point A on line AB, as shown in **Figure 8-20**.
2. Level the instrument before proceeding further. Recheck for plumb.
3. From point A (or station A), turn the telescope so the vertical crosshair is directly in line with the edge of a stake or rod held at point B. When using a transit, sight the telescope on the stake.
4. Use a measuring tape along line AB to locate distance to the corner. Drive a corner stake at this point.
5. Set the horizontal circle on the instrument at zero to align with the vernier zero and swing the instrument 90° (or any other required angle).
6. Position the rod or stake along line AC so it aligns with the crosshairs.
7. Locate the other corner along line AC using a measuring tape.
8. Move the instrument to point C, sight back to point A, and then turn 90° to locate the line of sight to point D.
9. Measure the distance to point D and place a stake.
10. Use a measuring tape to check the diagonal distances. If these are equal, the building line is square. If the resulting figure is a rectangle or square, you have completed the layout. However, you may want to move the instrument to point D to check your work.

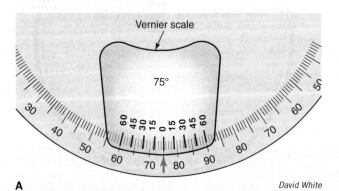

A   *David White*

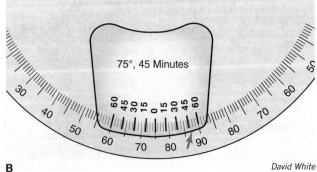

B   *David White*

Figure 8-19. Reading the horizontal circle of a transit. A—When the zero mark of the vernier is exactly on a degree mark, the reading is an even degree. In this case, the reading is 75°. B—When the zero mark is between degrees, read across the vernier to find the minute mark that aligns with a degree mark. This reading is 75°, 45′.

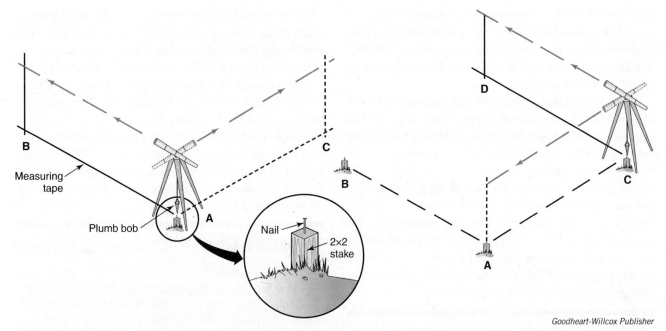

Figure 8-20. Steps for laying out building lines. Left—Locate the instrument over a stake marking a corner. Line up 0 on the instrument circle with the building line AB. Swing the instrument 90° to establish line AC. Right—Move the instrument to point C to establish point D. Rod must be used when the instrument is a builder's level. Rod must be held plumb using a level. A transit is a much better instrument in this operation since it is not necessary to use the rod. Simply swivel the transit telescope and sight on the corner stake.

In practice, you will find that it is difficult to locate a stake in a single operation. This is especially true when using a builder's level, where the line of sight must be "dropped" to ground level with a plumbed rod or straightedge. Usually, it is best to set a temporary stake, as in **Figure 8-21**. Mark it with a line sighted from the instrument. Then, with the measuring tape pulled taut and aligned with the mark, drive the permanent stake and locate the exact point as shown.

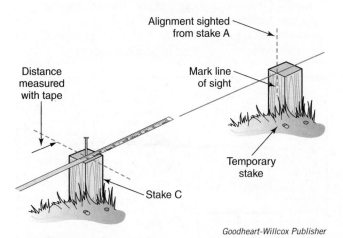

Figure 8-21. A temporary stake may be used to establish an exact point. First, set the temporary stake and mark the line of sight on it. Drive a second stake and transfer the mark from stake A.

All major rectangles and squares of a building line can be laid out using leveling instruments in the manner just described. After batter boards are set and lines attached, a carpenter's level and square can be used to locate stakes for small projections and irregular shapes.

8.5.3 Finding Grade Level

Many points on the building site need to be set at certain elevations or grade levels. These points might include the following:

- The depth of excavation, such as for a basement
- The finished height (elevation) of the foundation footings
- The height (elevation) of foundation walls
- The elevation of floors
- Site features, such as proper grading to ensure that surface water is directed away from the building
- Bearing elevation (point at which footings contact the earth) for footings in order to ensure adequate frost protection of the foundation
- Establishing rise and run of steps and walkways as part of the exterior of the structure

Not all building sites are level. Finding the difference in the grade level between several points or transferring the same level from one point to another is called *grade leveling*. This operation is immediately useful to the excavator, who must determine how much earth must be removed to excavate a basement or trench a foundation footing. Grade leveling is also useful to determine how much earth must be deposited in a particular area to achieve a desired height or elevation at that location.

When the leveling instrument has been set level, the line of sight will also be level. The readings can be used to calculate the difference in elevation, **Figure 8-22**. If the building site has a large slope, the instrument may need to be set up more than once between the points where you want to take readings. The first reading is taken with the rod in one position. Then, the instrument is carefully rotated to get the reading at a second rod position. Of course, if a laser level is being used, it will not be necessary to rotate the instrument. Simply move the rod to the second position and take that reading. This position may be higher or lower than the first position.

Pro Tip
From a practical standpoint, it is simpler to work from a higher point on the site than a lower point. Depending on the actual slope, this measurement can often be made with a single reading.

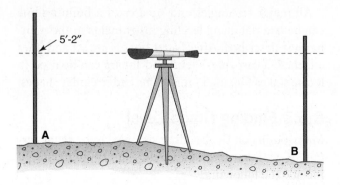

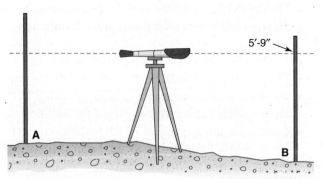

Goodheart-Willcox Publisher

Figure 8-22. Establishing a grade level and finding the difference between two points on a building lot. Top—Level the instrument and take a line of sight reading on point A. Mark down the rod reading. Bottom— If using a builder's level or transit, swing the telescope 180°, and take a line of sight reading on the rod. Compare the two elevations. In this example, point A is 7″ higher than point B.

The term *grade* means the level of the ground. *Elevation* refers to the major structural levels of the building. More specifically, these levels include the top of the footing, top of the foundation wall, and finish height of the first floor.

There should be a reference point (level) for all elevations established on the building site. This reference point is called the *benchmark*, also known as the datum or simply the beginning point. This point must remain undisturbed during the construction of the building project. A stake driven at one corner of the lot or building site, or even a mark chiseled into concrete curbing, often serves as the benchmark.

PROCEDURE
Checking Grade
Building sites are rarely perfectly level. All have high and low points. These highs and lows need to be determined before the height of the foundation is established. This job is easier when the site is fairly level, as in **Figure 8-22**.
1. Locate the instrument midway into the site, then level the instrument.
2. Take a line-of-sight reading on a rod held at one edge of the site.
3. From the position of the target on the rod, note the elevation (5′-2″ in **Figure 8-22**) and record it.
4. Take a line-of-sight reading on the rod located at the opposite side of the site.
5. Note the elevation at the target position on the rod (5′-9″ in **Figure 8-22**). Record it as before.
6. Subtract the lower elevation from the higher one to find the difference (7″). This number is the actual vertical increase or decrease from one known point to another.

When setting grade stakes for a footing or erecting batter boards, set the instrument in a central location on the site, as shown in **Figure 8-23**. The distances to the target (or rod) will be roughly equal. This will improve the accuracy of the readings taken for each corner. An elevation established at one corner can be quickly transferred to other corners or points in between.

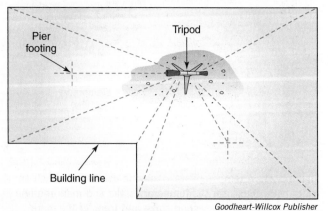

Goodheart-Willcox Publisher

Figure 8-23. A central location for the instrument will make finding and setting grade stakes easier and more accurate.

8.5.4 Setting Footing Stakes

Grade stakes for footings are usually first set to the approximate level "by eye." They are then carefully checked with the rod and level as they are driven deeper. The top of each stake should be driven to the required elevation.

Sometimes, reference lines are marked on construction members, stakes, or other objects near the work. A carpenter then transfers them to the formwork with a carpenter's level and rule as needed. This eliminates the need to repeatedly establish the same elevation.

There may be situations where the existing grade will not permit the setting of a stake or reference mark at the actual level of the grade. See **Figure 8-24.**

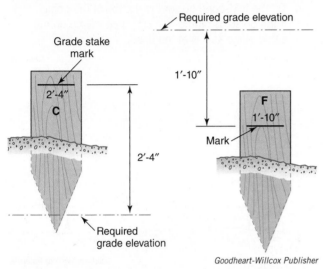

Figure 8-24. Cut and fill stakes are used to tell the excavator how much material must be removed or added to reach grade level. The letter C means "cut" (remove) and the letter F means "fill" (add).

Goodheart-Willcox Publisher

In such cases, a mark is made on the stake with the information on how much fill to add or remove. The letters C and F, standing for *cut* and *fill*, are generally used.

PROCEDURE
Using Multiple Readings

When laying out steeply sloped building plots or carrying a benchmark to the building site, it will likely be necessary to set up the instrument in several locations. **Figure 8-25** shows how reading from two positions is used to calculate, establish, or determine differences in grade at various locations on the plot.

1. Set up the instrument midway between two points on the plot. In **Figure 8-25**, these are identified as points A and B.
2. Take a line-of-sight reading at point A (or station A). Record the reading in the notebook. In this case, the reading is 5'-8".
3. With the rod at point B and the instrument still at setup #1, take a second reading. Record the reading in the notebook. In this instance, the reading is 2'-6".
4. Move the instrument to a point midway between points B and C. Level it as before.
5. Take a second reading on point B and record it. In this example, the reading is 6'-0".
6. With the rod located at point C, take a reading by rotating the telescope 180°. Note and record the elevation. In this case, the reading is 2'-4".
7. Calculate the differences in grade level for each pair of stakes, as shown in **Figure 8-25**.
8. Add the resulting distances and subtract the "minus" sum from the "plus" sum to find the difference in grade from one edge of the plot to the other. The result, in this example, is 6'-10".

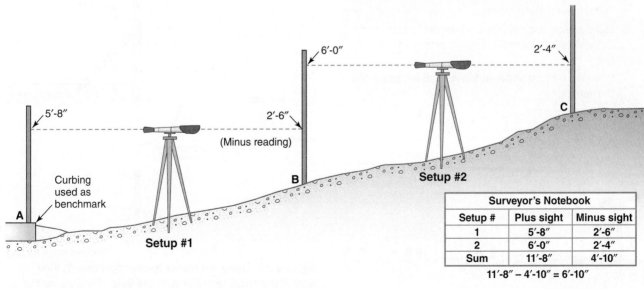

Surveyor's Notebook		
Setup #	Plus sight	Minus sight
1	5'-8"	2'-6"
2	6'-0"	2'-4"
Sum	11'-8"	4'-10"

11'-8" − 4'-10" = 6'-10"

Goodheart-Willcox Publisher

Figure 8-25. When there is a large slope on the property or when long distances are involved, the instrument will need to be set up in two or more locations.

8.5.5 Running Straight Lines with a Transit

Although the builder's level can be used to line up stakes, fence posts, poles, and roadways, the transit is more accurate for these tasks, especially when different elevations are involved. Set the instrument directly over the reference point. Level the instrument and then release the lock that holds the telescope in the level position. Swing the instrument to the required direction or until a stake is aligned with the vertical crosshair. Tighten the horizontal circle clamp so the telescope can move only in a vertical plane. Now, by pointing the telescope up or down, any number of points can be located in a perfectly straight line. See **Figure 8-26**.

8.5.6 Vertical Planes and Lines

Beyond the leveling tasks just mentioned, the transit is also a good instrument for these tasks:

- Measuring vertical angles above or below the line of sight
- Plumbing building walls, columns, and posts

PROCEDURE
Measuring Vertical Angles

1. Position the instrument near the structure. Level the instrument.
2. Release the lever that holds the telescope in a horizontal position.
3. Swing the instrument vertically.
4. Set the horizontal crosshair at the point you wish to measure.
5. Tighten the vertical clamp.
6. Make a final, fine adjustment with the tangent screw to locate the horizontal crosshair exactly on the point.
7. Read the vertical angle on the vertical arc scale and the vernier.

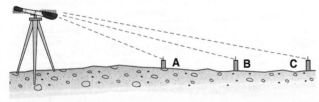

Goodheart-Willcox Publisher

Figure 8-26. How to use the transit to align a row of stakes.

PROCEDURE
Establishing Plumb Lines

Plumb lines can be checked or established by first operating the instrument as shown in **Figure 8-27**. As you tilt the telescope up and down, all of the sighted points are located in the same vertical plane. To plumb structures, such as posts or walls, follow these steps and refer to **Figure 8-28**:

1. Set up the transit at a distance from the object equal to at least the height of the object. Tilt the telescope to sight on the base.
2. Loosen the horizontal clamp and line up the vertical crosshair with the base of the object.
3. Tighten the horizontal clamp.
4. Tilt the telescope upward to the top of the object. If the object is plumb, the crosshair will be on the same plane as it is at the base.

Continued

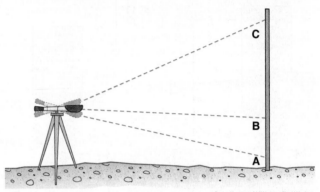

Goodheart-Willcox Publisher

Figure 8-27. A transit can be used to lay out or check points in a vertical plane.

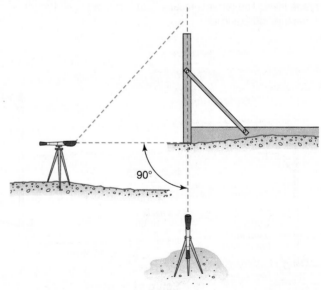

Goodheart-Willcox Publisher

Figure 8-28. Using the transit to plumb an object. First, align the vertical crosshair with the base. Then, swing the telescope to or near the top. Adjust the object for plumb as needed. Reposition the instrument at a 90° angle to the first line and repeat the process.

5. If object is not plumb, adjust the brace to bring the object into plumb.
6. Move the transit to a second position, preferably 90° either to the right or left, and repeat the procedure.

A plumb bob and line may often be the most practical way to check vertical planes and lines. For layouts inside a structure, where a regular builder's level or transit is impractical, use a plumb line.

8.6 Total Stations

A *total station* is similar to a transit in that it measures horizontal and vertical angles, but it also measures distances. The total station combines optical sensors with electronic circuitry to calculate angles and distances. **Figure 8-29** shows a total station. Total stations are expensive instruments, but they are accurate and can measure over any distance as long as line of sight is maintained. For these reasons they are used only on large construction projects and for surveying.

Trimble

Figure 8-29. A total station can measure distances and angles with great accuracy.

Construction Careers
Surveyor/Surveying Technician

Surveyors and surveying technicians establish official land boundaries and the exact location and extent of building sites and other land uses. Survey parties, usually consisting of several surveying technicians under the direction of a party chief, traditionally use a variety of instruments to establish property boundaries and elevations. A licensed surveyor uses this information, combined with research in legal records and other sources, to prepare an official map called a plat of survey that establishes the official location of the property. Such a survey is often required before a deed can be issued in a property transfer.

Surveyors are professionals who must meet educational and experience requirements and pass licensing examination. In the past, surveyors could be licensed by gaining experience on a survey party, then passing a licensing exam. However, most states today require a four-year degree in surveying or a related field, such as civil engineering, in addition to the exam and field experience.

Surveying technicians often have some postsecondary training, usually in a community college certificate or associate degree program. High school courses in algebra, geometry, trigonometry, and drafting are good preparation. Technicians are usually responsible for using optical, physical, and electronic tools to make the needed measurements in the field. While traditional tools such as transits and measuring tapes are still used, more and more survey parties are working with total stations and global positioning system receivers that use satellite data for precise location-finding. Advancement to party chief is possible with additional formal training and experience.

Celeborn/Shutterstock.com

Surveyors and surveyor technicians determine the exact land boundaries of the property. When all the information is gathered, they create a plat of survey to determine the official and legal location of the property. The survey is needed for the property to change ownership in a sale.

Working conditions for survey parties can be extreme, since they are exposed to all kinds of weather conditions and may have to carry equipment for long distances in rugged terrain. Licensed surveyors often work in the field, but also spend time indoors doing research, data analysis, and report writing.

Approximately 2/3 of all surveyors and surveying technicians are employed by architectural and engineering firms or companies providing related services. Most of the remaining employment is provided by government at all levels, ranging from federal agencies, such as the US Forest Service, to state and local planning departments and highway agencies. A relatively small number of surveyors are self-employed.

Chapter Review

Summary

- The correct location on the building site must be identified before the foundation or slab can be constructed.
- Laying out a building involves locating the outside corners of the foundation, driving stakes, and then stretching building lines between stakes to mark where the walls will be.
- Measuring tapes can be used for layout work, but leveling instruments are more precise. Leveling instruments are also used to establish grades and elevations.
- Laser levels are replacing conventional transits and levels in building site layout. Their biggest advantage is requiring only one person to take measurements, rather than two.

Know and Understand

Answer the following questions using the information in this chapter.

1. *True or False?* Zoning ordinances often regulate the minimum property setback.

2. When establishing a building line with a measuring tape, use the _____ method to establish a square corner.
   A. 1-2-3
   B. 6-8-10
   C. leveling
   D. squaring

3. In residential construction, it is important that building lines be accurately established in relation to _____ lines.
   A. lot
   B. building
   C. property
   D. zoning

4. What leveling instrument uses an internal compensator to ensure leveling accuracy?
   A. Builder's level.
   B. Transit.
   C. Auto level.
   D. Laser level.

5. *True or False?* Most builder's levels have a telescope with a power of about 50×.

6. When sighting through the telescope, you should adjust the _____ until the image is sharp and clear.
   A. barrel
   B. horizontal-motion clamp
   C. focusing knob
   D. tangent screw

7. *True or False?* Vernier scales are generally the same on all leveling instruments.

8. Staking a building starts with locating the _____.
   A. building lines
   B. setbacks
   C. property lines
   D. lot lines

9. Grade leveling is immediately useful to the _____, who must determine how much earth must be removed to dig a basement or trench a foundation footing.
   A. surveyor
   B. excavator
   C. builder
   D. architect

10. The reference point for all elevations established on a building site is referred to as the _____.
    A. grade
    B. stake
    C. benchmark
    D. ground level

11. The _____ is more accurate than the builder's level when lining up stakes, fence posts, poles, and roadways.
    A. laser level
    B. transit
    C. auto level
    D. plumb line

12. A _____ may often be the most practical way to check vertical planes and lines.
    A. plumb bob and line
    B. builder's level
    C. transit
    D. laser level

13. A total station differs from a transit in that it measures _____.
    A. height
    B. distance
    C. depth
    D. elevation

Apply and Analyze

1. Explain the process of establishing building lines.
2. Explain how to establish a square corner in a building line.
3. How do you level a builder's level to ensure accurate readings?
4. Explain how to take accurate readings with a leveling instrument when setting grade stakes for a footing or erecting batter boards.
5. When is it more convenient to use a plumb bob and line over a builder's level?

Critical Thinking

1. Why would it be wise to carefully consider the placement of a leveling instrument on the jobsite?
2. Explain what problems might result once a building is complete if the finished grade of a site is not sloped and contoured correctly.

Communicating about Carpentry

1. **Speaking.** Working in a group, brainstorm ideas for creating classroom tools (posters, flash cards, and/or games, for example) that will help your classmates learn and remember the different leveling tools. Choose the best idea(s), then delegate responsibilities to group members for constructing the tools and presenting the final products to the class.
2. **Speaking.** Pick 5–10 of the key terms that you practiced pronouncing. Write a brief scene in which those terms are used as you imagine them being used by carpenters in a real-life context. Then rewrite the dialogue using simpler sentences and transitions, as though an adult were describing the same scene to elementary or middle school students. Read both scenes to the class and ask for feedback on whether the two scenes were appropriate for their different audiences.

Foundations and Framing

Orange Line Media/Shutterstock.com

CHAPTER **9**

Footings and Foundations

OBJECTIVES

After studying this chapter, you will be able to:

- Lay out building lines and set up batter boards.
- Describe excavation procedures.
- Discuss the types of foundation systems used for residential buildings.
- Explain footing requirements and how to build footing forms.
- Describe the building, erecting, and use of forms for poured foundation walls.
- Describe concrete masonry units (CMUs) and how they are used.
- List steps and professional practices for laying up concrete block foundation walls.
- Explain foundation insulating and waterproofing procedures.
- Explain how slab-on-grade construction is used.
- Discuss design factors that apply to sidewalks and driveways.
- Explain how cold weather affects construction.
- Discuss protection against termites and other structure-destroying pests.
- Estimate concrete materials required for a specific area.

TECHNICAL TERMS

admixture
aggregate
anchor bolt
anchor strap
backfilling
batter board
buck
capillary action
cement
chair
closure block
concrete masonry unit (CMU)
control point
crawl space foundation
curing
dampproofing
fixed anchor

flat ICF wall
floating
form tie
grade beam foundation
head joint
header block
heaving
hydration
ledger board
lintel
mason's line
monolithic slab
mortar
nailing strip
permanent wood foundation
  (PWF)
pilaster
plain footing

post-and-beam ICF wall
rebar
reinforced footing
screeding
screen-grid ICF wall
slab foundation
slab-on-grade
spread footing
stepped footing
story pole
structurally supported
  slab
troweling
U-factor
waffle-grid ICF wall
waler
wall pocket
waterproofing

I n the construction of single-family dwellings and other structures, carpenters must work with many people in other construction trades. They also must work closely with an architect and owner in carrying out the total building plan.

On some jobs, carpenters may be required to lay out the building lines and supervise the excavation. They may also build forms for footings and poured foundation walls. Anyone working in the carpentry trade needs a working knowledge of standards and practices in concrete work. In this chapter, some of the material also deals with masonry. It is included because of the close relationship to carpentry.

9.1 Preparing the Site

Preparation of the building site may require removal of trees, boulders, and other obstructions, **Figure 9-1**. Grading may be needed before the building lines are laid out. This may require the placement of grade level stakes. Establishing grade properly is explained in Chapter 8, **Building Layout**. It usually requires the use of a transit or laser level.

Before any site work begins, check local ordinances regarding required permits. Many communities require environmental permits before the contour of the lot can be changed or trees removed. Environmental permits ensure that groundwater runoff is properly controlled and will not cause soil erosion or flooding of adjacent property. It is important that any groundwater not collect in pools or on neighboring property, and not run toward the building. Usually, an engineer or the owner obtains any environmental permits, but anyone changing the site should check to ensure the necessary permits have been secured.

If the property is wooded, use care in deciding which trees are to be removed. Much depends on the types of trees and where they are located. In general, evergreens should be used as protection against the cold winter winds. Deciduous (leaf-dropping) trees are best used as shade from the hot summer sun. Try to place the house to take advantage of the protection offered by trees already on the property. The architect who prepared the site drawings would normally specify where the house is to be located on the site and which trees should remain.

Trees that are to be taken down should be clearly marked so the crew responsible for their removal take the right ones. Avoid digging trenches through the root system of trees being retained because it could cause them to die. Likewise, trees usually will not tolerate more than a foot of additional fill over their root systems. The trees are cut down close to the ground and any usable logs are harvested. Then, the root systems of those trees are removed.

Once all unwanted trees have been removed, any useable topsoil is cleared and stockpiled for later reuse. If the topsoil contains many boulders or tree roots, it is probably not good enough to use for final grading, but it can be used where fill is needed to bring the site to the desired contour.

9.2 Laying Out Building Lines

After the site is prepared, the next step is to locate and mark lot lines. This must be done carefully and accurately. Errors in locating property boundaries can lead to locating the building too close to lot lines or even encroachment on adjoining property. To protect the owner and builder, establishing boundaries is best done by or with the help of a registered engineer or licensed surveyor, **Figure 9-2**. Such help may include

Goodheart-Willcox Publisher

Figure 9-1. Site preparation may involve removing trees and boulders.

Dmitry Kalinovsky/Shutterstock.com

Figure 9-2. Establishing property or lot lines is a task best left to an engineer or surveyor. This surveyor is using a total station for the job.

establishing building lines and grade levels, which can also be performed by a carpenter once property lines are located. Therefore, be familiar with the setback requirements of the authority having jurisdiction and the setbacks specified on the plot plan.

It is best to locate building lines with a builder's level, transit, or theodolite. Follow the procedures described in Chapter 8, **Building Layout**.

After locating all building lines, carefully check them. Measure their length and, even though they were laid out with an optical instrument, measure the diagonals of squares and rectangles, **Figure 9-3**. An out-of-square foundation causes problems throughout construction.

9.2.1 Batter Boards

Batter boards are temporary framework of stakes and ledger boards used to locate corners and building lines when laying out a foundation. They are set up well away from where each corner of the new building will be. Batter boards are usually built with 2×4s, but 2×6s or 2×2s may be used as well, **Figure 9-4**. It is important that the

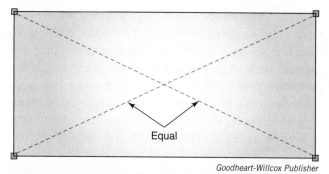

Goodheart-Willcox Publisher

Figure 9-3. The diagonals of a square or rectangle are equal. Always measure diagonals to be sure corners are square.

stakes be driven well into firm soil and that all nailing be tight and secure. If the batter boards move during construction, the building lines will be off. Add additional bracing to strengthen batter boards when necessary.

The horizontal boards are called *ledger boards*. Nail the ledger boards to the stakes. Check that they are level and at a convenient working height, preferably at or slightly above the top of the foundation wall. If practical, the batter boards should be level with each other.

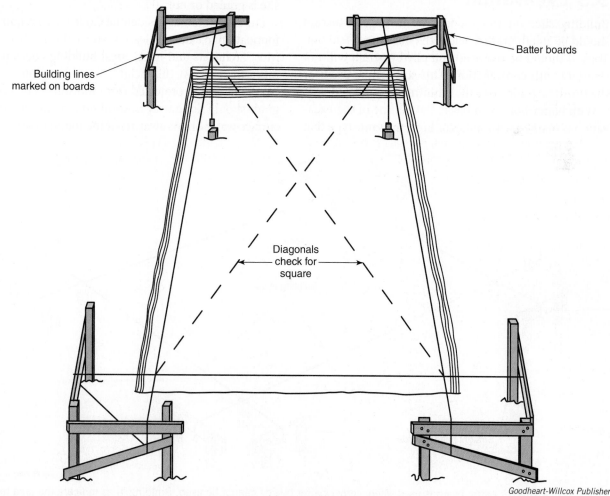

Goodheart-Willcox Publisher

Figure 9-4. Set up batter boards at least 4′ from building lines on all four corners. Allow more room if heavy equipment is likely to disturb boards. On loose soil, or when boards are more than 3′ off the ground, use braces.

Most excavations require equipment mobility and space to place the excavated soil. Therefore, batter boards must be set back far enough so the excavator does not disturb them. A garage with a slab-on-grade foundation might require 3′–4′ of clearance. A residence may need as much as 8′–10′ of clearance, and a commercial building much more. If it will be necessary to use heavy machinery between the batter boards and the building line, use straight batter boards set back farther from the corner. A straight batter board consists of two stakes and one ledger, with no corner. See **Figure 9-5**.

Using strong cord (such as mason's line) and a plumb bob or long level, locate the lines and verify they intersect over the layout stakes. Mark the tops of the ledger boards where the intersecting lines rest. Make a shallow saw kerf or drive a nail at this mark. Pull the lines tight and fasten them. If you are using saw kerfs, drive nails into the backs of the ledger boards to fasten the lines. You may prefer to wrap lines around the ledger several times, running them through the saw kerf.

9.3 Excavation

Building sites on steep slopes or in rugged terrain should be rough graded before the building is laid out. Topsoil should be removed and piled where it will not interfere with construction. This soil can be used for the finished grade after the building is completed.

With batter boards in place, the outline of the excavation is marked with stakes or lime, the same type that is used to mark a football field. The lines are then removed from the batter boards so they will not interfere with the excavation.

For regular basement foundations, the excavation should extend at least 2′ beyond the building lines to allow clearance for formwork. Foundations for structures with a slab floor or crawl space need little excavating beyond the trench for footings and walls.

The depth of the excavation can be calculated from a study of the vertical section views of the architectural plans. In cold climates, it is important to locate foundations below the frost line. Local building codes usually cover these requirements. If footings are set too shallow or are not insulated, moisture in the soil under the footing may freeze. This could force the foundation wall upward, a condition called *heaving*. Heaving can cause cracks and serious damage.

It is common practice to establish both the depth of the excavation and the height of the foundation by using the highest elevation on the perimeter of the excavation. This is known as the *control point*, or high point, **Figure 9-6**. This practice is followed whether the site is graded or not.

The International Residential Code (IRC) requires the foundation to extend at least 4″ above the finished grade. Most architects and many local building codes require more than 4″. At this height, wood framing and finish members are protected from moisture. The finished grade should slope away from all sides of the structure so surface water drains away from the foundation.

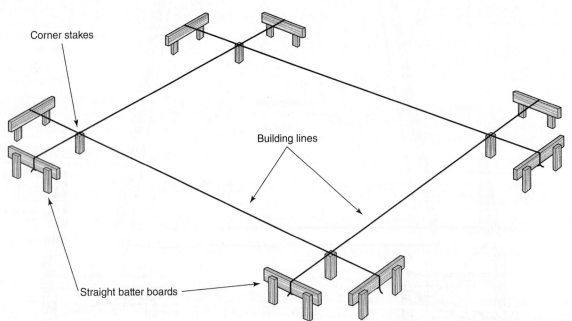

Goodheart-Willcox Publisher

Figure 9-5. A straight batter board is used where corner batter boards cannot be used. Building lines indicate the area to be excavated. The actual excavation should be at least 2′ outside of the building lines to provide working room for forming and placing the foundation.

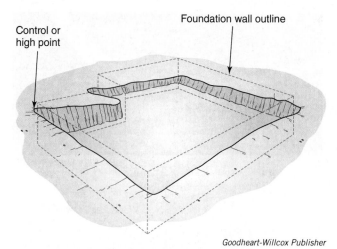

Control or high point

Foundation wall outline

Goodheart-Willcox Publisher

Figure 9-6. The control point is the highest elevation outside of the excavation and used to establish depth of excavation.

The depth of the excavation may also be affected by the elevation of the site. It may be higher or lower than the street or adjacent property. The level of sewer lines also has an effect. Normally, solving these problems is the responsibility of the architect. Information on grade, foundation, and floor levels is usually included in the construction drawings.

9.4 Foundation Systems

All structures settle. A properly designed and constructed foundation distributes the weight to the ground in such a way that the settling is negligible or at least uniform. **Figure 9-7** shows several foundation types in simplified form.

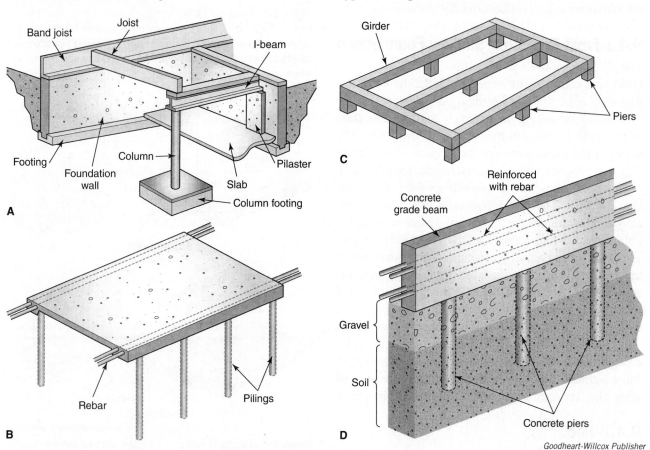

Goodheart-Willcox Publisher

Figure 9-7. Common foundation types. A—Spread footing with foundation wall. B—Slab-on-grade foundation. Piers (dashed lines) can be used as additional support in unstable soils. C—The pier-and-girder foundation is used in warm climates where water pipes and drainpipes are not in danger of freezing. D—Reinforced grade beam foundation.

For light construction, such as residential, *spread footing* is common. It transmits the load through the walls, pilasters, columns, or piers. These elements rest on a footing that transmits the load to the soil beneath. This type is often used when the owner needs a basement. *Crawl space foundations* are similar. The principal difference is that the foundation walls are not as high, usually only a few feet high.

The *slab-on-grade*, or ground-supported slab, is placed directly on the ground. This type of foundation is popular in warm climates because freezing of supply piping and drains is not a problem. It is also used for smaller structures, such as detached garages. Piling and girder foundations are also usually found in warm climates.

The *grade beam foundation* is used where soil has poor load-bearing qualities. This foundation is similar to the spread footing. It differs in that the footing is reinforced with *rebar* (steel reinforcing bars) and rests on concrete piers. A machine-operated auger can be used to dig the holes for the piers. In firm soil, concrete can be poured directly into the holes. In less stable or soft soil, round fiber forms are placed in the holes and filled with concrete. Rebar is placed in the holes or forms before the pour. This reinforcement extends above the ground and ties into other rebar laid horizontally in the beam.

9.4.1 Frost-Protected Shallow Foundations

The IRC allows the use of shallow foundations if the building is continuously heated and if the shallow foundation is frost-protected with rigid polystyrene. The method allows footings in extremely cold climates, such as in North Dakota, to be only 16″ below grade. The depth of the foundation and the amount of insulation required depends on the expected outside temperatures.

This method has been successfully applied in Scandinavian countries. The rigid insulation is placed both vertically and flat around the foundation perimeter along the slab edges and foundation corners. This insulation prevents frost heave even where footings do not extend below the frost line. Insulation is installed vertically to protect the edge of the foundation. Flashing is installed at the top of the foundation to prevent water from getting behind the insulation. Flashing is usually sheet copper or galvanized steel covering the top of the foundation and extending far enough outside the foundation to protect the insulation. More insulation, 12″–36″ wide, is placed flat around all sides. This is covered with several inches of soil, **Figure 9-8**.

9.4.2 Footings

Footings are used to support the load placed on the foundation and transfer this load to the soil beneath, **Figure 9-9**. Footings are especially helpful for preventing

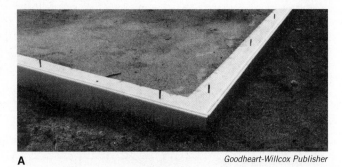

A   *Goodheart-Willcox Publisher*

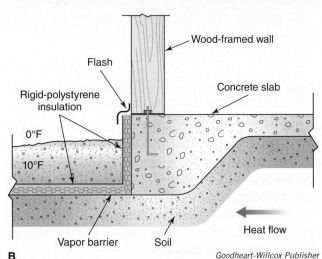

B   *Goodheart-Willcox Publisher*

Figure 9-8. Frost-protected shallow foundations. A—This type of foundation looks like any other except for the visible rigid insulation around the perimeter. B—Rigid polystyrene insulation placed around perimeter keeps the soil from freezing.

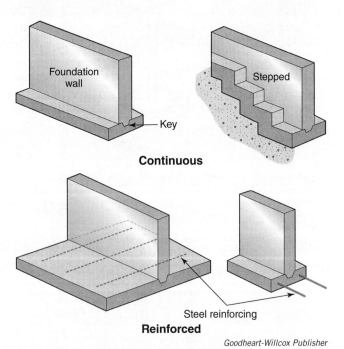

Goodheart-Willcox Publisher

Figure 9-9. Different types of footings are needed for different slope and soil conditions. Vertical runs of a stepped footing may not exceed 1/2 of the horizontal run between steps.

settling in areas with troublesome soil. *Plain footings* carry light loads and usually do not need reinforcing. *Reinforced footings* have steel rebar embedded in them for added strength against cracking. They are used when the load must be spread over a large area or bridged over a weak spot, such as excavations or sewer lines.

A *stepped footing* is one that changes grade levels at intervals to accommodate a sloping lot. Vertical sections should be at least 6″ thick. Horizontal sections between steps should be at least 2′ long. The height of a step should not be more than half the horizontal run. If masonry units are to be used over the footing, distances should fit standard brick or block modules.

Footings must be wide enough to spread the load over a sufficient area. Load-bearing capacities of soils vary considerably. See **Figure 9-10**. In residential and smaller building construction, a common practice is to make the footing twice as wide as the thickness of the foundation wall, **Figure 9-11**. It is typical for the footing to be as thick as the wall is wide. Footings that must support cast-in-place concrete walls may include a recess or groove forming a keyed joint. Steel dowels can also be installed to secure a poured wall. For larger structures, engineers take soil samples and determine the bearing capacity of the soil more precisely. While soil bearing capacity is important for residential and other light-frame structures, they do not place nearly the load on the soil as do larger structures with a lot of concrete. If the soil is high in clay or otherwise questionable, wise builders test soil samples even for smaller buildings. In such areas, the building code may require a soil analysis.

Footings under columns and posts carry heavy, concentrated loads and are usually from 2′ to 3′ square. The thickness should be about 1 1/2 times the distance from the face of the column to the edge of the footing. Reinforced footings are used in regions subject to earthquakes, and situations where the footings must extend over soils containing poor load-bearing material.

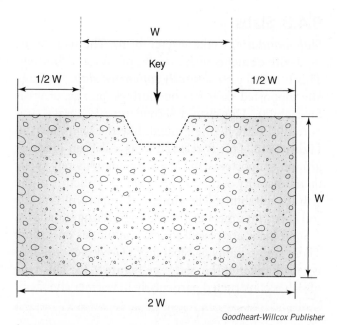

Goodheart-Willcox Publisher

Figure 9-11. Standard footing design for residential construction should be twice the width of the foundation wall that will rest on it. The thickness of the footing is typically the same as the wall width. Sometimes reinforcement is added. The key is designed to anchor the wall. It is often replaced with dowels of steel rebar.

Some structural designs may also require the use of reinforcement. The common practice is to use two No. 5 (5/8″) rebar for 12″ × 24″ footings. At least 3″ of concrete should cover the reinforcement at all points.

In a single-story dwelling where the chimney footings are independent of other footings, they should have a minimum projection of 4″ on each side. For a two-story house, chimney footings should have a minimum thickness of 12″ and a minimum projection of 6″ on each side. Exact dimensions will vary according to the weight of the chimney and the nature of the soil. Where chimneys are a part of outside walls or inside bearing walls, chimney footings should be constructed as part of the wall footing. Concrete for both chimney and wall footings should be placed at the same time.

Load-Bearing Capacities of Soil	
Soil Type	**Maximum Allowable Foundation Pressure (Tons per Sq Ft)**
Hard, sound rock	60
Soft, weathered rock	8
Medium gravel	6
Granular soils	3
Stiff clay	3
Medium density silt and silty soils	1.5

Goodheart-Willcox Publisher

Figure 9-10. The load-bearing capacity of different soil types varies considerably.

9.4.3 Slabs

Slab foundations take several forms. The slab can be used with elements such as walls, pilings, and footings. This is called a *structurally supported slab*. Structurally supported slabs rely on footings, pilings, or walls for stability. However, a second type is always supported directly by the ground, like the one shown in **Figure 9-12**. This is referred to as a slab-on-grade or ground-supported slab. These foundations rely solely on the stability of the soil base and the structure of the slab for durability. Therefore, it is important that the soil conditions meet applicable engineering standards.

Some slabs are placed in one continuous operation. There are no joints or separately placed sections. This is called a *monolithic slab*. See **Figure 9-13**. This type of slab is constructed particularly in warmer climates.

Pierce Construction, Ltd.

Figure 9-12. A slab-on-grade foundation "floats" on the soil under it. Reinforcing steel is often placed around the perimeter.

> **Safety Note**
> When using wire mesh for slab reinforcement, take precautions to prevent the ends of the unrolled mesh from springing up and causing injury. Secure both ends of the mesh or reverse the material so it curves downward as it is unrolled.

9.4.4 Forms for Footings

After the excavation is completed, carefully check the batter boards. They may have been disturbed by the excavating equipment. Make necessary adjustments before proceeding with layout of the footings and construction of the forms. Forms contain the concrete until it is cured. Replace the string lines on the batter boards. You need them to locate corners of the footing form. Drop a plumb bob from the intersection of the lines to the bottom of the excavation. Then proceed as follows.

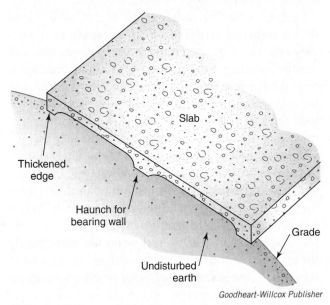

Goodheart-Willcox Publisher

Figure 9-13. A monolithic slab-on-grade often has thickened edges and may have interior thickened regions to support bearing walls.

PROCEDURE
Constructing Footing Forms

1. Drive stakes and establish points at the corners of the foundation walls.
2. Set up a laser level or a builder's level at a central point of the excavation. Drive a number of grade stakes level with the top of the footing along the footing line and at approximate points where column footings are required. Corner stakes can also be driven to the exact height of the top of the footing.
3. Connect the corner stakes with lines tied to nails in the top of the stakes.
4. Working from these building lines, construct the outside form for the footing. The form boards are located outside of the building lines by a distance equal to the footing extension beyond the building line. This varies according to the width of the

foundation wall. Footings should be twice the width of the wall, unless otherwise specified on the plans. For example, the footing for an 8″ foundation wall is usually 16″. Thus, it should extend 4″ on each side of the finished wall. The top edge of the form boards must be level with the grade stakes. See **Figure 9-14**.

5. With the outside form boards nailed in place, it is easy to set and level the inside form using a steel tape and a level. **Figure 9-15** illustrates this method.
6. Brace the forms as necessary. Wet concrete places a lot of force against the forms and it is difficult to readjust the forms after the concrete pushes them out of alignment.
7. When the footing forms are completely built, use a laser level or builder's level to recheck their height at several points.

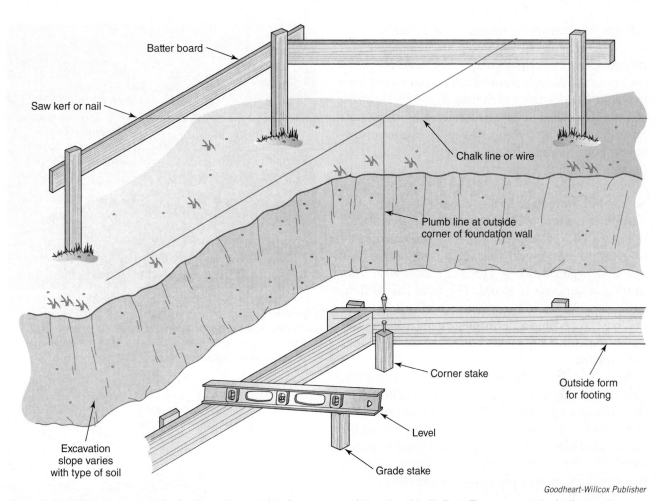

Batter board

Saw kerf or nail

Chalk line or wire

Plumb line at outside corner of foundation wall

Corner stake

Outside form for footing

Level

Grade stake

Excavation slope varies with type of soil

Goodheart-Willcox Publisher

Figure 9-14. Laying out forms for footings. The outside forms are positioned and built first. The corner stake indicates the building line. Footings should be twice the width of the wall. The top of the grade stake is the height of the footing. Place one end of a level on the grade stake and the other end on the outer form board. The form boards must be level with the grade stake.

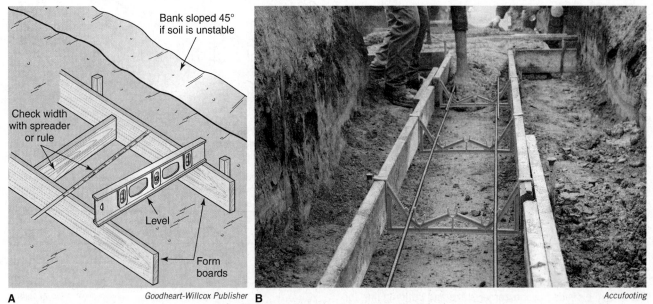

Bank sloped 45° if soil is unstable

Check width with spreader or rule

Level

Form boards

A   *Goodheart-Willcox Publisher* **B**   *Accufooting*

Figure 9-15. Setting forms. A—Measure the distance and set the inside form board. Use a rule and a level to space and level the inside form boards. B—Here, a plastic spreader is used to hold the forms in alignment. This type of spreader remains in the concrete permanently.

Normally, forms are constructed of 2″ lumber. Support stakes are placed at 4′ intervals. Temporary precut spreaders (also called spacers) may be held between the inside and outside form boards so you do not have to measure the distance each time, **Figure 9-16.** Strips of 1″ lumber called ties are nailed across the top of the form boards at intervals to strengthen the form during the pour. Remove the ties as you place the concrete. Bracing of footing forms may sometimes be desirable. Attach the brace to the top of a form stake and to the bottom of a brace stake located about 1′ away.

Another method of building footing forms is shown in **Figure 9-17.** Plastic channels with preformed metal spreaders can be quickly installed using only a level and an occasional stake to keep the hollow form from shifting as concrete is placed. The form has perforations along its outer edges. These perforations allow the form to become a permanent drain that removes water from around the foundation. The plastic form is not removed.

Stepped footings require some additional formwork. Vertical blocking must be nailed to the form to contain the concrete until it sets. See **Figure 9-18.** Be careful not to install any form members that will be trapped after the concrete has been placed. The soil should be able to contain the concrete at these points. All footings, except in rare situations, must be placed on undisturbed soil.

Goodheart-Willcox Publisher

Figure 9-17. Form-a-Drain® is a patented plastic footing form that also serves as a drain to collect and carry away water. The form remains in place as part of the foundation. Preformed metal spreaders also act as ties.

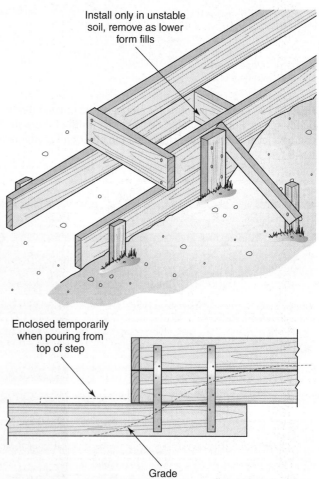

Goodheart-Willcox Publisher

Figure 9-18. Form constructed for a vertical section in a stepped footing. The lower level is placed first and usually allowed to set up slightly before the step is poured.

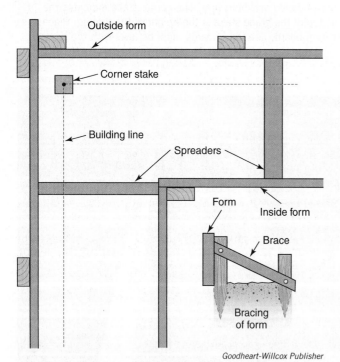

Goodheart-Willcox Publisher

Figure 9-16. Properly located footing form. Note that the outer form extends beyond corner stake and building line for a distance determined by the thickness of the foundation wall. Ties may be fastened across the tops of the forms to keep the forms from spreading.

Column footings are pads of concrete designed to support columns by distributing the load to the ground. They carry weight transmitted to the column from the beams, stringers, and joists of a building, **Figure 9-19**.

Forms for column footings are usually set after the wall footing forms are complete. These are located by taking measurements from the building lines. Forms are leveled with a leveling instrument or to previously set grade level stakes.

Some hand digging and leveling of the excavation will probably be necessary as forms are set. The top of the footing must be level. The bottom may vary as long as the minimum thickness is maintained.

Proper location of nails is necessary to avoid problems in form removal. It is important that they be only temporarily nailed (always from the outside) to stakes and to each other. Duplex (double-headed) nails may be used.

When the forms are completed, check for sturdiness and accuracy. Remove the line, line stakes, and grade stakes.

If the plans call for rebar in the footings, it should be put in place after the forms are constructed. Rebar should be tied with tie wire where it crosses another piece and where another piece is added to extend its

Goodheart-Willcox Publisher

Figure 9-20. Rebar is tied together with steel tie wire. It should be held off the earth at the bottom by rebar chairs.

length, **Figure 9-20**. The rebar should be supported off the ground, so it will be completely buried in the concrete. There are many styles of supports, often called chairs, for this purpose. If the rebar is to be turned up to extend into the foundation wall to be formed later, make sure it is placed where it will fall in the middle of that wall. With the forms built and checked for alignment and the rebar in place, the concrete can now be placed.

9.5 Erecting Wall Forms

Many different types of wall forming systems are available. There are certain basic considerations that should be understood and applied to all systems. For quality work, the forms must be tight, smooth, defect free, and properly aligned. Joints between form boards or panels should be tight. This prevents the cement paste from being lost, which tends to weaken the concrete and cause honeycombing.

Wall forms must be strong and well-braced to resist the side pressure created by the plastic concrete. This pressure increases tremendously as the height of the wall is increased. Regular concrete weighs about 150 lb per cu ft. If it is immediately placed into a form 8′ high, it creates a pressure of about 1200 lb per sq ft along the bottom edges of the form. In practice, this

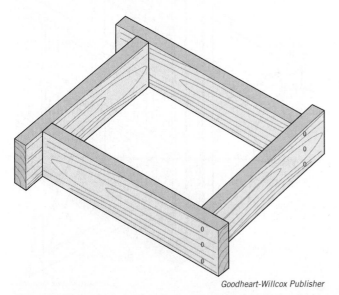

Goodheart-Willcox Publisher

Figure 9-19. Column footing form constructed of 2× lumber. There is no need to cut the pieces to length.

pressure is reduced through compaction and hardening of the concrete. It tends to support itself. Thus, the lateral pressure is related to these factors:

- The amount of concrete placed per hour
- The outside temperature
- The amount of mechanical vibration

9.5.1 Carpenter-Built Forms

Before setting up the outside foundation wall form, you will need to find and mark the building line on top of the footing. To do this, set up your lines on the batter boards once more. Then drop a plumb line from the intersections (corners) of the building lines to the footing. Mark the corners on the footing. Snap a chalk line from corner to corner on the footing. This line serves as a guide, marking the outside face of the foundation wall. As you set up the foundation forms, align the inner face of the outside form with the chalk line, **Figure 9-21**.

Low wall forms, up to about 3′ in height, can be assembled from 3/4″ plywood supported by 2×4 studs spaced 2′ apart. The height can be increased somewhat if the studs are closer together. See **Figure 9-22A**.

For walls over 4′ in height, the studs should be backed with *walers*, horizontal stiffeners added to provide greater strength to the form, **Figure 9-22B**. Walers are usually doubled 2×4s with space between them to run the ties. Spacing of the walers depends on the amount of pressure the concrete exerts on the form.

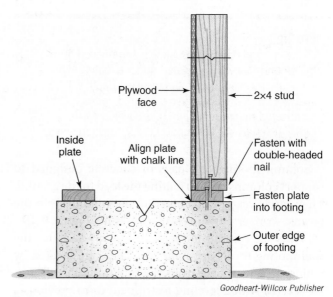

Figure 9-21. One method of locating and fastening an outside wall form to the footing. The plywood face of the form must align with the building line.

9.5.2 Form Hardware

Wire ties and wooden spreaders have been largely replaced with various manufactured devices. **Figure 9-23** shows three types of form ties. The rods go through small holes in the sheathing and studs. The tie holds the two sides of the form together at the right distance. Form ties, also called snap ties, are designed so that a portion of the tie rod remains in the wall.

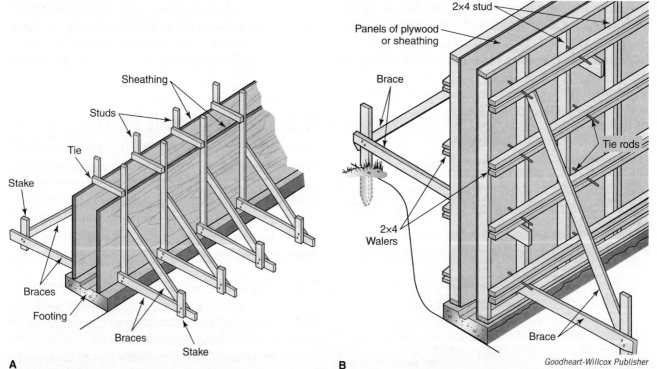

Figure 9-22. Wall forms. A—Design for wall forms up to 3′ high. Use plywood sheathing and space studs 2′ apart. B—Prefabricated wooden panels for a wall form. For walls over 4′ high, walers provide greater strength to the form. Add bracing to the top and bottom.

Wire Ties

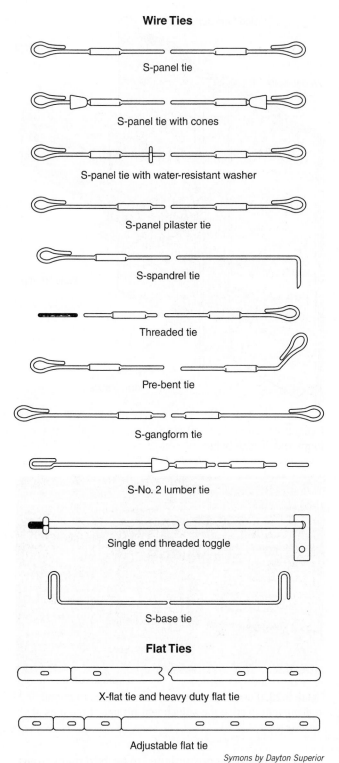

S-panel tie

S-panel tie with cones

S-panel tie with water-resistant washer

S-panel pilaster tie

S-spandrel tie

Threaded tie

Pre-bent tie

S-gangform tie

S-No. 2 lumber tie

Single end threaded toggle

S-base tie

Flat Ties

X-flat tie and heavy duty flat tie

Adjustable flat tie

Symons by Dayton Superior

Figure 9-23. There are many types of ties to fit every forming situation.

They are called snap ties because the exposed portions are snapped off when the forms are removed. To secure the corners, the walers are extended far enough beyond the corner to allow the use of vertical strongbacks. The strongbacks are nailed to the walers to keep the corner tight, **Figure 9-24**.

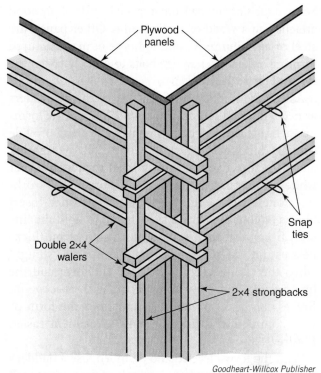

Goodheart-Willcox Publisher

Figure 9-24. Corners of wall forms must be carefully fastened so they can withstand the pressure from the poured concrete.

After the concrete has set, the clamps can be quickly removed and the forms stripped away. To break off the outer sections of the snap tie rod, a special wrench is used. The rod breaks at a small indentation located about 1″ beneath the concrete surface. The hole in the concrete is later patched with grout or mortar.

9.5.3 Panel Forms

Carpenters may build their own wooden panel forms using 3/4″ plywood and 2×4 studs to form 2′ × 8′ or 4′ × 8′ units. However, prefabricated panels are now used for most poured concrete wall forms, **Figure 9-25**.

Symons by Dayton Superior

Figure 9-25. Most wall forming is now done with manufactured forms that are quickly assembled and dismantled.

Some panels are made from a special grade of plywood attached to a wood or metal frame. Other panels are steel or aluminum. A common type of manufactured forming system uses wedge bolts to quickly fasten the forms, **Figure 9-26**.

As you are erecting the forms, be sure to add walers as needed for stiffening, **Figure 9-27**. Brace the forms on each side as needed to straighten them and provide support. Straightening can be done easily by sighting along the top edge of the form from corner to corner while another carpenter secures the bracing. If working alone, stretch a line from corner to corner. Insert blocks of equal thickness at each end to hold the line away from the form. Use a test block of the same thickness as the wall is straightened. The form is straight when the test block slips between the form and the line. Secure the form with a brace at each checkpoint. Braces can be 2×4s attached directly to the form or with commercial fittings to make adjustment easier, **Figure 9-28**.

With the forms leveled and plumbed, use a builder's level or laser level to mark the inside of the form at each corner indicating the level of the finished wall. Strike a chalk line between these marks all the way around the form. Using small nails, tack a level strip above the line so the worker placing the concrete can readily see the desired level.

Pro Tip
Carefully recheck formwork before placing the concrete. A form that fails as the concrete is placed and vibrated wastes material and causes extra work.

Symons by Dayton Superior

Figure 9-26. One common type of manufactured form system uses wedge bolts to tighten and hold the walers and strongbacks.

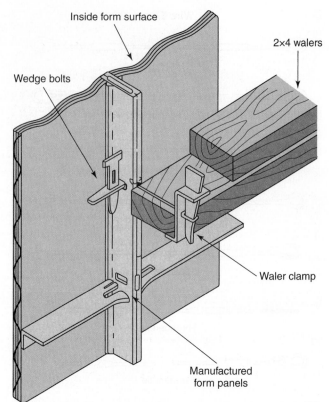

Symons by Dayton Superior

Figure 9-27. Here, 2×4 walers are tightened against the forms with wedge bolts.

Goodheart-Willcox Publisher

Figure 9-28. Forms must be braced to keep them plumb and in alignment. Notice the adjustment fittings at the tops of these braces.

Since panel forms are designed to be used many times, they should be treated to prevent the concrete from sticking to the surfaces. Use special form release coatings that are commercially available, **Figure 9-29**. After each removal, thoroughly clean form components. Carefully sort and stack them for the next job or storage.

Manufacturers have developed many forming systems to replace or supplement panel forms built by a carpenter. The panel units are light for easy handling and transporting from one building site to another.

Figure 9-29. Forms should be coated with a release agent before use. This reduces "worming" and surface voids in the concrete. After each use, forms should be carefully cleaned with a compound designed for that purpose.

Western Forms

Specially designed devices are used to quickly and accurately assemble and space the components. For residential work, these systems may be made from aluminum or from steel frames and exterior grade plywood panels. Lightweight polystyrene forms are designed to remain as insulation for the structure. Sometimes plywood forms are coated with a special plastic material to create a smooth finish on the concrete and prevent it from sticking to the surface.

Green Note

Form release agents with a petroleum or solvent base can cause health and environmental problems. Emission of volatile chemicals (VOCs) into the air can cause odor or health problems among workers. Water-based release agents can prevent concrete from sticking to wood forms without releasing VOCs into the air. Commercial water-based products are made from natural oils.

9.5.4 Curved Wall Forms

Forms for curved concrete walls can be made by bending plywood or other sheet material to the required radius. Sawing vertical kerfs on the inside of the form material may be needed to bend the sheet. Provide extra support along the curved form.

9.5.5 Wall Openings

Several procedures are followed to form openings in foundation walls for doors, windows, and other voids. In poured walls, frames called *bucks* are built and fastened into the regular forms. Beveled keys or *nailing strips* may be attached to the buck and cast into the concrete. Frames are then secured to these strips after the

bucks are removed. **Figure 9-30** shows two methods of framing openings. Use duplex nails to fasten the buck to the inside face of the form wall. Drive duplex nails through the outside of the form wall into the buck.

Special framing must also be attached inside the form for pipes or voids for carrying beams. As with windows and doors, formwork for these structures must be attached to either the inside or outside wall form before the other form is erected.

Tubes of fiber, plastic, or metal can be used for small openings. Hold them in place with wood blocks or plastic fasteners attached to the form. Larger forms made of wood can be attached with duplex nails driven through the form from the outside.

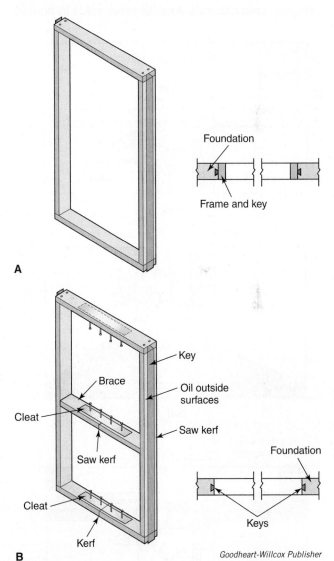

Goodheart-Willcox Publisher

Figure 9-30. Wood forms are used to frame openings in foundation walls. Carpenters often construct them onsite. A—Permanent buck that will be left in the wall. B—This frame is designed to be removed. Members are cut partway through for easy removal. Cleats and braces reinforce the members at saw cuts. Bucks and frames are nailed to the form with duplex nails driven from the outside.

In concrete block construction, door and window frames are first set in place. The masonry units are constructed around the sides. The outside surface of the frames has grooves into which the mortar flows, forming a key. Basement windows are often located level with the top of the foundation wall. **Figure 9-31** shows details of a basement window installation.

9.5.6 Forming a Wall Pocket

A *wall pocket* is a void in a concrete wall used to receive a large framing member that supports a floor. Normally, a pocket is required to receive a steel or wood beam. It should be large enough to provide 1/2″ of clearance on the sides and end so the beam can move without damaging the concrete wall. As with other voids in poured concrete walls, a buck must be installed on the inside form. Never allow the pocket to go all the way through the wall. This exposes the beam to weather and also causes air leakage. **Figure 9-32** shows installation of a buck on the face of the inside form. It should be held in place by duplex nails driven through the wooden form.

9.5.7 Pilasters

Long walls usually have pilasters. A *pilaster* is a thickened section of a concrete or masonry wall that strengthens the wall or provides extra support for beams. **Figure 9-33** shows a form setup for pouring a pilaster into a concrete foundation wall. Pilasters in foundation walls are placed on the inside of the wall, so they can resist the pressure of soil on the outside.

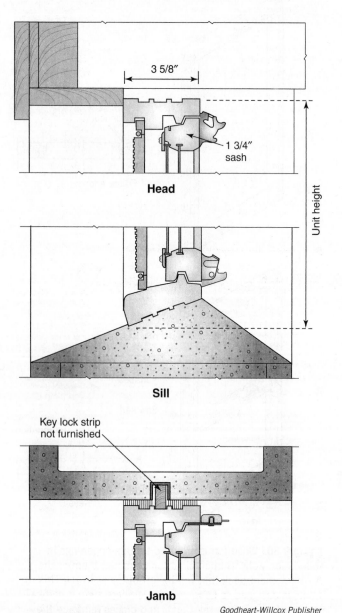

Goodheart-Willcox Publisher

Figure 9-31. Detail of basement window unit that can be placed in a concrete or masonry foundation wall.

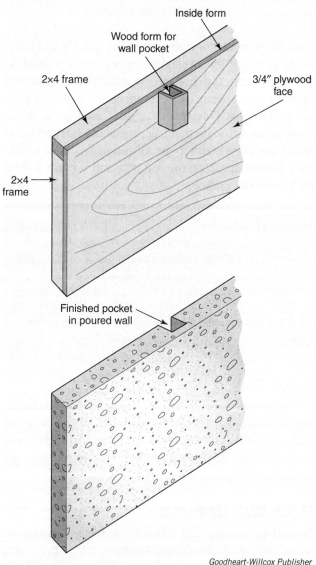

Goodheart-Willcox Publisher

Figure 9-32. Forming for a wall pocket. Attach the pocket form to the face of the inside wall form. Drive duplex nails through from the exterior so they can be removed when the concrete has cured.

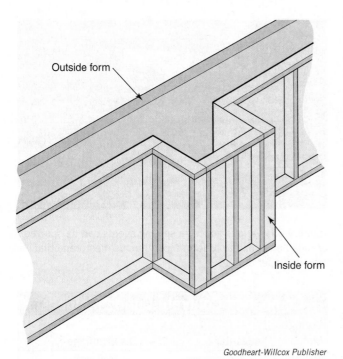

Goodheart-Willcox Publisher

Figure 9-33. Inside section of a foundation form setup for a pilaster. Often, the studs at the sides are omitted.

9.5.8 Column Forms

For standard square or rectangular columns, prefabricated forms save time. Those shown in **Figure 9-34A** and **9-34B** are quickly assembled and set up on a column footing. To prevent blowout, brace the bottom of the form.

Tubular fiber or paper forms for piers may also be used to save time. These are available in diameters from 6″ to 48″. For residential construction, sizes 12″ and 16″ are most often used. **Figure 9-34C** shows a pier formed using a fiber tube.

9.5.9 Beam Forms

Concrete beams and slabs that do not rest on the ground require forms beneath the concrete. Temporary shoring is constructed to support the bottom and sides of the forms. The shoring is removed when the forms are removed.

9.6 Concrete

Concrete is made by mixing carefully measured amounts of these items:

- Cement
- Fine aggregate (sand)
- Coarse aggregate (gravel or crushed stone)
- Water

First, the cement and aggregate are mixed dry. Then, water is added, causing a chemical change called *hydration* to take place. Hydration, not drying, causes hardening of the mixture.

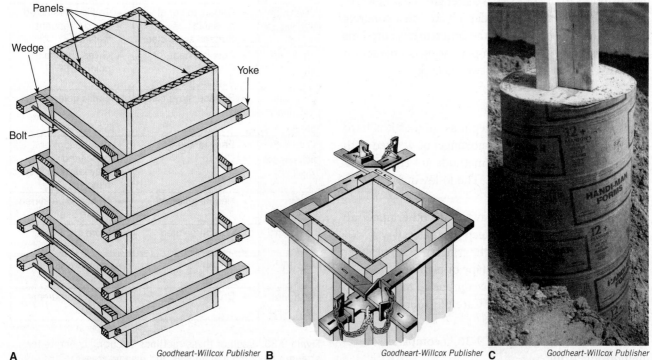

A *Goodheart-Willcox Publisher* **B**   *Goodheart-Willcox Publisher* **C**   *Goodheart-Willcox Publisher*

Figure 9-34. Prefabricated column forms save construction time. A—Yoke and wedge arrangement. B—Scissor clamps are adjustable. C—Fiber tube forms are easy to use below grade, but may not be rigid enough to stay straight when they are unsupported above grade.

9.6.1 Effect of Improper Amount of Water in Mix

The proper amount of water is important to the strength of the concrete. Generally, the ratio should be no more than 1/2 lb of water to 1 lb of cement. The moisture in the sand must always be a consideration in fixing the ratio.

Too much water in the mix will cause weak concrete. Excess water will rise to the surface and sit there. This is known as bleeding. Even though the water eventually evaporates, the concrete surface may craze, dust, and scale. Also, when the water evaporates it leaves pockets where water can later enter and freeze. With too little water in the mix, hydration will not take place. With too little mixing, water will not blend into the mix. Some ingredients will separate.

9.6.2 Proper Curing

Fresh concrete should be kept moist during the initial stage of hydration so it can properly cure. The curing process can go on for years after placement. However, most of the strength of the concrete develops during the first week or two. If allowed to dry out too soon, concrete may reach only 40% of its full strength.

Proper *curing* of concrete prevents loss of its moisture for a period of time. At seven days, with moisture, it reaches 50% of its eventual strength. After a month of curing, this strength is double. The strength of concrete usually refers to its compressive strength. That is, its resistance to being crushed under a load. Most concrete strength tests are done when the concrete has cured for 28 days. Another benefit of proper curing is a reduction in shrinkage, which also reduces cracking.

9.6.3 Curing Methods

Curing of the concrete can begin as soon as it is hard enough that the concrete can no longer be worked with a trowel. The easiest curing methods are flooding or keeping a sprinkler running. The following are other methods that are used:

- Covering with burlap, straw, or any other material that will hold water and keep the material wet.
- In hot weather, covering the concrete with plastic and sealing the laps with tape or planking.
- Use of a liquid-membrane compound to seal in moisture, **Figure 9-35**. Carefully follow the manufacturer's directions.

Refer to the table in **Figure 9-36**. It compares several methods of curing.

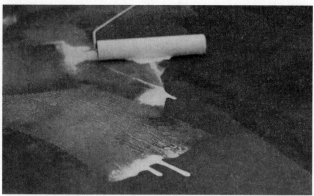

GCapture/Shutterstock.com

Figure 9-35. Various concrete sealing agents can be sprayed or rolled on concrete to form a continuous membrane that aids in proper curing.

Comparing Concrete Curing Methods		
Curing Method	Advantage	Disadvantage
Ponding	• Inexpensive • Needs little supervision • Surface stays clean • Good buffer for temperature • Controls quality best	• May leak • Poor method for cold weather • Traffic is restricted
Sprinkling/fogging	• Surface stays clean	• Involves more work/cost • Poor method for cold weather • Traffic is restricted
Cover with wet burlap	• Cheap; material reusable • Surface stays clean	• Traffic is restricted • Needs frequent wetting
Black/grey mix	• Absorbs heat • Cheap • Asphalt tile slicks	• Gives surface a color
Cover with waterproof paper	• Surface stays clean	• Traffic deteriorates it
Cover with polyethylene sheet	• Easy to use • Cheap	• Can be torn or punctured • May leave surface blotchy
White pigmented	• Heat is reflected	• Needs to be stirred before use
Clear resin base	• Can be painted	• More expensive
Wax resin base	• Cheap	• Leaves a gummy residue
Clear wax base	• Cheapest	• Leaves a gummy residue

Goodheart-Willcox Publisher

Figure 9-36. Each of these methods of curing concrete for strength has both advantages and disadvantages.

9.6.4 Slump Testing

Sometimes a slump test is made to assess the consistency, stiffness, and workability of fresh concrete. The amount of slump may be influenced by the amount of water in the mix. More water means more slump. However, there are other factors that affect slump:

- Type of aggregrate
- Admixtures
- Air in the mix
- Ambient temperature
- Proportions of the mix
- Mixing time
- Standing time

The test is made with a sheet metal cone 4″ in diameter at the top, 8″ in diameter at the bottom, and 12″ high.

9.6.5 Giving Concrete Tensile Strength

The compressive strength of concrete is high, but its tensile strength (stretching, bending, or twisting) is relatively low. For this reason, concrete used for beams, columns, and girders must be reinforced with steel rebar. Rebar is sized by numbers 3–18. For numbers 3–8, each number represents 1/8″ in diameter. So, a No. 3 rebar is 3/8″ and a No. 8 rebar is 1″. Bar sizes 9–18 do not follow the 1/8″ per number rule. They are based on the cross sectional area of the square bars that were

used to form them. A number 18 bar is actually just over 2 1/4″ in diameter. These larger sizes are rarely used in building construction. No. 4 rebar may be used in a footing to prevent settling at one or more locations. When concrete must resist compression forces only, reinforcement is usually not necessary. Slabs for floors or similar applications are often reinforced with welded wire fabric, **Figure 9-37**. The concrete can also be reinforced with high-strength fibers being added to the mix, **Figure 9-38**. These fibers yield nearly as much tensile strength as welded wire fabric and save time and money.

Goodheart-Willcox Publisher

Figure 9-38. Fibers to reinforce concrete may be glass, polypropylene, or nylon.

Welded Wire Fabric Reinforcement for Concrete Structures		
Type of Construction	**Recommended Style**	**Remarks**
Barbecue foundation slab	6×6-8/8 4×4-6/6	Use heavier style fabric for heavy, massive fireplaces or barbecue pits.
Basement floor	6×6-10/10 6×6-8/8 6×6-6/6	For small areas (15-foot maximum side dimension), use 6×6-10/10. As a general rule, the larger the area or the poorer the subsoil, the heavier the gauge.
Driveway	6×6-6/6	Continuous reinforcement between 25- to 30-foot contraction joints.
Foundation slab (residential only)	6×6-10/10	Use heavier gauge over poorly drained subsoil, or when maximum dimension is greater than 15 feet.
Garage floor	6×6-6/6	Position at midpoint of 5- or 6-inch thick slab.
Patio or terrace	6×6-10/10	Use 6×6-8/8 if subsoil is poorly drained.
Porch floor • 6-inch thick slab up to 6-foot span • 6-inch thick slab up to 8-foot span	6×6-6/6 4×4-4/4	Position 1 inch from bottom form to resist tensile stresses.
Sidewalk	6×6-10/10 6×6-8/8	Use heavier gauge over poorly drained subsoil. Construct 25- to 30-foot slabs as for driveways.
Steps (free span)	6×6-6/6	Use heavier gauge if more than five risers. Position fabric 1 inch from bottom form.
Steps (on ground)	6×6-8/8	Use 6×6-6/6 for unstable subsoil.

Goodheart-Willcox Publisher

Figure 9-37. Recommended styles of welded wire fabric reinforcement for concrete structures.

PROCEDURE
Making a Slump Test
Figure 9-39 shows the steps involved in making a slump test.

1. Secure a slump cone, wet it, and place it upright on a solid level base. The large-diameter end should be down.
2. Fill the cone with fresh concrete in three layers, each equal to about one-third of the volume of the cone. Rod (tamp) each layer 25 times before adding the next layer. See **Figure 9-39A**.
3. Strike off the top, then slowly and evenly remove the cone. See **Figure 9-39B**. This should take 5–12 seconds. Avoid jarring the mixture or tilting the cone during this step.
4. Turn the cone, small end down, and place it next to the concrete.
5. Lay the tamping rod across the top of the cone so it extends over the concrete. See **Figure 9-39C**.
6. Use a ruler to measure the amount of slump.

9.6.6 Cement

The "glue" that holds concrete together is Portland *cement*. It is manufactured from limestone mixed with shale, clay, or marl. Each sack of Portland cement holds 94 lb. This is equal to one cubic foot in volume. Cement should be a free-flowing powder. If it contains lumps that cannot be easily pulverized between thumb and fingers, it should not be used because it may alter the rate of hydration, which, in turn, will affect the rate of strength gain. Important properties of cement include a strong binding power, high hydration capability, and its admixture compatibility.

9.6.7 Aggregates

Aggregate may consist of sand, crushed stone, gravel, or lightweight materials such as expanded slag, clay, or shale. The large, coarse aggregate forms the basic structure of the concrete. This is seldom over 1 1/2″ in diameter. The voids between these particles are filled with smaller particles. The voids between these smaller particles are filled with still smaller particles, and so on. Together, they form a dense mass. With the recent interest in green construction and conservation of materials, old concrete is sometimes crushed and used as a coarse aggregate.

9.6.8 Admixtures

Modern concrete almost always contains additives to give the concrete certain properties. These additives are called *admixtures*. The most common admixtures are fillers added to decrease the cost of the mix without decreasing its strength. Hardening agents are added to

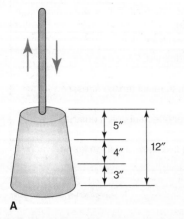

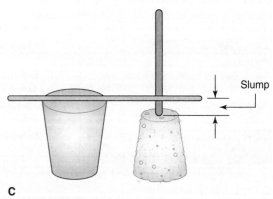

A   B   C

Goodheart-Willcox Publisher

Figure 9-39. Method of performing a slump test. It should be done whenever the consistency of the concrete is critical to the job. Testing can be performed at the jobsite or at the ready-mix plant.

make the concrete extra hard. Air-entraining agents introduce tiny air bubbles into the mix. These air bubbles act like tiny shock absorbers to relieve the stresses caused by freezing. Coloring agents may be added to produce concrete of a desired shade. Retarding admixtures cause the concrete to set more slowly.

9.7 Ordering Concrete

If there is a local ready-mix supplier, it is generally more convenient to have concrete delivered rather than mixed onsite. The mix is purchased by the cubic yard (27 cu ft). A minimum order is usually 1 cu yd, but there is an extra charge for any order under several yards. The amount varies from one supplier to the next. Ingredients are carefully measured, often in automated plants. Mixing takes place on the truck in route to the jobsite.

The plant personnel can help in selecting the right mix for the builder's needs. The mix proportions can be adjusted to give the strength needed for the type and location of the construction. Both climate and usage have a bearing on these proportions. If a specific slump is to be delivered, that information should be included in the order.

When placing an order, a carpenter should tell the plant how the concrete will be used (such as a footing, wall, slab-on-grade, or sidewalk). The amount needed should also be specified, as well as when it must be delivered. An order should be for about 5% more than the calculated amount to allow for consolidation, spillage, and form movement.

Since concrete must be placed as soon as possible after mixing, builders and suppliers must work together to avoid delays. Before the concrete truck arrives, forms should be checked for accuracy. The crew must be onsite, equipped with the proper tools, and ready to work the concrete into the forms.

Have a plan for what will be done with any leftover concrete. If the job will include outdoor fireplaces, masonry signs, or retaining walls, build the forms for their footings ahead of time, so extra concrete can be placed in them.

9.8 Placing Concrete

Usually, concrete can be placed directly into the forms from the ready-mix truck. To move the concrete to other areas not accessible to the truck, a wheelbarrow or a pump is generally used, **Figure 9-40**. When placing concrete, follow these general guidelines:

- Place concrete as near as possible to where it will rest. Never allow it to run or be worked over long horizontal distances.
- Never allow concrete to drop further than 4'. Doing so, or working over a long distance, could cause segregation. This is a condition in which large aggregates become separated from the cement paste and smaller aggregates.
- Promptly place concrete in forms after mixing.
- Concrete for walls should be placed in the forms in horizontal layers of uniform thickness not exceeding 1'–2'. As the concrete is placed, vibrate it enough to thoroughly compact the concrete. This produces a dense mass.
- Working the concrete next to the form tends to produce a smooth surface. It prevents honeycombing along the form faces. A spade or thin board may be used for this purpose. Large aggregates are forced away from the forms and any air trapped along the form face is released. Mechanical vibrators are effective in consolidating concrete.
- Mechanical vibrators also create added pressure on the forms. This factor must be considered in the form design. The vibrator should not be held

A   *Iomiso/Shutterstock.com*   B   *Bannafarsai_Stock/Shutterstock.com*   C   *Lakeview Images/Shutterstock.com*

Figure 9-40. Proper methods must be used to move concrete from the ready-mix truck to where it is to be placed. A—The spout from the ready-mix truck should be used wherever it will reach. B—Where accessibility makes other methods impossible, concrete can be placed with a concrete pump. C—For horizontal transport, a wheelbarrow can be useful.

in one location long enough to draw a pool of cement paste from the surrounding concrete.

- When pouring slabs, work in strips 4′–6′ wide. Avoid concentrating water at the ends, in corners, and along the face of the form.
- Use a baffle on slopes to prevent segregation and collecting of aggregate at the bottom of the slope. Better still, start placing concrete at the bottom and work uphill.

Safety Note

The chemical reaction that causes concrete to cure generates heat. If concrete is left in contact with bare skin for an extended time, it can cause serious chemical burns. Therefore, always wear safety glasses, rubber boots, and gloves when working with concrete.

9.8.1 Anchors

Wood plates are fastened to the top of foundation walls with 1/2″ *anchor bolts* or *anchor straps*. They are spaced not more than 4′ apart. See **Figure 9-41**. In concrete walls, anchors are set in place as soon as the pour is completed and leveled off.

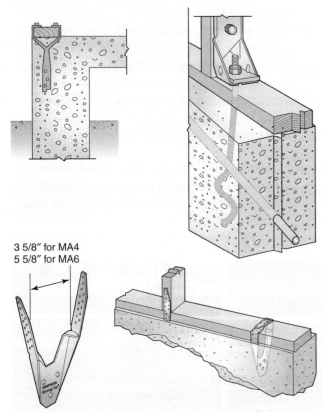

3 5/8″ for MA4
5 5/8″ for MA6

The Panel Clip Co.; ©Simpson Strong-Tie Company, Inc.

Figure 9-41. Patented anchor bolts and straps are embedded in the concrete. They secure the sills to the foundation.

9.8.2 Removing Forms

Wall forms help protect concrete from drying too fast. They should not be removed until the concrete is strong enough to carry the loads that will be placed on it. The material should be hard enough so the surface is not damaged by the stripping operation. Sufficient hardening of concrete will normally take a day or two.

9.9 Concrete Block Foundations

In some localities, *concrete masonry units (CMUs)*, commonly called concrete blocks, are used for foundation walls and other masonry construction. The standard CMU is made from a mixture of Portland cement and aggregates such as sand, fine gravel, or crushed stone. It weighs from 40 to 50 lb.

Lightweight units are made from Portland cement and natural or manufactured aggregates. Among these aggregates are volcanic cinders, pumice, and foundry slag. A lightweight unit weighs 25–35 lb. It usually has a much lower U-factor than a standard block. The *U-factor* is a measurement of the heat flow or heat transmission through materials. As a result, a lightweight CMU wall will have better insulating properties than a standard CMU wall.

Blocks should comply with specifications provided by the American Society for Testing Materials. ASTM specifications for a Grade A, load-bearing unit requires that the compression strength equal 1000 lb per square inch. Thus, an 8×8×16 block must withstand about 128,000 lb, or 64 tons.

9.9.1 Sizes and Shapes

CMUs are classified as solid or hollow. A solid unit is one in which the core or cell area (hollow portion) is 25% or less of the total cross-sectional area. Blocks are usually available in 4″, 6″, 8″, 10″, and 12″ widths and 4″ and 8″ heights. **Figure 9-42** shows some of the sizes and shapes. Sizes are actually 3/8″ shorter than their nominal (named) dimensions to allow for the mortar joint. For example: the 8×8×16 block is actually 7 5/8″ × 7 5/8″ × 15 5/8″. With a standard 3/8″ mortar joint, the laid-in-the-wall height is 8″ and the length 16″.

It is important to limit the moisture content of concrete block. Cover the blocks stockpiled at the jobsite to protect them from the elements. They should be secured on pallets or other supports to keep them off the ground. Never wet blocks before or while laying a wall.

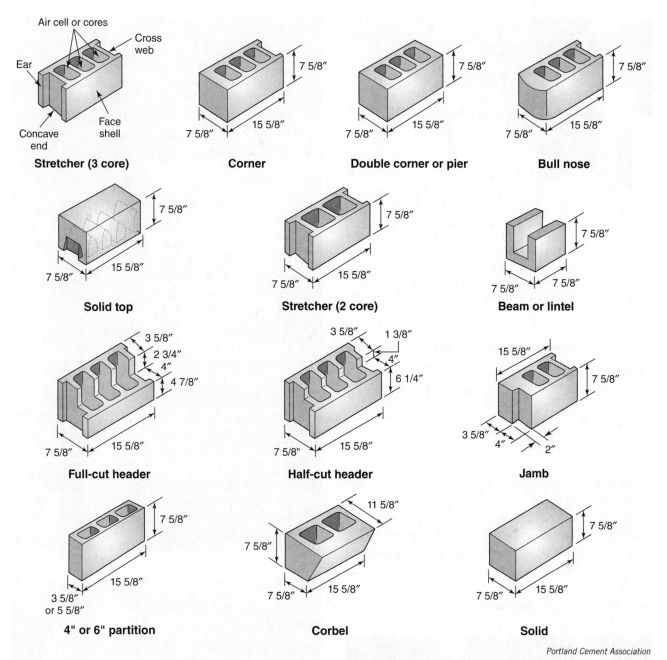

Figure 9-42. Concrete blocks are manufactured for many different purposes. These shapes are typical. Sizes are specified in this order: width, height, and length. Sizes are nominal, allowing for mortar joints.

Portland Cement Association

9.9.2 Mortar

To build a wall, blocks are bonded one to another with *mortar*. This is a mixture of Portland cement, lime, and masonry sand. These materials give the mortar its strength, ability to retain water, durability, and bonding capabilities.

When water is added just before use, the mortar retains its plastic properties long enough for a mason to lay up the block. Mortar can be made to resist different pressures, depending on where it is to be used. ASTM has set standards for different uses. Mortar used in foundations must withstand pressures from 1800 psi to 2500 psi when set. Mortar should be used within 1 1/2–2 hours after it has been mixed.

9.9.3 Laying Concrete Block

Foundation block should be laid on a good footing or concrete slab. A mason or carpenter first locates the corners on the footing and snaps a chalk line on all sides. This marks the building line and the outer edge of the blocks. The first course is then strung out dry (without mortar) to check the layout, as shown in **Figure 9-43**.

Portland Cement Association

Figure 9-43. After marking the building line on the footing with a chalk line, a mason always lays out the first course of block dry. This allows for adjustments that minimize cutting of block.

Pro Tip

The color of chalk in your chalk line, or chalk box, denotes its permanency. When snapping lines on a footer, use red chalk. Red is considered permanent. Blue chalk is less permanent, but should not be used on finish surfaces. For chalk that wipes away freely, use white or orange.

The next step is to lay down a mortar bed on the footing for the first course of blocks, **Figure 9-44**. Corner blocks are laid first. The *head joint* is the end of the block that will butt against the previously laid block. These are "buttered" with mortar before being laid.

The corners are usually built up first from four to six courses high, **Figure 9-45**. All corner blocks must be at the right height, in line, level, and plumb.

A

Goodheart-Willcox Publisher

Portland Cement Association

Figure 9-44. A full bed of mortar 1 1/2″ thick is laid down on the footing or slab for the first course of block.

B

Portland Cement Association

Figure 9-45. Checking for plumb and level. A—All corners are first laid up 4–6 courses high and carefully checked for plumb. B—A mason carefully checks for level as each course is laid.

A   *Portland Cement Association*

B   *Portland Cement Association*

Figure 9-47. Laying block. A—The head joints of the blocks are buttered before the blocks are laid. Blocks should be lightly bumped against the previously laid block for a good joint. B—A mason's line is used as a guide for laying block between corners. It maintains a straight line of block for each course.

Goodheart-Willcox Publisher

Figure 9-46. A story pole can be created from a straight piece of lumber and used to check the height of each course.

A mason uses a level to check this after laying every three or four blocks. Sometimes, a ***story pole*** is used to check the height of each course, **Figure 9-46**.

Each course at the corners is stepped back one-half of a block from the previous course. Lay the level on an angle across the ends of each unit. All the edges of the last block in each course should fall along the straight line formed by the level. If not, the corner must be rebuilt.

With all four corners built up, laying the wall between corners can begin, **Figure 9-47**. Note how the blocks are buttered before being laid. A ***mason's line*** stretched at each course acts as a reference for proper vertical and horizontal alignment. A mason's line is attached to the corner by a line holder. This can be a line block, line pin, or line stretcher.

As each block is laid, be careful that it does not touch the line. A mason's level is now used only to check the face of each block to keep it lined up with the face of the wall. Use the edge of the trowel to cut off mortar squeezed out of the joint. Use a lifting motion to capture the excess with the trowel. Do not allow it to fall off the trowel or smear the wall.

Some building codes require wire reinforcement between every third or fourth course of block. **Figure 9-48** shows wire reinforcement for masonry.

Hohmann and Barnard, Inc.

Figure 9-48. Wire reinforcement is embedded in the mortar between courses of blocks.

Story pole with mark at each course height

The last block to be placed in any course is called a *closure block*. Both ends of this block are buttered before the block is carefully placed, **Figure 9-49**.

Header blocks are concrete blocks that are dimensioned and shaped to form a shelf in a wall. They may be used when brick or stone veneer will be used as the exterior wall finish. **Figure 9-50** shows a more common method of providing a ledge to support brick veneer siding. Here, a mason places standard 8″ block atop 12″ block. The ledge that is formed easily supports a veneer facing of 4″ masonry.

Block foundation walls should be topped with anchors to secure the sill, **Figure 9-51**. These should be spaced 6′ on center and within 12″ of the end of any sill piece, or as specified on the plans and specifications.

Goodheart-Willcox Publisher

Figure 9-50. Forming a ledge by laying a standard 8×8×16 block on top of a 12″ wide block. The ledge that is formed will support a stone or brick veneer facing.

Before the top two courses are laid, put down a strip of metal lath wide enough to cover the core spaces. After the top course has been laid, fill the cores with mortar and install the anchor bolts a minimum of 7″ deep. Alternatively, anchor bolts can be placed in cores filled for vertical reinforcement. Anchor clips are installed in the same way. Anchors of either type should be about 18″ long. For areas with termite problems (discussed later in this chapter), top courses of stretcher block should be solid blocks or have all voids filled with concrete. This also helps distribute the load from floor joists.

Portland Cement Association

Figure 9-49. The closure block is buttered at both ends and then carefully placed in the wall.

> **Code Note**
> Section R404 of the International Residential Code requires vertical reinforcement in most CMU walls. Common practice is to fill cores with masonry grout and rebar every 48″, 56″, or 72″, depending on the width of the wall and the height of backfill. Some walls require even more reinforcement than this. Always check local codes to ensure compliance.

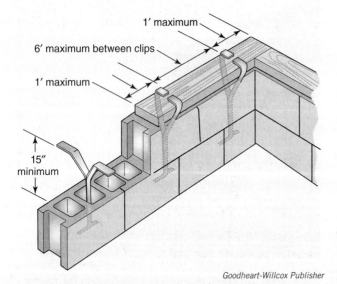

Goodheart-Willcox Publisher

Goodheart-Willcox Publisher

Figure 9-51. Install anchor straps or bolts in block walls to secure the sills. Always embed anchors in the second-from-top course. Place screening over voids beneath the second-from-top course to prevent mortar from falling away from the anchors.

9.9.4 Anchoring Intersecting Walls

Concrete block walls that intersect at 90° should use some type of connector to bind the intersecting walls together. Anchoring should be installed at every sixth course. There are several anchoring devices available, **Figure 9-52**. One is a metal strap known as a *fixed anchor*. The anchor ends are bent in opposite directions. It is placed in the wall between courses. One end projects upward into the hollow core of a block in the next course. The other end extends downward into a core from the intersecting wall. Both cores are then filled with mortar, concrete, or grout before the next course is laid. Note: Place metal lath or wire mesh across the corresponding cores in the course below to contain the mortar.

Hardware cloth or joint reinforcement is another device often used to connect two intersecting walls. It is also installed between courses and embedded in the mortar.

Another method uses a 1/4″ diameter tie with a rounded open loop on each end. The ends are hooked around embedded rebar.

9.9.5 Lintels

A structural unit called a *lintel* supports masonry located above openings for doors and windows, **Figure 9-53**. A lintel can be one of the following:

- A precast concrete unit that includes metal reinforcing bars (rebar).
- Standard blocks supported by steel angle iron.
- A course of lintel blocks across the opening, supported by a frame. Reinforcing bars are added and the blocks are filled with concrete.

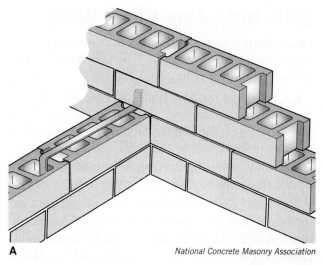

A *National Concrete Masonry Association*

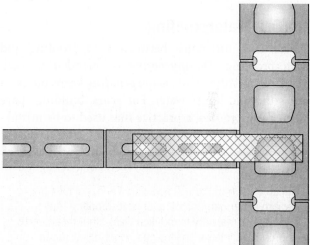

B *National Concrete Masonry Association*

Figure 9-52. Connecting intersecting block walls. A—Intersecting walls are often tied together with an anchor strap. B—Another method of tying together intersecting walls involves the use of hardware cloth.

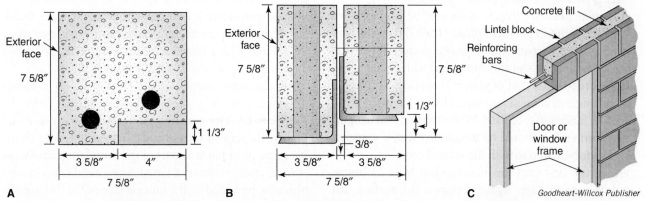

A   **B**   **C**   *Goodheart-Willcox Publisher*

Figure 9-53. Lintel construction methods. A—This is a cutaway view of a precast solid lintel for a block wall. Note the reinforcing rebar (black dots). B—This is a cutaway view of lintel made up of standard block units supported by steel angle iron. Notches allow space for installing window or door frames. C—Open-sided lintel blocks allow adding rebar and concrete. Blocks are laid over a supporting frame.

9.10 Insulating Block Walls

In northern climates, insulating the foundation walls may be necessary to maintain a warm basement. There are several ways to reduce the heat flow.

- Lightweight masonry units that have a lower U-factor may be used
- Adhering rigid insulation sheets to the exterior surface
- Cavity wall construction or the use of various forms of insulation applied to the interior or exterior surface

Building codes in some states do not permit the use of interior insulation on concrete or masonry foundation walls. Additional information about insulation is included in Chapter 18, **Thermal and Sound Insulation**.

9.10.1 Waterproofing

There is a difference between dampproofing and waterproofing. *Dampproofing* is intended to keep out soil moisture, while *waterproofing* keeps out both moisture and liquid water. For years, buildings have been dampproofed, a practice that used to be mistakenly referred to as waterproofing.

Code Note

Section R406 of the IRC specifies the conditions that require either dampproofing or waterproofing. Any concrete or masonry foundation walls "that retain earth and enclose interior spaces and floors below grade shall be dampproofed from the top of the footing to the finished grade." The IRC then provides a list of the permissible materials, which include bituminous coating and acrylic-modified cement. Waterproofing is only required by the IRC "in areas where a high water table or other severe soil-water conditions are known to exist."

Drain tile installed at the footing to take away ground water is part of the dampproofing system.

Concrete masonry (block) walls may be dampproofed by an application of cement plaster (called parging), followed by several coats of an asphaltic material. The wall surface should be clean and dampened with a water spray just before the first coat of plaster is applied. The plaster can be made of cement and sand (1 to 2 1/2 mix by volume), or mortar may be used. When the first coat has partially hardened, it should be roughened with a scratcher to provide better bond with the second coat.

After the first coat has hardened at least 24 hours, apply a second coat. Again, dampen the surface just before applying the plaster. Both coats should extend from about 6″ above the finished grade to the footing.

A cove of plaster should be formed between the footing and wall. This precaution prevents water from collecting and seeping through the joint. The second coat should be kept damp for at least 48 hours.

Where there is a high water table, causing the foundation to be subjected to excessive amounts of ground water, the foundation must be waterproofed from the top of the footing to the level of the finished grade. The IRC accepts eight materials for waterproofing:

- Two-ply hot-mopped felts
- Fifty-five-pound roll roofing
- Six-mil polyvinyl chloride
- Six-mil polyethylene
- Forty-mil polymer-modified asphalt
- Sixty-mil flexible polymer cement
- One-eighth-inch cement-based, fiber-reinforced, waterproof coating
- Sixty-mil solvent-free, liquid-applied synthetic rubber

Roll roofing, polyvinyl chloride, and polyethylene are applied by troweling a bituminous product directly on the concrete blocks. Then, while the coating is still wet, adhering the material to it. The others are applied directly to the cement parging.

As new building products are developed, they may be substituted for approved products or systems; however, these products must meet the standard set by the building code and must be approved for use. Peel-and-stick membranes, spray on membranes, and drain boards such as the one shown in **Figure 9-54** can replace or be added to a waterproofing or damp proofing system to improve performance.

Perimeter drains are installed alongside a footing. This drainage system carries water away from the foundation to an outlet that always remains open. In some localities, perimeter drains may be connected to a sump pump installed in the basement floor. The drain is usually installed at a slope of about 1″ in 20′. Perforated piping is lightweight and delivered in coils. The pipe is laid on a bed of at least 2″ of crushed stone or washed gravel and is covered with 6″ of the same material. Either the pipe or the crushed stone or gravel must be covered with a filter membrane to prevent soil from clogging the perforated pipe, **Figure 9-55**.

An alternate system called Form-a-Drain® builds drainage into a patented footing form. Refer to **Figure 9-17**. The system uses a plastic channel 4″–6″ high with perforations at regular intervals. The channel remains in place as part of the foundation drainage.

As an energy conservation measure, rigid foam insulation may be added to the outside face of the foundation wall. In such cases, no cement plastering or waterproofing should extend any higher than 2″ below the final grade. More information on this type of insulation is found in Chapter 18, **Thermal and Sound Insulation**.

The drainage plane provides a capillary break between the soil and the waterproofing membrane

Waterproofing

Water is given a clear path away from the foundation through gravel media and a perforated drainage pipe

A   *lusia599/Shutterstock.com*   B   *Dörken Systems Inc*

Figure 9-54. Waterproofing a foundation wall. A—A dimpled drainage plane can be installed over the membrane to give water an easy path of drainage. B—This section view shows a method of waterproofing basements that is especially effective in very wet soils.

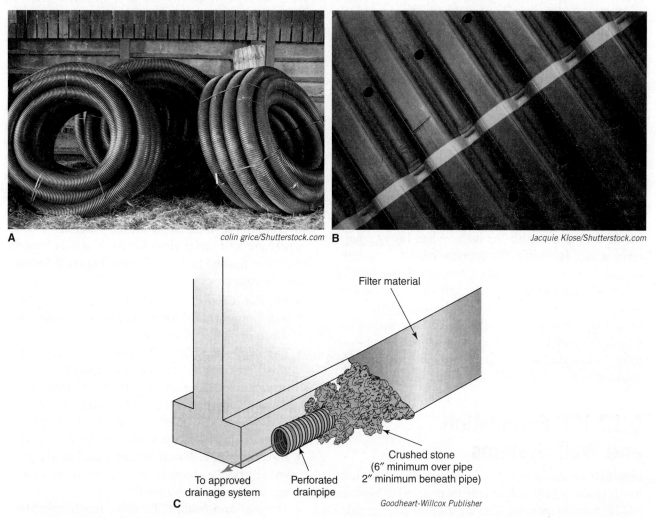

A   *colin grice/Shutterstock.com*   B   *Jacquie Klose/Shutterstock.com*

Filter material

To approved drainage system

Perforated drainpipe

Crushed stone (6″ minimum over pipe 2″ minimum beneath pipe)

C

Goodheart-Willcox Publisher

Figure 9-55. Installing footing drains. A—Perforated plastic drain tubing is delivered to the site in large coils. B—Small holes or slits in the pipe allow groundwater to enter the pipe. C—The openings in the drainpipe must be protected from being plugged with soil by filter material.

9.11 Backfilling

After the foundation has cured and waterproofing has been completed, the excavation outside the walls needs to be refilled with earth. This is known as *backfilling*. If not done properly, it can cause serious foundation problems. There are four important elements to consider when backfilling for the job to be done correctly:

- Protecting the new foundation from damage
- Using clean material (no debris)
- Compacting (tamping) the backfill
- Sloping the final grading to keep water away from the foundation

Since backfilling places considerable inward pressure on the walls of the foundation, the first floor framing and rough flooring are usually installed first. This is the best insurance against damage. However, many builders want to backfill as soon as the wall is cured. Doing so speeds up framing, because the foundation is more accessible. In such cases, bracing the walls becomes important. This is especially critical with concrete block foundations. If the foundation wall has windows that will be below grade, steel areaways must be installed to create window wells.

Backfill material should be equal to the amount that was taken out during excavation. However, if that soil is poor—such as heavy, moisture-laden clay—granular material should be substituted. Never bury construction lumber scraps or any other organic material against the foundation. It will only cause trouble later with uneven settling as the organic material decays. Also, such material may attract termites.

If a perimeter drain system has been installed, use care not to disturb or crush the lines. When heavy equipment is used to backfill, the operator must also be careful not to damage the walls. Compacting the backfill avoids problems caused by later settling. If the backfill is not compacted and settles later, the depression in the grade may cause water to collect and puddle against the foundation wall. Final grading should slope the soil away from the foundation. The finish grade should fall a minimum of 6″ for the first 10′ away from the foundation.

9.12 ICF Foundation and Wall Systems

Insulated concrete form (ICF) is a system that uses forms made of light polystyrene foam or other types of rigid insulating material. ICFs are used for both foundation walls and exterior walls of a building. Once forms are erected and braced, concrete is poured into them. The forms are left in place, resulting in a structural wall that is insulated on both sides. Depending on the specific system, it also provides surfaces for a carpenter to install both exterior and interior wall coverings.

> **Green Note**
> ICF foundations and even ICF above-grade walls are popular for green homes. Compared with conventional walls, ICF walls provide greater resistance to heat transfer. Thus, the use of ICFs reduce the amount of energy required for heating and cooling throughout the life of the structure. In addition, ICF construction produces less waste than that produced by conventional form construction.

9.12.1 Code Restrictions for ICFs

Builders planning to use an ICF foundation should first check the local building code. Some codes prohibit the use of ICF systems under certain conditions. The ban concerns areas of the country where termite infestation is heavy. The International Residential Code places restrictions on the use of ICFs below grade in heavily infested areas. Some exemptions apply. Again, carpenters should check local codes for details.

9.12.2 Types of ICFs

There are three basic types of ICFs, and each has different dimensions. One consists of insulating planks 8″–12″ high × 4′–8′ wide. These forms have interlocking ties and notches for stacking. A second shape has panels ranging from 1′ × 8′ to 4′ × 12′. The panels are held together and spaced with ties. Individual panels are glued or wired together. The third shape is a block ranging from 8″ × 16″ to 16″ × 4′. The blocks are hollowed out to provide a cavity for the concrete.

There are four categories of ICFs, based on how the concrete is formed inside the form. **Figure 9-56** illustrates the four types:

- The *flat ICF wall* system forms a solid concrete wall of uniform thickness. The form is made up of sheets or planks of insulating material.
- The *waffle-grid ICF wall* system forms a concrete web. Vertical parts of this web are spaced a maximum of 12″ apart OC (on center). Horizontal members are spaced at most 16″ apart OC. The thicker vertical members and thinner horizontal webs are similar in appearance to a breakfast waffle.
- A *screen-grid ICF wall* system is similar to the waffle-grid. Its appearance is nearer to that of a window screen than a waffle.
- The *post-and-beam ICF wall* system has vertical and horizontal concrete members spaced more than 12″ apart OC. This system looks more like a concrete frame than a monolithic concrete wall.

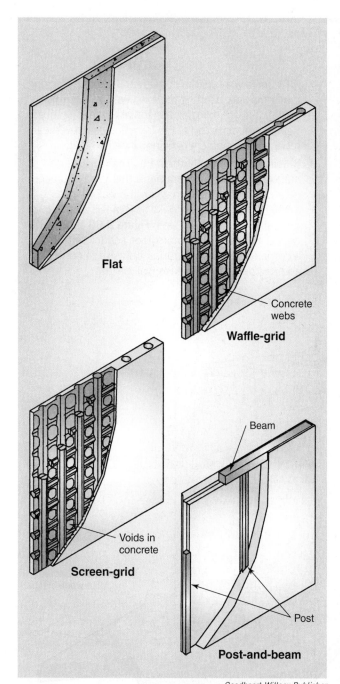

Flat

Concrete webs

Waffle-grid

Beam

Voids in concrete

Screen-grid

Post

Post-and-beam

Goodheart-Willcox Publisher

Figure 9-56. Insulated concrete forms (ICFs) are a lightweight alternative to conventional forms. They fall into the four general types shown.

Block-type ICFs are usually transported on pallets, **Figure 9-57.** Either the masons or carpenters will install them after the footings are placed and cured.

Pro Tip

Always follow the more detailed installation manual provided by the manufacturer of the ICF system being used. Procedures for various brands can differ. Check local codes for conformance, as well.

Goodheart-Willcox Publisher

Figure 9-57. ICFs are usually flat-packed when shipped to the construction site. This saves space and protects them from damage during shipping and handling. They are easily folded open for installation.

9.12.3 Placing the Concrete

Any method can be used to place concrete in the insulating form walls. For practical reasons, a concrete pump is often used. The height of the wall often makes other methods impractical. See **Figure 9-58**.

The concrete receives a "moist cure" because hydration continues over a longer period due to the insulating properties of the form. This increases the concrete's strength and reduces cracks and foundation leaks.

Normally, the delivery hose should be held horizontal over the form so the concrete falls naturally. If the wall is higher than 12′, it is advisable to make pours in stages. Place reinforcing steel between the cold seams of the pours. Some manufacturers warn that making single pours higher than 4′ may cause the insulating form to fail.

Lou Oates/Shutterstock.com

Figure 9-58. The use of a pump or conveyor is often the best choice for concrete placement in an ICF system.

PROCEDURE

Installing the Typical Flat ICF

Each manufacturer of ICFs supplies an installation manual for its specific system. The following steps are typical for setting up flat systems.

1. Footings should be wide enough to receive the forms with sufficient clearance to install 2×4 cleats along the outside edge, **Figure 9-59**. Securely nail cleats to the footing. This may be done with a powder-actuated tool. There should also be enough room on the inside edge for scrap 2×4s. These should be placed 4′–6′ apart. Footings should be keyed in the same way as in any other foundation or wall system.

2. Insert spreaders, called **form ties**, in every course to keep the width of the wall cavity uniform, unless the ICF comes with built-in spreaders. Continue stacking the forms and inserting the form ties to the proper height. If needed, install horizontal reinforcing rebar. Rest the rebar on top of the form ties. Insert vertical rebar as the concrete is being placed.

3. Bracing is extremely important. The ICF manufacturer may recommend that vertical braces be anchored to the form when the wall reaches a height of 4′. These should be 8′ apart along the outside of the form wall. **Figure 9-60** shows a typical system that also provides scaffolding where a carpenter can stand during pours or to continue adding courses to increase wall height.

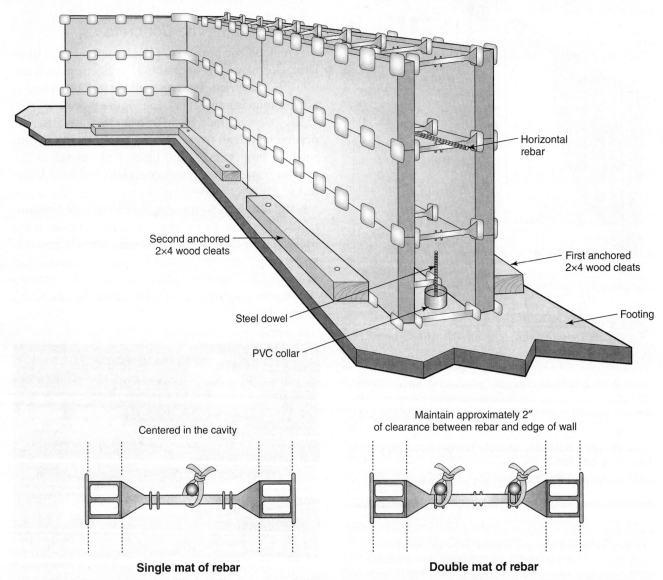

Horizontal rebar

Second anchored 2×4 wood cleats

First anchored 2×4 wood cleats

Footing

Steel dowel

PVC collar

Centered in the cavity

Single mat of rebar

Maintain approximately 2″ of clearance between rebar and edge of wall

Double mat of rebar

Lite-Form International

Figure 9-59. ICF planks are anchored by cleats fastened to footings. Note the placement of spacer ties and rebar.

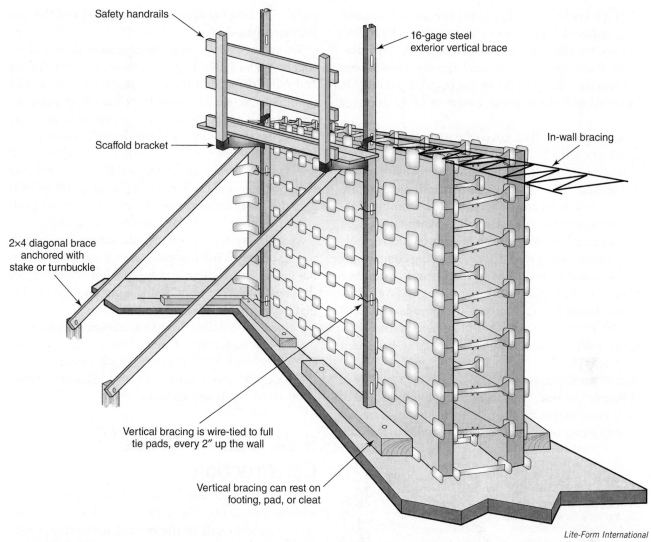

Safety handrails

16-gage steel exterior vertical brace

Scaffold bracket

In-wall bracing

2×4 diagonal brace anchored with stake or turnbuckle

Vertical bracing is wire-tied to full tie pads, every 2″ up the wall

Vertical bracing can rest on footing, pad, or cleat

Lite-Form International

Figure 9-60. Bracing wall forms. B—At 4′ high, the assembled insulating forms must be braced as shown. Place vertical and angled braces 8′ apart along one side of the form.

9.13 Wood Foundations

The *permanent wood foundation (PWF)* is a special building system that saves time because it can be installed in almost any weather, **Figure 9-61**. Moreover, it provides comfortable living space in basement areas because the stud walls can be fully insulated. All wood parts are pressure-treated with a solution of chemicals that makes the fibers useless as a food for insects and eliminates fungi that cause damage and decay.

Foundation sections of 2″ lumber and exterior plywood can be panelized in fabricating plants or constructed on the building site. Pressure-treated wood foundations have been approved by major code groups, including the IRC, and accepted by the US Federal Housing Authority (FHA), US Housing and Urban Development (HUD), and the Farmers Home Administration.

Southern Forest Products Association

Figure 9-61. When installing a wood foundation, all components are pressure-treated against rot and insect damage. Outside wall panels located below grade are waterproofed and protected by a moisture barrier and drainage system.

As for a conventional basement, the site is excavated to the required depth (below the frost line). Plumbing lines are installed and provisions are made for foundation drainage, following local requirements. Some soils require a sump (a pit for water collection) that is connected to a storm sewer, pump, or other drainage system.

The subgrade is then covered with a 4″–6″ layer of porous gravel or crushed stone and carefully leveled. Footing plates of 2×8 material are directly installed on this base. The wall sections of the foundation are then erected and securely fastened to the footing plate with noncorrosive fasteners. All fasteners should be made of stainless steel, silicon bronze, copper, or hot-dipped zinc-coated steel. Special caulking compounds must be used to seal all joints between sections and between footing plates and sole plates. Before pouring the basement floor, cover the porous gravel or crushed stone base with a 6-mil polyethylene film, **Figure 9-62**. Use a screed board after the basement concrete is placed.

Install the first floor frame on the double top plate of the foundation wall. Pay special attention to method of attachment, so that inward forces will be transferred to the floor structure. Where joists run parallel to the wall,

install blocking between the outside joist and the first interior joist.

Before backfilling, pour the basement floor and install first floor joists. This is necessary to avoid shifting and other damage to the foundation. Do not backfill until the concrete basement floor has cured. Also, before backfilling, apply a 6-mil polyethylene moisture barrier to sections of the outside wall below grade. Bond the top edge of the barrier to the wall at grade level with a special adhesive. Install a treated wood strip over this and caulk it. Later, the strip will serve as a guide for backfilling. Lap vertical joints in the polyethylene film at least 6″. Seal joints with adhesive.

As with any foundation system, satisfactory performance demands full compliance with standards covering design, fabrication, and installation. Standards for the construction of wood foundations are contained in manuals prepared by several associations.

Carpenters installing wood foundations should make certain that each piece of treated lumber and plywood carries the mark "AWPB-FDN." This assures them that the materials meet requirements of code organizations and federal regulatory agencies.

9.14 Slab-on-Grade Construction

Many commercial and residential structures are built without basements. The main floor is formed by placing concrete directly on the ground, referred to as slab-on-grade construction. Footings and foundations are similar to those for basements. However, they need to extend downward only to solid soil below the frost line. See **Figure 9-63**.

In slab-on-grade construction, insulation and moisture control are essential. The earth under the floor is called the subgrade and must be firm and completely free of sod, roots, and debris.

A coarse fill at least 4″ thick is placed over the finished subgrade. The fill should be brought to grade and thoroughly compacted. This granular fill may be slag, gravel, or crushed stone, preferably ranging from 1/2″ to 1″ in diameter. The material should be uniform, without fines, to ensure maximum air space in the fill. Air spaces add to the insulating qualities and reduce capillary attraction of subsoil moisture. *Capillary action* is the action by which moisture passes through fill or other building materials. Often, capillary action occurs against the force of gravity. In areas where the subsoil is not well drained, a line of drain tile may be required around the outside edge of the exterior wall footings.

Plywood may overlap field-applied
top plate for shear transfer
(flashing not required
if siding overlaps)

8″ minimum

Blocking

Face grain
direction

Floor joist

Caulk

Polyethylene
film

Floor joist

Cover plate

Optional
interior finish

4″ concrete slab* over
vapor barrier over
gravel, coarse sand,
or crushed rock

Excavated
sump pit

To storm sewer
or daylight

*A wood basement floor system is available; write APA for details

APA-The Engineered Wood Association

Figure 9-62. A cutaway of a typical wood foundation. Note the drainage sump. It keeps the gravel base and subsoil dry around the foundation.

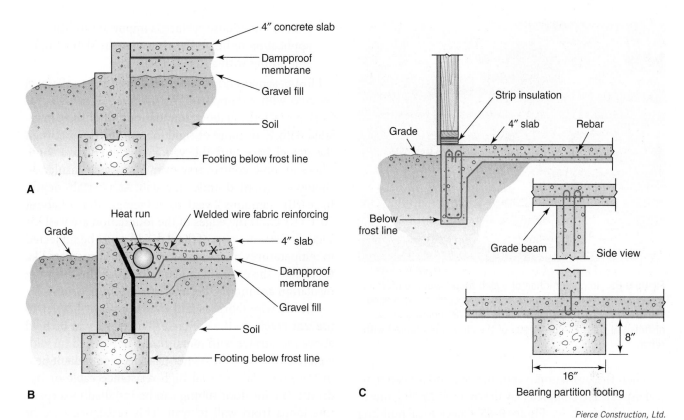

A

B

C

Bearing partition footing

Pierce Construction, Ltd.

Figure 9-63. Types of slab-on-grade. A—Unreinforced slab supported only by undisturbed soil beneath it. Used where soil is coarse and well drained. B—Slab is reinforced with welded wire fabric. The inside of the foundation wall is insulated because of the perimeter heat duct included in the slab. C—Monolithic slab. This type is used over problem soils. Loads are carried over a large area of the slab.

While preparing the subgrade and fill, make the various mechanical installations. Under-floor ducts are usually embedded in the granular fill. Water service supply lines, if placed under the floor slab, should be installed in trenches deep enough to avoid freezing. Connections to utilities should be brought above the finished floor level before placing the concrete.

After the fill has been compacted and brought to grade, a 6 mil polyethylene or other approved vapor barrier should be placed over the subbase. Its purpose is to stop the movement of water into the slab. Strips should be lapped 6″ and taped to form a complete seal. A vapor barrier is essential under every section of the floor. Instructions supplied by the manufacturer should be carefully followed.

Perimeter insulation is important. It reduces heat loss from the floor slab to the outside. The insulation material must be rigid and stable while in contact with wet concrete. **Figure 9-64** shows a foundation design with perimeter insulation typical of residential frame construction. Note how rigid insulation is applied to the exposed foundation exterior. The thickness of the insulation varies from 1″–2″, depending on the average outside temperature and type of heating. The insulation

can be either horizontally or vertically placed along the foundation wall. Refer to Chapter 18, **Thermal and Sound Insulation**, for additional information.

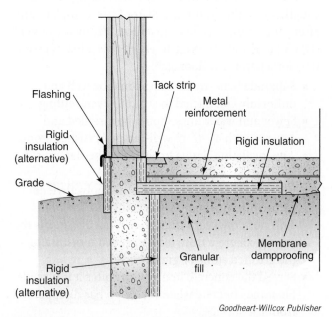

Goodheart-Willcox Publisher

Figure 9-64. Perimeter insulation is important when using slab-on-grade construction in cold climates.

Lamax/Shutterstock.com

Figure 9-65. In construction of a slab foundation, reinforcing mesh is laid on top of a vapor barrier. As concrete is placed, the mesh is lifted up so that it is 1″–1 1/2″ below surface of finished slab. The perimeter of the slab is reinforced with rebar.

When the insulation, vapor barrier, and all mechanical aspects are complete, lay down a reinforcing mesh, if it is to be used. See **Figure 9-65**. Check local building code requirements. Usually a 6″ × 6″ × 10-gage mesh is sufficient for residential work. This should be carefully located from 1″ to 1 1/2″ below the surface of the concrete.

9.15 Basement Floors

Many of the considerations previously listed for slab-on-grade floors also apply to basement floors. Basement floors are placed later in the building sequence, usually after the framing and roof are complete and after the plumber has installed drain/waste/vent (DWV) and water service lines. Here is a list of essentials for a basement floor slab:

- Subgrade beneath the slab should be well and uniformly compacted to avoid uneven settling.
- Sewer and water lines should be installed and covered with 4″–6″ of well-tamped crushed stone.
- Any insulation that is to be under the slab must be in place.
- Put down a vapor barrier of 6-mil polyethylene before placing the reinforcement and concrete. This not only keeps the slab dry, but also keeps the curing concrete from losing needed moisture to the crushed stone or gravel base.
- Screed the concrete immediately after it is placed. Float the concrete when the sheen has disappeared from the surface.

- If a smooth, dense surface is important for the application of finish flooring, give the slab a final smoothing with a steel trowel.

Most basement floors are placed over a base of gravel covered with a dampproof membrane. Floor slabs for frost-protected, shallow foundations often are insulated with 2″ or more of rigid foam insulation between the gravel base and the vapor barrier, **Figure 9-66**. There is some controversy over how much insulation should be installed under the slab. At a certain depth, the earth maintains a year-round temperature of about 50°F (10°C). If the edges of the foundation are well insulated, the inner part of the slab will never be subjected to temperatures much below 50°. Some carpenters like to insulate the entire slab just to be safe. However, the insulation around the perimeter is most important.

Welded wire fabric is laid over the vapor barrier to prevent cracks in the concrete. The fabric can be held above the surface with metal *chairs* or blocked up with concrete bricks. If the owner has specified radiant heat in the floor slab, special high-resistance cable or hydronic radiant floor tubing can be installed in serpentine loops from wall to wall. This resistance wire or tubing is fastened at intervals to the metal reinforcing bars with plastic ties. This work must be completed before the ready-mix concrete arrives.

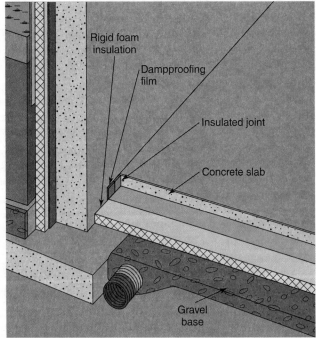

Goodheart-Willcox Publisher

Figure 9-66. Rigid insulation 2″ thick is sometimes used under a basement floor. Seams in the insulation are sealed with tape. A dampproofing film is rolled out over the rigid insulation.

The concrete is delivered to the basement through a window, stairway, or an open section of the outside wall. The rough opening for a fireplace located on an outside wall may also provide easy access. Concrete should not be moved any great distance once placed in the form. A wheelbarrow is often used to carry it from the ready-mix truck's chute to the form.

Working from one side of the form to the other, workers screed the concrete as fast as the form fills. See **Figure 9-67**. Power equipment may also be used for screeding. The chief purpose of *screeding* is to level the surface by striking off (removing) the excess concrete. Striking off starts as soon as the concrete is placed in the form. The screed rides on the edges of the form or previously poured sections of the slab. Low spots found behind the screeded concrete are filled and struck off once more. Two people move the screed across the form. They use a sawing motion as they move the screed along.

Screeding leaves a level, but coarse, surface. *Floating* provides a smoother finish. Workers use a tool, called a float, to fill up the hollows and compact the concrete. Hand floats may be made of magnesium or wood. The most common tool for floating slabs is the long-handled bull float, **Figure 9-68**. Floating is done when the concrete has begun to set. It must be firm enough so that a person's weight on it produces a footprint no more than 1/4″ deep. There should be no bleed water present on the surface. Floating too soon brings too many fines (fine

Kasten-Weiler Construction

Figure 9-67. Screeding levels the concrete as the form is filled. This operation may also be done with a power screed.

aggregates, such as fine sand) and more water to the surface. This produces fine cracks in the surface of the cured concrete.

Troweling produces a denser and smoother surface than floating. Trowels are made of metal and have a smooth finish on them. Troweling requires a firmer set than floating. It can proceed when working the surface does not bring up more water. Troweling can be done by hand for smaller jobs, but large slabs are usually finished with a power trowel, as shown in **Figure 9-68B**.

A — *Photo Courtesy of Mike Hess*   B — *Aisyaqilumaranas/Shutterstock.com*

Figure 9-68. Finishing the surface. A—Floating the finished slab by hand with a bull float. This tool can be made of wood, steel, or magnesium. B—Troweling a smooth finish with a power trowel.

9.16 Entrance Platforms and Stairs

Entrance platform foundations, such as porches, can be constructed as a part of the main foundation. If not, they should be firmly attached to the main foundation. Reinforcing bars placed in the wall when it is constructed can provide a solid connection.

The IRC specifies that stair risers may not be greater than 7 3/4″. This makes 2×8s a good choice for forming the risers. They are actually 7 1/4″ wide, falling well within the IRC requirement. If a riser of more than 7 1/4″ is needed, 2×10s can be "ripped" to the desired width. The IRC also specifies that treads must be at least 10″ wide.

> **Pro Tip**
> To maintain a comfortable steepness, the total of two risers and one tread should be between 24″ and 25″.

Steps may be formed up and poured as part of the platform, **Figure 9-69**. Use 2″ stock to keep the risers from bowing. Detailed information on building stairs is included in Chapter 21, **Stair Construction**. The 2×8 riser boards are set at an angle of about 15° to provide a slight overhang (nosing). Also, bevel the bottom edges of the boards so a mason can trowel the entire surface of the tread.

Some concrete steps must be placed against a wall or between two existing walls. A form can be constructed like the one shown in **Figure 9-70**.

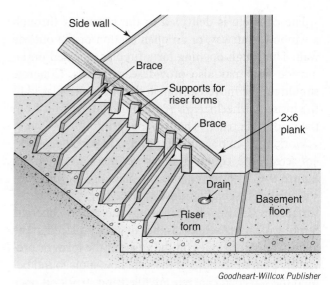

Goodheart-Willcox Publisher

Figure 9-70. A method of building forms when concrete steps must be poured between two existing walls.

9.17 Sidewalks, Steps, and Drives

Usually, sidewalks and drives are poured after the finished grading is completed. If there is extensive fill, wait until the grade has settled or tamp the soil.

Main walks leading to front entrances should be at least 4′ wide. Walks to secondary entrances may be 3′ or slightly less. In most areas, sidewalks are 4″ thick and the formwork is constructed with 2×4 lumber. Walks and drives are usually placed directly on the soil. If there is a moisture problem and frost action, a coarse granular fill should be put down first.

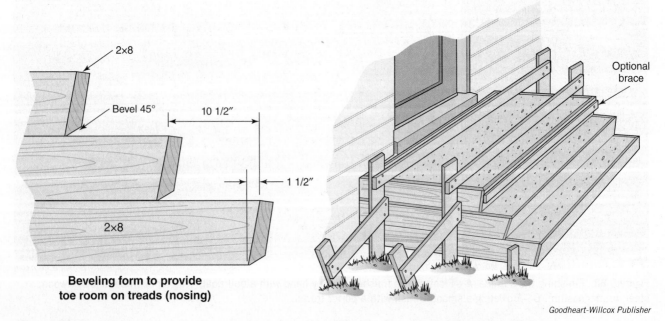

Goodheart-Willcox Publisher

Figure 9-69. Constructing forms for steps. Nosing is formed by tilting the form boards inward at the bottom.

When joining two levels of sidewalks with steps, it is usually best to pour the top level first. If retaining walls are used, these should be constructed next, along with a segment of the lower sidewalk. Finally, form up and pour the steps.

When setting sidewalk forms, provide a slope to one side of about 1/4″ per foot to keep water from pooling on the finished surface, **Figure 9-71**. Also, increase the thickness to 6″ or add reinforcement where there will be heavy vehicular traffic. The concrete should not be permitted to bond against foundation walls or entrance platforms. It should be permitted to "float" on the ground. A 1/2″ thick piece of beveled wood siding can be used to create the separation. It should be removed after the concrete has cured and the joint should be filled with a waterproof, compressible material, such as silicone caulk.

Driveways should be 5″ or 6″ thick with reinforcing mesh included. A single driveway should be at least 10′ wide and a double driveway a minimum of 16′ wide. The minimum crosswise slope should be 1/4″ per foot.

Always place the concrete between the forms close to its final position. Do not overwork the concrete while it is still plastic (can still be molded). This tends to bring excess water and fine material to the surface. It also causes scaling and dusting after the concrete has cured.

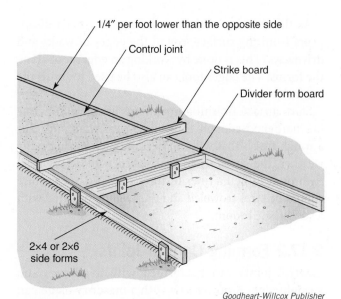

Goodheart-Willcox Publisher

Figure 9-71. The basic setup for constructing a sidewalk. Control joints should be 1/5 of the thickness of the concrete.

9.17.1 Finishing Walks and Drives

After the concrete is roughly spread between the forms, immediately screed it. After screeding, move a float over the surface. When skillfully performed, this operation removes high spots, fills depressions, and smoothes irregularities. See **Figure 9-72**.

A   *Goodheart-Willcox Publisher*   B   *Goodheart-Willcox Publisher*

Figure 9-72. Students in Alfred State College's Building Construction and Masonry programs learn the basics of finishing concrete flatwork. A—A manual screed is being used to flatten and shape concrete to grade. B—Floating the surface smoothes irregularities such as high spots and depressions.

As the concrete stiffens and the water sheen disappears from the surface, round the edges of walks and driveways. This is done by working an edger tool along the forms. The edger tool can also be used to finish the control joints.

Start surface finishing operations after the concrete has hardened enough to become somewhat stiff. For a rough finish that will not become slick during rainy weather, the surface can be stroked with a stiff bristle broom. When a finer texture is desired, the surface should be steel-troweled and then lightly brushed with a soft bristle broom.

9.17.2 Forming Control Joints

Control joints can be formed with a groover and straightedge. A power saw with a masonry blade can also be used. Start sawing 18–24 hours after the concrete is placed. These joints should extend to a depth of at least one-fifth the thickness of the concrete. For sidewalks and driveways, the distance between the control joints is usually about equal to the width of the slab.

For proper curing of concrete, protect it against losing moisture too quickly during the early stages of hardening (hydration). Refer to **Figure 9-36**, which describes several different methods of controlling curing. A convenient method is spraying the concrete with a plastic-based curing compound, which forms a continuous membrane over the surface.

9.18 Cold Weather Construction

Cold weather may call for some changes in the way concrete and masonry materials are handled and placed.

These steps may be necessary:
- Heat the materials
- Cover freshly placed concrete or masonry
- Erect an enclosure and keep the construction area heated

When temperatures fall below 40°F (5°C), concrete should have a temperature of 50°F–70°F (10°C–20°C) when it is placed. Since most concrete is ready-mix, only the building site needs to be sheltered and heated. However, when blocks are being placed, the materials also need shelter and heat.

Temporary shelter can often be arranged with scaffolding sections, lumber, and canvas or plastic tarpaulins. If a shelter cannot be built, the bagged materials and masonry units should be wrapped with tarpaulins when the temperature is below 40°F (5°C). Be sure the material is stored so that ground moisture cannot reach it. Follow recommendations listed in **Figure 9-73**.

Safety Note
When erecting temporary enclosures to heat workspaces or material storage spaces, use only nonflammable sheeting as a covering. The open flame of a kerosene or propane heater can ignite flammable coverings. The blue plastic tarps that are common on jobsites will combust and be completely consumed in a minute or two.

9.18.1 Protecting Concrete

Protect freshly placed concrete with a covering. Hydration creates heat. The covering helps hold the heat in while the concrete cures.

Avoid placing concrete on frozen ground. When the ground thaws, uneven settling may cause cracks. Before pouring, make sure that reinforcing metal, embedded fixtures, and the inside surfaces of forms are free of ice.

Cold Weather Construction		
Air Temperature Below	Material Protection	Masonry Protection
40°F	• Heat mixing water. • Keep mortar temperatures 40°F to 120°F.	• Cover walls and masonry materials, using canvas or plastic, to protect from moisture and freezing.
32°F	• Include the above protective recommendations. • Heat sand to thaw frozen clumps. • Heat wet masonry units to thaw ice.	• Provide windbreak for workers when wind speed is above 15 mph. • Cover walls and materials after workday to protect against moisture and freezing. • Keep masonry temperatures above 32°F using heaters or insulated blankets for 16 hours after placing units.
20°F	• Include all of the above protective recommendations. • Heat dry masonry units to 20°F	• Enclose structure and heat enclosure to keep temperature above 32°F for 24 hours after placing masonry units.

Goodheart-Willcox Publisher

Figure 9-73. Follow these recommendations from the Portland Cement Association for cold-weather construction.

9.18.2 Mortar Temperatures

The temperature of the materials used in mortar is important. Of the materials used in mortar, water is the easiest to heat. It can also store more heat and help bring cement and aggregate up to temperature. Generally, it should be cooler than 180°F (80°C). There is a danger that hotter water could cause the mortar to set instantly.

When the air temperature is lower than 32°F (0°C), sand should be heated to thaw frozen lumps. If desired, the sand temperature can be raised as high as 150°F (65°C). When heating sand, a 50-gallon drum open on one end or a metal pipe works well for containing flames. Heap the sand over and around the container.

9.19 Insect Protection

Wood-loving insects cause varying amounts of damage to structures throughout most areas of the country. Among these pests, the most destructive are termites. Other insects infesting and damaging wood structures include carpenter ants and several types of beetles. Control of these pests—especially termites—involves both following certain rules during construction and treating wood and soil with insecticides.

9.19.1 Termites

Two types of termites are found in temperate to tropical climates. Subterranean termites are the most destructive. They like warm, moist soil where wood or cellulose material is in good supply. In their search for wood, they build earthlike shelter tubes that give them passage over foundation walls, through cracks in walls, on pipes, or other supports leading into buildings.

Drywood termites are so-called because they do not need moisture or contact with the ground. They are airborne and bore into wood structures as well as furniture, particularly when the material is not protected by some type of finish. Though these termites are more common to the tropics, infestations are found along coastal areas from Virginia to northern California.

9.19.2 Termite Control

In new construction, prevention of termite damage begins with the design and construction of the building. The following measures are recommended. Always check for the requirements of local codes.

- Chemical treatment of the soil near the foundation walls.

- Chemical treatment (prior to construction) of soil under a slab-on-grade foundation.
- Removal of wood debris, stumps, and form boards from the building site during and after construction.
- Sealing foundation walls with material, such as damp-proofing, that prevents termites from entering through cracks that may develop.
- Maintaining a minimum of 8″ between grade and the wood sills and siding.
- Capping of foundation walls with termite shields and placing similar shielding around pipes that run from the soil into the building. Placement of shields is discussed in Chapter 10, **Floor Framing**. A shielding method developed by one manufacturer for its ICF foundation system is shown in **Figure 9-74**.

Control of termites in older buildings is also possible, **Figure 9-75**. Studies have shown that appropriate chemicals added to soil around and under existing buildings prevent or control infestation for many years. Treatment is best left to a professional exterminator.

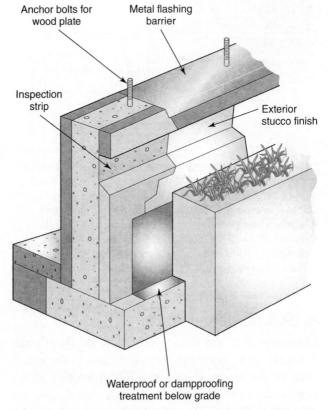

Lite-Form International

Figure 9-74. This detail shows a termite guard developed for use with an ICF foundation system.

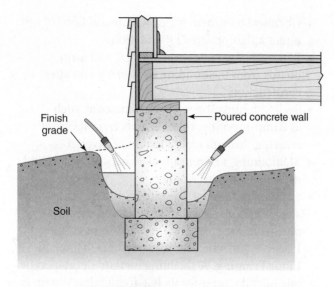

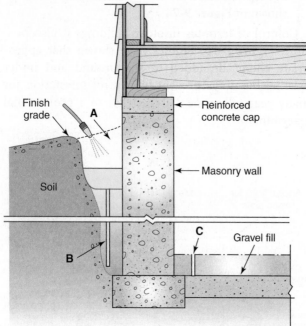

Goodheart-Willcox Publisher

Figure 9-75. With existing buildings, treat the soil as shown to control termites. Top—Trench around both exterior and interior perimeters of a crawl space foundation. Treat the soil as it is replaced. Bottom—Treating a full basement. A—Trench and treat as with the crawl space. B—Create rod holes down to the footings to aid distribution of the chemical. C—Drill holes in the concrete floor at 1′ intervals and fill with chemicals.

9.19.3 Carpenter Ants

Carpenter ants are found mostly in the Northeast, Midwest, and Northwest. They only nest in the wood, they do not eat it. Their tunneling in wooden structures causes the damage. Once in the wood, they are difficult to control, since the treatment must get into the nests. A 2′ wide band of chemical treatment around the building discourages ants from entering.

9.20 Estimating Materials

Concrete is measured and sold by the cubic yard. A cubic yard is 3′ square and 3′ high. It contains 27 cu ft ($3 \times 3 \times 3 = 27$). To determine the number of cubic yards needed for any square or rectangular area, use the following formula. All dimensions should be converted to feet and fractions of feet.

$$\text{Cubic yards} = \frac{\text{width} \times \text{length} \times \text{thickness}}{27}$$

When using this formula, you are first calculating the volume of concrete in cubic feet (width × length × thickness) and then converting the cubic feet to cubic yards by dividing by 27.

For example, to find the concrete needed to pour a basement floor that is $30′ \times 42′ \times 4″$:

$$\text{Cubic yards} = \frac{30 \times 42 \times 1/3}{27}$$

$$= \frac{\overset{10}{\cancel{30}} \times \overset{14}{\cancel{42}} \times 1/3}{\underset{9}{\cancel{27}} \, \underset{1}{\cancel{3}}} = \frac{140}{9}$$

$$= 15.56 \text{ cubic yards}$$

Allow extra concrete for waste or slight variation in the cross sections of the form. Usually an additional 5%–10% is added. For the above example, order 17 cubic yards of concrete.

The number of concrete masonry units needed can be estimated by determining the number of units required in each course and then multiplying by the number of courses between the footing and plate. For example: find the number of 8×8×16 blocks required to construct a foundation wall with a total perimeter (distance around the outside) of 144′. The wall will be laid 11 courses high.

$$\text{Total number} = \frac{\text{perimeter}}{\text{unit length}} \times \text{number of courses}$$

$$= \frac{144′}{16″} \times 11$$

$$= \frac{144′}{(4/3)′} \times 11$$

$$= \frac{144 \times 3}{4} \times 11$$

$$= \frac{\overset{36}{\cancel{144}} \times 3}{\underset{1}{\cancel{4}}} \times 11$$

$$= 108 \times 11$$

$$= 1188 \text{ blocks}$$

Another method of estimating block and mortar quantities is based on the amount of material needed for 100 sq ft of wall area. For example, the table in **Figure 9-76** shows that 100 sq ft of wall area contains 112.5 8×8×16 blocks. Thus, to estimate the number of blocks needed for a wall, you can determine the area of the wall in sq ft and divide by 100. (This determines the number of 100 sq ft "units" in the wall.) Then multiple this value by 112.5 to calculate the number of blocks required, or multiple by 2.6 to calculate the cubic feet of mortar required.

Quantities of Concrete Block and Mortar

Wall Thickness	For 100 Sq Ft of Wall		Mortar* for 100 Concrete Blocks
	Number of Blocks	Mortar*	
8″	112.5	2.6 cu ft	2.3 cu ft
12″	112.5	3.9 cu ft	3.5 cu ft

Based on block having an exposed face of 7 5/8″ x 15 5/8″ and laid up with 3/8″ mortar joints.
*With face shell mortar bedding – 10 percent waste included.

Portland Cement Association

Figure 9-76. Needed quantities of concrete block and mortar can be calculated with this chart.

Construction Careers
Cement Mason

Concrete is one of the most widely used building materials. In residential work, it is primarily used for foundations and basements, but in commercial and industrial construction, it often is a major above-ground structural material.

Photodisc/Photodisc/Thinkstock

Cement masons work with various tools when finishing concrete or laying concrete blocks. Masons must also know the proper mixtures for concrete and take into account weather and any other conditions that can affect curing.

On big construction projects, the foundation work may involve carpenters, steel workers, cement masons, laborers, and more. On small residential jobs, fewer people are involved. Usually, carpenters construct and strip concrete forms and place reinforcing steel, insulation, and vapor barriers. Cement masons do all concrete finishing and lay concrete blocks. Either cement masons or laborers place the concrete in the forms.

A cement mason must have a thorough knowledge of how concrete behaves when being placed in forms, how it cures, and what effects various tools, such as screeds and floats, have on the concrete surface. A mason must monitor temperature, wind, and moisture conditions and take appropriate actions to make sure that the concrete properly cures and attains maximum strength. A mason must also know the effects of various concrete additives and how to properly use them.

Working with concrete is physically demanding and fast paced, since concrete remains in a workable state for a limited time. Although concrete work is usually suspended in extremely wet or cold weather, a cement mason must be able to effectively work in a variety of weather conditions.

Most cement masons and concrete finishers work for general contractors or specialized concrete contractors. Only about 1 in 20 (the lowest percentage of any building trade) are self-employed.

Many cement masons begin as construction laborers, working as helpers and gradually learning the trade from experienced workers. They may also enter the trade through an apprenticeship program lasting three to four years. In addition to on-the-job experience, apprenticeship programs require a minimum of 144 hours per year of classroom work in such areas as blueprint reading, safety, and applied mathematics.

Chapter Review

Summary

- Carpenters are sometimes responsible for laying out building lines and building the concrete forms for building foundations.
- Batter boards consist of horizontal boards and stakes. They are set up well away from where each corner of the new building will be. Strong string stretched between the batter boards define the building outline.
- A control point is established on the site as a reference for the depth of excavation and height of the foundation.
- Footings must be placed around the building perimeter to support the foundation.
- Foundation types include slab-on-grade, spread footings (where a basement is required), piling and girder foundations, and grade beam foundations.
- Forms for concrete must be erected, reinforced, then removed after the concrete sets. In other cases, the foundation walls are constructed using concrete masonry units (blocks).
- Walls must be properly constructed and anchored to withstand stresses and insulated to prevent water damage and heating/cooling losses.
- The area around the foundation must be backfilled.
- In certain climates, permanent wood foundations are installed. Protection against pests that damage wood structures should be taken before construction begins.
- Usually, sidewalks and drives are laid after the finished grading is completed.

Know and Understand

Answer the following questions using the information in this chapter.

1. *True or False?* Grading may be needed before building lines are laid out.

2. It is best to locate building lines with a builder's level, transit, or _____.
   - A. auto level
   - B. laser level
   - C. theodolite
   - D. total station

3. A residence may need as much as 8′ to _____ of clearance for batter boards.
   - A. 10
   - B. 4
   - C. 3
   - D. 15

4. *True or False?* Building sites on steep slopes or rugged terrain should be rough-graded before the building is laid out.

5. The highest point on the perimeter of an excavation is known as the _____.
   - A. grade level
   - B. control point
   - C. grade point
   - D. building line

6. IRC now allows frost-protected shallow foundations that are only _____″ below grade.
   - A. 6
   - B. 12
   - C. 16
   - D. 36

7. A _____ footing is one that changes grade levels at intervals to accommodate a sloping lot.
   - A. spread
   - B. plain
   - C. reinforced
   - D. stepped

8. In residential construction, a common practice is to make the width of the footing _____ as wide as the foundation wall.
   - A. half
   - B. twice
   - C. one and a half times
   - D. equally

9. *True or False?* Concrete for chimney and wall footings must be placed at different times.

10. The spacing of walers depends on the amount of _____ the concrete exerts on the form.
    - A. temperature
    - B. friction
    - C. pressure
    - D. moisture

11. Which of the following materials is generally *not* used for small wall openings?
    - A. Plastic.
    - B. Fiber.
    - C. Concrete.
    - D. Metal.

12. Concrete is made by mixing _____, sand, gravel, and water in proper proportions.
    - A. cement
    - B. grout
    - C. mud
    - D. mortar

13. Which of the following methods is regarded as the easiest way to cure concrete?

    A. Covering with burlap or straw.
    B. Covering with plastic.
    C. Using a liquid-membrane compound.
    D. Flooding.

14. How long is a No. 5 rebar?

    A. 3/8″
    B. 1/2″
    C. 5/8″
    D. 3/4″

15. Most modern concrete contains additives called _____.

    A. admixtures
    B. slumps
    C. plastic
    D. aggregates

16. When placing concrete in forms, working the concrete next to the forms tends to produce a _____ surface along the form faces.

    A. rough
    B. smooth
    C. honeycomb
    D. thick

17. A lightweight CMU will have _____ insulating properties compared to a standard CMU wall.

    A. better
    B. worse
    C. equal
    D. consistent

18. When laying concrete blocks, what is used to check the height of each course?

    A. Mason's line.
    B. A story pole.
    C. A tape measure.
    D. Builder's level.

19. There must be no damage to the foundation, no debris, the backfill must be compacted, and _____ when backfilling a new foundation.

    A. the backfill must be done as soon as the wall is cured
    B. lumber scraps must be buried against the foundation
    C. the final grade must be sloped
    D. the final grade must be level with the foundation

20. Which ICF wall uses vertical and horizontal concrete members spaced more than 12″ apart OC?

    A. Flat.
    B. Waffle-grid.
    C. Screen-grid.
    D. Post-and-beam.

21. _____ is a common method used to build the main floor of many commercial and residential structures without basements.

    A. Slab-on construction
    B. Wood foundation
    C. Screeding
    D. Insulated concrete form (ICF)

22. What may occur if a slab is floated too soon?

    A. A smoother finished is produced.
    B. The concrete appears hollow.
    C. Fine cracks form in the concrete surface.
    D. The concrete's tensile strength increases.

23. A sidewalk should have a slope to one side of about _____″ per foot.

    A. 1/8
    B. 1/4
    C. 1/2
    D. 1

24. *True or False?* In cold weather, it is accepted practice to pour concrete over frozen ground.

25. *True or False?* Subterranean termites are the most destructive when it comes to wooden structures.

Apply and Analyze

1. When is slab-on-grade foundation generally used over other foundations?

2. What is the main difference between a grade beam foundation and a spread footing?

3. Explain the purpose of using a pilaster in a foundation wall.

4. Explain the process of making concrete.

5. Why should concrete never be dropped further than 4′ during placement?

6. What is the difference between dampproofing and waterproofing?

7. Explain why a slab is floated after it has been screeded.

8. Determine the amount of concrete (in cubic yards) needed to pour a slab with the following dimensions: $28′ \times 48′ \times 4″$. Add at least 10% for waste, and express your answer as a whole unit (that is, without a fraction of a cubic yard).

9. Determine the number of 8×8×16 concrete masonry units needed to construct a foundation wall 11 courses high with a 120′ perimeter. Include 10% for waste.

Critical Thinking

1. Many contractors save the wood lumber used for footing forms to reuse elsewhere. However, this wood is generally very dirty and covered in concrete residue after being stripped away from the poured concrete footings. What are some good ways to reuse this lumber where the dirt and concrete residue will not detract from the look or function?

2. Johan is prepping a new basement to pour the concrete floor. The prints call for 4″ of crushed stone as a subgrade under the slab. The stone must be brought in using a wheelbarrow. Johan filled the perimeter of the floor to grade but decided it was too much work to complete the job correctly. Thinking no one would notice, he decided to leave the entire center of the floor subgrade a little low. If no one catches this misstep before the concrete arrives what could be the consequences? How will this impact the concrete pour and the cost of the job?

3. Insulated concrete forms have gained popularity in recent years. How do you think this has disrupted or changed the workflow on new home construction sites?

Communicating about Carpentry

1. **Listening.** In small groups, discuss with your classmates—in basic, everyday language—your knowledge on footings and foundations. Try incorporating terms from the chapter in your descriptions. Conduct this discussion as though you had never read this chapter. Take notes on the observations expressed. Then review the points discussed, factoring in your new knowledge of foundations. Develop a summary of what you have learned about footings and foundations and present it to the class, using the terms that you have learned in this chapter.

2. **Speaking.** Working in groups of two or three students, create a three-dimensional building (either something you built or a drawing) and label the creation using the terms that you have learned in this chapter. Work with your group members to organize your project. Be creative in deciding what materials you will use to create your figure. Delegate responsibility to team members for bringing each of the needed materials. As much as possible, use materials that are readily available in your home (for example, boxes, string, balloons, plastic bottles, paper, tape, glue, or pipe cleaners). Do not use any sharp objects. Incorporate the terms that you learned in this chapter. You must be accurate in your placement of each term. You may use arrows, and you may create several layers on your building.

3. **Speaking.** Obtain small amounts of Portland cement, mason sand, crushed stone, or clean gravel. Mix three batches of concrete, varying the mix for each batch. Wear safety glasses and rubber or chemical resistant gloves while mixing the concrete. Record how much of each ingredient you used in each batch. Place your concrete in similar-sized cardboard boxes (smaller than a shoe box). Use a clean dowel, stick, or piece of rebar to rod the concrete (work the rod up and down several times), ensuring that there are no voids in the batch. Be careful not to tear your box form. After seven days, break the concrete with a sledgehammer, noting any differences in the strength of the batches. Make a presentation and report your findings to your class. Wear safety glasses, long pants, and sturdy shoes when using a sledgehammer to break concrete.

CHAPTER 10

Floor Framing

OBJECTIVES

After studying this chapter, you will be able to:

- Explain the difference between platform, balloon, and post-and-beam framing.
- Identify the main parts of a platform frame.
- Calculate the load on girders and beams used in residential construction.
- Lay out and install sills on a foundation wall.
- Describe how layouts are made on a header joist.
- Explain the correct procedure to follow when correcting problems with floor frames.
- Identify the parts of a floor truss.
- Describe materials used for subflooring.
- Estimate materials (sizes and amounts) required to construct a specific floor frame.

TECHNICAL TERMS

balloon framing	I-joist	solid bridging
band joist	joist	span
bridging	ledger	steel post
cantilevered	platform framing	subfloor
chord	post anchor	tail joist
cross bridging	post-and-beam framing	trimmer
deflection	ribbon	underpinning
firestop	sill	web
girder	sill plate	wide flange
header	sill sealer	

When the building foundation is completed and the concrete or mortar has properly set up, floor framing may begin. Framing of the floor is often completed before the foundation is backfilled. Installing the floor frame first helps the foundation withstand the pressure placed on it by the backfill material.

On the other hand, backfilling may be done before the floor frame is installed. This offers two advantages: it makes delivery of lumber easier and allows convenient access for the start of framing. As described in Chapter 9, **Footings and Foundations**, a carpenter must make sure the foundation wall is adequately braced against damage or cave-in from the pressure of the backfilling. This is critical with concrete block walls. In any case, the site should be brought to rough-grade level after backfilling.

The building's basic design—such as one-story, two-story, or split level—often determines the type of framing used. Other factors may also influence the decision. In some parts of the country, for example, buildings must be constructed with special resistance to wind and rain. In other areas, earthquakes may be the greatest hazard. In cold climates, heavy loads of wet snow may require sturdier or special roof framing. All structures should be built to reduce the effects of lumber shrinkage and warping. They must also resist the hazard of fire.

10.1 Types of Framing

There are three basic types of framing used in residential construction:

- *Platform framing*, also called western framing, is the most popular for residential construction. See **Figure 10-1**.
- *Balloon framing* is no longer used in new construction, but carpenters who remodel or repair older homes should know about it. This type of framing is discussed in Chapter 29, **Remodeling, Renovating, and Repairing**.
- *Post-and-beam framing*, also called plank-and-beam framing, is different from the other two types. Its heavy structural members are at least 4″ thick. This kind of construction is covered in Chapter 26, **Post-and-Beam Construction**.

In both platform and balloon framing, the common structural members are joists, studs, plates, and rafters. Material with a nominal 2″ thickness is used. While sawn 2″ lumber is still used extensively for framing, engineered lumber often replaces sawn materials. These new materials include I-joists, open-web truss joists, laminated-veneer lumber, and laminated-strand lumber. They are lightweight, have consistent strength, can be purchased in long lengths, and use fewer natural resources than sawn lumber.

Green Note

Adaptions of all three framing types are used in green building. Framing methods such as advanced framing, trussed walls, and double-stud walls are all variations that have evolved from the three basic framing types.

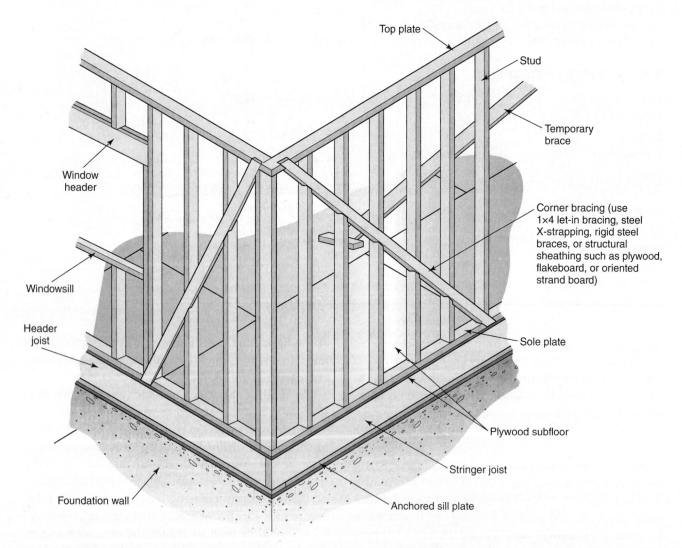

Goodheart-Willcox Publisher

Figure 10-1. In platform framing, a frame of joists provides support for the walls of the first level of the building. Ceiling joists or floor joists for the second level rest on the top plate.

10.1.1 Platform Framing

In platform framing, the first floor is built on top of the foundation wall as a platform. It provides a work area for safely and accurately assembling and raising wall sections. Wall sections are one story high. Outside walls and interior partitions support rafters or upper stories. Each floor is separately framed. Platform framing is satisfactory for both one-story and multistory structures. Settlement due to lumber shrinkage occurs in an even and uniform manner throughout the structure.

Figure 10-2 shows the first-floor platform of a platform frame. Usually, the first-floor platform rests on a *sill*. This is a wood member that is attached by bolts or other metal fasteners embedded in the foundation. Sometimes, the platform rests on an *underpinning*. This is a wall made of short studs that extends above a low foundation. It is usually used on a house that is to have a full basement. Sometimes an underpinning is called a cripple wall.

Figure 10-3 illustrates typical construction methods used at the first- and second-floor levels. A *firestop* is

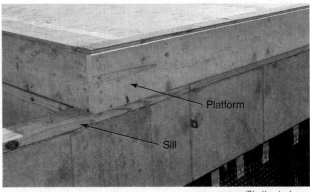

Figure 10-2. An example of platform framing. The first floor platform rests on a sill that is anchored to the foundation.

used to prevent the spread of a fire. Firestops can be solid wood blocking that closes the opening between framing members or it can be fire-retardant insulation. The only firestop needed is built into the floor frame at the second floor level. It prevents the spread of fire in a horizontal direction. It also serves as solid bridging, holding the joists in a plumb (vertical) position.

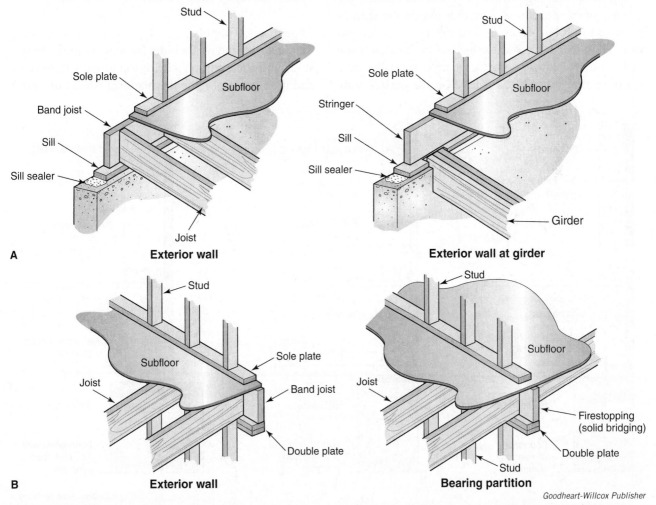

A   **Exterior wall**

Exterior wall at girder

B   **Exterior wall**

Bearing partition

Figure 10-3. Platform framing details. A—First floor. Sill sealer installed under the sill prevents air infiltration. B—Second floor framing is similar to first.

Referring to the illustrations in **Figure 10-3A**, note that a carpenter usually places a strip of insulation, called a *sill sealer*, between the sill plate and the foundation. This prevents heat from being lost at that point.

Architectural plans specify the type of framing. There will be section views of floors, walls, and ceilings. A typical detail drawing of first floor framing includes not only the type of construction, but also the size and spacing of the various members. See **Figure 10-4**. Chapter 5 of the International Residential Code (IRC) covers floor framing.

10.1.2 Balloon Framing

In balloon framing, the studs are continuous from the sill to the rafter plate. A *ribbon* let into the studs supports the second floor joists. They are spiked to the stud as well, **Figure 10-5**. Firestop must be added to the space between the studs. This stud space, which also occurs in load-bearing partitions, permits easy installation of mechanical system components, such as plumbing, wiring, and air ducts.

In balloon framing, shrinkage is reduced because the amount of cross-sectional lumber is low. Wood shrinks across its width, but practically no shrinkage occurs lengthwise. Thus, the high vertical stability of the balloon frame makes it adaptable to two-story structures, especially where masonry veneer or stucco is used on the outside wall.

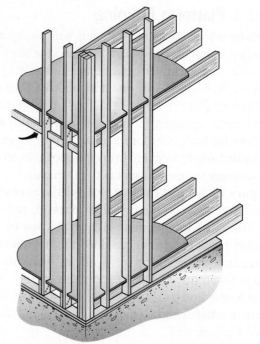

Goodheart-Willcox Publisher

Figure 10-5. Balloon framing. The second floor joists rest on a ribbon (arrow) let into the studs.

Though once popular, this type of framing is rarely, if ever, used today. Balloon framing requires longer studs and the addition of firestop at each floor level, and it does not provide the platform (floor) on which

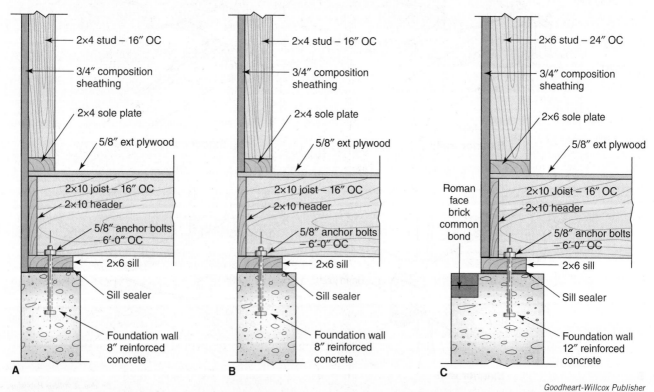

Goodheart-Willcox Publisher

Figure 10-4. Architectural detail drawings show methods of construction as well as the materials to use. A—Sheathing brought to the foundation. B—Sheathing brought to the sole plate. C—Brick veneer construction.

to assemble the wall framing. However, many balloon frames still exist and carpenters may encounter this design in remodeling or repair projects on older buildings.

10.2 Girders and Beams

Joists are the horizontal members of the floor frame. They rest on top of foundation walls, girders, and the top plates of second stories. The *span* is the distance between the walls. Usually, the span is so great that additional support must be provided. *Girders*, also called beams, rest on the foundation walls and on posts or columns. They support the joists at their midpoint. Girders may be solid timbers, built-up lumber, engineered lumber, or steel beams. Sometimes, a load-bearing partition replaces a girder or beam.

Laminated-veneer lumber (LVL) or laminated-strand lumber (LSL) is frequently used for girders for the following reasons:

- Dimensionally stable
- Consistent strength properties
- Eliminates the labor of constructing a built-up beam
- Available in lengths to span most houses

LVL and LSL manufacturers publish span tables and installation instructions for girders. **Figure 10-6** shows one manufacturer's span table for LVL girders.

Built-up girders can be made of three or four pieces of 2″ lumber nailed together with 10d nails. See **Figure 10-7**. Modern engineered lumber, such as laminated beams, may be used for girders. These are factory-made by gluing together thin strips of wood or veneer. Such beams can be manufactured in many different sizes and lengths. Joints should rest over columns or posts.

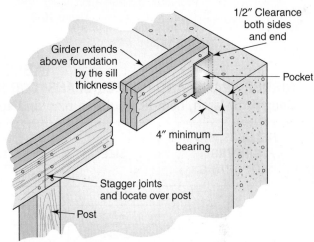

Goodheart-Willcox Publisher

Figure 10-7. When built-up wood girders are fabricated on the job, nails should be spaced no farther apart than 32″ along the top and bottom edges. A metal bearing plate should be placed under the girder at the foundation wall. Clearance in the pocket allows air circulation and prevents rot from moisture.

2.0E GP Lam LVL Floor Beams

This table shows the size (e.g.: 2–11¼″ = 2 plies of 1¾″x11¼″) of beams needed to support loads of one floor only, i.e., a second story floor or one story floor over a basement. (See drawing at right.)

When floor joists span continuously from wall to wall (not cut at beam) this table requires that "B" be not less than 45%, or greater than 55% of "A".

Example: If "A" = 328, "B" must be between 14.48 (322.45) and 17.68 (322.55)

For non-conforming situations, use FASTBeam® analysis and selection software or contact Georgia-Pacific.

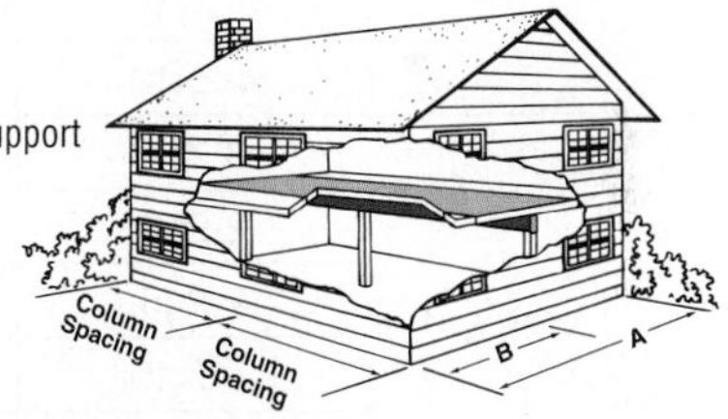

Total Floor Joist Span "A"		Column or Support Spacing (center-to-center)									
		11′	12′	13′	14′	15′	16′	17′	18′	19′	20′
	24′	2-11¼″ / 3-9¼″	2-11¼″ / 3-9½″	2-11⅞″ / 3-11¼″	2-14″ / 3-11¼″	2-14″+ / 3-11⅞″	2-16″+ / 3-14″	2-16″+ / 3-14″	2-18″+ / 3-16″	2-18″+ / 3-16″	2-18″+ / 3-16″
	28′	2-11¼″ / 3-9¼″	2-11⅞″ / 3-11¼″	2-14″+ / 3-11¼″	2-14″+ / 3-11⅞″	2-16″+ / 3-14″	2-16″+ / 3-14″	2-16″+ / 3-14″	2-18″+ / 3-16″	2-18″+ / 3-16″	3-16″
	32′	2-11¼″	2-14″+ / 3-11¼″	2-14″+ / 3-11⅞″	2-14″+	2-16″+ / 3-14″	2-16″+ / 3-14″	2-18″+ / 3-16″	2-18″+ / 3-16″	3-16″+	3-18″+
	36′	2-11⅞″+ / 3-11¼″	2-14″+ / 3-11¼″	2-14″+ / 3-11⅞″	2-16″+ / 3-14″	2-16″+ / 3-14″	2-18″+ / 3-14″	3-16″+	3-16″+	3-18″+	3-18″+
	40′	2-11⅞″+ / 3-11¼″	2-14″+ / 3-11¼″	2-14″+	2-16″+ / 3-14″	2-16″+ / 3-14″	3-16″+	3-16″+	3-16″+	3-18″+	3-18″+

+ See note 2.

NOTES:

1. Table is based on continuous floor joist span and simple or continuous beam span conditions. If floor joists are not continuous above the beam, take the sum of the joist spans then multiply by 0.8. This is the total floor joist span to consider.
2. Required end bearing length (based on 565 psi) is 3.0″ unless the subscript + is shown. In that case, 4.5″ is required.
3. At intermediate supports of continuous spans, use the following guidelines or refer to page 40.
   – 7½″ bearing length for beams requiring 3″ bearing at the beam ends
   – 10½″ bearing length for beams requiring 4½″ bearing at the beam ends
4. All headers require full-width bearing support, e.g., 2x6 for 5¼″, 3-ply members. The adequacy of supporting columns to be verified by others.
5. Table is based on residential floor loading of 40 psf live load and 12 psf dead load.
6. Live load reductions have been applied per IBC section 1607.9.1.
7. Deflection is limited to L/360 at live load and L/240 at total load.
8. For other uniform load conditions refer to pages 42–43.
9. A single 3½″ thick ply can be substituted for any two 1¾″ thick plies.
10. For multiple ply fasteners, see pages 51–53.

Georgia-Pacific Corp.

Figure 10-6. Span table for laminated-veneer lumber (LVL) girders.

PROCEDURE

Sizing Built-Up Girders

1. Determine the number of stories of the building.
2. Find the width of the building. That is, find the distance between exterior walls.
3. Select the proper girder size according to the local building code in your area. Built-up wood girder sizes that comply with the IRC are shown in **Figure 10-8**.
4. Determine the span of the girder by measuring the distance between columns or piers supporting the girder.
5. If engineered lumber is used, consult the manufacturer's span table to determine the specifications for girder sizes.

TABLE 602.7(2)
GIRDER SPANS[a] AND HEADER SPANS[a] FOR INTERIOR BEARING WALLS (Maximum spans for Douglas fir-larch, hem-fir, southern pine and spruce-pine-fir[b] and required number of jack studs)

HEADERS AND GIRDERS SUPPORTING	SIZE	BUILDING Width[c] (feet)					
		12		24		36	
		Span[e]	NJ[d]	Span[e]	NJ[d]	Span[e]	NJ[d]
One floor only	2-2×4	4-1	1	2-10	1	2-4	1
	2-2×6	6-1	1	4-4	1	3-6	1
	2-2×8	7-9	1	5-5	1	4-5	2
	2-2×10	9-2	1	6-6	2	5-3	2
	2-2×12	10-9	1	7-7	2	6-3	2
	3-2×8	9-8	1	6-10	1	5-7	1
	3-2×10	11-5	1	8-1	1	6-7	2
	3-2×12	13-6	1	9-6	2	7-9	2
	4-2×8	11-2	1	7-11	1	6-5	1
	4-2×10	13-3	1	9-4	1	7-8	1
	4-2×12	15-7	1	11-0	1	9-0	2
Two floors	2-2×4	2-7	1	1-11	1	1-7	1
	2-2×6	3-11	1	2-11	2	2-5	2
	2-2×8	5-0	1	3-8	2	3-1	2
	2-2×10	5-11	2	4-4	2	3-7	2
	2-2×12	6-11	2	5-2	2	4-3	3
	3-2×8	6-3	1	4-7	2	3-10	2
	3-2×10	7-5	1	5-6	2	4-6	2
	3-2×12	8-8	2	6-5	2	5-4	2
	4-2×8	7-2	1	5-4	1	4-5	2
	4-2×10	8-6	1	6-4	2	5-3	2
	4-2×12	10-1	1	7-5	2	6-2	2

For SI: 1 inch = 25.4 mm, 1 foot = 304.8 mm

a. Spans are given in feet and inches.
b. Spans are based on minimum design properties for No. 2 grade lumber of Douglas fir-larch, Southern pine, and spruce-pine-fir.
c. Building width is measured perpendicular to the ridge. For widths between those shown, spans are permitted to be interpolated.
d. NJ = Number of jack studs required to support each end. Where the number of required jack studs equals one, the header is permitted to be supported by an approved framing anchor attached to the full-height wall stud and to the header.
e. Spans are calculated assuming the top of the header or girder is laterally braced by perpendicular framing. Where the top of the header or girder is not laterally braced (for example, cripple studs bearing on the header), tabulated spans for headers consisting of 2×8, 2×10, or 2×12 sizes shall be multiplied by 0.70 or the header or girder shall be designed.

Table R602.7(2) Excerpted from the 2018 International Residential Code; Copyright 2017; Washington, D.C.: International Code Council. Reproduced with permission. All rights reserved. www.ICCSAFE.org

Figure 10-8. Spans for built-up wood girders.

10.2.1 Steel Beams

In many localities, steel beams are used instead of wood girders. The W-beam (*wide flange*) is generally used in residential construction. Common structural steel shapes and their designations are shown in **Figure 10-9**.

Wood beams vary in depth, width, species, and grade. Steel beams vary in depth, width of flange, and weight.

Sizes of steel beams depend on the load. **Figure 10-10** shows how the load is calculated. After the approximate load on a steel beam has been determined, the correct

	Structural Steel Shapes				
Shape	**Letter Designation**	**Size Designation**	**Shape**	**Letter Designation**	**Size Designation**
Wide Flange	W	Depth × Weight Per Lineal Foot	Structural Tube	HSS	Depth × Width × Wall Thickness
Standard	S	Depth × Weight Per Lineal Foot	Tee	T	Depth × Width Per Lineal Foot
Channel	C	Depth × Flange × Weight Per Lineal Foot	Angle	L	Long Leg × Short Leg × Thickness

Goodheart-Willcox Publisher

Figure 10-9. Several structural steel shapes may be used in the construction of a house.

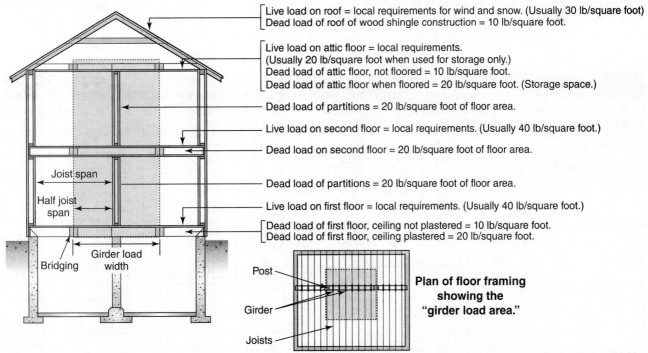

Live load on roof = local requirements for wind and snow. (Usually 30 lb/square foot)
Dead load of roof of wood shingle construction = 10 lb/square foot.

Live load on attic floor = local requirements.
(Usually 20 lb/square foot when used for storage only.)
Dead load of attic floor, not floored = 10 lb/square foot.
Dead load of attic floor when floored = 20 lb/square foot. (Storage space.)

Dead load of partitions = 20 lb/square foot of floor area.

Live load on second floor = local requirements. (Usually 40 lb/square foot.)

Dead load on second floor = 20 lb/square foot of floor area.

Dead load of partitions = 20 lb/square foot of floor area.

Live load on first floor = local requirements. (Usually 40 lb/square foot.)

Dead load of first floor, ceiling not plastered = 10 lb/square foot.
Dead load of first floor, ceiling plastered = 20 lb/square foot.

Joist span

Half joist span

Bridging   Girder load width

Post

Girder

Joists

Plan of floor framing showing the "girder load area."

Goodheart-Willcox Publisher

Figure 10-10. This diagram shows the method of figuring loads for the frame of a two-story home.

size can be selected from the manufacturer's span table, such as the one shown in **Figure 10-11**. This table lists a selected group of steel beams commonly used in residential structures. For example, if the total, evenly distributed load on the beam is 15,000 lb and the span between supports is 16'-0", then a W8×18 beam should be used. This specifies an 8" deep beam weighing 18 lb per lineal foot. The width of the flange is 5 1/4".

10.2.2 Posts and Columns

For girder and beam support, *steel posts* are most popular. The post should be capped with a steel plate to provide a good bearing area. A steel post designed specifically for this purpose has a threaded hole at one end. A rod attached to a plate is threaded into this hole. This arrangement allows for adjustment as the wooden beam and other structural members shrink.

For wood posts shorter than 9', it is safe to assume that a post with its greater dimension equal to the width of the girder it supports will carry the girder load. For example, a 6×6 post is suitable for a girder 6" wide. For a girder 8" wide, a 6×8 or 8×8 post should be used.

Adequate footings must be provided for girder posts and columns. Wood posts should be supported on footings that extend above the floor level, as shown in **Figure 10-12**. To make sure the posts will not move off their footings, pieces of 1/2" diameter reinforcing rod or iron bolts should be embedded into the footings before the concrete sets. They should project about 3" into holes bored in the bottoms of the posts. A *post anchor* can be used to securely hold the wood post, **Figure 10-13**. It supports the bottom of the post above the floor, protecting the wood from dampness.

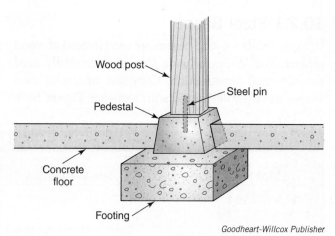

Goodheart-Willcox Publisher

Figure 10-12. For moisture protection, footings for columns must extend above the floor level.

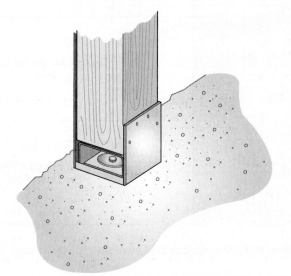

Goodheart-Willcox Publisher

Figure 10-13. Steel post anchor permits lateral adjustment. Various designs are available.

Allowable Loads for Wide Flange Steel Beams													
Designation Weight/Foot	Nominal Size Depth × Width	Span in Feet											
		8'	10'	12'	14'	16'	18'	20'	22'	24'	26'
W8×10	8×4	15.6	12.5	10.4	8.9	7.8	6.9	—	—	—	—
W8×13	8×4	19.9	15.9	13.3	11.4	9.9	8.8	—	—	—	—
W8×15	8×4	23.6	18.9	15.8	13.5	11.8	10.5	—	—	—	—
W8×18	8×5 1/4	30.4	24.3	20.3	17.4	15.2	13.5	—	—	—	—
W8×21	8×5 1/4	36.4	29.1	24.3	20.8	18.2	16.2	—	—	—	—
W8×24	8×6 1/2	41.8	33.4	27.8	23.9	20.9	18.6	—	—	—	—
W8×28	8×6 1/2	48.6	38.9	32.4	27.8	24.3	21.6	—	—	—	—
W10×22	10×5 3/4	—	—	30.9	26.5	23.2	20.6	18.6	16.9	—	—
W10×26	10×5 3/4	—	—	37.2	31.9	27.9	24.8	22.3	20.3	—	—
W10×30	10×5 3/4	—	—	43.2	37.0	32.4	28.8	25.9	23.6	—	—
W12×26	12×6 1/2	—	—	—	—	33.4	29.7	26.7	24.3	22.3	20.5
W12×30	12×6 1/2	—	—	—	—	38.6	34.3	30.9	28.1	25.8	23.8
W12×35	12×6 1/2	—	—	—	—	45.6	40.6	36.5	33.2	30.4	28.1

Goodheart-Willcox Publisher

Figure 10-11. Allowable uniform loads for wide flange steel beams. Loads are given in kips (1 kip = 1000 lb).

The bracket can be adjusted for plumb if the anchor bolt was improperly located.

When a wood column supports a steel girder, consider fitting the end of the column with a metal cap. If wood supports wood, install a metal cap to give an even bearing surface. The metal also prevents the end grain of the post from crushing the horizontal grain of the wood girder.

A built-up wood post may be made by nailing together three 2×6s. The pieces should be free from defects and securely nailed together. Otherwise, excessive loading may cause the members to buckle away from each other.

> **Pro Tip**
> Be sure the tops of posts and columns and the pockets in foundation walls are flat. This ensures the girder or beam is well supported with its sides plumb.

10.2.3 Framing over Girders and Beams

One method of framing joists over girders and beams is shown in **Figure 10-14**. The steel beam is set level with the top of the foundation wall. The 2″ wood pad then brings the joists level with the sill. When a wooden girder is used, it is usually set so the top is level with the sill.

If the ceiling height under joists needs to be lowered, the joists can be notched and carried on a *ledger*, **Figure 10-15**. This arrangement is seldom used in new construction. When it is necessary for the underside of the girder to be flush with the joists to provide an unbroken ceiling surface, the joists should be supported with hangers or stirrups. See **Figure 10-16**.

Framing joists to steel beams at various levels can be accomplished with special hangers in somewhat the same manner as for wood girders. With solid wood joists, you must make allowances for the fact that joists will likely shrink, while the steel beam will remain the same size. For average work with a 2×10 joist, an allowance of 3/8″ above the top flange of the steel girder or beam is usually sufficient. **Figure 10-17** shows two butted-joist methods of framing over girders.

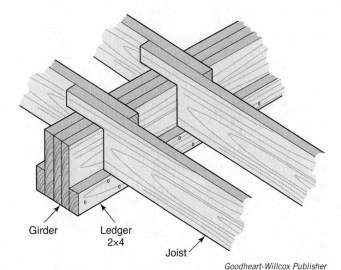

Girder   Ledger 2×4   Joist

Goodheart-Willcox Publisher

Figure 10-15. In this arrangement, the joists rest on the ledger strip, not on top of the girder.

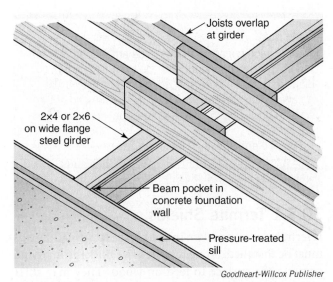

Joists overlap at girder

2×4 or 2×6 on wide flange steel girder

Beam pocket in concrete foundation wall

Pressure-treated sill

Goodheart-Willcox Publisher

Figure 10-14. Joists supported on top of a steel beam. The top of the beam is set flush with the top of the foundation wall.

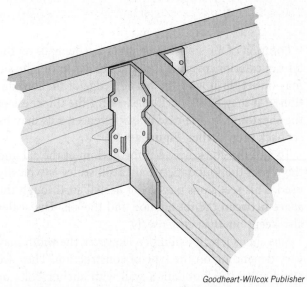

Goodheart-Willcox Publisher

Figure 10-16. Hangers are often used to attach joists to girders and headers.

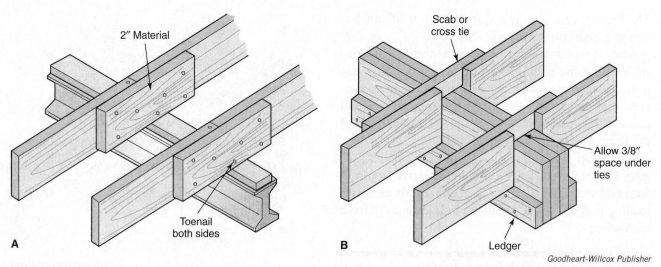

Goodheart-Willcox Publisher

Figure 10-17. Some carpenters like to butt joists over the girder. A—Joists are butted over a steel I-beam and wood 2×4 and secured with toenailing and a short scab of 2″ material. B—Joists are tied together with a scab or cross tie.

10.3 Sill Construction

After girders and beams are in place, the next step is to attach the sill to the foundation wall. The sill, also called the *sill plate* or mudsill, is the part of the side walls or floor frame that horizontally rests on the foundation. It supports the frame of the building and ties the frame to the foundation. The term mudsill originated from the process of correcting irregularities in masonry work by embedding the sill in a layer of fresh mortar or grout.

> **Code Note**
> Building codes specify that a sill must have a minimum level of protection against termite and decay damage. The IRC requires that all wood in contact with concrete or masonry be pressure-treated with a preservative to prevent decay and insect damage.

Location of the sill on the foundation depends on the type of exterior covering the building will have. The sill may be set flush with the outside edge of the foundation or it may be set back to allow for the thickness of the sheathing. In the case of brick veneer, sills may be set back even farther, **Figure 10-18**.

To seal the crack between the bottom of the sill and the top of the foundation wall, a sill sealer is typically used. This seal stops passage of heated air through the space between the foundation and the sill. The sealer also keeps out dust and insects.

Sills are usually 2×6 lumber. However, the width may vary depending on the type of construction. They are attached to the foundation wall with anchor bolts or straps.

Goodheart-Willcox Publisher

Figure 10-18. This sill has been set back 4″ from the face of the foundation to allow for masonry to be applied later.

> **Code Note**
> The IRC requires anchor bolts to be embedded at least 7″ into the concrete foundation, to have a minimum diameter of 1/2″, to be spaced no further than 6′ OC. An anchor bolt is required within 12″ of the end of the sill, and each board in the sill must contain at least two anchor bolts.
> The size and spacing of anchors, however, is a matter of concern to each community. Always check the specifications in local building codes.

10.3.1 Termite Shields

If termites are a problem in your locality, special shields must be installed. Some termites live underground and come to the surface to feed on wood. They may enter through cracks in masonry or build earthen tubes on the sides of masonry walls to reach the wood.

The wood sill should be at least 8″ above the ground. Also, a protective metal shield should extend out over the foundation wall, as shown in **Figure 10-19**. The shield should be not less than 26 gauge.

Figure 10-20 is a map of the United States that identifies various levels of termite infestation. Canada and Alaska are considered to be in region IV. Hawaii and Puerto Rico are in region I. In areas where termite damage is great, additional measures should be taken. Sometimes it is necessary to use chemically treated lumber for lower framing members. Also, the soil around the foundation and under the structure can be treated.

10.3.2 Installing Sills

Figure 10-21 shows two types of sill anchor–anchor straps and anchor bolts. With straps, position the sill and attach the straps with nails. Some types must be bent over the top of the sill. Others are nailed on the sides.

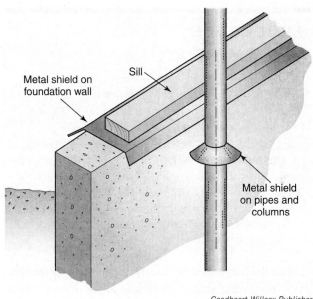

Goodheart-Willcox Publisher

Figure 10-19. Termite shields. Use galvanized sheet steel or other suitable metal that resists rusting.

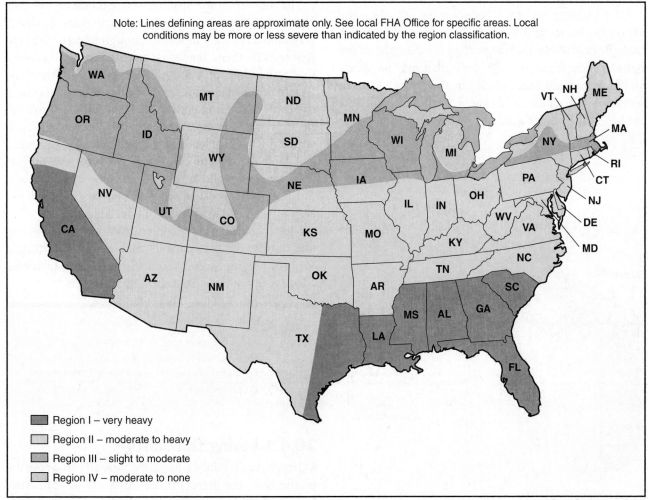

Forest Products Laboratory

Figure 10-20. Nearly all parts of the United States have some risk of termite infestation. Termites can damage buildings.

Using Anchor Bolts

1. Remove the washers and nuts.
2. Lay the sill along the foundation wall.
3. Using a square, draw lines across the sill on each side of the bolts, as shown in **Figure 10-21**.
4. Measure the distance from the center of the bolt to the outside of the foundation. If the sheathing or brick veneer is to be flush with the foundation, subtract the thickness of the sheathing or the brick veneer.
5. Use this distance to locate the bolt holes. You will probably need to make separate measurements for each anchor bolt.

Since the foundation walls may not be perfectly straight, many carpenters snap a chalk line along the top where the outside edge of the sill should be located. This ensures an accurate floor frame, which is basic to all additional construction. Variations between the outside surface of the sheathing and foundation wall can be shimmed when siding is installed.

After marking all of the holes on the sill, place the sill on sawhorses and bore the holes. Most carpenters bore the hole about 1/4″ larger than the diameter of the bolts to allow for some adjustment that may be necessary due to slight inaccuracies in the layout. As each section is laid out and holes bored, position the sill over the bolts to check for accuracy.

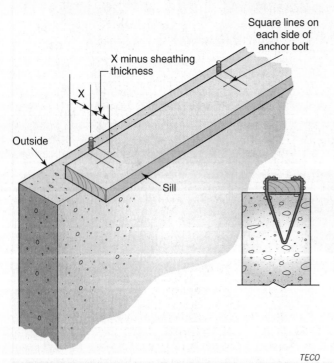

TECO

Figure 10-21. Laying out anchor bolt holes. An anchor strap (right) needs no layout. The sill is positioned and the strap is nailed to the sill.

After fitting all sill sections, remove them from the anchor bolts. Install the sill sealer and then replace the sill sections. Install washers and nuts. Before the nuts are tightened, ensure that the sills are properly aligned. If there is a setback for sheathing or brick veneer, check the distance from the edge of the foundation wall.

Now, check that the sill is level and straight. Low spots can be shimmed with wooden wedges. However, it is better to use grout or mortar.

10.4 Joists

Floor joists are framing members that carry the weight of the floor between the sills and girders. In residential construction, floor joists may come in one of these forms:

- I-joists
- Nominal 2″ lumber placed on edge
- Open-web truss joists

In heavier construction, steel bar joists and reinforced concrete joists are used.

The most common spacing of wood joists is 16″ OC. However, 12″, 19.2″, and 24″ OC spacings are also used. Joists must not only be strong enough to carry the load that rests on them, they must be stiff enough to prevent undue bending or vibration. **Deflection** is when the weight of the load causes the joist to bend downward at the center. Building codes usually specify that the deflection of a joist must not exceed 1/360th of the span with a normal live load. This is 1/2″ for a 15′-0″ span.

A table in Appendix B, **Technical Information** lists safe spans for solid wood joists under average loads. For floors, this is usually figured on a basis of 50 lb per sq ft (10-lb dead load and 40-lb live load).

I-joists are manufactured with varying sizes of top and bottom flanges and web thick-nesses, resulting in joists of varying strength and stiffness. Span tables for I-joists are available from the manufacturer.

Code Note
Building codes typically specify allowable span distances for floor joists based on the size of the joist and the species and grade of wood. Always verify that the joist sizes for the project are acceptable to the applicable building code.

10.4.1 Laying Out Joists

When laying out joists, begin by carefully studying the plans. Note the direction the joists are to run. Also, become familiar with the location of posts, columns, and supporting partitions. The plans may also show the centerlines of girders.

The position of the floor joists can be directly laid out on the sill, **Figure 10-22**. In platform construction, the joist spacing is usually laid out on the band joist rather than the sill. ***Band joists*** are those joists that sit on the sill and to which other floor joists are butted and attached. They are also called joist headers or rim joists. The position of an intersecting framing member may be laid out by marking a single line and then placing an X to indicate the side of the line the joist is to be installed, **Figure 10-23**. When using solid wood joists, the crown of the joist should be turned upward. The crown, also called a crook, is a slight warping.

Rather than having to measure each individual joist space with a steel tape, some carpenters make a master layout (called a rod) on a strip of wood. Use it to transfer the layout to band joists or the sill. The same rod is then used to make the joist layout on girders and the opposite wall. When the joists are lapped at the girder, the X (joist location) is marked on the other side of the layout line for the opposite wall. In this case, the spacing between the stringer and first joist is different from the regular spacing, **Figure 10-24**.

As mentioned earlier, some plans call for the sills and band joists to be set back from the edge of the foundation to leave a ledge around the outside of the sill. This ledge may be 1″ or less to accommodate the sheathing, or several inches to accommodate masonry veneer. When the sheathing is applied, it is flush with the foundation wall. In **Figure 10-24B** and **Figure 10-24C**, the sill is flush to the edge of the wall, but the band joists are offset.

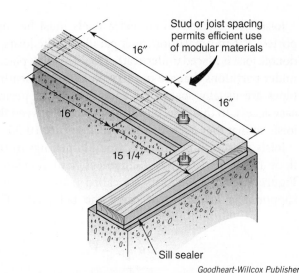

Goodheart-Willcox Publisher

Figure 10-22. After the sill is attached to the foundation, locations for studs or joists may be marked using a square.

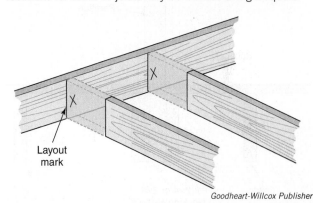

Goodheart-Willcox Publisher

Figure 10-23. How to mark the actual location of framing members being attached to a band joist. Layout marks show where the edge of a joist should be. The X indicates which side of the line the joist should be positioned.

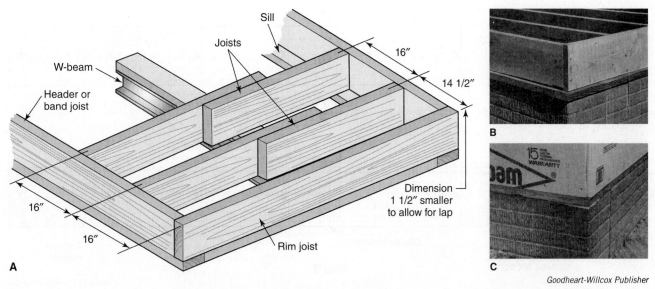

Goodheart-Willcox Publisher

Figure 10-24. A—When joists lap over the center girder, spacing between the first joist and the rim joist is different. B—Some carpenters set back the stringer and headers, leaving a ledge on the sill. The ledge is the same width as the thickness of the sheathing. C—The sheathing is flush with the foundation wall.

Joists are doubled where extra loads must be supported. When a partition runs parallel to the joists, a **double joist** is placed underneath. Doubled joists placed **under** partitions that are to carry plumbing or heating **pipes** are usually spaced far enough apart to permit **easy access**. In this case, blocking is placed between the joists so that they act as a single piece, **Figure 10-25**.

Joists must also be doubled around openings in the **floor** frame for stairways, chimneys, and fireplaces, **Figure 10-26**. These joists are called *trimmers*. They support the *headers* that carry the tail joists. The *tail joists* are short joists that run from the band joist to the header of the opening. A carpenter must become thoroughly familiar with the plans at each floor level so adequate support can be provided.

Select straight lumber for the band joist and lay out the standard spacing along its entire length, **Figure 10-27**. If an LVL rim board is used, it is manufactured straight. Add a line to mark the position for any doubled joists and trimmer joists that are required along openings. Where regular joists will become tail joists, change the X mark to a T, as shown in **Figure 10-27**.

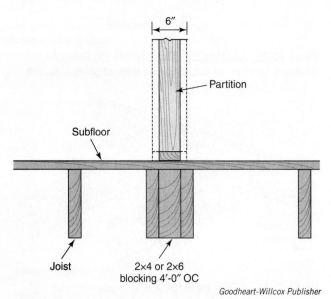

Goodheart-Willcox Publisher

Figure 10-25. Joists under partitions are doubled and spaced to allow access for heating or plumbing runs. If the wall must hold a plumbing stack (vent to roof), the wall is framed with 2×6s.

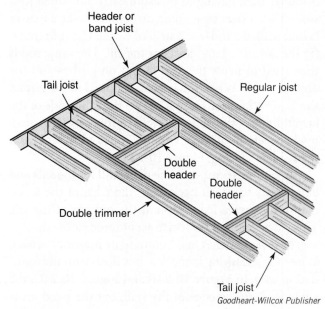

Goodheart-Willcox Publisher

Figure 10-26. Framing members are doubled around floor openings.

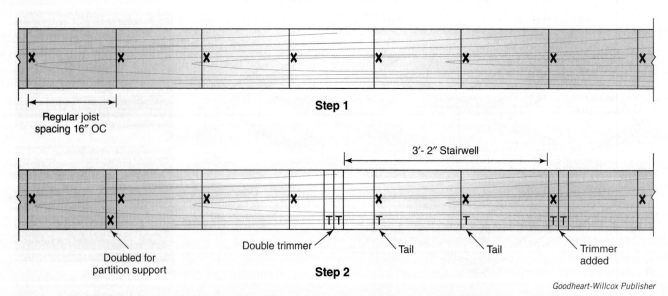

Goodheart-Willcox Publisher

Figure 10-27. Laying out a band joist with joist positions. In step 1, a rod is used to mark regular spacing. In step 2, double and trimmer joist positions have been added.

10.4.2 I-Joists

I-joists are engineered joists with top and bottom flanges that have replaced dimensional lumber joists on many jobs. The flanges are made from Douglas fir in solid lumber or LVL. The *web* is the material between the flanges. It is usually made of oriented strand board (OSB).

The web is glued into grooves cut in the flanges, **Figure 10-28**. No nails are used. These manufactured joists are available in such a wide range of depths and lengths that it is possible to span the width of a building with a single joist. They are not prone to shrinking or warping. This reduces the occurrence of squeaking floors caused by drying, shrinking lumber. Special techniques must be used to fasten I-joists to other frame components. I-joists are often attached to joist headers and girders with steel hangers, **Figure 10-29**.

Weyerhaeuser

Figure 10-28. I-joists supported by a glulam girder. I-joists are constructed with solid wood chords and plywood or OSB webs. They must be placed upright during storage and transport to avoid damage.

Code Note
For fire protection, IRC R501.3 requires that manufactured I-joists be covered with 1/2″ gypsum wallboard or equivalent, unless protected by an automatic sprinkler system.

Installing I-Joists

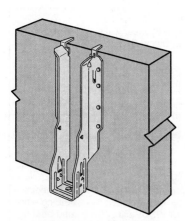

1. Attach hanger to header or girder.

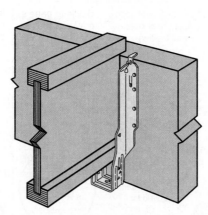

2. Slide I-joist into hanger.

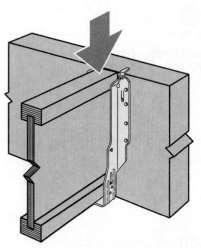

3. Firmly push or snap I-joist into seat of hanger.

Simpson Strong-tie

Figure 10-29. Special hangers are available for installing I-joists.

10.4.3 Nailing I-Joists

Manufacturers of I-joists have specific requirements for nailing, reinforcing, and cutting holes in their joists. The special construction techniques detailed in **Figure 10-30** are typical. Special nailing requirements might include the following:

- To reduce splitting, nail joists at bearing points with two 8d nails, one on either side, no closer than 1 1/2″ from the end. The size of the nails required may vary according to the thickness of the flanges. Generally, the specifications provide for about 2″ of the nail to penetrate the second piece.
- Nail rim joist that is 1 3/4″ or thinner to the wood I-beam using two 8d nails, one each at top and bottom flange.
- Attach 2×4 or wider "squash" plates to the top and bottom flanges to help support bearing walls. See details A2 and B2 in **Figure 10-30**.

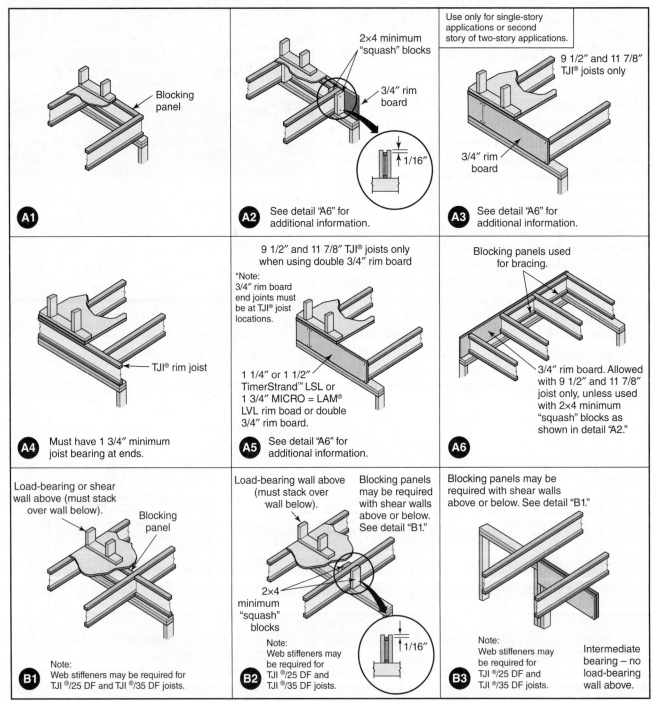

TrusJoist MacMillan

Figure 10-30. I-joists are light, strong, and, if correctly installed, provide a rigid, quiet floor.

10.4.4 Rim Boards

Rim boards are the framing members between the sill and the bottom of the wall framing. In framing with sawn wood, the rim board may be called the rim joist, band joist, or joist header. Engineered rim boards are designed for use with I-joists. Rim boards may be made with LVL, LSL, or OSB. I-joists do not expand and contract with changes in their moisture content, so it is necessary to use rim boards that have similar properties. Engineered rim boards are generally available in 1″, 1 1/8″, 1 1/4″, and 1 1/2″ thicknesses and in the same depths as I-joists.

Safety Note
Use great care in working with I-joists. They are unstable until properly braced. Failing to observe the following precautions and practices could cause accidents and damage:
- Never allow workers to walk on joists until the joists are braced.
- Do not stack building materials on unsheathed joists. Rather, place them only over beams or walls.
- Subflooring must be completely attached to each joist before additional loading of the floor.
- Install and nail all blocking, hangers, and rim boards at the ends of the joists.
- To temporarily brace joists, provide a rigid structure such as a braced wall or permanent deck (sheathing) nailed to the first 4′ of the joists.
- Attach 1×4 struts from the braced area across the tops of the rest of the joists. Unless this is done, sideways buckling or a rollover is likely.
- When joists are cantilevered, struts must be attached to both the top and bottom flanges.
- Keep flanges straight with a tolerance of no more than 1/2″ of true alignment.

10.4.5 Installing Solid Wood Joists

After the band joists are laid out, toenail them to the sill. Position all full-length joists with the crown turned up. Tightly hold the end against the band joist and along the layout line so the sides of the joist are plumb. Butt the joists against the band joist and fasten them with 16d nails.

Next, fasten the joists along the opposite wall. If the joists butt at the girder, they should be connected with a scarf or metal fastener. If they lap, they can be nailed together using 10d nails. Use 8d nails to toenail the joists to the girder.

Nail doubled joists together using 10d nails spaced about 1′ along the top and bottom edges. First, check that the crowns of the joists are at the same height. If not, toenail through the higher one to bring them to the

same level. Next, drive several nails straight through to tightly pull the two surfaces together. Clinch the protruding ends. Finish the nailing pattern, driving the nails at a slight angle.

Some carpenters lay a bead of caulk along the joint formed by the band joist and the sill to keep out air and dust. A sill sealer serves the same purpose.

Green Note
The development of modern engineered wood products has done much to support the movement toward green construction. According to the APA—The Engineered Wood Association, I-joists use 50% less wood than old-style sawn 2× joists. LVL, frequently used for girders and rim boards, also conserves wood by providing the required strength characteristics while using less wood. LSL further conserves resources because it can be manufactured from small-diameter, misshapen trees that would otherwise be unusable. Because engineered wood products can be purchased in any length, the waste at the construction site is greatly reduced.

10.4.6 Framing Openings

When installing header and tail joists for the opening, first, nail in the trimmer joists. As described earlier, a trimmer is a full-length joist or a stud that reinforces a rough opening. Sometimes, a regular joist is located where it can serve as the first trimmer. **Figure 10-31** is a plan view of a finished assembly.

The length of the headers can be determined from the layout on the band joist. Cut headers and tail joists to length. Make the cuts square and true. Considerable strength will be lost in the finished assembly if the members do not tightly fit together. Lay out the position of the tail joists on the headers by transferring the marks from the band joist.

Pro Tip
Be accurate in laying out and cutting floor framing members. The strength of the assembly depends on all of the parts tightly fitting together.

After the headers are installed, the tail joists are attached. One of the tail joists can be temporarily nailed to the trimmer on each side of the opening to accurately locate the header and hold it while it is being nailed.

Figure 10-32 illustrates the procedure for fastening tail joists, headers, and trimmers. After the first header and tail joists are in position between the first trimmers, nail the second or double header in place.

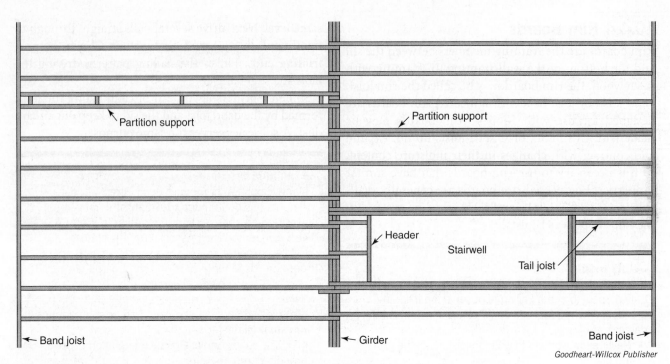

Figure 10-31. This is a partial plan view of a floor framed for a stair opening and partition supports.

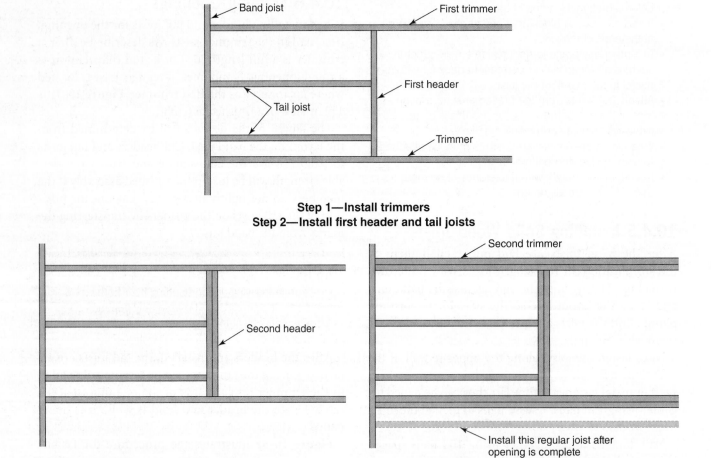

Figure 10-32. These steps can be followed in assembling frames for floor openings. In step 4, waiting to install the regular joist provides clearance for nailing the second trimmer.

Be sure to nail through the first trimmer into the second header. Use three 16d nails at each end. Finally, nail the second trimmer to the first trimmer.

A good nailing pattern for the entire assembly is shown in **Figure 10-33**. To allow more room for nailing, do not install joists adjacent to trimmers until the trimmers are doubled and nailed. This nailing pattern will support a concentrated load of 300 lb at any point on the floor. It will also hold a uniformly distributed load of 50 lb per sq ft with any spacing and span of tail joists ordinarily used in residential construction. This is true only if the long dimension of the floor opening is parallel to the joist. If the long way of the opening is at a right angle to the joists, excessive loading may be carried to the junction of headers with trimmers. Anticipated loads should be checked and more nails or additional supports should be provided at these junctions when needed.

Metal joist hangers are often used to assemble headers, trimmers, and tail joists, **Figure 10-34**. They are manufactured from 18 gauge, zinc-coated sheet steel. The National Forest Products Association recommends the use of joist hangers or ledger strips to support tail joists that are over 12′ long.

10.4.7 Bridging

Bridging is wood or metal pieces fitted in pairs from the bottom of one floor joist to the top of adjacent joist and used to distribute the floor load. It is installed between joists to hold joists vertical and help transfer

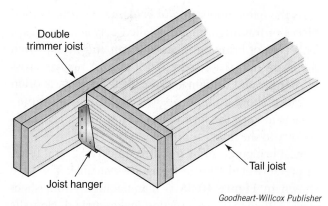

Goodheart-Willcox Publisher

Figure 10-34. Joist hangers are often used for assembling floor framing members.

loads from one joist to the next. Studies have shown that bridging may be eliminated if joists are properly secured at the ends and subflooring is adequate and carefully nailed. This illustrates the importance of quality work practices. Even so, many local building codes include bridging requirements, and general standards suggest that bridging be installed at intervals of no more than 8′.

Regular bridging, sometimes called herringbone bridging or *cross bridging*, is composed of pieces of 1×3 or 2×2 lumber diagonally set between the joists to form an X. Its purpose is to keep the joists in a vertical position and to transfer the load from one joist to the next.

Figure 10-35 shows how a carpenter's framing square can be used to lay out a pattern for bridging. Pieces can quickly be cut using a power miter saw.

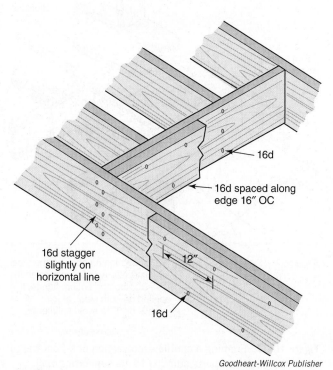

Goodheart-Willcox Publisher

Figure 10-33. Nailing pattern for attaching floor opening members.

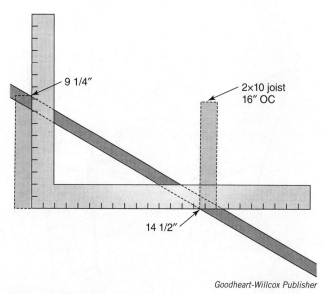

Goodheart-Willcox Publisher

Figure 10-35. A carpenter's framing square can be used to lay out bridging. The line for the lower cut can be found by shifting the tongue of the square to the 14 1/2″ mark on the stock.

As the name implies, **solid bridging** consists of solid pieces of framing lumber installed between the joists. This type of bridging is easier to cut and install when there are odd-sized spaces in a run of regular cross bridging. Solid bridging, also called blocking, is often installed above a supporting beam where its chief purpose is to keep the joist vertical. However, it also adds rigidity to the floor.

Steel bridging can be installed quickly. Several types of prefabricated steel bridging are available. The type shown in **Figure 10-36** is manufactured from sheet steel. A V-shaped cross section makes it rigid. No nails are required and it is driven into place with a regular hammer.

After the bridging is installed, the floor frame should be carefully checked to see that nailing patterns have been completed in all members. After this is done, the frame is ready to receive the subflooring. **Subflooring** is a covering of boards or panels, nailed to the joists, over which a finish floor is laid.

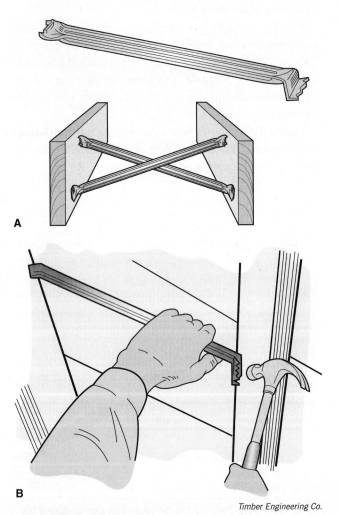

Figure 10-36. Using steel bridging. A—One type of steel bridging. B—Installation method.

Timber Engineering Co.

PROCEDURE
Installing Wood Bridging

1. Snap a chalk line across the tops of the joists at the center of their span.
2. Use two 8d nails to attach the top of the bridging to each side of every joist. Alternate the positioning of the bridging, first on one side of the chalk line and then on the other.
3. After the subflooring is complete or before the under surfaces of the floor are enclosed, the lower ends of the bridging are nailed to the joist.

10.5 Special Framing Problems

A building's design may include a section of floor that overhangs a lower floor or basement level. When the floor joists run at right angles to the walls, use longer joists, **Figure 10-37**. If the run of joists is parallel to the supporting wall, extend **cantilevered** (extending horizontally beyond a supporting surface) joists inward two to three times the length of the overhang, **Figure 10-38**. The exact spacing between joists and length of the members depends on the weight of the outside wall. This information is covered in IRC Section R502.3.3.

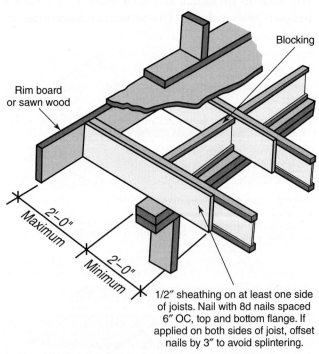

Blocking

Rim board or sawn wood

2'-0" Maximum

2'-0" Minimum

1/2" sheathing on at least one side of joists. Nail with 8d nails spaced 6" OC, top and bottom flange. If applied on both sides of joist, offset nails by 3" to avoid splintering.

Goodheart-Willcox Publisher

Figure 10-37. Framing a cantilevered section of a floor frame when the joists run perpendicular to the supporting wall. Blocking holds the joists vertical, adds rigidity, and closes up the space to provide a firestop.

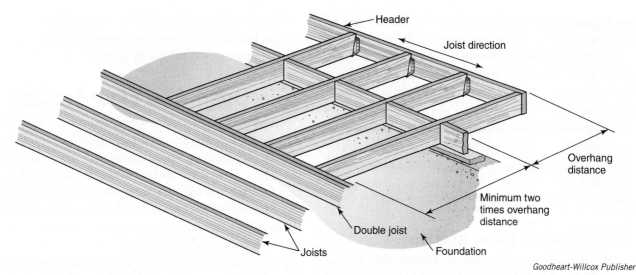

Goodheart-Willcox Publisher

Figure 10-38. When cantilevered joists run parallel to the supporting wall, extend the overhanging joists inward two to three times the length of the overhang. Always check local codes for exact requirements.

Entrance halls, bathrooms, and other areas are often finished with tile or stone that is installed on a concrete base. To provide room for this base, the floor frame must be lowered. When the area is not large, using doubled joists of a smaller dimension is acceptable, **Figure 10-39.** Reducing the distance between joists provides additional support. When the area is large, add steel or wood girders and posts.

Bathrooms must support unusually heavy loads—heavy fixtures like bathtubs and often the additional weight of a tile floor. The fixed dead load imposed by a tile floor averages around 30 lb per sq ft. The load from bathroom fixtures adds from 10 to 20 lb per sq ft, for a total of 40–50 lb dead load. In addition, it is frequently necessary to cut joists to bring in water service and waste drains. Therefore, take special precautions in framing bathroom floors to provide adequate support.

10.5.1 Cutting Openings in Solid Wood Floor Joists

Before cutting joists to install plumbing, it is useful to know how stress affects floor joists. This knowledge will help you determine where to make holes and cut notches. When the top of a joist is in compression and the bottom in tension, there is a point at which the stresses change from one to the other. At this point, there is neither tension nor compression. In the usual rectangular joist, this point is assumed to be midway between the top and bottom, **Figure 10-40.** Variations in the quality of lumber and other conditions may shift the point slightly. Still, this assumption is accurate enough. Since there is no compression or tension at the center, a hole has little effect on the joist's strength, provided the hole is not larger than one-third of the total depth of the joist.

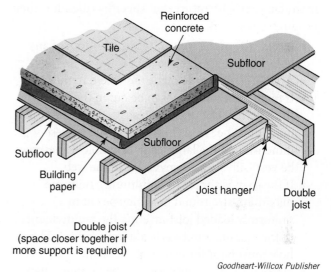

Goodheart-Willcox Publisher

Figure 10-39. Smaller, doubled joists are used when a concrete base is needed for tile or stone surfaces.

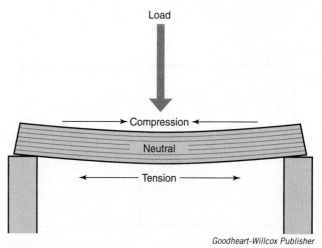

Goodheart-Willcox Publisher

Figure 10-40. When a load is placed on a joist or beam, the top is placed under compression and the bottom is placed under tension. The center of the joist is under little stress.

See **Figure 10-41**. The IRC requires that no holes may be made within 2″ of the top or bottom of the joist.

Weight produces the greatest bend if it is at the center of the span. Therefore, a hole is more likely to reduce the strength of a joist or beam if it is near the center of the span.

> **Pro Tip**
> When possible, cut holes at or close to the vertical middle of a joist. If the opening is limited to 1/3 of the total joist width, the reduction in strength is insignificant.

10.5.2 Cutting Holes in I-Joists

Most I-joists have perforated knockouts that can be removed to provide holes for pipes and wiring. Special rules apply for cutting through I-joists to run pipes, ducts, and electrical conduit or cable. Check the joist manufacturer's literature for specific rules for their product. In general:

- Leave 1/8″ of the web on top and bottom of any hole. Do not cut flanges.
- A 1 1/2″ hole can be made anywhere in the web.
- If more than one hole is to be cut in the web, the length of the uncut web section between the holes must be twice the length of the longest dimension of the largest adjacent hole. Holes may be vertically located anywhere in the web.
- If the span is simple (5′ minimum), only one maximum size round hole may be cut in a uniformly loaded joist meeting the requirements of the manufacturer (as listed in its guide). The hole must be at the center of the span.
- Holes through a cantilever can be no more than 1 1/2″ in diameter.
- Provide at least 1 1/2″ between a hole and a bearing surface.

- In determining allowable spacing of field-cut holes, 1 1/2″ knockouts provided by the manufacturer need not be considered.

10.6 Open-Web Floor Trusses

Open-web floor trusses are sometimes used in new construction. They are designed with the aid of computers and factory-built to specifications for their intended use. These designs ensure that loading requirements are met through the use of a minimum amount of material. Engineered jigs are used in the assembly to build in the proper camber (bend) in each unit. Open-web floor trusses offer several advantages:

- Since they are manufactured in a variety of lengths and depths, they are suited to many different loading conditions.
- The trusses provide long, clear spans with a minimum of depth (14″ and 16″ are most common).
- The open webs make them lighter and easier to handle.
- Installing plumbing pipes, heating ducts, and electrical systems through the web requires no cutting of material. Cutting is time-consuming and often weakens traditional lumber joists.
- Sound transmission is reduced.

Most trusses are fabricated with wooden *chords* (top and bottom members of a truss) connected by galvanized steel webs. The webs have metal teeth that are pressed into the sides of the chords. They also have a reinforcing rib that withstands both tension and compression forces. See **Figure 10-42**.

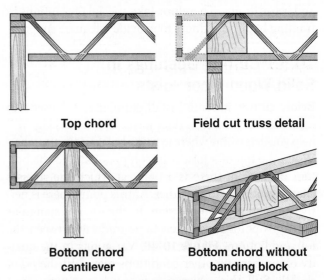

Top chord   Field cut truss detail

Bottom chord cantilever   Bottom chord without banding block

TrusWal Systems Corp.

Figure 10-42. Truss construction details. Chords are made of lumber. Webbing is galvanized steel. Trusses provide a wide nailing surface because the chord is laid flat.

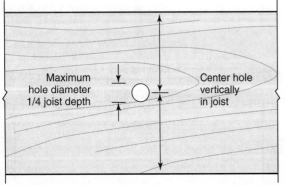

Maximum hole diameter 1/4 joist depth   Center hole vertically in joist

Goodheart-Willcox Publisher

Figure 10-41. Holes for plumbing pipes should be vertically centered in the joist and have a diameter no more than 1/4 of the depth of the joist.

10.7 Subfloors

The laying of the *subfloor*, or rough flooring, is the final step in completing the floor frame. Panel materials are used for this purpose. OSB and plywood are the materials of choice. These products are discussed in Chapter 3, **Building Materials**. The subfloor serves three purposes:

- Adds rigidity to the structure
- Provides a base for finish flooring material
- Furnishes a work surface where a carpenter can lay out and construct additional framing

10.7.1 OSB

OSB is a popular choice for subflooring because, unlike plywood, it does not delaminate when it is subjected to moisture. OSB does, however, absorb moisture, causing the edges to swell and the surface to become rougher. To combat this problem, a modified type of OSB has been developed, using waterproof resin to impregnate the fibers and bind them. One manufacturer of these panels guarantees them against water damage for 500 days, which is more than enough time to enclose the building. OSB panels for flooring are available in thicknesses from 19/32″ to 1 1/8″ with tongue-and-groove edges.

OSB should be installed with the long edges running perpendicular to the joists. The boards should be fastened using urethane-based adhesive and with nails or screws spaced every 6″ on supported panel ends and every 12″ in the centers of the panels. Fasteners should penetrate the wood joist 1″. All plywood and OSB expands slightly as it absorbs moisture, so a 1/8″ gap should be left between edges and ends of sheets. A 10d box nail makes a good spacer. Most OSB flooring products have self-spacing tongue-and-groove edges, so no additional space is required. Joints should be staggered in successive courses, **Figure 10-43**.

Stiffness is increased significantly when panels are glued to the joists in addition to using fasteners. This system ensures squeak-free construction, eliminates nail popping, and reduces labor costs.

Before each panel is placed, a 3/8″ bead of construction adhesive (glue) is applied to the joists, as shown in **Figure 10-44**. Spread only enough adhesive to lay no more than two panels. Two beads of adhesive are applied on joists where panel ends butt together. All screwing or nailing must be completed before the adhesive sets.

When laying tongue-and-groove panels, apply adhesive along the groove. Use a 1/8″ bead so that excessive squeeze-out is avoided. The bead can be either continuous or spaced. Drive sheets into the groove of previously laid subflooring by using a framing hammer and a protective piece of scrap lumber. Leave a 1/8″ space

Photo courtesy of Huber Engineered Woods, ©2020 Huber Engineered Woods, LLC.

Figure 10-44. Applying construction adhesive for a glued floor system. A 1/4″ wide bead of adhesive applied to the top of the joist is sufficient. Two beads are applied where panel ends butt.

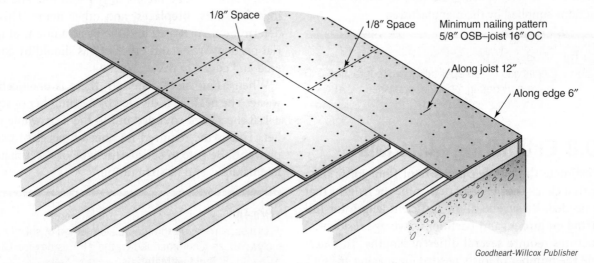

Goodheart-Willcox Publisher

Figure 10-43. Properly installing OSB subfloor. Note the nailing pattern and how joints are staggered for added strength.

TFoxFoto/Shutterstock.com

Figure 10-45. Fasten the glued subflooring with mechanical fasteners before the adhesive sets.

at all end and edge joints. Immediately fasten the subfloor with screws or nails. Use a nail gun or an autofeed screw gun to work more quickly, **Figure 10-45**.

10.7.2 Plywood

Plywood was once considered the best product for subflooring. It has been replaced by OSB as the most common subflooring material. Both provide a smooth, even base and act as a horizontal diaphragm that adds strength to the building.

5/8″ or 3/4″ plywood subflooring is applied much like OSB, with the long dimension of the sheet perpendicular to the joists and the end joints staggered. Use 8d nails or 1 5/8″ screws spaced 6″ along edges and 10″ along intermediate members. Plywood intended for subflooring has tongue-and-groove edges, similar to that found on OSB flooring.

Subfloor underlayment panels are available for joists or beam spacing of 16″, 20″, 24″, or 48″. Maximum support spacing is stamped on each panel. A 3/4″ thickness is used for 24″ OC spacing. Be sure to follow the instructions supplied by the manufacturer.

Pro Tip
Material specifications and application procedures for subflooring are provided in a booklet published by APA—The Engineered Wood Association.

10.8 Estimating Materials

To estimate the number and size of floor joists on a job, first check the plan and determine the lengths that are needed. Be sure to allow sufficient length for full bearing on girders and partitions. Average residential structures require several different lengths. To determine the number of joists needed for spacing 16″ OC,

multiply the length of the wall that carries the joists by 3/4, then add one. For 24″ OC spacing, multiply the length of the wall by 1/2. Also, add extra pieces for doubled joists under partitions and trimmer joists at openings. If solid wood joists are being used, add extra pieces for headers at openings. If the floor is to be framed with I-joists, add LVL according to the plans for headers at openings. Note the following example:

No. of joists = length of wall × 3/4 + 1 + extras

The sizes and numbers of joist headers will be the lengths of the foundation sections on which they rest. The depth of the band joists will be the same as the adjoining floor joists. Your estimate should include the cross-sectional size and the number of pieces of each length. For example, a complete estimate might be one of the following:

Required floor joists:
  20 pcs—1 3/4″ × 9 1/2″ × 28′-0″ I-joists
  12 pcs—1 3/4″ × 9 1/2″ × 32′-0″ I-joists
  2 pcs—1 3/4″ × 9 1/2″ × 14′-0″ I-joists

Required joist headers:
  2 pcs—1 1/4″ × 9 1/2″ × 20′-0″ LVL rim board
  2 pcs—1 1/4″ × 9 1/2″ × 12′-0″ LVL rim board
  1 pc—1 1/4″ × 9 1/2″ × 8′-0″ LVL rim board for header

For solid wood:
  40 pcs—2×10 × 16′-0″
  36 pcs—2×10 × 14′-0″

Required band joists:
  4 pcs—2×10 × 16′-0″
  2 pcs—2×10 × 12′-0″

Procedures for estimating the subflooring vary, depending on the type of material used. Usually the area is calculated by multiplying the overall length and width and then subtracting major areas that are not to be covered. These include breaks in the wall line and openings for stairs, fireplaces, and other items. This will determine the net area and the basic amount of material needed. Waste and other extras should be considered and added to this.

When using sheet materials, there is practically no waste. The net area is divided by 32 (number of sq ft in a 4×8 sheet) and rounded out to the next whole number. This is the required number of pieces of OSB or plywood. Be sure to specify the type of sheet material, its thickness, and its span rating.

Pro Tip
An alternative and more complete method of specifying plywood, as recommended by the APA, is provided in Chapter 3, **Building Materials**.

Chapter Review

Summary

- There are three basic types of framing used in residential construction: platform (western) framing, balloon framing, and post-and-beam framing.
- Platform framing is by far the most common type of framing. The first floor is constructed on top of the foundation wall and used as a work area for constructing the remaining wall, floor, and roof framing components.
- Floor framing consists of joists that rest on the foundation walls or girders. The joist ends rest on a sill that is fastened to the foundation wall and are held in position by a joist header fastened around the perimeter.
- Bridging may be installed between joists to hold joists vertical and help transfer loads from one joist to the next.
- Special framing must be used for floor areas cantilevered beyond the foundation walls.
- Some building plans call for open-web trusses for floor joists. They are light, strong, and provide space for electrical, piping, and HVAC installations.
- Subfloors of OSB or plywood are installed on top of the joists, making the whole floor structure more rigid.
- Proper fastening of the subfloor to the joists is important for strength, smoothness, and durability.

Know and Understand

Answer the following questions using the information in this chapter.

1. The type of framing used in most residential construction is _____ framing.
   - A. platform
   - B. balloon
   - C. post-and-beam
   - D. plank-and-beam

2. The studs of a balloon-type frame run continuously from the _____ to the rafter plate.
   - A. insulation
   - B. ribbon
   - C. sill
   - D. joists

3. _____ may be solid timbers, built-up lumber, engineered lumber, or steel beams.
   - A. Ribbons
   - B. Joists
   - C. Sills
   - D. Girders

4. Steel beams vary in depth, width of flange, and _____.
   - A. weight
   - B. grade
   - C. species
   - D. height

5. _____ are most popular for girder and beam support.
   - A. Solid timbers
   - B. Built-up lumber
   - C. Steel posts
   - D. Engineered lumber

6. *True or False?* The term mudsill originated from the process of correcting irregularities in masonry work by embedding sill in a layer of fresh mud.

7. Which of the following is *not* typically used as floor joists in residential construction?
   - A. I-joists.
   - B. Nominal 2″ lumber placed on edge.
   - C. Steel bar joists.
   - D. Open-web truss joists.

8. The most common spacing of wood joists is _____″ OC.
   - A. 12
   - B. 16
   - C. 19.2
   - D. 24

9. In platform construction, the joist spacing is usually laid out on the _____ joist rather than the sill.
   - A. band
   - B. floor
   - C. tail
   - D. I-

10. *True or False?* Regular bridging is designed to keep joists in vertical position and transfer the load from one joist to the next.

11. *True or False?* Steel bridging is installed with nails and a nail gun.

12. Before cutting joists to install plumbing, it is useful to know how _____ affects floor joists.
    A. material   C. temperature
    B. stress   D. piping

13. Most trusses are fabricated with wooden _____ connected by galvanized steel webs.
    A. knockouts   C. chords
    B. webs   D. bars

14. OSB is often preferred for subflooring over plywood because it does not delaminate when it is exposed to _____.
    A. heat   C. heavy loads
    B. termites   D. moisture

15. All plywood and OSB expands slightly, so when installing, leave a _____″ gap between edges and end of sheets.
    A. 1/8   C. 1 1/8
    B. 3/8   D. 2

For Questions 16–19, use the following simplified sketch of a floor frame outline to answer the questions.

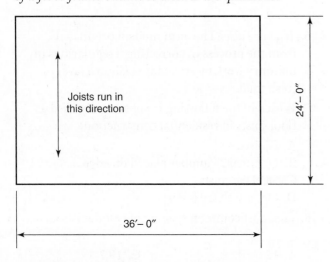

16. If the joist spacing is 16″ OC and there are no floor openings, how many joists are needed?
    A. 28 joists.   C. 58 joists.
    B. 31 joists.   D. 65 joists.

17. If the joist spacing is 24″ OC and there are no floor openings, how many joists are needed?
    A. 12 joists.   C. 19 joists.
    B. 15 joists.   D. 24 joists.

18. What is the total length of header or rim joist needed for the floor frame?
    A. 36′   C. 61′
    B. 54′   D. 72′

19. How many sheets of 4′×8′ sheathing are needed for the subfloor? (Assume no floor openings and no extra sheathing for waste.)
    A. 11 sheets.   C. 27 sheets.
    B. 19 sheets.   D. 44 sheets.

Apply and Analyze

1. How do you size a built-up girder?

2. Why is it important to install a metal cap when a wood column supports a wooden girder?

3. How can you avoid measuring every individual joist space when laying out joists?

4. When is it best to use solid bridging, or blocking, over regular cross bridging?

5. How much space should be left between holes horizontally cut into the web of an I-joist?

6. Explain the purpose of subflooring and underlayment.

7. Describe how to perform an estimate for floor framing materials.

Critical Thinking

1. When floor systems are framed with offset joists, such as in **Figures 10-14** and **10-24**, the sheathing pattern is impacted. Explain how and give a method that can be used to transition from one side of the girder to the other.

2. Research the strengths and weaknesses of using manufactured wood I-joists as opposed to using traditional framing lumber. When building your own home which would you choose and why?

Communicating about Carpentry

1. **Speaking and Listening.** Divide into groups of four or five students. Using your textbook as a starting point, research your topic and prepare a report on how to install platform floor framing. As a group, deliver your presentation to the rest of the class. Take notes while other students give their reports. Ask questions about any details that you would like clarified.

2. **Speaking.** Pick a figure in this chapter. Working with a partner, tell and retell the important information being conveyed by that figure.

Through your collaboration, develop what you and your partner believe is the most interesting verbal description of the importance of the chosen figure. Present your narration to the class.

3. **Speaking.** Obtain (or draw) a floor framing assembly. Label the areas on the assembly where these types of components are located: joists, sills, anchor straps, anchor bolts, joists, openings in the framing, and bridging. Be prepared to explain the purpose of each component.

4. **Speaking and Reading.** From the local building code in your area, find the requirements for floor framing. Prepare a list of the requirements along with sketches to clarify complicated written descriptions. Make an oral report to your class and pass around your sketches.

5. **Drawing.** Obtain a set of architectural plans for a house with a conventional basement. Study the methods of construction specified in the section and detailed drawings. Then, prepare a first-floor framing plan. Start by tracing the foundation walls and supports shown in the basement or foundation plan and then add all joists, headers, and other framing members. Your drawing should be similar to the partial drawing shown in **Figure 10-31**.

CHAPTER **11**

Wall and Ceiling Framing

OBJECTIVES

After studying this chapter, you will be able to:

- Identify the main parts of a wall frame.
- Explain methods of forming the outside corners and partition intersections of wall frames.
- Show how rough openings are handled in wall construction.
- Explain plate and stud layout.
- Describe the construction and erection of wall sections and partitions.
- List the materials commonly used for sheathing.
- Demonstrate the process of ceiling frame construction.
- Explain the benefits of using advanced framing.
- Explain what SIPs are and how they are erected.
- Estimate materials required for wall frames, ceiling frames, sheathing, studs, and headers.

TECHNICAL TERMS

advanced framing
ceiling frame
ceiling joist
cripple stud
header
housewrap
metal-strap bracing

nailer
partition
rough opening
sheathing
soffit
sole plate
story pole

strongback
structural insulated panel
  (SIP)
stud
top plate
trimmer stud

Wall framing is assembling the vertical and horizontal members that support the outside and inside walls of a structure. This frame also supports upper floors, ceilings, and the roof. It serves as a nailing base for inside and outside wall-covering materials. Inside walls that do not carry any structural load are called *partitions*.

There are references to 16d common nails and 8d common nails throughout this chapter. Wherever penny sizes are referenced, the equivalent size nail gun nail is acceptable.

11.1 Parts of the Wall Frame

Anyone planning to become a carpenter needs to know the correct names for the parts of a house frame. The wall-framing members used in conventional platform construction include sole plates, top plates, studs, headers, and sheathing. *Studs* are the vertical members of the wall frame. Except where interrupted by openings for windows and doors, they run full length from sole plate to top plate. The *sole plate* is the horizontal member at the bottom of the wall frame, supported by a wood subfloor, concrete slab, or other closed surface. The *top plate* is the horizontal member at the top of a wall frame. All of these components covered on the exterior by boards or prefabricated panels known as *sheathing*.

Short studs, known as *cripple studs*, are used above and below wall openings. Full-length studs become cripple studs when they end due to an opening. See **Figure 11-1** and **Figure 11-2**. Note that extra studs are used at the corners, at the sides of the rough openings for doors and windows, and where an interior wall meets an outside wall.

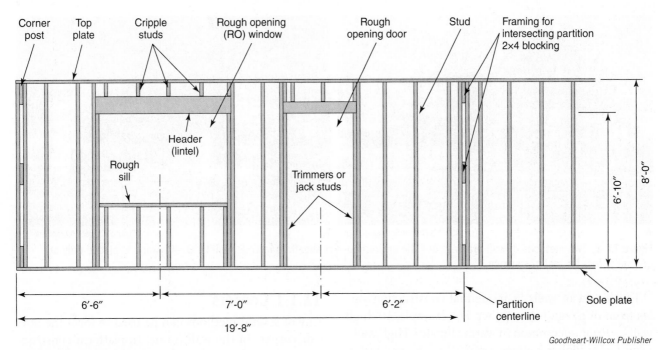

Figure 11-1. This is a drawing of a typical wall frame section with its members named. An extra stud and 2×4 blocking at an intersecting partition provide a nailing surface for an inside corner.

Goodheart-Willcox Publisher

Figure 11-2. This typical framing shows a rough opening for a window. Cripple studs and trimmers are in place.

Trimmer studs, or jack studs, are shortened studs that stiffen the sides of rough openings. They bear the direct weight of a header. Normally, carpenters install doors and windows so the tops are at the same height. This practice not only improves appearance, but also allows all trimmers to be cut the same length. All of the trimmers can be cut at one time, speeding up production.

Studs and plates are made from 2×4 or 2×6 lumber. Headers usually require heavier material.

It is common practice to sheath the exterior walls with structural OSB or plywood. These panels provide rigidity to the wall frame to keep it from racking (shifting out of square). If the wall is not sheathed with approved panels, bracing made of 1×4 stock or steel strips must be let into the wall frame at corners. An alternative method of bracing uses structural panels at the corners and covers the rest of the wall with nonstructural material, **Figure 11-3**.

In one-story structures, studs are sometimes placed 24″ OC (on center). However, 16″ OC spacing is more common. This spacing has evolved from years of established practice. It is based more on accommodating the wall-covering materials than on the actual calculation of imposed loads.

The stud spacing remains consistent, regardless of interruptions by openings or intersecting walls. This is done so that 4′ × 8′ panels can be installed with minimum cutting or waste.

A
Cynthia Farmer/Shutterstock.com

B
Cynthia Farmer/Shutterstock.com

Figure 11-3. Two methods of wall bracing. A—The entire house is sheathed in OSB panels. B—Plywood is placed only at critical points for rack resistance.

The height of walls in residential construction varies from one region to another. For example, 10′ high walls are not uncommon in warm climates. High walls, resulting in higher ceilings, are used to create a feel of spaciousness in a room. In colder climates, 8′ high walls are more common to help minimize heating costs.

Wall-framing lumber must be strong and straight with good nail-holding power. Warped lumber is not acceptable, especially if the interior finish is drywall. Stud grade, also called No. 3 grade, is approved and used throughout the country. Species such as Douglas fir, larch, hemlock, yellow pine, and spruce are satisfactory. Where straightness is especially critical, such as kitchen walls where cabinets will be installed, engineered lumber studs can be used. See Chapter 3, **Building Materials**, for additional information.

Pro Tip
Wall studs can be purchased precut to 92 5/8″ for a standard wall height of just over 8′. Buying precut studs saves cutting time on the job.

11.1.1 Corners

Any of several methods can be used to form the outside corners of the wall frame. In platform construction, the wall frame is usually assembled in sections on the rough floor and then tilted up into place. Corners are formed when a sidewall and end wall are joined.

One common method of corner construction uses a stud spaced a distance equal to the wall thickness from the end of the wall. Three or four short pieces of blocking are nailed in the space between the two studs. The adjoining wall is butted against the double stud, forming a surface for fastening interior wall finish on both walls, **Figure 11-4A**. Another method is to turn the extra stud 90°, as shown in **Figure 11-4B**.

Select only straight studs for corners. Assemble the wall with 10d nails spaced 12″ apart. Stagger the nails from one edge to the other. Attach the filler blocks with nails as well.

In climates that require the house to be well insulated, 2×6 studs are commonly used for exterior walls. This allows thicker insulation to be installed. **Figure 11-5** shows typical corner construction for 2×6 framing.

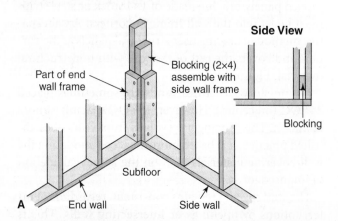

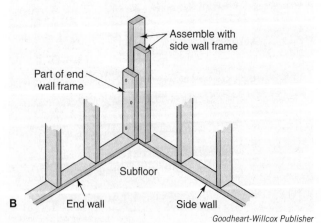

Goodheart-Willcox Publisher

Figure 11-4. Placement of studs to form corners in platform construction. A—Corner built from three full studs and blocking. B—Corner built with three full studs and no blocking.

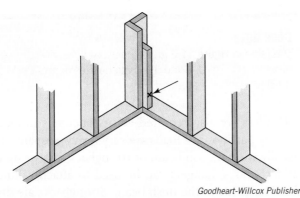

Goodheart-Willcox Publisher

Figure 11-5. Typical corner construction in 2×6 framing. The inside corner may be formed with a 2×4 (arrow).

11.1.2 Partition Intersections

Partitions should be solidly fastened to the outside walls. This requires extra framing on the outside wall. The framing must also provide a nailing surface on inside corners for wall covering, such as drywall. Several methods can be used to accomplish these purposes:

- Install extra studs in the outside wall and attach the partition to them.
- Insert blocking and nailers between the regular studs.
- Use blocking between the regular studs and attach nailers or backup clips to support inside wall coverings at all inside corners.

A *nailer* is lumber added as a backing at inside corners. **Figure 11-6** shows various methods of framing partitions.

11.1.3 Rough Openings

Study the house plans to learn the size and location of the *rough openings*, openings formed by framing members to receive and support windows or doors. A rough opening is often referred to as RO on drawings. Plan views have dimension lines. Usually, the measurement is taken from corners or intersecting partitions to the centerlines of the openings. Heights of rough openings are given in elevation and section views. Sizes of rough openings are listed in a table called a door and window schedule.

Headers support the weight of the building across door and window openings. Door and window headers are a different use of the same principle as headers in a floor frame. One way to make a header is to cut and nail together two or three framing members. Insert 1/2″ plywood spacers between the pieces to make the header the same thickness as the wall, **Figure 11-7**. Use 12d or 16d nails and stagger them 16″ on center. Fasten the header in the rough opening using 16d nails driven through the studs into the ends of the header.

The header length is equal to the rough opening plus the width of two trimmers (3″). The width of the lumber to be used in the header depends on the span of the opening, the load that the building will place on the header, and the expected snow load. Local building codes or the building plan usually include requirements for headers.

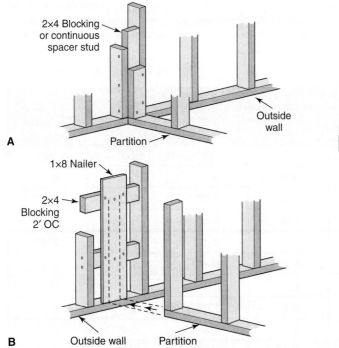

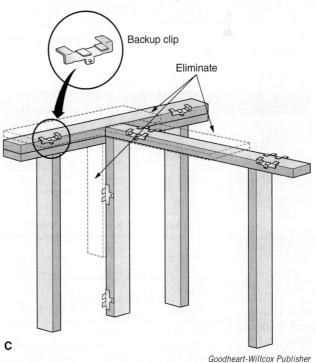

Goodheart-Willcox Publisher

Figure 11-6. Framing details where partitions intersect outside walls. A—Using extra studs. B—Blocking installed between studs. C—Backup clips are sometimes used and take the place of some framing studs.

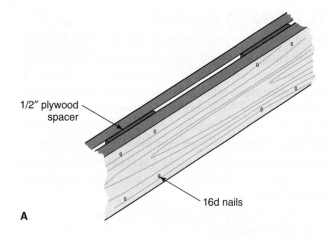

1/2″ plywood spacer

16d nails

A

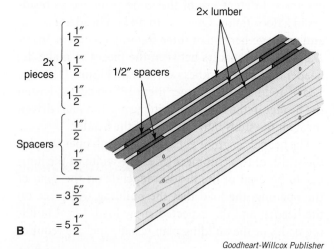

2× lumber

$1\frac{1}{2}″$
2x pieces $1\frac{1}{2}″$
$1\frac{1}{2}″$

1/2″ spacers

Spacers $\frac{1}{2}″$
$\frac{1}{2}″$

$= 3\frac{5}{2}″$

$= 5\frac{1}{2}″$

B

Goodheart-Willcox Publisher

Figure 11-7. Headers may be built up of several members. Place plywood spacers 16″–24″ apart on centers. A—A header for a 2×4 stud wall requires two members with 1/2″ spacers to equal the width of the studs. B—A 2×6 stud wall requires three members with 1/2″ plywood spacers between each member.

Code Note
The table in **Figure 11-8** gives the allowable spans of headers according to the International Residential Code (IRC) for various load conditions.

Headers are also required across openings in load-bearing partitions. If loads are very heavy or spans unusually wide, a flush beam or strongback may be used. In such cases, hangers can be used to attach ceiling-for-floor joists to the flush beam. Strongbacks are discussed later in this chapter. **Figure 11-9** shows a truss joist being attached to a microlam header flush beam.

Kasten-Weiler Construction

Figure 11-9. When open-web joists are used for a second-story floor frame, an LVL header may be used to support a cripple joist.

Allowable Header Spans

Header Supporting	Size	Span with 30 lb per sq ft Snow Load		Span with 50 lb per sq ft Snow Load		Span with 70 lb per sq ft Snow Load	
		20′	28′	20′	28′	20′	28′
Roof and Ceiling	2-2×4	3′-6″	3′-2″	3′-2″	3′-6″	2′-10″	2′-6″
	2-2×10	8′-5″	7′-3″	7′-3″	6′-3″	6′-6″	5′-7″
	3-2×8	8′-4″	7′-5″	7′-5″	6′-5″	6′-8″	5′-9″
Roof, Ceiling, and One Clear-Span Floor	2-2×4	3′-1″	2′-9″	2′-9″	2′-5″	2′-7″	2′-3″
	2-2×10	7′-0″	6′-2″	6′-4″	5′-6″	5′-9″	5′-1″
	3-2×8	7′-2″	6′-3″	6′-5″	5′-8″	5′-11″	5′-2″
Roof, Ceiling, and Two Clear-Span Floors	2-2×4	2′-1″	1′-8″	2′-0″	1′-8″	2′-0″	1′-8″
	2-2×10	4′-9″	4′-1″	4′-8″	4′-0″	4′-7″	4′-0″
	3-2×8	4′-10″	4′-2″	4′-9″	4′-1″	5′-5″	4′-8″

Goodheart-Willcox Publisher

Figure 11-8. The allowable header spans for exterior walls under various loads according to the IRC.

In conventional framing, extra studs are included around rough openings, as shown in the assembly in **Figure 11-10**. The studs and trimmers support the header and provide a nailing surface for window and door casing.

11.1.4 Alternate Header Construction

In large window openings, the size of the header may reduce the length of the upper cripple studs to a point where they cannot be easily assembled. In this case, the cripple studs can be replaced with flat blocking. Another solution is to increase the header size to completely fill the space to the plate. Some builders follow this practice and extend it to include all openings, regardless of the span. The cost of labor required to cut and fit the cripple studs is usually greater than the cost of the larger headers. A disadvantage of such construction is extra shrinkage. Shrinkage may cause cracks above doors and windows unless special precautions are taken when applying the interior wall finish.

Green Note
As described in the previous paragraph, headers are often sized to fill the space above the rough opening to the wall plate. Without any added insulation, this creates a tremendous thermal bridge. To cut heat loss and build greener headers, size headers properly and add a layer of rigid foam insulation between a built-up header's layers or to the exterior face.

11.2 Plate Layout

Sole plates and top plates are the same size as the studs, typically 2×4 or 2×6. Use only straight stock for plates. Select two pieces of equal length and lay them side by side along the location of the outside wall. The length is determined by what can be easily lifted off of the floor and into a vertical position after it is assembled. Remember that the weight may include all of the framing for rough openings, bracing, and sheathing. If wall jacks or a forklift are available for lifting, sections can be made larger. Where they must be lifted by hand, attach sheathing after the wall is up. Always locate joints over a stud. The centerlines of rough openings are marked first.

Pro Tip
Carefully check over your rough opening layouts for errors. Do the math before cutting and framing.

11.2.1 Laying Out the Second Exterior Stud Wall

Laying out the second exterior wall follows the same procedure as the first outside wall, with one exception. If sheet material is used for rough siding, then the location of the first stud from the corner post must allow for the edge of the panel to be flush with the outside edge of the siding. If the siding is 3/4″ thick and the studs are 16″ OC, lay out the first stud 15 1/4″ from the end of the plate.

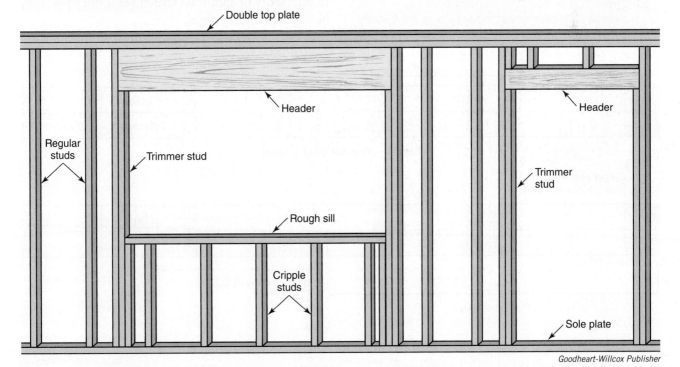

Goodheart-Willcox Publisher

Figure 11-10. Framing door and window openings. Notice how the wide header above the window eliminates cripple studs between the header and the top plate.

Then, when the first siding panel is installed, the first edge will be even with the outside of the siding and the second edge is centered on a stud.

PROCEDURE

Laying Out Plates for the First Outside Wall

1. On the rough floor, mark the width of the plates on all sides from outside in. Snap a chalk line to mark inside edge of the wall frame. Place sole and top plates along main sidewalls. Align the ends with floor frame and mark regular stud spacing along both plates, **Figure 11-11**.

2. Study the architectural plans and lay out the centerline for the rough opening of each door and window.

3. Measure and mark off half of the width of the opening on each side of the centerline.

4. Mark the plate for trimmer studs outside of these points. On each side of the trimmer stud, include marks for a full-length stud. Identify positions with the letter T for trimmer studs and X for full-length studs.

5. Mark all stud spaces located between the trimmers with the letter C. This designates them as cripple studs.

6. Lay out the centerlines where intersecting partitions butt. Add full-length studs at these points, if required by the method of construction.

7. When blocking is used between regular studs, the centerline is needed as a guide for positioning the backing strip.

8. Carefully plan the layout of wall corners so they correctly fit together when the wall sections are erected.

11.2.2 Story Pole

A *story pole* is a long measuring stick created and used by a carpenter on the job. It represents the actual wall frame with markings made at the proper height for every horizontal member of the wall frame—sole plate, rough windowsill, headers, and top plates. See **Figure 11-12**. Its use saves time that would be spent checking the drawings for these dimensions.

Since the pole must be light and easy for a carpenter to handle, it is usually a strip of 1×2 or 1×4 lumber. It must be long enough to reach from the rough floor to the underside of the ceiling or floor joists above.

When marking a story pole, transfer all of the heights for horizontal members from the drawings to the pole at one time. Measurements must be accurate and lines must be square across the pole. With this guide, there is no need to consult the plans time and again to find the lengths of studs, trimmers, and cripple studs. All this information is conveniently listed on the pole. A story pole is particularly useful in split-level construction, multistory buildings, or where stub walls are needed. A stub wall is a short wall that might be used, for example, to separate a breakfast nook from the main kitchen area. Stub walls are also common in framing split-level houses, where one level is not a full story above the one beside it.

When the header height of the doors is different from that of the windows, mark the height on the other side of the story pole. This keeps the two heights separate. In multistory or split-level structures, a story pole may be required for each level.

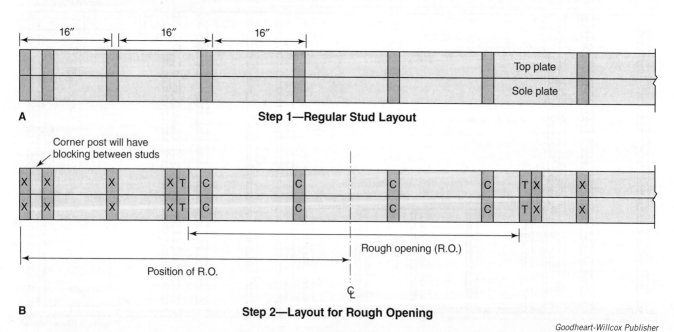

A   **Step 1—Regular Stud Layout**

B   **Step 2—Layout for Rough Opening**

Goodheart-Willcox Publisher

Figure 11-11. Layout of sole and top plates. A—Regular stud spacing has been marked. B—Layout is converted for a window opening. Stud type is marked

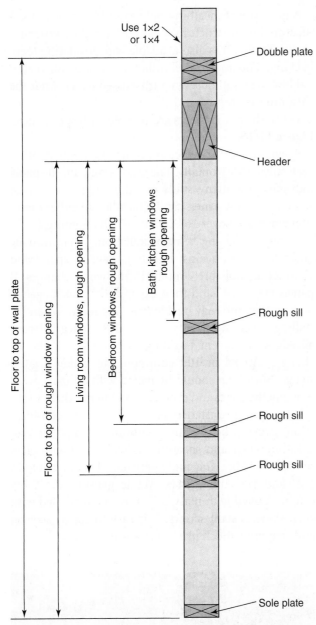

Use 1×2
or 1×4

Double plate

Header

Floor to top of wall plate

Floor to top of rough window opening

Living room windows, rough opening

Bedroom windows, rough opening

Bath, kitchen windows rough opening

Rough sill

Rough sill

Rough sill

Sole plate

Goodheart-Willcox Publisher

Figure 11-12. A story pole is a handy guide that marks the height of every horizontal member and the length of every vertical member of the wall frame. It usually extends only one story, but may include more.

11.3 Wall Sections

Wall sections are assembled on their edges on the rough flooring. All plates, studs, headers, and rough sills are nailed in place.

Wall sheathing is often applied to the frame before it is raised. Make certain that the framework is square before starting the application. Diagonal measurements across the corners must be equal. To keep the frame square while the sheathing is being applied, fasten a diagonal brace across one corner.

PROCEDURE
Constructing a Wall Section

After the layout of the studs has been marked on the sole plate and top plate, use the following procedure to construct the wall:

1. Working from the plans or a story pole, cut the various stud lengths. It is seldom necessary to cut standard full-length studs. These are usually precision end trimmed (P.E.T.) at the mill and delivered to the construction site ready to assemble.
2. Cut the headers and rough sills. Take their lengths directly from the plate layout. Assemble the headers.
3. Move the top plate away from the sole plate about a stud length. Turn both plates on edge with the layout marks inward. Place a full-length stud, crown up, at each position marked on the top plate and sole plate. See **Figure 11-13**.
4. Nail the top plate and the sole plate to the full-length studs using two 16d nails or 3 1/4″ nails if a nail gun is being used.
5. Set the trimmer studs in place on the sole plate and nail them to the full-length studs.
6. Place the header so it is tight against the ends of the trimmers. Nail through the full-length stud into the header using 3 1/4″ nails, **Figure 11-14**.
7. The upper cripple studs, if used, can be installed after the header is installed.
8. For window openings, transfer marks for the cripple studs from the sole plate to the rough sill and assemble the cripple studs with 3 1/4″ nails, **Figure 11-15**. If the wall section is erected before installing the cripples, toenail the lower ends of the cripple studs to the sole plate. Install the rough windowsill.
9. Add studs or blocking at positions where partitions will intersect.
10. Install any wall bracing that may be required for special installations. Remember, the inside of the wall is face down.
11. If the sheathing is to be applied before the wall is erected, it is the last component added after all framing is in place.

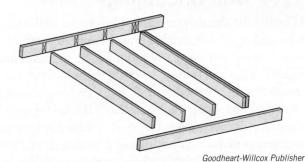

Goodheart-Willcox Publisher

Figure 11-13. Assemble the full-length studs and plates as shown. Place a stud at every position marked on the plate. Turn crowns upward and nail through the plates into the ends of the studs.

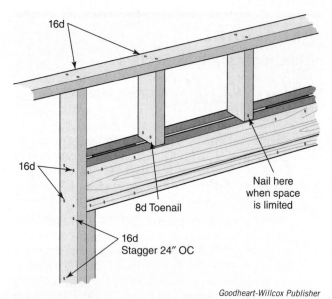

Figure 11-14. How to assemble and fasten headers, cripple studs, and trimmers.

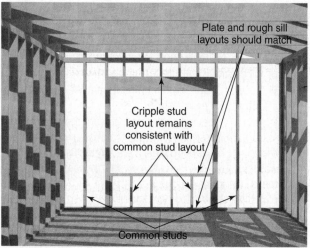

Figure 11-15. Cripple studs can be installed before the wall section is erected. If the wall sections are first erected, toenail the cripple studs to the sole plate.

11.4 Wall Sheathing

Whether the sheathing is applied before the wall frame is erected or after, it should always be applied before the roof framing has begun. Sheathing adds strength and rigidity to the wall frame. Most light frame buildings are sheathed with plywood or OSB. The sheathing material should carry a stamp indicating that it is rated for use as sheathing. If sheathing grade panels are used at least on the corners of the building, diagonal bracing should not be required. Some plans call for rigid insulation to be used as sheathing. In this case, either diagonal bracing or structural panels must be used at the corners.

A new type of sheathing, sold as ZIPSytem®, is manufactured with structural OSB coated with a water-resistive barrier. It is also available with insulating foam backing. This material eliminates the need for diagonal bracing and housewrap (discussed later). After the sheathing is applied and the walls are erected, seams between sheathing pieces are covered with special tape, **Figure 11-16**.

Plywood and OSB sheathing may be applied either vertically or horizontally. The grade stamp on the panel indicates how many studs or other supports it must span. Generally, panels can be installed in either direction when used over 16″ OC or 24″ OC framing.

During panel sheathing installation, steps must be taken to protect against buckling. Panels expand and contract as moisture content in the wood changes. If panels are "tight" and then expand, they push against one another and may buckle from the force. When installing sheathing panels, leave a 1/8″ space between all edges to prevent buckling as humidity in the air changes. An 8d or 10d common nail makes a good spacer. Sheathing should be nailed with 8d nails every 6″ along the edges and every 12″ on intermediate studs.

Where fire resistance is an important consideration, gypsum boards are a common choice for wall sheathing. Gypsum sheathing is reinforced with glass fibers and is fire and mold resistant. It is available in 1/2″ and 5/8″ thicknesses. While gypsum sheathing can be applied with nails, it is most often applied with corrosion-resistant screws. It is easy to cut by scoring and snapping, much like interior wallboard.

Figure 11-16. Coated OSB sheathing keeps water out, is structurally strong, and improves the heating and cooling performance of the building. Joints are sealed with tape made for that purpose.

11.5 Erecting Wall Sections

Most one-story wall sections can be raised by hand, **Figure 11-17**. Larger sections or structures require the use of a crane or other equipment. When raising sections to which sheathing has not been applied, it is good practice to install temporary diagonal bracing if regular bracing is not included. Some carpenters attach temporary blocking to the edge of the floor frame before raising wall sections. The blocking keeps the wall section from sliding off the platform as it is raised.

DJ Kinder/Shutterstock.com

Figure 11-18. The walls are temporarily braced to keep them plumb until the ceiling or roof framing is attached.

Safety Note

Before raising a section, be sure it is in the correct location. Have bracing at hand and ready to be attached. If the section is large, have extra help available. Make sure each worker knows what to do.

Immediately after the wall section is up, secure it with braces attached near the top and running to the subfloor at about a 45° angle, as shown in **Figure 11-18**. Next, make final adjustments in the position of the sole plate. Be sure it is straight. Then, nail it to the floor frame using 16d nails driven through the subfloor and into the joists.

Loosen the braces one at a time and plumb the corners and midpoint along the wall. On one-story construction, a carpenter's level is generally used for this. A plumb line can also be used. Braces temporarily attached to square a wall section can be removed when the permanent braces and sheathing are installed.

After one section of the wall is in place, proceed to other sections. The sequence to be followed is not generally important. Most carpenters prefer to erect main sidewalls first and then tie in end walls and smaller projections. The sequence for each individual project can be decided on as the wall framing is begun. Design and construction methods help determine how to proceed.

Pro Tip

When plumbing a wall with a carpenter's level, hold the level so you can look straight in at the bubble. If the wall framing member or surface is warped, you should hold the level against a long straightedge that has a spacer lug at the top and bottom. Some carpenters use an extendable level that allows you to plumb from the top plate to the bottom plate of a wall.

A   *North Bennett Street School, Boston*   **B**   *North Bennett Street School, Boston*

Figure 11-17. Tilting up wall sections. A—Raising a section of wall constructed from 2×6s. B—The front wall of garage is being raised by student carpenters. Note the angle and cross bracing to keep the center from racking.

11.6 Partitions

When the outside wall frame is complete, partitions are built and erected. At this stage, it is important to enclose the structure and make the roof watertight. Only bearing partitions are installed at this time. Bearing partitions are those that support the ceiling and roof. Roof and floor trusses require no other support than the outside walls, **Figure 11-19**. Erection of nonbearing partitions can wait until after the building is enclosed.

Establish the centerlines of the partitions from studying the plans. Mark the centerlines on the floor with a chalk line. Lay out the plates, studs, and headers. Cut the headers, then assemble and erect the partitions in the same way as outside walls. Erect long partitions first, then cross partitions. Finally, build and install short partitions that form closets, wardrobes, and alcoves.

The corners and intersections are constructed the same way as outside walls. Refer to **Figure 11-6**. The size and amount of blocking, however, may be reduced. The chief concern is to provide nailing surfaces at inside and outside corners for wall covering material.

11.6.1 Nonbearing Partitions

Nonbearing partitions do not require headers above doorways and other openings. Many rough openings can be framed with single pieces of 2×4 or 2×6 lumber since there is virtually no load on them. Trimmers may be added for rigidity. They also provide added framework for attaching casing and trim. Door openings in partitions and outside walls are framed with the sole plate included at the bottom of the opening. After the

framework is erected, the sole plate in the door opening is cut out with a handsaw. Rough door openings are generally made 2″–2 1/2″ wider than the finished door size.

The partitions between noisy areas and quiet areas are often soundproofed. This may require a special method of framing. See Chapter 16, **Windows and Exterior Doors**, for information on insulation.

Small alcoves, wardrobes, and partitions in closets may be framed with 2×2 material or by turning 2×4 stock sideways, thus saving space. This is satisfactory when the thinner constructions are short and intersect regular walls. Snap a chalk line across the rough floor to mark the position of partitions, **Figure 11-20**.

During wall and partition framing, add various important details. At this stage, add basic provisions to the framing for recessed and surface-hung cabinets, tissue-roll holders, and similar items. Architectural plans usually provide information concerning their size and location.

Openings for the installation of heating ducts are easily cut and framed at this time, **Figure 11-21**. Bathtubs and wall-mounted toilets require extra support, **Figure 11-22**. Wall backing for drapery brackets, towel bars, shower curtains, and wall-mounted plumbing valves should also be added, **Figure 11-23**. Plumbing fixture rough-in drawings are helpful in locating the backing, **Figure 11-24**. For most items, 1″ nominal backing material provides adequate support.

George_GL/Shutterstock.com

Figure 11-19. The open-web trusses used here are supported by the outside walls alone. Partitions do not need to be installed until the roof is on and the building is closed in.

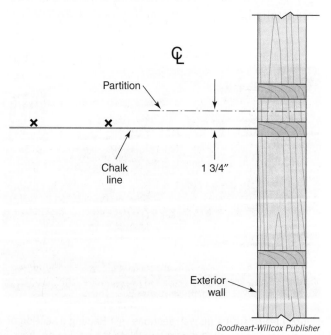

Goodheart-Willcox Publisher

Figure 11-20. Snap a chalk line on the floor to mark the position of partitions.

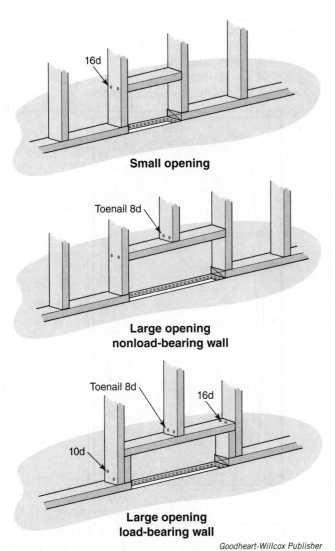

Small opening

**Large opening
nonload-bearing wall**

**Large opening
load-bearing wall**

Goodheart-Willcox Publisher

Figure 11-21. Special framing for heat ducts. Cripple studs should be added for large openings in load-bearing walls.

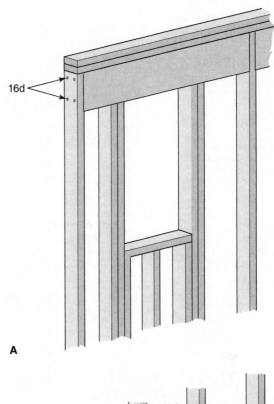

A

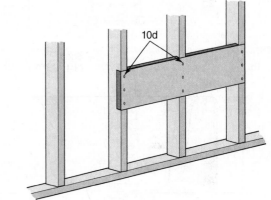

B

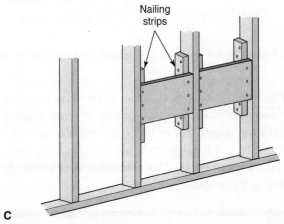

C

Goodheart-Willcox Publisher

Figure 11-23. Backing for mounting various fixtures and appliances. A—Extending the header over windows provides a base for attaching drapery rod brackets. B—Backing let into studs. Never cut back more than 25% of the stud width on bearing walls or partitions. C—Backing attached to nailing strips.

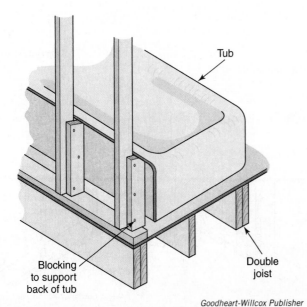

Goodheart-Willcox Publisher

Figure 11-22. Extra joists and blocking are needed to support tubs.

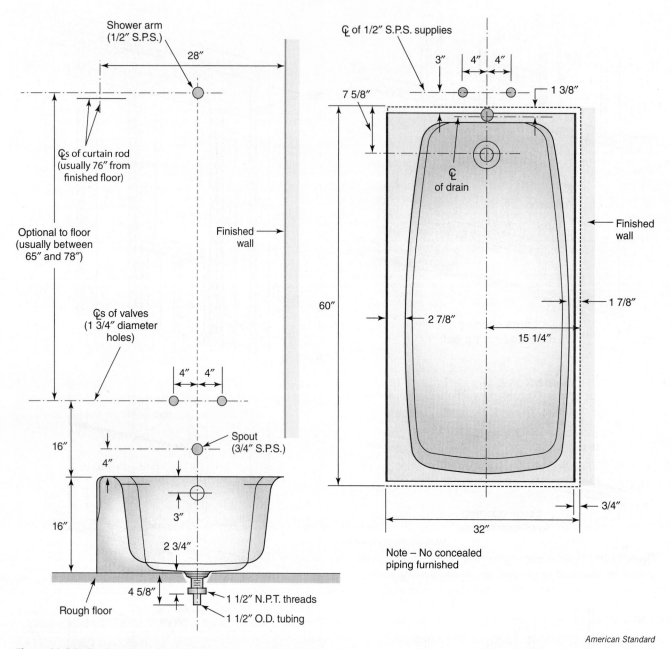

American Standard

Figure 11-24. Typical rough-in dimensions are helpful when locating special backing.

Small items that are not critical to the structure can often increase efficiency and quality of work during the finishing stages. For example, corner blocks make it possible to nail baseboards some distance back from their end, **Figure 11-25**. This eliminates the possibility of splitting the wood.

Pro Tip

Proper planning and monitoring of the job are important to avoid any unnecessary delays. A carpenter must continually study and plan the sequence of the job so that neither the weather nor work of other tradespeople will cause slowdowns or bottlenecks.

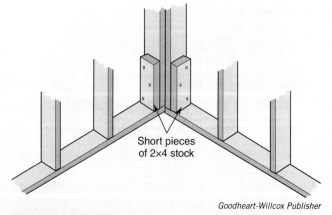

Goodheart-Willcox Publisher

Figure 11-25. Adding blocking in corners provides a better nailing surface for attaching baseboards.

11.6.2 Plumbing in Walls

Where plumbing is run through walls, special construction may be required. Depending on the size of the drain and venting pipes, a partition may have to be made wider. Usually a 6″ frame is sufficient. **Figure 11-26** shows two methods of construction.

Lateral (horizontal) runs of pipe require drilling holes or notching the studs. Sometimes, a wooden block is used to bridge a notch cut for plumbing, but usually plumbers and electricians drill holes in the center of the stud. The edge of a hole for a pipe and wiring may not be closer to the edge of the stud than 1 1/4″. That is more to protect the wiring or plumbing from nails and screws than to protect the strength of the stud.

11.6.3 Bracing

Exterior walls usually need some type of bracing to resist lateral (sideways) loads. Some types of material, such as plywood or OSB sheathing, provide sufficient rigidity. In these cases, additional bracing can be eliminated. Always check the exact requirements of the local building code.

Metal-strap bracing is widely used. It is made of 18 or 20 gauge galvanized steel and is 2″ wide. One type includes a 3/8″ center rib, **Figure 11-27**.

To install the ribbed strap bracing, snap a chalk line across the erected wall frame. Set a portable circular saw to cut just deep enough to completely accept the rib on the bracing. Make a cut on the chalk line on each member. Drive two 8d nails through the rib into each framing member. The let-in corner bracing shown in **Figure 11-28** is no longer permitted by building codes in areas subject to earthquakes. In such areas, plywood or OSB shear panels must be used for greater strength and rigidity.

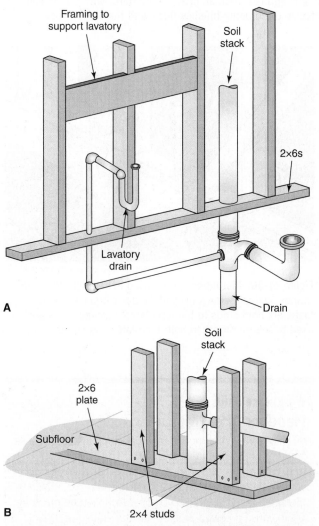

Goodheart-Willcox Publisher

Figure 11-26. A partition containing plumbing pipes may need different construction to provide room for the plumbing. A—A wall constructed of 2×6 studs and plate. B—The use of 2×4 studs on a 2×6 plate eliminates the need to notch or bore studs for lateral runs.

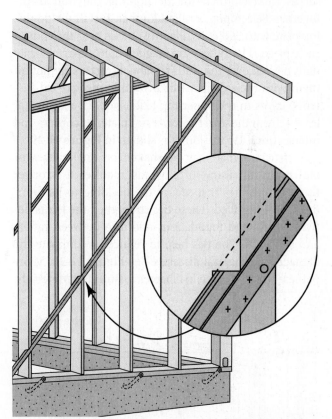

Goodheart-Willcox Publisher

Figure 11-27. Metal-strap bracing. The brace extends from the top plate to the sole plate. It is easily installed by cutting an angled saw kerf into the studs, top plate, and sole plate. The rib is designed to slip into a saw kerf at each member of the frame and is fastened with nails.

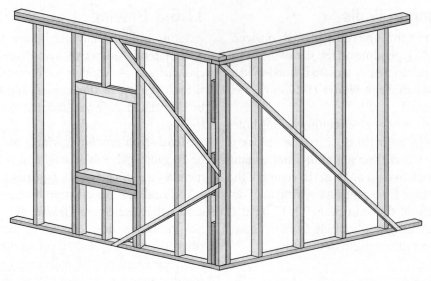

Goodheart-Willcox Publisher

Figure 11-28. Bracing a wall with 1×4 lumber. Note how the bracing is applied when an opening interferes with the diagonal run. Let-in bracing requires a recess be cut into the studs and plates so the bracing finishes flush with the wall surface.

11.6.4 Double Top Plate

In traditional framing, to add support under ceiling joists and rafters, the top plate is doubled. This also serves to strengthen the tie holding the wall frame together. Select long, straight lumber. Install the double top plate with 16d nails. Place two nails near the ends of each piece. The others are staggered 16″ apart. Locate nails near or over the studs so subcontractors cannot hit them with their drills while installing mechanical systems. Joints in the upper top plate should be located at least 4′ from those in the lower top plate. At corners and intersections, lap the joints as shown in **Figure 11-29**.

A structure that may be shaken by earthquakes or blown by hurricane-force winds requires a stronger frame. Metal ties that secure the building to its foundation are installed. These ties strengthen the joints between walls and foundations and also between rafters and sidewalls. The ties help to reduce or eliminate the damage from natural disasters. See **Figure 11-30**. Many of these ties are shown in Chapter 3, **Building Materials**.

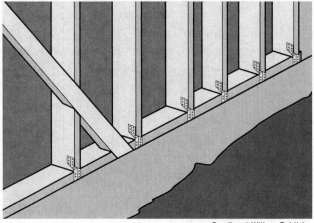

Goodheart-Willcox Publisher

Figure 11-30. Metal ties can be used to strengthen building frames against damaging winds and earthquakes. Metal fasteners hold studs to the top plates. Fasteners can also be used to secure rafters to both top plates and studs.

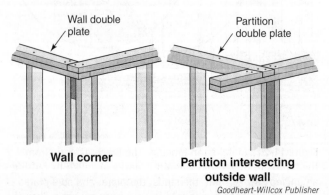

Wall double plate

Partition double plate

Wall corner

Partition intersecting outside wall

Goodheart-Willcox Publisher

Figure 11-29. Double plates are lap-jointed for strength wherever they intersect.

Workplace Skills
Handling the Pressure

As with many construction tasks, framing can get intense. It seems like everyone is in a hurry. The pressure to perform can be overwhelming. This is especially true if you are the new person on a crew. Mistakes can raise the aggravation levels of everyone. Don't let the pressure get to you. Perceived judgements can cause us to lack confidence, which can lead to poor decisions. Rise to the occasion. A fast-paced work environment can be fun if you can relax, see the tempo as a challenge, and take control of your reactions. As a result, your skill and speed will improve. Before you know it, you will be challenging others to pick up the pace!

11.6.5 Straightening Walls

At this point, corner posts are plumb and top plates are doubled. Now, the wall needs to be straightened and braced between the corners. Obtain three blocks of the same thickness. Tack one at each end of the wall. Tightly stretch a line between the two blocks. Using the third block as a gauge, align the wall and brace it at intervals. Be careful that the gauge comes up to, but does not touch, the line. If it touches the line, the line is no longer straight!

This procedure is a two-person job as described. It is difficult for one person to hold the wall while securing the brace. Some carpenters use a manufactured brace that can be adjusted to align the wall. Others fashion spring braces on the job.

11.7 Tri-Level and Split-Level Framing

Tri-level and split-level housing presents special challenges in wall framing. Generally, a platform type of construction is used. However, the floor joists for upper levels may be carried on ribbons let into the studs. The plans prescribe the type of construction. They should also include calculations of distances between floor levels. When working with split-level designs, a good carpenter prepares accurate story poles that show full-size layouts of these vertical distances for all levels of the building.

11.8 Special Framing

Framing carpenters are sometimes asked to build structures with special features. It helps if these features are carefully engineered and detailed by an architect. In such cases, construction details are included in the plans. When they are not, a carpenter must develop the plan.

Most bay windows are prefabricated by the window manufacturer. However, a carpenter may be asked to build a bay window. When asked to do so, the carpenter must visualize the details of construction, lay out and construct the floor frame to carry the project, and build the wall and roof frame. See **Figure 11-31**.

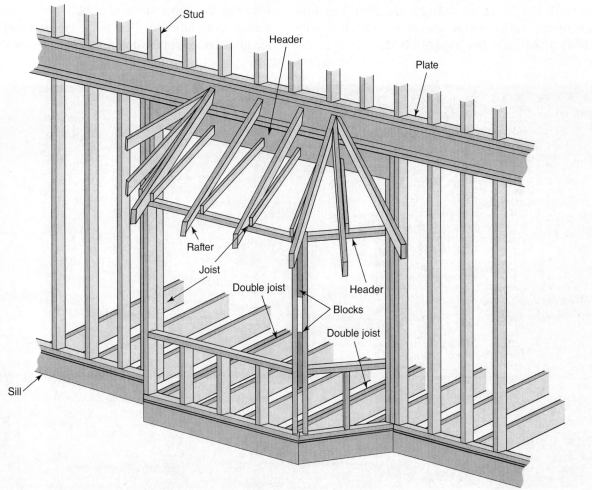

National Forest Products Assn.

Figure 11-31. Traditional framing for a bay window. To illustrate how joists are extended over the sill, the subfloor is not shown here. Joists at either side of the extension should be doubled.

Framing of the floor extension is best done by extending the floor joists, cantilevering them over the foundation. To help carry the weight of the structure, double the joists forming each side of the bay. A header over the opening in the wall carries the weight of the structure above. It also provides a nailing surface for the window's ceiling joists.

Another special framing project might be a cabinet *soffit* that closes in the space between the ceiling and the tops of cabinets. Soffits are most often found in kitchens, although bathrooms frequently have them, too.

Construction of a soffit begins by snapping a chalk line on the studs at the bottom edge of the soffit. This is usually 84″ from the finished floor. Next, nail a 2×2 along this line using one 10d nail at each stud. Then, snap a second chalk line snapped along the ceiling at a distance that is at least the depth of the cabinets. Nail another 2×2 onto the ceiling along this line. If the ceiling joists run parallel to the soffit, 2×4 blocking is used to bridge between two joists and provide support for the 2×2. Next, cut a 2×4 or 2×2 for the lower front edge of the soffit and blocking to frame the front and bottom of the soffit. These pieces are attached with screws or 8d nails. See **Figure 11-32**.

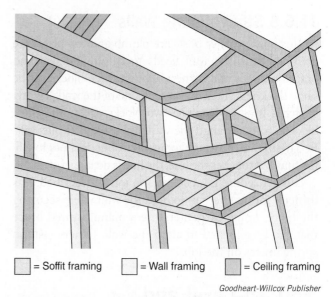

☐ = Soffit framing   ☐ = Wall framing   ☐ = Ceiling framing

Goodheart-Willcox Publisher

Figure 11-32. A framed soffit before the firestops are installed. Note the 2×2 frame against studs and ceiling joists. Special blocking is used at the corner.

11.9 Multistory Floor Framing

Framing the upper floor of a multistory dwelling is similar to framing the first floor, **Figure 11-33**. Joists are placed on top of the double plate along with headers

Goodheart-Willcox Publisher

Figure 11-33. In platform framing, the first story wall framing is independent from the second story wall framing. The floor and ceiling joists create a platform on which the second story is formed.

(band joists). Joists may be fastened to the plate with steel anchors or toenailed. See **Figure 11-34**. Some joists need the support of a bearing wall or girder, **Figure 11-35**.

11.10 Ceiling Framing

A *ceiling frame*, as the name suggests, is the system of support for all components of the ceiling. This frame may be the underside of the floor joists for the next story, the bottom chords of roof trusses, or ceiling joists just below the roof. See **Figure 11-36**. The basic construction of an assembly just below the roof is similar to floor framing. The main difference is that lighter joists are used and headers are not included around the outside. When trusses are used to form the roof frame, no additional ceiling frame is required.

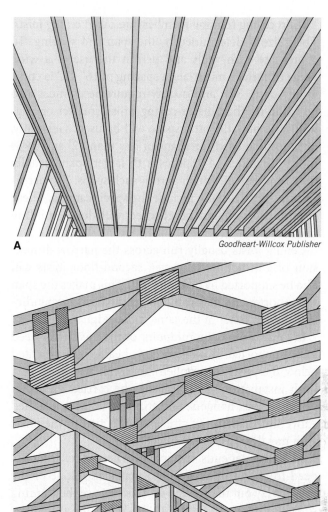

A *Goodheart-Willcox Publisher*

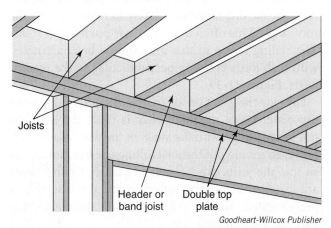

Goodheart-Willcox Publisher

Figure 11-34. Upper story framing. The floor frame is fastened to the double top plate of the previous story. Joists are toenailed to the plate.

Figure 11-36. Ceiling framing. A—Ceiling joists for a flat roof. B—Second floor trusses. Ceiling covering is fastened to the underside and subfloor sheathing to the top side.

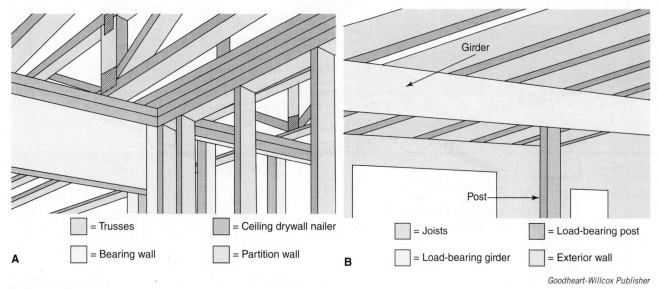

Goodheart-Willcox Publisher

Figure 11-35. For wide spans, joists may be supported on a bearing wall (A) or girder, as well as the exterior walls (B).

Main ceiling framing members are called *ceiling joists*. Their size is determined by the span and spacing. To coordinate with walls and permit the use of a wide range of surface materials, a spacing of 16″ OC is commonly used. Size and quality requirements must also be based on the type of ceiling finish (plaster or drywall) and how the attic space will be used. Generally, 2×6s may be used for spans of less than 12′ and 2×8s for spans under 14′. Spaces with larger spans usually have roof trusses, which permit such spans. The architectural plans usually include the specifications. These requirements should be checked with local building codes.

Ceiling joists usually run across the narrow dimension of a structure. However, second-floor joists can also be supported by bearing walls. This makes the span even shorter and may result in ceiling joists perpendicular to each other in the same frame, **Figure 11-37**.

In large rooms, the midpoint of the joists may need to be supported by a beam. This beam can be located below the joists or installed flush with the joists. In a flush installation, the joists may be carried on a ledger or by joist hangers. Sometimes a beam is installed above the joists in the attic area. It is tied to the joists with metal straps.

At their outer ends, the upper corners of ceiling joists need to be cut at an angle to match the slope of the roof. To lay out the pattern for this cut, use the framing square as shown in **Figure 11-38**. When the amount of stock to be removed is small, the cuts can be made after the joists and rafters are in place.

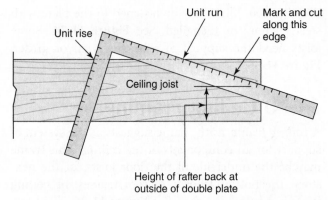

Goodheart-Willcox Publisher

Figure 11-38. A framing square is used to lay out the trim cut on the end of a ceiling joist. Using the rise and run of the building, the cut matches the slope of the rafters.

Parallel Ceiling Joists

When ceiling joists run parallel to the edge of the roof, the outside member is likely to interfere with the roof slope. This often occurs in low-pitched hip roofs. The ceiling frame in this area should be constructed with stub joists running perpendicular to the regular joists, **Figure 11-39**.

Lay out the position of the ceiling joists along the top plate. When a double plate is used, the joists do not need to align with the studs in the wall. The layout, however, should put the joists alongside the roof rafters so that the joists can be nailed to them. Ceiling joists are installed before the rafters unless the assembly is prefabricated. Toenail ceiling joists to the plate using two 10d nails on each side.

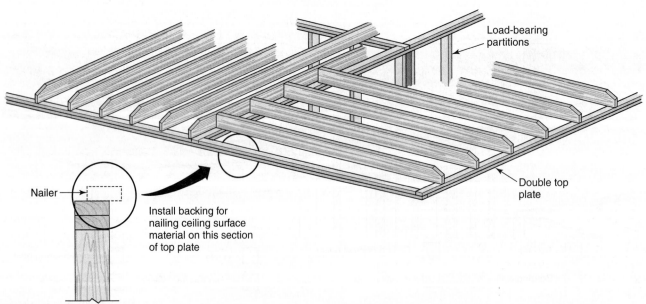

Goodheart-Willcox Publisher

Figure 11-37. A ceiling frame. The joists in the foreground are turned at right angles to reduce the span.

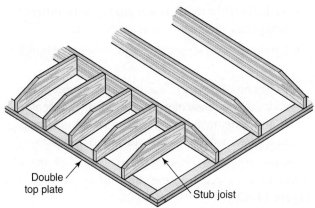

Goodheart-Willcox Publisher

Figure 11-39. Stub ceiling joists butted to a full-length joist. The stub joists are required for a low-pitched hip roof due to the lack of space near the edge.

Partitions or walls that run parallel to the joists must be fastened to the ceiling frame. A nailing strip or drywall clip to carry the ceiling material must also be installed. Various size materials can be installed in a number of ways. The chief requirement is that they provide adequate support for interior wall coverings. **Figure 11-40** shows a typical method of making such an installation. **Figure 11-6C** shows backup clip installation for a ceiling.

An access hole, also called a scuttle, must be included in the ceiling frame to provide entrance to the attic area. Fire regulations and building codes usually list minimum size requirements. The building plans generally indicate the size of the scuttle and show where it should be located. The opening is framed following the procedure used for openings in the floor.

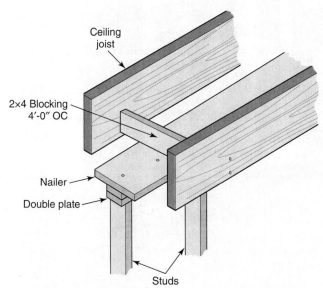

Goodheart-Willcox Publisher

Figure 11-40. Use special blocking to anchor partitions to the ceiling frame when they run parallel to the joists.

If the size of the opening is small (2′–3′ square), doubling of joists and headers is not required.

11.11 Strongbacks

Long spans of ceiling joists may require a *strongback*. This is an L-shaped support constructed of 2″ lumber. It is attached across the tops of joists to strengthen them and maintain the space between them. It also evens up the bottom edges of the joists so the ceiling is not wavy after the drywall is applied.

To construct a strongback, first mark the proper joist spacing (16″ or 24″ OC) on a 2×4. Position the 2×4 across the tops of the ceiling joists and fasten it with two 16d nails at each joist. Apply pressure against the joists as needed to maintain proper spacing.

Select a straight 2×6 or 2×8 for the second member. Place it on edge against either side of the 2×4 just attached to the joists. Attach one end to the 2×4 with a 16d nail. Work across the full-length of the strongback, aligning and nailing it. Stepping on either the 2×4 or the member on edge helps align each joist. Nail the vertical member to each joist and to the 2×4, **Figure 11-41**.

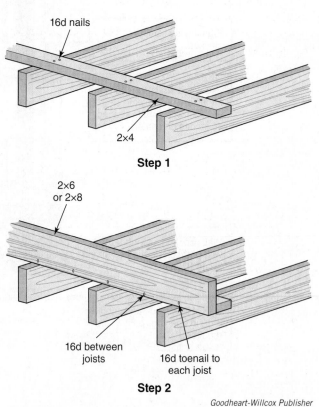

Goodheart-Willcox Publisher

Figure 11-41. Building a strongback. Step 1—Nail a 2×4 to the joists. Step 2—Turn a 2×6 or 2×8 on edge and nail it to the 2×4 and the joists.

11.12 Advanced Framing

Advanced framing is the term APA-The Engineered Wood Association uses for what was first called optimum value engineering (OVE). In recent years, advanced framing has become the more common name for this system. It is a system of framing using 2×6 studs spaced 24″ OC and using structural OSB or plywood as sheathing. Using this system brings these benefits:

- Reduces material usage. Board footage of framing lumber can be up to 10% less than with conventional framing

- Reduces labor costs. Fewer studs, joists, rafters, and plates equate to less assembly time

- Complies with structural requirements of building codes

- Builds more energy efficient homes to meet energy codes. An increase of wall cavity insulation content, from approximately 75% with conventional framing, to approximately 85% with advanced framing

Some carpenters choose to use some of the elements of advanced framing, but not all of them. See **Figure 11-42**.

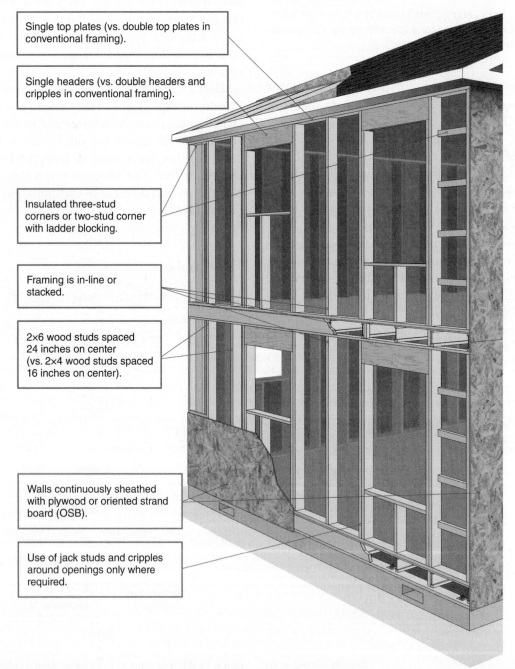

Single top plates (vs. double top plates in conventional framing).

Single headers (vs. double headers and cripples in conventional framing).

Insulated three-stud corners or two-stud corner with ladder blocking.

Framing is in-line or stacked.

2×6 wood studs spaced 24 inches on center (vs. 2×4 wood studs spaced 16 inches on center).

Walls continuously sheathed with plywood or oriented strand board (OSB).

Use of jack studs and cripples around openings only where required.

APA-The Engineered Wood Association

Figure 11-42. Advanced framing techniques may include some or all of the details shown here.

If all of the framing members are in line vertically, the top plate does not need to be doubled. The studs must be placed directly over the floor joists at each level and trusses or rafters must be placed directly over the studs. **Figure 11-43** shows how the framing members align.

Advanced framing replaces wood with insulation wherever that can be done without sacrificing structural integrity. Headers are built with a single ply of glulam or dimensional lumber, **Figure 11-44**. Headers are eliminated on non-load bearing walls and replaced by a single plate and cripple studs, just like the sill framing of a window.

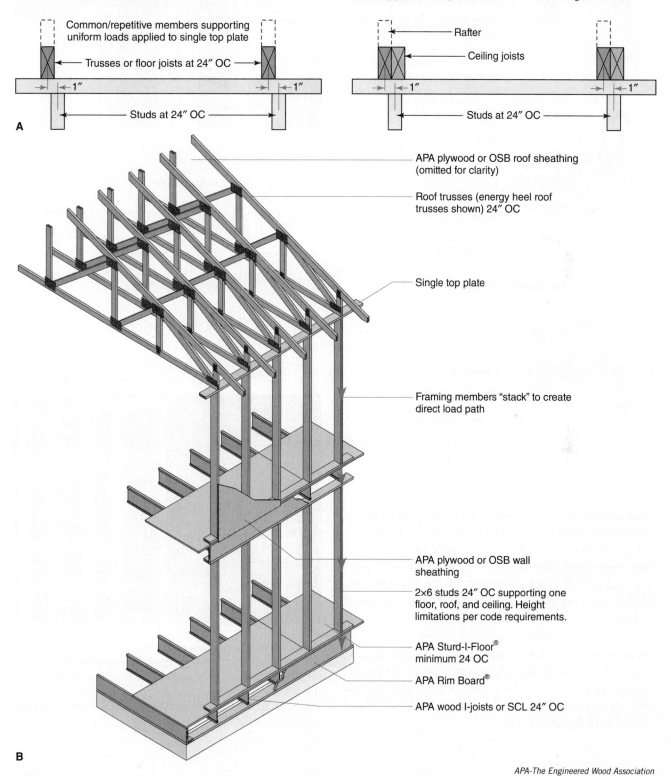

Figure 11-43. In-line framing 24″ on center (OC). A—Framing members must be centered over the member below within 1″. B—Aligned framing members create a straight load path from the roof to the foundation.

Single-Ply Header

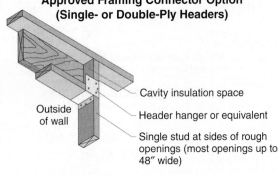

Outside of wall

Top plate

Single-ply load-bearing header (flush outer face of header with outer edge of studs)

Cavity insulation space (to stud depth less single header thickness)

Header bottom plate (to complete rough opening at header)

For many one-story builtings, single studs at sides of rough openings may be adequate.

Jack stud or approved framing connector

Approved Framing Connector Option (Single- or Double-Ply Headers)

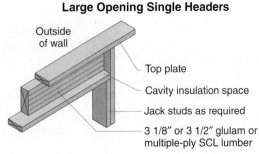

Outside of wall

Cavity insulation space

Header hanger or equivalent

Single stud at sides of rough openings (most openings up to 48″ wide)

Large Opening Single Headers

Outside of wall

Top plate

Cavity insulation space

Jack studs as required

3 1/8″ or 3 1/2″ glulam or multiple-ply SCL lumber

APA-The Engineered Wood Association

Figure 11-44. A single-ply header can replace a built-up header, leaving room to add insulation.

This leaves space for insulation where the extra lumber would have been used. As can be seen in the figure, another savings can be had by using header hangers instead of trimmer studs. The hanger is nailed to the rough opening stud and the header slips into the top of the hanger. Steel connecting plates can be used on single top plates to make splices and attach adjoining walls.

To gain maximum insulation, corners are constructed so they can be fully insulated. This can be accomplished with the three-stud corner shown in **Figure 11-5** or with a two-stud corner using drywall clips to attach the interior drywall. A two-stud corner is shown in **Figure 11-45**.

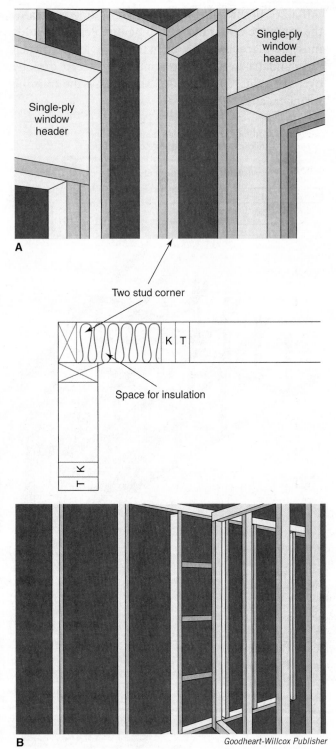

Single-ply window header

Single-ply window header

A

Two stud corner

K T

Space for insulation

T K

B

Goodheart-Willcox Publisher

Figure 11-45. Corners are constructed so that insulation can be added later. A—A two-stud corner. Clips will be installed to provide an attachment surface for the interior drywall. B—A ladder corner can be constructed or a full length 1× nailer may be installed for interior partitions and for exterior walls.

11.13 Structural Insulated Panels

Structural insulated panels (SIPs) are panels made with OSB or plywood sheathing on the exterior side, an insulating foam core, a structural panel, and usually OSB on the inside, **Figure 11-46**. SIPs may be manufactured with an interior wall finish, but OSB is more common. SIPs are available in sizes ranging from 4′ × 6′ to 8′ × 24′. They are ordered cut to specific sizes and shapes for the job, with each panel numbered to correspond with a number on the plans. Openings, such as for windows, are usually cut out before the SIP is shipped to the jobsite.

SIPs provide several advantages. Although they cost more than the materials for a conventionally framed wall, the cost of building a house with SIPs is about the same because of the savings in labor. They are especially energy efficient because they have few, if any, openings in the insulated foam and tight-fitting joints in the interior and exterior sheathing. SIPs also save on materials. The National Association of Home Builders has estimated that construction of a conventional 2000 square-foot house produces about 7000 pounds of waste. With SIPs there is practically no waste during construction.

SIPs can be used for walls and roofs. To erect a SIP wall, a sole plate is nailed to the floor platform, similar to what is done for a stick-built wall. The plate is set back from the edge of the floor a distance equal to the thickness of the exterior sheathing on the SIP. A heated tool is used to cut the foam in the SIP back far enough along the bottom edge to accommodate the sole plate, **Figure 11-47**. The SIP is then set down on the sole plate with the sheathing slipping over each side of the plate. The sheathing is then nailed to the plate. Where panels must fit side by side, the foam along the side of the SIP is routed out a distance of half the thickness of a 2×6 (assuming the wall has 5 1/2″ of foam) on the edge of each panel. A 2×6 is inserted half its thickness into the space created and nailed in place. That panel is erected as previously described. The second panel is slid up next to the first, so the 2×6 is completely enclosed in the two panels. The second panel is then nailed to the 2×6, forming a strong, tight joint. Caulking is applied to all joints before they are closed, so the wall will be air tight and well insulated. Whole walls can be assembled on the ground, then hoisted into place, **Figure 11-48**.

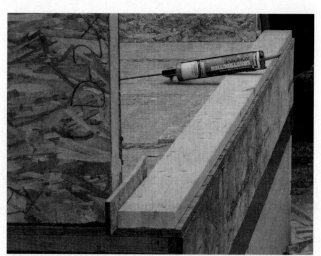
Goodheart-Willcox Publisher

Figure 11-47. The sole plate is set back from the edge of the floor to accommodate the sheathing on the SIP. Notice that a scrap piece of sheathing is used as a spacer to provide a place for the inner sheathing to slip behind the adjoining wall.

Goodheart-Willcox Publisher

Figure 11-46. A SIP has structural panels on the outside and a foam core on the inside.

brizmaker/Shutterstock.com

SV Production/Shutterstock.com

Figure 11-48. This SIP wall is being lifted into place with a crane. It has all of the door and window openings already prepared.

Green Note

The thermal envelope is a protective shell that curbs unwanted air leakage in or out of a building. It allows a building to conserve energy and reduce heating costs by passively preventing heat loss through use of insulation, air barriers, and advanced framing techniques (AFTs). Advanced framing techniques (AFTs) minimize waste by eliminating redundant lumber use without weakening the structure. Since heat escapes faster through wood than insulation, using less lumber and more insulation strengthens the building's thermal envelope.

11.14 Housewrap

Housewrap is a thin, tough, plastic sheet material that is applied to exterior walls. It prevents movement of air into or out of a building. See **Figure 11-49**. Manufactured under such trade names as Tyvek® and Typar®,

it comes in rolls of various widths up to 10′. Housewrap can have numerous functions depending on its type and its installation as part of the building envelope. It is efficient at preventing air infiltration. At the same time, it is water resistant. Housewrap is designed to prevent bulk water from entering the structure, yet allows interior water vapor to pass through and escape. This property is known as vapor permeability.

The typical housewrap is made from high-density, spun, polyethylene fibers. These fibers are virtually tearproof. Housewrap is stapled to the sheathing before exterior doors and windows are installed.

Housewrap can also be applied to concrete block walls as an air- or weather-resistant barrier. Usually, block masonry is then faced with stucco or brick. Caulk is used at 6″ intervals as the preferred fastening method. Follow the manufacturer's instructions. For more on the functions and installation of housewrap, see Chapter 15, **Building Envelope and Control Layers**.

11.15 Estimating Materials

To estimate wall and ceiling framing materials, first determine the total lineal feet of the walls. Add the length of each wall and partition. The plans include the dimensions of outside walls. These can be added together. Very short partitions may not be dimensioned. In this case, use a scale on the drawing. It is a good idea to place a colored pencil check mark on each wall and partition as its length is added to the list.

For plates, multiply the total figure by three (one sole plate plus two top plates) if you are using a double top plate or by two if you are using a single top plate. Add about 10% for waste. Order this number of linear feet of lumber in random lengths or convert to the number

Ken Wolter/Shutterstock.com

Figure 11-49. Housewrap is installed over the sheathing on exterior walls. The wrap is a tough plastic fabric that prevents air infiltration.

of pieces of a specific length. Each lineal foot of 2×4 equals 2/3 bd ft. Each lineal foot of 2×6 equals 1 bd ft. Thus, the total lineal feet of walls and partitions can be quickly converted to board measure, if required. For example:

Wall and partition length = 240

$$
\begin{aligned}
\text{Total plate material} &= \text{length} \times 3 + 10\% \\
&= 240 \times 3 + 10\% \\
&= 792 \text{ lineal ft} \\
&= \text{or } 57 \text{ pcs} \\
&= 2'' \times 4'' \times 14'\text{-}0''
\end{aligned}
$$

11.15.1 Estimating Studs

The total length of all walls and partitions is also used to estimate the number of studs required. When studs are spaced 16″ OC, multiply the total length (in feet) by 3/4. If the framing is 24″ OC, divide the total length by 2. Next, add two more studs for each corner, intersection, and opening. Using 240′ for the wall length, assume there are 12 corners, 10 intersections, 20 openings, and the framing is 16″ OC. Find the number of studs needed, including outside walls and partitions:

$$
\begin{aligned}
\text{Total studs} &= \text{total length} \times 3/4 + 2 \\
&\quad (\text{corners} + \text{intersection} + \text{openings} \\
&= \frac{240 \times 3}{4} + 2\,(12 + 10 + 20) \\
&= \frac{\overset{60}{\cancel{240}} \times 3}{\cancel{4}} + 2\,(42) \\
&= 60 \times 3 + 84 \\
&= 264
\end{aligned}
$$

PROCEDURE

Installing Housewrap

To install housewrap on erected walls, follow these steps:
1. Start from the bottom at a corner. Leave about a foot extra to wrap around the corner. Make sure the roll is perfectly vertical. Otherwise, the wrap will run either uphill or downhill on the wall. The bottom edge should run along the top edge of the foundation.
2. Secure the wrap with staples.
3. Start at a corner and continue to unroll a few feet at a time. Keep the sheet straight with no wrinkles. Check that you are following the foundation line.
4. Secure the wrap every 12″–18″ on a stud line.
5. Roll over window and door openings. Cut these openings later. See Chapter 16, **Windows and Exterior Doors**, for details.

Continued

PROCEDURE *(continued)*

6. Overlap every horizontal joint at least 6″. Upper layers should overlap lower layers. Overlap vertical joints 12″. Continue all around the building. Use a 1×4 to push the wrap into inside corners.
7. Secure all laps with housewrap tape.

Where the building has more than one floor, wrap can be installed vertically:
1. Working from the top plate of the top story, staple a 2×2 to the end of the roll and lower it down the wall to the band joist of the lowest floor.
2. Be sure to allow a flap at the beginning vertical corner. Then, secure the wrap at the band joist. Remove the 2×2.
3. Fold the flap around the corner and fasten it.
4. Staple 12″ apart along each stud.
5. Overlap each succeeding strip at least 12″. Immediately tape the joints to prevent wind from getting under the wrap.

Many carpenters estimate the number of studs by simply counting one stud (spaced 16″ OC) for each lineal foot of wall space. The overrun on spacing provides the extras needed for corners and openings. This method is rapid and fairly accurate. However, it does not estimate enough material for a small house divided into many rooms. On the other hand, too many studs will likely be estimated for a large house with wide windows and open interiors.

Material for headers must be calculated by analyzing the requirements for each opening. Use the rough opening width plus the thickness of the trimmers.

Ceiling joists are estimated by the same method used for floor joists. Since ceilings are relatively free of openings, no extras or waste needs to be included. Because of this, the short method should not be applied to ceiling joists. Use the following formula and include the size of the joists required:

Number of ceiling joists = wall length × 3/4 + 1

11.15.2 Estimating Headers

To estimate the amount of header material required, begin by identifying which headers, if any, will be doubled or built-up. If header hangers are being used without trimmer studs, the header length is the actual rough opening width. If trimmers are being used, add 3″ (1 1/2″ for the trimmer at each side) of the opening. If LVL or other engineered lumber is to be used, add all of the header lengths together and add several inches for saw kerfs. If dimensional lumber will be used, add all of the lengths and divide by 10 to determine the number of 10′ pieces required. The same calculation can be made to use 8′ pieces (divide by 8).

11.15.3 Estimating Wall Sheathing

To estimate the amount of wall sheathing, first find the total perimeter of the structure, **Appendix B**. Multiply this figure by the wall height. When the sheathing extends over the sill construction, measure the height from the top of the foundation. The product is the gross square footage of the wall surface. Do not allow for window and door openings unless they are very large. Most carpenters sheath over the opening then cut it out later.

Divide the net area by the square footage per sheet to determine the number of pieces required. For example, if the net area to be sheathed is 1060 sq ft and the sheets selected are 4′ × 8′, calculate the quantity needed as follows:

$$
\begin{aligned}
\text{Net area} &= 1060 \text{ sq ft} \\
\text{Fiberboard sheet size} &= 4' \times 8' \\
&= 32' \text{ sq ft} \\
\text{No. of sheets needed} &= 1060 \div 32 \\
&= 33.1 \text{ (round up to 34)}
\end{aligned}
$$

Pro Tip

Problems in estimating, although based on simple formulas, usually contain so many variables that good judgment must be applied to their solution. This judgment is acquired through experience.

Construction Careers
Framing Carpenter

Most of the carpenters employed in residential and light commercial construction are framing carpenters. They are sometimes referred to as rough carpenters to distinguish them from the finish carpenters who do trim and specialty work. Framing carpenters primarily work with wood as a structural material. They fabricate the floor structures, wall sections, and roof framing of the building. In recent years, framing carpenters have become increasingly involved with the cutting, fitting, and fastening of light steel structural members (joists and studs), especially in commercial construction.

Approximately 1/3 of all carpenters are self-employed. Another 1/3 work for general contractors. The remaining 1/3 are spread among specialty contractors, manufacturing firms, government agencies, and other employers. The growth of manufactured housing has created jobs for framing carpenters in firms that prefabricate building sections or entire structures.

Basic skills for framing carpenters are the ability to make use of measuring devices and to read and accurately follow construction drawings. They must be able to efficiently and safely use a variety of hand and power tools. Framing carpentry is strenuous work, involving lifting and carrying, standing or kneeling for long periods, and working in situations where the danger of injury from falls or other accidents is present. Most of the work is done outdoors, often in rainy, dusty, cold, or hot weather conditions. Wearing appropriate clothing and proper personal protective gear is important.

Framing carpenters acquire their skills in a variety of ways. Many learn through informal on-the-job training working with more experienced carpenters. Others enter the field from vocational school programs or take part in formal apprenticeship training. A carpentry apprenticeship typically is 3–4 years in length and combines classroom training with practical experience and instruction on the jobsite. Because they are exposed to most aspects of the construction process, carpenters who work for general contracting firms have opportunities for advancement to positions such as foreman, carpentry supervisor, or general construction supervisor.

Lakeview Images/Shutterstock

Framing carpenters work mostly with structural materials made of wood. They work on all sections of a building, from top to bottom. They are trained in the use of hand tools and power tools, as well as measuring devices.

Chapter Review

Summary

- Wall framing consists of the vertical and horizontal members that form the outside of a structure.
- Wall frames support upper floors, ceilings, and the roof.
- Framing members include the vertical studs, horizontal members called top and sole plates, headers, and sheathing.
- Studs and plates are made from 2×4 or 2×6 lumber, usually placed 16″ or 24″ on center OC. Wall sections are assembled on the rough flooring, and then raised into place.
- Once the outside wall frame is completed, the inside walls (partitions) are built and erected.
- Many inside partitions are nonbearing, which means they do not carry the weight of the structure.
- Tri-level and split-level housing presents special challenges in wall framing.
- Ceiling framing may be the underside of the floor joists for the next story or an assembly just below the roof.
- Advanced framing reduces material usage and labor costs, complies with structural requirements of building codes, and is energy efficient.
- After walls are sheathed, plastic housewrap is usually installed to prevent air infiltration.

Know and Understand

Answer the following questions using the information in this chapter.

1. Full-length studs become _____ when they end due to an opening.
   A. trimmer studs
   B. cripple studs
   C. headers
   D. top plates

2. *True or False?* 10′ high walls are common in colder climates, while 8′ high walls are common in warmer climates.

3. The header length is equal to the rough opening plus the width of two _____.
   A. trimmers
   B. top plates
   C. partitions
   D. plywood spacers

4. Wall sheathing should *always* be applied _____.
   A. before the wall frame is raised
   B. before roof framing
   C. after the wall frame is raised
   D. after roof framing

5. How much space should be left between edges of sheathing panels to account for panel expansion and contraction?
   A. 1/8″
   B. 1/2″
   C. 5/8″
   D. 6″

6. Immediately after a wall section is up, secure it with _____.
   A. studs
   B. nails
   C. braces
   D. metal ties

7. Bearing partitions support the ceiling and _____.
   A. floor trusses
   B. end walls
   C. main sidewalls
   D. roof

8. Rough door openings are generally made _____ to _____ wider than the finished door size.
   A. 1″, 1 1/2″
   B. 2″, 2 1/2″
   C. 2″, 4″
   D. 3″, 3 1/2″

9. The top plate is _____ in traditional framing.
   A. omitted
   B. doubled
   C. tripled
   D. braced

10. Prefabricated bay windows are an example of _____ framing.
    A. ceiling
    B. multistory
    C. special
    D. split-level

11. What room in the house is a soffit most often found?
    A. Kitchen.
    B. Bedroom.
    C. Utility room.
    D. Basement.

12. In a multistory dwelling, joists are placed on top of the _____ along the headers.
    A. ceiling
    B. trusses
    C. double plate
    D. door framing

13. *True or False?* When trusses are used to form the roof frame, no additional ceiling frame is required.

14. A _____ must be included in the ceiling frame to provide an entrance to the attic area.
    A. partition
    B. header
    C. scuttle
    D. door frame

15. A strongback is used for all of the following, *except* _____.
    A. strengthening a long span of ceiling joists
    B. maintaining proper spacing between ceiling joists
    C. keeping joists even along their lower edges
    D. aligning the height of headers

16. When compared to a conventionally frame wall, the cost of building a house with SIPs is roughly _____.
    A. double
    B. half
    C. the same
    D. dependent on the size of the house

17. *True or False?* A house constructed with SIPs produces practically no waste during construction.

18. What is generally the first step in estimating wall and ceiling framing materials?
    A. Determine the amount of plates needed.
    B. Determine the number of studs required.
    C. Determine the total lineal feet of the walls.
    D. Determine the amount of header material required.

19. The amount of ceiling joists needed is estimated by the same method used for _____.
    A. studs
    B. floor joists
    C. headers
    D. wall sheathing

20. How many wall sheathing sheets are needed for a net area of 1460 sq. ft. and the selected sheets are 4′ × 8′?
    A. 45 sheets.
    B. 46 sheets.
    C. 121 sheets.
    D. 122 sheets.

Apply and Analyze

1. Give two reasons why carpenters install doors and windows with their tops at the same height.

2. What are two alternative methods to construct a header for large window openings without decreasing the length of the cripple studs?

3. Why is a story pole often used when laying out wall framing?

4. After corner posts are plumb and top plates are doubled, how is a wall straightened?

5. What are four major benefits of advanced framing?

6. Name two different uses for housewrap.

7. Using 120′ for the wall length, assume there are 6 corners, 5 intersections, 10 openings, and the framing is 24″ OC. Find the number of studs needed, including outside walls and partitions.

Critical Thinking

1. Obtain a set of plans for a one-story house. Study the details of construction, especially typical wall sections. Prepare a scale drawing of the framing required for the front walls. Be sure the rough openings are the correct size and in the proper location. Your drawing should look similar to **Figure 11-1**.

2. Working from the set of plans used in #1, develop an estimated list of materials for the wall frame and sheathing. Include the number and length of studs, number and size of lumber for headers, material for plates, and type and amount of sheathing. Obtain prices from a local supplier and figure the total cost of the materials.

3. Obtain literature about OSB, plywood, foamed plastic, and gypsum sheathing. Download this material from the Internet, or obtain directly from manufacturers. Also, study books and other reference materials. Prepare a report for the class, based on the information you obtain. Include grades, manufacturing processes, characteristics, and application requirements. Discuss current prices and purchasing information. Be prepared to discuss specifications. Relate these specs to your local code.

Communicating about Carpentry

1. **Speaking.** Write a brief scene in which 5–10 terms are used as you imagine them being used by carpenters in a real-life context. Then rewrite the dialogue using simpler sentences and transitions, as though an adult were describing the same scene to elementary or middle-school students. Read both scenes to the class and ask for feedback on whether the two scenes were appropriate for their different audiences.

2. **Reading.** With a partner, create flash cards for the key terms in the chapter. On the front of the card, write the term. Practice reading aloud the terms, clarifying pronunciations where needed. (You may also use a dictionary.)

3. **Listening and Speaking.** Working with a partner, create a simple scale model of a typical wall, floor, or ceiling framing. Identify the following components, making sure to place them properly: bridging, ceiling joists, crown, girder, hanger, headers, housewrap, joists, ledger, partitions, ribbon, rough openings, sill, soffit, studs, subflooring, top plates, trimmer. Additional components may be included, depending on the type of framing being used. Drawings may be substituted for the physical model if necessary. Be creative in selecting materials for the various structures. Options include balsa wood, toothpicks, popsicle sticks, or materials from a model or craft store.

4. **Reading and Writing.** Conduct library or internet research to learn about the history and development of the balloon framing method of residential construction. Identify the parts of the country where it was most extensively used and what role it played in the expansion of Midwestern and Western cities. Try to determine when and why it was replaced by the platform framing method. If facilities are available, develop your information into a PowerPoint presentation.

CHAPTER **12**

Roof Framing

OBJECTIVES

After studying this chapter, you will be able to:

- List and describe the various types of roofs.
- Identify the parts of a roof frame.
- Define the terms *slope* and *pitch*.
- Use a framing square, speed square, and rafter tables.
- Lay out common rafters.
- Describe the layout and erection of a gable roof.
- Find the length of a hip rafter.
- Explain the design and erection of trusses.
- Describe the procedure for sheathing a roof.
- Estimate roofing materials.

TECHNICAL TERMS

bird's mouth	gambrel roof	ridge
camber	gussets	rise
collar tie	hip jack	roof truss
common difference	hip jack rafter	run
common rafter	hip rafter	scissors truss
cornice	hip roof	shed roof
cripple jack	hypotenuse	skip sheathing
dead load	king post truss	slope
dormer	live load	speed square
extended rake	lookout	truss
fascia	mansard roof	valley jack
Fink truss	pitch	valley rafters
flat roof	purlin	
gable roof	rafter	

Roof framing provides a base to which the roofing materials will be attached. The frame must be strong and rigid. In addition, a carefully designed and well-proportioned roof can add a distinctive and decorative character to the structure.

Roofs must be strong enough to withstand the weight and stress of snow and high winds. This must be kept in mind in the design of the roof components. In the conventional rafter system, these components include not only the rafters, but ceiling joists, collar ties, and purlins.

In roof trusses, roof components include the chords, webs, and connector plates or gussets.

In the design of roof framing systems, both live loads and dead loads must be considered. The **live load** on a roof is the total of all moving and variable loads, such as the weight of snow and the pressure from wind. This load varies from one locality to another. The **dead load** is the weight of permanent, stationary construction and equipment included in a building. The rafters, decking, and roof coverings are all part of the dead load.

12.1 Roof Types

Roof types vary widely, as illustrated in **Figure 12-1**. Most of them can be grouped as follows:

- *Gable roof*—Two surfaces slope from the centerline (*ridge*) of the structure. This forms two triangular ends called gables. Because of their simple design and low cost, gable roofs are often used for homes.
- *Hip roof*—All four sides slope from a central point or ridge. The angles created where two sides meet are called hips. An advantage of this roof type is its strength and the protective overhangs formed over end walls and sidewalls.
- *Gambrel roof*—In this variation of the gable roof, each slope is broken at midspan. This style is used on two-story construction. It permits extra space for more efficient use of the second floor. Dormers are usually included. The gambrel roof is a traditional style typical of the American colonial period and the period immediately following.
- *Flat roof*—This roof is supported on joists that also carry ceiling material on the underside. It may have a slight pitch (slope) to provide drainage.
- *Shed roof*—This simplest of pitched roofs is sometimes called a lean-to roof. The name comes from its frequent use on additions to a larger structure. It is often used in contemporary designs where the ceiling is attached directly to the roof frame.
- *Mansard roof*—Like the hip roof, the mansard has four sloping sides. However, each of the four sides has a double slope. The lower, outside slope is

nearly vertical. The upper slope is slightly pitched. Like the gambrel roof, the main advantages are the additional space gained in the rooms on the upper level and making the house look lower than it is. The name comes from its originator, architect Francois Mansart (1598–1666).

There are two basic systems used in framing the roof of platform structures: conventional, stick-built rafters and truss rafters. A carpenter builds conventional roof frames onsite. The ceiling joists and rafters are laid out, cut, and installed one at a time.

> **Green Note**
> Although not considered framework, SIPs can also be used to construct a roof structure. SIP systems are engineered for the individual project. Consult the working drawings for assembly requirements.

Rafters are sloped framing members that run downward from the peak of the roof to the plates of the outside walls. They are the supports for the roof load. As you learned in Chapter 11, **Wall and Ceiling Framing**, the ceiling joists tie the outside walls together and support the ceiling materials for the rooms below. They also secure the bottom ends of the rafters, preventing them from pushing outward when loaded.

Roof trusses are engineered and prefabricated assemblies. They combine rafter and ceiling joist in one. Usually, trusses are factory-built and delivered to the building site, **Figure 12-2**. The lower chords also serve as ceiling joists and support the ceiling coverings.

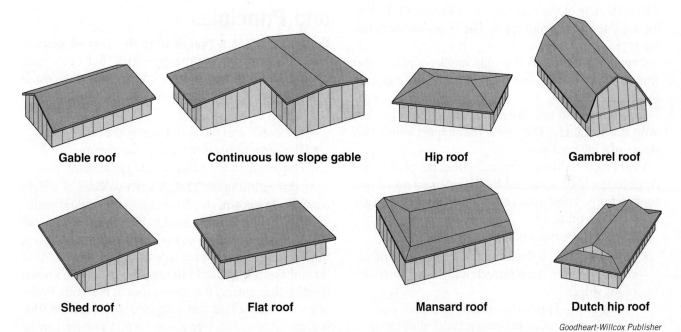

Gable roof   Continuous low slope gable   Hip roof   Gambrel roof

Shed roof   Flat roof   Mansard roof   Dutch hip roof

Goodheart-Willcox Publisher

Figure 12-1. Common types of roofs used in residential construction. The gable is most often used.

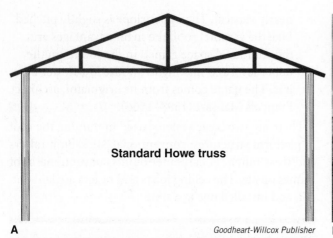

Standard Howe truss

A   *Goodheart-Willcox Publisher*

B   *alveyhighway/Shutterstock.com*

Figure 12-2. Roof trusses. A—The standard roof truss efficiently supports heavy loads. B—This set of prefabricated trusses is getting loaded on a truck for delivery.

Trussed roof framing has become more common than stick-framed roofs, but every framing carpenter encounters conventional rafter framing from time to time. If you know how to lay out and frame a conventional roof frame, learning how to erect trusses is fairly easy.

12.2 Roof Supports

Roofs are supported by one or all of the following systems, depending on the type of rafter design:

- Outside walls
- Ceiling joists (beams that hold the ceiling materials)
- Interior bearing walls (partitions that support structures above)

12.3 Parts of a Roof Frame

The plan view of the roof shown in **Figure 12-3** combines gable and hip roof types. The view also identifies the types of rafters.

Common rafters run at a right angle (in plan view) from the wall plate to the ridge. A gable roof has only this type of rafters.

Hip rafters also run from the plate to the ridge, but only at a 45° angle. They form the support where two slopes of a hip roof meet.

Valley rafters extend diagonally from the plate to the ridge in the hollow formed by the intersection of two roof sections. These roof sections are usually at a right angle to each other.

There are three types of jack rafters:

- *Hip jack*—This is the same as the lower part of a common rafter, but intersects a hip rafter instead of the ridge.
- *Valley jack*—This is the same as the upper end of a common rafter, but intersects a valley rafter instead of the plate.

- *Cripple jack*—Intersects neither the plate nor the ridge and is terminated at each end by hip and valley rafters. The cripple jack rafter is also called a cripple rafter, hip-valley cripple jack, or valley cripple jack.

Rafters are formed by laying them out and making various cuts. **Figure 12-4** shows the cuts for a common rafter and the sections formed. The ridge cut allows the upper end to tightly fit against the ridge. The *bird's mouth* is formed by a seat cut and plumb (vertical) cut when the rafter extends beyond the plate. This extension is called the overhang or tail. When there is no overhang, the bottom of the rafter is ended with a seat cut and a plumb cut.

12.4 Layout Terms and Principles

Roof framing is a practical application of geometry. This is an area of mathematics that deals with the relationships of points, lines, and surfaces. It is based largely on the properties of the right (90°) triangle:

- The horizontal distance (run) is the base.
- The vertical distance (rise) is the altitude.
- The length of the rafter is the *hypotenuse*.

As shown in **Figure 12-5**, if any two sides of a right triangle are known, the third side can be found mathematically. The formula used is $H^2 = A^2 + B^2$. *A* is the altitude or height, *B* is the base, and *H* (hypotenuse) is the line forming the third side of the triangle. To find the unknown length, add the squares of the two known lengths, then extract the square root of that sum. Without a calculator that has a square root function, this is time-consuming. The same correct answer can be found using other methods.

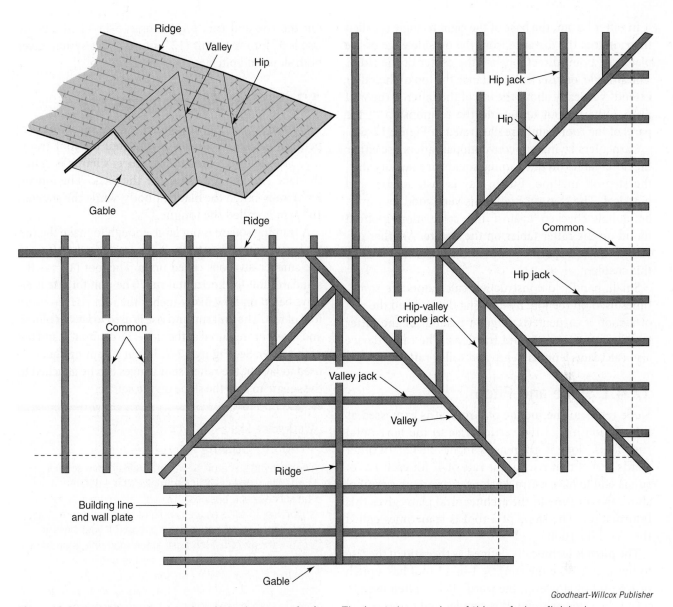

Goodheart-Willcox Publisher

Figure 12-3. A roof frame in plan view. Note the types of rafters. The inset shows a view of this roof when finished.

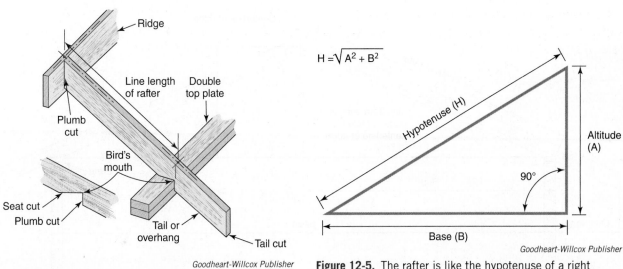

Goodheart-Willcox Publisher

Figure 12-4. Rafter parts. Various cuts and surfaces are important because rafters must have a snug fit at all joints.

$$H = \sqrt{A^2 + B^2}$$

Goodheart-Willcox Publisher

Figure 12-5. The rafter is like the hypotenuse of a right triangle. If you square the rise and run of a roof, then add the two figures, you have the square of the rafter's length.

In rafter layout, the base of the right triangle is called the **run**. It is the distance from the outside edge of the plate to a point directly below the center of the ridge. The altitude, or **rise**, is the distance the top of the rafter extends vertically above the top of the rafter at the wall plate. Other layout terms and the relationship of the parts of the roof frame are illustrated in **Figure 12-6**.

Carpenters laying out conventional rafters use either the rafter tables on the framing square or a layout called the step-off method. Either way is fast, simple, and practical. The step-off method is very good, but must be double-checked against the calculation method found on the rafter tables on the square. Another option is to use a speed square. This is discussed later in this chapter.

Small, handheld construction calculators are sometimes used on the job. By entering the rise and the run of a roof, a carpenter can get a readout of the rafter length in feet, inches, and fractions. The same device also calculates lengths of hip and valley rafters.

12.4.1 Slope and Pitch

Slope refers to the incline of a roof. It is expressed as the relationship of the vertical rise to the horizontal run. Thus, slope is given as "x inches-in-12." In other words, a roof that rises at the rate of 4″ for each 12″ of run is said to have a 4-in-12 slope. A triangular symbol above the roofline in the architectural plans gives this information. The slope of a roof is sometimes called the cut of the roof.

The **pitch** is technically defined as the ratio of the rise to the span (twice the run). A 4-in-12 roof has a pitch of 1/6 (4/24). However, the word pitch is often used to mean the same as slope. It is simpler and easier to call

out the rise and run, for example, 5-in-12 means the rise is 5″ for every unit (12″) of run. This system gives both slope and pitch.

12.5 Unit Measurements

The framing square, also called a steel square or carpenter's square, is a basic layout tool for roof framing. The side with the manufacturer's name is called the face and the opposite side the back. The longer, 24″ arm is called the blade or body, while the shorter, 16″ arm is called the tongue.

A framing square is not large enough to make the rafter layout with one setting. It is necessary, therefore, to use smaller divisions called units. The foot (12″) is the standard unit for horizontal run. The unit for rise is always based on how many inches the roof rises in every foot of run. The unit run and rise is used to lay out plumb and level cuts required at the ridge, bird's mouth, and tail of the rafter. See **Figure 12-7**. When a framing square is used to lay out the rafter, stair gauges can be attached to the square to save the slope on the square.

Workplace Skills
Problem Solving

The ability to see a problem and brainstorm a solution is a highly valued skill. It can make all the difference in the success of a business. Do not rely on your boss or supervisor to solve all the problems. When you are stuck, think it through. What are your options? If your boss is not on-site, or is just too busy at the moment, figure it out. Do not sit around and wait for them to solve the problem for you. Waiting on someone else looks bad for you and kills production. Developing your problem solving skills is key to your success in any career.

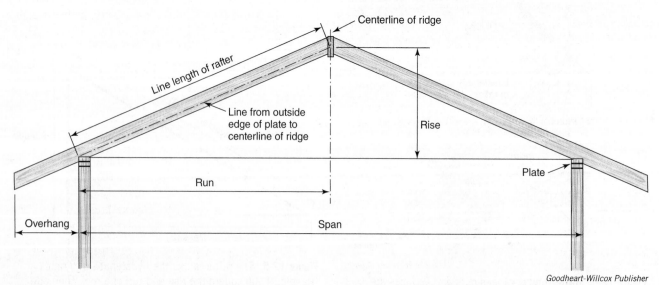

Goodheart-Willcox Publisher

Figure 12-6. These are basic terms in rafter layout. Do you see how the rise, run, and rafter length form a right triangle?

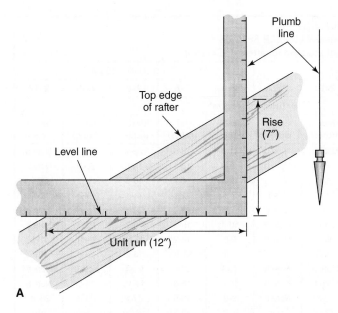

A

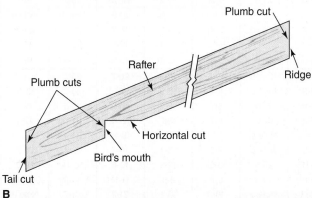

B

C

Goodheart-Willcox Publisher

Figure 12-7. A—How to position a square for marking one unit of measure on a rafter. The run is measured along the blade. The rise (in inches per foot) is measured on the tongue. B—This same position is used to lay out and mark the bird's mouth and plumb cuts. C—Stair gauges are a convenient way to save the slope on a framing square.

12.6 Framing Plans

When working with simple designs, a carpenter can easily visualize the roof framing. Since the wall framing is already erected, this is the only additional information needed:

- Rise and run of the roof
- Amount of required overhang
- Subfascia and type of fascia
- Type of ridge
- Any other information relating to roof construction

These items are included in the house plans. When the structure has a complex roof design, an architect often includes a roof framing plan. If a framing plan is not included, a carpenter should prepare one. To do this, make a scaled drawing on tracing paper laid directly over the floor plans. Include ridges, overhang, and every rafter. Make the drawing similar to the one shown in **Figure 12-3**. However, you can use a single line to represent each framing member. In your drawing, maintain accurate spacing between rafters. Draw hips and valleys at 45° angles and make jack rafters parallel to the common rafters.

Pro Tip
Dual-pitched hips are not at a typical 45° angle. They may not intersect over the outside corner, but slightly to one side or the other.

12.7 Rafter Sizes

As in floor and ceiling framing, the strength (size) of the rafter is determined by the species and grade of lumber, spacing, span or length, and the load to be placed on the roof. Local building codes must be consulted for their specifications. **Figure 12-8** shows one of several rafter tables included in the IRC. As this table illustrates, determining the size of a rafter can be quite involved. Always rely on the building code in effect to make this determination. Stock for hips, valleys, and ridges is usually larger than for other roof framing members.

Code Note
The IRC includes rafter span tables for the following conditions:

- Roof live load of 20 psf or ground snow load of 30 psf, 50 psf, or 70 psf
- Ceiling not attached to rafters or ceiling attached to rafters
- Dead load of 10 psf or 20 psf
- Rafter spacing of 12″, 16″, 19.2″, or 24″
- Wood species and grade
- Nominal rafter size: 2×4, 2×6, 2×8, 2×10, or 2×12

Table R802.4.1(1)
RAFTER SPANS FOR COMMON LUMBER SPECIES
(Roof live load = 20 psf, ceiling not attached to rafters, L/Δ = 180)

RAFTER SPACING (Inches)	SPECIES AND GRADE		DEAD LOAD = 10 psf					DEAD LOAD = 20 psf				
			2×4	2×6	2×8	2×10	2×12	2×4	2×6	2×8	2×10	2×12
			Maximum rafter spans[a]									
			(feet – inches)	(feet – inches)	(feet – inches)	(feet – inches)	(feet – inches)	(feet – inches)	(feet – inches)	(feet – inches)	(feet – inches)	(feet – inches)
12	Douglas fir-larch	SS	11-6	18-0	23-9	Note b	Note b	11-6	18-0	23-9	Note b	Note b
	Douglas fir-larch	#1	11-1	17-4	22-5	Note b	Note b	10-6	15-4	19-5	23-9	Note b
	Douglas fir-larch	#2	10-10	16-10	21-4	26-0	Note b	10-0	14-7	18-5	22-6	26-0
	Douglas fir-larch	#3	8-9	12-10	16-3	19-10	23-0	7-7	11-1	14-1	17-2	19-11
	Hem-fir	SS	10-10	17-0	22-5	Note b	Note b	10-10	17-0	22-5	Note b	Note b
	Hem-fir	#1	10-7	16-8	22-0	Note b	Note b	10-4	15-2	19-2	23-5	Note b
	Hem-fir	#2	10-1	15-11	20-8	25-3	Note b	9-8	14-2	17-11	21-11	25-5
	Hem-fir	#3	8-7	12-6	15-10	19-5	22-6	7-5	10-10	13-9	16-9	19-6
	Southern pine	SS	11-3	17-8	23-4	Note b	Note b	11-3	17-8	23-4	Note b	Note b
	Southern pine	#1	10-10	17-0	22-5	Note b	Note b	10-6	15-8	19-10	23-2	Note b
	Southern pine	#2	10-4	15-7	19-8	23-5	Note b	9-0	13-6	17-1	20-3	23-10
	Southern pine	#3	8-0	11-9	14-10	18-0	21-4	6-11	10-2	12-10	15-7	18-6
	Spruce pine-fir	SS	10-7	16-8	21-11	Note b	Note b	10-7	16-8	21-9	Note b	Note b
	Spruce pine-fir	#1	10-4	16-3	21-0	25-8	Note b	9-10	14-4	18-2	23-3	25-9
	Spruce pine-fir	#2	10-4	16-3	21-0	25-8	Note b	9-10	14-4	18-2	23-3	25-9
	Spruce pine-fir	#3	8-7	12-6	15-10	19-5	22-6	7-5	10-10	13-9	16-9	19-6
16	Douglas fir-larch	SS	10-5	16-4	21-7	Note b	Note b	10-5	16-3	20-7	25-2	Note b
	Douglas fir-larch	#1	10-0	15-4	19-5	23-9	Note b	9-1	13-3	16-10	20-7	23-10
	Douglas fir-larch	#2	9-10	14-7	18-5	22-6	26-0	8-7	12-7	16-0	19-6	22-7
	Douglas fir-larch	#3	7-7	11-1	14-1	17-2	19-11	6-7	9-8	12-12	14-11	17-3
	Hem-fir	SS	9-10	15-6	20-5	Note b	Note b	9-10	15-6	19-11	24-4	Note b
	Hem-fir	#1	9-8	15-2	19-2	23-5	Note b	9-0	13-1	16-7	20-4	23-7
	Hem-fir	#2	9-2	14-2	17-11	21-11	25-5	8-5	12-3	15-6	18-11	22-0
	Hem-fir	#3	7-5	10-10	13-9	16-9	19-6	6-5	9-5	11-11	14-6	16-10
	Southern pine	SS	10-3	16-1	21-2	Note b	Note b	10-3	16-1	21-2	25-7	Note b
	Southern pine	#1	9-10	15-6	19-10	23-2	Note b	9-1	13-7	17-2	20-1	23-10
	Southern pine	#2	9-0	13-6	17-1	20-3	23-10	7-9	11-8	14-9	17-6	20-8

Check sources for availability of lumber in lengths greater than 20 feet.

For SI: 1 inch = 25.4 mm, 1 foot = 304.8 mm, 1 pound per square foot = 0.0479 kPa.

a. The tabulated rafter spans assume that ceiling joists are located at the bottom of the attic space or that some other method of resisting the outward push of the rafters on the bearing walls, such as rafter ties, is provided at that location. Where ceiling joists or rafter ties are located higher in the attic space, the rafter spans shall be multiplied by the following factors:

H_C/H_R	Rafter Span Adjustment Factor
1/3	0.67
1/4	0.76
1/5	0.83
1/6	0.90
1/7.5 or less	1.00

where:

H_C = Height of ceiling joists or rafter ties measured vertically above the top of the rafter support walls.
H_R = Height of roof ridge measured vertically above the top of the rafter support walls.

b. Span exceeds 26 feet in lenght.

Portions of this publication reproduce excerpts from the 2015 International Residential Code for One- and Two-Family Dwellings, International Code Council, Inc., Washington, D.C. Reproduced with permission. All rights reserved.

Figure 12-8. This is part of one of eight rafter tables found in the IRC. Refer to your local code for the tables that apply in your area.

PROCEDURE

Laying Out Rafters (Step-off Method)

In this example, the total run of the rafter (ridge line to building line) is 6'-8" and the rise will be 5-in-12. There is a 1'-10" overhang.

1. Place the framing square on the stock and align the numbers for the unit run (12") and rise with the top edge of the rafter. The unit run is located on the blade and the unit rise on the tongue.

2. Set the rise and run on the square using stair gauges. To ensure accuracy, the correct marks on the square must be exactly positioned over the edge of the stock each time a line is marked. Be sure to use a sharp pencil to make the layout lines.

3. Starting at the top of the rafter, position the square and draw the ridge line along the edge of the tongue. While still holding the square, mark the length of the odd unit (8" in this example).

4. Shift the square along the edge of the stock until the tongue setting is even with the 8" mark. Draw a line along the tongue and mark the 12" point on the blade for a full unit, **Figure 12-9**.

5. Move the square to the 12" point just marked and repeat the marking procedure. Continue until the correct number of full units is laid out. This number is the same as the number of feet in the total run (6 in this example).

6. Form the bird's mouth by drawing a horizontal line (seat cut) to meet the building line so the surface is about equal to the width of the plate. Next, draw a horizontal line (seat cut) to the lower edge of the rafter. The size of the bird's mouth may vary depending on the design of the overhang, but it should provide a full bearing surface on the top plate. In **Figure 12-9**, note that the square has been turned over to mark these cuts and also to lay out the overhang. This may or may not be necessary, depending on the length of the rafter blank.

7. To lay out the overhang, start with the plumb cut of the bird's mouth and mark full units first. Then, add any odd unit that remains. The tail cut may be plumb, square, or a combination of plumb and level. Check the cornice details shown in the architectural plans for exact requirements.

8. The final step in the layout consists of shortening the rafter at the ridge. With the square in position, draw a new plumb line back from the ridge line 1/2 of the thickness of the ridge board, **Figure 12-9**.

9. Make the ridge, bird's mouth, and tail cuts.

10. Label the rafter as a pattern, indicating the roof section to which it belongs.

12.8 Laying Out Common Rafters

Rafters can be laid out by the step-off method, the rafter table on the framing square, or with a construction calculator. Any of these methods gives the correct rafter length. The step-off method is easy to use and the layout can be double-checked with the rafter tables.

Using any one of the methods, a carpenter first lays out, checks, and cuts a pattern rafter. The pattern is used to mark other rafters of the same size and type. The pattern will be used as one of the rafters after all of the rafters have been laid out. For pattern layout, select a piece of lumber that is straight and true.

To cut the pattern, place the blank on a pair of sawhorses. Usually, a carpenter stands on the crowned side (top edge) of the rafter. It is easier to handle the framing square from that position. However, in order to describe rafter layouts that can be quickly understood,

Figure 12-9 shows this position in reverse. The rafter in this illustration is shown in its installed position. As you lay out rafters, try to visualize how each will appear when it is set in the completed roof frame. Forming the habit of visualizing the rafter in its proper place helps to eliminate errors.

Pro Tip
The bird's mouth can be started with a circular saw, but you will need to use a handsaw to finish the plumb cut and seat cut.

12.8.1 Using the Rafter Table

You can also calculate the length of a common rafter using the table on the framing square. This table is on the face side of the square. It is the one marked *Length Common Rafters Per Foot Run*. You need to know the slope and run before starting.

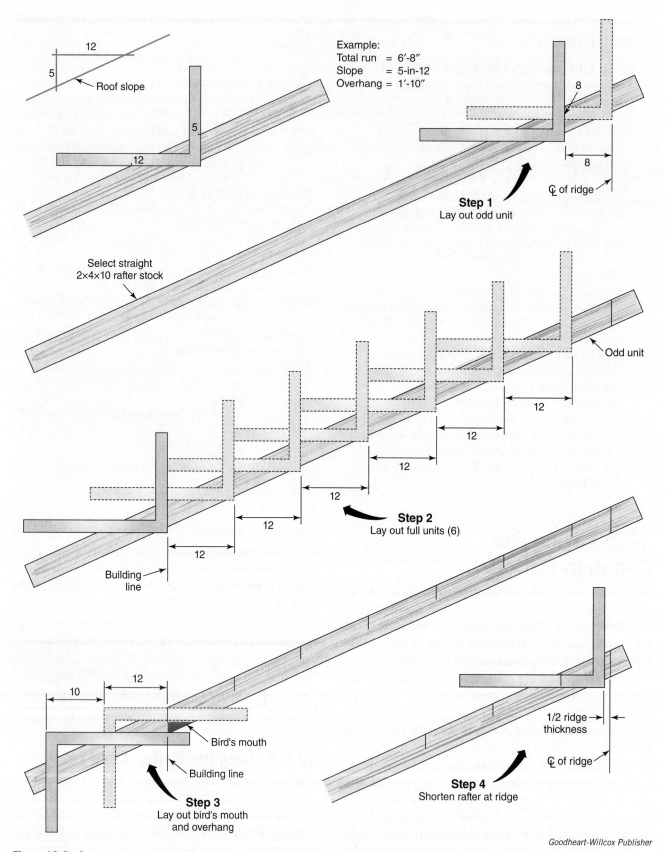

Roof slope

Example:
Total run　= 6′-8″
Slope　　　= 5-in-12
Overhang = 1′-10″

Step 1
Lay out odd unit

₵ of ridge

Select straight
2×4×10 rafter stock

Odd unit

Step 2
Lay out full units (6)

Building
line

Bird's mouth

Building line

Step 3
Lay out bird's mouth
and overhang

1/2 ridge
thickness

₵ of ridge

Step 4
Shorten rafter at ridge

Goodheart-Willcox Publisher

Figure 12-9. Carpenters can use this procedure for laying out a common rafter. It is known as the step-off method because each foot of run requires another step of the square.

To see how this works, suppose the building is 13'-4" wide, making the run of a rafter 6'-8" and the rise per foot 5". Find the inch mark for 5" on the blade of the square. See **Figure 12-10**. Under that number, on the top line of the table, look for the number nearest to the 5" mark. This is the length of the rafter in inches for one foot of run. To find the length of the rafter from the building line to the center of the ridge, multiply the units of run by the figure from the rafter table, as in the following examples.

Example 1:

$$Run = 6'\text{-}8'', Slope = 5\text{-}in\text{-}12$$
$$Run = 6'\text{-}8''$$
$$= 6\ 2/3\ units$$
$$Rise = 5'', number\ from\ rafter\ table = 13$$
$$Rafter\ length = 6\ 2/3 \times 13$$
$$= 86.67''$$
$$= 7'\text{-}2\ 11/16''$$

Example 2:

$$Run = 12', Slope = 4\text{-}in\text{-}12$$
$$Run = 12\ units$$

$$Rise = 4'', number\ from\ the\ rafter\ table = 12.65$$
$$Rafter\ length = 12 \times 12.65$$
$$= 151.8''$$
$$= 12.65'$$
$$= 12'\text{-}7\ 13/16''$$

These calculations result in the line length of the rafter running from the center of the ridge to the outside of the plate. To make the pattern layout, add the overhang and subtract half the thickness of the ridge board from the length.

12.8.2 Using the Speed Square

Sometimes a carpenter uses a *speed square* to find rafter lengths and determine the angle of cuts. This triangular measuring tool is smaller and easier to carry than a framing square. Some carpenters use a larger version of the speed square, called a *rafter square*, to lay out rafters. To use a speed square, you need to know the pitch of the roof. This is given in the plans or it can be determined by some simple math. Rafter charts for every pitch are supplied by the square's manufacturer.

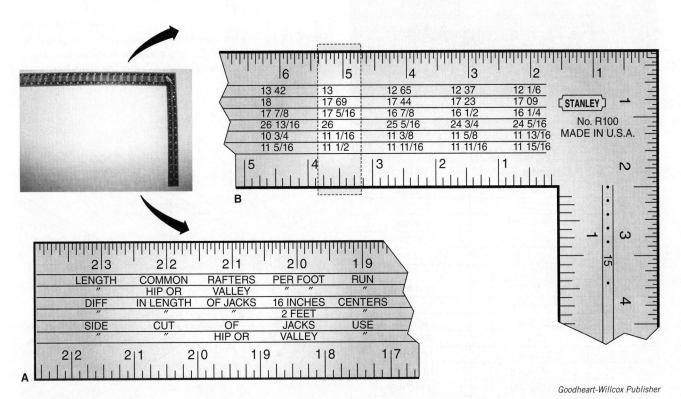

Goodheart-Willcox Publisher

Figure 12-10. The rafter table is printed on the face side of the framing square. A—The blade of the square carries tables for figuring lengths of rafters. B—If the unit rise is 5", the rafter length for 12" of run is 13".

PROCEDURE

Laying Out Rafters (Speed Square Method)

1. Once you know the rise per foot of run, refer to the chart for that rise. Read down the chart to the correct run to find the length of the rafter.
2. Mark the length on the rafter.
3. Place the pivot point of the speed square on the length mark while lining up the rise on another scale. See **Figure 12-11**. This step gives the angle for marking the plumb cut at the ridge.
4. Using the rafter pattern, cut the number required.

A
Goodheart-Willcox Publisher

Speed Square Chart		
8-12 Pitch		
Feet of Run	**Common Rafter**	**Hip or Valley Rafters**
1	1'-2 3/8"	1'-6 3/4"
2	2'-4 7/8"	3'-1 1/2"
3	3'-7 1/4"	4'-8 1/4"
4	4'-9 3/4"	6'-3"
5	6'-1/8"	7'-9 3/4"
6	7'-2 1/2"	9'-4 5/8"
7	8'-5"	10'-11 3/8"
8	9'-7 3/8"	12'-6 1/8"
9	10'-9 3/4"	14'-7/8"
10	12'-1/4"	15'-7 5/8"
11	13'-2 5/8"	17'-2 3/8"
12	14'-5 1/8"	18'-9 1/8"
14	16'-9 7/8"	21'-10 5/8"
16	19'-2 3/4"	25'-1/8"

B
Goodheart-Willcox Publisher

Figure 12-11. A—Using a speed square to mark the plumb cut on a common rafter. B—A sample speed square chart for figuring common rafters and valley or hip rafters when pitch of roof is 8-in-12.

12.9 Erecting a Gable Roof

Lay out the rafter spacing along the wall plate as the ceiling joists are laid out. When the rafter spacing is the same as ceiling joist spacing, every rafter is nailed to a joist. When rafters are spaced 24″ OC and ceiling joists 16″ OC, the layout is arranged as in **Figure 12-12**. The plate layout is important, so carefully follow the roof framing plan. To allow more room over the wall plate for additional insulation, some carpenters attach rafters to a 2×4 nailed on top of the ceiling joists, **Figure 12-13**.

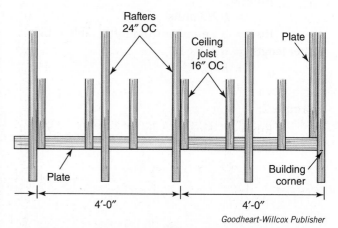

Goodheart-Willcox Publisher

Figure 12-12. A plan view of ceiling joist and rafter layout for 24″ OC rafters and 16″ OC joists. A joist is nailed to every other rafter, acting as a tie beam to keep the walls from spreading.

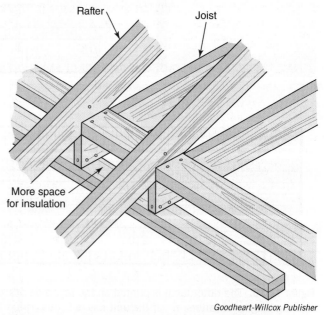

Goodheart-Willcox Publisher

Figure 12-13. This alternate framing method has rafters resting on a 2×4 plate placed on top of the ceiling joists. This allows room to place thicker insulation batts over the wall plate without restricting airflow to the soffit vents.

Select straight pieces of ridge stock and lay out the rafter spacing by transferring the marking directly from the plate. Joints in the ridge should occur at the center of a rafter. Cut the lengths that will make up the ridge and lay them across the ceiling joists close to where they will be assembled with the rafters.

To make the initial assembly, some carpenters prefer to first attach the ridge to several rafters on one side. This assembly is then raised and supported while the rafters are nailed to the plate. Then, several rafters are installed on the opposite side.

Another method of supporting a ridge board may be used. Mark two 2×4s for the height of the rise and attach one to either end of the ridge board. The edge of the ridge board should align with the mark. Lift up the assembly and tack the base of each 2×4 to the plate of the bearing wall. Plumb and secure the vertical 2×4s with braces. This method requires fewer carpenters.

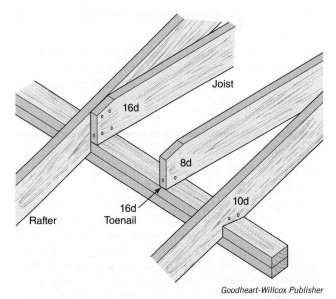

Goodheart-Willcox Publisher

Figure 12-14. The nailing pattern for joists and rafters at the wall plate.

PROCEDURE

Erecting a Site-Built Roof

1. Select straight rafters for the gable end and nail one in place at the plate.
2. Install a rafter on the opposite side with a worker at the ridge supporting both rafters.
3. Place the ridge board between the two rafters and temporarily nail it in place.
4. Move about five rafter spaces from the end and install another pair of rafters.
5. Plumb and brace the assembly. Make any adjustments necessary in the nailing of the first rafters. **Figure 12-14** shows the assembly and nailing pattern at the plate. Special framing anchors are often used, especially in areas where hurricanes are possible, **Figure 12-15**.
6. After aligning the ridge, install the remaining rafters. First, nail the rafter at the plate and then at the ridge. For a carpenter working alone or on a small crew, temporary supports hold the ridge board until rafters are installed, **Figure 12-16**.
7. Drive 16d nails through the ridge board into the rafter. The rafters on the opposite side of the ridge are toenailed. Install only a few rafters on one side before placing matching rafters on the opposite side. This practice makes it easier to keep the ridge straight.
8. Periodically check the ridge to ensure it remains straight and level.
9. Continue to add sections of ridge and assemble the rafters. Add bracing when required. Always install rafters with the crown (curve) turned upward.

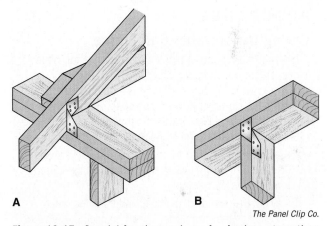

The Panel Clip Co.

Figure 12-15. Special framing anchors. A—Anchors strengthen the joint between the rafter and wall. B—The same type of clip can be used to attach the plate to studs.

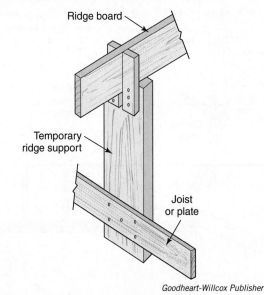

Goodheart-Willcox Publisher

Figure 12-16. A temporary support, built onsite, holds the ridge board level until a few rafters are installed.

Safety Note

When framing a roof, use extra care to prevent a fall. Fall protection devices must be used. Erect solid scaffolding wherever it is helpful. Avoid working directly above another person.

Code Note

The connection between the top plate and the rafter provides resistance to uplift. Uplift can occur when strong winds push against the side of a building and roof. In some cases, additional tie-connections may be needed based on the rafter spacing, roof span, and design wind speed. In the United States, these additional tie-down connections are most commonly needed along the gulf coast and Atlantic coast regions. Be sure to check the requirements in your local building code.

PROCEDURE

Laying Out a Gable End Frame

1. Using a level and straightedge, mark the top plate to show the location of a stud directly below the ridge, if possible. Often, a ventilator is located in the center of the end frame. If this is the case, mark a distance 1/2 of the ventilator width to either side of the plumb line. This mark is the location of the first stud.
2. Position the stud on the top plate and use a level to recheck for plumb. Mark the location of the rafter on the side of the stud.
3. Lay out the next stud, plumb it, and mark the rafter location on the stud's side. The difference between the two stud lengths is the *common difference*, **Figure 12-17**.
4. Using the common difference, continue marking, cutting, and placing studs until the edge of the frame is reached. If the rafters on both sides have the same rise and run, the end frame should be symmetrical (both sides identical).

PROCEDURE

Finding the Common Difference with the Framing Square

The common difference can also be obtained using the framing square. See **Figure 12-18**.
1. Set the square on a stud. Align the unit run on the blade with the edge of the stud. Then align the unit rise on the tongue to the same edge. (This is the same as the step method used in determining the length of a rafter.)
2. Mark a line on the stud along the outside edge of the blade.
3. Slide the square away from you. Keep the outer edge of the blade on this line. Watch the inch marks on the blade. Stop when the inch number for the stud spacing (16″ or 24″) aligns with the edge of the stud.
4. Read the inch mark on the tongue of the square that aligns with the edge of the stud. This is the common difference.

12.10 Gable End Frame

The end frames of a gable should be assembled after the rafters and ridge have been installed. The end frame consists of vertical studs running from the top plate of the bearing walls to the end rafters. The framing of the gable end is most easily done while it lies flat over the ceiling joists. Overhang, brick racks, frieze trim, louvers, housewrap, and siding can all be installed at this time. After it is erected, the end gable should be well braced before installing the ridge board and rafters. The following procedure can be used to determine the length and location of studs.

Roof designs often include an *extended rake* (gable overhang). Typical framing is illustrated in **Figure 12-19**. It requires the gable end frame be constructed before the roof frame is completed.

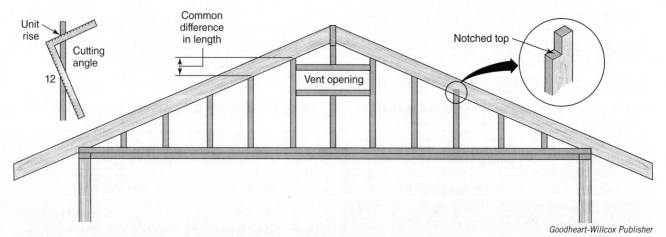

Goodheart-Willcox Publisher

Figure 12-17. Lay out studs for a gable end as shown. The change in length from one stud to another is the common difference.

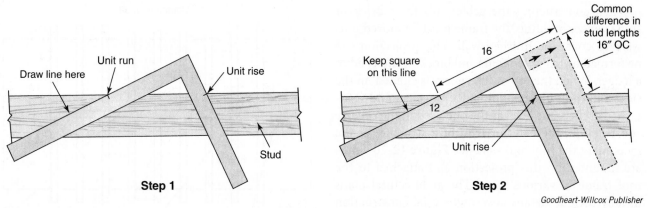

Goodheart-Willcox Publisher

Figure 12-18. Using a framing square to find the common difference for gable end studs. Step 1—Line up the run and rise on the edge of a stud. Step 2—Slide the square in the direction shown until the 16″ (or 24″, depending on the OC spacing you are using) mark on the blade aligns with the edge of the stud.

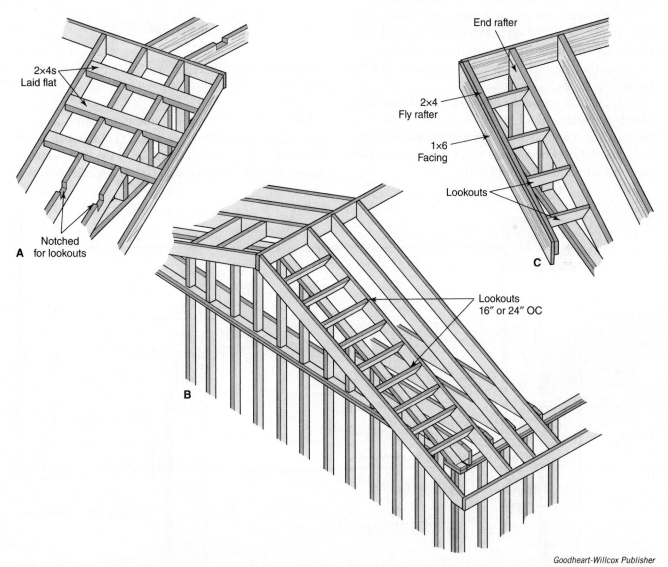

Goodheart-Willcox Publisher

Figure 12-19. Methods of framing a gable overhang. A—The 2×4 lookouts are laid flat over notched rafters. B—The plate atop the gable end studs supports lookouts laid on edge. Make sure the top of the plate lines up with the bottoms of the rafters. C—A small overhang with short lookouts supports a 2×4 fly rafter and a fascia. This method should not be used for overhangs of 1′ or more.

When constructing the gable ends for a brick or stone veneer building, the frame must be moved outward to cover the finished wall. This projection can be formed by using *lookouts* and blocking attached to a ledger. When the top of the veneer is aligned on the sides and ends of the building, the ledger should be attached at the same level as the one used in the *cornice* construction, the exterior trim of a structure at the meeting of the roof and wall. See **Figure 12-20**. Studs are mounted on this projection and attached to the roof frame in various ways. The architectural plans usually include details covering special construction features of this type.

12.11 Hip and Valley Rafters

Hip roofs or intersecting gable roofs have some or all of the following rafters: common, hip, valley, jack, and cripple. Two roof surfaces slanting upward from adjoining walls meet on a sloping line called a hip. The rafter supporting this intersection is known as a hip rafter.

First, cut and frame the common rafters and ridge boards. The ridge of a hip roof is cut to the length of the building, minus twice the run of the common rafters, plus the thickness of the rafter stock. It intersects the common rafters, **Figure 12-21**.

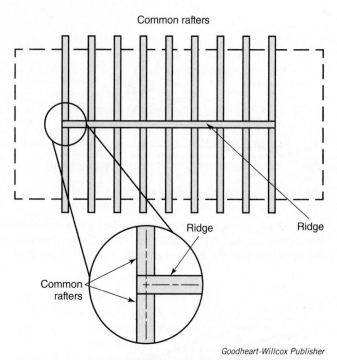

Goodheart-Willcox Publisher

Figure 12-21. The first step in framing a hip or intersecting roof is to install the ridge boards and common rafters.

From each corner of the building, measure along the side wall a distance that is equal to 1/2 of the span. Mark the point, which is the centerline of the first common rafter. All other rafters are laid out from this position.

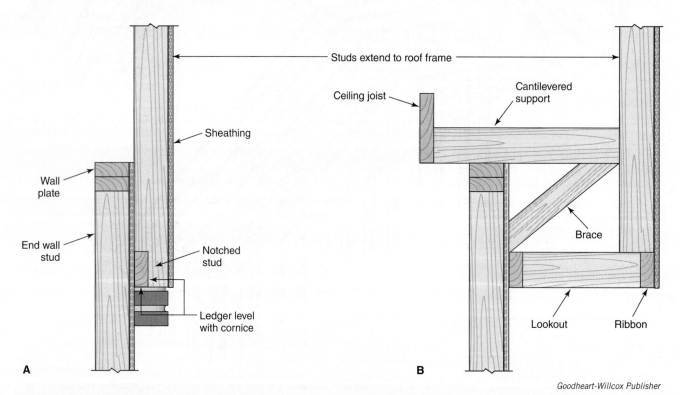

Goodheart-Willcox Publisher

Figure 12-20. Gable end framing for masonry veneered buildings. A—Build out the framing to cover the brick or stone veneer. B—An alternate framing that extends the gable end farther. Such framing might also be used over doors and for special features.

The length of a hip rafter can be determined by using the rafter table found on most framing squares. You must first know the run of the common rafters and the unit rise of the roof. The length of a hip or valley rafter per foot of run is usually the second line on the framing square, below the line for the length of common rafters per foot of run. See **Figure 12-10**.

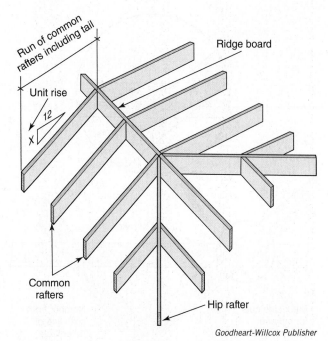

Figure 12-22. Use the run of common rafters, including the run of the overhang, to calculate the length of a hip rafter, including its overhang.

PROCEDURE
Finding the Length of a Hip Rafter

Use the rafter table on a framing square or a rafter table from another reference source.

1. Find the unit rise (number of inches the roof rises for every horizontal foot) at the top of the table. These are the regular inch graduations at the edge of the square.
2. In the column under that number, find the length of the hip rafter per foot of run of the common rafters. Add the overhang at the edge of the roof to the run of the common rafters, **Figure 12-22**.
3. Multiply the length of the hip rafter per foot of run (the number found in Step 2) by the number of feet of run of a common rafter (1/2 the width of the building).
4. Reduce the length by 1/2 the 45° thickness of the ridge board. The 45° thickness is the length of a 45° line drawn across the edge of the board. The 45° thickness of a 2× (1 1/2″ actually) is 2 1/8″, so half of that is 1 1/16″.

Example:

$$\text{Run} = 12' + 8''\text{overhang}$$
$$= 12'\text{-}8''$$

Length of hip rafter (from table) = 17.69″

$$\text{Calculated length of hip rafter} = 17.69'' \times 12.67$$
$$= 224.13''$$
$$= 18.68'$$
$$= 18'\text{-}8\ 3/16''$$

Length of the hip rafter
(minus ridge board)   = 18′-8 3/16″ − 1 1/16″
= 18′-7 1/8″

This is the length of the hip rafter including 8″ overhang.

Hip and valley rafters must be shortened at the ridge by a distance equal to 1/2 of the 45° horizontal thickness of the ridge, **Figure 12-23**. The side cuts are then laid out as shown in the illustration. Another method is to use the numbers from the sixth line of the rafter table. For example, consider that the slope is 5-in-12. The number on the sixth line below the 5″ mark is 11 1/2. All of the numbers in the table are based on or related to 12, so line up 12 on the blade and 11 1/2 on the tongue along the edge of the rafter, **Figure 12-24**.

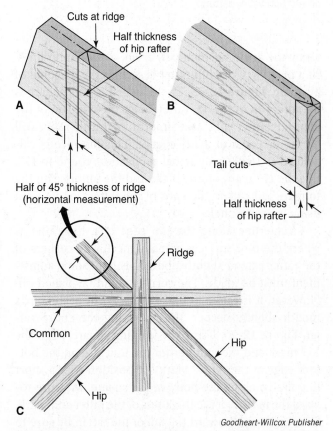

Figure 12-23. Shortening the hip rafter and making side cuts. A—Shorten the top of the hip rafter by 1/2 of the thickness of the ridge and mark. Then, measure back 1/2 of the thickness of the rafter. Mark both sides for angle cuts. B—Note how the tail is marked for cutting. C—This drawing explains why angled cuts must be 1/2 of the thickness of the ridge.

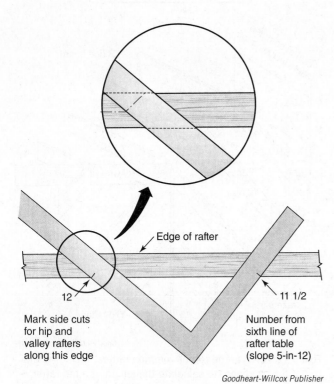

Figure 12-24. Using the framing square to lay out side cuts on hip and valley rafters.

Draw the angle for each cut. Then, draw plumb lines. In a plan view, the hip rafter is seen as the diagonal of a square, **Figure 12-25**. The diagonal of this square is the total run of the hip rafter. Since the unit run of the common rafter is 12″, the unit run of the hip rafter will be the diagonal of a 12″ square. This distance is calculated as 16.97″. For actual application, round to 17″. Use the 17″ mark on the blade and the unit rise on the tongue. Tail cuts at the ends of the rafters are laid out using the same angle.

A centerline along the top edge of a hip rafter is where the roof surfaces actually meet. The corners of the rafter extend slightly above this line. Some adjustment must be made. The corners could be planed off. However, it is easier to make the seat cut of the bird's mouth slightly deeper, thus lowering the entire rafter, **Figure 12-26**. Using the framing square, align the 17″ mark and the number for the unit rise on the bottom edge of the rafter. Mark the position with a short line drawn along the body of the square. Measure toward the tail 1/2 of the thickness of the rafter and mark. Shift the square toward the tail of the rafter. Be sure to maintain the alignment of 17″ and the unit rise. When the square aligns with the previous mark, mark the seat of the bird's mouth. Do not mark the plumb cut until you have read the next paragraph.

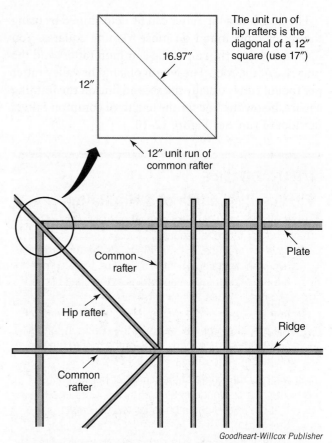

Figure 12-25. The length of a hip rafter is the diagonal of a square formed by the walls and two common rafters. Use the 17″ mark on the blade of the square for the run.

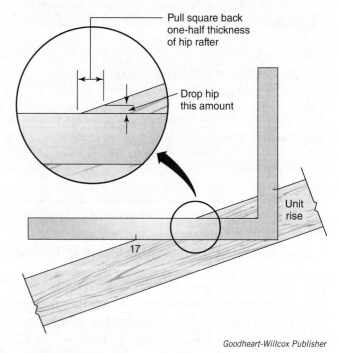

Figure 12-26. Using the square to find and mark the distance to drop a hip rafter. The bottom of the rafter is facing up.

The plumb (vertical) cut of the bird's mouth for valley rafters must be trimmed so it fits into the corner formed by the intersecting walls. Although side cuts of approximately 45° could be made, it is more practical to move the plumb cut toward the tail of the rafter by a horizontal distance equal to 1/2 of the 45° thickness of the rafter.

The tail of the hip rafter is actually the diagonal of a square formed by extending the line of each of the walls the length of the jack rafters, **Figure 12-27**. You can find its length by constructing a full-size square on paper. Then, measure the diagonal. Mark the plumb cut using the run (17″) and the rise (in inches) on the square. The tail must form a nailing surface for intersecting fascia boards. Refer to **Figure 12-23** for directions on how to make the tail cut.

A valley is where two pitched roofs meet to form an inside corner. The rafter at the center of the valley is a valley rafter. Provided both roofs have the same slope, which they usually do, the valley rafter is at a 45° angle, the same as a hip rafter. The pitch and run of valley rafters are the same as that of hip rafters, so the same table can be used for both. The procedure for finding the length of a hip rafter can be used to find the length of a valley rafter. The one difference is that valley rafters usually do not have any overhang or tail. As seen in **Figure 12-28**, the tails of the common rafters are often close enough, so no tail is needed on the valley rafter.

After hip and valley rafters are laid out and cut, they are installed on the roof. **Figure 12-29** is a rafter plan showing their positions.

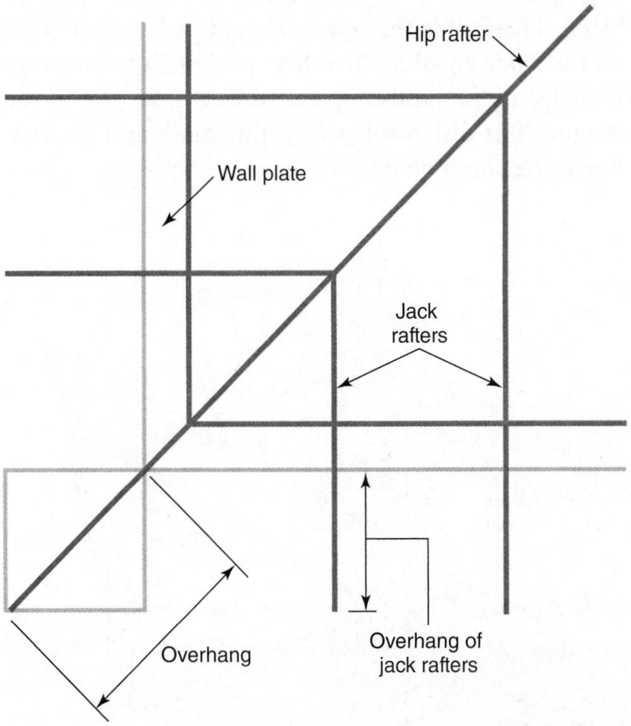

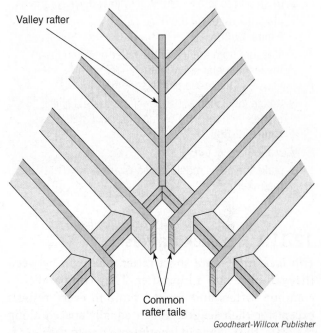

Goodheart-Willcox Publisher

Figure 12-28. When the common rafter tails meet or almost meet, there is no need for a tail on the valley rafter.

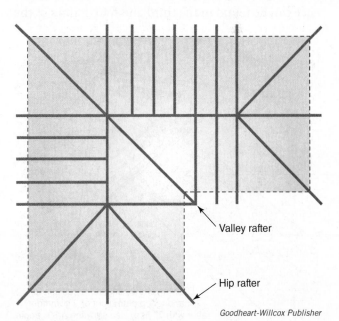

Goodheart-Willcox Publisher

Figure 12-29. This rafter plan shows hip and valley rafters along with common rafters. In actual construction, jack rafters are cut and installed last.

Goodheart-Willcox Publisher

Figure 12-27. Finding the length of the hip rafter overhang. You can construct the square full size on paper, then measure the length of the diagonal. A rafter table on the square also gives this information.

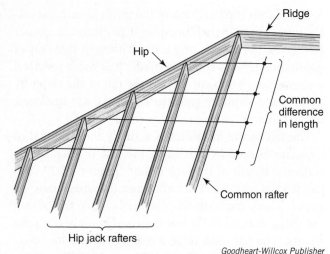

Figure 12-31. When evenly spaced, hip jack rafters have the same difference in length from one to the next. By using this information, all rafters can be cut without measuring each distance.

PROCEDURE

Laying Out Hip and Valley Rafters

The following is the procedure for laying out hip and valley rafters:

1. Find the rafter length from the table corresponding to the rise per foot of run.
2. Read down the table to the correct run.
3. Read across to the column for hip and valley rafters.
4. Mark the rafter for the ridge plumb cut, **Figure 12-30A**.
5. Measure from the top of the rafter down and mark the length obtained from the chart. Use the speed square to make the plumb seat mark.
6. Rotate the speed square until the dashed line is aligned with the seat plumb mark just made. See **Figure 12-30B**.
7. Mark the seat cut.
8. Make the adjustments previously described and cut the rafter.

12.11.1 Hip Jack Rafters

Hip jack rafters are short rafters that run between the wall plate and a hip rafter. They run parallel to common rafters and are the same in every respect except for their length. When equally spaced along the plate, the change in length from one to the next is always the same. This consistent change is the common difference, **Figure 12-31**. The common difference can be found in the third and fourth lines of the

rafter table. For a roof slope of 5-in-12 with rafters 24″ OC, the figure from the table on the square is 26″. See **Figure 12-32**.

The common difference can also be found with the layout method illustrated in **Figure 12-33**. Using the square, align the unit run and rise of the roof with the edge of a smooth piece of lumber. Draw a line along the blade. Then, slide the square along this line to a point on the blade equal to the rafter spacing. Put a mark on the edge of the lumber against the outside edge of the tongue. The distance between this mark and the first line is the common difference.

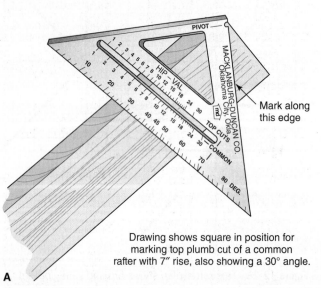

A

Drawing shows square in position for marking top plumb cut of a common rafter with 7″ rise, also showing a 30° angle.

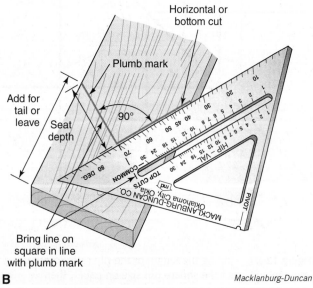

B

Figure 12-30. Using a speed square for plum and bird's mouth cuts. A—The plumb cut is set up for a hip or common rafter. Extra allowance must be marked for the miter cut. B—Marking the seat cut for the bird's mouth.

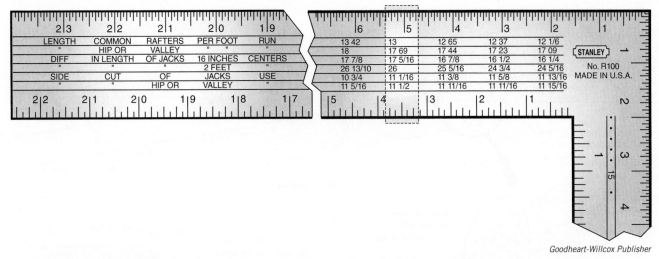

Figure 12-32. The common difference of jack rafters spaced at 24″ OC can be found on the fourth line of the rafter table. Simply read down from the unit rise per foot.

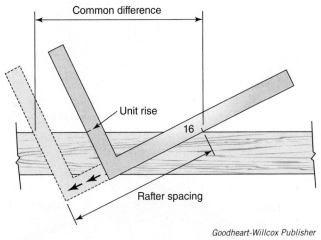

Figure 12-33. Using the framing square layout method of determining the common difference of jack rafters. Step 1—Place the square on the rafter as you would for the rise and run of the rafter. Draw a line along the edge of the blade. Step 2—Now, for a spacing of 16″ OC, slide the square in the direction of the arrows until the 16″ mark on the blade aligns with the mark made on the rafter.

To lay out jack rafters, select a piece of straight lumber. Lay out the bird's mouth and overhang from the common rafter pattern. Next, lay out the line length of a common rafter. This is the distance from the plumb cut of the bird's mouth to the centerline of the ridge. For the first jack rafter down from the ridge, lay out the common difference in length.

Now, take off 1/2 of the 45° thickness of the hip, **Figure 12-34.** Square this line across the top of the rafter and mark the center point. Through this point, lay out the side cut as shown in the illustration. Another method is to use a square and the number from the fifth line of the rafter table, **Figure 12-35.** Mark the plumb lines to follow when the cut is made.

For the next hip jack, move down the rafter the distance of the common difference and mark the cutting line. A sliding T-bevel is a good tool to use. Continue until all jacks are laid out. Then, use the T-bevel to mark the jack rafters as required.

Each hip in the roof assembly requires one set of jack rafters made up of matching pairs. A pair consists of two rafters of the same length with the side cuts made in opposite directions.

12.11.2 Valley Jacks

Similar procedures are followed in laying out valley jack rafters. For these, however, it is usually best to start the layout at the building line and move toward the ridge, **Figure 12-36.** The longest valley jack is the same as a common rafter except for the side (angle) cut at the bottom.

Use the common rafter pattern and extend the plumb cut of the bird's mouth to the top edge. Lay out the side cut by marking (horizontally) 1/2 of the thickness of the rafter. If you prefer, use the framing square and apply the numbers located in the fifth line of the rafter table.

The common difference for valley jack rafters is obtained by the same procedure used for hip jacks. Lay out this distance from the longest valley jack to the next. Continue along the pattern until all lengths are marked. Now, use this pattern to cut all the valley rafters. They are cut in pairs in the same way as hip jacks.

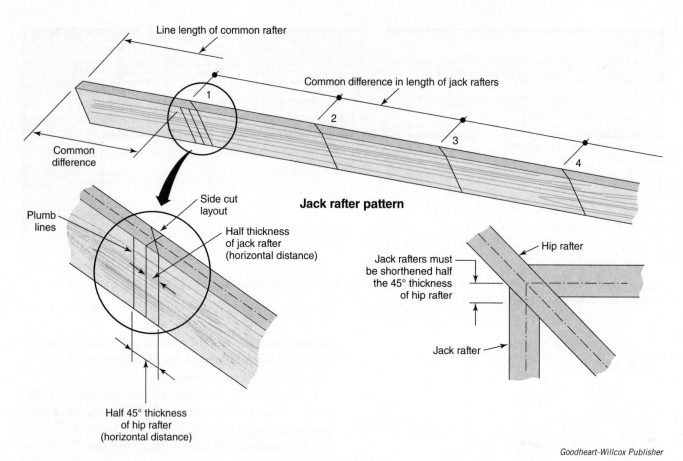

Jack rafter pattern

Goodheart-Willcox Publisher

Figure 12-34. Laying out jack rafters. A master pattern may be made (top).

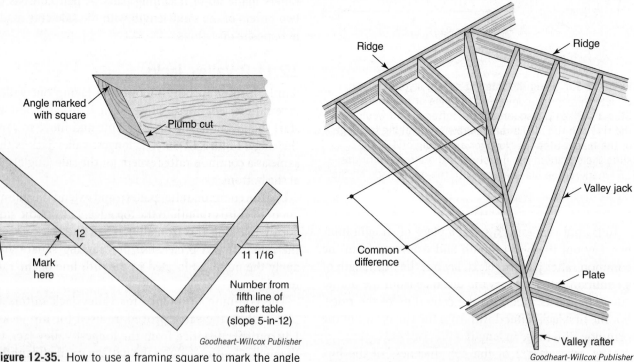

Goodheart-Willcox Publisher

Figure 12-35. How to use a framing square to mark the angle cut on the top of the hip jack rafter. The mark is always made on the top of the rafter. Be sure to mark the plumb cut as well.

Goodheart-Willcox Publisher

Figure 12-36. Valley jack rafters as assembled. Find the common difference in the same manner as for the hip jacks, except begin at the plate.

When all jack rafters are laid out, they should be carefully cut. Use either a portable electric saw with an adjustable base and guide or a compound miter saw.

> **Pro Tip**
> The strength of a roof frame depends a great deal on the quality of the joints. Use special care on the side cuts of jack rafters so the joining surfaces tightly fit together.

12.12 Erecting Jack Rafters

When all jack rafters are cut, assemble them into the roof frame, **Figure 12-37**. The nailing patterns depend on the size of the various members. Use 10d nails. Space the nails so they are near the heel of the side cut as they go from the jack into the hip or valley rafter.

Jack rafters should be erected in opposing pairs. This prevents the hip and valley rafters from being pushed out of line. It is good practice to first place a pair about halfway between the plate and ridge. Carefully sight along the hip or valley to make sure that it is straight and true. Temporary bracing could also be used for this purpose. Be sure that the outside walls running parallel to the ceiling joists are securely tied into the ceiling frame before hip jack rafters are installed. These rafters tend to push outward.

After rafters have been erected and securely nailed, carefully check over the frame. If some rafters are bowed sideways, they can be held straight with a strip of lumber located across the center of the span. Sight each rafter and move as needed. Drive a nail through the strip into the rafter to hold it in place. When the roof has been sheathed to this point, remove the spacer strip.

The last material to be attached to the roof frame before the decking is the *fascia*, **Figure 12-38**. This is the main trim member and is attached to the plumb-cut ends of the rafters. It conceals the rafter ends, provides a finished appearance, and furnishes a surface to which gutters can be attached.

Before attaching the fascia, however, stretch a chalk line along the rafter ends to ensure the rafter ends are the same length. Trim any ends that are too long so the eave line is straight.

The fascia may be directly attached to the rafter ends. Some carpenters prefer to install a sub fascia of 2× lumber first to even up the ends of the rafters and provide a solid nailing surface for the 1″ fascia board. Cut the upper edge of the fascia board at an angle to match the slope of the roof. Corners should be mitered and carefully fitted.

12.13 Special Problems

For intersecting roofs where the spans of the two sections are not equal, the ridges will not meet. To support the ridge of the narrow section, one of the valley rafters is continued to the main ridge. See **Figure 12-3**. The length of this extended or supporting valley is found by the same method used in the layout of a hip rafter.

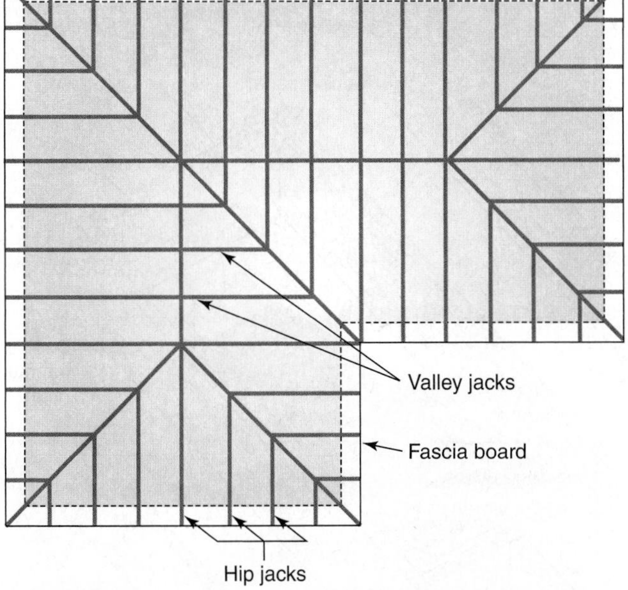

Valley jacks

Fascia board

Hip jacks

Goodheart-Willcox Publisher

Figure 12-37. A plan view of a complete roof frame. All hip and valley jacks are in place and the fascia board has been installed.

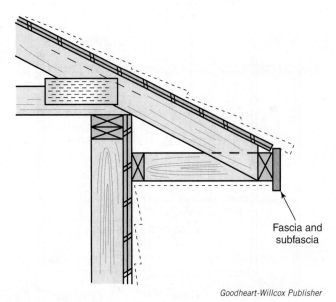

Fascia and subfascia

Goodheart-Willcox Publisher

Figure 12-38. The fascia board is fastened to the rafter or truss tails before installing the sheathing.

It is shortened at the ridge, just like the hip, but only a single side cut is required. The other valley rafter is fastened to the supporting valley with a square, plumb cut.

A rafter framed between the two valley rafters is called a valley cripple jack. The angle of the side cut at the top is the reverse of the side cut at the lower end. The run of the valley cripple is one side of a square, **Figure 12-39**. This run is equal to twice the distance from the centerline of the valley cripple jack to the intersection of the centerlines of the two valley rafters. Lay out the length of the cripple by the same method used for a jack rafter. Shorten each end 1/2 of the 45° thickness of the valley rafter stock. Make the side cuts in the same way as for regular jack rafters.

Rafters running between hips and valleys are called hip-valley cripple jacks. They require side cuts on each end. Since hip and valley rafters are parallel to each other, all cripple rafters running between them in a given roof section are the same length. The run of a hip-valley cripple jack rafter is equal to the side of a square, **Figure 12-40**. The size of the square is determined by the length of the plate between the hip and valley rafter. Use this distance and lay out the cripple in the same manner as a common rafter. Shorten each end by an amount equal to 1/2 of the 45° thickness of the hip and valley rafter stock. Now, lay out and mark the side cuts. Follow the same steps used for hip and valley jacks. The side cuts required on each end form parallel planes.

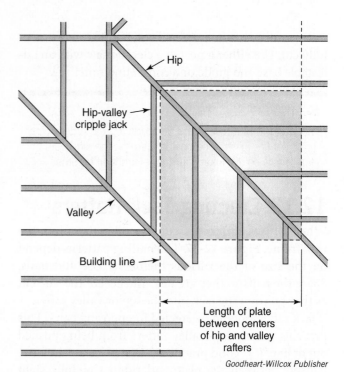

Goodheart-Willcox Publisher

Figure 12-40. The run of a hip-valley cripple jack is equal to the length of the wall plate from the valley rafter to the hip rafter.

12.14 Roof Openings

Some openings may be required in the roof for chimneys, skylights, and other structures. To frame large openings, follow the same procedure used in floor framing. Frame a chimney opening as shown in **Figure 12-41**.

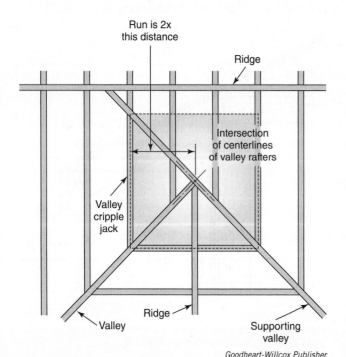

Goodheart-Willcox Publisher

Figure 12-39. Calculating the run of a valley cripple jack.

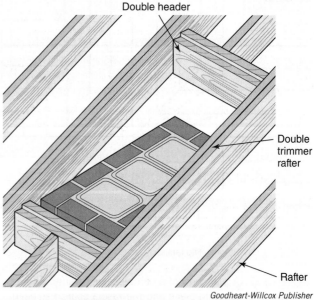

Goodheart-Willcox Publisher

Figure 12-41. Framing an opening in the roof for a chimney. Allow 2″ clearance all around the opening. The headers above and below the chimney are plumb.

To construct small openings, first complete the entire framework. Then, lay out and frame the opening.

Use a plumb line to locate the opening on the rafter from openings already formed in the ceiling or floor frame. Nail a temporary wooden strip across the top of the rafters to be cut. The supporting strip should be long enough to extend across two additional rafters on each side of the opening. This will support the ends of the cut rafters while the opening is being formed. Now, cut the rafters and nail in the headers. If the size of the opening is large, double the headers and add a trimmer rafter to each side.

12.15 Roof Anchorage

Rafters usually rest only on the outside walls of a structure. Rafters also lean against each other at the ridge, thus providing mutual support. This causes an outward thrust along the top plate that must be considered in the framing design.

Sidewalls are normally well secured by the ceiling joists, which are also tied to some of the rafters. End walls, however, are parallel to the joists. They need extra support, especially when located under a hip roof. Stub ceiling joists and metal straps are one method for reinforcing such walls. See **Figure 12-42**. Framing anchors can be substituted for the metal straps, especially when subflooring is part of the assembly.

12.16 Collar Ties

Collar ties tie together two rafters on opposite sides of a roof, **Figure 12-43**. They do not support the roof, but provide bracing and stiffening to hold the ridge and rafters together. Collar ties may be 2×6 or 1×6 boards installed at every third or fourth pair of rafters.

Purlins

Additional support must be provided when the rafter span exceeds the maximum allowed. A purlin provides this support. A **purlin** is a horizontal member that spans several rafters. A 2×4 purlin is used under the rafters to provide support for long rafters. See **Figure 12-44**.

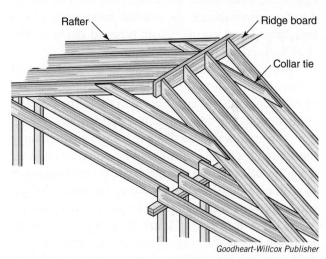

Goodheart-Willcox Publisher

Figure 12-43. Collar beams tie the rafters and ridge together, thus reinforcing the roof frame. Together with the ceiling joists, they secure the frame against spreading outward.

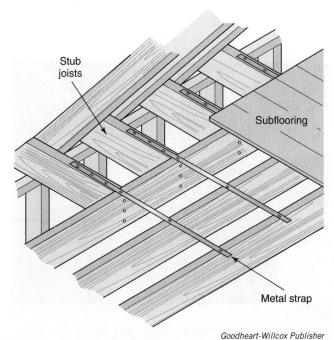

Goodheart-Willcox Publisher

Figure 12-42. Metal strapping and stub joists can be used to tie hip roofs to the end walls.

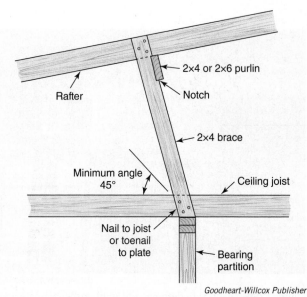

Goodheart-Willcox Publisher

Figure 12-44. The purlin in this illustration supports long runs of rafters at their midpoint. Weight is transferred to a bearing partition through 2×4 braces.

The purlin is supported by bracing and 2×4 stock resting on a plate located over a supporting partition. Bracing under the purlin may be placed at an angle not greater than 45° to transfer loads from the midpoint of the rafter to the support below. Purlins are also used over truss rafters when they are spaced farther apart than 24″. In this case, roof sheathing is fastened to these purlins.

12.17 Dormers

A *dormer* is a framed structure projecting above a sloping roof surface. It normally contains a vertical window unit. Although their chief purpose is to provide light, ventilation, and additional interior space, dormers also enhance the exterior appearance of the structure. There are two basic types of dormers: the shed dormer and the gable dormer.

The width of a shed dormer is not restricted by its roof design, **Figure 12-45**. It is used where a large amount of additional interior space is required. In the simplest construction, the front wall is extended straight up from the main wall plate. Double trimmer rafters carry the sidewall. The rise of the roof is sometimes figured from the top of the dormer plate to the main roof ridge. In such cases, the run is the same as the main roof. Be sure to provide sufficient slope for the dormer roof.

The gable dormers in **Figure 12-46** are tied into the roof well below the ridge. They are designed to provide openings for windows. Gable dormers can be located at various positions between the plate and ridge of the main roof. The inner end of a gable dormer is

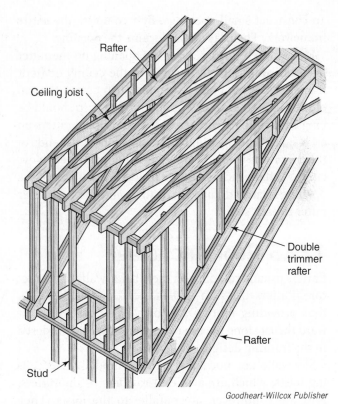

Goodheart-Willcox Publisher

Figure 12-45. Typical framing for a shed dormer. A nailer strip (not shown) is added along the double trimmer to support roof sheathing.

framed against a header, similar to the headers used for other roof openings, **Figure 12-47**. The valley formed where the dormer roof meets the main roof is framed with a valley rafter, the same as for other intersecting roofs.

A   Goodheart-Willcox Publisher

B   Goodheart-Willcox Publisher

Figure 12-46. A—Gable dormers add light and space, allowing the attic to be used as a living area. B—The studs in this gable dormer are 2×6 and rest on a sole plate. Note that they extend through the roof to their own wall plate.

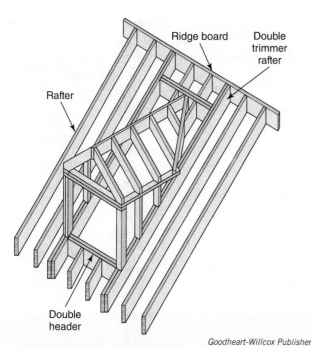

Goodheart-Willcox Publisher

Figure 12-47. A gable dormer is framed much like any other intersecting roof, except the top of the valley rafter rests on a header instead of the ridge board.

12.18 Framing Flat Roofs

Flat roofs provide a long, low appearance to the building. Sometimes, ceiling and roof members in a flat roof are combined as one system. The interior surfaces can be used as the finished décor or a place to fasten the finishing materials. When the underside of the roof is to be left exposed, rigid insulation is installed between the decking and the built-up roofing materials.

Methods and procedures used to frame flat roofs are about the same as those followed in constructing a floor. Most designs require an overhang with the ends of the joists tied together by a header or band. Cantilevered lookout rafters are tied to doubled roof joists, **Figure 12-48**. Corners can be formed as shown in the figure or carried on a longer diagonal joist that intersects the double joist.

In mild climates, carpenters may use nominal 2″ planking as sheathing over widely spaced beams. Both can be exposed on the underside.

Builders in the southwestern United States sometimes build traditional homes in the Pueblo or Territorial style. This style of home includes a flat roof supported by round poles called vigas. The vigas are spaced anywhere from 16″ to 30″, depending on the span. The poles rest on the wall plates and extend through the exterior wall about 2′–3′. A through-bolt secures them to the plate. Usually, the outside wall extends 2′–3′ above the roof as an architectural feature called a parapet wall. **Figure 12-49** shows a traditional southwestern home under construction.

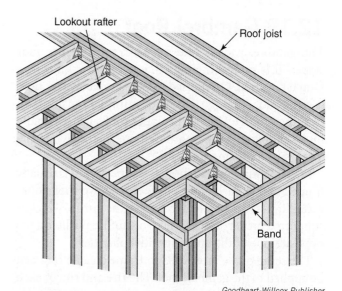

Goodheart-Willcox Publisher

Figure 12-48. Lookout rafters are cantilevered over the outside wall to form an overhang for this flat roof.

A   Pierce Construction, Ltd.

B   Pierce Construction, Ltd.

Figure 12-49. Flat roofs are a feature of some traditional home styles built in southwestern United States. A—Looking up at the ceiling of a Pueblo style house. Above the viga (pole) is a roof framed in 2×12s. Note the solid bridging. B—This exterior view shows the walls extending above the flat roof.

12.19 Gambrel Roof

The gambrel roof is like a gable roof, but has four slopes. It is typical in an architectural style known as Dutch Colonial. Often used in two-story construction, it gives added living space with minimal outside wall framing, **Figure 12-50**. The upper roof surface usually forms a 20° angle with a horizontal plane, while the lower surface forms about a 70° angle.

In residential construction, this type of roof is usually framed with a purlin located where the two roof slopes meet. Rafters are notched to receive the purlin, which is supported on partitions or tied to another purlin on the opposite side of the building with collar ties.

The same procedures used to frame a gable roof can be applied to the gambrel roof. The rise and run of each surface is found on the architectural plans. The two sets of rafters are laid out in the same way as previously described for a common rafter.

It may help to make a full-size sectional drawing at the intersection of the two slopes. This makes it easier to visualize and proportion the end cuts of the rafters. **Figure 12-51** shows basic gambrel roof framing for a small building, such as a garden house or toolshed. Note the use of gussets to join the rafter segments. A *gusset* is plywood or steel plate fastened to the outside of a joint to give the joint strength Also note the simple framing over the plate that is used to form the roof overhang.

12.20 Mansard Roof

Figure 12-52 shows a mansard-roofed dwelling. Like the gambrel roof, it has two slopes with the lower slope being steeper. The mansard roof, however, extends around all four sides of the building. Second floor joists extend beyond the first floor wall frame. This extension provides support for the lower rafters.

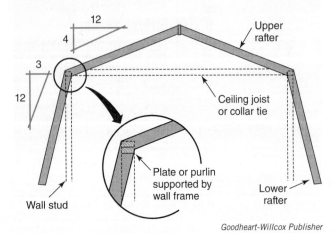

Goodheart-Willcox Publisher

Figure 12-50. Basic framing for a gambrel roof. The gambrel roof provides extra living space on the second story with minimal wall framing.

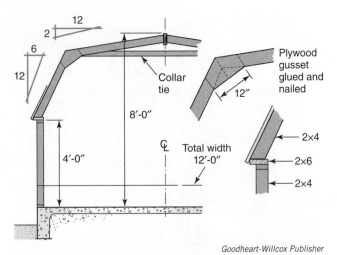

Goodheart-Willcox Publisher

Figure 12-51. A gambrel roof for a small building is easy to design and build.

Spiroview Inc/Shutterstock.com

Figure 12-52. An example of the mansard roof style. The main characteristics of the style are the two slopes and the overhang on all sides.

12.21 Special Framing

Figure 12-53 shows upper floor framing for a 1 1/2-story structure. Generally, this framing results in some saving of material, since walls and ceilings can be made a part of the roof frame. Knee walls are usually about 5' high. The ceiling height is typically 7'-6".

Low-sloping roofs such as the one shown in **Figure 12-54** usually need support at several points. The locations of those supports depends on the roof design and should be specified by an engineer. The strength derived from the triangular shape of regular-pitched roofs is greatly reduced in this design. Thus, carefully prepared architectural plans are essential for this type of roof structure.

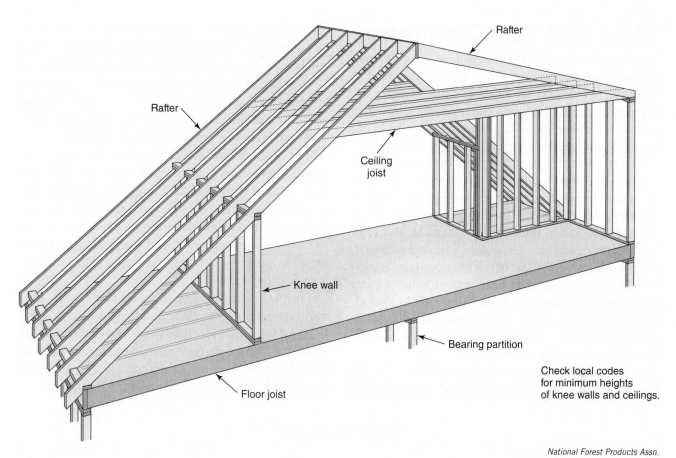

Figure 12-53. Wall, ceiling, and roof framing for a 1 1/2 story structure.

National Forest Products Assn.

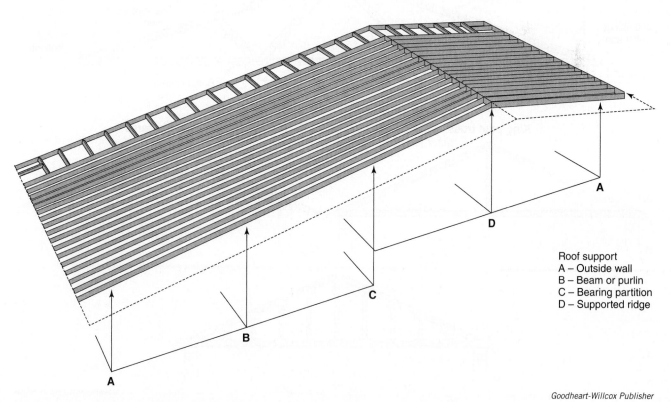

Roof support
A – Outside wall
B – Beam or purlin
C – Bearing partition
D – Supported ridge

Goodheart-Willcox Publisher

Figure 12-54. This low-slope roof is for a split-level design. The arrows indicate support points provided by the outside walls, bearing partitions, and purlins or beams.

12.22 Roof Truss Construction

A *truss* is a framework designed to carry a load between two or more supports. The principle used in its design is based on the rigidity of the triangle. Triangular shapes are built into the frame in such a way that the stresses of the various parts are parallel to the members making up the structure.

Roof trusses are frames that carry the roof and ceiling surfaces. They rest on the exterior walls and span the entire width of the structure. Due to their many advantages, roof trusses are used more extensively than conventional rafter framing. For example, since no load-bearing partitions are required, more freedom in the planning and division of the interior space is possible. Roof trusses permit larger rooms without extra beams and supports. The use of roof trusses also allows surface materials to be applied to outside walls, ceilings, and floors before partitions are constructed. Trussed roofs are also faster to erect.

There are many types and shapes of roof trusses, **Figure 12-55.** One commonly used in residential construction is the W truss, or *Fink truss*. This type can be used for spans up to 50′. The *king post truss* has top and bottom chords and a vertical center post. It is used on shorter spans up to 24′. The *scissors truss* is used for buildings having a sloping ceiling. In general, the slope of the bottom chord is 1/2 of the slope of the top chord. Truss rafters can be designed for spans up to 50′.

Most roof trusses are factory built to engineered specifications. Architectural drawings provide the manufacturer with all of the design specifications. The truss designer must know the span, roof slope, live and dead loads, and wind loads the roof must withstand. In most localities, engineering documents from a truss engineer must be submitted in order to obtain a certificate of occupancy.

If carpenters choose to build their own trusses, they must use carefully developed and engineered designs. Both APA-The Engineered Wood Association and the Truss Plate Institute have such designs and instructions. Refer to the different roof truss designs shown in Appendix B, Technical Information.

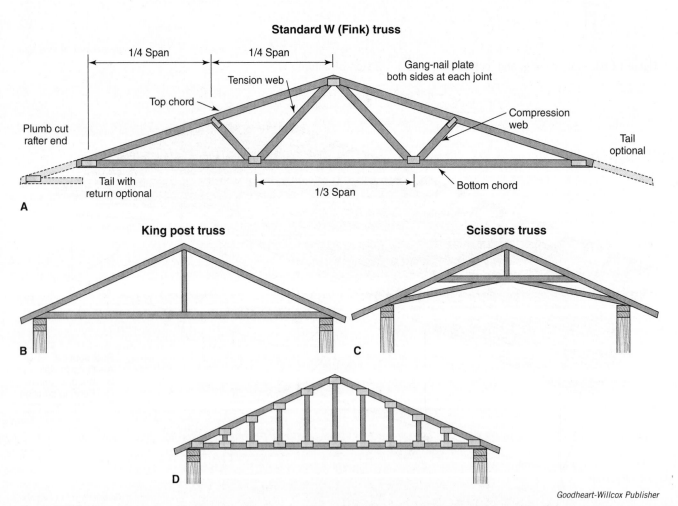

Goodheart-Willcox Publisher

Figure 12-55. Four types of roof trusses. A—The standard W, or Fink, truss is commonly used in residential construction. B—The king post truss. C—The scissors truss permits cathedral ceilings. D—Gable end truss.

Roof trusses must be made of structurally sound 2×4 lumber. Joints must be carefully fitted and tightly fastened with metal connectors or gussets on both sides. A carpenter is seldom required to determine the sizes of truss members or the type of joints used. Site-built trusses are precut and assembled at ground level. Spacing of 24″ OC is common. However, 16″ OC or other spacing may be required in some designs.

When the truss is in position and loaded, there is a slight sag. To compensate for this, the lower member (bottom chord) is slightly arched during fabrication of the unit. This adjustment is called *camber*. Camber is measured at the midpoint of the span. A standard 24′ truss usually requires about 1/2″ of camber.

In truss construction, it is essential that joint slippage be held to a minimum. Regular nailing patterns are usually not satisfactory. Special connectors must be used. Various types are available, **Figure 12-56**. All of them securely hold the joint and are easy to apply. Factory-built trusses are usually assembled with truss plates that are attached by being pressed into place with large rollers. When plywood gussets are used, they must be applied with glue and nails to both sides of the joint.

Trusses can be laid out and constructed on any clear floor area. Make a full-size layout on the floor. Snap chalk lines for long line lengths. Use straightedges to draw shorter lines.

Trusses for residential structures can normally be erected without special equipment. Each truss is simply placed upside down on the walls at or near the point of installation. Then, the peak of the rafter is swung upward by hand. Use a pole as needed. On multistory

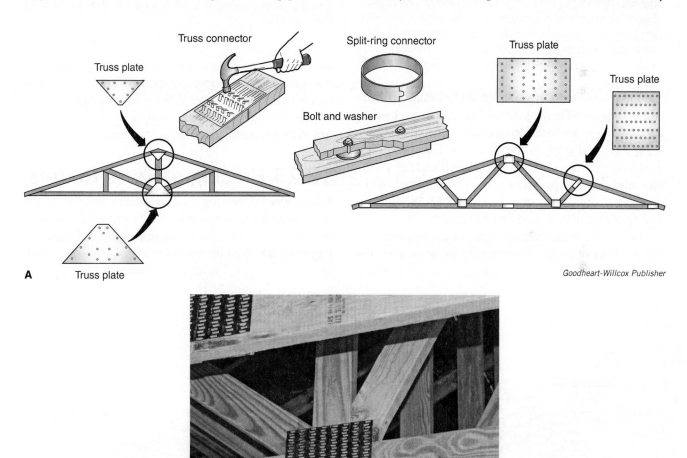

Goodheart-Willcox Publisher

Figure 12-56. Plates and connectors for roof trusses. A—Truss plates are made in many sizes, shapes, and types. Some are perforated for nails, some require no nails. Split-ring connectors fit into recesses bored into mating joints. B—Truss plate on factory-built truss. The yellow tag has a number that corresponds with the truss drawings.

dwellings or buildings with larger trusses, installation by hand may be impossible or impractical. In such cases, a crane must be used, **Figure 12-57**. Always attach lifting chains at or near joints and use a tag line to keep the truss from spinning or swinging.

Pro Tip
Use extra care when raising roof trusses. The first truss should be held with bracing and all succeeding trusses carefully braced to prevent collapse.

Safety Note
If using a personal fall arrest system as fall protection, never use the trusses as an anchorage point until a series of trusses have been installed and properly braced. Anchorage points must be capable of supporting 5000 lb per person. If subjected to a fall, trusses that are used for anchorage and not properly braced are likely to fall and collapse. An alternative fall protection system must be used until trusses are braced.

Roofs framed with trusses need not be limited to gable types. Today, a wide range of configurations can be produced. Designs are based on carefully prepared data covering load and lumber specifications. Computers are used to apply these data to develop specific designs. Further efficiency results from the use of specialized methods, machines, and fasteners. **Figure 12-58** shows a variety of roof trusses used on a hip roof.

Always refer to the framing plans before installing roof trusses. If basic design requirements are to be met, the final erection and bracing of a roof system must be carried

Pamela Au/Shutterstock.com

Figure 12-57. Hoisting a large truss into place with a crane. A worker on the ground is holding a tag line to keep the truss from spinning.

out according to plans and specifications. The installer is responsible for the proper storage, handling, and installation of truss rafters delivered to the jobsite, **Figure 12-59**. Trusses should not be stored on rough terrain or uneven surfaces that could cause damage to them.

12.23 Bracing of Truss Rafters
Proper ground bracing and temporary bracing of the truss rafter during erection is vital. Be prepared to ground brace the first truss erected using either method shown in **Figure 12-60**. As additional truss rafters are placed, install lateral and diagonal bracing as shown in **Figure 12-61**.

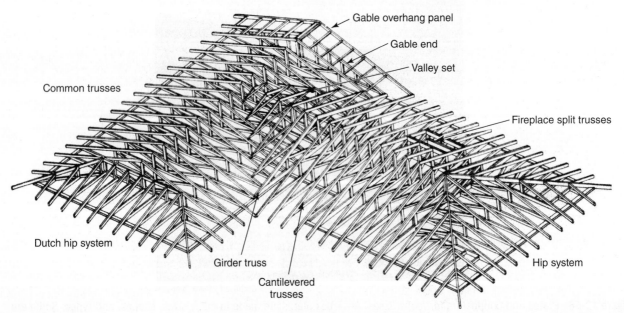

Common trusses

Gable overhang panel

Gable end

Valley set

Fireplace split trusses

Dutch hip system

Girder truss

Cantilevered trusses

Hip system

TrusWal Systems Corp.

Figure 12-58. This roof frame is constructed with many different prefabricated truss units.

CAUTION: The builder, building contractor, licensed contractor, erector, or erection contractor is advised to obtain and read the entire booklet "Commentary and Recommendations for Handling, Installing & Bracing Metal Plate Connected Wood Trusses, HIB-91" from the Truss Plate Institute.

CAUTION: All temporary bracing should be no less than 2×4 grade marked lumber. All connections should be made with minimum of two 16d nails. All trusses assumed 2' on-center or less. All multiply trusses should be connected together in accordance with design drawings prior to installation.

TRUSS STORAGE

CAUTION: Trusses should not be unloaded on rough terrain or uneven surfaces that could cause damage to the truss.

8'-10'   8'-10'

Trusses stored horizontally should be supported on blocking to prevent excessive lateral bending and lessen moisture gain.

Trusses stored vertically should be braced to prevent toppling or tipping.

WARNING: Do not break banding until installation begins or lift bundled trusses by the bands.

WARNING: Do not use damaged trusses.

DANGER: Do not store bundles upright unless properly braced.

DANGER: Walking on trusses that are lying flat is extremely dangerous and should be strictly prohibited.

Reproduced from HIB-91, Courtesy of Truss Plate Institute

Figure 12-59. Observe these warnings for proper storage of trusses on the jobsite.

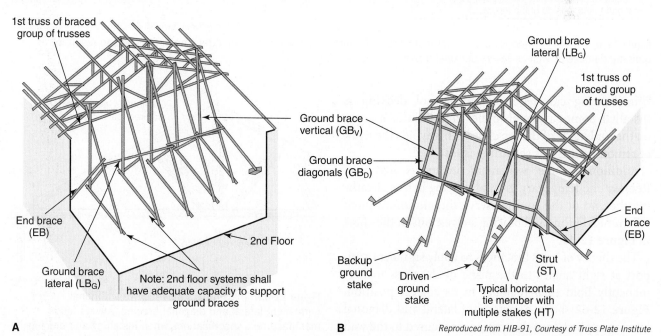

1st truss of braced group of trusses

Ground brace vertical (GB$_V$)

Ground brace diagonals (GB$_D$)

End brace (EB)

Ground brace lateral (LB$_G$)

2nd Floor

Note: 2nd floor systems shall have adequate capacity to support ground braces

Ground brace lateral (LB$_G$)

1st truss of braced group of trusses

End brace (EB)

Backup ground stake

Driven ground stake

Strut (ST)

Typical horizontal tie member with multiple stakes (HT)

A

B

Reproduced from HIB-91, Courtesy of Truss Plate Institute

Figure 12-60. Temporary ground bracing of truss rafters. Consult a registered professional truss engineer for exact bracing. Bracing should be erected before raising the first truss. A—Ground bracing installed in the building's interior. B—Bracing installed outside the building.

Span	Minimum pitch	Top chord lateral brace spacing (LB$_s$)	Top chord diagonal brace (DB$_s$) [# trusses]	
Up to 32′	4/12	8′	20	15
Over 32′ - 48′	4/12	6′	10	7
Over 48′ - 60′	4/12	5′	6	4
Over 60′	See a registered professional engineer			

DF = Douglas Fir-Larch   SP = Southern Pine
HF = Hem-Fir   SPF = Spruce-Pine-Fir

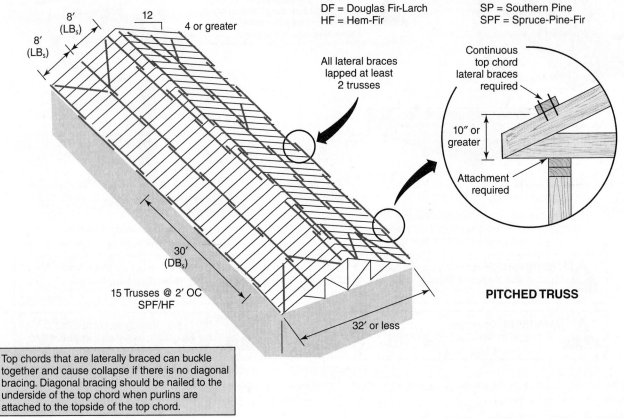

8′ (LB$_s$)
8′ (LB$_s$)
12
4 or greater

All lateral braces lapped at least 2 trusses

Continuous top chord lateral braces required

10″ or greater

Attachment required

PITCHED TRUSS

30′ (DB$_s$)

15 Trusses @ 2′ OC SPF/HF

32′ or less

Top chords that are laterally braced can buckle together and cause collapse if there is no diagonal bracing. Diagonal bracing should be nailed to the underside of the top chord when purlins are attached to the topside of the top chord.

Reproduced from HIB-91, Courtesy of Truss Plate Institute

Figure 12-61. Proper temporary bracing for roof trusses. Failure to follow these bracing recommendations can result in building damage as well as severe personal injury.

This bracing can be removed as roof decking is installed. Some carpenters prefer to install this bracing on the underside of the top chord. This avoids the work of removing them as sheathing is installed.

Additional inside permanent bracing is advisable. Remove the temporary ground bracing before sheathing the end of the roof. Observe the minimum pitch recommendations and bracing suggestions in the chart in **Figure 12-61**.

The theory of truss bracing is to apply sufficient support at right angles to the plane of the truss to permanently hold each member in its correct position. **Figure 12-62** shows a method of lateral and diagonal bracing applied once the trusses are secured to the wall plate.

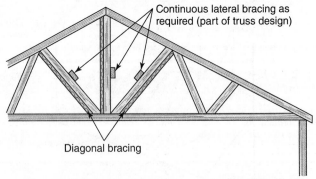

Continuous lateral bracing as required (part of truss design)

Diagonal bracing

Goodheart-Willcox Publisher

Figure 12-62. Truss rafters may require additional, permanent lateral and diagonal bracing. Always follow manufacturer's specifications when installing and bracing trusses.

Fasten truss rafters to the doubled wall plate as you would conventional rafters. Use toenailing or patented connectors. Sometimes both are used, **Figure 12-63**. Never toenail the lower chord of truss rafters to non-bearing walls. The trusses flex under snow loads and the bottom chord moves vertically. This tends to lift the wall, causing cracks and gaps at the junction of the wall and the floor. **Figure 12-64** shows a clip that can be used to attach the truss so the truss can move up and down without disturbing the wall. When the roof frame is complete, check it carefully to see that all members are securely fastened and that nailing patterns are adequate.

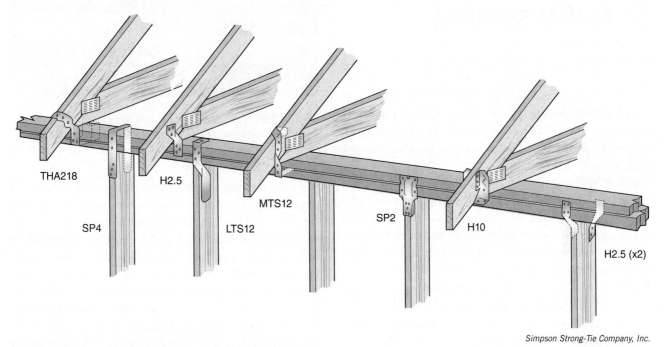

Simpson Strong-Tie Company, Inc.

Figure 12-63. Attaching trusses to the wall plate. Trusses can be attached using metal tie-downs. Trusses can also be fastened with 10d nails, two on each side.

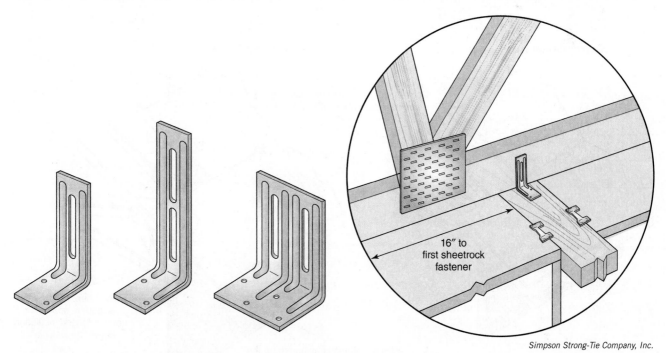

Simpson Strong-Tie Company, Inc.

Figure 12-64. A roof truss clip keeps the truss and nonbearing wall in alignment, but allows for vertical movement of the lower truss chord.

12.24 Roof Sheathing

Sheathing (also called decking) provides a nailing base for the roof covering and adds strength and rigidity to the roof frame. The most common sheathing material is structural panels of plywood or oriented strand board (OSB), **Figure 12-66**. Other special materials are also available. For example, one product consists of panels formed with solid wood boards, bonded together with heavy paper.

Before applying sheathing to a roof, check for a level nailing surface. Use a long level or a straight piece of lumber 6′–10′ long. Shim low trusses or rafters as needed. Blocking can be used to straighten bowed or warped top chords of trusses or rafters. Install the sheathing panels with screened surfaces or skid-resistant coatings facing up. To avoid bows in the sheathing, only stand where framing is supporting the panel as it is fastened.

Goodheart-Willcox Publisher

Figure 12-66. Oriented strand board has become popular as sheathing material. The textured side should be installed facing up to provide better footing.

Before starting the sheathing work, erect the needed scaffolding. The scaffolding should make it easy and safe to install the boards or panels along the lower edge of the roof. If asphalt shingles or other composition materials are used for the finished roof, there should be no gaps in the sheathing, except the 1/8″ that is recommended by the panel manufacturer. For wood shingles, metal sheets, or metal tile, *skip sheathing* should be used. In this method, board sheathing is spaced so there are voids (gaps). This allows wood shingles to dry out, thus preventing rot. The spacing must be arranged according to the length of the shingles. Nails must fall

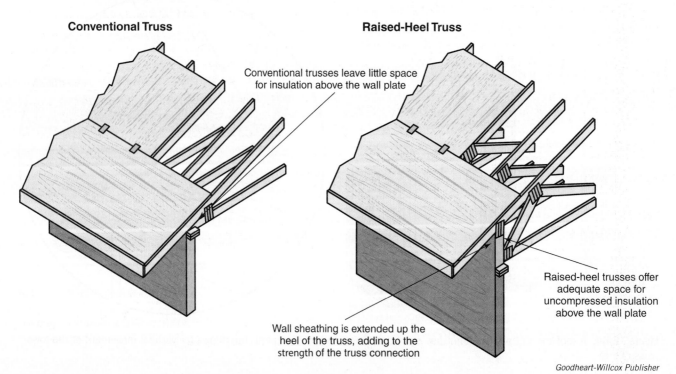

Conventional Truss

Conventional trusses leave little space for insulation above the wall plate

Raised-Heel Truss

Raised-heel trusses offer adequate space for uncompressed insulation above the wall plate

Wall sheathing is extended up the heel of the truss, adding to the strength of the truss connection

Goodheart-Willcox Publisher

Figure 12-65. Raised-heel trusses give more space for insulation in the attic at the eaves.

over wood, rather than voids. Boards should be attached with two 8d nails at each rafter. Joints must be located over the center of the rafter. For greatest rigidity, use long boards, particularly at the rake.

Carefully fit sheathing at valleys and hips and securely nail it. This ensures a solid, smooth base for the installation of flashing materials. Around chimney openings, allow a 1/2″ clearance from the masonry. Framing members must have a 2″ clearance. Always securely nail sheathing around roof openings.

12.24.1 Structural Panels (Sheets)

Structural panels are an ideal material for roof sheathing. They can be rapidly installed, hold nails well, and resist swelling and shrinkage. Because the panels are large, they add considerable rigidity to the roof frame. Always install panels with the long edge perpendicular to the rafters. Locate end joints directly over the center of the rafter. Small pieces can be used, but they should always cover at least two rafter spaces.

The required thickness of the panels varies with the roofing material and rafter spacing. Plywood or OSB must be rated for roof sheathing, as indicated on the grade stamp. See Chapter 3, **Building Materials**.

Panels should be nailed to rafters with 8d nails spaced 6″ apart on edges and 12″ apart elsewhere. If wood shingles are used and the sheathing is less than 1/2″ thick, 1×2 nailing strips spaced according to shingle exposure should be affixed to the sheathing.

If there are several carpenters on the job, sheets may be slid up a ladder. A special rack can also be used, **Figure 12-67**. This saves steps for a carpenter as it keeps

Goodheart-Willcox Publisher

Figure 12-67. A rack secured to the fascia holds sheathing panels where carpenters working on the roof can easily reach them.

sheathing panels close by. It is also a handy way to store panels until carpenters are ready to install them. On large construction jobs, a power panel elevator or tele-handler may be used to save time.

12.24.2 Installing Sheathing

Start installing the sheathing at the eaves and work up toward the ridge. If necessary, drive temporary fasteners at the corners to square panels on the rafters. Fasten one edge and then install intermediate fasteners, working inward from the edges. Lay down rows of panels from one edge of the roof to the other. Stagger the end joints between rows of panels so there are no continuous joints from the eave to the ridge. Maintain the same fastening sequence for each panel. This procedure avoids internal stress in the panels. It is advisable to snap a chalk line on the sheathing to mark the center of rafters or trusses. Carpenters can do this on the ground. The factory usually places such lines on OSB sheathing to indicate 16″ and 24″ OC spans.

When using power nailers or staplers, stand on the panel over the framing to ensure contact with the framing as the fastener is driven. Fastener heads must be driven flush with the panel surface. Fastener heads that are not flush with the surface will protrude into the roofing and cause it to fail prematurely. Maintain a 1/8″ space between the edges and ends of panels. Most roof sheathing has a tongue and groove milled along the long edge. Ensure the joint is engaged by protecting the top edge with a sacrificial block and hitting it with a framing hammer. If using square-edge sheathing, panel clips should be installed. A 10d nail can be used to gauge the spacing along end joints. It is important to center each panel end on a rafter. Trim panel ends, if necessary.

After all sheathing is installed, trim the extra overhang at the ends of a gable roof. The cutting line should be carefully marked, preferably with a chalk line. It is important for the appearance of the finished roof that this edge is straight.

12.24.3 Panel Clips

Clips are manufactured to strengthen roof sheathing panels between rafters. See **Figure 12-68**. These clips are sometimes called *H clips* because of their shape. They eliminate the need for blocking on long truss or rafter spans. The clips are slipped onto the panel edges midway between the rafter or truss spans. Two clips should be used where supports are 48″ OC. The clips are manufactured to fit five panel thicknesses: 3/8″, 7/16″, 1/2″, 5/8″, and 3/4″. An average house requires 250 panel clips.

Safety Note

Follow these safety rules when working on a roof.

- Work from the scaffold to install the fascia and to lay the first row of roof sheathing.
- Never place tools, materials, or debris where they can slide off of the roof. Never throw any discarded material off of the roof. Doing so can injure someone below.
- If debris of any kind must be cleared from a roof, assign a worker on the ground to secure and rope off the impact area. Maintain proper communication between workers on the roof and workers on the ground during this entire procedure.
- Fall protection: Familiarize yourself with OSHA rules that require the use of fall protection, guardrails on scaffolding, or hard hats in certain situations. The following are the basic fall protection rules:
- Any height 6′ or more above a lower level (like a roof surface) requires fall protection. This can be guardrails, safety nets, or a fall-arrest system, such as ropes and harnesses. If you are working from scaffolding, the maximum height is 10′ without fall protection.
- Use of body belts as a fall arrest is prohibited because the belts restrict breathing after a fall.

12.25 Estimating Materials

The number of rafters required for a plain gable roof is easy to figure. Simply multiply the length of the building by 3/4 for spacing 16″ OC, 3/5 for 20″, or 1/2 for 24″, round up, and add one more. Double this figure to include the other side of the roof. To determine the length of the rafter, use the scale on the framing square as previously described in this chapter. This example shows how to estimate rafters for the following building.

Example:

Building size 28′ × 40′. Roof slope 4-in-12. Overhang 2′-0″. Rafter spacing 24″ OC.

$$\text{Total rafter run} = 16'\text{-}0''$$
$$\text{Total rafter length} = 16'\text{-}11''$$
$$\text{Nearest standard length} = 18'\text{-}0''$$
$$\text{Number of pieces} = 2 \, (\text{Length of wall} \times 1/2 + 1)$$
$$= 2 \, (40 \times 1/2 + 1)$$
$$= 2 \times 21$$
$$= 42$$
$$\text{Rafter estimate} = 42 \text{ pcs } 2'' \times 8'' \times 18'\text{-}0''$$

When estimating a hip roof, it is not necessary to figure each jack rafter. The number of jack rafters required for one side of a hip is counted to obtain the number of pieces of common rafter stock. This normally supplies sufficient rafter material for the other side of the hip. For a short method on a plain hip roof, proceed as if it is a gable roof. Add one extra common rafter for each hip. Then, figure and add the hip rafters required.

For complex roof frames, it is best to work from a complete framing plan. Apply the methods described for plain roofs to the various sections. Make check

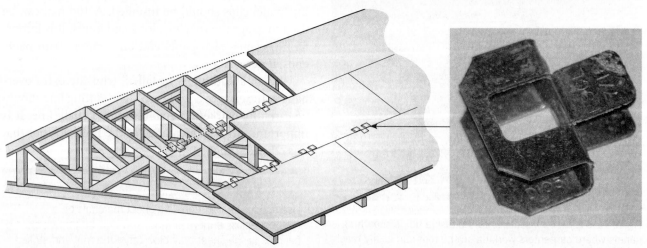

The Panel Clip Co.

Figure 12-68. Panel clips eliminate blocking on long truss or rafter spans.

marks on the rafters as they are figured so you will not double up or skip members. In estimating material for the total roof frame, remember to include material for ridges, collar ties, and bracing.

To estimate the roof sheathing, first calculate the total surface. Then apply the same procedures as used for subflooring and wall sheathing. Since the total area of the roof surface is also needed to estimate shingles, asphalt felt, or other roof surface materials, it is worth the extra time to accurately figure the area.

To figure sheathing for a plain gable roof, multiply the length of the ridge by the length of a common rafter, then double the amount. Calculate a plain hip roof as though it is a gable roof. However, instead of multiplying the length of a common rafter by the ridge, multiply it by the length of the building plus twice the overhang.

When estimating sheathing for complex roof plans and intersecting roof lines, first determine the main roof areas. Multiply the common rafter length by the length of the ridge, then multiply that product by two. Now, add the triangles that make up the other sections located over jack rafters. The area of a triangle equals 1/2 of the base times the altitude. The altitude of most triangular roof areas is the length of a common rafter located in or near the perimeter of the triangle. A plan view of the roof lines is helpful since all horizontal lines (roof edges and ridges) are seen true length and can be scaled. Always add an extra percentage for

waste when estimating sheathing requirements for roofs that are broken up by an unusually large number of valleys and hips.

12.26 Model and Small-Scale Construction

Students of carpentry can often gain worthwhile experience through construction of small portable buildings or scaled-down models. Working with models requires much time. It is often best to construct only part of a building. Small buildings, such as play houses or storage sheds, can be sold to recover the cost of materials.

A scale of 1 1/2″ = 1′-0″ usually makes it possible to apply regular framing procedures to the construction of a model that is not too large to handle and store. Cut framing members to their nominal size. For example, a 2 × 4 cut to this scale actually measures 1/4″ × 1/2″, while a 2×10 measures 1/4″ × 1 1/4″.

Make all framing materials from clear white pine or sugar pine. Both have sufficient strength and are easy to work with. Use small boards and fast-setting glue for assembly.

Materials other than wood can often be simulated from a wide range of items. For example, foundation work can be built of rigid foamed plastic and then brushed with a creamy mixture of Portland cement and water.

Chapter Review

Summary

- Roof framing must be well designed and strong enough to withstand the stress of high winds and snow loads.
- There are various types of roofs, but gable and hip roofs are most common.
- Gable roofs and gable end frames are the simplest to frame and erect.
- Roof framing may be of the traditional stick-built rafter design or prefabricated roof trusses.
- Rafters used to construct roof frames are classified as common, hip, valley, or jack rafters, depending on their function and location.
- Roof framing is a practical application of geometry, requiring careful measurement during the layout process.
- Carpenters laying out common rafters may use the step-off method, a framing table on the framing square, or a construction calculator to determine the correct length.
- Hip roofs require more calculation and special cuts for the hip, valley, and jack rafters.
- Openings must be framed in roofs for chimneys, skylights, or other structures. Such openings are framed using methods similar to floor framing.
- Shed and gable dormers project from the roof. They require both wall and roof framing techniques.
- Specific framing techniques are required for flat, gambrel, and mansard roofs.
- Although roof trusses may be built onsite, most are factory built. When being erected, they must be carefully braced to prevent collapse.
- The roof is structurally completed by applying sheathing to the rafters.
- The most common sheathing material is structural panels of plywood or OSB.

Know and Understand

Answer the following questions using the information in this chapter.

1. The _____ load is the external stresses placed on a roof, such as wind and snow.
   A. dead
   B. live
   C. external
   D. extra

2. A _____ roof is often used in contemporary designs where the ceiling is attached directly to the roof frame.
   A. gable
   B. hip
   C. shed
   D. mansard

3. *True or False?* The two basic systems used in framing the roof of platform structures are conventional and stick-built rafters and valley rafters.

4. A common rafter is found on a _____ roof.
   A. hip
   B. flat
   C. mansard
   D. gable

5. The _____ jack intersects neither the plate nor the ridge and is terminated at each end by hip and valley rafters.
   A. hip
   B. valley
   C. cripple
   D. common

6. *True or False?* The length of the rafter is known as the run.

7. Which of the following is *not* a method used to lay out rafters?
   A. Step-off method.
   B. Rafter table on the framing square.
   C. Calculating slope and pitch.
   D. Using a construction calculator.

8. To use a speed square, you need to know the _____ of the roof.
   A. hypotenuse
   B. pitch
   C. bird's mouth
   D. common difference

9. Gable ends for a brick or stone veneer building use _____ and blocking attached to a ledger to move the frame over the finished wall.
   A. extended rakes
   B. special joists
   C. lookouts
   D. brick racks

10. *True or False?* The ridge of a hip roof is cut to the length of the building, minus twice the run of the common rafters, plus the thickness of the rafter stock.

11. A _____ is where two pitched roofs meet to form an inside corner.
   A. valley
   B. hip
   C. tail
   D. cornice

12. The longest valley jack is the same as a common rafter except for the _____ cut at the bottom.
   A. ridge
   B. side
   C. seat
   D. plumb

13. Jack rafters should be erected in _____ to keep the hip or valley rafters straight.
   A. line with the common rafters
   B. opposing pairs
   C. matching pairs
   D. parallel patterns

14. What is the *fascia*?
   A. A temporary bracing.
   B. A strip of lumber used as a spacer.
   C. A rafter framed between two valley rafters.
   D. A main trim member attached to ends of rafters.

15. _____ tie together two rafters on opposite sides of a roof.
   A. Gussets
   B. Metal bracings
   C. Collar ties
   D. Purlins

16. A dormer provides light, ventilation, additional interior space, and _____.
   A. supports long rafters
   B. enhances the exterior appearance of a structure
   C. carries the sidewall
   D. reduces roof material waste

17. *True or False?* The principle used in a truss's design is based on the rigidity of the triangle.

18. *True or False?* A fink truss is generally used for buildings with a sloping ceiling.

19. The theory of truss bracing is to apply sufficient support at _____ angles to the plane of the truss to hold each member in position.
   A. equal
   B. side
   C. right
   D. opposite

20. Which of the following is *not* a panel thickness a panel clip is manufactured to fit?
   A. 3/8″
   B. 1/2″
   C. 5/8″
   D. 7/8″

21. To calculate the sheathing for a plain gable roof, multiply the length of the _____ by the length of a(n) _____, and then double the product.
   A. common rafter, rafter spacing
   B. ridge, common rafter
   C. building, rafter
   D. ridge, hip

22. Use a scale of _____ when constructing a model.
   A. 1 1/2″ = 1′-0″
   B. 1/2″ = 1′-0″
   C. 2 1/2″ = 1′-0″
   D. 3 1/2″ = 2′-0″

Apply and Analyze

1. How does a carpenter find the slope of a roof?

2. Name three ways rafters can be laid out.

3. How do you find the common difference with a framing square?

4. Explain how to find the length of a valley rafter.

5. What must be done to keep the first truss installed on a roof plumb before additional trusses are installed?

6. Describe how skip sheathing is beneficial for wood shingles.

7. What two factors determine the thickness of structural panels required for a sheathing application?

8. How many rafters are required for a gable roof for a building 42′ long with rafters spaced 16″ OC, an overhang of 1′-0″ at the rake, and a slope of 4-in-12?

Critical Thinking

1. Obtain a set of house plans where the design includes a hip roof and intersecting sections. It should not have a roof framing plan. Study the elevations and detail sections. Then, prepare a roof framing plan. Overlay the floor plan with a sheet of tracing paper. Trace the walls and draw all roof framing members to accurate scale. Be sure to include openings for chimneys and other items.

2. Prepare an estimate of the framing materials required for the roof used in Question 1. Include the dimensions for all needed lumber. Refer to the detail drawings or specifications to find lumber size requirements. If this information is not included in the plans, find it in the local building code.

3. Using the same plans you used in Activities 1 and 2, if you were to substitute conventional framing with a raised-heel truss system, how would the look of the house change? If a homeowner requested this change before construction, what steps should the builder take to ensure the house meets expectations from an aesthetic point of view?

Communicating about Carpentry

1. **Mathematics and Speaking.** Create an informational pamphlet on the different opportunities available as a roofer. Describe how much time is required to become employed full-time in these positions. Include images in your pamphlet. Present your pamphlet to the class.

2. **Speaking.** Working in a small group, discuss the similarities and differences between roofing and flooring. Discuss for each the materials needed, the career paths, the difficulty of work, length of schooling and expertise for certification, and prestige of the career.

3. **Reading.** Before you read the chapter, read the table and photo captions. What do you know about the material covered in this chapter just from reading the captions? After reading the chapter, determine if the knowledge you retained just from reading the captions is accurate.

CHAPTER 13

Framing with Steel

OBJECTIVES

After studying this chapter, you will be able to:

- List the advantages and disadvantages of steel framing.
- Describe the fastening methods used with steel framing components.
- Explain how wood and steel structural components are combined in floor framing.
- Demonstrate construction of walls using metal studs.
- Describe the safety precautions that must be used when working with steel framing components.
- Explain the use of a jig for fabricating steel roof trusses on the jobsite.

TECHNICAL TERMS

metal stud
self-tapping drywall screw
steel framing member

structural sheathing
track
weld bead

X-bracing

For years, steel framing has been used extensively in light commercial construction. Today, mostly for reasons of cost, it is increasingly used in residential construction. With some additional equipment, carpenters have been able to readily adapt to the different construction methods involved in steel framing.

13.1 Steel Framing

Steel framing members are manufactured in various widths and gauges. They are used as studs, joists, and truss rafters, **Figure 13-1**. Most manufacturers use a color code to prevent the different gauges from being mixed at the construction site. Prices are based on thousands of lineal feet.

Dimensions of Steel Framing for Studs and Joists		
Weight/Gauges of Light Gauge Steel Studs and Joists		
Mil	Gauge	Decimal Inch
16	26	0.0163
33	20	0.0329
43	18	0.0428
54	16	0.0538
68	14	0.0677
97	12	0.0966
Web Sizes: 2 ½"–14" Flange Widths: 1 3/8"–3"		

Goodheart-Willcox Publisher

Figure 13-1. Steel framing for studs and joists comes in several dimensions.

Studs, joists, and track are manufactured by brake forming and punching galvanized coil and sheet stock. Strength of the steel varies from one manufacturer to another. Steel components used for structural framing are coated to resist rust and corrosion. They are manufactured in a range of sizes and thicknesses, **Figure 13-2**. The four types of framing members are stud, track, U channel, and furring.

Metal stud systems are most often used for nonload-bearing walls and partitions. Size and spacing are first taken from the architectural drawings. The studs are attached to base and ceiling channels, called *tracks*, with screws or clips. A track is a U-shaped channel that is used at the top and bottom of a wall. The steel studs fit into the track and are secured by screws or welds. A typical stud consists of a metal channel with openings through which electrical lines, plumbing lines, and bracing can be installed. See **Figure 13-3**. Wall surface material, such as drywall, is attached to the metal studs with *self-tapping drywall screws*.

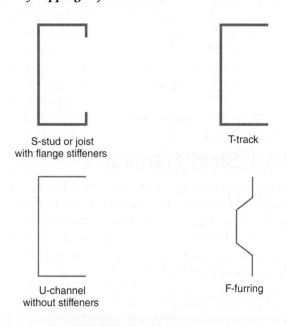

S-stud or joist
with flange stiffeners

T-track

U-channel
without stiffeners

F-furring

SSMA Designation

600S162-54

Stud

54 mil thick

6″

1.62″ (1 5/8″)
flange

Goodheart-Willcox Publisher

Figure 13-2. Steel Stud Manufacturers Association (SSMA) nomenclature.

Goodheart-Willcox Publisher

Figure 13-3. Metal studs are stamped and prepunched to accommodate installation of mechanical systems, such as electrical wiring and plumbing. Here, a plastic grommet has been inserted to protect the electrical cable from the sharp edges of the hole.

Pro Tip

It is important that pre-punched knockouts are aligned for mechanicals, especially conduit and rigid pipe, even flexible wire suffers from an unprofessional appearance when installed through misaligned studs. See **Figure 13-4**.

Framing with steel requires some special tools. These include a variable speed drill and screw gun, hearing protectors, clamping pliers, metal snips, metal punch, metal cutoff blade for a portable saw, magnetic level, metal cutoff saw (chop saw), and a right-angle drill. In some cases, welding equipment is needed.

Safety Note

Wear gloves when handling light gauge metal. Cut edges can be very sharp. Also wear safety glasses and hearing protection when cutting metal framing.

Lorraine Cline/Shutterstock.com

Figure 13-4. Here one stud was installed upside-down, disrupting the alignment of the pre-punched knockouts.

13.1.1 Advantages of Steel Framing

Steel members are lighter than their wood counterparts. This makes it easier for a carpenter to assemble and erect wall assemblies, **Figure 13-5**. Studs and joists are normally straight and consistent in size. They do not shrink, warp, swell, or have knots and other imperfections that affect the quality of the construction. There is also little limitation in the length of steel framing. Joists can be manufactured in lengths up to 40′.

Steel framing is noncombustible, does not absorb moisture, is impervious to insect destruction, and does not support mold or fungus. Steel framing can be designed to withstand the destructive forces of high winds, tornados, hurricanes, and earthquakes. Its strength and ductility (flexibility) properties allow it to easily meet the wind and seismic ratings established by the International Residential Code.

Green Note
Steel framing contains at least 25% recycled materials and all metal scrap is recyclable. This amounts to a savings for the builder and contributes to green construction efforts.

Industry sources claim that scrap and waste amounts to only 2% when using steel framing, compared to 20% for lumber, **Figure 13-6**. Steel scrap is fully recyclable, while wood is not. On average, new steel manufactured today contains 25% scrap steel. This reduces the pressure on finite resources and the load on valuable landfill space.

When steel framing is used in commercial buildings, it has a distinct advantage over wood due to the different construction methods used. In commercial structures, the framework is typically made up of beams and columns of steel or reinforced concrete. This framework supports the floors, which are installed before the walls. The floors have a camber; they are not level. Further, mechanical systems are also installed before the walls.

Taking into consideration all of these factors, it is necessary to install the walls piece by piece. First, a carpenter installs the plates and then the studs, fastening them around the mechanical systems in the wall. See **Figure 13-7**.

Susan Law Cain/Shutterstock.com

Figure 13-6. The scrap resulting from cutting steel studs to size is recyclable. Note the PPE that this carpenter is wearing. The face shield protects his skin from sparks. Gloves protect against cuts. Earplugs (or earmuffs) are required for hearing protection. Safety glasses protect the eyes from flying objects and impact. In this case, he is also protected from falling objects by the hard hat.

H.L. Stud Corp.

Figure 13-5. A wall section combining steel and wood. The wall section is so light one carpenter can carry it.

Figure 13-7. Installing steel framing in a commercial building. A—Plates are fastened in place where the walls will be located. The channels are fitted over or around mechanical systems, such as the electrical conduit in the foreground. B—Studs are installed into the floor and wall channels and fastened with screws or welding. These extra-long studs allow space for HVAC ducts and other systems above a suspended ceiling.

If wood is used, each stud must be accurately measured and cut—a time-consuming, labor-intensive job. Steel studs, by contrast, can be cut in bundles of ten, nested in the channels of the plates, and fastened with screws.

Therefore, cost comparisons between wood and steel framing vary depending on the application. Cost for framing is a factor that depends on more than the cost of materials only. The current market rates for raw materials, and the labor expended on the job both play a role. A large commercial job, for instance, may have a number of partition walls that an experienced crew can frame very quickly with steel. In this example, a steel-framed system would result in significant labor savings over a wood-framed system.

13.1.2 Disadvantages of Steel Framing

Steel has disadvantages that impact safety and cost. The standard stud for nonloadbearing walls has a metal thickness of 26 gauge and is flimsy. For comparison, 20-gauge framing is approximately 23% heavier and is much sturdier than 26-gauge framing. Care must be taken when handling them. Steel edges are sharp and may cause injury, even when hemmed components are used, **Figure 13-8**.

Even though steel prices are competitive with wood prices, steel can have higher engineering and labor costs. It takes longer to drive screws into metal than to drive nails into wood. Also, installing blocking, sheathing, and siding takes longer with steel studs than with wooden studs.

Track Options

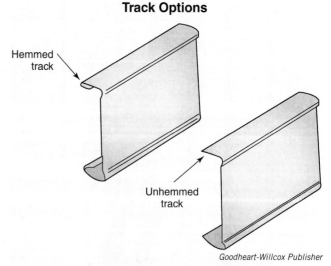

Hemmed track

Unhemmed track

Figure 13-8. Steel framing comes hemmed or unhemmed. A hemmed channel reduces the chance of injuries from sharp steel edges.

In cold climates, there is an added cost for the thermal breaks needed to conserve energy. Steel also suffers a cost penalty in extra insulation for exterior walls. Some carpenters do not use steel studs on outside walls because they are poor insulators. The R-value (heat resistance) of a wall can be reduced by as much as 50% when steel members are used.

13.2 Framing Floors

The methods for using steel in floor construction vary. Steel framing members can be used exclusively or combined with wood members. Depending on the materials, various fastening methods are used. **Figure 13-9** shows various methods of attaching floor joists.

Self-tapping screws are often used to fasten steel to steel. Special nails may be used to attach steel members to wood framing members. Pneumatic or electric drills are used to drive screws. Powder-actuated tools are usually used to attach the bottom track to a concrete slab. They may also be used to attach subflooring, such as oriented strand board (OSB), to steel joists. **Refer to** Chapter 5, **Power Tools**, for additional information on power tools. Subflooring can be attached also to joists with adhesives and self-tapping screws.

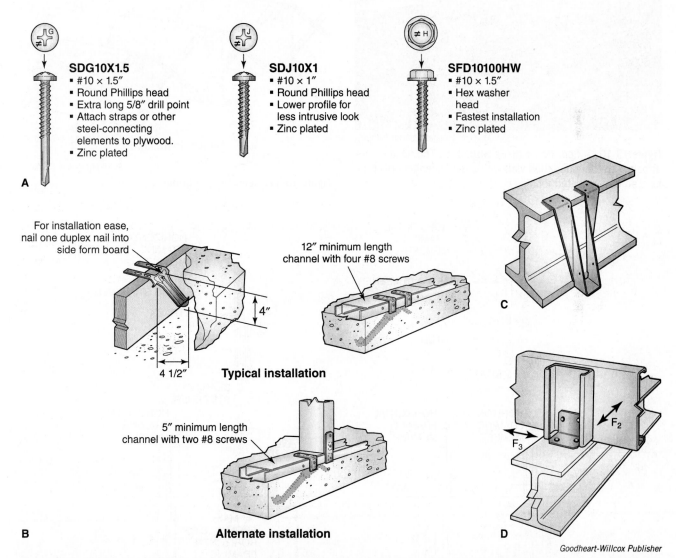

SDG10X1.5
- #10 × 1.5"
- Round Phillips head
- Extra long 5/8" drill point
- Attach straps or other steel-connecting elements to plywood.
- Zinc plated

SDJ10X1
- #10 × 1"
- Round Phillips head
- Lower profile for less intrusive look
- Zinc plated

SFD10100HW
- #10 × 1.5"
- Hex washer head
- Fastest installation
- Zinc plated

A

For installation ease, nail one duplex nail into side form board

4"

4 1/2"

Typical installation

12" minimum length channel with four #8 screws

C

5" minimum length channel with two #8 screws

F_2

F_3

B

Alternate installation

D

Goodheart-Willcox Publisher

Figure 13-9. Fasteners and joints for metal floor framing. A—Screw fasteners designed for light gauge steel construction. Long drill points allow for installation through several layers of steel before the threads engage. Note the markings on the screw heads: G has an extra long point; J has rounded head for lower profile; H has a hex head for faster installation. B—A mudsill anchor embedded in concrete. C—A hanger that can be attached by welding or with screws. D—This angle bracket attaches the joist to the I-beam.

Figure 13-10 shows a floor framed entirely in steel. The headers (band joists) are fastened to the foundation with steel pins. A powder-actuated tool drives the pins through a flange on the header into the concrete wall. Joists spaced 16″ OC are first fastened to the headers with screws, then welded. Where floor framing is supported by load-bearing walls, the joists must be reinforced with stiffeners and must be aligned with the studs in the bearing wall, **Figure 13-11**. Light gauge steel framing does not resist lateral (sideways) forces well, so bridging is used to stiffen the floor, **Figure 13-12**.

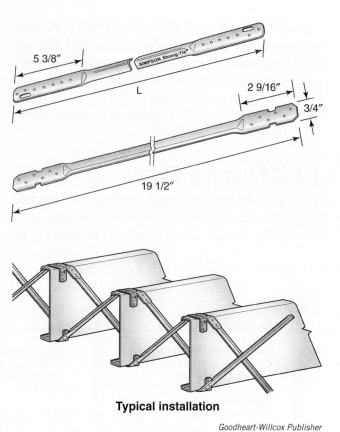

Typical installation

Goodheart-Willcox Publisher

Goodheart-Willcox Publisher

Figure 13-10. A floor frame made with steel. The headers are attached to the foundation wall with steel fasteners driven in by a powder-actuated tool.

Figure 13-12. Tension-type bridging.

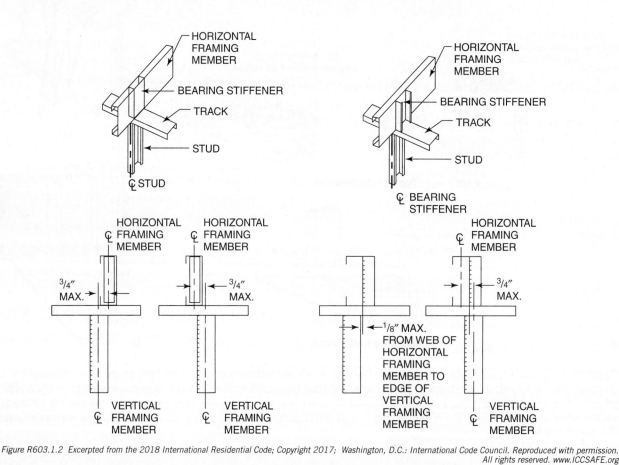

Figure R603.1.2 Excerpted from the 2018 International Residential Code; Copyright 2017; Washington, D.C.: International Code Council. Reproduced with permission. All rights reserved. www.ICCSAFE.org

Figure 13-11. Inline framing of floor and bearing wall. Note the stiffener used to reinforce the joists.

Another method involves attaching a metal pan across the joists and pouring a 2 1/2″–3″ thick, fiber-reinforced concrete floor on top of the joists, **Figure 13-13**.

13.3 Framing Walls and Ceilings

As noted earlier, steel-framed residential construction is increasing in popularity, **Figure 13-14**. *Metal studs* can be used with either metal plates or wood plates. Joints are fastened with No. 8 self-tapping screws.

As with wood construction, headers must be installed over openings in load-bearing walls. **Figure 13-15** shows three methods of constructing a header. As with wood wall framing, proper fastening and bracing is important.

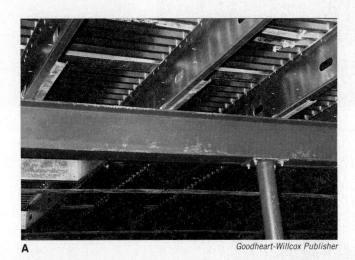

A   *Goodheart-Willcox Publisher*

B   *Goodheart-Willcox Publisher*

Figure 13-13. A steel-framed concrete floor. A—The view from the underside. Joists are one piece from wall to wall. The steel I-beam provides support. The floor pan has been attached to the joists with screws in preparation for pouring a concrete floor. B—A thin concrete base has been poured over the floor joists. The concrete is reinforced with fibers to prevent cracks.

Figure 13-14. A—Entire homes can be framed from light gauge steel. B—Steel studs are frequently used as partition walls on commercial buildings.

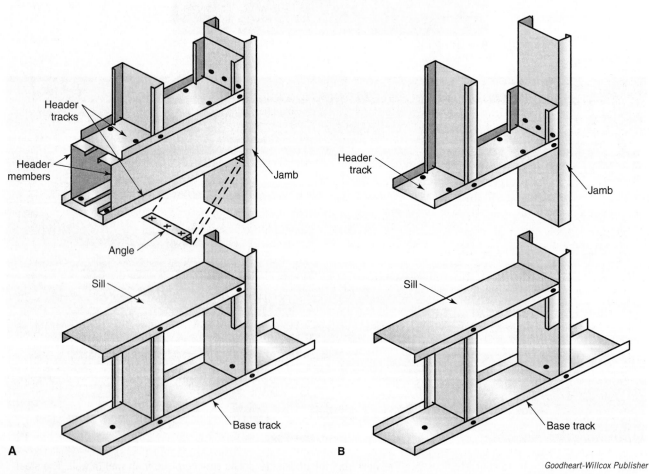

Figure 13-15. Wall framing. A—Load bearing headers must be constructed to specifications to ensure capacity. B—Non-load-bearing headers simply use a track.

Wall sections should have horizontal bridging to space studs, **Figure 13-16**. Temporary diagonal bracing of sections may be required to keep the walls square and plumb until permanent bracing can be installed. Permanent bracing of load-bearing walls is needed to withstand shear forces that could cause racking.

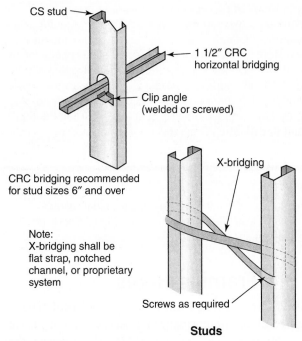

CS stud

1 1/2″ CRC horizontal bridging

Clip angle (welded or screwed)

CRC bridging recommended for stud sizes 6″ and over

X-bridging

Note:
X-bridging shall be flat strap, notched channel, or proprietary system

Screws as required

Studs

Solid blocking:
• Locate blocking at each end of wall, adjacent to openings, and as required
• For track:
Where blocking material thickness allows, notch and bend flanges 90° or anchor to verticals w/clip angles

Screw as required at each stud

Flat straps, notched channel, X-bridging, or proprietary bridging (each side) lap splice straps minimum 4″ (10 cm)

Screws as required (each side)

Note:
Number of rows of bridging as required by design

CEMCO and Dale Industries Inc.

Figure 13-16. Details for spacing and bracing steel wall frames.

Structural sheathing is normally applied to both sides of the wall frame. Type 2 plywood or OSB is rated for wind speeds of 100 mph. Sheets should be applied with the long side parallel to the studs. If plywood is used, it should extend from the top track to the bottom track. Shear strength can also be ensured by using *X-bracing*. The X-braces are steel straps that diagonally extend from the top of the wall to the bottom. They are attached to the tracks and studs with screws. Unlike diagonal bracing used in wood framing, the X-braces rest entirely on the surface of the wall frame. That is, they are not let-in.

Construction of steel-framed partitions is similar to wood construction. They can be framed in place or assembled on the deck and stood up.

Plumbing and electrical systems make use of prepunched or specially cut openings in the studs. Drywall or other surfacing material is attached to the studs with self-tapping screws. See **Figure 13-17**.

A
Goodheart-Willcox Publisher

B
Goodheart-Willcox Publisher

Figure 13-17. Steel-framed partitions. A—Water and waste piping is installed through openings cut in the steel studs. Protective bushings are used where flexible piping passes through a stud. B—Rigid electrical conduit is routed through openings in studs and connected to junction boxes that are attached to the studs with self-tapping screws.

A thermal break or insulation should be applied to the outside face of exterior walls, **Figure 13-18**. This improves the R-value of the outside wall by preventing the transfer of heat through the metal studs and joists.

Installation of drywall, sometimes called gypboard, is similar to installation over wood framing. Install drywall with its length parallel to the studs and joists. If a second layer is to be used, install it horizontally. Space the fasteners 24″ OC for the base layer and 16″ OC for the face layer, **Figure 13-19**.

Jack Klasey

Figure 13-18. Thermal insulation is installed between the steel studs and a concrete exterior wall to control heat transfer. This electrician is installing conduit in the wall.

Goodheart-Willcox Publisher

Figure 13-19. Screws are used to mount drywall on the steel studs.

Pro Tip
Additional information and technical guidelines are available from the International Residential Code and the Steel Framing Alliance.

13.4 Framing Roofs

Like their wooden counterparts, steel roof trusses, **Figure 13-20,** are normally engineered and constructed by companies specializing in this work. However, they can be fabricated onsite using the same methods and calculations as for wooden members. In such cases, be sure to follow engineering specifications.

Truss members can be cut with a portable electric saw or a chop saw. Next, the chords and webs are placed in a jig (a temporary arrangement of blocks to align the parts of the truss quickly and accurately during assembly) set up on the ground. The lapped and butt joints are fastened with self-tapping screws, a *weld bead*, or both.

Erection of the steel trusses follows the same procedure as for wood trusses. After they are fastened to the wall plate, permanent bracing is welded to the webs. **Figure 13-21** shows prefabricated trusses in place, fastened to the plate, and permanently braced. Plywood or OSB sheathing is then fastened to the rafters with sheet metal screws.

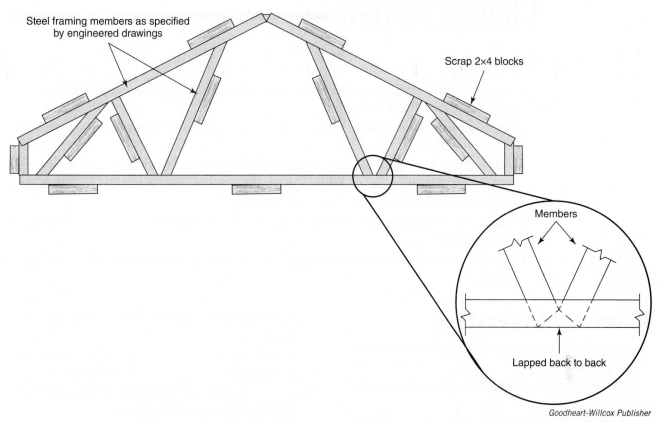

Figure 13-20. A wooden truss jig can be built from scrap framing lumber on a flat floor deck. Joints can be lapped and screwed, welded, fastened with plates made from sheet steel, or assembled with prefabricated connectors. All connections must be made to comply with engineering specifications.

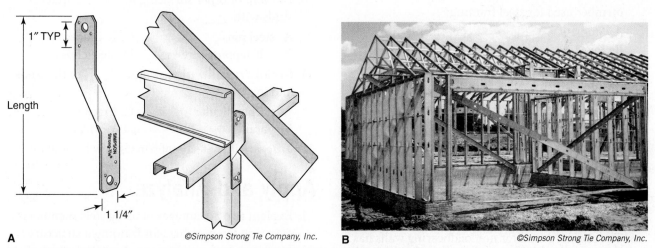

Figure 13-21. Steel truss installation. A—This all-purpose tie is suitable for securing steel trusses to the top plate, as shown. B—Rafter installation is complete when bracing (lighter-colored pieces) has been installed.

Chapter Review

Summary

- Steel framing has long been used in commercial construction, but is becoming more widespread in residential work.
- Framing members are used as studs, joists, trusses, and rafters and fastened in place with self-tapping screws.
- Steel framing has a number of advantages and disadvantages when compared to conventional wood construction.
- Some structures combine wood and steel framing.
- Extra caution must be used when handling steel framing materials to avoid injury caused by sharp edges.
- Construction and erection methods used for steel framing are similar to those used for wood.

Know and Understand

Answer the following questions using the information in this chapter.

1. Pricing for steel construction materials is based on _____.
   A. hundreds of lineal feet
   B. thousands of lineal feet
   C. weight of the materials in pounds
   D. overall size of the materials

2. Which of the following is *not* a type of framing member used for steel framing?
   A. Studs.
   B. Joists.
   C. Tracks.
   D. Drywall.

3. A stud consists of a metal channel with openings through which electrical and plumbing lines, and _____ can be installed.
   A. bracing
   B. partitions
   C. walls
   D. rafters

4. *True or False?* Framing in steel requires the same tools as wood framing.

5. *True or False?* Steel framing produces less waste than wood framing.

6. The standard stud for nonloadbearing walls has a metal thickness of _____ gauge.
   A. 16
   B. 20
   C. 26
   D. 32

7. *True or False?* Steel and wood framing members can be combined in the same structure.

8. Steel band joists are often fastened to foundations using _____.
   A. weld beads
   B. steel pins
   C. self-tapping screws
   D. adhesives

9. _____ are steel straps that diagonally extend from the top of the wall to the bottom.
   A. X-braces
   B. Structural sheathing
   C. Weld beads
   D. Metal studs

10. Drywall or other surfacing material is attached to studs with _____.
    A. steel pins
    B. self-tapping screws
    C. X-braces
    D. steel clips

11. Install drywall with its length _____ to the studs and joists.
    A. perpendicular
    B. parallel
    C. braced
    D. vertical

12. *True or False?* Steel roof trusses may be built onsite.

Apply and Analyze

1. Explain two advantages of using steel members over wooden members for framing a structure.

2. Explain two disadvantages of using steel members over wooden members for framing a structure.

3. How does X-bracing ensure shear strength?

4. How are steel roof trusses constructed?

5. What is a jig?

Critical Thinking

1. Wood framing members support and transfer live loads and dead loads directly through the framing members. This makes details like stud lengths and bearing points, such as trimmer studs, on load-bearing walls very important. When working with light-gauge structural steel framing, how are the live and dead loads transferred? What is the importance of engineered drawings on such a job?

2. Light-gauge steel framing is often used to build interior partition walls in commercial buildings. What problems does this solve over wood framing? What trade-offs need to be made to use steel instead of wood?

3. In northern climates, large, steel-framed buildings with low-sloped or flat roofs are designed to flex with snowfalls. This means the roof structure could droop in excess of an inch or more over a long span when there is a heavy snowfall. What could happen if a steel stud and gypsum board interior partition wall were built under one of these long spans? How do designers and builders solve this problem? Research this subject and report back to find out.

Communicating about Carpentry

1. **Listening.** With a group of other students, determine materials needed for using steel framing on a building. Each student should go to a building supply store or its website to investigate and calculate how much these materials would cost. Determine the cost and quality differences between different types of framing. Then each group presents its findings to the class, and the other class members take notes on the presentation.

2. **Speaking and Listening.** Create a collage that identifies different types of steel and wood framing. Show and discuss your collage in a group of 4–5 classmates. Are the other members of your group able to determine the types of framing that you tried to represent?

3. **Listening and Speaking.** Working with a partner, compare and contrast the use of wood framing versus steel framing. Consider the perspective of the construction company, the construction crew, and the owner of the building. In what situations would one product be preferable to the other? Record the key points of your discussion. Hold a class discussion. Compare your responses to those of your peers.

Closing In

Mark Byer/Shutterstock.com

CHAPTER **14**

Roofing Materials and Methods

OBJECTIVES

After studying this chapter, you will be able to:

- List the covering materials commonly used for sloping roofs.
- Define roofing terms.
- Describe how to prepare the roof deck.
- Describe reroofing procedures for both asphalt and wood shingles.
- Demonstrate correct nailing patterns.
- Select appropriate roofing materials for various slopes and conditions.
- Describe the application procedure for a built-up roof.
- Describe the use and application of wood shingles in residential construction.
- Explain how various roofing products are applied.
- Describe the different types of roofing tiles and how they are made.
- Demonstrate the proper positioning of gutters.
- Estimate materials needed for a specific roofing job.

TECHNICAL TERMS

asphalt shingle	flashing	selvage
base flashing	gravel stop	shingle butt
batten	gutter	side lap
built-up roof	head lap	six-inch method
cant strip	ice-and-water barrier	square
cap flashing	nailer board	starter strip
closed-cut valley	open valley	sweat sheet
coverage	plies	underlayment
drip edge	roll roofing	wood shake
eaves flashing strip	roofing tile	woven valley
eaves trough	saddle	
exposure	saturated felt	

R oofing materials protect the structure and its contents from the sun, rain, snow, wind, and dust. In addition to weather protection, a roof should offer some measure of fire resistance and be extremely durable. Since a large amount of surface is usually visible, especially on sloping roofs, the roofing materials can contribute to the attractiveness of the building. Roofing materials can add color, texture, and pattern, **Figure 14-1**.

planet5D LLC/Shutterstock.com

Figure 14-1. The roofing material not only protects the structure from weather, it also adds to the overall appearance.

Preparing a roof deck for its protective covering involves a number of operations. Most of these operations must follow a definite sequence. All items that project through the roof should be built or installed before roofing begins. These include chimneys, vent pipes, and special facilities for electrical and communications service. Performing any of this work after the finished roof is applied could damage the roof covering.

14.1 Types of Materials

Materials used for pitched (sloping) roofs include the following:

- Asphalt, wood, metal, and mineral fiber shingles
- Slate
- Tile made from clay or cement
- Sheet materials, such as rubberized single-ply membrane, roll roofing, galvanized iron, aluminum, tin, and copper

For flat or low-slope roofs, a membrane system is used. It consists of a continuous, watertight surface usually obtained by using a built-up roof, rubberized membrane, or seamed-metal sheets.

Built-up roofs are fabricated on the job. Roofing felts are laminated (stuck together) with asphalt or coal tar pitch. Then, hot asphalt is mopped over the felt layers. Finally, this surface is coated with crushed stone or gravel.

Rubber roofing is applied directly to the sheathing with adhesive. Seams are taped with special tape. Rubber roofing is sold in rolls up to 100′ wide, so taping is required only on large commercial buildings.

Flat metal roofs are assembled from metal sheet stock. Seams are soldered or sealed with special compounds to ensure that the roof is watertight.

When selecting roofing materials, it is important to consider these factors:

- Initial cost
- Maintenance costs
- Durability
- Appearance

The pitch of the roof also determines the selection. Low-sloped roofs require a more watertight system than steep roofs, **Figure 14-2**. Materials such as tile and slate require heavier roof frames.

Local building codes may prohibit the use of certain materials. In some cases, the materials may be a fire hazard. Certain localities may require materials that resist high winds or other elements.

14.2 Roofing Terminology

There are a number of specialized terms used in the roofing trade. Refer to **Figure 14-3**. Two of these, slope and pitch, are defined in Chapter 12, **Roof Framing**. Another widely used term is the *square*. This is the unit

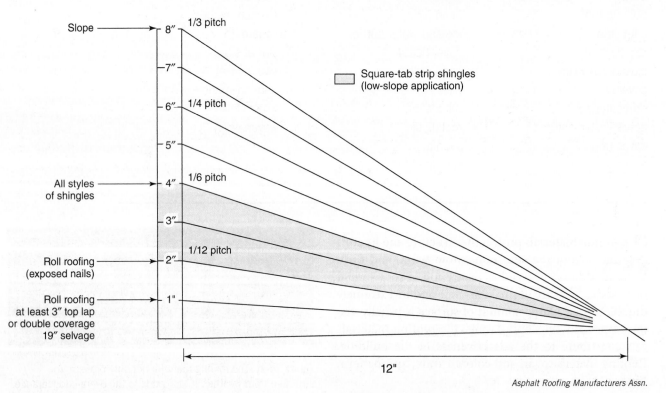

Asphalt Roofing Manufacturers Assn.

Figure 14-2. Roofing manufacturers list minimum slope requirements for various asphalt roofing products.

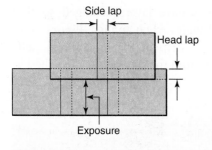
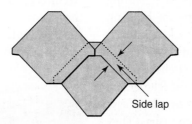
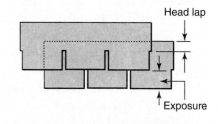

Goodheart-Willcox Publisher

Figure 14-3. Exposure, head lap, and side lap are important terms used in the application of roofing materials.

of measure for estimating and purchasing shingles. It is the amount of roofing material needed to provide 100 sq ft of finished (shingled) roof surface.

Coverage is the amount of weather protection provided by the overlapping of shingles. Depending on the material and method of application, shingles may furnish one, two, or even three thicknesses of material on the roof. This is referred to as single coverage, double coverage, or triple coverage.

Exposure is the distance between the edge of one course of shingles and the edge of the next higher course. The term also refers to the exposure of lapped siding. The distance from the lower edge of an overlapping shingle or sheet to the top edge of the shingle or sheet beneath is called *head lap*. The overlap length for side-by-side elements of roofing is *side lap*. A *shingle butt* is the lower, exposed edge of a shingle.

14.3 Preparing the Roof Deck

The roof sheathing should be smooth and securely attached to the frame. It must provide an adequate base to receive and hold the roofing nails and fasteners. In areas subject to hurricane-force winds, ring-shank nails should be used to install sheathing because of their greater holding power.

Inspect the roof deck to see that nailing patterns are complete and that there are no nails sticking up. Joints should be smooth and free of sharp edges that might cut roofing materials. Repair large knotholes over 1″ diameter by covering with a piece of sheet metal. Clean the roof surface of chips or other scrap material.

All types of shingles can be applied over solid sheathing. However, skip sheathing or application of shingle breather underlayment is a better choice for wood shingles or shakes **Figure 14-4**. Wood absorbs a certain amount of moisture. Either application method provides ventilation and promotes faster drying after a rainfall. Shingle breather underlayment is discussed later in this chapter. Solid boards have often been used for skip sheathing. Boards over 6″ wide should not be

Alfred State College

Figure 14-4. Furring allows for a secure nailing surface and space for ventilation between the wood shingles or shakes and the underlayment.

used as sheathing, however, because they are more liable to warp and cup than are narrower boards. It is important for the sheathing to provide a flat surface for the shingles.

Attics should be properly ventilated to remove moisture. Moisture vapor originating in occupied spaces may sometimes enter the attic. If the vapor becomes chilled below the dew point (produces droplets of water), it will condense on the underside of the roof deck. This causes the sheathing (especially sheet materials) to warp and buckle. To avoid this condition, install openings in locations that provide adequate ventilation, such as in the soffits and gable ends. Many homes have ventilated ridge caps to allow humid air to escape the attic. Vents or louvers should provide 1/2 sq in of opening per square foot of attic space.

14.4 Asphalt Roofing Products

Asphalt roofing products are widely used. These products fall into three broad groups: saturated felts, roll roofing, and shingles.

Saturated felts are used under shingles as sheathing paper. They are also used as laminations for built-up roofs. Saturated felts are made of dry felt soaked with asphalt. They are available in different weights, with 15 lb the most common for use as underlayment beneath shingles. This number indicates the weight of the felt needed to cover 100 sq ft.of the roof deck with a single layer.

Roll roofing and shingles are outer roof coverings. They must be weather resistant. Their base material is synthetic fibers, organic felt, or fiberglass. This base is saturated and then coated with a special asphalt that resists weathering. A surface of ceramic-coated, opaque mineral granules is then applied. The mineral granules shield the asphalt coating from damaging ultraviolet light, add color, and provide protection against fire. **Figure 14-5** shows weights and specifications for several asphalt roofing products.

Asphalt shingles are the most common type of roofing material used on houses. They are manufactured as strip shingles, interlocking shingles, and large individual shingles. Architectural asphalt shingles are the most popular roof covering used on homes today.

Architectural shingles have layers arranged so that they mimic wood shingles when applied, **Figure 14-6**. Most shingles have a strip of factory-applied, self-sealing adhesive. Heat from the sun softens the adhesive.

Goodheart-Willcox Publisher

Figure 14-6. Architectural asphalt shingles do not have tabs, but are layered to give them a random textured appearance.

Installation Data for Common Asphalt Roofing Products

Product	Approximate Shipping Weight		Square Per Package	Length	Width	Side or End Lap	Top lap	Exposure	Underwriters Laboratories Listing*
	Per roll	Per square							
Mineral Surface Roll 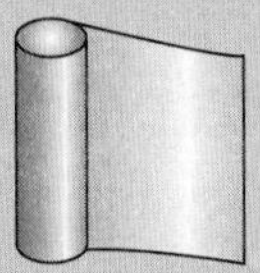	75# to 90#	75# to 90#	1	36′ to 38′	36″	6″	2″ to 4″	32″ to 34″	C
	Available in some areas in 9/10 or 3/4 square rolls.								
Mineral Surface Roll (Double Coverage) 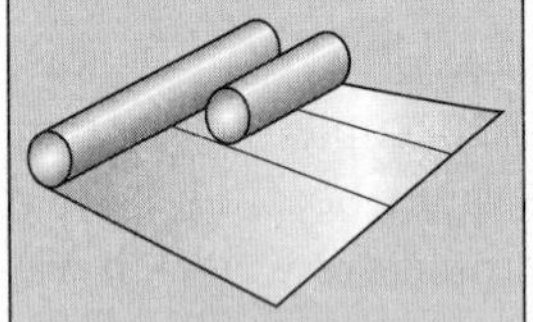	55# to 70#	110# to 140#	1/2	36′	36″	6″	19″	17″	C
Smooth Surface Roll 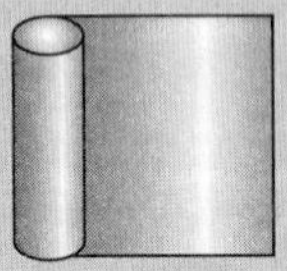	40# to 65#	40# to 65#	1	36′	36″	6″	2″	34″	None
Saturated Felt (Nonperforated) 	60#	15# to 30#	2 to 4	72′ to 144′	36″	4″ to 6″	2″ to 19″	17″ to 34″	None

Asphalt Roofing Manufacturers Assn.

Figure 14-5. This chart provides installation data for common asphalt roofing products. *UL rating at time of publication. Refer to the manufacturer's product literature at the time of purchase.

This results in a bond between each shingle and the shingle below. This bond prevents shingles from being raised by wind. The self-sealing action usually takes place within a few days during warm weather. In winter, the sealing time can be considerably longer, depending on the climate. **Figure 14-7** shows installation data for common asphalt shingles.

Safety Note

Safety considerations are very important in roofing work. Be sure to erect a secure scaffold that will support workers and is at a waist-high level with the eaves. Study Chapter 6, **Scaffolds, Ladders, and Rigging**, for more information.

14.4.1 Underlayment

A roof ***underlayment*** is a thin cover of asphalt-saturated felt or other material. It has a low vapor resistance. This underlayment serves these purposes:

- Protects the sheathing from moisture until the shingles are laid
- Provides additional weather protection by preventing the entrance of wind-driven rain and snow
- Prevents direct contact between shingles and resinous areas in the sheathing

Materials such as coated sheets or heavy felts should not be used. They may act as a vapor barrier and allow moisture and frost to gather between the covering and the roof deck. Although 15 lb roofer's felt is most

Installation Data for Common Asphalt Shingles

Product	Configuration	Per Square: Approximate shipping weight	Per Square: Shingles	Per Square: Bundles	Size: Width	Size: Length	Exposure	Underwriters Laboratories Listing
Self-Sealing Random-Tab Strip Shingle. 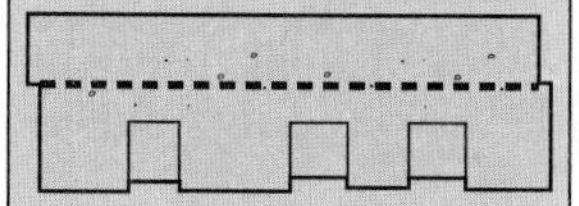 Laminates	Various edge, surface texture and application treatments	285# to 390#	66 to 90	3 to 5	11 1/2″ to 14″	36″	4″ to 6″	A or C Many wind resistant
Self-Sealing Random-Tab Strip Shingle 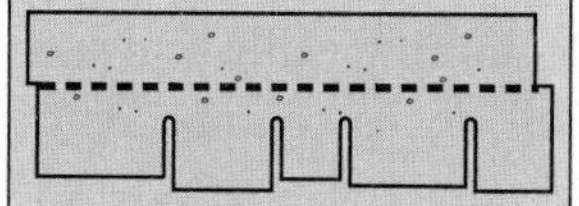 Single-Thickness	Various edge, surface texture and application treatments	250# to 300#	66 to 80	3 or 4	12″ to 13 1/4″	39.37	5″ to 5 5/8″	A or C Many wind resistant
Self-Sealing Square-Tab Strip Shingle 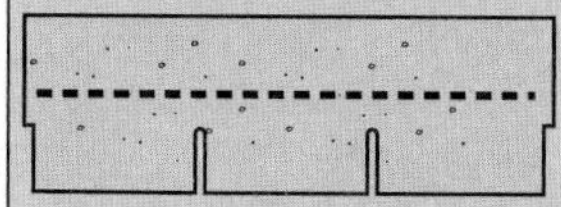 Three-Tab	Two-tab or Four-tab	215# to 325#	66 to 80	3 or 4	12″ to 13 1/4″	39.37	5″ to 5 5/8″	A or C All wind resistant
	Three-tab	215# to 300#	66 to 80	3 or 4	12″ to 13 1/4″	39.37	5″ to 5 5/8″	
Self-Sealing Square-Tab Strip Shingle 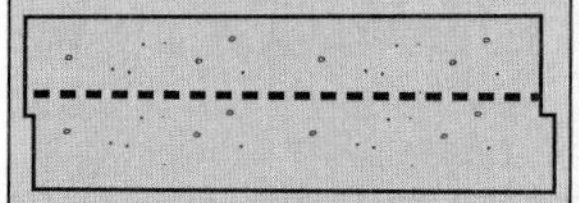 No-Cutout	Various edge and surface texture treatments	215# to 290#	66 to 81	3 or 4	12″ to 13 1/4″	39.37	5″ to 5 5/8″	A or C All wind resistant
Individual Interlocking Shingle 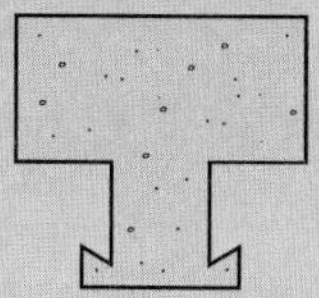 Basic Design	Several design variations	180# to 250#	72 to 120	3 or 4	18″ or 22 1/4″	20″ to 22 1/2″	—	A or C Many wind resistant

Asphalt Roofing Manufacturers Assn.

Figure 14-7. Installation data for common asphalt shingles.

commonly used as an underlayment, requirements vary depending on the kind of shingles and the roof slope.

General application standards for underlayment suggest a 2″ head lap at all horizontal joints and a 4″ side lap at all end joints. Lap at least 6″ on each side of the centerline of hips and valleys.

> **Pro Tip**
> Do not put down underlayment on a damp roof. It may trap moisture and damage the roof.

14.4.2 Drip Edge

The roof edges along the eaves and rake (the inclined edge of a gable roof) should have a metal *drip edge*, **Figure 14-8**. Various shapes formed from galvanized steel or aluminum are available. They extend back about 3″ from the roof edge and are bent at the factory

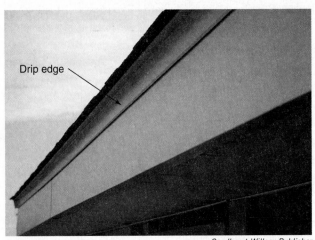

Drip edge

A   *Goodheart-Willcox Publisher*

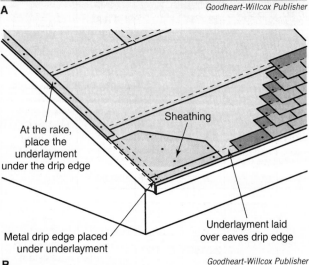

At the rake, place the underlayment under the drip edge

Sheathing

Metal drip edge placed under underlayment

Underlayment laid over eaves drip edge

B   *Goodheart-Willcox Publisher*

Figure 14-8. A—Drip edge is made of sheet metal bent to cause water to drip off the edge, not run back under the eave. B—Proper application of underlayment and metal drip edge.

to go downward over the edge. This causes the water to drip free of the underlying cornice construction. It also keeps the shingles from drooping over the edge when heated by the sun.

Install the drip edge with shingle nails 8″–10″ apart. Keep the nails near the upper or inner edge. Lap joints in the drip edge about 2″. At the eaves, the underlayment should be laid over the drip edge. At the rake, place the underlayment under the drip edge.

14.4.3 Ice-and-Water Barrier

It is recommended that an *ice-and-water barrier* be installed at the eaves for buildings in cold climates. Installed on new construction or during reroofing, it prevents leak-through from ice dams or wind-blown rain. These materials are self-sealing around nails and deck joints.

Available in 36″ wide rolls, ice-and-water barrier has an adhesive backing and some are reinforced with fiberglass.

The lower edge of the ice-and-water barrier should be placed even with the lip of the drip edge at the eve. Start the strip parallel with the drip edge. The barrier should cover the sheathing and drip edge from the roof's edge to at least 24″ inside the building's outside wall line. See **Figure 14-9**.

14.5 Flashing

Intersections with other roofs, adjoining walls, and such projections as chimneys and soil stacks complicate the installation of roof coverings. Making these areas watertight requires a special building material called flashing.

Owens Corning

Figure 14-9. In cold climates, put down an ice-and-water barrier and extend it up the roof slope beyond the building line. It is also advisable to use it on valleys and under other flashing on the roof.

Flashing is water-resistant sheet material designed to keep joints in the roof watertight. Materials used for flashing include zinc-coated (galvanized) metal, copper, lead, aluminum, ice-and-water barrier, and roll roofing.

14.5.1 Valley Flashing

A valley is the junction of two sloping roofs. The slopes direct water into this area. Thus, water drainage is heavier at this point, so it is important that the area is protected against leakage. Flashing is a method of sealing valleys against leakage. Proper installation of flashing for the type of valley shingle application used is critical to producing a roof that does not leak.

All valley flashing begins with applying a 36″ layer of ice-and-water shield to the valley. Sweep the deck to ensure it is clean. Check to make sure there are no nails or other objects that may cause a blemish in the membrane. Strike a chalk line 18″ from one side of the valley. Cut the membrane to length and place it in the valley. Being careful not to let it move, fold back one side and remove the masking from the back to expose the adhesive. Smooth the membrane into place, working on a small area at a time. When one side is completely in place, fold back the other side of the membrane and remove the masking from that side. Working from the valley toward the edge, smooth the ice-and-water barrier into place.

Further preparation depends on how you plan to shingle the valley:

- Open valley
- Woven valley
- Closed-cut valley

Whatever the style of flashing, make certain that it is smooth, has no obstructions, has the capacity to quickly move water away, and can handle occasional water backup.

14.5.2 Installing Open Valley Flashing

For valley flashing under asphalt shingles, use aluminum, copper, or galvanized sheet metal. Install *open valley* flashing as shown in **Figure 14-10**. Fasten the sheet metal flashing without puncturing the surface, **Figure 14-11**. All underlying joints are lapped at least 12″.

Before applying the shingles, snap a chalk centerline in the valley. Also, snap a line on each side of the center to mark the width of the waterway. This should be 6″ wide at the ridge and widened. The lines should move away from the valley at the rate of 1/8″ for every foot as they approach the eave. A valley 8′ long would be 7″ wide at the eave. When a course of shingles meets the valley, the outside chalk lines serve as guides in trimming the last unit. After the shingle is trimmed,

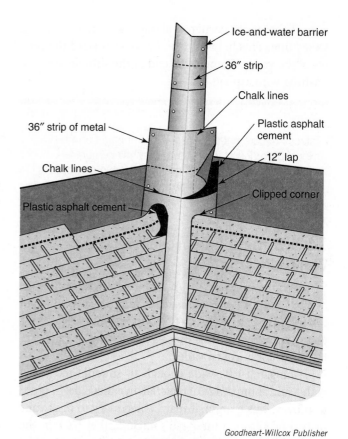

Goodheart-Willcox Publisher

Figure 14-10. Open valley flashing is laid down before shingling begins. The bottom is trimmed to match the eave line. Shingles should be sealed with asphalt cement where they lap over the flashing.

Goodheart-Willcox Publisher

Figure 14-11. Use the edge of the roofing nail head to hold the metal flashing in place. This allows the metal to move as it expands in the heat and contracts in the cold.

cut off the upper corner at about a 45° angle with the valley line. This helps direct any water toward the center of the valley. Cement the end of the shingle over the flashing with asphalt cement.

Pro Tip

Take extra care not to let asphalt cement get on the exposed portion of the flashing. Asphalt cement is very difficult to remove if it gets on a visible part of the flashing. Solvents that will dissolve asphalt cement will also dissolve the asphalt in the shingles.

14.5.3 Installing Woven Valley Flashing

Some roofers prefer to run shingles across valleys to create a *woven valley*. This method is often used when reroofing. See **Figure 14-12**. Only asphalt strip shingles may be applied this way. Architectural shingles are too thick for a woven valley.

Flashing must be wide enough to straddle the valley with a minimum of 12″ of material on either side. In order to provide this margin, it may be necessary to insert one or two tabs from a three-tab shingle farther back in the course. Fasteners must be kept at least 6″ away from the valley centerline.

When laying shingles across the valley, firmly press them into place to avoid bridging the valley. Keep fasteners at least 6″ away from either side of the centerline. Use two nails at the end of each terminal strip.

PROCEDURE

Installing a Woven Valley

1. Apply a 36″ wide ice-and-water barrier across the valley.
2. Lay the first course of shingles along the eave of one roof surface.
3. Extend one strip at least 12″ across the valley.
4. Lay the first course on the intersecting roof and extend it across the valley over the previously applied shingle.
5. Alternate succeeding courses, first along one roof surface and then the other. Refer again to **Figure 14-12**.

14.5.4 Installing Closed-Cut Valley Flashing

Another method of flashing a valley is the *closed-cut valley* method. In this method, the two intersecting roof surfaces (except the first course) are individually shingled. Refer to **Figure 14-13**.

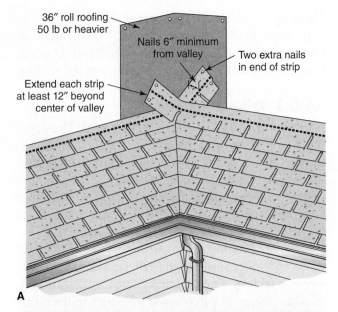

Goodheart-Willcox Publisher

Figure 14-12. Woven valley. A—Method of laying woven valley shingles. B—An example of woven valley shingling.

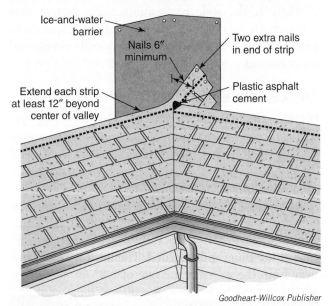

Goodheart-Willcox Publisher

Figure 14-13. Details of construction for a closed-cut valley. Shingles on the right are cut along a line 2″ back from the valley centerline.

When applying a new roof over the top of old shingles, build up the trough in an open valley to the average level of the roof surface. This can usually be done with strips of beveled wood. Then, flash the valley as described in the text.

14.5.5 Flashing at a Wall

Where the sloping part of a roof abuts a vertical wall, step flashing should always be used. This flashing should be 10″ wide and 2″ longer than the exposed face of the regular shingles. Bend the 10″ width at a right angle so that it extends 5″ over the roof and 5″ up the wall, as shown in **Figure 14-14**.

Install metal flashing as each course of shingles is laid. Nail it to the roof at the top edge, as shown. Do not nail flashing to the wall, since settling of the roof frame could damage the seal.

Install wall siding after the roof is completed. The siding then serves as cap flashing, **Figure 14-15**. Position the siding just above the roof surface. Allow enough clearance to paint the lower edges.

PROCEDURE
Installing a Closed-Cut Valley

1. Install ice-and-water barrier.
2. Apply the first course on both surfaces. At the valley, interweave this course only.
3. Apply all of the shingles on one roof surface. Carry each course across the valley and onto the adjoining roof at least 12″. If a shingle should fall short, insert a one- or two-tab section well away from the valley.
4. Apply remaining shingles in the same way, extending each course across the valley. Press shingles firmly into the valley. Drive two nails into the end of the valley shingle.
5. When the first roof surface is complete, snap a chalk line 2″ from the centerline of the valley on the second, unshingled surface.
6. Apply the second course of shingles along the eaves of the intersecting roof.
7. Where each course meets the valley, trim the shingle where it falls on the chalk line.
8. Trim off 1″ of the upper corner of the shingle at a 45° angle to prevent water from running back along the top edge.
9. Embed the end of the shingle in a 3″ wide strip of plastic asphalt cement.
10. Apply and complete succeeding courses in the same way.

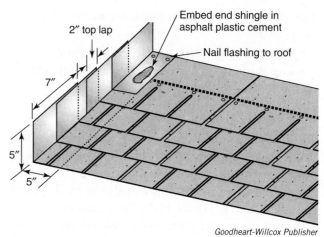

Goodheart-Willcox Publisher

Figure 14-14. Apply metal step flashing against vertical siding with each course of shingles. The vertical portion goes under the siding.

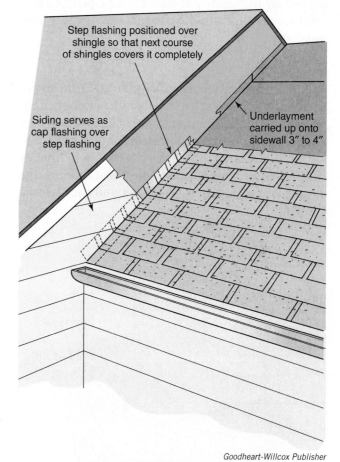

Goodheart-Willcox Publisher

Figure 14-15. Metal step flashing is applied to waterproof joints between the sloping roof and the walls. Note that the siding serves as cap flashing.

14.5.6 Chimney Flashing

Flashing around a masonry chimney must allow for some movement caused by settling or shrinkage of the building framework. This must be done in such a way that movement causes no damage to the water seal.

The flashing has two parts that move independently of each other:

- *Base flashing*, which is attached to the roof
- *Cap flashing* (also called counterflashing), which is attached to the chimney

Base Flashing

Sheet metal is usually preferred for base flashing. It should be applied by the step method previously described in the section on wall flashing. However, mineral-surfaced roll roofing can also be used to form the base flashing, **Figure 14-16**. First, cement the front unit into place and then attach the sidepieces. The flashing on the high (back) side is installed last. All sections are cemented together as they are applied.

Housings for prefabricated chimneys require flashing similar to masonry chimneys. Some prefabricated chimney units have flashing flanges that simplify their installation.

Cap Flashing

Cap flashing is sheet metal shaped to cover the top of the base flashing. When used around a masonry chimney, it is set into the mortar joints and bent down over the base flashing. In new construction, the cap flashing is mortared into the joints when the brick chimney is built. If the chimney is laid up without flashing, the mortar joints must be chiseled out and the flashing

forced in. Finally, the joints must be filled or pointed with good mortar. The metal is set into the joints 1 1/2″. Cap flashing on the front of the chimney may be one continuous piece. Sides must be stepped in sections because of the roof slope, **Figure 14-17**.

14.5.7 Chimney Saddle

Large chimneys on sloping roofs generally require an auxiliary roof deck on the high side. This structure is called a *saddle* or cricket. It diverts water from behind the chimney, preventing a buildup of ice and snow. If water, ice, or snow is allowed to collect behind the chimney, roof leaks could result.

Figure 14-18 shows a chimney saddle framing design. The frame is nailed to the roof deck and then sheathed. A small saddle does not require framing. It could be constructed from triangular pieces of 3/4″ exterior plywood.

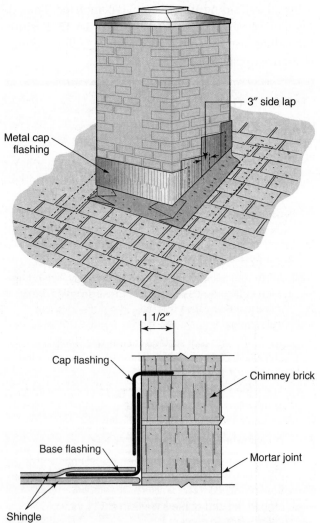

Goodheart-Willcox Publisher

Figure 14-17. Metal cap flashing is set into the mortar joints as the chimney is built. Cap flashing must go over the top of the base flashing as shown in the cross section.

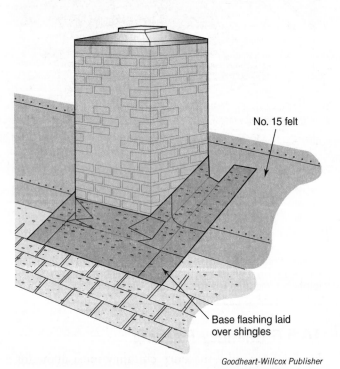

Goodheart-Willcox Publisher

Figure 14-16. Base flashing seals the joints between the chimney and the roof. Mineral-surfaced roll roofing is cut and cemented into place. The installation is ready for cap flashing, which covers the base flashing.

Suggested saddle frame. Nail to roof deck and then apply sheathing. Slope is equal to that of the main roof.

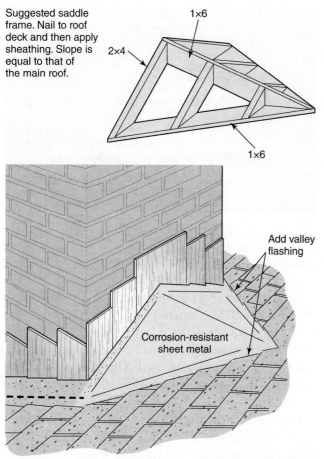

1×6

2×4

1×6

Add valley flashing

Corrosion-resistant sheet metal

Goodheart-Willcox Publisher

Figure 14-18. A chimney saddle, also called a cricket, diverts water away from the upper side of a chimney. Small saddles need not be framed. They can be formed from triangular pieces of 3/4″ exterior plywood.

Saddles are usually covered with corrosion-resistant sheet metal. However, mineral-surfaced roll roofing can be used. Valleys formed by the saddle and main roof should be carefully flashed in the same way as for regular roof valleys.

14.5.8 Vent Stack and Skylight Flashing

Pipes and skylights in the roof must also be carefully flashed, **Figure 14-19**. The roofing must be laid up to the protrusion. Cut and fit the shingles around the protrusion, then carefully cement a flange in place and lay shingles over the top. The flange must be large enough to extend along the roof surface at least 4″ below, 8″ above, and 6″ on each side of the stack. Most vent stack flashing is fitted with a neoprene boot that seals out moisture, as shown in **Figure 14-19**.

One-piece plastic skylights have a one-piece flashing that must be fitted under shingles at the top and sides and over shingles at the bottom. A wood curb is sometimes used to raise the skylight above the roof. The flashing procedure is similar to flashing a chimney.

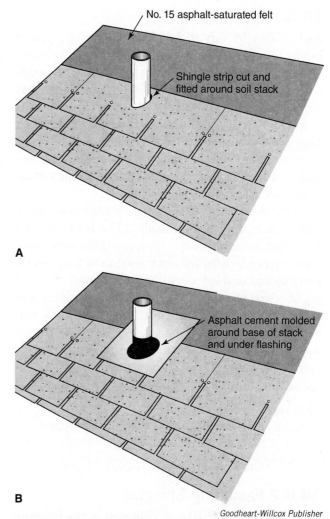

No. 15 asphalt-saturated felt

Shingle strip cut and fitted around soil stack

A

Asphalt cement molded around base of stack and under flashing

B

Goodheart-Willcox Publisher

Figure 14-19. Flashing stacks. A—Lay shingles up to the stack and fit the last course around it. B—Install a flange and apply shingles over the upper side of the flange.

Green Note
When recycling material on or off the jobsite, reuse it as directly as possible. When building materials are 'recycled,' they are often 'down-cycled' into lower-quality products. Lumber becomes wood chips and mulch. Concrete is ground down into low value aggregate. When a material is reused directly, without recycling, no quality is lost.

14.6 Asphalt Shingles

Generally, strip shingles are laid starting on one end of the roof and working to the other. However, when installing three-tab shingles on longer roofs, it may be best to start at the center and work both ways. In this case, snap a chalk line at a right angle to the ridge from the eaves to the peak. This line establishes the starting point.

14.6.1 Chalk Lines

Asphalt shingles vary slightly in length, usually plus or minus 1/4″ in a 36″ strip. There may be some variations in width, as well. Therefore, chalk lines should be used to achieve the proper placement so shingles are accurately aligned both horizontally and vertically. See **Figure 14-20**. Snap a number of chalk lines between the eaves and ridge. They serve as reference marks for starting each course. Space them according to the type of shingle and layout pattern. Full or cut shingles are aligned with the vertical chalk lines to form the desired pattern.

Chalk lines parallel to the ridge are used to maintain straight horizontal lines. Usually, a line is snapped for every fifth or sixth course. However, if the roofers are inexperienced, chalk lines may need to be snapped for every second or third course. To do this, mark the spacing of courses (using the shingle exposure) all the way to the ridge and snap chalk lines as needed. Some carpenters snap a horizontal chalk line every 10″ all the way to the ridge. Assuming a 5″ exposure, the top of every other course then falls on a chalk line.

Pro Tip
To get the best performance from any roofing material, always study and follow the manufacturer's directions. They are usually included on the packaging material.

14.6.2 Fastening Shingles

When roofing materials are delivered to the building site, handle them with care and protect them from damage. If a lift is available, use it to hoist the materials to the roof, **Figure 14-21**. Try to avoid handling asphalt shingles in extreme heat or cold. In high heat they become soft and tear easily. In extreme cold they become brittle and may break.

Christina Richards/Shutterstock.com

Figure 14-20. Snap chalk lines occasionally to keep shingles in alignment.

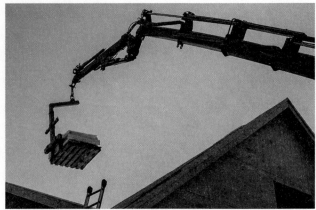

Goodheart-Willcox Publisher

Figure 14-21. A hoist saves the labor of carrying roof coverings to the roof. Some roofers use an elevator or conveyor to bring shingles to the point of application.

Nails used to apply asphalt roofing must have large heads (3/8″–7/16″ diameter) and sharp points. The International Residential Code (IRC) requires at least 12 gauge, galvanized-steel nails with barbed shanks. Aluminum nails are also used. The length should be sufficient to penetrate nearly the full thickness of the sheathing.

The number of nails and correct placement are both vital factors in proper application of shingles. For three-tab, square-butt shingles and for laminated architectural shingles, use a minimum of four nails per strip, as shown in **Figure 14-22**.

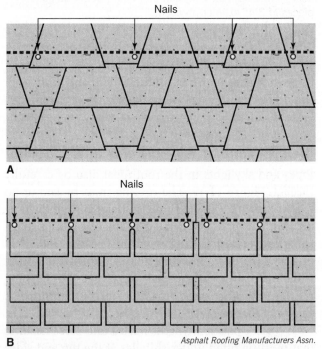

Asphalt Roofing Manufacturers Assn.

Figure 14-22. An approved nailing pattern for asphalt strip shingles. This placement catches the tops of the preceding course, providing additional holding power. A—Laminated architectural shingles. B—Three-tab shingles.

The building code may call for as many as six nails per shingle. Generally speaking, fasteners should be placed 5/8″ below the adhesive band found on the shingle strip. Carefully align each shingle and start the nailing from the end, next to the one previously laid. Proceed across the shingle. This will avoid buckling the shingle. Drive nails straight so the edge of the head does not cut into the shingle. The nail head should be driven flush with, not sunk into, the surface. If for some reason the nail fails to hit solid sheathing, drive another nail in a slightly different location.

Pneumatic-powered nail guns are often used to install asphalt shingles, **Figure 14-23**. Nail guns are loaded with coils of shingle nails. Always follow the manufacturer's recommendations for staples and special power nailing equipment.

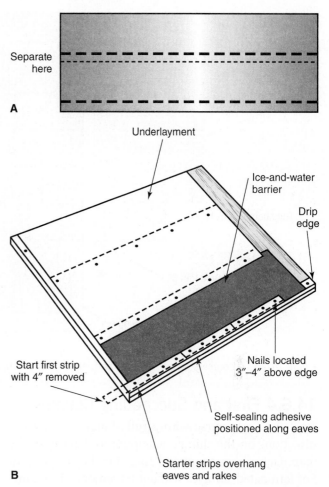

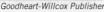

Goodheart-Willcox Publisher

Figure 14-23. Pneumatic tools are widely used to install asphalt shingles. A—A pneumatic nail gun that feeds roofing nails from the round magazine located below the handle. B— Nails are manufactured in coils that fit into the nail gun's magazine.

Goodheart-Willcox Publisher

Figure 14-24. A—Starter strips come bundled in pairs like shingles or in a roll. B—Starter strips are installed at the bottom of the eaves for weather protection and to securely seal the first course.

Pro Tip
If a fastener must be removed from a shingle, repair the hole with asphalt cement applied according to the manufacturer's directions.

14.6.3 Starter Strip

A *starter strip* is a strip of mineral-surfaced material that seals and secures the bottom edge of the first course of shingles and covers the gap between the tabs of three-tab shingles, **Figure 14-24**. Failure to install it will result in damage to the exposed underlayment and sheathing.

Let the strip slightly overhang the drip edge. Secure it with nails spaced 3″–4″ above the edge. Space the nails so they are not exposed at the cutouts between the tabs of the first course of shingles.

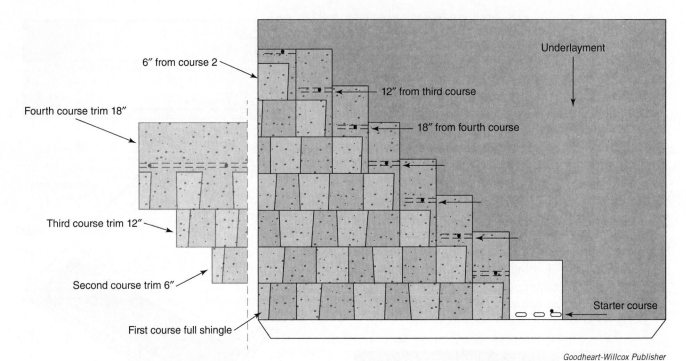

Goodheart-Willcox Publisher

Figure 14-25. A course starting set uses both ends of cut shingles to create a stair-step pattern.

14.6.4 First and Succeeding Courses

Start the first course with a full shingle. Follow the directions on the shingle wrapping to determine the manufacturer's recommendations for the cut pattern. For laminated shingles without tab cutouts, it is common to cut approximately 5″–7″ from the outer end of the second course. Cut 11″–13″ from the end of the third course and 17″–19″ from the end of the fourth course. Exact length of the cut depends on the length of a full shingle. This creates a stair-step installation pattern, **Figure 14-25.** Start over by using a full shingle on the fifth course. A utility knife and straightedge is the easiest way to cut asphalt shingles.

Three-tab shingles are laid so the cutouts are centered on the middle of the tab in the course directly below. Thus, the cutouts in every other course are exactly aligned. This is called the *six-inch method*.

For this pattern, cut 6″ from a strip to start the second course. The third course is started with one full tab (12″) removed from the strip. The fourth course is started with half of the strip (18″) removed. Continue reducing the strip lengths in the same sequence for subsequent courses.

Using an approved nailing pattern for asphalt shingles is important in securing both the best appearance and full weather protection. Manufacturers recommend that four nails be used, as shown in **Figure 14-22.** When shingles are applied with an exposure of 5″, nails should be placed 5/8″ above tops of cutouts. Locate one nail above each cutout and one nail in 1″ from each end.

Pro Tip
When laying asphalt shingles, soft-soled shoes that will not damage the surface and edge of the shingles should be worn. Asphalt products are easy to damage when worked at high temperatures. Try to avoid laying these materials on extremely hot days.

14.7 Hips and Ridges

Special hip and ridge shingles are sometimes available from the manufacturer. These shingles can also be easily made onsite with three-tab shingles. Laminated shingles cannot be cut apart to make hip and ridge cap shingles. Metal ridge roll is not recommended for asphalt shingles. Corrosion may discolor the roof.

PROCEDURE
Making and Installing Ridge Shingles
1. Cut three-tab shingles into three equal pieces.
2. Taper the lap (covered) portion of each ridge shingle with a utility knife. See **Figure 14-26A**.
3. After the ridge shingles are cut, bend them lengthwise in the centerline, **Figure 14-26B**. In cold weather, the shingle should be warmed before bending to prevent cracks and breaks.
4. Beginning at the bottom of the hips or at one end of the ridge, lap the units to provide a 5″ exposure, as illustrated in **Figure 14-26C**. Secure with one nail on each side, 5 1/2″ back from the exposed end, and 1″ from the edge.

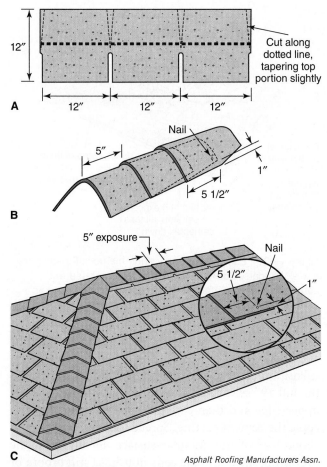

Figure 14-26. Hip and ridge shingles. A—Making cap shingles for hips and ridges. B—Properly placed nails catch the lap end of the previous cap. C—Nail hip and ridge shingles 5 1/2″ back from their edges. Use one nail on each side.

Asphalt Roofing Manufacturers Assn.

14.8 Wind Protection

The factory-applied adhesive on shingles prevents wind from lifting and damaging tabs. Only a few warm days are needed to seal tabs to the course underneath. In cold weather, the sealing adhesive may not seal until warmer weather arrives. If dirt and dust are blown under the shingles during cold weather, the shingles may never seal. For this reason, asphalt roofing should not be applied during cold winter months. In high-wind areas, wind-resistant shingles should be used. In the past it was common practice to use extra nails in high-wind areas, but the IRC now calls for wind-resistant shingles with normal nailing, as specified by the manufacturer. Wind-resistant shingles have stronger adhesive to prevent the wind from lifting the shingle.

Self-sealing shingles are satisfactory for roofs with slopes up to about 60°. For very steep slopes, like those used on mansard roofs, special application steps must be followed. You may have to seal them in place with quick-setting asphalt cement. Follow the recommendations provided by the roofing manufacturer.

If shingles without a sealing strip are used, the tabs can be cemented. Apply a spot of special tab cement, about 1″ square, with a putty knife or caulking gun and then press the tab down. Avoid lifting the tab any more than necessary while applying the cement.

Interlocking shingles are designed to resist strong winds. Details of the interlocking devices and methods of application vary considerably by manufacturer. Always study and follow the manufacturer's directions when installing all types of shingles.

14.9 Low-Slope Roofs

Roofs with slopes as low as 2-in-12 can be made watertight and wind resistant with asphalt shingles. However, for slopes less than 4-in-12, certain additional procedures should be followed. First, use two layers of felt underlayment. Lap each course of felt over the preceding one by 19″. In areas where the daily average temperature in January is 25°F (–5°C) or colder, cement the two felt layers together from the eaves up the roof to 24″ inside of the interior wall line of the building. As an alternative, an ice-and-water barrier can be used. **Figure 14-27** demonstrates how the barrier should be installed.

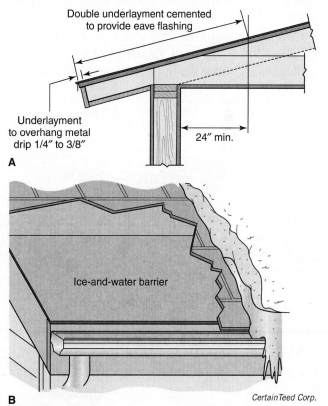

CertainTeed Corp.

Figure 14-27. Pay special attention to underlayment materials for low-slope roofs. A—Two plies are cemented together for a watertight eave flashing. B—This eave flashing is a single layer of a thick, polymer-modified asphalt reinforced with a fiberglass mat. This membrane must extend at least 24″ upward from the side walls.

Shingles provided with factory-applied adhesive and manufactured to conform to the Underwriters Laboratories *Standard for Class "C" Wind-Resistant Shingles* should then be installed. See **Figure 14-28** for special application methods.

14.10 Roll Roofing

Asphalt roll roofing is manufactured in a variety of weights, surfaces, and colors. Rolled roofing has a much shorter lifespan that strip shingles, typically 5–10 years. Because of its shorter lifespan, it should not be used as a flashing material, especially in open-valley applications.

For best results, roll roofing should be installed at temperatures of 45°F (10°C) or above. It can be used on slopes as low as 1″ rise per foot. The 36″ width roofing includes a granular-surfaced area that is 17″ wide and a smooth surface, called a *selvage*, that is 19″ wide.

Although double-coverage roll roofing can be applied parallel to the rake, it is usually applied parallel to the eaves, as shown in **Figure 14-29**. The starter strip can be made by cutting off the granular-surfaced portion. Use two rows of nails to install the starter strip—one 4 3/4″ below the upper edge and the other 1″ above the lower edge.

Coat the entire starter strip with asphalt cement and overlay a full-width sheet. Attach the sheet with a row of nails 4 3/4″ from the upper edge and a second row

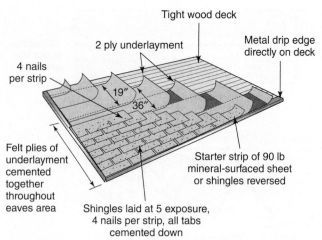

Goodheart-Willcox Publisher

Figure 14-28. Use special application methods for shingling low-slope roofs. Double plies of asphalt-saturated felt cemented together. Ice-and-water barrier can be used as eave flashing.

8 1/2″ below the first row. The nail interval should be about 12″.

Position each succeeding course so that it overlaps the full 19″ selvage area. Nail the sheet in place at the upper edge and then carefully turn the sheet back to apply the cement coating. Spread the cement to within about 1/4″ of the granular surface. Firmly press the overlaying sheet into the cement using a stiff broom or roller. Avoid excessive use of cement. Be sure to follow the manufacturer's recommendations.

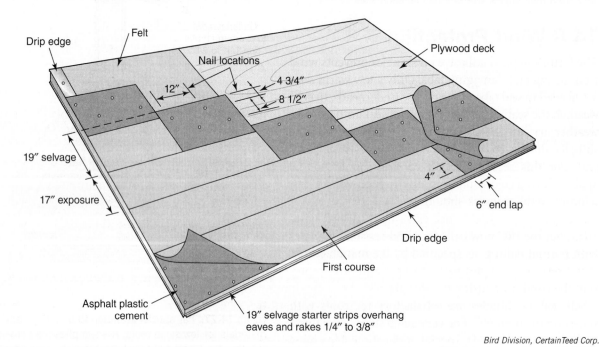

Bird Division, CertainTeed Corp.

Figure 14-29. Double-coverage roll roofing applied parallel to the eaves. Ice-and-water barrier can be used as underlayment along the eaves.

14.11 Reroofing

When reroofing, choose between removing the old roofing or leaving it in place. It is usually not necessary to remove wood shingles, asphalt shingles, or roll roofing before putting on a new asphalt roof provided these conditions are met:

- The strength of the existing deck and framing is adequate to support the weight of workers and the additional roofing, snow loads, and wind loads.
- The existing deck is sound and will provide good anchorage for the nails used in applying new roofing.
- The additional layer of shingles is allowed by the governing building code.
- The requirements are satisfied to meet the manufacturer's warranty of the newly applied shingles .

When putting on new roofing over old wood shingles, remove all loose or protruding nails. Renail the shingles into new nail locations. Renail loose, warped, and split shingles. Replace missing shingles. At the eaves and rakes, cut back the shingles far enough to allow the application of 4″–6″ wide, nominal 1″ thick strips. These strips should be nailed in place with their outside edges projecting beyond the roof deck the same distance as the old wood shingles.

When the old roof consists of square-butt asphalt shingles with a 5″ exposure, new self-sealing strip shingles can be applied as shown in **Figure 14-30**. This application pattern ensures a smooth, even appearance. In addition, it establishes a new nailing pattern about 2″ below the old one.

The joint between a vertical wall and roof surface should be sealed when reroofing. First, apply a strip of smooth roll roofing about 8″ wide. Firmly nail each edge, spacing the nails about 4″ OC. As the shingles are applied, spread asphalt cement on the strip. Thoroughly bed the shingles. To ensure a tight joint, use a caulking gun to apply a final bead of cement between the edges of the shingles and the siding.

Sometimes the old shingles are to be removed before applying a new roof. A roofing shovel or standard shovel is used. Work from the bottom up, sliding the shovel under the shingles and prying upward. Clear away all old materials down to the sheathing. Check the surface for any protrusions, such as old fasteners, that might puncture the new coverings. Inspect the sheathing and replace any areas that are damaged. Sweep the roof to remove any remaining debris. Then proceed as you would for installing a new roof.

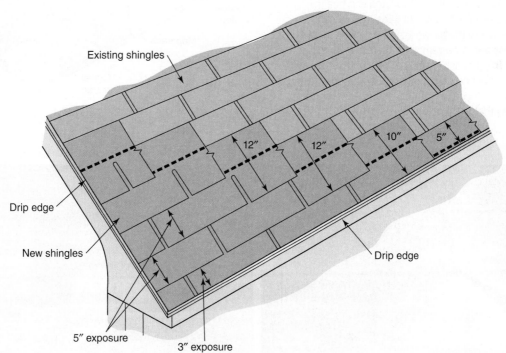

Asphalt Roofing Manufacturer's Assn.

Figure 14-30. Reroofing over old asphalt shingles. Lay down a starter strip the same width as the exposure. Then trim the first course to butt against the third course of old shingles. Butt full-width strip shingles against the next old course. If using three-tab shingles, offset cutouts so they do not fall over cutouts in the old shingles.

PROCEDURE

Installing Asphalt Shingles over Asphalt Shingles

1. For the starter course, remove the tabs from the new shingles and also remove 2″ from the tops. The remaining portion should now be equal to the exposure of the old shingles.

2. Cut off 3″ from the rake end of the first starter strip. This ensures that all cutouts of the old shingles will be covered.

3. Overhang the starter course at the eaves and rake edges about 1/4″. Nail it in place.

4. Cut 2″ from the butts for the first course. All other courses are full depth. Align the cut edge with the butts of the old shingles.

5. Start by removing 6″ from the width on the starting strip. Let the top edge butt against the butts of the old shingles in the next course. While this will reduce the exposure of the preceding course, the guttering will conceal the difference.

6. Increase the cutoff of the starting strip by an additional 6″ through the sixth course. Start the seventh with a full-width shingle.

Safety Note

Cement asbestos shingles were installed on many buildings through the 1960s. Asbestos is a health hazard. Removal of asbestos shingles may release asbestos fibers into the air where they can be inhaled. If these shingles are to be removed, the job must be done by crews certified in asbestos abatement. Improper asbestos removal endangers workers and may result in significant fines. If you are uncertain of asbestos content, there are labs that will test samples for you.

14.12 Built-Up Roofing

A flat roof or a roof with very little slope must be covered with a watertight system. Many flat roofs are covered with built-up roofing. Such a roof is very durable. Companies that manufacture the components provide detailed specifications and instructions for making the installation. The components are mainly saturated felt and asphalt.

On a clean wood deck, first lay down a heavy layer of saturated felt. Nail this down with galvanized nails, **Figure 14-31**. Nails must have a large head or be driven through tin caps. Mop each succeeding layer in place with hot asphalt. Built-up roofs for residential structures normally have three or four *plies* (layers of asphalt-saturated felts). When the felts are all in place, they are coated with hot asphalt and covered with slag, gravel, crushed stone, or marble chips. See **Figure 14-32**. These materials provide a weathering surface and improve the roof's appearance. Three to four hundred pounds of the mineral covering are used on a 100 sq ft section of roof.

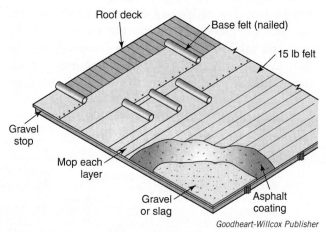

Goodheart-Willcox Publisher

Figure 14-31. Constructing a built-up roof.

A   *mikeledray/Shutterstock.com*

B   *mikeledray/Shutterstock.com*

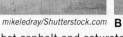

Figure 14-32. Built-up roofing. A—After hot asphalt and saturated felt layers are laid down, more hot asphalt is spread over the felt underlayment to bind the gravel. The gravel stop is applied after the base felt is laid. B—Applying washed gravel, 400 lb per square ft section of roof.

The asphalt used between each layer and to bed the surface coating is a petroleum product. It begins to flow, very slowly, at a fairly low temperature. This results in a self-healing property that is essential for flat roofs, since water is likely to stand on the roof. A special low-temperature asphalt known as dead-flat asphalt is used for flat roofs. For sloping roofs, steep asphalt (an asphalt with a high melting point) is used. In hot climates, use only steep asphalt.

A *gravel stop* is attached to the roof deck to serve as a trim member. This is a type of drip edge usually fabricated from galvanized sheet metal. It helps keep the mineral surface and asphalt in place, **Figure 14-33**. Gravel stops are installed after the base felt has been laid. Joints between sections of gravel stop are bedded in a special mastic that permits expansion and contraction in the metal.

Flashings around chimneys and vents or where the roof joins a wall must be constructed with special care. Leaks are most likely to occur at these locations. The best flashing materials include lead jackets, sheet copper, and special flashing cement.

Basic flashing construction is shown in **Figure 14-34**. The *cant strip* provides support for the felt layers as they curve from a horizontal to a vertical attitude.

Bare spots on a built-up roof should be repaired. First, clean the area. Then, apply a heavy coating of hot asphalt and spread more gravel or slag.

Cut away felts that have fallen apart and replace them with new felts. The new felts should be mopped in place, allowing at least one additional layer of felt to extend not less than 15″ beyond the other layers.

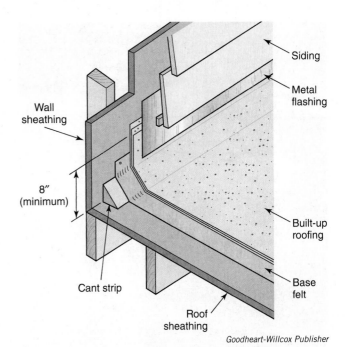

Goodheart-Willcox Publisher

Figure 14-34. Basic flashing construction details where a flat roof meets a wall. The cant strip is a triangular wooden strip that provides a gently curving base for layers of roofing felt or rubberized roof membrane.

14.13 Rubber Roofing

Many tire manufacturers recycle unused tires to make rubber roofing. The first rubber roofing system was developed by Firestone in 1980. In addition to its durability, rubber roofing can be installed quicker and easier than most other roofing systems. Rubber roofing is available in great widths, usually eliminating the need for seams. This makes it a good choice for flat roofs.

Before applying rubber roofing, it is important that the plywood or oriented strand board (OSB) roof deck be swept clean and all nails driven flush with the surface. Most manufacturer warranties will not cover their product if the roof deck is not clean. With the roof stripped and clean, measure and cut for any openings, such as chimneys and vent pipes. Allow at least 12″ of membrane to extend up any walls that adjoin the roof. Lay the roofing sheet in place and check for any necessary trimming. Fold half the sheet back over itself, exposing the sheathing under half the sheet, **Figure 14-35**. Apply the manufacturer's specified adhesive for the roofing system you are using. Most adhesives require some drying time before the membrane is put in place. Next, roll out the sheet of roofing a little at a time. Check for air bubbles as you unroll the membrane into the adhesive. If air bubbles appear, work them toward the edge before you unroll more roofing. Trim the roofing as necessary.

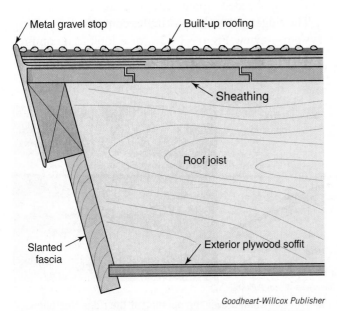

Goodheart-Willcox Publisher

Figure 14-33. This section view through the edge of a flat roof overhang shows the metal gravel stop installation.

Johns Manville

Figure 14-35. The rubber roofing membrane is folded back so adhesive can be applied.

Johns Manville

Figure 14-36. These carpenters specialize in roofing. They are applying 20' rolls of rubber roofing.

With the roof covered, it is time to flash any penetrations. Use rubber flashing intended for use with the brand of roofing you are using. Clean the area around the penetration with primer made for that purpose. Apply rubber flashing cement according to the printed instructions. Finally, wrap the flashing around the vertical object (pipe, vent, etc.) and fold it down onto the primed surface of the roofing. The flashing cannot be moved or adjusted once it touches the flashing cement. **Figure 14-36** shows a carpenter installing a rubber roof.

14.14 Ridge Vents for Asphalt Roofing

Moisture buildup and high temperatures can be problems in the attic space during certain seasons. Venting at the eaves, gable ends, and at or near the ridge helps dissipate heat and moisture. **Figure 14-37** shows two types of ridge vent. One type is rolled over the ridge opening. This opening is a gap of about 2″ or 3″ left in the sheathing at the ridge to allow air to pass through. Shingles are secured over the vent to keep out weather. The second type is rigid plastic. It is fastened over the vent opening with large-headed asphalt shingle nails.

The ridge vent and cap shingles complete the roofing system. Review **Figure 14-38** for a look at a comprehensive asphalt roofing system.

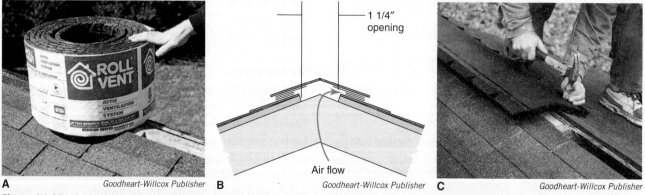

A   *Goodheart-Willcox Publisher*   B   *Goodheart-Willcox Publisher*   C   *Goodheart-Willcox Publisher*

Figure 14-37. A ventilated ridge provides attic ventilation. A—A roll vent is installed over the opening at the ridge that allows free passage of air. B—When a ventilated ridge cap is used, an opening is left in the sheathing and all other roofing is cut away to keep the opening clear. C—This rigid plastic vent is attached to the ridge with roofing nails.

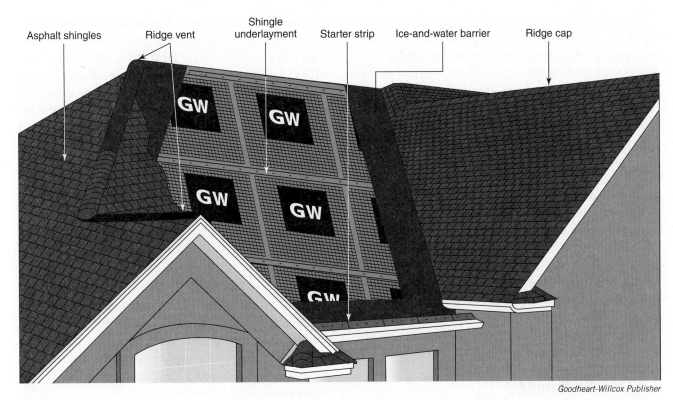

Asphalt shingles   Ridge vent   Shingle underlayment   Starter strip   Ice-and-water barrier   Ridge cap

GW GW GW GW GW

Goodheart-Willcox Publisher

Figure 14-38. Each component of an asphalt roof plays an important role in creating a weather-tight roof system that will last for decades.

14.15 Wood Shingles

Wood shingles are a traditional material used in residential construction, **Figure 14-39**. They are available with or without a polymer fire-retardant treatment.

Building codes often prohibit untreated wood roofing materials. Since wood weathers to a mellow color after exposure, wood shingles provide an appearance that is desired by many homeowners. When properly installed, they also provide a very durable roof.

Brady Adams

Figure 14-39. Wood shingles are both beautiful and functional when installed correctly. The most common usage is on rustic and historical buildings.

Wood shingles are made from western red cedar, redwood, or cypress. All of these woods are highly resistant to decay. The shingles are taper sawed and graded No. 1, No. 2, No. 3, and No. 4 (a utility grade). The best grade is cut in such a way that the annular rings are perpendicular to the surface. Butt ends vary in thickness, as shown in **Figure 14-40**. Wood shingles are manufactured in random widths and in lengths of 16″, 18″, and 24″. They are packaged in bundles. Four bundles contain enough shingles to cover 100 sq ft of roof using a standard application.

The exposure of wood shingles depends on the slope of the roof. When the slope is 5-in-12 or greater, standard exposures of 5″, 5 1/2″, and 7 1/2″ are used for 16″, 18″, and 24″ sizes, respectively. On roofs with lower slopes, the exposure should be reduced to 3 3/4″, 4 1/4″, and 5 3/4″, respectively. This provides a minimum of four layers of shingles over the entire roof area. In any type of construction, there should be a minimum of three layers at any given point to ensure complete protection against heavy, wind-driven rain.

14.15.1 Sheathing

Solid sheathing for wood shingles may consist of matched or unmatched 1″ boards, but usually it is plywood or OSB. Open, skip, or spaced sheathing is sometimes used because it costs less and permits

shingles to dry out quickly, **Figure 14-41A**. One reason for using solid sheathing is to gain the added insulation and resistance to air infiltration that such a deck offers. If applying wood shingles over solid sheathing, it is advisable to use an underlayment that allows the shingles to breathe. One such product on the market is known as Cedar Breather®. Made up in rolls of stiff fiber, it maintains a thin airspace between the decking and the shingles. The breather material allows drying of shingles after a rainfall. Refer to **Figure 14-41B**.

One method of applying skip roof sheathing is to space 1×3, 1×4, or 1×6 boards the same distance apart as the anticipated shingle exposure. Each course of shingles is nailed to a separate board.

14.15.2 Underlayment

Normally, an underlayment is not used for wood shingles, except when applied over solid sheathing. If roofing felt is used to prevent air infiltration, rosin-sized building paper or dry unsaturated felts are suitable. Saturated paper is usually not recommended. It may cause condensation problems.

To prepare a solid roof deck for application of wood shingles, first install 30 lb roofing felt. Allow 1/4″ overhang at the eaves. Allow 4″ of overlap as you work toward the ridge.

Grades and Specifications of Wood Shingles					
Grade	Length	Thickness (at Butt)	No. of Courses per Bundle	Bdls/Cartons per Square	Description
No. 1 Blue label	16″ (Fivex)	0.40″	20/20	4 bdls	The premium grade of shingles for roofs and sidewalls. These top-grade shingles are 100% heartwood. 100% clear and 100% edge-grain.
	18″ (Perfections)	0.45″	18/18	4 bdls	
	24″ (Royals)	0.50″	13/14	4 bdls	
No. 2 Red label	16″ (Fivex)	0.40″	20/20	4 bdls	A good grade for many applications. Not less than 10″ clear on 16″ shingles. 11″ clear on 18″ shingles and 16″ clear on 24″ shingles. Flat grain and limited sapwood are permitted in this grade.
	18″ (Perfections)	0.45″	18/18	4 bdls	
	24″ (Royals)	0.50″	13/14	4 bdls	
No. 3 Black label	16″ (Fivex)	0.40″	20/20	4 bdls	A utility grade for economy applications and secondary buildings. Not less than 6″ clear on 16″ and 18″ shingles, 10″ clear on 24″ shingles.
	18″ (Perfections)	0.45″	18/18	4 bdls	
	24″ (Royals)	0.50″	13/14	4 bdls	
No. 4 Undercoursing	16″ (Fivex)	0.40″	14/14 or	2 bdls	A utility grade for undercoursing on double-coursed sidewall applications or for interior accent walls.
			20/20	2 bdls	
	18″ (Perfections)	0.45″	14/14 or	2 bdls	
			18/18	2 bdls	
No. 1 or No. 2 Rebutted-rejointed	16″ (Fivex)	0.40″	33/33	1 carton	Same specifications as above for No. 1 and No. 2 grades but machine trimmed for parallel edges with butts sawn at right angles. For sidewall application where tightly fitting joints are desired. Also available with smooth sanded face.
	18″ (Perfections)	0.45″	28/28	1 carton	
	24″ (Royals)	0.50″	13/14	4 bdls	

Cedar Shake and Shingle Bureau

Figure 14-40. Wood shingles are made in several grades and to certain specifications for various applications.

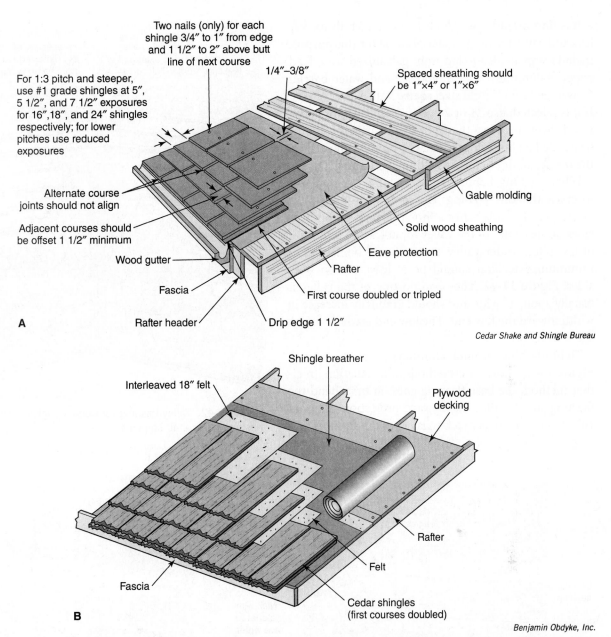

For 1:3 pitch and steeper, use #1 grade shingles at 5″, 5 1/2″, and 7 1/2″ exposures for 16″,18″, and 24″ shingles respectively; for lower pitches use reduced exposures

Two nails (only) for each shingle 3/4″ to 1″ from edge and 1 1/2″ to 2″ above butt line of next course

1/4″–3/8″

Spaced sheathing should be 1″×4″ or 1″×6″

Alternate course joints should not align

Adjacent courses should be offset 1 1/2″ minimum

Wood gutter

Fascia

Rafter header

Drip edge 1 1/2″

First course doubled or tripled

Rafter

Eave protection

Solid wood sheathing

Gable molding

A

Cedar Shake and Shingle Bureau

Interleaved 18″ felt

Shingle breather

Plywood decking

Fascia

Cedar shingles (first courses doubled)

Felt

Rafter

B

Benjamin Obdyke, Inc.

Figure 14-41. General application details for wood shingles. A—Install double or triple layers at eaves and allow the butts of the first course to project beyond the fascia by 1 1/2″. Note the use of both tight and spaced solid wood decking. B—To allow shingles to dry over solid wood decking, a layer of shingle breather is first installed. It provides an airspace between the deck and the shingles.

Next, tack on the breather, placing a nail or tack every three sq ft. Butt each course against the previous course.

Begin applying shingles after the first course of breather. This is to avoid walking on the breather. Follow the manufacturer's directions when applying shingles. Use a nail length that gives a 3/4″ penetration into the deck.

14.15.3 Fire Resistance

Recent tests have shown that flame-spread and burn-through rates for wood shingles and shakes can be reduced. This is achieved by pressure-treating the shingles with fire retardants. Flame-penetration time can also be increased by using 1/2″, Type X gypsum board under solid or spaced sheathing. For more information, see Chapter 9 of the IRC.

14.15.4 Flashing

In areas where outside temperatures drop to 0°F (–15°C) or colder, there is a possibility of ice forming along the eaves. An *eaves flashing strip* or ice-and-water barrier is required. The installation procedure is identical to that used for asphalt shingles.

It is important to use good-quality materials for valleys and eaves flashing. Materials used for this purpose include tinplate, lead-clad iron, galvanized iron, lead, copper, aluminum sheets, and ice-and-water barrier. Galvanized iron is mild steel coated with a layer of zinc. If it is selected, use 24 or 26 gauge metal. Tin or galvanized sheets with less than 2 oz of zinc per sq ft should be painted on both sides with a rustproof primer. Allow the primer to dry before installing the flashing.

When making bends in the flashing, use care not to crack the zinc coating. On roofs of 6-in-12 pitch or steeper, the valley sheets should extend up on both sides of the center of the valley for at least 7″. On roofs of less pitch, wider valley sheets should be used. The minimum extension should be at least 10″ on both sides, **Figure 14-42**. The open portion of the valley is usually about 4″ wide and should gradually increase in width toward the low end. The low end is where drainage is heaviest.

Tight flashing around chimneys is also essential. **Figure 14-43** shows two methods of installation. In either method, the base flashing goes on first. Bend the flashing to 90°, allowing upward projection of about 10″. Allow at least as much to lay over the sheathing.

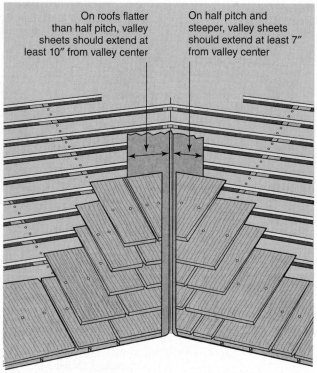

On roofs flatter than half pitch, valley sheets should extend at least 10″ from valley center

On half pitch and steeper, valley sheets should extend at least 7″ from valley center

Cedar Shake and Shingle Bureau

Figure 14-42. Valley flashing for wood shingles is similar to flashing for asphalt shingles.

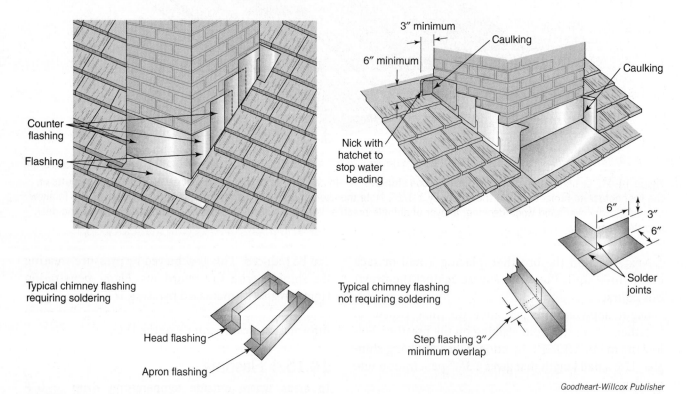

Counter flashing

Flashing

Typical chimney flashing requiring soldering

Head flashing

Apron flashing

3″ minimum

6″ minimum

Caulking

Caulking

Nick with hatchet to stop water beading

Typical chimney flashing not requiring soldering

Step flashing 3″ minimum overlap

6″

3″

6″

Solder joints

Goodheart-Willcox Publisher

Figure 14-43. Two methods of flashing around a brick chimney. In either method, cap flashing should be mortared into the joints as the chimney is being built. For an existing chimney, it is necessary to carefully remove old mortar to install the flashing.

14.15.5 Nails

Only rust-resistant nails should be used with wood shingles. Hot-dipped, zinc-coated nails have the strength of steel and the corrosion resistance of zinc. These types of nails are recommended. **Figure 14-44** shows sizes of nails for various jobs.

Most carpenters prefer to use a shingler's or lather's hatchet to lay wood shingles, **Figure 14-45**. This tool has a blade for splitting and trimming. Most have a gauge for spacing the weather exposure.

14.15.6 Applying Shingles

The first course of shingles at the eaves should be doubled or tripled. Horizontally space all shingles 1/4″–3/8″ apart. Because wood shingles absorb moisture, this spacing allows expansion when they become rain soaked.

Use only two nails to attach each shingle. The proper placing of these two nails is of considerable importance. The nails should be near the butt line of the shingles in the next course. Under no circumstances should nails be driven below this line. This would expose the nails to the weather. Driving the nails 1″–1 1/2″ above

Tru Value Hardware, Ashland, WI

Figure 14-45. A shingler's hatchet is especially suited for laying wood shingles and shakes. Most have adjustable gauges for measuring weather exposure.

the butt line is good practice. Two inches above is an allowable maximum. Place nails not more than 3/4″ from the edge of the shingle at each side. When so nailed, the shingles lie flat and give good service.

The second layer of shingles in the first course should be nailed over the first layer so the joints in each course are at least 1 1/2″ apart. A good shingler uses care in breaking the joints in successive courses so they do not line up in three successive courses. Joints in adjacent courses should be at least 1 1/2″ apart.

Rust-Resistant Nails					
For New Roof Construction			**Over-Roofing Construction**		**Double Coursing**
3d	3d	4d	5d	6d	5d
For 16″ and 18″ shingles		For 24″ shingles	For 16″ and 18″ shingles	For 24″ shingles	For all shingles
1 1/4″ long	1 1/4″ long 14# 1/2 gauge	1 1/2″ long #14 gauge	1 3/4″ long #14 gauge	2″ long #13 gauge	1 3/4″ long #14 gauge
Approx. 376 nails to lb	Approx. 515 nails to lb	Approx. 382 nails to lb	Approx. 310 nails to lb	Approx. 220 nails to lb	Approx. 380 nails to lb

Goodheart-Willcox Publisher

Figure 14-44. Use only rust-resistant nails of the size recommended for your particular application.

For shingles containing both flat and vertical grain, joints should not be aligned with the centerline of the heart grain. Split flat grain shingles in two before nailing. Treat knots and other defects at the edge of the shingle.

It is good practice to use a board as a straightedge to line up courses of shingles, **Figure 14-46**. Temporarily tack the board in place to hold the shingles until they are nailed. Two shinglers often work together. One distributes and lays the shingles along the straightedge while another nails them in place. As shingling progresses, check the alignment every five or six courses with a chalk line. Measure down from the ridge occasionally to be sure shingle courses are parallel to the ridge.

On a roof section where one end terminates at a valley, carefully cut valley shingles to the proper angle at the butts. Use wide shingles. Nail the shingles in place along the valley first.

Dripping from gables can be prevented by installing a wedge. Use a piece of 6″ bevel siding along the edge and parallel to the end rafter. Place the wedge under the shingles as shown in **Figure 14-47**.

> **Pro Tip**
> Use care when nailing wood shingles. The wood is soft. It can be easily crushed and damaged under the nail heads. Drive the nail just flush with the surface.

14.15.7 Shingled Hips and Ridges

Carefully and tightly cover ridges and hips to avoid roof leaks. In the best type of hip construction, nails should not be exposed to the weather, **Figure 14-48**. Select shingles of approximately the same width as the roof exposure. Snap lines onto the shingled roof, one on each side of the ridge. Mark the lines the correct distance back from the centerline of the ridge. These lines indicate the edges of the cap shingles. On small houses, hip caps may be made narrower.

Factory-assembled hip and ridge units are available. Weather exposure should be the same as that used for the regular shingles. Be sure to use longer nails that will penetrate well into the sheathing.

14.15.8 Reroofing with Wood Shingles

Wood shingles may be applied to old as well as new roofs, **Figure 14-49**. If the old wood-shingle roofing is in reasonably good shape, it need not be removed. In reroofing houses covered with composition material, whether in the form of roll roofing or asphalt shingles, it is usually best to strip off the old material. Otherwise, moisture may condense on the roof deck below. Decay of sheathing could follow.

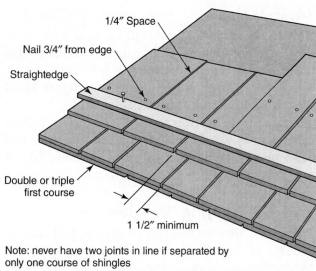

Note: never have two joints in line if separated by only one course of shingles

Goodheart-Willcox Publisher

Figure 14-46. Use a wooden straightedge as a guide in laying wood shingles. It keeps the courses straight and maintains proper shingle exposure.

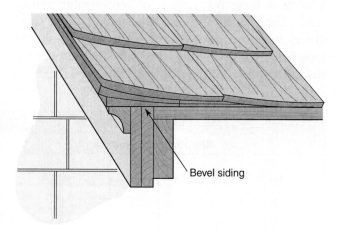

Goodheart-Willcox Publisher

Figure 14-47. Bevel siding installed along the rake tilts shingles inward to prevent water from dripping off the edge.

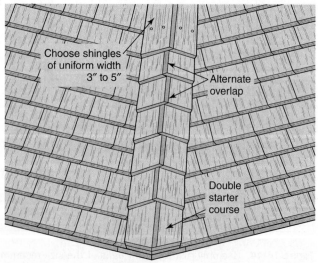

Goodheart-Willcox Publisher

Figure 14-48. Installing wood shingles at hips and ridges.

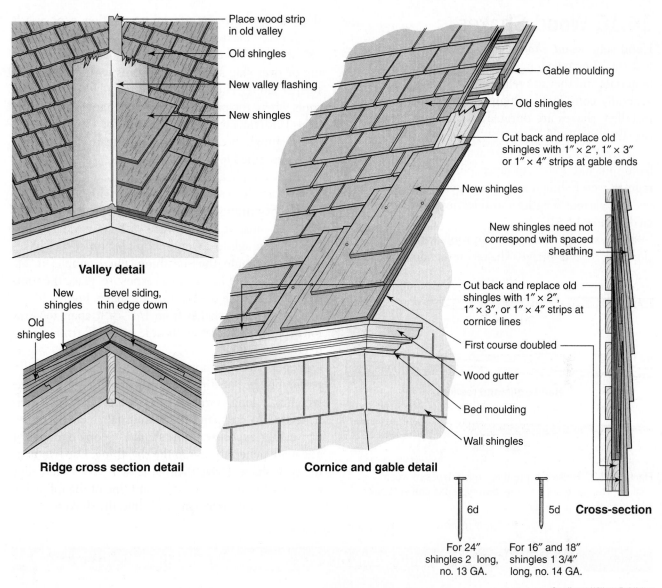

Valley detail

Ridge cross section detail

Cornice and gable detail

Place wood strip in old valley

Old shingles

New valley flashing

New shingles

New shingles
Bevel siding, thin edge down
Old shingles

Gable moulding

Old shingles

Cut back and replace old shingles with 1″ × 2″, 1″ × 3″ or 1″ × 4″ strips at gable ends

New shingles

New shingles need not correspond with spaced sheathing

Cut back and replace old shingles with 1″ × 2″, 1″ × 3″, or 1″ × 4″ strips at cornice lines

First course doubled

Wood gutter

Bed moulding

Wall shingles

Cross-section

6d

5d

For 24″ shingles 2 long, no. 13 GA.

For 16″ and 18″ shingles 1 3/4″ long, no. 14 GA.

Goodheart-Willcox Publisher

Figure 14-49. Detail drawings for reroofing with wood shingles. Note how valleys and ridges are built up.

Before applying new shingles, renail or replace all warped, split, and decayed shingles. To finish the edges of the roof, cut off the exposed portion of the first two rows of old shingles along the eaves using a sharp hatchet or portable circular saw. Nail a 1″ wood strip in this space. Place the outer edge flush with the eave line. Prepare shingle edges along the gable ends in a similar manner.

Raise the level of the valleys by applying wood strips. Install new flashings over the strips. Remove old hip and ridge caps to provide a more even base for new shingles. As new shingles are applied, space them 1/4″ apart to allow for their expansion in wet weather. Let them project 1/2″–3/4″ beyond the edge of the eaves.

Now, follow the same procedure described for new roofs. However, longer nails are needed. For 16″ and

18″ shingles, rust-resistant or zinc-clad 5d box nails or special 1 3/4″ long, 14 gauge over-roofing nails should be used. A 6d, 13 gauge rust-resistant nail is used for 24″ shingles.

Usually, no particular attention needs to be given to how the nails penetrate the old roof. It does not matter whether they strike the sheathing strips or not because complete penetration is obtained through the old shingles with the longer nails. An adequate number of nails needed to anchor all of the shingles of the new roof will strike sheathing or nailing strips.

Place new flashings around chimneys, but do not remove the old flashing. Liberally use high-grade, non-drying mastics to get a watertight seal between the brick and metal.

14.16 Wood Shakes

Hand-split *wood shakes* provide a pleasing surface texture. They take longer to install than most roofing materials and they are relatively expensive, so they are generally considered a high-end product. If properly installed, shakes are durable and may last as long as the structure itself.

Generally, wood shakes are available as straight split, hand split and resawn, or taper split, **Figure 14-50**. Like regular wood shingles, they are available in random widths. Various lengths and thicknesses are standardized, **Figure 14-51**.

Shakes should not be used on roofs with insufficient slope for good drainage. The recommended minimum slope is 4-in-12. The maximum weather exposure is 13″ for 32″ shakes, 10″ for 24″ shakes, and 8 1/2″ for 18″ shakes. Shakes can be applied over solid sheathing or over skip sheathing on roofs with a 4-in-12 or steeper slope. They can also be installed on mansard roofs where the pitch is not greater than 20° from vertical. Underlayment is the same as for wood shingles. Use a shingle breather over solid sheathing, as previously described for wood shingles.

Start the application by laying down an ice-and-water barrier or a 36″ strip of 30 lb felt along the eaves. Double the beginning course of shakes. A narrow strip of 30 lb felt must be laid between each course. This must be wide enough to cover the top portion of the shakes and extend onto the sheathing. For example, if 24″ shakes are being laid at a 10″ exposure, place the roofing felt 20″ above the butts of the shake, **Figure 14-52**. Space individual shakes from 1/4″–3/8″ apart to allow for expansion. Offset joints at least 1 1/2″ from course to course.

Proper nailing is important. Use rust-resistant nails, preferably the hot-dipped zinc-coated type. The 6d size, which is 2″ long, normally is adequate. Use longer nails if needed to compensate for unusual shake thickness or weather exposure. Nails should be long enough for adequate penetration into the sheathing. Use two nails for each shake. Drive them at least 1″ from each edge and about 1″ or 2″ above the butt line of the following course. Do not drive nail heads into the shakes.

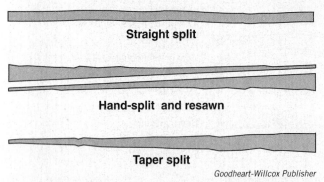

Straight split

Hand-split and resawn

Taper split

Goodheart-Willcox Publisher

Figure 14-50. These are the three basic types of wood shakes. The hand-split and resawn type has one side sawn flat.

Grades and Specifications of Four Styles of Wood Shakes

| Grade | Length and Thickness | 18″ Pack** | | Description |
		Courses per bdl	Bdl per sq	
No. 1 Hand-split & resawn	15″ Starter-finish	9/9	5	These shakes have split faces and sawn backs. Cedar logs are first cut into desired lengths. Blanks or boards of proper thickness are split and then run diagonally through a bandsaw to produce two tapered shakes from each blank.
	18″ × 1/2″ Mediums	9/9	5	
	18″ × 3/4″ Heavies	9/9	5	
	24″ × 3/8″	9/9	5	
	24″ × 1/2″ Mediums	9/9	5	
	24″ × 3/4″ Heavies	9/9	5	
No. 1 Taper-sawn	24″ × 5/8″	9/9	5	These shakes are sawn both sides.
	18″ × 5/8″	9/9	5	
No. Taper-split	24″ × 1/2″	9/9	5	Produced largely by hand using a sharp-bladed steel froe and wooden mallet. The natural shingle-like taper is achieved by reversing the block, end-for-end, with each split.
No. 1 Straight split	18″ × 3/8″ True Edge*	20″ Pack		Produced in the same manner as taper-split shakes except that by splitting from the same end of the block, the shakes acquire the same thickness throughout.
		14 Straight	4	
	18″ × 3/8″	19 Straight	5	
	24″ × 3/8″	16 Straight	5	

Note: *Exclusively sidewall product, with parallel edges.
**Pack used for majority of shakes.

Cedar Shake and Shingle Bureau

Figure 14-51. Wood shakes are manufactured in four styles with various specifications and sizes.

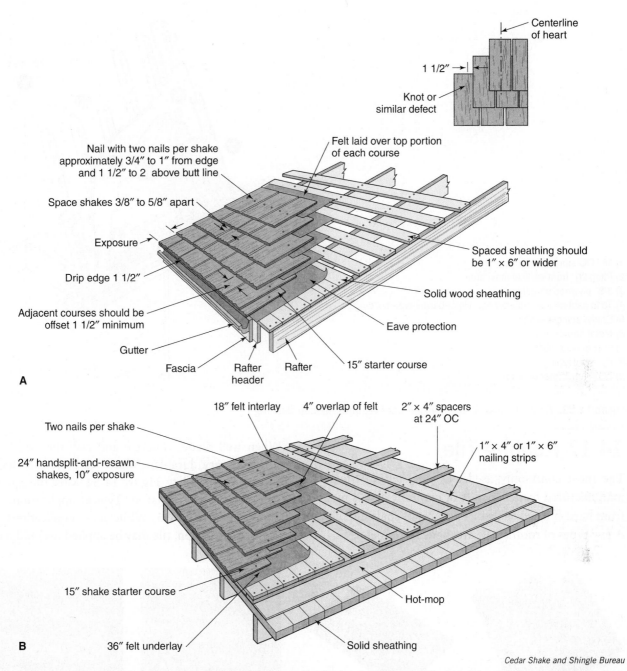

Nail with two nails per shake approximately 3/4″ to 1″ from edge and 1 1/2″ to 2 above butt line

Space shakes 3/8″ to 5/8″ apart

Exposure

Drip edge 1 1/2″

Adjacent courses should be offset 1 1/2″ minimum

Gutter

Fascia

Rafter header

Rafter

Centerline of heart

1 1/2″

Knot or similar defect

Felt laid over top portion of each course

Spaced sheathing should be 1″ × 6″ or wider

Solid wood sheathing

Eave protection

15″ starter course

A

Two nails per shake

24″ handsplit-and-resawn shakes, 10″ exposure

15″ shake starter course

36″ felt underlay

18″ felt interlay

4″ overlap of felt

2″ × 4″ spacers at 24″ OC

1″ × 4″ or 1″ × 6″ nailing strips

Hot-mop

Solid sheathing

B

Cedar Shake and Shingle Bureau

Figure 14-52. Shake application. A—Protect eaves with ice-and-water barrier. Lay an 18″-wide strip of 30 lb asphalt-saturated felt between each course. Straight-split shakes should be laid with the froe-end (end from which the shake has been split) toward the ridge. B—Recommended method for applying shakes to low-slope roofs. The lattice framework is embedded in a bituminous surface coating.

Valleys are laid as recommended for regular wood shingles. Underlay all valleys with 30 lb roofing felt or ice-and-water barrier. Metal valley flashing must be at least 20″ wide.

For the final course at the ridge line, try to select shakes of uniform size of about 6″. Trim off the ends so they meet evenly. Carefully apply a strip of 30 lb felt lengthwise along all ridges and hips. Nail them in place following the procedure described for regular wood shingles. Prefabricated hip-and-ridge units

are available. Their use will save time and provide uniformity.

Chimneys or other structures that project through the roof must be flashed and counterflashed on all sides. Flashing should extend at least 6″ under the shakes. Flashing should be applied at the top and bottom of the roof, at the inside and outside corners, and around any openings. Inside and outside corners of wood, vinyl, or aluminum are required. This material must be properly flashed and caulked. See **Figure 14-53.**

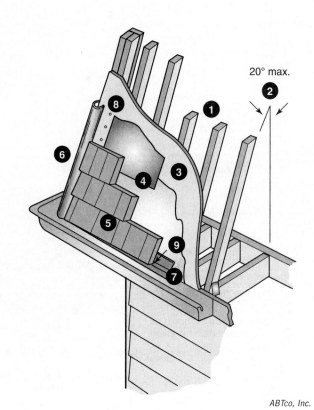

1) 16″ OC stud spacing
2) Pitch 20° from vertical maximum
3) 3/8″ plywood sheathing
4) 15 lb asphalt impregnated non-vapor-barrier-type building paper
5) Shake shingle siding
6) Metal corner post
7) 6″ drip edge flashing
8) Corner flashing
9) 3/8″×1 1/2″ starter strip

Figure 14-53. Applying shakes to a mansard roof requires special flashing.

14.17 Roofing Tile

The most commonly used types of *roofing tile* are manufactured from concrete or clay. The clay is made from hard-burned shale or mixtures of shale and clay. A few types of roofing tile are made from metal.

When well made, concrete and clay tiles are hard, dense, and very durable. Colors, textures, and shapes come in a great variety, **Figure 14-54.** Most tile is made of concrete; some is glazed. Typical applications are shown in **Figure 14-55.** While most applications are on new construction, tile may be applied over old roofs

Figure 14-54. Roof tiles come in curved, barrel, and flat shapes and in many colors. Special tile shapes are manufactured for ridges, hips, and rakes.

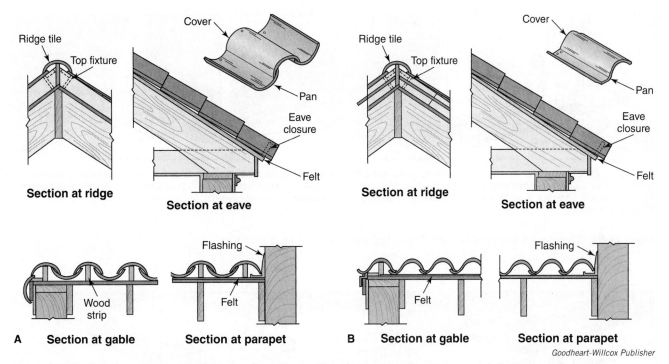

Figure 14-55. Typical application methods for tile roofs. A—Two-piece pan and cover commonly known as mission tile. B—Spanish tile.

provided the old covering is in reasonably good condition, and the roof framing and sheathing is strong enough to support the added weight. Typically, tile weighs from 5.8–10.25 lb per sq ft.

On new construction, roof trusses are engineered to support the tile. Additional roof framing or bracing may be required on existing construction. Sheathing should be at least a nominal 1″ thick and bridge no more than 24″.

14.17.1 Installing Tile Roofing Units

Requirements for underlayment depend on the roof pitch and local climate. Below a 3-in-12 pitch, the tile roofing is considered decorative. Therefore, a minimum of two plies of 15 lb felt, hot-mopped between layers, is recommended. This is topped with vertical battens spaced 24″ apart from the eave to the ridge that are also mopped with asphalt. Then, horizontal battens are laid and fastened at intersections with the vertical battens. *Battens* are strips of wood installed on the roof to hold the tiles in place. Normally, 1×2 battens are used over solid sheathing, spaced 24″ apart, **Figure 14-56**. The first row of battens should be placed so the tiles overhang the eaves by 1 1/2″.

Prior to laying tile, valleys must be prepared. **Figure 14-57** shows proper installation of flashing and underlayment. Then, tile roofing is laid from right to left, beginning at the right rake, **Figure 14-58**.

Monier Group

Figure 14-56. Tiles are attached to battens spaced according to the size of the tiles.

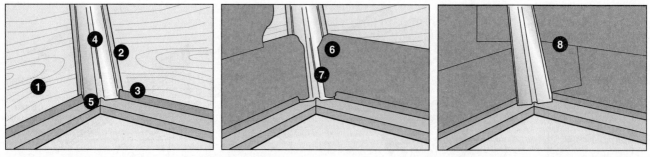

1) Decking. 2) 36" sweat sheet under valley flashing is recommended. 3) Cut off top of eave riser strip to permit valley drainage. 4) Standard galvanized iron (G.I.) valley flashing with crimped edges. 5) Extend valley flashing beyond eave riser strip. 6) Cut top corner of underlayment to ensure proper diversion of water into valley flashing. 7) Overlap valley flashing with underlayment. 8) Optional weaved underlayment treatment of valley.

Lifetile

Figure 14-57. Flashing a valley for a tile roof. Preformed standard flashing is available.

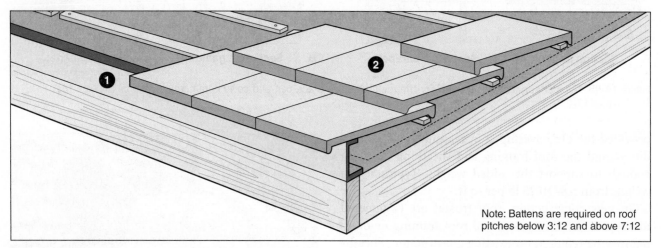

Note: Battens are required on roof pitches below 3:12 and above 7:12

1) Eave riser strip. 2) Shingle-lap vertical courses.

A

Profile Tiles

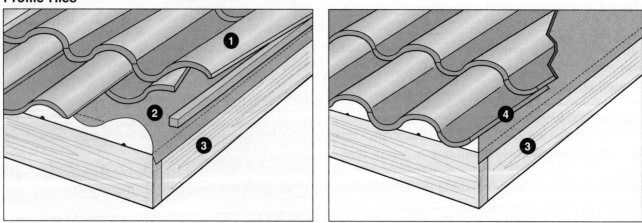

Right rake 2″×2″ wood strip starter   **Right rake pan section starter**

1) When full tile is used, edge of tile must not extend beyond center of nominal 2" x 2" starter. 2) Recess nominal 2" x 2" wood starter 4" to 6" up from eave line and 3/4" in from edge of barge rafter. 3) Barge rafter. 4) Start half-tiles at least 1" from edge of barge rafter.

B

Lifetile

Figure 14-58. Starting to lay tile roofing. A—Flat tiles have joints staggered between courses like shingles. B—Profiled (curved) tiles have joints aligned from the eave to the ridge.

Lugs located on the underside of each tile hook over the battens. Flat tiles are usually laid with each course overlapping the previous one. **Figure 14-59** shows the proper method of laying tile up to the valley. A circular saw fitted with a masonry cutting wheel is used to make angle cuts on tile.

Pro Tip
Be careful while moving about on the roof. Step only on the lower one-third of the tiles on the overlapped area.

14.17.2 Hips, Ridges, and Rakes

Nailer boards are wood strips installed on edge at hips and ridges to support trim tiles. The height of the ridge board is 2″–6″. The board must be high enough to maintain an even plane of trim tile. Trim tiles are attached with one corrosion-resistant 10d nail. Nose ends should be set in a bead of roofer's mastic that also covers the nail head. See **Figure 14-60**.

Fasten rake tile with two nails. The joints between the field tile and trim tile should be weatherproofed with a bed of mortar or an approved dry ridge-hip system.

Tile is extremely heavy. Use a mechanical method—conveyor or lift truck—to deliver it to the roof. This should be done after the deck has been prepared with flashing, underlayment, and battens. Be sure to include flashing around chimneys, pipes, and vents. **Figure 14-61** shows proper installation of vent flashing for tile roofing.

Severe weather conditions and taller structures require extra construction steps when installing roofing tile.

Where wind velocities may exceed 80 mph or where the roof is more than 40′ above the ground, observe the following guidelines. Be sure to consult building codes in these cases.

- Nail the head of every tile.
- Fasten the noses of eave courses with special clips.
- Secure rake tiles with two nails.
- Set noses of ridge, hip, and rake tiles in a bead of appropriate, approved roofer's mastic.
- Tiles cut too small for nailing should be set in an approved mastic or secured to the roof with wire.

PROCEDURE
Preparing a Tile Valley

1. Lay down a 36″-wide *sweat sheet*, dividing the width across the valley. A sweat sheet is a strip of felt or ice-and-water barrier.
2. Install the flashing. Valley flashing should be 26 gauge corrosion-resistant metal extending at least 11″ each way. In the center of the flashing there should be a diverter rib not less than 1″ high. Ends should be lapped 6″.
3. Cut off the top of the eave riser strip and extend the metal valley flashing slightly over the eave and upward the length of the valley. Edges along the length of the flashing should be turned up 1/2″ by 30°. Standard flashing usually comes preformed, ready for installation.
4. Put down underlayment, lapping it over the flashing. Another method involves putting down the underlayment and weaving it across the valley before installing the flashing.

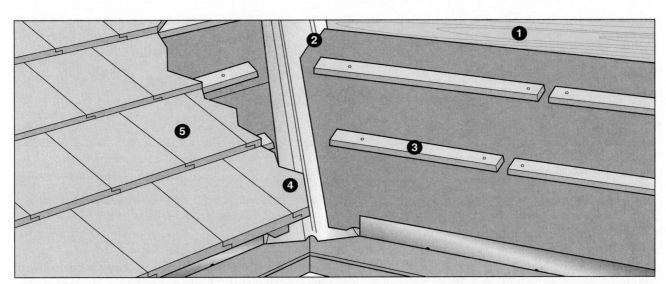

1) Decking. 2) Cut top corner of underlayment to ensure proper diversion of water into valley flashing. 3) Battens are optional on roof pitches between 3:12 and 7:12. 4) Remove lugs under tiles that rest on valley flashing. 5) Shingle-lap tile courses on flat tile installation.

Lifetile

Figure 14-59. Proper tile installation at a valley. Lugs must be removed when the top of the tile rests on the flashing.

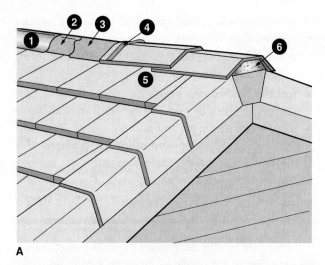

1) Ridge nailer to be of sufficient height to maintain even plane of ridge tiles. 2) Underlayment carried over or under ridge nailer. 3) Optional second layer of felt over nailer board as weatherblock. 4) Apply continuous bead of approved roofers' mastic at overlapping areas and over nail holes. 5) Provide minimum 3" headlap. 6) Optional mortar fill end treatment.

Note: Use one 10-penny nail per ridge tile.

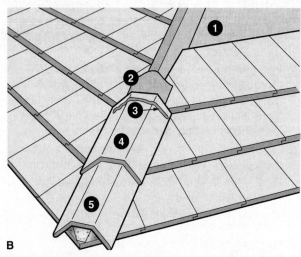

1) Underlayment carried over or under hip nailer. 2) Optional second layer of felt over nailer board as weatherblock. 3) Apply continuous bead of approved roofers' mastic at overlapping areas and over nail holes. 4) Provide minimum 3" headlap. 5) Hold back hip nailer 6" from eave edge.

Lifetile

Figure 14-60. Details of ridge and hip trim tile installation. A—Typical ridge design. B—Typical trim tile installation at a hip.

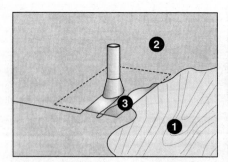 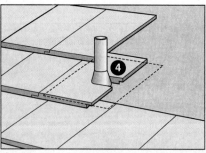 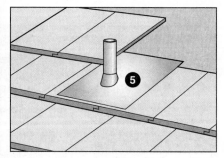

1) Decking. 2) Underlayment. 3) Standard galvanized iron (G.I.) base flashing. 4) Notch tile to accept flashing. 5) Standard galvanized iron (G.I.) top flashing. Seal with approved roofers' mastic.

Lifetile

Figure 14-61. Vent flashing for tile roofing is similar to the method used for asphalt or wood shingles. With some high-profile tiles, lead or other flexible flashing is used.

14.18 Metal Roofing

Structures in areas where heavy snow occurs may be given a roof covering of metal sheeting. This is because snow tends to slide off of the metal roof before building to a deep accumulation. Sheets heavily coated with zinc galvanizing (2 oz per sq ft) or aluminum-zinc alloy are approved for permanent structures. Sheets with lighter coatings of zinc are less durable and are likely to require painting every few years.

14.18.1 Zinc-Aluminum Coated Steel Roofing

Some steel roofing has a zinc-aluminum coating. A typical zinc-aluminum coating is 55% aluminum, 43% zinc, and 2% silicon. This coating is applied by a hot-dip process. Next, a chromate treatment adds corrosion resistance. This is followed by a primer and top coatings of polyesters, silicone, fluorocarbons, or plastisols. The top coats are available in a large selection of colors.

Before any coatings are applied, the flat, 26 gauge or 29 gauge sheets are passed through forming rolls to add ribs to the sheets. These ribs stiffen and strengthen the sheets. See **Figure 14-62**.

Apply zinc-aluminum-coated roofing sheets over solid decking. The recommended underlayment is 30 lb asphalt-saturated roofing felt. Other suitable barrier materials may be substituted. An ice-and-water barrier should be applied at the eaves in cold climates.

If it is being used for reroofing, remove the old roof covering. If this is not practical, hot mop a layer of underlayment and then put down 2×2 vertical battens at regular intervals. Mop more asphalt over this construction. If additional roof insulation is needed, sheet insulation can be applied between the battens. Cross battens are laid horizontally across the roof and roofing sheets are fastened to the cross battens. Use only approved, self-penetrating, self-tapping screws. Do not overdrive screws. This can cause panel distortion. In some panel designs, screws are concealed by attaching the next panel.

14.18.2 Galvanized, Corrugated Roofing

Sheets are generally 24 or 26 gauge. The heavier gauge has no particular advantage besides its added strength. A zinc coating for durability is more important than strength for this type of roofing. On temporary buildings and in cases where the most economical construction is required, lighter metal can be used.

14.18.3 Slope and Laps

Metal roofing sheets may be laid on slopes as low as 3-in-12 (1/8 pitch). If more than one sheet is required to reach the ridge, ends should lap no less than 8″. When the roof pitch is 1/4 or more, 4″ of end lap is satisfactory.

To make a tight roof, lap sides of corrugated sheets by 1 1/2 corrugations, **Figure 14-63**. If only a single-corrugation lap joint is used, wind-driven rain will likely be forced through the joint. When using roofing 27 1/2″ wide with 2 1/2″ corrugations and 1 1/2 corrugation laps, each sheet covers a net width of 24″ on the roof.

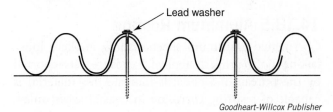

Goodheart-Willcox Publisher

Figure 14-63. Applying corrugated sheet metal roofing. Sheets properly laid with 1 1/2 corrugation laps.

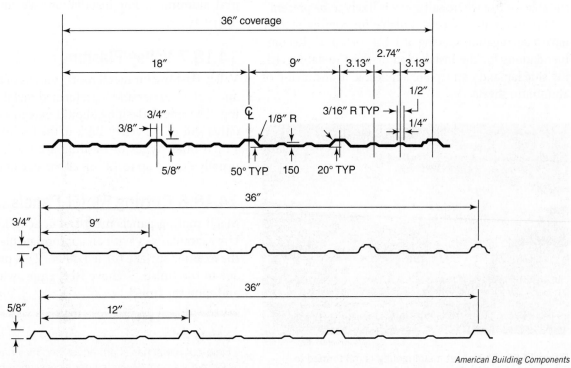

American Building Components

Figure 14-62. The ribs in metal roofing help stiffen the sheets and provide for leak-proof overlapping.

Sheet metal roofing is also available in a form that looks like clay tile. See **Figure 14-64**. Computer-controlled roll-forming machines produce a continual stepped-panel with tile forms about 7″ wide and 12″ long. Standard sheets are 36″ wide and measure from 2′–20′ long. The product is available in a variety of colors.

14.18.4 Fasteners

Corrugated or ribbed sheet metal roofing should be supported every 24″, but always consult and follow the manufacturer's installation instructions. Most carpenters use self-drilling screws with attached neoprene washers to fasten the sheets. Screws should be driven through the tops of the corrugations or ribs. Typically, No. 10 × 1″ hex washer-head screws are installed spaced 18″–24″ OC. In areas of high winds, closer spacing may be required. Consult local codes. Use a driving tool with a variable speed of 200–2500 rpm and a depth-sensing nosepiece.

Various accessories are available for closures and fittings. Their application is shown in **Figure 14-65**.

14.18.5 Aluminum Roofing

Corrugated aluminum sheets usually make a long-lasting roof, if properly applied. Exposure tests reported by the Bureau of Standards indicate this material is capable of resisting corrosion in coastal areas unless subjected to direct contact with saltwater spray. Where this is likely to happen, aluminum roofing is not recommended. Asphalt, tile, or rubber roofing would be suitable for use where salt spray is likely to be present.

Aluminum alloy sheets available for roofing usually have a corrugation spacing of 1 1/4″ or 2 1/2″. Recommendations for the installation of sheet metal regarding side lap and end lap are applicable to the laying of aluminum sheets.

Mr Twister/Shutterstock.com

Figure 14-64. This sheet metal roofing is roll formed to resemble mission tile. It is available in a variety of colors.

Aluminum is soft and the sheets used for roofing are relatively thin. Thus, they should be laid on tight sheathing or on decks with openings no more than 6″ wide. Aluminum roofing should be nailed with no less than 90 nails to a square, or about one nail for each square foot. Use aluminum alloy nails and place nonmetallic washers between nail heads and the roofing.

If desired, the sheathing may be covered with water-resistant building paper or asphalt-impregnated felt. Paper that absorbs and holds water should never be used.

To avoid corrosion, aluminum sheets should be stored so that air will have free access to all sides. Otherwise, a white deposit will form. This deposit can quickly create pinholes.

14.18.6 Aluminum Shakes

Aluminum shakes are manufactured of an aluminum-magnesium alloy with a nominal .019″ thickness. They are available in brown, red, dark gray, white, and natural aluminum. For installation, see manufacturer's instructions.

14.18.7 Valley Flashing

Valley flashing for metal roofs consists of a layer of ice-and-water barrier and a preformed metal valley flashing. The metal flashing should extend at least 9″ on either side of the valley. Ends of the flashing are lapped 12″. If other types of flashing are used, they should usually extend up to 20″ on either side of the valley.

14.18.8 Cutting Metal Panels

Metal roofing panels must be cut at valleys and hips. Use a circular saw with an appropriate blade. Cut with the exterior surface turned down. This prevents damage to the finish. Remove steel chips, which can rust and spoil the finish.

1) Ridge/Hip

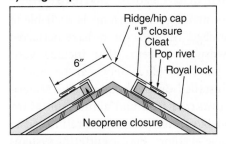

2) "W" Formed Valley

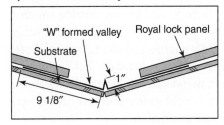

3) Gable Trim

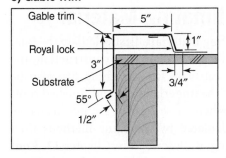

4) Peak Cap

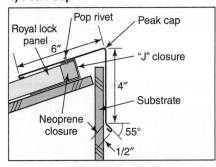

5) Endwall Flash

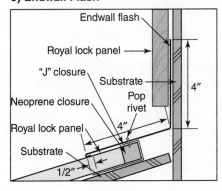

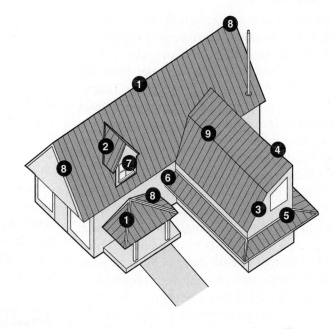

6) Sidewall Flash

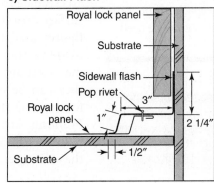

8a/b) Start/Finish Flash

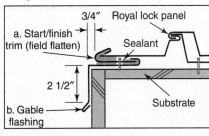

8a/c) Rake Wall

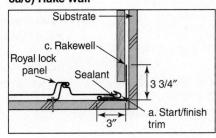

7) Drip Edge

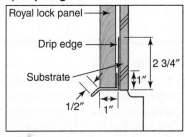

9) Transition

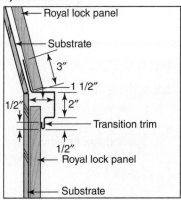

American Building Components

Figure 14-65. Various accessories are available for proper installation of metal roofing.

14.19 Gutters

The term *gutter* refers to a separate unit that is attached to the eave. The term *eaves trough* usually applies to a waterway built into the roof surface over the cornice. Gutters or eaves troughs collect rainwater from the edge of the roof and carry it to downspouts. Downspouts are vertical pipes that direct water away from the foundation or into a drainage system. For proper drainage, a gutter should slope 1/4″ for every 4′ in the direction of the downspout.

For best results, gutters must be sized to suit the roof areas from which they receive water. For roof areas up to 750 sq ft, a 4″-wide trough is suitable. For areas between 750 and 1400 sq ft, 5″ troughs should be used. For larger areas, a 6″ trough is recommended. Quality of gutters, like that of flashing, should correspond to the durability of the roof covering. If galvanized steel guttering is used, it should have a heavy zinc coating.

The size of downspouts or conductor pipes also depends on the roof area. For roofs up to 1000 sq ft, downspouts of 3″ diameter have sufficient capacity, if properly spaced. For larger roofs, 4″ downspouts should be used.

Pro Tip

An eaves trough must be carefully designed and built since any leakage will penetrate the structure. Because of the added time required and the extra cost of construction, they are seldom used in new construction or refurbishing work. Separate metal gutters are usually used in place of eaves

14.19.1 Metal and Plastic Gutters

A wide variety of metal gutter systems is available to control roof drainage. Manufacturers have perfected gutter and downspout systems that include various component parts. Whole systems can be quickly assembled and installed on the building site. Materials consist of galvanized iron and aluminum. Many systems are available either primed or prefinished to match a wide range of colors. Plastic guttering systems are also used extensively.

Gutter systems include inside and outside mitered corners, joint connectors, pipes, brackets, and other items. All are carefully engineered and fabricated. Parts easily slip together and are generally held with soft pop-rivets or sheet metal screws. **Figure 14-66** shows standard parts of a typical gutter and downspout system and how they are assembled.

14.20 Estimating Material

To estimate roofing materials, first calculate the total surface area to be covered. In new construction, the figures used to estimate the sheathing can also be used to estimate the underlayment and finished roofing materials. When these figures are not available, they can be calculated by the same methods used for roof sheathing, as described in Chapter 12, **Roof Framing**. Another method used to estimate roof area is to determine the total footprint (ground area) of the structure. Include all eave and cornice overhang.

PROCEDURE

Installing Gutters

1. At the high end of the eaves, attach a chalk line 3/4″ below the shingles.
2. Establish the proper slope at the other end and snap a line. The upper edge of the gutter will be located on this line.
3. Attach an end cap to the end of the gutter. The design of this assembly varies, depending on the gutter material and the manufacturer. In some cases, mastic is used where parts join. In others, gaskets are provided.
4. Begin attaching the gutter at the end opposite the downspout. Fasten the gutter at 2′ or 3′ intervals using the fastening devices provided. In some cases,

spike and ferrule fasteners are provided. In others, brackets or hangers are used.
5. Carefully check that the assembly is maintaining the proper slope.
6. If the gutter must wrap around corners, install miters, making sure that joints are made watertight.
7. Attach downspout elbows and downspouts. Each section has a large end and a crimped end. These are installed with the large end up so that debris does not catch on the crimped end.
8. Attach downspouts to the wall using downspout bands. Use two on each section.
9. If leaf guards are specified, install them.

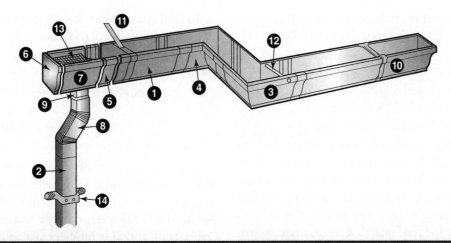

Components of a Metal Gutter System			
Key	**Description**	**Key**	**Description**
1	5″ K gutter	9	K Outlet tube (with flange)
2	3″ Square corrugated downspout	10	5″ K Fascia hanger
3	5″ Miter (outside)	11	5″ K Strap hanger
4	5″ Miter (inside)	12	7″ Spike (aluminum) 5″ Ferrule (aluminum)
5	5″ K Slip joint connector	13	5″ K Strainer
6	5″ K End cap left or right	14	3″ Pipe band (ornamental)
7	5″×3″ K End section with outlet tube		Touch-up paint
8	3″ Square corrugated 75° elbow or 60° elbow style a and b		Gutter seal (tube or cartridge)

Crown Aluminum Industries

Figure 14-66. A metal gutter system. The parts can be assembled on the jobsite.

Convert the ground area to roof area by adding a percentage determined by the roof slope, as follows:

- Slope of 3-in-12, add 3% of the area
- Slope of 4-in-12, add 5 1/2% of the area
- Slope of 5-in-12, add 8 1/2% of the area
- Slope of 6-in-12, add 12% of the area
- Slope of 8-in-12, add 20% of the area

There is a simple method for determining the roof pitch when it is not known. You can estimate it from the ground with the help of a folding carpenter's rule. Stand some distance away from the building and fold the rule into a triangle. Hold the folded rule at arm's length and frame the roof inside of the triangle. Adjust the triangle until the slope of its sides lines up with the roof, as in **Figure 14-67**. Be sure the base of the triangle is level. Read off the dimensions on the base of the rule that is marked in **Figure 14-67** as the "reading point." Then, refer to the chart in **Figure 14-68** to locate the proper pitch and slope.

To find the number of squares to be covered, divide the total square feet of roof surface by 100. For example, if the total ground area, including overhang, is found to be 1560 and the slope of the roof is 4-in-12:

$$Roof\ Area = 1560 + (1560 \times 5\ 1/2\%)$$
$$= 1560 + (1560 \times 0.055)$$
$$= 1560 + 85.80\ (or\ 86) = 1646$$
$$Number\ of\ squares = 16.46\ or\ 16\ 1/2$$

After the number of squares is established, additional amounts must be added. For asphalt shingles, it is generally recommended that 10% be added for waste. This, however, may be too much for a plain gable roof and too little for a complex intersecting roof. Certain allowances must also be added for reduced exposure on low-sloping roofs. Usually, the 10% waste figure can be reduced if allowance is made for hips, valleys, and other extras. For asphalt or wood shingles, one square is usually added for each 100 lineal feet of hips and valleys.

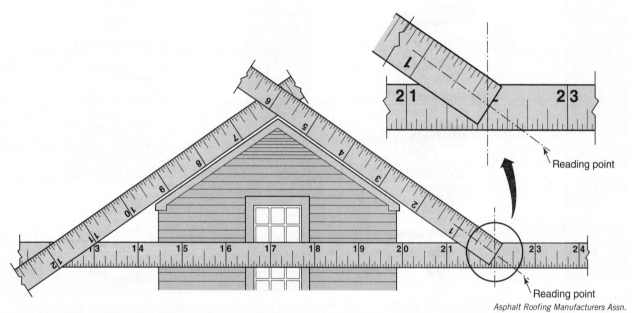

Asphalt Roofing Manufacturers Assn.

Figure 14-67. Determining the roof slope. Stand a sufficient distance from the gable end and frame the roof inside of a folded carpenter's rule held at arm's length. Adjust the end of the rule to get a proper reading at the reading point.

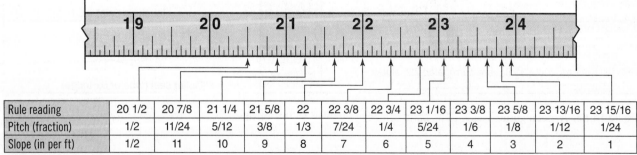

Rule reading	20 1/2	20 7/8	21 1/4	21 5/8	22	22 3/8	22 3/4	23 1/16	23 3/8	23 5/8	23 13/16	23 15/16
Pitch (fraction)	1/2	11/24	5/12	3/8	1/3	7/24	1/4	5/24	1/6	1/8	1/12	1/24
Slope (in per ft)	1/2	11	10	9	8	7	6	5	4	3	2	1

Asphalt Roofing Manufacturers Assn.

Figure 14-68. Reading point conversions. Locate the reading point on the chart and read downward to find the pitch and slope.

Quantities of starter strips, eaves flashing, valley flashing, and ridge shingles must be added to the total shingle requirements. All of these are figured on lineal measurements of the eaves, ridge, hips, and valleys. For a complex structure, a plan view of the roof is helpful in adding together these materials. All ridges and eave lines are seen as true length and can simply be measured with a scale to find their length. Hips and valleys are not seen as true length and a small amount must be added. Using the percentage listed for converting ground area to roof area usually provides sufficient accuracy.

Safety Note

Worker safety is important everywhere on a construction site. Working on a pitched roof can be made safer by observing these precautions:

- Wear boots or shoes with rubber or crepe soles. Soles and heels must be in good condition. If the soles are worn, discard the shoes or have them repaired.
- Workers must wear safety harnesses with lines tied off to a solid anchoring point.
- Rain, frost, and snow make a roof slippery. Wait until the surface is dry before working.
- Keep a broom or brush handy to sweep the roof clear of sawdust, loose debris, and dirt.
- Install shingle underlayment as soon as possible. Such material reduces the danger of slipping.
- Install temporary 2×4 cleats as toe holds. They can be removed as shingles are installed.
- Remove unused tools, cords, and other loose items from the roof; they can be serious hazards.

In addition to these precautions, check local, state, and OSHA requirements. Be alert to other potential hazards and practice common sense. Taking chances often leads to injuries.

Construction Careers
Roofer

All types of buildings—residential, commercial, or industrial—must have a weather-tight roof or the buildings will rapidly become unusable. The roofer is the tradesperson who installs roofing material on new buildings and repairs or replaces roofs on existing structures.

The type of roofing installed on a building depends on the roof structure. Flat or slightly sloped roofs are primarily used on industrial, commercial, and multiple unit residential (apartment) buildings. These roofs have either a built-up roofing system or a rubber roof. Built-up roofing consists of several layers of felt impregnated with a waterproofing material laid down and sealed with a molten, tar-like material called *bitumen*. The combination provides a waterproof and quite durable covering for the roof. Usually, bitumen and gravel are applied as a top layer to provide wear resistance. A newer technology used for flat roofs is a thin sheet of rubber or thermoplastic material that is rolled out onto the roof surface in long strips. The seams between adjoining strips are sealed so the entire roof is covered with a single waterproof layer. Various fastening and covering methods are used to hold the sheet in place.

Residential buildings typically have pitched (sloping) roofs and are covered with shingles or tiles. The most common form of shingles is made of asphalt-impregnated fiberglass felt covered with a layer of fine stone granules for durability. Slate, rot-resistant wood, metal, tile, and other materials are also used for shingles. The shingles are laid in overlapping rows, starting at the lowest edge of the roof and proceeding up the ridge or peak, to form a weatherproof covering.

Two out of three roofers are employed by contractors who perform both types of roofing work, either as new construction or repair/replacement. The remaining one-third own small roofing companies, typically specializing in residential installations.

The work is hot, dirty, and strenuous. Roofers must constantly be aware of their surroundings because of the danger of injuries from hot bitumen or falls. Roofers must be able to do heavy lifting and work under adverse and extreme weather conditions.

People often enter the field as helpers to experienced roofers, gradually acquiring skills on the job. Some participate in a three-year apprenticeship program that requires 144 hours of classroom time and 2000 hours of on-the-job experience each year.

Huntstock/Thinkstock

Roofers must be prepared to work in all types of weather. Roofers must possess knowledge of different materials used for various building structures, whether it is installing a new roof or repairing an existing roof. Because of the inherent danger from working so high off the ground, roofers must take great caution to protect against falling.

Chapter Review

Summary

- The roofing materials applied to a structure protect it from sun, wind, rain, and snow. They should also provide some degree of fire protection and add to the attractiveness of the building.
- The most common roof surfacing material is asphalt shingles, but mineral fiber shingles, sheet materials, wood shingles and shakes, and clay and concrete tiles are also used.
- Sheet materials include roll roofing, rubberized membranes, and metals such as galvanized iron, steel, aluminum, tin, and copper.
- In cold climates, an ice-and-water barrier is applied along roof edges and valleys before the surfacing material is put in place.
- Rubber roofing eliminates the need for seams, which makes it a good choice for flat roofs.
- Spaced or skip sheathing may be used for wood shingles and is typical for tile or slate roofs.
- Valley flashing occurs at the junction of two sloping roofs.
- Shingle application begins at the eaves and proceeds upward, course by course, to the ridge.
- Many flat roofs are covered with built-up roofing, which is durable and watertight.
- Since all roofing work is done at a height, safety practices must be carefully observed to prevent injury.

Know and Understand

Answer the following questions using the information in this chapter.

1. *True or False?* Low-slope roofs require a more watertight system than steep roofs.

2. The _____ is the unit of measure for estimating and purchasing shingles.
   - A. square
   - B. slope
   - C. lineal foot
   - D. pitch

3. In shingle application, the distance between the bottom edge of one course and the bottom edge of the next higher course is called _____.
   - A. coverage
   - B. exposure
   - C. head lap
   - D. side lap

4. Boards over _____″ should not be used as sheathing for roof decks.
   - A. 1
   - B. 2
   - C. 6
   - D. 12

5. The most commonly used roofing material for houses are _____.
   - A. saturated felts
   - B. roll shingles
   - C. flat metal sheets
   - D. asphalt shingles

6. Which of the following materials should *not* be used when valley flashing under asphalt shingles?
   - A. Aluminum.
   - B. Copper.
   - C. Saturated felt.
   - D. Galvanized sheet metal.

7. A valley 10′ long would be _____″ wide at the eave.
   - A. 1
   - B. 9
   - C. 15
   - D. 105

8. *True or False?* Only architectural shingles should be used in woven valley flashing.

9. Where the sloping part of a roof abuts a vertical wall, _____ flashing should always be used.
   - A. step
   - B. open valley
   - C. woven valley
   - D. closed-cut valley

10. What purpose does a neoprene boot serve when installed on vent stack flashing?
    - A. It protects wooden materials from termites.
    - B. It seals out moisture.
    - C. It adds strength to the flashing material.
    - D. All of the above.

11. For laminated architectural shingles, use a minimum of _____ nails per strip.
    - A. two
    - B. four
    - C. six
    - D. eight

12. For laminated shingles without tab cutouts, it is common to cut approximately _____″ to _____″ from the outer end of the third course.
    - A. 3, 4
    - B. 5, 7
    - C. 11, 13
    - D. 17, 19

13. *True or False?* Asphalt roofing should not be applied during cold winter months.

14. *True or False?* Roofs with slops as low as 1-in-12 can be made watertight and wind resistant with asphalt shingles.

15. Because of its shorter lifespan, _____ should not be used as flashing material with shingles, especially in open-valley applications.
    A. rolled roofing
    B. flat metal sheet
    C. asphalt
    D. metal ridge rolls

16. To remove old shingles, use a _____.
    A. utility knife
    B. straightedge
    C. roofing shovel
    D. All of the above.

17. *True or False?* In hot climates, use only dead-flat asphalt.

18. What is attached to a built-up roof deck and serves as a trim member?
    A. Asphalt.
    B. A gravel stop.
    C. A cant strip.
    D. An extra plie.

19. In the best grade of wood shingles, the annular rings run _____ to the surface.
    A. perpendicular
    B. parallel
    C. steep
    D. flat

20. In any type of construction, there should be a minimum of _____ layers at any given point to ensure complete protection against heavy, wind-driven rain.
    A. two
    B. three
    C. four
    D. six

21. The installation of an eaves flashing strip is identical to that used for _____.
    A. wood shingles
    B. wood sheathing
    C. wood shakes
    D. asphalt shingles

22. A narrow strip of 30 lb roofing felt is placed between each course when applying _____.
    A. asphalt shingles
    B. wood shingles
    C. wood shakes
    D. roofing tile

23. *True or False?* Most roofing tile is made of metal.

24. When preparing a tile valley, use a(n) _____ to divide the width across the valley.
    A. special clip
    B. batten
    C. eaves flashing strip
    D. sweat sheet

25. *True or False?* When installing corrugated metal roofing, the joints should be lapped one corrugation.

26. *True or False?* Aluminum roofing is not recommended for use in coastal areas because it can corrode when subjected to saltwater spray.

27. The vertical pipes of a gutter system are called _____.
    A. eaves troughs
    B. downspouts
    C. joint connectors
    D. brackets

28. *True or False?* To determine the number of squares of asphalt shingles needed for a roof, divide the total square feet of roof area by 100.

Apply and Analyze

1. Explain three reasons why an underlayment is required for a roof.
2. Describe the process of applying flashing to a valley.
3. Name the two parts in flashing that move independently of each other.
4. Why are chimney saddles generally required on large chimneys?
5. How are three-tab asphalt shingles laid out on a long roof?
6. What is the six-inch method?
7. What conditions must be met to put new asphalt roofing over existing roofing?
8. What is the purpose of a ridge vent?
9. What is the main advantage of a metal roof in cold climates?
10. What is the difference between a gutter and an eaves trough?

Critical Thinking

1. The IRC allows roof recovering with some exceptions. One of the exceptions is when a roof already has two or more layers of roofing. What are the possible ramifications of adding three or more layers of roofing to a home? How could this impact the structure over time?
2. List the strengths and weaknesses of each type of roofing for a home in your climate zone:
    A. Asphalt shingles
    B. Wood shingles
    C. Slate
    D. Clay tile
    E. Metal roofing
3. If you are stripping and reroofing a house, what are the advantages of having shingles delivered with a hoist that lifts them to the roof? What are the disadvantages?

Communicating about Carpentry

1. **Speaking.** Working in small groups, create flash cards for the key terms in this chapter. Each student chooses some terms and makes flash cards for those terms. On the front of the card, write the term. On the back of the card, write the pronunciation and a brief definition. Use your textbook and a dictionary for guidance. Then take turns quizzing one another on the pronunciations and definitions of the key terms.

2. **Speaking and Listening.** Divide into groups of four or five students. Each group should choose one of the following types of roofing: asphalt, wood, tile, and aluminum roofing. Using your textbook as a starting point, research your topic and prepare a report on the similarities and differences. Include the situations that may call for one to be used instead of the others, determine materials needed and their cost, and the application procedures. As a group, deliver your presentation to the rest of the class. Take notes while other students give their reports. Ask questions about any details that you would like clarified.

CHAPTER 15

Building Envelope and Control Layers

LEARNING OBJECTIVES

After studying this chapter, you will be able to:

- Describe the principles of environmental separation.
- List the four control layers in a structure.
- Give examples of materials commonly used as a water-resistive barrier (WRB).
- List the factors that affect the dew point.
- Describe what happens when air reaches its dew point.
- Explain vapor drive.
- Demonstrate how vapor drive changes from heating months to cooling months.
- Describe how exterior ridged insulation board can reduce condensation in a wall system.
- Explain the importance of proper installation for control layer components.
- Explain the challenges associated with high-performance building envelopes.

TECHNICAL TERMS

building envelope
building science
condensation
control layer
dew point
diffusion
expanded polystyrene (EPS)
  foam

extruded polystyrene (XPS)
  foam
hydrophobic
mineral wool
perm rating
polyiso
pressure differential
rainscreen

rigid insulation board
R-value
vapor barrier
vapor drive
vapor retarder
water vapor
water-restive barrier (WRB)
wind-washing

Control layers work together to create the ***building envelope***, the separation between the outside environment and the inside environment of an enclosure. Siding, roofing, flashing, insulation, and housewrap are just a few of the components that can make up these layers, **Figure 15-1**. Building science helps us understand the function of control layers. ***Building science*** is the application of knowledge of the physical world to building processes and materials.

The failure to realize the importance of control layers can lead to faulty design or installation. Serious structural issues for a building, or health issues for its occupants, can result, **Figure 15-2**.

Carpenters must recognize their role in the proper installation of these systems. Some materials can act as a control layer for more than one element.

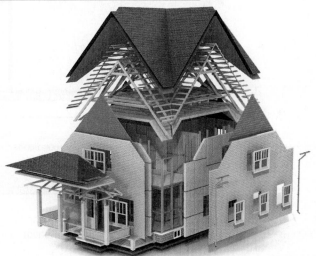

studiovin/Shutterstock.com

Figure 15-1. A building envelope is comprised of all the different components separating the interior of a home from the outside environment.

Copyright Goodheart-Willcox Co., Inc.

373

Cynthia Farmer/Shutterstock.com

Figure 15-2. A faulty window installation can quickly lead to serious damage on a building.

It is important that we first understand the functions of control layers to avoid installation mistakes. Studying building science helps us recognize how environmental changes inside and outside of an enclosure impact the systems that separate them.

Green Note

One of the most important factors of efficiency is the building envelope. A home can utilize the latest energy-efficient materials and still perform poorly if they are not assembled correctly. The building envelope must be both well-designed and assembled properly in order to ensure performance. It is pointless for an owner to pay top-dollar for a high-performance window if the builder installs it poorly.

15.1 Environmental Separation

The building envelope is the physical separation between the climate-controlled space inside a building and the unconditioned environment outside, **Figure 15-3**. Carpenters should understand the impact of environmental changes on both sides of the wall system. Major environmental factors that affect a structure include ultraviolet radiation, water, air pressure, water vapor, and temperature. *Control layers* work together to mitigate the impact that these elements have on a building. If a building is to remain structurally sound, energy-efficient, and a healthy place of habitation, it must maintain four control layers. In order of relative importance, these layers are the water control layer, the air control layer, the vapor control layer, and the thermal control layer.

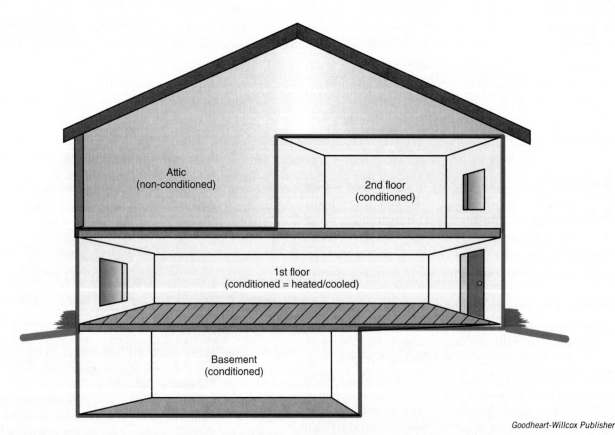

Goodheart-Willcox Publisher

Figure 15-3. The building envelope separates conditioned space from unconditioned space.

15.1.1 Climate Zones

Climate conditions vary from location to location. A good design will consider all of these factors and will vary location to location. For instance, colder climates require more insulation on the exterior side of the wall, marine climates have extra need for waterproofing details, and resolutions for issues like condensation are dealt with differently from zone to zone.

Code Note

The International Residential Code (IRC) contains a climate map, **Figure 15-4**, which divides the United States into eight distinct climate zones, each with varying levels of temperature and humidity that pose their own challenges. These zones are further designated each as Moist (A), Dry (B), and Marine (C).

15.1.2 Ultraviolet Radiation

Ultraviolet (UV) radiation from the sun's rays affects all surfaces. Fading, cracking, and general breakdown of materials often results. Materials such as shingles and siding are designed to resist the effects of UV radiation. However, there is no specific control layer to combat UV radiation. The goal is not necessarily to keep UV rays out. It is important, however, to protect building materials from harmful UV rays if they are not designed to resist them. For instance, housewrap and many exterior insulation products should only be exposed to UV radiation for a limited time. They are designed to be covered by an exterior cladding and will begin to break down if not protected.

15.2 Water Control

Bulk water can be introduced to the building envelope in a variety of ways. The most obvious method is rain, with wind-driven rain being especially persistent. Homes must be designed and built to protect vulnerable materials from water exposure. Water can also infiltrate a structure through ice. In some cases, water can flow into narrow or porous spaces, often resisting the force of gravity. Consider how a sponge draws water up from a pool of water against the force of gravity. This same principle is at work in construction materials.

Regardless of how water gets into a structure, water leakage into a building is a tremendous threat to a home's structural integrity and the health of its occupants. A building's first line of defense against bulk water is the roofing and siding. It is important to understand that siding should never be considered waterproof. In fact, some systems, such as vinyl siding, allow water to pass through at a variety of junctures. Corners, windows, and doors tend to be weak points for cladding systems.

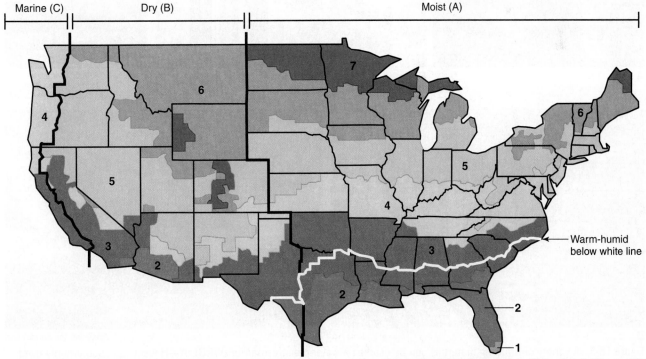

Map: CarryLove/Shutterstock.com; Data: US Department of Energy—DOE

Figure 15-4. The International Code Council has designated a variety of climate zones for the United States. Control Layer requirements can vary from zone to zone. Not shown: Guam, Puerto Rico and the Virgin Islands are all Zone 1. Alaska is in Zone 7 and Zone 8.

Vinyl siding and some other cladding systems rely heavily on the water-resistive barrier to keep bulk water from infiltrating. A builder should always assume that some water will leak past the siding. The water-resistive barrier must be capable of managing water leakage.

Water-restive barriers (WRBs), sometimes called weather-resistive barriers, are installed between the wall sheathing and the cladding. In the event that structural wall sheathing is not used, the WRB is applied directly to the studs. WRBs can be made up of a variety of materials and can serve the function of one or more of the control layers. The material used and its installation determine the function(s) of the WRB, though all WRBs act as a water control layer.

Examples of commonly used WRBs include 15 lb felt paper, building paper, housewrap, rigid foam board insulation (polystyrene or polyisocyanurate), coated sheathing systems, and liquid-applied membranes,

Figure 15-5. These materials can vary in function and installation, depending upon the design of the building envelope system. It is critical for the installer to understand the design and intended function of the system and its components in order to ensure longevity of the building.

Code Note
Building codes require the exterior wall envelope to be designed and built in a manner that prevents accumulation of water in the wall assembly by providing a WRB behind the exterior cladding. There is also a requirement that the WRB has a way to drain water to the outside of the cladding.

The WRB is the final line of defense against water. It must be designed and constructed in a way that ensures a watertight installation. Windows and doors are

A

Typar B

C
Photo courtesy of Huber Engineered Woods, ©2020 Huber Engineered Woods, LLC.

D
Dörken Systems Inc. E
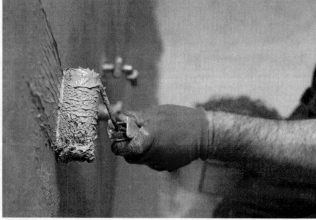
fotodrobik/Shutterstock.com

Figure 15-5. A variety of building materials can be used as a water-resistive barrier (WRB). A—Housewrap is commonly used on wall systems. B—Special attention must be given to flashing details when using rigid foam board insulation. C—Coated sheathings provide a surface that is impenetrable to water, though nailing and taping details are critical on these systems. D—Peel and stick membranes adhere directly to the sheathing. E—Liquid-applied membranes are used primarily on commercial projects.

flashed into the WRB system to keep water out of the framing, **Figure 15-6**. See Chapter 16, **Windows and Exterior Doors** for flashing details. Splices in material should be detailed according to manufacturer's instructions, with a path for water drainage.

Some housewraps, called drainable housewraps, are specially designed to encourage drainage behind materials that might trap water. They have a textured surface and a *hydrophobic* (water-repelling) coating to release surface tension and encourage drainage by gravity. **Figure 15-7**.

Behind rigid sidings, such as wood or fiber cement, a rainscreen can be built to encourage drainage. A *rainscreen* is a gap between the WRB and the cladding that allows water to drain freely, **Figure 15-8**. This space is created by using furring strips over the WRB. Siding is fastened directly to the furring strips. The bottom of the rainscreen is covered with an insect screen to keep insects from getting in. Rainscreens bring the added benefit of ventilation. Ventilation behind the siding allows residual water to evaporate and escape. The bottom acts as an air inlet for the ventilation air. The top of the rainscreen can be tied into the eave to allow ventilation to continue into the attic space. This air movement keeps the siding dry on the wall side, which reduces cupping in wood siding and ensures longer paint life.

Exterior foam board insulation can be an effective water control layer, if all flashing details are constructed in a way that forces bulk water to the outside face of the foam. All seams must be watertight. Be sure to consult manufacturer installation instructions for the material being used. Most foam insulations require specific types of flashing tape to comply with long-term adherence requirements. Using products that are not compatible with one another or are not installed properly will result in failed assemblies that allow water infiltration, resulting in a liability for the contractor. This is true of all WRB systems.

Goodheart-Willcox Publisher

Figure 15-6. This window, installed in a wall using a housewrap as a WRB, is properly flashed to keep water out.

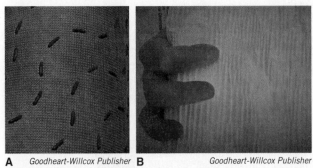

A *Goodheart-Willcox Publisher* B *Goodheart-Willcox Publisher*

Figure 15-7. Drainable housewrap encourages water drainage behind ridged sidings. Shown here are two types of drainable housewraps.

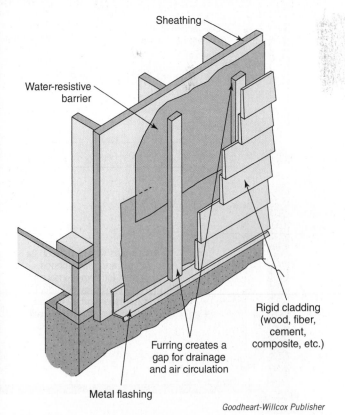

Goodheart-Willcox Publisher

Figure 15-8. A rainscreen behind this siding provides a space for water to freely drain and for ventilation, allowing the backside of the siding to dry.

15.3 Air Control

A *pressure differential* is a difference in air pressure between the inside and outside of a building envelope. Pressure differentials affect a home by forcing air into or out of the structure through any leak in the air barrier. This air may bring with it water (either bulk or vapor) or contaminants, such as dust, pollen, or mold. Additionally, air movement through the building envelope significantly reduces the efficiency of the structure's ability to maintain the temperature and comfort level inside the building.

A pressure differential can occur because of wind or systems inside a house that pull air out, **Figure 15-9.**

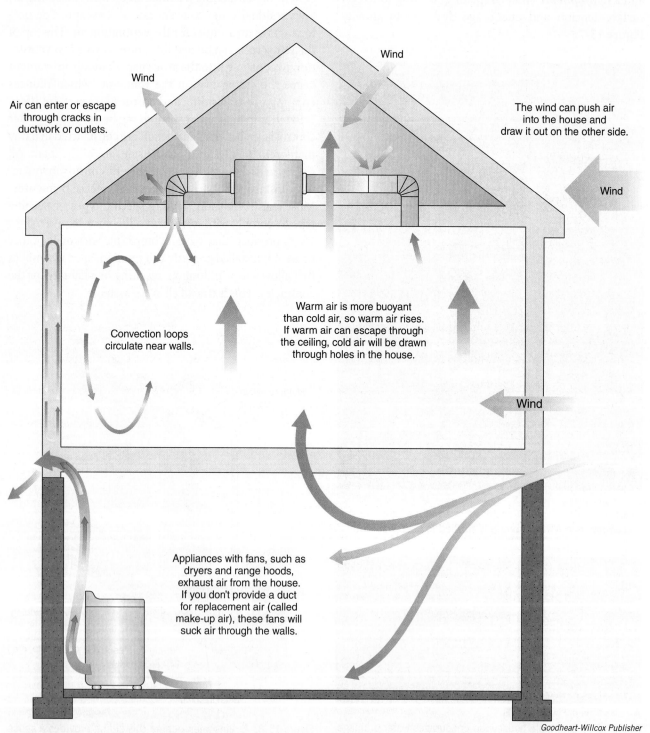

Goodheart-Willcox Publisher

Figure 15-9. Pressure differentials create a suction effect that moves air through the building envelope. This image shows forces at work that can create pressure differentials and common air leaks in a home.

Appliances in a home such as the range-exhaust hood, bathroom ventilation fans, and some gas-fired appliances can draw inside air out of the house. Some homes have wood-burning fireplaces or wood stoves that require combustion air. Any of these examples can create a low-pressure atmosphere inside the home. Unless a make-up air system is part of the design, air is drawn in through gaps and leaks in the building envelope to balance out this differential. This replacement air can pull moisture and contaminants into the wall cavities and the controlled environment of the building. A house that experiences the opposite pressure differential—higher pressure inside than outside—can also occur. A higher pressure inside the building will force air out of the house through leaks in the air barrier.

The air control layer separates the air inside a building from the air outside the building. The performance of the air barrier also impacts the thermal and vapor control of the house. A properly detailed air barrier will keep a house from feeling drafty and will stop the movement of moisture-laden air through wall cavities where condensation can occur. *Condensation* is the conversion of water vapor into liquid.

Warm air can hold more water vapor than cold air. When air cools, it loses its ability to hold the moisture it contains. When this happens, water droplets form. Condensation occurs when air reaches its *dew point*. The dew point varies depending on the temperature and relative humidity level of the air. If you live in a cold climate—any region where the average January temperature is 25°F (–5°C) or colder—the dew point can occur inside the wall structure or even in the insulation itself.

The importance of air barriers has increased as building codes have required more thermal insulation. A good air barrier improves the efficiency of fibrous insulations and reduces condensation inside wall and attic assemblies.

Air barriers are made up of multiple components or materials that are joined together to stop air movement. It is acceptable to have the air barrier installed on the inside or the outside of the wall. However, to comply with building codes, it must be incorporated as part of the building's thermal envelope and it must be continuous. The air barrier encapsulates the entire conditioned space, including the floors and ceilings.

Code Note

International Residential Code (IRC) says that air barriers must be continuous. This means that an air barrier for the wall system must be coupled to the air barrier on the ceiling, as well as the foundation walls and floor.

A house with a properly installed air barrier is considered a *tight* enclosure. Tight enclosures tend to hold moisture if it becomes trapped in the wall cavities, slowing the drying process considerably. This adds to the importance of properly detailing other control layers. A well-designed and well-installed system will prevent moisture from accumulating in the layers of the building envelope, but also will allow moisture to migrate out of the wall.

Traditionally, housewrap has been the choice material for air barriers on walls. While housewrap is a viable option, it is only considered a true air barrier if all edges of the housewrap are taped, **Figure 15-10**. Taping the housewrap at the bottom of the wall can have an adverse impact on drainage. If bulk water makes its way behind the housewrap, it is advantageous to leave the bottom untaped to provide a path for water to escape. Architects and builders must take this into account when designing the air barrier system. Without the seams taped, housewrap can still work as a water barrier, but without taping all seams and edges, including at the top and bottom of the wall, it is not a true air barrier.

Goodheart-Willcox Publisher

Figure 15-10. For housewrap to function as an air barrier, it should have no holes or tears and must be taped at all seams.

To maintain continuity, housewrap must be tied into the interior air barrier on the ceiling, whether directly, or through a combination of approved materials, **Figure 15-11**. These details are often overlooked or compromised. Other areas such as pinholes, cuts or tears, or even fastener penetrations can weaken the effectiveness of the air barrier. If housewrap is used as an air barrier, a trained carpenter must carefully install it. Wall sheathing or gypsum board can also be installed as an air barrier if detailed to function as such, **Figure 15-12**.

An air control layer that uses the sheathing as the air barrier will allow conditioned interior air to contact the inside of the sheathing. In this case, heated structures in cold climates may experience condensation on the inside of the sheathing unless the temperature of the sheathing is kept above the dew point. Conversely, an air control layer that uses the interior gypsum board as an air barrier will experience wind-washing inside the wall cavity, which will reduce the performance of the insulation. **Wind-washing** is the movement of unconditioned air through the insulation. Wind-washing is not a problem in dense insulations such as foam board or spray foam.

Using either the sheathing or gypsum board as an air barrier will require the framing attachments at the top and bottom plate to be continuously caulked, **Figure 15-12**. Joints in sheathing must also be taped with sheathing tape or caulked. All penetrations must be sealed, **Figure 15-13**. Gypsum board will need to be sealed around electrical boxes. To accomplish this, sealant must be used to block all gaps in and around the box, or gasketed electrical boxes can be used. All wire penetrations through the boxes must be sealed as well.

Coated sheathings, such as ZIP system®, can double as an air and water control layer if properly installed. Special sealant tapes and proper fastening procedures must be used. Consult manufacturer recommendations for proper installation and sealing requirements when using these systems.

Other WRB materials can also double as air barriers. Foam board insulation acts as an air barrier if all seams are taped. Spray foam insulation is also an effective air barrier. Fiberglass batts and loose fill insulations, such as cellulose, are not materials that can be used as air barriers.

On the foundation, masonry walls and concrete floors act as an air barrier. The key with all air barriers is that they are continuous, meaning all gaps, cuts, joints, and connections are sealed. Weak points in a faulty air barrier are generally at connection points, such as the mud sill, or around electrical fixtures, penetrations, chimneys and vents, and windows and doors, **Figure 15-9**.

A blower door test can measure how airtight the building is. Blower door tests are generally conducted when the interior finishes are nearing completion, but can be performed earlier in construction to identify trouble spots. For more on blower door testing, see Chapter 18, **Thermal and Sound Insulation**.

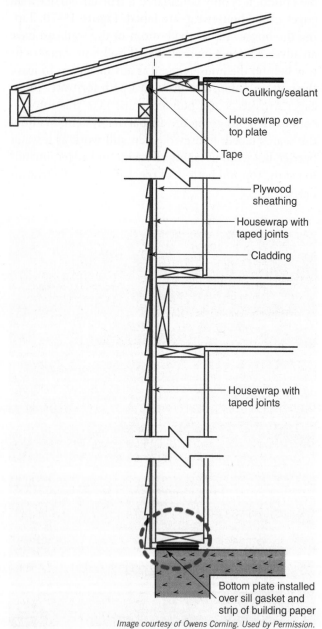

Caulking/sealant

Housewrap over top plate

Tape

Plywood sheathing

Housewrap with taped joints

Cladding

Housewrap with taped joints

Bottom plate installed over sill gasket and strip of building paper

Image courtesy of Owens Corning. Used by Permission.

Figure 15-11. A continuous air barrier that maintains contact with the thermal envelope is required by code. It can be made of various materials connected in an approved method, or direct attachment.

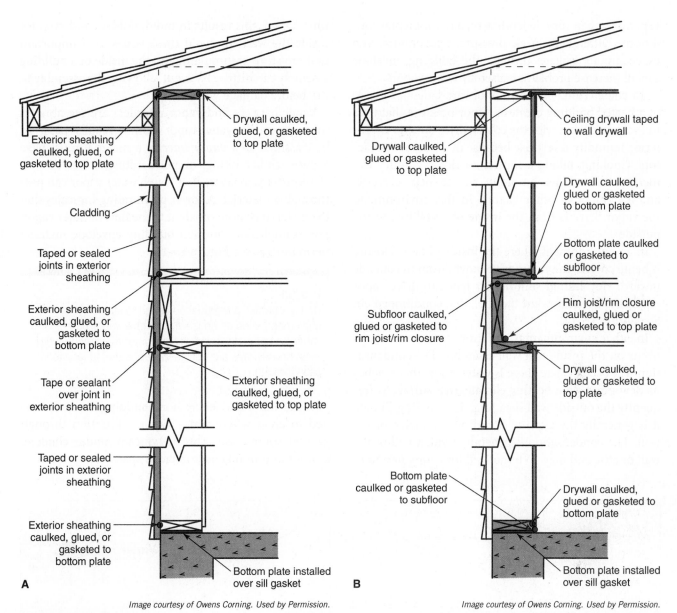

Image courtesy of Owens Corning. Used by Permission.   Image courtesy of Owens Corning. Used by Permission.

Figure 15-12. If detailed properly, the air barrier can be created on the inside or outside of the wall. A—Wall sheathing on the outside of the wall. B—Gypsum board on the inside of the wall. Take special note of the use of sealants.

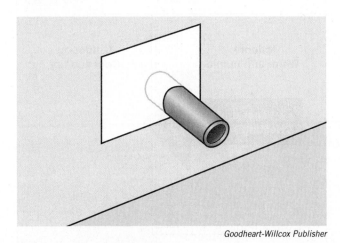

Goodheart-Willcox Publisher

Figure 15-13. Here, all-weather tape has been used to air seal around a pipe penetration.

15.4 Vapor Control

Water vapor is a gaseous form of water always present in the air. It is able to penetrate wood, stone, concrete, and most other building materials. Water vapor can be a threat to the health of a building and its occupants. High moisture content in organic materials can lead to mold, mildew, and rot.

Vapor migrates slowly through building materials. It moves from a higher humidity level to a lower humidity level; this tendency is called *diffusion*. Diffusion is accelerated in highly porous materials and is slowed by materials with low porosity. Diffusion is the reason for vapor drive. *Vapor drive* is the direction in which diffusion is occurring at a given time. The intensity of the vapor drive can vary depending on environmental conditions. Differences in air pressure, temperature, and especially air leakage through the building envelope can all increase problems caused by water vapor.

In colder climates, warm, moisture-laden air inside of a heated building creates a vapor pressure that tries to escape and mix with the colder, drier outside air. Interior humidity levels rise because of a number of factors. Cooking, taking a shower, or doing laundry can increase the relative humidity, so can pets, exercise, and a number of other factors. In this environment, the vapor drive is from the inside of a building to the outside.

In an environment where the inside of an enclosure is being cooled, the vapor drive moves from the outside inward. Hot, humid summer air tends to drive vapor inside the house, toward the cool, dry, conditioned air. See **Figure 15-14**.

In either environmental condition, condensation can occur on the inside of a wall assembly. The condensation will occur where there is water vapor that reaches its dew point. In a heating climate, that surface is frequently the outside wall sheathing. In a cooling climate it is generally the gypsum board on the inside of the wall. The condensation is generally hidden within the wall or attic and may go unnoticed for a long period of time. Often, this results in mold, mildew, and even rot inside the wall. To avoid these issues, it is important that moisture accumulating on the inside of a building assembly can diffuse back out, allowing the assembly to dry before damage occurs.

Vapor barriers and vapor retarders are membranes through which water vapor cannot readily pass. To be classified as a *vapor barrier* a material must have a permeability rating, or perm rating, lower than 1. *Perm ratings* measure how much water vapor can pass through a material. A lower perm rating indicates that the material is more resistant to the flow of water vapor. For examples of common building envelope material perm ratings, see **Figure 15-15**.

Pro Tip

Do not confuse a material designed to stop air movement, such as housewrap, with a vapor barrier. A vapor barrier is designed to stop water movement, while most housewraps are designed to breathe, to let vapor pass through.

Use of vapor barriers in homebuilding is mostly limited to areas where water vapor infiltration through ground sources is a threat, typically under concrete slabs or as ground cover in crawl spaces.

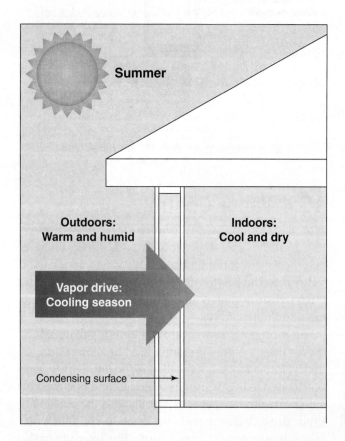
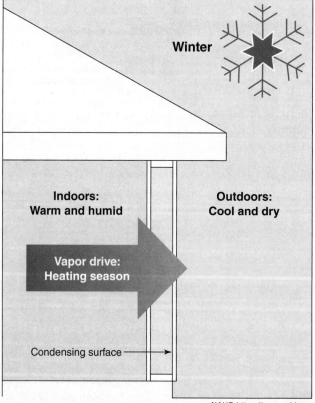

NAHB https://www.nahb.org

Figure 15-14. Vapor drive reverses from heating to cooling months.

Building Envelope Perm Ratings

Material	Perm Rating	Vapor Barrier	Common Use
Glass	0	Class I	Windows, doors, skylights
Aluminum foil	0.05	Class I	Foil faced insulation
Polyethylene sheeting	0.06	Class I	Capillary break under concrete slab
Foil faced polyisocyanurate insulation (PIR)	0.1	Class I	Exterior insulation
Extruded polystyrene (XPS)	1	Class II	Exterior insulation
Kraft faced insulation	1	Class II	Wall insulation
OSB Sheathing	Approx. 2	Class III	Wall sheathing
Expanded polystyrene insulation (EPS)	5	Class III	Exterior insulation
15 lb felt paper	5	Class III	WRB/Shingle underlayment
Standard latex paint	Approx. 7.5	Class III	Interior wall protection
Plywood sheathing	Approx. 10	Class III	Wall sheathing
Wood siding	Approx. 10	Class III	Siding
Housewrap	Range between 5 and 50	Class III or vapor open	WRB/ Air barrier
1/2″ Drywall	Approx. 50	Vapor open	Interior wall covering

Goodheart-Willcox Publisher

Figure 15-15. Vapor barriers have a perm rating of 1 or less. Vapor retarders have a perm rating between 1 and 10. Materials with a perm rating above 10 are considered permeable.

In the past, vapor barriers were placed on the predominately warm side of framed walls. If the home was built in a climate that was dominated by heating months, the vapor barrier, usually 6 mil plastic sheathing, was installed behind the gypsum board. The logic behind this system was that the vapor barrier would stop the air with a high moisture content from ever reaching a surface that was below the dew point. Essentially, the goal was to stop the vapor drive through the wall. This type of assembly relies heavily on the integrity of the vapor barrier and its proper installation. A weak point in the vapor barrier, such as poor detailing around electrical boxes, can cause condensation in a concentrated area.

As the standard of living continues to rise, more homeowners are using air conditioning systems rather than fans. Most climates now experience both heating and cooling for a number of months. In this case, the vapor drive changes throughout the seasons, **Figure 15-14**. Vapor barriers placed on the inside of a cooled space in the summer can cause condensation to occur on the vapor barrier itself. Except in the coldest climates, true vapor barriers are no longer used on the inside of conditioned spaces. A vapor barrier should never be used on both the inside and outside of an enclosure. Doing so would trap any rogue moisture in the assembly indefinitely.

The International Residential Code (IRC) now relies more heavily on *vapor retarders*, rather than vapor barriers. Vapor barriers have a perm rating of 0.1 or less. Vapor retarders have a perm rating greater than 0.1 but not more than 1. Materials with a perm rating above 1 are considered permeable, allowing vapor to freely pass.

Vapor retarders allow some moisture vapor migration from one side to the other. This slows vapor diffusion into the wall, but still allows assemblies to dry from the inside out. Building materials, such as kraft-faced fiberglass insulation, some perforated foil-faced products, and some rigid insulations act as vapor retarders, see **Figure 15-15**. Along with vapor retarders, a properly installed air barrier is also helpful in preventing condensation.

Like other control layers, proper understanding and installation of the function and purpose of the vapor control layer is critical in the performance of the building envelope.

15.5 Thermal Control

Temperature is the primary, though not the only, environmental comfort factor for an enclosure. Regulating the temperature of the enclosure at a comfortable level requires more than just insulation and a heating or air conditioning system. Factors such as relative humidity, solar radiation, type of heat source, and air movement through and in the enclosure all contribute to the comfort of a space.

The thermal control layer is made up of cavity insulation inside ceilings, walls, and floors, as well as any

Code Note

Climate zones are an important factor in determining both what classification of vapor barrier or vapor retarder to use and where it is placed in the wall assembly.

exterior insulation such as rigid foam board placed on the outside of walls, or under a concrete floor slab. The thermal control layer manages heat loss in a structure (or heat gain in cooling months). A material's resistance to the passage of heat is measured as an ***R-value***. When a building material has a low R-value, it means that the material has little resistance to the passage of heat. Without proper water, air, and vapor control, the R-value of the insulation can be reduced and the thermal control layer will be compromised. This chapter will focus on how the thermal control layer relates to the water-resistive barrier. For further information on thermal dynamics, types of insulation, and installation practices, refer to Chapter 18, **Thermal and Sound Insulation.**

Code Note

In climate zones 6 and 7, the International Residential Code (IRC) prescribes exterior wall insulation of R-5 or greater. Climate zones 3, 4, and 5 are required to use at least R-5 exterior insulation and R-13 cavity insulation unless R-20, or higher, cavity insulation is used. See **Figure 15-16**.

Thermal control plays a role in preventing condensation inside wall and ceiling assemblies and on basement floor slabs. As mentioned earlier, exterior insulation can keep sheathing temperature above the dew point, preventing condensation. Use of exterior insulation is becoming standard on many wall and slab assemblies, and is even being used on roof decks in some homes with vaulted ceilings.

Pro Tip

Although International Residential Code (IRC) prescribes certain levels of insulation in floors, walls, or ceilings, the codes do not always have the final say. Architects and builders may be able to reduce thermal performance in one area by adding to another. This is accomplished by analyzing the building envelope as a whole using computer software. Values are entered into the program for each variable in the thermal layer. The software generates a report signifying whether the building envelope meets minimum energy compliance. This allows architects and builders to incorporate designs such as cathedral ceilings, which would not otherwise meet minimum code compliance.

TABLE N1102.1.2 (R402.1.2)
INSULATION AND FENESTRATION REQUIREMENTS BY COMPONENT[a]

CLIMATE ZONE	FENESTRATION U-FACTOR[b]	SKYLIGHT[b] U-FACTOR	GLAZED FENESTRATION SHGC[b, e]	CEILING R-VALUE	WOOD FRAME WALL R-VALUE	MASS WALL R-VALUE[i]	FLOOR R-VALUE	BASEMENT[c] WALL R-VALUE	SLAB[d] R-VALUE & DEPTH	CRAWL SPACE[c] WALL R-VALUE
1	NR	0.75	0.25	30	13	3/4	13	0	0	0
2	0.40	0.65	0.25	38	13	4/6	13	0	0	0
3	0.32	0.55	0.25	38	20 or 13 + 5[h]	8/13	19	5/13[f]	0	5/13
4 except Marine	0.32	0.55	0.40	49	20 or 13 + 5[h]	8/13	19	10/13	10, 2 ft	10/13
5 and Marine 4	0.30	0.55	NR	49	20 or 13 + 5[h]	13/17	30[g]	15/19	10, 2 ft	15/19
6	0.30	0.55	NR	49	20 + 5[h] or 13 + 10[h]	15/50	30[g]	15/19	10, 4 ft	15/19
7 and 8	0.30	0.55	NR	49	20 + 5[h] or 13 + 10[h]	19/21	38[g]	15/19	10, 4 ft	15/19

For SI: 1 foot = 304.8 mm.

NR = Not Required.

    a. *R*-values are minimums. *U*-factors and SHGC are maximums. Where insulation is installed in a cavity that is less than the label or design thickness of the insulation, the installed *R*-value of the insulation shall be not less than the *R*-value specified in the table.

    b. The fenestration *U*-factor column excludes skylights. The SHGC column applies to all glazed fenestration.

    Exception: In Climate Zones 1 through 3, skylights shall be permitted to be excluded from glazed fenestration SHGC requirements provided that the SHGC for such skylights does not exceed 0.30.

    c. "10/13" means R-10 continuous insulation on the interior or exterior of the home or R-13 cavity insulation on the interior of the basement wall."15/19" means R-15 continuous insulation on the interior or exterior of the home or R-19 cavity insulation on the interior of the basement wall. Alternatively, compliance with "15/19" shall be R-13 cavity insulation on the interior of the basement wall plus R-5 continuous insulation on the interior or exterior of the home.

    d. R-5 insulation shall be provided under the full slab area of a heated slab in addition to the required slab edge insulation *R*-value for slabs as indicated in the table. The slab edge insulation for heated slabs shall not be required to extend below the slab.

    e. There are no SHGC requirements in the Marine Zone.

    f. Basement wall insulation shall not be required in warm-humid locations as defined by Figure N1101.7 and Table N1101.7.

    g. Alternatively, insulation sufficient to fill the framing cavity providing not less than an *R*-value of R-19.

    h. The first value is cavity insulation, the second value is continuous insulation. Therefore, as an example, "13+5" means R-13 cavity insulation plus R-5 continuous insulation.

    i. Mass walls shall be in accordance with Section N1102.2.5. The second *R*-value applies where more than half of the insulation is on the interior of the mass wall.

Table N1102.1.2 Excerpted from the 2018 International Residential Code; Copyright 2017; Washington, D.C.: International Code Council. Reproduced with permission. All rights reserved. www.ICCSAFE.org

Figure 15-16. This table shows the prescriptive code requirements for insulation values in a given climate.

When installed on walls, *rigid insulation board* acts as part of the thermal control layer, depending on the design and material used, it may also serve as the water, air, and/or vapor control layer in an exterior building envelope. Rigid insulation board used on exteriors comes in a variety of materials with differing characteristics, **Figure 15-17**. Which material is used depends

A   *Goodheart-Willcox Publisher*

B   *Goodheart-Willcox Publisher*

C   *Goodheart-Willcox Publisher*

Figure 15-17. Rigid insulation board comes in a number of different varieties. A—Extruded polystyrene foam. B—Expanded polystyrene foam. C—Polyisocyanurate.

on its intended function with regard to the building envelope.

Extruded polystyrene (XPS) foam—typically blue, pink, or green in color—has an insulation value of R-5 per inch. It is rated for ground contact installation and will not absorb water. XPS has a perm rating of 1 and is classified as a strong vapor retarder.

Expanded polystyrene (EPS) foam is similar to the material used in white Styrofoam® coolers and cups. EPS is a petroleum product that uses recycled polystyrene. It is made up of small beads formed during manufacturing. The beads are blown up like tiny balloons using pentane gas, a blowing agent that is safe for the ozone layer.

There are two types of EPS foam typically used in homebuilding:

- Standard EPS foam absorbs water and should not be used on ground contact installations. Insulation value is R-3.6 per inch. EPS has a perm rating of 5.
- Type IX EPS is rated for ground contact and therefore does not absorb water readily. Insulation values range from R-4 to R-5 per inch and can vary depending on the temperature. Type IX has a perm rating of 2.

Polyisocyanurate (PIR), more commonly known as ***polyiso***, is yellow in color and is often foil-faced but can be unfaced or faced with a permeable material. Polyiso is a moisture-resistant insulation material. It uses a blowing agent that slightly depletes the ozone. Its R-value is higher than other rigid foam products. Polyiso will absorb water if kept in contact with moisture continuously; therefore, it cannot be used for ground-contact installations. The R-value is approximately R-6 per inch, but can change with moisture content and temperature. The R-value of polyiso also slightly reduces over time. Permeability ratings depend on the facing of the material. Foil-faced polyiso has a perm rating approaching 0, which classifies it as a vapor barrier.

Mineral wool board insulation is made from natural and recycled materials. It is usually brown and has a fibrous appearance, **Figure 15-18**. ***Mineral wool*** is a material that is spun into fibers from molten rock. It is water-resistant and allows vapor to pass through freely. This makes it a good choice as an insulation over housewrap. Mineral wool insulation board can be faced with foil for use as an air and vapor barrier. Mineral wool insulation has R-values ranging from R-3.7 to R-4.2 per inch. Unlike foam, which is toxic when it burns, mineral wool is non-combustible, providing a measure of fire-resistance to the wall.

Serhii Krot/Shutterstock.com

Figure 15-18. Mineral wool rigid board insulation is gaining popularity in the homebuilding industry as an exterior insulation.

Using rigid exterior insulation board creates challenges for the installer. Of particular importance is how the insulation relates to the water control layer and its flashing details around windows and doors. How these details are established depends on where the water control layer is. The system may be designed so that bulk water drains to the inside of the insulation board or to the outside of the insulation board. A good system will ensure that the water control layer is continuous and well built.

15.6 High-Performance Building Envelopes

As building codes and standards recommend and require higher efficiency, builders must recognize the importance of their role in this process. It is vital for carpenters and other tradespeople to educate themselves on the function and science behind the products they are installing.

Consider a high-performance building envelope as a high-performance automobile. In an old work truck, a poor alignment or an engine misfire might not even be noticeable; however, in a racecar a poor alignment or misfire could be catastrophic. High-performance vehicles are often more sensitive and less forgiving than an old farm truck. In a similar fashion, high-performance buildings are also more sensitive than the houses and buildings constructed decades ago. In older buildings, a leak in the WRB might still allow water in, but that water was likely to dry because there may have

been little or no insulation in the wall. Air movement through the wall would also be likely, drying out any moisture that was trapped. These buildings were inefficient and often uncomfortable in poor weather.

Today's high-performance building envelopes are both efficient and comfortable. However, because they are built so tight, air quality can suffer. Air that is trapped inside the building can become stagnant. Some building materials, such as new paint and carpet, can emit volatile organic compounds (VOCs); see Chapter 24, **Painting, Finishing, and Decorating**. Other contaminants inside the house can build up as well. Therefore, building codes require mechanical ventilation systems to control the air exchanges in a building. Systems such as heat-recovery ventilators (HRV) or energy-recovery ventilators (ERV) use heat exchangers to scavenge heat from the exhaust air and transfer it to the intake air and vice versa during the cooling months, **Figure 15-19**. They are also equipped with an air filter to remove external contaminants such as pollen and mold spores. Through mechanical ventilation, the rate of air coming into and out of the building is controlled and conditioned, rather than leaking through the building envelope. See more on HRV and ERV systems in Chapter 33, **Heating, Ventilation, and Air Conditioning**.

If there is an air pressure differential between the inside and outside of the house, air will be drawn through any breach in the air barrier. Imagine a balloon that has a pinhole in it. All of the air in the balloon will pass through that single pinhole to equalize the pressure. Research shows, as much as 30 quarts of water can pass through a $1'' \times 1''$ hole in a drywall air barrier during one heating season, **Figure 15-20**. This moisture will condense and collect in concentrated areas as it cools below the dew point. Because there is no air movement inside the wall cavity of a high-performance envelope, drying is limited to diffusion. The same scenario can play out if a window or door is not properly flashed, allowing bulk water to enter a wall. Since the drying ability of a tight enclosure is limited, these issues continue to compound. Very quickly, damage or harmful mold can result.

The building envelope control layers must work together to properly control the water, air, vapor, and thermal constraints put on a house. The combinations of acceptable systems and materials is vast. However, equally vast are opportunities to assemble a system that will not perform in an acceptable manner. Tremendous care should be taken to ensure the function and integrity of control layers.

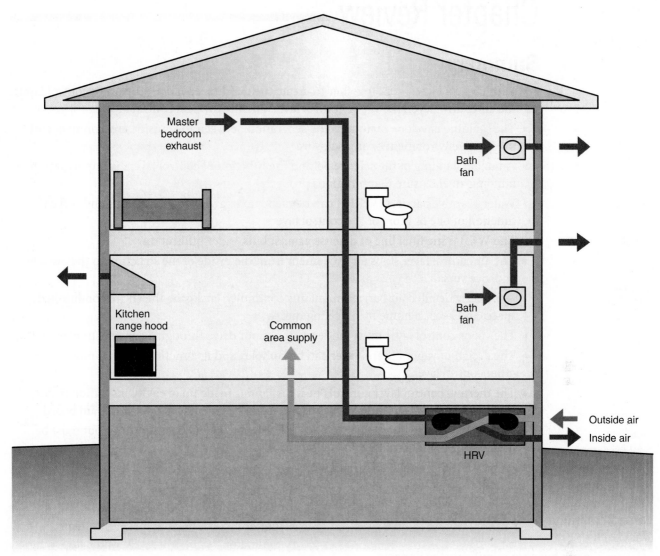

Figure 15-19. Heat recovery ventilators and energy recovery ventilators scavenge heat from outgoing air, or vice versa in a cooling climate, while bringing fresh, filtered air into the building.

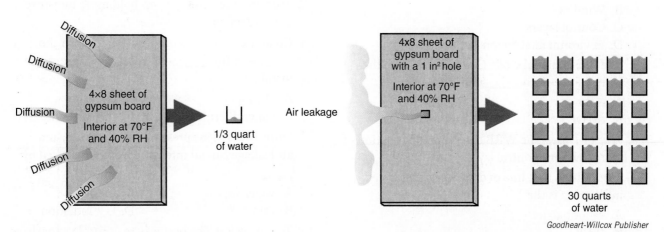

Figure 15-20. A leak in the air barrier of a building permits far more water vapor to travel through the building envelope than diffusion.

Chapter Review

Summary

- Four control layers work together to create the building envelope: the water control layer, the air control layer, the vapor control layer, and the thermal control layer.
- The building envelope maintains the separation between the outside environment and the inside environment of an enclosure.
- An understanding of the science behind the function of control layers is important when building an enclosure that functions properly.
- Water-resistive barriers (WRBs) can be made up of a variety of materials and serve the function of one or more of the control layers.
- The WRB is the final line of defense against bulk water infiltration.
- The air control layer stops air movement from the inside of the structure to the outside and vice versa.
- A properly detailed air barrier maintains continuity, enclosing the entire conditioned space, and keeps a home from feeling drafty.
- The vapor control layer must moderate the vapor drive through a construction assembly.
- The design of vapor control layer can be complex and its function may be impacted by other control layers.
- The thermal control layer controls heat loss and is made up of cavity insulation inside ceilings, walls, and floors, as well as any exterior insulation such as rigid foam board.
- High-performance building envelopes are both efficient and comfortable but must be built to the highest standard to avoid any failure in the control layers.

Know and Understand

Answer the following questions using information in this chapter.

1. _____ work together to mitigate the impact that ultraviolet radiation, water, air, vapor, and temperature have on a building.
   A. Shingles and siding
   B. Windows
   C. Control layers
   D. Environmental factors

2. The International Code Council breaks up the United States into _____ climate zones.
   A. eight   C. six
   B. twelve   D. three

3. *True or False?* The WRB can serve the function of more than one control layer.

4. A building's first line of defense against water infiltration is the _____.
   A. vapor barrier
   B. WRB
   C. flashing
   D. roofing and siding

5. A _____ is a gap between the WRB and the cladding that allows water to drain freely.
   A. gutter   C. rainscreen
   B. ventilator   D. water stop

6. _____ occurs when air reaches its dew point.
   A. Pressure difference   C. Precipitation
   B. Evaporation   D. Condensation

7. *True or False?* Cool air can hold more moisture than warm air.

8. Coated sheathings can function as both the _____ and the _____ control layers if properly installed.
   A. air, vapor   C. air, water
   B. vapor, thermal   D. thermal, air

9. Differences in air pressure, temperature, and air leakage can all intensify problems caused by _____.
   A. water vapor   C. rot
   B. rain   D. UV radiation

10. *True or False?* The best way to control vapor drive is to place a vapor barrier on both the inside and the outside of an enclosure.

11. The use of _____ insulation is becoming standard on many wall and slab assemblies.
    A. exterior
    B. interior
    C. fibrous
    D. rigid

12. Depending on the type and the installation, _____ can be used to serve as the water, air, vapor, and thermal control for a building.
    A. siding
    B. rigid insulation board
    C. sheathing
    D. housewrap

13. Because high-performance building envelopes are built so tight, _____ can suffer.
    A. siding integrity
    B. temperature levels
    C. air quality
    D. vapor levels

Apply and Analyze

1. List the four control layers in order of importance.
2. What is the difference between standard housewrap and drainable housewrap?

3. What is R-value?
4. XPS is the trade abbreviation for what kind of insulation?
5. A material's perm rating measures what?
6. Explain why mineral wool is a good choice for insulation over housewrap.

Critical Thinking

1. Design and draw a wall system that will function in a cold climate. The wall system must meet the following requirements:
   - R-20 minimum
   - Include and identify materials to be used for all four control layers
   - Identify important details that will impact the function of the control layers such as flashing, caulk, tape, etc.

2. Explain why it is so important to ensure the function and integrity of control layers.

CHAPTER 16

Windows and Exterior Doors

OBJECTIVES

After studying this chapter, you will be able to:

- Discuss standards for window and door fabrication.
- Identify the various types of windows.
- Explain how energy efficient windows affect heat resistance.
- Interpret a window schedule.
- Explain how window frames are adjusted for wall thickness.
- Summarize procedures for installing a standard window.
- Describe procedures for installing a replacement window.
- Describe the procedure for installing a bow or box bay window unit.
- Calculate rough openings.
- Prepare a rough opening for installation of a door frame.
- Describe the procedure for sliding glass door installation.
- Explain the correct construction of garage door frames.
- Select appropriate garage door hardware.

TECHNICAL TERMS

awning window	glass block	mullion
casement window	hopper window	muntin
double-hung window	horizontal sliding window	panel door
drip cap	jalousie window	sash
emissivity	jamb	storm sash
fixed window	jamb extension	U-factor
flush door	light tube	window and door schedule

Windows and doors are an important part of a structure. Building codes dictate requirements for egress (a means of escape), ventilation, and natural light for many rooms. A carpenter should have a basic understanding of these codes, as well as the various types, sizes, and standards of window and door construction. Window and door quality can range from economy-grade to high-end luxury. In order to maintain a quality external building envelope, it is critical that windows and doors are installed correctly. Installation can vary depending on whether a job is on new construction or an existing structure. Faulty installation can lead to leaks in the water or air control layer. Therefore, it is important for the installer to follow best practices to ensure proper installation.

Placement of the outside doors and the windows starts after the sheathing and weather-resistive barrier (WRB) has been installed.

16.1 Manufacture

Windows and doors are built in large millwork plants. They arrive at the building site as completed units ready to be installed in the openings of the structure.

16.1.1 Materials

Windows used in residences are made of wood, aluminum, steel, vinyl, fiberglass, or composite materials. Many are made of a combination of these materials. Wood is the original window construction material. Since wood does not transmit heat as readily as metal, there is less tendency for wooden window frames to condense humid air inside the residence. Ponderosa pine is commonly used in the fabrication of windows. It is carefully selected and kiln dried to a moisture content of 6%–12%.

Wood will decay under certain conditions and must be treated with preservatives. Exposed surfaces must be painted or clad with metal or plastic. Metal is stronger than wood and thus permits the use of smaller frame members around the glass. To eliminate the problem of heat transfer, better quality metal frames have thermal stops built into the frames.

16.1.2 Manufacturing Standards

Standards for the manufacture of windows and doors have been established by the Window & Door Manufacturers Association (WDMA), the American Architectural Manufacturers Association (AAMA), and the Canadian Standards Association (CSA). The International Building Code and the International Residential Code include the following requirement:

"Exterior windows and sliding doors shall be tested by an approved independent laboratory, and bear a label identifying manufacturer, performance characteristics and approved inspection agency to indicate compliance with AAMA/WDMA/CSA 101/I.S.2/A440. Exterior side-hinged doors shall be tested and labeled as conforming to AAMA/WDMA/CSA 101/I.S.2/A440..."

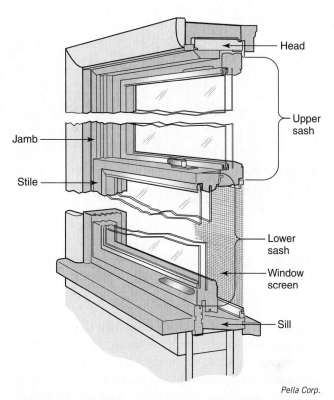

Pella Corp.

Figure 16-1. Standard sections used to show details of all types of windows. Note the part names.

16.2 Parts of Windows

As shipped from the factory, windows are complete and assembled, except for the interior trim. A carpenter must know the names of all the window parts, **Figure 16-1**, their functions, and where in the assembly they are located. In most cases, window terminology is the same, regardless of window type or the material used in manufacture. Knowing all this, a carpenter will understand or be able to give instructions about them.

A *sash* is a frame that holds many window parts. Essentially, it contains and holds the glass in place and forms a tight seal with the frame. The sides of the sash are called stiles. The other parts of the frame are the top rail and the bottom rail.

The sash is encased in another frame that is firmly secured to the wall frame. The sides of this frame are called *jambs*. The top is known as the head and the bottom is the sill.

Manufacturer's literature usually contains section views and details of window construction. However, because the window units are delivered to the construction site complete, with all parts except the interior trim in place, these drawings are often not included on the plans.

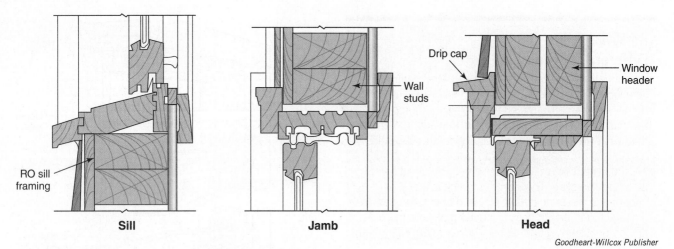

Sill **Jamb** **Head**

Goodheart-Willcox Publisher

Figure 16-2. Typical detail drawing shows a section through a window. Such details are not usually included with the house plans, but are available from the window manufacturer.

Architectural drawings also include wall framing members and surface materials, as shown in **Figure 16-2**. A *drip cap*, shown in the head section of **Figure 16-2**, is designed to carry rainwater out over the window casing. When the window is protected by a wide cornice, this element may not be included.

A *mullion* is formed by the window jambs when two units are joined together. The section view of a typical mullion in **Figure 16-3** shows how window units fit together. Windows can be joined together like this from the factory. These are referred to as mulled window units.

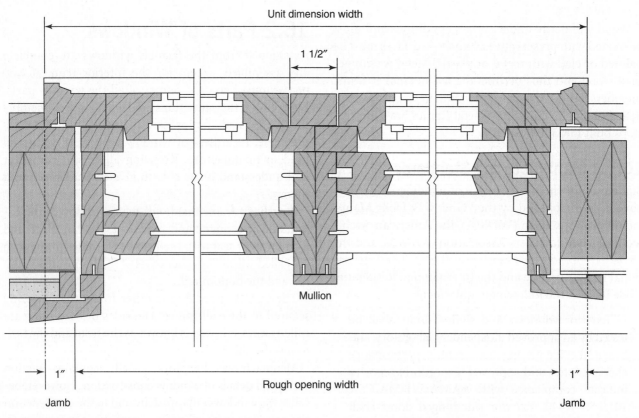

Rock Island Millwork

Figure 16-3. Section drawing of a mullion in plan view. This is where two window jambs are joined.

Years ago, window glass was available only in small sheets. By using rabbeted strips called *muntins*, a number of small panes of glass could be used to fill large sashes, **Figure 16-4**. Today, muntins are applied between the layers of thermal pane glass, or as an overlay on a large sheet of glass. See **Figure 16-5**. They do not actually separate or support small panes of glass. Made of wood or plastic and available in various patterns, overlays can snap in and out of the sash for easier painting and cleaning.

16.3 Types of Windows

In general, windows can be grouped under one or a combination of three basic types:

- Sliding
- Swinging
- Fixed

Each of these includes a variety of designs or methods of operation. Sliding types include the double-hung window and horizontal sliding window. See **Figure 16-6**. Swinging units that are hinged on a vertical plane are called casement windows, while those hinged on a horizontal plane can be either awning windows or hopper windows.

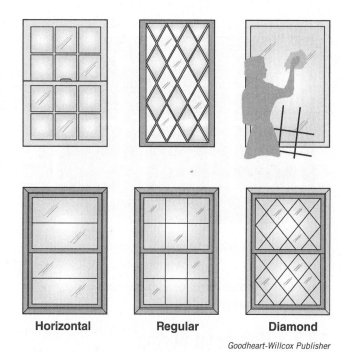

Goodheart-Willcox Publisher

Figure 16-5. Standard muntin patterns. Muntin assemblies in modern windows easily detach for cleaning and painting.

| Horizontal | Regular | Diamond |

A   *Marvin Windows*   B   *Marvin Windows*

Figure 16-6. A—Double-hung sliding window. B—Horizontal sliding window.

16.3.1 Double Hung

A *double-hung window* consists of two sashes that slide up and down past each other in the window frame. These are held in any vertical position by a friction fit against the frame or by springs and various balancing devices. Double-hung windows are widely used because of their economy, simplicity of operation, and adaptability to many architectural designs.

Jacek Wojnarowski/Shutterstock.com

Figure 16-4. In traditional wooden window units, muntins separated individual panes of glass.

Muntin

Pane

Noel V. Baebler/Shutterstock.com

Figure 16-7. Three mulled double-hung units.

Figure 16-7 shows a mulled window unit consisting of three double-hung windows.

16.3.2 Horizontal Sliding

Horizontal sliding windows have two or more sashes. At least one of them can horizontally move within the window frame, **Figure 16-8**. The most common design consists of two sashes, both of which are movable. In units with three sashes, the center one is usually fixed (not movable).

16.3.3 Casement

A *casement window* has a sash that is hinged on the side and swings outward. Some units consist of two or more windows separated by mullions. Sashes are operated by a cranking mechanism or a push bar mounted on the frame, **Figure 16-9**. Latches are used to close, lock,

Suti Stock Photo/Shutterstock.com

Figure 16-8. A horizontal sliding window has at least one moveable sash and may have one fixed sash.

4045/Shutterstock.com

Figure 16-9. Casement windows are hinged on the side and are operated by a crank or push bar. This photo shows a double casement window (center), flanked by fixed windows. Note the slight difference in appearance.

and tightly hold the sash against the weather stripping. Crank operators make it easy to open and close windows located above kitchen cabinets or other built-in fixtures.

The swing sash of a casement window permits full opening of the window. This provides good ventilation. Frequently, where a row of windows is desirable, fixed units are combined with operating units. Screen and storm sashes are attached to the inside of standard casement windows.

16.3.4 Awning

Awning windows have one or more sashes that are hinged at the top and swing out at the bottom, **Figure 16-10**. They are often combined with fixed units. Several operating sashes can be vertically stacked in such a way that they close on themselves or on rails that separate the units.

Marvin Windows

Figure 16-10. Awning windows have one or more sashes that are hinged at the top.

Most awning windows have a "projected action" in which the top rail moves down as the bottom of the sash swings out. Crank and push bar operators are similar to those on casement windows. Screens and storm sashes are mounted on the inside.

Awning windows are sometimes installed side by side to form a "ribbon" effect (a narrow band of windows, resembling a ribbon) at the top of a wall. Such an installation provides privacy for bedroom areas and also permits greater flexibility in furniture arrangements along outside walls. Often, the ribbon is placed above fixed windows to provide ventilation.

> **Pro Tip**
> Consideration of outside clearance must be given to both casement and awning windows. When open, they may interfere with movement on porches, patios, or walkways that are located adjacent to outside walls.

16.3.5 Hopper

The *hopper window* has a sash that is hinged along the bottom and swings inward, similar to an upside down awning window. It is operated by a locking handle located in the top rail of the sash. Hopper windows are easy to wash and maintain. However, their operation may interfere with drapes, curtains, and the use of inside space near the window.

16.3.6 Jalousies

A *jalousie window* is a series of horizontal glass slats held and controlled at each end by a movable metal frame. The metal frames are attached to each other by levers. The slats tilt together in about the same manner as a mini or venetian blind. Jalousie windows provide excellent ventilation. However, they are not very weathertight. Their use in northern climates is usually limited to three-season rooms (excluding winter, when more heat is required) and breezeways.

16.3.7 Fixed

The *fixed window* can be used in combination with any of the other units. Its main purpose is to provide daylight and a view of the outdoors. When used in this type of installation, the glass is set in a fixed sash and mounted in a frame that matches the ventilating windows. Large sheets of plate glass are often separated from other windows to form window walls. They are usually set in a special frame formed in the wall opening.

16.4 Window Heights

One of the important functions of a window is to provide a view of the outdoors. An architect must be aware of the dimensions shown in **Figure 16-11**. These dimensions may need to be adjusted when designing for persons with disabilities.

In residential construction, the standard height from the bottom of the window head to the finished floor is 6'-8". When this dimension is used, the heights of window and door openings are the same. If inside and outside trim must align, 1/2"-3/4" must be added to this height for thresholds and door clearances. Window manufacturers usually provide exact dimensions for their standard units. See **Figure 16-12**.

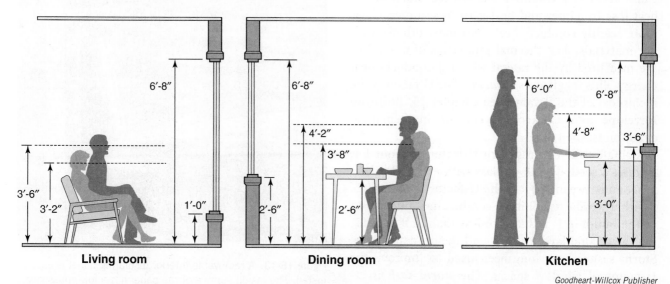

Goodheart-Willcox Publisher

Figure 16-11. Architectural drawings should take into account the standards for window heights. Avoid placing horizontal framework at the eye levels shown. Designs may need to be adjusted for people with disabilities.

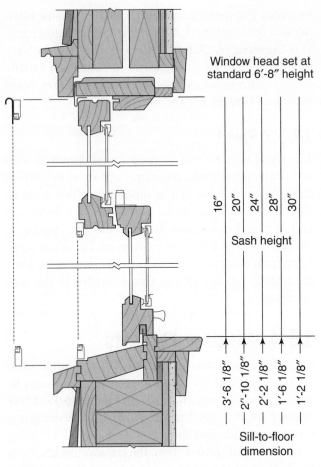

Window head set at standard 6'-8" height

Sash height

16"
20"
24"
28"
30"

3'-6 1/8"
2"-10 1/8"
2'-2 1/8"
1'-6 1/8"
1'-2 1/8"

Sill-to-floor dimension

Pella Corp.

Figure 16-12. Consult manufacturer's product literature for window head and sill heights.

16.5 Energy Efficiency of Windows

Glass areas of a dwelling account for much of the heat loss in winter and heat gain in summer. Glass more readily conducts heat than most other building materials. The thermal properties of a window are measured by the rate at which it conducts heat energy. This is called its *U-factor*. The U-factor is the reciprocal of the R-value, see Chapter 15, **Building Envelope and Control Layers**. Its mathematical equation is: U= 1/R.

A single pane of glass has a U-factor of about 1.1. Adding a second pane of glass with a 1/2" dead air space in between reduces the U-factor to about 0.5. In other words, its ability to conduct heat into or out of the building is reduced to less than half by adding the second layer of glass and the dead air space. Storm sashes have long been used to improve the U-factor of window spaces. The *storm sash* is attached to the outside of the window frame. They are

double- or triple-track units with movable screens and glass panes. In warm months, the screen is moved into position. In cold months, the glass is moved into position.

Another variation of the storm sash is the storm panel, **Figure 16-13**. Panes of glass mounted in metal frames can be attached to either the inside or outside of the window sash. It is generally used on horizontal sliding and casement or other hinged sashes. When a lower U-factor is required, the storm panel can be equipped with sealed double glazing.

Modern windows no longer use storm sashes or panels. Instead, the sashes use thermal panes, which have two, or even three layers of glass with a space between each.

16.5.1 Thermal Panes

For movable sashes, two or three layers of 1/8" glass are fused together with a space between layers. Special seals are used to trap the air or other gas between panes, creating an insulating layer. Spaces are generally 1/4"–1/2" wide. When double glazing was first developed, the space between the panes was filled with dehydrated (moisture removed) air. Today, it is quite common for the space to be filled with argon, a stable gas that conducts heat even more slowly than dry air. Triple-glazed, argon-filled windows can achieve a U-factor below 0.2.

Pella Corp.

Figure 16-13. A removable interior insulating panel is easy to install. The inside surface of the panel has a low-emissivity coating that reflects radiant heat.

Figure 16-14 shows sealed triple glazing in a standard casement unit. Double and triple glazing offer the following advantages:

- Lower heat loss in cold weather
- Reduced downdrafts along the window surface
- Reduced heat penetration in summer months
- Decreased or eliminated sweating and fogging of windows in cold weather
- Less outside noise transmitted through the window

Green Note

Windows and glass doors can leak a significant amount of heat in winter. Installing high-efficiency, triple-glazed windows in cold climates can reduce heat loss by 15%–30%. Double-glazed windows will meet weather needs and code specifications in warmer, more moderate parts of the country.

16.5.2 Low-Emissivity Glazing

Emissivity is the relative ability of a material to transmit or absorb and emit heat. Research in the area of glazing technology has developed a method of lowering the U-factor of double-glazed windows.

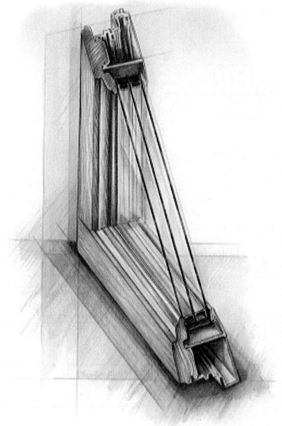

Pella Corp.

Figure 16-14. A triple-glazed unit has two air spaces.

Commonly referred to as low-e windows, these units have a clear outer pane, an air space, and a special coating on the air-gap side of the inner pane. The special, factory-applied coating consists of an extremely thin layer of metal oxide. It reflects infrared (heat wave) radiation, but allows regular light waves to pass through.

During winter months, warm surfaces within a room (wall, floor, furniture) radiate heat waves. When these waves strike the low-e surface of the window, they are reflected back into the room. This action reduces heat loss. During summer months, heat waves from walks, drives, and other outside surfaces are prevented from entering. This lowers air conditioner loads.

16.5.3 NFRC

The National Fenestration Rating Council (NFRC) is a nonprofit organization that establishes ratings for various categories of window, door, and skylight performance. Generally, an NFRC label is placed on every newly manufactured window. This label contains the performance rating information for that particular window as outlined in **Figure 16-15**. These ratings are used by the US Department of Energy to determine if the window, door, or skylight qualifies for the Energy Star certification. For more on ENERGY STAR, refer to Chapter 28, **Green Building and Certification Programs**.

National Fenestration Rating Council

Figure 16-15. The carpenter should be familiar with the performance factors set forth by the NFRC. Requirements for each rating vary by geographic region and intended function of the window, door, or skylight.

16.6 Screens

Windows opened for ventilation require screens to keep out insects. The mesh, usually made from fiberglass or aluminum, should have a minimum of 252 openings per square inch. Most screens have a light metal frame. Manufacturers have perfected many methods for mounting and storing screen panels.

16.7 Windows in Plan and Elevation Drawings

Carpenters study the plans and elevations of the working drawings to determine the types of windows called for and where they are to be located. It is common practice in frame construction to locate the horizontal position of windows and exterior doors by dimensioning to the center of the opening, as shown in **Figure 16-16**. In masonry construction, the dimension is to the edge of the opening.

Elevations show the type of window and may include glass size and heights, **Figure 16-17**. The position of the hinge line (point of dotted line) indicates the type of swinging window. Sliding windows require a note on the plan view to indicate they are not fixed units. Supporting mullions are indicated in the plans and elevations.

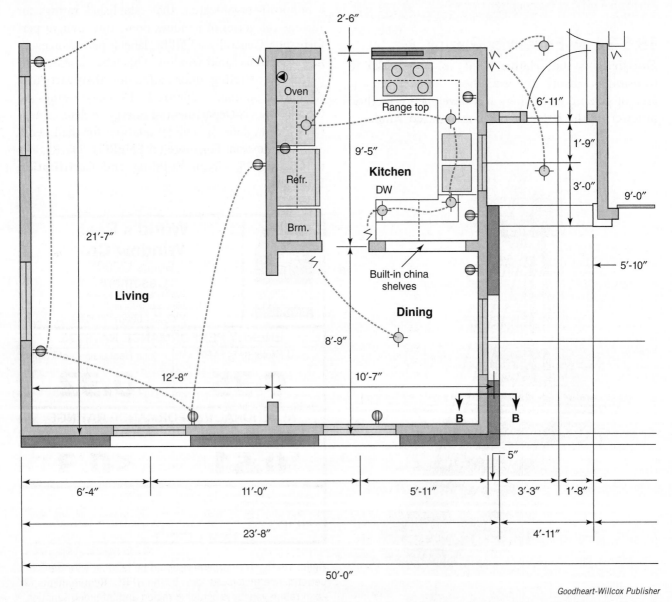

Goodheart-Willcox Publisher

Figure 16-16. Floor plans show locations of windows and doors. Can you locate the door shown on this partial floor plan?

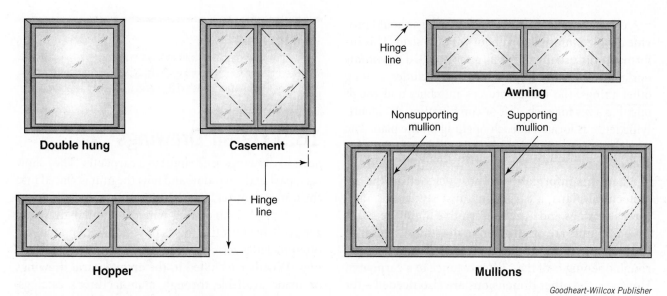

Goodheart-Willcox Publisher

Figure 16-17. Windows shown in elevation views. Horizontal sliding windows are noted to distinguish them from fixed units. Elevations show supporting and nonsupporting mullions.

16.8 Window Sizes

Besides the type and position of the window, a carpenter should know the size of each unit or combination of units. Window sizes may include the following:

- Glass size
- Sash size
- Rough frame opening
- Masonry or unit opening

Figure 16-18 shows the position of these measurements and approximately how they are figured from the glass size. Measurements slightly vary from one manufacturer to another.

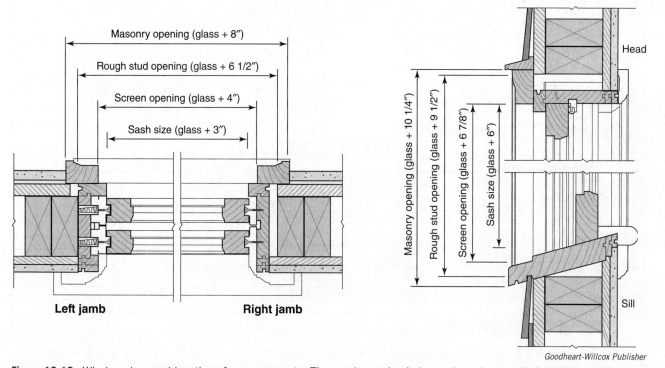

Goodheart-Willcox Publisher

Figure 16-18. Window sizes and location of measurements. The rough opening is larger than the overall size of the frame to permit alignment and leveling when the unit is installed.

A complete set of architectural drawings should provide detailed information about window sizes. This information is usually listed in a table called a *window and door schedule*, **Figure 16-19**. It includes, among other things, the manufacturer's numbers and rough opening sizes for each unit or combination. An identifying letter is located at each opening on the plan. The same letter is used in the schedule to identify the window or door.

When this information is not included in the architectural plans, a carpenter must study manufacturer's catalogs and other descriptive literature. Sizes for basic units are often given in diagrams, similar to the example found in **Figure 16-20**. The size of the rough opening is of major importance to a carpenter. However, the other dimensions are also needed—for example, the height of the rough opening above the floor.

Pro Tip

When the manufacturer lists window and door sizes, whether they consist of rough openings, sash size, or other items, the horizontal dimension is listed first and the height second.

16.9 Detail Drawings

Section drawings are helpful to a carpenter. They show each part of the window and how the unit is placed into the wall structure. **Figure 16-21** shows detail drawings for a typical double-hung window. Similar drawings are available for other types of windows. An architect often includes selected detail views in the set of drawings. Whether included in the architectural drawings or made available through manufacturer's catalogs, detail drawings such as these are essential in building the rough frame of the wall structure and in installing the window units. See **Figure 16-22**.

DOOR SCHEDULE

IND	SIZE	TYPE	ROUGH OPENING	REMARKS
D1	(2) 1-6x6-8	PASSAGE	3'-2" x 6'-10 7/8"	-
D2	2-0x6-8	PASSAGE	2'-2" x 6'-10 7/8"	-
D3	(2) 2-0x6-8	PASSAGE	4'-2" x 6'-10 7/8"	-
D4	2-4x6-8	PASSAGE	2'-6" x 6'-10 7/8"	-
D5	2-6x6-8	PASSAGE	5'-2" x 6'-10 7/8"	-
D6	2-8x6-8	PASSAGE	2'-10" x 6'-10 7/8"	-
D7	2-8x6-8	POCKET	2'-10" x 6'-10 7/8"	-
D8	5-0x6-8	BI-PASS	5'-2" x 6'-10 7/8"	-
ED1	2-8x6-8	EXTERIOR FIRE DOOR	2'-10" x 6'-10 7/8"	SELF CLOSING HINGES
ED2	3-0x6-8	EXTERIOR SERVICE DOOR	3'-2" x 6'-10 7/8"	SELF CLOSING HINGES
ED3	3-0x6-8	EXTERIOR	3'-2" x 6'-10 7/8"	W/(2) 15" SL & ARCH TRANSFORM
ED4	(2) 2-8x8-0	FRENCH	5'-4" x 6'-10 7/8"	TEMPERED GLASS
ED5	9-0x8-0	GARAGE-OVHGD	9'-0" x 8'-0"	OVERHEAD GARAGE DOOR
ED6	16-0x8-0	GARAGE-OVHGD	16'-0" x 8'-0"	OVERHEAD GARAGE DOOR
ED7	6-0x6-8	SLIDING GLASS DOOR	6'-2" x 6'-10 7/8"	-

WINDOW SCHEDULE

IND	MANUFACTURER	SIZE	TYPE	ROUGH OPENING	REMARKS
W1	ANDERSEN	TW2032	TILT WASH DOUBLE HUNG	2'-2 1/8" x 3'-5 1/4"	-
W2	ANDERSEN	TW2846	TILT WASH DOUBLE HUNG	2'-10 1/8" x 4'-9 1/4"	-
W3	ANDERSEN	TW2846-3	TILT WASH DOUBLE HUNG	8'-5 3/4" x 4'-9 1/4"	-
W4	ANDERSEN	TW2852	TILT WASH DOUBLE HUNG	2'-10 1/8" x 5'-5 1/4"	-
W5	ANDERSEN	TW2852-2	TILT WASH DOUBLE HUNG	5'-7" 15/16 x 5'-5 1/4"	-
W6	ANDERSEN	TW3052	TILT WASH DOUBLE HUNG	3'-2 1/8" x 5'-5 1/4"	-
W7	ANDERSEN	DHP41042	PICTURE	4'-11 7/8" x 4'-5 1/4"	-
W8	ANDERSEN	DHT2810	TRANSFORM	2'-10 1/8" x 1'-10 1/2"	-
W9	ANDERSEN	3-0 x 1-8	CUSTOM ARCH TRANSFORM	-	-
W10	ANDERSEN	5-10 x 1-8	CUSTOM ARCH TRANSFORM	-	TEMPERED

NOTE: ALL WINDOWS ARE SERIES 400

Goodheart-Willcox Publisher

Figure 16-19. Window and door schedules give detailed information about the size and type of window or door to be used. The location is denoted on the print by an alphanumeric code, which can be cross-referenced to the schedule.

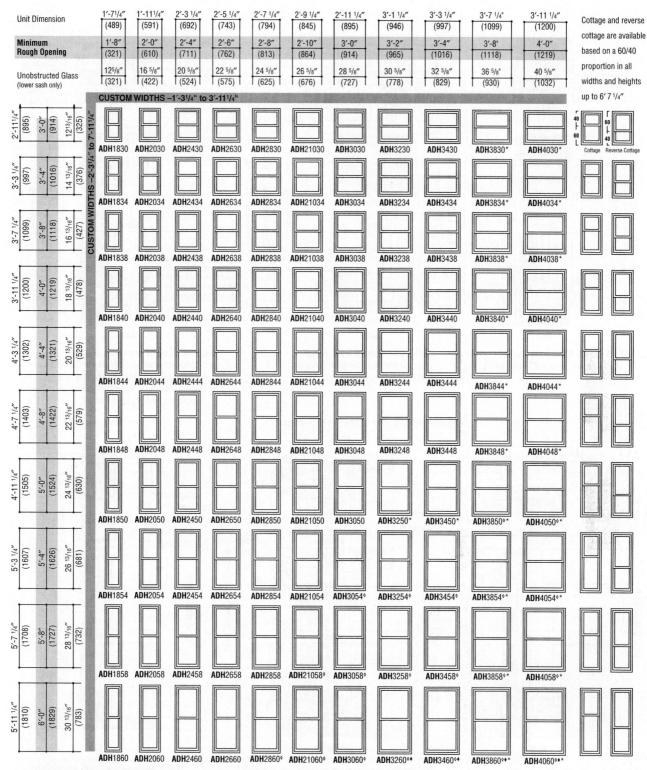

Unit Dimension	1'-7 1/4" (489)	1'-11 1/4" (591)	2'-3 1/4" (692)	2'-5 1/4" (743)	2'-7 1/4" (794)	2'-9 1/4" (845)	2'-11 1/4" (895)	3'-1 1/4" (946)	3'-3 1/4" (997)	3'-7 1/4" (1099)	3'-11 1/4" (1200)
Minimum Rough Opening	1'-8" (321)	2'-0" (610)	2'-4" (711)	2'-6" (762)	2'-8" (813)	2'-10" (864)	3'-0" (914)	3'-2" (965)	3'-4" (1016)	3'-8" (1118)	4'-0" (1219)
Unobstructed Glass (lower sash only)	12 5/8" (321)	16 5/8" (422)	20 5/8" (524)	22 5/8" (575)	24 5/8" (625)	26 5/8" (676)	28 5/8" (727)	30 5/8" (778)	32 5/8" (829)	36 5/8" (930)	40 5/8" (1032)

Cottage and reverse cottage are available based on a 60/40 proportion in all widths and heights up to 6' 7 1/4"

CUSTOM WIDTHS –1'-3 1/4" to 3'-11 1/4"

CUSTOM WIDTHS –2'-3 1/4" to 7'-11 1/4"

Height	Col1	Col2	Col3	Col4	Col5	Col6	Col7	Col8	Col9	Col10	Col11
2'-11 1/4" (895) / 3'-0" (914) / 12 13/16" (325)	ADH1830	ADH2030	ADH2430	ADH2630	ADH2830	ADH21030	ADH3030	ADH3230	ADH3430	ADH3830*	ADH4030*
3'-3 1/4" (997) / 3'-4" (1016) / 14 13/16" (376)	ADH1834	ADH2034	ADH2434	ADH2634	ADH2834	ADH21034	ADH3034	ADH3234	ADH3434	ADH3834*	ADH4034*
3'-7 1/4" (1099) / 3'-8" (1118) / 16 13/16" (427)	ADH1838	ADH2038	ADH2438	ADH2638	ADH2838	ADH21038	ADH3038	ADH3238	ADH3438	ADH3838*	ADH4038*
3'-11 1/4" (1200) / 4'-0" (1219) / 18 13/16" (478)	ADH1840	ADH2040	ADH2440	ADH2640	ADH2840	ADH21040	ADH3040	ADH3240	ADH3440	ADH3840*	ADH4040*
4'-3 1/4" (1302) / 4'-4" (1321) / 20 13/16" (529)	ADH1844	ADH2044	ADH2444	ADH2644	ADH2844	ADH21044	ADH3044	ADH3244	ADH3444	ADH3844*	ADH4044*
4'-7 1/4" (1403) / 4'-8" (1422) / 22 13/16" (579)	ADH1848	ADH2048	ADH2448	ADH2648	ADH2848	ADH21048	ADH3048	ADH3248	ADH3448	ADH3848*	ADH4048*
4'-11 1/4" (1505) / 5'-0" (1524) / 24 13/16" (630)	ADH1850	ADH2050	ADH2450	ADH2650	ADH2850	ADH21050	ADH3050	ADH3250*	ADH3450*	ADH3850◊*	ADH4050◊*
5'-3 1/4" (1607) / 5'-4" (1626) / 26 13/16" (681)	ADH1854	ADH2054	ADH2454	ADH2654	ADH2854	ADH21054	ADH3054◊	ADH3254◊	ADH3454◊	ADH3854◊*	ADH4054◊*
5'-7 1/4" (1708) / 5'-8" (1727) / 28 13/16" (732)	ADH1858	ADH2058	ADH2458	ADH2658	ADH2858	ADH21058◊	ADH3058◊	ADH3258◊	ADH3458◊	ADH3858◊*	ADH4058◊*
5'-11 1/4" (1810) / 6'-0" (1829) / 30 13/16" (783)	ADH1860	ADH2060	ADH2460	ADH2660	ADH2860◊	ADH21060◊	ADH3060◊	ADH3260◊*	ADH3460◊*	ADH3860◊*♦	ADH4060◊*♦

40 / 60 / 60 / 40   Cottage   Reverse Cottage

- "Window Dimension" always refers to outside frame to frame dimension.
- "MinimumRoughOpening" dimensions may need to be increased to allow for use of building wraps, flashing, sill panning, brackets, fasteners orother items.
- Dimensions in parenthesis are in millimeters.
- ◊ Sizes with equal sash that meet or exceed clear opening area of 5.7 sq. ft., clear opening width of 20" and clear opening height of 24".
- ♦ Sizes with cottage sash that meet or exceed clear opening area of 5.7 sq.ft., clear opening width of 20" and clear opening height of 24".
- *Two locks are standard.

Andersen Corp.

Figure 16-20. Websites, catalogs, and brochures display the variety of window units a manufacturer offers. This page shows some of the sizes of available double-hung window units in one series from a major window manufacturer.

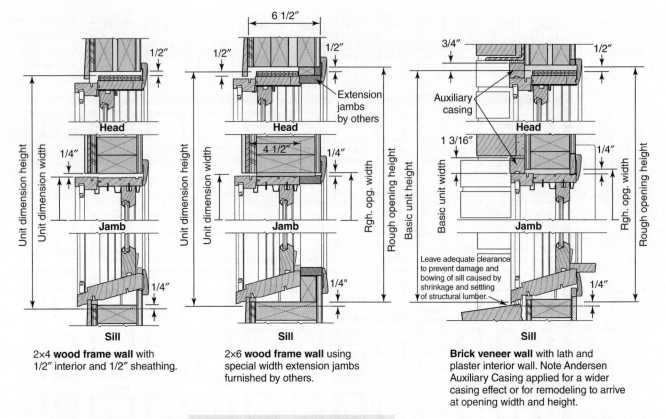

6 1/2″

1/2″

1/2″   1/2″

Extension
jambs
by others

Head

Head

4 1/2″

1/4″

Unit dimension height
Unit dimension width

Unit dimension height
Unit dimension width

Rgh. opg. width
Rough opening height

3/4″

1/2″

Auxiliary
casing

Head

1 3/16″

1/4″

Basic unit height
Basic unit width

Rgh. opg. width
Rough opening height

Jamb

Jamb

1/4″

1/4″

Jamb

Leave adequate clearance
to prevent damage and
bowing of sill caused by
shrinkage and settling
of structural lumber.

1/4″

Sill

Sill

Sill

2×4 **wood frame wall** with
1/2″ interior and 1/2″ sheathing.

2×6 **wood frame wall** using
special width extension jambs
furnished by others.

Brick veneer wall with lath and
plaster interior wall. Note Andersen
Auxiliary Casing applied for a wider
casing effect or for remodeling to arrive
at opening width and height.

Caulking is required under sill stop or stool.

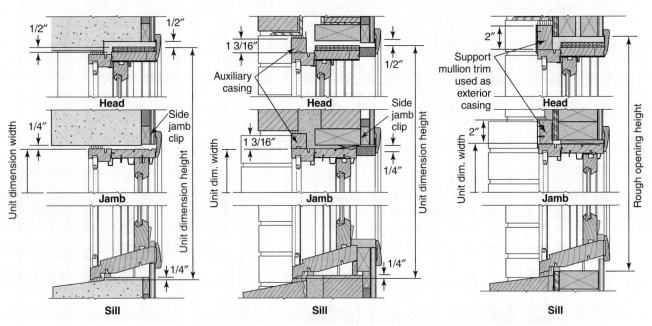

1/2″

1/2″

Head

Side
jamb
clip

1/4″

1 3/16″

1/2″

Auxiliary
casing

Head

Side
jamb
clip

2″

Support
mullion trim
used as
exterior
casing

Head

Unit dimension width

Unit dimension height

1 3/16″

1/4″

Unit dim. width

Unit dimension height

2″

Unit dim. width

Rough opening height

Jamb

Jamb

Jamb

1/4″

1/4″

Sill

Sill

Sill

Pre-cast wall with 1/2″ drywall
interior on furring strips. Unit secured by
side jamb clips attached to side jamb and
nailed into furring strips.

10″ cavity masonry wall with 1/2″ drywall
interior on furring strips. Note unit has Andersen
Auxiliary Casing applied for a wider casing effect
or for remodeling to arrive at large openings.

Brick veneer wall with Andersen
Support Mullion Trim used for wider
casing effect or for remodeling to arrive
at opening width and height.

Andersen Corp.

Figure 16-21. These detail drawings show a standard double-hung window installed in wall structures of different materials
and thicknesses.

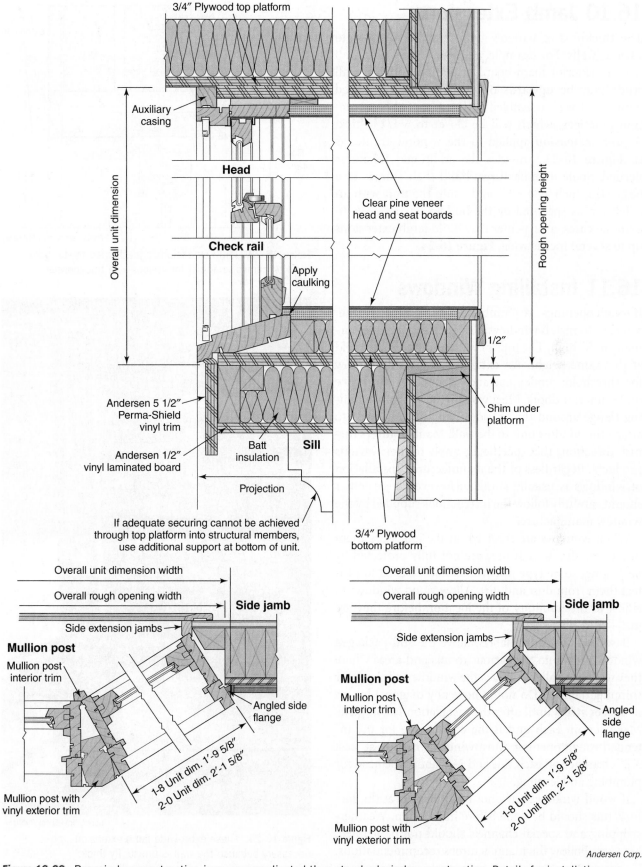

Figure 16-22. Bay window construction is more complicated than standard window construction. Details for installation are extremely important for a carpenter.

16.10 Jamb Extensions

The thickness of window units may be adjusted to various walls. For example, a cold-climate frame wall with 1″ exterior foam board insulation and 2×6 wall studs may be upwards of 7 1/2″ thick, compared with a 2×4 wall assembled with traditional construction practices, which will be closer to 5 1/2″ thick. A *jamb extension* applied to the window frame, as in **Figure 16-23**, provides the additional thickness needed. Some manufacturers build their frames to a basic size, such as 4 5/8″, and equip the unit with an extension as specified by the builder or architect. In extreme cases, a carpenter may build jamb extensions up to several inches wide, **Figure 16-24**.

16.11 Installing Windows

If rough openings are plumb, level, and correctly sized, it is easy to install windows. Residential doors are normally 6′-8″ high. The tops of windows are usually held at this same height. An extra 1/2″ is added to allow for thresholds under entrance doors and clearance under interior doors. Most wood windows have a nailing flange around the outside. This flange is used to attach the window unit to the wall. Manufacturers furnish directions that specifically apply to their various products. Regardless of the manufacturer, installation of windows is usually similar. However, a carpenter should carefully follow the instructions supplied by the window manufacturer.

When windows are received on the job, store them in a clean, dry area. If they are not fully protected by packaging, some type of cover should be used to protect them from dust and dirt. Allow wood windows to adjust to the humidity of the location before they are installed.

Based on the carton labels, move the still-packaged window units into the various rooms and areas where they will be installed. Unpack the window and check for shipping damage. Do not remove any diagonal braces or spacer strips until after the installation is complete. Make certain each unit is the correct size for the intended rough opening. Most windows require at least 3/8″ clearance on each side and 3/4″ above the head for plumbing and leveling.

If wood windows have not been primed at the factory, this should be done before installation. Weatherstripping and special channels should not be painted, however. Follow the manufacturer's recommendations.

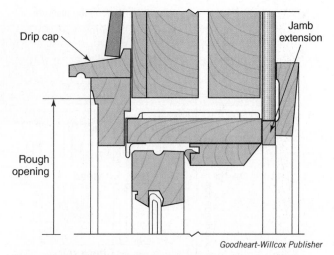

Figure 16-23. Jamb extensions are attached to standard window frames to adjust for various wall thicknesses.

Figure 16-24. These extra-wide jamb extensions are necessary because the wall is unusually thick.

Most window units are installed from the outside of the structure. See **Figure 16-25**. From inside the structure, the unit can be turned at an angle and moved to the outside through the rough opening. When handling the window entirely from the inside, one person should help with leveling and shimming while the other worker holds the window in place.

Most window units come with all hardware in place. This includes sash locks and keepers. The sash lock is often a rotating piece with a cam or inclined lip that tightens against the keeper. The keeper is the piece that attaches to the upper sash (in the case of a double-hung window) to mate with the sash lock as it is rotated. If the sash lock and keeper have not been installed at the factory, they are easily installed with screws. They must be positioned so that the lock releases from the keeper in the unlocked position and tightens securely as the lock is moved to the locked position.

Marvin Windows and Doors

Figure 16-25. Rest the window sill on the bottom of the rough opening and then swing the top into place.

PROCEDURE

Installing a Window with a Flange on a Home with Traditional Housewrap

Note: WRB and flashing details can vary widely. Always be sure to research and understand the methodology behind water management in the system you are using. Adjust the installation practices accordingly. Doing so will ensure a weathertight installation.

1. Apply housewrap to the wall, covering over window openings.
2. Cut the housewrap according to the example in **Figure 16-26**, being sure to fold the top flap up. The flap will be folded down later to create a shingle-lap at the top of the window.
3. Install sill flashing. A variety of sill flashing options, such as flexible flashing tape, liquid-applied membranes, or aluminum/plastic sill pans, are acceptable. It is advisable that the sill framing be sloped toward the outside of the wall to prevent water from infiltrating to the interior.
4. Apply sealant to the sides and top flange of the window. Alternatively, the sides and top of the opening can be sealed. Do not seal the bottom of the window. This leaves a path for water to escape if any were to get in.
5. Place the unit into the rough opening. The sash should be closed and locked. The flange should overlap the exterior sheathing.
6. While resting the window sill on the rough opening, tilt the top into place.
7. Shift the window as needed to horizontally center it in the rough opening.
8. Use shims (wedge blocks or shingles) under the jambs to level the frame. Place a level on the sill and adjust the shims until the sill is perfectly level. There may be a tendency for the sill to sag on mulled (multiple) units.
9. Temporarily secure one corner by tacking a roofing nail (1 1/2″ long) through the lower flange.
10. Double-check for level once more. Adjust as needed.
11. Secure the opposite corner, tacking a nail through the flange.
12. Plumb the side jambs with shims.
13. Check the corners with a framing square or measure the diagonals. If the unit is not square, adjust as needed.
14. Temporarily drive several nails into the top of the side flange.

Continued

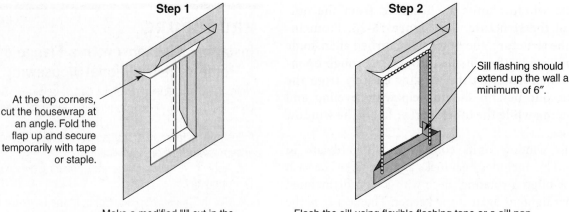

Step 1

At the top corners, cut the housewrap at an angle. Fold the flap up and secure temporarily with tape or staple.

Make a modified "I" cut in the housewrap. Fold the material into the opening and trim excess.

Step 2

Sill flashing should extend up the wall a minimum of 6″.

Flash the sill using flexible flashing tape or a sill pan. Install sealant along the sides and top of the opening. No sealant is used on the bottom.

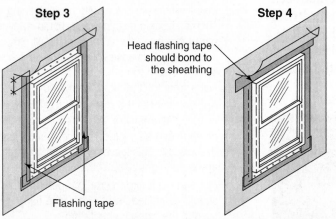
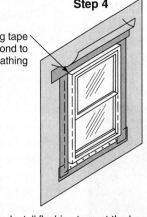
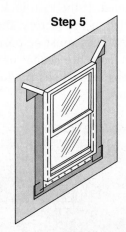

Step 3

Flashing tape

Install the window plumb, level, and square. Cover the side flanges with flashing tape, rolling the tape up onto the window frame slightly.

Step 4

Head flashing tape should bond to the sheathing

Install flashing tape at the head in the same manner as the sides.

Step 5

Fold top flap down and tape the edges.

Goodheart-Willcox Publisher

Figure 16-26. The sequence to flashing a window properly is important. All material laps should allow gravity drainage of water.

PROCEDURE *(continued)*

15. Check over the entire window again to see that it is square and level. Use additional shims if necessary. Be careful not to bow or distort the window frame with the shims.

16. Check the sash for ease of operation. Make sure the spacing is even between the sash and frame.

17. Permanently nail the flanged window in place with 1 3/4″ or longer galvanized roofing nails. Space the nails 12″–16″ OC, or as specified by the window label. See **Figure 16-27**.

18. Cover the side and top flanges with adhesive-backed flashing tape. Start with the sides, then tape the top flange so that the tape overlaps in a shingle fashion.

19. To seal the top flange against water penetration, fold the top flap of housewrap down over the head flashing tape and secure with housewrap tape.

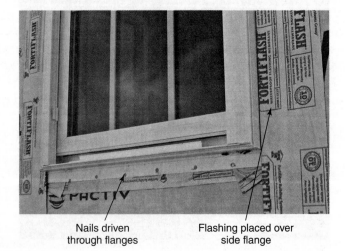

Nails driven through flanges

Flashing placed over side flange

Goodheart-Willcox Publisher

Figure 16-27. A double-hung window with a flange is held in place by nails driven through the flange and sheathing into the building frame. The head and side flanges here are covered with adhesive flashing.

Installing window units without flanges requires the same general procedure. The only difference is that the flashing tape should roll up onto the frame. Special flashing tape is available to make this job easier. Alternatively, the entire rough opening can be flashed in a manner similar to the sill, **Figure 16-28**. Securing the window is accomplished by nailing or screwing metal clips attached to the jambs of the window. Alternatively, some windows can be fastened through the jamb into the rough frame of the wall.

Pro Tip

When installing window units in masonry walls and walls with brick veneer, attach them to the wood bucks using the same procedure as explained for frame construction. When brick veneer is the exterior finish, allow enough clearance for caulking between the masonry and the window sill. In case of shrinkage and settling of the underlying wood framing, the window will be protected from damage.

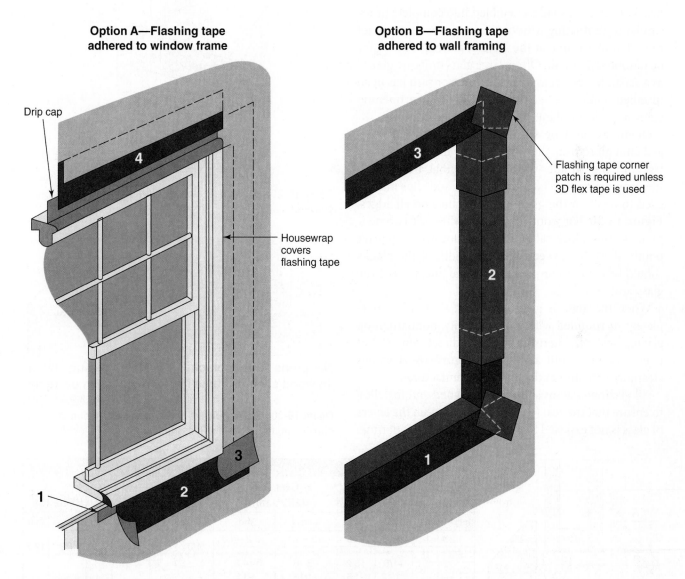

Option A—Flashing tape adhered to window frame

Drip cap

Housewrap covers flashing tape

Option B—Flashing tape adhered to wall framing

Flashing tape corner patch is required unless 3D flex tape is used

Flashing tape is installed in order from bottom to top with all joints overlapping:

1. Sill framing
2. Bottom of window
3. Sides of window
4. Head of window

1. Sill framing
2. Sides of opening
3. Head of opening

Goodheart-Willcox Publisher

Figure 16-28. Extra diligence must be taken when flashing a window without a weathertight nailing flange. Wrapping the rough opening with flexible flashing tape keeps water from damaging the framing. Additionally, another layer of flashing tape should be adhered to the window frame, bonding it to the wall like a flanged window.

16.12 Installing Fixed Units

When the fixed glass panel is of medium size, it is usually mounted in a sash and frame. **Figure 16-29** is a detail drawing of a fixed unit. Often such units are combined with matching ventilating units. The installation of fixed units is essentially the same as for regular windows. Fixed units are usually larger and heavier. Take extra precautions in handling the units and making the installation. Have extra help to prevent dropping the unit.

Large insulating units have 1/4″ thick glass and are heavy. They are usually assembled into complete units, similar to ventilating windows. Sometimes they are not mounted in a frame at the factory, although they may be mounted in a sash. Often, large glass units are glazed as a separate operation after the frame or sash has been installed. Glazing is setting the glass in the opening using a glazing sealant.

Openings must be square, free of twists, and rugged enough to bear the weight of the heavy glass unit. Use only high-grade wood materials that are dry and free from warp. Special setting blocks and clips may be used to position the glass with clearance on all edges, **Figure 16-30**. For wood frames or sashes, use two neoprene setting blocks at least 4″ long located at quarter points along the lower edge. The width of the blocks should be equal to or greater than the thickness of the glass unit.

When the glass is glazed, its edge should be completely surrounded with a high-quality, nonhardening glazing sealant. There must be no direct contact between the glass unit and frame. **Figure 16-31** shows clearances recommended by one manufacturer.

All glazing systems should be designed and installed to ensure that the seal (organic type) between the layers of glass is not exposed to water for long periods of time.

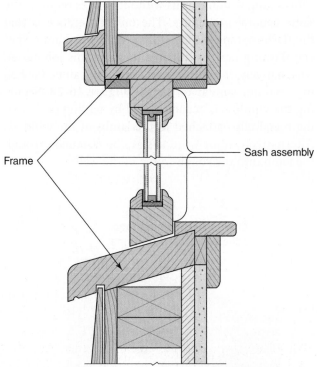

Goodheart-Willcox Publisher

Figure 16-29. Detail of 1″ double-glazed window in a conventional sash and frame.

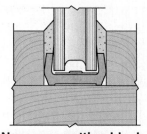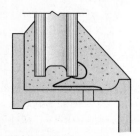

Neoprene setting block in wood sash or frame   **Metal glazing clip in metal sash or frame**

Goodheart-Willcox Publisher

Figure 16-30. Setting blocks and clips. The use of metal clips is limited to welded glass units.

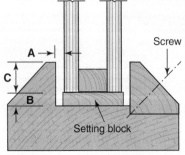

| Glass Thickness | | Dimensional Tolerance | | | | Minimum Clearance | | | | | |
| | | Up to 48″ (1220 mm) | | Over 48″ (1220 mm) | | Face (A) | | Edge (B) | | Bite (C) | |
inches	mm	inches	mm	inches	mm	inches	mm	inches	mm	inches	mm
1/2	12	±1/16	±1.6	+1/8 – 1/16	+3.2 – 1.6	1/8	3.2	1/8	3.2	1/2	12.7
5/8	15	+1/8 –1/16	+3.2 –1.6	+3/16 –1/16	+4.8 –1.6						
23/32	18										
3/4[1]	19										
3/4[2]	19	±1/16	±1.6	+1/8 – 1/16	+3.2 – 1.6	3/16	4.8	1/4	6.4	1/2	12.7
7/8	22	+1/8 –1/16	+3.2 –1.6	+3/16 –1/16	+4.8 –1.6						
31/32	24										
1	25										

[1] 1/4″ (6 mm) Air space
[2] 1/2″ (12 mm) Air space

Goodheart-Willcox Publisher

Figure 16-31. Recommended clearances for sealed insulating glass.

Weep holes (small holes that provide a way for moisture to drain) as recommended by the Sealed Insulating Glass Manufacturers Association (SIGMA) should be included.

Insulating glass units cannot be altered in any way on the building site. It is essential, then, that sash and frames be carefully designed and that specified dimensions are followed. Wooden window frames should be treated with a wood preservative.

Step-by-step procedures for installing insulating glass are given in Appendix B, **Technical Information**. Also provided are standard thicknesses of insulating glass for fixed window units. Details for the construction of a window wall are given as well.

16.13 Glass Blocks

A *glass block* is made of two formed pieces of glass that are fused together to leave an insulating air space between them. They come in different patterns and are usually available in three nominal sizes: 6″ × 6″, 8″ × 8″, and 12″ × 12″. See **Figure 16-32**. All are 3 7/8″ thick. Special shapes are available for turning corners and for building curved panels. The blocks come in both light-diffusing and light-directing types.

Glass blocks have good insulating properties. Used in outside walls, they provide light, help prevent drafts, muffle disturbing noises, cut off unpleasant views, and ensure privacy, **Figure 16-33**. Inside the home, partitions and screens of glass blocks add a pleasant touch to rooms they divide.

Installing glass blocks into a panel is not difficult. Regular masonry tools are used. Even though carpenters seldom make the actual installation, they are required to build the framework. Therefore, knowledge of the design requirements is helpful.

Goodheart-Willcox Publisher

Figure 16-33. This glass block window provides a high level of privacy in a bathroom.

16.13.1 Installing Small Glass Block Panels

Details for installing glass block panels of 25 sq ft or less are given in **Figure 16-34**. In such panels, the height should not exceed 7′, and the width should not exceed 5′. Panels may be supported by a mortar key at jambs in masonry construction or by wood members in frame construction. Wall anchors or wall ties are not required in the joints. Expansion space is only required at the head.

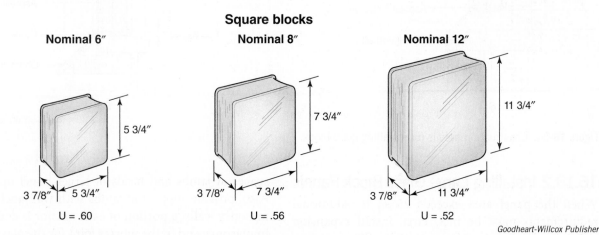

Square blocks

Nominal 6″   Nominal 8″   Nominal 12″

5 3/4″   7 3/4″   11 3/4″

3 7/8″ 5 3/4″   3 7/8″ 7 3/4″   3 7/8″ 11 3/4″

U = .60   U = .56   U = .52

Goodheart-Willcox Publisher

Figure 16-32. Nominal and actual sizes of commonly used glass block.

For panels 25 sq ft and less

1 – Preparation of opening

Keep finger space between brick wythes clear of mortar for depth 3/4″.

Where concrete block walls are used, sash block must be used at jambs of openings.

Apply a heavy brush coat of asphalt emulsion to sill and jambs.

2 – Laying procedure

Allow mortar to key in at jamb.

Use full mortar bed, do not furrow.

Initial cleaning to be done when joints are tooled.

Lay glass block on full mortar bed.

3 – Caulking and cleaning

Caulk at head both inside and outside.

Final cleaning to be done with fine wire brush after final set.

Typical installation details

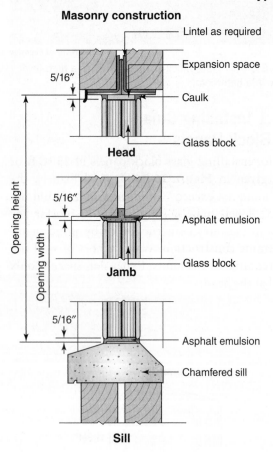

Masonry construction

Lintel as required
Expansion space
Caulk
Glass block
5/16″
Head

Asphalt emulsion
Glass block
5/16″
Jamb

Asphalt emulsion
Chamfered sill
5/16″
Sill

Opening height
Opening width

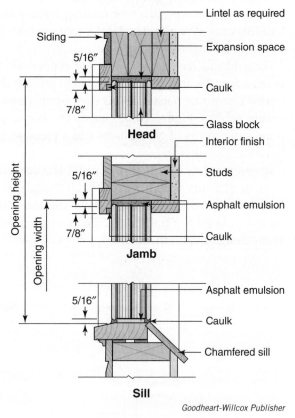

Wood frame construction

Siding
Lintel as required
Expansion space
Caulk
Glass block
5/16″
7/8″
Head

Interior finish
Studs
Asphalt emulsion
Caulk
5/16″
7/8″
Jamb

Asphalt emulsion
Caulk
Chamfered sill
5/16″
Sill

Opening height
Opening width

Goodheart-Willcox Publisher

Figure 16-34. Construction details for installing glass block.

16.13.2 Installing Large Glass Block Panels

When the panel area exceeds 25 sq ft, additional requirements must be met. First, install *expansion strips* (resilient material) to partially fill expansion spaces at jambs and heads of larger panel openings. Also, provide support at jambs with wall anchors. In masonry walls, a portion of each anchor is embedded in masonry and in the mortar joint for the glass block.

Anchors should be crimped within expansion spaces and spaced on 24″ centers to rest in the same joints as wall ties. Anchors are corrosion resistant and 2′ long × 1 3/4″ wide.

Wall ties should be installed on 24″ centers in horizontal mortar joints of larger panels. Lap ties not less than 6″ whenever it is necessary to use more than one length. Do not bridge expansion spaces. Wall ties are also corrosion resistant. They are 8′ long and 2″ wide.

Pro Tip
Glass block panels should never be larger than 10′ wide or 10′ high.

16.13.3 Openings for Glass Blocks

To determine the height or width of the opening required for a glass block panel, multiply the number of units by the nominal block size and then add 3/8″. For example, a panel that is four 8″ blocks wide by five 8″ blocks high requires an opening 32 3/8″ wide and 40 3/8″ high.

16.14 Replacing Windows

Older windows waste a tremendous amount of heating and cooling energy. Homeowners often replace these windows with new units that have excellent weatherstripping, double or triple glazing, and low-e glass. Manufacturers provide a wide range of sizes and a variety of special trim members that are helpful in making the replacement. They also provide detailed instructions for installing their products.

First consideration in window replacement is the type and size of the new units. To determine the size, it is best to remove the inside trim and measure the rough opening. If the trim members are to be reused, pry them off carefully. Remove nails by pulling them through the trim from the back side. Pulling nails through the face of trim may cause the wood to splinter at the surface.

Pro Tip
A type of window known as a replacement window is available. This unit fits inside of the jamb, stiles, and head of the existing unit. These units can be ordered in custom sizes in 1/8″ increments. In this case it is only necessary to discard the sashes, parting stops and weights. This common method makes installation much easier. However, it is critical to ensure accurate measurements are taken before ordering replacement units. This method requires tight tolerances in order for trim pieces to fit properly.

16.15 Skylights

Skylights can make a dramatic difference in the appearance and feeling of a room, **Figure 16-35**. Skylights in bathrooms provide good light and privacy. They may be the best way to secure natural light in stairwells and hallways. Skylights can convert attic space into living space at a fraction of the cost of dormers with regular windows.

In recent years, the *light tube* skylight has come into use. This type of skylight, typically about 12″ in diameter, consists of a clear dome and a tube with a reflective lining that concentrates the light. As shown in **Figure 16-36**, the tube can direct the light to areas not directly beneath the dome. Since only a small opening is made in the roof, light tube units are easier and faster to install than a traditional skylight.

Andersen Corp.
Figure 16-35. Skylights provide natural overhead lighting for this bedroom.

Solatube North America, Ltd.
Figure 16-36. Light tube skylights are often used to supply light to windowless bathrooms or interior hallways.

Rough openings for skylights are framed in about the same way as described for chimneys. A detail drawing of a typical installation is shown in **Figure 16-37**. Flashing flanges included in the skylight frame should be installed according to the manufacturer's directions. Skylights of this type should not be installed in roofs with less than a 3-in-12 slope.

In standard frame construction, a shaft is required to connect the ceiling opening with the roof. **Figure 16-38** shows several types of shafts. It is very important to construct the shaft as airtight as possible to prevent air infiltration and heat gain or loss. Attach to the sides of the shaft insulation that has the same thickness as the ceiling insulation.

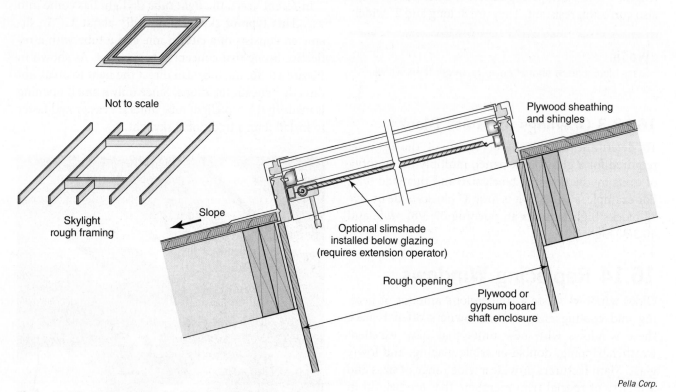

Pella Corp.

Figure 16-37. Cross-section detail for installation of a skylight.

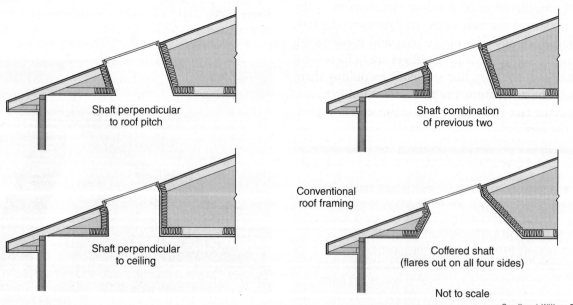

Goodheart-Willcox Publisher

Figure 16-38. A skylight shaft may take any of several shapes.

15.16 Installing Bow and Bay Windows

Bow and bay windows come assembled as a unit, ready for installation. Since these units are larger than standard windows, as well as projecting a considerable distance outward from the building wall, they require extra support. This support is important to prevent sagging of the window and reduce stress on the wall framing.

Cable supports are one way of providing bracing. Cables extend from the outer edge of the window's base to framing members above the window. The cables are concealed with wall covering materials.

A carpenter may choose any of several methods of attaching the cables to the building frame. In one method used for single-story houses, cables are attached to rafter tails in the soffit, **Figure 16-39**. It may be necessary to install bracing between two rafter tails. The cables are secured to cleats at the top and to T-nuts at the bottom. In another variation, the support might be provided by a horizontal or vertical member of the building's wall frame. In a gable installation, the cleats may be anchored to a lintel, studs, second floor band joist, or sill plate.

Extra support along the base is necessary for larger units. If box or angle bay units have a center sash of $5' \times 5'$ or larger, use a steel angle or channel or a properly sized wood member across the bottom, spanning the center sash. Drill holes at either end of the support member for the steel cables. See **Figure 16-40**.

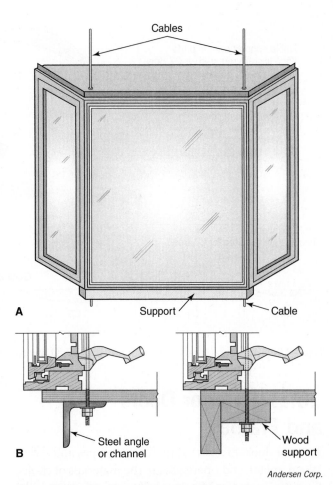

Andersen Corp.

Figure 16-40. Window support. A—Heavier box or angle bay units must have extra support at their bases. B—Channel steel, angle steel, or wood is recommended.

Kits are provided by the manufacturer along with instructions and suggestions for cable locations. See **Figure 16-41**. When preparations are completed, support the window unit with jacks, lever arms, or some other method. Thread the cable through the top and bottom platforms, allowing 1″ of threads below the bottom platform. Attach washers and nuts to each cable. Anchor the cables to the cleats previously attached overhead and draw the cables taut. Remove the temporary support and check the unit for plumb, level, and sash reveal. Readjust as necessary. Finally, tighten the upper nuts to prevent movement and then tighten the lock nut. Notch the head and seat boards and slide them into place. Always follow specific instructions provided by the manufacturer.

Green Note

Regardless of how energy efficient a window or door unit is, using faulty installation practices can leave a home drafty. Always consult manufacturer's recommendations and follow current Energy Star best practices for window and door installation.

Andersen Corp.

Figure 16-39. Supporting a bay window. Cables are attached either to rafter tails or to 2″ lumber spanning the distance between two rafters.

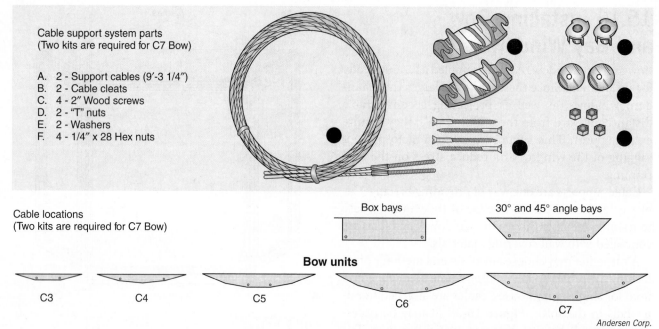

Cable support system parts
(Two kits are required for C7 Bow)

A. 2 - Support cables (9′-3 1/4″)
B. 2 - Cable cleats
C. 4 - 2″ Wood screws
D. 2 - "T" nuts
E. 2 - Washers
F. 4 - 1/4″ x 28 Hex nuts

Cable locations
(Two kits are required for C7 Bow)

Box bays

30° and 45° angle bays

Bow units

C3   C4   C5   C6   C7

Andersen Corp.

Figure 16-41. Manufacturers of prefabricated bow and bay windows have kits and instructions for all types of installations.

16.17 Exterior Doors and Frames

Exterior doors are offered in different types and styles. Wood, metal, and fiberglass are the materials of choice for exterior doors. Metal and fiberglass doors may be more common because wood doors tend to be more expensive.

Flush doors have a flat face applied to each side of a light framework. A core of the same thickness as the frame is inside the frame. Solid core construction consists of wood blocking or particleboard. Hollow core doors use spacing materials at intervals between the facing materials. Metal flush doors usually have a core of rigid insulation that reduces heat transfer through the entryway.

Panel doors have a solid wood framework of stiles (vertical members) and rails (horizontal members) that hold panels. Panels are thinner, often decorative, parts that fill in spaces between the rails and stiles. The panels may be glass, solid wood, or wood louvers.

Outside door frames and doors are installed at the same time as the windows following similar procedures. Secondary and service entrances usually have frames and trim members to match the windows. Main entrances, however, often contain additional elements that add an important decorative feature, **Figure 16-42**.

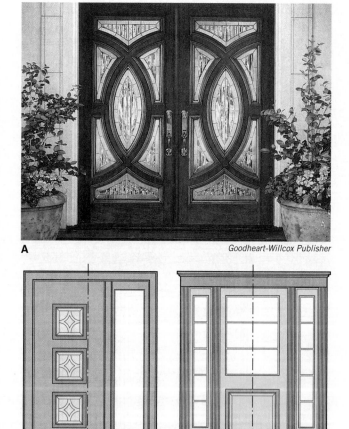

A

Goodheart-Willcox Publisher

B

Goodheart-Willcox Publisher

Figure 16-42. A—Main entry doors. B—Frames and doors for main entrances as they appear on elevation drawings.

Exterior doors in residential construction are nearly always 6′-8″ high, although 7′-0″ sizes are available. Main entrances usually are equipped with a single door that is 3′-0″ wide. Narrower (2′-8″ and 2′-6″) sizes are used for rear and service doors.

Outside door frames, like windows, have heads, jambs, and sills. The head and jambs are made of 5/4″ stock since they must carry not only the main door but also screen and storm doors.

Door frames are manufactured at a millwork plant and arrive at the building site either assembled and ready to install (prehung) or disassembled (knocked down, or KD). Sometimes KD units are assembled by the dealer or distributor. It is relatively easy to assemble door frames on the job when the joints are accurately machined and the parts are carefully packaged and marked. While details of a door frame may vary by manufacturer, the general construction is the same, **Figure 16-43**. The head and jambs are rabbeted, usually 1/2″ deep, to receive the door. In residential construction, outside doors swing inward and the rabbet must be located on the inside.

Stock door frames are designed for standard wall framing. However, they can also be adapted to stone or brick veneer construction, as shown in **Figure 16-44A**. Stock frames can also be fitted with extension strips that convert the frames to fit a greater wall thickness, **Figure 16-44B**. Extension strips are also used on window frames.

Doorsill designs vary considerably. However, the IRC requires the top of the sill to be no more than 1/2″ above the finished floor. Sills may be made of wood,

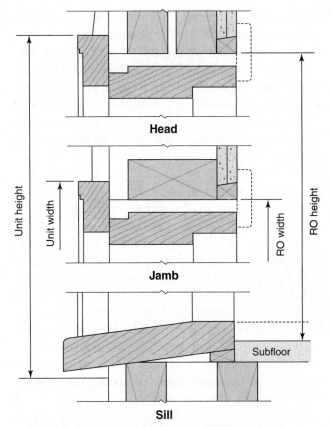

Head

Jamb

Subfloor

Sill

Calculations for determining frame and rough opening sizes:

Unit height = Door + 4 1/2″
Unit width = Door + 4″
RO height = Door + 2 1/2″
RO width = Door + 2 1/2″

Goodheart-Willcox Publisher

Figure 16-43. Always check drawings for exterior door frame details and sizes. Make sure rough openings are the correct sizes.

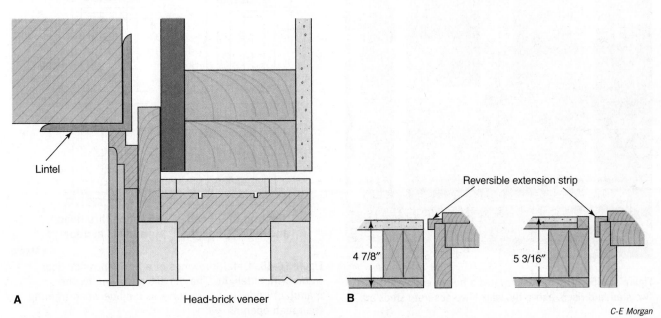

Lintel

A   Head-brick veneer

Reversible extension strip

4 7/8″   5 3/16″

B

C-E Morgan

Figure 16-44. A—Detail of how to install the head of a door frame in brick veneer construction. B—Stock frames can be made wider. These reversible extension strips convert 4 5/8″ jambs to either 4 7/8″ or 5 3/16″.

metal, stone, or concrete. The outside stoop at the entrance is placed just below the doorsill or it may be lowered by a standard riser height (7 1/2"), **Figure 16-45.**

Setting a doorsill so it will be level with the finished floor requires cutting away a section of the rough floor. Part of the top edge of the floor joist must also be cut away. This is done at the time the door frame is installed. The framing of the rough opening, however, comes earlier—before the door frame is delivered to the job. A carpenter must carefully check the working drawings. The size of the rough opening is usually included in the door and window schedule. The height of the opening is shown in detail sections. For standard construction, the rough opening can be calculated by adding about 2 1/2" to the door width and height. When this information is not included in the working drawings, consult the manufacturer's literature. This contains not only rough opening requirements, but also detail drawings, **Figure 16-46.** The latter are especially important for front entrances that consist of more complex structures or have fixed sidelight window units.

Setting the threshold and hanging the door are a part of the interior finishing operation. These operations are covered in Chapter 22, **Doors and Interior Trim.** After the installation is complete, lightly tack a piece of

1/4" or 3/8" plywood over the sill to protect it during further construction work. At this time, many builders prefer to hang a temporary combination door so that the interior of the structure can be secured, providing a place to store tools and materials.

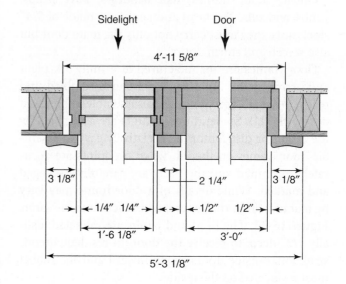

Plan

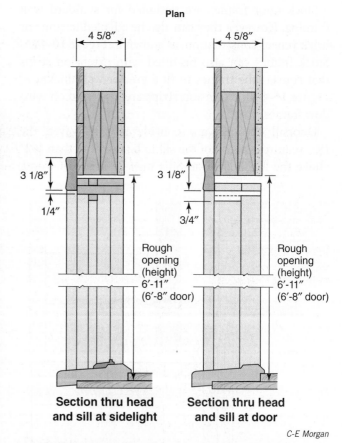

Section thru head and sill at sidelight   **Section thru head and sill at door**

C-E Morgan

Figure 16-46. Detail drawings of a front entrance door frame with sidelight. The dimensions given by the manufacturer's literature serve as a guide when framing the rough opening.

![Photo of a front entrance with double doors and sidelights]

Pease Industries, Inc.

Figure 16-45. This door is elevated a height equivalent to two standard risers above the walk. Two concrete steps are added.

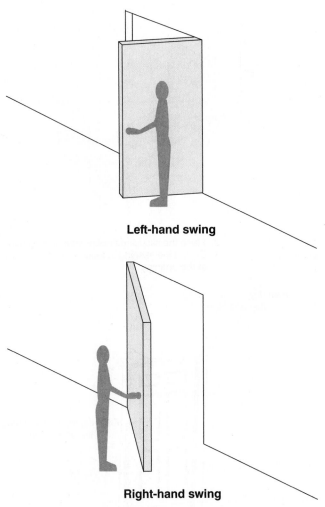

Left-hand swing

Right-hand swing

Goodheart-Willcox Publisher

Figure 16-47. Determining the hand of a door. Face the door from the room it opens into. The side the lockset is on as you are facing the door determines the hand of the door.

16.17.1 Determining the Hand of a Door

Swing-type doors are either left-hand or right-hand swing. There are a couple of ways to determine the hand of a door. The simplest is to stand in front of the closed door, ready to grasp the knob and pull it toward you. If the door knob is in front of your right hand, it is a right-hand door. If the knob is on your left, it is a left-hand door. See **Figure 16-47**.

Pro Tip

When setting door and window frames, make sure no adjustments are necessary before completing the final nailing process. Never drive any of the nails completely into the wood until all nails are in place and a final check has been made.

PROCEDURE

Installing a Prehung Entry Door

1. Check the dimensions of the rough opening. For proper clearance, the opening should be about 3/4″ wider and 1/2″ higher than the unit dimensions to allow for shims.
2. Check that the opening is plumb in both directions, square, and level.
3. Check for clearance between the bottom of the door and the finish floor.
4. Install sill pan or flash the sill with flexible flashing tape that extends up the studs about 6″. Wrap studs with flashing tape, ensuring the tape overlaps the sill flashing in a shingle fashion. Alternatively, tape can be adhered to the brick mold after installation. This method is similar to a flanged window installation. See **Figure 16-48**.
5. Apply a double bead of caulking compound to the sill flashing to seal under the door sill.
6. Lay the door unit flat and apply caulk to the underside of the sill. Set the unit in place.
7. Check the hinge-side jamb in both directions for plumb. If there are thin spacer shims located between the frame and door, do not remove them until the frame is firmly attached to the rough opening.
8. Slip shims into the space between the hinge-side jamb and the rough opening and tack to prevent movement.
9. Insert shims on the hinge-side jamb behind the hinges between the side jambs and rough opening. Install them at the top, bottom, and midpoint.
10. Place shims behind the hinges to maintain the spacing when the longer hinge screws that secure the hinges through the jamb and into the rough opening are installed. Manufacturers usually recommend that at least two of the screws in the top hinge be replaced with 2 1/4″ screws.
11. Shim the lock-side jamb top and bottom, making sure to maintain spacing between the jamb and door. Tack it temporarily and check for plumb and square.
12. Drive 12d casing nails through the jambs and shims and into the structural frame members.
13. If the unit has an adjustable threshold, adjust it for a smooth contact with the bottom edge of the door.
14. Install a drip cap flashing over the head of the door, tying the flashing into the WRB so that it is weathertight.

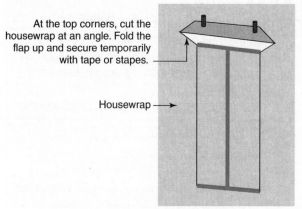

At the top corners, cut the housewrap at an angle. Fold the flap up and secure temporarily with tape or stapes.

Housewrap →

1. Make a modified "I" cut in the housewrap. Fold the material into the opening and trim excess.

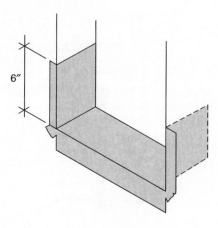

6″

2. Flash the sill using flexible tape or a sill pan. Extend the flashing at least 6″ up the sides of the opening.

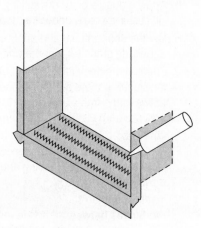

3. Apply sealant where the door threshold will rest per door manufacturer instructions.

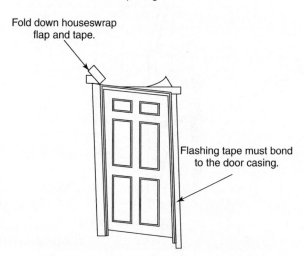

Fold down houseswrap flap and tape.

Flashing tape must bond to the door casing.

4. Install door by tilting it in place. Ensure it is plumb, level, and square. Install side flashing and then top flashing tape, rolling the tape onto the door frame slightly. Fold top flap down and tape the edges.

Goodheart-Willcox Publisher

Figure 16-48. Construction details like these differ from one manufacturer to the next. Be sure to study the manufacturer's information before installation.

Pro Tip
After a prehung door unit is installed, some builders remove the door from its hinges and carefully store it to avoid damage during continuing construction. However, for security reasons, it may be necessary to install a lock and leave the door in place.

16.18 Sliding Glass Doors

To accommodate outdoor living, patio doors are often included in residential designs. The French or casement type doors are also used for this purpose. Installation of French doors is similar to installation of single prehung doors, except the unit is wider and heavier.

A sliding glass door rides on nylon or stainless steel rollers. This makes it easy to operate. When equipped with quality weatherstripping and insulating glass, heat loss and condensation are limited to satisfactory levels, even in cold climates. A sliding glass door is available as a factory-assembled frame and door unit. Parts and installation details are similar to those for sliding windows.

Sliding glass door units contain at least one fixed and one sliding panel. Some may have three or four panels. The type of unit is commonly designated by the number and arrangement of the panels as viewed from the outside. See **Figure 16-49**. The letters *L* or *R* indicate which way the operating panel opens. Some manufacturers use the letter *X* to indicate operating panels and

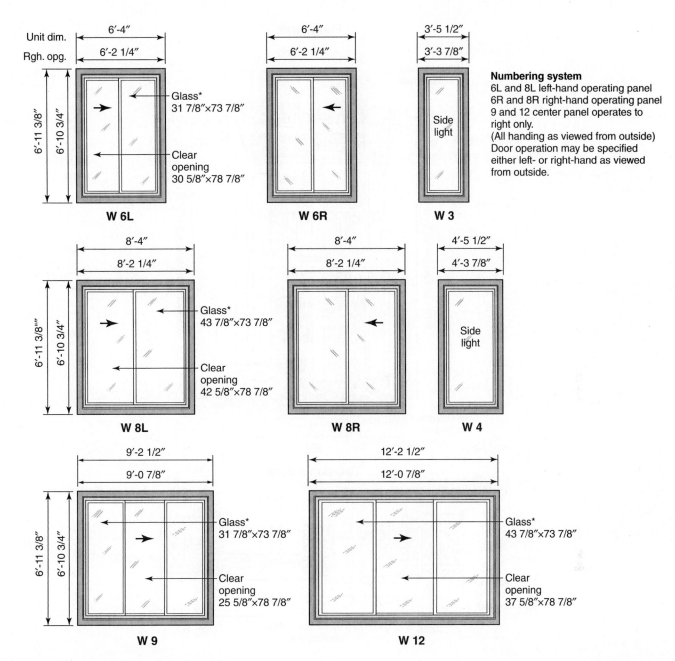

Figure 16-49. Like windows, doors must be flashed to keep water out of the framing.

O for fixed panels. Sliding glass door units are glazed with tempered (safety) glass.

Manufacturers of sliding glass doors provide detailed instructions for installing their particular product. Study this material carefully before proceeding with the work. Construction details vary from one manufacturer to another, as do unit sizes and rough opening requirements. See **Figure 16-50.** If detail drawings or rough opening sizes are not included in the architectural plans, obtain the information from the manufacturer's literature.

The installation of sliding door frames is similar to the procedure described for regular outside doors.

Before setting the frame in place, lay a bead of sealing compound across the opening to ensure a weathertight joint. If the heavy, glass doors are to properly slide, the sill must be level and straight.

Plumb side jambs and install wedges in the same way as you would install regular door frames. After careful checking, complete the installation of the frame by driving galvanized casing nails through the side and head casings into structural frame members.

At this point, many carpenters prefer to check the fit between the metal sill cover and doors and then, rather than installing them, carefully store these items.

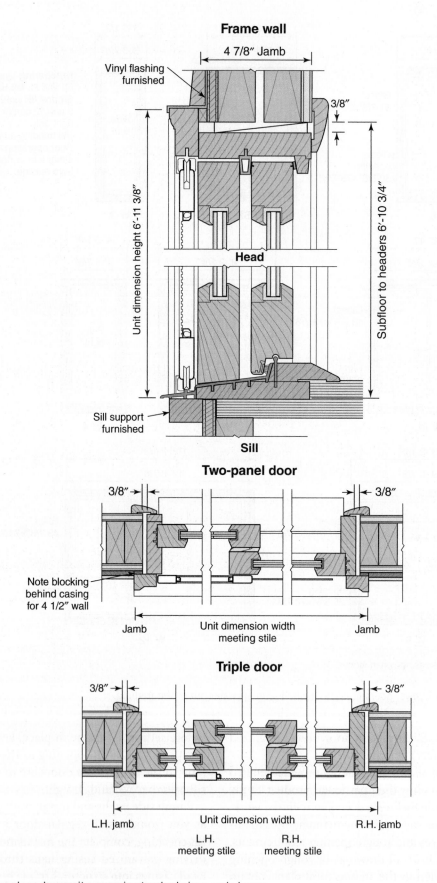

Figure 16-50. Sliding glass door units come in standard sizes and shapes.

Andersen Corp.

The opening is temporarily enclosed with plywood or polyethylene film attached to a frame. Then, during the finishing stages of construction, after inside and outside wall surfaces are completed, the sill cover, threshold, doors, and hardware are installed.

> **Pro Tip**
> After installing large glass units, it is considered good practice to place a large *X* on the glass while the building remains under construction. Use masking tape or washable paint to apply the *X*. This alerts workers to not walk into the glass or damage it with tools and materials.

16.19 Garage Doors

Basically, there are three types of garage doors—hinged or swinging, swing-up, and overhead doors. The overhead door has almost completely replaced the hinged and swing-up door types. See **Figure 16-51**.

Wooden garage doors are constructed much like exterior entry doors. Flush-type doors with foamed plastic cores are available for installations that require high levels of sound and thermal insulation. Garage doors are also manufactured from steel, aluminum, and fiberglass. Many designs are available to match contemporary or traditional architecture. Overhead doors are generally available in widths from 8′ to 20′ in 1′ increments and heights of 7′ or 8′. Greater heights can be special ordered.

For use in areas subject to severe storms with high winds, hurricane-proof garage doors have been developed. These doors feature heavier construction than conventional doors and are equipped with steel bracing posts and locking hardware to withstand hurricane-force winds. See **Figure 16-52**.

A   *Stanley Door Systems*

A   *Steve Olewinski*

B   *Steve Olewinski*

Figure 16-51. A—Roll-up garage doors are hinged in sections. Note how the sections move upward and tilt to horizontal position. B—Rollers attached to each section run in a special steel track to support the door.

B   *Stanley Door Systems*

Figure 16-52. Hurricane-proof garage door. A—Exterior view. B—The inside view shows steel posts and heavy-duty locking hardware.

16.19.1 Garage Door Frames

Frames for garage doors can include side jambs and a head like exterior entry doors or are built from framing lumber when the garage is framed. The size of the frame opening is usually the same size as the door. However, the manufacturer's specifications and details should be checked before placing the order. The door frame is often wrapped in metal or can be trimmed out to match the doors and windows. Custom frames can also be ordered.

Figure 16-53 shows typical garage doorjamb sections for both wood frame and masonry construction. Note the thickness (2″ nominal) of the heavy inside frame to which the track and hardware will be mounted. The width of this member should be at least 4″ with no projecting bolt or lag screw heads.

The rough opening width for frame construction is normally about 3″ greater than the door size. The height of the rough opening should be the door height plus about 1 1/2″, as measured from the finished floor.

Give careful consideration to the inside height. A minimum clearance between the top of the door and the ceiling must be provided for hardware, counterbalancing mechanisms, and the door itself when open. On some special, low-headroom designs, this distance may be as little as 6″. Always be sure to check the manufacturer's requirements for each door.

To replace an existing overhead garage door, take exact measurements of the opening before placing the order.

Dimensions should include both width and height in feet and inches. Check the old frame for damage. Replace it if necessary. Also, check that there is sufficient side room for attaching the vertical track. Normally, about 3 1/2″ is needed for standard extension or torsion bar springs. Clearance requirements should be checked with the supplier.

When installing garage door frames, follow the general procedure used for regular door frames. Make certain the jambs are plumb and the head is level. Before beginning the installation, set up several sawhorses to support the door sections as they are readied for installation. Manufacturers always supply detailed directions and procedures for the installation of their products. Carefully follow these printed materials. See **Figure 16-54**.

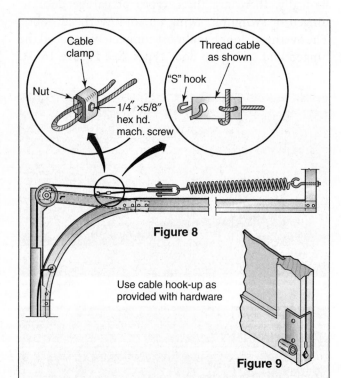

14. Install eye bolts at rear of horizontal tracks.
15. Assemble springs and sheave wheels. Hook springs to eye bolts as shown above.
16. Hook clip ends of cables to bottom bracket pins. See Figure 9.
17. Thread cables over sheaves. Put "S" hooks in about center of row of holes in top of horizontal channel. Pull cables thru "S" hooks until springs are stretched enough to hold door open. Install cable clamps as shown above.
18. Close door. Lengthen or shorten springs to correctly balance door by moving "S" hooks in channel holes or by moving cable clamps. This should always be done with door in open position.
19. Apply pull rope handles and lock assembly.
20. Final spring adjustment should be made after door has been glazed and finish painted.
21. Keep all hinges, rollers, and bearings well oiled.

Goodheart-Willcox Publisher

Figure 16-54. Partial sample of a manufacturer's installation instructions. These are generally well illustrated and easy to understand.

Goodheart-Willcox Publisher

Figure 16-53. Typical jamb construction for garage door frames when an overhead door is used.

16.19.2 Hardware and Counterbalances

Garage door hardware must be well designed so the door will operate easily. The track, hinges, and bolts should be made of galvanized steel. The metal must be heavy enough to last the life of the door. To offset the weight of the door, various counterbalancing devices are used. Two of the most common types are the extension spring and the torsion spring. The torsion spring and its mechanism are more expensive, but provide a smoother and more consistent action. It is especially recommended for wide doors and those that are heavier in construction. See **Figure 16-55**. Most garage doors in new construction are equipped with an electric-powered door opener, **Figure 16-56**.

Safety Note

Counterbalancing torsion springs used on garage door systems are under high tension and store considerable potential energy. They can cause injury and damage if suddenly released. Special tools must be used to adjust a torsion spring. Such work should only be done by a properly trained technician.

PROCEDURE

Installing an Overhead Garage Door

1. Temporarily tack door stops in place on the doorjambs.
2. Place the first door section on the sawhorses and complete any needed preparation. For example, some manufacturers provide a channel for weatherstripping. This should be installed before proceeding.
3. Position the section in the door opening, centering it and leveling it using shims.
4. Temporarily tack the section to the stops to maintain the position.
5. Assemble the door sections, one on top of the other, in the opening. As you assemble, attach hinges and other hardware.
6. Attach rollers to the door.
7. Slip vertical track sections onto the rollers and attach track sections to the jambs.
8. Mount horizontal track sections. This may involve nailing wood supports to bridge across joists.
9. Raise and prop door in the open position.
10. Attach the counterbalancing mechanism.
11. Open and close door and make necessary adjustments to the track.
12. Adjust stops for a smooth, tight fit.

A
Clopay Building Products

B
Clopay Building Products

Figure 16-55. A—Extension springs are mounted just above the horizontal track, perpendicular to the closed garage door. Extension springs get their lifting power by stretching. B—Torsion springs are mounted above the closed door, parallel and horizontal to the top section of the door as shown here. Torsion springs wind and unwind to provide lifting power.

karenfoleyphotography/Shutterstock.com

Figure 16-56. Electric-powered garage door openers are common. They allow remote-control opening and closing of the door without leaving the car.

Chapter Review

Summary

- To make a house weathertight and minimize air infiltration, windows and exterior doors must be properly installed.
- There are three basic types of windows: sliding (double hung and horizontal), swinging (casement, hopper, or awning), and fixed (usually used in combination with swinging or sliding units).
- To control heat gain and loss, energy-efficient windows are double or triple glazed and many have a low-e coating on the glass to reflect infrared rays (heat waves).
- Building drawings usually include tables called window and door schedules that list all of the sizes and types of doors and windows used in the project. This information is important to a carpenter, since it shows the rough opening size needed for each door or window.
- Skylights, bow windows, and bay windows have special installation requirements.
- Doors are made in flush and panel styles and are installed in much the same way as windows. Prehung doors simplify and speed up installation.
- French doors and sliding glass doors leading to patios or decks are widely used.
- The garage door is another specialized type of exterior door. These are sectional doors that roll upward on a track.

Review Questions

Answer the following questions using the information in this chapter.

1. _____ is the original window construction material.
   A. Steel
   B. Wood
   C. Fiberglass
   D. Aluminum

2. *True or False?* Sashes are used to fill the layers between thermal pane glass, or as an overlay on a large sheet of glass.

3. The side frame pieces of a window sash are _____ and the top and bottom pieces of the sash frame are the _____.
   A. jambs, mullions
   B. heads, sills
   C. stiles, rails
   D. stiles, sills

4. A window consisting of two sashes that slide up and down past each other in the window frame is called a _____ window.
   A. double-hung
   B. horizontal sliding
   C. casement
   D. hopper

5. *True or False?* A window that is hinged at the top and the bottom of the sash opens outward is an awning window.

6. A _____ window can be used in combination with any other window unit.
   A. hopper
   B. jalousie
   C. fixed
   D. casement

7. A single pane of glass in a window has a U-factor of about _____.
   A. 0.2
   B. .05
   C. 1
   D. 1.1

8. *True or False?* Window size information is generally found in the window and door schedule of an architectural drawing.

9. To adjust for various wall thicknesses, a(n) _____ is applied to standard window frames.
   A. flange
   B. air gap
   C. jamb extension
   D. special wall stud

10. *True or False?* Most window units are installed from the inside of the structure.

11. After a window unit has been temporarily set in the rough opening, the next step is to _____.
    A. fasten it securely in place
    B. align the top to the proper height
    C. plumb the side jamb
    D. level the sill

12. Large, insulating, fixed window units are made from glass that is _____″ thick.
    A. 1/8
    B. 1/4
    C. 1/2
    D. 4

13. _____ are used to partially fill expansion spaces at jambs and heads of larger glass block panel openings.
    A. Expansion strips
    B. Wall anchors
    C. Wall ties
    D. Small glass block panels

14. When figuring the size of the opening for glass block panels, multiply the number of units by the nominal block size and add _____.
   A. the area of the opening
   B. 3/8″
   C. 8″
   D. 25 sq ft

15. To determine the size for a window replacement, it is best to remove the inside trim and measure the _____.
   A. rough opening   C. head
   B. jambs   D. frame

16. _____ supports are one way of providing bracing for bow and bay windows.
   A. Frame   C. Cable
   B. Steel   D. Tie

17. A door that is made up of rails and stiles holding wood panels is called a(n) _____ door.
   A. flush   C. sliding
   B. panel   D. entry

18. The top of the doorsill sho0uld not be more than _____″ above the finished floor.
   A. 1/8   C. 1/2
   B. 1/4   D. 2 1/2

19. *True or False?* When a door swings toward you with the lockset on the right, it is a right-hand door.

20. *True or False?* The hinged garage door has almost completely replaced the swing-up and overhead door types.

21. The _____ is recommended for wide garage doors and those that are heavier in construction.
   A. extension spring
   B. torsion spring
   C. swing-up
   D. electric-powered door opener

Apply and Analyze

1. Give two reasons why casement windows are ideal for windows above kitchen cabinets.
2. How can you lower the U-factor of modern window spaces?
3. How do *low-e* windows help reduce heat loss in winter months?
4. What is the difference between installing window units with flanges versus without flanges?
5. Why are skylights beneficial additions to attic spaces?

6. Why is extra support important for bow and bay windows?
7. How do you install an overhead garage door?

Critical Thinking

1. A large replacement window was installed by a local contractor. It is winter and there is a lot of condensation developing on the glass. What factors might cause this to happen? What questions might you ask the homeowner to determine the cause?
2. While working on a remodeling job, you remove an entry door. Your experienced coworker mentions that the door was not flashed properly when it was originally installed. What are some visible signs she may have picked up on? What kind of problems could this issue cause over time?
3. Triple-pane windows are expensive. Research costs for comparable double- and triple-pane windows in your area by calling a supplier. Perform a cost-to-benefit analysis to determine what factors, such as length of home ownership or climate region, might justify an upgrade from double-pane to triple-pane glass when building a new home. For the sake of this exercise assume the home has 10 windows that are 36″ wide and 60″ tall.
4. While performing the exercise in question #3, what other factors did you determine might be important when making this decision?

Communicating about Carpentry

1. **Speaking and Listening.** Working in small groups, create a poster that illustrates the different windows and exterior doors described in this chapter. As you work with your group, discuss the meaning of each term. Afterward, display your posters in the classroom as a convenient reference aid for discussions and assignments.
2. **Speaking.** Describe the steps, from beginning to end, of removing a broken garage door and replacing it with a new one. Include details of cost of removal, new materials, and any other relevant information.

CHAPTER 17

Exterior Wall Finish

LEARNING OBJECTIVES

After studying this chapter, you will be able to:

- Identify the parts of a cornice and rake.
- Describe cornice and rake construction.
- Illustrate approved methods of flashing installation.
- Describe how wood siding and shingles are applied.
- Estimate the amount of siding or shingles required for a specific structure.
- Discuss the proper application of wood bevel siding.
- Discuss the proper application of fiber-cement siding.
- Explain proper installation practices of vinyl and aluminum cornice materials.
- Describe proper installation of vinyl siding and accessories.
- List the most common siding choices and their characteristics.
- Discuss exterior insulation and finish systems (EIFS) and their application.
- Describe the tools and materials used when finishing veneer walls.

TECHNICAL TERMS

bevel siding	exterior trim	receiving channel
board-and-batten siding	fascia	single coursing
board-on-board siding	fascia backer	snub cornice
box cornice	fiber-cement siding	soffit
brake	frieze	soft-coat system
buttlock	hard-coat system	starter track
channel rustic siding	heel	sub-fascia
cornice	ledger strip	toe
double coursing	lookout	top lock
double-channel lineal	mortar	utility trim
drop siding	open cornice	veneer wall
exterior finish	preacher	vinyl siding
exterior insulation and finish system (EIFS)	rake	wood shingle

The term *exterior finish* includes all exterior materials of a structure. It generally refers to the roofing materials, cornice trim boards, wall coverings (such as siding), and trim members around doors and windows. The installation of special architectural woodwork at entrances or the application of a ceiling to a porch or breezeway area is also included under this broad heading.

Previous chapters describe the application of the finished roof and the air, vapor, and water control layers, as well as installation of the trim around windows and outside doors. This chapter covers the construction and finish of cornice work and the materials and methods used to provide a suitable outside wall covering.

It is important to understand that the exterior wall finish is the first line of defense against water infiltration.

However, it is not the only line of defense. Before the installation of the exterior wall finish, it is critical that the weather-resistive barrier (WRB) has been installed properly, diverting any bulk water from infiltrating the wall system.

Wall covering and cornice system installations differ based on the materials and type of system that is used. For instance, wood, fiber cement, and composite materials are all installed using similar processes, while vinyl, aluminum, and steel use a different system that incorporates interlocking pieces and receiving channels. Additionally, stucco, EIFS, and masonry veneer systems use an altogether different process. For this reason, it is important to differentiate between the systems. Although the installation techniques of these systems may differ significantly, the goal is ultimately the same: to give the building an attractive finish that shields the WRB from wind, driving rain, and harmful ultraviolet radiation. This chapter begins by examining wood and similar systems, then moves to vinyl and aluminum, and finishes by looking at masonry systems.

17.1 Cornice Designs and Terms

Cornice is a term for all of the parts that enclose or finish off the overhang of a roof. It usually includes the fascia board, lookouts and a soffit for a closed cornice, along with any decorative moldings used to conceal joints. Highly decorative cornice systems were once commonplace, **Figure 17-1**. However, increased labor and maintenance costs have driven builders and architects to resort to simpler systems.

Cornices provide a finished connection between the wall and roof. The cornice at the eve often includes a gutter, or eaves-trough, to catch and divert water as it runs off the roof. If the building has a gable or gambrel end, that cornice may be referred to as raked or sloping. The cornice may have multiple functions on a building. Aside from adding architectural or aesthetic value, the cornice overhang protects the wall finish from water by directing runoff away from the wall and in many cases capturing it in a gutter. In some historic buildings, the cornice houses a hidden eaves-trough, as in **Figure 17-2**. Often, the underside of a cornice, called the *soffit*, is used as an air intake for attic ventilation. In homes built with energy efficiency in mind, the cornice can be a functional shading device, keeping the intense summer sun from coming through the windows.

During the construction process, the cornices of a home are generally finished off before the siding. This gives the wall finish a clean termination point. The style of the house determines the cornice design. There are three basic cornice styles: box, open, and snub. A fourth, which might be considered a variation of the box style, is used with truss rafters. See **Figure 17-3**.

The *box cornice* is the most common. It completely encloses the rafter and other parts of the overhang.

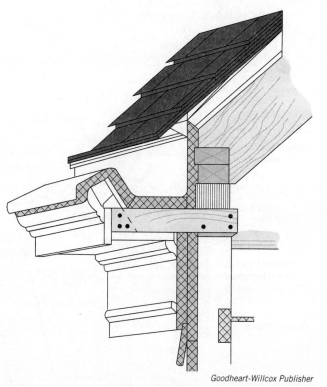

Goodheart-Willcox Publisher

Figure 17-2. Integrated eaves-troughs were typically lined with copper or sheet metal and are notorious for leaking. They have generally been retained for historic preservation purposes only.

Goodheart-Willcox Publisher

Figure 17-1. Decorative cornice assemblies, such as the one on this brick Victorian, were once commonplace in new homes.

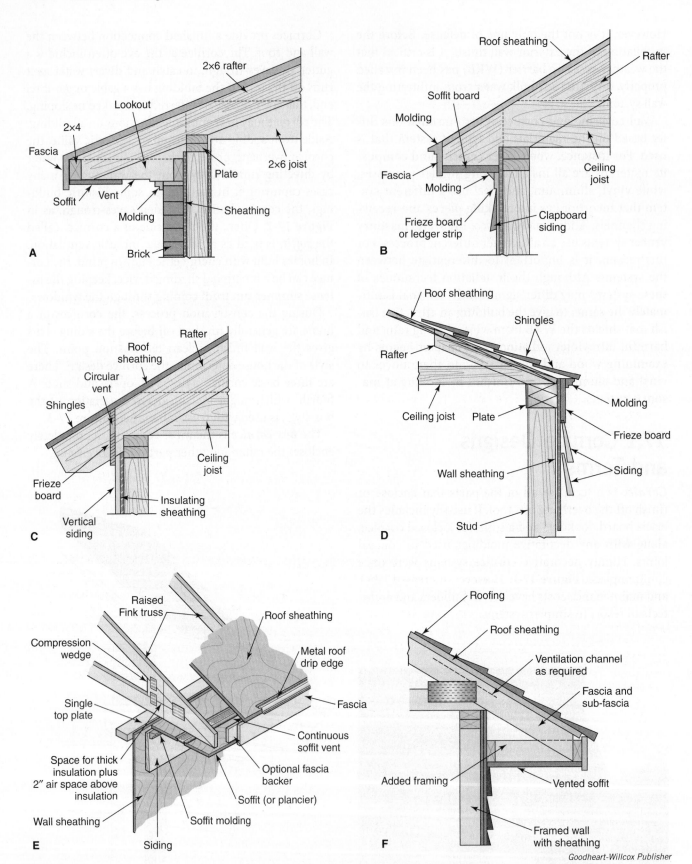

Figure 17-3. Typical cornice details for different architectural styles. A—Normal box cornice. B—Narrow box cornice. C—Open cornice. D—Snub cornice. E—Box cornice applied to truss rafters. F—Box cornice applied to truss with tails

Goodheart-Willcox Publisher

An *open cornice* may have no enclosing parts. There is no soffit covering the underside and often no fascia board concealing the rafter tails. The underside of the roof sheathing, since it is exposed at this point, must be attractive. Thus, tongue-and-groove, beaded, or V-grooved lumber is used. Open cornices are used when the overhang is very large or when rafters are large, laminated, or solid beams. They can also be used to add architectural interest.

A *snub cornice* has no overhang. The rafters end at the sidewall. This style does not provide walls or windows with any protection from the weather. It is sometimes used to save material and labor.

Roof trusses are more common than site-built rafters. Depending on the truss construction, additional framing is often needed to support a wooden cornice system. Therefore, vinyl or aluminum materials are commonly used. Some trusses, however, are designed for use with a snub cornice, eliminating the need for soffit altogether.

Figure 17-3 shows details of different cornice constructions. Several closed or box cornice designs are illustrated in **Figure 17-4**.

17.1.1 Parts of the Cornice

Figure 17-5 shows structural and trim parts of a box cornice that are constructed of wood or a similar material. It is important to note that variations can exist with all systems. Always check manufacturer recommendations when using a product such as PVC, hardboard, or composite.

The *fascia* is the main trim member. It is horizontally attached along the ends of the rafters. A *ledger strip* is a 1″ or 2″ ribbon horizontally nailed along the wall to support the lookouts. *Lookouts* are usually 2×4 pieces attached between the rafters and the wall. Their purpose is to provide support for soffit materials.

In box cornice systems, the soffit closes off the space between the wall and the fascia board. This system can be constructed from a range of materials, including plywood, hardboard, solid lumber, and fiber cement. If the soffit is designed to ventilate the attic space, vents are provided along its width or at intervals. Sometimes, a 1″ or 2″ wood nailing strip is attached to the rafter ends in a box cornice to support the fascia. This strip is called a *fascia backer*. When PVC and composite materials are used, a stronger sub-fascia board is generally required. *Sub-fascia* is a framing member behind the fascia trim that provides support and backing for fascia trim. Along with the lookouts and ledger, these components provide a frame to which the soffit material can be applied.

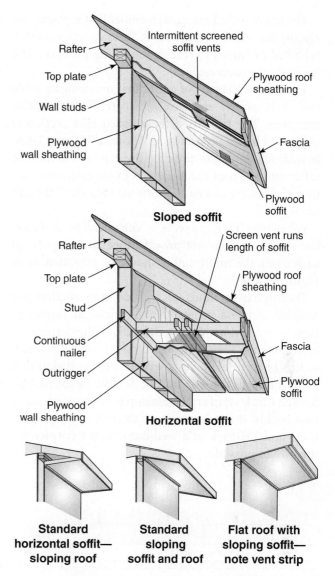

Council of Forest Industries of British Columbia

Figure 17-4. Three different box cornice designs.

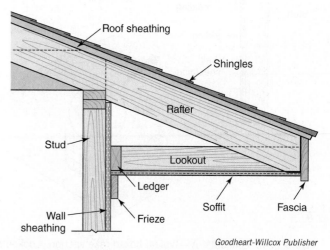

Goodheart-Willcox Publisher

Figure 17-5. Parts of a typical box cornice. Soffit materials are often prefabricated, usually from metal, plastic, or plywood.

The *frieze* is the horizontal member that is placed flat against the wall below the soffit. Its lower edge is often rabbeted or furred to receive the siding material. The frieze is often decorative.

The *rake* is the part of a roof that overhangs a gable end. It is usually enclosed with carefully fitted trim members. The trim used for a boxed rake section is supported by the projecting roof boards, **Figure 17-6A**. In addition, lookouts or nailers are fastened to the end rafter and the roof sheathing. As in cornice construction, these serve as a nailing base for the rake soffit and fascia.

When the rake projects a considerable distance, the sheathing does not provide adequate support. In such cases, the roof framing should be extended, as in **Figure 17-6B**.

Traditional methods of cornice construction are simpler and less ornate than in classical construction. Notice the architectural details of the cornice return shown in **Figure 17-7**. Incorporating details such as this in a project takes a level of time and skill that many builders are not willing or able to invest. Although this level of craftsmanship is not seen on many new builds, it is advantageous to mimic the proportions and the look of a well-constructed cornice return when possible.

Wermers Design and Architecture

Figure 17-7. A traditionally detailed and proportioned cornice adds architectural appeal to a building.

17.2 Wall Sheathing and Flashing

Siding can be applied over various sheathing materials. However, the finish on wood siding will last longer if applied over a ventilated rainscreen. When the sheathing is solid wood, plywood, or oriented strand board, the siding is directly nailed to the material at about 24″ intervals. End joints in the siding may occur

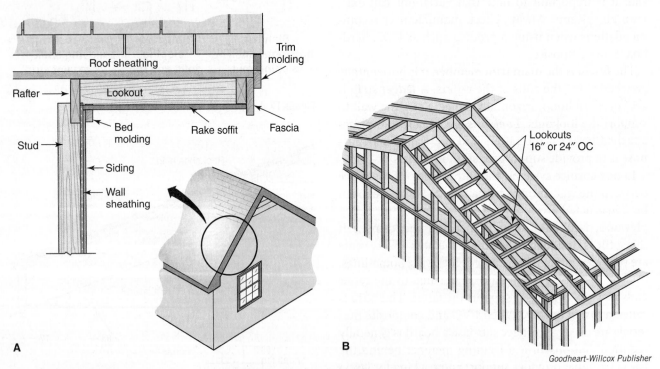

A

B

Goodheart-Willcox Publisher

Figure 17-6. Lookouts. A—Typical boxed rake section. Lookouts provide a nailing surface and support for the soffit. B—Sometimes, the lookouts are cantilevered to support a wide overhang on the gable.

between framing members. Non-structural sheathing cannot be used as a nailing base. In these cases, the siding is attached by nailing through the sheathing and into the frame. Additional information about wall sheathing can be found in Chapter 11, **Wall and Ceiling Framing**.

An insulation board of rigid polystyrene may be used in place of, or in addition to, sheathing. It adds substantial R-value to walls. Boards are usually available in 1/2″, 3/4″, and 1″ thicknesses. **Figure 17-8** shows a typical application. A WRB is applied to seal the walls against air and bulk water infiltration.

Before the application of siding, install flashing where it is required around openings. Metal flashing is usually installed over the drip caps of doors and windows, **Figure 17-9**. In areas not subjected to wind-driven rain, the head flashing may be omitted when the vertical distance between soffit and the top of the finished trim of the opening is equal to or less than 1/4 of the overhang width. For structures with unsheathed walls, flashing and a WRB must still be installed.

Pro Tip

Be sure the structural frame and sheathing are dry before applying the siding material. If excessive moisture is present during the application, later drying and shrinkage will likely cause the siding to buckle.

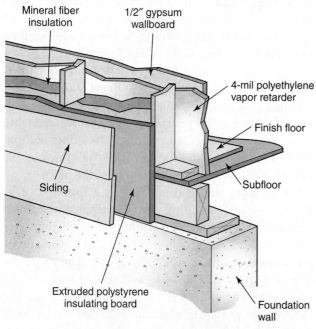

Amoco Foam Products Co.

Figure 17-8. Using rigid polystyrene insulation board on an outside wall. Fasten with galvanized roofing nails (3/8″ head) or 16 gauge staples (3/4″ crown).

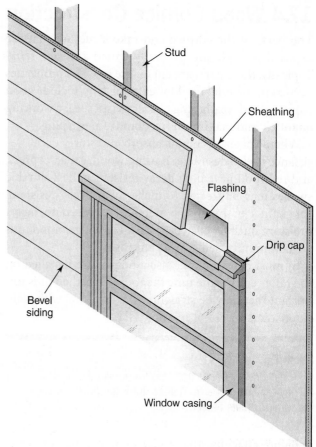

Goodheart-Willcox Publisher

Figure 17-9. Apply metal flashing over the drip caps above windows and doors that are not protected by the roof overhang.

17.3 Wood Wall Finishes

One of the most traditional materials used for the exterior finish of American homes is wood siding. Depending on the type, wood siding can give a variety of distinct looks that accentuate the architectural style of the house. From the ornate features of the classic Victorian, to the rustic appeal of a cabin in the woods, the versatility of wood siding is tremendous. The look of quality wood siding is so desirable that most manufactured siding materials are designed to look like wood.

Although the appearance of new wood siding may be preferred by property owners, it is not without drawbacks. The cost of wood siding materials, installation, and proper maintenance can be high when compared to other materials. If maintenance is overlooked, wood siding can quickly lose its attractive appearance. Additionally, decay, flammability, and vulnerability to insects can influence the decision to use wood or not.

17.4 Wood Cornice Construction

The parts of the cornice and rake structure that are exposed to view are generally called *exterior trim*. Typically, these parts are cut on the job. The properties needed in material used for exterior trim include good painting and weathering characteristics, easy working qualities, and maximum freedom from warping.

Where materials might absorb moisture, decay resistance is also desirable. If using wood, cedar, cypress, and redwood have high decay resistance. Less-durable species may be treated to make them decay-resistant. End joints or miters of members exposed to heavy moisture must be coated. Special caulking compounds are commonly used for this purpose.

In most construction, fascia boards are installed on the rafter ends at the time the roof is sheathed. It is important that they be straight, true, and level with well-fitted joints.

Pro Tip
Always use rust-resistant nails for outside finish work. They may be made from aluminum, galvanized steel, or cadmium-plated steel.

Before attaching the fascia, some builders prefer to nail on a 2×6 sub-fascia to provide structural support. This practice is required to add support if using some manufactured materials, such as PVC. The fascia trim, or fascia board, may be attached to this sub-fascia. Corners of fascia boards should be mitered. End joints should meet at a 45° angle.

The rake should be constructed to match the cornice. After the main trim members are installed, moldings can be set in corners to cover irregularities. However, the use of molding is minimal to maintain a smooth trim appearance. Weather protection, such as primer or stain, should be applied as soon as possible after installation.

Once the cornices are complete, siding installation can begin.

PROCEDURE

Framing a Wooden Box Cornice

1. If lookouts are to be used, install a ledger strip along the wall.
2. With a level, locate and mark points on the wall that are level with the bottom edges of the rafter tails. Snap a chalk line through these points.
3. Nail on the ledger strip (or a metal channel, depending on soffit material to be used).

Continued

PROCEDURE (continued)

4. Cut the lookouts. Lookouts are usually made from 2×4 stock. Locate them at each rafter.
5. Toenail one end of the lookout to the ledger. Check that the other end is flush with the bottom of the rafter. Then, nail the other end to the rafter.
6. Rip cut the top edge of the fascia to match the slope of the roof.
7. Install the fascia to the rafter ends.
8. Unless working on a hip roof, build cornice returns to cap the ends of each eve.
9. If using thin material for the soffit, attach a nailing strip along the inside of the fascia to provide a nailing surface. Sometimes the back of the fascia is grooved to receive soffit material.
10. Cut the soffit material to size.
11. Secure the soffit with rust-resistant nails or screws. When regular casing or finish nails are used, they must be countersunk and the holes filled with putty. Do this after the prime coat of paint. Installation of metal soffits is explained later.

17.5 Horizontal Wood Siding

Wood siding is usually applied over a base consisting of sheathing and housewrap. The sheathing may be oriented strand board, plywood, boards, or rigid insulation board. Where the type of sheathing does not provide sufficient strength to resist a racking load, the wall framing should be braced as described in Chapter 11, **Wall and Ceiling Framing**.

Edge views of a number of types of horizontal siding are shown in **Figure 17-10**. *Bevel siding* is most commonly used and available in various widths. It is made by sawing plain-surfaced boards at a diagonal to produce two wedge-shaped pieces. The siding is about 3/16″ thick at the thin edge and 1/2″–3/4″ thick on the other edge, depending on the width of the piece.

Wide bevel siding often has shiplapped or rabbeted joints. The siding lies flat against the studding instead of touching it only near the joints, as ordinary bevel siding does. This reduces the apparent thickness of the siding by 1/4″ but permits the use of extra nails in wide siding and reduces the chance of warping. It is also economical, since the rabbeted joint requires less lumber than the lap joint used with plain bevel siding. The rabbet, however, must be deep enough so that—when the siding is applied—the width of the boards can be adjusted upward or downward to meet windowsill, head casing, and eave lines.

Edge Views of Various Horizontal Siding

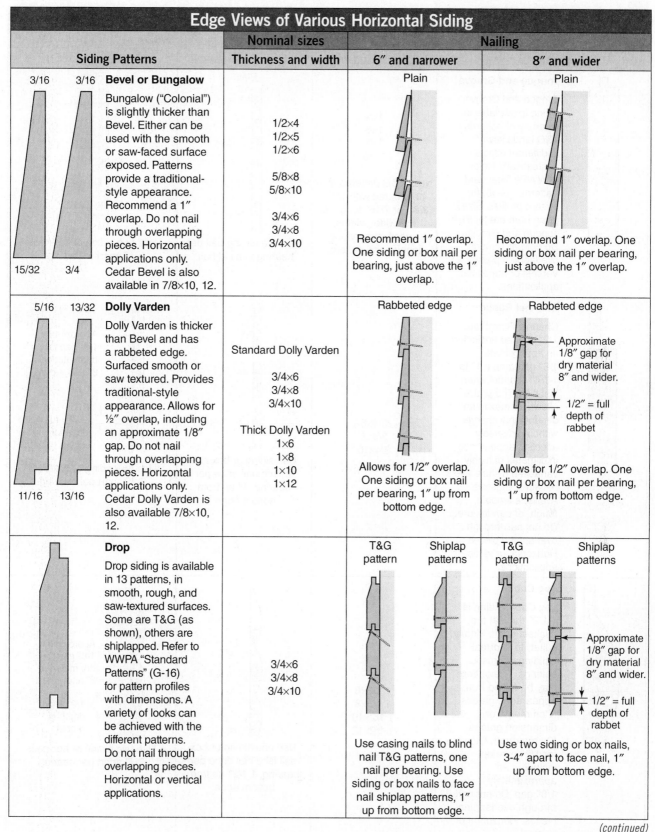

Siding Patterns	Nominal sizes	Nailing	
	Thickness and width	6″ and narrower	8″ and wider
Bevel or Bungalow — 3/16, 3/16, 15/32, 3/4 Bungalow ("Colonial") is slightly thicker than Bevel. Either can be used with the smooth or saw-faced surface exposed. Patterns provide a traditional-style appearance. Recommend a 1″ overlap. Do not nail through overlapping pieces. Horizontal applications only. Cedar Bevel is also available in 7/8×10, 12.	1/2×4 1/2×5 1/2×6 5/8×8 5/8×10 3/4×6 3/4×8 3/4×10	Plain Recommend 1″ overlap. One siding or box nail per bearing, just above the 1″ overlap.	Plain Recommend 1″ overlap. One siding or box nail per bearing, just above the 1″ overlap.
Dolly Varden — 5/16, 13/32, 11/16, 13/16 Dolly Varden is thicker than Bevel and has a rabbeted edge. Surfaced smooth or saw textured. Provides traditional-style appearance. Allows for ½″ overlap, including an approximate 1/8″ gap. Do not nail through overlapping pieces. Horizontal applications only. Cedar Dolly Varden is also available 7/8×10, 12.	Standard Dolly Varden 3/4×6 3/4×8 3/4×10 Thick Dolly Varden 1×6 1×8 1×10 1×12	Rabbeted edge Allows for 1/2″ overlap. One siding or box nail per bearing, 1″ up from bottom edge.	Rabbeted edge Approximate 1/8″ gap for dry material 8″ and wider. 1/2″ = full depth of rabbet Allows for 1/2″ overlap. One siding or box nail per bearing, 1″ up from bottom edge.
Drop Drop siding is available in 13 patterns, in smooth, rough, and saw-textured surfaces. Some are T&G (as shown), others are shiplapped. Refer to WWPA "Standard Patterns" (G-16) for pattern profiles with dimensions. A variety of looks can be achieved with the different patterns. Do not nail through overlapping pieces. Horizontal or vertical applications.	3/4×6 3/4×8 3/4×10	T&G pattern Shiplap patterns Use casing nails to blind nail T&G patterns, one nail per bearing. Use siding or box nails to face nail shiplap patterns, 1″ up from bottom edge.	T&G pattern Shiplap patterns Approximate 1/8″ gap for dry material 8″ and wider. 1/2″ = full depth of rabbet Use two siding or box nails, 3-4″ apart to face nail, 1″ up from bottom edge.

(continued)
Western Wood Products Assn.

Figure 17-10. Edge views of six different types of horizontal siding. Nominal sizes are used in figuring footage of siding. Note nailing suggestions.

Edge Views of Various Horizontal Siding

Siding Patterns	Nominal sizes Thickness and width	Nailing	
		6″ and narrower	8″ and wider
Tongue and Groove Tongue and Groove siding is available in a variety of patterns. T&G lends itself to different effects aesthetically. Refer to WWPA "Standard Patterns" (G-16) for pattern profiles. Sizes given here are for Plain Tongue and Groove. Do not nail through overlapping pieces. Vertical or horizontal applications.	1×4 1×6 1×8 1×10 Note: T&G patterns may be ordered with 1/4″, 3/8″, or 7/16″ tongues. For wider widths, specify the longer tongue and pattern.	Plain Use one casing nail per bearing to blind nail.	Plain Use two siding or box nails 3-4″ apart to face nail.
Channel Rustic Channel Rustic has 1/2″ overlap (including an approximate 1/8″ gap) and a 1″ to 1-1/4″ channel when installed. The profile allows for maximum dimensional change without adversely affecting appearance in climates of highly variable moisture levels between seasons. Available smooth, rough, or saw textured. Do not nail through overlapping pieces. Horizontal or vertical applications.	3/4×6 3/4×8 3/4×10	Use siding or box nail to face nail once per bearing, 1″ up from bottom edge.	Approximate 1/8″ gap for dry material 8″ and wider. 1/2″ = full depth of rabbet Use two siding or box nails 3-4″ apart per bearing.
Log Cabin Log Cabin siding is 1-1/2″ thick at the thickest point. Ideally suited to informal buildings in rustic settings. The pattern may be milled from appearance grades (Commons) or dimension grades (2× material). Allows for 1/2″ overlap, including an approximately 1/8″ gap. Do not nail through overlapping pieces. Horizontal or vertical.	1 1/2×6 1 1/2×8 1 1/2×10 1 1/2×12	Use one siding or box nail face nail once per bearing, 1 1/2″ up from bottom edge.	Approximate 1/8″ gap for dry material 8" and wider. 1/2″ = full depth of rabbet Use two siding or box nails 3-4″ apart per bearing.

Channel rustic siding and drop siding are usually 3/4″ thick and 6″, 8″, or 10″ wide. **Channel rustic siding** has shiplap-type joints. **Drop siding** usually has tongue-and-groove joints. Drop siding is heavier, has more structural strength, and makes tighter joints than bevel siding. Because of this, it is often used on garages and other buildings that are not sheathed.

Wood used for exterior siding should be a select grade, free of knots, pitch pockets, and other defects. The best grade has edge grain because it is less likely to warp than a flat grain. The moisture content at the time of application should be what it will reach in service. This is about 12%, except for the southwestern United States, where the moisture content should average about 9%.

Siding should be carefully handled when it is delivered to the building site. The wood from which it is made is usually quite soft. The surface can be easily damaged. Try to store siding inside the structure or keep it covered with a weatherproof material until it is applied.

17.5.1 Installation Procedures

Wood siding is precision-manufactured to standard sizes. It is easily cut and fitted. Plain beveled siding is lapped so it will shed water and provide a windproof and dustproof covering. A minimum lap of 1″ is used for 6″ widths, while 8″ and 10″ siding should lap 1 1/2″. With lap siding, it is an advantage to be able to vary the exposure. This allows you to have a full course below and above windows and over the top of doors.

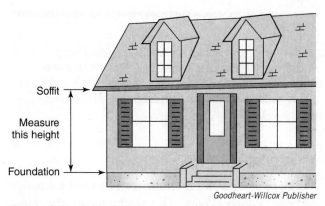
Goodheart-Willcox Publisher

Figure 17-11. Begin layout of the story pole by taking the total measurement from beneath the soffit or frieze board to about 1″ below the top of the foundation.

Goodheart-Willcox Publisher

Figure 17-12. Transfer the story pole layout to all inside and outside corners, as well as to door and window casings.

PROCEDURE

Preparing a Story Pole and the Layout

1. Measure the distance from the soffit or frieze board, if one is applied, to about 1″ below the top of the foundation, **Figure 17-11**.
2. Divide this distance into spaces equal to the width of the siding minus the lap.
3. Adjust the lap allowance (maintain minimum requirements) so the spaces are equal.
4. When possible, adjust the spacing so courses of siding are continuous above and below windows or other wall openings without notching.
5. Mark the position of the top of each siding board on the story pole.

6. Hold the story pole against the wall at each inside and outside corner of the structure. Transfer the layout marks to the wall as is shown in **Figure 17-12**. Also transfer the story pole layout marks along both sides of window and door casings.
7. Some carpenters prefer to tack nails at these layout points, since lines can be quickly attached to them and used to align the siding stock. Nails can also be useful for holding a chalk line if guidelines are laid out by snapping.
8. After making the layout, carefully check it.

17.5.2 Wood Corner Boards

A square piece of solid lumber can be used for inside corners, **Figure 17-13**. The thickness of the piece depends on the thickness of the siding. The siding should never project above the corner board.

Outside wood corners can be formed with two pieces of trim lumber, **Figure 17-14**. The thickness, again, depends on the siding. One of the two corner pieces should be narrower by the thickness of the other trim piece. When butted together, they will appear to have the same width. Attach these to the structure before installing the siding.

Corner boards may be plain or molded, depending on the architectural treatment required. After the corner boards are in place, installation of siding can begin.

Inside or outside corners can be capped with formed metal pieces, as shown in **Figure 17-15**. Although outside corners can be lapped or mitered, metal corners are used more often. They can be quickly installed and provide a neat, trim appearance.

rsooll/Shutterstock.com

Figure 17-14. The most common treatment for outside corners is to use corner boards made of trim lumber as shown on this weathered siding.

A

Goodheart-Willcox Publisher

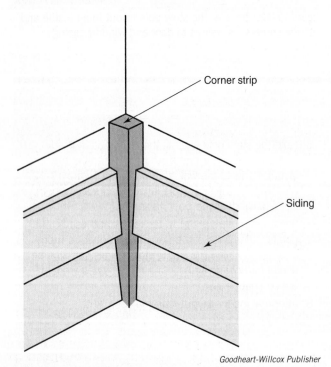

Goodheart-Willcox Publisher

Figure 17-13. Inside corners can be formed from solid lumber. When necessary, use two thicknesses to secure the required size.

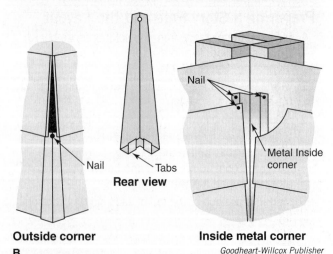

Outside corner

B

Inside metal corner

Goodheart-Willcox Publisher

Figure 17-15. Metal corners can be used for horizontal siding. A—A finished installation. B—Attachment details. All nails are covered.

Some carpenters mark cuts at corner boards and window and door casings using a *preacher*, sometimes called a siding gauge. As shown in **Figure 17-16**, a preacher is a small block made from 3/8″ or 1/2″ hardwood. It is notched to fit over the siding. To use the preacher, make certain the siding piece is properly located where it is to be installed. Position the notched piece over the siding and firmly hold it against the edge of the corner board or trim. Mark the siding where it is to be cut off.

For quality work, a carpenter first makes the cuts with a fine-tooth saw and then smoothes the ends with a few strokes of a block plane. Square butt joints are used between adjacent pieces of siding. Stagger the joints as widely as possible from one course to the next.

Consider giving wood siding a coat of water-repellent preservative before it is installed. However, you can also do this after the installation. Preservatives contain waxes, resins, and oils that protect the wood from weather and rot. In addition to this treatment, joints in siding may be bedded in a special caulking compound to make them watertight.

PROCEDURE

Installing Bevel Siding

1. Nail a spacer strip along the foundation line. Its thickness should be the same as the thin edge of the siding, **Figure 17-17**. This will give the first course the proper tilt.
2. Apply the first piece of siding. Allow the butt edge to extend below the spacer strip to form a drip edge.
3. Cut and fit horizontal wood siding tightly against corner boards, window and door casings, and adjoining boards. Cut the siding long enough so it has a hairline overlap. If then bowed slightly, it can be snapped in for a tight fit.
4. Apply the remaining rows, staggering joints.

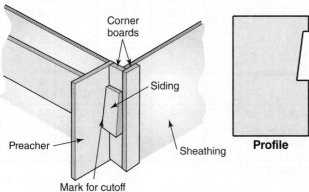

Goodheart-Willcox Publisher

Figure 17-16. Some carpenters use a guide called a preacher to accurately mark siding cuts.

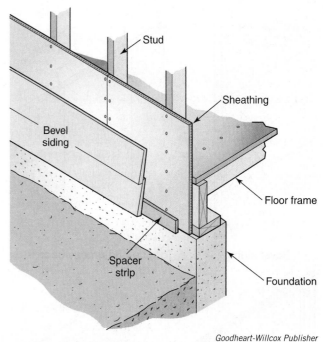

Goodheart-Willcox Publisher

Figure 17-17. Install a spacer strip under the first course of bevel siding to give it the proper tilt to match succeeding courses.

17.5.3 Nailing

To fasten siding, the types of nails recommended are stainless steel, high-tensile-strength aluminum, hot-dipped galvanized, or other noncorrosive nails. Avoid fasteners that might produce unsightly rust stains. Even small-headed plain steel nails, countersunk and puttied, are likely to rust. If there is a possibility the material will split when nailing end joints, drill holes for the nails. **Figure 17-18** shows wood siding nailing patterns for different types of siding. Also refer to **Figure 17-10**.

Face nail horizontal siding to each stud. Nails should penetrate into the wood at least 1 1/2″. For 1/2″ siding over wood or plywood sheathing, use 6d nails. Over non-structural or gypsum sheathing, use longer nails according to the thickness of the sheathing. Use 8d nails for 3/4″ siding over wood or plywood sheathing, and 10d nails over non-structural or gypsum sheathing. For 1″-thick siding, it is best to use 8d or 10d nails. Avoid using nails that are longer than necessary. They may interfere with electrical wiring or plumbing in the wall.

For narrow siding, the nail is generally placed about 1/2″ above the butt edge. In this location, the fastener passes through the upper edge of the lower course.

When applying wide (4″–6″) bevel siding, drive the nail through the butt edge 1″ above the lap so that it misses the thin edge of the piece of siding underneath.

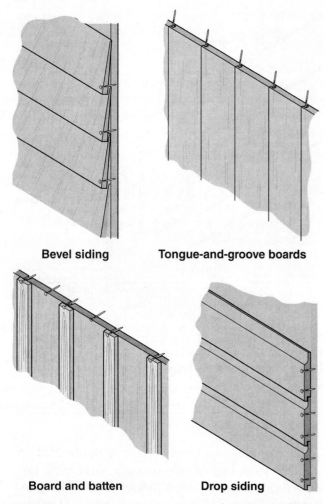

Bevel siding　　**Tongue-and-groove boards**

Board and batten　　**Drop siding**

Goodheart-Willcox Publisher

Figure 17-18. Different types of wood siding require different nailing patterns.

This permits expansion and contraction of the siding boards with seasonal changes in moisture content. It eliminates the tendency of the siding to cup or split when both edges are nailed. Since the amount of swelling and shrinking is proportional to the width of the material, vary the nail distance proportionally.

Pro Tip
Wood primarily expands and contracts across the grain due to changes in moisture content. Vinyl and aluminum materials primarily expand and contract along their length due to changes in temperature. Understanding material properties is an important part of proper installation and is useful when troubleshooting problems in existing structures.

17.5.4 Painting and Maintenance

Wood siding is subject to decay and weathering. Neither will occur if simple precautions are taken. Decay is the disintegration of wood caused by the growth

of fungi. These fungi grow in wood when the moisture content is too high. If the structure is built on a foundation that has been carried well above the ground and the construction is such that water runs off instead of into the walls, decay should not be a problem.

A wide range of finishing materials is on the market. Most fall into one of four general categories: clear water repellents, bleaching oils, stains, and paints. More will be said about each of these finishes in a later chapter. Certain factors need to be considered when selecting a finish. These may include desired appearance, preparation and maintenance requirements of the finish, location of the structure, climate, and current condition of the siding.

If the siding is to be painted, a priming coat should be put on as soon as possible. If an unexpected rain should wet unprimed wood siding, the first coat of paint should not be applied until the wood has dried and the primer applied.

To avoid future separation between primer and topcoat, apply the first topcoat of paint within two weeks of the primer. The second topcoat should be applied within two weeks of the first.

17.5.5 Estimating Siding

To determine how much siding to order, it is necessary to increase the footage to make up for the difference between nominal and finished sizes. More must also be added for waste resulting from the cutting of joints and the overlap in beveled siding. The table in **Figure 17-19** provides a factor. Multiply the net square

Estimating Horizontal Wood Siding			
Type	**Size (inches)**	**Lap (inches)**	**Multiply net wall surface by**
Bevel siding	1×4	3/4	1.45
	*1×5	7/8	1.38
	1×6	1	1.33
	1×8	1 1/4	1.33
	1×10	1 1/2	1.29
	1×12	1 1/2	1.23
Rustic and drop siding (shiplapped)	1×4	—	1.28
	*1×5	—	1.21
	1×6		1.19
	1×8		1.16
Rustic and drop siding (dressed and matched)	1×4	—	1.23
	*1×5	—	1.18
	1×6		1.16
	1×8		1.14

*Unusual sizes.

Goodheart-Willcox Publisher

Figure 17-19. When estimating horizontal wood siding, multiply the net wall surface to be covered by the factor in the right-hand column.

footage of the wall surface to be covered by this factor. The following example shows the steps.

$$\text{Wall material} = 1 \times 10 \text{ bevel siding with } 1\ 1/2'' \text{ lap}$$
$$\text{Wall height} = 8'$$
$$\text{Wall perimeter} = 160'$$
$$\text{Door and window area} = 240 \text{ sq ft}$$
$$\text{Total area to be covered} = (8 \times 160) - 240$$
$$= 1280 - 240$$
$$= 1040 \text{ sq ft}$$
$$\text{Siding needed} = 1040 \times 1.29$$
$$= 1342 \text{ sq ft or bd ft}$$

Calculate the area of gable ends by multiplying the height above the eaves by the width, then divide by two. Since considerable waste occurs in covering triangular areas, add at least 10% to this calculation. When the structure includes many corners due to projections and recesses in the wall line, add an additional 5% to the factors shown in **Figure 17-19**.

17.6 Vertical Siding

Vertical siding is commonly used to set off entrances or gable ends. It is also often used for the main wall areas, **Figure 17-20**. Vertical siding may be plain-surfaced, matched boards; pattern-matched boards; tongue-and-groove boards; or square-edge boards covered at the joint with a batten strip.

If the siding is directly applied to the building frame, backing blocks (called bearings) should be horizontally installed at 16″–24″ intervals between studs. These provide a good nailing surface. If structural panels, such as oriented strand board, are used as sheathing, the panels provide a satisfactory nailing surface.

Matched vertical siding made from solid lumber should be no more than 8″ wide. It should be installed with two 8d nails not more than 4′ apart. Backing blocks should be used between studs to provide a good nailing base. The bottoms of the boards are usually undercut to form a drip edge.

If the siding material is wide, such as 4′ wide panels, there is no need to install blocking between studs. Simply nail the sheets onto the studs.

Board-and-batten siding applications are designed around wide, square-edged boards spaced about 1/2″ apart. They are fastened at each bearing with one or two 8d nails. Use 10d nails to attach the battens, **Figure 17-21**. Locate nails in the center of the batten so the shanks will pass between the boards and into the bearing. This allows the boards to expand and contract independent of the battens.

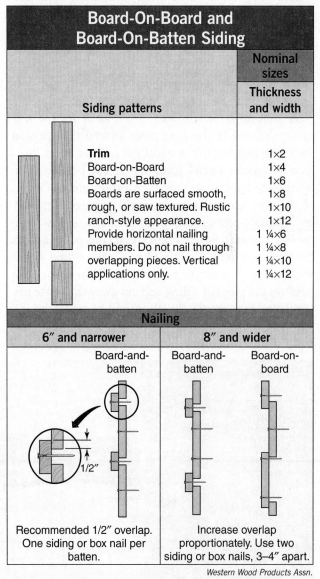

Board-On-Board and Board-On-Batten Siding

Siding patterns	Nominal sizes
	Thickness and width
Trim	1×2
Board-on-Board	1×4
Board-on-Batten	1×6
Boards are surfaced smooth,	1×8
rough, or saw textured. Rustic	1×10
ranch-style appearance.	1×12
Provide horizontal nailing	1 ¼×6
members. Do not nail through	1 ¼×8
overlapping pieces. Vertical	1 ¼×10
applications only.	1 ¼×12

Nailing	
6″ and narrower	**8″ and wider**
Board-and-batten	Board-and-batten / Board-on-board
Recommended 1/2″ overlap. One siding or box nail per batten.	Increase overlap proportionately. Use two siding or box nails, 3–4″ apart.

Western Wood Products Assn.

Figure 17-21. Board-on-board and board-and-batten siding give a building a rustic appearance. They also provide for expansion and contraction.

Artazum/Shutterstock.com

Figure 17-20. Vertical wood boards with battens make a durable and beautiful siding material.

Board-and-batten effects are also possible using large vertical sheets of exterior plywood or composition material. Simply attach vertical solid wood strips over the joints and at intervals between the joints.

A variation of the board-and-batten siding is the **board-on-board siding**. Refer to **Figure 17-21**. Note the nailing pattern.

Normally, corner boards are not used on vertical siding. If installing tongue-and-groove siding, the groove must be ripped from the corner piece. To mark the edge of the corner piece for ripping, place the board on the starting edge, using a level to ensure it is vertical. The bottom end should be about 1″ below the sheathing. The top end should butt against or be covered by the trim. Mark the groove edge for ripping and make the cut, slightly undercutting the edge.

Install the corner board using face nailing at the corner and blind nailing at the tongue. Next, temporarily tack another board at the opposite edge and stretch a string between them as a guide for the bottoms of the rest of the siding.

When nearing an opening in the wall, such as a window or door, cut and fit the siding piece just before the piece that must be trimmed. Set it aside for the moment. Cut and fit the next piece around the opening. Tack it in place of the one just cut and set aside. Use a level to transfer a mark from the top and bottom of the opening to the siding. Now, take a short piece of siding of the same width to use as a marking gauge. If the siding is tongue-and-groove, you will need to cut away the groove to avoid a false marking. Hold this gauge against the trim and draw a cutting line on the siding as you slide the piece from the top to the bottom of the trim. Remove the piece of siding and cut away the waste material along the marks. Finally, install both pieces.

17.7 Wood Shingles

Wood shingles are sometimes used for wall covering, **Figure 17-22**. There are many different types available. Some are especially designed for sidewall application with a grooved surface and factory-applied paint or stain.

Shingles are very durable and can be applied in various ways to provide a variety of architectural effects. Hand-split shingles are occasionally used, but they are expensive and difficult to install.

Most shingles are made in random widths. No. 1 grade shingles vary from 3″ to 14″ wide. Only a small number of the narrow width is permitted. Shingles of a uniform width, known as dimension shingles, are also available. For sidewall application, follow these recommendations for maximum exposure:

- For 16″ shingles, 7 1/2″ exposure
- For 18″ shingles, 8 1/2″ exposure
- For 24″ shingles, 11 1/2″ exposure

17.7.1 Application of Shingles to Sidewalls

Shingles can be applied to walls using one of two basic methods: single course or double course. **Single coursing** is similar to applying wood shingles to a roof. However, greater weather exposure of the shingle is allowed. In **double coursing**, a second layer is applied over the first. This method permits use of a lower grade shingle under the shingle exposed to the weather. Double coursing also allows greater exposure. The exposed shingle butt should extend about 1/2″ below the butt of the undercourse.

When double coursing shingle siding, secure the outer course with two 5d, small-head, rust-resistant

A   *Bernhard Richter/Shutterstock.com*

B   *Dolores M. Harvey/Shutterstock.com*

Figure 17-22. Wood shakes and shingles offer a variety of looks. A—The weathered shingles on this cabin give it a rustic look. B—This home appears ornate and refined.

nails at a point 2″ above the butt. This makes a greater weather exposure possible. Frequently, exposures of as much as 12″ for 16″ shingles, 14″ for 18″ shingles, and 16″ for 24″ shingles are satisfactory. **Figure 17-23** lists the sizes of standard sidewall shingles with a grooved surface. Approximate coverage for various lengths and exposures is included.

Spaced sheathing is satisfactory on implement sheds, garages, and other uninhabited structures. Here, protection from the elements is the principal consideration, not prevention of air infiltration.

To obtain the best effect and avoid unnecessary cutting of shingles, butt lines should be even with the upper edges of window openings. Likewise, they should line up with the lower edges of such openings. Steps for adjusting course spacing to accomplish this are explained later in this section.

Pro Tip

It is better to tack a straightedge to the wall to use as a guide for maintaining straight and level courses of shingles, rather than attempt to shingle to a chalk line.

17.7.2 Single Coursing of Sidewalls

The single-coursing method for sidewall shingle application is much like roof application. The major difference is in the exposures employed. In roof construction, maximum permissible exposures are slightly less than 1/3 of the shingle length. This produces a three-ply covering. Vertical sidewall surfaces present fewer weather-resistance problems than do roofs. Accordingly, a two-ply covering of shingles is usually adequate.

In single-coursed sidewalls with No. 1 grade shingles, weather exposure should never be greater than 1/2 of the length of the shingle minus 1″. Thus, two layers of wood are found at every point in the wall. For example, when 16″ shingles are used, the maximum exposure should be 1″ less than 8″, or 7 ″.

Single-coursed shingle sidewalls should have concealed nailing, **Figure 17-24**. This means that the nails must be driven about 1″ above the butt line of the next course.

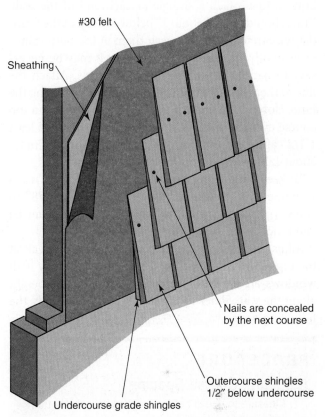

Goodheart-Willcox Publisher

Figure 17-24. Single-course method of applying shingles to sidewalls. Solid backing and nailing base is provided by wood sheathing.

Sizes for Sidewall Shingles

Grade	Length	Thickness (at butt)	No. of Courses per bdl/carton	Bdls/Cartons per square	Shipping weight	Description
No. 1	16″ (Fivex)	.40″	33/33	1 carton	60* lbs.	Same specifications as rebutted-rejointed shingles, except that shingle face has been given grain-like grooves. Natural color or variety of factory-applied colors. Also in 4′ and 8′ panels.
	18″ (Perfections)	.45″	28/28	1 carton	60* lbs.	
	24″ (Royals)	.50″	13/14	4 bdls.	192 lbs.	

Note: *70 lbs. when factory finished.

Coverage for Sidewall Shingles

Length and thickness	Approximate coverage of one square (4 bundles) of shingles based on following weather exposures																										
	3 1/2″	4″	4 1/2″	5″	5 1/2″	6″	6 1/2″	7″	7 1/2″	8″	8 1/2″	9″	9 1/2″	10″	10 1/2″	11″	11 1/2″	12″	12 1/2″	13″	13 1/2″	14″	14 1/2″	15″	15 1/2″	16″
16″ × 5/2″	70	80	90	100*	110	120	130	140	150**	160	170	180	190	200	210	220	230	240†	—	—	—	—	—	—	—	—
18″ × 5/2″ 1/4″	—	72 1/2	81 1/2	90 1/2	100*	109	118	127	136	145 1/2	154 1/2**	163 1/2	172 1/2	181 1/2	191	200	209	218	227	236	245 1/2	254 1/2†	—	—	—	—
24″ × 4/2″	—	—	—	—	—	80	86 1/2	93	100*	106 1/2	113	120	126 1/2	133	140	146 1/2	153**	160	166 1/2	173	180	186 1/2	193	200	206 1/2	231†

Note: *Maximum exposure recommended for roofs. **Maximum exposure recommended for single coursing on sidewalls. †Maximum exposure recommended for double coursing on sidewalls.

Goodheart-Willcox Publisher

Figure 17-23. Sizes and coverage for sidewall shingles. Shingle thickness is based on the number of butts required to equal a given measurement.

The shingles of this course then adequately cover the nails. Two nails should be driven into each shingle up to 10″ wide. Place each nail about 3/4″ from the edge of the shingle. On shingles wider than 10″, drive two more nails approximately 1″ apart in the center of the shingle at the same distance above the butt line as the other nails. Use rust-resistant, 3d box nails.

For obvious reasons, shingles are applied from the bottom up. Attach a shingle to each end of the wall. Butts should extend about 1″ below the top of the foundation. Stretch a line between them at the butts. Since even tightly stretched lines may sag, it may be necessary to nail on other shingles at intervals. Attach the line to them, as well. Install the starter course. Bring the butts close to the line without touching it. Apply a top course over the starter course, offsetting joints at least 1 1/2″ and keep the butts of the outer course shingles about 1/2″ below the butts of the starter course.

To keep succeeding courses straight and level, use a straightedge. Space untreated shingles at least 1/8″ to allow for swelling when wet. Treated shingles can be laid much closer together.

Adjust exposure of courses so they evenly break at the top of the window trim. This is practical only if all windows are set with their heads at the same height along the wall. Refer to **Figure 17-25** as you study the following steps.

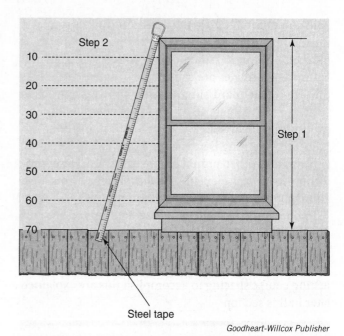

Goodheart-Willcox Publisher

Figure 17-25. Method for laying out shingle siding courses so they evenly break above and below windows. This can only be done if all windows are the same size.

PROCEDURE
Adjusting Course Spacing

1. Measure the height of the window from the bottom trim to the top trim.
2. Divide this measurement by the shingle exposure you are using and round to the nearest whole number. For example, suppose the height is 50″ and the shingle exposure 7 1/2″. Dividing the height by the exposure gives you 6.67″. Go to the closest whole number, which is 7″. This is the number of even courses that will bring the break to the top of the window trim.
3. To mark the wall for these courses, extend a steel tape to some large multiple of 7″ (for example, 70″).
4. Run the tape at a slant from the top of the window trim to a level line even with the bottom of the stool (if one is to be installed later). The 70″ mark on the tape should rest on this line.
5. Mark the sheathing at 10″, 20″, 30″, 40″, 50″, 60″, and 70″. These are the course lines that will break at the top of the window.
6. Mark these lines on a scrap of lumber so you can transfer them to the other side of the window and to all the windows as needed. If there is a second story, mark those windows, too.

17.7.3 Estimating Quantities

In estimating the quantity of shingles required for sidewalls, areas to be shingled should be calculated in square feet. Deduct window and door areas. Consult the table in **Figure 17-23**. The coverage of one square (4 bundles) at the exposure to be used should be divided into the wall area to be covered. The resulting figure is the number of squares needed. Add 5% to allow for waste in cutting and fitting around openings and for the double starter course.

17.7.4 Double Coursing of Sidewalls

In double coursing, a low-cost shingle is generally used for the bottom layer. This is covered with a No. 1 grade shingle or a processed shingle or shake. Many types of shingles are available for the outer course. Prestained shingles are available in attractive colors and are particularly suitable. Although wide exposures usually require the use of long shingles, this effect is obtained in double coursing by the application of doubled layers of regular 16″ or 18″ shingles. The maximum exposure to the weather of double-coursed 16″ shingles is 12″. For 18″ shingles it is 14″.

The proper application of shingles on a double-coursed sidewall is illustrated in **Figure 17-26**. Most procedures used for regular siding can be followed.

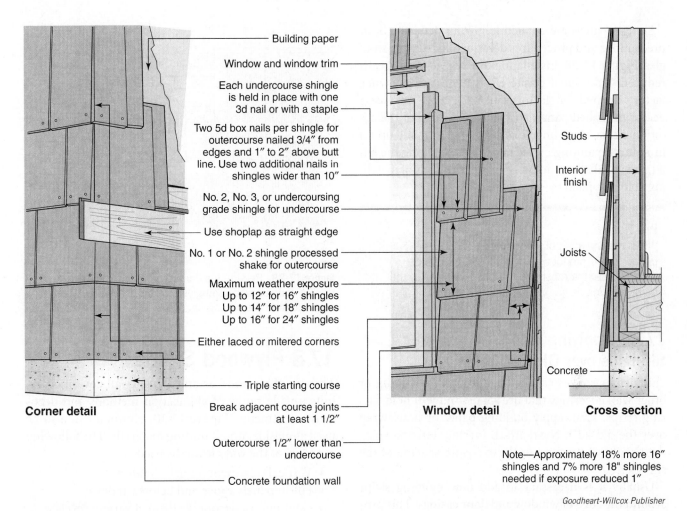

Corner detail

- Building paper
- Window and window trim
- Each undercourse shingle is held in place with one 3d nail or with a staple
- Two 5d box nails per shingle for outercourse nailed 3/4″ from edges and 1″ to 2″ above butt line. Use two additional nails in shingles wider than 10″
- No. 2, No. 3, or undercoursing grade shingle for undercourse
- Use shoplap as straight edge
- No. 1 or No. 2 shingle processed shake for outercourse
- Maximum weather exposure Up to 12″ for 16″ shingles Up to 14″ for 18″ shingles Up to 16″ for 24″ shingles
- Either laced or mitered corners
- Triple starting course
- Break adjacent course joints at least 1 1/2″
- Outercourse 1/2″ lower than undercourse
- Concrete foundation wall

Window detail

Cross section

- Studs
- Interior finish
- Joists
- Concrete

Note—Approximately 18% more 16″ shingles and 7% more 18″ shingles needed if exposure reduced 1″

Goodheart-Willcox Publisher

Figure 17-26. Double-coursed shingle siding.

There are variations to straight-line coursing. In staggered coursing, butts of alternating shingles are offset below the line. Offsets are never more than 1″ for 16″ and 18″ shingles or 1 1/2″ for 24″ shingles. In ribbon coursing, both layers are oriented in a straight line. The top course is raised roughly 1″ above the undercourse. Both courses are normally butt nailed.

17.7.5 Shingle and Shake Panels

Shingles and shakes for sidewall application are available in panel form. The panels consist of two courses of individual shingles (usually western red cedar) permanently bonded to a backing, **Figure 17-27.** Standard length panels are 8′. End shingles of each course are offset for staggered joints that match up with adjacent panels. Panels are available in various textures, either unstained or factory finished in a variety of colors. Special metal or mitered wood corners are also manufactured.

Shakertown Corp.

Figure 17-27. Panelized shakes or shingles can afford a look that is very similar to individual shakes or shingles, and can speed installation considerably. Panels are generally 8′ long and consist of two courses bonded to a base.

Shingle panels are applied following the same basic precautions and procedures described for regular shingles, **Figure 17-28**. Installation time, however, is greatly reduced. Additional onsite labor is also saved when factory-primed or factory-finished units are used. Factory-finished panels cost slightly more, but the additional cost of prefinishing is more than made up for in reduced painting costs later. When applying the latter, the installation should be made with nails of matching color supplied by the manufacturer.

> **Pro Tip**
> When making an application of a specialized or prefabricated product such as shingle panels, be sure to follow the recommendations furnished by the manufacturer.

17.7.6 Applying Wood Shingles over Old Siding

Shingles or shakes can be applied over old siding or other wall coverings that are sound and will hold furring strips. First, apply building paper or housewrap over the old wall. Next, attach furring. Furring spacing should correspond to the exposure spacing of the shingles.

Usually it is necessary to add new molding strips around the edge of window and door casings. This provides a new trim edge for the shingles. **Figure 17-29** illustrates how nailing strips are applied over an old stucco surface.

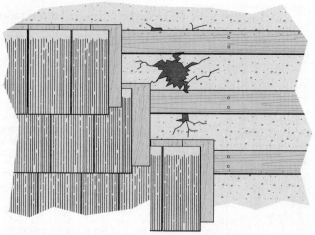

Goodheart-Willcox Publisher

Figure 17-29. Re-siding a damaged stucco wall with double-coursed wood shingles. Furring strips are spaced to correspond to the nailing lines.

17.8 Plywood Siding

The use of plywood as an exterior wall covering permits a wide range of application methods and decorative treatments, **Figure 17-30**. Plywood can also be used alongside other building materials. The following are a few of the ways it can be used:

- Vertical treatment for gable ends
- Fill-in panels above and below windows
- Continuous, decorative band at various levels along an entire wall

Siding panel applications

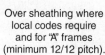

| Direct to studs (over felt) recommended for sidewalls and Mansards 60° and steeper. | Over sheathing where local codes require and for "A" frames (minimum 12/12 pitch). |

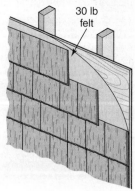

Shakertown Corp.

Figure 17-28. Proper application of panelized siding. Panels are self-aligning.

A   *APA-The Engineered Wood Association.*

B   *APA-The Engineered Wood Association.*

Figure 17-30. Examples of plywood siding styles.
A—Rough-sawn surface. B—Reverse board-and-batten.

Workplace Skills
Attention to Detail

Details matter. Regardless of what kind of exterior wall finish you are installing, or what job you are doing, details matter. Small things like consistent nail spacing on a fascia trim installation, or a careful caulk job on the gutters can make your work stand out above the rest. One day, long after the job has been wrapped-up, when the homeowner kicks back in a lawn chair while enjoying a barbeque with the neighbors, you want them to scan your work and comment on what a great job you did, not point out flaws. This is sure to win you a better reputation and more work down the road.

Artazum/Shutterstock.com

Figure 17-31. This grooved plywood is applied over a sheathed wall and has a rough-sawn surface.

All plywood siding must be made from exterior-type plywood. Douglas fir is the most commonly used species. However, cedar and redwood are also available. Panels come in either a sanded condition or with factory-applied sealer or stain. For information on grading standards, see Chapter 3, **Building Materials**.

Panel sizes are 48″ wide by 8′, 9′, or 10′ long. A 3/8″ thickness is normally used for direct-to-stud applications. A 5/16″ thickness may be used over approved sheathing. Thicker panels are required when the texture treatment consists of deep cuts. For unsheathed walls, plywood thickness should not be less than 3/8″ on 16″ stud spacing, 1/2″ for 20″ stud spacing, and 5/8″ for 24″ stud spacing.

Application of large sheets is generally made with the long dimension vertical. This eliminates the need for blocking to support horizontal joints. **Figure 17-31** shows a home sided with grooved vertical panels having shiplap edges.

In vertical installations, center the joints over studs. In horizontal installations, place solid blocking behind joints. Standard application requirements are given in **Figure 17-32**. Battens are an option, but should not be used with plywood siding that is textured.

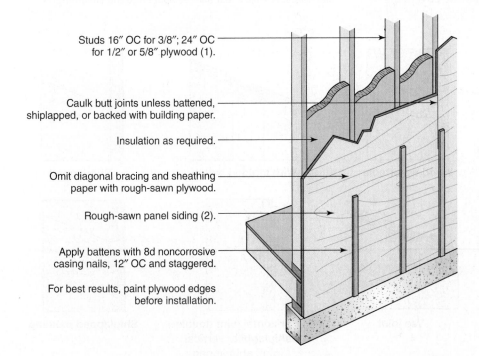

Studs 16″ OC for 3/8″; 24″ OC for 1/2″ or 5/8″ plywood (1).

Caulk butt joints unless battened, shiplapped, or backed with building paper.

Insulation as required.

Omit diagonal bracing and sheathing paper with rough-sawn plywood.

Rough-sawn panel siding (2).

Apply battens with 8d noncorrosive casing nails, 12″ OC and staggered.

For best results, paint plywood edges before installation.

Notes:
1. May use 3/8″ panel siding over 24″ OC supports; 5/16″ over 16″ OC.
2. Nail 6″ O.C. at panel edges and 12″ OC at intermediate supports. Use galvanized, aluminum, or other noncorrosive casing or siding nails — 6d for 3/8″ and 1/2″ panels; 8d for panels 5/8″ and thicker. Nail 3/8″ in from edges.

APA-The Engineered Wood Association

Figure 17-32. These recommendations are the standard for thickness of panel and method of installing vertical siding. Always check local codes to ensure compliance.

Figure 17-33 shows several ways to handle joints between plywood panels. All edges of plywood siding—whether butted, V-shaped, lapped, covered, or exposed—should be sealed with a heavy application of high-grade exterior primer, aluminum paint, or oil paint. Special caulking compounds are also recommended.

Installing vertical panel siding is not difficult, although handling the panels may be clumsy. Start from a corner with the edge flush, square, and plumbed to the corner. Make sure the trailing edge is centered on a stud. Use 6d siding nails for 1/2″ panels and 8d nails for thicker panels. Space nails 6″ around edges and 12″ when field nailing. Apply successive panels, keeping a straight line.

Plywood lapped siding may look the same as regular, beveled siding. Heavy shadow lines are secured by using spacer strips at the lapped edges. Application requirements are given in **Figure 17-34**. A bevel of at least 30° is recommended. The lap should be at least 1 1/2″.

Vertical joints of lapped siding should be butted over a shingle wedge and centered over a stud, unless structural sheathing is used. Nail siding to each of the studs along the bottom edge and not more than 4″ OC at vertical joints. Nails should penetrate studs or sheathing at least 1″.

If plywood lap siding is wider than 12″, a wooden tapered strip, such as a shingle, should be used at all studs, with nailing at alternate studs. Outside corners should butt against corner boards or be covered.

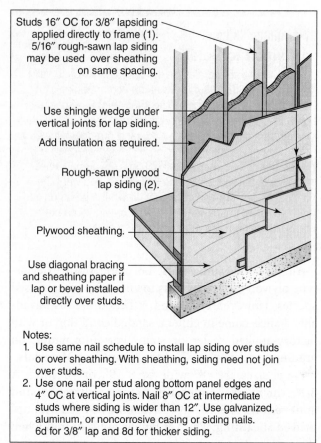

Studs 16″ OC for 3/8″ lapsiding applied directly to frame (1). 5/16″ rough-sawn lap siding may be used over sheathing on same spacing.

Use shingle wedge under vertical joints for lap siding.

Add insulation as required.

Rough-sawn plywood lap siding (2).

Plywood sheathing.

Use diagonal bracing and sheathing paper if lap or bevel installed directly over studs.

Notes:
1. Use same nail schedule to install lap siding over studs or over sheathing. With sheathing, siding need not join over studs.
2. Use one nail per stud along bottom panel edges and 4″ OC at vertical joints. Nail 8″ OC at intermediate studs where siding is wider than 12″. Use galvanized, aluminum, or noncorrosive casing or siding nails. 6d for 3/8″ lap and 8d for thicker siding.

Goodheart-Willcox Publisher

Figure 17-34. Application requirements for lapped plywood siding. If siding is directly applied to the frame, install corner bracing. Always check local codes for variance from these standards. A WRB must also be applied to the sheathing.

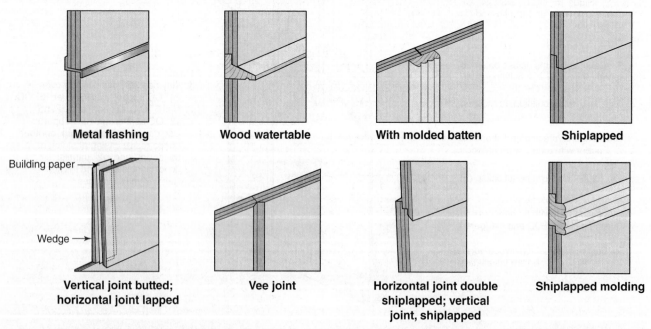

Metal flashing

Wood watertable

With molded batten

Shiplapped

Building paper

Wedge

Vertical joint butted; horizontal joint lapped

Vee joint

Horizontal joint double shiplapped; vertical joint, shiplapped

Shiplapped molding

APA-The Engineered Wood Association

Figure 17-33. Joint details for plywood siding. All edges should be sealed with paint before installing or with a special caulking compound at the time of installation.

17.9 Fiber-Cement Siding

In recent years, siding manufactured of fiber-reinforced cement has seen considerable growth in the residential market. Today, *fiber-cement siding* is used in new construction of homes as a substitute for wood siding, **Figure 17-35**. Each manufacturer has a slightly different formula, but basically fiber-cement is made of Portland cement, sand, and cellulose fibers. This siding makes a durable covering for a house and is not subject to damage by insects or ultraviolet light.

Green Note
Fiber-cement siding is considered a green material because there is very little waste in its manufacturing. Some manufacturers use fly ash, an industrial waste product, as the fine aggregate, thereby making the product more environmentally friendly.

Fiber-cement siding is made in various forms, including traditional lap siding, vertical panel siding, shingle panels and individual shingles, trim planks, and perforated soffit panels. Surface textures are woodgrain, stucco, or smooth and may be painted. Factory-finished siding is available in a number of colors. The material is strong, fire- and impact-resistant, and will not deteriorate from weather exposure.

17.9.1 Cutting Fiber-Cement

One major manufacturer of fiber-cement siding lists three methods of cutting its products:

- Good: Use a circular saw equipped with a special blade for fiber-cement products. Position the cutting station so the wind will carry any dust away from workers.

- Better: Use a circular saw equipped with a special blade for fiber-cement products and a HEPA (high-efficiency particulate absorption) dust collection system. Position the cutting station so the wind will carry any dust away from workers.
- Best: Score the piece with a utility knife and straightedge, then snap it on the scored line; or use shears, **Figure 17-36**. These are the only methods to be used when cutting indoors.

Safety Note
When cutting fiber-cement siding materials with a saw, wear an approved dust mask or respirator to prevent breathing crystalline silica dust. Excessive exposure to the dust can cause silicosis, a serious lung disease. Sawing should always be done outdoors so that air movement can carry the dust away from the person doing the cutting. Workers in the immediate vicinity should wear dust masks while cutting is being done.

Goodheart-Willcox Publisher

Figure 17-35. Fiber-cement siding is widely used for siding and trim on new homes.

A *Goodheart-Willcox Publisher* **B** *Goodheart-Willcox Publisher*

Figure 17-36. A—Fiber cement shears are available from a variety of manufacturers. B - Shearing is a preferred method for cutting fiber-cement siding since it does not create dangerous dust.

17.9.2 Installing Fiber-Cement Siding

While installation principles of fiber-cement lap siding are similar to those of wood lap siding, installation details differ. Proper selection and spacing of fasteners is more critical with fiber-cement siding. Also, extra precautions must be taken when cutting fiber cement. It can be installed directly onto braced studs spaced up to 24″ OC or on plywood or OSB panels at least 7/16″ thick. In either case, a WRB or rain screen should be applied directly beneath the siding. It is important to prevent water from getting behind the siding. Points where water might seep behind the siding should be flashed, as shown in **Figure 17-37**. End joints where two siding planks meet end-to-end are required by the IRC to have some sort of joint treatment. **Figure 17-38** shows end-joint flashing. Consult the manufacturer's specifications for the proper fastener size and type to meet local wind load requirements. Nailing is best done with a pneumatic nail gun. Set the air pressure on the gun so the nail is driven just snug to the face of the siding. If a nail is underdriven, drive it snug with a smooth-faced hammer.

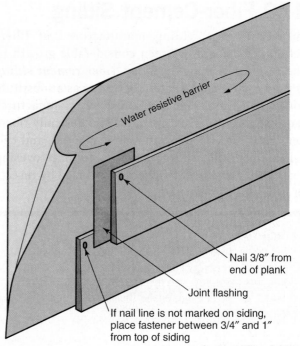

Goodheart-Willcox Publisher

Figure 17-38. Asphalt-saturated felt can be used to flash end joints.

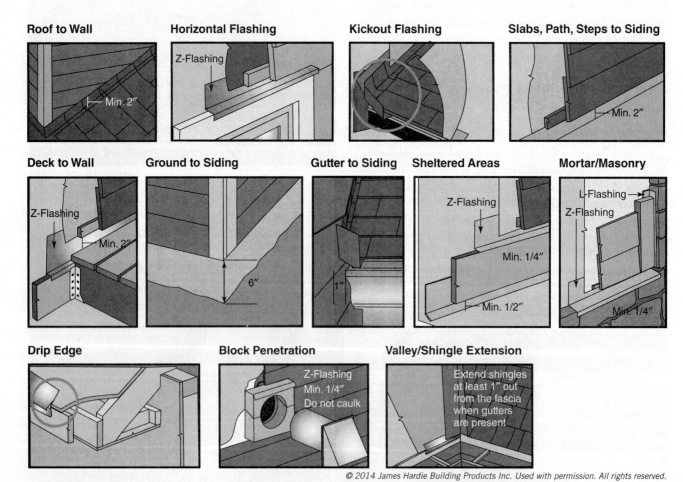

Figure 17-37. Fiber-cement siding must be properly flashed to prevent moisture from getting behind the siding.

Do not use aluminum fasteners, staples, or clipped-head nails. **Figure 17-39** shows how fiber-cement siding should be fastened.

17.9.3 Fiber-Cement Trim and Soffit

If the building is sided with fiber-cement horizontal siding, vertical siding, or shingles, the trim should also be fiber-cement. This ensures that the colors and textures will match and the trim will have the same long durability as the siding. Fiber-cement trim boards are available in 4/4 and 5/4 thicknesses and width comparable to what is available in wood trim boards. Fiber-cement soffit material is thinner and lighter in weight than siding or trim boards. Soffit material is available in vented and unvented styles. These materials can be cut with the same tools as fiber-cement siding, following the same safety precautions. The easiest way to fasten most trim is with a pneumatic nail gun, but it can also be hand-nailed. It may be necessary to predrill holes when hand-nailing to prevent corners from chipping out. Do not overdrive nails as this lessens their holding power.

17.10 Vinyl Siding

Vinyl siding has largely replaced aluminum siding as an inexpensive, quick-to-apply siding material. Vinyl is used in the construction of new homes more than any other siding. Vinyl siding is manufactured from rigid polyvinyl chloride. Both aluminum and vinyl siding have interlocking edges and are available for horizontal and vertical applications. The installation steps for aluminum are similar to those for vinyl.

Vinyl siding is made from a rigid polyvinyl chloride compound. It is tough, durable, and very common. The vinyl material is extruded into either horizontal or vertical siding units and accessories. The panel thickness is between 0.40″ and 0.45″. Panels are available in various widths.

Horizontal strips are manufactured to resemble wood drop siding and Dutch lap, **Figure 17-40**. Some strips are made to simulate 8″-wide lap siding. Some panels look like two courses of 4″ or 5″ siding. Several other variations exist, but these are the most common. The bottom hooked edge, or **buttlock**, hooks into the **top lock** of the course below and is nailed through slots in the top edge. Special inside and outside corner posts cover the ends.

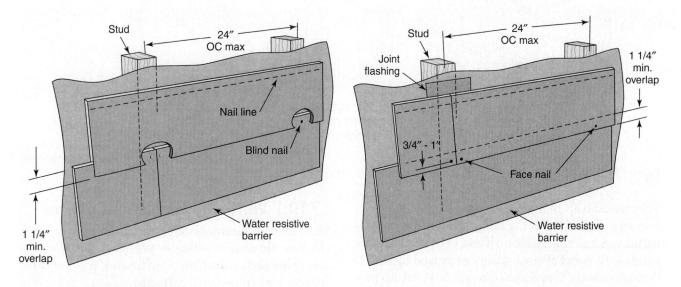

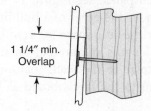

Minimum overlap for both face and blind nailing

1 1/4″ min. Overlap

* When face nailing to OSB, planks must be no greater than 9 1/4″ wide and fasteners must be 12″ OC or less.
** Also see General Fastening Requirements; and when considering alternative fastening options refer to James Hardie's Technical Bulletin USTB 17 - Fastening Tips for HardiePlank® Lap Siding.

Figure 17-39. Proper nailing of fiber-cement products.

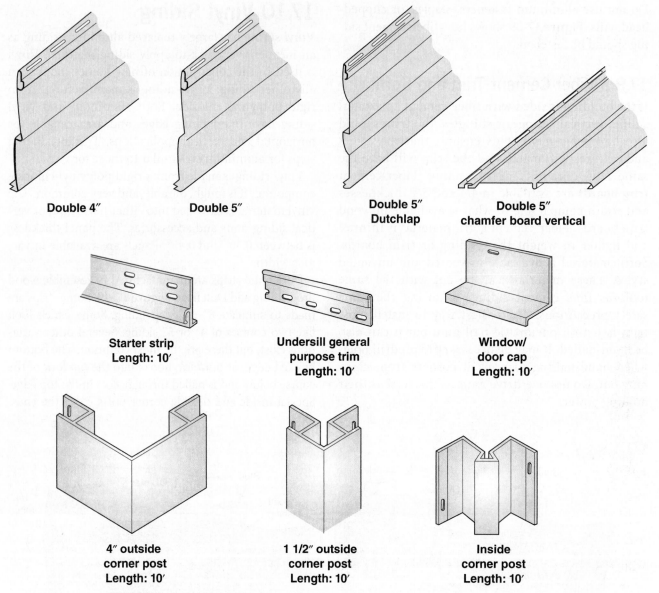

Double 4″

Double 5″

Double 5″
Dutchlap

Double 5″
chamfer board vertical

Starter strip
Length: 10′

Undersill general
purpose trim
Length: 10′

Window/
door cap
Length: 10′

4″ outside
corner post
Length: 10′

1 1/2″ outside
corner post
Length: 10′

Inside
corner post
Length: 10′

Bird Division, CertainTeed Corp.

Figure 17-40. Various accessories are made for use with vinyl siding.

They have a built-in *receiving channel* to permit expansion and contraction of the siding. The receiving channel accepts and conceals the cut end of a siding or soffit panel. A 12′ panel of vinyl siding can expand up to 1/2″ through seasonal temperature changes. It is vital that provisions are made for this movement by following proper installation procedures.

Another important characteristic of vinyl siding is that it is not designed to be the last line of defense against weather. While it is a buffer for the water-resistive barrier, it will not keep water out. Therefore, it is vital that the WRB is designed, installed, and flashed properly. If it is not, water infiltration can occur, damaging wall systems through mold, mildew, and rot.

17.10.1 Vinyl and Aluminum Cornices

Vinyl and aluminum systems are common even when vinyl or aluminum siding is not used. These products come prefinished, are easy to work with, and are attractive when installed correctly.

Vinyl and aluminum soffit panels are available in various widths and can be purchased as a solid, perforated, or semi-perforated panel, see **Figure 17-41**. Using perforated soffit under the eaves of a building allows for airflow to enter the attic space for ventilation. Panels come in 12′ lengths and are cut on site to be installed perpendicular to the wall, spanning the distance from the wall to the fascia board as in **Figure 17-42**.

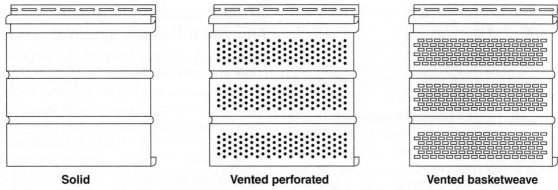

Solid **Vented perforated** **Vented basketweave**

Goodheart-Willcox Publisher

Figure 17-41. Various configurations are available for soffit panels. Generally, vented panels are used on eaves to allow an air inlet for attic ventilation.

Goodheart-Willcox Publisher

Figure 17-42. Typical new construction, closed box soffit and fascia trim installation at the eaves.

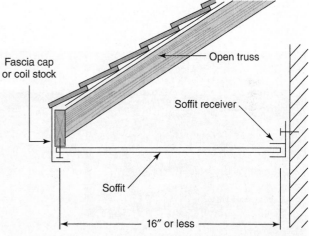

Goodheart-Willcox Publisher

Figure 17-43. A closed box soffit is simple and quick to install using vinyl and aluminum systems.

Vinyl and aluminum materials expand and contract with changes in temperature. These systems use a series of receiving channels. When installed correctly, these channels allow the installed product to expand and contract without causing damage or irregularities.

The most common configuration with new construction is to install a closed box soffit on an open eave. See **Figure 17-43**. On existing homes that are being re-sided, it is common to cover over old wood cornices with vinyl or aluminum. In either case, multiple configurations can be used with these systems.

Vinyl and aluminum systems can also be used on porch and carport ceilings. For these applications, it is important to understand that, unlike vinyl siding, soffit panels are not designed to be spliced end to end. For spans longer than the length of a full panel, or for a transition around a corner, a double-channel lineal, or back-to-back J-channel must be installed, **Figure 17-44**. A *double-channel lineal* is a vinyl siding trim that joins two soffit panels at a transition around a corner or at an end-to-end joint

Goodheart-Willcox Publisher

Figure 17-44. The mitered corner in this porch ceiling requires a transition point with a receiving channel on both sides. Here, builders used a double-channel lineal bent from coil stock. Pieces like this can be purchased or fabricated on site using a portable sheet metal brake.

Fascia trim pieces cover and protect the sub-fascia board as shown in **Figure 17-43**. This trim can be purchased preformed in aluminum or vinyl. Many contractors prefer to bend their own fascia trim, and even receiving channels on site with a portable field brake, **Figure 17-45**. A *brake* is a tool that is commonly used by siding contractors. Brakes can be outfitted with coil rollers, cutters, and measuring devices to make them efficient and easy to use. Aluminum coil stock can be purchased in a variety of colors to customize the finish trim on a project. Using a brake gives contractors the ability to build aluminum trim pieces in a variety of shapes and profiles. Bending trim, rather than buying it, can save money if an experienced tradesperson is doing the work.

17.10.2 Installing Vinyl and Aluminum Cornice Systems

Before vinyl siding is installed, the soffit and fascia trims are installed. Generally, vinyl and aluminum cornice construction begins with the installation of the soffit receiving channel(s) and panels. The process of installation is similar for both vinyl and aluminum. The method of installation depends on whether the work is on new construction or a re-side. Refer to **Figure 17-46** for acceptable configurations. Unless the job involves re-siding an existing structure, the cornice will be open. Even if the job is a re-side, the old soffit may need to be removed, creating an open cornice. Any time vinyl or aluminum soffit is installed, the finished product will be a closed soffit.

A   *Goodheart-Willcox Publisher* B   *Goodheart-Willcox Publisher*

C   *Goodheart-Willcox Publisher* D   *Goodheart-Willcox Publisher*

Figure 17-45. A—A sheet metal brake is used to bend profiles from aluminum coil stock. B—Material is clamped in the jaw of the brake and the hinge is lifted by the user to create the desired bend. C—Custom trim, such as unique window or door casing covers cannot be purchased and must be bent on site. D—Accessories like this coil roller and cutter can speed the process and improve quality considerably.

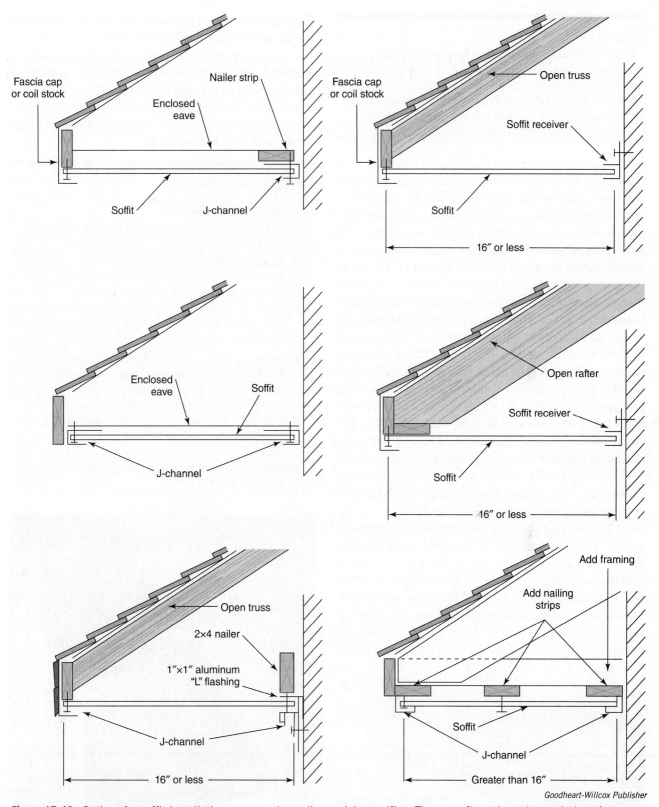

Figure 17-46. Options for soffit installation can vary depending on job specifics. These configurations show solutions for some of the challenges that the builder may face. It is important to understand the limitations of the system you are working with. Not every scenario will be presented in an instruction manual. A good understanding will give you the ability to think critically and overcome obstacles.

Goodheart-Willcox Publisher

Pro Tip

To cut vinyl soffit or siding, use a fine toothed blade installed backwards in a circular saw. Installing the blade backwards will reduce chipping and cracking of the vinyl. Aviation snips or siding shears may also be used. For rip cuts, score the material with a sharp utility knife, and then bend it to snap it off at the score mark.

Because these materials expand and contract with changes in temperature, it is very important to follow recommended installation procedures. If proper procedures are not followed, soffit panels can blow out during a storm, or buckle during the heat of summer. All receiving channels and panels are manufactured with a slotted nailing flange, as shown in **Figure 17-47**. These products are nailed loosely with aluminum or galvanized roofing or siding nails. The nails should be placed in the center of the nailing slot to allow for the material to expand or contract without restriction. Nails should be driven straight and a slight gap, 1/32″ or about the thickness of a dime, should be left between the head of the nail and the material. Nailing in this fashion is more like hanging a picture on a wall than pinning the material fast to it. Nails should be driven every 8″ to 12″ along all trim pieces. When finished nailing, the piece should be movable from side to side with a little effort.

Vinyl soffit panels should not be cut to fit tight. Because they need to have room to expand, the panels must be cut approximately 1/4″ short at each end. Always follow manufacturer instructions for product-specific guidelines. Panels are installed perpendicular to the wall. Typically, one end of the panel is placed into the receiving channel along the wall and the other end is nailed to the bottom of the sub-fascia board, **Figure 17-42**. If the span is greater than 16″, framing must be added as a nailer, **Figure 17-46**. Subsequent panels lock together and are nailed in the same fashion.

The exposed ends of the soffit material are covered by the fascia trim once soffit installation is complete.

It is often necessary for soffit to make a turn around an inside or outside corner. This occurs where there is a hip or a valley in the roof structure. When making this transition, a double-channel lineal or back-to-back J-channel must be installed to create a miter joint in the soffit panels. See **Figure 17-48**.

PROCEDURE
Hanging Vinyl Soffit

1. Install receiving channels, soffit receiver, F-channel, or J-channel.
2. Measure from the wall to the fascia board, subtracting 1/2″ (1/4″ for each end) to allow for expansion. Mark and cut this dimension on a soffit panel.
3. Insert the panel into the channel on the wall, and then secure it at the fascia board, making sure that it is square with the wall.
4. To turn a corner, measure from the channel at the wall corner to the channel at the corner of the fascia board. Subtract 1/4″ for expansion. Cut and install soffit double-channel lineal or back-to-back J-channel. If necessary, install nailing strips to provide backing for the lineal. Miter cut the corner soffit panels and install as described in Step 3.
5. After all soffit panels are installed, cut the metal or vinyl fascia cover to fit.

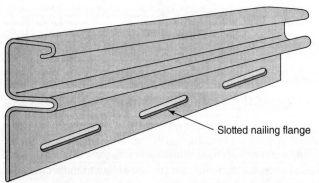

Goodheart-Willcox Publisher

Figure 17-47. Use the center of the nailing slot when attaching trim. Do not nail these trims tight. This allows material to expand and contract around the fasteners. The exception to this is the top nail of a vertical piece.

Slotted nailing flange

Radovan1/Shutterstock.com

Figure 17-48. Transitions often need to be made around a corner. Notice the double-channel lineal trim that is used to make this transition in the soffit.

After soffit installation is complete, fascia trim is installed. Fascia is an important component to the aesthetic of a home. Proper installation of the fascia trim will add to the curb appeal of a home, while a poor job can draw immediate attention to substandard work. Dents, scratches, kinks, and rippling in aluminum trim can reflect light in an unattractive manner, showing imperfections, even at a distance. See **Figure 17-49**. Because of this, it is important to handle aluminum trim with care. It should never be forced into place. Installers should follow proper installation techniques to avoid damage and rippling caused from material expansion.

Pro Tip

Aviation snips can be used to cut aluminum fascia; however, snips often leave a jagged edge when they cut. A better way is to use a sharp utility knife to score the material several times, and then bend until it snaps for a clean cut. Use a speed square as a straightedge to guide the blade while you score the material for a crosscut. A portable field brake can be used as a straightedge for lengthwise cuts. Be careful not to damage the brake, always start from each end and score toward the center to avoid nicking the brake's hinge.

Both aluminum and vinyl fascia trim are installed in a similar fashion. However, installation for vinyl may differ from manufacturer to manufacturer. Always consult the recommendations specified for the product or system being used.

Goodheart-Willcox Publisher

Figure 17-49. Oil canning, or buckling, is a common problem when working with aluminum or vinyl materials. The sheet material buckles and creates waves, rather than lying flat. Proper framing and installation can eliminate substandard work such as this.

PROCEDURE
Installing Aluminum Fascia Trim

1. The top of the fascia trim is held in place by the drip edge along the base of the roof. If no drip edge has been installed, a piece of vinyl or aluminum utility trim can be used. No nails are driven through the top edge of the fascia.

2. Measure from the lower side of the soffit panels to the top of the trim installed (drip edge or utility trim) on the upper side of the fascia board. Deduct approximately 1/4″ from this dimension and trim the fascia along its length.

3. Slip the top edge of the fascia into the drip edge (or other installed trim) and secure the fascia in place through the bottom flange with color-matched trim nails.

4. The fascia trim is primarily nailed along the bottom flange with matching painted trim nails. Nails should be spaced no more than 24″ apart and positioned at the V-groves of the soffit to avoid distorting the soffit panels.

   Note: It is recommended that holes be predrilled through the aluminum material to avoid denting it and to give the material some room to expand and contract. Holes should be larger than the shank of the trim nails, but smaller than the head. Use a hammer with a smooth face to avoid marring the material. Nails should not be driven tight.

5. Overlap the face of all splices by 1″. For inside and outside corners, bend a 1″ tab on the face of the trim piece to wrap around the unfinished corner. Cover the tab with the adjoining piece. Trim off the bottom flange where panels overlap.

17.10.3 Installing Vinyl Siding

Practices described here represent standard vinyl siding installation procedures. Although many vinyl siding systems use identical installation methods, always consult the manufacturer installation instructions before beginning the work. Not doing so could cause a mistake that may void the product warranty. Specialty systems such as insulated vinyl siding and polymer shakes or shingles also have very specific installation requirements.

Vinyl siding can generally be installed with the tools already in a carpenter's toolbox. The following tools are recommended by the Vinyl Siding Institute:

- Power saw, table or miter, with fine-tooth blade (12–16 teeth per inch)
- Other standard tools such as a hammer, square, chalk line, level, steel tape, and safety glasses
- Utility knife for cutting, trimming, and scoring

- Tin snips or aviation snips to speed cutting and shaping
- Snap-lock punch for cutting lugs on siding that has been trimmed for the top or finishing course under the eaves
- A siding unlocking tool

Other than a properly constructed WRB, no special preparation of the exterior walls is needed when using vinyl siding on new construction. Installation of horizontal and vertical siding is similar. However, there are some differences, as described in the following procedures. **Figure 17-50** shows before and after photos of a re-siding project. Trim for vinyl-sided buildings can be vinyl or aluminum. Various elements of vinyl siding and aluminum trim are shown in **Figure 17-51**. Special trim pieces that require intricate cuts and bends are created on a brake.

Pro Tip

There are three types of cuts for vinyl siding: short cuts, low volume cuts, and long cuts. For short cuts, such as cutting pieces to length, the quickest way is to use a portable circular saw. Use a fine-tooth blade and mount it on the saw backward, **Figure 17-52**. If only a small number of cuts must be made, they can be done with tin snips. For long cuts, such as to rip a piece to a narrower width, use a utility knife to score the surface, and then bend it at the score line to break it. The portable field brake can be helpful for these long cuts. Clamp the piece in the jaws of the brake and use it as a straight edge to guide your knife blade.

A *VANBECKS Roofing & Siding—Kalamazoo, MI*

B *VANBECKS Roofing & Siding—Kalamazoo, MI*

Figure 17-50. Done properly, re-siding enhances the appearance of a building. A—This siding is old and needs to be replaced. B—A vinyl recladding updates the look of the home and offers low maintenance.

Benjamin Obdyke, Inc.

Figure 17-51. The different elements of vinyl and aluminum exterior finish.

Figure 17-52. A simple jig saves time cutting siding units to length

The top course under the eave requires special trim and preparation of the panel. Every manufacturer provides a system for concealing and securing the top siding panel. Generally, this requires a **utility trim**, also known as an undersill trim. A utility trim is designed to grip the trimmed top edge of the panel.

Gable ends are covered in the same manner as the walls. Use two scrap pieces of panel to make a pattern for cutting the proper angle. See **Figure 17-53**. Lock one piece into the panel below. Hold the other piece against the gable. Mark a line across the bottom piece and cut. Follow the same procedure for the other side of the gable. This pattern can be cut and the angle transferred to full pieces for ensuring the correct cut will be made.

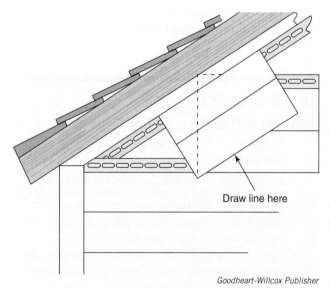

Draw line here

Figure 17-53. To transfer the angle to the panel being cut, make a pattern and use it as a stencil.

PROCEDURE

Installing Horizontal Siding

1. Snap a chalk line at the bottom of the wall. Carpenters usually start at the back of the building and work toward the front. Finish each side before starting the next. Always cover largest areas first. Short panels can be used to finish the end of a course and on smaller surfaces such as dormers.

2. Using the chalk line as a guide, nail on a starter strip, **Figure 17-54**. Allow space for inside and outside corner trim and J-channel.

3. Position and fasten all inside and outside corner posts, **Figure 17-55**, allowing 1/4″ space at the top. Place the first nail at the top of the top slot – this keeps the corner from sliding down. Nail in the center of all other slots, spacing nails 8″–12″ apart. Do not nail tight. Leave a 1/32″ gap (about the thickness of a dime) between the nail head and the material.

4. If splicing of corner posts is necessary, cut away 1″ from all but the outer face of the top of the lower corner post. Lap 3/4″ of the upper post over the lower. This allows 1/4″ for expansion.

5. Install J-channel around windows and doors, **Figure 17-56**. Cut side J-channels slightly longer than the trim. Notch the top ends. Cut top J-channel longer to provide a tab. Miter the faces of the channel pieces. Bend down the tabs on the top J-channel to act as a flashing over the side pieces.

6. Place the first panel in the starter strip. Lock it in place.

7. Hold the bottom edge of the panel and drive nails into the center of the nailing slots. Secure the rest of the piece by nailing every 12″–16″. Be sure to drive nails straight and leave a 1/32″ gap between the nail head and the material. Do not pull taut or stretch the panel upward while nailing. This deforms the siding and causes a bad lap with the panel beneath.

8. Factory ends of siding panels have a special notch designed to mate for freedom of movement during expansion and contraction. These factory cuts should be maintained at end splices. Overlap each panel 1″, facing the butt edge away from the main traffic areas. Stagger laps a minimum of 24″ horizontally or three courses vertically. Always allow a 1/4″ expansion gap where siding meets accessories. If you are installing it in cold weather, allow 3/8″ for expansion.

   Note: Do *not* use cut panels in the field of a wall (in between other pieces) always use full-length panels. Failure to do so can cause the end-joint splice pattern to be disrupted on subsequent courses.

9. Check every five or six courses for alignment with eaves.

10. Install the top course and utility trim according to manufacturer directions.

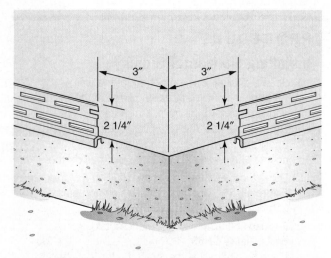

Goodheart-Willcox Publisher

Figure 17-54. Installing the starter strips. A 3″ setback at corners allows room for installing corner posts at inside and outside corners. Allow 1/4″ clearance for expansion between the siding and the corner post.

Goodheart-Willcox Publisher

Figure 17-55. Corner posts are held 3/4″ below the bottom of the starter strip. This keeps the bottom of the siding covered even when the corner shrinks in cold temperatures.

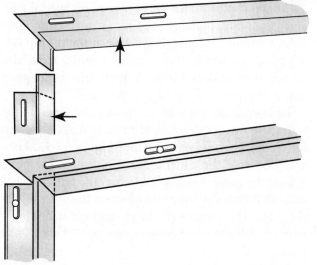

CertainTeed Corp.

Figure 17-56. Install J-channel around windows and doors to provide a finished appearance to siding and to seal joints against water. Cut and shape the flange and miter the corners as shown.

PROCEDURE
Typical Installation of Top Course

1. Install utility trim or J-channel with utility trim against the overhang.
2. To determine how much of the panel to cut off, measure the distance between the top, inside slot of the utility trim and the top lock of the last installed siding panel. Subtract 1/4″ for expansion. Cut the top panel lengthwise to this dimension.
3. Punch the top panel with a snap-lock punch to 1/4″ below the cut edge. This creates raised points that snap into the double J-channel.
4. Install the panel, locking the modified top edge into the utility trim.

PROCEDURE

Installing Vertical Siding

1. Install corner posts as described for horizontal applications. Be careful not to distort the post in any way. Make sure posts are plumb, since this affects installation of the panels.
2. Install J-channel at the top and bottom of the wall as you would for horizontal siding.
3. Carefully measure the distance from the inside of the top J-channel to the inside of the bottom J-channel. Subtract 1/2″ from the total length to allow for panel expansion. Cut the panels to this dimension.
4. Install a starter strip into the corner post channel, **Figure 17-57**. Check that there is sufficient clearance inside of the post channel for the panel to lock onto the starter strip. Drive a nail at the top of the first slot, then check for plumb and finish nailing.
5. Cut, shape, and fit J-channel around windows and doors, as in horizontal installations.
6. Trim and fit vertical panels around doors and windows, **Figure 17-58**.
7. Ending panels at corners are trimmed and fitted into a piece of utility trim inserted under the corner post.

PROCEDURE

Finishing Gable Ends for Vertical Siding

1. Use a level to establish and draw a vertical line in the center of the gable.
2. Attach two J-channels back-to-back on this line.
3. Install starter strips inside each of these J-channels.
4. Install J-channels along the rake and base of the gable, as shown in **Figure 17-59**.
5. Make a pattern for angle cuts. Measure and cut panels.
6. Install panels, starting at the center and working out to each side.

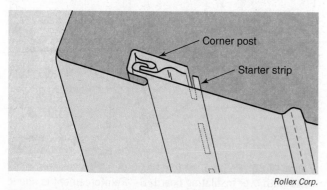

Rollex Corp.

Figure 17-57. In a vertical application, use a starter strip to lock in the first panel at the starting edge of each corner.

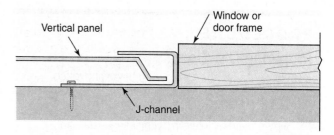

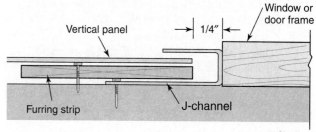

Rollex Corp.

Figure 17-58. Fitting trimmed panels into J-channels. If the trim cut is not made in the vee of the panel, install a furring strip to produce a snug fit in the J-channel.

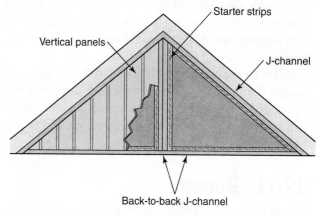

Rollex Corp.

Figure 17-59. Finishing a gable end with vertical siding. Install J-channel on all sides along with starter strips.

17.10.4 Re-Siding over Old Siding

New vinyl siding is often installed without removing the old siding. However, preparation involves several steps. Usually, windowsills, heads, and casings are clad with aluminum to reduce maintenance. Material for the trim cover is cut from prefinished aluminum coil stock and formed on a brake. This trim cover should be installed before the new siding.

Vinyl siding requires a smooth and continuous surface as its base. Therefore, the old siding is covered with new sheathing or exterior board insulation. A new WRB must be established and properly flashed to protect the old siding from any water that may become trapped between layers.

PROCEDURE
Preparing Old Siding

1. Nail loose boards and trim. Replace any rotted boards.
2. Remove downspouts, lighting fixtures, and moldings where they will interfere with siding installation.
3. Tie back shrubbery and trees that are close enough to be damaged or interfere with work.
4. A smooth surface must be in place for the new siding. Install new sheathing or rigid foam board insulation over the old siding.
5. Proceed with installation as if it were new construction.

17.10.5 Estimating Aluminum and Vinyl Siding

Panels are sold by the square (100 sq ft). Find the area of one side and one end wall of the building. Add the areas and multiply by two to get total area. Subtract the area of openings. Add 10% for waste. Divide by 100 to find the number of squares needed.

Order accessories after taking careful linear measurements. Carefully determine where each accessory will be needed as you measure. Accessories include starter strips, inside and outside corner posts, moldings, and J-channels.

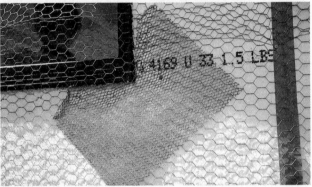

17.11 Stucco

When properly applied, stucco provides an exterior wall finish that is virtually maintenance-free. The finish coat may be tinted by adding coloring or the surface may be painted with a suitable material. Where stucco is used on houses more than one story high, the use of balloon framing for the outside walls is desirable. If stucco is applied to walls built with platform framing, shrinkage at band joists may cause distortion or cracks in the stucco surface.

Stucco is usually applied in three coats over concrete block, wood sheathing, or cement board. The three layers include a base or scratch coat 3/8″–1/2″ thick, brown coat 1/4″–3/8″ thick, and 3/8″ color or finish coat.

Stucco can be directly applied to a clean concrete block wall. On wood structures, "D" or better building paper, roofing felt, or housewrap and metal lath are necessary as a substrate, **Figure 17-60**. Type D building paper is similar to asphalt-saturated felt, except that instead of rag fiber or recycled cellulose fibers the base material is kraft paper made with new wood fibers.

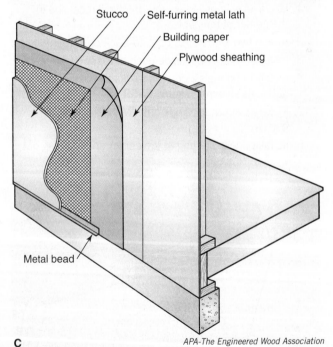

Stucco   Self-furring metal lath

Building paper

Plywood sheathing

Metal bead

Figure 17-60. Preparation for stucco. A—A worker applies metal lath over insulating board. B—Reinforcement is added at window corners. C—Construction details for stucco finish. Sheathing paper or housewrap is recommended over both interior and exterior plywood sheathing.

Felt is made with rag fibers or recycled cardboard fibers and is usually heavier than building paper. Because both kraft paper and felt can absorb some moisture, two layers are usually used under stucco. The metal lath should be heavily galvanized. Space it at least 1/4″ away from the sheathing so the scratch coat can be easily forced through, becoming thoroughly keyed into the metal lath, **Figure 17-61**. Metal or wood molding with a groove that also keys the stucco is applied at edges and around openings.

Galvanized furring nails, metal furring strips, and self-furring wire mesh are available. Nails should penetrate the sheathing at least 3/4″. When fiberboard or gypsum sheathing is used, nailing with adequate penetration should occur over studs.

A metal rake is employed to create scratch lines in the still-wet first coat. The brown coat is applied when the first has dried. It is usually given a smoother surface using straightedges and wooden floats. This produces an adobe-like appearance, as is found in the southwest. Workers use a steel trowel when the brown coat is the final coat.

The first two coats should cure for one or two weeks before the final coat goes on. Usually, the final coat is given a smoother, pebbled finish, but can also have a raised texture, **Figure 17-62**. If color is preferred, iron oxide pigments are added to the finish material. Typically, the finished stucco is 7/8″ thick.

Pro Tip

Stucco plaster is usually one part Portland cement, three parts sand, and hydrated lime that is 10% less by volume than the cement. Once applied, keep it wet for three days. Never apply stucco in temperatures below 40°F (5°C). Stucco needs warm temperatures for the chemical reaction that results in curing.

Lou Oates/Shutterstock.com

Figure 17-61. The base coat is going on over metal lath. Sufficient pressure is applied to push some of the stucco through the lath.

serato/Shutterstock.com

Figure 17-62. The topcoat is being applied. When dry, it will be painted.

17.12 Exterior Insulation and Finish Systems

Exterior insulation and finish systems (EIFS) are similar in appearance to stucco and are often referred to as synthetic stucco. Several manufacturers produce these systems, which offer high R-values, low maintenance, design flexibility, and a wide variety of colors.

EIFS are available either as polymer based (PB) or polymer modified (PM). The PB type, or *soft-coat systems*, are typically thin (1/8″), flexible, and attached with adhesives. The PM type, or *hard-coat systems*, are thicker (about 1/4″) and are mechanically attached. EIFS are offered in different textures.

Earlier systems have been redesigned to handle problems with water penetration and retention. The systems currently being installed drain away any water penetration before it can cause structural damage. **Figure 17-63** shows two typical EIFS applications in cutaway. Water management is accomplished in several ways:

- Application of housewrap or other water barrier materials to the sheathing
- Placing grooves on one surface of the foam panel to channel incidental water that has entered the wall to the outside through weep holes
- Installation of a drainage mat between the exterior insulation board and the water barrier
- Careful application of flashing at points where leakage is likely to occur

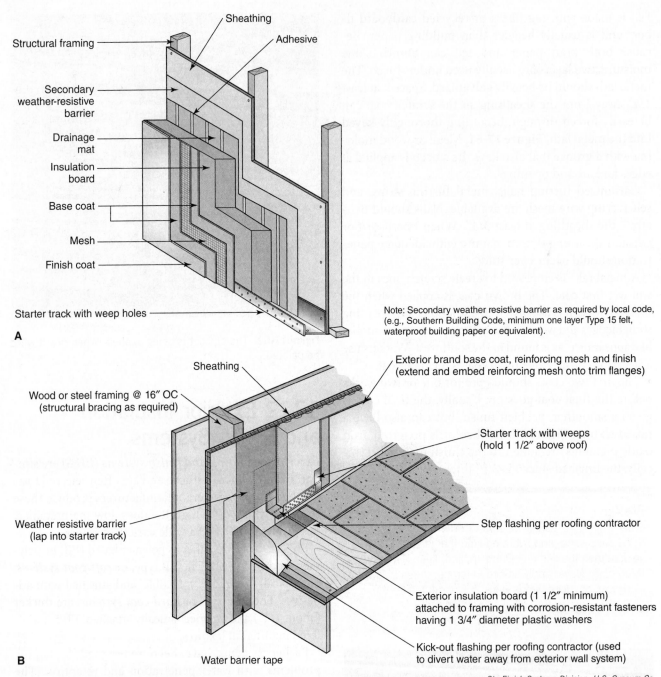

Structural framing

Sheathing

Adhesive

Secondary weather-resistive barrier

Drainage mat

Insulation board

Base coat

Mesh

Finish coat

Starter track with weep holes

Note: Secondary weather resistive barrier as required by local code, (e.g., Southern Building Code, minimum one layer Type 15 felt, waterproof building paper or equivalent).

A

Sheathing

Wood or steel framing @ 16″ OC (structural bracing as required)

Exterior brand base coat, reinforcing mesh and finish (extend and embed reinforcing mesh onto trim flanges)

Starter track with weeps (hold 1 1/2″ above roof)

Step flashing per roofing contractor

Weather resistive barrier (lap into starter track)

Exterior insulation board (1 1/2″ minimum) attached to framing with corrosion-resistant fasteners having 1 3/4″ diameter plastic washers

Water barrier tape

B

Kick-out flashing per roofing contractor (used to divert water away from exterior wall system)

Sto Finish Systems Division, U.S. Gypsum Co.

Figure 17-63. Typical EIFS installations. A—This cutaway shows the components of an EIFS-sided structure that incorporates a drainage cavity and water-resistive barrier. B—This cutaway illustrates a water-managed EIFS that incorporates grooves in the insulation board to channel any incidental moisture to weep holes.

Water-managed systems require accessories, many designed expressly for preventing water damage. See **Figure 17-64**.

Installation should be attempted only after carefully studying manufacturer's instructions.

An exterior insulation finish system can be installed over gypsum sheathing, glass mat–reinforced gypsum, plywood, oriented strand board, concrete, and concrete block. The surface of the substrate must be flat before beginning installation.

EIFS Manufacturer Supplies

Sloped-Sill Wedge

Description: Exterior grade rigid polyvinyl chloride (PVC) trim component used to provide a positive slope over the sill to direct intruding water to the outside face of weather-resistive barrier for drainage out of system. May require increased rough opening size.

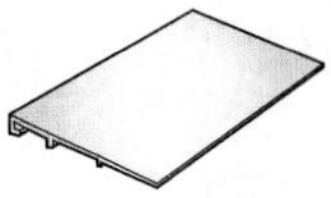

Size: 1/4″ high, 3 7/8″ and 5 7/8″ deep.

Length: 10′

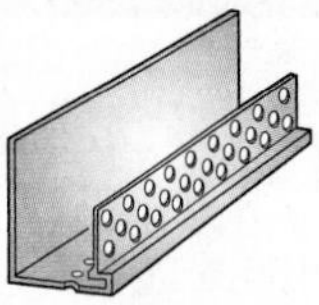

Starter Track

Description: Exterior grade rigid polyvinyl chloride (PVC) trim component with weep holes used at all horizontal terminations (foundations, bottom of wall/roof intersections, bottom of chimney/roof intersections, etc.).

Size: Sized to match system thickness specified.

Length: 10′

Water-Barrier Tape

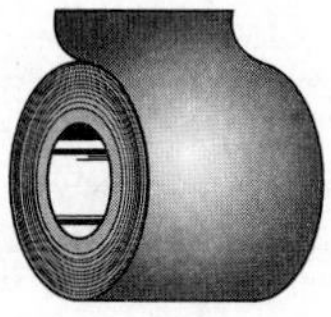

Description: A polyethylene-film coated, self-sealing tape that is primarily used to seal window sills to the building paper prior to the installation of the windows. The tape is self-adhesive with a disposable silicone-treated release sheet.

Thickness: 30 mils.

Width: 9″

Color: Black

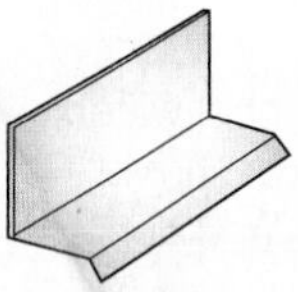

Drip Flashing

Description: Exterior grade rigid polyvinyl chloride (PVC) trim component used to provide flashing and drainage of the system above windows, doors, and other penetrations.

Size: Sized to match system thickness specified.

Length: 10′

Weather-Resistant Barrier

Description: Spunbonded olefin material is engineered to improve the effectiveness of wall systems that require a weather-resistive barrier. Features include: engineered surface design, superior water resistance, high permeance value, excellent tear resistance, pliability, light weight.

Size: 5″ wide, 200 long rolls, 0.22 oz./sq. ft.

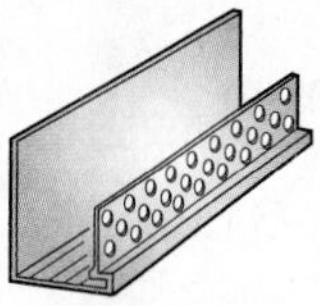

Casing Bead

Description: Exterior grade rigid polyvinyl chloride (PVC) trim component used to terminate system at edges of all windows and doors to create a space for sealant application when foam or panel bands are used for insulation board.

Size: Sized to match system thickness specified.

Wood and Steel Screws

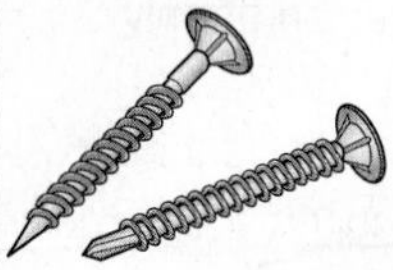

Description: Water head design provides greater holding power. Special coating provides corrosion resistance. For 14 to 20 ga.

Steel Framing: 1 1/4″ and 1 5/8″ steel screws.

For Wood Framing: 1 1/4″, 1 5/8″, or 2 1/4″ wood screws.

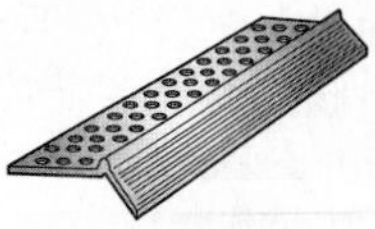

45° Bead

Description: Exterior grade polyvinyl chloride (PVC) trim component used to finish around window and door penetrations as a backing for perimeter sealant application.

Size: 1/2″ long, 45° angled return, 10′ length.

U. S. Gypsum Co.

Figure 17-64. A typical list of accessories supplied by an EIFS manufacturer.

The following is a general method of installing EIFS using a grooved insulation board. As always, carefully read and follow the detailed instructions supplied by the manufacturers.

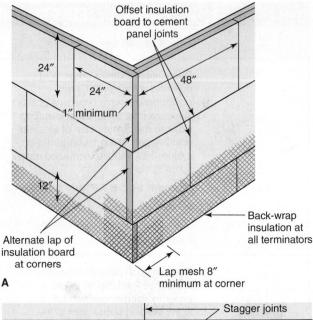

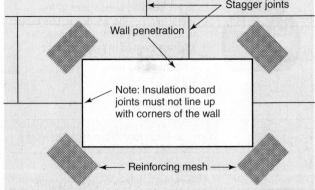

U.S. Gypsum Co.

Figure 17-65. Insulation board alignment. A—Alternate the lap of insulation board at outside corners and offset the vertical joints. B—Avoid lining up joints, either horizontal or vertical, with window or door openings. Apply reinforcement at the corners of wall penetrations.

PROCEDURE
Installing the Insulation Board

1. Install *starter track* at the bottom of the walls, over the tops of windows, and at roof/wall intersections.

2. Horizontally staple a weather-resistive barrier over the sheathing, working from the bottom up. Lap the first course over the flange of the starter track. Lap succeeding courses 4″–6″.

3. Prepare window and door openings by cutting the barrier and securing it at these openings as when applying housewrap.

4. Install flashing at top of openings to channel water to the outside in case of penetration. This should be done at the time you install windows and doors. Flashing should be installed through a slit made above the window head. Secure it with nails. Apply water barrier tape to provide a seal. Since windows vary from one manufacturer to the next, consultation between window manufacturer, EIFS manufacturer, and architect is critical for proper installation of flashing.

5. Install vertical trim around windows and doors.

6. Install control joints following the architect's drawings.

7. Install insulation board so grooves are vertical. Allow 1/16″–3/16″ clearance between the insulation board and the starter strip. Attach to every stud with approved fasteners. Some manufacturers use a drainage mat to direct incidental water to weep holes. Follow their instructions for installing this mat and the insulation board.

8. Precut L-shaped pieces to fit around openings. Avoid having joints at these locations.

9. Joints in the insulation board should be offset between sheathing and window openings. See **Figure 17-65**. Attach insulation board with fasteners no more than 8″ apart in the field, remaining 3″–5″ away from all edges. Fasteners should be long enough to penetrate 3/4″ into studs. Gaps 1/8″ or larger should be filled with slivers of insulation board cut to fit. Do not use adhesive.

10. Sand or rasp the insulation board to improve the bond with the base coat or added architectural features.

PROCEDURE
Applying the EIFS

1. Apply the base coat (an adhesive) to any foam pieces used as architectural features and firmly press them in place, **Figure 17-66**.

2. Coat fastener heads and allow them to dry four hours.

3. Using a steel trowel, apply the base coat to the sanded surface. Uniformly apply the base coat (adhesive).

4. Embed the reinforcing mesh immediately into the wet base coat. Smooth it with a trowel. At this time the base coat should be no thicker than the mesh. Lap the mesh 2″ on all sides.

5. When the base coat has firmed, trowel on a second coat and smooth to 3/32″ thickness.

6. When the base coat has dried a minimum of 24 hours, it is ready for primer or a finish coat.

Figure 17-66. Insulation board being applied in preparation for a new plaster finish.

PROCEDURE

Applying Primer and Finish Material to EIFS

1. Inspect the base coat for defects and repair before continuing.
2. Mix the primer following the manufacturer's instructions. If desired, tint or shade it to the finish color.
3. Apply the primer with a napped roller or quality latex paintbrush. Allow to completely dry before applying finish coats. Work only in temperatures above 40°F (5°C) to ensure proper drying and curing. Protect from rain.
4. Mix the finish material and apply with a trowel to a minimum 1/16″ thickness, **Figure 17-67**. Work rapidly, maintaining a wet edge.
5. Allow the finish to dry for 24 hours. During drying, protect it it from adverse weather, dust, or physical contact.

A   B
Goodheart-Willcox Publisher

Figure 17-67. Applying the final finish to an EIFS. A—A worker mixes colored stucco finishing materials. B—Other construction workers apply the final finish.

17.13 Brick or Stone Veneer

A *veneer wall* is usually not referred to as a masonry wall, although it is. It is a wall framed in wood or metal to which stone, brick, or concrete block is attached rather than siding. However, the weight of the veneer is supported directly by the foundation, **Figure 17-68**.

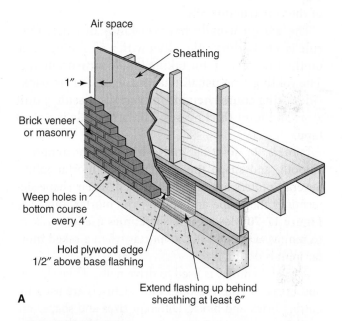

A

B
APA-The Engineered Wood Association

Figure 17-68. A—General construction details for brick veneer siding with plywood sheathing. One tie should be anchored to a stud for every 2 sq. ft. of brick area. B—A brick veneer being installed.

The foundation must be wide enough to provide a base for the masonry units. A base flashing of noncorroding metal should extend from the outside face of the wall, over the top of the ledge, and at least 6″ up behind the sheathing.

17.13.1 Brick Tools

A trowel is the most-used tool in the mason's tool kit. See **Figure 17-69**. The pointed end of the trowel is the *toe* or point; the wide end is the *heel*. Another important tool is the mason's level. A mason's level should have both vertical and horizontal vials that can be read from either side. Some levels are aluminum, while others are wood with metal edges designed to withstand the rough environment of the construction site.

The mason usually has two kinds of rules. One rule is a 6′ folding rule, often with a 6″ sliding scale on the first section for taking inside measurements. The folding rule usually has markings on its backside giving course heights for various masonry unit sizes and joint thicknesses. The other rule is a steel tape.

Jointers, or jointing tools, are used to compress, smooth, and shape the surface of the mortar joints. Several jointers are commonly used. Their shape determines the shape and style of the mortar joint. See **Figure 17-70** Masons also sometimes use joint rakers to remove a portion of the mortar when a raked mortar joint is desired.

A brick hammer is used to drive nails, strike chisels, and break or chip masonry units. Chisels are used for cutting brick and block. Different sizes and shapes are available.

Dmitry Kalinovsky/Shutterstock.com

Figure 17-70. A jointing tool shapes mortar joints before the mortar sets up. The bricks shown here are thin veneer bricks that are placed on the wall like tiles, then the joints are struck in using this thin trowel.

A mason's line is strung across a wall as a guide to keep courses level and walls straight. The line is secured at each end by line holders. For more accurate and faster cutting of brick or block, masons may use a power-driven masonry saw.

17.13.2 Masonry Materials

Bricks are structural units made to several sizes from clay or shale. This material is mixed with water and then fired (baked at a high temperature) in large kilns.

Brick comes in various sizes, some of them modular and some nonmodular. Modular sizes are based on 4″. This includes dimensions of one-half of 4″ as well as two or three times 4″. See **Figure 17-71**.

Sizes include allowance for the thickness of the mortar joint. Thus, the nominal size is smaller than the actual space the brick will occupy in a wall.

A   *bogdanhoda/Shutterstock.com*

B   *khawfangenvi16/Shutterstock.com*

Figure 17-69. Masonry tools. A—Mason's trowel is used to handle mortar and strike off mortar joints. B—A mason's level is used to check masonry walls for plumb and level.

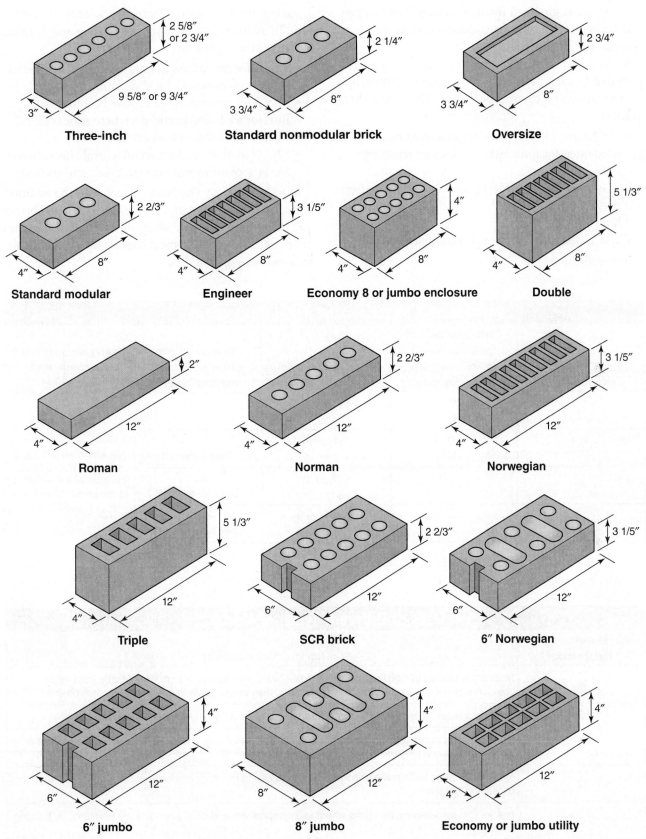

Figure 17-71. Nominal dimensions of brick.

Bricks are engineered for various uses. The two general types are building or common and facing. Each type has different grades.

Building or common brick is a strong, general-purpose brick intended for use where strength is more important than appearance. There are three grades:

- *SW (severe weather) grade* resists freezing and is used often for foundation courses or retaining walls.
- *MW (moderate weather) grade* is used where there may be exposure to below-freezing temperatures, but in dry locations.
- *NW (no weather) grade* is used to back up interior masonry.

Facing brick is used where appearance is important. In addition, the brick must be strong and durable. There are three grades:

- *FBX* is for general use in exposed interior or exterior walls or partitions. Color and size are uniform.
- *FBS* is for general use in exposed exterior and interior walls and partitions where wider color variations and sizes are permitted.
- *FBA* is used to produce architectural effects from the lack of uniformity in size, color, and texture.

Mortar is mostly Portland cement with the addition of hydrated lime and sand. It is designed to bond bricks and blocks into a strong, waterproof wall. Various types have been developed to meet the needs of different situations. See **Figure 17-72**.

Mortar Materials and Their Proportions				
Mortar type	Parts by volume of Portland cement or Portland blast furnace slag cement	Parts by volume of masonry cement	Parts by volume of hydrated lime or lime putty	Aggregate, measured in a damp, loose condition
M	1 1	1 (Type II) —	— 1/4	Not less than 2 1/2 and not more than 3 times the sum of the volumes of the cements and lime used.
S	1/2 1	1 (Type II) —	— Over 1/4 to 1/2	
N	— 1	1 (Type II) —	— Over 1/2 to 1 1/4	
O	— 1	1 (Type I or II) —	— Over 1 1/4 to 1 1/2	
K	1	—	Over 2 1/2 to 4	

A

Uses of Different Types of Mortar	
ASTM mortar type designation	Construction suitability
M	Masonry subjected to high compressive loads, severe frost action, or high lateral loads from earth pressures, hurricane winds, or earthquakes. Structures below grade, manholes, and catch basins.
S	Structures requiring high flexural bond strength, but subject only to normal compressive loads.
N	General use in above grade masonry. Residential basement construction, interior walls and partitions. Concrete masonry veneers applied to frame construction.
O	Non-load-bearing walls and partitions. Solid load bearing masonry of allowable compressive strength not exceeding 100 psi.
K	Interior non-load-bearing partitions where low compressive and bond strengths are permitted by building codes.

B

Goodheart-Willcox Publisher

Figure 17-72. A—This chart shows the materials in mortar and their proportions. B—Uses of different types of mortar.

Corrosion-resistant metal ties are used to secure the veneer to the framework. These are usually horizontally spaced 32″ apart and vertically spaced 15″ apart. Where something other than wood sheathing is used, the ties should be secured to the studs.

Weep holes are small openings in the bottom course. They permit the escape of any water or moisture that may penetrate the wall. Space weep holes about 4′ apart.

Select a type of brick suitable for exposure to the weather. Such brick is hard and low in water absorption. Sandstone and limestone are most commonly used for stone veneer. These materials vary widely in quality. Be sure to select materials locally known to be durable. Procedure for laying brick and stone are essentially the same as laying concrete block. **Figure 17-73** shows typical patterns for both brick and stone veneer.

17.14 Shutters

Some architectural designs may require the installation of shutters at the sides of window units. These consist of frame assemblies with solid panels or louvers. In the early days of our country, shutters served an important practical function. They could be closed over the window to protect the glass. Closing and locking the shutters also provided some security to the inhabitants.

Today, except where hurricanes are likely to occur, shutters and blinds are decorative. They tend to extend the width of windows, stressing horizontal lines of the structure. Hinges are seldom used. Shutters are usually attached to the exterior wall with screws so they can be easily removed for painting and maintenance. Stock sizes include various heights to fit standard window units. Widths range from 14″ to 20″ in 2″ increments.

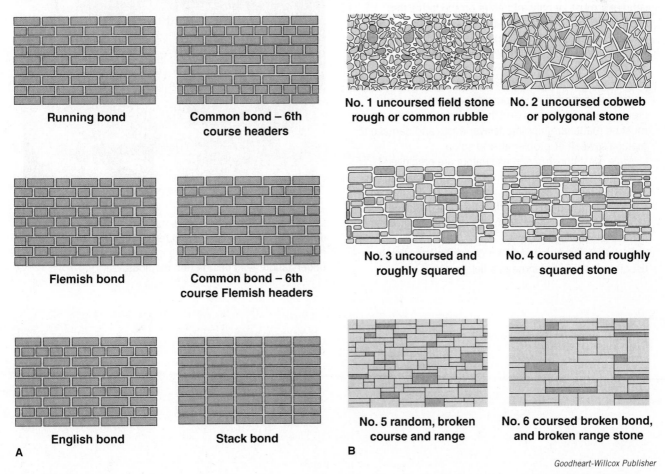

Goodheart-Willcox Publisher

Figure 17-73. These are common patterns for brick and stone veneer siding. A—Basic structure bonds for brick. B—Common stone patterns.

Construction Careers
Estimators

Accurate estimates of cost are vital to the success of a construction project. Developing such estimates in detail is the responsibility of the construction estimator. Usually employed by a general contractor, the construction estimator factors in everything required to complete a job, including the price of lumber, the cost of renting a large crane for a day, the labor pay rates to insurance coverage, and expected loss of time resulting from bad weather. The resulting estimate is used as the basis for the contractor to competitively bid on the construction contract. For firms building single homes or residential developments, such an estimate is used to set the selling price of the homes.

Construction estimators usually begin with a set of detailed building plans that provide dimensions and other information needed to begin the estimate. In most cases, the estimator also visits the site. This visit provides information on terrain, site access, drainage, and the availability of utilities.

Next, careful analyses of material and labor requirements are done. These are called takeoffs and are entered on standard estimating forms. The estimator must identify and enter on the forms the exact quantities of each item that the contractor must provide. If portions of the work (such as plumbing or electrical) are subcontracted, the estimator must analyze the bids submitted by the subcontractors. Decisions involving sequencing of operations must be made, since improperly scheduled equipment, materials, or work crews may cause delays and increased costs. Allowances must be made for shipping delays, waste and damage to materials, weather problems, and so on.

While most construction estimators are employed by building contractors, some work for large architectural or engineering firms. Estimators may also operate their own consulting businesses, providing fee-based services to contractors, government agencies, and building owners or financing organizations.

Educational requirements for construction estimators usually consist of a degree in a field such as construction management, engineering, or architecture. A strong aptitude for mathematics and proficiency with computer software are also valuable assets. Many estimators also have practical experience in construction work, giving them familiarity with materials, job practices, and the various specialty trades. Additional on-the-job training is usually provided, since each company has its own specific way of preparing and presenting estimates.

Professional certification is voluntary, but beneficial. Such certification is administered by two professional organizations—the Association for the Advancement of Cost Engineering and the Society of Cost Estimating and Analysis. The requirements vary, but include a number of years of professional experience and passing an examination.

Marcin Balcerzak/Hemera/Thinkstock

Construction estimators review construction drawings to determine the quantities of materials required for the project. Estimators have strong math skills and a good understanding of construction methods.

Chapter Review

Summary

- The exterior finish of a structure includes installation of siding or other exterior wall finish materials, cornices, and the trim around windows and doors.
- A cornice is the finished connection between the exterior wall and the roof. It includes the fascia board, lookouts, soffit, and trim molding.
- A box cornice conceals the rafter ends with a fascia board and soffits. An open cornice does not use soffits and may not have a fascia board covering the rafter tails.
- If the rafters do not extend past the sidewall, the cornice is called a snub cornice.
- Siding styles include horizontal, vertical, and shingles.
- Horizontal siding—wood, fiber-cement, or vinyl—is overlapped to resist water infiltration and has an attractive three-dimensional appearance due to its beveled cross section.
- Vinyl siding expands and contracts with changes in temperature. Because of this, it is important that it is installed in a manner that allows unrestricted movement.
- Shingles are traditionally made of rot-resistant wood, but are also available in a fiber-cement material.
- Vertical siding may be individual boards, panels, or large sheets of material such as plywood.
- Fiber-cement siding is being used as a substitute for wood siding on new builds. It is durable and requires less maintenance than wood.
- Vinyl siding has largely replaced aluminum siding and is used in new construction of homes more than any other siding product.
- Exterior walls may be covered with a layer of cement-like material called stucco or with an applied exterior insulation and finish system (EIFS).
- An EIFS application can mimic stucco, stone, or other materials.
- Some buildings use a thin layer of brick or stone, called a veneer, as the exterior finish.

Review Questions

Answer the following questions using the information in this chapter.

1. Which one of the following is *not* considered part of the exterior finish of a house?
   A. An entry door.
   B. Ceiling of a porch or breezeway.
   C. Cornice.
   D. Exterior wall coverings.

2. The _____ protects the wall finish from water by directing runoff way from the wall and often captures it in a gutter.
   A. soffit
   B. cornice overhang
   C. eaves-trough
   D. WRB

3. *True or False?* The open cornice is the most commonly used cornice design.

4. The wood nailing strip nailed to the rafter ends in a box cornice is called a _____.
   A. ledger strip   C. sub-fascia
   B. lookout   D. fascia backer

5. The _____ is the part of a roof that overhangs a gable end.
   A. fascia   C. rake
   B. frieze   D. lookout

6. In areas not subject to wind-driven rain, head flashing can usually be omitted over an opening if the vertical distance to the soffit is less than or equal to _____ the width of the overhang.
   A. 1/4   C. double
   B. 1/2   D. triple

7. Of the various types of horizontal wood siding, _____ siding usually has the most strength and tightest joints.

A. bevel   C. drop
B. channel rustic   D. wide bevel

8. The best grade of solid wood siding is made from _____-grain material.

A. horizontal   C. flat
B. vertical   D. edge

9. Plain beveled siding in a 10″ width should be lapped about _____ ″.

A. 1   C. 2
B. 1 1/2   D. 3 1/2

10. Which of the following nails should *not* be used to fasten siding?

A. Stainless steel.
B. Hot-dipped galvanized.
C. High-tensile-strength aluminum.
D. Small-headed plain steel.

11. To avoid future separation between primer and topcoat, apply the first topcoat of paint within _____ weeks of the primer.

A. two   C. four
B. three   D. six

12. *True or False?* Vertical siding can be used to set off entrances, gable ends, and main wall areas.

13. What is the maximum exposure for 18″ wide shingles?

A. 7 1/2″   C. 11 1/2″
B. 8 1/2″   D. 14″

14. Shingles should always be applied to sidewalls from _____.

A. the bottom up
B. the top down
C. left-to-right
D. largest to smallest

15. In _____ coursing, butts of alternating shingles are offset below the line.

A. single   C. staggered
B. double   D. ribbon

16. _____ is the most commonly used species for plywood siding.

A. Douglas fir   C. Redwood
B. Cedar   D. Pine

17. Which of the following is not typically an ingredient for making fiber-cement?

A. Fiberglass.   C. Sand
B. Portland cement.   D. Cellulose fibers.

18. What is placed behind end joints in horizontal fiber-cement siding?

A. Vented soffits.
B. Flashing.
C. A WRB.
D. Special calk compound.

19. Vinyl siding has largely replaced _____ siding.

A. wood   C. aluminum
B. fiber-cement   D. steel

20. *True or False?* Brakes are commonly used to build aluminum and vinyl trim pieces.

21. What tool can be used to help speed cutting and shaping of vinyl siding?

A. Utility knife.   C. Snap-lock punch.
B. Aviation snips.   D. Hand saw.

22. When installing horizontal siding, carpenters usually start from the _____ and work to the _____ of the building.

A. top, bottom   C. back, front
B. bottom, top   D. front, back

23. *True or False?* When re-siding with vinyl siding, the old siding must be removed.

24. Vinyl siding panels are sold by the square, which is _____ square feet.

A. 25   C. 100
B. 50   D. 250

25. The weight of the veneer in a veneer wall is supported by the _____.

A. foundation   C. concrete block
B. siding   D. joints

26. A _____ is the most-used tool in the mason's tool kit.

A. mason's level   C. brick hammer
B. trowel   D. steel tape

27. *True or False?* Modern-day shutters are only used for decorative purposes, except where hurricanes are likely to occur.

Apply and Analyze

1. Why is a snub cornice design sometimes cheaper than other cornice designs?

2. Explain two methods of protecting wood siding from weathering.

3. What is *double coursing*?

4. What is considered by one manufacturer to be the best method for cutting fiber-cement indoors?

5. Explain why vinyl siding has almost all but replaced aluminum siding.

6. How do you install aluminum fascia trim?

7. Explain how to apply the three coats of stucco over a concrete block wall.

Critical Thinking

1. Research various styles of cornice returns on the internet. Compare your findings with the styles used on the home in **Figure 17-35**. Note that this home already features two differing styles of cornice return. Express your opinion of the look of this home as built. Do you think the right cornice choices were made? If given the choice, would you do anything differently? Why or why not?

2. Oil canning (**Figure 17-49**) is a common defect on vinyl siding when it is installed improperly. Consider what forces contribute to oil canning. Name three mistakes that could cause vinyl siding to oil can and explain what proper installation techniques are used to prevent each.

Communicating about Carpentry

1. **Listening.** In small groups, discuss with your classmates—in basic, everyday language—the exterior wall finish on your home.

Conduct this discussion as though you had never read this chapter. Take notes on the observations expressed. Then review the points discussed, factoring in your new knowledge of exterior wall finish. Develop a summary of what you have learned about exterior wall finish and present it to the class, using the terms that you have learned in this chapter.

2. **Speaking and Reading.** Research the products and services available for roofing materials, exterior doors, and exterior walls. Collect promotional materials for a variety of products and services from product manufacturers. Analyze the data in these materials based on the knowledge gained from this lesson. With your group, review the list for words that can be used in the subject area of roofing, exterior doors, and exterior walls. Practice pronouncing the word, and discuss its meaning. As a fun challenge, work together to compose a creative narrative using as many words as you can from your new list.

3. **Speaking and Writing.** Compare and contrast the types of exterior wall finishes available. Include the pros and cons of each type of material. Prepare an oral or written presentation.

SECTION 4 | Finishing

Javani LLC/Shutterstock.com

CHAPTER 18

Thermal and Sound Insulation

OBJECTIVES

After studying this chapter, you will be able to:

- Summarize the principles of conduction, convection, and radiation.
- Define technical terms relating to thermal properties of construction materials.
- Interpret thermal ratings charts.
- Describe the types of insulation.
- Select appropriate areas for insulation in a given structure.
- List general procedures for installing batt and blanket, loose, foamed, and rigid insulation.
- List ways to increase sound and noise control.
- Describe methods of construction that raise STC ratings in desired areas.
- Describe how acoustical materials work.

TECHNICAL TERMS

batt insulation
blanket insulation
blower door test
British thermal unit (btu)
coefficient of thermal
   conductivity (k)
conductance (C)
conduction
convection
convection current
crawl space

decibels (dB)
degree day
draft excluder
flexible insulation
foamed-in-place insulation
frequency
impact sounds
loose-fill insulation
masking sounds
noise

noise-reduction coefficient
   (NRC)
radiation
reflective insulation
resistance (R)
sound transmission class
   (STC)
sound transmission loss (STL)
threshold of pain
total heat transmission (U)

In construction, insulation includes materials that do not readily transmit energy in the form of heat, electricity, or sound. This chapter primarily deals with using insulation to prevent the transfer of heat either into or out of a building. To a lesser extent, the use of insulation to prevent sound transmission is also discussed. Acoustical treatments and sound control use many of the same materials used for thermal insulation. They are, therefore, both described in this chapter.

Insulation requirements of homes and other buildings have radically changed as people have become more concerned with energy conservation. As energy costs have increased, so has the amount of insulation being placed in walls, floors, and ceilings, **Figure 18-1**.

Owens-Corning

Figure 18-1. Blown cellulose insulation is added over batt insulation to reduce a home's heat loss through the attic.

At the same time, the insulating qualities of doors and windows have been upgraded to hold down energy costs. The amount and quality of insulation in a home can play a huge role in its green rating. Today, ductwork is likely to be sealed and insulated if it is not placed in a heated space.

Insulation is necessary in buildings where the temperature of the interior space must be controlled. In cold climates, the major concern is retaining heat in the building. In warmer regions, insulation keeps external heat from entering the building.

A wide range of insulation materials is available to fill energy-efficiency requirements. The materials are engineered for efficient installation and come in convenient packages that are easy to handle and store.

18.1 Building Sequence

While carpenters are completing the exterior finish of a structure, other tradespeople are installing mechanical systems on the inside. These systems must be completed before interior walls are insulated and closed in with drywall or plaster. Ductwork for heating and air conditioning is installed in the floors, walls, and ceilings. An electrician installs wiring for electrical circuits, attaching conduit or cable to the building's framework. The boxes that enclose connections for convenience outlets, switches, and lighting fixtures are also located and attached.

A plumber installs the water supply piping and drainage/waste/venting (DWV) piping, **Figure 18-2**. These pipes supply fresh water and drain away wastewater. Tubs and shower bases are also usually installed at this time. They must be carefully covered to protect them during the interior finish operations.

During the heating and plumbing rough-in, tradespeople often need to cut through structural framework. A carpenter should check this work. Framing members may have been seriously weakened and must be reinforced before interior walls are covered.

Goodheart-Willcox Publisher

Figure 18-2. Before thermal and sound insulation materials are installed, other construction trades must make their installations. In this structure, the electrical wiring, outlet boxes, and rough plumbing have been installed.

18.2 How Heat Is Transmitted

Although carpenters are seldom required to design buildings or figure heat losses, they should have some knowledge and understanding of the theory of heat transfer and the factors involved.

Heat seeks a balance with surrounding areas. When the inside temperature is controlled within a given comfort range, there is some flow of heat. Heat moves from the inside to the outside in winter and from the outside to the inside during hot summer weather.

Heat is transferred through walls, floors, ceilings, windows, and doors at a rate directly related to the difference in temperature and the resistance to heat flow provided by intervening materials. Heat moves from one place to another by one or a combination of three methods, **Figure 18-3**:

- Conduction
- Convection
- Radiation

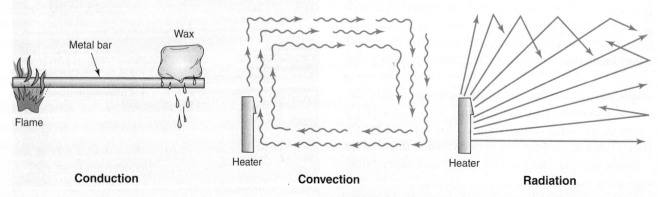

Goodheart-Willcox Publisher

Figure 18-3. How heat is transferred. Conduction takes place only in solids. Convection uses air or water to move heat to a cooler space. Radiation makes use of wave motion.

Actually, heat transmission through walls, ceilings, and floors is a result of all three transmission methods. In addition to this, some heat is moved by convection as air flows through cracks around doors, windows, and other openings in the structure.

18.2.1 Conduction

Conduction is transmission of heat from one molecule to another within a material, or from one material to another when they are held in direct contact. Dense materials, such as metal or stone, more rapidly conduct heat than porous materials, such as wood and fiber products. Any material conducts some heat when a temperature difference exists between its surfaces.

18.2.2 Convection

Convection is the transfer of heat by another agent, such as air or water. In large spaces, molecules of air can carry heat from warm surfaces to cold surfaces. When air is heated, it becomes lighter and rises. Thus, airflow (called *convection currents*) is created within the space. Air is a good insulator when it is confined to a small space (cavity) in which convection flow is limited or absent. In walls and ceilings, trapped air restricts convection currents and will reduce the flow of heat.

18.2.3 Radiation

Heat can be transmitted by wave motion in about the same manner as light. This process is called *radiation* because it represents radiant energy.

Heat received from the sun is a good example of radiant heat. The waves do not heat the space through which they move. However, when they come in contact with a conductive surface, some of the energy is absorbed, while some may be reflected.

Effective resistance to radiation heating comes about through reflection. Shiny surfaces, such as aluminum foil, are often used to provide this type of insulation.

18.3 Thermal Insulation

All building materials resist the flow of heat to some degree. The amount of resistance depends on their porosity (how porous) or density. Air is an excellent insulator when confined to the tiny spaces or cells inside of a porous material. Dense materials such as masonry or glass contain few, if any, air spaces and are poor thermal insulators.

Fibrous materials are generally good insulators, not only because of the porosity in the fibers themselves,

but also because of the thin film of air that surrounds each individual fiber.

Commercial insulation materials are made of glass fibers, glass foam, mineral fibers, organic fibers, and foamed plastic. Most insulation relies on pockets of trapped air to resist the movement of heat through the material. If the air pockets absorb excessive amounts of moisture, their insulating value is greatly reduced. In addition, insulation should be resistant to any physical change, such as settling or being compacted, that would reduce its effectiveness in preventing heat flow. Selection is based on initial cost, effectiveness, durability, purpose and the adaptation of the insulating material's form to that of the construction and installation methods.

18.4 Terminology

A coefficient is a number that serves as a measure of a property—in this case, the ability to transfer heat. The thermal properties of common building materials and insulation materials are known, or can be accurately measured. Heat transmission (the amount of heat flow) through any combination of these materials can be calculated. First, it is necessary to understand certain terms.

A *British thermal unit (Btu)* is a measure of heat. One Btu is the amount of heat needed to raise the temperature of 1 lb of water 1°F.

$$\text{Btu} = \text{weight} \times \text{temperature difference}$$

The *coefficient of thermal conductivity (k)* is the amount of heat (Btu) transferred in one hour through 1 sq ft of a given material that is 1″ thick and has a temperature difference between its surfaces of 1°F.

The *conductance (C)* of a material is the amount of heat (Btu) that will flow through the material in one hour per sq ft of surface with 1°F of temperature difference. The thickness of the material is not a factor. For example, the C-value for an average hollow concrete block is 0.53.

The *resistance (R)* is the reciprocal (inverse) of conductivity or conductance. All insulation has an R-value listed by the manufacturer. A good insulation material has a high R-value.

$$R = \frac{1}{k} \; or \; \frac{1}{C}$$

The *total heat transmission (U)* is measured in Btu per sq ft per hour with 1°F temperature difference for a structure (wall, ceiling, and floor). The structure may consist of several materials or spaces. A standard frame wall with composition sheathing, gypsum lath and plaster, and a 1″ blanket insulation has a U-value of about 0.11. To calculate the U-value where the R-values are known, apply this formula:

$$U = \frac{1}{R_1 + R_2 + R_3 + \ldots R_n}$$

The harshness of climate is measured in degree days. The higher the number of degree days, the colder the climate and the more insulation needed. A *degree day* is the product of one day and the number of degrees Fahrenheit that the mean temperature is below 65°F. Figures are usually quoted for a full year. Degree days are used by a heating engineer to determine the design and size of the heating system. Use this formula to determine a degree day if the high is 60°F and the low is 30°F:

$$\text{Degree day} = 60 + \frac{30}{2}$$
$$= 65 - \frac{90}{2}$$
$$= 65 - 45$$
$$= 20$$

Figure 18-4 shows how insulation reduces the U-value for a conventional frame wall. Note that a 5 1/2″ thick blanket reduces the U-value from 0.29 to 0.053. This is about an 82% reduction. Actually, the U-value for the total wall structure is slightly higher because the wood studs have a lower R-value than the blanket insulation.

R-values of wall structures with various types and amounts of insulation are shown in **Figure 18-5**. The R-value provides a convenient measure for comparing heat loss in materials and structural designs. However, to determine the total heat loss (or gain) through a wall, ceiling, or floor, you must convert R-values to U-values. The total of these U-values is needed to calculate the size of heating and cooling equipment. Refer to Appendix B, **Technical Information**.

It is important to understand that heat transmission decreases as the insulation thickness is increased, but not in a direct relationship. This can be noted through a study of U- and R-values for various materials and structures. For example, in a frame wall with a U-value of 0.24, the addition of 1″ of insulation reduces the heat loss by 46%, to U-0.13. A second inch of insulation reduces the loss by 16% to U-0.09. A third inch reduces the loss by 10% to U-0.065. Additional thicknesses continue to lower the U-value, but at a still-lower percentage. At some point, it becomes useless to add more insulation.

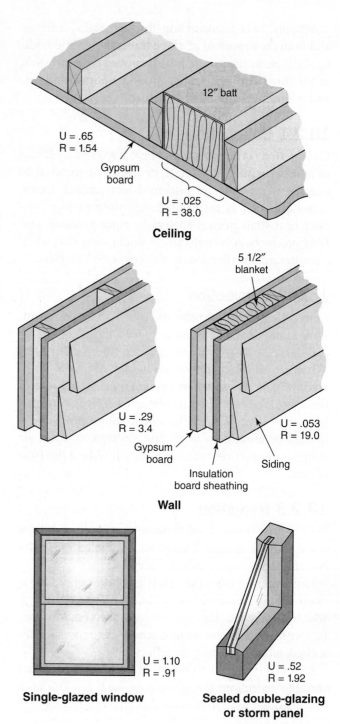

Goodheart-Willcox Publisher

Figure 18-4. Insulation reduces flow of heat. Double-glazed windows enclose trapped air, which is a good insulator.

Windows provide another example of how U-values decrease. The heat loss through a sash with a single pane of glass is reduced from U-1.10 to about U-0.52 when a second pane is added. This provides a reduction of about 50%. A third pane (triple glazing) reduces the heat loss to U-0.35. Windows with three layers of glass are expensive. However, in northern climates, the resulting lower heating costs and added comfort often justify this extra expense.

Types of Wall Construction and R-Values

Uninsulated 2×4 Stud Wall

Air films	R = 0.9
3/4″ wood exterior siding	1.0
1/2″ insulation board	1.2
Air space	1.2
Vapor barrier	0.0
1/2″ gypsum board	0.5
	4.8 total R

2×4 Stud Wall with Batt Insulation

Air films	R = 0.9
3/4″ wood exterior siding	1.0
1/2″ insulation board	1.2
3 1/2″ batt or blanket insulation	11.0
Vapor barrier	0.0
1/2″ gypsum board	0.5
	14.6 total R

2×4 Stud Wall with Rigid Board

Air films	R = 0.9
3/4″ wood exterior siding	1.0
1″ polystyrene rigid board	5.0
3 1/2″ batt or blanket insulation	11.0
Vapor barrier	0.0
1/2″ gypsum board	0.5
	18.4 total R

Improved Insulated 2×4 Stud Wall

Air films	R = 0.9
3/4″ wood exterior siding	1.0
3/4″ insulation board	2.0
3 5/8″ batt insulation	13.0
Vapor barrier	0.0
5/8″ urethane insulation board	5.0
1/2″ gypsum board	0.5
	22.4 total R

2×6 Insulated Stud Wall

Air films	R = 0.9
3/4″ wood exterior siding	1.0
3/4″ insulation board	2.0
5 1/2″ insulating blanket	19.0
Vapor barrier	0.0
1/2″ gypsum board	0.5
	23.4 total R

Improved 2×6 Insulated Stud Wall

Air films	R = 0.9
3/4″ wood exterior siding	1.0
3/4″ insulation board	2.0
5 1/2″ batt or blanket insulation	19.0
Vapor barrier	0.0
5/8″ urethane insulation	5.0
1/2″ gypsum board	0.6
	28.5 total R

Iowa Energy Policy Council

Figure 18-5. Types of wall construction and their R-values. Materials are listed in order from the outside in. "Air films" refer to the inside and outside film of stagnant air that forms on any surface.

R-values for commonly used insulation and building materials are listed in **Figure 18-6**. The original source of most data on this subject is the American Society of Heating, Refrigerating, and Air Conditioning Engineers (ASHRAE). R-values can be converted to U-values by calculating the reciprocal (dividing the value into 1).

Commonly Used R-Values for Construction and Insulation Materials

Material	Kind	Insulation value
Masonry	Concrete, sand, and gravel, 1″	R-0.08
	Concrete blocks (three core)	
	Sand and gravel aggregate, 4″	R-0.71
	Sand and gravel aggregate, 8″	R-1.11
	Lightweight aggregate, 4″	R-1.50
	Lightweight aggregate, 8″	R-2.00
	Brick	
	Face, 4″	R-0.44
	Common, 4″	R-0.80
	Stone, lime, sand, 1″	R-0.08
	Stucco, 1″	R-0.20
Wood	Fir, pine, other softwoods, 3/4″	R-0.94
	Fir, pine, other softwoods, 1 1/2″	R-1.89
	Fir, pine, other softwoods, 3 1/2″	R-4.35
	Maple, oak, other hardwoods, 1″	R-0-.91
Manufactured Wood Products	Plywood, softwood, 1/4″	R-0.31
	Plywood, softwood, 1/2″	R-0.62
	Plywood, softwood, 5/8″	R-0.78
	Plywood, softwood, 3/4″	R-0-.93
	Hardboard, tempered, 1/4″	R-0.25
	Hardboard, underlayment, 1/4″	R-0.31
	Particleboard, underlayment, 5/8″	R-0.82
	Mineral fiber, 1/4″	R-0.21
	Gypsum board, 1/2″	R-0.45
	Gypsum board, 5/8″	R-0.56
	Insulation board sheathing, 1/2″	R-1.32
	Insulation board sheathing, 25/32″	R-2.06
Siding and Roofing	Building paper, permeable felt, 15 lb.	R-0.06
	Wood bevel siding, 1/2″	R-0.81
	Wood bevel siding, 3/4″	R-1.05
	Aluminum, hollow-back siding	R-0.61
	Wood siding shingles, 7 1/2″ exp.	R-0.87
	Wood roofing shingles, standard	R-0.94
	Asphalt roofing shingles	R-0.44
Insulation	Cellular or foam glass, 1″	R-2.50
	Glass fiber, batt, 1″	R-3.13
	Expanded perlite, 1″	R-2.78
	Expanded polystyrene bead board, 1″	R-3.85
	Expanded polystyrene extruded smooth, 1″	R-5.00
	Expanded polyurethane, 1″	R-6.00
	Mineral fiber with binder, 1″	R-3.45
Inside Finish	Cement plaster, sand aggregate, 1″	R-0.20
	Gypsum plaster, light wt. aggregate, 1/2″	R-0.32
	Hardwood finished floor, 3/4″	R-0.68
	Vinyl floor, 1/8″	R-0.05
	Carpet and fibrous pad	R-2.05
Glass	(see Chapter 16 Windows and Exterior Doors)	

Goodheart-Willcox Publisher

Figure 18-6. R-values for commonly used construction and insulation materials.

Values of R-18 or higher in residential wall construction can be obtained two ways:

- Using 2×6 studs, which provide a thicker wall cavity for insulation, **Figure 18-7**.
- Using 2×4 studs sheathed with thick, rigid insulation panels made from foamed polystyrene plastic.

18.5 Amount of Insulation

Insulation is required in any building where a temperature above or below the outdoor temperature must be maintained inside of the building. Comfort, health, economy, and environmental impact are considerations for thermal insulation.

Comfortable and healthy indoor temperatures depend not only on the temperature of the air, but also on the temperature of the wall, ceiling, and floor surfaces. It is possible to heat the air in a room to the correct comfort level and still have cold room surfaces. The human body will lose heat by radiation (or conduction, if there is direct contact) to these colder surfaces.

The amount of insulation for a given structure must be based not only on comfort standards, but also on factors such as insulation costs (material and labor), probable fuel costs in the future, and local climatic conditions. **Figure 18-8** shows a map of the continental

Goodheart-Willcox Publisher

Figure 18-7. One of the easiest ways to secure R-18+ values in outside walls is to use 2×6 studs and install insulation in the stud space.

United States and a chart of recommended insulation for each area according to winter/summer temperatures. In warmer climates, smaller amounts of insulation are needed, not to protect against cold, but to keep out summer heat.

Department of Energy R-Value Recommendations

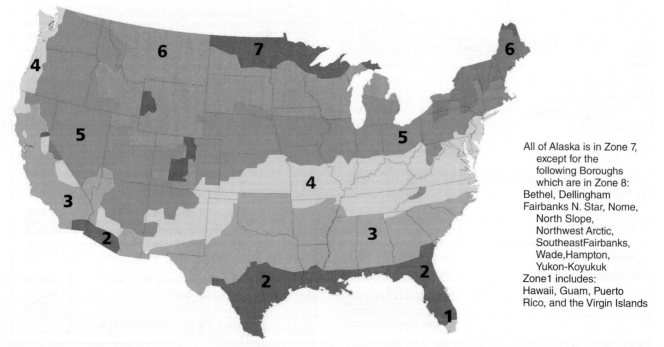

All of Alaska is in Zone 7, except for the following Boroughs which are in Zone 8: Bethel, Dellingham Fairbanks N. Star, Nome, North Slope, Northwest Arctic, SoutheastFairbanks, Wade,Hampton, Yukon-Koyukuk
Zone1 includes: Hawaii, Guam, Puerto Rico, and the Virgin Islands

U.S. Department of Energy

Figure 18-8. North America is divided into zones according to the climate in each area. The R-values listed are recommended by the U.S. Department of Energy. *(continued)*

New Wood-Framed Houses

| Zone | Heating System | Attic | Cathedral Ceiling | Wall | | Floor |
				Cavity	Insulation Sheathing	
1	All	R30 to R49	R22 to R38	R13 to R15	None	R13
2	Gas, oil, heat pump	R30 to R60	R22 to R38	R13 to R15	None	R13
	Electric furnace					R19 to R25
3	Gas, oil, heat pump	R30 to R60	R22 to R38	R13 to R15	None	R25
	Electric furnace				R2.5 to R5	
4	Gas, oil, heat pump	R38 to R60	R30 to R38	R13 to R15	R2.5 to R6	R25 to R30
	Electric furnace				R5 to R6	
5	Gas, oil, heat pump	R38 to R60	R30 to R38	R13 to R15	R2.5 to R6	R25 to R30
	Electric furnace				R5 to R6	
6	All	R49 to R60	R30 to R38	R13 to R15	R5 to R6	R25 to R30
7	All	R49 to R60	R30 to R38	R13 to R15	R5 to R6	R25 to R30
8	All	R49 to R60	R30 to R38	R13 to R15	R5 to R6	R25 to R30

Existing Wood-Framed Houses

| Zone | Wall | | Floor |
	Uninsulated Attic	Existing 3 to 4 Inches of Insulation	
1	R30 to R49	R25 to R30	R13
2	R30 to R60	R25 to R38	R3 to R19
3	R30 to R60	R25 to R38	R19 to R25
4	R30 to R60	R38	R25 to R30
5–8	R49 to R60	R38 to R49	R25 to R30

Wall Insulation-Whenever exterior siding is removed on an:

Uninsulated wood-frame wall:	Insulated wood-frame wall:
▪ Drill holes in the sheathing and blow insulation into the empty wall cavity before installing the new siding ▪ Insulative sheathing may be used as necessary to meet R-value requirements	▪ For Zones 4-8, add R5 insulative sheathing before installing the new siding Reference: Department of Energy/CE-01802008, Insulation Fact Sheet

Figure 18-8. *(continued)*

18.6 Types of Insulation

Thermal insulation is made in many forms, **Figure 18-9**. It may be grouped into five broad classifications:

- Flexible
- Loose fill
- Rigid
- Foamed-in-place
- Reflective

Flexible insulation is manufactured in two types: blanket and batt. *Blanket insulation* is generally furnished in rolls or strips of convenient length and in various widths suited to standard stud and joist spacing. It comes in thicknesses of 3/4″–12″.

The body of the blanket is made of loosely felted mats of mineral or vegetable fibers, such as glass fiber, rock wool, slag, wood fiber, and cotton. Organic fiber mats are usually chemically treated to make them resistant to fire, decay, insects, and vermin. Blanket insulation is often enclosed in paper with tabs on the side for attachment to framing members. The covering materials may be treated on one side to serve as a vapor retarder. In some cases, the covering is surfaced with aluminum foil or other reflective insulation.

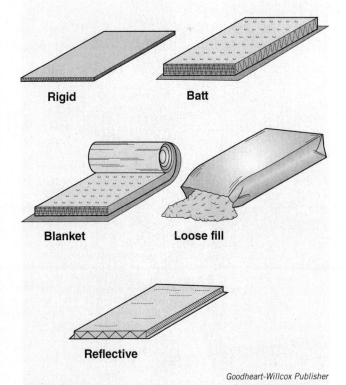

Rigid   Batt

Blanket   Loose fill

Reflective

Goodheart-Willcox Publisher

Figure 18-9. Insulation is made in these basic forms plus foamed-in-place. Batt and blanket are both types of flexible insulation.

Figure 18-10 shows an unfaced blanket that is easy to install between wall studs. Friction holds it in place. After all of the insulation is in place, a vapor barrier (explained later in this chapter) should be applied over the entire surface of the wall frame, since the blanket is unfaced. It prevents water vapor inside the building from entering the insulation.

Batt insulation is made of the same fibrous material as blankets. Thickness can be greater in this form and may range from 3 1/2″ to 12″, **Figure 18-11**. Batts are generally available in widths of 15″ and 23″ and in 24″ and 48″ lengths. They are available with a single-flanged cover or with both sides uncovered.

Loose-fill insulation is made from such materials as rock wool, glass, slag wool, wood fibers, shredded redwood bark, shredded paper, granulated cork, ground or macerated (softened by soaking in a liquid) wood pulp products, vermiculite, perlite, (vermiculite and perlite are both naturally occurring materials that greatly expand, causing them to hold trapped air) powdered gypsum, sawdust, and wood shavings. Loose-fill insulation is used in bulk form and supplied in bags or bales, **Figure 18-12**. Loose insulation may be poured or blown. It is commonly used to insulate spaces between studs or to build up any desired thickness on a flat surface, such as above ceilings.

Goodheart-Willcox Publisher

Figure 18-11. Batts are sold in various thicknesses. Thicker batts are designed to give high insulation values for ceilings and attics.

Kurteev Gennadii/Shutterstock.com

Figure 18-12. Loose insulation is often used over ceilings in attic areas. Coverage and R-values are listed on the bag.

One of the chief advantages of this type is that when insulating an older structure, only a few boards need to be removed in order to blow the material into the walls. **Figure 18-13** shows typical coverage of loose fiberglass insulation blown into walls and attics.

Rigid insulation is manufactured in sheets and other inflexible forms. These foam panels have higher R-values per inch than other forms of insulation. They are also more expensive. Their durability and rigidity allow them to be used in conditions where other types of insulation would not be suitable. The uses and properties of rigid foam insulation are covered in Chapter 15, **Building Envelope and Control Layers.**

Owens-Corning

Figure 18-10. Unfaced blanket insulation must snugly fit between studs as it is held there by friction.

Sidewall Coverage Information

Nominal R-value	Thickness(framing timber)	Density	Minimum weight per sq ft	Maximum coverage per bag**
To obtain a thermal resistance (R) of:	Installed insulation should not be less than (inches):	Pounds per cubic foot:	Weight per sq ft of installed insulation should not be less than (lb):	Contents of bag should not cover more than (sq ft):
R-14	3.5" (2×4)	1.8	0.525	55
R-22	5.5" (2×6)		0.825	35
R-29	7.25" (2×8)		1.088	26
R-37	9.25" (2×10)		1.387	21
R-15	3.5" (2×4)	2.3	0.670	43
R-23	5.5" (2×6)		1.054	27
R-31	7.25" (2×8)		1.389	21
R-39	9.25" (2×10)		1.773	16

Attic/Open Blow Coverage Information

R-value	Bags per 1000 sqft.*	Maximum sq ft per bag*	Minimum weight per sq ft	Minimum thickness
To obtain a thermal resistance (R) of:	Bags per 1000 sq ft of net area:	Contents of bag should not cover more than (sq ft):	Weight per sq ft of installed insulation should not be less than (lb):	Installed insulation should not be less than (inches):
R-60	43.5	23.0	1.307	20.50"
R-50	35.2	28.4	1.057	17.50"
R-44	31.0	32.3	0.928	15.75"
R-38	26.7	37.4	0.803	14.00"
R-30	20.2	49.5	0.606	11.00"
R-26	18.2	54.8	0.547	10.00"
R-22	15.0	66.5	0.451	8.50"
R-19	13.1	76.1	0.394	7.50"
R-13	9.3	107.5	0.279	5.50"
R-11	7.9	126.6	0.237	4.75"

*Coverage per bag does not include framing members, which will increase coverage, depending on 16" or 24" OC.
**For sidewall applications—To compensate for wall framing, the net coverage per bag should be increased. To calculate for conventional wood stud framing, when framing is 16" OC multiply coverage value shown by 1.14. When framing is 24" OC multiply coverage value shown by 1.11.

Ark Seal, Inc.

Figure 18-13. These R-values and coverage charts are for a blown-in type of insulation.

18.6.1 Foamed-in-Place Insulation

Foamed-in-place insulation is usually polyisocyanurate (PIR) that is pumped into a spray nozzle where it is mixed with a blowing agent and a curing agent. It is sprayed onto the surface to be insulated. It quickly foams up to several times its original volume, then cures into either semiflexible foam or a rigid foam, **Figure 18-14**.

Foamed-in-place insulation requires special equipment and specially trained personnel, so it is generally more expensive to install than other types of insulation. It offers some advantages over batts, blankets, and loose fill in that it does an excellent job of filling all voids and sealing any possible air leaks. It is also practical to fill voids in concrete blocks.

CertainTeed Corporation

Figure 18-14. Foamed-in-place insulation.

Spray-foam insulation is also sold in aerosol cans for use in small quantities. It is available in several formulations:

- Low expanding—Fills small gaps and cracks less than 1″ wide.
- High expanding—Fills large voids.
- Window and door—Insulates between window or door frames and the wall framing. It is more flexible than other spray-foam products, so it does not bow the door or window frame as it expands, **Figure 18-15**.
- Fireblock—Seals around pipes and ducts to prevent flame spread during a fire.
- Pestblock—Openings are filled with bitter-tasting foam that repels rodents and other pests.

18.6.2 Reflective Insulation

Reflective insulation is usually a metal foil or foil-surfaced material. It differs from other insulating materials in that the number of reflecting surfaces, not the thickness of the material, determines its insulating value. In order to be effective, the metal foil must be exposed to an air space, preferably 3/4″ or more in depth.

Aluminum foil is available in sheets or corrugations supported on paper. It is often mounted on the back of gypsum lath. One effective form of reflective insulation has multiple-spaced sheets.

18.6.3 Other Types of Insulation

There are insulation materials that do not fit the other classifications. Some examples include a confetti-like material mixed with adhesive and sprayed onto the surface to be insulated, multiple layers of corrugated paper, and lightweight aggregates like vermiculite and perlite used in plaster to reduce heat transmission.

Lightweight aggregates made from blast furnace slag, burned clay products, and cinders are commonly used in concrete and concrete blocks. They improve the insulating qualities of these materials.

18.7 Where to Insulate

Heated areas should be surrounded with insulation by placing it in the walls, ceiling, and floors, **Figure 18-16**. Insulation placed in these areas should function in conjunction with any exterior insulation that may have been installed as part of the building envelope, such as rigid insulation under the siding. Together, these systems become the thermal envelope.

In most cases it is best to place the insulation as close to the heated space as possible. For example, if an attic is unused, the insulation should be placed in the attic floor rather than in the roof structure. However, the type and properties of the insulation being used, and a thorough understanding of control layers, condensation, and ventilation play a key role in how the thermal envelope functions and performs.

Goodheart-Willcox Publisher

Figure 18-15. The space between door and window frames and the wall framing should be filled with flexible spray foam.

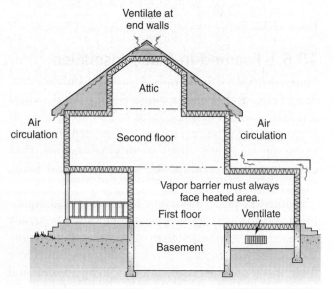

Goodheart-Willcox Publisher

Figure 18-16. Where to insulate residential structures. In many cases, basement walls and crawl space walls are also insulated.

If the attic space or certain portions of it must be heated, insulate the walls and overhead. If the insulation is placed between the rafters, it may be necessary to allow space between the insulation and sheathing for free air circulation. The floors of rooms above unheated garages or porches also require insulation for maximum comfort.

Where a basement is used as a living or recreation area, masonry walls should be insulated with EPS or XPS rigid board insulation. Then, walls may be framed with studs or furred out and drywalled. Studded walls can also be insulated as in **Figure 18-17**. Be sure to check local codes; some place restrictions on materials used in basement living spaces. Insulating basements is highly recommended as an energy conservation measure. Not only does it save heat and provide comfort, it has the added benefit of giving the basement better acoustical qualities. More on this subject is covered later in the chapter.

Even if basement walls are not insulated, insulating material should be installed over the band joists and headers, **Figure 18-18**. This area has little protection against heat loss.

18.7.1 Insulating Crawl Spaces

Floors over unheated space directly above the ground require a similar degree of insulation as walls in the same climate zone, **Figure 18-19**. This area is called a *crawl space*. It is usually enclosed by foundation walls. Such areas require ventilation and, therefore, have about the same temperature as the outside.

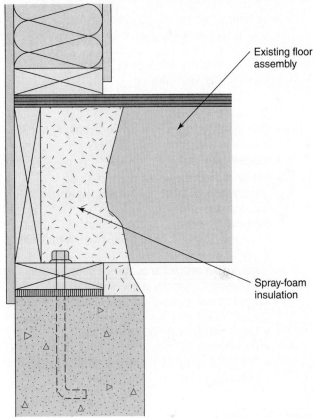

Goodheart-Willcox Publisher

Figure 18-18. Insulate band joists atop basement walls, as there is little protection from loss of heat. The band joist must be air-sealed as well as insulated. Spray-foam insulation is commonly used.

Owens-Corning

Figure 18-17. Insulating a basement wall with R-11 unfaced fiberglass. A framework of 2×4 studs has been installed against the masonry wall and provides support for insulation and drywall or paneling.

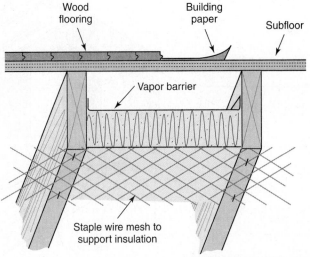

Manville Building Products, Inc.

Figure 18-19. Basic construction and insulation requirements in a crawl space. Insulation can be held in place with bowed wire, chicken wire, or fishing line.

Code Note

The International Residential Code gives three options to manage underfloor space in residential construction.

1. Vented crawl space without a vapor retarder
   - For each 150 sq ft of area, 1 sq ft must be allowed for venting through the foundation or exterior walls.
   - One ventilated opening must be located within 3′ of each corner of the crawl space.
2. Vented crawl space with a Class I vapor retarder covering the ground surface
   - For each 1500 sq ft of area, 1 sq ft must be allowed for venting.
   - One ventilated opening must be located within 3′ of each corner of the crawl space.
3. Unvented crawl space
   - In this case, the air in the crawl space is considered conditioned, requiring the walls and rim joists to be insulated in accordance with basement foundation wall requirements.
   - All exposed ground surface must be covered with a Class I vapor retarder
   - Joints must overlap 6″ and must be sealed or taped.
   - The edges of the vapor retarder must extend 6″ up the foundation wall and be attached and sealed to the wall.
   - One of the following is provided in the crawl space
     - Continuously operated exhaust fan
     - Conditioned air supply
     - The crawl space air is used as an HVAC plenum
     - Dehumidification

Alternatively, **Figure 18-20** shows an unvented crawl space. In this case, insulation must be moved to the foundation wall. A Class I vapor retarder keeps ground moisture from being introduced into the living space, and conditioned air or mechanical ventilation moves the air to keep it from becoming trapped and stagnant. Covering the soil surface in the crawl space with 6-mil polyethylene plastic film can control moisture coming up through the ground. Lay the material on the soil with edges overlapping at least 6″. If the ground is rough, it is best to put down a layer of sand or fine gravel before placing the vapor retarder. Even in vented crawl spaces, covers of this kind greatly restrict the evaporation of ground water, so less ventilation is needed. Refer to the Code Note.

The grade level beneath the building should be higher than the outside grade if water is likely to migrate inside the foundation wall. A ground cover in the crawl space is especially valuable where the water table is near the surface or if the soil easily absorbs water. Be sure the covering extends well up along the foundation wall.

18.7.2 Insulating Slab Foundations

Many homes, as well as other structures, are built on concrete slab floors. Such floors should be insulated and a vapor barrier laid down to prevent heat loss and vapor penetration. Very little heat is lost into the ground

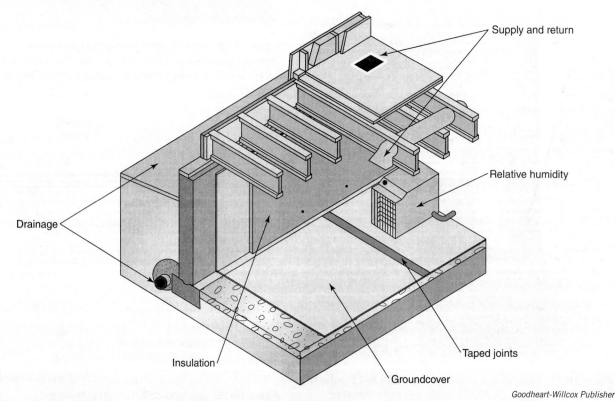

Supply and return

Relative humidity

Drainage

Insulation

Taped joints

Groundcover

Goodheart-Willcox Publisher

Figure 18-20. This example of a conditioned crawlspace eliminates the need for exterior ventilation. Conditioned air is supplied through the HVAC system. Humidity must also be managed to avoid moisture issues.

under the central area of the floor. Only the perimeter needs to be insulated. The vapor barrier, however, must be continuous under the entire floor. The insulation can be horizontally installed extending about 2′ under the floor. It can also be vertically installed along the foundation walls, as shown in **Figure 18-21**. See Chapter 9, **Footings and Foundations**, for more information on insulating concrete slab foundations.

18.7.3 Insulating Existing Foundations

Insulating walls and ceilings of existing structures is fairly simple. However, insulating floors or foundations of structures without basements can be a difficult task. Crawl spaces may not be easy to enter or may not provide enough room to work. Attaching a rigid insulation board to the outside surface of the foundation is usually the best solution.

Figure 18-22 shows extruded polystyrene applied to the exterior wall of a crawl space. Dig a trench along the wall. Clean the wall surface and repair cracks. The insulation board and its protective cover will extend outward beyond the wall siding. To waterproof this joint, install a metal flashing. Extend the flashing at least 1″ upward behind the siding.

Cut the polystyrene panels to size and then attach them to the wall with 3/8″ beads of mastic. Follow the manufacturer's directions. Only a few minutes are required for the mastic to make an initial set.

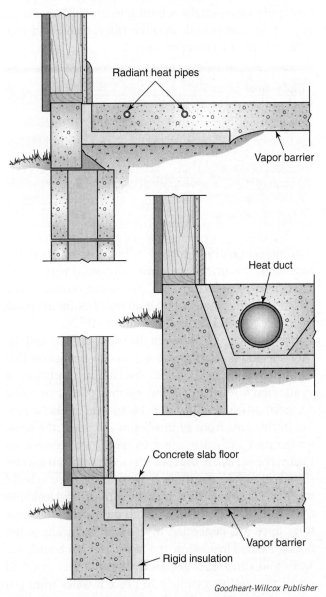

Goodheart-Willcox Publisher

Figure 18-21. Slab floors need insulation along their perimeters. The vapor barrier should be continuous under the entire floor.

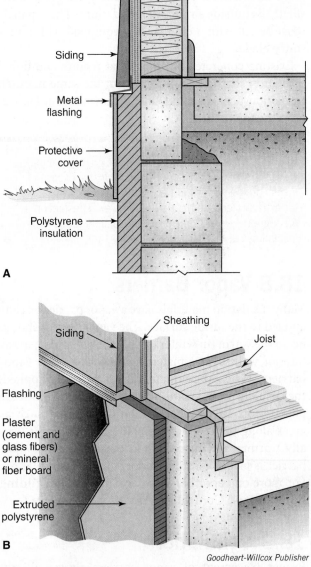

Goodheart-Willcox Publisher

Figure 18-22. Extruded polystyrene can be installed on the outside surface to provide insulation for all types of existing foundations. A—Slab foundation. B—Crawl space or basement foundation.

Polystyrene panels must be protected against ultraviolet light, wear, and impact forces. An approved method is to cover the surface with special plaster made from cement, lime, glass fibers, and a water-resistant agent. This mixture is troweled onto the surface 1/8″–1/4″ thick. It should extend downward to about 4″ below finished grade. The buried portion of the panel does not require protection.

Cement board panels, exterior plywood, or other durable weatherproof panels can also be used to protect the insulation. These panels can usually be attached with an approved mastic. Their use is practical when only a small area of the insulation board projects above the finished grade.

Polystyrene insulation is available in panels that have a weatherproof coating of fiber-reinforced cement on one side. They are especially designed to insulate the outside surface of foundations. These panels must be cut with a circular saw equipped with a masonry blade.

Existing concrete slab floors that require additional insulation can be handled in about the same manner. Insulation panels should extend downward at least 2′ or to the top of the footing.

Green Note
Never throw away unused insulation. Install excess insulation in interior walls and on top of attic insulation for extra protection, but do not force extra insulation in a confined space. Compressing insulation reduces its insulating value.

18.8 Vapor Barriers

Many insulation materials have a vapor barrier already applied to the inside surface. The vapor barrier should be on the warm (in winter) side of the wall. Also, many interior wall surface materials are backed with vapor barriers. When these materials are properly applied, they usually provide satisfactory resistance to moisture penetration. In very hot and humid regions, such as areas along the coast of the Gulf of Mexico, it is usually warmer and more humid outside, so the vapor barrier is placed on the exterior side of the insulation. For more on vapor barriers, see Chapter 15, **Building Envelope and Control Layers**.

18.9 Ventilation

Proper placement of vapor barriers alone will not protect the structure against moisture. Steps must be taken to provide good ventilation where appropriate.

The cold side (outside) of walls should be weathertight, but still permit the wall to breathe. Building paper or housewrap can be used over wood sheathing to reduce air infiltration while still permitting moisture in the wall to escape.

Attics can be constructed to be conditioned or unconditioned spaces. **Figure 18-23** shows both configurations. In a conditioned attic, the use of closed-cell spray foam between the rafters acts as an air barrier and a vapor retarder, keeping moisture from reaching a condensing surface. Therefore, attic ventilation is not required for this building configuration. However, it is especially important to ventilate an unheated attic or the space directly under a low-pitched or flat roof that is not insulated in this way. **Figure 18-23** also shows the most common systems for ventilating these spaces. Generally, an air intake is built into the soffit overhang and an exhaust is built into the ridge. Gable and roof surface vents are sometimes used.

Code Note
The International Residential Code sets certain requirements for ventilating attic spaces. There shall be a 1″ space minimum between the roof sheathing and insulation. Net free ventilation shall not be less than 1/150th of the area of the space. Ventilators in the upper portion of the space being ventilated must provide half of the requirement. The remainder must be located in the eaves or cornices.

To prevent entry of insects or birds, openings to the outside of the attic space must be covered with perforated vinyl, hardware cloth, corrosion-resistant wire cloth screening, or similar material. Minimum mesh opening shall be 1/8″ and maximum 1/4″.

The limited space between the top wall plate and the roof sheathing restricts the amount of insulation that can be installed in this area. Insufficient insulation or ventilation directly under low-pitched roofs may cause a special problem, **Figure 18-24**. In winter, heat escaping in this area from rooms below may melt the snow on the roof and cause water to run down the roof. At the overhang, where the roof surface is colder, the water may freeze again, causing a ledge or dam of ice to build up. Water may back up under the shingles and leak into the building. This leakage can go undetected for some time since it is concealed in the attic and walls of the house. Ruined insulation, damaged gypsum board, rot, and mold can result. Ice and water barriers installed under the roofing can help to prevent water from entering the attic; however, the root cause of poor insulation and ventilation should be addressed.

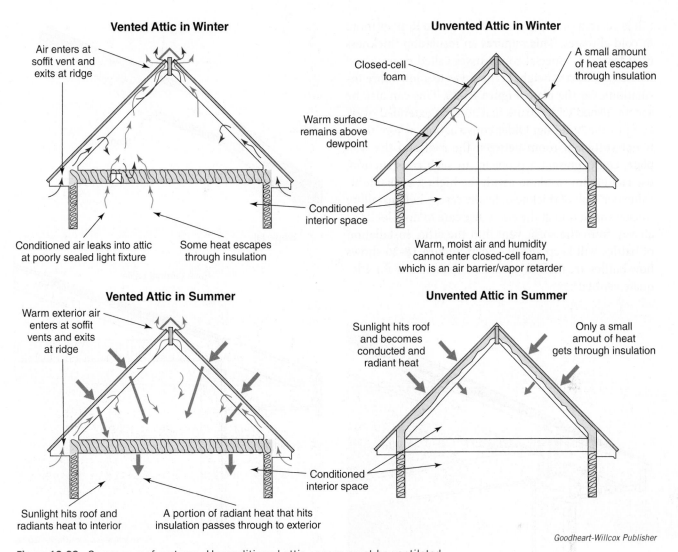

Vented Attic in Winter

Air enters at soffit vent and exits at ridge

Conditioned air leaks into attic at poorly sealed light fixture

Some heat escapes through insulation

Unvented Attic in Winter

Closed-cell foam

A small amount of heat escapes through insulation

Warm surface remains above dewpoint

Conditioned interior space

Warm, moist air and humidity cannot enter closed-cell foam, which is an air barrier/vapor retarder

Vented Attic in Summer

Warm exterior air enters at soffit vents and exits at ridge

Sunlight hits roof and radiants heat to interior

A portion of radiant heat that hits insulation passes through to exterior

Conditioned interior space

Unvented Attic in Summer

Sunlight hits roof and becomes conducted and radiant heat

Only a small amout of heat gets through insulation

Goodheart-Willcox Publisher

Figure 18-23. Common roof systems. Unconditioned attic spaces must be ventilated.

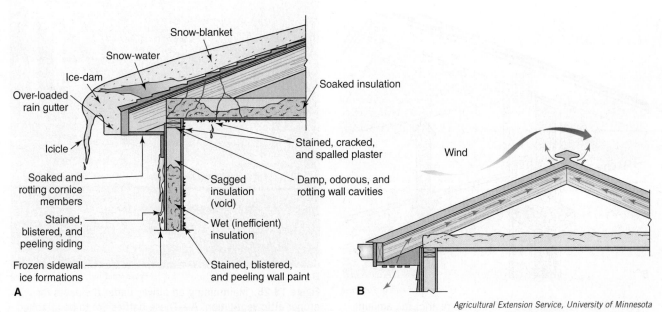

Snow-blanket

Snow-water

Ice-dam

Over-loaded rain gutter

Icicle

Soaked and rotting cornice members

Stained, blistered, and peeling siding

Frozen sidewall ice formations

Soaked insulation

Stained, cracked, and spalled plaster

Damp, odorous, and rotting wall cavities

Sagged insulation (void)

Wet (inefficient) insulation

Stained, blistered, and peeling wall paint

Wind

A

B

Agricultural Extension Service, University of Minnesota

Figure 18-24. Ice dams form on roofs with too little insulation and not enough attic ventilation. A—Escaping heat melts the snow, causing runoff. Water freezes in the overloaded gutter, damming up water. B—Insulation over the outside wall and good ventilation should solve the problem of ice dams.

It is current practice to insulate attics to R-49 or more in cold climates. This requires an insulation thickness of at least 14″. Special roof trusses called energy-heel trusses allot extra height at the eaves to allow more insulation over the plate, **Figure 18-25**. This can also be accomplished with a truss that adds an extended overhang to the building. Older construction, where there is not sufficient room between the roof and the wall plate, could present a problem. In such construction, use rigid foam insulation with the highest available R-value per inch of thickness. In any case, when installing thicker insulation at the eaves, use care to not block the airway from the soffit vent into the attic. Installation of baffles will keep airways open. **Figure 18-26** shows how baffles are used to ensure an open airway for adequate ventilation.

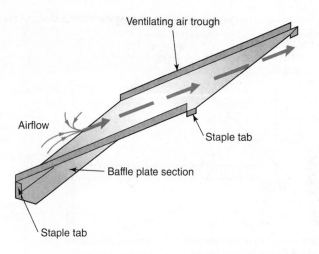

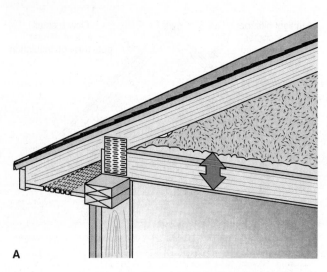

A

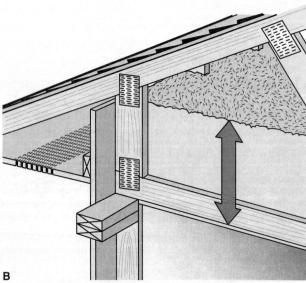

B

Goodheart-Willcox Publisher

Figure 18-25. A—Standard roof trusses restrict the amount of attic insulation that can be installed at the eaves of a building. B–Raised-heel trusses extend the wall surface to allow for additional insulation.

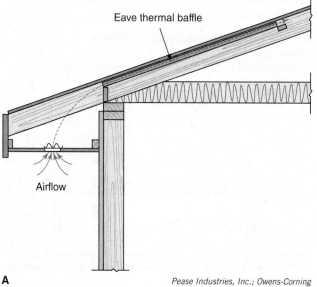

A

Pease Industries, Inc.; Owens-Corning

B

Pease Industries, Inc.; Owens-Corning

Figure 18-26. Maintaining an airway under the eaves for proper attic ventilation. A—These baffles are to be attached to the roof between rafters. They prevent insulation from shutting off airflow. B—This view of the attic shows the baffles installed at the eaves.

Gable roofs often have louvered ventilators in the gable ends. **Figure 18-27** shows two models. There are many sizes available.

Ventilators located on the roof slope may leak if not properly installed. Whenever possible, these ventilators should be installed on a section of the roof that slopes to the rear. This hides them from view on the front of the house.

Sometimes, it may be possible to use a false flue (looks like a chimney, but actually only ventilates the attic) or a section of a chimney for attic ventilation, **Figure 18-28**. In most cases, it is best to install continuous ridge venting. An example of this method is discussed and illustrated in Chapter 14, **Roofing Materials and Methods**.

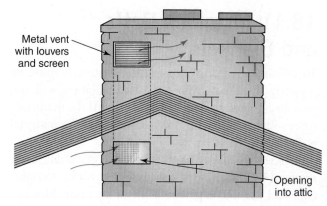

Goodheart-Willcox Publisher

Figure 18-28. A section of a chimney can sometimes be used for attic ventilation.

18.10 Safety with Insulation

Installing insulation is not particularly hazardous. The Occupational Safety and Health Administration (OSHA) does not have specific recommendations for working with fiberglass. The American Conference of Governmental Hygienists publishes a book of suggested limits.

A   Ideal Co.

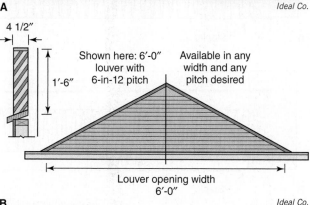

B   Ideal Co.

Figure 18-27. Gable-end ventilation. A—Prefabricated metal gable-end vent. B—Vents can also be constructed from wood and are available in a variety of shapes.

18.11 Installing Batts and Blankets

To efficiently perform, insulation materials must be properly installed. Even the best insulation will not provide its rated resistance to heat flow if the manufacturer's instructions are not followed or if the material is damaged.

Blankets or batts can be cut with shears or a utility knife, **Figure 18-29**. Compress the insulation with a straightedge and use it as a guide for the utility knife. Measure the space and then cut the insulation 2″–3″ longer. On kraft-paper-faced batts, remove a portion of the insulation from each end so that you will have a flange of the backing or vapor barrier to staple to the framing. When working with blanket insulation, it is usually best to mark the required length on the floor, unroll the blanket, align it with the marks, and cut the pieces.

For wall installation of blankets, first staple the top end to the plate and then staple through the flanges down along the studs, aligning the blanket carefully, **Figure 18-30**. Finally, secure the bottom edge to the sole plate.

To install batts in a wall section, place the batt at the bottom of the stud space and press it into place. Start the second batt at the top, pressing it tight against the plate. Sections can be joined at the midpoint by butting them together. The vapor retarder should be overlapped at least 1″, unless a separate one is installed.

Owens-Corning

Figure 18-30. Install blanket insulation by stapling it to the studs, working from top to bottom.

Some batts are designed without covers or flanges and are held in place by friction between the batt and the studs.

Flanges are common to most blankets or batts. These are stapled to the face or side of the framing members, **Figure 18-31**. Pull the flange smooth and space the staples no more than 12″ apart. The interior finish serves to further seal the flange in place when it is fastened to the stud face. In drywall construction, specifications may require that the faces of the studs be left uncovered. Fit the flanges of the insulation smoothly along the sides of the framing and space the staples 6″ or closer.

Goodheart-Willcox Publisher

Figure 18-29. Use a utility knife to cut fiberglass insulation to the required length. To save time, lay out and cut several pieces at once.

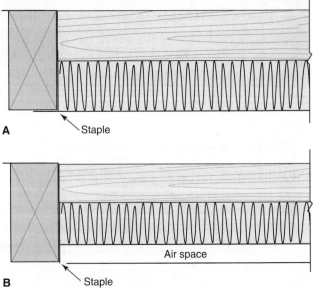

Goodheart-Willcox Publisher

Figure 18-31. Two methods of installing blankets and batts. A—Flush with the inside surface of the stud. B—With the flange stapled along the side of the stud to form an air space.

Be sure there are no gaps or fish mouths (wrinkles). Fully lap joints and avoid making perforations.

New construction usually requires that plumbing be installed in interior walls. In cold climates, water supply pipes should never be located in outside walls. In older dwellings, you must carefully thread the insulation behind any pipes that might be located in the outer wall. If water supply pipes are present, add a separate vapor barrier between the pipe and the interior surface to prevent condensation on the cold pipe. Insulation should also be carefully fitted around and behind electrical boxes, **Figure 18-32**.

Ceiling insulation can be installed from below or, if attic space is accessible, from above. When batts are used, they are usually installed from below. Follow the same general procedure recommended for walls. Snugly butt pieces together at their ends and carry insulation over the outside wall, as shown in **Figure 18-33**. In cold climates, extra thicknesses are recommended up to R-60. When constructing stick-built rafters, it may be necessary to place a 2×4 on top of the ceiling joists above the outside wall. Fasten the rafters to it, rather than to the wall plate below. See **Figure 18-34**.

Goodheart-Willcox Publisher

Figure 18-32. Carefully fit insulation around plumbing located in outside walls, as well as around and behind electrical boxes, to eliminate insulation voids.

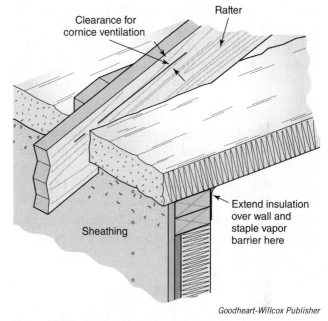

Goodheart-Willcox Publisher

Figure 18-33. Ceiling insulation should extend over the top of the wall plate to help avoid ice dams on the roof. Be sure to leave an airway between the cornice and attic for adequate ventilation.

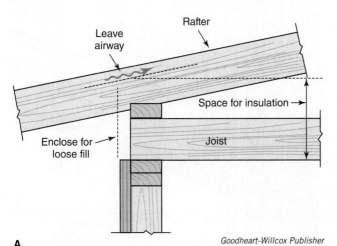

A

Goodheart-Willcox Publisher

B

Goodheart-Willcox Publisher

Figure 18-34. Allowing for thicker insulation. A—One method of framing a low-pitched roof to get the extra space needed for ceiling insulation. Rafters rest on a 2×4 added over the top of the ceiling joists. B—An alternate truss rafter design for insulation/airway clearance.

In multistory construction, the floor frame should be insulated at the band joist, **Figure 18-35**. Insulation should also be installed in the perimeter of the first floor. Cut and fit pieces so they snugly fit between the joists and against the header.

Insulate large wall and ceiling areas first. Then, insulate the odd-sized, smaller spaces above and below windows. Small cuttings remaining from the main areas can be used. Take the time to carefully apply the insulation around electrical outlets and other wall openings. Be careful that you do not cover outlet boxes or they may be missed when the wall coverings are applied.

For a thorough insulation job, all voids must be filled. Be sure to include the voids between window and door frames and the rough framing, as shown in **Figure 18-36**. It is often faster and more effective to insulate these spaces with spray-foam insulation made for that purpose. Use scraps left over from larger spaces. Loosely push the insulation into small voids and cracks with a stick or screwdriver. Be careful not to compress the insulation, since this will reduce its insulating qualities. Cover the area with a vapor retarder, **Figure 18-37**.

Insulate cantilevered floor projections that carry a chimney chase or bay window unit, or extend a room over an outside wall. See **Figure 18-38**. To seal against air infiltration, the sheathing must be tight and the insulation flange and vapor barrier must be carefully stapled to the sides of the joist as shown.

Bullard-Haven Technical School

Figure 18-36. Insulate window and door frames by carefully stuffing pieces of batt insulation into the cavities around the frames. Try not to compress the material too much.

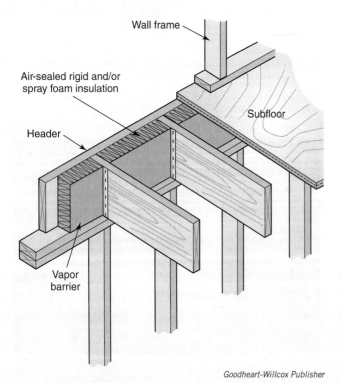
Goodheart-Willcox Publisher

Figure 18-35. Always insulate the perimeter of the floor frame, as shown, for each floor of the building.

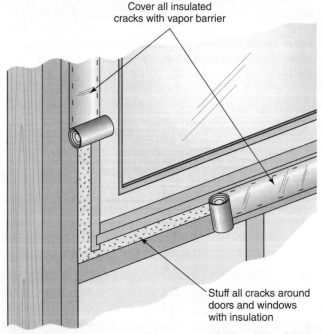
Goodheart-Willcox Publisher

Figure 18-37. Cover insulated cracks around wall openings with vapor barrier to seal against air infiltration.

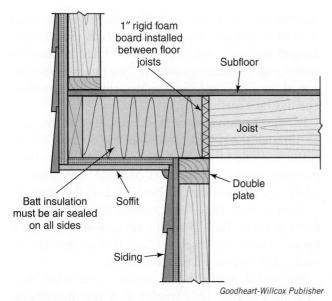

Figure 18-38. Floors that are cantilevered over an outside wall should be insulated to at least R-19.

If weather conditions permit, this segment of insulation can be installed before the subfloor is laid. This simplifies installation since the work can be done from above.

18.12 Installing Loose and Foamed Insulation

Loose insulation is most used above ceilings, where it can be placed by pouring or blowing, **Figure 18-39**. It can be directly poured from bags into the spaces between joists. Mineral fibers made from rock, slag, or glass are widely used. Blown-in fill insulation is also made from cellulose fibers.

Owens-Corning

Figure 18-39. Blowing loose-fill insulation into an attic area. Wear a cap, face mask, goggles, and gloves.

When pouring or blowing loose insulation in an attic, contain the fill around the soffit area with batt insulation. This is required since breezes tend to blow the loose materials away from the eave area. Either install thick batts next to outside walls or install baffles that will direct the incoming air upward away from the insulation.

Code Note
Be careful when insulating around lights or exhaust fans recessed into the ceiling. As a safeguard against fire, the National Electrical Code requires a 3″ space around any heat-producing device, unless it is rated for insulation contact (IC). See **Figure 18-40**.

It is easy to blow loose insulation into ceilings or walls, whether in new or remodeled structures. Special blowers are needed for this method. Usually, they can be rented from the supplier of the insulation or from home improvement stores.

One blown-in insulation system mixes a thin coating of binder adhesive into fiberglass fibers and then sprays the fiberglass into cavities behind netting. The adhesive eliminates settling problems that account for voids in the insulating blanket. The "blown-in-blanket" material is mixed on-site. The system uses a fine-mesh

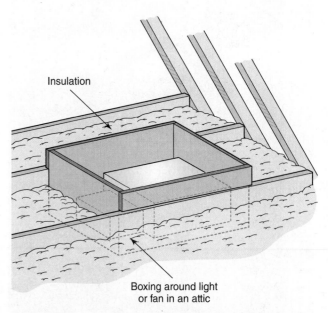

Goodheart-Willcox Publisher

Figure 18-40. Before adding loose-fill insulation, box in fans and light fixtures that project into ceilings. The National Electrical Code (NEC) requires a 3″ air space on all sides. This box should be capped and insulated over or fixtures that are rated for insulation contact (IC) should be used.

nylon netting that is glued or stapled to the building studs, **Figure 18-41**. The netting restrains the bonded fibers injected into each wall cavity, **Figure 18-42**.

Figure 18-41. Fine-mesh nylon netting has been attached to studs prior to insulating with adhesive-coated loose insulation. Drywall will be installed on the opposite side before the coated insulation is blown in.

Figure 18-42. Installing blown-in-blanket insulation. The red hose delivers thinned adhesive to insulating fibers. Netting holds the insulation in the cavity.

A properly filled cavity has a slight bulge. A wide roller is used to bring it flush with the studs. Small voids around windows and doors can be filled by hand or sealed with a foamed insulation. Two methods are suggested for filling the cavities: two-three hole method and one hole method.

For the *two-three hole method*, insert the tip of the nozzle through the netting 2'–3' from the bottom. Fill the cavity to within a foot above the point of insertion. Then, pointing the nozzle tip upward, continue filling until the cavity is about one-half full. Next, reinsert the nozzle tip 2' from the top. With the tip aimed downward, fill the lower portion. Then, point the tip upward to complete the fill. If necessary, insert the tip at remaining voids and underfilled areas to fill the entire cavity.

The *one hole method* is only used for cavities 3 1/2" thick by 8' high and 16" OC. Attach a 5'–8' length of flexible hose to the end of the nozzle. Insert the hose into the center of the cavity and push it in until it is about 2' from the bottom. As the cavity fills, pull the hose upward. When the cavity is filled halfway, reinsert the hose upward to within about 2' of the top, slowly retracting the hose as the remaining cavity fills.

Blown-in insulation can also be used to fill cavities in concrete block walls. Thermal resistance is greatly increased. For example, the R-value of a standard concrete block (R-1.9) is increased to R-2.8 when the cores are filled with insulation. A lightweight 8" block will be increased from R-3.0 to R-5.9. Attach a 5' length of 2 1/2" PVC pipe to a flexible hose. Lower the pipe into each cavity and slowly remove it as the cavity fills. For retrofit of existing walls, 1 1/2" holes are drilled at intervals for access with the nozzle tip.

The same method can also be used to retrofit walls in old buildings. Working from the outside, drill holes in the sheathing after removing sections of siding. Be careful not to damage the siding, since it must be reinstalled later.

The installed R-value for fill insulation varies depending on the installation method (pouring or blowing) and the insulating material used. Manufacturers include these figures on the bag or bale labels. R-values for a 20 lb bag of poured mineral wool insulation are listed in **Figure 18-43**.

As described earlier, foamed-in-place insulation is sometimes used, especially in new construction. Some of these materials can be installed only in open cavities, since they expand in volume 100 times beyond the applied thickness. In such cases, they cannot be used in existing walls unless the drywall or plaster is first removed. Foamed-in-place products seal up gaps and cracks and are nearly impermeable to moisture.

R-Values and Coverage for Mineral Wood Pouring Insulation			
R-value	Minimum Thickness	Maximum Net Coverage Area (sq ft)	Minimum Weight per Sq Ft (lb)
R-38	11 1/4″	8.5	2.344
R-33	9 3/4″	9.8	2.031
R-30	8 7/8″	10.8	1.849
R-26	7 3/4″	12.4	1.615
R-22	6 1/2″	14.8	1.354
R-19	5 5/8″	17.1	1.172
R-11	3 1/4″	29.5	0.677

Goodheart-Willcox Publisher

Figure 18-43. R-values and coverage for 20 lb of mineral wool pouring insulation.

18.13 Installing Rigid Insulation

Insulating board is widely used for exterior walls. Its application is covered in Chapter 15, **The Building Envelope and Control Layers**. Insulating board is commonly used in the following ways:

- As an insulating material installed over the roof deck
- As exterior wall insulation applied over sheathing
- As a base for application of synthetic stucco exterior coverings
- As insulation under a concrete floor slab
- Foundation insulation on the interior or exterior (if rated for ground contact) insulation

Plastic foam (polystyrene or polyisocyanurate) insulation is widely accepted as a rigid insulating material and has been successfully applied to a wide variety of construction types. The installation methods for slab or block insulation vary with the type of product. Always study the manufacturer's specifications.

A number of products are especially designed to insulate concrete slab floors, such as sheets of rigid polystyrene laid down inside the perimeter, before pouring a concrete slab foundation. Sometimes a builder lays down sheets of the same material outside of the slab perimeter. The insulating value of these extra sheets makes deep, below–frost line footings unnecessary.

Figure 18-44 shows a section view of plastic foam insulation board applied to the interior of a masonry wall. It is bonded to the wall surface with a special mastic. This provides a permanent insulation and vapor barrier. After the boards are installed, conventional plaster coats can be applied to the surface.

Their value also lies in the superior insulating qualities they provide.

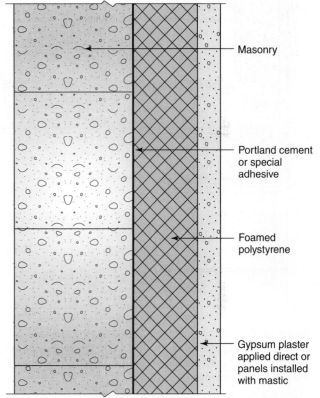

Masonry

Portland cement or special adhesive

Foamed polystyrene

Gypsum plaster applied direct or panels installed with mastic

Goodheart-Willcox Publisher

Figure 18-44. Cross section of a masonry wall. The warm side has been insulated with rigid foamed polystyrene.

18.14 Insulating Basement Walls

When basements are to be used as living space, exterior walls should be insulated. Check local codes for any restrictions on insulating basement walls. The outside surface of concrete or masonry walls should be waterproofed below grade and should include a footing perimeter drain. See Chapter 9, **Footings and Foundations**.

The inside surface may be finished by using studs or furring strips to form a cavity for the insulation and provide a nailing base for surface coverings. In cold climates, a framework of 2×4 studs spaced 16″ OC is best. Use concrete nails, screws, or mastic to secure the sole plate and fasten the top plate to the joists. **Figure 18-45** shows one method of insulating the band joist in a basement.

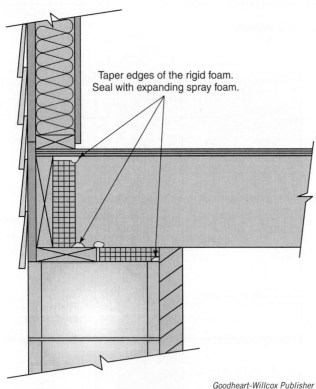

Goodheart-Willcox Publisher

Figure 18-45. Approved method for insulating band joist between the basement wall and first floor. Fiberglass blanket insulation may be added but should be air-sealed to achieve maximum performance.

Taper edges of the rigid foam. Seal with expanding spray foam.

18.15 Stopping Air Infiltration

A continuous air barrier that maintains contact with the thermal envelope must be maintained around the conditioned space. For more on air barriers, see Chapter 15, **Building Envelope and Control Layers**. Likewise, all fibrous insulations should maintain contact with an air barrier on all six sides to maintain performance. Otherwise, wind-washing will reduce the thermal performance.

Pay special attention to areas around windows and doors and other small openings in the structure. In the construction process, air infiltration can be reduced by properly assembling materials and by sealing joints. Caulking and sealing is usually required at these locations:

- Joints between the sill and foundation
- Joints around door and window frames
- Intersections of sheathing with the chimney and other masonry work
- Cracks between drip caps and siding
- Openings between masonry work and siding

Be sure to seal around the electrical service entrance and hose bibbs. The preferred method to seal these penetrations is to use flexible flashing tape that is incorporated into the water-resistant barrier to both air-seal and water-seal around the pipe, **Figure 18-46**.

Inside of the structure, give special attention to recessed light fixtures and any built-in units located in outside walls. Also, seal electrical conduit and plumbing pipes that run from the attic into walls and partitions located in the living space. Seal conduit where it enters electrical boxes and seal the boxes to the inside wall surface.

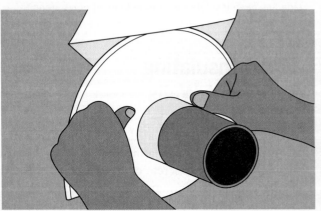
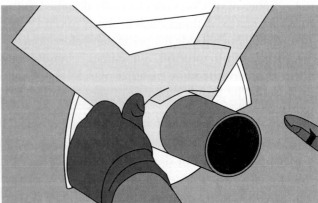

Goodheart-Willcox Publisher

Figure 18-46. Wall penetrations should be flashed and air-sealed into the WRB using flexible flashing tape.

18.16 Estimating Thermal Insulation Materials

The amounts of insulating materials are figured on the basis of area (square feet). The thickness is then specified as separate data. The size of packages varies, depending on the type and thickness. For example, one manufacturer packages 1 1/2″ blankets in rolls of 140 sq ft. A 3″ blanket in the same width comes in rolls of 70 sq ft. Batts are furnished in packages (sometimes called tubes) that contain as much as 100 sq ft. A 6″ batt usually contains 50 sq ft.

To determine the amount of insulation needed for exterior walls, first calculate the total perimeter of the structure. Then, multiply that number by the ceiling height. Deduct from the total the area of doors and windows. Many carpenters deduct only large windows or window-walls and disregard doors and smaller openings. This extra allowance makes up for loss in cutting and fitting and also provides for additional material needed around plumbing pipes, recessed lighting boxes, and other items. Here is an example.

$$\text{Perimeter} = 30' + 40' + 36' + 20' + 6' + 20'$$
$$= 152'$$
$$\text{Area} = 152' \times 8'$$
$$= 1216 \text{ sq ft}$$
$$\text{Window wall} = 12' \times 8'$$
$$= 96 \text{ sq ft}$$
$$\text{Net area} = 1216 - 96$$
$$= 1120 \text{ sq ft}$$

The same procedure can be used to estimate reflective insulation. Make a greater allowance for cutting and waste, especially when using the accordion type. This type cannot be effectively spliced. Rolls of reflective insulation hold from 250 to 500 sq ft.

Rigid insulation is also estimated on the basis of area. It can be calculated from dimensions shown on the working drawings.

When estimating the amount for floors and ceilings, use the same figures calculated for the subfloor area. Stairwell openings and openings for large fireplaces can

be deducted. However, not deducting these amounts may provide the extra insulation needed for waste and special packing around fixtures.

For a perimeter insulation strip in a concrete slab floor, multiply the perimeter of the building by the width of the strip. There is little waste on such an installation, since even small pieces can be used.

Fill insulation comes in bags that usually contain 3 or 4 cu ft. This example shows how the required amount of cu ft can be calculated:

$$\text{Area} = 1200 \text{ sq ft}$$
$$\text{Thickness} = 4''$$
$$= 1/3'$$
$$\text{Cu ft required} = 1200 \times 1/3$$
$$= 400$$
$$\text{Less 10\% allowance for joists 16'' OC} = 400 - 40$$
$$\text{Net amount} = 360 \text{ cu ft}$$
$$\text{Number of bags (4 cu ft per bag)} = 90$$

Manufacturer's directions and specifications usually have tables that provide a direct reading of the number of bags required for a certain thickness of application. These are especially helpful in estimating the amount needed for such items as filling the cores of concrete blocks.

18.17 Blower Door Testing

All buildings, no matter how tightly they are constructed, have some air leaks. Excessive leaks can cause one part of the house to feel drafty while the rest of the house may feel comfortable. A **blower door test** measures the tightness of a building by mounting a large fan in an otherwise sealed doorway. Often, energy efficiency incentive programs such as Energy Star require a blower door test. There are three main components for a blower door test:

- A frame and a flexible panel designed to temporarily fill a doorway
- A powerful variable-speed fan that is attached to the blower-door frame
- At least two pressure gauges: one (the air pressure gauge) to measure the pressure difference between the home's interior and the outdoors, and another (the airflow manometer) that measures the fan's airflow

The tighter the building, the greater the difference will be between outside and inside air pressure. **Figure 18-47** shows a blower door in place in the doorway of a house. With higher pressure outside, air is drawn into the building through any cracks or openings. A smoke pencil can be used to detect where the air is flowing into the building.

The Energy Conservancy

Figure 18-47. A blower door consists of an adjustable frame, a flexible covering, and a powerful fan.

Follow these steps to prepare for a blower door test:

- Close all exterior doors and windows.
- Open all interior doors.
- Close any wood-burning fireplaces or stove dampers.
- Disable heating equipment and nonelectric water heaters by turning their thermostats down (and in some cases by shutting off electrical power to the equipment).
- Fill all plumbing traps with water.
- Close openings for continuous ventilation systems and heat recovery ventilators.

The blower door technician will slowly increase the fan speed. It is generally a good idea to walk through the house when the fan is turned up part way to make sure nothing unexpected is happening, such as an unlocked door opening or a missed vent discharging air. The test usually takes about an hour.

18.18 Acoustics and Sound Control

Noise is unwanted sound. Noise is a nuisance and unpleasant. It reduces human efficiency and can cause undue fatigue. A by-product of our modern world, it has reached a magnitude that usually requires some measure of sound control in every home and building.

The wall and roof structures of houses, especially those built in cold climates, are heavy enough to block average outside noises. Therefore, noise or sound control mainly applies to interior partitions, floors, and surface finishes.

Sounds inside an average home are generated by conversation, television, radios, stereos, computers, and musical instruments. Vacuum cleaners, washing machines, food mixers, and garbage disposals are examples of mechanical equipment that create considerable noise. Plumbing, heating, and air conditioning systems may be a source of excessive noise if poorly designed and installed. The activities in playrooms and workshops may create sounds that are undesirable if transmitted to relaxing and sleeping areas.

The solution to problems of sound and noise control can be separated into three parts:

- Reducing the source
- Controlling sound within a given area or room
- Controlling sound transmission to other rooms, **Figure 18-48**

A carpenter should have some understanding of how the latter two can be accomplished and be familiar with sound conditioning materials and construction techniques. As in the case of thermal insulation, a carpenter must appreciate the importance of careful work and proper installation methods.

Bilanol/Shutterstock.com

Figure 18-48. Installing batts of acoustical insulation between partition studs reduces room-to-room sound transmission.

18.18.1 Sound Intensity

The number of *decibels (dB)* indicates the loudness or intensity of the sound. Refer to **Figure 18-49**. Notice that the rustle of leaves or a low whisper is on the threshold of audibility. That is, the sound is barely heard by the human ear. At the top of the scale are painfully loud sounds of over 130 dB, which is often referred to as the *threshold of pain*.

There is a logarithmic relation, on the decibel scale, to the amount of sound energy involved. If a given sound level is 10 dB greater than another, its intensity is 10^1, or ten times, greater than the first level. If the sounds differ by 20 dB, the ratio of their intensities is 10^2 or 100 times greater; if by 30 dB, the ratio is 10^3 or 1000 times greater, and so on up the scale.

18.18.2 Sound Transmission

When sound is generated within a room, the sound waves strike the walls, floor, and ceiling. Much of this sound energy is reflected back into the room. The rest is absorbed by the surfaces. If there are cracks or holes through the wall (no matter how tiny), part of these sound waves travel through as airborne sounds.

The sound waves striking the wall also cause it to vibrate as a diaphragm, reproducing these waves on the other side of the wall. Sound transmission through theoretically airtight partitions is the result of such a diaphragm action. These are the sounds carried through a building by the vibrations of the structural materials themselves. Footsteps heard through the floors of a structure are an example of impact sounds, explained in further detail later in this chapter. The sound insulation value of such substances is almost entirely a matter of their relative weight, thickness, and area. In partitions of normal dimensions, this value depends mostly on weight.

As the sound moves through any type of wall or barrier, its intensity is reduced. This reduction is called *sound transmission loss (STL)* and is expressed in decibels. A wall with a STL of 30 dB will reduce the loudness level of sound passing through it from, for example, 65 dB to 35 dB. See **Figure 18-50**. The transmission loss of any floor or wall is determined by the materials, design, and quality of the construction techniques.

Although transmission loss rated in decibels is still used, a system that rates the sound-blocking efficiency is widely accepted. It is called the *sound transmission class (STC)* system. Standards have been established through extensive research by such organizations as the National Institute of Standards and Technology.

STC numbers have been adopted by acoustical engineers as a measure of the resistance to sound transmission of a building element. Like the resistance in thermal insulation (R), the higher the number, the better the sound barrier. **Figure 18-51** shows how a composite STC rating is applied to a given wall construction through a specified range of frequencies. **Figure 18-52** further describes these STC ratings by making a simple application to a wall separating two apartments.

Decibel Levels for Various Sounds		
Common Description	**Decibels**	**Threshold of Feeling**
Threshold of pain	130 120	Space shuttle (180+) (lift-off) Boeing 747
Deafening	110 100	Thunder, artillery Nearby riveter Elevated train Boiler factory
Very loud	90 80	Loud street noise Noisy factory Truck unmuffled Police whistle
Loud	70 60	Noisy office Average street noise Average radio Average factory
Moderate	50 40	Noisy home Average office Average conversation Quiet radio
Faint	30 20	Quiet home or private office Average auditorium Quiet conversation
Very faint	10 0	Rustle of leaves Whisper Soundproof room Threshold of audibility

Goodheart-Willcox Publisher

Figure 18-49. Decibel levels for a wide range of sounds. Those above 90 decibels are disagreeable and can even be painful or damaging to the hearing.

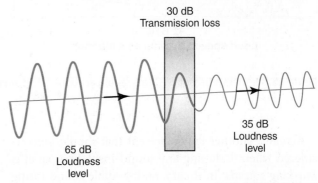

Goodheart-Willcox Publisher

Figure 18-50. Sound transmission losses occur as sound waves travel through a wall. Values vary, depending on the frequency of the sound waves.

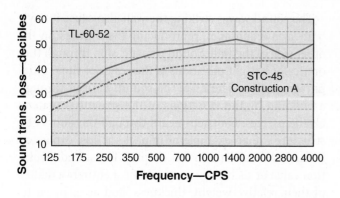
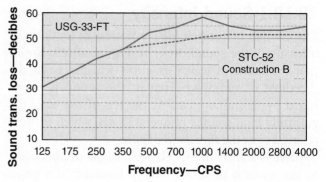

Figure 18-51. These graphs show the transmission loss values in decibels at various frequencies for two STC-rated constructions. The dotted lines represent these losses.

30 STC
Loud speech can be understood fairly well

35 STC
Loud speech audible but not intelligible

42 STC
Loud speech audible as a murmur

50 STC
Loud speech not audible

Figure 18-52. How different STC ratings of partitions affect transmission of noise between two apartments.

Actually, another factor present that should be considered when designing any sound-insulating panel is ***masking sounds***. In theory, an inaudible sound rating of zero on the decibel scale is for a perfectly quiet room. Since there are noises in every habitable room that tend to mask the sound entering, it is only necessary to reduce sound below the ambient (existing) sound level within the space to be insulated. Assume that there is a radio playing soft music in the listening room (about 30 dB). Thus, sound of less than 30 dB entering the room is completely masked by the ambient sound in the room.

18.18.3 Wall Construction

How high must an STC rating be for a given wall? This largely depends on the types of areas it separates. For example, partitions between bedrooms in an average home usually do not require special soundproofing, while those between bedrooms and activity or living rooms should have a high STC rating. Partitions surrounding a bathroom should also have a high STC number. Extra attention should be given to the placement of insulation around pipes.

Figure 18-53 shows a number of practical constructions for partitions and their STC ratings. Sound-deadening board can be used in conjunction with 1/2″ or 5/8″ drywall to increase the STC rating of a wall. These lightweight, low-density panels dampen sound transmissions in higher frequencies within our hearing range and can reduce sound transmission through interior or exterior walls. Sound-deadening board is applied to the framing with screws or nails and drywall is applied over the sound-deadening board.

To secure high STC ratings in a wall structure when using a sound-deadening board, carefully follow manufacturers' application details. Where nails are used, the size, type, and application patterns are critical. In drywall construction, the joints should be taped and finished, and the entire perimeter sealed. Openings in the wall, such as for convenience outlets and medicine cabinets, require special consideration. For example, electrical outlets on opposite faces of the partition should not be located in the same stud space.

18.18.4 Double Walls

Partitions between apartments are often constructed to form two separate walls. Standard blanket insulation is installed in about the same manner as for thermal insulation purposes. It should be stapled to only one row of the framing members.

Methods of Dampening Sound Levels in Interior Walls		
Description	Approximate STC Rating	Wall Assembly
2×4 studs with one layer of 5/8″ gypsum on each side. No insulation.	35	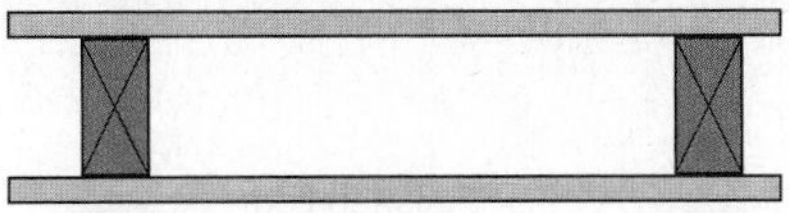
2×4 studs with one layer of 5/8″ gypsum on each side; batt insulation in cavity.	38	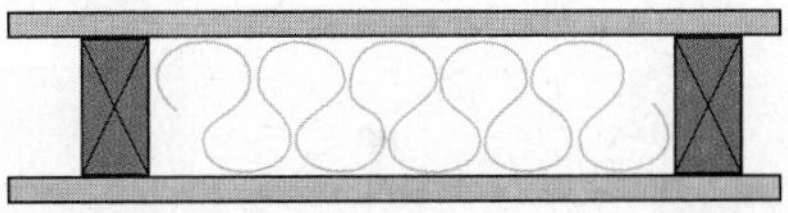
2×4 studs with two layers of 5/8″ gypsum on each side; batt insulation in cavity.	45	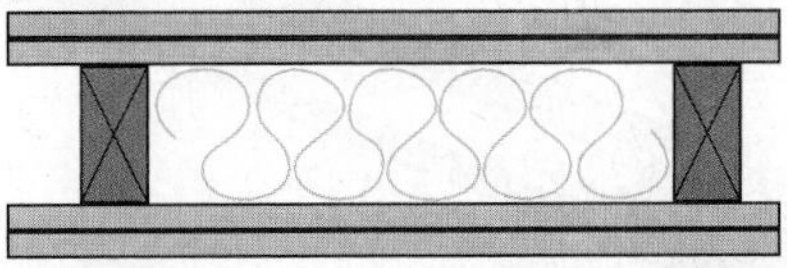
Staggered 2×4 studs with one layer of 5/8″ gypsum on each side; batt insulation in cavity.	48	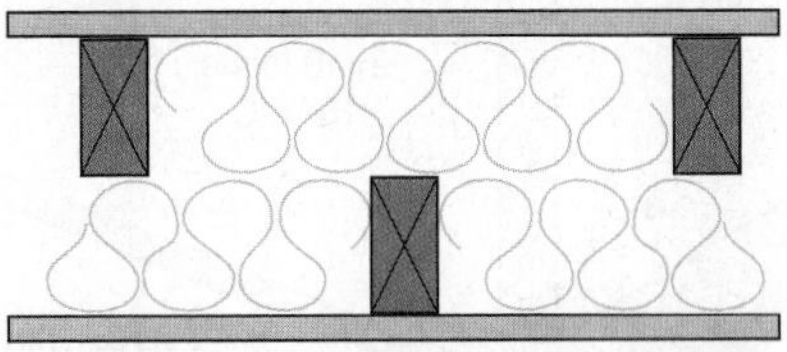
Staggered 2×4 studs with two layers of 5/8″ gypsum on each side; batt insulation in cavity.	54	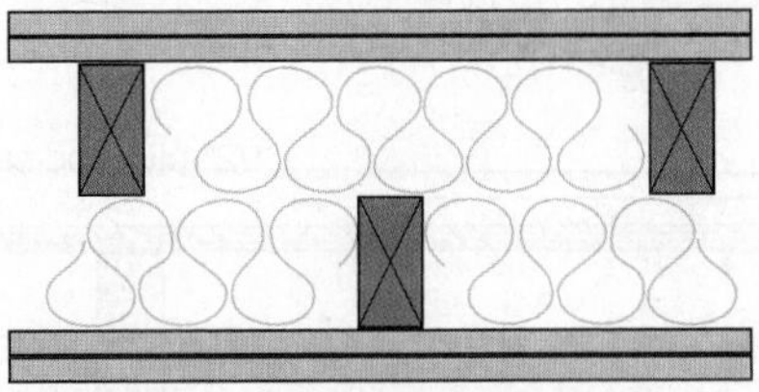
3 5/8″ steel studs with two layers of 5/8″ gypsum on each side; batt insulation in cavity.	44	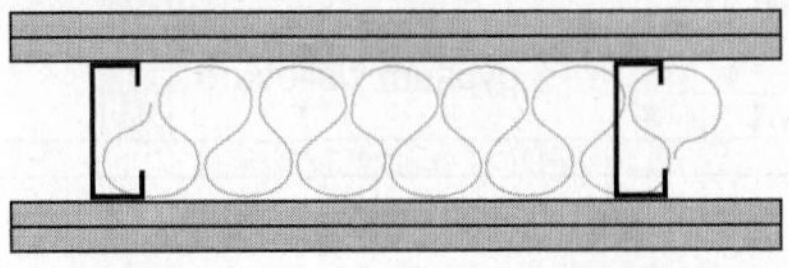

Goodheart-Willcox Publisher

Figure 18-53. Several different methods of constructing interior walls to dampen sound levels.

For economical and space-saving construction, strips of special, resilient channel are nailed to standard stud frames. See **Figure 18-54**. The base layer of gypsum board is attached with screws. The surface layer is bonded with an adhesive. Since laminated systems like this minimize the use of metal fasteners, they result in a finer appearance along with better sound and fire resistance.

18.18.5 Floors and Ceilings

In general, the considerations for soundproofing that were used for walls apply to floors and ceilings as well. Floors are subjected to *impact sounds*. These are the noises from activities such as walking, moving furniture, or operating vacuum cleaners and other equipment. Sound control through floors is somewhat more difficult. Often, an impact sound causes more annoyance in the room below than it does in the room where it is generated. The addition of carpeting or similar material to a regular hardwood floor is effective in minimizing impact sounds.

Properly installed sound-deadening board increases the STC rating of the floor or ceiling, **Figure 18-55**.

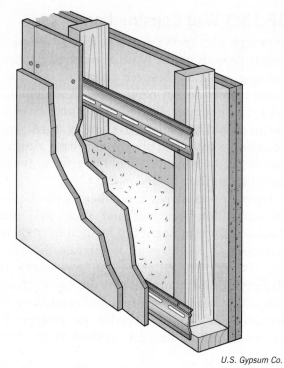

U.S. Gypsum Co.

Figure 18-54. Resilient channels can be added to wood studs on one side, with a double drywall layer and 3″ insulation batt for an STC rating of 50.

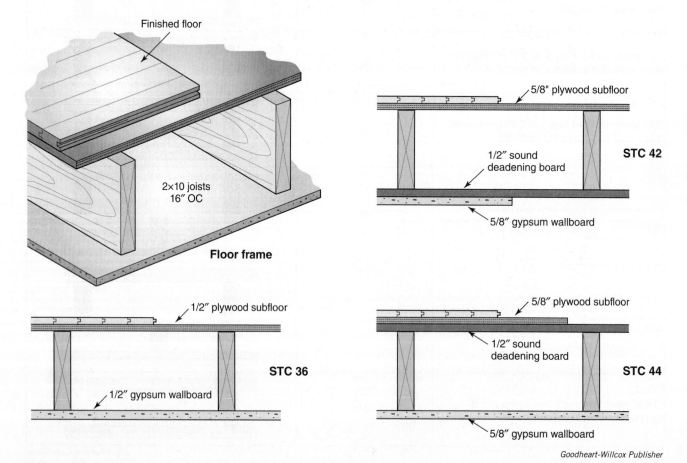

Goodheart-Willcox Publisher

Figure 18-55. Floor and ceiling soundproofing constructions. *(continued)*

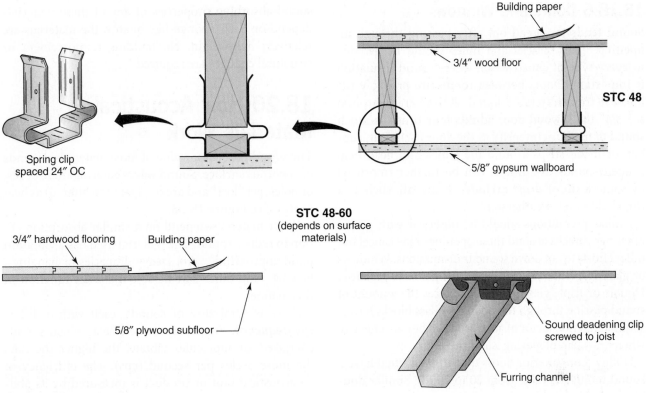

STC 48

Building paper

3/4″ wood floor

5/8″ gypsum wallboard

Spring clip
spaced 24″ OC

STC 48-60
(depends on surface
materials)

3/4″ hardwood flooring   Building paper

5/8″ plywood subfloor

Sound deadening clip
screwed to joist

Furring channel

Figure 18-55. *(continued)*

The use of metal clips to attach the ceiling material is a practical solution. Various suspended ceiling systems also provide high levels of sound control.

Figure 18-56 shows the installation of a floor/ceiling system with an STC rating over 52. It consists of 2×10 joists placed 16″ OC, with a standard wood subfloor and finished floor. The floor is covered with carpet and pad. Resilient metal channels are attached to the joists with 1 1/4″ screws. Nails must not be used. Gypsum panels are attached to the channels with screws. The system includes a 3″ insulation blanket. The STC rating would be slightly higher if the floor is framed with I-joists.

An existing floor can be soundproofed using the method shown in **Figure 18-57**. Sleepers of 2×3 wood are laid over a glass wool blanket, but not nailed to the old floor. When the new floor is laid, be sure the nails do not go all the way through the sleepers. The only contact between the new and old floor is the glass wool blanket. It compresses to about 1/4″ under the sleepers. The system makes the floor resilient, in addition to reducing sound transmission.

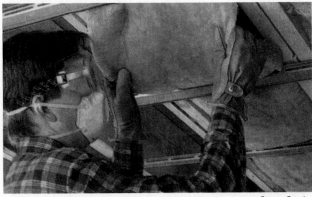

Owens-Corning

Figure 18-56. Soundproofing the ceiling above a basement room. The resilient channels will carry the ceiling panels. When the floor above is covered with a carpet and pad, an STC of 52 can be attained.

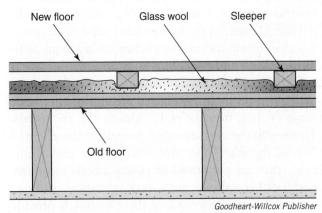

New floor   Glass wool   Sleeper

Old floor

Goodheart-Willcox Publisher

Figure 18-57. Sleepers that are floated over insulation provide significant soundproofing to an existing floor.

18.18.6 Doors and Windows

Sound tends to spread out after passing through an opening. Thus, cracks and holes should be avoided in every type of construction where sound insulation is important. Doors between rooms are probably the greatest transmitters of sound. A 1/4″ crack around a 1 3/4″ thick wood door admits four times as much sound of medium intensity as the door itself. Felt, rubber, or metal strips around the jambs and head help deaden sound. Conditions can be further improved by some form of *draft excluder* at the sill, such as a threshold or felt weatherstop.

Similar precautions should be observed with glazed openings. Cracks around these openings may cancel out other efforts to cut down sound transmissions. Windows or glazed openings should be as airtight as is practical. Double or triple glazing greatly reduces the amount of sound passing through the opening. Glass blocks have a sound reduction factor of about 40 dB. They are effective where transparent glazing is not needed.

Hollow-core interior doors that are well fitted have a sound reduction value from 20 to 25 dB. Similar double doors hung with at least 6″ air space between have a sound reduction factor as high as 40 dB. Using felt or rubber strips around the stops increases this factor.

For special installations, soundproof doors are available. They can be built to suit almost any condition. Special hardware is used on this type of door to prevent sound transmission through the doorknobs.

18.19 Noise Reduction within a Space

While it is important to design walls and floors that reduce sound transmission between spaces, it is also advisable to treat the enclosure so that sound is trapped or reduced at its source. Reducing the noise level within the room not only cuts sound transmission to other rooms, but improves living conditions within the room. Areas in the home where noise reduction is most important include the kitchen, utility room, family room, and hallways.

There are a number of different types of acoustical material available to the builder. These come in a wide range of sizes, from 12″ × 12″ tiles to 4′ × 16′ boards. Those with the best acoustical properties absorb up to 70% of the sound that strikes them. The most common types are perforated or porous fiberboard units, perforated metal pan units, cork acoustical material, and acoustical plaster. All of these materials provide high absorption qualities and (except for acoustical plaster) have a factory-applied finish. Since the sound-absorbing properties of any of these materials depend on their sponge-like quality, the materials are relatively lightweight. No building reinforcement or structural changes are required.

18.20 How Acoustical Materials Work

The sound-absorbing value of most materials depends on a porous surface. Sound waves entering these pores, or holes, get "lost" and are dissipated (scattered) as heat energy. See **Figure 18-58**.

Other materials depend on a similar absorption action to reduce sound. The material used has a vibration point approaching zero. Heavy draperies or hangings, hair felt, and other soft flexible materials function in this manner.

Noise is a mixture of sounds, each with a different *frequency*. Frequency is the rate at which sound-energized air molecules vibrate; the higher the rate, the more cycles per second (cps). The efficiency of an acoustical unit or product is measured by its ability to absorb sound waves. This efficiency is measured by the *noise-reduction coefficient (NRC)* of a material for the average middle range of sounds. For most installations, the NRC can be used to compare the values of one material over another. However, some materials are designed to do a better job for either high- or low-frequency sounds. For special cases, such as music studios, auditoriums, and theaters, an acoustical engineer should be consulted. Many manufacturers of acoustical materials furnish this service.

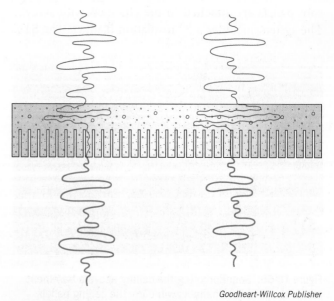

Goodheart-Willcox Publisher

Figure 18-58. Porous materials absorb sound waves by capturing the waves in the material's many voids.

Perforated fiberboard acoustical materials have a low-density, fibrous composition. They are formed into tile shapes. Holes of various sizes are drilled almost through the tile. The sounds that strike these units are trapped in the holes. Since the walls that the holes are drilled into are relatively soft and fibrous in nature, they form tiny pockets that absorb the sound.

There are also some fibrous products that are not drilled. These depend on the surface porosity and the low vibration point of the material to absorb the sound. Most have a relatively smooth surface that provides a high degree of light reflection without glare. This type of product is usually installed on the ceiling or upper wall surface by gluing or stapling. Where the surface is in poor condition, the material can be nailed or glued to furring strips. Each manufacturer has specifications for installation.

The perforated-metal, pan-type material performs in somewhat the same manner as the other materials. The sound enters through the holes in the surface of the pans and is trapped by the backing material. The backing material is soft and resilient. The perforated-metal, pan-type material is commonly used in institutional and commercial buildings where ease of maintenance and fire resistance are important factors.

18.20.1 Installation of Acoustical Materials

Carefully follow the manufacturer's recommendations when applying acoustical materials. If they are not properly installed, the sound-deadening materials may not do the job. In many cases, the amount of air space in back of the material is a factor in its sound absorption qualities.

Because most acoustical materials are soft, they are usually installed on the ceiling or upper portion of sidewalls. For sounds originating in the average room, the ceiling offers a sufficient area for sound absorption. Directions concerning the methods and procedures for installing ceiling tile are included in Chapter 19, **Interior Wall and Ceiling Finish**.

18.20.2 Suspended Ceilings

The panels of a suspended ceiling are installed on metal runners that form a grid hung below the actual ceiling, **Figure 18-59**. This system allows the large panels to simply be dropped into place. Suspended ceiling systems are used to conceal pipes, electrical wiring, and structural beams. The entire ceiling, or any part of it, may be removed and relocated without damage to the material.

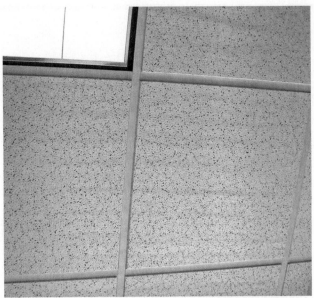
Goodheart-Willcox Publisher

Figure 18-59. Removable 2′ × 2′ panels rest on a metal grid in this suspended ceiling. They can be removed easily to provide access to pipes, electrical wiring, or HVAC ducts above the ceiling. Lighting fixtures also rest on the grid.

The panels are made from ground cork, glass fibers, or other types of porous materials. The material is pressed into acoustical panels of various sizes and thicknesses. Because of the resistance of both cork and glass fiber to moisture, panels made from them are ideal for use as ceiling material in indoor swimming pools, commercial kitchens, or any place where humidity is a problem. Additional information on suspended ceilings is included in Chapter 19, **Interior Wall and Ceiling Finish**.

18.20.3 Maintenance

If properly done, painting of perforated insulation material usually does not lower the material's efficiency. If improperly applied, however, paint will soon fill the pores of the material and destroy its efficiency. Spray painting is usually preferable to brush painting for these materials. The thinner mixture is less likely to clog the pores of the material.

Any dirt clogging the pores of the material should be first removed. This may be done with a vacuum cleaner or a soft-bristle brush. Some acoustical material can be cleaned by washing.

Pro Tip
Manufacturers of acoustical tiles and other products have prepared detailed instructions for installation and maintenance. Be sure to carefully follow them.

Construction Careers
Insulation Installer

Increased emphasis on energy efficiency in homes and commercial structures has led to increased job opportunities for insulation installers. Installers work with many kinds of insulation, including fiberglass batts, loose-fill materials, rigid sheets of foam plastic, and expanding foam sprays. More complex installation techniques are typically needed for industrial plants and some commercial applications than for residential buildings.

Most insulation workers are employed by building finishing contractors and typically work on larger residential developments and commercial/industrial buildings in urban areas. Smaller building contractors (primarily residential) usually do their own insulation work. Carpenters, drywallers, or HVAC installers usually handle the insulating tasks.

Insulation installers typically work indoors, but working conditions often are dirty and dusty. The conditions can also be uncomfortable during hot, humid weather. Fine particles from fiberglass and other insulating materials can cause skin, eye, or respiratory system irritation. For this reason, following safety guidelines, ensuring adequate ventilation, and wearing proper personal protective gear are very important.

Installation skills are most often learned on the job by working with an experienced installer. High school or vocational school classes in construction, woodworking, and blueprint reading are useful preparation for this field. A formal apprenticeship program combining classroom work and four years of on-the-job training is available in some localities. Apprenticeship programs help prepare workers for more complex types of installation work in industrial plants and similar settings and can lead to advancement into supervisory roles.

nightfrost/iStock/Thinkstock

Good building insulation leads to increased energy savings. Insulation may be installed by carpenters, drywallers, or HVAC installers. Installers much take precautions to avoid skin contact or inhaling any insulation particles.

Chapter Review

Summary

- Insulation materials do not readily transmit energy in the form of heat, electricity, or sound.
- In construction work, much of the applied insulation is intended to prevent the transmission of heat energy—keeping heat out in the summer and holding it in during the winter.
- Heat travels in three ways: conduction, convection, and radiation.
- Insulation is used inside walls, in attics, under floors, and around foundations. Insulation placed in these areas should function in conjunction with any exterior insulation that may have been installed as part of the building envelope.
- The efficiency of insulation is determined by its R-value. The higher the R-value, the more the material resists transmitting heat.
- Types of insulation are flexible, loose fill, rigid, foamed-in-place, and reflective.
- Attics and crawl spaces can be configured as conditioned or unconditioned spaces. It is important to carefully follow building codes for the configuration being used.
- Proper attic ventilation is important to protect a home from ice dams in cold climates.
- Some safety precautions must be taken when working with fiberglass insulation to avoid possible skin or eye irritation.
- A blower door test is used to measure a building's air leaks.
- Insulation can be used to control sound transmission by reducing the vibrations transmitted through structural materials.

Know and Understand

Answer the following questions using the information in this chapter.

1. *True or False?* Requirements of sound control and insulating against heat loss are so different that the same materials cannot be used for both.

2. When heat moves from one molecule to another within a given material, the method of heat transmission is called _____.
   A. conduction
   B. convection
   C. radiation
   D. heat flow

3. _____ is a good insulator when it is confined to a small space when convection flow is limited or absent.
   A. Water
   B. Air
   C. Ductwork
   D. Flexible blanket insulation

4. Which of the following is considered a poor thermal insulator?
   A. Air.
   B. Aluminum foil.
   C. Fibrous blanket.
   D. Glass.

5. The _____ of a material is the amount of heat (Btu) that will flow through the material in one hour per sq ft of surface with 1°F of temperature difference.
   A. coefficient of thermal conductivity
   B. conductance
   C. resistance
   D. total heat transmission

6. *True or False?* A good insulation material has a high R-value.

7. Which of the following is typically *not* a consideration when choosing thermal insulation?
   A. Comfort.
   B. Health.
   C. Appearance.
   D. Environmental impact.

8. _____ insulation has foam panels with higher R-values per inch than other forms of insulation.
   A. Flexible
   B. Loose-fill
   C. Rigid
   D. Foamed-in-place

9. What type of insulation is generally the *least* cost efficient?
   A. Flexible.
   B. Foamed-in-place.
   C. Rigid.
   D. Reflective.

10. The thermal envelope is made up of _____ and _____ insulation.
   A. flexible, rigid
   B. interior, exterior
   C. flashing, exterior
   D. floor, wall

11. _____ are practical for protecting insulation only when a small area of the insulation board projects above the finished grade.
   A. Ventilators
   B. Polystyrene panels
   C. Cementitious panels
   D. Vapor barriers

12. _____ are used to keep airways open to ensure adequate ventilation when insulating an attic space.
   A. Baffles
   B. Vapor barriers
   C. Energy-heel trusses
   D. Ventilators

13. In drywall construction, blanket or batt insulation with flanges should be stapled to stud faces at intervals of no more than _____".
   A. 1
   B. 2
   C. 6
   D. 12

14. When pouring or blowing loose insulation in an attic, contain the fill around the soffit area with _____ insulation.
   A. loose
   B. batt
   C. blanket
   D. foamed-in-place

15. *True or False?* If using the basement as a living space, interior walls should be insulated.

16. The preferred method of sealing penetrations around electrical service entrances and hose bibbs is to use _____ that is incorporated into the water-resistant barrier.
   A. caulk
   B. a foam insulation board
   C. flexible flashing tape
   D. a vapor barrier

17. The total perimeter of a structure is 200′, the ceiling height is 10′, and the total area of the doors and windows is 150 sq ft. What is the net area of the insulation needed for the exterior walls?
   A. 60 sq ft.
   B. 360 sq ft.
   C. 1850 sq ft.
   D. 2150 sq ft.

18. Which of the following is *not* a main component used for a blower door test?
   A. A smoke pencil.
   B. A temporary frame and panel.
   C. A variable-speed fan.
   D. Pressure gauges.

19. The unit of measure used to indicate the loudness or intensity of a sound is called the _____.
   A. noise
   B. decibel
   C. threshold of pain
   D. volume

20. A wall with an STL of 30 dB reduces the loudness of a given sound traveling through it from 65 dB to _____ dB.
   A. 60
   B. 35
   C. 45
   D. 50

21. A method of rating the sound-blocking efficiency of a wall, floor, or ceiling structure is called the _____ system.
   A. sound transmission loss (STL)
   B. decibel
   C. sound transmissions class (STC)
   D. diaphragm action

22. A(n) _____ can be used to dampen sound transmissions in higher frequencies within our hearing range and reduce sound transmission through interior or exterior walls.
   A. insulation board
   B. double wall
   C. insulation blanket
   D. sound-deadening board

23. *True or False?* To ensure a high STC rating, electrical outlets on opposite faces of a partition should not be located in the same stud space.

24. The efficiency of an acoustical unit or product is measured by the _____ of a material for the average middle range of sounds.
   A. frequency
   B. diaphragm action
   C. noise-reduction coefficient (NRC)
   D. sound transmission loss (STL)

Apply and Analyze

1. Explain the three methods of heat transmission.
2. How is a degree day used?
3. Which of the five broad classifications of thermal insulation is best used to insulate spaces between studs in an old building?
4. Give two reasons why a basement should be insulated.
5. How are vapor barriers installed differently in colder climate regions versus hot, humid regions?
6. Describe the *two-three hole method*.
7. Name three ways an insulating board can be used.
8. What steps must be followed to prepare for a blower door test?

9. What are *impact sounds*?

10. Why is spray painting preferred to brushing when refinishing acoustical ceiling tile?

Critical Thinking

1. What kind of impact can rodents, such as mice, rats, and squirrels, have on a building? Left unchecked, how can these animals damage the thermal performance of a home?

2. Working from a set of blueprints supplied by your instructor or downloaded from the internet, identify the points in the building that will require special attention when insulating. Write out what you feel is the best solution for each of these areas.

3. Imagine you are a contractor who is bidding a job to build a small home that meets high-efficiency standards. What systems and components do you think will be important to achieve compliance in a project such as this? What tests will be used when the job is finished to ensure compliance?

Who can you contact for help to ensure compliance before and during construction?

Communicating about Carpentry

1. **Speaking.** Working in a group, brainstorm ideas for creating classroom tools (posters, flash cards, and/or games, for example) that will help your classmates learn and remember the different thermal and sound insulation materials. Choose the best idea(s), then delegate responsibilities to group members for constructing the tools and presenting the final products to the class.

2. **Art.** After reading this chapter, you should have a good understanding of thermal and sound insulation. Create a model or draw a picture with labels describing thermal and sound insulation. In your model or picture, compare buildings without insulation with buildings that have insulation protection.

CHAPTER **19**

Interior Wall and Ceiling Finish

OBJECTIVES

After studying this chapter, you will be able to:

- Explain wall and ceiling covering materials.
- Describe wallboard cutting, nailing, and adhesive techniques.
- Illustrate how double layer and predecorated wallboard are applied.
- Describe the characteristics of gypsum plaster.
- Explain how gypsum lath and expanded metal lath are applied.
- Illustrate the use of plaster grounds.
- Describe plastering methods.
- Lay out ceiling tile and install furring strips.
- Describe methods for leveling and installing a suspended ceiling.
- Estimate quantities of lath, wallboard, and ceiling tiles for a specific interior.

TECHNICAL TERMS

banjo
brown coat
cement board
chevron paneling
darby
diagonal paneling
expanded metal lath
field

finish coat
grounds
gypsum lath
gypsum wallboard
hardboard
herringbone paneling
insulating fiberboard lath
interior finishing

mold-resistant wallboard
plaster
scratch coat
single-layer construction
suspended ceiling
wood lath
working time

The installation of cover materials to walls and ceilings is part of *interior finishing*. This stage of construction can start after the mechanical systems (plumbing, heating, and electrical wiring) and insulation are installed and inspected. Exterior doors must also have been hung and windows installed. They will protect the finishing materials from weather damage. Interior walls can be covered with any of a number of materials, **Figure 19-1**:

- Gypsum wallboard—Commonly called drywall or sheetrock, *gypsum wallboard* is a laminated material with a gypsum core and paper covering on either side. It usually comes in sheets 4′ wide × 8′, 9′, 10′, 12′, or 14′ long. It can be special ordered in 54″ widths. It is sold in several

thicknesses: 1/4″, 5/16″, 3/8″, 1/2″, and 5/8″. Gypsum wallboard is used on both walls and ceilings.

- Gypsum lath for plaster—This is a base of 16″ × 48″ to 4′ × 8′ gypsum board usually 3/8″ or 1/2″ thick. It is applied as a backing for final covering with plaster.
- Predecorated gypsum paneling—This is the same as gypsum wallboard. However, decorative vinyl finishes have been applied and edges have received special treatment so that no other finishing work need be done after the panels have been installed. The finishes are tough and easily cleaned.
- Plywood and particleboard—Plywood is fabricated in 4′ widths. Lengths include 8′, 9′, and 10′.

A   *archideaphoto/Shutterstock.com*   B   *YKvisual/Shutterstock.com*   C   *fotoplan/Shutterstock.com*

Figure 19-1. Various materials are used as wall coverings. A—Gypsum wallboard is the most widely used material. It accepts many different surfacing materials, such as wallpaper and paint. B—Wood is used in solid form, as shown here, or in various engineered panels. C—Ceramic tile is used extensively for walls in kitchens and bathrooms.

Panels are manufactured in thicknesses of 1/4″, 3/8″, 7/16″, 1/2″, 5/8″, and 3/4″. Usually, the sheets are prefinished in a variety of colors and patterns. The surface material may be either a hardwood or softwood. Surfaces can be embossed, stained, or color toned.

- Hardboard and fiberboard—These are produced from wood fibers in sizes and thicknesses similar to plywood. The face finish is simulated to look like wood. Other decorative patterns are also applied. Sheets may be embossed and grooved to take on the look of random planking, leather, or wallpaper. Surfaces may also be coated with plastic. Variations of fiberboard are used as ceiling coverings.

- Solid-wood paneling—These are boards or pieces of solid wood. Widths of boards vary from 2″ to 12″ and thicknesses are a nominal 1″ or 2″. Faces may be rough-sawed, plain, or molded in a variety of patterns. Lengths vary from 4′ to 10′. Shingles, usually considered a siding or roofing material, are occasionally used on interior walls.

- Plaster—For many years, *plaster* was the most popular wall covering. It is made of powdered gypsum to which other materials are added to improve drying time. A plastered wall system includes a base support, such as metal or gypsum lath, over which coats of wet plaster are applied.

- Cement board—Available under several different brand names, **cement board** is a versatile fiber-reinforced cement panel material. It is not considered a finishing material itself, but serves as a base (underlayment) for finishing materials on floors, countertops, and exterior or interior walls. It is fireproof, not damaged by water, and resists impact. Some cement board products are lightweight with fiberglass-reinforced matting and silicone-treated cores. **Figure 19-2** shows a sample of a typical cement board product.

James Hardie Building Products

Figure 19-2. A sample of a fiber-reinforced cement underlayment. It can be used as backing for ceramic tile, marble, and plastic laminates. The dot at the edge is a guide for nailing.

- Special finishes—These include a variety of products and materials such as brick, stone, glazed tile, plastic tile, and plastic laminates. They are used either as accent materials or to provide a wear-resistant surface. They are often found in kitchens and bathrooms.

Ceilings can be covered with many of the same materials used for walls. Composition tiles are especially suitable because they are easy to install.

Before beginning the wall and ceiling application, check the framework. Be certain that sufficient backing has been installed for fixtures and appliances. Refer to Chapter 11, **Wall and Ceiling Framing**. Nailers provide a surface for fastening wall coverings at all intersections of wall and ceiling surfaces. They must be included at all vertical corners and at all intersections between walls and ceilings. Special or built-in equipment, such as a prefabricated fireplace, must be installed before the interior wall surface is applied.

19.1 Drywall Construction

Drywall materials are the most common wall coverings used in construction. Most builders prefer to use drywall because it saves time. Regular plaster requires considerable drying time and introduces a lot of moisture into the building. Either type of finish presents advantages and disadvantages. Drywall construction, for example, requires that studs and ceiling joists be perfectly straight and true; otherwise, the wall surface will be uneven. This can sometimes be corrected in ceiling joists by installing a strongback. Where steel or engineered wood joists and studs are used in framing, this is not a problem. The wood framing material must also have a moisture content very near to what it will eventually reach in service. This helps prevent nail pops and joint cracks. Using screws to fasten drywall also helps with this problem.

19.1.1 Handling and Storage

Drywall should be delivered only a few days before it is to be installed. If stored on the jobsite for too long, it may be damaged. Likewise, joint compound and veneer plaster finishes have a short shelf life and should not be stored for long periods.

Handle drywall as carefully as you would millwork. It can be transported to the point of use manually or by machine, **Figure 19-3**. In multistory construction, use of a machine may be the most practical and time-saving way to deliver the material to the point where it will be installed.

Stack drywall flat on a clean floor in the center of the largest rooms. Unless the floor is a slab-on-grade, do not stack all of the drywall in one room. The weight of a large amount of drywall may be more than the framing can withstand in a single area. Place those sheets to be used on the ceiling on the top of the pile. This is because the ceilings will be drywalled before the walls. Never stack longer sheets on top of shorter ones—the overhanging portion could crack.

Store metal corner beads, casing beads, and trim where they will not be bent.

19.1.2 Types, Thicknesses, and Styles

There are several types of gypsum wallboard. Regular wallboard is used where a special type is not specified. Type X has an improved, thicker core that does not crumble as quickly as other boards in a fire. It is specified for fire-rated construction, such as to separate an attached garage from living spaces or to separate attached dwelling units, as in an apartment building. Mold-resistant board incorporates various methods of improving resistance to mold and mildew. **Figure 19-4** lists a variety of thicknesses, edge joint designs, and types. **Figure 19-5** illustrates several standard edge designs for gypsum wallboard. The tapered edges form a shallow depression between adjacent sheets. This depression is brought level with drywall tape and filler. The result is a smooth, uninterrupted surface. Drywalling requires some special tools for marking, cutting, installing, and finishing. See **Figure 19-6**.

Main Types of Gypsum Wallboard		
Type	Thickness (inches)	Edges
Regular (ASTM C36, FS SSL30d)	1/4 5/16 3/8 1/2 5/8	Tapered Square Square Tapered Bevel
Fire resistant type "X" wallboard	1/2 5/8	Square Tapered Bevel
Insulating wallboard (aluminum foil on back surface)	3/8 1/2 5/8	Square T & G Tapered Round
Regular backing board (ASTM C442, FS SSL30d)	1/4 3/8 1/2	Square Square T & G
Foil-backed backing board	3/8	Square
Fire resistant type "X" backing board	1/2 5/8	T & G
Coreboard (Homogeneous or laminated)	3/4 1	Square T & G Ship lap
Pre-decorated	3/8	Bevel Round Square
	1/2 5/8	Bevel Square

Goodheart-Willcox Publisher

Figure 19-4. Main types of gypsum wallboard.

Kent E Roberts/Shutterstock.com

Figure 19-3. Transporting drywall into a new residence using power equipment.

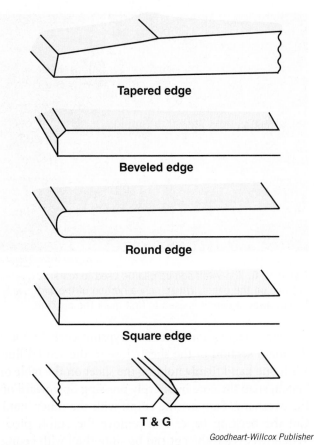

Tapered edge

Beveled edge

Round edge

Square edge

T & G

Goodheart-Willcox Publisher

Figure 19-5. Gypsum wallboard is manufactured with several different edge styles.

19.2 Single-Layer Construction

Single-layer construction is used where economy, fast installation, and fire resistance are important. It is well suited to remodeling jobs, and for resurfacing damaged or cracked plaster walls. For both new construction and remodeling, use 1/2″ or 5/8″ gypsum wallboard. The 5/8″ sheet is preferred for high-quality construction. Cover ceilings first, then the walls.

There are two methods of arranging the drywall sheets:

- Parallel—Long edges of panels run in the same direction as studs and joists.
- Perpendicular—Long edges of panels are at right angles (90°) to studs and joists.

The second method is generally preferred for several reasons:

- There are fewer feet of joints to be finished, thus saving time and reducing costs.
- Panels bridge more studs and joists, making the building's frame stronger.
- The strongest dimension of the panel runs across the frame, so that each piece is fastened to the greatest number of studs possible.
- There are fewer problems with irregularities in alignment and spacing of the frame.
- Horizontal joints are easier to treat because they are lower on the walls.

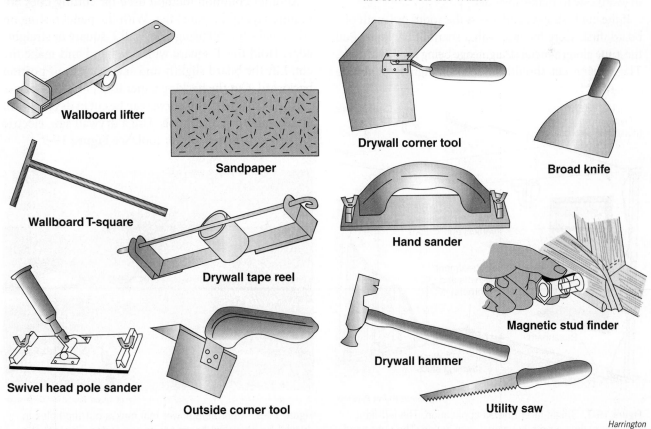

Wallboard lifter

Sandpaper

Wallboard T-square

Drywall tape reel

Swivel head pole sander

Outside corner tool

Drywall corner tool

Broad knife

Hand sander

Magnetic stud finder

Drywall hammer

Utility saw

Harrington

Figure 19-6. These tools are used for preparing and installing drywall. Note the names.

In either method, vertical wall joints must be centered on studs for proper fastening. **Figure 19-7** shows both parallel (vertical) and perpendicular (horizontal) applications. As a general rule, use whichever method results in the fewest joints. Stagger end joints and locate them as far away from the center of walls and ceilings as possible.

Loosely butt wall panels against the ceiling panels. In a parallel application, use a wallboard lifter to raise the panel to the ceiling. A lifter is a foot-operated lever device. Stepping on it while one end is under the drywall raises the panel enough to press it against the ceiling. Refer to **Figure 19-6**. In horizontal applications, the top wall panels are installed first. This is done so that any gaps or cut edges come at the floor where trim will cover them.

19.2.1 Measuring and Cutting

All measurements should be carefully taken from the spot where the wallboard will be installed. Usually, it is best to make two readings, one for each side of the panel. Following this procedure eliminates errors. It also corrects for openings and framing that are not plumb or square. Use a steel tape to take measurements. Transfer these measurements to the drywall panel. Where the cut will not be entirely across the panel, draw a line along the straightedge, as in **Figure 19-8**. Then, use a drywall saw to make any cuts.

If the cut is straight and across the width or length of a board, first, score the face with a sharp utility knife. Pull the knife along a metal straightedge being used as a guide. The scoring cut should be deep enough to penetrate

Goodheart-Willcox Publisher

Figure 19-8. A drywall square can be used to mark a square cut line on the panel. Where only a portion of the width will be cut away, a pencil is used to first mark the cut line.

the paper facing and enter the gypsum core. Support the main section of the sheet close to the scored line. With one hand firmly holding the sheet on the table or bench, snap the core by sharply pressing downward on the overhang. Support the cutoff with the other hand. Cut the backing paper and remove the cutoff piece. When necessary, the cut can be smoothed with coarse sandpaper mounted on a block of wood.

Another common method used for cutting does not require laying the panel flat. With the panel resting on edge, make the cut using a drywall T-square or straightedge. Hold the T-square with one hand and make the cut. Lift the board slightly and snap the cutoff portion backward. Cut the backing paper to separate the parts.

Irregular shapes and curves can be cut in drywall with either hand or power tools. Use a drywall saw, electric saber saw, or rotary power tool. See **Figure 19-9**.

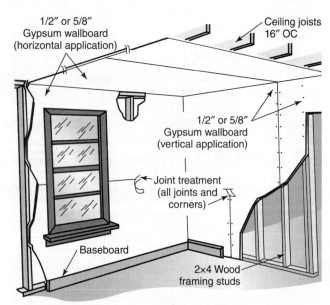

Goodheart-Willcox Publisher

Figure 19-7. Single-layer drywall application. The left-hand wall and ceiling have a horizontal application. The right-hand wall has a vertical application.

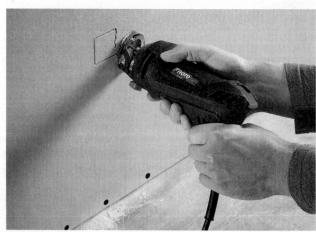

RotoZip, a division of Robert Bosch Tool Corporation

Figure 19-9. A portable power tool makes cutting holes in drywall for electrical boxes easier and faster. The tool can also be used to cut circles or irregular shapes.

19.2.2 Screw Fastening

Screw application requires a screw gun. This is a positive-clutch electric power tool designed for attaching drywall. It uses a Phillips bit and has an adjustable head that controls screw depth. To adjust for proper screw depth, the control head is rotated. The screw gun is designed so that it does not operate until the screw and gun tip are pressed against the drywall.

Using screws to fasten gypsum board is more common than using nails. **Figure 19-10** shows the common types of drywall fasteners. Most drywall mechanics prefer screws because they provide a firm, tight attachment to wood or metal framing and are less likely to detach. Special, self-tapping screws are used for metal-framed wallboard systems. Fasteners of this type must be driven so the screw head rests in a slight dimple formed by the driving tool, **Figure 19-11**. The paper face of the drywall should not be cut, nor should the gypsum core be fractured. Since screws hold the wallboard more securely than nails, ceiling spacing can be extended to 12″ and side walls to 16″.

Braces can be used to support large panels while they are fastened to a ceiling. Start fastening ceiling panels at the abutting edge. Next, screw or nail the *field* (area between edges), working away from the abutted edge. Then, finish with the opposite side. If the perimeter is fastened first, panels will sag and not draw up tight to joists. After the ceiling is completed, drywall can be applied to the walls, **Figure 19-12**.

Hilti Tools

Figure 19-11. An electric screw gun with a special depth-adjusting clutch is used to drive drywall screws into steel studs. The clutch disengages when the nose strikes the panel surface.

Goodheart-Willcox Publisher

Figure 19-12. After ceilings are drywalled, apply drywall to walls. Try to get a close fit in the corners.

19.2.3 Nail Fastening

Nail spacing varies depending on the materials being used. For single-layer construction, space nails no farther apart than 7″ on ceilings and 8″ on walls. Keep nails at least 3/8″ from ends and edges. Annular ring nails that are 1 1/4″ long with a 1/4″ diameter head

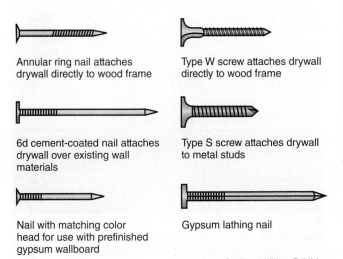

Annular ring nail attaches drywall directly to wood frame

Type W screw attaches drywall directly to wood frame

6d cement-coated nail attaches drywall over existing wall materials

Type S screw attaches drywall to metal studs

Nail with matching color head for use with prefinished gypsum wallboard

Gypsum lathing nail

Goodheart-Willcox Publisher

Figure 19-10. Gypsum wallboard and gypsum lath fasteners. Others are available.

are generally recommended. Tightly draw the drywall against the framing so there can be no movement of the board on the nail shank as it is being driven. Press the board tightly against the stud or joist to avoid breaking through the wallboard face. Drive nails straight and true. An angled nail will tear the face paper as it is countersunk. Use extra care during the final strokes so the nail head rests in a slight dimple formed by the crowned head of the hammer or stapler. Be careful not to break the paper face of the board.

19.2.4 Adhesive Fastening

Drywall may also be fastened with a special adhesive that is sold in cartridges. Apply it in a continuous bead over the frame surface using a hand or powered gun. One main advantage of adhesive is there are only a few depressions from mechanical fasteners that need to be filled later. Adhesives also produce a sturdier wall that is more resistant to impact sounds.

Safety Note
Some adhesives have flammable solvents. Do not use flammable adhesives where open flames are present.

PROCEDURE
Applying Adhesives
1. Select the proper adhesive and read the manufacturer's directions for use.
2. Apply adhesives only when temperatures are between 50°F and 100°F (10°C–40°C). Keep adhesive containers closed. Evaporation of the solvent can affect the adhesive's performance and ability to bond.
3. Check all surfaces. They must be free of dirt, oil, or other contaminants.
4. Observe manufacturer's open time for the adhesive. If exceeded, a poor bond is sure to result. As a general rule, apply no more adhesive than can be covered within 15 minutes.
5. Apply a continuous bead of adhesive to the center of all studs, joists, or furring.
6. Where two pieces of wallboard join on a framing member, use a zigzag bead pattern. The bead should be 1/4″–3/8″ wide. Then, when the board is in place, it will be held by a band at least 1″ wide and 1/16″ thick.
7. Use temporary screws or bracing to ensure full contact between the adhesive and the drywall. Go over each surface, applying hand pressure to force the panel into the adhesive. All of these precautions help the adhesive to develop proper bonding strength.

19.2.5 Concealing Joints and Fastener Heads

Mechanical taping tools are often used to apply the compound and tape. **Figure 19-13** illustrates the use of a *banjo* to apply compound and tape at the same time. This tool has a reservoir to hold the compound and a reel to hold the tape. Manual pressure is applied to spread the compound and apply the tape as the tool is moved along the joint. After the tape and compound are applied, the joints are smoothed with a broad knife.

The introduction of pressure-sensitive, glass-fiber tape has reduced the time required to conceal and reinforce joints and interior angles. It has an open weave of 100 meshes per square inch. This mesh provides excellent reinforcing and keying of plaster or compound coats. Simply use hand pressure to attach it to the wall,

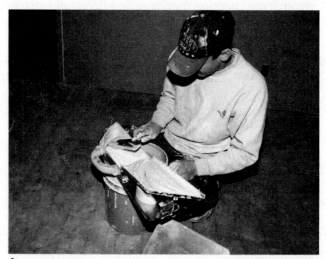
A   *Loren LaPointe Drywalling*

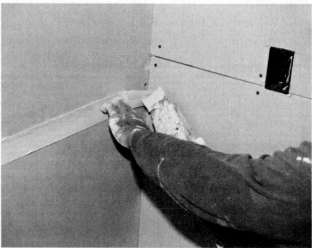
B   *Loren LaPointe Drywalling*

Figure 19-13. Using a banjo for applying tape to drywall joints. A—The reservoir is being filled with compound. The roll of tape is visible under the reservoir cover. B—The operator squeezes out compound as the tape is pulled across the joint.

then bond it by pulling a finishing knife or trowel along its length. The tape is also easily applied to inside corners.

Next, apply two coats of fast-setting joint compound over the tape. Care should be taken on every coat of joint compound to finish it as smooth as possible as it is applied. Expert drywall finishers need to do only light sanding of the final coat. If the surface is to be covered with a texture paint, the joints can be finished with a single coat of compound. When applying joint compound to ceilings and the upper walls, it will be necessary to stand on a platform to reach the work. Often, stilts are used to reach high areas, **Figure 19-14**.

Figure 19-14. Stilts are very helpful when working on ceilings and upper walls. Many stilts are adjustable in height, making it possible to work on high ceilings and tall walls.

A wide variety of joint compounds is available for drywall construction. Some are in a powder form ready to be mixed with water. Others are ready to use from the containers. Always read and follow the manufacturer's recommendations.

PROCEDURE
Concealing Joints and Fastener Heads

1. Apply a bedding coat of joint compound into the depression formed at all butt joints by the tapered edges of the board. Use a 5″ or 6″ joint knife, **Figure 19-15**.
2. Center the reinforcing tape over the joint. Press tape into compound by drawing the knife along the joint with enough pressure to remove excess compound. Smooth it out to avoid wrinkling or buckling. Apply a skim coat over the tape.
3. After the embedding coat is completely dry, apply a second coat over the tape. Feather the edges at least 1/2″–3/4″ beyond the edges of the first coating.
4. When this coat is completely dry, apply a third coat with the edges feathered out about 2″ beyond the second coat.
5. After the last coat is dry, sand lightly, if necessary. Fasteners are also concealed with compound, each coat being applied at the same time the joints are covered.

19.2.6 Corners

Outside corners are reinforced with a metal corner bead, which is made in various styles. The bead is installed after horizontal and vertical joints have been taped. Fasten the bead by driving screws or nails through the wallboard and into the framing, **Figure 19-16**.

A

B

C

Figure 19-15. Taping wallboard joints with taping knives. A—First apply compound to the channel at the joint. B—Then, embed the tape. Be sure it is centered over the joint. C—Immediately apply a skim coat over the tape and smooth the edges. Use a broader knife for this step.

Goodheart-Willcox Publisher

Figure 19-16. Attaching metal corner bead to an outside corner.

After installation, the bead is concealed with joint compound in about the same manner as regular joints. To finish and reinforce edges around doors, windows, and other openings, metal-channel trim is available.

At inside corners, reinforcing tape is used. First, apply a bedding coat of joint compound to both sides of the corner. Then, fold the tape along the centerline and smooth it into place. Remove excess compound and finish surfaces along with the other joints.

19.2.7 Mixing Compound

Place a quantity of compound in a clean container. Mix in only clean water according to instructions on the bag. Use care that all compound becomes uniformly damp. Avoid contaminating the mixture with dirty water or previously mixed compound. To do so will affect setting time. Mix only what can be used up within the *working time* indicated on the bag. Adding water to retemper compound once it begins to set is useless—it will not prevent setting or increase working time.

Ready-mixed compound should be used as it comes from the container. Add cool water in half-pint increments for a thinner mix. Use a potato-masher type

mixer and lightly remix after each addition of water. Test the compound after each addition to avoid the mixture from becoming too thin.

19.2.8 Attaching Drywall to Steel Framing

Single-layer application to steel framing members is similar to wood frame application. Arrange panels either parallel or perpendicular to framing and attach them with 1″ long Type S screws, **Figure 19-17**. The leading ends or edges of panels must first be attached to the open edge of the stud or joist flange.

19.2.9 Air Sealing Drywall

Drywall is an effective air barrier when sealed properly. See Chapter 15, **Building Envelope and Control Layers**, for information regarding air barriers and their function in an enclosure. When using the drywall as an air barrier, all penetrations, such as electrical boxes, pipe penetrations, and fixtures, must be sealed from air leakage. This can be accomplished by using gasketed electrical boxes or appropriate sealant where there is the possibility of any air movement through the drywall. It is also important for all drywall to be in contact with the top and bottom plates on the wall framing. A continuous bead of sealant must connect the drywall to the wall plates. Additionally, no drywall joints can be left unfinished.

Goodheart-Willcox Publisher

Figure 19-17. Screws are used to attach drywall to steel framing in this office building project.

19.3 Double-Layer Construction

Double-layer (also called two-ply) gypsum board applications over wood framing ensure a strong wall surface, **Figure 19-18**. Double-layer drywall is often specified to be Type X, fire-rated wallboard. It is also often 5/8″ wallboard to increase the fire and sound protection it provides. This method is adaptable to the use of either predecorated panels or standard beveled drywall with treated joints.

For areas where there is likely to be moisture coming into direct contact with the wall, there are highly water-resistant backing boards, including some that are impervious to water. Their use is especially recommended in shower areas as a base for tile and other protective coverings.

Sound deadening backing board is sometimes used for the base of double layered walls. It is specified where high sound transmission class (STC) ratings are needed.

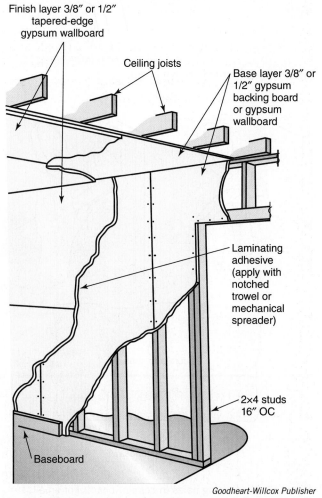

Finish layer 3/8″ or 1/2″ tapered-edge gypsum wallboard

Ceiling joists

Base layer 3/8″ or 1/2″ gypsum backing board or gypsum wallboard

Laminating adhesive (apply with notched trowel or mechanical spreader)

2×4 studs 16″ OC

Baseboard

Goodheart-Willcox Publisher

Figure 19-18. A cutaway of double-layer gypsum wallboard construction. The finish layer may be applied at a right angle to the base (as shown) or running parallel.

19.3.1 Attaching the Layers

Apply backing board to framing with staples, nails, or screws. The finish layer is laminated to the base layer with an adhesive or joint compound. Joints of the finish layer should be offset at least 10″ from the joints of the base layer. Finish layers can be applied parallel to the base layer or at right angles to it. If the two-ply application is used to comply with a fire rating, the joint compound must be fire rated.

19.3.2 Applying Adhesive

First, cut and fit drywall sheets. Adhesive is usually applied to the entire surface. However, strip lamination is used in some applications. This method uses strips with ribbons of adhesive spaced at regular intervals.

Many methods of applying adhesive are acceptable. Trowels and powered devices are available. Whatever method is used, the spacing and size of the bead of adhesive must provide the required spread when panels are pressed into position. A notched spreader is often used to apply adhesive to the entire back surface of a finish-layer panel. Strip lamination is frequently used for sidewall panels. The application can be made either on the base surface or on the face panel. Temporary bracing may be used to hold panels in position until bonding has taken place. There are several rules to follow when using recommended fasteners:

- On-ceiling applications when using a laminating adhesive—Space fasteners 16″ OC along ends and edges. At mid-width, use one fastener for every framing member.
- When laminating with compound—Provide permanent supplementary fastening or temporary fastening (usually overnight) until the compound has dried.

When nails are used, they should provide a minimum penetration of 3/4″ into the wood framing members. Consult a nail chart.

19.3.3 Finishing Double-Layer Wallboard

If the wallboard is to receive other covering material, joints should be taped, nails concealed, and corners finished in the same way as for single-layer construction. If a veneer of plaster is to be applied, use reinforcing tape and a single bedding coat of compound over joints. If the construction is to be fire rated with Type X drywall, use fire-resistant compound. Tape inside corners and apply a special bead to outside corners to form grounds for the veneer coat. Always use a single length of corner bead to extend from the floor to the ceiling.

19.4 Special Backing

Special backing is available as a base for tile in areas where walls are frequently wet, such as showers. One type is known as cement board. It is manufactured from a slurry of Portland cement reinforced with polymer-coated, fiberglass mesh embedded in both sides. Some backing board products are rigid, while others are somewhat flexible.

These materials, depending on the brand, can be used in many different applications on both interior and exterior surfaces, **Figure 19-19**. Applications include the following:

- Floor underlayment for tile, resilient coverings, carpeting, or thin brick
- Floor or wall heat shields for stoves
- Backing for tiles on walls in shower stalls or tub surrounds
- Base for countertops
- Base for exterior finishes such as ceramic tile, thin brick, or synthetic stucco

19.4.1 Working with Cement Board

Cement board can be worked with ordinary carpenter's tools. Panels can be fastened with nails, screws, or staples. Common panel thicknesses are 1/4″, 7/16″, and 1/2″. Panel dimensions are 3′ × 5′ and 4′ × 8′.

Lindasj22/Shutterstock.com

Figure 19-19. Cement board is used as backing for ceramic tile on walls, countertops, and other locations that are frequently exposed to water. A construction laser is being used here for tile alignment.

To cut cement board, score it several times with a tungsten-tipped knife. Use a straightedge or drywall square as a guide, **Figure 19-20**. Use the straightedge to apply topside pressure at the score and snap the panel upward. Rough edges can be smoothed with a rasp or coarse sandpaper. Small holes should be outlined with a series of drilled holes. Use a tungsten carbide-tipped masonry bit. On larger holes, score all sides and then make a diagonal score across the opening. In either case, break out the hole from the face side with a hammer. See **Figure 19-21**.

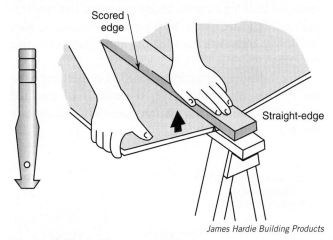

James Hardie Building Products

Figure 19-20. Cutting cement board.

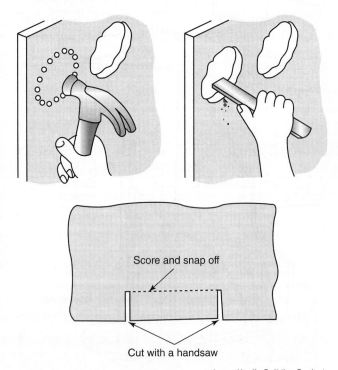

James Hardie Building Products

Figure 19-21. Cutting holes in cement board. For small openings, make a series of small holes and break out the waste with a hammer. Score larger openings and saw where possible.

Cement board is installed in about the same way as conventional wallboard. **Figure 19-22** shows a detail of the construction at the edge of a tub. Stud spacing should not be greater than 16″ OC. Note the furring strip. It ensures alignment between the tub lip and the wallboard. Also note the 1/4″ space that should be maintained along the tub edge.

19.5 Mold-Resistant Wallboard

Mold-resistant wallboard is a type of gypsum wallboard processed to withstand the effects of moisture and high humidity. The covering is treated to be mold and mildew resistant. Its facing paper is light green so it can be easily identified. Because of this coloring, it is often referred to as greenboard. Mold-resistant wallboard is not as water-resistant as cement board, but is still extensively used. Standard thicknesses include 1/2″ and 5/8″. Standard width is 4′ and lengths range from 8′ through 12′. This product is not used as a base under ceramic tile or other nonabsorbent finishing materials in showers and tub alcoves. However, it can be used in other areas of the bathroom.

19.6 Veneer Plaster

Veneer plaster is a high-strength material applied as a coat less than 1/8″ thick. Because of the composition and thinness of the coat, it dries rapidly.

Trim and decoration work may proceed after a minimum drying time of 24 hours.

A special gypsum board is used for the base. Its face surface consists of several layers of paper. The outer layer rapidly absorbs moisture and makes it easier to apply the plaster coat. The inner layer keeps the gypsum core dry and rigid. To identify the face surface, note that the outer layer is rolled over the long edge. Other than this, the materials and methods are nearly the same as those for regular drywall construction.

Veneer plaster can be applied as a one- or two-coat system. Either system can be given a smooth or a textured surface. Corner bead, trim, and grounds must be carefully set for a 1/16″ thickness in one-coat applications and 3/32″ for two-coat applications. **Figure 19-23** shows a two-coat application of veneer plaster.

The manufacturer's directions should be carefully followed for best performance and workability of veneer plaster. Proper mixing is especially important. This is usually accomplished with a cage-type paddle mounted in an electric drill.

19.7 Predecorated Wallboard

A variety of predecorated gypsum wallboard is available. This type of wallboard is usually vertically applied because of the difficulty involved in successfully matching and finishing butt joints.

Wall surfaces must be dry before installation can begin. Panels should be unpacked and stood on their long edges, exposing both sides to room air, for 24–48 hours before being attached.

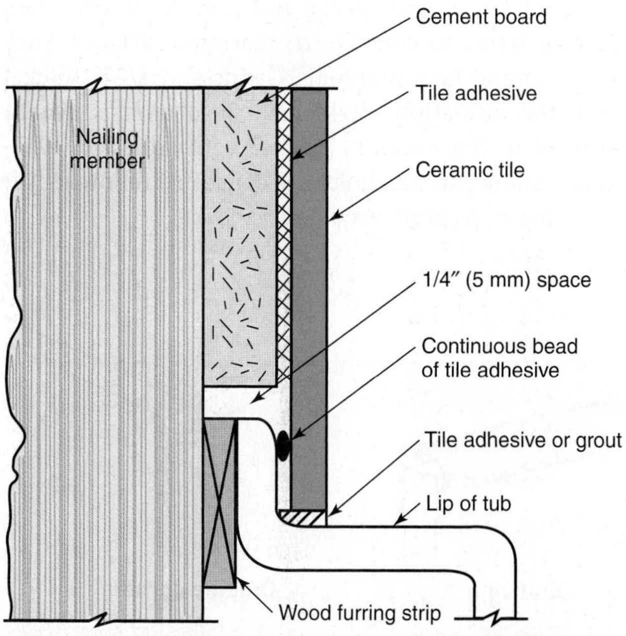

Cement board

Tile adhesive

Ceramic tile

Nailing member

1/4″ (5 mm) space

Continuous bead of tile adhesive

Tile adhesive or grout

Lip of tub

Wood furring strip

Gold Bond Building Products

Figure 19-22. Installation of cement board around a tub.

StockphotoVideo/Shutterstock.com

Figure 19-23. Application of a two-coat veneer plaster system. The first coat is being applied with a standard trowel.

The panels can be attached to furring strips, studs, or other solid surfaces. On remodeling jobs, first remove wallpaper and loose paint. Also, repair any damaged plaster. The use of an adhesive to bond the panels to a base layer is common practice. However, color-matched nails are available, as well. For best results, manufacturers recommend both gluing and nailing for 1/4″ panels. To avoid damaging the finish, be sure to drive colored nails with a plastic-headed hammer, rawhide mallet, or a special cover placed over the face of a regular hammer. Nails should be spaced 8″ apart and should never be closer than 3/8″ to the ends or edges of the wallboard. Avoid a tight fit at floors and ceilings. An expansion space of 1/16″ should be allowed to avoid buckling.

When adhesives alone are used for fastening, check each panel after about 15–30 minutes to ensure that the adhesive is set. Firmly press along the edges and framing members. Use a rubber mallet or cover a block of wood with a soft cloth and tap along all areas where adhesive was applied. Adhesive on the decorated surface must be immediately removed with a soft cloth and mineral spirits.

To trim edges and joints of predecorated panels, you can use aluminum or plastic trim made to match the finished surface of the wallboard, **Figure 19-24**. Cut the trim with a hacksaw. Attach it with flat-head wire nails spaced 8″–10″ apart. When attaching divider strips, first place the trim on one panel that is carefully aligned and nail the exposed flange in place. Then insert the next panel. **Figure 19-25** shows a completed panel installation.

19.8 Wallboard on Masonry Walls

Gypsum wallboard can be installed over metal or wood furring strips attached to a masonry wall. Where the structure consists of an interior wall that is straight

ABTco, Inc.

Figure 19-25. This bathroom has walls with prefinished paneling.

and true, the panels can be directly laminated to the masonry surface with a special adhesive.

Exterior walls must be thoroughly waterproofed. Insulation should be included if the structure is located in a cold climate. **Figure 19-26** shows a masonry wall application made with furring strips. The insulation may be rigid foam or batts. It is best to use a powder-actuated nailer and special fasteners to attach wood furring strips to concrete or masonry surfaces. Such strips should be a nominal 2″ wide and 1/32″ thicker than the insulation. Rigid foam insulation is usually bonded to the masonry surface with adhesive. Wallboard joints and nail holes are concealed following the finishing steps previously described.

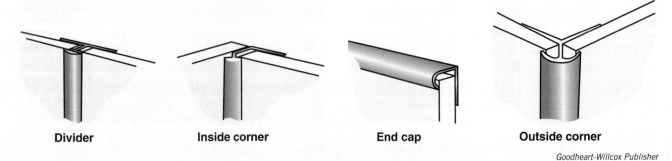

Divider   Inside corner   End cap   Outside corner

Goodheart-Willcox Publisher

Figure 19-24. Trim can be used to cover raw edges of predecorated gypsum wallboard.

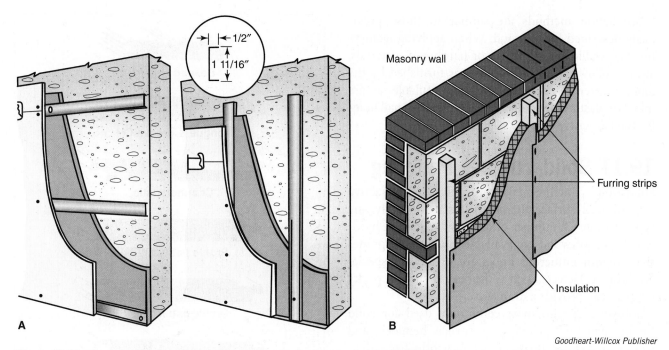

Goodheart-Willcox Publisher

Figure 19-26. Two methods of preparing masonry walls for interior finish. A—Wallboard can be attached to metal furring channels. Rigid insulation is used. B—Wood furring strips and blanket insulation.

Installations of wallboard on interior masonry walls, especially those below grade, must be carefully done. Be sure to follow recommendations furnished by manufacturers.

19.9 Installing Furring

When the wall framing is irregular or when framing members are spaced too far apart for normal attachment of wallboard, furring can be attached to the wall or ceiling frame to provide a good structure to carry the wallboard. Furring is generally low-grade softwood in narrow strips.

Horizontally nail 1×3 or 1×4 furring strips across the studs. Start at the floor line and continue up the wall. Spacing depends on the panel thickness. Thin panels need more support. Install vertical strips every 4′ along the wall to support panel edges. Level low areas by shimming behind the furring strips. Use a line stretched from corner to corner to locate low spots.

19.10 Hardboard

Through special processing, *hardboard* (also called fiberboard) can be manufactured with a low moisture-absorption rate. The face is often scored to form a bead-board pattern for use as wainscoting. Originally, wainscoting was wood panels or boards covering the bottom portion of walls to protect the plaster.

Today, wainscoting is used to create a certain architectural look. **Figure 19-27** shows decorative hardboard used for wainscoting. Panels for wall application are usually 1/4″ thick.

Since hardboard is made from wood fibers, the panels slightly expand and contract with changes in humidity. Panels should be installed when they are at their maximum size. If installed when moisture content is low, there will be a tendency for panels to buckle between the studs or attachment points. Manufacturers of prefinished hardboard panels recommend the panels be unwrapped and then separately placed around the room for at least 48 hours before installation.

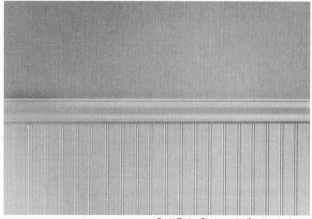

Brett Taylor Photography/Shutterstock.com

Figure 19-27. This prefinished hardboard wainscoting is made to look like pine.

Installation methods are similar to those previously described for plywood. When applying factory-finished wallboard, plywood, or hardboard materials, always follow the recommendations furnished by the manufacturer. Special adhesives are available, as are metal or plastic trim in matching colors. Drill nail holes for the harder types.

19.11 Solid-Lumber Paneling

Solid-wood paneling makes a durable and attractive interior wall surface. It may be used in nearly any type of room. A number of different species of hardwood and softwood are available. Sometimes, grades that contain numerous knots are used for a special appearance. Defects, such as the deep fissures in pecky cypress, can provide a dramatic effect.

Softwood species most commonly used for solid lumber paneling include pine, spruce, hemlock, and western red cedar. Boards range in widths from 4″ to 12″ (nominal size) and are dressed to 3/4″ thickness. Board-and-batten or shiplap joints are sometimes used, but tongue-and-groove (T & G) joints combined with shaped edges and surfaces are more popular.

Paneling patterns are often reversible, offering two choices in a single panel. Some patterns are smooth on one side and saw-textured on the other. **Figure 19-28** shows several patterns offered by one manufacturer. Dozens of variations are possible, varying by species of wood, texture, finish, and how the paneling is applied to a wall.

As with plywood paneling, allow the boards to adjust to the temperature and humidity of the room before installation. Stand them around the room. At the same time, match the boards for color and grain. If tongue-and-groove boards are to be stained or finished later, apply the same finish to the tongues. Unless this is done, later shrinkage of the wood will expose the unfinished surface of the tongue.

When solid-wood paneling is horizontally applied, furring strips or blocking is not required. The boards are directly nailed to the studs. Inside corners are formed by butting the paneling units flush with the other walls.

Vertical installations require furring strips at the top and bottom of the wall and at various intermediate spaces. Sometimes, 2×4 blocking is installed between the studs to serve as a nailing base. Even when heavy tongue-and-groove boards are used, these nailing members should not be spaced more than 48″ apart.

Narrow widths (4″–6″) of tongue-and-groove paneling are blind nailed. This eliminates the need for countersinking and filling nail holes. It also provides a

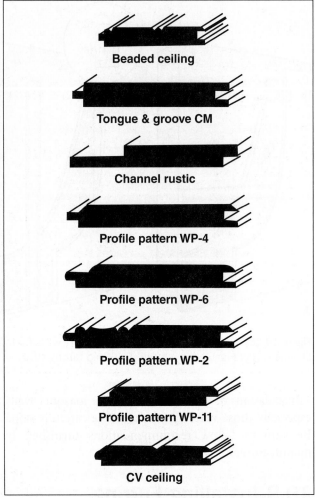

Beaded ceiling

Tongue & groove CM

Channel rustic

Profile pattern WP-4

Profile pattern WP-6

Profile pattern WP-2

Profile pattern WP-11

CV ceiling

Western Wood Products Assn.

Figure 19-28. Smooth-surfaced solid-wood paneling is available in several pattern profiles.

smooth, blemish-free surface. This is especially important when clear finishes are used.

Exterior wall constructions where the interior surface consists of solid-wood paneling should include a tight application of building paper or housewrap. This will prevent the infiltration of wind and dust through the joints. In cold climates, insulation and vapor barriers are important.

Pro Tip
If random widths are used, boards on adjacent walls must match and be accurately aligned.

19.11.1 Installing Solid Paneling at an Angle

Though more difficult and time-consuming to install, angled paneling is attractive and goes well with informal designs. *Diagonal paneling* angles in only one direction.

Angles of 22 1/2°, 30°, and 45° are most popular. *Chevron paneling* is installed in a V or inverted V pattern. *Herringbone paneling* alternates the direction of the angle at regular intervals. All three types are shown in **Figure 19-29** along with tables on coverage.

To determine the proper angle for diagonal paneling, first establish a vertical line in the middle of the wall using a plumb line. Draw an intersecting horizontal line about one foot off the floor. Mark 3′ from the intersection in each direction (horizontal and vertical).

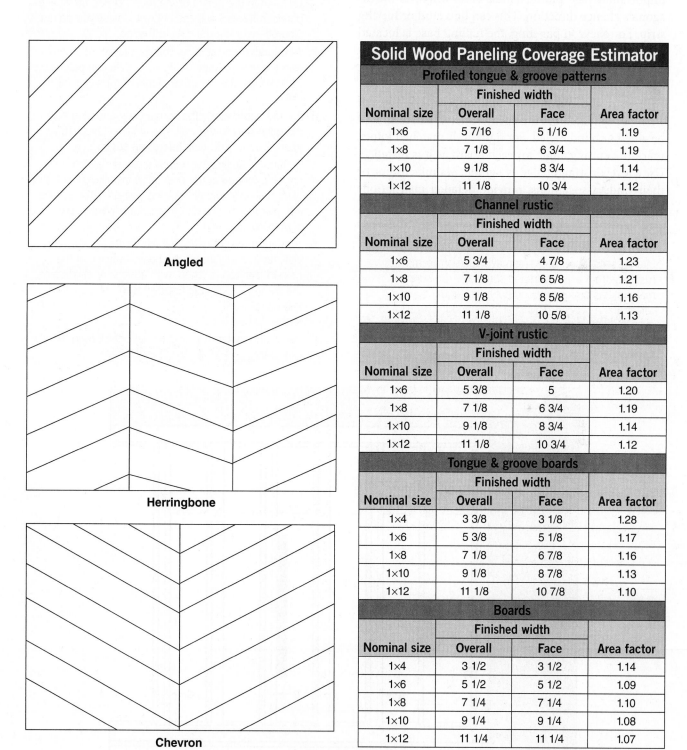

Angled

Herringbone

Chevron

Solid Wood Paneling Coverage Estimator

Profiled tongue & groove patterns

Nominal size	Overall	Face	Area factor
1×6	5 7/16	5 1/16	1.19
1×8	7 1/8	6 3/4	1.19
1×10	9 1/8	8 3/4	1.14
1×12	11 1/8	10 3/4	1.12

Channel rustic

Nominal size	Overall	Face	Area factor
1×6	5 3/4	4 7/8	1.23
1×8	7 1/8	6 5/8	1.21
1×10	9 1/8	8 5/8	1.16
1×12	11 1/8	10 5/8	1.13

V-joint rustic

Nominal size	Overall	Face	Area factor
1×6	5 3/8	5	1.20
1×8	7 1/8	6 3/4	1.19
1×10	9 1/8	8 3/4	1.14
1×12	11 1/8	10 3/4	1.12

Tongue & groove boards

Nominal size	Overall	Face	Area factor
1×4	3 3/8	3 1/8	1.28
1×6	5 3/8	5 1/8	1.17
1×8	7 1/8	6 7/8	1.16
1×10	9 1/8	8 7/8	1.13
1×12	11 1/8	10 7/8	1.10

Boards

Nominal size	Overall	Face	Area factor
1×4	3 1/2	3 1/2	1.14
1×6	5 1/2	5 1/2	1.09
1×8	7 1/4	7 1/4	1.10
1×10	9 1/4	9 1/4	1.08
1×12	11 1/4	11 1/4	1.07

Western Wood Products Assn.

Figure 19-29. Three patterns for angled solid paneling. Tables such as those at the right are used for estimating coverage of various patterns of solid paneling.

Draw a diagonal line from each point on the horizontal line through the point marked on the vertical line. The result is two 45° angles. For other angles, use a protractor to establish the diagonal lines.

For chevron and herringbone panel application, it is important to have plumb nailing bases wherever the diagonals change direction. This can be a stud or furring strip. For chevron paneling, the nailing base is located midway across the wall. For herringbone paneling, locate a plumbed vertical nailer every 36″, **Figure 19-30**. Center an additional nailing surface at 18″.

To start chevron and herringbone patterns, install triangles of paneling at the centerline(s). These should be glued or blind nailed. For tight, well-matched joints, mark the cutting angle for each board as it is installed. Use a level or straightedge as a guide. Manufacturers, or their trade associations, provide detailed instructions for such installations.

Pro Tip
Interior solid-wood paneling may cause problems resulting from expansion and shrinkage. Be sure the material has a moisture content about equal to what it will attain in service. This should be about 8%–10% for most regions of the United States. Prior to installation, store the paneling in the room and allow room air to reach both sides of the wood.

PROCEDURE
Installing Vertical Solid Paneling
1. Start installation at a corner. Choose a straight board and cut it slightly shorter—normally about 1/4″—than the height of the room. Divide the space between the top and the bottom, unless you do not plan on installing molding at the top.
2. With tongue-and-groove boards, plumb the board with the groove edge to the wall. Tack it in place while you scribe the edge to match the adjoining wall.
3. Rip the board along the scribed line and face nail the edge next to the wall. Use finishing nails 16″ apart. Then, blind nail the trailing edge.
4. Continue installing boards by slipping the groove onto the tongue of the previously installed board and blind nail only. Slightly warped boards can be pried into place with a wood chisel. Badly warped pieces should not be used.
5. To mark the last board so it tightly fits the adjoining wall, temporarily install the board in place of the next-to-last board. Then, with a block or dividers of the same width as the finished face of the next-to-last board, scribe a line. Remove the board and cut to the scribed line.
6. Install and fasten the next-to-last board, then the last board.

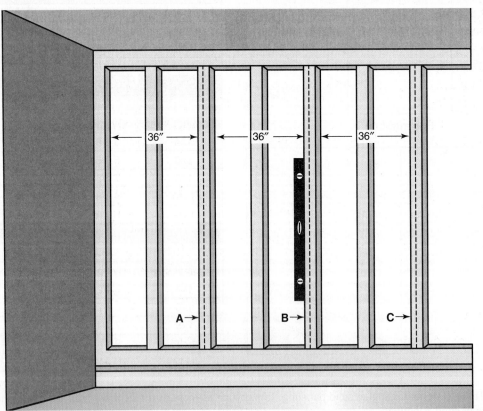

Western Wood Products Assn.

Figure 19-30. Herringbone paneling requires furring strips every 18″ OC. Plumb lines should be drawn every 36″ OC.

19.12 Plaster

Through the years, gypsum plaster has provided desirable qualities in a wall and ceiling finish such as beauty, durability, economy, fire protection, structural rigidity, and resistance to sound transmission. It is often used to achieve desired fire protection. It is also highly adaptable, since it can be readily applied to curved or irregular surfaces.

When plaster is used for the interior wall and ceiling surfaces, a carpenter usually installs grounds that serve as guides for the plasterer. *Grounds* are strips of wood or metal placed along floors and around wall openings as a thickness guide. A carpenter also applies the lath that forms the plaster base. For this reason, it is important for a carpenter to have a general knowledge of how the plaster coats are applied.

Pro Tip
Plaster will set up (harden) under water. For this reason, leftover plaster should never be discarded in a sink or toilet.

19.12.1 Plaster Base

A plaster finish requires some type of base. For many years, *wood lath* was used for this purpose. The lath consisted of thin, narrow strips of soft wood that were nailed to studs. The strips were spaced a short distance apart, so that the wet plaster could be forced through the gaps. As the wet plaster was pushed through, it slumped downward to form a key. When dry, this key firmly held the plaster to the lath. Other materials have generally replaced wood lath, but carpenters may encounter it in remodeling jobs on older buildings. Today, commonly used plaster bases include gypsum lath and expanded metal lath, **Figure 19-31**. Plaster is also sometimes applied directly to masonry surfaces or special gypsum block units.

Gypsum lath consists of a rigid gypsum filler with a special paper cover. Panels measure 16″ × 48″, 16″ × 96″, and 48″ × 96″ and are horizontally applied to the framing members of the structure. For a stud or ceiling joist spacing of 16″ OC, 3/8″ gypsum lath is used. For a 24″ OC spacing, 1/2″ gypsum lath is required. For maximum fire protection, 5/8″ lath may be used. Gypsum lath is also made with a backing of aluminum foil vapor barrier. Lath with perforations improves the plaster bond and extends the time the wall surface remains intact when exposed to fire. Some building codes specify this type.

Expanded metal lath consists of a copper alloy steel sheet that is slit and expanded to form openings for keying the plaster. The two most common types are diamond mesh and flat rib. Standard-size pieces are 27″ wide by 96″ long. To improve its rust-resistance, metal lath is usually galvanized or dipped in black asphaltum paint.

Insulating fiberboard lath is also used as a plaster base. It comes in a 3/8″ and 1/2″ thickness with a shiplap edge. Fiberboard lath has considerable insulation value and is sometimes used on ceilings or walls adjoining exterior or unheated areas. **Figure 19-32** shows the available sizes.

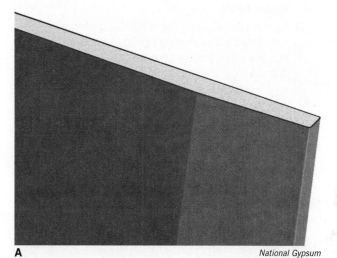

A   *National Gypsum*

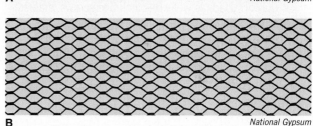

B   *National Gypsum*

Figure 19-31. Plaster base materials. A—Gypsum lath. B—Expanded metal lath.

Gypsum and Insulating Lath Sizes

Type	Thickness (inches)	Width (inches)	Length (inches)
Plain	3/8 1/2	16 16	48 or 96 48
Perforated	3/8 1/2	16 16	48 or 96 48
Insulating	3/8 3/8 or 1/2	16 24	48 or 96 as requested to 12′
Long length	1/2	24	as requested to 12′

Goodheart-Willcox Publisher

Figure 19-32. Gypsum and insulating lath are available in these sizes.

PROCEDURE

Installing Gypsum Lath

Before lath is applied, inspect the framing for proper spacing and alignment. Check to see that corners, ceiling lines, and openings have nailers to support the ends of the lath. Insulating lath is installed in the same manner as gypsum lath, except that 13 gauge 1 1/4″ blued nails should be used.

1. Install the ceiling and then the walls. Work from the top down on walls. Apply with the long dimension at right angles to the framing.

2. Stagger the end joints between adjacent courses, **Figure 19-33**.

3. Gypsum lath is easily cut to size by scoring one or both sides with a pointed or edge tool. Break the lath along the line. Be sure to make neatly fitted cutouts for plumbing pipes and electrical outlets.

4. Turn the folded or lapped paper edges of gypsum lath toward the framing. Edges and ends of lath should be in moderate contact.

5. Apply 3/8″ thick gypsum lath with 13 gauge gypsum lathing nails 1 1/8″ long. Nail sizes for other installations are given in **Figure 19-34**. The lath must be nailed at each stud or joist. Begin nailing from the center and work toward outer edges. As you nail, firmly press the lath against the frame. Drive nails flush without cutting the face paper. Keep nails 3/8″ away from ends and edges. Other types of fasteners may be used, as well. This includes screws, staples, and special clips. Staples should be flattened wire. Drive them so the crown lies in the same direction as (parallel to) the wood framing. Screws must always be used when attaching to steel frame members.

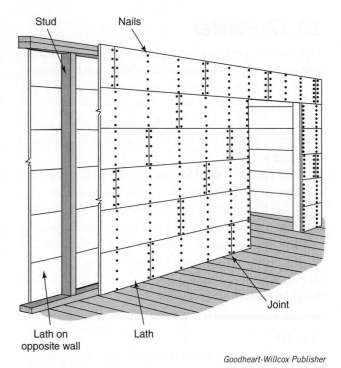

Goodheart-Willcox Publisher

Figure 19-33. Gypsum lath over wood studs. Stagger joints.

19.12.2 Metal Lath

In commercial construction, metal lath is common. In residential work, it is often used only in shower stalls or tub alcoves, **Figure 19-35**. When the construction is properly designed, the metal lath provides a rigid wall surface.

Metal lath contains many small voids. These openings allow proper keying of the plaster. Produced in different weights, each bundle has some identifying mark for weight. For example, one manufacturer uses a color code.

Fastening Requirements for Gypsum Lath								
Type framing	**Base thickness**		**Fastener**	**Max. frame spacing**		**Max. fastener spacing**		
Wood	inch	mm	Nails, 13 gauge, 1 1/8″ long, 19/64″ flat head, blued	inch	cm	inch	cm	
	3/8	10	Staples—16 gauge galvanized, flattened wire flat crown 7/16″ wide, 7/8″ divergent legs	16	40	5	13	
	1/2	13	Nails—13 gauge, 1 1/4″ long, 19/64″ flat head, blued	24	60	4	10	
			Staples—16 gauge galvanized, flattened wire flat crown 7/16″ wide, 1″ divergent legs					
USG steel stud	3/8	10	1″ Type S screws	16 24	40 60	12	30	
TRUSTEEL stud	3/8	10	Clips	16	40	16	40	

Goodheart-Willcox Publisher

Figure 19-34. Follow these fastening requirements for gypsum lath.

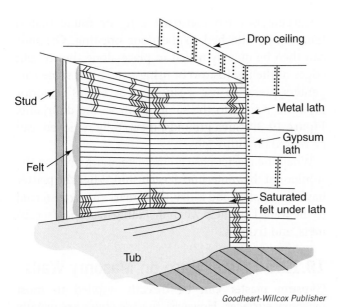

Figure 19-35. Metal lath is installed over a layer of waterproof felt paper in areas subject to moisture. The paper can be eliminated if cement board is used under the lath.

Ends of bundles are spray painted, 3.4 lb in red, 2.5 and 2.75 lb in white. Diamond mesh 1.75 lb is not painted.

Metal lath must be of the proper type and weight for the support spacing. Sides and ends are lapped and corners are returned (overlapped). Studs are usually first covered with a 15 lb asphalt-saturated felt. Portland cement plaster is often used as the first coat when the surface will be finished with ceramic tile. Gypsum plaster is used for top coats.

19.12.3 Reinforcing

Since some drying always occurs in wood frame structures, shrinkage can be expected. This is likely to cause plaster to crack around openings, in corners, or wherever there is a concentration of cross-grained wood.

To minimize cracking, expanded metal lath is often used in key positions over the plaster base. Strips 8″ wide are applied at an angle over the corners of doors and windows as shown in **Figure 19-36**. Lightly tack or staple these into place so they become a part of the plaster base only. If nailed securely to the framing, warping, shrinking, and twisting of the frame will be transmitted into the plaster and cause cracks.

Metal lath should also be used under and around wood beams that will be covered with plaster. Be sure to extend the edges of the reinforcing well beyond the structural element being covered.

Inside corners may be reinforced with a specially formed metal lath or wire fabric sometimes called Cornerite™ **Figure 19-37**. Minimum widths should be 5″ so that there is 2 1/2″ on each surface of the internal angle.

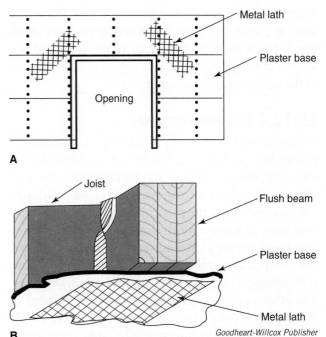

A

B

Figure 19-36. Reinforcing the plaster base. A—Place metal lath around the jamb area of large openings, especially where large headers are used. B—Provide reinforcing under flush beams.

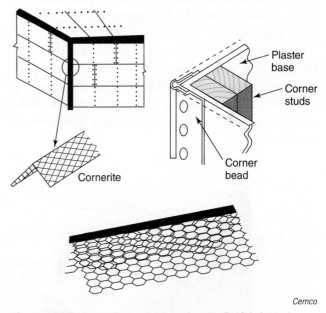

Figure 19-37. Cornerite and corner beads. Prefabricated reinforcing is made for both inside and outside corners.

In some plaster base systems that employ a special attachment clip, Cornerite is not recommended.

Outside corners where no wood trim will be applied are reinforced with metal corner beads. They must be carefully applied and plumbed or leveled. This is because they serve not only as a reinforcement for the plaster, but as a ground (guide) for its application.

When applying corner bead, use a straightedge and level or plumb line. Also, use a spacer block to check the distance between the corner of the bead and the surface of the plaster base. This distance must be equal to the thickness of the plaster coats.

19.12.4 Plaster Grounds

Plaster grounds are usually wood strips as thick as the plaster base and plaster. For average residential construction, this is 3/8″ (for the plaster base) plus 1/2″ (for the plaster) for a total thickness of 7/8″. Grounds are installed before the plaster is applied. The plasterer uses them as a gauge for thickness of the plaster. They also help keep the plaster surface level and even. Later, the grounds may become a nailing base for attaching trim members. They are used around doors, windows, and other openings. Sometimes, they are included at the bottom of walls along the floor line, **Figure 19-38**. In some wall systems—especially large commercial and institutional buildings—metal edges and strips serve as grounds.

Windows are usually equipped with jamb extensions that are adjusted for the various thicknesses of materials used in the wall structure. These jamb extensions serve as plaster grounds. A carpenter seldom needs to make any changes.

Grounds around some openings are removed after the plastering is complete. Those used at door openings must be carefully set (plumbed) and conform to the width of the doorjamb to ensure a good fit of the casing. The width of standard interior doorjambs is usually 5 1/4″.

Carpenters often construct a jig or frame that is temporarily attached to the door opening. Grounds can then be quickly nailed in place along the straight edges of the jig. Instead of using two strips, some carpenters prefer to use a single piece of 3/8″ exterior plywood ripped to the same width as the doorjamb, **Figure 19-39**. Such grounds, if carefully removed, can be reused on future jobs.

When the plaster base is complete, mark lines on the subfloor at the centerline of each stud. After the plaster has been applied and is dry, these marks are transferred to the wall. They help when installing baseboards, cabinets, and fixtures.

19.12.5 Plaster Base on Masonry Walls

Gypsum plaster can be directly applied to most masonry surfaces. However, furring strips are usually installed on outside walls to prevent excessive heat loss through the walls. These carry the plaster base materials and provide an air space. Strips are attached in one of the following ways:

- Driving case-hardened nails into the mortar joints or using a powder-actuated nail gun
- Nailing into metal or wood plugs placed in the wall during construction
- Applying adhesive

When heat loss needs to be further reduced, blanket insulation can be installed. Some carpenters use rigid polystyrene panels. These are directly attached to the wall with adhesives. If furring strips are not used, the plastic foam can serve as a plaster base. Follow manufacturer's recommendations when making an installation of this kind.

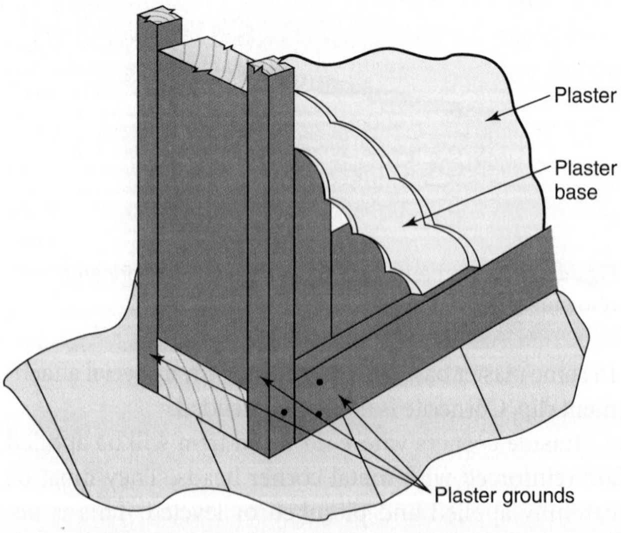

Figure 19-38. Plaster grounds are installed around openings and sometimes along the floor.

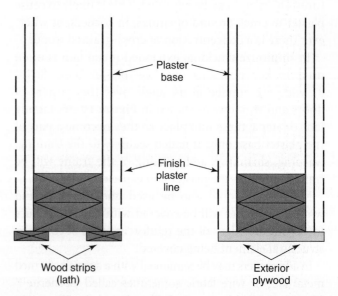

Figure 19-39. Two methods of installing removable grounds at door openings. These are top views.

19.12.6 Plastering Materials and Methods

Plaster is applied in two or three coats. The base coats are prepared by mixing gypsum with an aggregate, either at the gypsum plant or on the job. The aggregate may be wood fibers, sand, perlite, or vermiculite. Sand is the most commonly used aggregate.

Discard plaster that has started to set. Wash out the mixer with clean water after each batch has been prepared. Keep tools and equipment clean.

In three-coat work, the first application, called the **scratch coat**, is directly applied to the plaster base. It is cross-raked, or scratched, after having "taken up" (stiffened), **Figure 19-40**. The scratch coat is then allowed to set and partially dry. The second application, or **brown coat**, is then applied and leveled with the grounds and screeds. A long flat tool called a **darby** and a rod (straightedge) are used. When the brown coat has set and is somewhat dry, it is time to apply the third or **finish coat**. This coat is about 1/16″ thick.

In two-coat work, the scratch coat and brown coat are applied at almost the same time. The cross raking of the scratch coat is omitted. The brown coat of plaster is usually applied (doubled-back) within a few minutes. This application method is the one most frequently used over gypsum or insulating lath plaster bases in residential construction.

The minimum plaster thickness for all coats is 1/2″ when applied to regular gypsum or insulating lath bases. A 5/8″ thickness is usually required over brick, tile, or masonry. When plaster is applied to metal lath, it should measure 3/4″ in thickness from the backside of the lath.

A plaster job should be constantly inspected for base coat thickness. Unless grounds and screeds are used on ceilings or large wall areas, it is extremely difficult to keep the thickness uniform. Should the thickness be reduced, the possibility of checks and cracks is much greater. A 1/2″ thickness of plaster possesses almost twice the resistance to bending and breaking as a 3/8″ thickness.

The final, or finish, coat consists of two general types: sand-float (textured) and putty (smooth). In the sand-float finish, special sand is mixed with gypsum or lime and cement. After the plaster is applied to the surface, it is smoothed with a float to produce various effects. The final finish depends on the floating method and the coarseness of the sand. A smooth finish is produced by applying a putty-like material consisting of lime and gypsum or cement. It is troweled perfectly smooth, like concrete, **Figure 19-41**.

National Gypsum Co.

Figure 19-41. A putty-like mixture of lime and gypsum or cement is used for the finish coat of plaster. It is troweled to an even thickness with a smooth finish.

National Gypsum Co.

Figure 19-40. Cross-raking a scratch coat of plaster. The raking helps the brown coat adhere to the scratch coat.

Pro Tip
Sometimes, a plastering machine is used instead of hand troweling. It not only saves a great deal of labor but also improves the quality of the plaster application. The lapsed time between mixing and application is shortened. The machine also makes possible the control of the plaster coat's density.

19.13 Tiles

Ceramic tile and stone are often used on walls and floors in bathrooms and on countertops and backsplashes in kitchens. These materials can make a waterproof, long-wearing, attractive surface, **Figure 19-42**.

There are four kinds of tiles, including the aforementioned ceramic, that are widely used in both residential and commercial buildings:

- Ceramic and porcelain—These tiles are made of clay, minerals, and water and are fired in a kiln to make a strong ceramic material. They may be glazed, covered with a glass-like coating that is fired to make an attractive, water-repellant surface. They may also be left unglazed for a more natural appearance. The difference between porcelain and regular ceramic tile is primarily in the temperature to which it is fired in manufacturing. Firing porcelain at a higher temperature makes a denser, more water-resistant material. Porcelain is generally more expensive than regular ceramic title. Ceramic and porcelain tile are available with a wide range of patterns and colors. They come in sizes from $1'' \times 1''$ to $24'' \times 24''$ and irregular shapes that fit together to form interesting patterns.

- Quarry—This tile is made from shale and natural clays, is typically reddish-brown and unglazed, and is used in high-traffic settings. Although reddish-brown is the most common color, it is also available in tan, grey, and deeper brown. Many commercial buildings use quarry tile because it hides stains well and is very durable.

- Glass—Glass tile is just that, glass made to be used as wall tiles. It is waterproof and generally translucent in color. This creates a very different appearance from other tile products. Glass tile is often used for backsplashes and wall tile, but it is not suitable for floors because it is not as tough as other tile products.

Arsel Ozgurdal/Shutterstock.com

Figure 19-42. Ceramic tile makes an attractive, water-resistant, durable surface.

- Natural stone—There are many varieties of natural stone that can be produced as tiles. The most common are marble and granite. Each variety of stone has different properties. Marble, for example, is attractive but does not wear as well as some others.

19.13.1 Preparing for Tile Installation

Carpenters do not usually lay tile, but they are often called upon to prepare a surface for tiling. Even the toughest tiles will not flex. If tile adheres well to a surface that tends to flex as it is walked on or as equipment sends vibrations through the structure, the tile is likely to crack or separate from its backing. The framing under a tile floor must be sufficient to support the floor and prevent such flexing. Plywood and OSB tend to expand and contract as their moisture content changes. To prevent such problems, cement board or

other tile backing is applied to the surface to be tiled, **Figure 19-43**. Waterproofing and uncoupling membranes can be applied directly to gypsum and subflooring. They are practical and effective alternatives to cement board.

Cement board can be cut by scoring the face with a carbide-tipped scoring knife, **Figure 19-44**, and a straightedge, then snapping the board onto the scored line. Make several passes with the scoring knife to create a pronounced score line. For cutouts and small

Goodheart-Willcox Publisher

Figure 19-44. A scoring knife has a carbide cutting edge. A utility knife can be used to score cement board, but it will dull quickly.

Goodheart-Willcox Publisher

Figure 19-43. Tile must have a strong, stable base. A—Cement board or other backing intended for use as tile backing makes a stable base for tile. B—Waterproofing tile membranes can be applied directly over gypsum board. C—Decoupling membranes can be used on floors.

holes, score around the cutout and break out from the face side with a hammer.

Safety Note

Do not cut cement board indoors with a power saw. The dust from sawing cement board contains silica dust, which is harmful if inhaled. Wear a dust mask when cutting cement board.

19.14 Ceiling Tile

Ceiling tiles are found in both old and new construction. They can be installed over engineered metal strips, wood furring strips, solid plaster, drywall, or any smooth, continuous surface. There is a considerable range in material used to make ceiling tile. Some types are fiberboard, mineral, perforated metal, and glass-fiber. When selecting a product, consider its appearance, light reflection, fire resistance, sound absorption, maintenance, cost, and ease of installation.

A standard size tile is 12″ × 12″. However, the tiles are available in larger sizes; for example, 24″ × 24″ and 16″ × 32″. A wide range of surface patterns and textures is manufactured.

Figure 19-45 illustrates an overlapping, tongue-and-groove edge that provides a wide flange to receive staples. This type of joint also permits efficient installation when an adhesive is used to hold the tile in place.

PROCEDURE

Layout Procedure

1. Measure the two short walls and locate the midpoint of each one.
2. Snap a chalk line to establish the centerline. In the middle of this line, establish a chalk line that is at right angles to the long centerline. All tiles are installed with their edges parallel to these lines.
3. For an even appearance, the border courses along opposite walls should be the same width. For example, in a room 10′-8″ wide, use nine full tiles and two border tiles trimmed to 10″ each.
4. After determining the width of the border tile, snap chalk lines parallel to the centerlines to provide a guide for the installation of the border tiles or the furring strips that will support them.

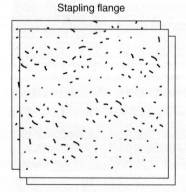
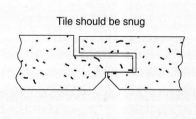
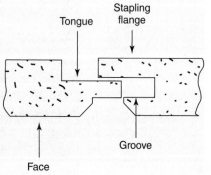

Slide tile into position to make snug joint; do not force or hammer

Stapling flange

Tile should be snug

Tongue / Stapling flange / Groove / Face

Figure 19-45. The tongue-and-groove joint used on standard ceiling tile. Flanges receive staples, while the groove holds one edge of the next tile.

19.14.1 Furring

Furring should be directly nailed to the ceiling joists on new construction. If applied over an old ceiling surface, be sure to locate and mark the joists before attaching the strips. Place the first furring strip flush against the wall at a right angle to the joists. Nail it with two 8d nails at each joist. Nail the second strip in place so that it will be centered over the edge of the border tile. Use the line previously laid out. All other strips are then installed on center. For 12″ × 12″ tile, locate strips every 12″. **Figure 19-46** shows completed furring.

Some carpenters prefer to start from the center and work each side toward the walls. If the furring strips are uniform in width, a spacer jig may speed up the job. Double-check the position of the furring strips from time to time. Make sure they will be centered over the tile joints.

It is essential that the faces of the furring strips be level with each other. Check alignment with a carpenter's level, straightedge, or line stretched across the strips. To align strips, drive tapered shims between the strips and the joist. If only one or two joists extend below the plane of the others, it may be best to notch these joists before installing the furring strips.

If pipes or electrical conduit are located below the ceiling joists, it may be necessary to double the furring strips. The first course is spaced 24″–32″ OC and attached directly to the joists. The second course is then applied at a right angle to the first. Space strips according to the width of the tiles, as previously described. Large pipes or ducts that project below the ceiling joists should be boxed in with furring strips before the tile is installed. Wood or metal trim can be used to finish corners and edges.

There is an alternative to furring. It consists of installing gypsum wallboard to the joists or an existing ceiling surface. The tile is then attached to the wallboard either with adhesive or special staples.

Pro Tip

Most manufacturers recommend that fiberboard tile be unpacked in the area where it will be installed at least 24 hours before application. This will allow the tile to adjust to room temperature and humidity.

PROCEDURE

Installing Ceiling Tile

1. Carefully check over the furring.
2. Snap chalk lines on the strips to provide a guide for setting the border tile. Do this along each wall. Be certain the lines are correct. Double-check to see that the chalk lines form 90° angles at the corners.
3. To cut the border tile, first deeply score the face with a knife drawn along a straightedge. Break it along this line by placing it over a sharp edge. Make any irregular cuts around light fixtures or other projections with a coping or compass saw. Power tools can also be used. Some tiles are made of mineral fiber, which rapidly dulls regular cutting edges.
4. Start the installation with a corner tile and then set border tile out in each direction. Fill in full-size tile.
5. When you reach the opposite wall, trim the border tile to size. If border tiles are full-width, remove the stapling flanges and face nail the tile. Locate these nails close to the wall so they will be covered by the trim.

19.14.2 Adhering Ceiling Tile

Figure 19-47 shows tile being attached with a stapler. Be sure the tile is correctly aligned. Firmly hold it while setting the staples. For 12″ × 12″ tile, use three staples along each flanged edge. Use four staples for 16″ × 16″ tile. Staples should be at least 9/16″ long.

Goodheart-Willcox Publisher

Figure 19-46. A completed furring installation.

Lakeview Images/Shutterstock.com

Figure 19-47. Staples can be used to attach ceiling tile to furring strips.

There are several types of adhesive designed especially for installing ceiling tile. The thick-putty type is applied in daubs about the size of a walnut. Apply adhesive on the back at each corner of 12″ × 12″ tile, about 1 1/2″ away from each edge. Position the tile, then slide it back against the adjoining tile. This motion, along with firm pressure, spreads the adhesive so the daubs are about 1/8″ thick. Some adhesives are thinner and applied with a brush.

In remodeling work, be sure the old ceiling is clean and that any paint or wallpaper is adhering well. Always follow the recommendations and directions provided by the manufacturer of the products being used.

Pro Tip
Be sure to keep your hands clean while handling ceiling tile.

19.14.3 Metal Track System

Another ceiling tile system uses 4′-long metal tracks to replace the wood furring strip. The track is nailed or screwed to the old ceiling or to joists at 12″ OC intervals, **Figure 19-48**. Tongue-and-groove panels are slipped into place. A clip snapped into the track slides over the tile lip. No other fasteners are used, **Figure 19-49**.

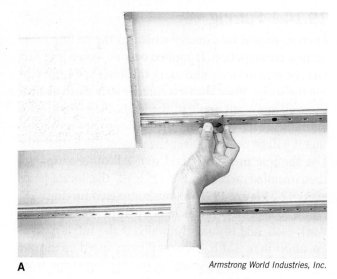
A   *Armstrong World Industries, Inc.*

B   *Armstrong World Industries, Inc.*

Figure 19-49. This track system uses clips to hold the ceiling tiles. Use one clip for each tile. Border tiles must have two clips. A—Snap the clip into the track. B—Snugly slide the clip against the tile.

Armstrong World Industries, Inc.

Figure 19-48. Attach the predrilled track to the ceiling at 12″ intervals.

19.15 Suspended Ceilings

When heating ducts and plumbing lines interfere with the application of a finished surface, a *suspended ceiling* is practical. In other instances, it provides a simple way of lowering high ceilings. The installation consists of a metal framework designed to support tile or panels, **Figure 19-50**. The framework is mainly supported by wires tied to the building structure directly above.

The height of the suspended ceiling must first be determined. Then, trim must be attached to the perimeter of the room. Use a level chalk line or a laser level as a guide, **Figure 19-51**.

Carefully calculate room dimensions, lay out positions of the main runners, and then install screw eyes 4′ OC in the existing ceiling structure. The panels next

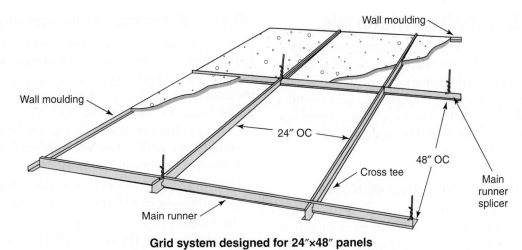

Grid system designed for 24″×48″ panels

U.S. Gypsum Co.

Figure 19-50. This sketch is typical of the framework used for suspended ceilings.

Spectra-Physics Laserplane, Inc.

Figure 19-51. This installer is using a laser level to plumb the suspended ceiling framework.

to the wall (border panels) may need to be reduced in width to provide an arrangement that is symmetrical (the same on both sides of room). Plan the layout as previously described for regular ceiling tile.

Install the main runners by resting them on the wall trim and attaching them to wires tied to the screw eyes. Use chalk lines or string stretched between the wall trim to ensure a level assembly. Sections of runners are easily spliced. Odd lengths can be cut with a fine-tooth hacksaw or aviation snips.

After all main runners are in place, recheck the level and adjust the wires as necessary. Next, check the required spacing for tiles or panels and install the cross tees. Simply insert the end tab of the cross tee into the runner slot.

When the suspension framework is complete, panels can be installed. Each panel is tilted upward and turned slightly on edge so it will "thread" through the opening.

After the entire panel is above the framework, turn it flat and lower it onto the grid flanges.

A concealed suspension system is shown in **Figure 19-52**. The tongue-and-groove joint is similar to the joint on regular ceiling tile. Joints are interlocked with the flange of the special runner when the installation is made.

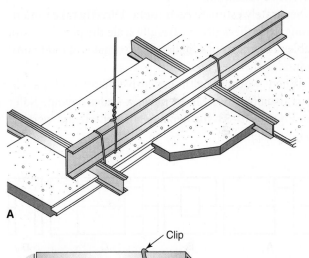

A

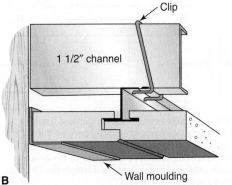

B

Goodheart-Willcox Publisher

Figure 19-52. A concealed suspension system for ceiling tile. A—Channels of the suspension system are concealed by the ceiling panels. B—Tongue-and-groove joint detail for the panels.

19.16 Estimating Materials

The amount of material required to cover walls or ceilings is determined by first calculating the total number of square feet of wall or ceiling area. Regular-size door and window openings are disregarded. This provides an allowance for waste. However, larger openings, such as window walls or picture windows, should be subtracted.

19.16.1 Determining Area of Rooms

The ceiling area is usually the same as the floor area. It is much easier to take floor dimensions and multiply the length times the width. To find wall area, add all of the wall lengths and multiply by the wall height.

19.16.2 Sheet Materials

When ordering sheet materials such as wallboard or paneling, be sure to specify the length. Always plan to use the longest practical sheet. This will keep the number of butt joints to a minimum or may even eliminate them. Divide the total length of the walls by the width of the sheets to find the number of sheets needed for a vertical application.

Separately estimate each room. Directly take dimensions from the walls or carefully scale the plan. Consult tables and charts of suppliers for estimates of joint compound, adhesives, and nails.

When working with expensive hardwood panels, it is usually advisable to make a scaled layout. Refer to **Figure 19-53** for typical horizontal and vertical arrangements.

19.16.3 Estimating Solid Paneling

Estimates for solid paneling are based on its nominal (unfinished) size. Seasoning and planing bring it to its dressed size. Forming of joints may further reduce the actual face widths. For example, a 1×4 tongue-and-groove board has a face width of 3 1/8″.

First, calculate the square footage of the wall to be covered. Then, multiply by factors taken from lumber tables such as the one in **Figure 19-32**:

- For 1×4 tongue-and-groove boards, use 1.28.
- For 1×6 tongue-and-groove boards, use 1.17.

On standard vertical applications, add 5% for waste when required lengths can be selected or when the lumber is end-matched.

19.16.4 Estimating Gypsum Lath

Since gypsum lath is produced in sections smaller than full sheets, you will need to use a different method to figure the needed quantity. First, determine the area of the ceiling and add to this the area of the walls. For example:

$$\begin{aligned}
\text{Ceiling area} & \\
\text{(same as floor area)} &= 1250 \text{ sq ft} \\
\text{Total length of walls} &= 14' + 12' + 14' + 12' + 10' \\
&= 62' \\
\text{Wall area} &= 62' \times 8' \\
&= 496 \text{ sq ft} \\
\text{Total area} &= 496 + 1250 \\
&= 1746 \text{ sq ft}
\end{aligned}$$

Divide this by the number of square feet in a sheet of the lath you are using. If you are using 16″ × 48″ (1 1/3′ × 4″)

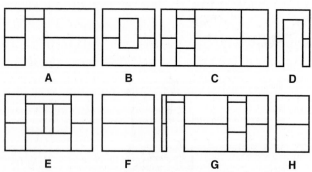

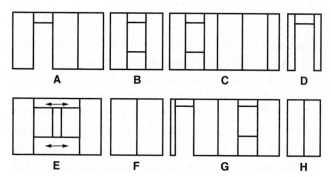

In design 1, a simple two-panel horizontal arrangement has the single continuous horizontal joint placed midway between the floor and ceiling. The panel design is defined by the vertical joints at openings, such as those in elevations A, C, E, and G. However, key panels may be omitted where the panel and length exceed the wall element width, as in elevations B and D.

In design 2, a vertical-arrangement panel design involves lining up vertical joints with wall openings. Then, the plain wall space is vertically divided in widths proportionate to the openings. When the width of a door or window opening exceeds the panel width, panels may be placed horizontally, as shown by arrows in elevation E.

Goodheart-Willcox Publisher

Figure 19-53. Two commonly used arrangements for laying out wall panels. The basic rule is to "work from the openings." First, line up vertical joints above doors and above and below windows. Divide the remaining plain wall space into an orderly pattern as stud location allows.

lath, each piece has an area of 5.3 sq ft. The job will require 329.4 or 330 pieces of lath.

Plasterers usually base their prices and estimates on the number of square yards to be covered. To convert square feet to square yards, divide by 9 (1 sq yd = 9 sq ft).

19.16.5 Estimating Ceiling Tile

Quantities of ceiling tile are estimated by figuring the area (square footage) to be covered. Round off any fractional parts of a foot in width and length to the next larger full unit when making this calculation.

Add extra units when it is necessary to balance the installation pattern with border tile along each wall. When using 12″ × 12″ tile, the number required equals the square footage plus the extra allowance described. Standard 12″ × 12″ ceiling tile are packaged 64 sq ft to a carton.

Construction Careers
Plasterer

Plastering is one of the oldest building trades, dating back at least to the ancient Romans. The plasterer applies interior wall finish material in several coats to form a smooth, hard finish. The finish material is a mixture of water and lime- or gypsum-based plaster. The plaster is applied over a solid surface, such as concrete block, or a perforated metal material called lath that is attached to wall studs. Plasterers also may use gypsum wallboard (drywall) as a base, applying a thin "skim coat" of lime-based plaster as a smooth, abrasion-resistant finish.

Stucco masons are members of a related trade. They apply exterior finish coats to buildings in much the same way that plasterers cover interior walls. Their material, however, is a weather-resistant mixture of Portland cement, lime, and water. Plasterers and stucco masons also apply another type of exterior finish, known as an exterior insulated finishing system (EIFS), that has achieved wide acceptance in recent years. The EIFS consists of rigid foam insulation board and a reinforcing mesh attached to building walls, then covered with polymer-based finish similar to stucco. The polymer coat can be finished to mimic various types of stone or other traditional wall materials.

Plasterers and stucco masons must be in good physical condition. Their work involves standing for long periods and engaging in fairly strenuous physical activity, especially involving arm and shoulder muscles. Working conditions may be dusty and the materials used can cause skin, eye, or respiratory irritation. Personal protective equipment should be worn where appropriate.

Most plasterers and stucco masons work for small-to-medium-sized specialty contractors. Fewer than 10% are self-employed.

While some apprenticeship programs are available, most plasterers and stucco masons learn their skills on the job. They start as helpers, carrying and mixing materials, setting up scaffolding, and cleaning tools and the work site. They learn their job skills from experienced tradespeople, gradually assuming more responsibility. Apprentice training, where available, typically lasts 2–3 years and involves at least 144 hours of classroom work in addition to on-the-job training.

Revel Pix LLC/iStock/Thinkstock

Before repairing a ceiling in an older home, this plasterer is removing the existing plaster to expose the wood lath.

Construction Careers
Drywall Installer

Gypsum wallboard, generally referred to as drywall, has replaced plaster as a wall and ceiling finish in many applications. The wide acceptance of drywall, with the resulting employment opportunities for drywall installers, is due to its lower cost and easier, faster installation process.

Drywall installers are responsible for measuring and cutting to size the standard-size drywall sheets, then fastening them to walls and ceilings. The installation involves fitting material around doors, windows, and openings such as electrical boxes, plumbing pipes, and HVAC ducts. While nailing was once extensively used for fastening to wood studs and ceiling joists, screws are now more likely to be used. Screws are used with both wood and metal studs.

On large construction projects, finishers called *tapers* may handle the next installation step. These workers apply joint compound and paper or mesh tape to cover the joints between drywall sheets. Compound is also used to cover screw heads and other imperfections. Several coats of compound are applied and sanded. A skilled taper makes joints virtually invisible. On smaller jobs, the same person often does both installation and taping.

A related occupation to drywall installation is ceiling tile installation. A ceiling tile installer may mount acoustical tiles directly on drywall surfaces or may install a grid of metal channels for a suspended ceiling. Acoustical panels fit into the channels.

Another related occupation is the *lather*. These workers fasten metal mesh or rockboard (a more strong and rigid variation of drywall) to studs and ceiling joists as a base for traditional plaster application. As the popularity of drywall has increased, the number of lathers has declined.

Drywall installers work in a protected environment, since the building must be closed in before their work can begin. Sheets of drywall are heavy and often handled with aid from a helper or mechanical devices. Large amounts of fine dust are generated during finishing work. Respiratory protection is required.

Although apprenticeship programs do exist, most drywall installers learn their skills on the job, starting as a helper. Because of the large amount of measuring involved, drywall installers must have good math skills. Approximately 20% of drywall installers are self-employed. Business skills are also useful for those installers.

Christina Richards/iStock/Thinkstock

A drywall installer using a lift to support a large piece of gypsum board being installed on a ceiling.

Chapter Review

Summary

- Interior finish work—installing cover materials for walls and ceilings—cannot begin until all exterior windows and doors have been installed.
- Many different covering materials are used to finish interior walls and ceilings, including gypsum wallboard, gypsum lath for plaster, predecorated gypsum paneling, plywood and particleboard, hardboard and fiberboard, solid-wood paneling, plaster, cement board, and special finishes.
- The most popular in interior finish material is gypsum wallboard, which is a laminated material with a gypsum core and paper covering on either side.
- Drywall application may be in a single layer, though, double-layer applications provide greater sound insulation and fire resistance.
- Special water-resistant gypsum boards are used in potentially humid areas, such as bathrooms.
- Cement board is used as a base for tile application in showers and tub areas, as is mold-resistant wallboard.
- Paneling is usually in the form of individual boards or planks. Plaster is used as fire protection.
- Ceilings are often finished in the same way as walls, but may have acoustical tiles installed to aid in sound control. Some tile is attached to furring strips; other types are placed in a suspended metal grid.

Know and Understand

Answer the following questions using the information in this chapter.

1. _____ was once considered the most popular wall covering.
   A. Gypsum wallboard
   B. Plywood
   C. Solid-wood paneling
   D. Plaster

2. Which of the following interior wall covering materials is fireproof and resists water damage?
   A. Gypsum wallboard.
   B. Plaster.
   C. Cement board.
   D. Brick.

3. *True or False?* To store drywall materials, stack drywall flat on a clean floor in the center of the largest rooms.

4. Most drywall mechanics prefer _____ because they provide a firm, tight attachment to wood or metal framing.
   A. nails   C. braces
   B. screws   D. adhesives

5. When applying compound and tape, what tool allows application of these materials simultaneously?
   A. Banjo.   C. Knife.
   B. Screw gun.   D. Trowel.

6. *True or False?* Adding more water to retemper joint compound will help extend its working time.

7. A continuous bead of _____ must connect the drywall to the top and bottom plates on the wall framing.
   A. metal   C. fasteners
   B. sealant   D. joint compound

8. Which of the following statements is *true* for Type-X gypsum wallboard?
   A. It is the type used for most interior walls and ceilings.
   B. It is made to reduce moisture.
   C. It offers fire and sound protection.
   D. All of the above.

9. To cut cement board, score it several times with a _____-tipped knife.
   A. silver   C. titanium
   B. steel   D. tungsten

10. *True or False?* A special plaster is used for the base of veneer plaster walls.

11. The face of hardboard is often scored to form a bead-board pattern for use as _____.
    A. wainscoting
    B. plaster protection
    C. textured wall surfaces
    D. furring

12. Solid lumber paneling patterns are often _____.
    A. limited   C. expensive
    B. reversible   D. textured

13. _____ paneling is installed in a pattern that alternates the direction of the angle at regular intervals.
    A. Diagonal   C. Herringbone
    B. Chevron   D. Solid lumber

14. Strips of wood or metal installed at the edge of openings to provide a guide for the plasterer are called _____.
    A. furring   C. plaster studs
    B. laths   D. grounds

15. _____ lath is commonly used in commercial construction.
    A. Wood
    B. Metal
    C. Gypsum
    D. Insulating fiberboard

16. *True or False?* Inside corners of wood-framed structures may be reinforced with Cornerite™.

17. _____ is the most commonly used aggregate with plaster base mixtures.
    A. Wood fiber   C. Sand
    B. Perlite   D. Vermiculite

18. The minimum plaster thickness for all coats is _____ when applied to brick, tile, or masonry.
    A. 1/2″   C. 3/4″
    B. 5/8″   D. 1″

19. _____ tiles are made from shale and natural clays, and is used in high-traffic settings.
    A. Quarry   C. Porcelain
    B. Ceramic   D. Natural stone

20. *True or False?* Furring should be directly nailed to the ceiling joists on new construction.

21. The metal framework of a suspended ceiling is mainly supported by _____ tied to the building structure directly above.
    A. joists   C. wires
    B. beams   D. runners

22. To find wall area, add all of the wall lengths and multiply by the _____.
    A. ceiling height   C. floor area
    B. wall height   D. floor length

Apply and Analyze

1. Why is the perpendicular method preferred when arranging drywall sheets in single-layer construction?

2. What kind of indoor wall construction is used where high sound transmission class (STC) ratings are needed?

3. How do you determine the proper angle for installing diagonal solid paneling?

4. Explain how carpenters minimize plaster cracking in wood frame structures.

5. Describe three-coat work with plaster.

6. When are suspended ceilings typically used?

7. When ordering sheet materials for walls and ceilings, why must you always plan to use the longest practical sheet?

Critical Thinking

1. Create a material list to cover all ceilings and walls with gypsum wallboard in the room where you are sitting. It may be helpful to draw a 2D sketch of each wall and the ceiling complete with dimensions. Research what lengths and widths of gypsum board are available to you locally (often 52″ widths are available). Justify what size and quantity of materials you would purchase for this job, considering factors such as the number of joints, minimizing waste, linear feet of joints to finish, delivery access, and material handling restrictions.

2. In commercial construction, suspended ceilings in rooms are common. It is important that gypsum board continues all the way to the top of the studs, even where it is concealed above the suspended ceiling. What reasons can you think of that make this an important detail? When might these areas also need compound and tape?

3. Carpenters building a tile shower have a number of options for tile backing. In the past, gypsum wallboard was used as a backing (or base) for wall tile; this is no longer an approved method. What are the issues with using gypsum wallboard as a tile backer in a shower? How do you think these issues manifest themselves over time?

Communicating about Carpentry

1. **Speaking and Writing.** Working with a partner, create a model of interior walls and ceilings both before and after finish has been applied. Create the model so that the parts can be removed. Use your model to demonstrate to the class the different options available when choosing which type of finish to use.

2. **Speaking and Listening.** With two classmates, role-play a situation in which you are a carpenter explaining to a customer why you need to apply interior wall and ceiling finish to the customer's house. Explain the procedures you will be using, and why. Adjust your vocabulary as necessary while responding to their questions and clarifying information. Then switch roles.

3. **Speaking and Writing.** Without referring to your textbook, make a list of materials used to cover walls and ceilings as part of the interior finishing. After you have completed your list, check the materials covered in your textbook and perform additional research on the materials. Create a chart with images of each type of material. Taking turns with your classmates, label each type of material on the chart and explain its purpose, when it is commonly used, and any other details you can recall.

CHAPTER **20**

Finish Flooring

OBJECTIVES

After studying this chapter, you will be able to:

- Describe strip, engineered, plank, and block wood flooring.
- Lay out and install strip flooring on concrete or plywood subfloors.
- Describe the procedure for installing laminate flooring.
- Describe the procedure for applying hardboard, particleboard, and plywood underlayment.
- Outline the basic steps for installing resilient flooring.
- Explain the different types of floor tiling and how they are installed.

TECHNICAL TERMS

bisque	lugged tile	sanded grout
blind nailed	mosaic tile	self-adhering tile
block flooring	paver tile	setup line
cove base	perimeter-bonded flooring	side-and-end matched
engineered wood flooring	permeability	sleeper
face nailing	plank flooring	spline
finish flooring	pull bar	strip flooring
full-adhesive-bonded flooring	quarry tile	undercut
grout	racking	

In construction, *finish flooring* is any material used as the final surface of a floor. A wide selection of materials is manufactured for this purpose. Many new and improved materials are being used in the production of composition (resilient) flooring. Notable among these are the vinyl plastics, which have largely replaced asphalt tile, rubber tile, and linoleum.

Flagstone, slate, brick, and ceramic tile are frequently selected for special areas. These areas include entrances, bathrooms, kitchens, and multipurpose rooms. Floor structures usually need to be designed to carry the greater weight of this type of flooring material.

The finish flooring is usually laid after wall and ceiling surfaces are completed, but before interior door frames and other trim are added. The floor surface should be covered to protect it during other inside finish work.

Sanding and finishing of a wood floor surface is the last major operation as the interior is completed.

20.1 Wood Flooring

Wood is popular as flooring in residential structures. Wood flooring, especially hardwood, has the strength and durability to withstand wear, while providing an attractive appearance, **Figure 20-1**. Hardwood and softwood flooring is available in a variety of widths and thicknesses, and as random-width planks or unit blocks.

Oak is a widely used species. However, maple, birch, beech, and other hardwoods also have desirable qualities. Softwood flooring includes such species as pine and fir. These are fairly durable when produced with an edge-grain surface.

Artazum/Shutterstock.com

Figure 20-1. Strip hardwood flooring is very durable and has a natural beauty that is brought out by a clear finish. A wide variety of hardwood flooring with a factory-applied finish is available.

20.1.1 Types of Wood Flooring

The following are general types of wood flooring used in residential structures:

- Strip
- Plank
- Block (parquet)
- Engineered

As the name implies, *strip flooring* consists of pieces of solid wood cut into narrow strips. It is the most widely used wood flooring and is laid in a random pattern of end joints. Most strip flooring is tongue-and-groove on the sides and ends. This design is also referred to as *side-and-end matched*. Another feature of strip floor-ing is the *undercut*, as shown in **Figure 20-2**. This is a wide groove on the bottom of each piece that enables it to lie flat and stable even when the subfloor surface is slightly uneven.

Unfinished strip flooring is manufactured with sharp, square corners on the face edges. After the floor is laid, the surface is sanded to remove unevenness. Prefinished strip flooring is sanded, finished, and waxed before it leaves the factory. Both edges of the face are chamfered. When the strips are laid, the chamfered edges form a vee joint. This makes any unevenness unnoticeable.

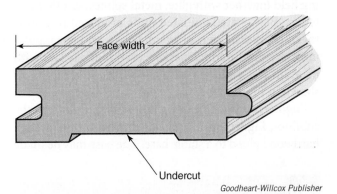

Goodheart-Willcox Publisher

Figure 20-2. Typical section of strip flooring. The ends are also tongue-and-groove. The undercut helps compensate for slightly uneven subfloors.

Plank flooring is also solid wood, **Figure 20-3A**. It gives an informal appearance to the floor. This is particularly appropriate for colonial and ranch-style homes. Plank floors are usually laid in random widths. The pieces are bored and plugged in the millwork shop to simulate the wooden pegs traditionally used to fasten the flooring in place. Today, this type of flooring has tongue-and-groove edges. It is laid in about the same manner as regular strip floors. Widths of the strips vary in different combinations.

A   *Robbins/Sykes, Tibbias Flooring Inc.* B   *alexander zveiger/Shutterstock.com*

Figure 20-3. Two types of wood floors. A—Softwood planks with walnut plugs. B—Unit block parquet flooring.

Block flooring looks like conventional parquetry. Often, this type is referred to as parquet flooring. Each block has an elaborate design formed by smaller wood blocks. The block unit consists of short lengths of flooring held together with glue, metal splines, or other fasteners. Blocks are laminated units produced by gluing together several layers of wood. Square and rectangular units are produced. Generally, each block is laid with its grain at a right angle to the surrounding units, as shown in **Figure 20-3B**.

Engineered wood flooring is designed to float on the subfloor, **Figure 20-4**. This flooring has a thin layer of hardwood glued to a stable base. The base may be made

of hardboard or other composite. It may also be made of veneer lumber. The top ply is prefinished. Individual pieces are 3/8″–3/4″ thick, 4″–8″ wide, and about 3′ long. The face veneer is made of two or three hardwood strips tightly edge-joined to resemble solid strip flooring. The thickness of the veneer and the joining are so precise that chamfering edges is not necessary.

20.1.2 Sizes and Grades

Hardwood strip flooring is generally available in widths ranging from 1 1/2″ to 3 1/4″. Standard thicknesses include 3/8″, 1/2″, and 3/4″. Solid planks are usually 3/4″ thick. However, greater thicknesses are available. Widths range from 3″ to 9″ in multiples of 1″. Unit blocks are also commonly produced in a 3/4″ thickness. Dimensions (width and length) are in multiples of the width of the strips from which they are made. For example, squares assembled from 2 1/4″ strips will be 6 3/4″ × 6 3/4″, 9″ × 9″, or 11 1/4″ × 11 1/4″.

Uniform grading rules are established by manufacturers working with the US Bureau of Standards and such organizations as the National Oak Flooring Manufacturers Association and the Maple Flooring Manufacturers Association. Grading is largely based on appearance. Consideration is given to knots, streaks, color, pinworm holes, and sapwood. The percentage of long and short pieces is also a consideration.

Damian Palus/Shutterstock.com

Figure 20-4. Engineered wood strip flooring has become very popular with homeowners and is available in many wood grain patterns.

Oak, for example, is separated into two grades of quarter-sawed stock and five grades of plain-sawed stock. This list shows the plain-sawed grades in descending order:

- Clear
- Select and better
- Select
- No. 1 common
- No. 2 common

White and red oak species are ordinarily separated in the highest grades. A chart on grades for hardwood flooring is included in Appendix B, **Technical Information**.

20.1.3 Delivery

It is important for wood flooring to have as close to the same moisture content as the air in the building in which it will be installed. If the moisture content of the flooring is higher than its surroundings, it will shrink as it dries, leaving cracks between the individual strips. If the moisture content of the flooring is lower than its surroundings, it will expand as it picks up moisture, causing the floor to buckle. Manufacturers recommend that wood flooring be delivered four or five days before installation. This period of conditioning permits the wood to match its moisture content with that in the building. Avoid transporting flooring in rain or snow. In damp or foggy conditions, protect the lumber with a tarp. Loosely pile the flooring throughout the structure. Do not store where floors are less than 18″ above the ground or where there is poor air circulation under the floor. A minimum inside temperature of 70°F (20°C) should be maintained. Avoid damp buildings. Make sure concrete work and plaster are thoroughly dry before bringing in flooring. The dryness of concrete can be checked by taping a 2′ × 2′ sheet of polyethylene to the concrete. Leave the polyethylene in place for a day or two, then remove it and look at the concrete. If the concrete is not completely dry, there will be a damp looking dark square where the polyethylene was. It generally takes more than a month for concrete to be dry enough to bring hardwood flooring into the building.

20.1.4 Subfloors

In conventional joist construction, most building codes specify a sound and rigid subfloor. It adds considerable strength to the structure and serves as a base for attaching the finish flooring.

For regular strip or plank flooring, the subfloor should consist of good-quality plywood or oriented strand board with a grade stamp indicating it is intended for structural use. It must be installed according to recognized standards. Refer to Chapter 10, **Floor Framing** for sizes and installation methods. Inadequate or improper nailing of subfloors usually results in squeaky floors.

20.2 Installing Wood Strip Flooring

Normally, laying the flooring should be the last building operation in the house, except for interior trim. Mechanical systems and wall and ceiling coverings should have been completed.

Before installing wood strip floors, carefully check the subfloor to make certain it is clean and that nailing patterns are complete. Put down a good quality (usually 15 lb) building paper at a right angle to the long dimension of the finish flooring. Extend it from wall to wall with a 4″ lap, as shown in **Figure 20-5**. Joist locations should be marked with a chalk line on the paper. Over the heating plant or hot air ducts, it is advisable to use a double-weight building paper. Insulation may also be directly attached to the underside of the subfloor. This extra precaution will prevent excessive heat from reaching the finish flooring, where it can cause cracks and open joints.

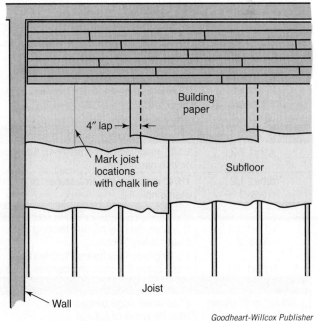

Goodheart-Willcox Publisher

Figure 20-5. This cutaway view shows the various layers of material under wood strip flooring. The subfloor is usually plywood with 15 lb building paper overlaid.

Lay strip flooring at a right angle to the floor joists. Strip flooring looks best when its direction aligns with the longest dimension of a rectangular room. Since floor joists normally span the shortest dimension of the living room, this will establish the direction the flooring runs in other rooms.

Stagger end joints so that several are not grouped in one small area. Shorter pieces should be used up in closets and in areas where there is the least traffic. Long strips should be used at entrances and for starting and finishing off a room.

Pro Tip

The moisture content usually recommended for flooring at the time of installation is 6% for the dry, southwestern states, 10% for the more-humid, southern states, and about 7% or 8% for the remainder of the country.

If you are installing new wood flooring in an existing home, place the flooring materials in the room a few days before installing. This allows the moisture content in the new flooring materials to reach equilibrium with the existing room. If the flooring is installed with a moisture content different from that of the room, excessive shrinkage or expansion of the flooring may occur.

20.2.1 Nailing

Floor squeaks or creaks are caused by the movement of one board against the other. The problem may be in either the subfloor or the finished floor. Using enough nails of the proper size will reduce these undesirable noises. When possible, the nails used for the finish floor should go through the subfloor and into the joist. Place a nail at each joist and one in between. Nail sizes are specified for different types of floors in **Figure 20-6**.

Start the installation by laying the first strip along either sidewall of the room. Select long pieces. If the wall is not perfectly straight and true, set the first strip along a chalk line. Place the groove edge next to the wall, staying at least 1/2″ away from the wall to allow for expansion of the flooring. This space will be covered later by the baseboard and base shoe.

Make sure the first strip is perfectly aligned. Then, face nail it as illustrated in **Figure 20-7**. In *face nailing*, the nail head must be set (sunk below face) and the hole filled. Predrilling holes will prevent splitting.

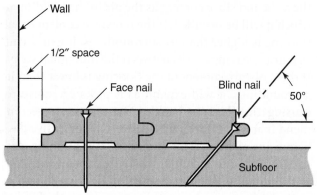

Goodheart-Willcox Publisher

Figure 20-7. Nail the starter strip through the face. Succeeding strips are blind nailed through the tongue.

Fastener Sizes and Spacing for Strip and Plank Flooring		
Flooring nominal size, inches	Size of fasteners	Spacing of fasteners
3/4×1 1/2 3/4×2 1/4 3/4×3 1/4	2″ machine-driven fasteners; 7d or 8d screw or cut nail.	10″ – 12″ apart
Following flooring must be laid on a subfloor.		
1/2×1 1/2 1/2×2	1 1/2″ machine-driven fastener; 5d screw, cut steel or wire casing nail.	10″ apart
3/8×1 1/2 3/8×2	1 1/4″ machine-driven fastener, or 4d bright wire casing nail.	8″ apart
Square-edge flooring as follows, face-nailed – through top face.		
5/16×1 1/2 5/16×2	1″, 15 gauge fully barbed flooring brad. 2 nails every 7 inches.	
5/16×1 1/3	1″, 15 gauge fully barbed flooring brad. 1 nail every 5 inches on alternate sides of strip.	
Plank flooring		
3/4×3″ to 8″ plank	2″ serrated-edge barbed nails. 7d or 8d screw or cut nail. Use 1 1/2″ length with 3/4″ plywood subfloor or slab.	8″ apart into and between joists

Goodheart-Willcox Publisher

Figure 20-6. This chart suggests fastener size and spacing for application of strip and plank flooring. Always follow the flooring manufacturer's instructions, which may vary from brand to brand.

Succeeding strips are **blind nailed**. The nail penetrates the flooring where the tongue joins the shoulder. Drive the nail at an angle of about 50°. Countersink all nails. Use care so that the edge of the flooring strips are not damaged. Special flooring nail guns are available for fast and consistent nailing. See **Figure 20-8**.

Each strip should fit tightly against the preceding one. When it is necessary to drive strips into position, use a piece of scrap flooring as a driving block.

Lay out seven or eight rows of flooring at a time. Stagger the pattern, cutting and fitting as you go. This step is known in the trade as **racking** the floor. Use different strip lengths. Match the strips so they will extend wall to wall with about 1/2″ clearance at each end.

End joints in successive courses should be 6″ or more from each other. Try to arrange the pieces so the joints are well distributed. Blend color and grain for a pleasing pattern. Lengths cut from the end of a course should be carried back to the opposite wall to start the next course. This is the only place it can be used since its cut end has no tongue or groove.

Flooring strips should run uninterrupted through doorways and into adjoining rooms. When there is a projection into the room, such as a wall or partition, follow the procedure shown in **Figure 20-9**.

When strip or plank flooring a large area, it sometimes helps to set up a starter strip at or near the center. Use a **spline** in the groove of the starter strip to create a tongue. Be sure to accurately measure and align the starter strip with the walls on each side of the room.

When the flooring has been brought to within 2′–3′ of the far wall, check the distance to the wall at each end to see if the strips are parallel to the wall. If not, slightly dress the grooved edges at one end until the strips have been adjusted to run parallel. This is necessary so that the last piece may be aligned with the baseboard.

Michael Pettigrew/Shutterstock.com

Figure 20-8. Strip floor installation using a portable nail gun. Never attempt to set nails with a hammer. You will damage the flooring.

PROCEDURE
Laying Flooring around Projections

1. Lay the main floor area to a point even with the projection.
2. Extend the next course all the way across the room. Set the extended strip to a chalk line and face nail it in place.
3. Form a tongue on the grooved edge by inserting a spline.
4. Install the flooring in the second room in both directions from this strip.

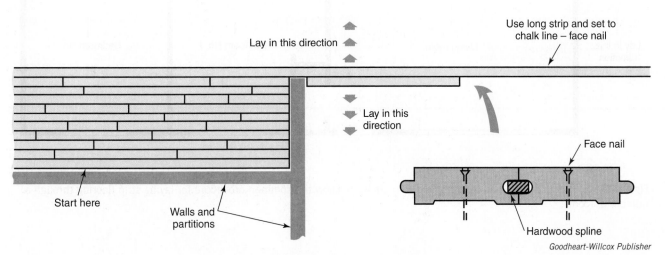

Goodheart-Willcox Publisher

Figure 20-9. Follow this procedure when laying strip floors around a wall or partition.

20.2.2 Multiroom Layout

Sometimes, a strip flooring installation is carried through all or a major part of a building. Before beginning, study the floor plan to determine the most efficient procedure to follow.

Figure 20-10 shows a typical residential floor plan with a method for installing strip flooring. Following this plan, a carpenter first snaps a chalk line the width of two flooring strips (plus 1/2″) away from the partition as shown. Carpenters may call this a *setup line*. Align the starter courses with this setup line before face nailing. Continue installation across the living room, across the hall, and into and through bedrooms No. 1 and No. 2.

Next, set a spline in the groove of the starter strip and lay the floor from this point into the dining room and bedroom No. 3. In bedroom No. 2, use a spline to lay the floor back into the closet. In small closet areas, such as shown in bedroom No. 1, it is usually impractical to reverse the direction of laying with a splined groove. Simply face nail the pieces in place.

The last strip laid along the wall of a room usually needs to be ripped so it has the required clearance (1/2″ minimum). Due to lack of clearance for blind nailing, the last several strips must be face nailed. Do not place ripped strips where they might detract from the floor's appearance. It is recommended that full-width strips always be used around entrances and across doorways.

> **Pro Tip**
> In general, plank flooring is installed following the same procedures used for strip flooring. In addition to the regular blind nailing, screws are set and concealed in the face of wide boards.

20.2.3 Wood Flooring over Concrete

Finished wood flooring systems can be successfully installed over a concrete slab. When the concrete floor is suspended with an air space below, a moisture barrier is usually not required. Below-grade installation over concrete is not recommended.

New concrete is heavy with moisture. On-grade slabs must be properly installed with a vapor retarder, such as 6-mil polyethylene film, between the 6″ gravel fill and the slab. This keeps ground moisture from entering the slab, but it also slows down curing of the slab. Always test for dryness—even though the slab may have been in place for two years or more.

For slab-on-grade installation, put down *sleepers*. These are wood strips attached to or embedded in the

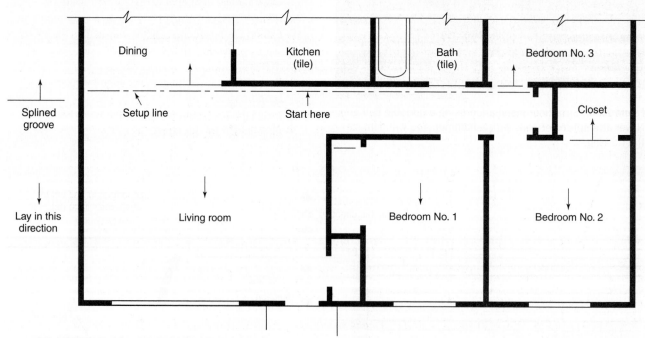

Goodheart-Willcox Publisher

Figure 20-10. This floor plan has been marked with arrows to show the common procedure for laying strip flooring through a number of rooms.

concrete surface. The sleepers serve as a nailing base for the flooring material.

Beware of damp slab-on-grade concrete floors. If a vapor barrier was not laid down under the concrete during construction, moisture can work up to the surface. This dampness will damage wood flooring.

Whenever concrete is directly placed on grade, either at or below ground level, an approved membrane moisture barrier must be installed between the flooring and the concrete. Evenly apply a coat of asphalt mastic thinned with an appropriate solvent over the concrete surface. Cover this with polyethylene film. Next, apply a coat of a different, special asphalt mastic. Then set wood sleepers into the mastic. The sleepers should be 2×4 lumber about 30″ long, laid flat and running at a right angle to the flooring. Lap them at least 3″ for 2 1/4″ flooring and 4″ for 3 1/4″ flooring. Spacing between centers is usually 12″–16″.

A newer method of laying strip floors over a concrete slab-on-grade offers savings over older methods. A double layer of 1×2 wood sleepers is nailed together with a moisture barrier of 4-mil polyethylene film placed between them, **Figure 20-11**. This is accepted by the US Federal Housing Authority. An alternative system described by the Wood Flooring Manufacturers Association (NOFMA) uses 3/4″ or thicker sheathing-grade plywood in place of sleepers.

Pro Tip
Installation of wood flooring over concrete must be done carefully. Secure detailed specifications from manufacturers of the products used. Organizations such as NOFMA can also provide instructional materials and advice.

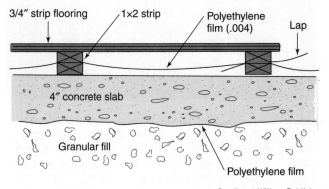

3/4″ strip flooring — 1×2 strip — Polyethylene film (.004) — Lap

4″ concrete slab

Granular fill

Polyethylene film

Goodheart-Willcox Publisher

Figure 20-11. Strip flooring over a concrete slab. The sleepers are doubled 1×2 strips.

PROCEDURE
Laying Wood Strip Flooring over Concrete
1. Clean and prime the floor.
2. Snap chalk lines 16″ apart.
3. Put down bands (rivers) of adhesive along the layout lines.
4. Embed treated wood sleepers in the adhesive.
5. Secure the sleepers with 1 1/2″ concrete nails or special concrete screws about 24″ apart. If screws are used, drill holes down through the sleepers into the concrete. Use a carbide-tipped masonry drill and a hole shooter. Strips can also be attached with powder-actuated equipment.
6. Place a layer of 4-mil–6-mil polyethylene film over the strips.
7. Join the polyethylene sheets by forming a lap over a sleeper.
8. Nail a second layer of untreated strips to the bottom sleepers, sandwiching the barrier. Use 4d nails about 16″ apart.
9. Install strip or plank flooring as previously described. No two adjoining flooring strips should break joints between the same two sleepers.

PROCEDURE
Laying Wood Strip Flooring Using a Plywood Subfloor over Concrete Slab
1. Put down an appropriate vapor retarder, such as 6-mil polyethylene.
2. Lay 3/4″ plywood loosely over the entire floor. Place the plywood at a diagonal to the direction of the strip flooring to help prevent cracks associated with panel edges. Stagger plywood joints every 4′. Allow 3/4″ spacing at walls and 1/4″–1/2″ between panels.
3. Cut to fit around doorjambs and other obstructions allowing 1/8″ gap.
4. Fasten the plywood to the slab. Use a powder-actuated concrete nail gun or a hammer and concrete nails. Start at the center of the panel and work outward. Use at least nine nails on each panel.
5. You may also glue the plywood to the vapor retarder. When using this method, cut 3/4″ plywood into 4′ squares and score the back 3/8″ deep on a 12″ grid pattern. Apply cut-back (reduced) mastic to the vapor barrier with a 1/4″ × 1/4″ notched trowel and lay down the plywood squares, pressing them into the mastic.

20.3 Estimating Strip Flooring

To determine the number of board feet of strip floor-ing needed to cover a given area, first calculate the area in square feet. Then, add the percentage listed in **Figure 20-12** for the particular size being used. The figures listed are based on laying flooring straight across the room. They provide an allowance for side matching plus 5% for end matching. For example:

$$
\begin{aligned}
\text{Total area} &= 900 \text{ sq ft} \\
\text{Flooring size} &= 3/4'' \times 2\ 1/4'' \\
\text{Bd ft of flooring} &= 900 + 38\% \text{ of } 900 \\
&= 900 + 342 \\
&= 1242 \\
\text{Number of bundles} &= 52 \ (24 \text{ bd ft to bundle})
\end{aligned}
$$

There are 24 bd ft in a standard bundle of wood strip flooring. Where there are many breaks and projections in the wall line, allow additional amounts.

20.4 Wood Block (Parquet) Flooring

Block or parquet flooring is made up of squares of wood in various patterns. It requires application meth-ods different from strip flooring. Flooring blocks are usually squares $12'' \times 12''$, but other sizes are available. They are produced in two different ways:

- Unit blocks are glued up from several short lengths of flooring.
- Laminated blocks are made by bonding three plies of hardwood with moisture-resistant glue.

Parquet flooring may use intricate patterns of small pieces in producing the squares. Each block is tongue-and-groove, as shown in **Figure 20-13**.

Styles and types of block, as well as installation meth-ods, vary somewhat among manufacturers. Detailed instructions usually come with the product. If not, they can be secured from the manufacturer or distributor.

Estimating Strip Flooring

Flooring board size	Additional percentage
$3/4'' \times 1\ 1/2''$	55%
$3/4'' \times 2''$	43%
$3/4'' \times 2\ 1/4''$	38%
$3/4'' \times 3\ 1/4''$	29%
$3/8'' \times 1\ 1/2''$	38%
$3/8'' \times 2''$	30%
$1/2'' \times 1\ 1/2''$	38%
$1/2'' \times 2''$	30%

Goodheart-Willcox Publisher

Figure 20-12. When estimating strip flooring, add the listed percentage to total area to allow for waste.

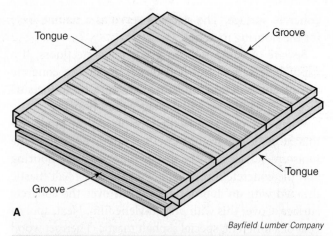

A

Bayfield Lumber Company

B

Bayfield Lumber Company

C

Bayfield Lumber Company

Figure 20-13. Examples of typical block or parquet flooring. Edges are always tongue-and-groove. A—One style of strip flooring. B—In this style, squares run perpendicular to each other. C—The wood strips in B are held together with metal strips mortised into the reverse side.

NOFMA recommends that installation be made over a double layer of subflooring. Block flooring is nearly always laid in a mastic. The mastic is evenly spread over the subfloor to a thickness of 3/32″. Some installations may require a layer of 30 lb asphalt-saturated felt. Lay the felt in a mastic coat and then coat the top surface with mastic to receive the flooring. Lay blocks in this coat. If installation is over a concrete slab, use a moisture barrier as is used for installation of strip flooring.

Allow some clearance on unit blocks for expansion. Rubber strips are sometimes used for this purpose. Generally, it is recommended that unit blocks be installed with a 1/2″ space at the wall.

> **Pro Tip**
> If blocks are to be nailed rather than set in mastic, blind nail them through the tongued edges as with strip or plank floors.

20.4.1 Installation Patterns

Installation can be made either on a square or diagonal pattern. When laying on a square pattern, never use a wall as a starting line. Walls are not always truly straight. Snap chalk lines equal distances from the sidewalls at a 90° angle to each other so that they cross in the center of the room. See **Figure 20-14**. Begin laying blocks from the center point where the chalk lines intersect. When using a diagonal pattern, start in a corner and measure equal distances along each of the walls. Snap a chalk line between these two points. This is your base line. Next, run a test line to intersect the base line at 90°. Begin layout at the intersection of the base line and the test line.

Layout procedures are similar to those described for resilient flooring materials later in this chapter. Block and parquet flooring and adhesive materials vary from one manufacturer to another. Secure detailed information concerning installation procedures from the manufacturer of the product.

20.5 Prefinished Wood Flooring

Most types of wood flooring materials are available with a factory-applied finish, **Figure 20-15**. These finishes are generally considered to be superior to those applied on the job. Another advantage of prefinished flooring is that it speeds up construction. Floors are immediately ready for service.

A disadvantage is that the installation must be made with greater care to avoid damaging the finish. Face nailing must be held to a minimum when working with prefinished materials. Although the holes created by face nailing can be covered with a special filler material, hammer marks caused by careless nailing are extremely difficult to repair.

A prefinished floor must be the last step in the interior finish sequence. Other trim that abuts the finished

peter s/Shutterstock.com

Figure 20-15. Prefinished flooring. The edges of each unit have a slight bevel.

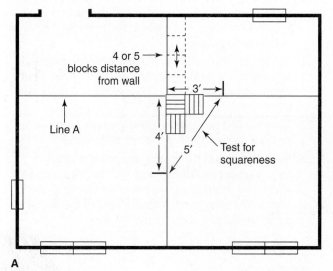

A

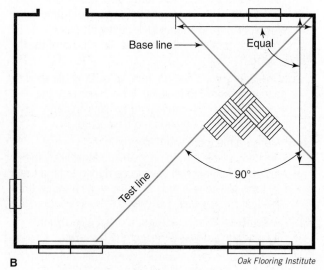

Oak Flooring Institute

B

Figure 20-14. Laying out chalk lines for parquet flooring patterns. A—Square pattern. B—Diagonal pattern.

floor must be set beforehand with proper clearances. For example, interior doorjambs and casing require spacer blocks equal to the floor thickness. Place these under the units while they are being installed. They are removed before installing the finish flooring. Unless prefinished baseboards are installed after the floor is laid, they must also be placed with the necessary clearance. Other special provisions need to be made around built-in cabinets and at stairways and entrances.

20.6 Laminate Flooring

Laminate flooring consists of a 1/4″ core of medium- or high-density fiberboard. This is sandwiched between layers of melamine paper and a protective topcoat. The top layer carries the wood-grained pattern. Sizes vary, but most planks are 46″–50″ long and 8″ wide.

The laminate is not fastened to a subfloor. Instead, it floats over the subfloor. An underlayment of foam takes care of cushioning and absorbs sound. Tongue-and-groove construction allows the planks to be snapped together as they are installed. Although the first types introduced required gluing of the joints, most manufacturers now offer flooring in both glue-less- and glued-joint types. Glued joints are used for installations where moisture could be present, such as bathrooms and kitchens.

Follow the manufacturer's recommendations on allowing the laminated flooring to acclimate to the building's humidity level. Some types require storing the unopened cartons in the room for two days before installation, but others can be installed without this waiting period.

Before installation, inspect each plank for defects. Since they are very thin, the strips can be installed over almost any kind of subfloor—even old floors. However, carpeting usually must be removed. Each manufacturer supplies instructions for installation of its product. Read and follow these instructions carefully.

Pro Tip
Normally, radiant heating creates no problem for laminate flooring and requires no precautions. But, if in doubt, consult the distributor.

PROCEDURE
Installing Laminate Flooring

1. Clean the floor.
2. If the subfloor is concrete, lay down a vapor barrier—usually a 6-mil polyethylene film. Overlap edges 8″. This step is important even when the concrete is covered with other types of floor coverings.
3. Put down a foam underlayment as recommended by the manufacturer.
4. Determine which direction to lay the planks. As a rule, run them in the same direction as the incoming light. There are other considerations. Small or narrow rooms look best with planks running parallel to the longest dimension of the room.
5. On existing construction, remove existing baseboards and quarter-round molding. When replaced, the quarter-round will conceal the expansion gap between the flooring and the baseboards.
6. Measure the room width at a right angle to the direction planks are to be installed. Most likely the last row of planks will have to be ripped to fit. If the last row will be less than 2″ wide, reduce the width of the first row, as well.
7. Assemble the first two rows. If using glued-joint flooring, do a dry assembly of the two rows.
8. Start at the left-hand corner of the longest, straightest wall. Lay down the first row as shown in **Figure 20-16**. The groove should be toward the wall. Be sure to use spacers between the wall and the planking. This should maintain a gap for expansion as directed by the manufacturer.
9. Mark and cut the end plank. Work from the underside.
10. Use the leftover piece to start the next row. If the cut piece is less than 8″ long, start with a new plank cut in half. Check that end joints are offset at least 8″ from adjacent rows.
11. The double row will allow you to see whether the wall is straight. If it is uneven, use a spacer to trace a cut line. Number the planks consecutively and make the cuts.
12. Allow the first two rows of glued-joint planks to set for an hour before installing the rest of the floor.
13. Assemble planks row by row, snapping or gluing them together as instructed by the manufacturer. For glued-joint flooring, apply a ribbon of glue to grooves, completely filling them. A groove against the wall or other fixed object does not require glue.

Continued

PROCEDURE (continued)

14. Tightly fit all joints. Use a tapping block and hammer or mallet to tighten planks, following the manufacturer's instructions. On glued-joint installations, some glue should ooze to the top of the joint for a good seal. Wipe off excess with a damp cloth.

15. Cut the last row to fit. Place a full row of planks to be marked directly on top of the second-to-last row that has already been installed. Then, using a piece of full-width scrap plank, place the tongue edge against the wall. Hold a marking pencil against the opposite edge and mark a line as the scrap is pulled along. Placing the tongue edge against the wall allows for the expansion room needed next to the wall. Trim the planks along the pencil line, then install the final row.

16. Use a tapping block to tap planks into place. See **Figure 20-17**. Where tight quarters do not permit use of the tapping block, use a **pull bar**.

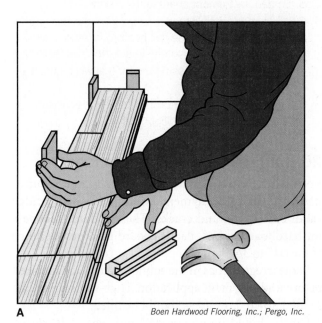

A   Boen Hardwood Flooring, Inc.; Pergo, Inc.

A   Boen Hardwood Flooring, Inc.; Pergo, Inc.

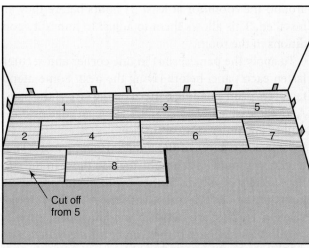

B   Boen Hardwood Flooring, Inc.; Pergo, Inc.

Figure 20-16. Begin installing wood laminate flooring at the left-hand wall with the groove to the wall. A—Use spacers to keep the flooring 1/2″ away from wall. B—Lay the planks in the order shown here. If the cutoff from the first row (#5) is longer than 8″, it can be used to start the third row.

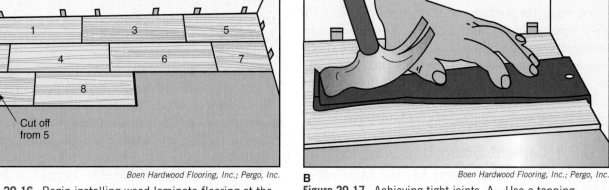

B   Boen Hardwood Flooring, Inc.; Pergo, Inc.

Figure 20-17. Achieving tight joints. A—Use a tapping block to drive tongues into grooves. B—When driving planks together in tight quarters, such as near a wall, use a pull bar and hammer.

20.7 Engineered Wood Flooring

Engineered wood flooring is similar to laminate, except that the materials are wood. Engineered wood flooring is prefinished with a durable finish. Most manufacturers warranty their finish for 25 years or more. Edges and ends are manufactured with tongue-and-groove joints that snap together tightly. Some manufacturers recommend gluing the joints. Apply a full bead of glue in the groove, snap the pieces together, and then hold them in place until the glue is dry. Apply a couple of pieces of masking tape applied across the joints. Installation of the flooring is the same as for laminate flooring.

Green Note
Eliminate waste created through theft and vandalism with green security. Keep accurate records to catch material shortage. At the end of the workday, store all materials and tools securely. Use solar-powered security cameras to monitor for thieves and vandals while the jobsite is closed.

20.8 Underlayment for Nonwood Floors

If resilient flooring materials are laid over rough or irregular surfaces, such as existing flooring or wood subflooring, they will eventually telegraph (show on their own surface) even the slightest irregularities. An underlayment is required to provide a satisfactory surface for resilient materials. This thin sheet material is laid down to provide a smooth, even surface for the finish flooring. The material can be plywood, hardboard, cement board, or other manufactured product. On concrete floors, a special mastic material is sometimes used when the surface condition does not meet the requirements of the finish flooring. An underlayment also prevents the finish flooring materials from checking or cracking when slight movement takes place in a wood subfloor. When used under carpeting and resilient materials, the underlayment is usually installed as soon as wall and ceiling surfaces are complete.

20.8.1 Cement Board Underlayment

Cement board underlayment has the advantage of being dimensionally stable. It is not necessary to leave expansion room between panels or at walls. Further, it is impervious to water and particularly suited for use under ceramic tile floors. Additional information about cement board is given in Chapter 19, **Interior Wall and Ceiling Finish**.

PROCEDURE
Installing Cement Board on Floors

1. Fasten subfloor of 5/8″ exterior grade plywood to joists no more than 24″ OC.
2. Determine the layout of the backing so that no joints in the backing board fall over joints in the plywood. Do not allow four corners of backing to meet at one point.
3. Allow 1/8″ between sheets of cement board.
4. Apply a supporting bed of mortar or modified thinset to subfloor using a 1/4″ square-notched trowel.
5. Embed the cement board in the mortar.
6. Use approved nails or screws to fasten the backing board every 8″. Set the nail heads or screws flush with the surface. Be careful to not overdrive them.
7. Fill all joints with the same mortar that will be used to set the tiles.
8. Embed 2″-wide alkali-resistant tape in the mortar and level with a joint knife or trowel.

20.8.2 Hardboard and Particleboard

Both hardboard and particleboard meet the requirements of an underlayment board. They will bridge small cups, gaps, and cracks. The standard thickness for hardboard is 1/4″. Particleboard thicknesses range from 1/4″ to 3/4″.

Large irregularities in the subfloor should be repaired before underlayment application. High spots should be sanded down. Low areas should be filled.

Panels should be unwrapped and separately placed around the room for at least 24 hours before they are installed. This allows them to adjust to humidity conditions in the room.

To apply the panels, start in one corner and securely fasten each panel before laying the next. Some manufacturers print a nailing pattern on the face of the panel. Allow at least a 1/8″–3/8″ space along an edge next to a wall or any other vertical surface.

Stagger the joints of the underlayment panels. Their direction should be at a right angle to those in the subfloor. Be especially careful to avoid alignment of any joints in the underlayment with those in the subfloor. Leave a 1/32″ space between the joints of hardboard panels. Particleboard panels can be lightly butted.

Underlayment panels should be attached to the subfloor with approved fasteners. These fasteners include ring-grooved and screw-shank nails. Spacing of nails for particleboard varies with the material's thickness. For 3/8″ panels, use 3d nails spaced 3″ apart around the edges and 6″ each way throughout the field of the panel. For 1/2″ or 5/8″ panels, use 4d nails spaced 6″ apart around the edges and 10″ apart in the field of the panel. Be sure to drive nail heads flush with the surface. When fastening underlayment with staples, use a type that is etched or galvanized and at least 7/8″ long. Space staples not over 4″ apart along panel edges.

Special adhesives can also be used to bond underlayment to subfloors. An advantage of using adhesive is that it eliminates the possibility of nail "popping" under resilient floors. A nail pop will eventually damage the resilient flooring.

20.8.3 Plywood

Install plywood underlayment with the smooth side up. Since a range of plywood thicknesses is available, vertical alignment of the surfaces of various finish flooring materials is easy, **Figure 20-18**.

Follow the installation procedures described for hardboard. Turn the grain of the face ply to run at a right angle to the joists. Stagger end joints. When applying 1/4″ plywood, use 3d ring-grooved or screw-shank nails. Nails should not be more than 3″ from the edges. Field spacing of nails should be 6″ each way.

20.9 Resilient Floor Tile

Surfaces of the underlayment must be smooth and joints level. Double-check that no fasteners are sticking above the surface. Fill cracks and open areas with a floor-leveling compound. Use a sander to remove rough edges and high spots. With the underlayment securely fastened and leveled, carefully sweep and vacuum the surface.

Smoothness is extremely important, especially under the more pliable materials, such as vinyl, rubber, and linoleum. Over a period of time, these materials will telegraph even the slightest irregularities such as the texture of wood grain or rough surfaces. Linoleum, either in tile or sheet form, is especially susceptible to

telegraphing. For this reason, a base layer of felt is often applied over the underlayment when this material is to be installed.

Underlayment panels should resist denting and punctures from concentrated loads. In addition, look out for any substances that might stain vinyl. These include patching compounds, marking inks, paints, solvents, adhesives, asphalt, and dyes. Some fasteners are coated with resin, rosin, or cement containing a dye that could discolor vinyl. They should only be used if they are known to be safe.

Pro Tip

There are many resilient flooring materials on the market. Make the application according to the recommendations and instructions furnished by the manufacturer of the product.

PROCEDURE

Preparing the Floor for Resilient Tile

1. Locate the center of the end walls of the room. Disregard any breaks or irregularities in the contour.
2. Establish a main centerline on the floor by snapping a chalk line between the two center points. When snapping long lines, hold the line at various intervals and snap only short sections.
3. Lay out a centerline at a right angle to the main one. Use a carpenter's square or set up a right triangle (base 4′, altitude 3′, hypotenuse 5′) and snap a chalk line.
4. With the centerlines established, make a trial layout of tile along the centerlines, as shown in **Figure 20-19**.

Continued

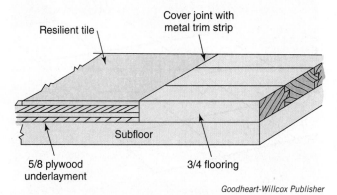

Goodheart-Willcox Publisher

Figure 20-18. Alignment of resilient floor tile with strip flooring is easy since plywood is available in different thicknesses.

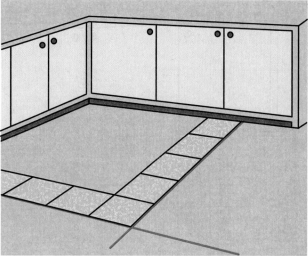

Goodheart-Willcox Publisher

Figure 20-19. Lay down tiles along both centerlines to the walls.

PROCEDURE (continued)

5. Measure the distance between the wall and last tile, **Figure 20-20**. If the distance is less than 2″ or more than 8″, move the centerline closer to the wall by half of the tile's dimension. This adjustment eliminates the need to install border tiles that are too narrow. Since the original centerline is moved exactly half the tile size, the border tile width will remain uniform on opposite sides of the room.

6. Check the layout along the other centerline in the same way.

7. Remove the loose tile put down in the trial layout.

8. Clean the floor surface again.

9. Using the type of spreader (trowel or brush) recommended by the manufacturer, spread the adhesive over one-quarter of the total area. Bring the adhesive right up to the chalk line, but do not cover it.

10. The next step is to lay the tile.

Pro Tip

Use care in spreading the adhesive. If it is spread too thin, the tile will not properly adhere. If applied too heavily, the adhesive will squeeze up between the joints.

20.9.1 Laying Tile

Allow the adhesive to take an initial set before any tile is laid. The set time will vary upward from a minimum of about 15 minutes, depending on the type of adhesive used. Test the surface with your thumb. It should feel slightly tacky but should not stick to your skin.

Start laying the tile at the center of the room. Make sure the edges of the tile align with the chalk line. If the chalk line is partially obscured by the adhesive, snap a new line or tack down a thin, straight strip of wood to act as a guide in placing the tile.

Carefully lay each tile in place. Do not slide the tile—sliding can cause the adhesive to work up between the joints and prevent a tight fit. Squarely butt each tile to the adjoining one with the corners in line, **Figure 20-21**. Take the time to see that each tile is correctly positioned. There is usually no hurry, since most adhesives can be "worked" over a period of several hours.

Rubber, vinyl, and linoleum are usually rolled after a section of the floor is laid. Asphalt tile does not need to be rolled. Be sure to follow the manufacturer recommendations.

After the main area is complete, set the border tile as a separate operation. To lay out a border tile, place a loose tile that will be cut and used over the last tile in the outside row. Then, take another tile and place it against the wall. Mark a line on the first tile with a sharp pencil, **Figure 20-22**. Cut the tile along the marked line using heavy-duty household shears or tin snips. Some types of tile require a special cutter, while others may be scribed and snapped. Asphalt tile can be readily cut with snips, if the tile is heated.

Various trim and feature strips are available to customize a tile installation. They are laid by following the general procedures previously described for regular tile.

After all four sections of the floor have been completed, *cove base* can be installed along the wall and around fixtures. A special adhesive is made for this operation. Cut the proper lengths and make a trial fit. Apply the adhesive to the cove base and press the base into place.

Carefully check over the completed installation. Remove any spots of adhesive on the floor or cove base. Carefully work using cleaners and procedures approved by the manufacturer.

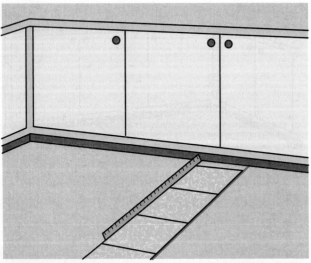

Goodheart-Willcox Publisher

Figure 20-20. Measure the width of the border tile and adjust the centerline, if necessary.

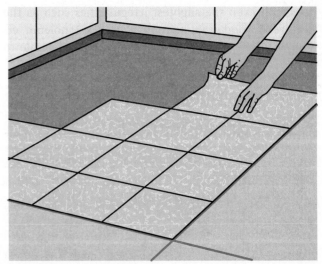

Goodheart-Willcox Publisher

Figure 20-21. When laying tile, align joining edges first, then lower the rest of the tile.

A

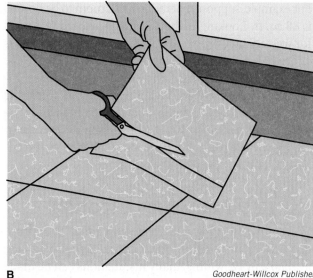

B

Goodheart-Willcox Publisher

Figure 20-22. Finishing off the borders. A—Mark the border tile. B—Cut the tile to fit using shears.

20.9.2 Installing Resilient Tile over Concrete

Before installing resilient tile over concrete, make sure the surface is properly prepared. It must be dry, smooth, and structurally sound. There must be no depressions, pits, scale, or foreign deposits. Remove any paint, varnish, oil, or wax. Trisodium phosphate mixed with hot water will remove most paints, except those with a chlorinated rubber or resin base. These must be removed by grinding.

Fill joints and cracks with a latex underlayment or crack filler. Follow the manufacturer's instructions for on-grade or subgrade applications. Cover dusty or chalked surfaces on suspended concrete floors with a single coat of primer before spreading adhesive. Chalking or dusty floors on or below grade may be a sign of leaching (water seeping through the concrete and carrying minerals from the concrete to the surface). Perform a bond and moisture test before going ahead. To make a bond and moisture test, install 3′ squares of the flooring material at different places. If they are securely bonded after 72 hours, the floor is dry enough for application of the tile.

20.10 Self-Adhering Tiles

Self-adhering tiles are similar to standard floor tile, except the adhesive is applied to the tile back at the factory. These tiles are easy to install. Remove the paper from the back of the tile, position the tile on the floor, and press it down. Follow the same layout procedure as for standard tiles, except adhesive is not spread on the underlayment.

Floors must be dry, smooth, and completely free of wax, grease, and dirt. Generally, self-adhering tiles can be laid over smooth-faced resilient floors. However, embossed tile, urethane finish, or cushioned floors should be removed.

Tiles should be kept at a room temperature of at least 65°F (20°C) for 24 hours before installation. The room should be kept warm for one week after installation. This ensures a firm bond to the subfloor surface. Always study and follow the manufacturer's directions.

20.11 Estimating Tile Flooring and Adhesive

To estimate the number of tiles needed, first find the area of the room. Multiply the length in feet and inches by the width in feet and inches. If the room has an irregular shape, section the room into squares or rectangles. Then, figure each area and add the results.

There must be an allowance for waste. Multiply the square footage by the factor shown in **Figure 20-23**.

Percentage of Tile Waste for Various Areas	
Area (square feet)	**Percentage**
Up to 50	14
50–100	10
100–200	8
200–300	7
300–1000	5
1000–5000	3
5000–10,000	2–3
Over 10,000	1–2

Goodheart-Willcox Publisher

Figure 20-23. Tile flooring waste percentages for 12″ × 12″ floor tile.

For example, suppose that the area of a room to be floored is 88 sq. ft. Looking at the chart, 10% should be added for waste. Thus, 88 × 1.10 = 96.8 or 97 sq ft. If one tile is 12″ × 12″, you will need about 97 tiles to complete the job. Add 1/3 more when laying down 9″ × 9″ tiles.

Estimate adhesive for laying resilient tile according to square yardage or square footage. Do not include the extra added for the waste. Divide the square yardage or footage by the per-gallon spreading capacity of the adhesive. This is shown on the container.

20.12 Sheet Vinyl Flooring

Sheet vinyl flooring is extremely flexible. This property makes installation much easier. Since sheets are available in 12′ widths, many installations can be made free of seams. This material usually can be installed over concrete, plywood underlayment, or noncushioned vinyl flooring.

There are two types of sheet vinyl flooring: full-adhesive bonded and perimeter bonded. *Full-adhesive-bonded flooring* must be fastened to the substrate with a coat of adhesive covering the entire surface. *Perimeter-bonded flooring* is fastened down only around the edges and at seams.

Full-adhesive-bonded flooring is normally cut to fit the space using a template before installation. It is then laid down to test the fit. If a seam is needed, the two pieces are pattern matched and slightly overlapped. One-half of the material is then folded back and adhesive is applied to the substrate. The folded portion is then laid in the adhesive. The procedure is repeated with the other half. If there is a seam, the process may be done in quarters instead of halves. The overlapped seam is cut through both pieces of vinyl and the scraps removed, **Figure 20-24.**

As the final step, a heavy roller is used to tightly bond the seam and the rest of the flooring.

To install perimeter-bonded flooring, smoothly spread the sheet over the floor. Let excess material turn up around the edges of the room. When there are seams, carefully match the pattern and fasten the two sections to the floor with adhesive.

Trim edges to size. After all edges are trimmed and fitted, secure them with a staple gun or use a band of double-faced tape. Always carefully study the manufacturer's directions before starting the work.

20.13 Vinyl Plank Flooring

Vinyl plank flooring is an attractive alternative to sheet vinyl or laminate flooring. It is a more affordable option than wood but maintains a wood look. Vinyl planks come as a glue-down or a snap-together floating product, **Figure 20-25.** Both types are cut easily using a utility knife.

A   *Maleo/Shutterstock.com*

B   *Alex Verrone/Shutterstock.com*

Figure 20-25. Vinyl plank flooring. A—Glue-down vinyl planks. B—Snap-together floating vinyl planks.

SSS55/Shutterstock.com

Figure 20-24. Where seams overlap, trim flexible vinyl sheet linoleum with a utility knife drawn along a straightedge. Be sure the straightedge is parallel to the wall.

Glue-down vinyl planks offer the advantages of a full-adhesive-bonded floor without the challenges associated with making an accurate cutting template. Glue-down planks withstand heavy foot traffic.

Snap-together floating vinyl planks are similar to laminate floors but have a surface that is slightly softer. Vinyl planks are more resistant to water than laminates and can be cleaned with a wet mop. Vinyl surfaces can tear under excessive forces. Laminates tend to chip along the edges when stressed, whereas vinyl will not. Installation of the two products is nearly identical. Always consult the manufacturer installation instructions for product-specific details.

20.14 Floor Tile

Tiles are made from a mixture of clay and other materials that may include ground shale, gypsum, talc, vermiculite, and sand. Sometimes, tiles are made from pure clay. During manufacture, the dry ingredients are mixed with water and other materials and shaped into a *bisque*. This is the body of the tile. The bisque may be formed by extrusion or die pressing, cut from a sheet, or formed by hand. The bisques are allowed to air dry for a short time before being placed in a kiln and fired at temperatures ranging from 1900°F to 2200°F (1000°C–1200°C). Tiles are available in a variety of colors and shapes.

20.14.1 Tile Types

The types of tile are grouped according to their permeability. *Permeability* is the tendency of the tile to absorb water. Permeable and impermeable tile are used on floors, walls, and countertops, **Figure 20-26**.

The following list is organized from most to least permeable:

- Nonvitreous
- Semivitreous
- Vitreous
- Impervious

Tile can also be organized by their characteristics. These are the most common:

- Paver
- Quarry
- Mosaic
- Lugged

Paver tile is either glazed or unglazed and at least 1/2″ thick. The smaller ones are 4″ × 6″ and larger ones are 12″ square. Paver tiles are typically used for flooring and walkways. *Quarry tile* is very dense and always unglazed. They are ideal for flooring and countertops. Generally, quarry tile is semivitreous or vitreous and ranges in thickness from 1/2″ to 3/4″. *Mosaic tile* is any tile 2″ square or smaller. They are usually vitreous and range from 3/32″ to 1/4″ square. See **Figure 20-27**. *Lugged tile* is any kind of tile that has protrusions around it to ensure exact spacing. The lugs are eventually concealed by grout. **Figure 20-28** shows several examples of tile designed and manufactured specifically for floors, walls, and counters.

20.14.2 Tile Adhesives

Tiles are attached with various types of adhesives. The most-used adhesives are mastics, dry-set, and latex-Portland cement mortars.

images and videos/Shutterstock.com

Figure 20-26. Ceramic tile comes in a variety of colors and textures. This versatile product provides durable and attractive surfaces not only for floors, but for walls, backsplashes, and countertops.

Studio Light and Shade/Shutterstock.com

Figure 20-27. This kitchen wall backsplash is laid with mosaic tile.

A   *sevi gadea/Shutterstock.com* **B**   *August_0802/Shutterstock.com* **C**   *sebos/Shutterstock.com*

Figure 20-28. Various flooring applications with tile. A—Ceramic tiles come in a range of shapes, colors, and textures. Here, tiles resemble wood planks. B—Quarry stone in granite and marble. These materials vary in hardness, veining, color, and finish. C—Glass mosaic tiles have become a popular choice as an accent. They are available in a wide range of colors and shapes.

Mastics are the least expensive tile adhesive and come ready to use. However, they do not have the strength or flexibility of mortar. Further, they should not be used where the tile will be subjected to heat (for example, around a fireplace). It is also thought that mortars are a better choice of adhesive where tile will be exposed to water.

Mortar mixes are basically mixtures of sand and Portland cement, but also contain additives that improve bonding. Dry-set can be mixed with plain water or a liquid modified with latex, acrylic, or epoxy resins. While expensive, epoxy dry-set mortar has high bonding strength and resistance to impact. Furthermore, it can be applied to nearly any surface—plastic laminate, plywood, and steel.

20.14.3 Estimating Tile

To determine the quantity of tile needed, first calculate the square footage of the area to be covered. Measure the length and width and multiply the two figures. Add an additional 10% to compensate for breakage, cuts, and extras for future repairs. Manufacturer's coverage charts are available where tile is sold. A sample is shown in **Figure 20-29**.

20.14.4 Installing Tile

Surfaces to be tiled should be tight and even. It may be necessary to check the nailing pattern of the substrate or to add an underlayment such as cement board or another type of backing board. Cement backing board sheets should be installed in a "brick" pattern. Do not allow four corners to meet at one point. Keep panels 1/8″ away from walls. Installation of cement board backing is covered in Chapter 19, **Interior Wall and Ceiling Finish**.

	Square Feet of Coverage	
	Tile size	
	4 1/4″ × 4 1/4″ 6″ × 6″	4′ × 4′ 8′ × 8′ 10′ × 10′ 12′ × 12′
	Carton size	
Number of cartons	**12.5 square feet per carton**	**11.1 square feet per carton**
1	12	11
2	25	22
3	37	33
4	50	44
5	62	55
6	75	66
7	87	77
8	100	88
9	112	99
10	125	111

American Olean Tile Co.

Figure 20-29. A coverage chart aids in determining the amount of tile needed.

It is important for good adhesion of the tiles that the mortar or mastic completely covers the back of the tile. A notched trowel should be used to apply a full coat to the underlayment. Use the notched edge to evenly spread the material.

Two methods of mortar installation are used: thick bed and thin set. In thick bed, the mortar is 3/4″–1 3/4″ thick. The mortar is allowed to set and cure before tiles are adhered with a dry-set adhesive. This method produces a surface that is strong and unaffected by frequent and prolonged wetting. In thin set, the adhesive (mortar) is applied in a thin layer to the substrate.

Thickness may vary from 1/32″ to 1/8″. Thin-setting mortar is less expensive, easier to install, and results in a quicker job.

The spaces between individual tiles must be filled with a form of mortar called *grout*. This step prevents moisture and dirt from getting between the tiles. Grout comes in dry form and can be mixed with water or latex- or acrylic-modified liquids.

Plain grout is cement with additives that give it a smooth consistency. It is used when spaces between tiles are less than 1/16″ wide. *Sanded grout* is cement with sand added to it. It is used where joints are wider than 1/16″.

Allow ceramic or clay tile installations to dry for 24–36 hours before grouting. Grout is applied with a grouting trowel. It has a cushioned surface that allows spreading of grout over the tiles without damaging them. First, the grout is spread over the surface, **Figure 20-30**. Then, it is worked into the joints. The final operation is to use a rough cloth or sponge to clean off grout from the tiles before it sets.

Goodheart-Willcox Publisher

Figure 20-30. Applying grout. Allow ceramic or clay tile installations to dry for 24–36 hours before grouting.

Construction Careers
Flooring Installer

Residential and commercial buildings, and some areas of industrial structures, make use of decorative or protective floor coverings. Depending on the application, these range from resilient tiles and "sheet goods" to wood or laminate flooring, tile products, or various types of carpeting.

The general term flooring installer covers several specialized trades, all dealing with some form of floor covering. Floor installers, or floor layers, work with wood, vinyl, and other types of resilient tile. They also install

Anne Kitzman/Shutterstock.com

These flooring installers are installing a wood floor. Installers also work with vinyl, tile, carpeting, and sheet goods. Floor installation requires a lot of bending and kneeling, and exposes the installer to dust particles and fumes.

sheet goods. Sheet goods are wide lengths of resilient flooring material that can often cover an entire room without the need for seams. Floor sanders and finishers specialize in finishing or refinishing wood floors. Carpet installers or layers work with different types of carpeting, using various installation methods. Some of these methods are stretched, glued down, or loose laid. Tile installers, or tile setters, cover floors with ceramic tiles or natural stone materials, such as marble.

Flooring installers generally work in fairly comfortable conditions, since their work is done after buildings are closed in and heated and construction debris has been removed. They must be in good physical condition, since the work requires a considerable amount of bending, kneeling, and carrying. Flooring installers encounter few safety hazards, but should wear personal protective gear as needed. For example, floor sanding generates considerable dust, adhesives, and other chemicals that can produce irritating fumes.

Carpet installers make up more than half of the flooring installer workforce. While many flooring installers work for flooring contractors or retail stores, more than 40% are self-employed. This is more than double the rate of self-employment for the construction trades as a whole.

While some apprenticeship programs exist, most flooring installers learn their trades through informal on-the-job training. Trainees start as helpers and are gradually given additional responsibility as their skills develop. Since flooring installation involves a great deal of measuring and calculation, good mathematical ability is an asset.

Chapter Review

Summary

- Finish flooring is available in many materials, ranging from flagstone, slate, brick, and ceramic tile through resilient tiles and sheet vinyl to various wood products.
- Finish flooring materials are usually installed after all wall and ceiling surfaces are completed.
- Strip flooring is fitted together with tongue-and-groove joints and is blind nailed through the groove.
- Parquet flooring is usually laid in a mastic.
- Laminate flooring floats over the subfloor. Tongue-and-groove joints between strips snap together to firmly hold the pieces together.
- Plywood, hardboard, or cement board are used as underlayments.
- Resilient flooring, such as tile or sheet vinyl, requires a smooth underlayment that will not telegraph irregularities through to the flooring.
- The most-used adhesives are mastics, dry-set, and latex-Portland cement mortars. Some tiles are supplied with adhesive already applied and can merely be pressed in place on the underlayment.
- Tiles are grouped according to their permeability.
- Spaces between tiles are filled with a mortar-like material called grout.

Know and Understand

Answer the following questions using the information in this chapter.

1. _____ have largely replaced asphalt tile, rubber tile, and linoleum as a finish flooring material.
   - A. Slates
   - B. Brick
   - C. Ceramic tiles
   - D. Vinyl plastics

2. _____ flooring the most widely used wood flooring.
   - A. Strip
   - B. Plank
   - C. Block
   - D. Engineered

3. *True or False?* Standard thicknesses of hardwood strip flooring include 3/8″, 1/2″, and 3/4″.

4. *True or False?* Wood flooring should be delivered on the day it is to be installed.

5. When the starter strip is located in the center of an area, install a _____ in the groove to create a tongue.
   - A. sleeper
   - B. setup line
   - C. spline
   - D. succeeding strip

6. What is the first step in laying wood strip flooring over concrete?
   - A. Snap chalk lines 16″ apart.
   - B. Clean and prime the floor.
   - C. Join polyethylene sheets by forming a lap over a sleeper.
   - D. Embed treated wood sleepers in bands of adhesive.

7. How many board feet are there in a standard bundle of hardwood flooring?
   - A. 12
   - B. 24
   - C. 48
   - D. 52

8. Laminate flooring consists of a fiberboard core, melamine paper, and a _____.
   - A. series of glued joints
   - B. mastic
   - C. protective adhesive
   - D. protective topcoat

9. *True or False?* Engineered wood strip flooring is attached to a plywood subfloor.

10. What can be used on underlayment to prevent nail "popping" under resilient floors?
    - A. Special fasteners.
    - B. A special nailing pattern.
    - C. Special adhesives.
    - D. Floor-leveling compound.

11. A resilient flooring material that is especially susceptible to telegraphing small irregularities in the base is _____.
    - A. linoleum
    - B. vinyl
    - C. rubber
    - D. asphalt

12. *True or False?* When laying resilient tiles, tiling is started at the entrance to the room.

13. The area of a room to be floored with self-adhering tiles is 105 sq ft and you are using 12″ × 12″ tiles. Accounting for 10% waste, how many tiles will you need to complete the job?
    - A. 89
    - B. 116
    - C. 132
    - D. 144

14. Which of the following is *not* a method of forming a bisque for floor tile?
   A. Extrusion.
   B. Die pressing.
   C. Cut from a sheet.
   D. Baking in a kiln.

15. _____ tile is very dense, always unglazed, and ideal for floors.
   A. Paver
   B. Quarry
   C. Mosaic
   D. Lugged

16. _____ is cement with sand added to it.
   A. Sanded grout
   B. Plain grout
   C. Mortar
   D. Mastic

Apply and Analyze

1. When is finish flooring generally laid in construction?
2. Explain the four types of wood flooring used in residential structures.
3. What is "racking the floor"?
4. Name one advantage and one disadvantage to using prefinished wood flooring.
5. Why does cement board underlayment require no expansion gap between panels?
6. How is installation of self-adhering resilient tiles different from that of standard resilient tiles?
7. What are two advantages that vinyl plank flooring offers when compared to laminate flooring?

Critical Thinking

1. Research options for flooring. Narrow your search to your top two choices for your kitchen. Make a material list to lay both types of flooring. Visit or call your local supplier to determine the cost to purchase all materials needed to lay both. Which material will you use and why? What are the advantages and disadvantages of each type of flooring?

2. Building practices vary from one geographic location to another. For instance, in the southern United States it is common to build a home on a slab foundation, while in northern states, homes usually have a basement or a crawlspace. What impact will this make on flooring choices and installation practices? What other factors based on geographic region or climate might impact flooring choices or installation practices? Explain.

Communicating about Carpentry

1. **Speaking.** With a partner, role-play the following situation: A carpenter must explain the different options for finish flooring to be installed in a customer's home. Use your own words to describe strip, plank, and unit block wood flooring, and explain the procedure for applying hardboard, particleboard, waferboard, and plywood underlayment. The customer should ask questions if the explanation is unclear. Switch roles and repeat the activity.

2. **Reading and Writing.** Working with a partner, research the benefits that carpenters can enjoy from joining a union. Determine how much dues must be paid to be a union member, and the benefits union members enjoy. Develop a report for the class on the benefits of being a member of the carpenters' union.

CHAPTER **21**

Stair Construction

OBJECTIVES

After studying this chapter, you will be able to:

- Identify the various types of stairs.
- Define basic stair parts and terms.
- Calculate the rise-run ratio, number and size of risers, and stairwell length.
- Lay out stringers for a given stair rise and run.
- Prepare sketches of the types of stringers.
- List commonly available prefabricated stair parts.

TECHNICAL TERMS

baluster	nosing	total run
balustrade	platform	tread
built-up stringer	riser	unit rise
cut-out stringer	run of stairs	unit run
handrail	stairwell	wall rail
headroom	straight run	winder
housed stringer	stringer	winding stairs
newel	total rise	

A stair is a series of steps, each elevated a measured distance, leading from one level of a structure to another. When the series is a continuous section without breaks formed by landings or other constructions, the terms flight of stairs or *run of stairs* are often used. Other terms that can be properly used include stairway and staircase.

Because of European influence, main stairs have often been the chief architectural feature in an entrance hallway or other area. However, in new construction, public rooms are usually on the first floor. Due to this, there is a trend to move the stairs to a less conspicuous location.

Stair construction requires a high degree of skill. The quality of the work should compare with that found in fine cabinetwork, **Figure 21-1**. The parts for main stairways are usually made in millwork plants and then assembled on the job. Even so, the assembly work must be performed by a skillful carpenter who understands the basic principles of stair design and knows layout and construction procedures.

Main stairways are usually not built or installed until after interior wall surfaces are complete and finish flooring or underlayment has been laid. Basement stairs should not be installed until the concrete floor has been placed.

Carpenters build temporary stairs from framing lumber to provide access until the permanent stairs are installed. These are often designed as a detachable unit so they can be moved from one project to another. Sometimes, permanent carriages are installed during the rough framing and temporary treads are attached. Carriages, or *stringers*, are the inclined supports that carry the treads and risers. In this case, after the interior is finished, the temporary treads are replaced with finished parts.

Figure 21-1. An attractive main set of stairs is a desirable architectural feature in a residence. Stair design and construction has long been considered one of the highest forms of joinery.

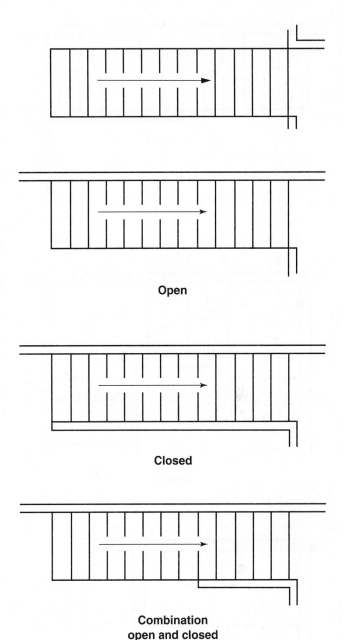

Open

Closed

**Combination
open and closed**

Goodheart-Willcox Publisher

Figure 21-2. Open and closed stairways. A stair is called open even if one side is enclosed by a wall.

21.1 Types of Stairs

Basically, there are two stair categories: service stairs and main stairs. Either of these may be closed, open, or a combination of open and closed. See **Figure 21-2**. In addition, the type of stairs may be platform, straight run, or winding.

The *platform* type includes landings where the direction of the stair run is usually changed. Such descriptive terms as L-type (long L and wide L), double L-type, and U-type (wide U and narrow U) are commonly used. See **Figure 21-3**.

In split-level houses, platform stairs with short and generally straight runs are used. Usually, stair runs of this type are located so that the stair run directly above automatically provides headroom, **Figure 21-4**.

The *straight run* stairway is continuous from one floor level to another without landings or turns. It is the easiest to build. Standard multistory designs require a long *stairwell* in the floor above to provide headroom.

This often presents a problem in smaller structures. A long run of 12–16 steps also has the disadvantage of being tiring. It offers no chance for a rest during ascent.

Winding stairs, also called geometrical, are circular or elliptical. They gradually change directions as they ascend from one level to another. These often require curved wall surfaces that are difficult to build. Because of their expense, winding stairs are usually only found in high-end homes.

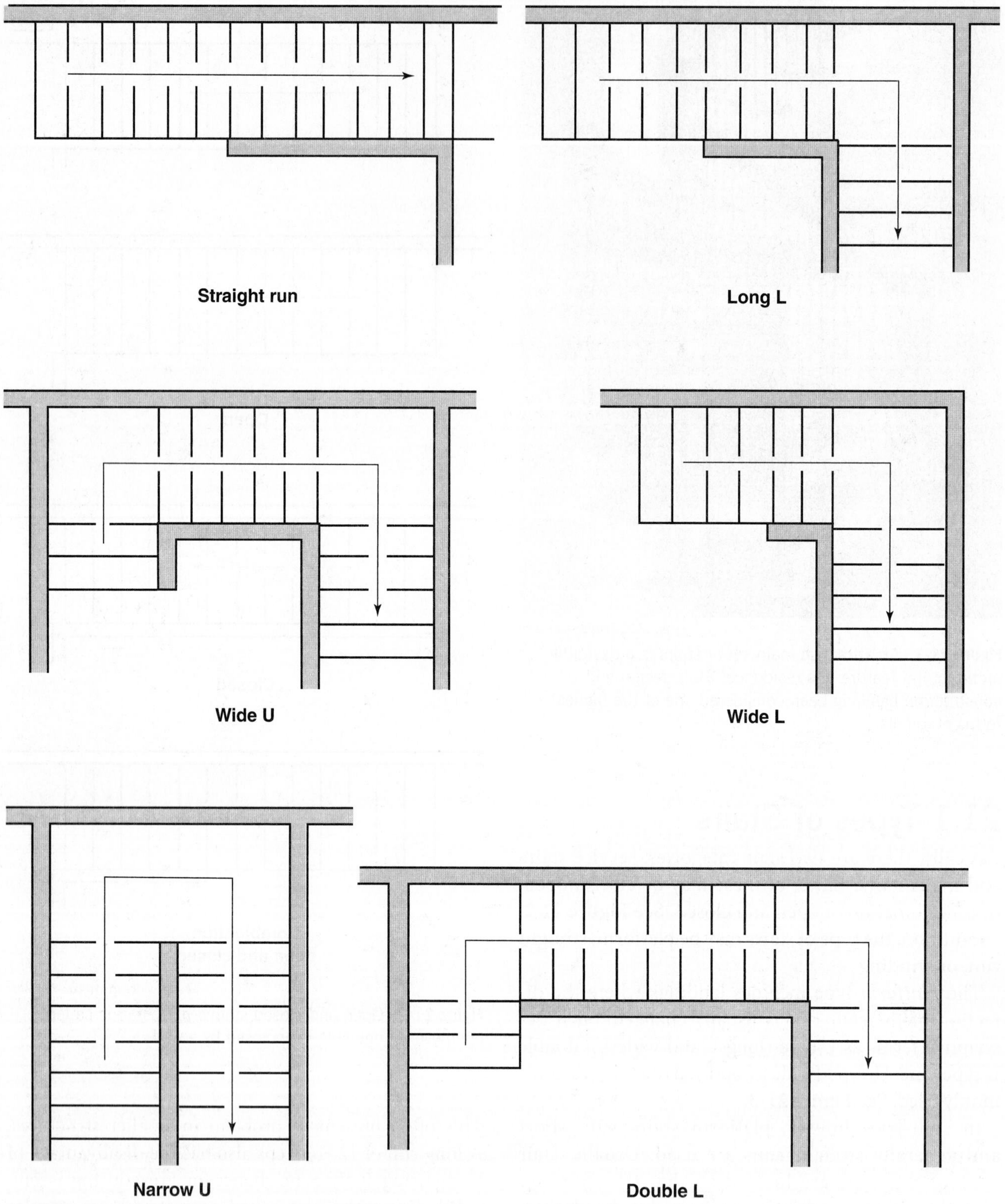

Straight run

Long L

Wide U

Wide L

Narrow U

Double L

Figure 21-3. Terms used to define various stair types.

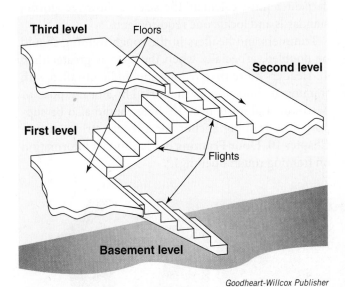

Figure 21-4. Stair runs are often made one above the other to gain headroom. This one is designed for a split-level home.

21.2 Stair Parts

Stairs are basically sets of *risers* (vertical stair members between consecutive stair treads) and *treads* (horizontal walking surface of a stair) supported by stringers, **Figure 21-5**. The relationship between the riser height (*unit rise*) and the tread width (*unit run*) determines how easily the stairs may be negotiated. Research has indicated that the ideal riser height is 7″, while the ideal tread width is 11″.

Code Note

Headroom is measured from a line along the front edges of the treads to the enclosed surface or header above. The minimum distance specified by the International Residential Code (IRC) is 6′-8″, but local building codes may be different. Refer again to **Figure 21-5**.

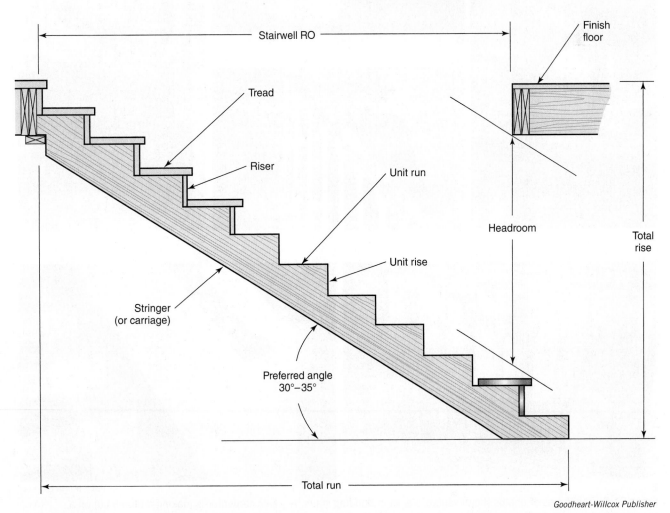

Figure 21-5. Basic stair parts and terms. The total number of risers is always one greater than the total number of treads.

21.3 Stairwell Framing

Methods of stair building differ from one locality to another. One carpenter may cut and install a stringer (carriage) during the wall and floor framing. Another may put off all stair work until the interior finishing stages. **Figure 21-6** shows several stages of stair building.

Regardless of procedures followed, the rough openings for the stairwell must be carefully laid out and constructed. If the architectural drawings do not include dimensions and details of the stair installation, then a carpenter must calculate the sizes. Follow recognized standards and local code requirements.

Trimmers and headers in the rough framing should be doubled, especially when the span is greater than 4′. Headers more than 6′ long should be installed with framing anchors, unless supported by a beam, post, or partition. Tail joists over 12′ long should also be supported by framing anchors or a ledger strip. Refer to Chapter 10, **Floor Framing**, for additional information on framing rough openings.

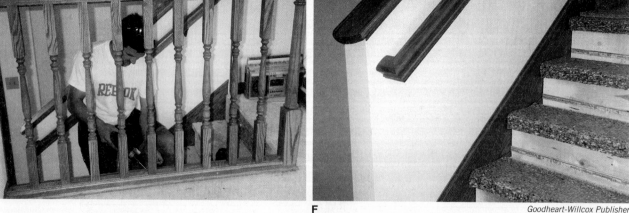

Goodheart-Willcox Publisher

Figure 21-6. This series of photos shows various stages in building stairs. A—This carpenter is making a plumb cut on a housed (closed) stringer. Grooves are cut in the stringer to receive the treads and risers. B—This stringer is a cut-out type. The 2×4 spacer (arrow) gives clearance for installation of the wall finish. C—Newel post is being installed. A carpenter is checking for plumb. D—The handrail has been installed and a carpenter is cutting and placing the prefinished balusters. E—This carpenter is fastening the lower rail of a banister to the floor. Since the banister is made of oak, it is necessary to drill pilot holes for the nails. Glue is also applied to each joint. F—These stairs are completed except for the installation of carpeting.

Providing adequate headroom is often a problem, especially in smaller structures. Installing an auxiliary header close to the main header permits a slight extension in the floor area above a stairway, **Figure 21-7**. When a closet is located directly above the stairway, the closet floor is sometimes raised for additional headroom.

21.4 Stair Design

Most important in stair design is the mathematical relationship between the riser and tread. There are three generally accepted rules for calculating the rise-run or riser-tread ratio:

- The sum of two risers and one tread should equal 24″–25″.
- The sum of one riser and one tread should equal 17″–18″.
- The height of the riser times the width of the tread should equal 70″–75″.

According to the first rule, a riser 7 1/2″ high requires a tread of 10″. A 6 1/2″ riser requires a 12″ tread. Additionally, the IRC has four rules concerning tread width and riser height:

- The maximum riser height is 7 3/4″.
- All risers in a flight of stairs must be within 3/8″ of the same height.
- The minimum tread width is 10″.
- All treads in a flight of stairs must be within 3/8″ of the same width.

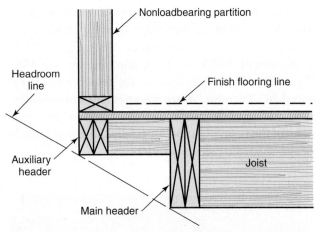

Goodheart-Willcox Publisher

Figure 21-7. Extending the upper floor area with a shallow auxiliary header to provide headroom. The partition over the auxiliary header must not be load bearing.

In residential structures, treads (excluding nosing) are seldom more than 12″. Nosing is a small extension of the tread. In a given run of stairs, it is extremely important that all of the treads be the same size. The same is true of the risers. A person tends to subconsciously measure the first few steps and will probably trip if subsequent risers are not the same.

When the rise-run combination is wrong, climbing the stairs will be tiring and cause extra strain on leg muscles. Further, the toe may kick the riser if the tread is too narrow. A unit rise of 7″–7 5/8″ with an appropriate tread width provides both comfort and safety. As stair rise is increased, the run must be decreased. See **Figure 21-8**.

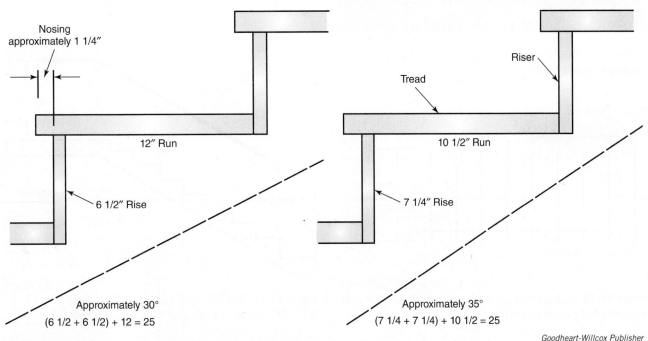

Goodheart-Willcox Publisher

Figure 21-8. Be careful about rise-run relationships in stair design.

A stair should be wide enough to allow two people to pass without contact. Further, it should provide space so furniture can be moved up or down, **Figure 21-9**. A minimum width of 3′ is required by the IRC, **Figure 21-10**. On spiral stairs, the requirement is reduced 26″. Furniture moving is an important consideration and extra clearance should be provided in closed stairs of the L- and U-type, especially those that include wedge-shaped treads, or *winders*.

Stairs must have a continuous rail along the side for safety and convenience. A *handrail* is a pole installed above and parallel to stair steps to act as a support for persons using the stairs. Also called a stair rail, a handrail is used on open stairways that are constructed with a low partition or banister. In closed stairs, the support rail is called a *wall rail*. It is attached to the wall with special metal brackets. Except for very wide stairs, a rail on only one side is sufficient. **Figure 21-11** illustrates the correct height for a rail.

A complete set of architectural plans should include detail drawings of main stairs, especially when the design includes any unusual features. For example, the stair layout in **Figure 21-12** shows a split-level entrance with open-riser stairs leading to upper and lower floors. An exact description of tread mountings, overlap, nosing requirements, and height of the handrail is not included. These items of construction are the responsibility of a carpenter, who must have a thorough understanding of basic stair design and how to lay out and make the installation.

All stairs, whether main or service, are shown on the floor plans. When details of the stair design are not included in the complete set of plans, the architect usually specifies on the plan view the number and width of the treads for each stair run. Sometimes, the number of risers and the riser height are also included.

Pro Tip

In a given run of stairs, be sure to make all of the risers the same height and all of the treads the same width. An unequal riser, especially one that is too high, may cause a fall.

pics721/Shutterstock.com

Figure 21-9. This run of stairs is wide enough for moving furniture up and down.

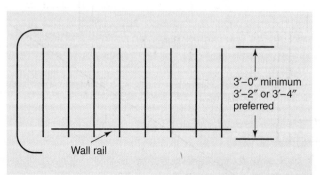

Goodheart-Willcox Publisher

Figure 21-10. A main stair should be at least 3′ wide for easy movement of people and furniture.

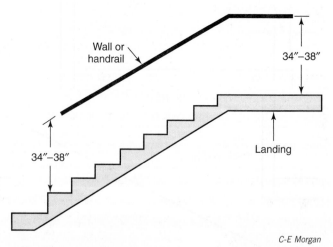

C-E Morgan

Figure 21-11. The IRC specifies a handrail height of 34″–38″. Always consult your local building codes.

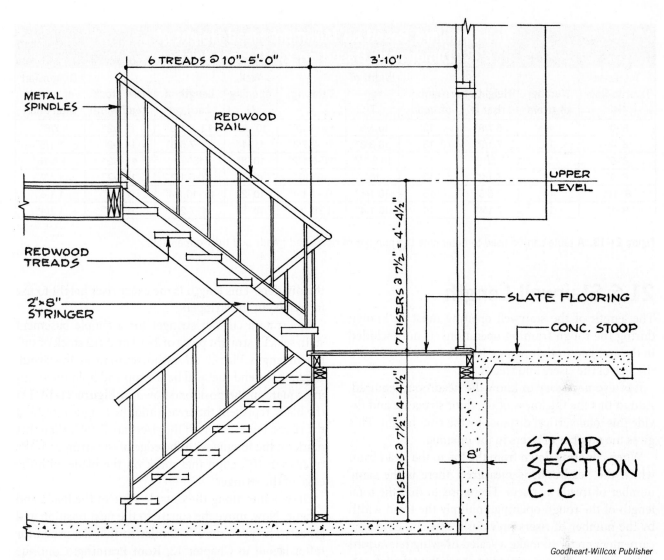

Figure 21-12. Architectural drawings show stair layouts. Note the information given for riser-tread ratios.

Goodheart-Willcox Publisher

21.5 Stair Calculations

To calculate the number and size of risers and treads (less nosing) for a given stair run, first divide the ***total rise*** (vertical distance from one floor to another) by 7 to determine the number of risers. Some carpenters divide by 8. Either number is accurate enough. For example, if the total rise for a basement stairway is 7′-10″ (94″), dividing by 7 yields 13.43. Since there must be a whole number of risers, round 13.43 to 13. Divide the total rise by that number to determine the unit rise:

$$\text{Unit rise} = 94″ \div 13$$
$$= 7.23, \text{round to 7 1/4″}$$
$$\text{Number of risers} = 13$$
$$\text{Riser height} = 7\ 1/4″$$

In any stair run, the number of treads is one less than the number of risers. A 10 1/2″ tread is correct for this example. The horizontal distance occupied by the stairs, measured from the foot of the stairs to a point directly beneath where the stairs rest on a floor or landing above is called the ***total run***. It is calculated as follows:

$$\text{Number of treads} = 12$$
$$\text{Total run} = 10\ 1/2″ \times 12$$
$$= 126″$$
$$= 10′\text{-}6″$$

The stairs in this example will have 13 risers 7 1/4″ high, 12 treads 10 1/2″ wide, and a total run of 10′-6″.

Some manufacturers supply tables for determining rise and run, riser, and tread ratios. See **Figure 21-13**.

Well Openings Based on Minimum Head Height of 6'-8" Dimensions Based on 2×10 Floor Joist									
Total rise floor to floor H	Number of risers	Height of riser R	Number of treads	Width of run T	Total run L	Well opening U	Length of carriage	Use stock tread width	Dimension of nosing projection
8'-0"	14	6 7/8"	13	10 5/8"	11'-6 1/8"	10'-10"	13'-8 1/2"	11 1/2"	7/8"
8'-4"	14	7 1/8"	13	10 3/8"	11'-2 7/8"	11'-1"	13'-7 5/8"	11 1/2"	1 1/8"
8'-6"	14	7 5/16"	13	10 3/16"	11'-0 1/2"	10'-8"	13'-7"	11 1/2"	1 5/16"
8'-9"	14	7 1/2"	13	10"	10'-10"	10'-1"	13'-6 1/2"	11 1/2"	1 1/2"
8'-11"	14	7 5/8"	13	10 1/4"	11'-1 1/4"	10'-2"	13'-10 1/4"	11 1/2"	1 1/4"
9'-1"	15	7 1/4"	14	10 1/4"	11'-11 1/2"	10'-8"	14'-7 3/4"	11 1/2"	1 1/4"

C-E Morgan

Figure 21-13. A table can be used to determine the number of risers and treads and their dimensions.

21.6 Stairwell Length

The length of the stairwell opening must be known during the rough framing operations. If not included in the architectural drawings, it can be calculated from the size of the risers and treads.

It is also necessary to know the headroom required. Add to this the thickness of the floor structure and divide this total vertical distance by the riser height. This gives the number of risers in the opening.

When counting down from the top to the tread from which the headroom is measured, there is the same number of treads as risers. Therefore, to find the total length of the rough opening, multiply the tread width by the number of risers previously determined. Some carpenters prefer to make a scaled drawing (elevation) of the stairs and floor section to check the calculations.

21.7 Stringer Layout

To lay out the stair stringer, first determine the riser height. Place a story pole in a plumb position from the finished floor below through the rough stair opening above. On the pole, mark the height of the top of the finished floor above.

Set a pair of dividers to the calculated riser height and step off the distances on the story pole. There will likely be a slight error in the first layout, so adjust the setting and try again. Continue adjusting the dividers and stepping off the distance on the story pole until the last space is equal to all of the others. Measure the setting of

the dividers. This length is the exact riser height to use in laying out the stringers.

To create a cut-out stringer for a simple basement stair, select a straight piece of 2×10 or 2×12 stock of sufficient length. Place it on sawhorses to make the layout. Begin at the end that will be the top and hold the framing square in the position shown in **Figure 21-14**. Let the blade represent the treads and the tongue represent the risers. For example, if the risers are 7 3/4", align that mark on the tongue with the edge of the stringer. If the treads are 10", align that mark on the blade with the edge of the stringer.

Draw a line along the outside edge of the blade and tongue. Now, move the square to the next position and repeat. The procedure is similar to that described for rafter layout in Chapter 12, **Roof Framing**. Continue stepping off with the square until the required number of risers and treads have been drawn, **Figure 21-15**.

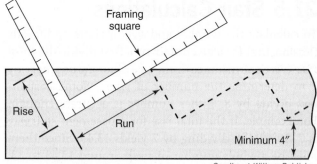

Goodheart-Willcox Publisher

Figure 21-14. Using a framing square to lay out a stringer.

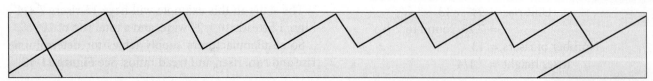

Goodheart-Willcox Publisher

Figure 21-15. A completed stringer layout will look something like this.

The stair begins with a riser at the bottom, so extend the last tread line to the back edge of the stringer. At the top, extend the last tread and riser line to the back edge.

One other adjustment must be made before the stringer is cut. Earlier calculations that gave the height of the riser did not take into account the thickness of the tread. Therefore, the total rise of the stringer must be shortened by one tread thickness. Otherwise, the top tread will be too high. The bottom of the stringer must be trimmed, as shown in **Figure 21-16**.

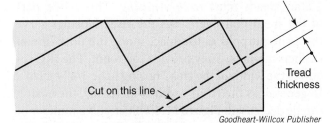

Goodheart-Willcox Publisher

Figure 21-16. Trim the bottom end of the stringer to adjust the riser height for the tread thickness.

Pro Tip

Extreme accuracy is required in laying out the stringer. Be sure to use a sharp pencil or knife and make the lines meet on the edge of the stock. Accuracy can be ensured in this layout by using stair gauges or by clamping a strip of wood to the blade and tongue. **Figure 21-17** shows how stair gauges are used.

21.8 Treads and Risers

The thickness of a main stair tread is generally 1 1/6″ or 1 1/8″. Hardwood or softwoods may be used. The US Federal Housing Administration requires that stair treads be hardwood, vertical-grain softwoods, or flat-grain softwoods covered with a suitable finish flooring material.

Lumber for risers is usually 3/4″ thick and should match the tread material. This is especially important when the stairs are not covered. In most construction, the riser drops behind the tread, making it possible to reinforce the joint with nails or screws driven from the back side of the stairs. **Figure 21-18** shows basic types of riser designs. A sloping riser is sometimes used in concrete steps since it provides an easy way to form a nosing. The IRC requires sloping risers to be within 30° of vertical.

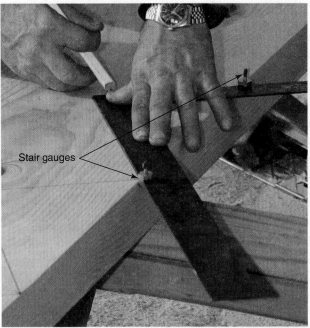

Dwight Smith/Shutterstock.com

Figure 21-17. Stair gauges are clamped to the edge of a framing square to make it easy to duplicate angles.

Where the top edge of the riser meets the tread, glue blocks are sometimes used. A rabbeted edge of the riser may fit into a groove in the tread. A rabbet and groove joint may also be used where the back edge of the tread meets the riser.

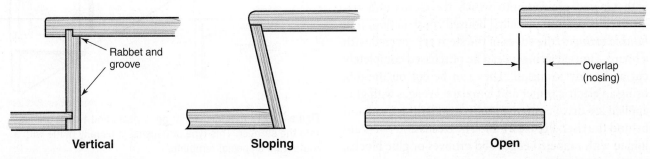

Goodheart-Willcox Publisher

Figure 21-18. Basic stair riser shapes. For the open riser, the tread should overlap the riser at least 2″.

Stair treads must have a *nosing*. This is the part of the tread that overhangs the riser. Nosings serve the same purpose as toe space along the floor line of kitchen cabinets. They provide toe room. The IRC requires the width of the tread nosing to be 3/4″– 1 1/4″. In general, as the tread width is increased, the nosing can be decreased. **Figure 21-19** illustrates a number of nosing forms. Cove molding may be used under the nosing to cover the joint between riser and tread and conceal nails used to attach the riser to the stringer or carriage.

Basement stairs may be constructed with an open riser (no riser board installed). Sometimes an open riser design is built into a main stair to provide a special effect. Open risers cannot be completely "open." The space where the riser would normally be placed must include a structure that would prevent a 4″ diameter sphere (ball) from passing through. Various methods of support or suspension may be used. Often, custom-made metal brackets or other devices are needed.

21.9 Types of Stringers

Treads and risers are supported by stringers that are solidly fixed to the wall or framework of the building. For wide stairs, a third stringer is installed in the middle to add support.

The simplest type of stringer is the *built-up stringer*. It is a stringer to which blocking has been added to form a base for adding treads and risers. It is formed by attaching cleats on which the tread can rest. *Cut-out stringers* are stringers into which the rise and run are cut. They are commonly constructed for either main or service stairs. This is the type created in the earlier layout description. Prefabricated treads and risers are often used with this type of stringer. An adaptation of the cut-out stringer, called semihoused construction, is shown in **Figure 21-20**. The cut-out stringer and backing stringer may be assembled and then installed as a unit or each part may be separately installed.

A popular type of stair construction has a stringer with tapered grooves into which the treads and risers fit. It is commonly called housed construction. In a *housed stringer*, the edges of the steps are covered with a board. Housed stringers can be purchased completely cut and ready to install. They can be cut on the job, using an electric router and template. Wedges with glue applied are driven into the grooves under the tread and behind the riser, **Figure 21-21**. The treads and risers are joined with rabbeted edges and grooves or glue blocks.

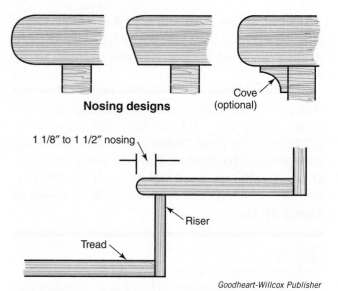

Nosing designs

Cove (optional)

1 1/8″ to 1 1/2″ nosing

Riser

Tread

Goodheart-Willcox Publisher

Figure 21-19. Common tread nosings.

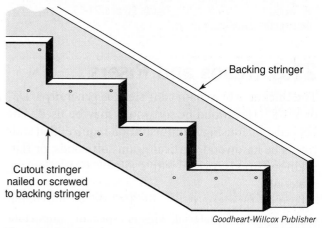

Backing stringer

Cutout stringer nailed or screwed to backing stringer

Goodheart-Willcox Publisher

Figure 21-20. This is a semihoused stringer, an adaptation of the cut-out stringer.

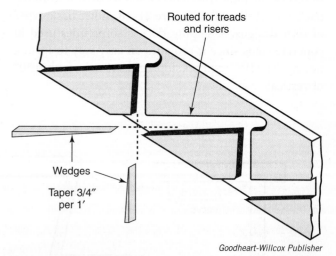

Routed for treads and risers

Wedges

Taper 3/4″ per 1′

Goodheart-Willcox Publisher

Figure 21-21. In a housed stringer, risers and treads are let into the stringer. This type of housing is usually made with a router and a special template.

To assemble the stairs, the housed stringer is spiked to the wall surface and into the wall frame. The treads and risers are then set into place. Work is done from the top downward.

Housed construction produces a stair that is strong and dust tight. It seldom develops squeaks. Housed stringers show above the profiles of the treads and risers and provide a finish strip along the wall. The design should permit a smooth joint where it meets the baseboard of the upper and lower levels.

Stairs are often built at a millwork shop and transported to the construction site as a fully-assembled unit, **Figure 21-22**. The entire stair assembly, minus the banister, is set in place in the building frame. This means that the framing must be done accurately, so the opening for the stair fits the prebuilt stair assembly.

21.10 Winder Stairs

Winder stairs present stair conditions that are frequently regarded as undesirable. In fact, some localities do not allow them. Check local building codes to see if this type of stairs is allowed. The use of winder stairs may sometimes be necessary, however, where space is limited. When used, it is important to maintain a

Top Tread Stairways

Figure 21-22. This stair is assembled and ready to be transported to the building site.

winder-tread width along the line of travel that is equal to the tread width in the straight run. The IRC specifies that the narrowest end of winder treads must be at least 6″ and the width at the line of travel must be at least 10″. The line of travel is a line 12″ from the inside edge of the stair, **Figure 21-23**. Before starting the construction of this type of stairs, a carpenter should make a full-size or carefully scaled layout in the plan view. The best radius for the line of travel can then be determined.

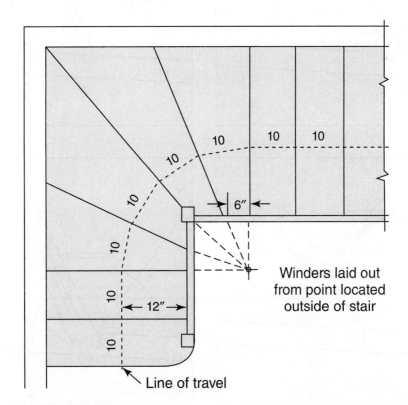

Goodheart-Willcox Publisher

Figure 21-23. Laying out winder stairs with lines representing the tread nosings converging outside of the construction.

PROCEDURE

Splitting Angles for Miter Cuts

A carpenter sometimes faces odd angles that must be accurately split to make a miter cut. This can be mathematically calculated, but the following method avoids the math and ensures great accuracy.

1. Select a plywood scrap about 6″ wide and 1′ long with a factory edge to use as a storyboard.
2. Draw a line near to and parallel with the factory edge.
3. Use a T-bevel to find the angle to be mitered and transfer the angle to the storyboard.
4. Draw a line along the blade of the T-bevel, as shown in **Figure 21-24A**.

5. Open a pencil compass or scribe about 3″ or 4″. Place the point of the instrument at the intersection of lines AB and AC and draw arcs of equal length across both lines, **Figure 21-24B**.
6. Swing arcs of equal distance from points B and C to create point D, **Figure 21-24C**. You may need to open up the compass or dividers more to create this point.
7. Draw a line to connect points A and D, **Figure 21-24D**. This is the miter angle.
8. Adjust the T-bevel to this angle and use it to set the miter saw.
9. Make a test cut on scraps to verify the accuracy of the angle.

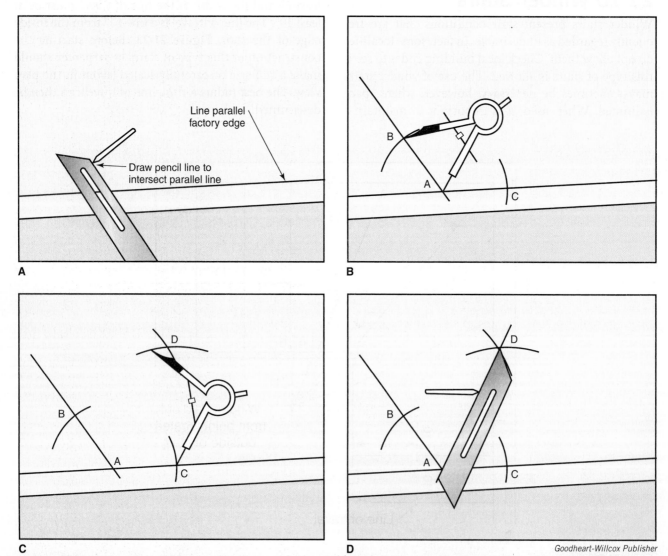

A

B

C

D

Goodheart-Willcox Publisher

Figure 21-24. Splitting an angle for a miter cut. A—After drawing a line parallel to the factory edge, transfer the miter angle to the board. B—Draw arcs of equal lengths from the intersection of lines AB and AC. C—Swing arcs to create point D. D—Draw line AD. This is the miter angle.

21.11 Open Stairs

Stairs that are open on one or both sides require some type of decorative enclosure and support for a handrail. Typical designs consist of an assembly of parts called a **balustrade**, **Figure 21-25**, consisting of a railing resting on a series of balusters that, in turn, rest on a base, usually the treads. The principal members of a balustrade are newels, balusters, and rails. They are usually made in a factory and assembled on the job by a carpenter.

A **newel** is the main post at the start of a stair and the stiffening post at the landing. The starting newel must be securely anchored either to the starter step or carried down through the floor and attached to a floor joist. **Balusters** are the spindles that support the handrail on open stairs. They are joined to the stair treads or shoe rail using either a round or square mortise. Two or three may be mounted on each tread.

Code Note

The main purpose of balusters is to prevent anyone, children especially, from slipping under the railings and falling to the floor below. The IRC and most local codes require baluster spacing of no more than 4″.

21.12 Using Stock Stair Parts

While many parts of a main staircase could be cut and shaped on the job, the usual practice is to use factory-made parts. These are available in a wide range of stock sizes and can be selected to fill requirements for most standard stair designs. See **Figure 21-26**. Stair parts are ordered through lumber and millwork dealers. They are shipped to the building site in heavy, protective cartons along with directions for fitting and assembly.

Figure 21-27 shows a balustrade assembled using stock parts. Hardware especially designed for stair work is illustrated in **Figure 21-28**.

21.13 Spiral Stairways

Metal spiral stairways eliminate framing and save space. See **Figure 21-29**. Units are available in aluminum or steel and in a variety of designs to fit requirements up to 30 steps and heights up to 22′-6″. Use of spiral stairs is often restricted by building codes.

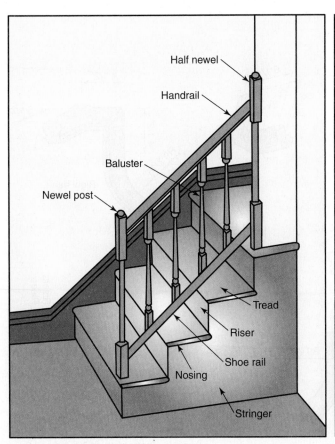
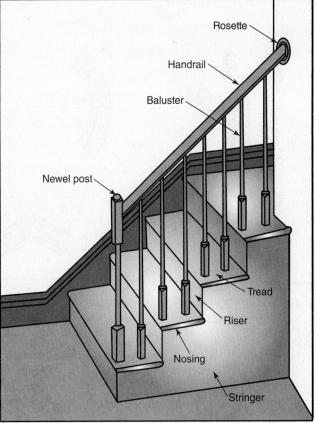

Goodheart-Willcox Publisher

Figure 21-25. Parts of an open stair. An assembly including a newel, balusters, and rail is called a balustrade.

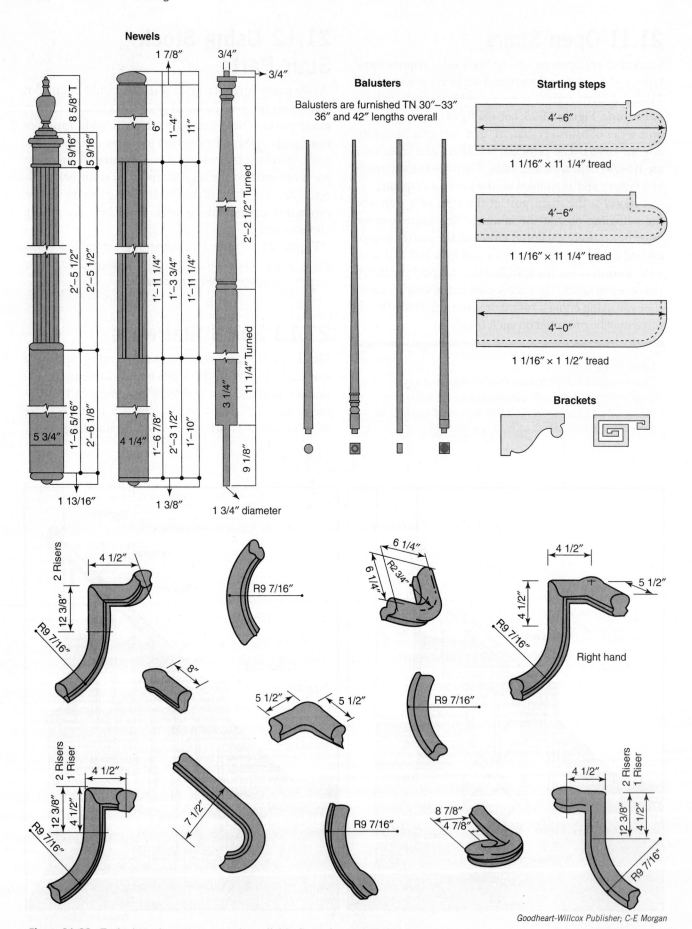

Figure 21-26. Typical stock parts commonly available for stair construction.

Goodheart-Willcox Publisher; C-E Morgan

Goodheart-Willcox Publisher

Figure 21-27. Balustrade assembly produced from stock parts.

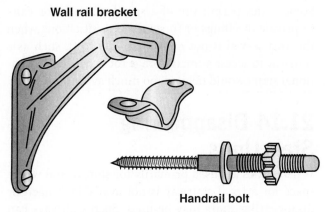

Wall rail bracket

Handrail bolt

Goodheart-Willcox Publisher

Figure 21-28. Hardware for stair rails. The handrail bolt is concealed in the center of a joint. The nut is accessible from below and can be adjusted with a screwdriver or a hammer and nail set.

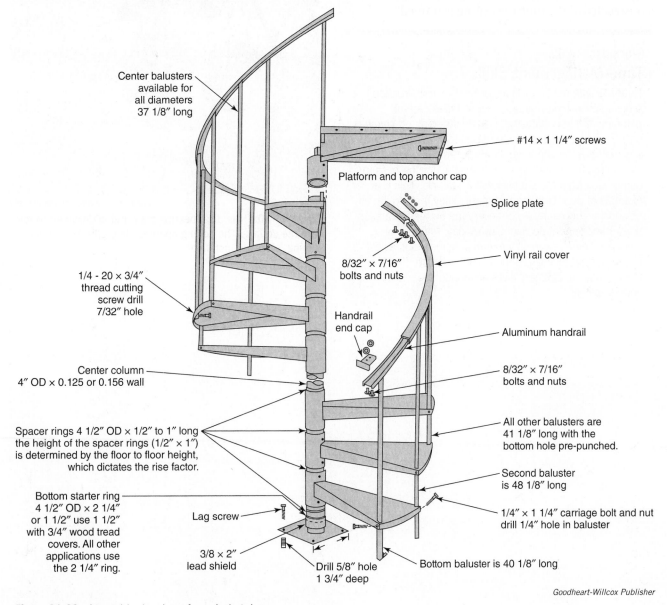

Center balusters available for all diameters 37 1/8″ long

#14 × 1 1/4″ screws

Platform and top anchor cap

Splice plate

Vinyl rail cover

8/32″ × 7/16″ bolts and nuts

1/4 - 20 × 3/4″ thread cutting screw drill 7/32″ hole

Handrail end cap

Aluminum handrail

8/32″ × 7/16″ bolts and nuts

Center column 4″ OD × 0.125 or 0.156 wall

Spacer rings 4 1/2″ OD × 1/2″ to 1″ long the height of the spacer rings (1/2″ × 1″) is determined by the floor to floor height, which dictates the rise factor.

All other balusters are 41 1/8″ long with the bottom hole pre-punched.

Second baluster is 48 1/8″ long

Bottom starter ring 4 1/2″ OD × 2 1/4″ or 1 1/2″ use 1 1/2″ with 3/4″ wood tread covers. All other applications use the 2 1/4″ ring.

Lag screw

3/8 × 2″ lead shield

Drill 5/8″ hole 1 3/4″ deep

1/4″ × 1 1/4″ carriage bolt and nut drill 1/4″ hole in baluster

Bottom baluster is 40 1/8″ long

Goodheart-Willcox Publisher

Figure 21-29. Assembly drawing of a spiral stairway.

Some codes permit use of a spiral stairway for exits in private dwellings or in some other situations when the area served is not more than 400 sq ft, such as a stairway to access a small storage room, where conventional stairs would take up too much available space.

21.14 Disappearing Stair Units

Where attics are used primarily for storage and where space for a fixed stairway is not available, hinged or disappearing stairs may be used. Such stairways can be purchased ready to install. They operate through an opening in the ceiling and swing up into the attic space when not in use, **Figure 21-30**. Where such stairs are to be provided, the attic floor should be designed for regular floor loading and the rough opening should be constructed at the time the ceiling is framed.

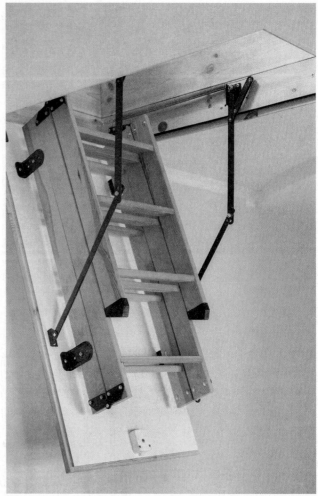

Olexandr Panchenko/Shutterstock.com

Figure 21-30. This disappearing stair unit is designed to fold into the ceiling. The ceiling opening should be framed as the ceiling joists are installed.

Chapter Review

Summary

- A stairway is a series of steps, each elevated a measured distance, leading from one level of a structure to another.
- Stairways are either platform, straight run, or winding (circular or elliptical).
- Stair parts include risers, treads, and stringers
- All treads must be the same size to ensure safe use. Risers must also be the same size.
- Stairwell framing procedures vary but involve precise lay out and construction.
- Stairs must have a continuous rail along the side for safety and convenience.
- The number of steps and risers is calculated using the total rise (vertical height) and total run (horizontal distance) occupied by the stairs.
- The length of a stairwell opening and required headroom must be known during the rough framing operations.
- Stringer layout starts by finding the riser height through a series of calculations.
- Types of stringers are built-up and cut-out. A type of cut-out stringer is the semihoused stringer.
- Housed stringers are a type of stringer with tapered grooves that treads and risers fit into.
- Winder stairs are used where space is limited.
- Stairs that are open on one or both sides require a balustrade.
- Many staircase parts, such as balusters and rails, can be purchased factory-made.
- Spiral and disappearing stair are often used where space is at a premium.

Know and Understand

Answer the following questions using the information in this chapter.

1. The platform type of a stairway includes _____, where the direction of the stair run is usually changed.
   A. stringers
   B. landings
   C. stairwells
   D. risers

2. _____ stairs are usually only found in high-end homes.
   A. Platform type
   B. Straight run
   C. Winding
   D. Main

3. One of the rules used to calculate riser-tread relationship states that the sum of one riser and one tread should equal _____"–_____".
   A. 7 1/2, 10
   B. 17, 18
   C. 24, 25
   D. 70, 75

4. A stair should be wide enough to allow _____ people to pass without contact.
   A. two
   B. three
   C. four
   D. five

5. A set of stairs in a split-level home has six risers with a tread width of 11". The total run of the stairs is _____".
   A. 44
   B. 55
   C. 66
   D. 72

6. To lay out the stair stringer, first determine the _____.
   A. riser height
   B. tread width
   C. total run
   D. total rise

7. A _____ is the part of the tread that overhangs the riser.
   A. stringer
   B. nosing
   C. toe space
   D. wall rail

8. _____ stringers have a stringer with tapered grooves where the treads and risers fit.
   A. Built-up
   B. Cut-out
   C. Semihoused
   D. Housed

9. *True or False?* Winder stairs are allowed by all building codes.

10. *True or False?* Open stairs require both a decorative enclosure and support for a handrail.

11. When a disappearing stair unit is used to provide attic access, the attic floor should be designed for _____ floor loading.
    A. regular
    B. heavy
    C. light
    D. rough

Apply and Analyze

1. Why is it important in a run of stairs for all of the treads and risers to be the same size?
2. What can open riser designs in stairs be used for?
3. Explain the process of splitting angles for miter cuts.
4. What are the three principal members of a balustrade?
5. When might spiral stairways be used?

Critical Thinking

1. Carefully study the stair requirements and then prepare a detail drawing similar to **Figure 21-12** for a stair to fit the following:
    - First floor to second floor dimension is 8'-11".
    - Dimensions of opening in upper floor are 3'-3" × 11'-8".
    - It is not necessary to include a landing. Use a scale of 1/2" equals 1'. Carefully select and calculate the riser-tread ratio. Be sure the number and size of risers is correct for the distance between the two levels. Check the headroom requirements against your local building code or the IRC. Submit the completed drawing and size specifications to your instructor.
2. Study a millwork catalog or millwork suppliers on the internet and become familiar with the stock parts shown for a main stairway. Working from a set of architectural plans or a stair detail that you may have drawn, prepare a list of all of the stair parts needed to construct the stairway. Include the number of each part and its size, quality, material, and catalog number. Obtain a cost estimate for the materials. Be prepared to discuss the materials and costs with your instructor and class.
3. Construct a small stair unit with a temporary handrail. Follow the guidelines provided by your instructor. Apply the stair design practices outlined in this chapter.

Communicating about Carpentry

1. **Speaking.** Write a brief scene in which 5–10 terms are used as you imagine them being used by carpenters in a real-life context. Then rewrite the dialogue using simpler sentences and transitions, as though you were describing the same scene to elementary or middle-school students. Read both scenes to the class and ask for feedback on whether the two scenes were appropriate for their different audiences.
2. **Speaking.** Pick a figure in this chapter. Working with a partner, tell and retell the important information being conveyed by that figure. Through your collaboration, develop what you and your partner believe is the most interesting verbal description of the importance of the chosen figure. Present your narration to the class.
3. **Reading and Listening.** In small groups, discuss the main topics in the chapter. Ask questions of other group members to clarify concepts or terms as needed. Prepare a presentation to visually enhance student learning.

CHAPTER 22

Doors and Interior Trim

OBJECTIVES

After studying this chapter, you will be able to:

- Describe how door frames and casings are installed.
- List the steps for installing a prehung door.
- Explain the difference between panel- and flush-type doors.
- List the steps for hanging a door.
- Name different types of locks and describe typical installation procedures.
- Compare the pocket and bypass types of sliding doors.
- Outline the order in which window trim members should be applied.
- Cut, fit, and install crown molding in a rectangular room.
- Cut, fit, and nail baseboard trim.

TECHNICAL TERMS

base shoe	doorstop	prehung door unit
baseboard	flush door	rail
coped joint	head jamb	reveal
core	hinge mortise	scarf joint
crown molding	molding	side jamb
dead bolt	panel door	stile
door casing	plinth block	stool
door frame	pocket door	threshold

This chapter deals with the methods and materials of an important part of interior finish, including the following:

- Installing door frames
- Hanging doors
- Fitting trim around openings
- Fitting trim at intersections of walls, floors, and ceilings

This aspect of carpentry requires great skill and accuracy. Well-fitted trim greatly improves the appearance and desirability of a home.

22.1 Interior Door Frames

The interior *door frame* is an assembly of wood parts that form an enclosure and support for a door and its hardware. The door frame covers the unfinished edges of the door opening. It also supports the trim pieces that are attached after installation.

The frame consists of two side jambs and a head jamb, **Figure 22-1**. The *side jamb* is the part of a door frame that is dadoed to receive the head jamb. The *head jamb* is the part of a door frame that fits between the side jambs, forming the top of the frame.

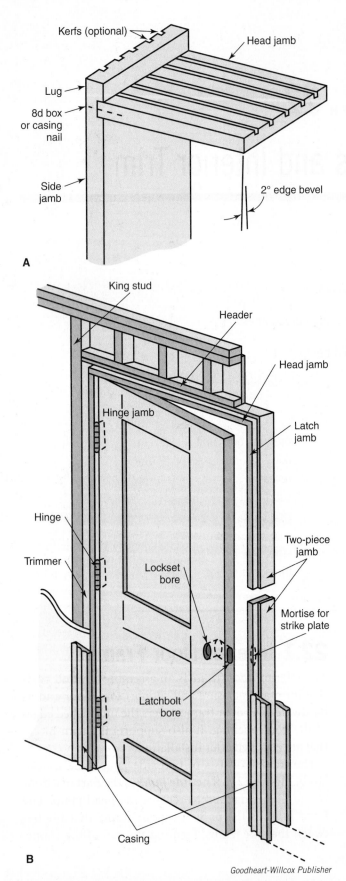

A

B

Goodheart-Willcox Publisher

Figure 22-1. A—Section view of an interior door frame. Parts are listed. Note that edges of all jambs may be beveled slightly so trim will fit snugly. B—Parts of a standard inside passage door frame.

Interior door frames are simpler than those for exterior doors. The jambs are not rabbeted and there is no doorsill. Refer to Chapter 16, **Windows and Exterior Doors**.

Standard jambs for regular 2×4 stud partitions are made from nominal 1″ material and are 4 5/8″ wide. The back side of the jamb is usually kerfed to reduce the tendency of the material to cup (warp). The edge of the jamb may be slightly beveled so the casing snugly fits against it with no visible crack.

Side jambs are dadoed at the top to receive the head jamb. The side jambs for residential doorways are 6′–9″ long (measured to the head jamb). This provides clearance at the bottom of the door for flooring materials, while allowing a 6′–8″ opening.

Interior doorjambs are sometimes adjustable. These are designed to fit walls of different thicknesses. The three-piece type depends on a rabbet joint and a concealing doorstop, **Figure 22-2**. Another type is made in two pieces.

Door frames are usually cut, sanded, and fitted (prehung) in millwork plants. This allows quick assembly on the job. Door frames should receive the same care in storage and handling as other finished woodwork.

A *prehung door unit* consists of a door frame with the door already installed. See **Figure 22-3**. The frame usually includes both sides of casing. A lock is not installed, although machining for its installation has usually been completed. Quite often, the door is prefinished. See **Figure 22-4**.

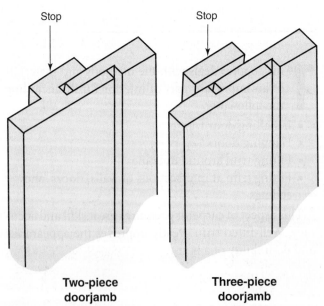

Two-piece doorjamb   **Three-piece doorjamb**

Goodheart-Willcox Publisher

Figure 22-2. Adjustable doorjambs fit any thickness of wall. The stop conceals the joints.

General Procedure for Installing a Prehung Interior Door

1. Remove door unit from carton and check for damage. Separate the two sections. Place tongue side (not attached to door) outside of the room.

2. Slide frame section that includes the door into the opening. Side jambs should rest on finished floor or spacer blocks.

3. Carefully plumb door frame and nail casing to wall structure. Be sure all spacer blocks are in place between the jambs and door.

4. Move to the other side of the wall and install shims between side jambs and rough opening. Shims should be located where spacer blocks make contact with door edges. Nail through jambs and shims.

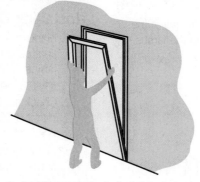

5. Install remaining hall of door frame. Insert the tongue edge into the grooved section already in place. Nail casing to wall structure.

6. Nail through slope into jambs. Remove spacer blocks and check door operation. Make any adjustments required. Drive extra nails where shims are located. Install lock set.

Frank Paxton Lumber Co.

Figure 22-3. General procedure for installing a prehung interior door.

Goodheart-Willcox Publisher

Figure 22-4. In this photo, a carpenter is applying the casing just before hanging the door. Installing casing on one side of the door before placing it in the rough opening gives the unit a positive stop against the finished wall surface.

PROCEDURE

Installing a Prehung Door

1. Check the length of the head and side jamb against the rough opening to determine if the door is correct for the opening.
2. Place the unit into the opening, **Figure 22-3**. Let the side jambs rest on the finish flooring or on spacer blocks of the right thickness. Spacers are needed only if the final flooring surface has not been laid.
3. Level the head jamb, shimming a side jamb as required.
4. Plumb the jambs side-to-side in the frame with a straightedge and level or a long carpenter's level. Make adjustments using doubled shims between the frame and the stud.
5. Plumb the frame front-to-back in the opening.
6. Temporarily fasten the top and bottom of each side jamb with an 8d casing nail.

Continued

Pro Tip

When setting a door frame, do not fully drive any of the nails until all shims have been adjusted and the jambs are straight and plumb.

22.1.1 Door Casing

Trim, known also as *door casing*, is applied to each side of the door frame to cover the space between the jambs and the wall surface. This secures the frame to the wall structure and stiffens the jambs so they can carry the door. **Figure 22-6** shows a section view through a doorjamb. The casing covers the shims and is attached to both the jamb and wall surface. It is usually installed with a *reveal*. This is a setback from the inside edge of the jamb of about 1/4″ or 3/8″. The setback allows room for hinges and striker plates, while improving the appearance of the trim.

PROCEDURE

Installing Casing

1. Mark the reveal. Some carpenters prefer to draw a light pencil line on the edge of the side and head jambs. Use a combination square with the blade extended to the reveal as a marking guide.

2. Check the bottom end of each side casing to see that it is square on what will be the bottom end. This is necessary for a tight fit against the finished floor.

3. Install either the side casings or the head casing. Many carpenters prefer to do the head casing first.

4. Determine the length of the head casing, being sure to include the width of the side casings.

5. Make miter cuts. Use a miter box or a power miter saw to make accurate cuts, **Figure 22-7**.

6. Align the head casing with the pencil marks for reveal and install it.

7. Mark the length of the side casings. Carefully place each piece against the jamb, aligned with the reveal setback. Mark the location of the miter joint at the top.

Continued

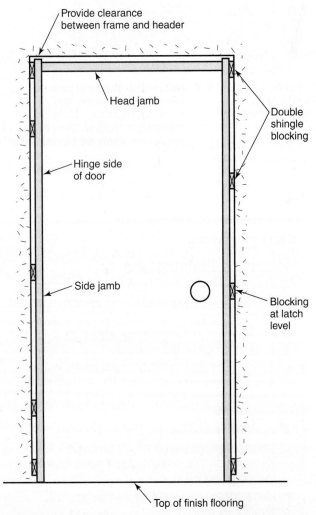

Goodheart-Willcox Publisher

Figure 22-5. Set the door into the opening and add double shims.

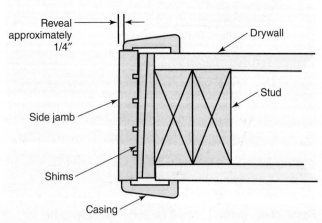

Goodheart-Willcox Publisher

Figure 22-6. This section view shows the position of casing, which is attached to the doorjamb.

Goodheart-Willcox Publisher

Figure 22-7. A power miter saw makes cutting accurate, clean miter cuts easy.

PROCEDURE (continued)

8. Make the miter cuts.
9. If the miters do not properly fit, trim them with a block plane.
10. Temporarily nail casing with finish nails.
11. When the fit is satisfactory, fully drive the nails and complete the nailing pattern. Use 4d or 6d nails along the jamb edge and drive 8d nails through the outer edge into the studs making up the rough opening. Each pair of nails should be spaced about 16″ (OC).
12. If you did not use a nail gun that sets the nails below the surface, use a nail set to sink the nails below the surface, **Figure 22-8**.

Goodheart-Willcox Publisher

Figure 22-8. When installing casing, use a nail set to drive casing nails below surface. The resulting holes will be filled before finishing.

Pro Tip

To avoid making hammer dents in finish lumber that must have a fine appearance, cut a saw kerf in the butt end of a shim shingle. Start the finish nail, and then slip the kerfed shim around the nail. Drive the nail flush with the shim shingle and finish by using a nail set. Any missed hammer blows will be absorbed by the shim, leaving the face of the finish trim unblemished.

22.1.2 Using Plinth Blocks

A *plinth block* is a decorative carved block trim set mainly at the corners of door frames and window frames. It allows a finish carpenter to make butt joints instead of mitered corners. Installation of the casing for plinth blocks is slightly different from that used with mitered corners. Usually, the head casing and plinths are installed first. Sometimes plinth blocks are also installed at the floor level on door frames. The side casings are typically fitted, marked, cut, and installed last. See **Figure 22-9**.

22.2 Panel Doors

There are two general door styles: panel and flush. The panel door is also referred to as a stile and rail door. This type of construction is used in sash, louver, storm, screen, and combination doors. Sash doors are similar to panel doors in appearance and construction, but have one or more glass lights in place of the wood panels.

Goodheart-Willcox Publisher

Figure 22-9. Plinth blocks are decorative corners.

A *panel door* has a frame with separate panels of plywood, hardboard, or steel set into the frame, **Figure 22-10**. Each door consists of an outside vertical member called a *stile*, and a horizontal member called a *rail*. The rails and stiles are usually made of solid material. However, some are veneer applied over a lumber core. In some cases, the doors are molded from a wood fiber, **Figure 22-11**. A variety of designs is formed by changing the number, size, and shape of the panels.

Special effects are secured by installing raised panels, which add line and texture. This panel is formed of thick material that is reduced around the edges where it fits into the grooves in the stiles and rails.

A   *Masonite Corp.*

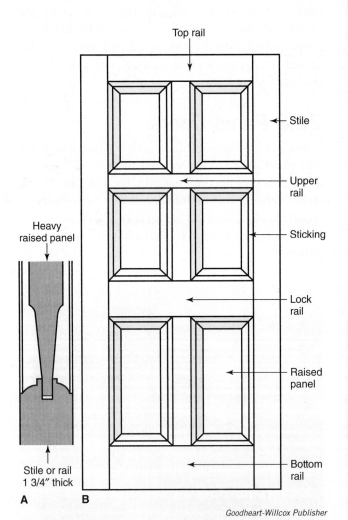

Goodheart-Willcox Publisher

Figure 22-10. Panel door construction. A—This section through the stile and panel shows the joint detail. B—The parts of a panel door.

B   *TruStile Doors*

Figure 22-11. A—Molded doors are made to simulate panel doors. A section of the steel skin on this exterior door has been removed to show the hardboard base. B—A molded door looks like a standard panel door when installed. This interior prehung double door is molded from wood fibers. The surface is primed before leaving the factory, then painted on the job.

22.3 Flush Doors

A *flush door* fits into the opening and does not project outward beyond the frame. It consists of a wood frame with thin, flat sheets of material applied to both faces. These doors are strong and durable. Flush doors account for a high percentage of wood doors used in homes and commercial structures.

The face panels are also called skins. These are commonly made of 1/8″ plywood. However, hardboard, plastic laminates, fiberglass, and metal are also used.

Flush doors are made with either solid or hollow cores. The *core* is the innermost layer(s) of plywood between the face panels. Cores of wood or various composition materials are used in solid (or slab) construction, **Figure 22-12**. The frame inside of the door is usually made of softwood that matches the color of the face veneers. The most common type of solid core construction uses wood blocks bonded together with the end joints staggered.

Flush doors with hollow cores are widely used for interior doors. They may also be used for exterior doors, if made with waterproof adhesives. Hollow core flush doors do not ordinarily provide as much thermal (heat) and sound insulation as solid core doors. Usually their fire resistance rating is lower as well. Some flush exterior doors have compression-molded fiberglass face panels. They are attached to a wooden frame and can be formed to reproduce various traditional and contemporary designs. The core is a high-density polyurethane foam that has a high R-value. Unlike steel-faced doors, the fiberglass door can be trimmed for a precision fit. The surface is textured like wood and can be stained or painted.

22.4 Sizes and Grades

Interior door widths vary with the installation. Widths of 2′-6″, 2′-8″, and 3′-0″ are most commonly used for interior doors. Most codes do not specify minimum sizes for interior doors, but common practice is 2′-8″ for bedroom doors, 2′-6″ for bathroom doors, and 2′-0″ for closets. Standard thickness for interior passage doors is 1 3/8″. Interior door sizes and patterns are illustrated in **Figure 22-13**.

Interior residential doors have a standard height of 6′-8″. A 7′-0″ door is sometimes used for entrances or special interior installations. A 7′-0″ height is usually considered standard for commercial buildings.

Grades and manufacturing requirements for doors are listed in industry standards developed by the Window and Door Manufacturers Association and the Fir and Hemlock Door Association (FHDA). The purpose of these standards is to establish nationally recognized dimensions, designs, and quality specifications for materials and work.

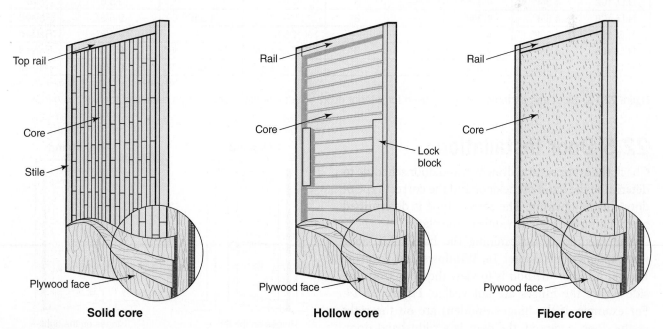

Solid core   **Hollow core**   **Fiber core**

Figure 22-12. Hollow core doors have a thin skin with honeycombed paper, wood blocking, or plastic foam filling the space between the skins. Frames are usually made of softwood.

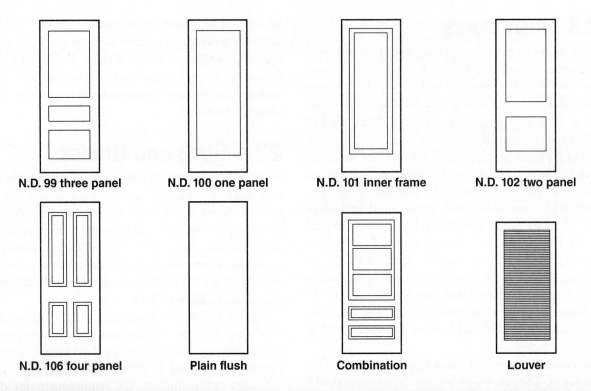

N.D. 99 three panel   N.D. 100 one panel   N.D. 101 inner frame   N.D. 102 two panel

N.D. 106 four panel   Plain flush   Combination   Louver

Construction Details								
Design No.	Stiles	Top rail	Cross rail	Lock rail	Intermediate rails	Mullions or muntins	Bottom rail	Panels
N.D. 99	4 3/4″	4 3/4″	4 5/8	4 5/8″	—	—	9 5/8″	Flat
N.D. 100	4 3/4″	4 3/4″	—	—	—	—	9 5/8″	Flat
N.D. 101	4 1/4″ Face	4 1/4″ Face	—	—	—	—	9 1/4″ or 1/2″ Face	Flat
N.D. 102	4 3/4″	4 3/4″	—	8″	—	—	9 5/8″	Flat
N.D. 106	4 3/4″	4 3/4″	—	8″	—	4 5/8″	9 5/8″	Raised
N.D. 107	4 3/4″	4 3/4″	—	—	4 5/8″	—	9 5/8″	Raised
N.D. 108	4 3/4″	4 3/4″	—	8″	3 7/8″ or 4 5/8″	3 7/8″ or 4 5/8″	9 5/8″	Raised
N.D. 111	4 3/4″	4 3/4″	—	8″	3 7/8″ or 4 5/8″	3 7/8″ or 4 5/8″	9 5/8″	Raised

National Woodwork Manufacturers Association

Figure 22-13. Sizes and patterns commonly used for interior doors. The table lists construction details of the various parts.

22.5 Door Installation

Check the architectural drawings and door schedule to determine the hand of the door and the correct type of door for the opening. The correct type is determined by checking the door schedule in the plans.

One method of determining the hand of a door was explained in Chapter 16, **Windows and Exterior Doors.** Another method is to view the door from the side where the hinges are not visible, **Figure 22-14.** For example, if the hinges (hidden) are on the right as the door is viewed, the door is a right-hand door. Conversely, if the hidden hinges are on the left of the viewer, it is a left-hand door.

Doors may be trimmed somewhat so they fit with enough clearance on each side to prevent the door from binding, even with some swelling during humid weather.

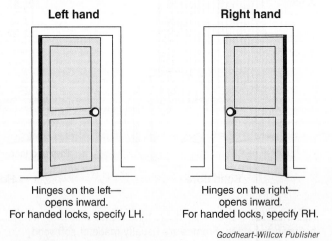

Left hand   Right hand

Hinges on the left—
opens inward.
For handed locks, specify LH.

Hinges on the right—
opens inward.
For handed locks, specify RH.

Goodheart-Willcox Publisher

Figure 22-14. To determine the hand of a door, imagine viewing it with hinges concealed. When ordering, use the abbreviations LH or RH.

Never attempt to cut down a door to fit a smaller opening. If a large amount of material is removed from its edges, the structural balance of the door may be disturbed. Warping may result and the door is weakened.

Cutouts for glass inserts in flush doors should never be more than 40% of the face area. The opening should not be within 5″ of the edge.

Carefully handle doors. Avoid soiling unfinished doors. If they are to be stored for more than a few days, stack them horizontally on a clean level surface. Cover them to keep them clean. Before installing doors, place them in the room for several days so they will reach the average prevailing moisture content before being hung or finished.

22.5.1 Fitting the Door

If the door is not prehung, it must be fitted to the frame. Mark the doorjamb that will receive the hinges. Also mark the edge of the side jamb where they will be mounted. This is opposite of the edge viewed to determine hand. Next, check measurements of both the door and the door opening.

Trim the door to fit the opening. Most doors are carefully sized at the millwork plant, leaving only a slight amount of on-the-job fitting and adjustment. The small amount of material can be removed by planing. Door trimming can be done with a hand plane or a power plane, **Figure 22-15**. While the door is being planed, it may be securely held on edge either by clamping it to sawhorses or using a special door holder.

Clearances should be 3/32″ on the lock side and 1/16″ on the hinge side, **Figure 22-16**. A clearance of 1/16″ at the top and 5/8″ at the bottom is generally satisfactory.

Bosch Power Tool Corp.

Figure 22-15. A power plane speeds up the task of planing a door. It has an edge guide fence that provides consistently accurate planing across the door.

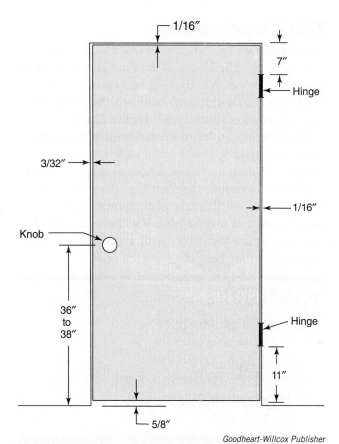

Goodheart-Willcox Publisher

Figure 22-16. Recommended clearances around interior doors. Some carpenters use a quarter (25-cent) coin to check the clearance at the top and lock side.

If the door is to swing across heavy carpeting, increase the bottom clearance. A *threshold* is a beveled or tapered piece attached to each side of a door. It is used to close the space between the bottom of a door and the sill or floor underneath. Thresholds are generally used only under exterior doors. The threshold may be installed before or after the door is hung.

After the door is brought to the correct size, plane a bevel on the lock side to provide clearance for the edge when it swings open. This bevel should be about 1/8″–2″ (approximately 3 1/2°). Narrow doors require a greater bevel than wide doors since the arc of swing is smaller. The type of hinge and the position of the pins should be considered in determining the exact bevel required. After the bevel is cut and the fit of the door is checked, use a block plane to soften (round) corners on all edges of the door. Smooth with sandpaper.

Pro Tip
Millwork plants can furnish prefitted doors that are machined to the size specified, with the lock edge beveled and corners slightly rounded. Doors can also be furnished with gains cut for hinges and holes bored for lock installation.

22.5.2 Installing Hinges

Hinge mortises are the recesses cut into the edge of a door to receive the hinges. If the door is not prehung, these must be cut. Mortises are best cut with a router. Used with a door-and-jamb template, the router saves time and ensures accuracy. See **Figure 22-17**.

Adjustments can be made for various door thicknesses and heights, as well as for different sizes of butt hinges. The design of most templates makes it nearly impossible to mount them on the wrong side of the door or jamb. For this type of equipment, hinges with rounded corners may be used. It will save the time required to square the corner with a wood chisel. Once the mortises have been cut, hang the door.

PROCEDURE
Cutting Hinge Mortises
1. Position the template on the door.
2. Make the cuts with the router.
3. Attach the template to the doorjamb.
4. Cut the matching mortises.

A   *Stanley/Black & Decker*

B   *Infinity Cutting Tools*

Figure 22-17. A—Hinge templates can be adjusted for different door sizes as well as for different hinge sizes. B—A router and template are being used to cut a gain for a round-cornered hinge.

22.5.3 Security Hinges

Many door hinges consist of two leafs connected by a pin. The leafs pivot on the pin, allowing the door to swing open and closed. Often, the pin can be removed by using a nail or small punch and a hammer to drive the pin up and out of the leafs. If an exterior door swings outward, the hinge pins are on the outside of the door, making it possible for anyone to drive the pins out and remove the door. There are several types of security hinges that prevent such unauthorized entry. A common type of security hinge has a protrusion on one leaf that fits into an opening in the opposite leaf, **Figure 22-18**. With the door closed, the protrusion prevents the door from being slid out of the opening, even with the hinge pin removed. The hinge is made additionally secure by replacing one or two of the screws on the jamb side with longer screws that will reach through the jamb and into the stud.

PROCEDURE
Hanging a Door
1. Place the hinge in the gain so the head of the removable pin will be on top when the door is hung.
2. Drive the first screw in slightly toward the back edge to tightly draw the leaf of the hinge into the gain.
3. Repeat steps 1 and 2 to attach the free leaf of the hinge to the jamb. Set only one or two screws in each hinge leaf.
4. Check the fit and then install the remaining screws.
5. After all hinges are installed, hang the door and check clearance on all edges.
6. Make required corrections by planing the door edges or adjusting the depth of the hinge gains.
7. Minor adjustments can be made by applying cardboard or metal shims behind the hinge leaf, **Figure 22-19**.

Goodheart-Willcox Publisher

Figure 22-18. The protrusions on this security hinge fit into the openings on the opposite leaf when the door is closed. This prevents the door from sliding out of the opening if the hinge pin is removed.

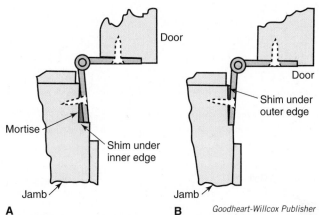

Figure 22-19. Cardboard or metal shims can sometimes be used to make minor adjustments in door clearance. A—Providing more clearance along the lock jamb. B—Closing the space along the lock jamb.

22.5.3 Doorstops

A *doorstop* is a molding nailed to the faces of the door frame jambs. Its purpose is to stop the door as the door is closed. Doorstops are usually the last trim members to be installed. If the door is prehung, the doorstop is installed at the factory. For doors that are not prehung, many carpenters cut the stops and tack them in place before installing the lock. Permanent nailing comes after the lock installation has been completed.

With the door closed, set the stop on the hinge jamb with a clearance of 1/16″. The stop on the lock side is set against the door except in the area around the lock. Here, allow a slight clearance for humidity changes and decorating. Set the stop on the head jamb so it aligns with the stops on the side jambs. Cut miter joints and attach the stop with 4d nails spaced 16″ OC.

22.6 Door Locks

Four types of passage door locks are illustrated in **Figure 22-20**. Cylindrical and tubular locks are used most often because they can be easily and quickly installed. Unit locks are installed in an open cutout in the edge of the door and need not be disassembled when installed. Such locks are commonly used on entrance doors for apartments and some commercial buildings where locks must be changed from time to time.

Cylindrical locks have a sturdy, heavy-duty mechanism that provides security for exterior doors, **Figure 22-21**. They require boring a large hole in the door face, a smaller hole in the edge, and a shallow mortise for the front plate. Often, doors come with these holes already bored at the factory. The tubular lock is similar, but requires a smaller hole in the door face.

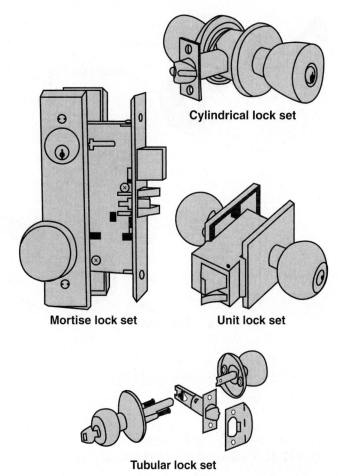

Figure 22-20. Four basic types of door lock sets. Mortise lock sets provide high security and are often found in main entrances to apartment buildings. Cylindrical and tubular locks are most often chosen for residential installations, with cylindrical being more secure for entrance doors. Unit lock sets are best for individual apartment units where locks are frequently changed.

When ordering door locks, it is sometimes necessary to describe the way in which the door swings. This is referred to as the "hand" of the door. Refer again to **Figure 22-14** for the procedure used to determine this specification.

Dead bolts (also called deadlocks or throwbolts) are special door security consisting of a hardened steel bolt and a lock. The lock is operated using a key on the outside and either a key or handle on the inside. Units are made with single-cylinder or double-cylinder action. Double-cylinder dead bolts require key use on both sides of the door. They should be used if the door contains a window that is positioned near the lock. A dead bolt prevents potential intruders from smashing the window from the outside and reaching through to unlock and open the door. However, there is a safety concern related to an emergency exit.

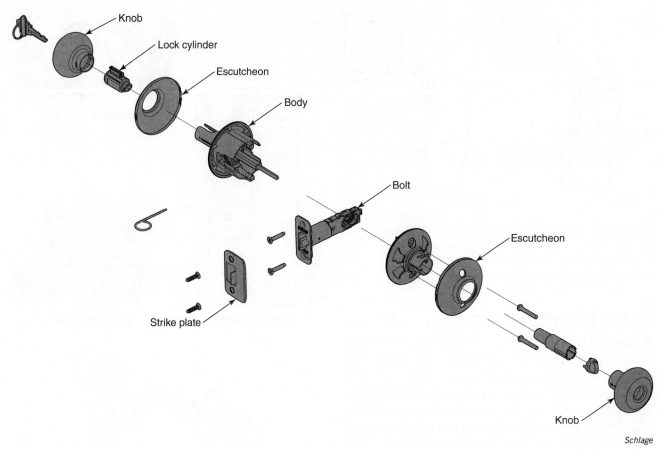

Figure 22-21. Parts of a cylinder lock set.

If an occupant needs to vacate the premises immediately and does not have a key for the dead bolt, the door is inaccessible. **Figure 22-22** shows two types of dead bolts.

A   **B**   *Weiser Lock*
Figure 22-22. Two types of dead bolts. A—Tubular dead bolt. B—A keyless dead bolt uses a numeric code.

PROCEDURE

Installing a Lock in a Nonpredrilled Door

1. Open the door to a convenient working position and block it with wedges placed underneath.
2. Measure up from the floor a distance of 38″ (35″, 36″, or 40″ are sometimes used) and mark a light horizontal line. This is the center of the lock.
3. Position the template furnished with the lock set on the face and edge of the door.
4. Lay out the centers of the holes, **Figure 22-23**.
5. Bore the holes according to the instructions provided with the lock set. Using a boring jig ensures accurate work, **Figure 22-24**.
6. Lay out and cut the shallow mortise on the door edge. A faceplate mortise marker, also called a marking chisel, is faster and more accurate than standard wood chisels.

Manufacturer's Instructions for Lock Sets

1. Mark door

Mark height line across edge of door. 38″ is the usual height above the floor. Fold template over edge of door, centering on height line. Mark centers of 7/8″ and 2 1/8″ holes.

2. Install latch unit

Bore 2 1/8″ hole through door, and 7/8″ hole into edge of door at points marked on template.

Cut out for latch front and install latch unit.

3. Install strike

Mark height line for strike on jamb. Mark vertical centerline on jamb. This centerline must be same distance from stop molding as latch case centerline is from edge of door that will hit stop molding.

Cut mortise in jamb for strike and box.

Insert box and strike and tighten screws securely.

4. Adjust lock

To adjust this lock for a 1 3/8″ door, unscrew outside rose plate 5/16″ from case cutout. To adjust for a 1 3/4″ door, unscrew outside rose plate to provide 1/2″ between rose plate and case cutout. To adjust for any thickness between 1 3/8″ and 1 3/4″, set the rose plate at a suitable intermediate position.

5/16″ For 1 3/8″ door
Case cutout

1/2″ For 1 3/4″ door
Case
Outside rose plate
Case cutout

5. Install lock

With latch case in place, insert lock assembly into 2 1/8″ hole, making sure that lock case hooks retainer legs and retractor hooks bolt tails. Do not force.

Retractor
Retainer leg
Bolt tail
Lock case

6. Install rose plate

Slip on rose plate and locate screw holes on vertical centerline with "Top" up. Insert machine screws and tighten alternately to obtain secure attachment.

See instructions above
Vertical

7. Install inside rose

Place inside rose over rose plate with notch in rose over spring retainer and snap rose down so rose is flush with door.

Spring
Notch

8. Install inside knob

Align lug on knob with narrow slot on side of spindle and push knob all the way in until retainer clicks into slot on knob.

9. Important

This lock is set for a right-hand door.

Right way	Wrong way

To change lock hand

Knob must be in unlocked position. Turn outside knob counterclockwise approximately 45° and insert small nail in hole of trim cap. Depress retainer and slide knob off. Turn knob 180° and replace knob.

Goodheart-Willcox Publisher

Figure 22-23. Manufacturers furnish detailed instructions for installation of their lock sets. Generally, cylindrical locks can be installed following similar instructions.

Porter-Cable Corp.

Figure 22-24. When installing locks, a boring jig for door hardware saves time and ensures accuracy.

22.6.1 Peepholes

A peephole, also called a door viewer, is an inexpensive security measure. Peepholes are available with a prism that allows you to see a wide area, in some cases 180°. To install a peephole, first determine the right height. That height should be approximately eye level for the shortest adult living in the home. Use a tape measure to mark this height in the center of the door. Most peepholes require a 1/2″ hole, but some require a 5/8″ hole. Read the instructions for the peephole or measure the diameter of the cylinder to determine what size drill bit to use. A spur bit or an auger bit with a pilot point works best on wood doors. Be careful to hold the drill perpendicular to the door and apply light pressure as you drill. A piece of masking tape on the opposite side of the door will help prevent the surface from splintering. Next, insert the outside part of the peephole into the drilled hole. Finally, insert the inside part and tighten it with a coin.

22.7 Sliding Pocket Doors

A *pocket door*, or recessed door, is suspended on a track that slides into a recess or pocket in a partition wall when opened. This space-saving feature is used where there is no clearance for a swinging door. The pocket door frame consists of a split side jamb attached to a framework built into the wall, **Figure 22-25**. The

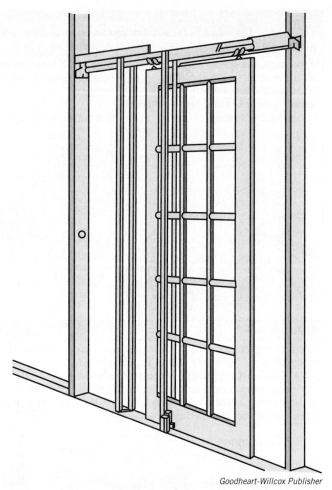

Goodheart-Willcox Publisher

Figure 22-25. The frame for a pocket door includes wall framing, into which the door will slide.

rough opening in the structural frame must be large enough to include the finished door opening and the pocket. The usual height of the rough opening is 6′-11 1/2″, as shown in **Figure 22-26**. The preassembled pocket framework and track is installed during the rough framing stage. Follow the manufacturer's instructions when framing up the interior partition and installing the pocket door.

Pocket-frame units are available from millwork plants in a number of standard sizes. When doing the rough-in, a carpenter should have the manufacturer's specifications for the door selected.

Manufacturers that specialize in builder's hardware have developed steel pocket door frames. They are easy to install and provide a firm base for wall surface materials.

Figure 22-27 shows a typical track and roller assembly for a pocket door. The hanger (wheel assembly) snaps into the plate attached to the top of the door. It can be easily adjusted up or down to plumb the door in the opening.

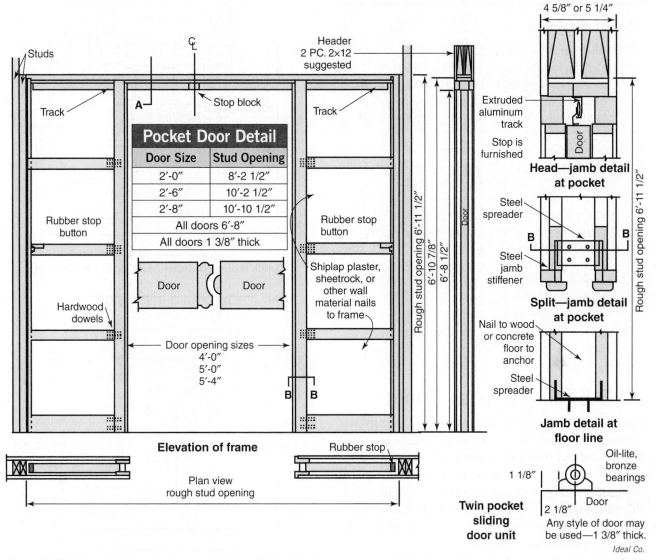

Figure 22-26. Structural details of a pocket door. The ladder-like framing on both sides of the pocket supports drywall. Drywall fasteners should not be so long as to penetrate this framework.

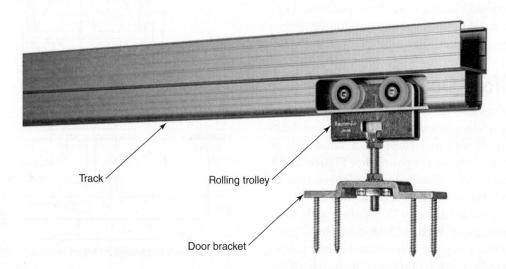

K.N. Crowder

Figure 22-27. A track and roller assembly for a pocket door. The threaded rod adjusts the door up and down.

PROCEDURE
Installing a Pocket Door

1. Install the header that will support the door, its tracking mechanisms, and the 1″ framing.
2. If provided, fasten a steel channel to the floor. The channel keeps the proper clearance needed in the pocket framework.
3. Attach the steel track to the header.
4. Install the hardware onto the door. Attach rollers to the top of the door where specified by the manufacturer. Pocket doors also require an edge pull and a recessed side pull for opening and closing.
5. Tilt the door outward to engage the rollers in the overhead track.
6. Slide the door into the floor channel. Slide it back and forth. It should easily move while evenly butting against side jambs.
7. Adjust the rollers as needed.
8. Apply wall coverings and trim to conceal the framing and track

22.8 Sliding Bypass Doors

Standard interior door frames can be used for a bypass sliding door installation. When the track is mounted below the head jamb, the height of a standard door must be reduced and a trim strip installed to conceal the hardware, **Figure 22-28**. Head jamb units are available with a recessed track that permits the doors to ride flush with the underside of the jamb. Hardware for sliding bypass doors is packaged complete with track, hangers (rollers), floor guide, screws, and instructions for making the installation.

A disadvantage of bypass sliding doors is that access to the total opening at one time is not possible. They are, however, easy to install and practical for wardrobes and many other interior wall openings.

22.9 Bifold Doors

Bifold doors consist of two doors that are hinged together. The folding action is guided by an overhead track. A complete unit may consist of a single pair or two or more pairs of doors or panels. See **Figure 22-29**. These door units are well suited to wardrobes, closets, pantries, and certain openings between rooms.

The opening for bifold units is trimmed with standard jambs and casing. **Figure 22-30** illustrates a hardware installation. The pivot brackets and center guides have self-lubricating, nylon bushings. The weight of the doors is supported by the pivot brackets and hinges between the doors, not by the overhead track and guide.

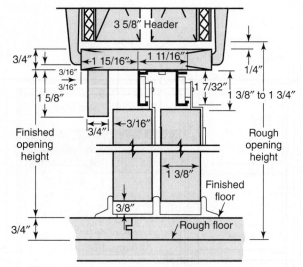
Track mounted under head jamb 1 3/8″ doors

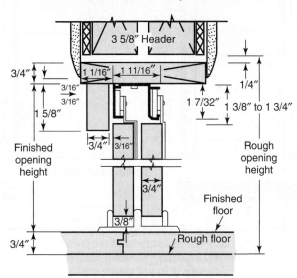
Track mounted under head jamb 3/4″ doors

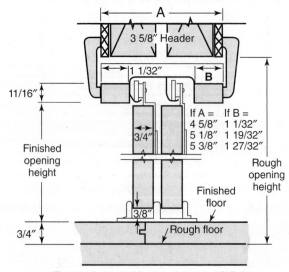
Track recessed in head jamb 3/4″ doors

Goodheart-Willcox Publisher

Figure 22-28. Drawings of typical bypass sliding door installations. Regular interior doors may be used.

Figure 22-29. Bifold doors are often used to close off wardrobe space.

tab62/Shutterstock.com

Two-door units generally range from 2′ to 3′ wide, while four-door units are available in widths from 3′ to 6′.

Folding door hardware comes in a package that includes hinges, pivots, guides, bumpers, aligners, nails, and screws. Instructions for the installation are also included. Millwork plants can supply matching doors with prefitted hardware.

When the total opening for a four-door unit is greater than 6′ or wider than 3′ for a two-door unit, it is usually necessary to install heavier hardware. Also, a supporting roller-hanger is used instead of a regular center guide.

22.10 Multifold Doors

Multifold doors are built from narrow panels with some type of hinge along the edges. See **Figure 22-31**. One manufacturer produces a design where the hinge action is provided by steel springs threaded through the panels. The entire door assembly is supported by nylon rollers located in an overhead metal track. The track is wood-trimmed to match the door.

Extruded aluminum track

Center guide

Top pivot bracket

Top pivot

Full mortised hinges

Door aligner

Bottom pivot

Bottom pivot bracket

Ideal Co.

Figure 22-30. Typical hardware used for folding door installation. The hinged units are supported by brackets. No weight is carried by the center guide.

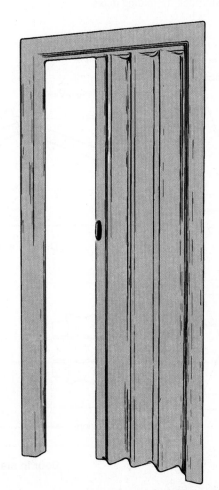

Goodheart-Willcox Publisher

Figure 22-31. Multifold doors can be used where space is limited.

Figure 22-32 shows how the track is installed in either a wood-framed or drywalled opening. The track for bifold doors is installed in the same way. Manufacturers furnish door units in a complete package that includes track, hardware, latches, and instructions for making the installation.

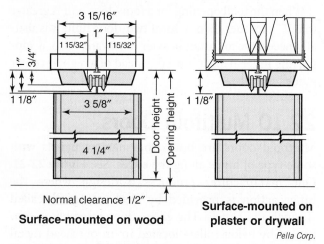

Surface-mounted on wood

Surface-mounted on plaster or drywall

Pella Corp.

Figure 22-32. Typical head sections show the installation of an overhead track for folding doors.

Door panel surfaces are available in a variety of materials and finishes. The best panel grades are made from wood veneers bonded to wood cores. Panels are also made of stabilized particleboard wrapped with wood-grain-embossed vinyl film.

An important advantage of this type of folding door is its space-saving feature. As the door is opened, the panels fold together forming a "stack" that requires little room space. For example, the stack dimension for an 8′ opening is only 11 1/2″. When it is desirable to clear the entire opening, the stack can often be housed in a special wall cavity. Bifold doors and regular passage doors must have swinging clearance in the room as they are opened and closed. **Figure 22-33** shows general details of construction with the door in both opened and closed positions.

Pro Tip
Folding doors are available in a variety of sizes. They can be used to separate room areas, laundry alcoves, and storage spaces.

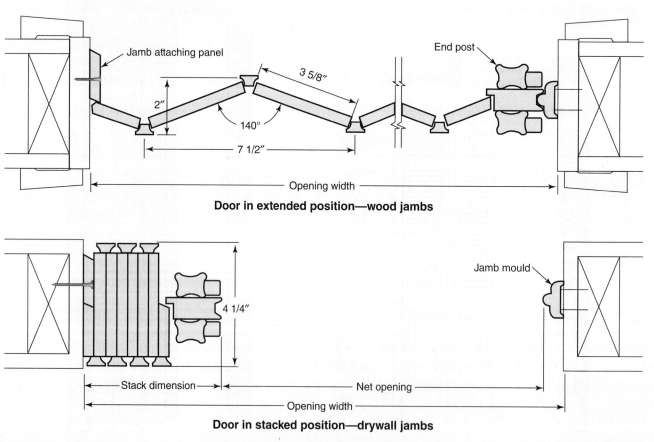

Door in extended position—wood jambs

Door in stacked position—drywall jambs

Pella Corp.

Figure 22-33. General details showing the operation of a folding door.

22.11 Moldings

Moldings are trim pieces that are relatively narrow wood or plastic strips. They have a curved profile through the length of the strip to accent and emphasize the ornamentation of a structure and to conceal surface or angle joints. They are designed to be functional as well as decorative. For example, window and door moldings cover the space between the jamb and the wall covering. They also make the installation more rigid.

A wide range of types, patterns, and sizes of moldings is used in residential construction. A few common shapes and typical uses are shown in **Figure 22-34**. There are many other standard profiles and each is usually available in several sizes. Many millwork plants make custom moldings in any shape and size desired. Of course, custom molding is much more expensive than stock profiles and there is usually a one-time set-up charge. Information on molding patterns along with a numbering system and grading rules are included in a manual that is available from Western Wood Products Association.

22.11.1 Classifying Moldings

Some moldings take their names from their shape. Full rounds, half rounds, and quarter rounds are examples. They may be used in many different locations to finish interior space. The following are a few examples:

- Full rounds are often installed as closet poles.
- Half rounds may serve to cover the raw edges of shelving.
- Baseboard, base shoe, and base cap are installed where the wall and floor meet.
- Quarter round used as a base cap dresses off the top of the base when the base's top edge is squared off.

Names of other moldings come from their use. Examples include casing, chair rail, and panel molding.

22.12 Window Trim

Interior window trim consists of the casing, *stool* (forms the interior sill cap of a window), apron, and stops, **Figure 22-35**. Millwork companies select and package the proper length trim members to finish a given unit or combination of units.

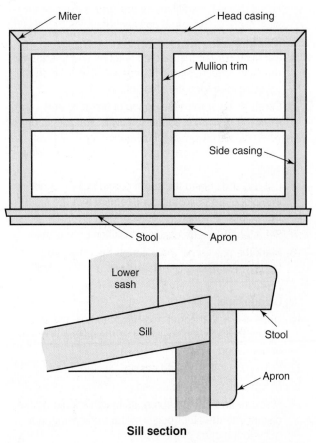

Goodheart-Willcox Publisher

Figure 22-35. Trim members used for a standard double-hung window.

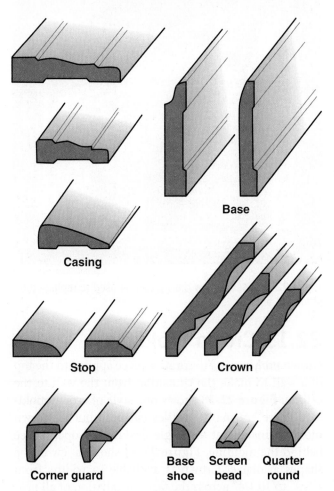

Goodheart-Willcox Publisher

Figure 22-34. Profiles of some typical molding patterns.

When the face of the apron is curved, the ends should be returned or coped so that the shapes of the ends are the same. The returned end is commonly used and formed with miter cuts as illustrated in **Figure 22-36**.

Sometimes, the stool and apron are eliminated. Instead, a piece of beveled sill liner is installed to match the window jamb. Regular casing is then applied around the entire window. See **Figure 22-37**. This is known as picture-frame trimming.

PROCEDURE

Installing Trim for Double-Hung Windows

1. To find the length of the stool, position a piece of side casing and measure beyond it by about 3/4″. Do this on both sides of the window and mark light lines on the wall.

2. Position the stool and mark the cutoff lines for the ends of the stool using the light lines on the wall.

3. Hold the stool level with the sill. Mark the inside edges of the side jambs. Also mark a line on the face of the stool where it will fit against the wall surface. For a standard double-hung window, this line is usually directly above the square edge of the stool rabbet. This is the notch where it fits over the top of the sill.

4. Carefully cut out the marked notches on the ends of the stool and check the fit. You will have to open the lower sash to slide the stool into position.

5. Carefully lower the sash on top of the stool and draw the cutoff line so the sash will clear the stool when closed. Allow slight clearance for final finish between the front edge of the stool and the window sash.

6. Cut the ends, sand the surface, and nail the stool into place. Some installations require that the stool be bedded in caulking compound.

7. Set a length of side casing in position on the stool.

8. Mark the position of the miter on the inside edge. Usually, the casing is installed with a 1/4″ or 3/8″ reveal.

9. Cut the miter.

10. Drive one nail at about mid-height to hold the side casing in place. Drill nail holes if the casing is made of hardwood.

11. Cut and install the head casing.

12. Finish nailing the side casing.

13. Measure and cut the apron. Its length should be the distance to the outer edges of the side casing, not quite to the ends of the stool.

14. Sand any visible saw cuts on the apron and nail it in place.

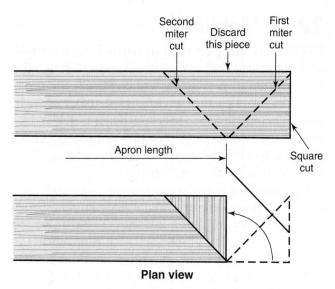

Plan view

Goodheart-Willcox Publisher

Figure 22-36. How to mark and cut a returned end on an apron. Use glue to attach the end piece.

Artazum/Shutterstock.com

Figure 22-37. Notice how casing can be used to replace the stool and apron.

22.13 Crown Molding

Crown molding is a decorative piece applied to the top of a wall to make the transition from the wall to the ceiling. **Figure 22-38** shows one style of crown molding. There are many profiles (shapes), but all crown molding installs at an angle, leaving an open space behind the molding. **Figure 22-34** shows a few of the shapes available. Crown is available in widths from 1 5/8″ to 10 1/4″ across the face. Installation of all profiles and sizes is basically the same.

Goodheart-Willcox Publisher

Figure 22-38. Crown molding is available in many patterns and sizes.

22.13.1 Installing Crown Molding

All crown molding touches the wall, where it can be nailed to studs. Where the ceiling joists run perpendicular to the wall, the top of the molding can be nailed into the joists. Where the joists run parallel to the wall, there may not be any framing available to nail the top of the molding, **Figure 22-39**. In these cases, triangular support blocks must be fastened to the wall to provide a nailing surface for the crown.

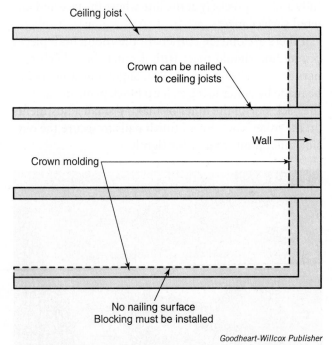

Goodheart-Willcox Publisher

Figure 22-39. Where ceiling joists are parallel to the wall, there may not be any framing to hold the top of the crown molding. Support blocks will be required to hold the crown.

The size of the support blocks is taken from measurements that are made with the molding. Hold a short piece of the molding in place near a corner, being sure that the top and bottom rest flat against the wall and the ceiling. Make a light pencil mark on the wall at the bottom of the molding, another at the top of the flat surface that is against the wall, and a third at the inside of the flat surface against the ceiling. The lower mark on the wall will be used to set the molding at the right angle when it is installed. The upper mark on the wall is one corner of the triangular block to be cut. The ceiling mark is the opposite corner of the triangular block.

Now, mark where the bottom edge of the crown molding will be on all of the walls. Measure down from the ceiling to the lower pencil mark made previously. Mark this measurement on both walls at each corner of the room. Snap a chalk line on each wall to show where the bottom of the molding will be. If you are working on finished walls, use white powdered chalk. It is easier to clean off walls than colored chalk. After the chalk lines are snapped, attach a support block near each corner on any wall that does not have a ceiling joist for nailing the top of the crown. Space additional support blocks every 16″ along the wall. Attach the blocks with construction adhesive and finishing nails into available framing, **Figure 22-40**. You may need to clip the 90° corner off the block to accommodate variations caused by drywall compound in the corner. Place a light pencil mark on the wall at the center of each stud or support block, below the crown, where it will be visible to aid in nailing the crown later.

To achieve tight-fitting joints, it is important to take accurate measurements. Take the measurements at the top of the wall where the molding will be installed. This measurement might be slightly different from one taken lower on the wall. It is easiest to get long measurements with two people; one to hold the end of the tape measure against the wall at one corner and another to read the measurement at the next corner.

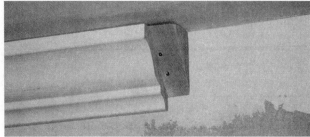

Goodheart-Willcox Publisher

Figure 22-40. Support blocks provide a nailing surface for the molding where no ceiling joists are present.

If a second person is not available, start in the middle of the wall and extend the end of the tape measure to the corner. Make a light pencil mark at a convenient measurement. Try to use round numbers. For example, it is easier to work with 6′–0″ than, say, 6′–2 3/8″. Next, measure from the opposite wall to the pencil mark and add the two measurements to find the total.

Start installing crown along a long wall. The first piece of crown molding should be cut square on both ends. If the wall is too long to finish with a single piece of molding, two pieces are joined with a mitered-lap joint, called a *scarf joint*, over a support block. Each piece is cut at a 45° angle and finish nails are driven through both pieces and into the support block, **Figure 22-41**. Install the left piece first to make fitting corners easier.

Carefully measure the next wall to the right, measuring to the adjacent wall, not the edge of the previously installed crown molding. Cut a miter on the end that will adjoin the previously cut piece. If the next corner is an inside corner, the right end of the molding will be cut square.

Mitering crown molding can be confusing because the molding must be turned upside down and swapped end-for-end. The molding is made so that one edge on the back rests against the wall and the other rests against the ceiling. To cut it, consider the base of the saw to be the ceiling and the saw fence to be the wall, **Figure 22-42**. A pencil mark on the saw fence indicates the proper angle to position the molding. The distance from the saw table (ceiling) to the mark on the fence (wall) is the same as the distance from the ceiling to the mark you previously made on the wall of the room. Cut what will be the right end of the molding (on the left as the piece is positioned on the saw) square to fit in the inside corner. Measure and mark the wall edge (the top as it rests on the saw) with the measurement you took from the wall. You will be measuring from the end you just cut square to the right (the end on the left when installed). Swing the saw to the left to cut a 45° miter outside the pencil mark. The ceiling edge of the molding will be shorter than the wall edge.

Goodheart-Willcox Publisher

Figure 22-42. When crown molding is placed on a power miter saw, the saw base substitutes for the ceiling and the saw fence substitutes for the wall. Tape is applied to the fence and marked with a pencil to show the proper position.

Cutting an inside miter on the molding reveals the profile where the mitered surface meets the molding face. Use a coping saw to cut the end of the molding to that shape. This is called coping the molding. **Figure 22-43** shows a cope being cut. Undercut the coped end slightly so the face will fit tightly against the adjacent piece of molding. Minor adjustments to the cope cut can be made with a file or sandpaper.

Continue around the room in this manner. The last piece will have to be coped on both ends. Measure carefully and cut precisely at the line where the face and mitered surface meet.

If there are outside corners in the room, both pieces of molding should meet with a miter joint. Such corners are not always exactly 90°. Slight adjustments can be made by either using a sharp block plane to cut back the back side of the miter or making small adjustments on the miter saw. Use 4d finish nails to secure the outside miter joint when it fits tightly.

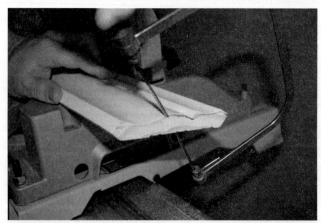

Goodheart-Willcox Publisher

Figure 22-43. Use a coping saw to follow the profile of the molding. Undercut the cope slightly.

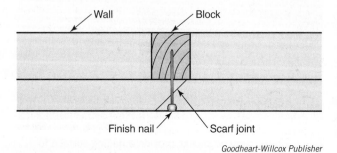

Goodheart-Willcox Publisher

Figure 22-41. Molding can be joined with a scarf joint, leaving only a slight line on the surface of the molding. The joint is held together with glue and finish nails.

22.14 Baseboard and Base Shoe

The *baseboard* is a finishing board that covers the joint where the wall intersects the finish flooring, **Figure 22-44**. It is among the last of the interior trim members to be installed, since it must be fitted to the door casings and cabinetwork and over the finish floor. Baseboards run around the room between door openings, cabinets, and built-ins. The joints at internal corners should be coped. Those at outside corners are mitered, **Figure 22-45**.

Base shoe is narrow molding used around the perimeter of a room where the baseboard meets the finish floor. It is used to seal the joint between the baseboard and the floor. It is usually fitted at the time the baseboard is installed, but not nailed in place until after any surface finishes have been applied. Base shoe is often used to cover the edge of resilient tile or carpet.

22.14.1 Installing Baseboard and Base Shoe

Baseboard and base shoe are installed in the same way. Select and place the baseboard material around the sides of the room. Sort the pieces so there will be the least amount of cutting and waste. Where a straight run of baseboard must be joined, use a scarf joint, as described in the section on crown molding. Be sure to locate the joint so it can be nailed over a stud.

Baseboard installation is much easier if stud locations have been marked. First, mark on the rough floor ahead of the plaster or drywall installation. Then, mark on the wall surface before installing the finish flooring or underlayment.

If the stud positions have not been marked, locate them with a stud finder. If a stud finder is not available, tap along the wall with a hammer until a solid sound is heard. Drive nails into the wall to locate the exact position and edges of the stud. Then mark the location of others by measuring the stud spacing (usually 16″ or 24″ OC).

Start installing the baseboard at an inside corner. Make a square cut on the end of the first piece of baseboard. The butt joint should be tight with the intersecting wall surface. Make scarf joints as needed to complete the run to the opposite inside corner. The next piece butts against the base just installed. This joint is coped. The *coped joint* is made by cutting the profile of the molding face on the end of one piece, so it fits snugly against the other piece. A coped joint takes longer to make than a plain miter, but it makes a better joint at an inside corner. When nailed into place, it will not open up. Also, if the wood shrinks after installation, no noticeable crack will appear.

All outside corners are joined with a miter joint. Hold the baseboard in position and mark at the back edge in line with the intersecting surface. Make the 45° cut using a miter saw, **Figure 22-46**.

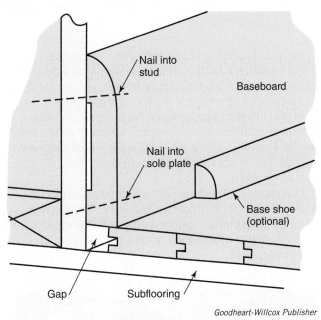

Goodheart-Willcox Publisher

Figure 22-44. Cutaway of baseboard and base shoe. They conceal the gap between the flooring and finished wall.

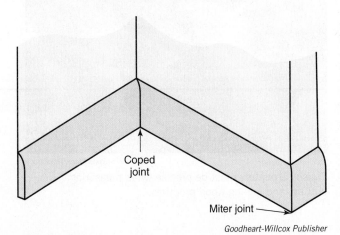

Goodheart-Willcox Publisher

Figure 22-45. Coped joints are suitable for inside corners. Use a mitered joint for outside corners.

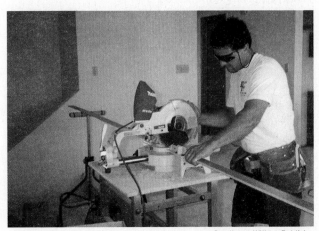

Goodheart-Willcox Publisher

Figure 22-46. Using a power miter box to cut a miter joint for outside corners.

Before installing each section of baseboard, check both ends to make sure the cut and fit is correct. To install, tightly hold the board against the floor. Fasten it with finishing nails long enough to penetrate well into the studs. The lower nail is angled slightly down so it is easier to drive and enter the sole plate. Set the nail heads.

Baseboards are normally butted against the door casing, as illustrated in **Figure 22-47**. The casing should be thick enough to accommodate the slightly thinner profile of the baseboard. Base shoe, if used, is ended at the casing with a miter cut as shown.

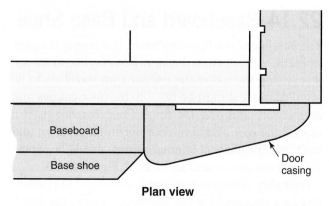

Plan view

Goodheart-Willcox Publisher

Figure 22-47. The baseboard is butt-jointed against door casing. The base shoe gets an outside miter cut.

Construction Careers
Finish Carpenter

A finish carpenter normally has all of the knowledge and skills of a framing carpenter. In addition, a finish carpenter has specialized knowledge and skills in such areas as interior and exterior trim work, stairway planning and building, door and window installation, and the installation of kitchen cabinets and countertops. A finish carpenter's skill set falls between that of a framing carpenter and a cabinetmaker.

Larger general contracting firms or housing development companies usually employ both framing and finish carpenters. In large residential developments, the carpenters often are organized into crews that move from one building to another at the appropriate construction stage. A finish carpenter crew, for example, normally completes all of the trim work and other finishing tasks in one structure, then moves on to the next structure and repeats the same tasks. Carpenters who work for smaller general contractors and remodeling companies or operate their own business usually perform both framing and finish carpentry work.

Finish carpenters must be skilled in measuring, cutting, and joining wood components. The work calls for a higher degree of precision, especially in making joints, than framing carpentry. Practical math skills and the ability to read prints are important skills for success in finish carpentry. Although their work may be somewhat less strenuous than that of framing carpenters, finish carpenters must still be in good physical condition. The work requires standing for long periods and can involve some heavy lifting. While most work is indoors, finish carpenters may be exposed to inclement weather. Some of their responsibilities often involve door and window installation and the application of a building's exterior trim.

Finish carpenters typically enter the trade by the same routes as framing carpenters—technical school programs, on-the-job training, or formal apprenticeship. Many began as framing carpenters and gradually acquired the skills needed to do finish carpentry work. Apprenticeship programs combine classroom instruction and 3–4 years of on-the-job experience. These programs prepare apprentices for both framing and finish carpentry work.

Like framing carpenters, finish carpenters are exposed to almost all aspects of construction work. This prepares them for advancement to supervisory positions within larger contracting firms and can also prepare them for self-employment.

Patrick Heagney/iStock/Thinkstock

Finish carpenters must be precise when measuring, cutting, and joining wood products.

Chapter Review

Summary

- Installing interior doors and trim requires great skill and accuracy.
- Interior door frames consist of two side jambs and a head jamb. They cover the unfinished edges of the door opening and provide support for the door and its hardware.
- Prehung door units come fully assembled. They are positioned in the door opening, adjusted, then nailed into place.
- Trim, known as door casing, is installed to trim the edges of the doorway where they meet the wall.
- Panel doors are built of stiles and rails that frame solid wood or plywood inserts.
- Flush doors have a smooth, thin sheet of material applied to both faces.
- Interior doors are available in a standard height and a variety of widths for different applications.
- Installation steps for traditional doors include cutting gains to install hinges, and boring the door to receive the lock mechanism.
- Cylindrical and tubular locks are popular due to ease of installation.
- Unit lock are installed in an open cutout on the edge of a door.
- Cylindrical locks have heavy-duty mechanisms that provide security for exterior doors.
- A pocket door is suspended on a track that slides into a recess or pocket in a partition wall when opened.
- Standard interior door frames can be used for a bypass sliding door installation.
- Multifold doors are built from narrow panels with some type of hinge along the edges.
- Moldings are available in different types, patterns, and sizes.
- Interior window trim members include casing, stool, apron, and stops.
- Crown molding is often installed at the joint between the walls and ceilings of some rooms.
- Molding used to trim the joint between the wall and floor is known as baseboard.

Know and Understand

Answer the following questions using the information in this chapter.

1. *True or False?* Interior door frames are simpler than those for exterior doors.

2. A _____ is applied to each of the door frames to cover the space between the jambs and the wall surface.
   - A. doorsill
   - B. head jamb
   - C. trim
   - D. plinth block

3. Where is a plinth block mainly used?
   - A. On door and window frames.
   - B. On flush doors only.
   - C. On panel doors only.
   - D. On prehung door units.

4. The two general types of doors are _____ and _____.
   - A. panel, sash
   - B. panel, flush
   - C. flush, combination
   - D. flush, sash

5. The face panels of flush doors, also called _____, are usually made of 1/8″ plywood.
   - A. stiles
   - B. rails
   - C. skins
   - D. cores

6. Common width for bedroom doors is to use a _____′-_____″.
   - A. 2, 6
   - B. 2, 8
   - C. 3, 0
   - D. 6, 8

7. The bottom clearance of a door should be increased if the door is to swing across _____.
   - A. wood floors
   - B. laminate floors
   - C. heavy carpeting
   - D. a corner

8. _____ are the recesses cut into the edge of a door to receive the hinges.
   A. Thresholds   C. Doorstops
   B. Hinge mortises   D. Peepholes

9. Doorstops should be set on the hinge jamb with a clearance of _____".
   A. 1/16   C. 1/8
   B. 1/12   D. 1/4

10. The two most commonly used types of residential door locks are cylindrical and _____.
    A. mortise   C. unit
    B. tubular   D. keyless

11. _____-cylinder dead bolts should be installed if the door contains a window.
    A. Keyless   C. Single
    B. Unit   D. Double

12. A piece of _____ on the opposite side of the door will help prevent splintering the surface when drilling for a peephole.
    A. wood   C. masking tape
    B. metal   D. plastic

13. The type of door that slides into a recess in a partition wall is a _____ door.
    A. pocket   C. bifold
    B. bypass   D. multifold

14. _____ doors are supported by pivot brackets and hinges between the doors.
    A. Pocket   C. Bifold
    B. Bypass   D. Multifold

15. *True or False?* The best panel grades for door surfaces are made from fiberglass and wood.

16. What type of molding may be used to cover the raw edges of shelving?
    A. Quarter rounds.
    B. Half rounds.
    C. Full rounds.
    D. Baseboard moldings.

17. What parts are eliminated from window trim to create picture-frame trim?
    A. casing, stool   C. stops, casing
    B. casing, apron   D. stool, apron

18. _____ are often used to cover the edge of resilient tile or carpet.
    A. Baseboards   C. Moldings
    B. Base shoes   D. Coped joints

19. *True or False?* All outside corners of baseboards are joined with a miter joint.

Apply and Analyze

1. What is the purpose of installing a reveal with the door casing?

2. If you are facing a door with the hinges on the right side and the door opens away from you, what is the hand of the door?

3. Why should you never attempt to cut down a door to fit a smaller opening?

4. Explain how to hang a door.

5. Which type of door is the most optimal for saving space and why?

6. Explain the phrase *coping the molding.*

Critical Thinking

1. Imagine you are working on a job as a finish carpenter. The outside corners have metal corner beads installed over the gypsum wallboard. The drywall finishers have coated the corners with drywall compound and it has made the corner slightly out of square. You recognized this only after cutting your baseboard trim pieces with the miter saw set at exactly 45°. This has left a gap in your joint. What solution(s) can you come up with to make the baseboard fit the corner while maintaining good joinery around the corner? Explain the steps you would take to carry out your solution.

2. When framing a rough opening for an interior door the framing crew crowned one of the king studs backwards. If no adjustments are made, what impact will this have on the door when it is placed?

3. In the scenario for question 2, what actions would you take to ensure the door is installed properly? Remember, at this point in the job, the wall finishes will have been applied.

Communicating about Carpentry

1. **Reading.** With a partner, create flash cards for the key terms in the chapter. On the front of the card, write the term. On the back of the card, write the phonetic spelling as written in the text. (You may also use a dictionary.) Practice reading the terms aloud, clarifying pronunciations where needed.

2. **Art.** Create a collage that identifies different types of doors and interior trim. Show and discuss your collage in a group of 4–5 classmates. Are the other members of your group able to determine the types of doors and interior trim that you tried to represent?

3. **Reading and Speaking.** Do some research on the types of interior doors available for residential buildings. Obtain images for as many types as possible. Explain how the doors are similar and how they are different. Point out specific qualities on each door in your explanations. Repeat the exercise with types of indoor trim available.

CHAPTER **23**

Cabinetry

OBJECTIVES

After studying this chapter, you will be able to:
- Identify the different types of cabinets used in homes.
- Identify cabinet locations by reading floor plans.
- Identify the dimensions of standard-sized cabinets.
- Install prefabricated base and wall cabinets.
- Explain the benefits and disadvantages of the most common countertop materials.
- Explain how to install a solid-surface countertop.
- Explain how carpenters work with plastic laminates.
- Explain how to install prefabricated countertops.
- Identify the different types of cabinet doors.
- Explain how to install euro-style hinges.

TECHNICAL TERMS

backing sheet	euro hinge	plastic laminate
base level line	granite	pull
build-up strip	hinge mortise	shelf
cabinetwork	lipped door	slip sheet
engineered quartz	overlay door	solid-surface

As used in the interior finish of a residence, *cabinetwork* refers to built-in kitchen and bathroom storage. In a general way, it also refers to such work as closet shelving, wardrobe fittings, desks, and bookcases. The term *built-in* emphasizes that the cabinet or unit is attached to the structure. The use of built-in cabinets and storage units was an important development in architecture and design.

In the kitchen, and in other areas of the home, storage units should be designed for the items that they will contain. Space must be carefully allocated, **Figure 23-1**. Drawers, shelves, and other elements should be proportioned to satisfy specific needs. These are three types of cabinetwork used in homes:

- Units that are built on the job by a carpenter— There are many places in a home where a carpenter may be required to build some shelving or casework. Even when most of the cabinets are factory produced, a carpenter is responsible for

the installation. This task requires skill and careful attention to detail.
- Custom-built units constructed in local cabinet shops or millwork plants—Custom cabinets in high-end homes are generally built in a cabinet shop and brought to the building site ready to be installed.
- Mass-produced cabinets from factories that specialize in this area of manufacturing—Except in high-end custom-built homes, most cabinets are made in a cabinet factory and shipped to the site ready for installation.

23.1 Factory-Built Cabinets

Most of the cabinetwork for residential and commercial buildings is constructed in factories that specialize in this work. Modern production machines and tools can save time and produce high-quality cabinetry. Mass-produced parts are assembled with the aid of jigs and fixtures.

Breadmaker/Shutterstock.com

Figure 23-1. These kitchen cabinets are attractive and provide efficient storage space in shelving and drawers. They also provide countertop working space for food preparation around sinks and cooktops.

Manufacturers offer a variety of shades and colors that are applied at the factory by experts, **Figure 23-2**. Because of the controlled conditions and special equipment, finishing materials that have high resistance to moisture, acids, and abrasion can be applied. Once the finishing process is complete and hardware has been installed, the units are carefully packaged and shipped to a distributor or directly to a construction site.

23.2 Drawings for Cabinetwork

For most houses with factory-produced cabinets, the floor plans show cabinet locations. Cabinet elevations provide additional information about cabinet positions and some dimensions. **Figure 23-3** shows typical cabinet elevations found in architectural drawings.

A *il21/Shutterstock.com* **B** *MintImages/Shutterstock.com*

Figure 23-2. Building cabinets in a factory can improve quality and yield a more economical price. A—Cabinet finishes are sprayed on for a consistent, quality finish. B—Premium-quality cabinet doors receive a coat of hand-rubbed stain from highly skilled wood finishers.

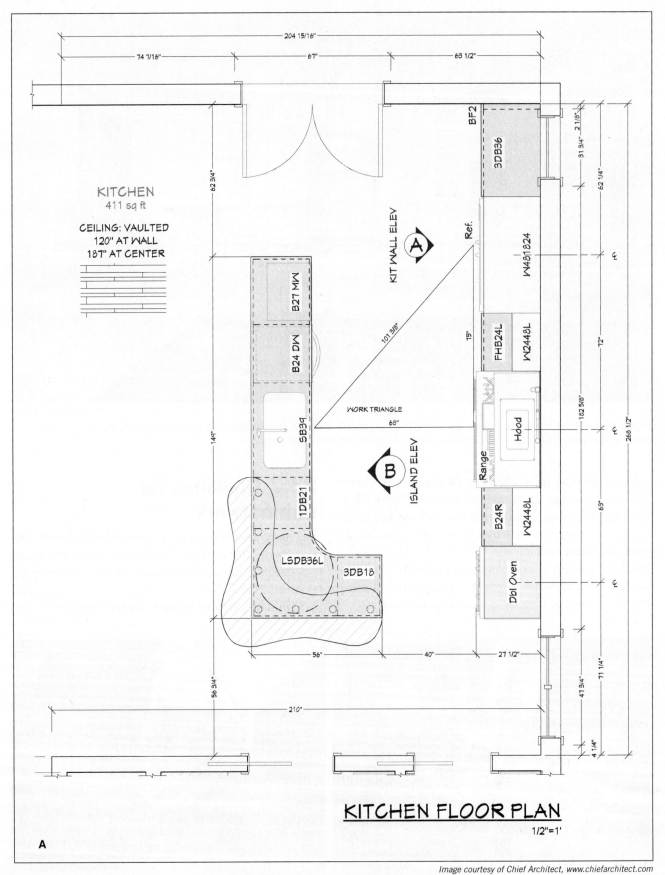

KITCHEN
411 sq ft

CEILING: VAULTED
120" AT WALL
187" AT CENTER

KITCHEN FLOOR PLAN
1/2"=1'

Figure 23-3. Kitchen cabinet layout. A—The floor plan shows cabinet locations. (Continued)

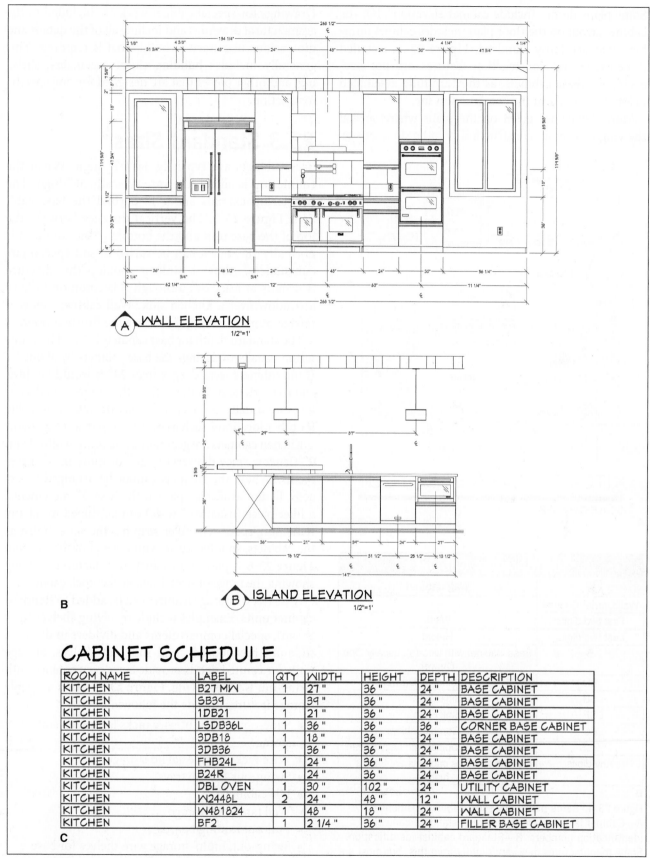

A WALL ELEVATION
1/2"=1'

B ISLAND ELEVATION
1/2"=1'

CABINET SCHEDULE

ROOM NAME	LABEL	QTY	WIDTH	HEIGHT	DEPTH	DESCRIPTION
KITCHEN	B27 MW	1	27 "	36 "	24 "	BASE CABINET
KITCHEN	SB39	1	39 "	36 "	24 "	BASE CABINET
KITCHEN	1DB21	1	21 "	36 "	24 "	BASE CABINET
KITCHEN	LSDB36L	1	36 "	36 "	36 "	CORNER BASE CABINET
KITCHEN	3DB18	1	18 "	36 "	24 "	BASE CABINET
KITCHEN	3DB36	1	36 "	36 "	24 "	BASE CABINET
KITCHEN	FHB24L	1	24 "	36 "	24 "	BASE CABINET
KITCHEN	B24R	1	24 "	36 "	24 "	BASE CABINET
KITCHEN	DBL OVEN	1	30 "	102 "	24 "	UTILITY CABINET
KITCHEN	W2448L	2	24 "	48 "	12 "	WALL CABINET
KITCHEN	W481824	1	48 "	18 "	24 "	WALL CABINET
KITCHEN	BF2	1	2 1/4 "	36 "	24 "	FILLER BASE CABINET

C

Figure 23-3 (Continued). Kitchen cabinet layout. B—Elevations show more detail. C—This set of plans includes a schedule showing all the cabinet sizes.

Some plans do not include cabinet elevations, but the cabinet layout on the floor plan includes cabinet numbers that are fairly standard, **Figure 23-4**. Detailed dimensions and construction features are not necessary for these cabinets, as they are fully assembled before they arrive at the construction site.

More details are given on the plans where special shelving or cabinet construction is to be done onsite.

Drawings for special cabinet work are included in the architectural drawings and include all of the details and dimensions necessary to build what is required. This generally includes building closet organizers, shelving, and other pieces that are designed for one specific installation.

23.3 Standard Sizes

Base cabinets are typically 34 1/2″ high. When the countertop is added, they are typically 36″ high. The countertop extends about 1″ beyond the base cabinets, **Figure 23-5**. The vertical distance between the top of the base unit and the bottom of the wall unit is normally 18″, but it can be varied to suit special circumstances. Above a sink or cooktop this distance is usually at least 24″, although it is common to have a window over a kitchen sink. Wall cabinets above a refrigerator are usually 72″ above the finished floor.

The standard depth for base cabinets is 24″. The countertop usually overhangs the base cabinets by about 1″. If the cabinets were deeper than 24″ it would be difficult to reach items in the back of the cabinets. Wall cabinets are usually 12″ deep, but other depths are available for other locations, such as over the top of a refrigerator.

Kitchen cabinets are generally available in widths from 9″ (narrow spice drawers) to 36″ or more in 3″ increments. Where the cabinets cannot be arranged to exactly fit the available space with these 3″ increments, a filler piece (usually 3″ wide) can be ripped to fill any remaining space. The filler strip has the same finish as the cabinets, so it blends in with the rest of the kitchen. **Figure 23-6** is part of a cabinet manufacturer's catalog showing the range of sizes available for stock cabinets.

A variety of storage features can be added to standard cabinet units. Examples include revolving shelves (lazy Susan), special compartments and dividers in drawers, slide-out breadboards, and slide-out shelves. Storage units for canned goods provide extra convenience with swing-out wood shelving, **Figure 23-7**. Other available storage features include the following:

- Slide-out or tilt-out wire racks for under-sink storage
- Wire racks for special lid storage
- Swing-out storage trays
- File cabinets
- Wall tambour storage
- Pull-out ironing boards
- Swing-out, multi-storage wire shelves for base cabinets

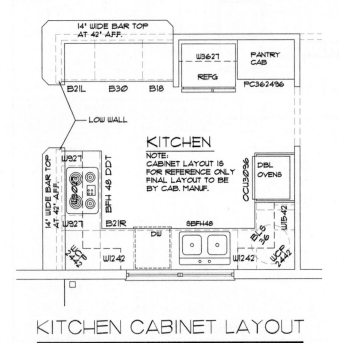

A   *The Garlinghouse Company*

Identification Cabinet

W	Wall cabinet
Wall Corner Cabinet	—
First two digits	Width
Last two digits	Height
B	Base Cabinet with Door(s), usually One Drawer
BD	Base Cabinet all Drawers
SB	Sink Base
SF	Sink Front
BC	Base Corner Cabinet
BLS	Base Corner Cabinet with Lazy Susan
Any digits	Width (all base cabinets are 34 1/2″ high)

B   *The Garlinghouse Company*

Figure 23-4. Some plans only show cabinet positions. A—Cabinet layout drawing including standard cabinet identification numbers. B—Standard cabinet identification. Some manufactures may vary slightly from this, but most are similar.

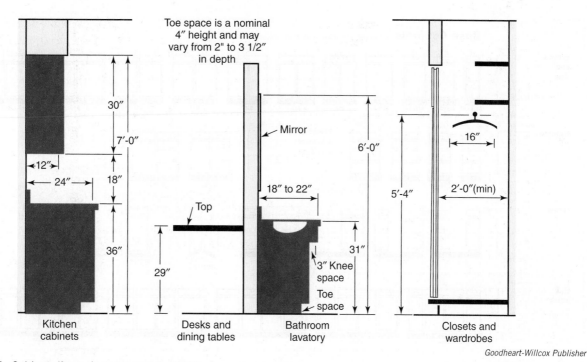

Figure 23-5. Cabinet dimensions are standardized.

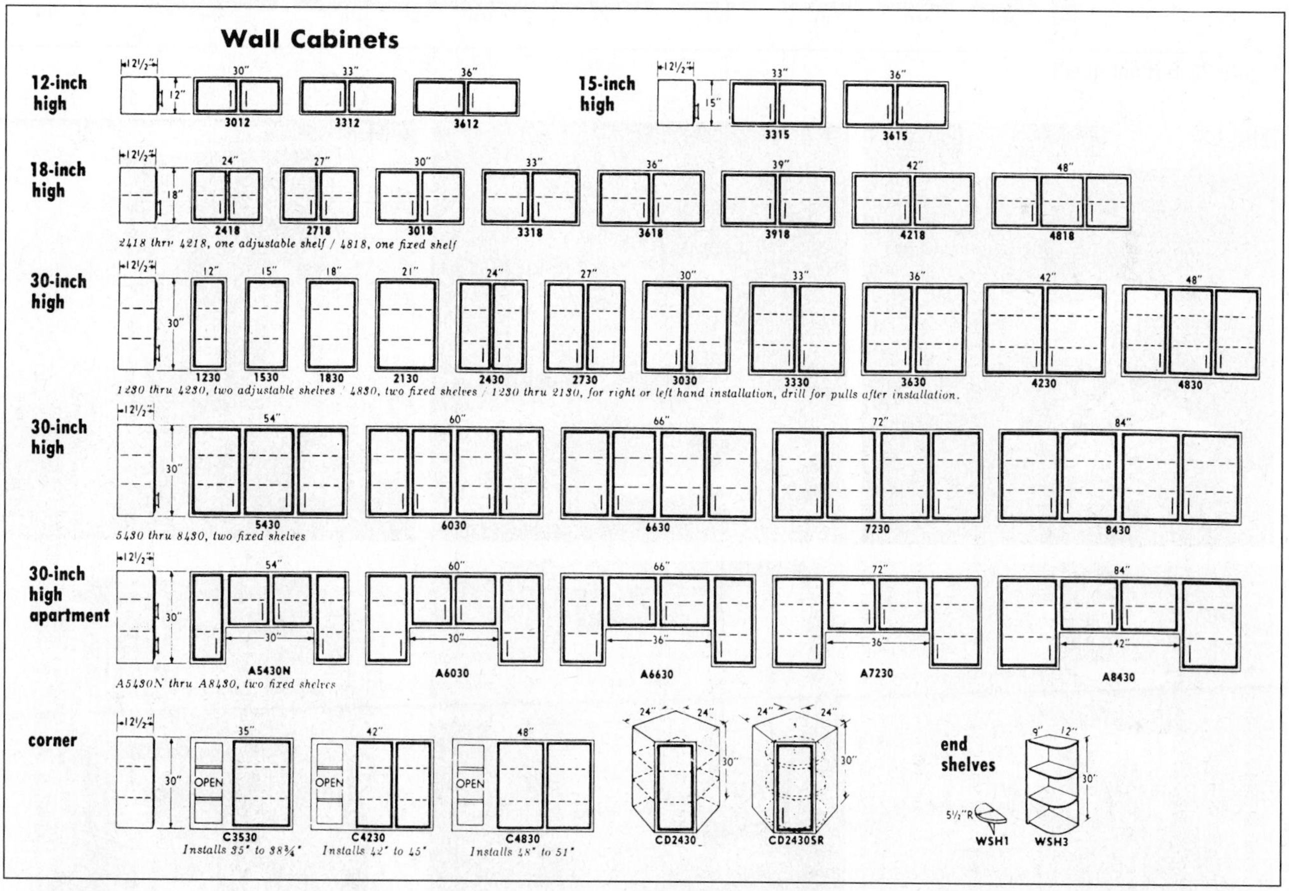

Goodheart-Willcox Publisher

Figure 23-6. Cabinet manufacturer's catalog sheet showing standard cabinet units and dimensions. (Continued)

I-XL Furniture Co.

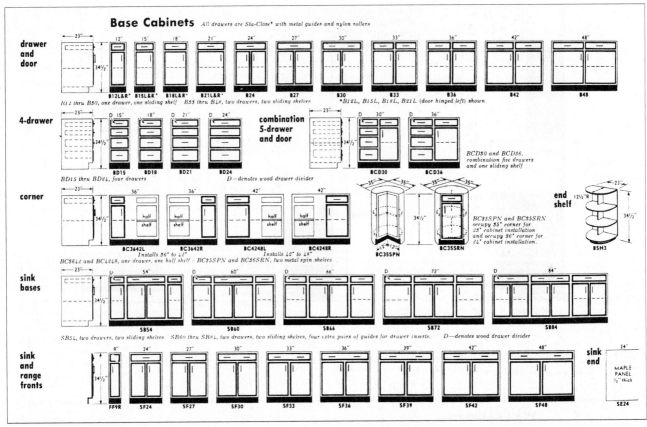

Figure 23-6 (Continued).

Rev-A-Shelf, Inc.

Figure 23-7. Cabinets can be ordered with special drawers, bins, and roll-out shelves.

In bathrooms, vanity cabinets with built-in lavatories are normally 31″ high, **Figure 23-8**. The depth depends on the type of fixture. The counter surface may be plastic laminate, stone, synthetic stone, or tile. Some lavatories provide knee room; others do not.

23.4 Cabinet Materials

A wide range of materials is used in factory-built cabinets. Low-priced cabinets are usually made from panels of particleboard with a vinyl film applied to exposed surfaces. The vinyl is printed with a wood grain pattern.

High-quality cabinets are made from veneers and solid hardwoods such as oak, birch, ash, and hickory.

Hardboard, particleboard, and waferboard may be used for certain interior panels and drawer bottoms. Frames are assembled with accurately made joints. Dovetail joints are generally used in drawer assemblies. See **Figure 23-9**. A completed kitchen installation is shown in **Figure 23-10**.

Breadmaker/Shutterstock.com

Figure 23-8. Bathroom lavatories are often built into a cabinet called a vanity.

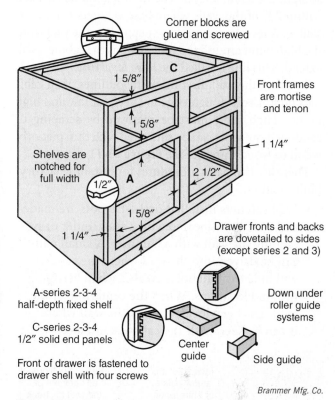

Corner blocks are glued and screwed

Front frames are mortise and tenon

1 5/8″

1 5/8″

1 1/4″

Shelves are notched for full width

1/2″

2 1/2″

1 5/8″

1 1/4″

Drawer fronts and backs are dovetailed to sides (except series 2 and 3)

A-series 2-3-4 half-depth fixed shelf

C-series 2-3-4 1/2″ solid end panels

Front of drawer is fastened to drawer shell with four screws

Down under roller guide systems

Center guide

Side guide

Brammer Mfg. Co.

Figure 23-9. Construction details of a factory-built cabinet.

Artazum/Shutterstock.com

Figure 23-10. This modern kitchen includes a large number of factory-built cabinets.

23.5 Cabinet Installation

Before cabinets can be installed, a carpenter must check walls and floors for uneven spots. After these spots are located, cabinets must be shimmed or scribed to make the installation plumb, true, and square.

Floors can be checked with a long straightedge or straight 2 × 4 and a level. The floor should be checked within 22″ of the walls where base cabinets will be installed. A level line should be snapped on the wall from the high point around the wall as far as the cabinets will extend. This line is called the *base level line*.

To check the wall, first mark the outlines of all cabinets on it. Use a straightedge to check for low and high points. High spots must be removed by scraping or sanding. Low spots can be shimmed with thin pieces of wood or wood shingles. See **Figure 23-11**.

There are two basic procedures for installing factory-built cabinets:

- Wall cabinets installed first—Layouts are made and wall studs located with a stud finder or by tapping the wall with a hammer. A small nail can be driven into the wall where the cabinet will hide the nail holes created when verifying the stud locations. When the centerline of the studs has been located, mark the wall where it can be seen with the cabinets in place.

Then, the wall units are lifted into position. They are held with a padded T-brace, also known as a T-stick or story stick, which allows the worker to stand close to the wall while making the installation. See **Figure 23-12**. After the wall cabinets are securely attached and checked, the base cabinets are moved into place, leveled, and secured.

- Base cabinets installed first—The tops of the base cabinets can then be used to support T-braces that hold the wall units in place. This procedure is illustrated in **Figure 23-13**.

Floors and walls are seldom exactly level and plumb. Shims and blocking must be used as necessary so the cabinets are not racked or twisted. Doors and drawers will not operate properly if the cabinet is distorted by improper installation.

Remove plaster at high points

Tack on shims at low points or shim when attaching cabinets to wall

Mark the outlines of all cabinets on the wall to check actual cabinet dimensions against your layout

84″

30″

19 1/2″

34 1/2″

Level   High spot

High point level

Strike level base line from high point of floor

Check space for tall unit by measuring up from high point level. Be sure tall unit will fit under soffit. It may be necessary to trim some material off the top of the tall units in order to fit under the soffit.

KraftMaid Cabinetry, Inc.

Figure 23-11. Before installing cabinets, mark the outline of every base and wall unit on the wall. Check for high and low spots, walls for plumb, and floors for level.

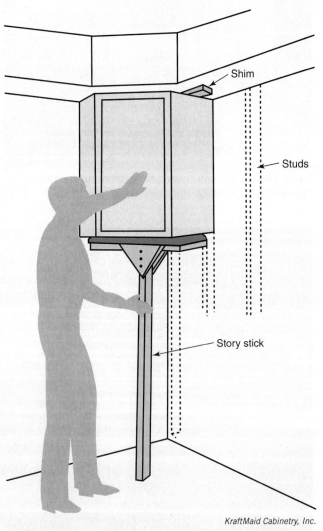

Shim

Studs

Story stick

KraftMaid Cabinetry, Inc.

Figure 23-12. Use a T-stick to support a wall cabinet. Mark the stud locations on the inside of the cabinet for mounting.

Basic Steps for Installing Factory-Built Cabinets

1. Locate the position of all wall studs where cabinets are to hang by tapping with a hammer. Mark their position where the marks can easily be seen when the cabinets are in position.

2. Find the highest point on the floor with a level. This is important for both base and wall cabinet installation later. Remove the baseboard from all walls where cabinets are to be installed. This will allow them to go flush against the walls.

3. Start the installation with the corner or end unit. Slide it into place then continue to slide the other base cabinets into the proper position.

4. When all base cabinets are in position, fasten the cabinets together. This is done by drilling a 1/4″ diameter hole through the face frames, and using the 3″ screws and T-nuts provided. To get the maximum holding power from the screw, one hole should be close to the top of the end stile and one should be close to the bottom.

5. Check the position of each cabinet with a spirit level, going from the front of the cabinet to the back of the cabinet. Next shim between the cabinet and the wall for a perfect base cabinet installation.

6. Starting at the high point in the floor, level the leading edges of the cabinets. Continue to shim between the cabinets and the floor until all the base cabinets have been brought to level.

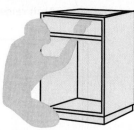

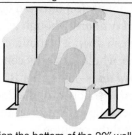

7. After the cabinets have been leveled, both front to back and across the front, fasten the cabinets to the wall at the stud locations. This is done by drilling a 3/32″ diameter hole 2 1/4″ deep through both the hanging strips for the 2 1/2″ × 8 screws that are provided.

8. Fit the counter top into position and attach it to the base cabinets by predrilling and screwing through the front corner blocks into the top. Use caution not to drill through the top. Cover the counter top for protection while all the cabinets are being installed.

9. Position the bottom of the 30″ wall cabinets 19″ from the top of the base cabinet, unless the cabinets are to be installed against a soffit. A brace can be made to help hold the wall cabinets in place while they are being fastened. Start the wall cabinet installation with a corner end cabinet. Use care in getting this cabinet installed plumb and level.

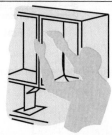
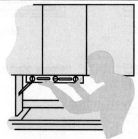
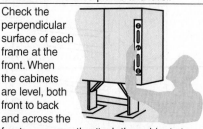

10. Temporarily secure the adjoining wall cabinets so that the leveling may be done without removing them. Drill through the end stiles of the cabinets and fasten them together as was done with the base cabinets.

11. Use a spirit level to check the horizontal surfaces. Shim between the cabinet and the wall until the cabinet is level. This is necessary if doors are to fit properly.

12. Check the perpendicular surface of each frame at the front. When the cabinets are level, both front to back and across the front, permanently attach the cabinets to the wall. This is done by predrilling a 3/32″ diameter hole 2 1/4″ deep through the hanging strip inside the top and below the bottom of the cabinets at the stud location. Enough number 8 screws should be used to fasten the cabinets securely to the wall.

Haas Cabinet Co., Inc.

Figure 23-13. Basic steps for installing factory-built cabinets.

Screws should go through the hanging strips (if used) and into the stud framing. Never use nails. Ideally, a framing carpenter will have installed horizontal blocking between the studs, allowing sufficient surface on which to mount the cabinets. Join units by first clamping them together and aligning them, then installing fasteners, **Figure 23-14**.

23.6 Cabinet Doors

Cabinet doors are sometimes shipped separate from the cabinets. In these cases, the carpenter who installs the cabinets must also install the doors. **Figure 23-15** shows six basic styles of cabinet doors.

Doors may be made of plywood, preferably with a lumber core, or they may consist of a frame with a panel insert. Fine cabinet doors often have a frame covered on each side with thin plywood. This construction is similar to a hollow core door. Because of its tendency to warp, solid stock is seldom used, except on very small doors.

Some manufacturers offer panel doors with glass inserts. These are useful for displaying items on cabinet shelves. Hinged doors for cabinets are of three basic types, depending on how they fit into or over the door opening. **Figure 23-16** shows the basic shapes of the three types:

- *Flush door*—The door fits into the opening and does not project outward beyond the frame.
- *Overlay door*—Though its edges are square like the flush door, it is mounted on the outside of the frame, wholly or partly concealing it.
- *Lipped door*—The door is rabbeted along all edges, so that part of the door is inside the door frame. A lip extends over the frame on all sides to conceal the opening.

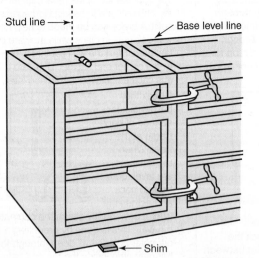

KraftMaid Cabinetry, Inc.

Figure 23-14. To install base cabinets, shim as needed at the floor to level the units and bring them up to the base level line. Clamp adjoining units, as shown, to hold them in alignment, then fasten cabinets to each other and to the wall studs.

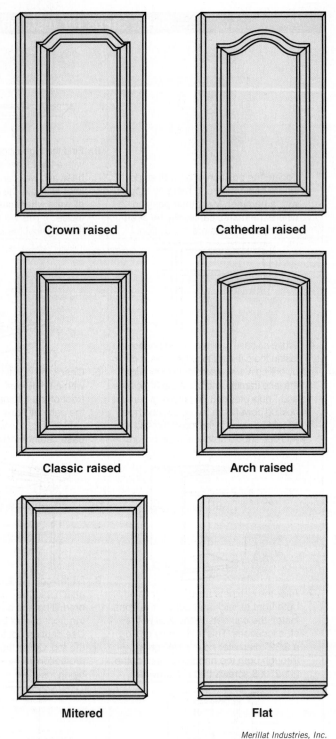

Merillat Industries, Inc.

Figure 23-15. Six basic styles used for cabinet doors.

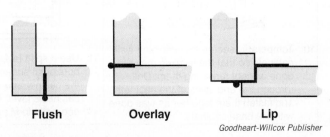

Goodheart-Willcox Publisher

Figure 23-16. Three basic types of cabinet doors.

23.6.1 Installing Doors

Flush doors are often installed with butt hinges. However, surface hinges, wraparound hinges, knife hinges, and euro hinges can be used. **Figure 23-17** shows several styles that are often used for cabinetwork.

The size of a hinge is determined by its length and width when open. Select the size to suit the size of the door. Large wardrobe or storage cabinet doors should have three hinges; smaller doors typically have two.

Flush doors must be carefully fitted to the opening with about 1/16″ clearance on each edge. *Hinge mortises* for hinges should be cut equally into the cabinet frame and the door. On large doors, it is practical to use a portable router for this operation.

Install the hinges on the door, then mount the door in the opening. Initially, use only one screw in each hinge leaf. Adjust the fit and then set the remaining screws. It may be necessary to plane a slight bevel on the door edge opposite of the hinges for better clearance.

Stops are set on the door frame so the door will be held flush with the surface of the opening when closed. The stops may be placed all around the opening or only on the side where the lock or catch is located.

Overlay doors provide an attractive appearance in some contemporary styles of cabinetwork. Overlay doors are usually installed with European-style, or euro, hinges, **Figure 23-18**. *Euro hinges* have standardized hole spacing, so the pocket for the hinge cup and screw holes can be located with a jig made for that purpose. Euro hinges are easy to install, completely concealed when the door is closed, and can be adjusted along three axes with a screwdriver. This makes adjusting the door a simple operation. They can be used with overlay doors, lipped doors, and flush doors. Euro hinges can be used with face-frame cabinets and with frameless cabinets. There are many sizes and types of euro hinges, so consult the manufacturer's specifications to choose the right hinges.

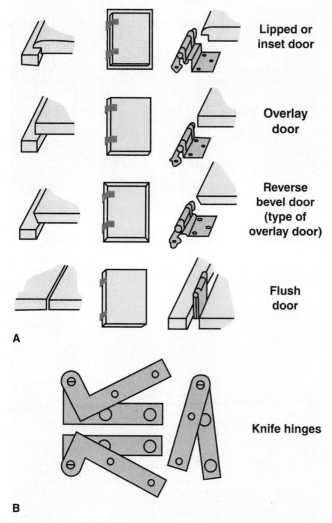

Lipped or inset door

Overlay door

Reverse bevel door (type of overlay door)

Flush door

A

Knife hinges

B

Baldwin Hardware Corp., Amerock Corp.

Figure 23-17. Hinge types used for cabinet doors. A—Common cabinet-door hinge types. B—Knife hinges.

Goodheart-Willcox Publisher

Figure 23-18. Euro-style hinges are easy to install and easy to adjust.

PROCEDURE

Installing Euro Hinges

1. Mark a line on the back of the door parallel with the hinge-side edge and 7/8″ in from the edge, **Figure 23-19A**.

2. Mark lines across the above line 3 1/2″ from the top and bottom of the door. This will be the center of the pocket hole for the hinge cup. Marking all of the hole locations can be simplified with a jig made for that purpose.

3. Using a 35 mm Forstner bit, bore a hole 1/2″ deep for the pocket holes, **Figure 23-19B**.

4. Press the hinge cups into the pocket holes, then use a square to ensure that the hinge is square with the edge of the door.

5. Use a 7/64″ drill bit to drill pilot holes through the hinge and into the door, **Figure 23-19C**.

6. Drive the screws that were supplied by the manufacturer (5/16″ wood screws) to hold the hinge in place.

7. Using a combination square, mark a vertical line inside the cabinet 2 1/4″ from the front edge.

8. Set the door in place and adjust for an even gap all the way around. The gap should be 1/16″.

9. Mark a line across this at 3 1/2″ from the bottom and top, plus 1/16″ for the gap.

10. Remove the hinge mounting plate and set it in the cabinet with the three screw holes aligned over the pencil marks.

11. Use a 7/64″ drill bit to drill pilot holes through the hinge plate and into the cabinet side.

12. Snap the hinges into the mounting plates, **Figure 23-19D**.

13. Make any necessary adjustments to the hinges.

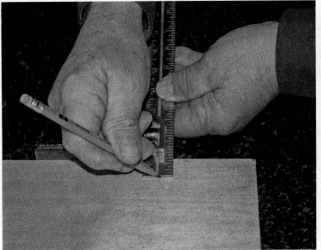
A
Goodheart-Willcox Publisher

B
Goodheart-Willcox Publisher

C
Goodheart-Willcox Publisher

D
Goodheart-Willcox Publisher

Figure 23-19. Installing a euro hinge. A—Marking the door for the pocket hole centerlines. B—Boring the pocket with a Forstner bit. C—Drilling screw holes. D—The installed hinge.

Lipped doors can be installed with euro hinges or a special offset hinge.

The finish hardware is not usually installed when the cabinets arrive at the site. There is a great variety of finish hardware available for cabinets. Each job may call for something different, **Figure 23-20**. It is usually best to install knobs, pulls, and handles after the doors are installed. There is less chance of a protruding piece of hardware scratching other surfaces.

Drawer *pulls* usually look best when they are located slightly above the centerline of the drawer front. They should be horizontally centered and level. Door pulls are convenient when located on the bottom third of wall cabinet doors and the top third of base cabinet doors.

Hinged doors require some type of catch to keep them closed. Several common types are illustrated in **Figure 23-21**. An important consideration in their selection is the noise they produce when the door is opened or closed. In general, the catch should be placed as near as is practical to the door pull. Carefully follow the instructions for installation of the catch that accompany the package.

23.7 Cabinets for Other Rooms

Some manufacturers of kitchen cabinets build storage units for other areas of the home. Built-in units provide drawers and cabinets that save space. They improve efficiency and also offer greater convenience. For example, many homes are being equipped with closet organizer systems for more efficient use of storage space, **Figure 23-22**. These systems consist of modular cabinet and shelving units that can be installed in various combinations, depending on the space available and specific storage needs.

In other rooms, carefully planned built-in cabinets provide efficient storage and display for items used in work, recreation, and hobby activities. Their custom-built appearance usually adds to room decor. They may even eliminate the need for some movable pieces of furniture.

23.7.1 Shelves

Shelves are widely used in cabinetwork, especially in wall units. If possible, make the shelves adjustable. This allows the storage space to be used for various purposes.

Figure 23-23 presents several methods of installing shelves. Carefully and accurately lay out the shelf support system so the shelves will be level. Usually, it is best to do the cutting and drilling before the cabinet is assembled.

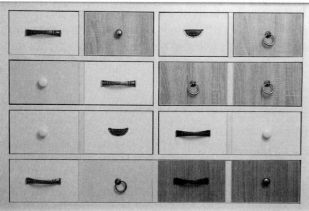

Amerock Corp.

Figure 23-20. Some examples of cabinet door and drawer hardware.

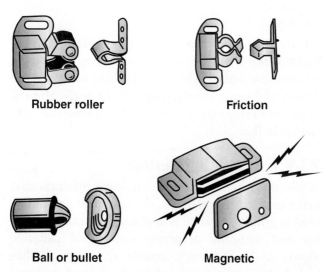

Rubber roller   **Friction**

Ball or bullet   **Magnetic**

Goodheart-Willcox Publisher

Figure 23-21. Various mechanisms are used by cabinet catches to hold doors shut.

Goodheart-Willcox Publisher

Figure 23-22. A combination of wall- and floor-mounted cabinetry is used in this walk-in closet to provide various types of storage.

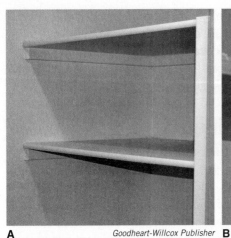

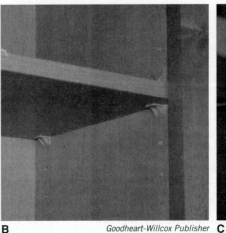

A   *Goodheart-Willcox Publisher*   B   *Goodheart-Willcox Publisher*   C   *Goodheart-Willcox Publisher*

Figure 23-23. Several methods of supporting shelves. A—Wood cleats attached to the wall or inside of the cabinet. B—Shelf pins fit into holes on the inside of the cabinet wall. C—Metal track holds brackets or clips that support shelf.

If shelf standards are of the type set in a groove, it is absolutely necessary to cut the groove before assembly. Some patented adjustable shelf supports are designed to mount on the surface.

Standard 3/4″ shelving should be supported every 42″ or at shorter intervals. This applies especially to shelves that will carry heavy loads. The front edge of plywood shelving should be overlaid with a strip of wood material that matches the cabinet wood. This may be solid stock or thin strips that are glued to the plywood edge.

23.8 Other Built-In Units

The information to this point in the chapter has been largely directed toward kitchen cabinets. The same general procedures can be applied to wardrobes, room dividers, and various built-in units for living rooms and family rooms. Built-in features are popular in other areas of the house for these reasons:

- Increased storage space
- Efficient organization and use of space
- Attractive, customized appearance of well-designed and carefully constructed units

Study the details of the construction provided in the architectural plans or develop a carefully prepared working drawing using the actual dimensions secured from the wall, floor, and ceiling surfaces. When drawings are provided, check the space available to make sure that it agrees with the drawings.

Wall and floor surfaces are seldom perfectly level or plumb. Slight adjustments will be needed as the cabinetwork is constructed. Room corners or the corners of alcoves designed for built-in units will likely not be square. Here, the cabinetwork will need to be trimmed or strips and wedges used to bring the work into proper alignment.

Keep a square and level constantly at hand, especially during the rough-in stages. Do not carry inaccuracies from the wall or floor into the cabinet framework. Doing so will make it difficult to keep the cabinet facing plumb and level. Errors will also be a source of annoyance during the hanging of doors and fitting of drawers.

23.9 Countertops

There is a wide variety of materials used for countertops:

- Ceramic tile
- Copper
- Glass
- Granite
- Marble
- Plastic laminate
- Soapstone
- Solid-surface (synthetic resin)
- Wood
- Zinc

Each has its own characteristics, benefits, and drawbacks. Whatever material is used, the countertop must be attractive, water-resistant, and able to withstand the bumps and bangs of everyday use. As with cabinets, most countertops are prefabricated in a shop that specializes in that work.

23.9.1 Stone Countertops

While there are several natural stone products that make excellent countertops, the one most commonly used is *granite*. Granite is available in a range of colors and naturally occurring swirls and patterns, **Figure 23-24**. It is very durable, resisting scratches and dents. It is not affected by the heat of cooking. Granite is, however, heavy and it can be stained by colored liquids. Granite, marble, and other natural stone

New Azul Aran   Aspen White   Black Fantasy Deep Brush   Mesquite   River Bordeaux Satin

Montana   Zion   Black Beauty   Delicatus Cream   Golden Ray

Arizona Tile

Figure 23-24. Granite from different locations around the world has different colors and grain patterns.

countertops require application of a special sealer to resist staining. The cost of granite varies, with exotic grain patterns and colors being most expensive. Granite suppliers often require that you purchase the entire slab, which means there can be considerable waste from unused portions. Special tools and skills are required to work with granite, so it is usually fabricated and installed by someone who specializes in stone countertops. During installation, silicone adhesive sealant is applied to the tops of the base cabinets and the granite is set in that adhesive. Where necessary, joints are sealed with two-part epoxy.

23.9.2 Engineered Quartz Countertops

Engineered quartz is a composite made of ground quartz, synthetic resin, and pigment. It is impervious to stains and spills. It is heat resistant and fairly scratch resistant. Because the color pigment is added as part of the manufacturing process, it can be produced in a wide variety of colors. Like granite, engineered quartz countertops require special tools and skills to fabricate, so they are usually installed by someone who specializes in stone countertops. It is expensive, but it is often the preferred material because of its durability. It is also favored by those who are striving for green-friendly materials because it is long-lasting and involves little waste. **Figure 23-25** shows a kitchen with an engineered quartz countertop. Engineered quartz

countertops are applied in the same manner as natural stone countertops.

23.9.3 Solid-Surface Countertops

Solid-surface counters are a blend of acrylic or polyester resins, powdered fillers, and pigments cast into slabs. This durable material is nonporous, so it does not stain easily. Because it is made of plastic resins, it scratches easily and can be affected by heat. It can be manufactured in almost any color and made to look like stone.

Cambria USA

Figure 23-25. Engineered quartz is durable, impervious to stains, and available in many colors and patterns.

Figure 23-26 shows a solid-surface countertop. Solid-surface is softer than stone, so it can be worked with woodworking tools outfitted with sharp, carbide-cutting edges. It is usually ordered cut to final size, ready for installation. It is adhered to base cabinets with silicone, and joints are glued with seam adhesive made for this purpose.

PROCEDURE

Installing Solid-Surface Countertops

Manufacturers of solid-surface countertops have specific instructions, including the adhesives to be used with their materials. They also manufacture sinks of the same material. These sinks can be bonded to the underside of the countertop using the manufacturers' instructions. Steel sinks can also be fastened to the underside of the countertop with fasteners that are glued into holes drilled partway into the bottom surface of the countertop.

1. Prepare the base cabinets. The countertop should be supported every 24″. Install additional pieces inside cabinet tops if necessary.
2. Apply 2″-3″-wide pieces of 1/2″ or 3/4″ particleboard or MDF (Medium Density Fiberboard) along the cabinet edges, **Figure 23-27**.
3. Surface edges of countertop pieces to be joined to ensure clean, tight-fitting joints. This is best done using a router with a straight bit and ball-bearing guide. The seam should fit tightly with only a faint visible line when dry fitted.
4. Clean surfaces to be joined with denatured alcohol.
5. Place the sheets to be joined face down and position the sheets to be joined about 3/16″ apart.
6. Attach wood clamping blocks to both pieces 1″ from each edge using hot-melt glue. These blocks will be used to clamp the two sheets together.
7. Fill the joint half full with the manufacturer's recommended seam adhesive.
8. Use spring clamps or bar clamps to hold the two sheets together while the glue cures, **Figure 23-28**. Do not overtighten the clamps, as this will force too much adhesive out of the joint, causing a *starved joint*.
9. Use denatured alcohol to soften the hot-melt glue, then tap the blocks lightly on their sides with a hammer to break them free from the glue.
10. Do not attempt to remove seam adhesive or marks from the hot-melt with a scraper. It can be removed and polished when the countertop is installed.
11. Apply a dime-size dab of silicone every 18″. No mechanical fasteners should be used to fasten the countertop. Never screw, staple, or nail into solid-surface.
12. Set the countertop in place on the silicone.
13. Using a random-orbit sander and starting with an abrasive equivalent to 180-grit, sand all seams to remove the remaining adhesive. Sand with increasingly finer abrasives until the desired finish is achieved. Clean all sanding dust from the surface between abrasive changes.

Goodheart-Willcox Publisher

Figure 23-26. Solid-surface countertops can be manufactured in almost any color.

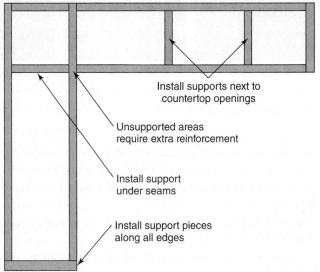

Install supports next to countertop openings

Unsupported areas require extra reinforcement

Install support under seams

Install support pieces along all edges

Wilsonart LLC

Figure 23-27. Apply particleboard or MDF around edges of cabinets to support the countertop.

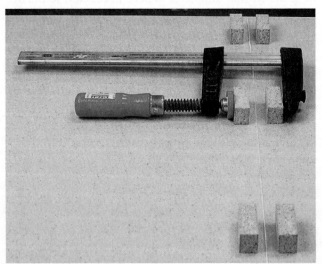

Wilsonart LLC

Figure 23-28. Use bar clamps or spring clamps to hold the two pieces together until the adhesive cures.

23.9.4 Plastic Laminate Countertops

Plastic laminate was once the most popular countertop material available, but it has lost some favor to the wide range of options available today. Plastic laminate is still an excellent choice and very attractive for many applications. It is easily worked with carpentry tools. Although most plastic laminate countertops are prefabricated and delivered to the site ready for installation, most carpenters work with plastic laminates at some point in their career.

Plastic laminate for countertops is usually approximately 1/16″ thick, offers high resistance to wear, and is unharmed by boiling water, alcohol, oil, grease, and ordinary household chemicals, **Figure 23-29**.

Although the laminate is very hard, it does not possess great strength and is serviceable only when bonded to plywood or particleboard. This base or core material must be smooth and dimensionally stable. However, some plywoods, especially fir, have a coarse grain texture that may telegraph (show through). Particleboard, which is less expensive than plywood, provides a smooth surface and adequate strength. The surface texture of oriented strand board (OSB) may telescope through the plastic laminate, so it is not used as a base.

When the core or base is free to move and is not supported by other parts of the structure, the laminated surface may warp. Bonding a backing sheet of the laminate to the second face can counteract this. It will minimize moisture gain or loss and provides a balanced unit with identical materials on either side of the core. For a premium grade of cabinetwork, Architectural Woodwork Institute standards specify that a **backing sheet** be used on any unsupported area exceeding 4 sq ft. Backing sheets are like the regular laminate without the decorative finish and are usually thinner. A standard thickness for use opposite a 0.060″ (1/16″) face laminate is 0.020″.

Working with Laminates

Plastic laminates can be cut to rough size with a handsaw, table saw, portable circular saw, or portable router, **Figure 23-30**. Use fine-tooth blades and support the material close to the cut. Laminates used on vertical surfaces, which are 1/32″ thick, can be cut with tin snips.

Formica Group North America

Formica Group North America

Figure 23-29. Plastic laminate countertops are hard and durable, resisting stains and wear. Laminates are available in a wide variety of colors and patterns.

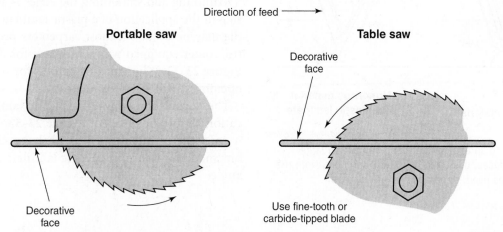
Goodheart-Willcox Publisher

Figure 23-30. Saws used to cut plastic laminate must have fine-toothed blades.

It is best to make the roughing cuts 1/8″–1/4″ oversize. Trim the edges after the laminate has been mounted. Handle large sheets carefully because they can be easily cracked or broken. Be careful not to scratch the decorative side.

Contact bond cement is used to apply the plastic laminate because no sustained pressure is required. It is applied with a spreader, roller, or brush to both surfaces being joined. On large horizontal surfaces, it is best to use a spreader, **Figure 23-31**.

For soft plywoods, particleboard, or other porous surfaces, the spreader is held with the serrated edge perpendicular to the surface. On hard, nonporous surfaces and on plastic laminates, hold the edge at a 45° angle. A single coat should be sufficient.

An animal hair or fiber brush may be used to apply the adhesive to small surfaces or those in a vertical position. Apply one coat, let it thoroughly dry, and then apply a second.

All of the surface should be completely covered with a glossy film. Dull spots, after drying, indicate that the application was too thin and that another coat should be applied.

Thoroughly stir the adhesive before using it and follow the manufacturer's recommendations. Usually, brushes and applicators must be cleaned in a special solvent.

After the adhesive has been applied to both surfaces, let it dry (usually for at least 15 minutes). You can test the dryness by lightly pressing a piece of paper against the coated surface. If no adhesive sticks to the paper, it is ready to be bonded. This bond can usually be made any time within an hour. Time varies with different manufacturers. If the assembly cannot be made within this time, the adhesive can be reactivated by applying another thin coat.

Safety Note
Some types of contact cement are extremely flammable. Nonflammable types may produce harmful vapors. Be sure the work area is well ventilated. Follow the manufacturer's directions and observe precautions.

Adhering Laminates

Once the adhesive is ready, bring the two surfaces together in the exact position required. The pieces cannot be shifted once contact is made. When joining large surfaces, place two sheets of heavy freezer paper or waxed paper, called *slip sheets*, over the base surface. Arrange the slip sheets so they overlap slightly near the middle of the surface. Then, slide the laminate into position. Withdraw one piece of the paper so one side of the laminate can be bonded and then remove the other slip sheet and apply pressure.

Total bond is secured by the application of momentary pressure. Hand rolling provides satisfactory results if the roller is small (3″ or less in length). Long rollers apply less pressure per square inch. Work from the center to the outside edges and be certain to roll the entire surface. In corners and areas that are hard to roll, hold a block of clean, soft wood on the surface and tap it with a rubber mallet.

Trimming and smoothing the edges is an important step in the application of a plastic laminate. A plane or file may be used, but most carpenters prefer an electric router equipped with a bit made for this purpose, **Figure 23-32**. When making cutouts for sinks or other openings, a saber saw is practical.

The corner and edges of a plastic laminate application should be beveled, **Figure 23-33**. This makes it smooth to touch and less likely to chip. Router bits made for trimming plastic laminate achieve this angle.

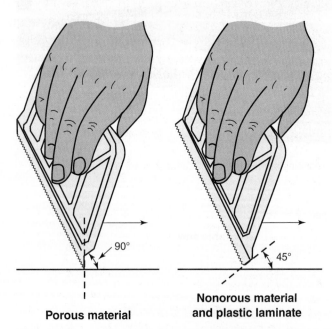

90°

45°

Porous material

Nonorous material and plastic laminate

Goodheart-Willcox Publisher

Figure 23-31. To spread contact bond cement on large horizontal surfaces, use a toothed metal spreader. Note the different angles used, depending on the material.

An alternative is to finish the edge of the counter with edge trim made to match the countertop. Edge trim has matching laminate factory-bonded to a fiber base or particleboard base. When properly applied with contact-bond cement, the seam between the edge trim and the top surface is difficult to see.

When working with plastic laminates, be especially careful that files, edge tools, or abrasive papers do not damage the finished surfaces. Such damage is difficult to repair.

Installing Laminate Countertops

Countertops are usually prefabricated in a millwork shop, but a carpenter may choose to build them to the proper measurements and specifications. Before beginning installation, double check the base cabinet tops. They must be level and even across the top. If necessary, shim them at the floor or base. ***Build-up strips*** are narrow wood pieces that are used to attach the countertop to the base cabinets. They are usually provided by the millwork shop. If not provided, a carpenter must cut them.

Figure 23-34 shows the installation of a mitered countertop. Miters must be accurately cut and fitted, a difficult procedure requiring great skill. The joint is sealed with silicone. Then it is secured and drawn tight from the bottom with bolts that are let into the substrate.

DeWalt

Figure 23-32. A small router, sometimes called a laminate trimmer, is used to finish the edge of the laminate.

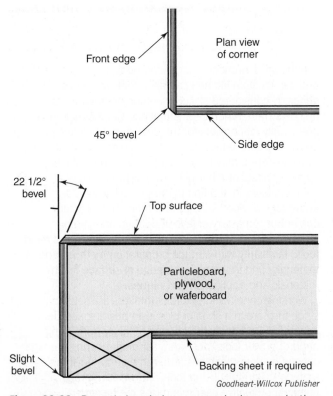

Goodheart-Willcox Publisher

Figure 23-33. Properly beveled corners and edges on plastic laminate countertops improve the appearance and durability of the installation. Note the angles shown in the two views.

Goodheart-Willcox Publisher

Figure 23-34. Properly fitting a mitered countertop is a job for a skilled carpenter or woodworker.

PROCEDURE

Installing Prefabricated Countertops

Normally, the prefabricated top will come with instructions for installation. In general, follow these steps:

1. If build-up strips are not provided, cut 3/4″ lumber into strips 2″ wide and 24″ long.
2. Space build-up strips 2′ apart, but be sure to place beginning and end strips 2″ in from the cabinet ends. This space allows room for end caps on the countertop.
3. To mount the build-up strips, drill screw holes at each end. Then, using 1 1/4″ drywall screws, attach one end to the top of the face frame and the opposite end to the back of the cabinet.
4. Position the countertop on top of the cabinets and tight against the wall.
5. Use a pencil to scribe a line transferring any irregularities in the wall to the backsplash.
6. Use a belt sander and medium-grit sandpaper to remove material from the backsplash up to the marked line. Always have the belt moving into the laminate to avoid chipping.
7. Check your work and repeat steps 5 and 6 until the fit is satisfactory.
8. If a sink or drop-in range is to be installed, check the manufacturer's instructions for size and location. Apply masking tape to the countertop and trace the cut lines on it. The tape helps to avoid chipping.
9. Drill a relief hole at each corner.
10. Install a sharp saw blade in a jigsaw and make the cut.
11. Position the countertop on the cabinet base and check the fit.
12. Working from underneath, drill pilot holes in the build-up strips.
13. Drive 1 1/4″ black drywall screws up through the build-up strips to anchor the countertop.
14. Place caulk at the joint between the backsplash and the wall.

Construction Careers

Cabinetmaker

Built-in shelving units, ornate wood paneling, furniture built to order, and custom-made cabinets for kitchens or other rooms are all products produced by a cabinetmaker. While some cabinetmakers are employed by high-end contractors and remodelers, most of those who work in the construction field are self-employed. Outside of the construction industry, cabinetmakers are employed by custom woodworking firms building furniture and cabinets to order.

Monkey Business Images/Shutterstock.com

Cabinetmakers must know the entire process of cabinetmaking from beginning to end. That means designing the cabinet, selecting the appropriate wood, and possessing proper cutting and fastening techniques and finishing skills.

Although a number of large companies produce stock and semicustom kitchen cabinets, their workers are generally considered skilled machine operators and finishers, rather than cabinetmakers. Cabinetmakers are usually responsible for the design and building of custom wood products to meet customer specifications. In some cases, they may execute a design produced by an architect or other person. Skills required of a cabinetmaker, in addition to design ability, include knowledge of wood species and their suitability for cabinet use; mastery of layout, cutting, and joining techniques; the ability to use all types of power and hand tools; familiarity with surface preparation methods and materials; and thorough knowledge of surface finish materials and application techniques.

Working conditions for cabinetmakers are seldom harsh, since they are in a shop or closed-in building site. Job hazards are primarily the chance of injury from using hand or power tools, dust from sanding, hearing damage caused by loud machines, and fumes from finishing materials. Carefully using tools and wearing proper personal protective gear minimizes the hazards.

In addition to vocational training in woodworking, persons wishing to enter the cabinetmaking field should take classes in math, design and drafting, and print reading. Most cabinetmakers acquire their skills over a period of years by working under the direction of an experienced cabinetmaker.

Chapter Review

Summary

- Cabinetwork refers to built-in kitchen and bathroom storage, but also closet shelving, wardrobe fittings, desks, and bookcases.
- Although most carpenters do not build cabinets, they should have an understanding of cabinet construction and the materials used.
- Cabinet locations, as well as where special shelving or cabinet construction is to be done are generally shown on floor plans.
- Cabinet doors may either be flush, overlay, or lipped. The lipped and overlay door styles overlap the door opening in the cabinet and conceal it.
- Euro hinges have standardized hole spacing, are easy to install, and can be used with all door types.
- In addition to kitchens, carpenters install cabinetry and shelving in other areas of the house to save space, improve efficiency, and offer greater convenience.
- Countertops may be made of any one of several materials, including ceramic tile, copper, glass, granite, marble, plastic laminate, soapstone, solid-surface, wood, and zinc.
- Most countertops are prefabricated and must be water resistant and durable.

Know and Understand

Answer the following questions using the information in this chapter.

1. *True or False?* Factory-built cabinets can be finished to resists moisture, acids, and abrasion.

2. Shelving dimensions for a specific cabinet installation can be found in the _____.
   A. floor plan of the house
   B. architectural drawings
   C. manufacturer's instructions
   D. cabinet layout

3. The base unit of a standard kitchen cabinet is _____" high.
   A. 24   C. 34 1/2
   B. 31 1/2   D. 36

4. *True or False?* Swing-out wood shelving is best used for convenient storage of canned goods?

5. Which of the following is *not* considered a high-quality cabinet material?
   A. Oak.   C. Particleboard.
   B. Birch.   D. Ash.

6. High spots on walls must removed by _____, while low spots are _____ with thin pieces of wood or wood shingles.
   A. shimming, scraped   C. scraping, nailed
   B. scraping, shimmed   D. leveling, scraped

7. A _____ door is rabbeted along all edges, so that part of the door is inside the door frame.
   A. hollow core   C. overlay
   B. flush   D. lipped

8. Smaller cabinet doors typically have _____ hinges.
   A. two   C. surface
   B. three   D. knife

9. *True or False?* It is best to install knobs, pulls, and handles before a cabinet door is installed.

10. Marking all of the hole locations for a euro hinge is conveniently done using a _____ made for that purpose.
    A. screw   C. drill bit
    B. jig   D. mounting plate

11. _____ is durable, impervious to stains, and available in many colors and patterns.
    A. Granite   C. Marble
    B. Plastic laminate   D. Engineered quartz

12. *True or False?* Plastic laminate is still the most popular countertop material available.

13. When making a plastic laminate installation, it is recommended that a _____ be used on any unsupported area greater than 4 sq ft.
    A. built-up strip   C. backing sheet
    B. slip sheet   D. special joint

14. _____ are narrow wood pieces used to attach laminate countertops to base cabinets.
    A. Slip sheets
    B. Trim strips
    C. Build-up strips
    D. Backing sheets

Apply and Analyze

1. Explain the three types of cabinetwork used in homes.
2. What must a carpenter do before cabinets can be installed?
3. What is the purpose of a stop on a flush cabinet door?
4. Where should a door pull be installed on a base cabinet door?
5. Name three reasons built-in cabinets are popular.
6. When adhering laminates, how should carpenters join large surfaces together?

Critical Thinking

1. When remodeling a kitchen, it is often necessary for homeowners to continue living in the house while carpenters remove the old kitchen and replace it. This process can take weeks. If you were a lead carpenter on such a job, how would you prepare the homeowners for the disruption in their daily lives before the job begins? What steps could you take to minimize stress on them while the project is taking place?
2. Put yourself in the role of the homeowner in question 1. During this project, what issues do you think will frustrate you about the disruptions of living with weeks of construction in your home? Of these issues, what would frustrate you the most? Does this perspective change how you feel about your answer to question 1?
3. When building cabinets in a shop, what do you think are the most important details of your work? If you do not pay attention to these details what are the consequences?

Communicating about Carpentry

1. **Listening.** With a group of three other students, determine materials needed for installing cabinetry. Each student should go to a hardware store or its website (Home Depot, Lowe's, etc.) to investigate and calculate how much these materials would cost. Determine the cost and quality differences between different types of cabinetry. Then each group presents their findings to the class, and the other class members take notes on the presentation.
2. **Listening, Writing, and Speaking.** Working with a partner, interview a veteran carpenter and ask how the profession has changed over the years. Among the questions to ask: Is the job fulfilling? What are the differences now than when you began your career? How has technology played a role in the tools you use? What words of advice would you have to someone who is thinking of entering this field as a career? Develop a report for the class on the responses to your interview.

CHAPTER 24

Painting, Finishing, and Decorating

OBJECTIVES

After studying this chapter, you will be able to:

- Cite safety rules that apply to painting and finishing.
- List tools and equipment and demonstrate their use.
- Select proper materials for various painting, finishing, and decorating jobs.
- Prepare exterior and interior surfaces for painting.
- Explain proper procedures for painting, finishing, and wallpaper hanging.
- Correct problems that may occur with previously painted exteriors.

TECHNICAL TERMS

binder	intermediate colors	siphon-feed spray gun
booking	pigment	tint
caulk	plug	value
coatings	pressure-feed spray gun	volatile organic compounds (VOCs)
complementary colors	primary colors	
cutting in	putty	wallpapering
filler	related colors	wet edge
gravity-feed spray gun	secondary color	wood bleaching
hue	shade	wood putty

Paint is more than an attractive covering—it is also a protective coating. Wood and some other covering materials used inside and outside a building require protection against soiling, rot, and other types of deterioration caused by the environment.

Exterior wood siding and trim, if left unprotected, will deteriorate from exposure to ultraviolet light, moisture, and microscopic organisms. Frequent wetting of wood, even for short periods, breaks down the lignins that bind the cellulose fibers. Then, rain washes away the loose fibers. Unfinished woods also can discolor, shrink, swell, check, and warp.

The term *coatings* is used to describe all types of finishes, whether they are designed for wood or for other materials, such as metals and drywall. Coatings can be clear or colored. They include paints, stains, varnishes, and various synthetic materials. Anodizing is a process used to color and preserve the surface of aluminum. It is also considered a coating.

24.1 Paint and Environmental Health Hazards

The element lead can affect almost every organ and system in your body. Ingestion or inhalation of lead can be dangerous. Children six years old and younger are most susceptible to the harmful effects of lead. People can breathe lead dust by spending time in areas where lead-based paint is deteriorating. Lead can also be inhaled during renovation or repair work that disturbs painted surfaces in older homes and buildings. According to the US Environmental Protection Agency (EPA), 87% of the homes built before 1940 and 24% of the homes built before 1978 contained lead-based paint, **Figure 24-1**. Test kits for determining if paint contains lead are available at paint and hardware stores. Anyone who paints or conducts renovations or repairs in pre-1978 or child-occupied facilities must be certified by the EPA or an authorized state agency. In addition to personal certification, the company doing the work must be registered with the EPA.

Although today's paints and finishing materials are formulated without lead, virtually all contain other harmful substances. Household paint contains up to 10,000 chemicals, of which about 300 are known toxins (harmful to human health) and 150 are known carcinogens (cause cancer). Some of the most harmful of these chemicals are ***volatile organic compounds (VOCs)***. VOCs are carbon-containing chemicals that readily evaporate into the air. When they enter the air, they react with other elements to produce ozone, which causes air pollution and a host of health issues including breathing problems, headache, nausea, and burning, watery eyes. Some VOCs have also been linked to not only cancer, but kidney and liver

damage as well. Painting and finishing materials are the second leading source of VOCs, after automobile exhaust.

Paint is typically made of pigment to give the paint color, a vehicle to help the pigment stay on the surface, and solvents to make the whole mixture more liquid. The vehicle is a film-forming liquid, such as acrylic polymer or alkyd oil. The solvent is part of the vehicle. Common solvents in paint are turpentine, mineral spirits, toluene, ketones, and water. The solvents are made to evaporate quickly and allow the paint to dry, which makes them the biggest source of VOCs. As the solvent evaporates into the atmosphere, the chemicals it contains are also released with it. Of the solvents most commonly used in paints, only water is low in VOCs. Getting rid of the oil-based solvents does not entirely rid the paint of VOCs. Water-based paints can contain VOCs in their pigments and vehicles. According to the EPA, latex and flat-finish paint with less than 250 grams of VOCs per liter of paint can be labeled as low-VOC paint. Paints with more gloss must be below 380 g/L to be labeled as low VOCs. Paints with less than 5 g/L can be labeled as no VOCs.

The VOCs in paint and finishing materials can be released even after the paint has dried. Low-VOC paint is healthier for both the painter and the eventual inhabitants of the building.

Green Note

Certain materials, such as paint, sealants, particleboards, and even carpeting, can off-gas chemicals, called volatile organic compounds (VOCs), and impact indoor air quality. Substitute high-VOC materials with low-VOC and VOC-free alternative products.

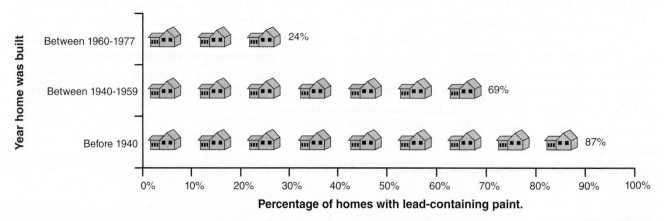

Goodheart-Willcox Publisher

Figure 24-1. Older homes are likely to contain lead-based paint.

Follow these safety rules when working with paints and finishes:

- Wear safety glasses when applying finishing materials.
- Wear rubber gloves, goggles, and a rubber apron when working with bleaches and acids.
- When working with or spraying thinners and reducers (such as naphtha, lacquer thinner, and enamel reducer), keep the work area well ventilated. The fumes are highly toxic.
- Store chemicals and soiled rags in safe containers.
- Wear an approved respirator when using toxic chemicals.
- Wear a respirator or mask and protective clothing when operating a paint sprayer, **Figure 24-2**.
- Thoroughly wash your hands after applying finishes to remove any toxic materials.
- Locate the nearest sink, shower, or eyewash station in case of a splash or spill that lands on your skin or in your eyes.
- Keep a fire extinguisher handy in the finishing area.
- When preparing a building built before 1978 for repainting, test for lead in the old paint before sanding or scraping. Breathing or contact with lead can cause serious health problems. Sanding of lead-containing paints requires special precautions to prevent breathing in the lead particles that would be released. Removal of paints containing lead should be done only by trained professionals.

24.2 Caulks and Sealants

Before any exterior surfaces are painted, any openings or cracks that might allow water or outside air to enter the building should be sealed with caulking compound, or caulk. *Caulk* is a material used to fill joints, cracks, or holes in construction. It is applied using a caulking gun, **Figure 24-3**. Caulk does not perform well where it may be stretched by expansion and contraction or other movement of the building parts. Sealants, on the other hand, have the ability to maintain a seal even when the base parts move slightly. When selecting caulks and sealants, their ability to adhere to the base material must be taken into account. Many manufacturers of building materials, such as window manufacturers and siding manufacturers, specify the type of caulk or sealant that should be used with their product.

Caulks and sealants are available with different chemical make-ups: silicone, acrylic, modified acrylic, and polyurethane, for example. Some caulk is paintable. Some is transparent, or translucent, and some is opaque and pigmented to blend with the substrate (base material). **Figure 24-4** shows which caulks and sealants work best for various applications.

Caulk is important to sealing all building penetrations (such as for pipes and wires) and joints between exterior surfaces. It also has interior uses. In virtually every building or home there are gaps where molding contacts the walls or pieces of molding join. Paint is usually not a good gap filler, so it is best to seal these gaps with caulk, which is easier to use than spackle or wood filler. Some examples where caulk can be used are around crown molding, chair rails, baseboards, door and window frames, gaps in mitered corners, and small wall cracks. These areas can be filled with acrylic (paintable) caulk.

Graco Inc.

Figure 24-2. Respiratory protection is important when operating a paint sprayer. Depending on the material being sprayed, wear a mask or approved respirator.

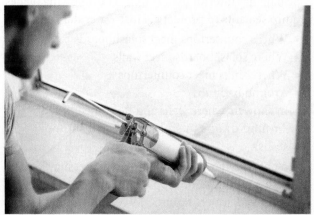

Africa Studio/Shutterstock.com

Figure 24-3. Calk is used around windows, ensuring a proper seal.

Caulking Applications

Caulking Compound	Recommended Uses	Cleanup	Shrinkage	Adhesion	Cost	Comments
Silicone	Seals most dissimilar building materials such as wood and stone, metal flashing, and brick.	Dry cloth if immediate; mineral spirits or naphtha.	Little or none	Good to excellent	High	Permits joints to stretch or compress. Silicones will stick to painted surfaces, but paint will not adhere to most cured silicones.
Polyurethane, expandable spray foam	Expands when curing; good for larger cracks indoors or outdoors. Use in nonfriction areas, as material can become dry and powdery over time.	Solvent such as lacquer thinner, if immediate.	None; expands quite a bit	Good to excellent	Moderate to high	Spray foam quickly expands to fit larger, irregular gaps. Flexible. Can be applied at variable temperatures. Must be painted for exterior use to protect from ultraviolet radiation. Manufacturing process produces greenhouse gases.
Water-based foam sealant	Around window and door frames in new construction; smaller cracks.	Water	None; expands only 25%	Good to excellent	High	Takes 24 hours to cure. Cures to soft consistency. Water-based foam production does not produce greenhouse gases. Will not over-expand to bend windows (new construction). Must be exposed to air to dry. Not useful for larger gaps, as curing becomes difficult.
Butyl rubber	Seals most dissimilar materials (glass, metal, plastic, wood, and concrete). Seals around windows and flashing, bonds loose shingles.	Mineral spirits or naphtha.	5%–30%	Good	Moderate to high	Durable 10 or more years; resilient, not brittle. Can be painted after one week curing. Variable shrinkage; may require two applications. Does not adhere well to painted surfaces. Toxic; follow label precautions.
Latex	Seals joints around tub and shower. Fills cracks in tile, plaster, glass, and plastic; fills nail holes.	Water	5%–10%	Good to excellent	Moderate	Easy to use. Seams can be trimmed or smoothed with moist finger or tool. Water resistant when dry. Can be sanded and painted. Less elastic than above materials. Varied durability, 2–10 years. Will not adhere to metal. Little flexibility once cured. Needs to be painted when used on exteriors.

US Department of Energy

Figure 24-4. Guidelines for selecting the best caulk for particular applications.

Sealant has interior uses as well. The following places require sealants to protect against water damage:

- Where countertops meet splashboards
- Where splashboards meet walls
- Where sinks meet countertops
- Around bathtubs
- In showers where walls and floor meet
- Around drains in sinks, tubs, and shower stalls

Sealants are generally not paintable, so it may be best to finish painting before applying sealants.

Caulks and sealants can be tooled with a wet finger. Do not apply more caulk than is necessary. Then, before the caulk starts to form a skin, run your wet finger along the seam or joint to remove excess and leave a smooth finish.

24.3 Painting and Finishing Tools

A variety of tools may be needed to tackle painting or finishing tasks. These include ladders, scaffolding, compressor and paint sprayer, sanders, brushes, rags, paint roller and pan, paint pads, tack rags, putty knives, and paint guards.

24.3.1 Brushes

Brushes are sold in many sizes and grades and are designed for various purposes. The most common brushes are shown in **Figure 24-5**.

A brush has several parts, as shown in **Figure 24-6**. The bristles used in brushes are either synthetic fibers or natural fibers (animal hair). Natural-fiber brushes are traditionally made from the hair of hogs, especially Chinese hogs, which grow long hair.

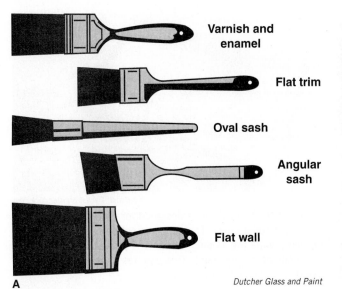

Varnish and enamel

Flat trim

Oval sash

Angular sash

Flat wall

A

Dutcher Glass and Paint

B

ZikG/Shutterstock.com

Figure 24-5. Brushes. A—Most painting or finishing tasks can be accomplished with these five kinds of brushes. B—Good painting and decorating stores carry a wide variety of brushes for all purposes.

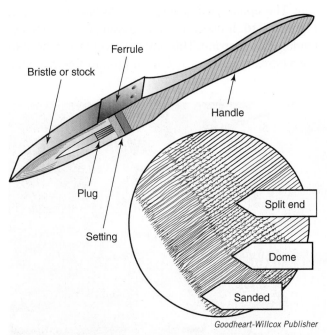

Ferrule

Bristle or stock

Handle

Plug

Setting

Split end

Dome

Sanded

Goodheart-Willcox Publisher

Figure 24-6. Parts of a brush. The inset shows the tip of a quality brush, pointing out its features.

Hog bristles are oval in cross section and have naturally tapered and flagged (split) ends. These flagged tips provide paint-holding ability through capillary action. Other natural fibers used for brushes include hair from horses, oxen, and fitch (a type of weasel) and tampico, which is derived from cactus.

The synthetic fibers of nylon or polyester are manufactured from petroleum products. They are widely used for certain types of brushes. Nylon and polyester bristles are flagged during manufacture. These bristles will outlast natural bristles. They are especially suited for water-based finishes.

Quality brushes have a smooth taper from the ferrule to the tip. These are formed by varying the length of the bristles. Inside the brush, completely surrounded by the bristles, is a plug made from wood, metal, fiber, or plastic. It creates a reservoir to hold the paint.

Wire brushes are useful for cleaning plaster off of subflooring and removing rust from metal. They are also used to remove old paint that is peeling or flaking, **Figure 24-7**.

Peipen/Shutterstock.com

Figure 24-7. Wire brushes are useful for removing rust and other unwanted foreign materials from a surface to be painted.

24.3.2 Rollers, Pans, and Pads

Paint rollers are sold in various widths. They are designed to hold tube-shaped pads that are available in many different types, naps, and sizes. **Figure 24-8** shows various rollers and roller covers. Rollers are ideal for painting large surfaces and can be used to apply many types of interior and exterior finishes. These include oil-based paints, water-based paints, floor and deck paints, masonry paints, and aluminum coatings. For smooth surfaces, short-nap roller covers are available. The rougher the surface to be painted, the longer the nap should be.

Pans for roller painting have a sloped bottom, with the lowest part designed to hold a supply of paint, **Figure 24-9**. The pan has a textured sloping surface for rolling off excess paint from the roller.

24.3.3 Mechanical Spraying Equipment

Nearly all types of paint, stain, and varnish can be applied with a compressor and spray gun. Spraying is the quickest and most practical method for painting large surfaces.

Benoit Daoust/Shutterstock.com

Figure 24-9. A roller pan is designed for even spreading of paint on a roller cover. The deepest section holds a supply of paint, while the sloping, upper part provides a surface for removing excess paint. This pan has a disposable plastic insert for easy cleanup.

Paint spray guns vary in how the paint is delivered to the atomizing nozzle: siphon, gravity, or pressure. The *siphon-feed spray gun* works like an atomizer, **Figure 24-10**. It depends on air passing over a tube extending into the paint. This action draws the paint up into the airstream, where it is broken up into fine particles.

Radovan1/Shutterstock.com

Figure 24-8. Paint rollers. Rollers and roller covers are made in different widths. Covers are designed in different naps for smooth to rough surfaces.

bodganhoda/Shutterstock.com

Figure 24-11. A gravity-feed spray gun relies on gravity to draw the material from the cup into the gun.

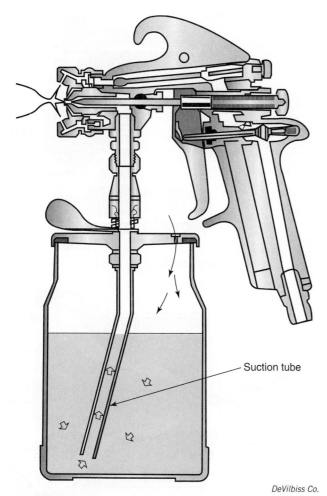

DeVilbiss Co.

Figure 24-10. A section view of an external-mix, siphon-feed spray gun.

Suction tube

A siphon-feed spray gun is designed for applying light-bodied materials such as shellacs, stains, varnishes, lacquers, and synthetic enamels. Siphon-feed guns are not normally used on a construction site.

A **gravity-feed spray gun** is similar to a siphon-feed, but the material (paint) cup is attached to the top of the gun, so the paint is drawn into the gun by gravity, **Figure 24-11.** They are used primarily for light stains, lacquers, and varnishes. Gravity guns use more of the paint than other spray guns. Most gravity-feed guns use a high volume of air delivered at a low pressure (HVLP). The air pressure used is 30 psi or less, so there is little overspray, with 70% or more of the material being delivered to the surface being sprayed.

In a **pressure-feed spray gun**, the paint is delivered to the gun under high pressure, often over 3,000 psi. The pressure created forces finishing material into the airstream. Pressure-feed guns are intended for spraying heavy-bodied paints. Pressure-feed guns used on construction sites are often fed by a separate pump unit that draws the material out of the paint bucket and through a hose to the gun, **Figure 24-12.**

A   *Graco Inc.*

B   *Graco Inc.*

Figure 24-12. Pressure feed sprayer. A—The pump draws paint from the container and delivers it under high pressure to the gun. B—The gun is attached to the pump by the feed hose. The high pressure atomizes the paint at the nozzle.

Small airless spray guns are completely self-contained, **Figure 24-13**. Handheld, airless spray units are convenient for small jobs, but would require frequent refilling for large jobs.

Spray guns may have either an internal-mix or external-mix spraying head, **Figure 24-14**. The external-mix head is used to apply fast-drying materials and provides greater control of the spray pattern. The internal-mix head is designed for slow-drying materials and is best used with low air pressure.

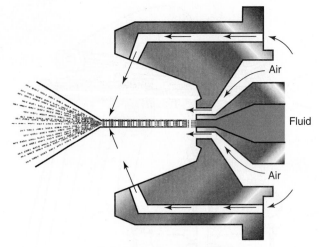

External-mix spray head

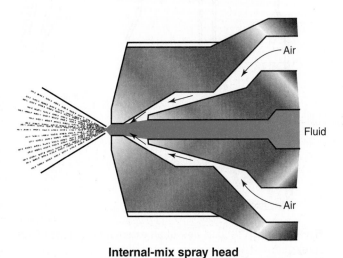

Internal-mix spray head

Goodheart-Willcox Publisher

Figure 24-14. The spray head, or nozzle, consists of an air cap and a fluid tip.

Graco Inc.

Figure 24-13. Airless sprayers can be self-contained. This one is cordless.

24.3.4 Ladders and Scaffolds

Both ladders and scaffolds are widely used in painting and decorating. They provide a safe method of reaching high places with painting tools. OSHA rules regarding safety in high places applies to painting. Review the information in Chapter 6, **Scaffolds, Ladders, and Rigging**.

24.3.5 Paint Removal Tools

An electric sander speeds the work of removing old paint, foreign materials, or roughness from a surface. A belt sander or a disc sander are useful for removing material rapidly, **Figure 24-15**. New sander shapes and sizes make most sanding locations accessible.

A heat gun, like the one shown in **Figure 24-16**, can be used to soften paint for removal with a hand scraper or putty knife. It is a much safer method for paint removal than the use of an open flame from a handheld propane torch. The heat gun can also be used to soften old putty before reglazing windows.

Porter-Cable

Figure 24-15. This electric disc sander is designed to remove paint from flat surfaces.

DIProduction/Shutterstock.com

Figure 24-16. A heat gun may be used to remove old paint. An open flame should never be used for paint removal.

24.4 Painting, Finishing, and Decorating Materials

There are various materials for painting, varnishing, and paperhanging. These include paints, vehicles or binders, solvents and thinners, varnishes, stains, fillers, sealers, bleaches, pumice, sandpaper, steel wool, wallpaper, and wallpaper paste.

24.4.1 Paints

Paints, varnishes, and stains are adhesive coatings, in a manner of speaking. They are spread in a thin, unbroken film that is designed to protect and beautify the surface. The term paint was once reserved for oil-based paints. However, it now includes coatings based on various resins such as acrylic latex. These are often referred to as water-based paints because they can be thinned or cleaned up with water.

The principal ingredients of paint are the ***pigment***, which provides color and opacity, and the vehicle. Paint stores and home improvement centers have small cards, often called paint chips that show the possible colors, **Figure 24-17**. All of these colors can be mixed by using different pigments. Vehicles are oils or resins that make the paint a fluid and allow it to be applied to, and spread over, a surface. The vehicle also includes a ***binder*** that remains behind after the oils or resins have dried. The binder then holds the particles of pigment together.

Andy Dean Photography/Shutterstock.com

Figure 24-17. Paint stores are usually well stocked with paints and aids that help buyers in selection of colors.

24.4.2 Clear Finishes

Shellacs, lacquers, and varnishes are clear coatings applied over wood. They not only protect the wood surface, but enhance the beauty of the grain and the wood color.

Shellac has been used as a wood finish for centuries. It is made by dissolving the excretions of the lac bug in alcohol. The alcohol evaporates after it is applied and left exposed to the air, leaving a hard, attractive finish. Shellac is available as a liquid in an alcohol base or as dry flakes to be dissolved before use. In its natural state it produces a warm, amber-colored finish called orange shellac. It is also available as white shellac, with the natural pigment removed, so it yields a nearly water-clear finish. Shellac produces an attractive finish, but it is easily damaged by small amounts of water and even more so by alcohol.

Lacquer is a synthetic finish that dries or cures by the evaporation of the solvent. Traditionally, that solvent is a volatile organic compound that evaporates very quickly. In recent years, lacquers have been developed that use water as the solvent base. In either case, lacquer is usually sprayed, not brushed. It yields a hard finish that can achieve high gloss. Colored lacquer is used to achieve the smooth, gloss finishes on automobiles.

Traditional varnish is an oil-based product with solids similar to what is used to make oil-based paint. Traditional varnish is seldom used since the introduction of polyurethane varnish, or polyurethane. Polyurethane is essentially a plastic in liquid form, until it cures through a chemical reaction. It does not dry. It is especially tough and can be formulated to produce gloss, semigloss, or satin finishes. Most polyurethane does not withstand the effects of the sun's ultraviolet rays very well, so unless it is specifically marked for exterior use, such as spar polyurethane, it should not be used on building exteriors. Polyurethane is available as either water-based or oil-based. Water-based polyurethane is clear and does not change the color of the substrate (base material). Oil-based polyurethane is slightly amber in color and produces a warmer color on wood. It also withstands heat a little better than does its water-based counterpart.

Sealers are used as an undercoat to keep stains and other clear coatings from being absorbed too deeply into porous wood. They also "tie down" stains and fillers that have been applied. Shellac is often used as a sealer to improve adhesion of topcoats and prevent bleeding of certain stains through topcoats. Special stain-blocking sealers are formulated to prevent knots and water stains from bleeding through the finish coating, **Figure 24-18**.

Goodheart-Willcox Publisher

Figure 24-18. Stain blocking sealers often have a shellac base. They prevent knots and other staining materials from bleeding through the paint.

Fillers are heavy-bodied liquids used to fill the pores of open-grain woods before application of stains and topcoats. *Putty* is a plastic, dough-like material used to fill holes and large depressions in wood surfaces. It is also used to fill nail holes before painting.

24.4.3 Stains

Many stains are designed for interior wood finishing, but stains for exterior uses on siding, decks, and fences are also popular. Interior stains are grouped as oil stains and dyes. Their use is generally limited to spray applications. Oil stains are either penetrating or pigmented. Penetrating stains are brushed on, allowed to soak into the wood, and the excess is removed with a dry cloth, **Figure 24-19**. Pigmented stains are designed to be applied with either a brush, roller, or spray. For deeper tones, the stain is brushed on and allowed to penetrate the wood and dry for a longer period of time before wiping.

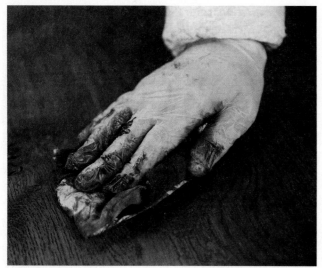

MintImages/Shutterstock.com

Figure 24-19. Stains can be applied with a cloth, pad, or brush.

Goodheart-Willcox Publisher

Figure 24-20. Paint can labels contain information on coverage, drying time, clean-up, safety, and other important facts.

Dyes are powder or concentrated liquid that are dissolved in either water or alcohol prior to use. When dye is dissolved in alcohol, it is sometimes called a spirit stain. Spirit stains rapidly set up and dry. Water stains dry more slowly but the water raises the grain of the wood, so sanding with very fine abrasive paper is necessary after the stain dries. Spirit stains and water stains yield brighter, more transparent results than oil stains, but they are more difficult to use, so their use is generally limited to furniture and factory-built cabinets.

Exterior stains are oil base and classified by the amount of pigment they contain. Semi-transparent stains are lightly pigmented. They color the wood, but allow the grain to show through. Solid stains are much like paint, with a heavy pigmentation that provides an opaque color finish. Exterior stains designed for use on decks and porches have a high abrasion resistance.

24.5 Container Labels

All finishing materials come in containers with labels. The label contains a lot of information that is important for the painter to know, **Figure 24-20**. The area of a surface that can be covered is usually given in square feet. The total area may vary depending on the roughness or porosity of the surface to be covered. Some finishing materials can be applied at lower temperatures than others. Most finishing materials pose a hazard if they are splashed in your eyes or ingested (swallowed), and others also can cause serious harm if they are absorbed through the skin. This information is contained on the label, as well as on a safety data sheet (SDS) that should be available with all materials. The painter will also want to know how the material can be thinned, if it can be thinned at all, and how to clean tools after using the material.

Pro Tip
A professional painter reads the labels on the finishing material being used before work begins and follows the manufacturer's instructions to achieve the best results.

24.6 Color Selection

Color selection is important when working with paints. A color wheel is useful in selecting colors that go well together, **Figure 24-21**. There are a few terms that are important to understand when working with color schemes.

- *Hue*—The class of a particular color, such as red or blue.
- *Primary colors*—Red, blue, and yellow. These colors can be combined to form all other colors.
- *Secondary colors*—Violet, green, and orange. A color that is produced when two primary colors are combined.

- *Intermediate colors*—Red-violet, blue-violet, blue-green, yellow-orange, red-orange, and yellow-green. A color combination of a primary color and a secondary color already containing that primary color.
- *Related colors*—Hues that are side-by-side on the color wheel.
- *Complementary colors*—Hues that are opposite each other on the color wheel.
- *Value*—The lightness or darkness of a color.
- *Tint*—Pure color with white added. Color nearer white in value.
- *Shade*—Pure color with black added. Color nearer black in value.

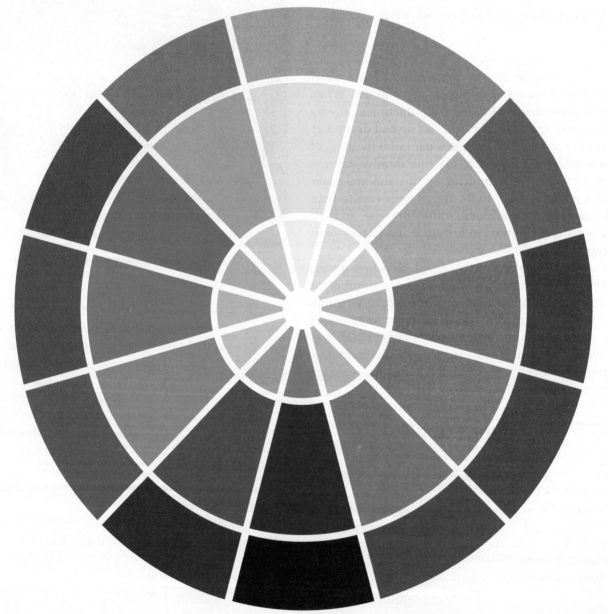

Goodheart-Willcox Publisher

Figure 24-21. A color wheel may be used to assist in picking color combinations that are pleasing and complementary.

Paints are available in factory-mixed colors or as bases to which color may be added at the paint store. For a custom-mixed color, the customer can select a color from an array of paint chips. Paint manufacturers supply instructions and proportions of pigments needed so the store can exactly match the chip. Many stores today also have color analyzers that allow them to mix paint to perfectly match a color sample from a fabric, picture, or other source.

24.7 Preparing Surfaces for Coating

A good paint or varnishing job is not possible without proper preparation of the surface. On new work, all that is necessary may be removal of any foreign matter, such as plaster, grease, dirt, and pitch, followed by a light sanding to get a smooth surface.

Sandpaper is used to smooth surfaces prior to painting or varnishing. Steel wool and pumice are used to smooth and enhance the surface of clear coatings once they have dried. Sandpaper may be used between coats to provide a "tooth" for better adhesion.

Renewal or recoating of old exterior finishes requires much more work. Remove loose paint by scraping, wire brushing, and sanding. Edges around bare spots must be feathered. Work with the grain while sanding. Sometimes, complete removal of the paint may be necessary because of extensive blistering, cracking, and alligatoring of the film. Fill cracks and nail holes with putty and then sand smooth. Before painting window frames, check the old finish and glazing. A heat gun will soften dried putty, making it easy to remove, **Figure 24-22**.

Coat all bare spots with primer before applying the finish coating. Primer is specially formulated to improve the adhesion to the substrate. In other words, it sticks to the wood better. Primer will also help prevent the wood, wallboard, or whatever is being painted from absorbing the solvent in the paint. Most paints are formulated to be toughest when the solvents evaporate into the atmosphere slowly. If the wood absorbs the solvent quickly, the paint coating will not be as durable. On extremely weathered surfaces, sand the surface down to "bright" wood. Dry, weathered wood will quickly draw the vehicle out of the paint, leaving a painted surface that quickly fails. To remedy this problem, many painters apply a 50/50 mixture of turpentine (or mineral spirits) and linseed oil. This is very important on heavily weathered surfaces if the new paint job is to last even one season. Allow this coating 48 hours of drying time before priming.

Greg McGill/Shutterstock.com

Figure 24-22. Removing dry, brittle glazing compound from windows can be difficult and cause damage to the muntins. Using a heat gun will soften the compound so it can be easily lifted off.

Before painting interior walls, repair any cracks in plaster or drywall that persist or reopen. Cover the crack with fiberglass tape and apply a compatible drywall compound. Sand and prime repaired areas before repainting.

24.8 Choosing Paint Applicators

Selecting an applicator for painting is important. The applicator must be well adapted to the surface and the coating being used. You should be familiar with the applicator and how to use it. Brushes have long been the primary applicator for painting, for a number of reasons:

- They work well on small, detailed surfaces. Control is accurate; you can put the paint exactly where needed. This reduces the need for masking.
- Since brushes do not produce overspray or spattering, there is less need to protect surrounding areas.

- Brushes produce better results on textured surfaces. This is important on open-grain, rough-sawn, or brushed siding. When rollers or sprayers are used on such surfaces, they are apt to leave voids.
- Brushes work well in conjunction with rollers. The brush is used on the detail work; the roller coats the larger surfaces.

24.9 Using Proper Brushing Technique

Gripping the brush the right way helps you do more accurate work. At the same time, a proper grip is less tiring. See **Figure 24-23**.

The pencil grip is usually best for small brushes used on detail work. This grip would be used, for example, to paint muntins in a window. The fingers wrap around the metal ferrule, giving a full range of motion.

The pinch grip gives good control of the brush. It is most comfortable for painting walls, siding, and trim. All of the fingers will fit on the ferrule of a 3″ or 4″ brush.

Do not grip the brush too tightly, or your hand may cramp from fatigue.

The wrap grip is preferred for overhead painting. In this grasp, the fingers and thumb wrap around the handle.

A brush is handy for cutting in the edges of boards. Several edges can be cut in at once, so long as the edges do not dry before the face of boards is painted.

24.10 Roller and Pad Application

Rollers are useful for coating large areas. They rapidly coat interior walls and exterior siding. Other advantages include the following:

- The roller's nap provides a uniform texture in the coating.
- Rollers are less expensive than good-quality brushes.
- By using an extension handle, the painter often can paint ceilings and walls without using ladders or scaffolds.
- Preparation and cleanup of rollers takes less time than brushes.

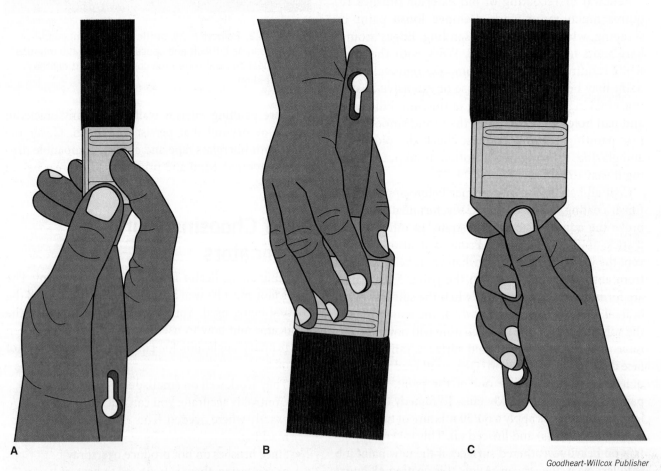

Goodheart-Willcox Publisher

Figure 24-23. Proper methods of holding a brush. A—Pencil grip is used with small brushes for control of paint on small areas. B—Painters prefer the pinch grip on vertical surfaces. C—Wrap grip is less tiring when used on overhead surfaces.

A   B   C

When painting siding, topcoats of a water-based exterior finish can be applied over an oil-base primer using a roller or paint pad. If using a roller, paint across one board at a time, working from top to bottom of the wall.

Paint pads have a napped fabric over a foam base. The pad is secured in a plastic holder. Special pans used with the pads hold about 1/2 gallon of paint. To fill the pad, the pan is tipped to flood a grid area with paint. The pad is pressed on the grid to load it. Three or four boards can be completed to a stopping point before moving to a new area.

24.11 Spray Painting

Spray painting is sometimes preferred to painting with brushes, rollers, or pads. It is most economical where there are large surfaces to be painted in one color. Areas that are to receive no paint must be carefully masked, as shown in **Figure 24-24**.

Proper mixing of paints is important for all types of paint application. Some painters box the paint by intermixing different containers of paint, **Figure 24-25**.

Greco Painting

Figure 24-25. When painting large surfaces, such as the outside of a house, it helps to box (intermix) the paint so that color will be uniform throughout the paint job.

This procedure ensures that the paint will be uniform in color and texture.

Safety Note
Safety precautions must be observed when spray painting to avoid the hazards of explosions, fire, and personal injury. Spraying should be done in a well-ventilated area. Flammable liquids should be kept in safety containers. Fumes and fine spray are toxic—always wear a respirator to avoid inhaling them. Wear protective clothing to keep paint spray off your skin. Always have a fire extinguisher handy.

24.12 Painting Wood Exteriors

Before painting new wood, seal all knots and pitch-containing areas with a shellac or knot sealer. Previously painted surfaces may have to be scraped or sanded to remove loose paint. Repair any imperfections and apply putty to holes.

The first coat of finish should be a pigmented primer or a sealer. Allow this coat to thoroughly dry before applying finish coats. Top coatings may be exterior oil paints, oleoresinous finishes, alkyd coatings, or water-based paints. The thickness and smoothness of a paint coat is controlled, to an extent, by the amount of paint carried in the brush. Do not dip the brush more than half of the length of the bristles.

Oil paints are applied in a way slightly different from other coatings. The paint is applied to a small area with diagonal strokes, then spread out with full-arm level strokes in a different direction. Each brush load should be brushed out with sweeping strokes. Use the tip for brushing and gradually lift the brush at the end of each stroke. This will give a thin feather edge at the end of each stroke.

belushi/Shutterstock.com

Figure 24-24. When using a sprayer to apply paint, it is important to mask off surfaces not being painted, such as windows and trim.

Alkyds and water-based paints require less brushing and dry rapidly. Thus, there is less time for brushing out than with oil paints. If too-rapid drying becomes a problem, inhibitors may be added; different types are used for latex paint, oil, and alkyd paints.

It is important to always paint to a *wet edge*. The paint band should be narrow enough so that the next band can be completed before the first has dried. This eliminates lap marks. Start painting at the highest point and move downward in horizontal bands. Never stop painting partway across a band. Always end a day's work at a window, door, edge of a board, or a corner.

Pro Tip

The lower edges of clapboard siding should be painted first. Rough surfaces require additional brushing in all directions. This will ensure that pores and crevices are well covered with paint. Poking at the surface with the brush tip will eventually destroy the flags on the bristle ends. If poking is necessary, use an old brush.

24.13 Problems with Coatings

Painters often experience problems with previously painted exteriors. When repainting previously painted exteriors, it is helpful to recognize certain defects in the old coatings. **Figure 24-26** illustrates some of these defects:

- Alligatoring and checking—Alligatoring is a severe case of checking. It is the result of applying relatively hard finishing coats over primers that are softer. To avoid this situation, apply progressively more flexible coats from primer to finishing coats. Allow more drying time between coats and follow the manufacturer's recommendations for thinning.

- Cracking and scaling—These conditions are usually the result of using paint lacking in elasticity. The coating becomes hard and brittle and unable to contract and expand with the wood. The conditions are aggravated by too-thick of an application. To avoid this condition, use high quality paint and avoid building up thick layers.

- Excessive chalking—Oils in the finish coat render a glossy finish. Ultraviolet rays cause the oils to break down, leaving loosely bound pigment on the surface. Some chalking is desirable to keep the paint surface clean. Heavy chalking, however, wears away the paint and leaves the surface unprotected. Using poor-quality paint is the major cause of excessive chalking. Before repainting, remove the chalked paint by scrubbing or wire brushing.

- Blistering and peeling—Blistering always precedes peeling. This condition is due to the pressure of water vapor under the coating pushing the paint film from the substrate. The only cure is to eliminate the source of the moisture. Moisture barriers, such as aluminum paint or Mylar film, will stop the migration of moisture-laden air from the interior to the exterior. Repair of external leaks may also be necessary. Install proper flashing and caulk joints and cracks. Make sure grading is correct to eliminate ground moisture coming in contact with siding. Proper ventilation of attics and crawl spaces will also reduce moisture buildup. Finally, allow surfaces to dry thoroughly before painting.

- Mildew—This is a form of plant life. It flourishes on soft paints. Eliminate mildew by washing the affected surface with a mild alkali such as washing soda, trisodium phosphate (TSP), or sodium metasilicate.

A   *Goodheart-Willcox Publisher*

B   *Goodheart-Willcox Publisher*

C   *Goodheart-Willcox Publisher*

Figure 24-26. Various types of paint failure. A great deal of prep work will be required before these surfaces can be repainted. A—Checking. B—Advanced alligatoring. C—Blistering and peeling.

24.13.1 Defects in Finishes

Certain defects in varnishes and other clear finishes are the result of improper application methods or application under unfavorable conditions.

- Crawling—The finish gathers up into small globules and does not cover the surface. This may be caused by applying the finish over a wet, cold, waxy, or greasy surface. Other causes are mixing of different kinds of varnish, using too much drier, or thinning with naphtha instead of turpentine or other approved thinner.
- Blistering—As with paints, blistering of clear finishes is caused by moisture vapor from below the surface.
- Blooming (clouding)—This condition may be caused by high humidity in the room where finishing is done. Another cause is the use of water and pumice instead of oil and pumice for rubbing the finish.
- Runs, curtains, and sags—Conditions caused by too-heavy of an application.
- Specks in surface—Results from hardened particles in the container of finish, dirty brushes, improperly prepared surfaces, or airborne dust in the room.

24.14 Interior Painting

Interior surfaces are seen more closely than exterior surfaces, so proper surface preparation is especially critical. Surfaces must be clean. Normally, a thorough dusting is enough. However, glossy surfaces or surfaces soiled with oily dirt will need to be washed with a strong washing compound such as trisodium phosphate to degrease and dull the surface. Rinse with clean, hot water. Pay particular attention to surfaces surrounding sinks and stoves, as well as frequently touched surfaces around switches, doorknobs, handrails, and the edges of doors. There are also special chemical deglossers that will prepare glossy painted surfaces for recoating. These are especially effective over latex paint.

Cover surfaces not being painted to protect them from paint drips or spatters. Spread drop cloths over floors and furniture. Mask off windows, trim, and edges that are not to be painted. Standard masking tape is available in different widths. Some tape is designed to cleanly peel off after being attached a week or more.

Remove all items attached to walls and trim that are not to be painted. This includes wall hangings, pictures, switch and receptacle plates, and hardware.

Assemble your tools. Thoroughly mix paints, even if they have recently been agitated on a paint store vibrator. Using a paddle or a power mixer, rapidly stir to suspend settled pigments.

Follow these steps in order when painting an entire room:

1. Paint the ceiling.
2. Paint the walls.
3. Paint the wood trim and doors.

Either rollers or brushes may be used; most painters prefer the roller because it allows the painting to progress more quickly.

Paint large surfaces in narrow strips so that you are always painting against a wet edge. Dry edges that are overlapped will show up as lap marks or have a different sheen when the paint is dry.

To paint into the corners formed by the ceiling and wall, a process known as *cutting in*, a small brush may be helpful. Cutting in provides a narrow painted strip of paint at the ceiling line, around doors and windows, and along baseboards. These edges cannot be reached with the roller without depositing paint where it is not wanted.

If painting with a brush, dip the bristles into the paint to one-third of their length. Tap the brush gently against the inside or edge of the paint container to release paint that might otherwise drip. Some painters make several small nail holes in the rim of the paint container. This allows paint collected there to drain back into the container. Paint stores have available a grid of expanded metal that hangs inside a five-gallon container for striking off excess paint from brush or roller. See **Figure 24-27**.

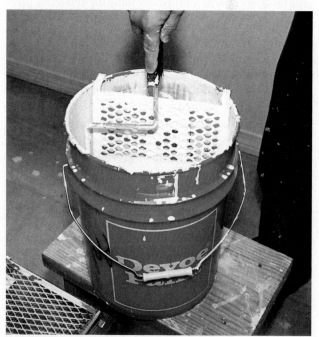

Goodheart-Willcox Publisher

Figure 24-27. A grid of expanded aluminum can be used in a five-gallon can to remove excess paint from a roller or brush.

When loading a roller, do it carefully to avoid dripping paint. If using a tray, pour a small amount of paint into the deep end. Dip the roller into the paint and work it into the nap by moving the roller back and forth on the slanted area of the tray while applying light pressure. This removes excess paint.

Start rolling in one corner of the wall or ceiling, painting a narrow strip. Overlap each stroke, working first in one direction and then in another, until the surface is covered. Gently work the roller avoiding quick, choppy strokes that cause paint spatters. Continue rolling until the newly painted section is covered and the roller needs reloading. Unless you are repainting with the same color, applying a second coat makes the end result look more professional than applying one coat.

Trim painting should be done only after walls are thoroughly dry. A 1 1/2″ brush is suitable for windows; use a 2″ brush for other trim. To avoid drips and runs, be careful not to overload the brush.

On multipaned windows, start painting on center mullions first, working outward in either direction. Trim should be painted last. Use masking tape to avoid spreading paint onto the window glass. Accidental spatters on the glass, however, can be removed with a razor blade when the paint has dried.

Paint doorjambs and casing before painting the door. To avoid laps and brush marks when painting paneled doors, paint the molded edges of the panels first, starting with the top panel. Then, fill in the panels using brush strokes with the grain. Next, paint the rails (horizontal cross boards). Finish by painting the stiles (vertical side boards). Refer to **Figure 24-28**. If the door swings away from the side being painted, paint the hinge edge the same color. If it swings toward the side being painted, paint the lock edge the same color. If both sides of the door are to have the same color, all edges should be painted.

Some painters prefer to finish multipaneled doors with a roller. First the recessed panels and decorative edges are brushed. Then stiles and rails are rolled.

Paint the baseboards last. A plastic, metal, or cardboard guard should be used to protect the finished floor or carpeting when painting the lower edge of the baseboard. As an alternative, painter's gummed masking paper can be used to protect hard-surfaced finish flooring.

Pro Tip

When painting a ceiling, an extension handle attached to the roller will be useful and easier than climbing up and down a stepladder or a platform.

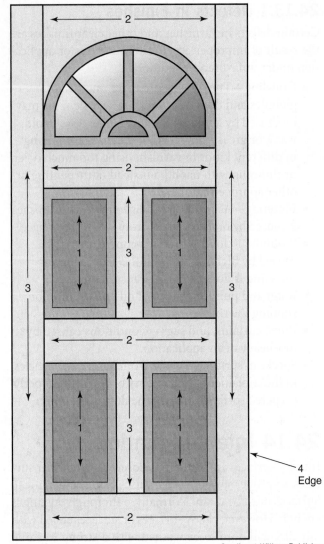

Goodheart-Willcox Publisher

Figure 24-28. Efficient method of painting a paneled door. Paint the panels first, then rails, then stiles, and finally, edges. Arrows indicate the direction of the brushing.

24.15 Working with Stains and Clear Finishes

Items to be finished should be smooth, free of dust, dirt, glue, and grease. Before applying stains or finishes, remove all hardware. Before staining and finishing a wood surface, remove or repair all surface dents. Minor defects such as machine marks, small scratches, and excess glue can be removed with light sanding. More serious conditions, such as dents and holes, require repairs.

Small dents can generally be removed by applying a drop of warm water to the spot. This causes the grain to swell and rise to or above the surface. Let it dry and then sand smooth.

Lift deeper dents with steam, using a damp cloth and a hot iron, **Figure 24-29**. Place the damp cloth over the dent and apply the hot iron to the cloth. Move the iron around to avoid burning the wood. Sand the dried surface smooth.

Holes include open joints, cracks, splits, and gouges. Fillers are needed to repair such defects. Basic fillers include plugs, Plastic wood®, wood putty, stick shellac, and wax sticks. *Plugs* are wooden pieces shaped to fit and forced into the defect. Plastic wood® is a combination of wood powder and a plastic hardener. *Wood putty* is a mixture of wood and adhesive in powder form. Stick shellac is hardened, colored shellac. It must be applied with a heated knife. Stick wax is a colored wax that closely resembles crayons. Stick wax is available in many colors to match various woods and stains, **Figure 24-30**. Apply the wax by rubbing the stick over the defect. This forces the wax into it.

Wood bleaching removes some of the natural color from the wood. Bleaches can be purchased ready to mix and usually consist of two solutions. Mixing is done in a glass or porcelain container. The solution is applied to the bare wood with a rubber sponge or a cotton rag. Wear gloves and a face shield for this operation.

Safety Note
Some wood bleach is caustic and some is acidic. Both can be dangerous, so it is especially important to follow the label directions.

Mohawk Finishing Products

Figure 24-30. Wax sticks to fill small holes are available in many colors.

Surfaces with all defects repaired are ready for staining and final finishing. While not necessary for every surface, stains are used to change the color of natural wood. When used on exterior surfaces, stains have a protective quality. They can be applied by wiping, brushing, or spraying, **Figure 24-31**. Wipe off excess, if any, and allow the stain to dry for 24 hours.

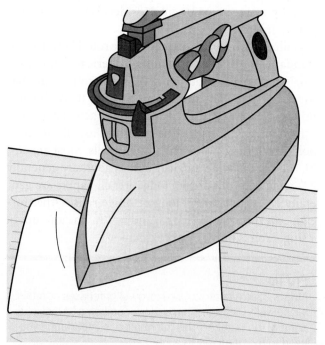

Goodheart-Willcox Publisher

Figure 24-29. Dents may be steamed out using a damp cloth and heat from an iron.

Goodheart-Willcox Publisher

Figure 24-31. Stain, varnish, and clear finishes can be applied to trim before it is installed.

Then, coat it with a sealer. The sealer "ties down" the stain and filler. Some fast-drying sealers are ready for final finishes in about 30 minutes. Follow directions on the container.

Clear finishes are brushed or sprayed, **Figure 24-32**. Brushes should be the finest available. Flow the finish on rather than brushing it out as with paint. Be sure not to leave holidays (bare spots). Do not apply too heavy of a coat because sags and drips will mar the finish.

Smooth the dried finish coats with fine steel wool and treat them to a rubbing with wax. Allow at least 48 hours of drying time before attempting to rub out the final coat.

PROCEDURE

Estimating Paint

1. Measure the height and width of the surface to be covered. Convert inches to feet, dividing the inches by 12. Carrying decimals two places to the right of the decimal point provides enough accuracy.
2. Multiply the two measurements to find total area in square feet.
3. Find the dimensions of windows, doorways, and any other spaces that do not require paint. Multiply height by width.
4. Deduct the resulting area from the area calculated in Step 2.
5. Check the label on the paint container to see how many square feet each container will cover.
6. Divide that number into the area to be covered. The resulting number indicates the number of containers required. If two coats are needed, double the number of containers.

24.16 Wallcoverings

Hanging wallcoverings, or *wallpapering*, is an important phase of decorating. Wallpaper was once the first choice for finishing interior walls. Today it is used on small areas or to accent a single wall. Doing professional-quality work demands a thorough knowledge of materials and procedures.

Most new wallcoverings are made from vinyl resins. The vinyl is made in a continuous, flexible film and applied to a paper or fabric backing. Such materials have the advantage of being long lasting and washable. Moreover, they can be made highly decorative. A more exotic type of wall hanging is embossed wallpaper simulating styles popular in the Victorian era.

Goodheart-Willcox Publisher

Figure 24-32. With proper masking, stains and finishes can be applied after trim has been installed.

One of the embossed products, Lincrusta, is a rigid product made from a linseed oil mixture with a deep relief. Anaglypta is a similar, lightweight embossed material made from cotton pulp. These materials are used by restorers or by interior decorators seeking a period look.

Wallcoverings are measured by the roll. There are two standards for dimensions. In the American standard, a single roll covers about 36 sq ft. Single rolls in the European system contain from 27 to 29 sq ft. With trimming, about 21–23 sq ft are available in actual coverage.

For convenience sake, the coverings usually come in bolts containing two or three single rolls. Most are 20″, 21 1/2″, or 22″ wide. English-made papers are generally 22″ wide and rolls are 36′ long. French papers are generally 18″ wide and 27′ long in single rolls. Other foreign coverings vary in length and width. The amount in a single roll is only an approximation.

Pro Tip

Always keep a record of wallcovering purchases or retain a label with the collection, pattern number, and dye lot or printing run number. This information is important if you need to reorder to finish a job.

Paperhangers require a number of hand tools, a paste table, and an expandable plank. Tools include shears for cutting paper, several sizes of rollers, smoothing and paste brushes, several knives, a razor blade and holder, a broad knife, a 6′ or 7′ straightedge, plumb bob, a chalk line, and chalk. See **Figure 24-33**.

24.16.1 Preparing Walls for Covering

Allow newly erected walls, whether plastered or drywalled, to thoroughly dry before covering. Treat plastered walls with a solution that neutralizes the lime and alkali in the plaster. Sizing is a glue solution that is brushed or rolled onto a newly plastered or drywalled surface to seal the pores and prevent the paste from being absorbed into the wall.

Old wallpaper is not a good foundation for new paper. Take the time to remove it. Some old coverings are "strippable" and can be easily removed. Cloth-backed vinyl coverings are usually strippable. However, much depends on whether the right adhesive was used and whether proper surface preparation was done. Use a putty knife to lift a bottom corner of the strip. Lift upward at an angle nearly parallel to the wall. Other methods of stripping old coverings include the following:

- Steaming—This requires a machine to produce a heated vapor that penetrates the covering to soften the adhesive. Scoring the old covering helps the steam get to the adhesive quicker. Special tools are available for scoring.
- Soaking—In this method, use a sponge or a garden sprayer to apply water. As in the steaming method, scoring makes it easier for the water to penetrate. On nonbreathable coverings like foil, vinyl, vinyl backed, or cloth-backed vinyl, scoring before soaking is necessary for penetration to the adhesive.
- Peeling—Some coverings have a vinyl outer layer that can be pulled away. This exposes a paper backing that is easily removed by soaking or another method.
- Chemical strippers—There are three types: wetting agents, solvent removers, and enzymes. These degrade the starch in the adhesive.

24.16.2 Preparing the Wallcovering

Before wallcoverings are cut to length, check to make sure that there is enough material to do the job. More will have to be purchased if there is too little to finish the job. Care must be taken that the new covering is from the same production run. Coverings from other runs may vary in color and ruin the job.

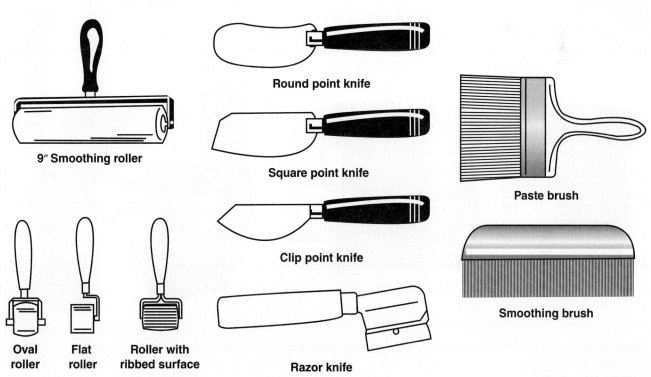

9″ Smoothing roller

Round point knife

Square point knife

Clip point knife

Oval roller   **Flat roller**   **Roller with ribbed surface**

Razor knife

Paste brush

Smoothing brush

Figure 24-33. These tools are recommended for wallpapering.

Wallcoverings are brittle. They must be softened before being applied to walls. On prepasted papers, place the prepared strip in a water tray for a time specified by the manufacturer. Then, place it on the wall.

There are also products that will activate the paste on prepasted papers. These watery products are rolled onto the back of the covering. One gallon will normally cover 10 single rolls.

Booking is a method of pasting and folding wallcoverings. This is an important step as it allows the paste to soak into the wallcovering. Allow the booked strip to "relax" for 3–5 minutes before hanging. Check the manufacturer's recommendations for the exact time.

When using coverings with a design, you must match the pattern of the strips on the paste table. Do this before cutting them. Do not cut the strips too short—allow about 2″ excess at both top and bottom. Trim strips to exact length after they are on the wall. Replace the razor blade after trimming each strip.

After hanging every three or four strips, go over the seams with a roller. Keep the roller clean and wipe away adhesive squeezed out of the seams. Do not press too hard—it will cause stretching or shrinking problems later. Never use seam rollers on flocked or embossed wallcoverings.

PROCEDURE
Booking Wallcovering Strips

1. Turn all cut, unpasted strips face down on the paste table.
2. Paste only one strip at a time. Paste two-thirds of the strip.
3. With clean hands, book the strip. To book, pick up the corners of the pasted end and fold it over only the pasted portion (pasted side in). Do not go beyond the pasted portion and do not crease the fold.
4. Paste and fold the other one-third. Edges must be even. If there is selvage on the covering, trim it at this time.
5. Loosely roll the booked strip and let it sit.

24.16.3 Hanging Wallcovering around Openings

Never attempt to precut a strip to fit around a window or door. Apply the paper over the opening and use a razor knife to trim the excess, **Figure 24-34**. Use this method when papering around light switches and receptacles, as well.

Safety Note
Make sure the circuit breaker has been turned off before working around any electrical device.

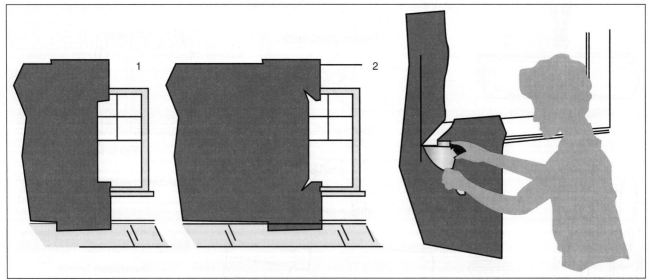

Figure 24-34. Proper procedure for papering around windows and doors. Before making final trim cuts, cut into all corners at a 45° angle.

PROCEDURE

Hanging Wallcovering

1. Using a plumb bob or a level, make a plumb starting line on the wall, as shown in **Figure 24-35**. This is used as a guide for hanging the first strip. It is better to start the first strip at a corner with a portion of the strip wrapping around the corner. The corner lap should be at least 1/2″ wide.

2. Overlap the top edge of the first strip 2″ onto the ceiling and align the edge with the plumb line.

3. Unfold the top one-third of the strip. Smooth the opposite side against the wall.

4. When the entire top portion is smoothed against the wall, as in **Figure 24-36**, unfold the second portion and align it with the plumb line.

5. Smooth out the strip from the center to the edges, working from the top down. Use brushing strokes in a sequence like that shown in **Figure 24-37**. Use care not to stretch the wet covering. When satisfied that the strip is properly aligned, smooth the entire surface again to remove remaining air bubbles.

6. Trim the ceiling and baseboard ends with a razor knife. Use a wall scraper or a broad knife as a guide. Tightly hold the scraper or broad knife against and parallel to the wall as you trim with the razor knife, **Figure 24-38**. Move the scraper or broad knife along the strip, but keep the razor knife in contact with the wall.

7. Rinse the surface of the strip to remove excess adhesive. Wipe away adhesive from the ceiling and baseboard at the same time. Frequently change the rinse water and use a good-quality, natural or synthetic sponge.

8. Match each succeeding strip to the previous strip. Butt the edge against the previous strip for a tight seam. Brush and smooth in the same way. Be sure to wipe off excess adhesive as before.

National Decorating Products Assn.

Figure 24-35. A plumb line is being set up using a level and a straightedge. This is usually done near a corner so the first roll of paper wraps around the corner at least 1/2″. A chalk line can be snapped for the plumb line.

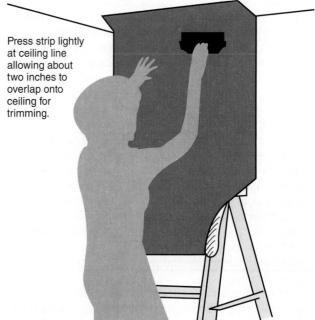

Press strip lightly at ceiling line allowing about two inches to overlap onto ceiling for trimming.

National Decorating Products Assn.

Figure 24-36. Attach the first strip of wallpaper as shown, aligning or centering it on the plumb line.

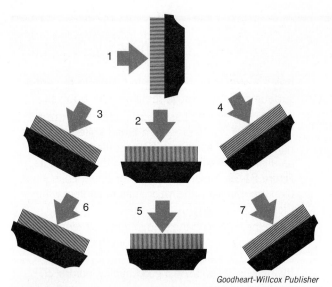

Goodheart-Willcox Publisher

Figure 24-37. Wallcovering should be brushed out in this sequence to remove air pockets and firmly adhere the strips to the wall.

Trim the excess wallcoverings at the ceiling and baseboard.

National Decorating Products Assn.

Figure 24-38. Trim off each strip at the ceiling and baseboard using a broad knife or scraper and a razor knife.

Construction Careers
Painter

Two out of five (40%) of all painters are self-employed—a percentage double that of the construction trades as a group. The nature of the trade lends itself to setting up a contracting business operated by the owner with possibly one or two employees. An approximately equal number of painters are employed by contractors who work on new construction, remodeling, and restoration. The remainder of workers in the trade are employed by building management firms, industrial plants, or government agencies and perform primarily maintenance work.

Painters are responsible for preparing the surface to be painted, selecting the right finishing material and applicator for the application, and applying the finishes. Painters often use ladders, scaffolds, or similar devices to work in areas that are sometimes high above the ground. In addition to possible falling hazards, painters may be exposed to toxic or dangerous materials when preparing surfaces or applying paint and varnish. Paying close attention to safety rules and wearing proper protective gear is important.

While many painters gain their skills on the job, apprenticeship programs exist and are recommended for continuing success in the field. Apprenticeship programs generally include 2–4 years of on-the-job training under the direction of experienced workers, and 144 hours of

classroom training each year. Advancement to supervisory or estimating jobs within an organization is possible. As noted earlier, many painters start their own contracting businesses.

Monkey Business Images/Shutterstock.com

Painters often work high above the ground, both indoors and outdoors. Precautions must be taken to prevent falls.

Chapter Review

Summary

- Paint is not only decorative, but protective. It keeps interior and exterior surfaces from being damaged by moisture and other environmental factors.
- Paint that contains lead must be handled carefully. The lead in lead-based paint has been found to cause brain damage and other severe health problems if it enters the human body.
- Many oil-based paints contain volatile organic compounds (VOCs) and can cause damage if inhaled or absorbed through the skin.
- Caulks and sealants are used to close cracks or openings that might allow water or outside air to enter the building.
- Finishing materials and supplies are chemicals that can cause skin, eye, and lung problems. They can also be flammable.
- Paints are made up of pigment, a vehicle that includes a binder, and solvent; varnishes and lacquers are clear coatings. Stains are usually transparent, but may include pigments.
- Proper surface preparation is vital to a good painting or varnishing job. This may involve cleaning, scraping and sanding, or even complete removal of the old finish, then priming the surface.
- Depending on the task, most coatings can be applied with brush, roller, pad, or spray gun.
- Clear finishes, such as shellacs, varnishes, and lacquers are applied over wood.
- Wallpaper is another category of surface finish. Most wallpapers are long lasting, washable, and made from vinyl resins.
- Before applying new wallpaper, the existing wallpaper must be removed by peeling, soaking, or steaming.
- Wallpaper is adhered to the wall with an adhesive. Careful attention must be paid to getting the first strip plumb and to edge-butting all succeeding strips.

Know and Understand

Answer the following questions using the information in this chapter.

1. Which of the following is *not* considered a coating?
   A. Paints.
   B. Stains.
   C. Sealants.
   D. Synthetic materials.

2. The leading source of VOCs is _____.
   A. paint
   B. finishing materials
   C. lead
   D. automobile exhaust

3. What level of VOCs is allowed in low-VOC flat-finish paint?
   A. 5 g/L
   B. 250 g/L
   C. 380 g/L
   D. 500 g/L

4. _____ is typically used to fill gaps between door and window frames.
   A. Caulk
   B. Sealant
   C. Paint
   D. All of the above.

5. _____ is typically used to protect areas around bathtubs from water damage?
   A. Caulk
   B. Sealant
   C. Paint
   D. All of the above.

6. *True or False?* Siphon-feed spray guns are used on construction sites by using a separate pump unit to draw material out of a bucket and through a hose to the gun.

7. A spray gun with an external-mix head is intended for spraying _____ materials.
   A. slow-drying
   B. fast-drying
   C. large
   D. small

8. A(n) _____ can be used to soften paint for removal with a hand scraper or putty knife.
   A. electric sander
   B. handheld propane torch
   C. spray gun
   D. heat gun

9. _____ is the ingredient that gives color to a paint.
   A. Pigment   C. Binder
   B. Vehicle   D. Acrylic latex

10. *True or False?* Fillers are used to fill the pores of open-grain wood surfaces.

11. Red, blue, and yellow are called the _____ colors.
    A. primary   C. related
    B. intermediate   D. complementary

12. The lightness or darkness of a color is known as its _____.
    A. hue   C. tint
    B. value   D. shade

13. _____ is used to smooth surfaces prior to painting or varnishing.
    A. Sandpaper   C. Pumice
    B. Steel wool   D. Primer

14. Which brush grip is most comfortable for painting walls, siding, and trim?
    A. Pencil.   C. Wrap.
    B. Pinch.   D. Standard.

15. _____ is most economical where there are large surfaces to be painted in one color.
    A. Roller painting C. Spray painting
    B. Brush painting   D. Staining

16. *True or False?* Alkyds and water-based paints require less brushing out than with oil paints.

17. Cracking and scaling are usually the result of using paint lacking in _____.
    A. oil   C. pigment
    B. elasticity   D. vehicles

18. _____ may be caused by applying a finish over a wet, cold, waxy, or greasy surface.
    A. Crawling   C. Blooming
    B. Blistering   D. Mildew

19. If painting with a brush, dip the bristles into the paint to _____ of their length.
    A. one-quarter   C. half
    B. one-third   D. three-quarters

20. _____ are wooden pieces shaped to fit and forced into a wooden defect.
    A. Plastic wood®   C. Wax sticks
    B. Plugs   D. Stick shellacs

21. A wall is 96″ high and 144″ wide. There is one 24″ × 24″ window on this wall. Given that one small container of paint can cover 60 sq ft, how many containers of paint are needed for two coats of paint on this wall? Round to the nearest whole number.
    A. 1 container.   C. 4 containers.
    B. 2 containers.   D. 6 containers.

22. Most new wallpaper is made of _____.
    A. lincrusta   C. vinyl resins
    B. anaglypta   D. plastic

23. _____ requires using a sponge or garden sprayer to help strip old wallcoverings.
    A. Steaming   C. Peeling
    B. Soaking   D. Sizing

24. *True or False?* Booking wallpaper is the process of ordering, or "booking", it from the supplier.

Apply and Analyze

1. List three factors in the environment that degrade unpainted wood siding.
2. Name the parts of a typical paintbrush.
3. What is polyurethane?
4. Give two reasons why brushes are used as the primary applicator for painting.
5. Why should you always paint to a wet edge?
6. When painting a room, what order should you paint the ceiling, walls, and wood trim and doors?
7. Explain how to use steam to remove a dent from wood.
8. Explain how to hang a wallcovering around an opening.

Critical Thinking

1. High quality paint brushes can be expensive. If you owned a painting company what system would you use to ensure paint brushes get properly cleaned and stored after use? How would you communicate the importance of your system to employees?

2. A customer in a remodeling job has asked you to paint over some existing wall paper to save the labor cost for stripping the it off the wall. When you apply latex primer to the wallpaper you find that large blisters of air form. What do you think is causing this to occur? Research possible solutions online. How would you go about solving the problem?

Communicating about Carpentry

1. **Speaking.** Working in a group, brainstorm ideas for creating classroom tools (posters, flash cards, and/or games, for example) that will help your classmates learn and remember the different painting, finishing, and decorating materials. Choose the best idea(s), then delegate responsibilities to group members for constructing the tools and presenting the final products to the class.

2. **Art.** After reading Chapter 24, you should have a good understanding of painting, finishing, and decorating. Create a model or draw a picture with labels describing paint, finish, and decorations. In your model or picture, compare buildings before and after the upgrades.

3. **Listening and Speaking.** Search online for local unions that may have workshops, guest speakers, or presentations on painting and finishing techniques and new materials. Also look online for demonstration and instruction videos. Your local library may also have various media covering painting and finishing techniques. Watch videos, attend a presentation, and take notes. Listen for terms that you have learned from your textbook. Write a short summary and be prepared to present your findings to the class.

anweber/Shutterstock.com

CHAPTER **25**

Chimneys and Fireplaces

OBJECTIVES

After studying this chapter, you will be able to:

- Explain how masonry chimneys are constructed around flue linings.
- Name the parts of a typical masonry fireplace.
- Describe procedures for the construction of the chimney, hearth, walls, and throat.
- Define the functions of the damper and smoke shelf.
- Calculate the flue area for a given fireplace.
- Describe the common types of factory-built chimneys and fireplaces.
- Install a factory-built chimney.
- List special considerations for installing factory-built fireplace units.
- Explain how glass enclosures installed on a fireplace can improve efficiency.

TECHNICAL TERMS

ash dump	downdraft	smoke chamber
chase	fireclay	smoke shelf
chimney	flue	splay
circulator	flue lining	termination top
corbel	hearth	throat
damper	refractory mortar	zero clearance

A *chimney* is a vertical, hollow, shaft that exhausts the smoke and gases from heating units, fireplaces, and incinerators. When properly designed and built, a chimney may also improve the outside appearance of a home, **Figure 25-1**.

The fireplace once was the only source of heat for dwellings. The fireplace remains popular, even though its efficiency as a source of heat is low compared with modern heating systems. The desire for a fireplace results from the cheerful, homelike atmosphere it creates and the value it adds to the home.

The rising cost of energy has led to many improvements in fireplace design. Masonry fireplaces are often equipped with glass doors to save heat. They may also include ductwork to circulate room air around the firebox and heat it. More importantly, they may have provisions to draw combustion air from outside of the house. There is a wide variety of factory-built fireplaces. Factory-built fireplaces are easy to install and usually less expensive than built-in-place masonry units.

Goodheart-Willcox Publisher

Figure 25-1. This brick chimney is an attractive feature of this home.

25.1 Masonry Chimneys

A masonry chimney must have its own footing. It is built in such a way that it neither provides support to, nor receives support from, the building frame. Footings should extend below the frost line, project at least 6″ beyond the sides of the chimney, and be at least 12″ thick. The walls of a chimney with a clay flue lining should be at least 4″ thick. Foundations for a chimney or fireplace, especially when located on an outside wall, may be combined with those used for the building structure.

The size of the chimney depends on the number, arrangement, and size of the *flues* (passage in a chimney through which smoke, gas, and fumes rise). The flue for a heating plant should have appropriate cross-sectional area and height to create a good draft. This permits the heating equipment to develop its rated output and allows it to draw all of the smoke out of a wood-burning fireplace. When deciding on flue sizes, always follow the recommendations of the heating equipment manufacturer.

Building codes require that chimneys be constructed high enough to avoid downdrafts caused by wind turbulence as it sweeps past nearby obstructions or over sloping roofs. Minimum heights required by the International Residential Code (IRC) are illustrated in **Figure 25-2**. The chimney should always extend at least 2′ above any roof ridge that is within a 10′ horizontal distance and 3′ above the highest point where it passes through a roof.

Combustible materials, such as wood framing members, must be located at least 2″ away from the chimney wall, **Figure 25-3**. The open spaces between the

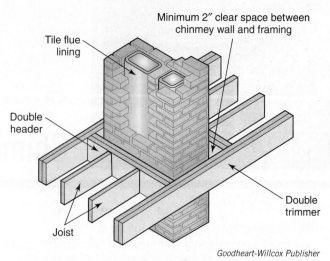

Goodheart-Willcox Publisher

Figure 25-3. Floor framing around a chimney. Combustible materials, such as wood framing, must be spaced at least 2″ from the chimney wall. When multiple flues are used in a chimney, stagger the joints in adjacent flue columns by at least 7″.

framework and chimney can be filled with mineral wool or other noncombustible material.

25.1.1 Flue Linings

The IRC and most local codes require that chimneys be lined. The *flue lining* in a chimney is a noncombustible sleeve, usually made of fireclay, that provides a smooth surface inside of a masonry chimney. Fireclay flue linings are recommended for wood-burning fireplaces. *Fireclay* is a heat-resistant clay material. Flue linings for other appliances such as gas fireplaces, furnaces, pellet stoves, and water heaters, may be made of clay or any lining that is approved by Underwriters Laboratories.

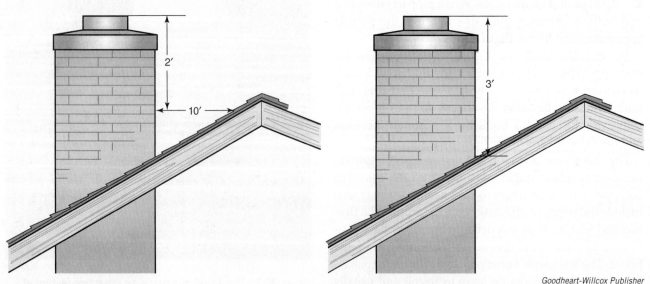

Goodheart-Willcox Publisher

Figure 25-2. Minimum chimney heights above the roof. Always check local building codes.

Such flue linings are often stainless steel. Linings are available in square, rectangular, and round shapes. Each shape is available in several sizes, as listed in **Figure 25-4**. Note that different methods of measurement are used for the three types:

- Outside dimensions for the old standard
- Nominal outside dimensions for the modular
- Inside diameter for the round linings

Small and medium flue linings are usually available in 2′ lengths. Flue rings are placed during the assembly of the chimney to provide a form for the masonry.

As a rule, connecting of a single gas-fired water heater to a furnace flue is often permitted. Do not combine larger flues.

25.1.2 Construction

Before constructing the chimney, allow its footing to cure for several days to give it proper strength. Use care in starting the masonry work. Make all lines level and plumb.

Each section of flue lining is set in place before the surrounding part of the chimney wall is built. The lining is a guide for the brick work. Joints in the flue lining are bedded in *refractory mortar*, a type of mortar made with fire clay that resist flames and high temperatures. Use care in placing the lining units so they are square and plumb. Brick work is carried up along the lining. Then, the next lining unit is placed.

Where offsets or bends are necessary in the chimney, miter the ends of the abutting sections of lining. This prevents reduction of the flue area. The angle of offset should be limited to 30° from vertical. The center

of gravity of the upper section must not fall beyond the centerline of the lower wall.

Chimneys are often corbelled to increase the size of the upper section. To *corbel* is to extend outward from the surface of a masonry wall, one or more courses, to form a supporting ledge. Corbelling is done just before the chimney projects through the roof, **Figure 25-5**. The larger exterior appearance is often more attractive. Also, breakage due to wind is less likely for thicker masonry.

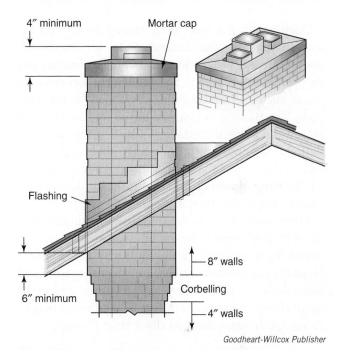

Goodheart-Willcox Publisher

Figure 25-5. Corbelling is used to enlarge the top section of the chimney. This helps it resist wind pressure and results in a more attractive appearance.

Flue Linings					
Standard Liners		Modular Liners		Round Liners	
Outside Dimensions (in)	Areas of Passage (sq in)	Nominal Outside Dimensions (in)	Areas of Passage (sq in)	Inside Diameter (in)	Areas of Passage (sq in)
8 1/2×8 1/2	52.56	8×12	57	8	50.26
		8×16	74		
8 1/2×13	80.50			10	78.54
		12×12	87		
8 1/2×18	109.69	12×16	120	12	113.00
13×13	126.56	16×16	162	15	176.70
13×13	182.84	16×20	208		
13×13	248.06	20×20	262	18	254.40
		20×24	320	20	314.10
		20×24	385	22	380.13
				24	452.30

Goodheart-Willcox Publisher

Figure 25-4. Flue liner dimensions and clear cross-sectional areas (areas of passage).

Corbelling should not exceed a 1″ projection in each course. The final size should be reached at least 6″ below the roof framing.

> ### Code Note
> The IRC requires the size and shape of a chimney wall or flue lining to remain unchanged as it passes through a floor, ceiling, or roof. Changes in size and shape are not allowed in a region from 6″ below to 6″ above the members being passed.

Openings in the roof frame should be formed before constructing the chimney. Refer to Chapter 12, **Roof Framing**. Water leakage around the chimney can be prevented by proper flashing. Corrosion-resistant metal, such as sheet copper, is often used for this purpose. The flashing is attached to the roof surface and extends up along the masonry. Cap flashing (counterflashing) is bonded into the mortar joints and is then lapped down over the base flashing. Refer to Chapter 14, **Roofing Materials and Methods**, for details concerning the flashing installation.

At the top of the chimney, the flue lining should project at least 4″ beyond the top brick course or cap. Surround the lining with cement mortar at least 2″ thick. Slope the cap so wind currents are directed upward and water drains away. When several flues are located in the same chimney, extend each flue to a different height. Horizontally space them no closer than 4″ apart.

25.2 Masonry Fireplaces

Figure 25-6 shows a cutaway view of a masonry fireplace. The chimney and fireplace are combined in a single unit. This unit often includes a flue for the regular heating equipment.

The *hearth* holds the fuel and contains the fire. It has two parts: front hearth (located in front) and back hearth (located below the fire area). The back hearth, side walls, and back wall are lined with firebrick that can withstand direct contact with flame. The side and back walls are sloped to reflect heat into the room.

A *damper* is a venting device located above the fire. The damper controls combustion, prevents heat loss, and redirects downdrafts.

The throat, smoke shelf, and smoke chamber are also important parts of the fireplace. They must be carefully designed to ensure good operation. Each of these parts serves a specific function in the proper

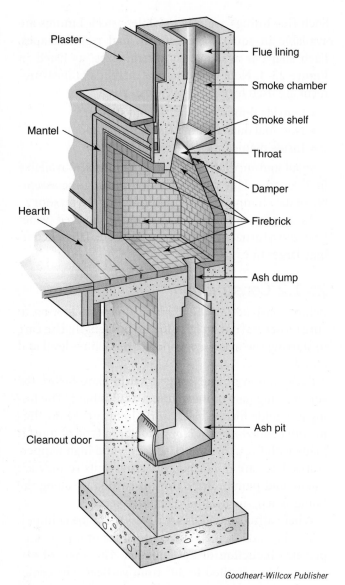

Goodheart-Willcox Publisher

Figure 25-6. Major components of a masonry fireplace.

operation of a fireplace. They will be discussed in greater detail later in this chapter.

Architectural plans often include details of the fireplace construction, **Figure 25-7**. The drawings include overall dimensions.

> ### Pro Tip
> Correct operation of a fireplace depends on an adequate flow of air into the building to replace the air exhausted through the flue. In older homes, air infiltration around doors and windows may be sufficient. However, to improve the energy efficiency of the house, combustion air is often piped to the fireplace from the outside.

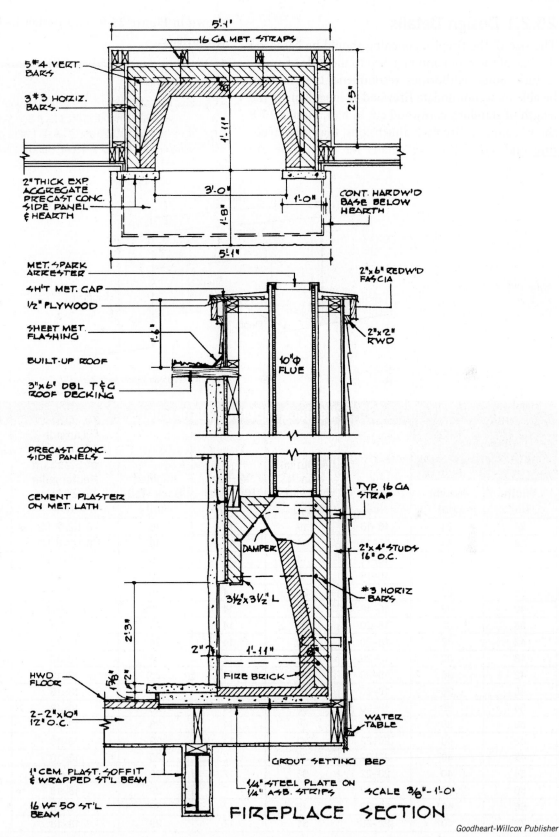

Figure 25-7. Detail drawings of fireplaces are included in architectural plans. Note the 2×4 framing on the outer edges in both views. A wooden chimney frame that juts out from a wall is called a chase.

25.2.1 Design Details

The size of the fireplace opening should be based on the size of the room and matched to the style of architecture. Some authorities recommend the fireplace be able to accommodate firewood 2′ long. This is the length of standard cordwood cut in half. **Figure 25-8** shows common fireplace dimensions. Masonry structure and wood framing details for a fireplace are shown in **Figure 25-9**. This design includes a flue for the furnace.

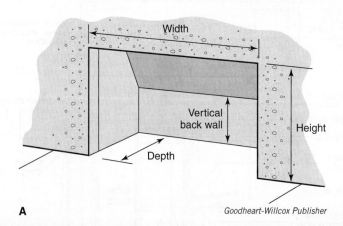

A

Goodheart-Willcox Publisher

Fireplace Opening Dimensions							
Fireplace Opening		Depth (in)	Minimum Back (horizontal) (in)	Vertical Back Wall (in)	Inclined Back Wall (in)	Outside Dimensions of Standard Rectangular Flue Lining (in)	Inside Diameter of Standard Round Flue Lining (in)
Width (in)	Height (in)						
24	24	16–18	14	14	16	8 1/2×8 1/2	10
28	24	16–18	14	14	16	8 1/2×8 1/2	10
24	28	16–18	14	14	20	8 1/2×8 1/2	10
30	28	16–18	16	14	20	8 1/2×13	10
36	28	16–18	22	14	20	8 1/2×13	12
42	28	16–18	28	14	20	8 1/2×18	12
36	32	18–20	20	14	24	8 1/2×18	12
42	32	18–20	26	14	24	13×13	12
48	32	18–20	32	14	24	13×13	15
42	36	18–20	26	14	28	13×13	15
48	36	18–20	32	14	28	13×18	15
54	36	18–20	38	14	28	13×18	15
60	36	18–20	44	14	28	13×18	15
42	40	20–22	24	17	29	13×13	15
48	40	20–22	30	17	29	13×18	15
54	40	20–22	36	17	29	13×18	15
60	40	20–22	42	17	29	18×18	18
66	40	20–22	48	17	29	18×18	18
72	40	22–28	51	17	29	18×18	18

B

Goodheart-Willcox Publisher

Figure 25-8. Recommended dimensions for a wide range of fireplace sizes.

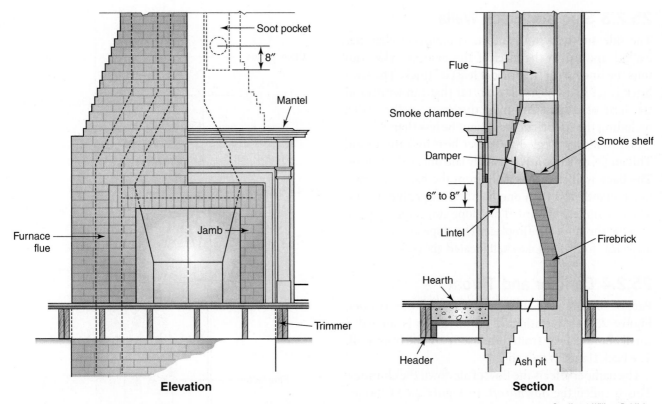

Elevation   **Section**

Figure 25-9. Elevation and section views show construction details for a typical masonry fireplace.

25.2.2 Hearth

The hearth, including the front section, must be completely supported by the chimney. The support is constructed by first building temporary supports and forms. Then, a minimum of 3 1/2″-thick concrete with reinforcing is poured. In the best construction, a cantilevered design is secured by recessing the back edge of the rough hearth into the rear wall of the chimney.

An *ash dump* may be provided in the rear hearth for clearing ashes, if there is space below for an ash pit. The dump consists of a metal frame with a pivoted cover. The basement ash pit should be of tight masonry and include a steel or cast iron clean-out door.

When the floor structure consists of a slab-on-grade, an ash pit may be formed by a raised hearth, as illustrated in **Figure 25-10**. This design may be used when the fireplace is located on an outside wall, but not if that wall is located where snow may block the outside access. In some designs, especially when no ash pit is included, the rear hearth is lowered several inches so ashes are contained in this area.

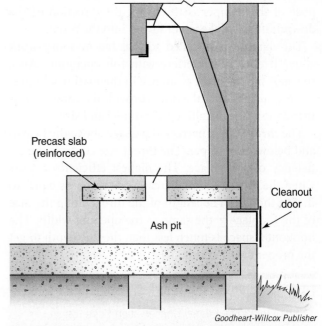

Figure 25-10. In slab-on-grade construction, a raised hearth provides space for an ash pit. This type of construction cannot be used in areas where snow may cover the cleanout door.

25.2.3 Side and Back Walls

The side and back walls of the combustion chamber extend upward to the level of the damper. The wall must be lined with firebrick at least 2″ thick. The firebrick is set in a special clay mortar that can withstand the heat of a fire. The total thickness of the walls, including the firebrick, should not be less than 8″.

Side walls are angled to reflect heat into the room. This angle, called *splay*, is usually laid out at 5″ per foot. The back wall rises vertically from the base for a distance slightly less than one-half of the opening height. Then, it slopes forward. This slope directs the smoke into the throat of the fireplace. The slope also keeps an area clear for the smoke shelf located above.

25.2.4 Damper and Throat

Part of the fireplace throat is formed by the damper, **Figure 25-11**. A stationary front flange is angled a small amount away from the masonry of the front wall. The back flange is movable.

The damper affects the flow of air down the chimney. This is called the *downdraft*. In **Figure 25-11**, notice the upward flow of hot air and smoke from the throat into the front side of the smoke chamber. The rapid upward passage of hot gases creates a downward current (downdraft) on the opposite side of the flue. One purpose of the damper is to change the direction of the downdraft so smoke is not forced into the room.

The damper is installed so that the masonry work above it does not interfere with full operation. Also, the ends have a slight clearance in the masonry to permit expansion. Manufacturers provide data and recommendations for installing each model and size.

The *throat* is the narrowed passage above the hearth and below the damper. The throat size controls the efficiency of a fireplace. The highest efficiency occurs when the cross-sectional area of the throat is equal to that of the flue. The length of the throat along the face of the fireplace is the same as the opening width. The horizontal space is much less than the flue depth to get the proper cross section.

Green Note

The International Energy Conservation Code requires tight-fitting dampers and outdoor combustion air for new wood-burning fireplaces. These measures reduce the amount of conditioned air from the home that is consumed by the fireplace. Replacing the consumed conditioned air requires energy, so following these guidelines reduces energy usage.

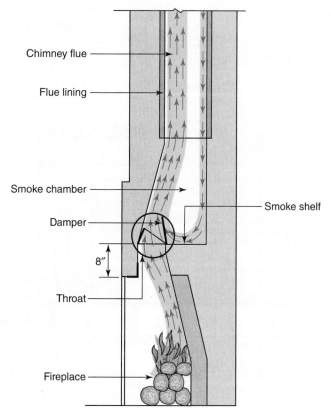

Goodheart-Willcox Publisher

Figure 25-11. Operating details of a fireplace. Arrows indicate smoke and air currents. Note the role of the damper in keeping the downward current from entering the room.

25.2.5 Smoke Shelf and Chamber

The *smoke shelf* is a horizontal shelf located adjacent to the damper in a fireplace. The smoke shelf helps the damper change the direction of the downdraft. The deeper the shelf behind the damper, the better the fireplace works. The depth may vary from 4″ to 12″, depending on the depth of the fireplace. Some smoke shelves are curved to reduce turbulence in the airflow. The length along the face for all types is equal to the full width of the throat.

The *smoke chamber* is the space extending from the top of the throat to the bottom of the flue. The area at the bottom of the chamber is quite large, since its depth includes that of the throat plus the depth of the smoke shelf. This space temporarily holds accumulated smoke if a gust of wind across the top of the chimney cuts off the updraft. Without this chamber, smoke would likely be forced out into the room. A smoke chamber also lessens the force of the downdraft by increasing the area through which it passes. Side walls are generally drawn inward one foot for each 18″ of rise. The surfaces of most smoke chambers are plastered with about a 1/2″ thickness of cement mortar.

25.2.6 Flue Size

The cross-sectional area of the flue is based on the area of the fireplace opening. As a general rule, the flue area should be at least 1/10 of the total opening. This applies to chimneys that are at least 20′ high. A somewhat larger flue may be needed to compensate for the lower chimney heights normally used in single-story construction. The upward movement of smoke in a low chimney does not reach a high velocity. Thus, a greater cross-sectional area is required for the flue.

One recommended method of calculating the area for a flue is to allow 13 sq in of area for the chimney flue to every sq ft of fireplace opening. For example, if the fireplace opening equaled 8.25 sq ft, then a flue area of at least 107 sq in is needed.

25.2.7 Construction Sequence

Masonry fireplaces are nearly always built in two stages. The first begins during the rough framing of the structure. Masonry work is carried up from the foundation and the main walls of the fireplace are formed. After a steel lintel is set above the opening, the damper is installed and the smoke chamber built. Then, the chimney is carried upward through the roof and the exterior masonry is completed. These steps usually occur before the roof deck is laid.

The second, finishing stage of the fireplace takes place during the application of interior trim after the finish wall surface is applied. Decorative brick or stone may be set over the exposed front face. The surface of the front hearth can be finished at this time, since the reinforced concrete base was placed during the rough masonry construction. The wood trim (mantel) is installed when masonry work is complete.

25.2.8 Special Designs

Some fireplace designs have openings on two or more sides. For these, follow the same planning principles as previously described for conventional designs. When calculating the flue area, the sum of the area of all faces must be used. **Figure 25-12** shows a corner fireplace design in which the flue area is based on the total of the front face opening plus the end face opening. In this particular construction, the side walls are not splayed. However, the rear wall is sloped in the usual manner.

Multiface fireplaces must incorporate a throat and damper with requirements similar to standard designs. Special dampers with square ends and sides are available for two-way fireplaces that serve adjoining rooms.

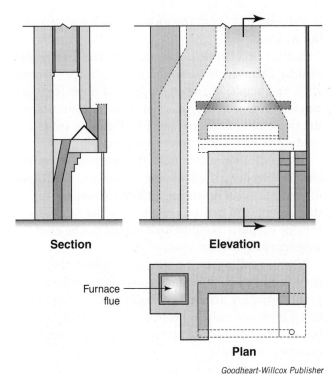

Section   **Elevation**

Furnace flue

Plan

Goodheart-Willcox Publisher

Figure 25-12. Design for a projecting corner fireplace.

25.2.9 Built-In Circulators

The heating efficiency of a fireplace can be increased by using a factory-built metal unit called a *circulator*, **Figure 25-13**. The sides and back are double walled, providing a space where air is heated. Cool air enters this chamber near the floor level. When the air is heated,

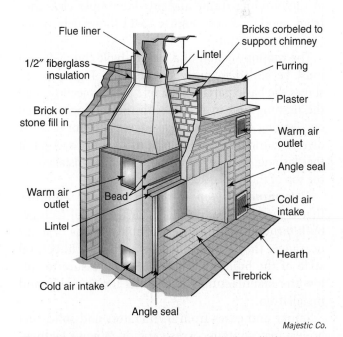

Flue liner — 1/2″ fiberglass insulation — Brick or stone fill in — Warm air outlet — Lintel — Cold air intake — Angle seal — Bead — Bricks corbeled to support chimney — Lintel — Furring — Plaster — Warm air outlet — Angle seal — Cold air intake — Hearth — Firebrick

Majestic Co.

Figure 25-13. Cutaway view of a circulator installation. Masonry is erected around the metal unit.

it rises and returns to the room through registers at a higher level.

Built-in circulators include not only the firebox and heating chamber, but also the throat, damper, smoke shelf, and smoke chamber. Since all of these parts are carefully engineered, proper flue draw is ensured when the flue size is adequate and installation is made according to the manufacturer's directions.

To install a circulator unit, first position it on the hearth, then build the brick and masonry work around the outside. Steel lintels are required across the top of the opening and may be required in other locations to provide support. The unit itself should not be used for support of any masonry work.

When installing a circulator unit, follow specifications furnished by the manufacturer. Some type of fire-proof insulating material is usually placed around the metal form not only to prevent the movement of heat but also to provide some expansion space between the metal and masonry.

25.3 Factory-Built Chimneys

Lightweight chimney units are available that require no masonry work. They provide flues for heating equipment or fireplaces and can be installed in single-story or multistory structures. Factory-built chimneys usually consist of double- or triple-walled sections of pipe that are assembled to form the flue. Special flanges and fittings are used to fasten the flue to the building frame and provide proper clearance from wood members. **Figure 25-14** shows details of a typical installation.

The rooftop section of a factory-built chimney unit must be sealed to prevent roof leaks. **Figure 25-15** shows the basic parts of a simple pipe projection. Bed the flashing unit in mastic with the roofing material overlapping the top and side edges. The storm collar diverts rainwater from the pipe to a conical section of the flashing. Some type of cap should be installed on the top of the flue to keep out rain or snow.

A ***termination top*** is the part of the chimney that extends above the roof. Many types of factory-built termination tops are available. For the best appearance, they should blend with the architectural style of the building, **Figure 25-16**. Be sure to follow the manufacturer's directions for assembly and installation.

Smoke and gases from incinerators and solid-fuel boilers may contain corrosive acids. When a factory-built chimney is used, the inside pipe should be made of stainless steel or have a porcelain-coated surface.

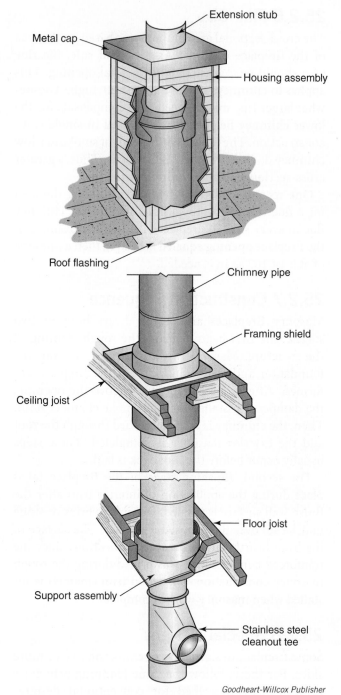

Goodheart-Willcox Publisher

Figure 25-14. Triple-wall metal pipe sections are used to erect this factory-built chimney. Depending on the application, inside diameter of the pipes can range from 6″ to 14″. A rain cap is installed on the extension stub.

Pro Tip

A factory-built chimney should have a label showing that it has been tested and listed by Underwriters Laboratories or another nationally recognized testing organization. It must also conform to local building codes. Be sure to make the installation in strict accordance with the manufacturer's instructions.

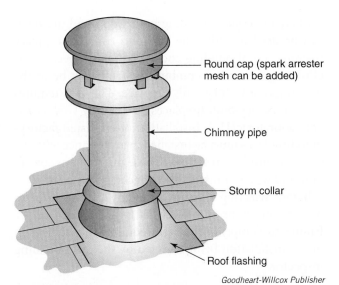

Goodheart-Willcox Publisher

Figure 25-15. The basic parts of a rooftop projection for a standard flue pipe.

Goodheart-Willcox Publisher

Figure 25-16. Termination tops and chimney housings are used with factory-built chimneys. This example has a wood framework covered with siding to match the exterior siding of the house.

25.4 Factory-Built Fireplaces

Many of the fireplaces in new homes are factory-built units. These units are generally less expensive than masonry units. Manufacturers of factory-built fireplaces provide a wide range of designs that are easy to install with ordinary tools. In some cases, multiple-steel-wall construction and special firebox linings permit *zero clearance*. This means that the outside of the

housing can rest directly on wood floors and close to wood framing members, **Figure 25-17**. Zero clearance does not necessarily mean that combustible materials like wood can be in direct contact with the fireplace.

Manufacturers provide detailed drawings, instructions, and recommendations for the installation of their fireplace units. These should be carefully studied and followed throughout installation. Other precautions should be taken if gas or electric starters are used or if gas logs are installed. Check local fire codes before starting work.

Safety Note
Improper fireplace installation or operation can be dangerous. A house fire, explosion, or carbon monoxide poison can result. Make sure you and the homeowner read and understand all installation and operating instructions before installing or operating a fireplace.

Pamela Au/Shutterstock.com

Figure 25-17. This zero-clearance fireplace is surrounded by wood framing. Its construction ensures that the outside remains cool even with a fire burning inside.

25.4.1 Operation and Construction

Factory-built fireplaces operate like a metal circulator built into masonry fireplaces. Room air enters intakes at floor level and flows through chambers around the firebox. As the air is heated, it rises. This motion carries the air through the grillwork at the top of the unit and back into the room, **Figure 25-18**. Units are sometimes equipped with circulating fans that are controlled by switches.

Some units can be attached to vertical ducts that carry the heated air to various locations within the room or to an adjoining room. Units may also have blowers that increase the flow of air and thus improve heating efficiency.

In new construction, the basic frame is usually built at the same time as exterior walls and partitions. The facing side of the frame should remain open until the fireplace is installed. If support is required above this opening, it should be framed with a header like that for a door. After the fireplace unit is installed, any front framing below the header can be added.

Many materials, including both wood and masonry, can be used to finish the wall and trim the fireplace opening. Local building codes frequently specify how close wood or other combustible trim may be to the opening of a factory-built fireplace. The IRC requires that all factory-built fireplaces be listed by Underwriters Laboratory (UL). Installation of UL-listed factory-built fireplaces must be installed in accordance with the manufacturer's instructions. Those instructions specify how and where combustible trim can be installed.

Before selecting and installing any type of factory-built fireplace, be sure to check local building codes. Failure to comply with these requirements can result in the installation being rejected by the local building inspector.

25.4.2 Chimneys for Factory-Built Fireplaces

Chimney (flue pipe) systems for factory-built fireplaces are designed for specific units and usually sold as a package along with the fireplace. **Figure 25-19** shows the typical assembly of standard components that run through the ceiling and roof structure.

Preway, Inc.

Figure 25-18. This cutaway view shows the operation of a factory-built fireplace. Room air (blue arrows) is drawn from the floor level into the heating chamber. The air is heated, then returned to the room (red arrows). Combustion air can also be drawn from the outside using a pipe connected to the inlet shown at lower left. Note the triple walls of the flue.

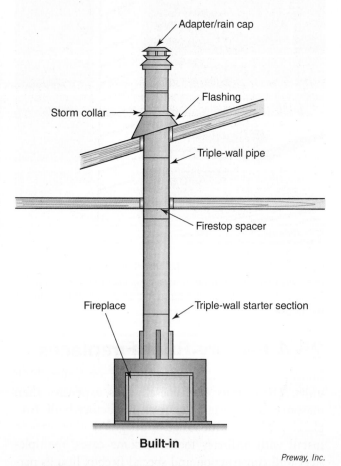

Built-in

Preway, Inc.

Figure 25-19. Chimney system used for factory-built fireplaces.

Special attention must be given to clearance and support. Openings through the ceiling and roof should be carefully framed to provide the correct size openings for firestop spacers and support boxes. **Figure 25-20** shows the installation of a firestop spacer in an attic area.

The chimney pipe must be supported at the roof. If not, the entire weight of the pipe will rest on the fireplace. By supporting the chimney pipe at the roof, its weight is better distributed and the pipe is more stable.

When installed on an exterior wall, the chimney and fireplace unit can be located in a special projection called a chase. A *chase* is a wood frame that juts from an outside wall and supports a factory-built chimney and fireplace. This box-like structure is built as a part of the floor, wall, or roof frame, **Figure 25-21**. When a chase is not used, the fireplace and chimney are located within the room. This reduces the amount of available living space. By using an exterior chase, the room area is not reduced by the presence of the fireplace.

The outer walls of the chase are usually finished to match the outer walls of the house, **Figure 25-22**. Insulate the walls of the chase and seal them against infiltration. Be sure insulation does not touch the flue pipe.

APA–The Engineered Wood Association

Figure 25-21. Typical wood framing for a chimney chase. The roof structure helps hold the chase steady on three sides.

Zadiraka Evgenii/Shutterstock.com

Figure 25-20. A firestop spacer is used to close off the opening around a triple-walled fireplace flue. To prevent the spread of fire, all vertical channels in a building frame should be closed off at least once every 10 vertical feet.

Goodheart-Willcox Publisher

Figure 25-22. A completed chase for a factory-built fireplace and chimney.

25.5 Glass Enclosures

Glass enclosures improve the efficiency of fireplaces. They can be installed on either masonry structures or factory-built units. They reduce the amount of heated room air that escapes up the chimney, even when there is no fire. Energy-efficiency is improved because the rate of burning can be controlled by adjusting draft vents and dampers.

Workplace Skills
Cultural Diversity and Equity

In construction, people from all walks of life work together. Men and women with diverse backgrounds regularly come together to achieve amazing results. Having a diverse crew brings with it the advantage of gaining various perspectives that can contribute to creativity and problem solving. It is important and beneficial for the entire crew to respect and appreciate each other's differences. Sometimes this requires extra effort and sensitivity toward a coworker's individuality. These issues can be complex and difficult. Focus on the strengths that others bring to the table. If you are struggling with these issues or are feeling discriminated against seek help.

Construction Careers
Mason

Workers who build walls, floors, walkways, or other structures from brick, concrete block, or stone are called masons. They often are referred to by the type of material they work with: brickmasons, blockmasons, or stonemasons. A special category is formed by the small number of masons who lay firebrick and heat-resistant refractory tile. These refractory masons work mostly in steel mills and iron foundries.

While brickmasons and blockmasons work on both residential and commercial structures, the majority of stonemasons work on structures such as office buildings and hotels. Stone installed in residential settings is primarily used for fireplaces, floors, and outdoor walkways or patios. The material used by brickmasons and blockmasons is supplied in standard sizes and shapes, while stonemasons often must cut and fit pieces of irregular shape and size.

Approximately 25% of all masons are self-employed. The rest work for general contracting firms or specialty businesses. A small number work for manufacturing firms, government agencies, and other employers. The US Bureau of Labor Statistics expects the job outlook for masons to grow faster than the average for all jobs. Masons must be able to accurately read and follow construction drawings. They make extensive use of measuring skills. Masons must be able to perform strenuous work that includes heavy lifting, often in unfavorable weather conditions. Another required skill is the ability to efficiently and safely use a variety of hand and power tools.

Most masons acquire their skills learning from experienced workers through informal on-the-job training. Some graduate from technical school programs. A small number—usually those working for large nonresidential contractors—complete a formal apprenticeship that is typically three years in length. The apprenticeship combines approximately 144 hours of classroom training with practical experience and instruction on the jobsite.

Masons who work for larger contracting firms have opportunities for advancement to supervisory positions. Some leave wage employment to open their own businesses. Opportunities also exist in related fields such as building inspection and construction management.

Digital Vision/Photodisc/Thinkstock

This figure shows masons working with bricks, but masons work with blocks and stones as well. Masons read construction drawings, possess measuring skills, and are able to use both hand and power tools.

Chapter Review

Summary

- Masonry chimneys are placed on their own footings.
- Chimney size depends on the number, arrangement, and size of flues needed for heating equipment and fireplaces.
- Chimneys must be constructed high enough to avoid downdrafts caused by wind turbulence.
- The IRC and most local codes require that chimneys be lined with a noncombustible sleeve.
- Each section of flue lining is set in place before the surrounding part of the chimney wall is built.
- The lining is a guide for the brick work.
- At the top of the chimney, the flue lining should project at least 4″ beyond the top brick course or cap.
- A fireplace consists of a hearth, damper and throat, smoke shelf, smoke chamber, and flue.
- The hearth must be completely supported by the chimney.
- The damper affects the flow of air down the chimney.
- The smoke shelf helps the damper change the direction of the downdraft.
- In general, the flue area should be at least 1/10 of the total opening.
- Masonry fireplaces are typically built in two stages: the first begins during the rough framing of the structure; the second, finishing stage takes place after the finish wall surface is applied.
- Built-in circulators increase the heating efficiency of a fireplace.
- Factory-built chimneys and fireplaces eliminate the need for masonry work.
- Factory-built fireplaces that have zero-clearance means they can rest directly on floors and close to wood framing members.
- A factory-built fireplace draws room air in at floor level, moves the air through chambers around the firebox, and expels the air at the top of the unit, back into the room.
- Chimneys for factory-built fireplaces must be carefully framed and properly supported at the roof.
- When installed on an exterior wall, factory-built chimney and fireplace units can be located in a chase.
- Glass enclosures on fireplaces improve efficiency by preventing heated air loss from the room, even when the fireplace is not being used.

Know and Understand

Answer the following questions using the information in this chapter.

1. The overall size of a chimney depends on the number, arrangement, and size of the _____.
   A. wood framing members
   B. flues
   C. fireplaces
   D. hearth

2. Wood framing should be located at least _____″ away from a masonry chimney.
   A. 2
   B. 3
   C. 6
   D. 10

3. The offset angle in a chimney should be limited to _____° from vertical.
   A. 15
   B. 30
   C. 45
   D. 90

4. When several flues are located in the same chimney, horizontally space them no closer than _____" apart.
   A. 2
   B. 3
   C. 4
   D. 8
5. To prevent heat loss from a room when the fireplace is not in operation, the fireplace must be equipped with a(n) _____.
   A. hearth
   B. damper
   C. throat
   D. ash dump
6. The hearth is completely supported by the _____.
   A. building frame
   B. fireplace
   C. chimney
   D. flue lining
7. The length of the throat along the face should be equal to the _____ of the fireplace opening.
   A. height
   B. depth
   C. area
   D. width
8. *True or False?* Downdraft occurs when a rapid upward passage of hot gases passes on the opposite side of the flue.
9. The _____ helps the damper change the direction of the downdraft.
   A. hearth
   B. smoke shelf
   C. smoke chamber
   D. flue
10. Which of the following is *not* generally included with built-in circulators?
    A. Firebox.
    B. Throat.
    C. Flue.
    D. Smoke shelf.
11. Factory-built chimneys consist of pipe sections with double or _____ walls.
    A. insulated
    B. triple
    C. multistory
    D. flashed
12. When a factory-built metal chimney is used for an incinerator, the inside pipe should be coated with porcelain or made of _____.
    A. copper
    B. cast-iron
    C. stainless steel
    D. silver
13. *True or False?* The IRC specifies how close wood or other combustible trim may be to the opening of a factory-built fireplace.
14. A _____ is a special projection on the outside of a house that contains a factory-built chimney.
    A. chase
    B. termination top
    C. circulator
    D. firebox

Apply and Analyze

1. Explain whether or not a masonry chimney provides support or receives support from the building frame.
2. What is corbelling?
3. Explain the purpose of providing an ash dump in the rear hearth.
4. Why is a larger flue sometimes needed in single-story constructions?
5. What is meant by permitting zero clearance for multiple-steel-wall construction and special firebox linings?

Critical Thinking

1. Wood and gas burning fires require oxygen to burn. As combustion occurs, smoke and waste gasses are vented through the chimney or exhaust vent. Research what options are available to supply make-up air for combustion. If no provision for make-up air is provided, what consequences could result?
2. The majority of fireplaces are not used as a primary heating source. What value do fireplaces create in a home? Why do you think they are so popular?
3. Research the costs of installing a masonry fireplace versus installing a factory-built vented gas fireplace. Compare the benefits of each. Report to your class on which option you would choose if building a new home for yourself. In your report, be sure to explain all of the factors that contributed to your decision and why those factors are important to you.

Communicating about Carpentry

1. **Speaking and Art.** Working with a partner, create a model of a chimney or fireplace. Create the model so the parts can be removed. Demonstrate the different options available when choosing a chimney or fireplace.
2. **Speaking and Reading.** Research the products and services available for chimneys and fireplaces. Collect promotional materials for a variety of products and services from product manufacturers. Analyze the data based on the knowledge gained from this lesson. With your group, review the list for words that can be used in this subject area. Break each word into its combining form, practice pronouncing the word, and discuss its meaning. As a fun challenge, work together to compose a creative narrative using as many words as you can from your new list.

CHAPTER **26**

Post-and-Beam Construction

OBJECTIVES

After studying this chapter, you will be able to:

- List the advantages and disadvantages of post-and-beam construction.
- Describe general specifications for supporting posts.
- Compare transverse and longitudinal beams.
- Describe how roof and floor planks should be selected and installed.
- Sketch basic construction details of structural insulated panels (SIPs) and box beams.

TECHNICAL TERMS

box beam
longitudinal beam

mortise-and-tenon joint
plank

post-and-beam construction
transverse beam

Residential and commercial framing can take many forms. Conventional stick-framing methods are the most popular in residential construction; however, homes can be also be framed using light-gauge steel framing, heavy-gauge steel framing, SIPs panels, or insulated concrete forms. *Post-and-beam construction*, or plank-and-beam construction, is a method of framing that utilizes large framing members—posts, beams, and planks. Because of their great strength, these members may be spaced farther apart than conventional framing members, **Figure 26-1**.

FOTOGRIN/Shutterstock.com

Figure 26-1. The posts and beams of this house frame will be left exposed as an architectural feature of the building.

Frames of this type are similar to mill construction, once used for barns and heavy-timbered buildings. It is often used today for upscale residential building, since it permits greater flexibility than conventional framing methods for contemporary and traditional designs. See **Figure 26-2** for a comparison of construction methods.

Post-and-beam construction is often combined with conventional framing, **Figure 26-3**. For example, the walls might be conventionally built and the roof framed with beams and planks. In such a case, the term plank-and-beam could only be applied to the roof structure. Similarly, it would be correct to refer to a heavy-timbered floor structure as a plank-and-beam system.

26.1 Advantages of Post-and-Beam Construction

The most obvious advantages of post-and-beam construction are the distinctive architectural effects created by the exposed beams in the ceiling and the added ceiling height. The underside of the roof planks may serve as the ceiling surface, thus providing savings in material, **Figure 26-4**.

Post-and-beam framing may also provide some savings in labor. The pieces are larger and fewer in number. They can usually be more rapidly assembled than conventional framing.

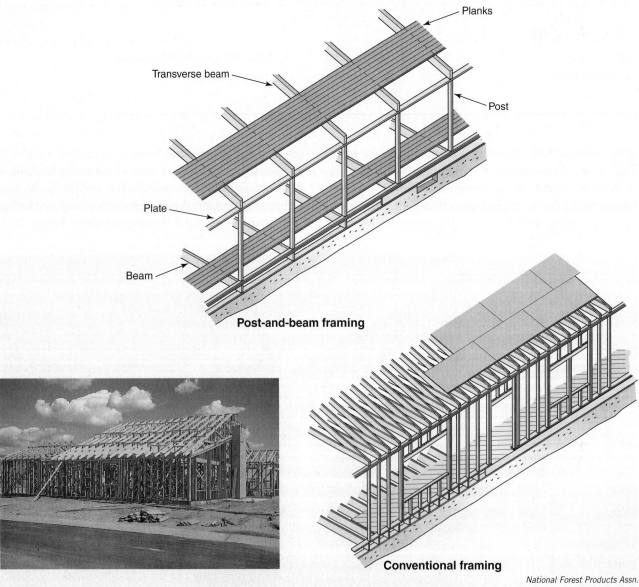

Post-and-beam framing

Conventional framing

National Forest Products Assn.

Figure 26-2. Post-and-beam framing compared to conventional framing. In post-and-beam construction, headers can be eliminated, which simplifies framing around windows and doors.

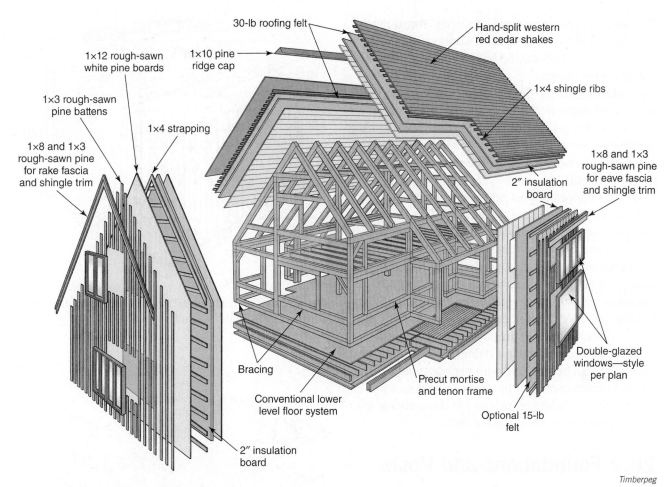

30-lb roofing felt

1×12 rough-sawn white pine boards

1×10 pine ridge cap

Hand-split western red cedar shakes

1×3 rough-sawn pine battens

1×4 strapping

1×4 shingle ribs

1×8 and 1×3 rough-sawn pine for rake fascia and shingle trim

1×8 and 1×3 rough-sawn pine for eave fascia and shingle trim

2″ insulation board

Double-glazed windows—style per plan

Bracing

Conventional lower level floor system

Precut mortise and tenon frame

Optional 15-lb felt

2″ insulation board

Timberpeg

Figure 26-3. A rustic variation of a post-and-beam house that follows traditional framing of colonial barns. On the first level, a conventional flooring system is used, while heavy beams are used on the second level. The pine board-and-batten siding helps stiffen the walls against racking.

Vermont Frames

Figure 26-4. The high ceilings and exposed beams of post-and-beam construction are attractive architectural features.

One of the chief structural advantages of this method is the simplicity of framing around door and window openings. Loads are carried by posts spaced at wide intervals in the walls. Large openings can be framed without the need for headers, **Figure 26-5**.

Vermont Frames

Figure 26-5. Wide expanses of windows are possible with post-and-beam construction because headers are not required.

Window walls can be formed by merely inserting window frames between the posts. Another advantage is that wide overhangs can be built by simply extending the heavy roof beams.

In addition to its flexibility in design, post-and-beam construction also provides high resistance to fire. Wood beams do not transmit heat, nor do they collapse in the manner of unprotected metal beams. Exposure of wood beams to flame results in a slow loss of strength.

Most limitations of post-and-beam construction can be resolved through careful planning. The absence of concealed spaces in outside walls and ceilings makes installation of electrical wiring, plumbing, and heating somewhat more difficult. Also, the plank floors, for example, are designed to carry moderate, uniform loads. Therefore, extra framing must be provided under load-bearing partitions, bathtubs, refrigerators, and other places where heavy loads are likely. Extra members may be needed to provide lateral stability to the frame and walls. This can be achieved with various types of bracing. It is more common, however, to enclose some of the wall area with large panels or use conventional stud construction as shown in **Figure 26-6**.

26.2 Foundations and Posts

Foundations for post-and-beam framing may consist of continuous walls or simple piers located under each post. Refer to Chapter 9, **Footings and Foundations**. Either continuous or pier foundations must rest on footings that meet local building code requirements.

Posts must be strong enough to support the load and large enough to provide full bearing surfaces for the ends of the beams. In general, posts should not be less than 4×4 nominal size. They may be made of solid stock or built up from 2″ pieces. Where the ends of beams are joined over a post, the bearing surface should be increased with bearing blocks as shown in **Figure 26-7**.

When posts extend to any great height without lateral bracing, a greater cross-sectional area is required to prevent buckling. Requirements are usually listed in local building codes through an l/d, or slenderness, ratio. The *l* represents the length in inches and the *d* stands for the smallest cross-sectional dimension (actual size). For example, a 4×4 piece 8′ long has a ratio of about 27. This is within the limits usually prescribed.

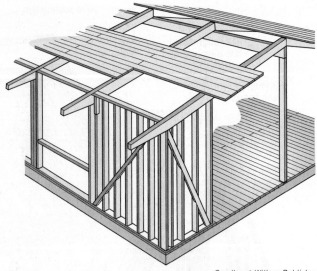

Goodheart-Willcox Publisher

Figure 26-6. To provide lateral stability in a post-and-beam structure, a conventional stud wall with let-in bracing can be used.

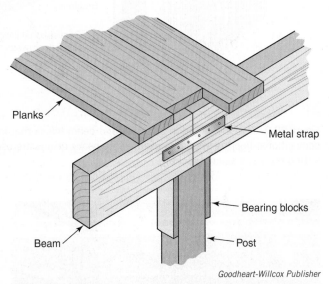

Goodheart-Willcox Publisher

Figure 26-7. Where beams are joined over a post, bearing blocks are used to ensure an adequate bearing surface. A heavy steel plate can be used in place of the bearing blocks.

The distance between posts is determined by the basic design of the structure. This spacing must be carefully engineered. Usually, posts are evenly spaced along the length of the building and within the allowable free span of the floor or roof planks. Due to modular dimensions, construction costs will be lower if post-and-beam positions occur at standard increments (intervals) of 16″, 24″, and 48″.

In single-story construction, a plate is attached to the top of the posts in about the same way as conventional framing. The roof beams are then positioned directly over the posts as shown in **Figure 26-8**.

26.3 Floor Beams

Beams for floor structures may be solid, glue-laminated, or built-up, **Figure 26-9**. Sometimes, the built-up beams are formed with spacer blocks between the main members.

Box beams may also be used. These are discussed later in this chapter. For single-story structures, where under-the-floor appearance is not critical, standard dimension lumber can be nailed together to form a beam of any desired size.

Design of sills for a plank-and-beam floor system can be similar to regular platform construction, **Figure 26-10**. When it is desirable to keep the silhouette of the structure low with the floor near grade level, the beams can be supported in pockets in the foundation wall.

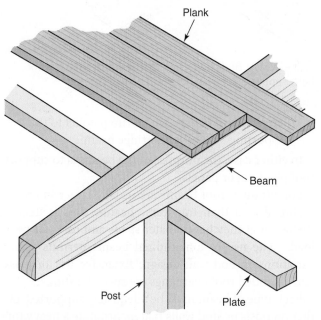

Goodheart-Willcox Publisher

Figure 26-8. To prevent sagging, roof beams must rest directly over supporting posts.

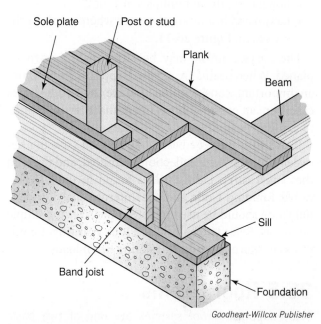

Goodheart-Willcox Publisher

Figure 26-10. Typical sill construction for a post-and-beam frame. If a post is not located over a beam, blocking must be used to provide proper support.

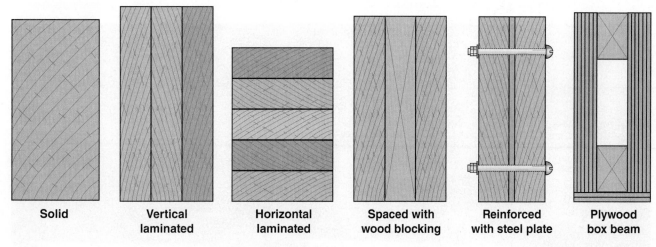

| Solid | Vertical laminated | Horizontal laminated | Spaced with wood blocking | Reinforced with steel plate | Plywood box beam |

Goodheart-Willcox Publisher

Figure 26-9. Beams can be solid or built-up from smaller dimension lumber.

26.4 Beam Descriptions

In general, it is best to use solid timbers when beam sizes are small or when a rustic architectural appearance is desired. Where high stress factors demand large sizes or a finished appearance is required, it is usually more economical to use laminated beams. Laminated beams are manufactured in a wide range of sizes and finishes. Solid timbers are available in various forms:

- A range of standard cross sections
- Lengths of 6′ and longer
- Longer lengths in multiples of 1′ or 2′
- Laminated beams are also used where great depth is needed, **Figure 26-11**.

The surface finish may be either rough sawn or planed. When beams are exposed, appearance becomes an important consideration. **Figure 26-12** illustrates casing-in, an on-the-job treatment that may be applied to exposed beams.

Beam sizes must be based on the span (spacing between supports), the deflection permitted, and the load they must carry. Design tables found in the International Building Code and the International Residential Code should be used to determine sizes for simple buildings. Refer to Chapter 10, **Floor Framing**, for additional information on calculating beam loads.

26.5 Roof Beams

Beam-supported roof systems are one of two basic types, **Figure 26-13**:

- *Transverse beams*, which run perpendicular to the ridge, and are similar to exposed rafters on wide spacings.

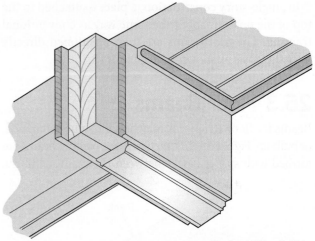

Goodheart-Willcox Publisher

Figure 26-12. Built-up beams can be covered (cased) for better appearance when they will be left exposed.

- *Longitudinal beams*, which run parallel to the supporting sidewalls and ridge beam.

In either case, the purpose of the beams is to support roof planking or panels.

Longitudinal roof beams are usually larger in cross-sectional area than transverse beams. This is due to the fact that they have greater spans and carry heavier loads. The use of longitudinal beams permits many variations in end-wall design. Extensive use of glass and extended roof overhangs are special features.

Each type of beam must be adequately supported, either on posts or stud walls that incorporate a heavy top plate. When supported on posts, the connection can be reinforced with a wide panel frame that extends to the top of the beam, **Figure 26-14**. A similar method used to support a ridge beam is shown in **Figure 26-15**.

StructureCraft

Figure 26-11. Laminated beams can be manufactured in any depth.

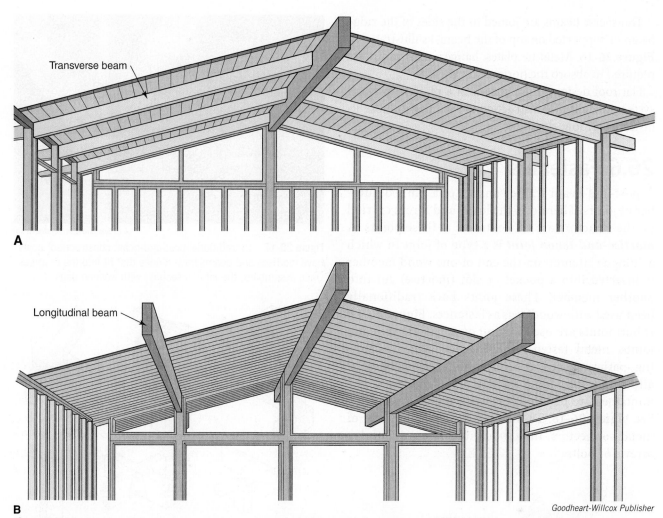

Goodheart-Willcox Publisher

Figure 26-13. Plank-and-beam roof construction methods. A—Transverse beam, with beams running perpendicular to the ridge. B—Longitudinal beam, with beams running parallel to the ridge.

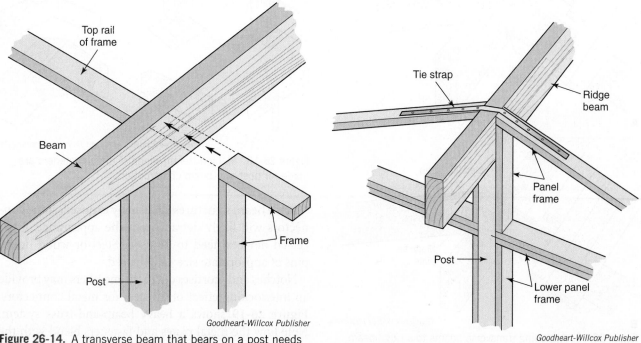

Goodheart-Willcox Publisher

Figure 26-14. A transverse beam that bears on a post needs support to prevent lateral (side-to-side) movement. The reinforcement is provided by the filler frame panel.

Goodheart-Willcox Publisher

Figure 26-15. A metal tie strap and panel frames prevent lateral movement of the ridge beam.

Transverse beams are joined to the sides of the ridge beam or supported on top of the beam, as illustrated in **Figure 26-16**. Metal tie plates, hangers, and straps are required to absorb the horizontal thrust.

Flat roof designs often consist of a plank-and-beam system. Details of construction are similar to those illustrated for low, sloping roofs.

26.6 Fasteners

A post-and-beam frame consists of a limited number of joints. Therefore, the loads and forces exerted on the structure are concentrated at these points. A *mortise-and-tenon joint* is a type of joint in which a "tongue" (tenon) on the end of one wood member is inserted into a pocket or slot (mortise) cut into another member. These joints have traditionally been used with wood pegs as fasteners, **Figure 26-17**. If butt joints are used instead of mortise-and-tenon joints, metal fasteners are needed. Regular nailing patterns used in conventional framing usually do not provide a satisfactory connection. The joints need to be reinforced with metal connectors. See **Figure 26-18**. To increase the holding power of metal connectors, they should be attached with lag screws or bolts.

A.L. Spangler/Shutterstock.com

Figure 26-17. In traditional post-and-beam construction, posts have mortises and beams have tenons that fit into the mortises. Once assembled, the joint is secured with wooden pegs.

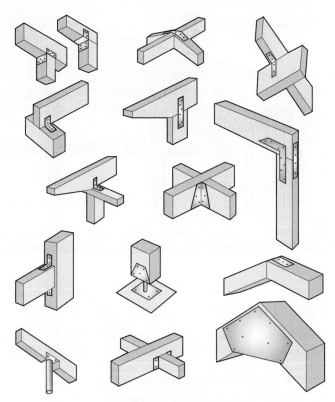
Western Wood Products Assn. and Timber Engineering Co.

Figure 26-18. Many different kinds of metal fasteners are made for post-and-beam construction.

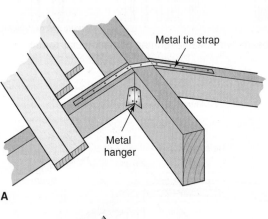

A

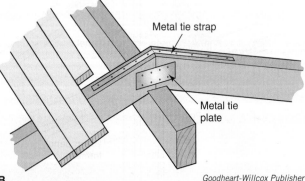

B

Goodheart-Willcox Publisher

Figure 26-16. Attaching transverse beams to a ridge beam. A—Beams attached to the side of the ridge. B—Beams supported on top of the ridge.

Since beam structures are usually exposed, some connectors will likely detract from the appearance. Concealed devices need to be used. Steel or wood dowel pins of appropriate size can be used.

Notches and mortises cut in the members may provide an interlocking effect or a recess for metal connectors. **Figure 26-19** shows a heavy beam-and-truss system. Note how the metal plates and fasteners blend with the rough surface of the structural members to provide a special architectural effect.

Figure 26-19. A heavy beam-and-truss system is used here to support roof beams at the midpoint and ridge. The rough-sawn surface finish helps mask joints in the laminated beams.

Boise-Cascade Corp.

Use extra care when assembling exposed posts and beams. Tool and hammer marks detract from the final appearance.

26.7 Partitions

Interior partitions are more difficult to construct under an exposed beam ceiling. Except for a load-bearing partition under a main ridge beam, it is usually best to make the installation after beams and planks are in place. Partitions running perpendicular to a sloping ceiling should have regular top plates with filler sections installed between the beams.

Partitions parallel to transverse beams have a sloping top plate. Some carpenters prefer to construct these partitions in two sections. First, build a conventional lower section the same height as the sidewalls. Then, add a triangular section above the lower section.

When nonbearing partitions run at a right angle to a plank floor, no special framing is necessary. However, when nonbearing partitions run parallel, additional support must be provided, **Figure 26-20**. Replace the sole plate with a small beam or add the beam below the plank flooring.

26.8 Planks

Planks for floor and roof decking can be anywhere from 2″ to 4″ thick, depending on the span. Edges are usually tongue-and-groove for a strong, tight surface. **Figure 26-21** illustrates standard tongue-and-groove designs. The identification numbers given in the figure are those listed by the Western Wood Products Association.

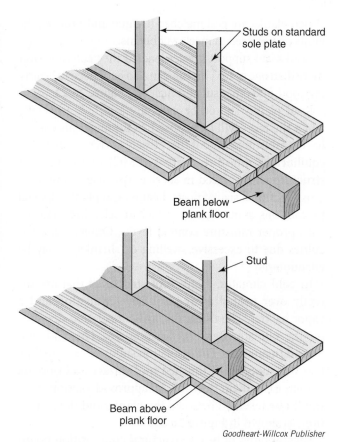

Goodheart-Willcox Publisher

Figure 26-20. Two methods used to support nonbearing partitions that run parallel to flooring planks.

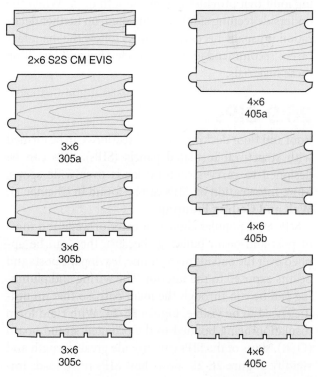

Goodheart-Willcox Publisher

Figure 26-21. Standard plank patterns in end view. Faces are machined flat, while edges are tongue-and-groove.

When planks are end matched (tongue-and-groove), the joints do not need to meet over beams.

Planks can support greater loads if they continue over more than one span. This rule can also be applied to beams, plywood, and other support material, **Figure 26-22**.

Roof planks should be carefully selected, especially when the faces will be exposed. Solid materials should have a moisture content that closely corresponds to the equilibrium moisture content (EMC) of the interior structure when placed in service. Because of the large cross-sectional size of posts, beams, and planks, special precautions should be observed in selecting material with proper moisture content levels. Otherwise, difficulties due to excessive swelling or shrinkage may be encountered.

In cold climates, plank roof structures that are directly over heated areas must have insulation and a vapor barrier. The thickness of the insulation depends on the climate. Refer to Chapter 18, **Thermal and Sound Insulation**. The insulation should be a rigid type that will support the finished roof surface and workers who must walk on the roof. An approved vapor barrier should be installed between the planks and the insulation, as shown in **Figure 26-23**.

Several types of heavy structural composition board are available for roof decks. These come in thicknesses of 2″–4″. The panel sizes are large and the material is lightweight. Edges usually have some type of interlocking joint that provides a tight, smooth deck. When the underside (ceiling side) is prefinished, no further decoration is usually necessary. Always follow the manufacturer's recommendations when selecting and installing these materials.

26.9 SIPs

Roof and floor decking and wall sections can be formed with structural insulated panels (SIPs). SIPs can be designed to carry structural loads over wide spans. More information on SIPs can be found in Chapter 11, **Wall and Ceiling Framing**.

SIPs are a popular choice for exterior walls and roofs of post-and-beam buildings because they can be applied over the outside of the frame, leaving the posts and beams exposed to the interior. The wiring and plumbing can be run through the foam core, so that no utilities need to be visible, **Figure 26-24**. With rigid foam insulation being bonded to the oriented strand board (OSB), skins of the SIPs can provide great strength and rigidity. **Figure 26-25** shows how SIPs resist loads imposed on a roof.

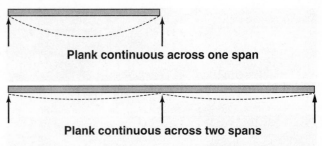

Plank continuous across one span

Plank continuous across two spans

Goodheart-Willcox Publisher

Figure 26-22. The stiffness of a plank is increased if it continuously extends over two or more spans.

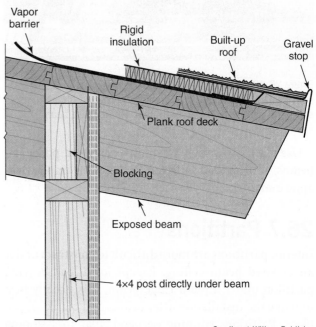

Vapor barrier

Rigid insulation

Built-up roof

Gravel stop

Plank roof deck

Blocking

Exposed beam

4×4 post directly under beam

Goodheart-Willcox Publisher

Figure 26-23. When a plank roof deck is directly over a heated space, a vapor barrier and insulation must be applied. This cross section shows the roof deck, vapor barrier, insulation, and built-up roofing.

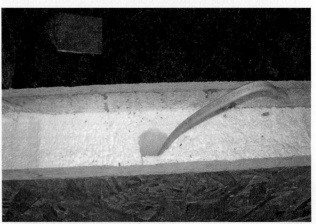

Goodheart-Willcox Publisher

Figure 26-24. It is common practice to run wiring through the foam cores of SIPs.

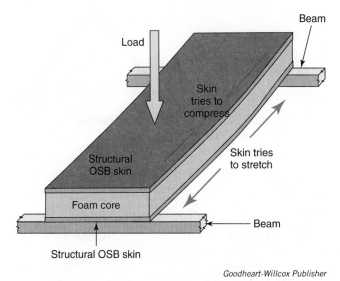

Goodheart-Willcox Publisher

Figure 26-25. This is how a stressed-skin panel provides support.

26.10 Box Beams

Box beams made of plywood webs offer a structural unit that can span distances up to 120′. The high strength-to-weight ratio offers a tremendous advantage in commercial structures where wide, unobstructed areas are required.

The basic plywood *box beam* consists of one or more vertical plywood webs that are laminated to seasoned lumber flanges, **Figure 26-26**. The flanges are separated at regular intervals by vertical spacers (stiffeners) that help distribute the load between the upper and lower flange. Spacers also prevent buckling of the plywood webs. The strength of the unit depends to a large extent on the quality of the glue bond between the various members. Plywood box beams must be carefully designed and fabricated under controlled conditions.

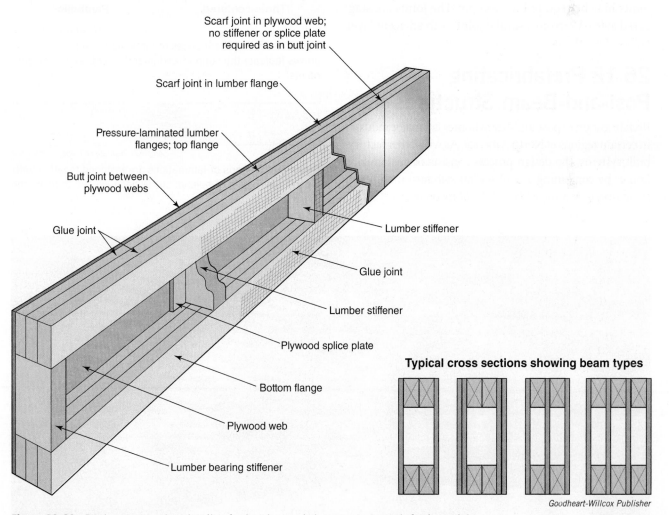

Goodheart-Willcox Publisher

Figure 26-26. Basic construction details of a box beam. It has great strength for its weight.

26.11 Laminated Beams and Arches

Laminated wood beams and arches are available in many shapes and sizes. They offer a great deal of flexibility in building design. In addition to its natural beauty, laminated wood offers strength, safety, economy, and permanence. Most laminated structural members are made of softwoods. They are manufactured and prefinished in industrial plants specializing in such production.

In residential work, beams are usually straight or tapered. In institutional and commercial buildings, however, they are often formed into curves, arches, and other complex shapes, **Figure 26-27**. Some of the basic curved forms are illustrated in **Figure 26-28**.

In the fabrication of beams and arches, lumber is carefully selected and machined to size. To get the required length, pieces must often be end joined. Since end grain is hard to join, a special finger joint may be used, as shown in **Figure 26-29**. A number of these joints may be required in each ply. The joints are staggered at least 2′ from a similar joint in an adjacent layer.

26.12 Prefabricating Post-and-Beam Structures

Prefabrication of post-and-beam homes is a major industry in certain regions of North America. As with other factory-built systems, the design process consists of customizing homes by combining any of several standard modules or structures to suit the wishes of the homeowner.

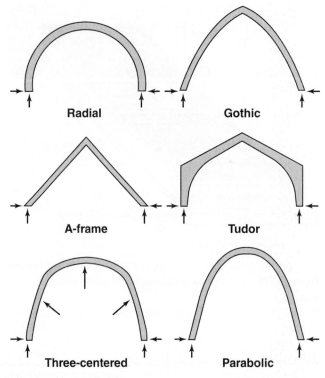

Radial   **Gothic**   **A-frame**   **Tudor**   **Three-centered**   **Parabolic**

Goodheart-Willcox Publisher

Figure 26-28. Different styles of laminated wood arches. The arrows indicate the support and lateral-thrust-reinforcement points.

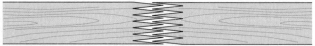

American Institute of Timber Construction

Figure 26-29. Ends of laminations are fastened together with glued finger joints. A special cutter head is used to make the joints.

APA–The Engineered Wood Association

Figure 26-27. Gracefully curved beams of laminated wood support the roof of this hockey rink.

The client reviews standardized structures working with architects and engineers. They work with models, combining them in ways that produce a plan to match their wishes and lifestyle. They also select the type of beams—planed, rough sawn, stained, and so on. See **Figure 26-30**.

Timbers are selected from the factory stock. These are then measured, marked, and cut. Large production machines are used to cut mortises and tenons, **Figure 26-31**.

Wall and roof sections are fabricated into large panels on the factory floor. Rigid insulation is cut and installed in the sections. Windows, exterior siding, and interior wall coverings are also installed, **Figure 26-32**. Completed panels are moved with overhead cranes, wrapped to protect them from weather, and loaded onto trailers for transport to the building site, as shown in **Figure 26-33**. Onsite, builders erect the frame. Then, wall and roof panels are installed, **Figure 26-34**.

Yankee Barn Homes, Inc.

Figure 26-31. Cutting a mortise in a beam using a large saw.

A

A   *Yankee Barn Homes, Inc.*

B   *Yankee Barn Homes, Inc.*

Figure 26-30. Finishing timbers for prefabricated post-and-beam buildings. A—If a client wants a smooth finish, beams can be planed. B—Stains can be applied to meet the client's wishes.

B   *Yankee Barn Homes, Inc.*

Figure 26-32. Factory assembly. A—Trimming and placing rigid insulation in a wall panel. B—Applying vertical wood siding to the exterior of a panelized gable end.

A

Yankee Barn Homes, Inc.

B

Yankee Barn Homes, Inc.

C

Yankee Barn Homes, Inc.

Figure 26-33. Delivering components for a building. A—The finished panel is moved from the production floor to the shipping area. B—Loading roof panels on a flatbed trailer truck. C—Weatherproof wrapping protects the panels in transit.

A

Yankee Barn Homes, Inc.

B

Yankee Barn Homes, Inc.

Figure 26-34. Closing in. A—Once the frame is complete, wall panels are put in place. B—Roof panels complete the assembly.

Chapter Review

Summary

- Because of their larger size and strength, post-and-beam framing members can be spaced farther apart than conventional framing members, increasing design flexibility.
- Post-and-beam construction advantages include the ability to create distinctive architectural effects, labor savings, simplicity of framing around door and window openings, and high fire resistance.
- Foundations may be continuous or may consist of piers that carry the load of each post.
- Beams for floor structures may be solid, glue laminated, or built-up.
- Use solid timbers when beam sizes are small or when a rustic architectural appearance is desired. Where high stress factors demand large sizes or a finished appearance is required, laminated beams are more economical.
- Roof beam systems may be transverse or longitudinal
- Pegged mortise-and-tenon joints are traditional in post-and-beam construction, but metal fasteners are sometimes used where appearance is less important.
- Except for a load-bearing partition under a main ridge beam, it is usually best to install partitions after beams and planks are in place.
- Floors and roof decking are made from 2″ to 4″ thick planks, usually with tongue-and-groove joints.
- SIPs are designed to bridge wide spans. They are often used for exterior walls and the roof on post and beam structures.
- Plywood box beams have a high strength-to-weight ratio and can span distances up to 120′.
- Laminated beams and arches offer many design possibilities.
- Prefabricated post-and-beam buildings are built in a factory, then shipped to the building site for erection.

Know and Understand

Answer the following questions using the information in this chapter.

1. Which of the following are *not* used in post-and-beam construction?
   - A. Posts.
   - B. Beams.
   - C. Studs.
   - D. Planks.

2. In post-and-beam construction, loads are carried by _____ spaced at wide intervals in the walls.
   - A. posts
   - B. beams
   - C. planks
   - D. braces

3. *True or False?* Most limitations of post-and-beam construction can be resolved through careful planning.

4. In general, posts should *not* be less than _____ nominal size.
   - A. 2×2
   - B. 4×4
   - C. 8×8
   - D. 16×16

5. The slenderness ratio of a post compares the total _____ in inches with the smallest _____ dimension.
   - A. area, post
   - B. width, cross-sectional
   - C. length, post
   - D. length, cross-sectional

6. Beams used for floor structures can be _____.
   - A. glue-laminated
   - B. solid
   - C. built-up
   - D. All of the above.

7. _____ are more economical to use where high stress factors demand large sizes or a finished appearance is required.
   - A. Solid timbers
   - B. Laminated beams
   - C. Sills
   - D. Built-up beams

8. *True or False?* Transverse beams have greater spans and carry heavier loads than longitudinal beams.

9. Traditionally, _____ joints have been used with wood pegs as fasteners in post-and-beam framing.
   A. butt
   B. mortise-and-tenon
   C. slip
   D. tongue-and-groove

10. Partitions running perpendicular to a sloping ceiling should have regular _____ with filler sections installed between the beams.
    A. top plates   C. dividers
    B. studs   D. planks

11. When planks are _____, the joints do not need to meet over beams.
    A. mortise-and-tenon designs
    B. one continuous span
    C. end matched
    D. insulated

12. *True or False?* SIPs will not support structural loads.

13. SIPs resist loads imposed on a roof through _____.
    A. framing lumber between panels
    B. compression/stretching resistance of the skins
    C. the foam core
    D. the glue bond

14. Box beams can span distances up to _____′.
    A. 12   C. 120
    B. 60   D. 200

15. Laminated wood beams are available in many _____.
    A. shapes and sizes
    B. customizable patterns
    C. hardwood options
    D. All the above.

16. *True or False?* Factory-built post-and-beam buildings offer the owner little choice in customizing their home.

Apply and Analyze

1. List three advantages of post-and-beam construction.

2. Describe the two types of beam-supported roof systems.

3. Why is it best to ensure that the moisture content of roof planks closely corresponds to the EMC of the interior structure?

4. Why are SIPs a popular choice for exterior walls and roofs of post-and-beam buildings?

5. Explain how a basic plywood box beam is used to offer support to a structure.

Critical Thinking

1. Research options for becoming an experienced timber framer in your geographic area (classes offered, associations, etc.). If you were to decide that this is a direction you want to take your career, what would be a viable option for the training needed to become an accomplished timber framer?

2. Imagine you are a contractor who was approached by a client interested in building a timber-framed home. How would you advise them?

3. Using what you have learned about timber-frame structures, what construction challenges might you face building a timber-framed structure over a conventional one?

Communicating about Carpentry

1. **Speaking and Listening.** Working in small groups, create a poster that illustrates the difference between post-and-beam framing and conventional framing. As you work with your group, discuss the meaning of each term. Afterward, display your posters in the classroom as a convenient reference aid for discussions and assignments.

2. **Speaking and Reading.** Research the products and services available for post-and-beam construction. Collect promotional materials from product manufacturers. Analyze the data in these materials based on the knowledge gained from this lesson. With your group, review the list for words that can be used in the subject area of post-and-beam construction. Practice pronouncing the word, and discuss its meaning. As a fun challenge, work together to compose a creative narrative using as many words as you can from your new list.

3. **Listening and Speaking.** In small groups, review the key terms listed at the beginning of the chapter. For each term, discuss the meaning of term and describe the term in simple, everyday language. Record your group's initial description, and then make suggestions to improve your description. Compare your descriptions with those of the other groups in a classroom discussion.

4. **Designing.** Using balsa wood or other material your instructor suggests and hot-melt glue or small wire nails, build a mock-up of a simple, one-story, post-and-beam house or barn. Use a scale of 1/8″=1′. Use a piece of 1/4″ plywood as a base to start your model. SIPs can be simulated with corrugated cardboard. Discuss and compare your mock-up with others in your group.

5. **Building.** Build a mockup of a SIP in which you can experiment with the design and size of the various parts. Use a scale of about 3″=1′. Skins can be made of 1/8″ hardboard. Share your design sketches with your instructor.

6. **Mathematics.** Working with one other person, refer to the following drawing. Estimate the weight per square foot of the 3″ × 6″ floor planks. This is the dead load on a floor. Multiply this by the total square feet of floor area to find the total dead load on the floor. Add your estimate of the weight of the furniture that might be found in a bedroom (queen-size bed, dresser, and chest of drawers). Add the weight of two occupants and a visitor (175 lb each) to the weight of the furniture. This is the live load on the floor. How much weight must be supported by each mortise-and-tenon joint in the drawing?

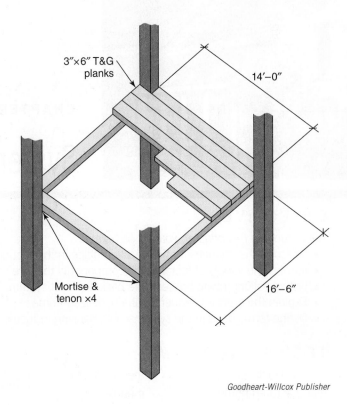

3″×6″ T&G planks

14′–0″

Mortise & tenon ×4

16′–6″

Goodheart-Willcox Publisher

CHAPTER **27**

Systems-Built Housing

OBJECTIVES

After studying this chapter, you will be able to:

- Describe the changes that have taken place in the technology of systems-built housing.
- Identify the variety of factory-built components that are utilized in a systems-built home.
- List and differentiate between the basic types of systems-built structures.
- Explain the erection sequence of a panelized home.
- Define terms used in the systems-built housing industry.

TECHNICAL TERMS

closed panel	modular home	precut home
manufactured home	module	systems-built house
mechanical core	panelized home	

A *systems-built house* refers to a house built from components assembled in a factory following precise design specifications. At one time, the term prefabrication was used to describe such construction. Either term indicates a process in which parts are cut and assembled into sections, modules, or entire homes in factories. Then, the units are shipped to the building site for final assembly and erection. Once erected, a systems-built house cannot easily be distinguished from a conventionally constructed (or stick-built) home, **Figure 27-1**.

Systems- or factory-built housing also includes *modules*. These are three-dimensional units that are fully assembled before they leave the manufacturing plant. Manufacturers of these modular structures also build commercial buildings such as banks, schools, office buildings, motels, and hotels, **Figure 27-2**.

At one time, prefabricated or factory-built housing limited the buyer to a few styles and plans offered by the manufacturer. All that has changed, partly because computer-assisted design can quickly adapt architectural drawings to the buyer's needs and lifestyle. Thus,

architects and engineers can quickly produce any style of home and easily incorporate custom designs. Systems-built homes include geodesic domes and log homes, as well as traditional and classic styles.

To make quality components, many tasks must be carried out under controlled assembly-floor conditions using special tools. Overhead cranes lift heavy assemblies and power tools are used for fabrication. SIPs and plywood box beams require accurate glue applications, special presses, and handling equipment. Often, the simplest prefabrication process in modern plants is done on large production equipment. The machine reduces the amount of human energy required and increases production. Efficiency is increased both in the plant and on the building site.

Green Note

Selecting materials, cutting, and assembly are all done with the aid of computers, so there is less waste and fewer trips to the site. This reduces waste in local landfills and emissions from excessive truck traffic.

Artazum/Shutterstock.com

Figure 27-1. A systems-built home cannot be distinguished from a stick-built home. This home was erected from factory-built modules.

Modular Genius

Figure 27-2. This school building was built from modules assembled in a factory.

Designs are completed with the assistance of computers. During the manufacturing process, the components move from one workstation to another where all building trade activities are represented. A quality control process ensures that the work is in conformity with state and local building codes.

Pro Tip
Modular construction is efficient. A module can normally be built in 14 working days and erected in one or two days.

27.1 Factory-Built Components

In home construction today, many factory-built components are used in place of jobsite finish work and stick-built framing work. Some of these components are windows, door units, soffit systems, stairs, and built-in cabinetwork.

One major framing component is the roof truss, **Figure 27-3**. These are used in all types of residential and light commercial construction. Manufacturing plants can usually furnish any type of trusses, including matched units for a complex roof. Roof trusses must be carefully designed by structural engineers and built to exact specifications, **Figure 27-4**. High-production saws are used to cut truss members to length. Special presses and jig tables are used to fasten the assembly with gang-nail connectors.

A

SR Sloan, Inc.

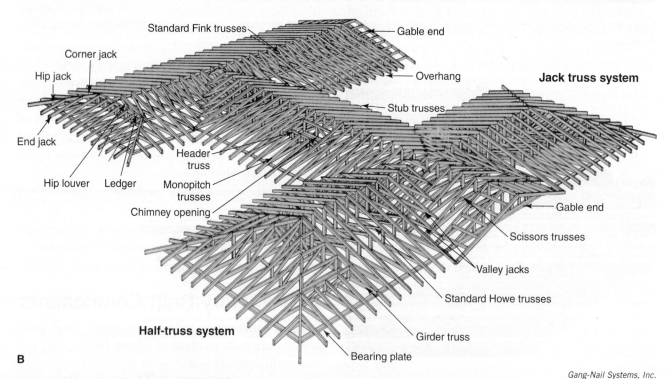

B

Gang-Nail Systems, Inc.

Figure 27-3. Factory-built truss rafters are used in all types of residential and light commercial structures. A—Trusses are laid out on a flat surface, then a roller presses the metal truss plates into the wood members. B—Prefabricated roof framing includes all the types of roof trusses used to form hips, valleys, overhangs, and gable ends.

Goodheart-Willcox Publisher

Figure 27-4. Green alignment lasers help assure precision assembly on this truss jug before the attachment plates are pressed together.

BuildSMART, LLC

Figure 27-6. Workers install a factory-assembled panelized wall.

The floor truss is another example of a prefabricated component, **Figure 27-5**. It permits a wide, unsupported span and uses a minimum of material. It also has openings for heating and air conditioning ducts, electrical circuitry, and plumbing lines.

Wall and roof sections or panels are made in various shapes and sizes. Panels provide structural strength in addition to forming inside and outside surfaces. Some panel systems are constructed of conventional framing with 2″ lumber, standard insulation, and coverings. **Figure 27-6** shows a wall panel being installed onsite.

Another type of panel is a SIP. These are described in Chapter 11, **Wall and Ceiling Framing** and in Chapter 26, **Post-and-Beam Construction**. They can be made to various heights and widths, depending on the design of the building. SIPs made with a rigid insulation core provide enough strength for walls and partitions. See **Figure 27-7**. These lightweight panels have oriented strand board (OSB) outer and inner skins. They have openings prebored through which wiring may be passed. Interior and exterior finish coverings may be attached directly to the OSB skin.

arturasker/Shutterstock.com

Figure 27-5. This floor truss system provides ample space for installation of plumbing lines.

Enercept, Inc.

Figure 27-7. This building is being constructed with SIPs for the floors and walls.

A building in various stages of completion can be prepared in a factory, then shipped as a "package" to the building site. Most such buildings are single-family homes. However, small commercial buildings, farm structures, and multi-family homes are also produced and moved this way.

27.2 Types of Systems-Built Homes

The Building Systems Councils include manufacturers in several categories of systems-built buildings. The Councils group systems-built housing in one of four categories:

- Concrete homes
- Modular homes
- Panelized homes
- Log homes

Other organizations have additional categories. The North Carolina Manufactured Housing Institute adds two other categories:

- Precut homes
- Manufactured homes (formerly called mobile homes)

Regardless of the type of housing chosen by the owner, all offer rapid construction, quality materials, lower building costs, and customizing.

27.3 Concrete Homes

All builders have used concrete for decades in such things as foundations, driveways, and walks. Concrete is becoming an increasingly popular building material for the basic structure of houses, including walls and even roofs. Concrete homes can be more stable than wood-framed houses in areas that are subjected to violent weather and earthquakes. Concrete absorbs heat from the sun and radiates it slowly into the house, making it attractive for passive-solar heating. The continuous insulation barrier that concrete offers reduces air infiltration and drafts. Concrete's durability is also a plus for green home credits. Many insurance companies offer lower premiums for fire-resistant concrete homes.

27.4 Modular Homes

According to the Modular Home Builders Association (MHBA), approximately 13,200 modular homes were sold in the United States during 2013. *Modular homes* are made up of two or more three-dimensional units. These units are called modules, sometimes referred to as mods. The modules are produced on a factory assembly line, **Figure 27-8**. Many modular homes are single-story.

Figure 27-8. Modular homes are constructed in a controlled-environment factory. A—The floor frame is constructed on a raised platform for convenience. B—As soon as the walls and interior partitions are in place, wiring can be installed and drywall can be hung. C—The ceiling frame is about to be lifted into place with a crane. D—The roof covering is being applied. Notice the safety nets and safety chains. E—The modules are being wrapped for transportation to the site. F—A completed home may be two or three stories.

By stacking modules on each other, however, two- and three-story homes are possible. Modules also may be smaller units, such as a bathroom or a kitchen, **Figure 27-9**.

Each module is 12′ or 14′ wide and may be up to 60′ long. Such units are nearly completed at the factory. Some include cabinets and plumbing fixtures. At the building site, crews set the module on a foundation with a crane, **Figure 27-10**. Many manufacturers of modules have their own setup crews. When the module is in position, it is fastened with bolts.

Wausau Homes, Inc.

Figure 27-9. Some modules are small, consisting of only one room. Such modules usually include all plumbing, wiring, and fixtures.

Innovative Building Systems

Figure 27-10. A modular home is being placed on its foundation with a crane. The siding will be applied to this home after it is set.

In this type of prefabrication, entire sections (modules) of the structure are built and finished in manufacturing plants. The sections or modules are then hauled to the site where they are assembled. The width of a module seldom exceeds 14′. Trucks and roads cannot handle wider units.

An advantage of modular construction is that nearly all of the detailed finish work can be done at the factory. Kitchen cabinets can be attached to the walls and other built-in features can be installed. Also, wall, floor, and ceiling surfaces can be applied and finished. Electrical wiring chases, heating and air conditioning ducts, plumbing lines, and even plumbing fixtures can be installed under close control in manufacturing plants using labor-saving tools. A section that has a group of plumbing and heating facilities and includes most of the utility hookups is called a *mechanical core*. It is often included with other sections that consist mainly of panels.

Mechanical cores group the kitchen, bath, and utilities in one unit that requires only three connections at the site. A core with a bathroom and kitchen on opposite sides of the same wall is typical. Some units are designed to include heating and air conditioning equipment, as well as electrical and plumbing equipment.

27.5 Panelized Homes

While most homes today have some prebuilt parts, such as roof trusses and engineered lumber, *panelized homes* leave the factory as a series of floor, wall, and roof panels. Exterior walls typically include the framing, sheathing, vapor barrier, siding, windows, and exterior doors installed. In some cases, the exterior walls may be SIPs. Interior walls are usually framed with openings and headers in place. When prefabricated panels are finished on both the inside and outside surfaces, they are called *closed panels*. "Closed" means that no additional material must be installed once the panels are erected. However, drywall and mechanical systems may be added later. Most panelizers cut door and window openings, hang windows, and predrill studs for electrical wiring and plumbing runs. Channels may be formed in rigid insulation for running electrical wiring. A few producers make panelized ceilings and floors in the same way.

Panels arrive on the building site numbered according to their location in the house. Erection of the building can begin as the panels are unloaded from the truck.

In panelized prefabrication, flat sections of the structure are built on assembly lines. Large woodworking machines cut framing members to the desired length and angle. Parts are stored and delivered to the assembly stations as needed. Wall and floor frames are formed by placing the various members in positioning jigs on the production line. The parts are then fastened with pneumatic nailers, **Figure 27-11**. Electrical wiring or other mechanical facilities may be installed while the frame is being built.

As the completed frames move along conveyor lines, wall surface materials are placed in position and nailed. Nailing is done with powered gang nailers or a computerized track nailer, **Figure 27-12**. For some jobs, this may be the final step before panels are sent to the site for delivery. For completely panelized homes, wall panels are developed further with mechanicals, such as electrical wiring, insulation, windows, doors, siding, and even gypsum on interior surfaces.

Goodheart-Willcox Publisher

Figure 27-11. Wall panels are assembled by workers on a jig, then moved on to a sheathing station.

Goodheart-Willcox Publisher

Figure 27-12. An operator uses this automated pneumatic nailer to nail sheathing with accuracy and precise spacing. The nailer rides along the track in a wheeled carriage. The operator uses a laser to align the track with a wall stud and computer software assigns the spacing between each nail.

With all panelized homes, a good deal of construction will still need to be accomplished on site. In other areas of the plant, roof units are prepared. For post-and-beam structures, panels that form both the roof and ceiling surfaces are common. To avoid painting in high places after erection, the ceiling side is painted at the factory.

Although most panelized prefabricated houses use a first floor deck built by conventional methods, some manufacturers design and build floor panels. Full-length joists are assembled with headers. Long plywood sheets are formed using scarf joints. These are then glued in place. The resulting stressed-skin construction is rigid and strong. Use of these panels prevents nail pops and squeaks.

As the panels near the end of the production line, they receive a final inspection. Each panel is marked with a number for easy assembly, then is moved to storage or loaded onto a truck. When all materials and millwork are added to complete the "package," it goes to the building site.

27.6 Precut Homes

For *precut homes*, lumber is cut, shaped, and labeled. Then, it is shipped to the jobsite. This reduces labor and saves time on the building site. Manufacturers of this type of house include materials needed to form the outside and inside surfaces. Also shipped are such millwork items as windows, doors, stairs, and cabinets. Optional items include electrical, plumbing, and heating equipment. Kitchen cabinets and other built-in units are usually made in plants specializing in these items. The units may be shipped either to the home fabricator or directly to the building site.

> **Pro Tip**
> At one time, the terms *manufactured* and *industrialized* were used when referring to prefabricated housing.

27.7 Log Homes

Log homes are essentially a type of precut home. Walls of log homes are built by stacking precut and machined logs on top of each other, rather than by framing with studs. This is the only difference between a log home and conventional stick-built homes or homes that are of modular or panelized construction.

Logs are combined in the factory with other modern building materials to produce any style of home for the buyer. They can be customized as the buyers wish. High-speed machines mill the logs to uniform shapes and lengths. At the same time, the logs receive the tongues, grooves, notches, and splines that hold them together.

27.8 Onsite Building Erection

Preparation for onsite building erection begins at the factory. Workers load the prefabricated units onto semitrailer trucks. Each unit is carefully marked and placed on the truck so it can be removed in the order it is to be assembled onsite, **Figure 27-13**.

At the building site, the owner's contractor prepares a foundation for the house. It is built according to the dimensions given on the foundation plan supplied by the building's manufacturer.

Many systems-built manufacturers send along their own setup crews as an efficiency measure. Usually the crews and trucks arrive at the site early in the morning.

A — *Goodheart-Willcox Publisher*

B — *Goodheart-Willcox Publisher*

C — *NextGen Building components–Christa Construction LLC*

Figure 27-13. Onsite building erection. A—On the home site, a crane lifts prefabricated panels in the order they are assembled. This speeds up the erection process. B—For larger homes, some factories send their own setup crews. After setup, the customer's contractor takes over. This crew completed closing in of the structure in a single day. C—Panelization allows for large projects, such as this apartment complex, to be built in a controlled environment off-site, saving valuable space on-site.

Since prefabricated units are heavy, a crane is used to lift the units from the trailers and set them in place on the foundation.

The floor deck is built in the standard way or is assembled from panels. Then, mechanical core units are set in place, walls and partitions are joined, ceiling-roof units are installed, and roof panels close the structure. **Figure 27-14** shows the construction of a systems-built home. The shell of a home can be finished in one day.

One disadvantage of prefabrication is soon discovered at the building site: weather may prevent work. The builder must choose a day without wind or rain. Wind makes it hard to unload and position large panels. Rain will damage inside wall materials applied at the factory. Rain may also cause the problem of soft ground. It limits the size of the crane that can be used. Unless the reach of the crane is great, there is no way to safely bring large units into wet areas.

In many ways, however, the weather is less important for systems-built construction than for conventional construction. A skilled crew can erect and enclose a systems-built single-family home in less than one day. A few weeks are usually required to complete the wiring, heating, plumbing, and decorating. With systems-built structures, the inside work begins on dry structures.

A   *Wausau Homes, Inc.*

B   *Wausau Homes, Inc.*

C   *Wausau Homes, Inc.*

D   *Wausau Homes, Inc.*

Figure 27-14. This sequence shows onsite erection of a prefabricated home. A—Floor panels have been laid onto the foundation and a mechanical core for the kitchen is being lowered into place. Note the bath module already in place. B—The front wall panel for the living room is being installed. An outside surface has not been applied since this section of the house will be finished with brick veneer. C—A crane lowers ceiling-roof panels into place over the bedroom area. The combined unit is hinged and closed for shipment. The unit is opened for fitting at the building site. D—Installing the final roof panel on the garage. No ceiling panels were required. Note that carpenters have started shingle work over the bedroom area. This factory-built home was enclosed in less than a day.

In conventional construction, the inside often gets wet before the building is closed in. Therefore, the finish work goes faster on a systems-built home. For this and other reasons, prefabrication competes well with other building methods.

27.9 Assembling a Panelized Home

Typically, a panelized structure is assembled onsite. Often a crane is used to lift and move the panels, **Figure 27-15**. The manufacturer of the panelized home will provide detailed instructions for joining the parts. These instructions will specify fastener types and sizes, adhesives and where to use them, and often the sequence in which to assemble the home. In multistory buildings, the instructions may allow the second-floor joists to be hung on joist hangers, or they may rest on top of the first-floor wall, **Figure 27-16**.

Ceiling panels can be attached to walls and ridge beams in two ways. **Figure 27-17** shows metal clips or ties being used as a means of securing the panels. Also shown is an attachment with roof spikes. These

Goodheart-Willcox Publisher

Figure 27-15. Complete wall panels are heavy, so they are set on the sill with the help of a crane.

are thin-shank, threaded nails 1 1/2″–2″ longer than the thickness of the panels. Asphalt or fiberglass shingles can be applied directly over structural roof panels. When wood, tile, or slate shingles are used, it is customary to first apply horizontal wood battens.

The following assembly procedure is for SIPs. These steps are typical of this type of construction.

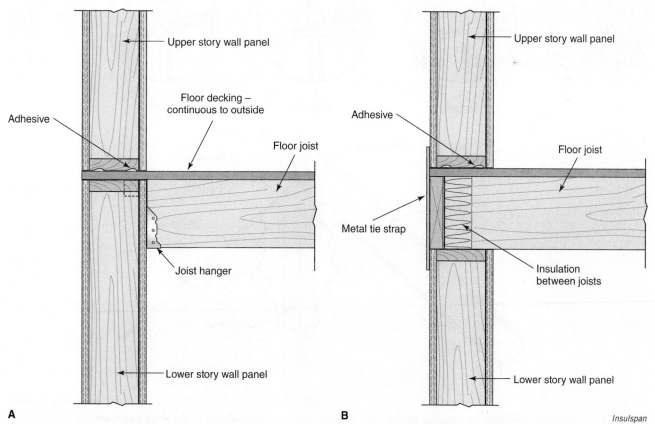

A

B

Insulspan

Figure 27-16. Methods vary for securing and supporting second-floor joists. A—Using joist hangers secured to top of plate. B—Resting second-floor joists on top of the wall plate.

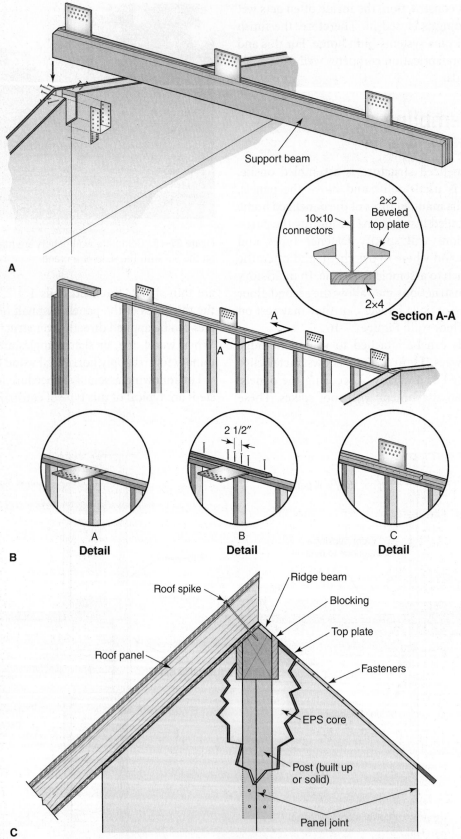

Enercept, Inc.; Insulspan

Figure 27-17. Alternate methods for securing roof panels. A—Steel connectors attached to the ridge beam secure the tops of roof panels. B—Clips attached to the wall plate secure the bottoms of roof panels. C—End view of a roof panel, ridge beam, and built-up king post. The beam must be solidly attached to the king post.

PROCEDURE

Erecting SIPs

1. Lay down a strip of polypropylene sill sealer along the outer edge of the subfloor.
2. Install 2×6 plates all around the perimeter with the outside edge extending 1/2″ beyond the edge of the subfloor, as shown in **Figure 27-18**.
3. Exterior wall panels are numbered in the order of their placement. Install them over the plate, **Figure 27-19**. Check each panel for vertical electrical wiring channels and drill a hole through the plate and subflooring in these locations.
4. Lay down a wavy line of caulk on the plate and vertical edges of each panel.
5. Carefully tip each panel into place. Avoid disturbing the caulk.
6. Plumb each panel and nail panels to the plate and to each other (on each side). Use 8d nails spaced 6″ apart.
7. Slide lintel panels into position, but do not immediately nail. Door and window panels are shown in **Figure 27-20**. Window rough openings may be cut in the field of a panel or have a lower panel and a lintel (upper) panel. Door openings have only a lintel panel.
8. Make height adjustments to lintels and then nail them in place.
9. If there is a king post on gable ends, toenail it to the plate at the centerline of the building, **Figure 27-21**.
10. Mark the panels that have vertical wiring cores. Place the mark on the inside panel.
11. Double up the wall plate. Stagger-nail the plates every 6″.
12. Drill through both plates into the wiring cores.

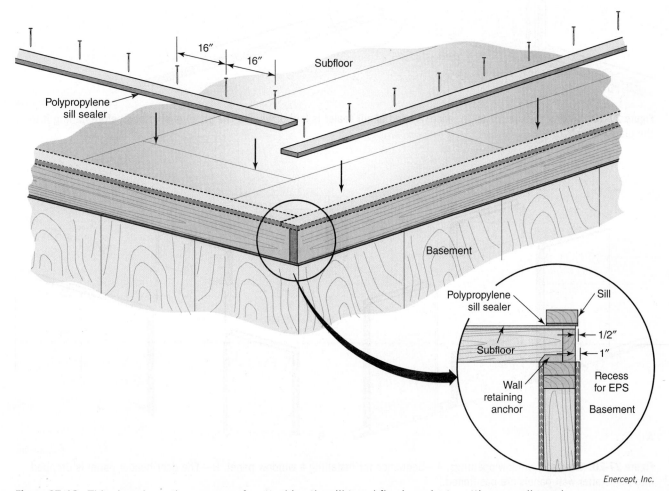

Enercept, Inc.

Figure 27-18. This view shows the sequence for attaching the sill to subflooring prior to setting up wall panels.

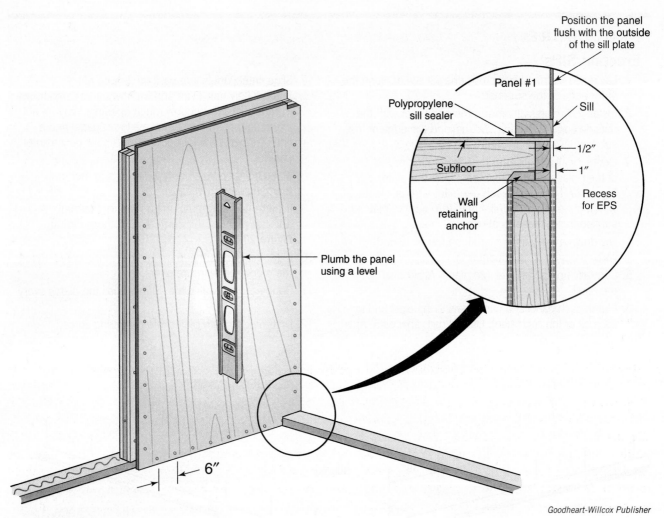

Position the panel flush with the outside of the sill plate

Panel #1

Polypropylene sill sealer

Sill

Subfloor

1/2"

1"

Wall retaining anchor

Recess for EPS

Plumb the panel using a level

6"

Goodheart-Willcox Publisher

Figure 27-19. Setting the first panel in place. A bead of sill sealer is laid down and the panel is plumbed before nailing it to the plate.

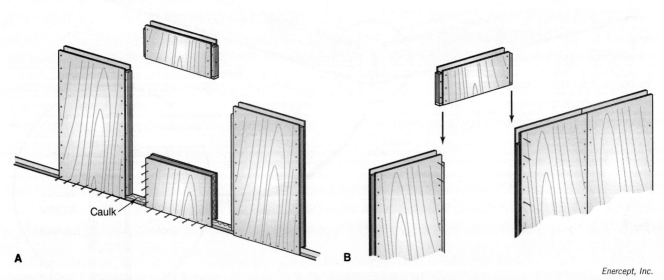

Caulk

A

B

Enercept, Inc.

Figure 27-20. Door and window openings. A—Sequence for installing a window panel. B—The door header panel is dropped into place after wall panels are positioned.

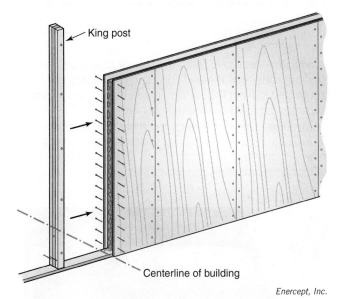

Enercept, Inc.

Figure 27-21. A king post may be located at the center of the wall directly under the gable. It must align with the center of the building. The post must be toenailed to the plate with 16d nails.

27.10 Manufactured Homes

The term *manufactured home* in the United States is defined and codified by federal law (Code of Federal Regulations, 24 CFR 3280): "Manufactured homes are built as dwelling units of at least 320 square feet (30 m²) in size with a permanent chassis to assure the initial and continued transportability of the home." The requirement to have a wheeled chassis permanently attached differentiates *manufactured housing* from other types of prefabricated homes, such as modular homes. The most common width is between 12′–14′, but two units may

be attached to result in a home up to 28′ wide. Lengths may be greater than 68′. The variety of choices available to the buyer has grown. What used to be a market of metal-clad, single-wide units now includes multisectional units with brick siding and marble countertops. The average size of units has grown to over 1400 sq ft.

There have also been significant changes in terms of public image. At one time, homes that were trucked to a site on their own wheels were called "mobile" homes or "trailer" homes. However, a survey done in the 1990s reported that 98% of such homes were never moved from their original site. To emphasize this and improve public image, the industry has replaced the term mobile home with *manufactured home*.

Production methods used in building manufactured homes are like those used for standard factory-built housing. Some of the structures, however, are different. For example, the floor must be one rigid unit. It usually consists of a wood frame attached to a welded steel chassis.

Designs sometimes include an expandable feature that permits a wider living room. A slide-out section is carried inside while the home is being transported. It is then expanded to the side after the home is on the site.

A manufactured home is finished and fully equipped at the factory. It is then pulled to the dealer or directly to the home site. Since manufactured homes are mounted on a chassis, they do not need a permanent foundation, although some are placed on foundations. Utility hook-ups are ready to be made at the site. Interior furnishings include all major appliances, carpeting, drapes, furniture, and lamps. **Figure 27-22** shows a new double-wide manufactured home being installed.

A   *Mark William Richardson/Shutterstock.com*   **B**   *Goodheart-Willcox Publisher*

Figure 27-22. A double-wide manufactured home is being moved onto a site. A—The two sections are separately towed in and then positioned on the lot ready to be joined. B—A carpenter prepares to move the units together. The tongue jack rests on a roller that rides in a U-channel. A come-along winch operated by a carpenter or a helper slides the unit up against the mating unit.

Construction Careers
Equipment Operator/Operating Engineer

Operators of heavy equipment on construction sites are responsible for such tasks as excavation and the movement of building materials around the site. They operate equipment such as bulldozers, graders, lifts, loaders, excavators, and cranes. Persons who are qualified to operate a number of different kinds of heavy equipment are classified as operating engineers.

Heavy equipment is typically controlled by using such devices as levers, switches, and dials. More complex equipment makes use of computerized controls. In addition to operating equipment, operators and operating engineers are often responsible for setting up and inspecting equipment, maintaining and adjusting machines, and sometimes performing minor repairs.

Most operators learn through on-the-job training, but formal training is available through apprenticeship programs and private technical schools. Apprenticeship programs for operating engineers are three years in length and involve 1600 hours of jobsite training and more than 400 hours of classroom work. Employers generally prefer to hire high school graduates. Experience in large vehicle operation, such as trucks, tractors, or military vehicles, is helpful in obtaining a job in this field.

The demand for operators and operating engineers is strong and expected to remain so well into the future. Continuing demand for housing and other structures, as well as highway and bridge construction and repair, should generate a large number of jobs in this field.

Almost one-half of a million workers are employed as equipment operators or operating engineers. About three-fourths of them work in building construction and related fields. Most of the remaining jobs are in highway and street maintenance (paving and surfacing).

Working conditions can be harsh, since construction equipment operators are exposed to all types of weather conditions—heat, cold, rain, wind, snow, and sleet. Working conditions are also noisy and many types of equipment subject their operators to vibration and jarring impacts as they run over rough ground.

Pay for operating engineers is comparable to other skilled members of the building trades. Annual earnings, however, may be reduced by severe weather conditions that prevent operating equipment on the jobsite.

michaeljung/Shutterstock

Heavy equipment operators, also known as operating engineers, move and excavate material around a jobsite that is too heavy to be done by hand. Operators must be careful when using the equipment. Proper communication with other members of the crew is essential to avoid potential injury. Operating engineers must also maintain proper care of the equipment when it is not in use. The work of an operating engineer sometimes depends on being able to excavate onsite. Some operating engineers' work is seasonal, with little work during the winter.

Chapter Review

Summary

- Houses built from components assembled in a factory are known as systems-built housing. Some manufacturers also assemble complete modules that are then attached to each other to form the finished structure.
- Factory-built components include roof and floor trusses and wall and roof panels.
- Complete "packages" are shipped to the building site for erection.
- Modular houses are built as two or more large finished sections that are then fastened together onsite.
- Panelized homes leave the factory as a series of floor, wall, and roof panels.
- Precut homes are delivered with all lumber cut to size and labeled.
- Conventional stick-built homes and log homes are both offered in precut form.
- Manufactured homes are complete structures built on a chassis equipped with wheels.

Know and Understand

Answer the following questions using the information in this chapter.

1. _____ are three-dimensional units that are fully assembled before leaving the manufacturing plant.
   A. SIPs
   B. Prefabricated houses
   C. Modules
   D. Panelized houses

2. *True or False?* SIPs made with a rigid insulation core, OSB, and inner skins provide enough strength for walls and partitions.

3. *True or False?* The Building Systems Councils categorizes precut and manufactured homes as systems-built homes.

4. An advantage of modular construction is that nearly all of the detailed finish work can be done _____.
   A. at the factory
   B. on the jobsite
   C. after the module is placed
   D. in transit to the jobsite

5. A structural unit that contains plumbing, heating, and electrical systems is called a(n) _____ core.
   A. modular   C. panelized
   B. mechanical   D. insulated

6. In _____ prefabrication, flat sections of the structure are built on assembly lines.
   A. precut   C. panelized
   B. modular   D. concrete

7. *True or False?* A log home is a factory-built structure of the panelized type.

8. At the building site, the foundation for the house is built according to the dimensions given on the foundation plan supplied by the _____.
   A. carpenter
   B. architect
   C. building's manufacturer
   D. building's owner

9. *True or False?* Prefabricated construction is not affected by weather conditions during erection.

10. The time it takes a skilled crew to erect and enclose a systems-built single-family home is roughly _____.
    A. less than one day   C. one week
    B. two days   D. two weeks

11. The term "mobile home" has long since been replaced with _____ home.
    A. modular
    B. prefabricated
    C. manufactured
    D. factory-built

Apply and Analyze

1. What innovation made it possible to offer great variety and customization to systems-built homes?
2. Which major factory-built framing component is used in all types of residential and light commercial construction?
3. What is meant by "closed" in a closed panel?
4. How is a panelized home typically assembled?
5. What is required of a manufactured home that differentiates it from a modular home?

Critical Thinking

1. Where do you see advantages of systems-built housing? Who benefits from these advantages?
2. Where do you see disadvantages of systems-built housing? Who suffers the consequences of these disadvantages?
3. As a carpenter speaking to someone considering building a new home, how would you counsel them with regards to modular, panelized, or traditionally framed homes?

Communicating about Carpentry

1. **Speaking.** Working in small groups, create flash cards for the key terms in this chapter. Each student chooses some terms and makes flash cards for those terms. On the front of the card, write the term. On the back of the card, write the pronunciation and a brief definition. Use your textbook and a dictionary for guidance. Then take turns quizzing one another on the pronunciations and definitions of the key terms.

2. **Listening.** In small groups, discuss with your classmates—in basic, everyday language—the different types of systems-built housing. Conduct this discussion as though you had never read this chapter. Review the points discussed, factoring in your new knowledge of systems-built housing. Develop a summary of what you have learned about systems-built housing and present it to the class, using the terms that you have learned in this chapter.

CHAPTER 28

Green Building and Certification Programs

OBJECTIVES

After studying this chapter, you will be able to:

- Name the building codes that impact the construction industry with regard to green building.
- Describe green certification programs and their differences.
- List the essential members of a green home building team.
- Explain why site selection and lot planning are important to a green home.
- List several factors to be considered in selection of green home products.
- Describe energy consumption considerations when making a green home plan.
- Identify sources of indoor pollution.
- Explain what is meant by best construction practices.

TECHNICAL TERMS

DOE Zero Energy Ready
   Home Program
ENERGY STAR
gray water
green certification program
green rater
green verifier

indoor air quality
International Energy
   Conservation Code(IECC)
International Green
   Construction Code (IgCC)
Leadership in Energy and
   Environmental Design
   (LEED®)

LEED® for Homes Provider
National Green Building
   Standard (NGBS)
passive house
radon renewable energy
WaterSense

Green building is a building system in which all building parts work together to lessen the impact on the environment, reduce energy consumption, and extend the usable life of the building with minimum maintenance. Green building must achieve these benchmarks while maintaining a comfortable and healthy living environment for a building's occupants. Standards are met by using a holistic approach to building design and construction. Simply using bamboo flooring or photovoltaic solar collectors on the roof does not make a home green.

Building a green home requires that the design team, the construction team, and the owner all work together from the beginning. The design of the home must be comfortable for the owner and it must make the best use of the resources, such as energy and water, that the home consumes. At the center of green building is energy conservation. Besides cost savings to the owner in

the way of lower utility bills, conservation brings the added benefits of slowing the use of finite natural resources and reducing atmospheric pollution.

28.1 Green Building Considerations

Opportunities to incorporate green building principles are vast, ranging from low-tech, common sense methods, such as adding more insulation, to high-tech methods, like utilizing occupancy sensors and smart thermostats to control electronics and HVAC equipment. This type of construction is not limited to the building itself. Green building should include a visionary approach, with an eye toward how the building fits into the surrounding environment, whether that environment is a pristine, secluded woodland lot, a suburban housing development, or an urban setting.

Copyright Goodheart-Willcox Co., Inc.

715

The approach should also consider how the project interacts within the social context. For instance, if a developer is planning a new housing development:

- In what ways can the natural ecology of the site be preserved or restored?
- Does the layout of the streets and sidewalks promote efficient and safe travel?
- Does the design of the development contribute to an increase or decrease in vehicular traffic?
- Does the design encourage the best use of green space, pedestrian traffic, and an active lifestyle?
- Are there charging accommodations to promote the use of alternative forms of transportation such as electric cars, public transport, and bicycles?
- Have considerations been made regarding light and sound pollution?

Paying attention to questions such as these in the planning stage will affect the quality of life for the entire community. Projects that are built with sustainability in mind should not only save natural resources but should enhance the quality of life for the residents. Encouraging transportation methods such as walking or biking promotes a healthy lifestyle while reducing carbon emissions, noise, and light pollution. Easy access to public transportation and carpooling accommodations also play a role.

As you can see, a great deal of thought should go into the overall vision of a green building project. If a builder is only constructing one building, the scope of these factors is reduced. However, in a broader context, the environmental benefits can be intensified if these principles are respected. Key considerations that should be given to any green building project include:

- Overall site development
- Controlling water runoff during and after construction
- Water use and efficiency for the home
- Energy use and efficiency for the home
- Energy sources and generation
- Material selection, sourcing, use, and waste disposal
- Indoor environmental control and air quality

28.2 Development of the Green Home Market

In the early 1970s, the world experienced a severe decrease in the availability of petroleum products. This caused a sharp increase in the cost of electricity and home heating fuels. The building construction industry responded by developing ways to conserve the energy that buildings used. That was the birth of green home building.

Code Note

Green building practices often lead to updates and changes in building codes. Research and development in the green building industry can prove to be beneficial to the health and safety of a home, as well as good for the environment. The inverse can also be true. The failure of a green building practice can reveal where a system is weak and in need of improvement. For instance, our understanding of condensation inside wall and roof systems has been improved because of experimentation with green building practices. As a result, building codes have been modified.

Since that time, we have learned that not only can we find ways to use less energy, but we can save water, consume fewer natural resources in the production of building materials, and make homes safer, more comfortable, and more durable. Using the methods and materials that have been developed over the years, a green home need not be more expensive than a non-green home. What makes it green is early planning, efficient use of the right materials, quality workmanship, and educating the homeowner or renter. While construction costs may be slightly higher in a green home, the savings in energy, waste reduction, and durability more than make up the difference in the long term. Today, the green home market is a big part of the total market for new homes and it is growing at an accelerating pace.

28.3 Green Building Codes, Standards, and Programs

In order to build a green home, we must first have an understanding of what defines green building and what sets the standard for the industry. What is the difference between an ordinary home and a green home? What features make a green home *green*? One way to differentiate a green home from an ordinary home is through a green certification program. A *green certification program* is a set of standards for green building practices that are followed by the building team and verified by an independent third party. Program standards address many of the parts of the green building system described earlier, such as energy efficiency, water conservation, and homeowner education. When a building team participates in a green home certification program, they generally receive a rating to indicate the degree to which green principles have been followed. The greener the house is, the higher the rating. There are a variety of green building certification programs around the United States.

Some are locally based and cover just a single city or town, while others are nationwide programs open to any homebuilder. Many of these programs have some overlap and others build from a lesser standard, adding additional requirements in order to meet certification requirements. It is important for owners, designers, and builders to understand the goals of a project before choosing which program(s) will be used.

28.3.1 International Energy Conservation Code

Unlike NGBS and LEED which are voluntary and discussed later in this chapter, the *International Energy Conservation Code (IECC)*, **Figure 28-1**, is a model code that covers minimum requirements for energy efficiency in new residential and commercial buildings. This code has been widely adopted by municipalities and sets requirements for exterior envelope insulation, window and door U-factors and solar heat gain coefficient ratings, HVAC duct insulation, lighting and power efficiency, and water distribution insulation.

While not explicitly a green building code, IECC is a code that has been used to improve energy efficiency. It is a model code that raises the bar to achieve a higher standard than the International Residential Code and International Commercial Code.

28.3.2 International Green Construction Code

The *International Green Construction Code (IgCC)*, **Figure 28-2**, is another model code that can be adopted by municipalities to mandate certain green building standards for construction. This code covers minimum requirements for improving the environmental and health performance of buildings and their sites. The IgCC builds on the IECC by creating more comprehensive and higher-level requirements for green building. Just as with other model codes, the government of the jurisdiction is free to adopt the IgCC in its entirety or with amendments. However, it is up to the jurisdiction to determine if it will adopt the code at all.

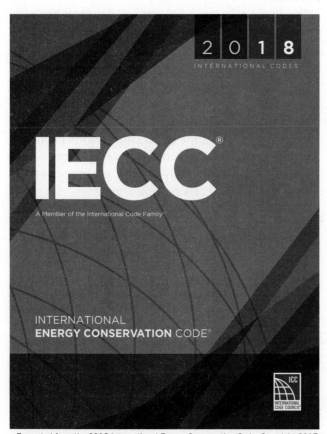

Figure 28-1. The IECC is a model code that covers minimum requirements for energy efficiency in new residential and commercial buildings.

Figure 28-2. The IgCC is a model code that covers minimum requirements for improving the environmental and health performance of buildings and their sites.

The IgCC is the first model code to integrate the sustainability measures of a site into the construction project. It provides standards for energy efficiency, resource conservation, water safety, land use, site development, indoor environmental quality, and building performance. When written, this code was crafted in a way that uses a whole-systems approach to the design, construction, and operation of buildings. This means it determines requirements with respect to the entire building and its environment, addressing the broader context in which a building is situated. IgCC accounts for factors such as neighborhood connections and improved pedestrian access.

28.3.3 National Green Building Standard

The *National Green Building Standard (NGBS)* is a building rating and certification program developed jointly by the International Code Council (ICC) and the National Association of Home Builders (NAHB). NGBS is one of the leading green programs that is used nationwide. The program is administered by the Home Innovation Labs, formerly known as the NAHB Research Foundation. Under this program, points are awarded for certain green practices and materials. Points are awarded by an independent *green verifier*, a green building expert who is certified by Home Innovation Labs to score green homes for the NGBS. The verifier reviews the plans for the house before construction, then inspects the building as it is being constructed. To be authorized by the Home Innovation Research Labs as a verifier, a person must have experience in green building, complete a training course, and pass a written test.

The verifier scores the home and awards one of four levels of green certification: bronze, silver, gold, or emerald. NGBS includes checklists of items that may receive points, **Figure 28-3**. Only homes that have been scored by an authorized green verifier can receive a certified rating.

28.3.4 Leadership in Energy and Environmental Design (LEED®)

Leadership in Energy and Environmental Design (LEED®) is the green building program of the US Green Building Council (USGBC®). Like NGBS, LEED provides independent verification of buildings aspiring to be certified as green. Prerequisites must be met and points are earned for buildings looking to receive different levels of certification. Both LEED and NGBS have items in every category of the rating system that

are mandatory. Those requirements must be met to achieve any level of certification. The mandatory items are usually not much more stringent than the requirements of national model codes.

LEED certifies both commercial and residential projects. The LEED program for residential certification is *LEED for Homes Provider*.

The LEED rating system evaluates the construction process in the following categories:

- Location and transportation
- Materials and resources
- Water efficiency
- Energy and atmosphere
- Sustainable sites
- Indoor environmental quality
- Innovation
- Regional priority credits

Each of these categories includes mandatory requirements and optional elements. A project receives credits for each optional element integrated into the project. The number of credits earned by a project determines its LEED certification level: certified, silver, gold, or platinum.

Involvement by two additional parties are required as part of the LEED certification process:

- *LEED® for Homes Providers* are green construction experts who are certified by LEED to assist green construction teams with designing and constructing green homes. The provider is involved in the project from start to finish. The provider reviews the initial construction plans and the complete construction process. This person or team also works closely with the green rater and provides a final certification review.
- *Green raters* are green experts who are certified by LEED to score each aspect of a building for LEED certification. Some green raters are employed by providers. Green raters perform inspections and tests throughout the construction process. A green rater must be certified by the USGBC (United States Green Building Council).

A typical LEED for Homes project requires the following:

1. Certification and inspection by the provider before the design is finalized.
2. An inspection by the green rater just before the drywall is installed.
3. Final inspection and testing by the green rater when construction is complete.
4. Final certification by the provider. See **Figure 28-4**.

GREEN BUILDING PRACTICES	POINTS
605 **RECYCLED CONSTRUCTION WASTE**	
605.0 Intent. Waste generated during construction is recycled. All waste classified as hazardous is properly handled and disposed of. **(Points not awarded for hazardous waste removal.)**	
605.1 Construction waste management plan. A construction waste management plan is developed, posted at the jobsite, and implemented with a goal of recycling or salvaging a minimum of 50 percent (by weight) of construction waste.	6
605.2 On-site recycling. On-site recycling measures following applicable regulations and codes are implemented, such as the following:	7
(a) Materials are ground or otherwise safely applied on-site as soil amendment or fill. A minimum of 50 percent (by weight) of construction and land-clearing waste is diverted from landfill.	
(b) Alternative compliance methods approved by the Adopting Entity.	
(c) Compatible untreated biomass material (lumber, posts, beams, etc.) are set aside for combustion if a solid fuel-burning appliance per Section 901.2.1(2) will be available for on-site renewable energy.	
605.3 Recycled construction materials. Construction materials (e.g., wood, cardboard, metals, drywall, plastic, asphalt roofing shingles, or concrete) are recycled offsite.	6 Max
(1) a minimum of two types of materials are recycled	3
(2) for each additional recycled material type	1
606 **RENEWABLE MATERIALS**	
606.0 Intent. Building materials derived from renewable resources are used.	
606.1 Biobased products. The following biobased products are used:	8 Max
(a) certified solid wood in accordance with Section 606.2	
(b) engineered wood	
(c) bamboo	
(d) cotton	
(e) cork	
(f) straw	
(g) natural fiber products make from crops (soy-based, corn-based)	
(h) products with the minimum biobased contents of the USDA 7 CFR Part 2902	
(i) other biobased materials with a minimum of 50 percent biobased content (by weight or volume)	

International Code Council

Figure 28-3. This is a sample page from the National Green Building Standard. The points for each item are shown on the right.

Water Efficiency (WE)	PRE-REQUISITES	POINT FLOOR	POSSIBLE POINTS	YES	MAYBE	NO
	NA	3	15	12	1	2

Water conservation begins with measures to reduce water use, both indoors and out. It can also include rainwater or water from nonpotable sources for some applications. This category has a minimum requirement or "point floor" of at least three points.

Credits

WE 1 Water Reuse	STATUS	POSSIBLE POINTS	YES	MAYBE	NO
1.1 Rainwater Harvesting System	☑ SCORED	4	4	0	0
1.2 Graywater Reuse System	☑ SCORED	1	1	0	0
1.3 Use of Municipal Recycled Water System		3	0	0	0

WE 2 Irrigation System	STATUS	POSSIBLE POINTS	YES	MAYBE	NO
2.1 High-Efficiency Irrigation System	☑ NOT AVAILABLE	3	0	0	0
2.2 Third-Party Inspection	☑ NOT AVAILABLE	1	0	0	0
2.3 Reduce Overall Irrigation Demand by at Least 45%	☑ SCORED	4	2	1	1

WE 3 Indoor Water Use	STATUS	POSSIBLE POINTS	YES	MAYBE	NO
3.1–3.2 High-Efficiency Fixtures and Fittings	☑ SCORED	6	5	0	1

PROJECT PROGRESS

Project Progress will reach 100% once every credit is scored. Quick Scored credits must be confirmed.

SELF-SCORED ACHIEVEMENT

ABOUT YOUR SCORE

POINTS (12)   MAYBES (1)

OVERALL PRE-REQUISITES	OVERALL POINT FLOORS
Not Met	**Not Met**

US Green Building Council

Figure 28-4. A sample checklist for a LEED® home. This checklist will be updated as the construction progresses.

28.3.5 ENERGY STAR® Certified Homes

ENERGY STAR is a labeling program created to reduce the energy consumption of devices. It was created and is administered as a joint effort between the US Environmental Protection Agency (EPA) and the US Department of Energy (DOE). The program certifies products such as home appliances, lighting, electronics, windows, doors, and roofing materials. In order to receive the ENERGY STAR label (**Figure 28-5**) a product must meet certain efficiency and testing standards. ENERGY STAR has continually expanded the range of products that it certifies and now includes remodeled existing homes, new homes, commercial buildings, and industrial plants as part of the program. ENERGY STAR certified homes are required to be at least 10% more efficient than homes build to code and are required to be built with improved quality, performance, and comfort. In order to achieve the ENERGY STAR label, a home must be constructed by a builder who has been designated as an ENERGY STAR partner and must

The US Environmental Protection Agency's Energy Star Program

Figure 28-5. The ENERGY STAR label identifies products that meet defined energy performance requirements.

meet all program requirements set forth by a design review checklist and verified by an ENERGY STAR rater. Since the program's inception, over two million ENERGY STAR certified homes have been built.

28.3.6 Passive House Standard

Imagine a home or building that does not require a conventional heating or air conditioning system to remain comfortable. This is one of the goals of the passive house construction concept, **Figure 28-6**. A *passive house* is built to a high standard of efficiency, which results in ultra-low energy consumption while maintaining a comfortable interior living environment. The passive house standard is a voluntary standard that aims to achieve maximum efficiency and comfort in a building. Training and certification for passive house construction is administered by Passive House Institute US, Inc.(PHIUS) , a nonprofit 501(c)(3) organization committed to making high-performance passive building the mainstream market standard.

Though not easy to achieve, passive houses are designed to make best use of heat collected from the sun's rays and to retain heat emitted by appliances and other sources, including body heat, during the heating months. They take advantage of sources such as shade structures and deciduous trees in the cooling months to maintain temperatures in the home. Furthermore, passive houses attempt to achieve this while retaining a high level of comfort, quality, and affordability. These objectives are met through five basic green building principles:

1. Continuous insulation—An unbroken envelope of high quality insulation helps the building maintain a consistent temperature.
2. High efficiency windows—Triple pane insulated glass and insulated frames allow solar heat in and reduce heat transfer out during heating months. These windows also act to eliminate overheating during the cooling months.
3. Continuous air barrier—Ensuring a continuous air barrier keeps conditioned air, heat, and moisture from leaking through the envelope.
4. Eliminate thermal bridges—The thermal envelope needs to be uniform.
5. High quality mechanical ventilation—A high efficiency heat or energy recovery ventilator maintains good air quality inside the home while minimizing energy loss.

You may have recognized some of these principles as standard building practices. Many of the efficiency principles used in traditional construction codes started as ideas that were used in voluntary standards such as the passive house standard. These building principles apply to all building types. However, passive house standards take these principles to a more stringent level than standard construction. It may be helpful to think of passive house as a kind of research-and-development venture. Successful passive house strategies are often adopted into standard building codes.

A *PHIUS* **B** *PHIUS*

Figure 28-6. A—Passive house under construction. B—With good design, passive houses can prove to be both efficient and attractive.

However, when used in practice, many production builders take shortcuts or use minimal compliance efforts to save time or money on a build. Passive house construction leaves no room for shortcuts. To be considered a passive house the building must meet specific criteria after completion, **Figure 28-7**.

Building to the passive house standard can be expensive; however, in the long-term much of the cost can be recuperated through lower utility bills. A passive house building can realize up to 75% savings on heating- and cooling-related energy bills compared to the average new building. These buildings are designed to provide a comfortable living environment in all weather conditions, high indoor air quality, a resilient structure that will withstand the test of time, and very low utility bills, while minimizing the impact on the environment.

28.3.7 DOE Zero Energy Ready Homes

In addition to the ENERGY STAR Certified Homes program, the US Department of Energy (DOE) administers a higher-level certification program called the *DOE Zero Energy Ready Home Program*. This green-home building program achieves energy efficiency that is 40%-50% better than a typical new home. It builds on the ENERGY STAR program. Like the ENERGY STAR program, the builder is required to register as a Zero Energy Ready Home partner in order to participate. To achieve certification under this program the home must meet specific checklist and performance requirements that include compliance with a number of other codes and programs, including:

- ENERGY STAR
- IECC insulation requirements
- Energy and water conservation and distribution requirements
- HVAC duct installation optimization
- Indoor air quality certification
- Preparations to ensure the home is solar-ready (in some climates)

Additionally, the program encourages builders to:

- Commit to building 100% of their homes to meet program requirements
- Meet the requirements of the passive house standard
- Reduce water consumption of plumbing fixtures by participating in the Environmental Protection Agency's (EPA) *WaterSense* conservation program
- Take specific measures to build more disaster resistant homes based on regionally specific natural hazards
- Apply comprehensive quality management practices
- Implement practices that ensure homes are solar hot water-ready and take advantage of solar thermal systems
- Urge buyers to share utility bill data with the DOE Zero Ready Home program

28.4 Green Building System

After determining the goals of a green build, what does it take to achieve the desired rating? Green building is a system that requires many parts to work together. The system includes tangible tasks, such as selecting materials made from recycled or renewable resources. It also includes understanding and communicating concepts to ensure the design and construction teams work together to guarantee that all aspects of the building work as a system. Having all parts of the building working together is the most important aspect of a green home. If a member of the team (an employee of a subcontractor, for instance) fails to recognize their role in the success of the project, the outcome of the build may be compromised.

In green home building, the house functions like an organism. A house is not merely an assembly of pieces and parts built by workers of various specialties

Passive House Standards

Building Demand	Standard
Primary Energy Demand	Not to exceed 120 kWh annually for all domestic applications (heating, cooling, hot water, and domestic electricity) per square meter of usable living space.
Airtightness	Maximum of 0.6 air changes per hour at 50 pascals pressure (as verified with an onsite pressure test in both pressurized and depressurized states).
Thermal Comfort	Thermal comfort must be met for all living areas year-round, with not more than 10% of the hours in any given year over 77°F (25°C)
Space Heating Demand	Not to exceed 15kWh annually or 10W (peak demand) per square meter of usable living space.
Space Cooling Demand	Roughly matches the heat demand with an additional, climate-dependent allowance for dehumidification.

Goodheart-Willcox Publisher

Figure 28-7. Passive houses must meet specific criteria after completion to be certified passive.

performing different tasks without consideration for the rest of the building. In order for the systems to function in concert, every material and subsystem must be planned, installed properly, and connected with care. Only when all components are working together is the home functioning optimally. This is what constitutes a genuine green building system.

28.4.1 The Green Building Team

Green building begins when the designers begin work on the home design. The owner and the architect meet with the general contractor, subcontractors (framing carpenters, plumber, HVAC technician, electrician, etc.), and possibly a land developer to discuss the plan. It is not necessary for everyone to attend every meeting, but the lines of communication between everyone working on the project must be open in order to coordinate efforts and ensure the green goals are achieved. Good team planning and communication is one of the most effective ways to ensure the project minimizes environmental impacts and maximizes the benefits that can be realized through each aspect of the build.

Teamwork is necessary to ensure all the components and systems are designed and installed properly. Even the smallest detail, such as a hole through an outside wall for a wire or pipe, has to be considered and communicated. Otherwise, the hole might not be sealed and can leak water and air, **Figure 28-8**. No one trade can complete its work without involving the work of other trades. Everyone on the jobsite must cooperate, work together, and look out for one another so the house is built to the highest standard and achieves its green goals.

At first, building a green home might be challenging. It requires working with new materials and building methods, **Figure 28-9**. With willingness to learn and an understanding of why things are done as they are, green building will become as familiar as conventional building.

> **Pro Tip**
> Construction materials and methods are changing faster than ever. To be a valuable member of the industry, carpenters must be prepared to learn as much as possible about the newest materials and latest methods.

28.4.2 Site Selection, Lot Planning, and Preparation

Part of green housing is green development. In other words, where the house is built and how the community surrounding the house is designed has a lot to do with reducing the impact on the environment and improving the lives of those living in the house. For example, a house built near public transportation may reduce the miles traveled in personal automobiles. A development located within walking distance of shopping will reduce vehicle traffic even more.

The objective when planning, selecting a lot, and doing site work for a green-built house is to minimize its impact on the site and the environment. Before construction begins, a plan is made outlining how the site work and construction will be conducted.

Goodheart-Willcox Publisher

Figure 28-8. If the opening for this electrical box is not properly sealed, air and water will enter the building at this point.

brizmaker/Shutterstock.com

Figure 28-9. SIPs are often used in green construction because they provide excellent insulation, are airtight, and conserve resources by creating little waste.

The plan addresses how to protect existing trees and shrubs, manage water runoff and control erosion, and minimize overall disturbance to the site.

The orientation of the house on the lot is also considered. Often, a green home is positioned on the lot to take advantage of sun for natural daylight and heat in winter, **Figure 28-10**. Deciduous trees, which lose their leaves in the winter, can act as a solar buffer, providing shade to a home in the summer and allowing the benefit of the sun's rays to come through in the winter. In warm climates where air conditioning is the major energy use, windows, roof overhangs, and landscaping are planned to avoid direct sunlight and provide shading to help keep the house naturally cooler, **Figure 28-11**.

SUSAN LEGGETT/Shutterstock.com

Figure 28-10. The large windows in this home face south where they will admit the sun's rays during the winter.

28.4.3 House Design and Planning

As mentioned earlier, the entire team—owner, architect, general contractor, and subcontractors—is involved in designing and planning for a green home. The architect's preliminary house plans are reviewed by those who will build the home, each looking for ways to improve energy and water efficiency, make the house more durable, simplify utility system installation, and make the house more comfortable. There are often group meetings where the designer, owner, contractor, and key trades people discuss the house plans, examine the impact of each recommendation, and how the work will be carried out. The general contractor and trade contractors can also recommend green building materials and equipment that best suit the project.

A building project that is planned by all who will work on it is generally easier to manage because everyone is familiar with the design, the green goals, and with each other. Costs are also controlled better than for a traditionally built home because the building team has already worked out how the details of each trade will affect the others. This eliminates the need to buy additional supplies that were not part of the original budget plan.

28.4.4 Resource Efficiency

Green home construction uses resources very efficiently, causing as little impact on the environment as possible. Efficient use of natural resources can result in more functional products. For example, engineered lumber makes use of nearly the entire tree and produces far less

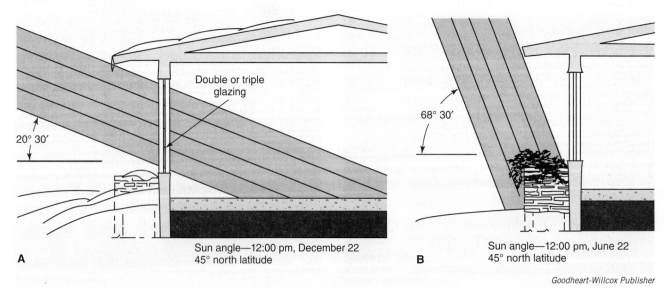

Goodheart-Willcox Publisher

Figure 28-11. The roof overhang can control solar heating. A—In the winter, when the sun remains closer to the horizon, the windows allow for passive solar heating. B—In the summer, the sun rises higher in the sky, so the roof overhang shades the windows, reducing the solar heat gain in the building.

waste than sawing conventional lumber. Engineered lumber can also be longer, stiffer, and more dimensionally stable than sawn lumber. See **Figure 28-12**.

The designer and the contractor should consider the following when choosing materials for a green building:

- Energy savings—Insulation, photovoltaic panels, and ENERGY STAR certified appliances, lighting, electronics, roofing, windows, and doors are all considered green because they conserve energy.
- Product longevity—Durable products need to be replaced less often than poorly built ones. This means that resources and energy used to make longer-lasting products have less impact on the environment. Installing materials correctly using manufacturer's suggested methods also increases product longevity.

- *Indoor air quality*—The amount of pollutants in the air inside buildings is a key concern to building occupants. The most common causes of poor indoor air quality are paints and carpeting that have high volatile organic compounds (VOCs), heating and ventilation devices that circulate stagnant or foul air, and lack of ventilation causing excessive stale air build-up.
- Production location—Products and materials harvested, mined, or manufactured relatively close to the jobsite require less energy to transport.
- Resource consumption during production–– Many manufacturers have made their production processes more energy- and water-efficient. By using less energy and water during production, the products have a smaller impact on the environment.

Green products are also chosen for their recyclability. Is the product made from recycled material? Can the material be recycled in the future when it is removed from the house, perhaps during a remodel, or will it end up as waste in a landfill? Many common materials used in construction (wood, steel, some plastics) can easily be recycled into new materials and therefore have less impact on the environment, **Figure 28-13**.

Weyerhaeuser

Figure 28-12. This engineered lumber utilizes 50% more of a log than conventional framing lumber.

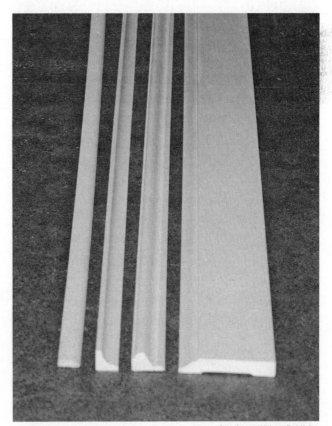

Goodheart-Willcox Publisher

Figure 28-13. This cellular vinyl molding can contain up to 90% recycled material.

These products have a lower environmental impact because their raw materials would have been disposed of as waste had they not been reused.

Designers and builders must consider the environmental impact of producing and disposing of the materials a project requires. Choosing products that are locally sourced and are made with readily renewable materials reduces the environmental impact of construction. Reusing and recycling materials helps reduce demand and the effects associated with waste disposal and unnecessary production. Green building is a mindset that requires builders to think big with long-term sustainability aspirations in mind. Ideally, even the end of the building's life will be considered. A good sustainable design will give thought to how a new building will be deconstructed and its materials might be reused or recycled.

28.4.5 Energy Efficiency

Energy consumption has far-reaching environmental impacts, from the mining of fossil fuels to the environmental emissions that come from burning nonrenewable energy sources. A home consumes energy month after month and year after year, so the impacts of energy use build up over time. For this reason, green home programs place great emphasis on energy efficiency. Energy is consumed not only during the operation of a home, but it is consumed as the house is built, as the raw materials are mined or harvested, and as the building materials are transported to the site, **Figure 28-14**.

About 21% of the energy consumed in the United States is used in homes. At one time, most of that energy went to heating the home. In recent years, however, appliances have consumed a greater percentage of energy than heating equipment. Homes have gotten larger, but due to the focus on insulation and weather tightness, less energy is needed to heat larger homes. At the same time, the average size of a refrigerator has increased dramatically; more, larger televisions are in use; an increasing number of homes have clothes dryers and dishwashers; and the list goes on. The goal of green construction is to allow for these conveniences while reducing the amount of energy the home consumes.

To combat this energy consumption, some green homes use *renewable energy*, which comes from sources that are naturally replenished, such as sun, wind, and water flow. Using solar heating and photovoltaic and wind-powered generators, these homes produce a portion of, or even all, the energy they need to operate, **Figure 28-15**.

There are several ways a green home is made more energy-efficient than an ordinary home. The appliances and lighting selected are the most energy efficient available. The energy-using products in a green home most often have the ENERGY STAR label, indicating that they have been tested and proven to be energy efficient. Devices carrying the ENERGY STAR service label, such as computer products, office equipment, kitchen appliances, and water heaters, generally use 20%–30% less energy than standard models.

The most effective way to reduce energy bills involves reducing heat transfer through the building envelope (exterior building surfaces). A green home is designed and

Goodheart-Willcox Publisher

Figure 28-14. Large amounts of energy are consumed transporting building materials. Green home planners look for materials as close to the site as possible to reduce the amount of fuel used in transport.

Solar water heater

Diyana Dimitrova/Shutterstock.com

Figure 28-15. Homes outfitted with photovoltaic solar panels are able to produce much of their own electricity; some produce enough energy to sell a portion back to the electrical grid. This home is also outfitted with a solar water heater, which reduces energy consumed by a traditional water heater.

built in a manner that maximizes the insulation in the floor, walls, and ceiling. It is also built to be as airtight as possible, reducing energy lost due to air leaks. The windows and doors are selected for energy-efficient glazing and excellent air seals. Performance gains in these areas can be accomplished in a variety of ways but will usually mean adjustments for the carpenters framing the home. Double stud walls, insulated roof decks, and structural insulated panels (SIPs) are just a few examples of systems used to increase insulation values, **Figure 28-16**.

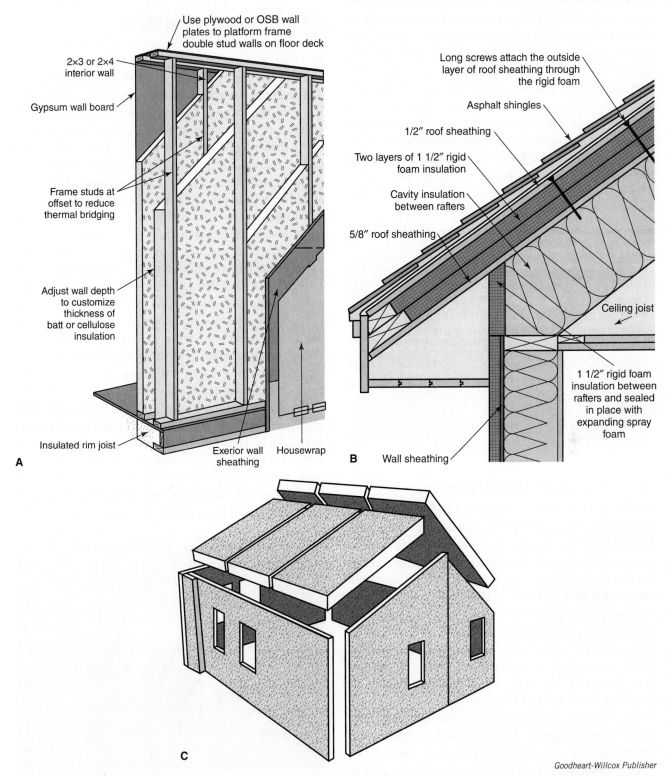

A—

Use plywood or OSB wall plates to platform frame double stud walls on floor deck

2×3 or 2×4 interior wall

Gypsum wall board

Frame studs at offset to reduce thermal bridging

Adjust wall depth to customize thickness of batt or cellulose insulation

Insulated rim joist

Exerior wall sheathing

Housewrap

B—

Long screws attach the outside layer of roof sheathing through the rigid foam

Asphalt shingles

1/2″ roof sheathing

Two layers of 1 1/2″ rigid foam insulation

Cavity insulation between rafters

5/8″ roof sheathing

Ceiling joist

1 1/2″ rigid foam insulation between rafters and sealed in place with expanding spray foam

Wall sheathing

C—

Goodheart-Willcox Publisher

Figure 28-16. This illustration shows one of a variety of methods that can be used for their construction. A—Double stud walls allow for additional insulation in exterior walls. B—An insulated roof deck can improve efficiency and reduce condensation issues if built properly. C—SIPs are available in different thicknesses and can be used for walls, ceilings and floors.

Reducing thermal bridging might be accomplished through use of advanced framing practices (see Chapter 11, **Wall and Ceiling Framing**) or the addition of exterior insulation.

Because of a green home's airtight characteristics, any gases given off by the components of the home will remain in the home. This means that materials must be selected that emit the lowest possible amount of undesirable gases. It is important to achieve a healthy and comfortable environment inside the building. Efficient ventilation systems are used to control fresh air intake and exhaust of stale air that would otherwise be trapped inside. The home should use efficient heating and cooling methods. Options like geothermal systems use a constant underground temperature to achieve high efficiency. Natural sources of heating and cooling such as passive solar heat and shade trees should also be used when available.

Code Note

The International Energy Conservation Code (IECC) defines specific requirements for commercial and residential construction. Energy efficiency is the focus of most of these requirements. The IECC is adopted as part of the building code in many states and municipalities. On the other hand, International Green Construction Code (IgCC) is adopted by municipalities who have determined to raise the bar on green building. IECC establishes minimums for energy conservation, IgCC establishes minimums for green building. It is up to each municipality to adopt all, or parts of the code that they wish to enforce.

28.4.6 Weather Resistance

Recall that the exterior surfaces of a building are called the *envelope*. The building envelope is comprised of four control layers: water control, air control, vapor control, and thermal control. The most important function of the building envelope is to keep out the weather—cold air in the winter, warm air in the summer, and water at any time. In fact, one of the biggest factors in reducing the building performance is failure of the envelope to keep water and unconditioned air out of the building. For more on the building envelope, control layers, and how they relate to energy efficiency, see Chapter 15, **Building Envelope and Control Layers**.

Green homes are carefully built to maintain high quality control layers.. Water is shed from the roof using carefully installed flashing to keep water out of seams between the roof and vertical surfaces such as chimneys, plumbing vents, and adjoining walls. Wherever the envelope is penetrated by a pipe, conduit, or wires, the seam must be sealed with flashing and sealant, **Figure 28-17**. The opening should be sealed by the worker making the opening, but every member of the green building team should be on the lookout for such things.

28.4.7 Water Conservation

Green homes strive to use less water for the occupants and for irrigation than ordinary homes. The plumbing fixtures use less water without reducing performance. The average American flushes a toilet five times per day and the average home has 3.2 occupants. Using toilets that flush completely with 1.28 gallons per flush (gpf) instead of those that use 3.5 gpf, saves 12,965 gallons of water in a year. That conserves water and reduces wastewater disposal requirements. It has the added benefit of lowering water utility bills.

Showering accounts for about 17% of the indoor water use of a typical family. The WaterSense program can help families reduce water use. By using WaterSense-labeled shower heads, the average family could save up to 2,900 gallons of water per year.

The landscaped area around the house might reduce the need for irrigation by using native drought-resistant plants and reducing lawn area, **Figure 28-18**. This is known as xeriscaping. Innovative systems such as rainwater collection and gray water reuse are sometimes incorporated into green homes to minimize the demand on potable (drinkable) water systems. *Gray water* is all washwater that has been used in the home, except water from toilets. Even the site plan of a green home may address water conservation by incorporating elements such as pavement that allows water to pass through and drywells to help rainwater soak into the ground and avoid run-off. When rainwater soaks into the ground

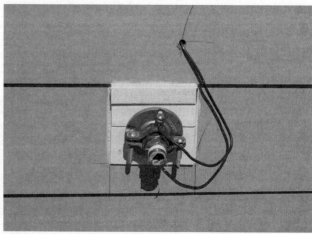

Figure 28-17. Seams and cracks must be sealed where pipes and wires pass through the building envelope.

Kathryn Roach/Shutterstock.com

Figure 28-18. This landscaping is done entirely with plants that are native to the area and require little water. The steps to the side of the planted area are permeable, allowing surface water to pass through the earth below.

rather than running off to a storm drain system, it helps recharge the groundwater aquifer, the layer of water beneath the surface of the earth from which wells draw their water.

28.4.8 Indoor Air Quality

The air inside a home can be as much as three times as polluted as the air outside, even in industrial areas. Indoor air pollution may come from a variety of sources, most of which can be eliminated or greatly controlled during design and construction.

Radon

Radium is a metallic element found in trace amounts in the ground. Radium produces *radon*, an invisible, radioactive gas given off by some rocks and soil. Radon can seep into a home through openings in the foundation system. Since it is heavier than air, radon sinks to the lowest level of a house, where it can accumulate and build up to potentially dangerous levels. Radon is known to cause cancer.

Radon tests are used to identify potentially dangerous buildups, **Figure 28-19**. Methods to correct unacceptable radon levels may include the following:

- Sealing cracks or other openings in foundation floors and walls to prevent radon entry
- Increasing ventilation in the area of buildup to remove radon and prevent buildup

- Installing vapor barriers, such as polyethylene membranes, between the foundation and the ground to prevent radon entry

Pro Tip

Applying these radon prevention methods in the construction of a new home can prevent radon issues throughout the life of the home.

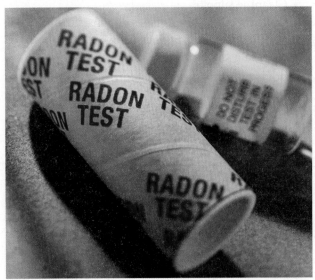

stevecoleimages/iStock/Getty Images Plus

Figure 28-19. This radon test kit is easy to use and readily available in hardware stores and home improvement centers.

Mold

Mold is a fuzzy fungus that can release spores into the air. These spores can pose health risks, such as respiratory problems and facial irritations.

Mold can grow on most surfaces in the presence of moisture. Mold cannot grow without moisture, so the best method of preventing mold growth is avoiding unwanted moisture buildups. Potential causes of mold-causing moisture include:

- Leaking roofs, walls, windows, or doors
- Leaking plumbing systems
- Moisture from the ground entering the building
- Inadequate ventilation in areas where water vapor is present, such as kitchens and bathrooms

If mold is discovered in a home, treatment includes two steps. First, the existing mold must be removed. Removal methods vary based on the type of mold. Second, the moisture problem that allowed the mold growth must be corrected. If this moisture problem is not addressed, the mold will return.

Volatile Organic Compounds (VOCs)

Volatile organic compounds (VOCs) are carbon-containing chemicals that readily evaporate into the air. When they enter the air, they react with other elements to produce ozone. Ozone causes air pollution and a host of health issues including breathing problems, headache, nausea, and burning, watery eyes. Some VOCs have also been linked to not only cancer, but kidney and liver damage as well.

Many products in a home can emit VOCs, including the following:

- Paints, lacquers, and similar products
- Cleaning supplies
- Craft supplies, such as glues and markers
- Building materials and furnishings, such as carpeting and furniture

Indoor air quality can be improved by using low-VOC products and no-VOC products. VOC buildup is reduced in homes with ventilation systems that exchange indoor and outdoor air more frequently.

Fuel-burning appliances such as furnaces, water heaters, gas fireplaces, wood stoves, and gas ranges all produce carbon dioxide, carbon monoxide, and water vapor when they operate. When the exhaust from these appliances leaks into the house, the results can lead to unhealthy air or, in the case of carbon monoxide, become deadly.

The first step to keep the indoor environmental quality good is to reduce the pollutants that enter the house. Radon and water vapor from the soil can be prevented from entering a home by sealing the house. Green building materials that have very low or no VOCs do not add to the pollutant load in the house, **Figure 28-20**. Exhaust venting of all combustion appliances is critical. The exhaust must lead to the outside of the house and ideally be routed through a sealed exhaust pipe so there is no chance for exhaust leakage inside.

Education is the best way to affect homeowners' behavior. They must understand that the things they bring into a house can pose a threat to their health. Homeowners must know about alternatives, such as low- or no-VOC carpeting. With knowledge of the alternatives, homeowners can make choices for furnishings, clothing, and cleaning products that minimize pollutants inside the house.

The second step to improving indoor air quality is to provide proper ventilation. Green homes are built tight. That is, they do not have cracks and seams that would allow outside air and indoor air to be exchanged on their own. Green homes are designed with ventilation to replace stale indoor air with fresh outdoor air. Often, this ventilation is accomplished with equipment built into the HVAC system.

Goodheart-Willcox Publisher

Figure 28-20. In green construction, products with low or zero VOCs are used to improve indoor air quality. Water-based paint with less than 250 grams of VOCs per liter of paint is considered low-VOC paint.

28.4.9 Best Construction Practices

There are many ways to frame a house, install cabinets, or shingle a roof that will meet the minimum requirements of building codes. Green building goes beyond the building codes and raises the bar for construction practices to a greater level of quality, **Figure 28-21**. Materials that are installed just well enough to meet the code might not perform efficiently or last a long time. For example, house wrap that is stapled to the sheathing in a hurry might have spots where the seams just barely overlap or do not overlap at all. When winter winds hit that wall, there is a good chance that some of that cold air will enter the building. In wind-driven rain, water might get behind the house wrap and allow mold to grow when the wall temperature rises. Flashing is another example. Roof flashing is critical to keep water from leaking into the house. Flashing that is up to code but also sized too small or installed with a large gap off the roof surface may leak during a windy rainstorm and lead to rot inside the attic. Best practices call for metal roof flashing to be large enough for the best overlaps with the roofing and siding, and installed properly so, regardless of the weather, water will not leak into the house. Using best practices not only improves the quality of the work being done, but it also lengthens the lifecycle of materials and the home itself. This is an effective means of making any project more sustainable while improving the reputation of the builder as an added benefit.

Involving all members of the team, owner, designer, general contractor, subcontractors, and workers in the whole process of building a home is a best practice. Often, the workers who are actually installing materials are the ones who will first have an idea that might lead to more durable construction or less waste. To employ this best practice, not only do the contractors and foremen need to accept the idea of whole-team involvement, but the carpenters, plumbers, and laborers need to accept that concept as well. Every member of the green building team has a responsibility to work toward the green goals of the project.

Best construction practices change over time as building researchers, manufacturers, and construction workers discover better ways of building. It is up to the entire building team to investigate the many sources of information about how to construct each detail of a house. This way, when the home is built, it is constructed using the best practices and to the highest quality standards.

Pro Tip
It costs no more to use best practices. All workers on a construction project should strive to follow best practices, even if the building is not planned for green certification.

Simpson Strong-tie

Figure 28-21. These carpenters are installing a steel shear panel that exceeds code requirements.

28.4.10 Operation, Maintenance, and Homeowner Education

A green home will not perform well if it is not operated and maintained as designed. Homeowners play the biggest role in how a green home performs over time. Their daily activities impact how much energy and water are used. Before homeowners move into a new green home, the building team must teach them how to operate and maintain the home and how their behavior affects the home's environmental impact. The owners need to know, for example, how to program the HVAC system thermostat and clean or replace the filter to get the highest efficiency from the equipment. They need to understand the importance of cleaning and maintaining gutters and downspouts so that all water is diverted at least 5′ from the foundation. Without this knowledge and understanding, the owners can defeat many of the green features built into a house without realizing it. Each trade has a hand in educating the homeowners either directly by leading a training session or indirectly by teaching the general contractor or designer, who then trains the owners.

There is too much information about the systems and materials in a house for owners to absorb it all during a few hours of training. Therefore, the building team must compile a reference manual for every green home. The manual details how to operate and optimally set the various systems and appliances. The manual also contains valuable information for future owners or tradespeople who may work on the house during routine maintenance and renovation work. The reference manual includes a list of all the materials and products used to build the house as well as manufacturer-printed literature. It often includes photographs of key elements of the house taken during installation and before the wall finishes were applied. This enables future workers to see what is inside the walls, making their work easier and safer. The manual also includes lifestyle impact information about using each of the home's systems, such as how the amount of time taken for showers affects utility costs. Today, many builders use project management software for estimating, bidding, scheduling, and documenting all aspects of the build. Many of these software programs allow the builder to share select portions of this data with the home buyer. They also have the capability of creating the reference manual digitally, opening up another avenue for communication between the builder and the client after the home is complete. When used effectively, these programs compile a more complete picture of the project from start to finish.

Workplace Skills
Self-Assessment Inventory

Carpentry allows for a wide variety of career choices. From framing or remodeling to real-estate or estimating, there is a place for people of nearly all skill sets and personality types. Determining where you fit best as a carpenter may be easier if you reflect on your strengths and weaknesses. Do you like interacting with new people or do you prefer a tight-knit group? Maybe you prefer to work solo? A self-assessment should include taking inventory of your interests, values, personality, and skills. It should include what type of work environment you prefer or thrive in, and what kind of work interactions and activities are important to you. It is also important to understand how you learn best, such as: are written instructions best for you, or is it better to observe someone else performing a task before you attempt it? Knowing this type of information about yourself can give you a lot of direction when choosing the path that best suits you.

Chapter Review

Summary

- Green building lessens the impact on the environment, reduces energy consumption, and extends the usable life of the building with a minimum of maintenance.
- Green building should include a visionary approach with an eye toward how the building fits into the surrounding environment and how the project interacts within the broader social context.
- The building construction industry responded to petroleum shortages in the early 1970s by developing ways to conserve the energy that buildings used. That was the birth of green home building.
- A green certification program is a set of standards for green building practices that are followed by the building team and verified by an independent third party.
- The International Energy Conservation Code (IECC) and the International Green Construction Code (IgCC) are model codes that cover minimum requirements for energy and green building.
- The National Green Building Standard (NGBS) is a building rating and certification program that awards point for certain green practices and materials.
- Green building programs, such as LEED®, ENERGY STAR, and others are used throughout the United States and other countries. Programs such as these define a level of performance or compliance for green building.
- Passive house construction creates homes or buildings that do not require conventional heating or air conditioning systems to remain comfortable.
- The DOE Zero Energy Ready Home Program the DOE Zero Energy Ready Home Program. This green-home building program achieves energy efficiency that is 40%-50% better than a typical new home.
- Green building is a system that requires many parts to work together. The system includes tangible tasks, as well as understanding and communicating concepts to ensure the design and construction teams work together.
- It is important for owners, designers, and builders to understand the goals of a project before choosing which program(s) are to be used.
- The objective when planning, selecting a lot, and doing site work for a green-built house is to minimize its impact on the site and the environment.
- When choosing materials, you must consider energy savings, product longevity, indoor air quality, volatile organic compounds (VOCs), production location, and resource consumption.
- Better insulation has allowed homes to use less energy, while appliances have gotten bigger and, as a result, use more energy. Energy Star is a program that certifies energy-using devices as being particularly efficient.
- Green homes are planned and constructed with attention to using materials that impact the environment as little as possible as they are mined or produced, assembled into the house, used during operation of the house, and disposed of at the ends of their lives.
- Green homes are built with an additional focus on quality. This is especially important with regard to the control layers of the building envelope.
- Indoor air quality can be greatly improved with proper green construction. Three sources of indoor pollution are radon, mold, and volatile organic compounds (VOCs).
- Best construction practices, such as proper house wrapping and house flashing, are followed during construction of a green home so that all parts of the building system will function together as well as possible.
- Energy conservation is an important consideration in green homes.

Know and Understand

Answer the following questions using the information in this chapter.

1. *True or False?* A home with photovoltaic solar collectors on the roof is an indication of a green home.

2. Which of the following factors does *not* contribute to making a home *green*?
   A. Efficient use of materials.
   B. Early planning.
   C. Longer construction time.
   D. Quality workmanship.

3. The _____ is a model code that can be adopted by a municipality to mandate construction practices that regulate energy conservation requirements for new residential and commercial buildings.
   A. IECC   C. LEED
   B. IgCC   D. NGBS

4. Under the NGBS program, an independent green _____ reviews plans for a home before construction, and inspects it as it is being constructed.
   A. carpenter   C. verifier
   B. contractor   D. inspector

5. ENERGY STAR began as a labeling program jointly created by the _____ and the Department of Energy.
   A. International Code Council
   B. Environment Protection Agency
   C. National Green Building Standard
   D. Leadership in Energy and Environmental Design

6. A(n) _____ house does not require a conventional heating or cooling system to remain comfortable.
   A. traditional   C. passive
   B. two-story   D. underground

7. When does the green building team usually meet for the first time?
   A. During the design stage, before any construction begins.
   B. As soon as the foundation is in place.
   C. After the materials are bought by the owner.
   D. When the general contractor hires the subcontractors.

8. _____ lumber makes use of nearly the entire tree and produces far less waste than sawing conventional lumber.
   A. Green   C. Deciduous
   B. Resilient   D. Engineered

9. *True or False?* Where a product is produced should be considered when deciding what materials to use for a green building.

10. *True or False?* About 21% of energy consumed in the United States is used in homes.

11. Selecting energy-efficient appliances often involves selecting products with a(n) _____ label.
    A. IECC   C. ENERGY STAR
    B. NGBS   D. EPA

12. One of the most energy-depleting events for building performance is _____ getting through the building envelope.
    A. air   C. appliances
    B. water   D. Both A and B.

13. _____ can be reused to minimize the demand on potable water systems.
    A. Gray water   C. Well water
    B. Filters   D. Groundwater

14. _____ is a harmful gas that can be present in trace amounts in the ground and builds up near the lower floors of a building.
    A. Mold   C. Carbon monoxide
    B. Radon   D. Chlorine

15. _____ play the biggest role in how a green home performs over time.
    A. Carpenters   C. Green verifiers
    B. Homeowners   D. EPA workers

Apply and Analyze

1. What is one key difference between a green home and an ordinary home?

2. What are the responsibilities of an LEED® Homes Provider?

3. Explain the five basic green building principles that passive houses maintain to avoid the use of conventional heating or air-conditioning systems?

4. What are the most common causes of poor indoor air quality?

5. How is xeriscaping an efficient, water-saving process?

6. Describe two ways to improve air quality and reduce VOC buildup in a home.

Critical Thinking

1. Search the internet for additional information regarding challenges faced by governments implementing the various green building codes discussed in this chapter. Based on the political, environmental, and economic climate of your local municipality, which model code do you think is appropriate for residential construction in your region? Support your answer.

2. Passive houses carefully manage solar energy throughout the year. Explain how the principles, design characteristics, and products that make this possible work together.

3. Based on the home certification programs discussed in this chapter (NGBS, LEED, ENERGY STAR, PHIUS, DOE Energy Ready Homes), compare the major advantages and disadvantages of each.

Communicating about Carpentry

1. **Speaking and Listening.** With two classmates, role-play a situation in which you are a carpenter explaining to a customer the benefits of converting a building to solar energy from conventional energy. Explain why green energy is here to stay. Your explanation should include environmental and cost consequences of this conversion. Adjust your vocabulary as necessary while responding to their questions and clarifying information. Then switch roles.

2. **Speaking and Listening.** Divide into groups of four or five students. Each group should choose one of the following topics: conduction, convection, and radiation. Using your textbook as a starting point, research your topic and prepare a report on how heat travels. As a group, deliver your presentation to the rest of the class. Take notes while other students give their reports. Ask questions about any details that you would like clarified.

3. **Reading and Listening.** In small groups, discuss the main topics in the chapter. Ask questions of other group members to clarify concepts or terms as needed.

4. **Reading and Speaking.** As a class, create a list of green construction topics. Divide into groups of two to four students and choose one or two of the topics to research. Gather relevant data and prepare a debate with one or two group members on the pro side, and one or two on the con side of the subject. Include topics such as cost, sustainability, accessibility, and dependability as part of your reasoning. Conduct a debate for the class. Be prepared to answer questions from your peers.

CHAPTER 29

Remodeling, Renovating, and Repairing

OBJECTIVES

After studying this chapter, you will be able to:

- Identify different types of residential construction by visual inspection.
- Set up a proper sequence of renovation or repair for the interior and exterior of a house.
- Repair and replace deteriorated components and systems.
- Remove parts of the structure without damaging the total structure.
- List steps for removal of wall sections prior to remodeling.
- Identify bearing walls.
- List steps for installing shoring.
- Install and support headers, concealed headers, and saddle beams.
- Follow accepted methods in replacing all types of doors.
- Make repairs to wood and asphalt shingles.

TECHNICAL TERMS

concealed header	king stud	shoring
firestop	ribbon	sistering
jack post	saddle beam	
jamb clip	shear wall	

In many ways, remodeling and renovation work are more painstaking than new construction. To successfully perform remodeling work the carpenter must be able to remain flexible with respect to planning. Often jobs do not go exactly as planned because of unforeseen challenges that arise as the existing building components are dismantled. This requires effective critical thinking and problem-solving abilities. It is more difficult to work with structures that may have defects such as sagging floors, rotting framework, and walls that are no longer plumb. Sections or components of the old building often must be dismantled without destroying what is to be retained or reused. Frequently, a carpenter must be able to visualize how a structure was built so that no damage is done to it or any of its systems during demolition and removal of old walls.

Remodeling and renovating are well worth the effort, **Figure 29-1**. Remodeling brings greater comfort and usability for the owner. Further, remodeling and repairing have other benefits, such as forestalling the structure's deterioration or enhancing its value.

The International Residential Code (IRC) covers repair, renovation, alteration, and reconstruction in Appendix J. Many of the provisions of that appendix refer to the sections of the code covering new construction, but several parts of Appendix J allow exceptions from the new construction requirements. It is wise to review this appendix before any work is done on an existing building.

Code Note

Often, building permits must be obtained before remodeling and renovation work can begin. Building inspectors may need to inspect the work at various stages of the project. Always check with your local municipality to determine if permits and inspections are required. The permit and inspection requirements will be based on the scope of the project.

Front Exterior—Before Renovation

Front Exterior—After Renovation

Side Porch—Before Renovation

Side Porch—After Renovation

Ceiling—During Renovation

Ceiling—After Renovation

Goodheart-Willcox Publisher

Figure 29-1. Before, during, and after photos of a full renovation project that saved this historic brick home. Transformations like these give remodeling contractors a sense of satisfaction in their work.

29.1 What Comes First?

In a long-neglected home, halting further deterioration with temporary repairs may be the first order of business. Leaks should be remedied before they cause further damage to the structure. Replace missing shingles and seal leaks with roofing cement. If the shingles and roof sheathing are beyond repair, replace the rotted sheathing and reshingle the roof. See **Figure 29-2**. However, if structural changes are planned that would change rooflines, the roof should be repaired after those changes.

Structural repairs may be necessary for personal safety. Before attempting to repair or replace wall coverings, doors, and windows, examine the house to see if there are structural repairs that need to be made. Such items could be seriously affected by jacking or replacement of foundations, sills, studs, or other frame members.

Start repairs with the foundation and sills and work your way up. Do not fix a structural problem involving the roof, then jack up the frame. Everything will have shifted, leaving you to make additional adjustments.

29.1.1 Sequence of Exterior Renovation

Renovating the exterior of the building should be the first order of business. Not all of the following steps will need to be taken. However, whatever steps are needed should be taken in order.

A   *Goodheart-Willcox Publisher*

B   *Goodheart-Willcox Publisher*

Figure 29-2. Roofing repairs are the most common renovation project. A—Old shingles being removed. B—Sheathing may need to be installed or replaced.

PROCEDURE

Renovating an Exterior

1. List what needs to be done.
2. Demolish whatever must be eliminated and clear away the debris.
3. Perform all needed structural work. Work from the bottom up. This step should include chimneys and masonry or concrete work. Take measures to protect open areas from the weather.
4. Complete the site work. Regrade and provide drainage as needed.
5. Repair or replace the roof, including flashing, gutters, and roof vents, **Figure 29-3**.
6. Scrape or strip paint.
7. Repair masonry and tuckpoint.
8. Repair or replace windows and outside doors, **Figure 29-4**.
9. Repair or replace siding, **Figure 29-5**.
10. Stain or prime wood siding and trim.
11. Caulk, glaze, and putty.
12. Paint.

A   *Goodheart-Willcox Publisher*

B   *Goodheart-Willcox Publisher*

Figure 29-3. Reroofing a home can add to the function, value, and appearance of a home. A—An old metal roof has been recoated multiple times, making the home look neglected. B—Stripping the old roofing and replacing it with shingles adds curb appeal.

Des Moines, Iowa, Public Schools

Figure 29-4. Replacing exterior windows and doors. A—The new, energy-efficient windows installed in this house are smaller than the units they replaced. One is located where a door was removed. B—Carpentry students check the operation of a replacement window after installing it. C—A new doorway is being framed in place of a window in this remodeling job. The siding at the left was removed to provide access to an electrical receptacle that had to be relocated.

Handyman Construction

Figure 29-5. Replacing siding and soffits on a wood-shingled home.

29.1.2 Interior Renovation

Interior work can proceed while exterior work is underway. If the building is unoccupied, it is usually best to bring all phases of the work along at the same rate. All of the demolition, mechanical system repair, drywalling, and stripping and painting can be done in sequence.

If the house is occupied, consideration must be given to the comfort of the occupants. In this situation, there are two choices:

- Completing a single room or area to the point where the space is livable, but without finishes or decoration.
- Zone-by-zone completion. In this approach, a floor or wing is completely finished before going on to the next floor or area.

If the house was built before 1978, it is likely that some or all of the paint and varnish may contain lead. Paint removal, whether scraping or sanding, can release fine particles containing lead into the atmosphere. The Environmental Protection Agency's Lead Renovation, Repair and Painting Rule (RRP Rule) requires that contractors performing renovation, repair, and painting projects that disturb lead-based paint in homes, child care facilities, pre-schools, and kindergartens built before 1978 must be certified by EPA (or an EPA-authorized state), use certified renovators who are trained by EPA-approved training providers, and follow lead-safe work practices.

PROCEDURE
Renovating an Interior

1. Demolish what is to be eliminated and remove the debris, **Figure 29-6**. This includes removal of structural elements such as studs. Provide temporary shoring as needed.
2. Perform structural work, such as alterations to bearing walls, **Figure 29-7**.
3. Insulate exterior walls and ceiling.
4. Make changes to or remove nonbearing partitions; install or remove soffits or pipe chases; repair subfloors; install nailers for built-ins, plumbing fixtures, and light fixtures.
5. Rough-in plumbing and electrical wiring.
6. Install drywall, repair lath and plaster, tape drywall joints, apply a skim coat of plaster. See **Figure 29-8**.
7. Put down underlayment for new flooring or tile.
8. Repair or install ceramic tile.
9. Install fixtures for plumbing and heating. Set fixtures before any finishing to avoid damage to floors, walls, or trim.
10. Install light switches and receptacles.
11. Install unfinished floor coverings.

Continued

PROCEDURE *(continued)*

12. Repair or install woodwork or trim. Refinish old woodwork and cabinetry, if in good repair, **Figure 29-9**.
13. Prime and paint walls.
14. Apply finish to floors, if unfinished.
15. Install prefinished flooring materials.
16. Touch up painted and clear-finished surfaces.
17. Install light fixtures and hardware such as cover plates and handrails.
18. Clean finished areas and wash windows.

Safety Note

When refinishing floors or woodwork that has been varnished, be sure to wear a dust mask or respirator and use a sander with a HEPA filter attachment to capture as much dust as possible.

A   *Des Moines, Iowa, Public Schools*

B   *Des Moines, Iowa, Public Schools*

Figure 29-6. Demolition work. A—Using a hammer drill to remove a masonry wall. B—A dumpster is usually located on a construction site to collect debris and scrap materials. When material is being discarded from upper stories, a chute is used to direct it into the dumpster.

Goodheart-Willcox Publisher

Figure 29-7. This carpenter is fastening one end of a large laminated beam that will serve as a header for an enlarged opening between an entryway and family room. The post in the foreground is part of temporary shoring used to support the structure during changes to the bearing wall.

Suz Waldron/Shutterstock.com

Figure 29-8. Drywall is installed after plumbing and electrical rough-in work is completed.

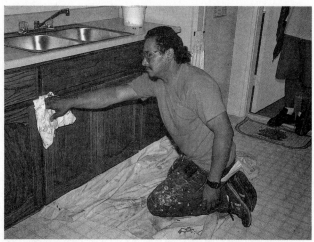

T.W. Lewis Construction Company

Figure 29-9 Renovation work often involves refinishing woodwork or cabinetry.

Code Note

Generally, when remodeling, all new work is required to meet the latest building code standards. Building codes may define multiple classifications of work, with each classification having different code requirements. For example, the IRC provides different requirements for four classes of work:

- Repairs—Restoring, replacing, or maintaining existing materials or equipment.
- Renovations—Extensive repair, modification of load-bearing elements, equipment upgrades.
- Alterations—Reconfiguring a space, adding or removing a door or window, reconfiguring or extending any system, installation of additional equipment.
- Reconstruction—Reconfiguring a space and affecting an exit, a renovation or alteration in which the work area cannot be occupied due to lack of means-of-egress, extensive alterations.

29.2 Design of Old Structures

Many older homes were built using balloon framing. This method was popular in the United States from about 1850–1930s. The main feature of this construction is long studs that run from the sill all the way to the plate on which the rafters rest. These are called building height studs. They may be spaced anywhere from 12″ to 24″ OC.

In balloon framing, the second floor joists rest on a narrow, horizontal member called a *ribbon*. This member is let into the studs. It is usually a 1×6 board. *Firestops* are short pieces of 2″ thick blocking installed between studs and joists at each floor level. Their purpose is to prevent spread of fire from one part of the dwelling to another.

Figure 29-10 shows details of balloon frame construction. Additional information on housing designs is located in Chapter 10, **Floor Framing**; Chapter 11, **Wall and Ceiling Framing**; Chapter 12, **Roof Framing**; and Chapter 26, **Post-and-Beam Construction**.

Figure 29-11 shows details of post-and-beam construction. This type of construction was popular in colonial times. Some new homes are constructed with this method.

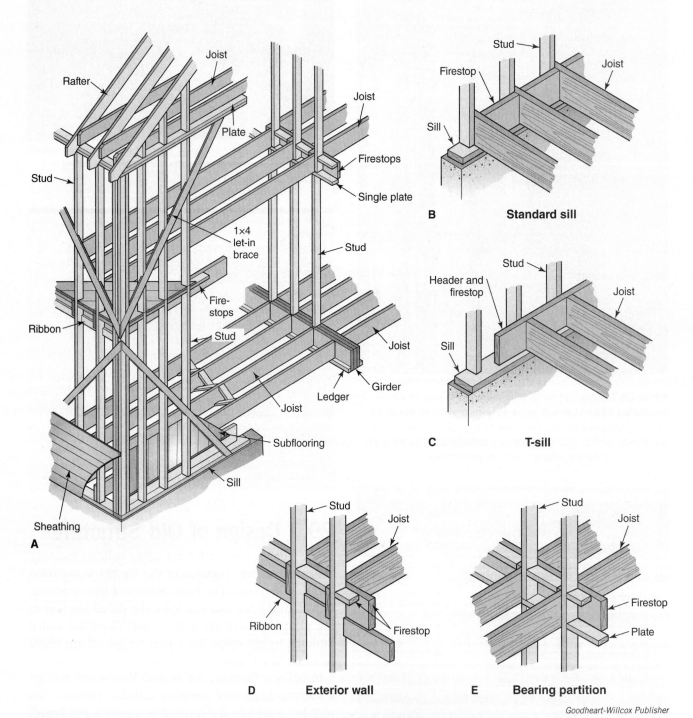

Figure 29-10. Balloon framing. A—Major features of balloon framing include studs running the full height of the building and ribbons used to support floor joists for the second story. B—Detail of a standard sill. C—Detail of a T-sill. D—Detail showing how joists are attached to the ribbon for a second story. E—Detail of framing used for a bearing partition.

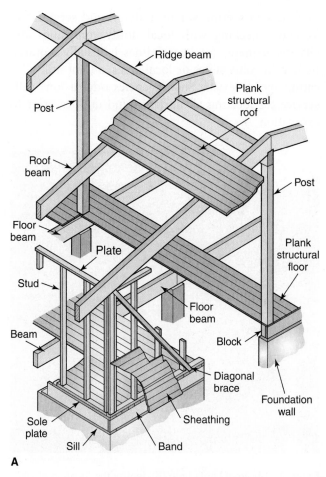

A

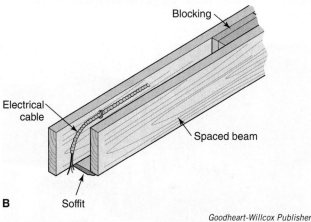

B

Goodheart-Willcox Publisher

Figure 29-11. Post-and-beam construction. A—General construction detail. B—Spaced beams used in this construction method allow electrical wires and plumbing lines to be hidden.

29.3 Replacing Rotted Sills

Rotted sills and other structural members are not uncommon in older homes. Often, leaks and other problems have not been corrected in these homes.

Replacing a sill in platform construction requires lifting and supporting the weight of the building while the old sill is removed and replaced with a new, pressure-treated sill. See **Figure 29-12**.

Reinforcing rotted studs usually does not require temporary support, unless the rot has caused the wall to settle and sag. Reinforcing is best done by *sistering*. This is accomplished by nailing another stud alongside the rotted member. Installing the sister requires removal of the interior wall covering or exterior siding.

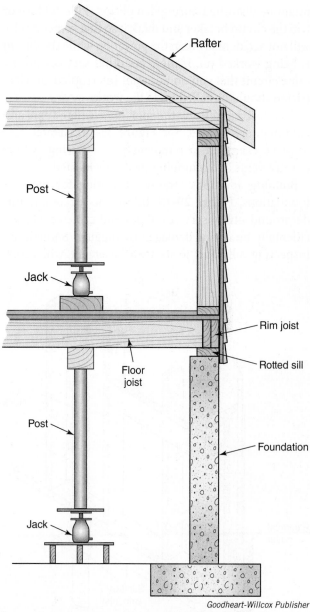

Goodheart-Willcox Publisher

Figure 29-12. Using shoring at two levels to raise and support the structure when replacing a rotted sill.

29.4 Hidden Structural Details

Before demolition of any kind is attempted for inside changes, study the building to determine what type of construction was used. This will make removal of structural elements much easier and safer.

If the house has a basement or crawl space, examine the areas beneath first-floor walls. Try to determine if there are gas or fuel lines, electric conductors, or plumbing running through the walls or floors where you will be working. See **Figure 29-13**. Locate and close valves to stop gas or water flow in service lines. As a precaution, cut power to any electrical circuit that might be disturbed during demolition. Pull the fuse or trip the circuit breaker and mark it so that someone else will not accidentally turn the power on while the circuit is being worked on. If there are other services on the same circuit that you do not want interrupted, an electrician should be called to do the work.

Go outside and study the roof. Note where vents and chimneys are located. See **Figure 29-14**. Greater care must be exercised in removing wall coverings where exhaust vents and plumbing stacks are located.

Building additions require excavating for piers or foundations, **Figure 29-15**. Be sure to locate any underground utility services. Pipes and cables could accidentally be cut or damaged by digging. Sometimes, inspecting sills and joists in a basement will reveal markings indicating where underground services are located. Checking with local utility companies and city departments for buried lines is usually required by law. In some states or local areas, a joint utility locating agency exists. Such agencies offer a "one-call" service for locating all underground utilities prior to excavation.

Goodheart-Willcox Publisher

Figure 29-14. Roof vents help to identify the location of DWV piping.

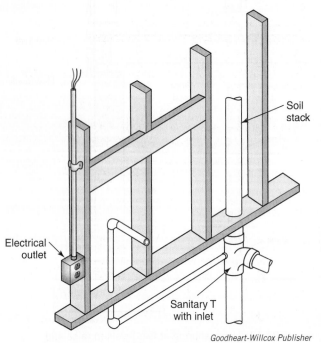

Goodheart-Willcox Publisher

Figure 29-13. Electrical, water supply, and DWV lines are usually concealed within walls. Use care in removing wall coverings to avoid damaging these lines.

Jack Klasey

Figure 29-15. Excavating for a building addition must be done with care to avoid damaging utility lines. Water, gas, and sewer lines must be located before beginning to dig.

29.5 Removing Old Walls

The first step in opening up or tearing out a wall is to remove trim such as baseboards, cove strips at the ceiling, and window and door trim. If these materials are to be reused, carefully remove them.

Next, remove the wall covering to expose the framing, **Figure 29-16**. Be sure to wear a mask to avoid breathing in dust and particles of construction materials. Most construction done during the past 40 years used drywall. In homes older than that, wood lath and plaster were used.

Tools useful for demolition include a sledgehammer, ripping or wrecking bar, hammer or hatchet, and rip chisel, **Figure 29-17**. The rip chisel is essential if any of the trim is to be saved and reused. A reciprocating saw is also useful for cutting framing.

Des Moines, Iowa, Public Schools

Figure 29-17. A sledgehammer is sometimes used in demolition work. In this case, a partition made from hollow masonry is being removed.

Safety Note

Before using any cutting tools or brute force to tear into a wall, thoroughly investigate what might be concealed behind the surface. It can be very dangerous to cut into a wall with live electric or gas service.

Pro Tip

Many popular television programs show workers smashing cabinets and walls with sledgehammers. Most cabinets can be removed just as quickly by removing the screws that hold them to the wall. There are also usually screws through the face frame to hold one cabinet to another. Using a sledgehammer to remove plaster and drywall runs the risk of damaging pipes and electrical equipment that is hidden in the wall cavity.

Many carpenters use a flat-bladed garden spade to remove lath and drywall. This tool will also remove drywall and lathing nails as it is run over the studs. If the wall is a partition, removal of the covering is all that is necessary to strip the framing. In outside walls, insulation and siding might also need to be removed to strip the framing.

For good housekeeping, it is often useful to rent a large trash container like the one shown in **Figure 29-18**.

Goodheart-Willcox Publisher

Figure 29-16. Removing wall coverings exposes the framing and utilities running inside the walls. To avoid unnecessary damage to systems or electrical shock, always be cautious when opening up existing walls. In this older home being renovated, the large pipes at left supply the radiators of the hot-water heating system. Also visible are electrical conduit and water supply pipes.

Steven Belanger/Shutterstock.com

Figure 29-18. A portable trash container, or dumpster, is rented and positioned at the jobsite to help keep it clean. When full, the dumpster is loaded onto a truck and transported to a disposal site.

Then, debris can be immediately cleared away so that it does not clutter up the jobsite. Loose debris is an eyesore as well as a safety hazard.

Pro Tip

To salvage interior trim for reuse, do not attempt to remove the nails by driving them back through the face of the trim. Instead, use a small wrecking bar or nippers to pull them through the back. This avoids splintering that would spoil the exposed face.

29.5.1 Recognizing Bearing Walls

Before removing a wall or a portion of it, determine whether or not it is a bearing wall. Bearing walls need shoring (temporary support) while the wall is being removed. Permanent supports of some kind must be in place before the remodeling is completed.

Determining which walls are bearing walls is quite simple with outside walls; interior partitions are another matter. All outside walls support some weight of the structure. End walls of gabled, single-story houses are the sole exception. They carry little weight except that of the wall section from the peak of the roof to the ceiling level. Outside walls running parallel to the ridge of the roof carry the most load since they provide almost all of the support for the roof, if the roof is framed with conventional rafters. Roof trusses are capable of spanning the entire width of the building so they are not supported at their centers. When the roof is framed with trusses, it is not necessary for any interior partitions to support the roof. Even with a trussed roof, first-floor walls may support both spliced ceiling joists and second-floor walls that are directly above. Identifying which partitions (interior walls) support weight from structures above is not always simple. The following conditions are usually an indication of a bearing wall. See **Figure 29-19**.

- The wall runs down the middle of the length of the house.
- Ceiling joists run perpendicular to the wall.
- Overhead joists are spliced over the wall, indicating they depend on the wall for support.
- The wall runs at right angles to overhead joists and breaks up a long span. The joists may not be spliced over the wall. Check span tables for load-bearing ability of the joist.
- The wall is directly below a parallel wall on the upper level. If the walls are parallel to the run of the joists, the joists may have been doubled or tripled to carry the load of the upper wall. The only way to be certain is to break through the ceiling for a visual inspection.

Familiarize yourself with the framing principles used for the type of structure you are remodeling. Study the framing chapters in this book for details.

29.6 Providing Shoring

Shoring is temporary support of a building's frame. It must be installed to support ceilings or upper floors before all or a portion of a bearing wall is removed. **Figure 29-20** shows one type. It is simply a short wall of 2×4 plates and studs. The shoring wall can be assembled on the floor and lifted into place. To distribute the load, use double plates on top and bottom.

In some cases, it may be necessary to place a second shoring on the other side of the bearing wall. Usually, the joists are lapped and spiked well enough to provide stiffness to the joint so a second shoring is not necessary.

A second shoring method is used by many carpenters. Larger dimension lumber—4×4s, for example—are used for plates. Adjustable steel posts, or *jack posts*, are used in place of the studs, **Figure 29-21**.

Be sure that the shoring is resting on adequate support. That is, it should rest across several joists. If shoring is running parallel to floor joists, lay down planking to distribute weight across several joists. Do the same at the ceiling.

Leave the shoring in place until a permanent support has been framed in to take the place of the bearing wall. This may be a lintel or a concealed header.

Once shoring is in place, remove the old bearing wall. Remove studs first, being careful not to disturb plaster or drywall at the ceiling. Try to salvage the studs for future use. Use a reciprocating saw to cut the studs away from the sole plate if needed. If the studs are not being recycled, use a sledge at the bottom to loosen them from the sole plate.

PROCEDURE

Installing Shoring

1. Measure the distance from floor to ceiling.
2. Cut the studs shorter than the floor-to-ceiling distance to allow 1/4"–1/2" clearance.
3. Nail the studs to one top and one bottom plate.
4. Attach the second top plate. Leave the second bottom (or sole) plate off to allow clearance when raising the shoring.
5. Square up the shoring wall and attach a 1×4 diagonal brace.
6. Lift the shoring in place. If necessary, get a helper to do some of the lifting. Keep the shoring far enough from the partition so you will have room to work.
7. Slide the second plate under the bottom plate. Use wood shingle shims at regular intervals along the bottom of the wall to bring the temporary wall to full height.

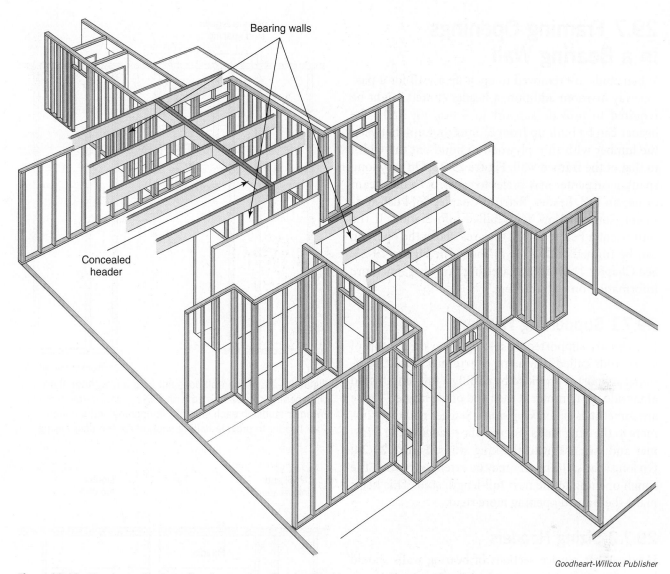

Goodheart-Willcox Publisher

Figure 29-19. Bearing walls can often be most easily identified by examining an unfloored attic space. Walls running the length of the building center usually support the joists.

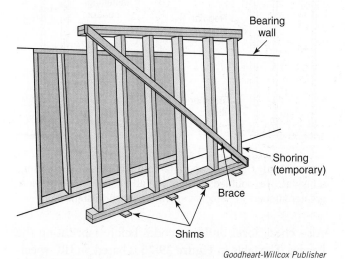

Goodheart-Willcox Publisher

Figure 29-20. Shoring must be used to support the weight of the structure when a bearing wall is being removed or altered. The shoring is positioned next to and parallel with the bearing wall.

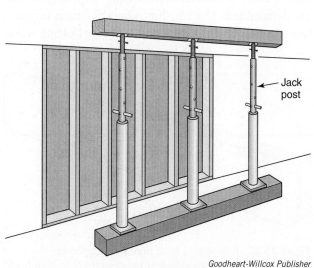

Goodheart-Willcox Publisher

Figure 29-21. An alternate shoring method uses heavy timbers and adjustable steel pillars called jack posts.

29.7 Framing Openings in a Bearing Wall

When studs are removed to open up a wall for a passageway or room addition, a header (lintel) might be required to provide support in a bearing wall. The header can be built up from 2″ lumber. Shim between the lumber with thin plywood to build out the width to that of the framed wall, **Figure 29-22**. On very long spans, a carpenter may prefer to construct a box beam or use an LVL header. Refer to Chapter 26, **Post-and-Beam Construction**, for detailed information. If the wall is not a bearing wall, the area above the opening can be framed with 2×4s as shown in **Figure 29-23**. See Chapter 11, **Wall and Ceiling Framing**, for more information on wall framing.

29.7.1 Supporting Headers

Headers are supported by resting them on the top of short studs called trimmers, or trimmer studs. Additional support is provided by nailing a full-length stud alongside the trimmer at each end of the header. These are sometimes called *king studs*. Stagger nail the trimmers to the king studs. Toenail the header to the trimmer and end nail it to the king studs, **Figure 29-24**. On long spans, the header may be extended beyond the rough opening to the next full-length stud. This helps make the framed opening more rigid.

29.7.2 Sizing Headers

Headers that replace sections of bearing walls should be sized according to recommended standards for load-bearing ability. The size is specified in the architectural drawings for the remodeling job. A carpenter should also be aware of what the standards are and how the sizes are determined. The IRC lists allowable spans for various size headers for exterior and interior bearing walls.

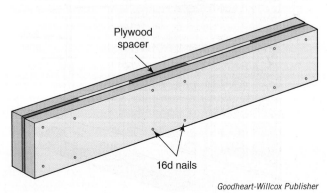

Figure 29-22. A header is built from two lengths of 2″ lumber with a plywood spacer between them to match finished width to the wall studs.

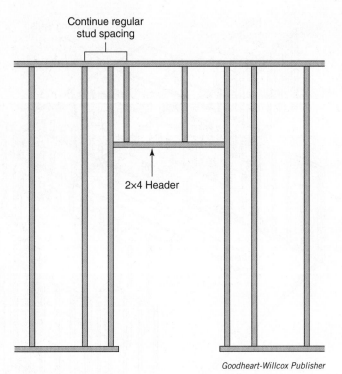

Figure 29-23. If the wall does not bear a load from above, the opening can be framed with 2×4s. Some carpenters use trimmer studs on each side of the opening and a double 2×4 header to provide nailing support for the door casing.

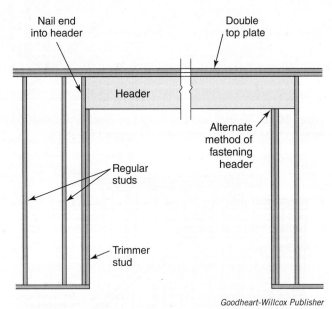

Figure 29-24. Standard and alternate methods of supporting a header. The doubled studs of the alternate method provide greater support for headers spanning longer distances.

Also check local building codes before fabricating the header. The table in **Figure 29-25** is based on IRC specifications but covers only a few possible opening conditions. Laminated-veneer lumber (LVL) is often used for headers.

Allowable Header Spans						
		Span with 30 lb per sq ft Snow Load		Span with 50 lb per sq ft Snow Load		Span with 70 lb per sq ft Snow Load	
Header Supporting	Size	Building width		Building width		Building width	
		20′	28′	20′	28′	20′	28′
Roof and ceiling	2 - 2×4	3′ - 6″	3′ - 2″	3′ - 2″	2′ - 9″	2′ - 10″	2′ - 6″
	2 - 2×10	8′ - 5″	7′ - 3″	7′ - 3″	6′ - 3″	6′ - 6″	5′ - 7″
	3 - 2×8	8′ - 4″	7′ - 5″	7′ - 5″	6′ - 5″	6′ - 8″	5′ - 9″
Roof, ceiling, and one clear-span floor	2 - 2×4	3′ - 1″	2′ - 9″	2′ - 9″	2′ - 5″	2′ - 7″	2′ - 3″
	2 - 2×10	7′ - 0″	6′ - 2″	6′ - 4″	5′ - 6″	5′ - 9″	5′ - 1″
	3 - 2×8	7′ - 2″	6′ - 3″	6′ - 5″	5′ - 8″	5′ - 11″	5′ - 2″
Roof, ceiling, and one clear-span floor	2 - 2×4	2′ - 1″	1′ - 8″	2′ - 0″	1′ - 8″	2′ - 0″	1′ - 8″
	2 - 2×10	4′ - 9″	4′ - 1″	4′ - 8″	4′ - 0″	4′ - 7″	4′ - 0″
	3 - 2×8	4′ - 10″	4′ - 2″	4′ - 9″	4′ - 1″	5′ - 5″	4′ - 8″

Goodheart-Willcox Publisher

Figure 29-25. This table lists headers for a few opening sizes and loading conditions. Complete tables are found in the IRC.

29.7.3 Concealed Headers and Saddle Beams

There are times when it is not desirable to have a header extending below the ceiling level. In such cases, the header must be at the same level or above the joists it supports. A *concealed header* is butted against the ends of the joists it is supporting. A *saddle beam* is positioned above the joists it supports. See **Figure 29-26**.

A concealed header is simple to add to an exterior wall in platform construction, **Figure 29-27**.

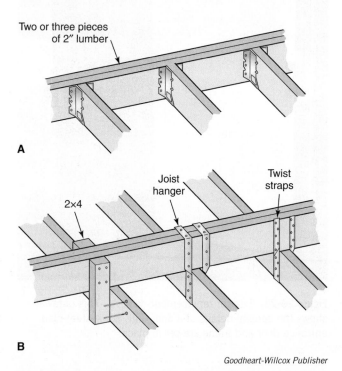

A—Two or three pieces of 2″ lumber

B—2×4, Joist hanger, Twist straps

Goodheart-Willcox Publisher

Figure 29-26. Hidden support methods using joist hangers. A—Concealed header. B—Saddle beam showing three attachment methods.

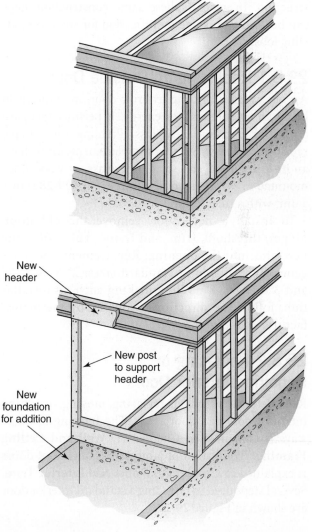

New header

New post to support header

New foundation for addition

Goodheart-Willcox Publisher

Figure 29-27. When remodeling a wall in platform construction, a header can often be constructed by adding to a band joist on a second floor.

Simply attach a second piece of 2″ lumber to the band joist. Add studs at each end to provide support for the header.

Because they are placed above the joists, saddle beams can only be used in the attic space. They should rest on top of joists supported by the remaining portions of the bearing wall.

> **Pro Tip**
> An architect or engineer should check the loads and specify the header sizes.

29.8 Small Remodeling and Repair Jobs

Most of the chapters in this book deal with new construction. In general, these same construction steps can be used with some modification for small remodeling jobs.

29.8.1 Replacing an Outside Door

An old, worn, ill-fitting exterior door, in addition to being an eyesore, wastes energy. It permits excessive air infiltration and heat loss. Prehung units make replacement easy. Standard replacement units have an insulated steel, steel-clad, fiberglass, or wood door mounted in a steel or wood frame. **Figure 29-28** shows a unit with a steel frame.

To install a new door, remove the old door, hinges, threshold, trim, and frame. Then, slide the new unit into the opening. Replacement doors are generally available in standard sizes: 2′-6″ × 6′-8″ and 3′-0″ × 6′-8″. When making such an installation, follow the directions provided by the manufacturer of the unit.

29.8.2 Replacing or Repairing Interior Doors

Remodeling may include replacing, moving, or adding interior doors and doorjambs. The framing of doorways is explained in Chapter 11, **Wall and Ceiling Framing**. Installation of jambs and hanging of doors is explained in Chapter 22, **Doors and Interior Trim**. Several steps used in replacing a standard interior door are shown in **Figure 29-29**.

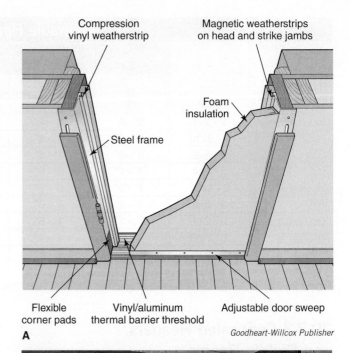

Goodheart-Willcox Publisher

B   *General Products, Inc.*

Figure 29-28. Exterior replacement door. A—This cutaway shows the general details of a door unit. B—A steel-clad entrance door and frame are set in place.

A

C

D

B

McDaniels Construction Co., Inc.

Figure 29-29. Preparing to replace an interior door. A—Measure the door opening after removing trim to determine the replacement door size needed. B—Removing the old door. C—Removing the old hinge leaves from the jamb. D—Removing the striker plate.

PROCEDURE

Hanging an Interior Door with Jamb Clips

1. Install a jamb clip directly behind the center of each hinge with the 1/2″ screws (provided). Make sure the jamb clip's clearance hole lines up directly behind one of the center screws of each hinge, **Figure 29-30**.

2. Install the other three jamb clips directly across from the hinges on the opposite side of the door.

3. Using a 6′ or 4′ level, draw a plumb line all the way down the drywall approximately 1/2″ in from the rough opening on the hinge side. This line must be perfectly plumb.

4. Place the door into the opening. Starting with the top bracket on the hinge side of the door, line up a reference notch on the bracket with the plumb line that was drawn in step 3. Screw the jamb clip into

place using a drywall screw. Use the same reference notch when screwing in the next two jamb clips on the hinge side. The door is now perfectly level.

5. Screw in the remaining three jamb clips on the latch side, starting on the top. Make sure the margin (gap between door and door frame) stays the same. Check door operation.

6. For added support, remove a center screw from each of the hinges and replace with a long set screw (2″ or longer) that goes through the hinge, bracket clearance hole, and into the wall stud.

7. Trim/casing can now be installed directly over the jamb clips to complete installation.

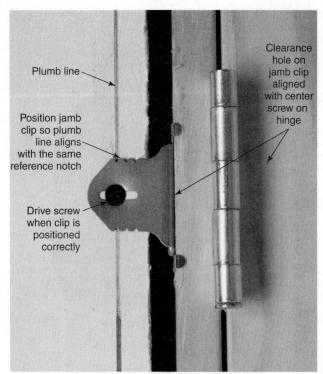

Figure 29-30. A jamb clip in place on the back of the jamb and fastened to the wall framing.

Use of a special steel clip, known as a *jamb clip*, presents another method of hanging an interior door, **Figure 29-31**. The steel clips eliminate the need to use wooden shims.

As a building ages, day-to-day usage, settling, and other factors cause problems with doors. Hinges wear and latches no longer work. Also, raising the floor level with new solid flooring material or carpeting makes it necessary to trim the door to clear the new flooring material. If the door itself is still in good condition, repairs are in order. Consider the following conditions and suggested repairs.

- If the door binds or will not close tightly, check for loose hinges. Reinstall the hinges. You may need to use longer screws or fill oversized screw holes and reinstall screws. If the door still binds, note where it is binding on the frame and mark the spot with a pencil, **Figure 29-32**. Then, use a hand plane to remove a small amount of material. Be careful not to remove too much wood.
- Installation of new flooring may make it necessary to remove 1/2″ or more from the bottom of the door. Remove the door from its hinges. Have a helper hold the door in the door opening in its normal position with about 1/16″ space at the top or with hinge leaves aligned. Mark the door bottom. If the door has veneered facings, make a

shallow cut on the line with a utility knife. Use a steel straight edge as a guide. Transfer the cut line to the opposite side and make a shallow cut there, as well. This will prevent splintering of the veneer when the excess is sawed or planed away.

- If the door binds on the threshold or floor, check for a loose upper hinge or worn hinges. Replace the hinges if worn. Check for sagging of the door by measuring diagonals. Sagging is confirmed if the diagonal from the top of the hinge side of the door is longer. Replace the door or plane the bottom where it is binding.
- Before replacing a seemingly defective door latch, check to see if it aligns with the striker. Reset the striker if alignment is the problem.

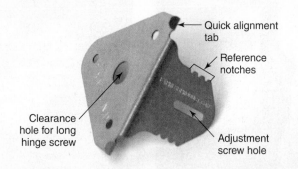

Figure 29-31. Jamb clips allow door installation without the use of shims.

Figure 29-32. Marking the point where a door is sticking provides a guide for removing the excess material with a plane.

29.8.3 Installing New Windows

Replacement windows are available in a variety of sizes and styles to fit rough openings of many old windows, **Figure 29-33**. Before ordering replacements, take careful measurements. Measure again before installing to ensure proper fit. Check around the opening for signs of rot or leakage. This must be fixed before installation, as even a new window will not solve this problem. Repair any problem before installing the new window.

Specific instructions for installation are provided by the window manufacturer. Carefully read them before beginning installation. Also, refer to Chapter 16, **Windows and Exterior Doors**.

29.8.4 Repairing Wood Shingles

Wood shingles are used for both roofing and exterior wall coverings. Over time, individual shingles may sustain damage and require repair or replacement.

Split shingles can be repaired. If the crack is small, fill it with asphalt roofing cement. If the shingle appears loose, secure it with small, galvanized nails. Cover the nail heads with asphalt roofing cement. If the crack is large or a piece of the shingle is missing, install a piece of flashing under the shingle. Make sure the flashing extends well above the top of the crack or that it extends under the next course. Secure the patch with galvanized nails. Use asphalt roofing cement over the patch.

Badly damaged wood shingles can be replaced without disturbing the surrounding shingles. Use a wood chisel or screwdriver to split the damaged shingle into narrow pieces. These can be pulled out with your fingers. Use a hacksaw blade to cut off the nails. Slide a new shingle of the same size into place, but leave it

Lisa Signori/Shutterstock.com

Figure 29-33. This vinyl replacement window fits the rough opening of the original window. The original trim has been kept in place and clad with aluminum in a color that suits the house.

about 1/4″ below adjoining shingles. Secure it with two galvanized nails driven at an angle against the upper course. Then, drive the new shingle upward until the nail heads are concealed.

29.8.5 Repairing Asphalt Shingles

Wind sometimes causes damage to asphalt shingles by flipping them up and even tearing them. If the shingle is not too badly damaged, apply a generous dab of asphalt cement to the underside. Press it down and secure it with broad-headed roofing nails. Apply asphalt cement to the nail heads.

Asphalt shingles are usually manufactured in two- or three-tab sections or strips. If a strip is badly damaged, it can be replaced without destroying the adjacent shingles. Lift the upper shingle and remove the nails from the damaged strip using a flat bar, or cut them off with a hacksaw blade. Remove the damaged strip, slip the new one into place, and secure it by blind nailing. As an alternative method, secure the new shingle by nailing through the upper shingle. Cover the nail heads with asphalt cement.

Pro Tip
Asphalt shingles are pliable when they are warm, but can be quite brittle when they are cold. If possible, wait for warm weather to make repairs to them.

29.9 Building Additions onto Homes

When building additions to older homes, check all dimensions of the existing construction carefully. In particular, check the following:

- Foundations—New masonry units may not match the size of older units. You may need to adjust the level of footings or alter the mortar joint thickness.
- Ceiling heights—In balloon framing, ceiling heights may vary from house to house. Study local codes to see if your changes are acceptable.
- Dimensions of framing members—The standards for dimension lumber (2×4s, 2×6s, etc.) have changed with time. Check the actual dimension of older lumber.

In addition, in recent years regions that experience serious damage from earthquakes, high winds, tornadoes, and hurricanes have revised building codes to reduce future damage. **Figure 29-34** shows how and where to install ties that anchor a wood frame building from the foundation to the rafters. Check local codes to see if such ties must be installed in an addition.

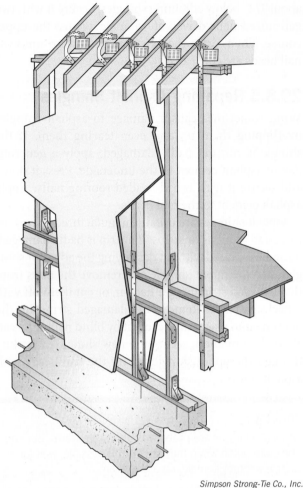

Building codes in these high-wind and earthquake zones also frequently have specific requirements for braced walls. This is covered in detail in the IRC. Walls with code-required bracing are called *shear walls*. Shear walls can be braced with let-in wood bracing or, in some cases, metal straps. When the wall has several openings for windows and doors, engineered shear panels may be required, **Figure 29-35**. Most building codes require the wall bracing to be designed by a licensed engineer.

29.10 Responsible Renovation

A major remodeling of a building may seem to generate a great deal of waste. Even so, it is usually far less damaging to the environment than new construction. Inevitably, some waste ends up in a landfill. This waste can be kept to a minimum by careful salvage and attention to what can be recycled.

Window trim, doors, wood paneling, hardwood flooring, tile, brick, and framing lumber can be set aside and reused. If not used in the remodeling, someone else may use it. Careful removal of recyclable materials is more time-consuming than demolition, however. Many carpenters and contractors find salvage too costly. This approach is most feasible for do-it-yourself projects.

Salvaged material should be carefully cleaned and stored out of the weather until it can be reused or sold. Some material will need to be given away. A "Free Material" sign on the lawn works well for giveaways. The savings in disposal costs may even pay for the extra time spent in careful removal.

Simpson Strong-Tie Co., Inc.

Figure 29-34. To help buildings withstand earthquakes and high winds, special metal ties are used to tie the various components of the building together and anchor the structure to its foundation.

Goodheart-Willcox Publisher

Figure 29-35. Steel shear panels are installed around this garage door to comply with bracing requirements listed in the code.

Inefficient fixtures should not be recycled. For example, old flush toilets are water wasters. Old windows are equally inefficient and waste energy. They should be used only for unheated garages or sheds.

Green Note

In selecting building materials for remodeling, choose those made from recycled material wherever possible. By doing so, you save natural resources and help to create markets for recycled products. This is a key requirement to every recycling program.

Safety Note

Dealing with hazardous materials is a concern in every remodeling project. In a house built before 1978, painted surfaces should be tested for lead content. Lead-based paint was banned in 1978. Kits for testing lead content in paint are available. Lead-contaminated materials must be disposed of by a person who has been trained and certified to work with materials containing lead. Other possible hazards include chlordane-contaminated lumber, lead solder in old pipes, and asbestos in pipe insulation, old flooring tile, and exterior siding.

Construction Careers
Asbestos Removal Technician

For many years, the mineral fiber asbestos was used for fireproofing, pipe insulation, and other applications in buildings. It is no longer used in construction since airborne particles of the material have been shown to cause cancer and diseases of the lungs. Many older buildings have large amounts of asbestos in them, which poses danger to workers involved in demolition or remodeling work. This material must be removed for the health and safety of workers and occupants. Asbestos contamination has been a particular concern in school buildings where young children can be exposed to asbestos particles.

These concerns have created a new building trades specialty, the asbestos removal technician. A related field is lead removal, since old lead-based paints are a health hazard as well. Many people working in this field are trained in both lead and asbestos removal. Older structures frequently have both types of material.

Since buildings usually cannot be occupied while asbestos is being removed, asbestos removal technicians typically work in teams to complete the work in as short a time as possible. The building or portion of the structure where the work is being done is sealed to prevent the escape of airborne asbestos particles. To protect their health, asbestos removal technicians (and, often, lead removal workers) wear full body suits with an air supply. They physically remove the asbestos materials from the structure, enclosing all debris in sealed containers for safe disposal. Specialized vacuum cleaners with highly efficient filters are used to clean up dust resulting from the removal activities.

Both asbestos and lead removal work can be dangerous, although strict protective measures are taken to prevent the breathing of toxic dust. As in most construction occupations, working with hand and power tools presents dangers, as do the necessary climbing, lifting, and other physical activities. Working inside full body suits is hot and uncomfortable and may be particularly difficult for people with claustrophobia. Work hours may often be long, since there are usually deadlines to complete the work. These specialty technicians are generally well-paid, however, ranking with the more well-compensated members of the traditional building trades.

Workers in both asbestos and lead removal must be licensed. Although formal education beyond high school is not needed, licensing requires successful completion of a formal 32–40 hour training program. Employment in these fields is expected to continue to grow as more and more aging buildings containing asbestos and lead are demolished or renovated.

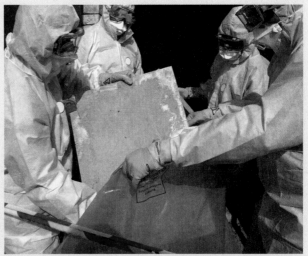

bermau/iStock/Thinkstock

Many older buildings contain asbestos, which, like lead-based paint, is harmful if inhaled. Asbestos removal technicians wear full body suits to prevent inhaling any toxic particles.

Chapter Review

Summary

- Remodeling and renovation is often more difficult than new construction. Existing structures may have hidden defects that must be addressed.
- A carpenter must be able to visualize how a structure was built so that no structural damage is done during demolition and removal of old walls.
- Work should be done in the proper sequence, beginning with structural repairs and changes working from the bottom up.
- Interior renovation work can proceed at the same time as exterior work.
- A major project in renovation is often replacing rotted building sills.
- If walls are to be removed, care must be taken to find hidden structural details, whether or not the wall is a bearing wall.
- When a bearing wall is removed or changed, proper shoring must be provided to support the weight of the structure.
- New openings in bearing walls require calculating header size to properly support loads above the openings.
- There are many small remodeling jobs that may be individually done or a part of a larger project. These include door and window replacement and repairs to roofing materials.
- Disposal of waste from remodeling projects should be done in an environmentally responsible manner.
- As in all construction activity, safe working practices and use of proper protective gear are vital.

Know and Understand

Answer the following questions using the information in this chapter.

1. *True or False?* Because of its high cost, remodeling a home is generally not worth the effort.

2. The first concern when planning the renovation of a long-neglected house is _____.
   A. condition of the siding
   B. condition of the foundation
   C. assuring that mechanical systems are working
   D. repairing any leaks found

3. The last step in renewing the exterior of a house is to _____.
   A. complete the site work
   B. caulk, glaze, and putty
   C. paint
   D. stain siding and trim

4. If a house was built before _____, it is likely that some or all of the paint and varnish may contain lead.
   A. 1978
   B. 1984
   C. 1990
   D. 2000

5. Balloon framing in older buildings used long studs known as _____ studs.
   A. ribbon
   B. building height
   C. balloon
   D. foundation

6. Before rotted sills can be repaired, _____.
   A. flooring in that area should be removed
   B. the building must be supported in some way to remove the load from the sills
   C. measurements should be taken of building's height
   D. a carpenter must obtain a license to perform this complicated task

7. Before starting excavation for an addition, determine the location of _____ utility services.
   A. unused
   B. multiple
   C. aboveground
   D. underground

8. A _____ is a useful tool to use if any trim removed is to be saved and reused.
   A. hatchet
   B. rip chisel
   C. reciprocating saw
   D. hammer

9. *True or False?* Drywall is a temporary wall installed to support a building's frame before all or a portion of a bearing wall is removed.

10. Headers are supported by resting them on top of short studs called _____.
    A. king studs
    B. lintels
    C. jack posts
    D. shoring

11. _____ are placed above the joists they support, meaning they can only be used in the attic space.
    A. King studs
    B. Concealed headers
    C. Saddle beams
    D. Studs

12. If a door binds or will not close tightly, check for _____.
    A. high flooring
    B. loose hinges
    C. damaged trim
    D. plumb

13. *True or False?* Split shingles can be repaired.

14. *True or False?* Building renovation is more harmful to the environment than new construction.

Apply and Analyze

1. What is the first step in a sequence of tasks for renovating the exterior?

2. What does the term *sistering* mean?

3. Before beginning a remodeling job, why is it wise to find out the type of construction used?

4. List five indications that a partition is a bearing wall.

5. A built-up wooden header must span 4′-1″ in the outside wall of a 28′ wide building with one floor, a ceiling, and a roof above with a 30 lb per square foot snow load. What size built-up headers should be used?

6. When installing new flooring, what repairs might need to be made to any doors on the same floor?

7. When building an addition in a high-wind area, what design feature must be included to resist the force of the wind on the walls?

Critical Thinking

1. Working on a renovation job requires excellent communication and sometimes conflict resolution skills. Imagine you have been hired to remodel a bathroom. When you pull up the old flooring you discover the sub-floor has rotted from a leaky shower. The homeowner insists that you cover the costs of replacing the rotten subfloor as part of the original contract. You believe a change-order should be generated and the homeowner should cover the costs since it was

an unforeseen condition. Outline how you would approach this situation in a way that will not jeopardize the reputation of your company.

2. As a follow-up to question #1, what steps could have been taken before the job began that may have safeguarded your company against the risks of unforeseen circumstances such as this one?

3. Kayla is considering having the old windows in her home replaced. She asks you for advice on what type and quality of replacement windows you would recommend. What questions would you ask her to help her make a decision?

Communicating about Carpentry

1. **Speaking.** With a partner, role-play the following situation: A carpenter must explain the sequence of renovation or repair for the interior and exterior of the house. The customer should ask questions if the explanation is unclear. Switch roles and repeat the activity.

2. **Writing and Speaking.** You are a building contractor. You have been hired to inspect a building that was recently renovated to make sure it is "up to Code." The company added new lighting, new cabinetry, as well as an additional room. Create your inspection plan in the form of a presentation. Detail what you checked and the specifications you have met to allow this building to open for business. Share the presentation with the class as though the class were the client who needs approval from the county inspection board.

3. **Reading and Writing.** The ability to read and interpret information is an important workplace skill. Presume you work for a local, well-known construction company. Your employer is considering bidding on a contract to restore a historic landmark. Your supervisor wants you to evaluate and interpret some research on restoring historic landmarks to period standards. Locate at least three reliable resources for the most current information on historic restorations. Obtain local codes and restrictions, as well as federal and state regulations, on historic restorations. Locate information on the landmark itself to learn its history and possibly obtain pictures of period furnishings. Write a report summarizing your findings in an organized manner.

CHAPTER 30

Building Decks and Porches

OBJECTIVES

After studying this chapter, you will be able to:
- Identify the different types of decks and porches.
- Describe the advantages and disadvantages of various structural and decking materials.
- Select and install the appropriate types of fasteners for deck construction.
- Prepare the site, then lay out and construct a deck.
- State the purposes of a pergola.
- Describe the differences between deck and porch construction.

TECHNICAL TERMS

alkaline copper quaternary (ACQ)
chromated copper arsenate (CCA)
closed porch
composite decking
copper azole (CA)
dado
disodium octaborate tetrahydrate (DOT)
elevated deck
galvanic corrosion
ground-level deck
multilevel deck
open porch
pergola
semi-enclosed porch

L ifestyles that value the outdoors have given rise to the construction of various types of decks. These decks become an extension of the main dwelling, functioning as outdoor rooms. There are two basic types of decks: elevated and ground-level. Either type may be a single-level or multilevel deck.

An *elevated deck*, **Figure 30-1**, has a supporting ledger physically attached to the building wall. The principles of construction for an elevated deck are similar to those used in building the main structure. It is supported by posts and at least one side is attached to the house. Height of the supporting posts vary, depending on the slope of the lot and the deck design. If soil conditions permit excavation for footings, they can be dug by hand or with a power post-hole auger. See **Figure 30-2**. Some augers are powered by a small gasoline engine mounted on the device itself or the auger may be attached to and controlled by a tractor or other mobile power source.

Trex Company

Figure 30-1. This multilevel deck is built on piers, posts, and a framing system composed of girders and joists.

Goodheart-Willcox Publisher

Figure 30-2. Power augers can be used to quickly excavate holes for foundation piers. This large power auger is mounted on a tractor.

A *ground-level deck*, as the name suggests, is built at ground level, so posts are not required, **Figure 30-3**. It does not require a railing for safety, although one may be used for appearance or to define the edges of the deck. A ground-level deck may be constructed with the joists resting on concrete piers. The piers may only project an inch or two above the finished grade.

A *multilevel deck* has connected sections built on more than one level, **Figure 30-4**. It may be elevated or built at ground level. On a sloping site, the upper level of the deck is typically attached to the building.

30.1 Structural Materials

Before construction begins, a carpenter must choose materials suited to the construction plan. Among the choices in lumber are the grade, species, and size. Engineered lumber is another choice for structural members. If a composite or plastic is the chosen material, then there are various materials and designs available. There are also choices to make in fasteners, connectors, anchors, and other hardware.

30.1.1 Lumber

Resistance to decay is a major concern when choosing lumber for a deck. Deck materials are always subject to weather conditions and may not be protected by paint or other coatings. Some woods naturally resist rot, but are more expensive than commonly used treated lumber.

Trex Company

Figure 30-3. A ground-level deck may be built with its frame resting directly on the ground or on very low concrete piers.

Javani LLC/Shutterstock.com

Figure 30-4. An example of a multilevel deck. Some multilevel decks have three or more levels.

The chart in **Figure 30-5** describes various species that are suitable for decks. As shown in the chart, certain woods are better suited than others for deck framing. Most prized for its resistance to rot and termite damage is redwood cut from the heartwood of the tree. As a rule of thumb, all heartwood grades have the word *heart* in the grade name.

30.1.2 Treated Lumber

Because of material cost, treated lumber is more often used than redwood, cedar, or tropical hardwoods. At one time, the most common preservative used for lumber was ***chromated copper arsenate (CCA)***. CCA-treated lumber is no longer available for residential use due to toxic emissions. Studies show that arsenic, a toxic substance, could leach out of the wood and contaminate the soil and groundwater. The Environmental Protection Agency (EPA) ordered production to end January 1, 2004. One special use of CCA treatment is still allowed—it can be applied to lumber used below ground in permanent wood foundations. For information on permanent wood foundations, refer to Chapter 9, **Footings and Foundations**.

Other preservatives have replaced arsenic as a wood treatment. Two widely used substitutes are ***alkaline copper quaternary (ACQ)*** and ***copper azole (CA)***. Both of these treatments are high in copper content, and usually leave the lumber with a brownish color. The high copper content also creates corrosion problems with some flashing and fastener materials, requiring the use of stainless steel or heavily galvanized fasteners.

Softwoods Suited for Decks	
Western Red Cedar	Highly resistant to warping and weathering, like redwood though coarser; the heartwood resists decay but not termites; easy to work, though weak and brittle; moderate ability to hold nails.
Douglas Fir Western Larch	Heavy, strong, stiff, holds nails well; not easily worked with hand tools; when pressure treated, resists decay and termites.
True Fir Eastern and Western Hemlock	Firs are mostly lightweight and moderately soft; hemlocks are moderately strong and stiff; firs are easy to work, but hemlocks more difficult; shrinkage could be a problem; resistant to termites and decay only when pressure treated.
Ponderosa Pine	Somewhat strong and stiff, but not as strong as southern pine; very resistant to warping; holds nails well; when pressure treated, resists termites and decay.
Red Pine	Strong and stiff, but not to the degree of southern pine; easier to work and holds nails well; resistant to termites and decay when pressure treated.
Southern Yellow Pine	Hard, very stiff, and strong with good nail-holding quality; moderately easy-working; decay and termite resistant when pressure treated.
Redwood	Heartwood famous for durability and resistance to decay and termites; moderately light, but limited in structural strength; workable, but brittle; moderate ability to hold nails.

Goodheart-Willcox Publisher

Figure 30-5. A basis for choosing woods for deck building. Refer to samples of woods shown in Chapter 3 of this text.

Lumber treated with a borate preservative is also available, although harder to find. ***Disodium octaborate tetrahydrate (DOT)*** is a borax-based wood preservative that is nontoxic and does not cause galvanic corrosion to fasteners. It is insect, mold, and rot resistant. It is basically nontoxic to humans and animals. DOT has been used primarily in locations protected from weather. This was due to concerns about borate leaching from the material and reducing its effectiveness as a preservative when the wood gets wet. Recent studies indicate that leaching is less of a problem than originally thought and new techniques have better bonded the borates to the wood. Wood treated with DOT has a bluish color.

30.1.3 Engineered Lumber

Engineered wood construction materials are derived from smaller pieces of wood that are either laminated or modified in some other way to make them stronger and eliminate warping and shrinking. Given the same dimensions, engineered structural units are stronger than sawn lumber. In deck applications, engineered lumber is used for beams and joists. Some engineered wood products are not suited for exterior use where they will be exposed to the weather.

Glue-laminated timber (glulam) is manufactured by stacking finger-jointed layers of standard lumber and adhering the layers with adhesives. Performance is improved by placing different grades of wood in the laminations. Glulam is a traditional choice for

exposed applications. Pressure treating glulams makes them resistant to rot or termite damage. The material can be curved during manufacture for use in special applications.

Parallel-strand lumber (PSL) is made up of long strips of wood running parallel to each other and saturated with adhesives. It is sold under the trade name Parallam®. For exterior use, pressure treatment renders Parallam resistant to both decay and termites.

30.2 Decking Materials

Although treated lumber, redwood, cedar, or tropical hardwoods such as mahogany or teak have been the material traditionally used for the decking, both composites and vinyl have captured part of the market in recent years. Additionally, heat treating methods have been developed that allow many common wood species become a suitable material for decking.

30.2.1 Composite Decking

Composite decking is a blend of 30%–50% plastic with wood fibers or sawdust, **Figure 30-6**. The plastic is usually recycled. The combination of wood and plastic produces a skidproof and paintable surface. Other advantages include low maintenance, no splintering, and easy workability. Screws sink into the surface and disappear. Some brands offer 10-year warranties; others as long as 20 years.

CPG Building Products—TimberTech ReliaBoard

Figure 30-6. Composite decking materials offer low maintenance and are easily worked with the same tools as conventional lumber. Surface finishes are available in wood grain or other textures. This installation makes use of a hidden fastener system.

There are a number of firms manufacturing composite decking boards in standard sizes. Standard thicknesses are 1 1/4″ (5/4″) and 2″. Standard widths are 4″, 6″, and 8″. Boards are either plain or tongue-and-groove and may be solid or hollow. Some have channels that allow electrical conductors for lights and electrical outlets to be concealed. Composites weather to a light gray and can be painted or stained for appearance, if desired.

Some composites may have an artificial (rather than wood-like) appearance. Over time, color may fade from exposure to sunlight. During construction, sawdust and shavings must be collected on a drop cloth, since the debris is not biodegradable.

30.2.2 Vinyl Decking

Polyvinyl chloride (PVC) deck systems made from extruded or molded vinyl plastic offer decking boards, railings, spindles, and fascia. The boards usually have a cellular core and are coated with a layer of solid PVC. Vinyl decking can also be hollow with internal reinforcement for stiffness. Vinyl is available in a variety of colors and resists the effects of ultraviolet rays. It is a relatively maintenance-free decking material.

30.2.3 Thermally Modified Lumber

Thermally modified wood is gaining popularity as a decking material. The process of thermal modification transforms common species such as yellow pine and ash into a more stable, rot and insect resistant product. This is accomplished by baking the wood at a high temperature in a kiln, while using steam to keep the outside surface from drying too quickly. The high temperatures burn off the sugars and resins that bacteria and other organisms feed on. This process modifies the cellular structure of the wood, making it less absorbent to moisture and more stable. Because the wood contains no food source, organisms that would cause rot and decay will not feed on it.

Thermally modified wood can be milled, and fastened just like traditional lumber but is slightly weaker than before treatment. It is lightweight and has a rich, attractive hue when new, but will fade and gray quickly when subjected to the elements if not treated with sealer or stain.

30.3 Fasteners and Connectors

Various types of fasteners and connectors are available for deck installation, **Figure 30-7**. All connectors, related hardware, and fasteners must be corrosion resistant when using ACQ or CA pressure-treated lumber.

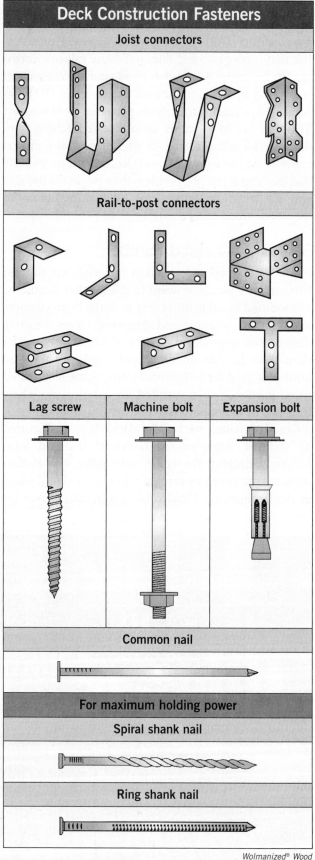

Deck Construction Fasteners

Joist connectors

Rail-to-post connectors

| Lag screw | Machine bolt | Expansion bolt |

Common nail

For maximum holding power

Spiral shank nail

Ring shank nail

Wolmanized® Wood

Figure 30-7. Fasteners typically used in deck construction. Joist connectors and rail-to-post connectors are formed out of sheet metal and designed to make deck frames stronger.

The copper-rich lumber treatment will set up galvanic corrosion with fasteners made from common steel or steel with a light galvanized coating. *Galvanic corrosion* is the eating away of metal fasteners and hardware due to a chemical reaction between dissimilar metals in a damp environment. Over time, this corrosion will weaken the deck structure to the point of possible collapse. Stainless steel fasteners and hardware are preferred, but the best grades of galvanized steel can be used. Hot-dipped galvanized coatings should have a G-185 rating. Electrogalvanized items should have a class rating of 40 or higher. Manufacturers of decking screws have developed ceramic coatings for their products that provide corrosion protection by serving as a barrier between the steel of the fastener and the copper in the preservative.

In addition to fasteners and other hardware, any flashing used at points where the decking adjoins the structure must be corrosion resistant. Stainless steel or copper flashing is preferred. Aluminum should never be used, since it is particularly susceptible to galvanic corrosion.

Special concealed fasteners that install on the underside of all types of deck boards are available. This method produces a deck surface unmarred by screw holes and eliminates a condition that causes rot in wood deck materials. Follow the manufacturer's instructions for ordering the quantity needed. Usually, the fasteners can be installed working from the top side of the deck.

PROCEDURE
Installing a Deck with Concealed Fasteners

1. Nail the fasteners to the top of all joists, alternating from side to side, **Figure 30-8**.
2. Begin attaching the deck boards, working from the top. Use two screws for each board. Joints should meet over a joist and a fastener must be installed on each side, **Figure 30-9**.
3. Continue attaching deck boards one at a time until the job is completed. **Figure 30-10** shows the underside of a completed deck. **Figure 30-11** shows a concealed fastening system that is used with composite decking.

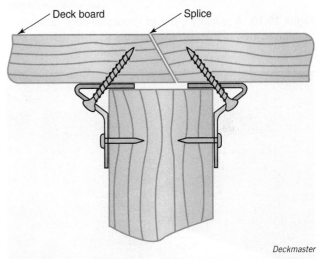

Deckmaster

Figure 30-9. This cutaway view shows the proper method for installing spliced deck boards using a concealed fastening system.

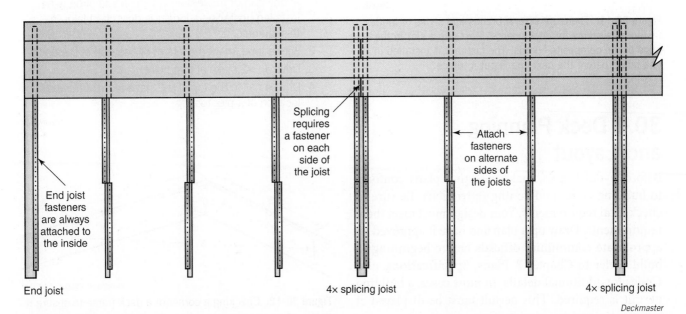

Deckmaster

Figure 30-8. This layout is a guide for one method for installing concealed deck fasteners.

Goodheart-Willcox Publisher

Figure 30-10. A properly installed concealed deck fastening system, as viewed from beneath the deck.

DeckWise

Figure 30-11. There are several brands and styles of hidden fastening systems for deck boards. Most grip a slot in the edge of the composite boards. The fastener is installed in one board before the next one is put in place.

30.4 Deck Planning and Layout

Before layout begins, make sure that plans conform to building codes and zoning restrictions. Be sure to check local requirements. Your design must meet these requirements. Draw up a plan and have it approved by appropriate community officials before beginning to build. Refer to Chapter 7, **Plans, Specifications, and Codes** for additional details. In most cases, a building permit is required. This permit must be displayed at the site during construction.

Before the deck is laid out, site preparation should be done. Site preparation involves any modification of the location where the deck will be built. This could involve several steps:

- Leveling the ground
- Providing necessary drainage
- Removing sod or laying down sheet material (black polyethylene or landscape fabric) to eliminate weed growth
- Excavating and installing piers

Layout is done most easily using an optical instrument such as a builder's level or a transit. If such an instrument is not available, it is possible to lay out the deck with measuring tapes.

After initial layout, you can check for square corners once more by installing a double rim joist hanger on one end of the ledger. Set a side rim joist in place and apply the 3-4-5 rule (a shorter-distance form of the 6-8-10 rule). Mark 3′ on the ledger and 4′ on the rim joist. Then, shift the rim joist back and forth until the diagonal measurement is 5′. Refer to **Figure 30-12**.

For a detached deck, the location will be determined by a carpenter after consulting with the homeowner. Layout is similar to the procedure for an attached deck, except that batter boards are needed at all four corners and at intermediate locations to locate support beams.

PROCEDURE
Laying out an Elevated Deck

1. Establish one corner of the deck. This is along one wall of the house. Usually, this point will be at one end of the ledger, which should already be installed. Installing a ledger is described later in this chapter.
2. With a tape, mark the length of one side of the deck.
3. Drive a 2×2 stake at that point.
4. Drive a nail in the top of the stake to mark the exact location of a pier footing.

Continued

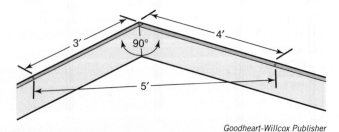

Goodheart-Willcox Publisher

Figure 30-12. Checking a corner of a deck frame to ensure it is indeed square.

PROCEDURE *(continued)*

5. Using the 6-8-10 rule for checking that corners are square, lay out all sides and drive stakes to mark each corner. Refer to Chapter 8, **Building Layout** for information on the 6-8-10 rule.

6. Install batter boards at each corner, **Figure 30-13**.

7. Stretch a mason's line from the nails driven at each end of the ledger to the batter boards. Refer to Chapter 9, **Footings and Foundations**.

8. Shift the lines at the batter boards as necessary until the diagonal measurements are equal.

9. Locate the positions of other piers needed to support beams.

30.5 Constructing the Deck

Like a house, a deck is built upward from a solid foundation. Its beams or girders must rest on firm ground, concrete piers, posts resting piers, or a ledger solidly attached to the house structure. Joists are installed to form the base for the decking surface. Railings, stairs, and other finishing touches are added to complete the deck.

30.5.1 Installing Piers

In cold climates or unstable soils, piers (concrete bases) supporting decks should extend below the region's frost line, **Figure 30-14**. The wooden posts that support the deck may be attached to the concrete piers with steel post bases, **Figure 30-15**.

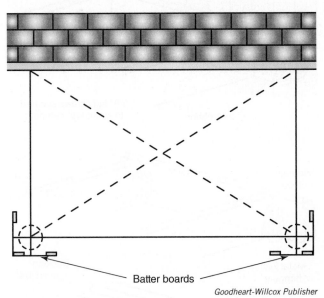

Batter boards

Goodheart-Willcox Publisher

Figure 30-13. Use batter boards and attach string to indicate the layout of the deck.

Goodheart-Willcox Publisher

Figure 30-15. An anchor bolt is set in the concrete before it cures, then the post base is slipped onto the anchor bolt to hold a wooden post.

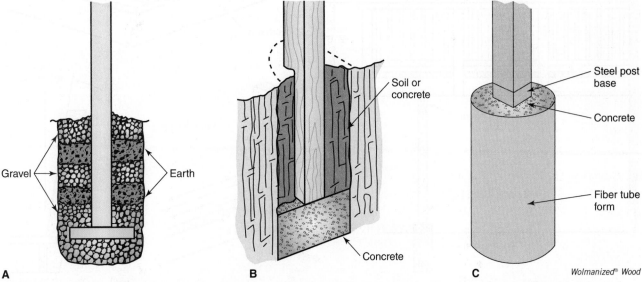

Gravel   Earth

Soil or concrete

Concrete

Steel post base

Concrete

Fiber tube form

A   B   C   *Wolmanized® Wood*

Figure 30-14. Three methods of setting posts/piers for the deck. It is important to get below the frost line in cold regions. A—Gravel and earth setting. B—Concrete setting. C—Concrete pier.

Location of the piers is important, since they must support the entire load of the deck itself as well as persons using it.

Piers support beams or girders that, in turn, support joists. Before locating the piers, be sure to check span tables for joists. The distance between beams is governed by the recommended span of the joist to be used. **Figure 30-16** shows design details for an elevated deck.

Piers are constructed of concrete. The underground portion should include or rest on a footing that extends several inches below the frost line. The aboveground portion is usually 8″ high. If poured, a supporting form, such as a Sonotube®, is required, **Figure 30-17**. (Because of its wide use, the brand name Sonotube is used generically by many people to describe all such supporting form tubes.) This is a round form made of either plastic or fiber. The tube is placed in the hole after the hole has been partly filled with concrete to form the footing. The tube is then filled with concrete. A metal fastener is embedded in the concrete to secure

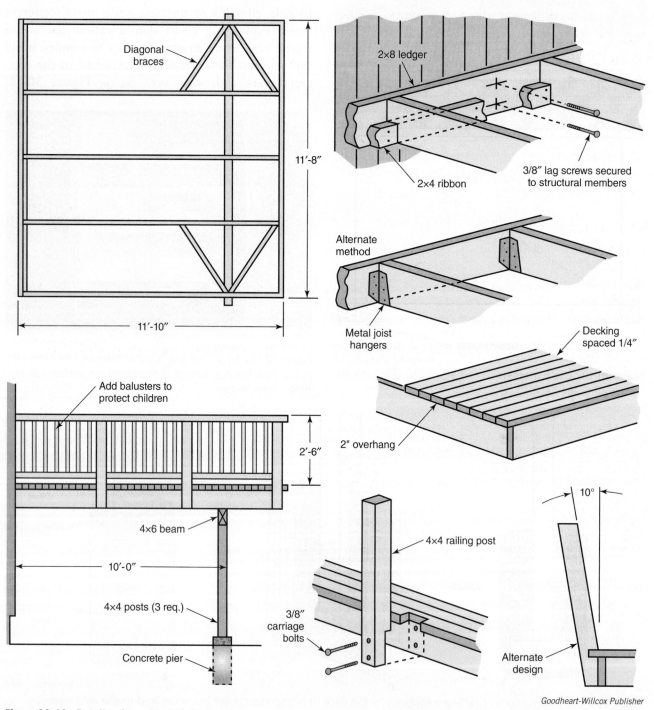

Goodheart-Willcox Publisher

Figure 30-16. Details of an elevated deck.

Goodheart-Willcox Publisher

Figure 30-17. Sonotubes are installed and ready for placing concrete.

either a post or beam to the pier, depending on the deck's height. **Figure 30-18** shows a post secured to a pier with a metal post base.

30.5.2 Installing Posts

Deck posts can be installed on top of piers or sunk below the ground on a footing. Usually, aboveground installation is preferred, since it protects the posts against rot and insect damage. Regardless of the installation method used, posts must be plumb (exactly vertical). Use a torpedo level or other short level on two adjacent sides to plumb each post. As shown in **Figure 30-19**, nail two temporary braces at right angles to each other to hold the post in position. After bracing, double-check to ensure that the post is still plumb.

Goodheart-Willcox Publisher

Figure 30-18. A metal post base is often used to secure a deck post to a pier. The air space below the fastener allows water to drain away, preventing the post from rotting.

Goodheart-Willcox Publisher

Figure 30-19. This carpenter is checking for plumb on this post. Bracing on two directions holds the post exactly vertical.

Pro Tip

Whenever pressure-treated wood is cut or drilled, the exposed, untreated area is vulnerable to rot and deterioration. Thoroughly treat all cut edges and bore holes in posts, beams, ledgers, joists, and deck boards with a generous coating of preservative solution.

30.5.3 Installing the Ledger

When one side of the deck is attached to a building, a ledger must be attached to the wall. It must be at least the same size as the deck joists (for example, 2×6 or 2×8) and run the full length of the deck.

Prepare the building wall by removing siding material so the ledger can be attached to the house band joist. The top of the ledger should be a comfortable distance below a doorway to the deck. In cases where the ledger must be attached to a concrete foundation or masonry wall, use expansion anchors in drilled holes.

To protect both the house structure and the ledger from water that could cause rot, install metal flashing and spacers, as shown in **Figure 30-20**. Use metal washers to space the ledger 1/4″–1/2″ away from the band joist to permit drainage. Bolt holes through the flashing should be caulked for watertightness. As noted earlier, only stainless steel or copper flashing should be used with copper-rich (ACQ or CA) treated lumber. Aluminum flashing will rapidly corrode.

> **Code Note**
>
> Building codes typically include requirements for ledger size and fastening. For example, the IRC requires ledgers to be at least 2×8 lumber and provides specific fastening requirements based on the span of the deck joists. Be sure to refer to the applicable building code requirements when constructing decks.

30.5.4 Placing Beams and Joists

With the ledger in place and posts plumbed and braced, the post tops must be cut down to the proper level. From the top of the ledger, measure down the depth of the beam to establish the cutoff line for the posts. It is general practice to slope the deck away from the building 1/8″ per foot. The cutoff point on the posts can be found with a laser level or a string. To use a string, stretch it from the mark on the ledger to the farthest post, **Figure 30-21**. Use a line level to level the string, then measure down and mark the calculated slope. For example, if the deck extends 20′ from the building, the mark on the farthest post should be 2 1/2″ lower than the mark on the ledger.

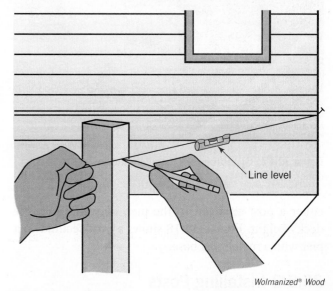

Wolmanized® Wood

Figure 30-21. Using string and a line level to determine the cutoff line for posts.

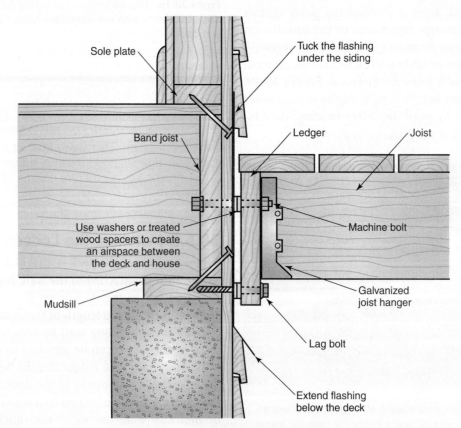

Goodheart-Willcox Publisher

Figure 30-20. Proper flashing and the use of spacers will help protect the house and deck ledger from water damage.

Drive a small nail into the post at the mark and attach the string. Mark all intermediate posts at the point where the string crosses them. Follow the same procedure with the remaining posts. Cut off the excess height with a portable circular saw.

Beams should be installed with the crown up. For maximum load-bearing ability, mount the beam on top of the posts. Attach the beam to the post with a metal post-and-beam connector. If splicing of the beam is necessary, locate the splice over a post. Install joists, crown up, resting them on top of the beam or hanging them between beams, flush with the top. Use hangers to attach the joists to the beam and ledger. If joists rest on top of the beam, fasten them down with framing anchors. For added strength, double the rim joists.

Safety Note
Since the ledger acts as a beam to support joists and other beams, it must be firmly attached to the building's structure with threaded fasteners (bolts, lag screws, or masonry anchors). Do not use nails.

30.5.5 Installing Deck Boards

As a general rule, attach decking perpendicular to the joists. Because sawed lumber is seldom exactly straight, it is best to snap a chalk line as a guide for laying the beginning row. Straighten boards as you go, **Figure 30-22**.

For redwood, cedar, or tropical hardwood decking, maintain a 1/4″ space between rows to allow drainage. If you are using treated lumber, no spacing is required. As the boards dry, the shrinkage takes care of the spacing. Spans for composites are essentially the same as for wood—5/4×6 deck boards for residential construction can span joists 16″ OC, while regular 2×6 boards can span 20″ OC joists.

Since sawed lumber is prone to cupping, it is good practice to turn the board so the bark side is up. You can determine this by checking the annular rings visible at the end of the board. The convex (upward-arching) side is the bark side. See **Figure 30-23**.

Goodheart-Willcox Publisher

Figure 30-23. To minimize cupping of deck boards, install them bark-side up. You can identify the bark side by observing the annular rings at the cut end of the board. Upward-arching (convex) rings point to the bark side.

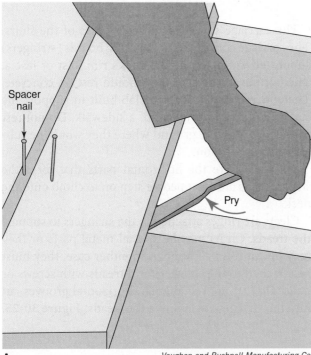

Spacer nail

Pry

A   *Vaughan and Bushnell Manufacturing Co.*

B   *Vaughan and Bushnell Manufacturing Co.*

Figure 30-22. Straightening a deck board. A—Insert the sharp point of tool into a joist, then pry against the board to close up spacing. To open space between boards, insert the tool between them and pry outward. B—A lever-type tool specifically designed for straightening deck boards.

Use either nails or screws as fasteners. Deck screws are often ceramic coated to prevent corrosion. Use hot-dipped galvanized nails. For greater holding power, always drive nails at an angle. Ring-shank nails have greater holding power and are less subject to popping, which exposes the nail heads over time. Composite decking is not subject to nail pops.

Pro Tip

On an attached deck, start installing the deck boards from the outer edge and work toward the building. A narrower board at the building is less noticeable than one at the outer edge of the deck.

30.5.6 Installing Railings

Design of railings must follow local code. Section R312.1 of the IRC requires railings to be at least 36″ high on any deck that is 2 1/2′ or more above the ground. The IRC also specifies the spacing of balusters. These are narrow, vertical strips of wood fastened between the top and bottom rails. They must be spaced close enough to prevent a 4″ ball from passing between them. This is the same requirement as for interior stair railings. The ability of the railing to withstand lateral loads may also be specified by the local code. Always check code requirements before building.

Railings require posts, top rails, bottom rails, and balusters. Some railings may also include an intermediate rail. Latticework replaces the balusters in some designs. Posts for railings are usually attached to the rim joist with through-bolts or lag screws. Posts are normally spaced about 4′ apart. Complete rails in many shapes for decks and steps may be purchased through home improvement centers and lumber yards, **Figure 30-24**.

30.5.7 Installing Stairs

Unless the deck is built at ground level, it will have stairs to provide access from the ground. Stairs may have as few as two treads. If the deck is at a second-floor level, the stairs may have a dozen or more treads. Stairs have three basic components:

- Carriage
- Treads
- Cleats or dadoes

Trex Company, Inc.

Figure 30-24. Factory-built railings, posts, and balusters are available from manufacturers.

The carriage supports all other parts of the stairs. The carriage consists of two sloping boards (stringers) supported at the top by the deck's rim joist or fascia. The bottom of the stringers should rest on concrete. The concrete base can be a slab built to support the stairs or it may be one end of a sidewalk. Do not rest the stringers on the ground where they would be subject to deterioration.

The treads are the horizontal parts that form the steps. Treads are what people step on to climb onto the deck.

Cleats are ridges attached to the stringers to support the treads. Cleats may be special metal parts or they may be cut from 1″ lumber. In either case, they must be attached to the stringers and treads with screws or nails. *Dadoes* are rectangular, horizontal grooves cut into the stringers to support the treads, **Figure 30-25**.

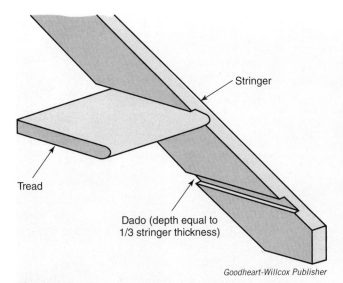

Stringer

Tread

Dado (depth equal to
1/3 stringer thickness)

Goodheart-Willcox Publisher

Figure 30-25. Dadoes may be cut in stringers to support treads.

Treads must be secured with waterproof glue and screws. Refer to Chapter 21, **Stair Construction** for more information on laying out and building stairs. The rules that apply to interior stairs for riser height and tread width also apply to deck stairs.

30.6 Pergolas

A *pergola* is a structure with an open canopy that provides partial shade from the sun. It is often built over all or a part of a deck, **Figure 30-26**. Posts supporting the structure may be extensions of the deck posts. One side may be fastened to the building. A pergola may also be built as a freestanding structure and is often used in gardens in this manner, **Figure 30-27**. Unlike a roof, the open canopy of the pergola is not intended to provide protection from the elements.

Santiago Cornejo/Shutterstock.com

Figure 30-26. This pergola is supported by posts fastened to horizontal frame members.

Figure 30-27. A freestanding pergola is frequently used in gardens. Many different designs are used for such structures.

30.7 Porches

A porch deck is structurally similar to an open deck, but its supporting members must be heavier to provide adequate support for the roof, **Figure 30-28**. As a rule, porches are narrower than decks. The porch width is often dictated by the extension of the house roof to cover the porch. There are three types of porches:

- *Open porch*—Has no railing, just a deck, roof, and supporting columns. The height of the porch deck above the ground may be governed by a local code.
- *Semi-enclosed porch*—Has either an ornamental balustrade or a low wall enclosure. The wall may have a wide sill. Some designs allow the addition of removable screens and storm panels to accommodate the change of seasons.

- *Closed porch*—Typically is framed with studs and sided like the rest of the house. It is more like an extension of the house than a transitional area.

Framing of the porch deck should allow a slope of 1″ in 8′ for proper drainage away from the building. Unlike deck boards, the floorboards of the porch are laid tight to one another. Often, the boards are tongue-and-groove. If not treated by the manufacturer, the boards can be soaked in preservative. Flooring should be painted on all faces and edges before installing, if possible. Each joint should be caulked before the board is nailed down. Since they have a number of exposed joints and faces that can collect water, porches are more susceptible to rot than other parts of the house. Be careful to design joints that shed water.

Southern Pine Council

Figure 30-28. A spacious, semi-enclosed porch with interesting architectural details. Semi-enclosed and open porches are especially popular in warmer climates.

Chapter Review

Summary

- Basic types of decks include elevated and ground-level structures. Elevated decks are supported by posts and a ledger beam fastened to the house.
- Either type of deck may have a single level or several levels.
- The supporting structure of the deck is typically wood—either conventional pressure-treated lumber or engineered lumber.
- The deck surface itself may be wood, composite, or vinyl.
- Copper-based preservative treatments are nontoxic, but must be used with stainless steel or specially treated fasteners that will not be weakened by galvanic corrosion.
- Concealed fastener systems are most commonly used to fasten deck boards.
- Layout of the deck must be done using the same techniques as a house foundation.
- The deck must be adequately supported, usually on concrete piers.
- A ledger beam is solidly attached to the house structure to support one end of the deck. Posts are located on the piers to support the rest of the structure.
- Beams or girders rest on the posts and ledger and, in turn, support the deck joists.
- Decking material, railings, and steps complete the structure.
- A porch is structurally like a deck with the addition of a roof. It may be open, semi-enclosed, or closed.

Know and Understand

Answer the following questions using the information in this chapter.

1. The two basic deck types are _____ and _____.
   A. elevated, single-level
   B. single-level, multilevel
   C. elevated, ground-level
   D. ground-level, multilevel

2. *True or False?* A ground-level deck does not require a railing for safety.

3. Redwood cut from the _____ is most resistant to rot and insect damage.
   A. heartwood
   B. southern pine
   C. red cedar
   D. red pine

4. _____-treated lumber is no longer produced for residential use.
   A. DOT
   B. CCA
   C. ACQ
   D. CA

5. _____ is manufactured by stacking finger-jointed layers of standard lumber and adhering the layers with adhesive.
   A. Parallel-strand lumber (PSL)
   B. Glue-laminated timber (Glulam)
   C. DOT-treated lumber
   D. Heat-treated lumber

6. Standard thickness of composite decking boards are 1 1/4″ and _____″.
   A. 2
   B. 4
   C. 6
   D. 8

7. Thermally modified wood can make common species such as yellow pine and ash more rot and _____ resistant.
   A. corrosion
   B. insect
   C. mold
   D. water

8. Fasteners and connectors used with ACQ or CA pressure-treated lumber must be _____ resistant.
   A. mold
   B. insect
   C. corrosion
   D. water

9. _____ corrosion occurs when dissimilar metals are in contact in a damp environment.
   A. Ceramic
   B. Steel
   C. Galvanic
   D. Copper

10. If a builder's level or transit is not available, it is possible to lay out a deck with _____.
    A. rulers
    B. joists
    C. any straightedge
    D. measuring tapes

11. *True or False?* It is not necessary to place footings or piers below the frost line.

12. Piers are constructed of _____.
    A. plastics
    B. concrete
    C. wood
    D. fibers

13. Beams should be installed with the crown
_____.
   A. up
   B. down
   C. facing the posts
   D. below the posts

14. *True or False?* A deck railing can be placed at whatever height seems comfortable for the homeowner.

15. _____ are slots cut into stringer to hold the treads.
   A. Cleats
   B. Treads
   C. Dadoes
   D. Carriages

16. A shade structure built with an open canopy is called a _____.
   A. roof
   B. patio
   C. shed
   D. pergola

17. Unlike deck boards, floorboards of a porch are laid _____ to one another.
   A. parallel
   B. perpendicular
   C. tight
   D. spaced evenly

Apply and Analyze

1. What is most commonly used to support elevated deck framing?

2. What are two advantages of composite deck boards?

3. Why should aluminum never be used as flashing for a deck?

4. What is a Sonotube?

5. Why is spacing not required when installing treated lumber deck boards?

6. Describe the three basic porch types.

Critical Thinking

1. Assume the height of the deck floor illustrated in **Figure 30-16** is 148″. Refer to *Section R311.7 Stairways* in the 2018 IRC. You can perform an internet search if you do not have a code book. Assume there are no horizontal limits that would restrict the tread width. Establish a stair design that meets all code requirements for this deck. Be sure to calculate the number of risers and the riser height. Make a sketch of your design.

2. Research products that are used for treating wood decks, such as paints, stains, and sealers. What do these products protect against? Based on information you found, which product would you recommend for a deck in your climate? Why?

Communicating about Carpentry

1. **Reading and Speaking.** Create an informational pamphlet on the different opportunities available as a carpenter who concentrates on special construction assignments such as remodels and renovations, and decks and porches. Describe how much time is required to become employed full-time in these positions. Include images in your pamphlet. Present your pamphlet to the class.

2. **Speaking and Art.** Working with a partner, create a model of a deck or porch. Create the model so that the parts can be removed. Use your model to demonstrate to the class the different options available when choosing which type of deck or porch to install.

Mechanical Systems

Steven White/iStock/Thinkstock

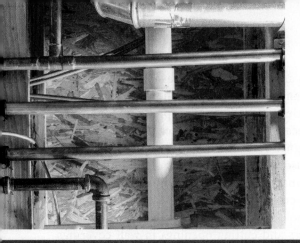

CHAPTER **31**

Electrical Wiring

OBJECTIVES

After studying this chapter, you will be able to:
- Define basic electrical terms.
- Explain what is included in an electrical wiring system.
- List the tools, devices, and materials required to do electrical wiring in a residential building.
- Demonstrate the proper use of tools and handling of materials.
- Demonstrate an understanding of basic circuit theory.
- Use approved methods for simple wiring installation tasks.
- Perform simple electrical troubleshooting.
- Describe the components of a home security alarm system.

TECHNICAL TERMS

alternating current (AC)
arc-fault circuit interrupter (AFCI)
branch circuit
circuit
circuit breaker
conductor
conduit
current

direct current (DC)
electric current
entrance panel
fish tape
ground-fault circuit interrupters (GFCI)
home run
induction
mechanical system

National Electrical Code (NEC)
receptacle
service
short circuit
structured wiring
switch
transformer

Once a building is constructed and closed in (roof, windows, and exterior doors installed), it is ready for the installation of electrical wiring. This is one of three installations, along with plumbing and heating/ventilating/air conditioning (HVAC), that are known as *mechanical systems*. Normally, these systems are the responsibility of persons skilled in building trades other than carpentry. These workers are generally hired by or work for subcontractors. Mechanical systems are installed before insulation and wall coverings because they must be placed in the frame of the building.

The electrical wiring system consists of conductors (wires), boxes, and various devices that control the distribution of electricity throughout the building. The system provides the *current* (flow of electrons) that powers lights, receptacles, heating and air conditioning units, and appliances.

Electricity can be extremely dangerous if it accidentally comes in contact with people, combustible materials, or metal that is not part of the intended electrical path. Current that accidentally contacts a person who is not properly insulated can cause serious injury and even death from electrical shock. An electrical short may cause a fire. For safety and proper operation of the building's electrical system, a qualified electrician must install electrical wiring according to national and local electrical codes.

The *National Electrical Code (NEC)* is a collection of rules and standards developed by experts to ensure the proper installation of electrical conductors and related devices. The NEC is a model code reviewed and revised every three years. A model code becomes a law—and is thereby enforceable—only when adopted by a local unit of government (such as a city, village, or county). Communities hire inspectors to examine new installations and enforce the local code when violations are found.

When doing any type of electrical work, it is necessary to be familiar with all local building code requirements. The NEC is the model electrical code used by most municipalities. Refer to Chapter 7, **Plans, Specifications, and Codes** for more information.

31.1 Tools and Equipment

Certain tools, equipment, and materials are essential when installing electrical wiring. **Figure 31-1** shows a basic tool list. Electricians must be comfortable using all carpentry tools, especially hammers and drills.

Basic Tool List	
Striking tools	**Cutting and sawing tools**
Lineman's or electricians hammer	Files
Drilling tools	Crosscut saw
	Jab saw
Right-angle drill	Hacksaw
Cordless drill	Circular saw, 7″
Drill bits, various sizes	Compound miter saw
▪ wood twist	Reciprocating saw
▪ metal	Pocketknife or electrician's
▪ masonry	knife
▪ spade	Cable cutters
▪ bit extenders	Wood chisel
Soldering and	Wire strippers
wire-joining tools	Cable strippers
Soldering iron	**Pliers**
Soldering gun	Slip joint pliers
Propane torch	Lineman's pliers
Solder, rosin core	Side cutting pliers
Soldering paste	Diagonal pliers
Crimping tool	Long nose pliers
Wire stripper	End cutting pliers
	Locking jaw pliers
	Special and
	miscellaneous tools
Fastening tools	Fish tape wire puller
Standard screwdriver	Wire pulling lubricant
Phillips screwdriver	Conduit or pipe cutter
Offset screwdriver	Reamer
Torque head screwdriver	Conduit bender (hickey)
Adjustable wrench	Fuse puller
Allen wrenches	Tap and die set
Socket/ratchet wrenches	Flashlight
Box end wrenches	Plumb bob
	Test light, continuity tester
Measuring tools	Level
Folding ruler	Conduit threader
Carpenter's extension ruler	Trouble light
Steel tape	Gas generator, about 1500 W
12″ ruler or meter stick	Portable space heater
Wire gage	Assorted wood or fiberglass
VOM	ladders
	Wire grips
	Chalk line
	Tool pouch

Goodheart-Willcox Publisher

Figure 31-1. An electrician's toolbox should be stocked with tools from this basic list.

Hammers are used for driving fasteners and staples, attaching hangers and electrical boxes, and striking chisels. An electrician's hammer is preferred for attaching boxes because it has a longer neck than a carpenter's hammer, **Figure 31-2**.

Saws are useful for making cuts in a building's frame. A handsaw may be used for notching or cutting studs. For heavier cuts, a power circular saw is more effective. A keyhole saw or jab saw is useful for sawing in tight quarters and cutting openings in wall coverings for electrical boxes, **Figure 31-3**. A reciprocating saw may also work wherever a jab saw is used; it is faster and can handle thicker materials. A hacksaw is used to cut metal such as conduit or cable.

Wood chisels are handy for trimming away small amounts of wood on framing members when mounting boxes and fixtures. They are the preferred tool for notching studs, joists, plaster, flooring, and old-style lath.

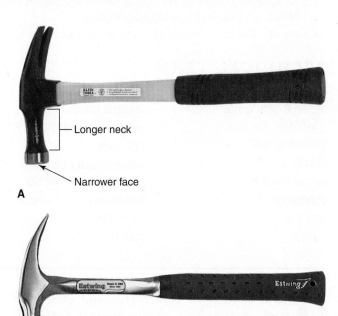

Klein Tools
Estwing Manufacturing

Figure 31-2. A—An electrician's hammer has a longer neck than the framing hammer used by carpenters, which are useful when installing outlet boxes or reaching into tight places. B—Carpenter's framing hammer.

Klein Tools

Figure 31-3. A jab saw is useful when cutting openings in drywall for electrical boxes. A jab saw is one type of drywall saw. The large triangular teeth cut fast and do not easily clog.

A number of different tools can be used for wire and cable cutting. Multipurpose tools cut, strip, and crimp, **Figure 31-4**. Some wire strippers are made for the single purpose of stripping wire, **Figure 31-5**.

Pliers are good for holding, shaping, and cutting. The basic types used are slip-joint pliers and lineman's pliers, **Figure 31-6**. Locking pliers are a versatile tool. They can serve as pliers, a lock wrench, an open-end wrench, or a pipe wrench.

The straight-blade screwdriver and the Phillips screwdriver are the fastening tools used most often. They are primarily used to attach wires to terminal screws and attach switches and receptacles to boxes.

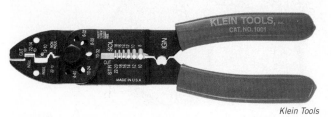

Klein Tools

Figure 31-4. This multipurpose tool is used to cut electrical wire, strip insulation from either solid or stranded wire, and crimp various types of connectors onto conductors.

Goodheart-Willcox Publisher

Figure 31-5. This wire stripper has openings for wire from 10 gauge down to 18 gauge.

Klein Tools

Figure 31-6. A side cutting, or lineman's, pliers is an essential tool for everyone doing electrical installations.

Drills are used to bore holes when running conductors or conduit through studs and joists. In tight quarters, a right-angle drill is used. For more detailed information on tools, refer to Chapter 4, **Hand Tools** and Chapter 5, **Power Tools**.

An electrician's tool kit may also contain soldering tools—usually an electric soldering iron. Even though the use of wire connectors has all but eliminated the need for soldering, there is still some use for soldered connections.

A steel tape measure is an important tool used for measuring. Typical tape measures are 25′ and 30′ long.

A fish tape is necessary for pulling wires through conduit, **Figure 31-7**. A *fish tape* is a long strip of metal, fiberglass, or nylon wound on a reel for ease of handling. The tape is pushed through installed electrical conduit and used to pull wires between boxes.

A circuit tester is used to determine if there is voltage at any point in a circuit. A simple tester may have only a pair of contact leads and a neon bulb that lights when voltage is detected. Most electricians also carry a multimeter, which will measure, not just detect, voltage. A multimeter also measures current flow (amps) and resistance (ohms). **Figure 31-8** shows a neon circuit tester and a multimeter.

Various other tools that round out an electrician's tool kit include a trouble light, pipe cutter, conduit bender, conductor bender, and a set of socket wrenches. For electrical work, use only wood or fiberglass ladders—neither is a good conductor of electricity.

Safety Note
Never use a metal ladder when doing electrical work. If the ladder accidentally contacts a current-carrying conductor, serious injury or even death could result.

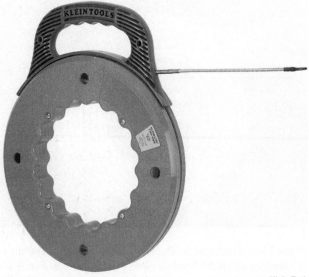

Klein Tools

Figure 31-7. One type of fish tape.

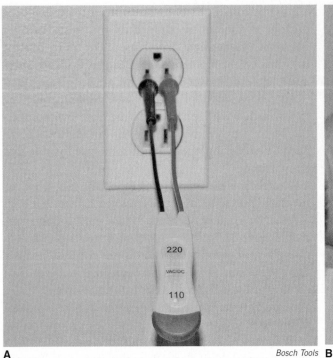

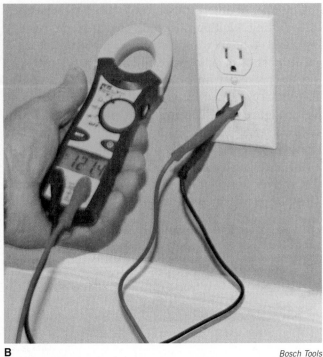

A   *Bosch Tools* **B**   *Bosch Tools*

Figure 31-8. Electrical test equipment. A—A neon circuit tester determines only if there is voltage present. B—A multimeter can be used to measure AC and DC circuits for voltage, amperage, and resistance.

31.2 Electrical System Components

An electrical system in a building is made up of the following:

- Conductors
- Boxes and box covers
- Circuit breakers or fuses
- Ground-fault circuit interrupters (GFCIs)
- Arc-fault circuit interrupters (AFCIs)
- Switches and receptacles
- Conduit or raceways
- Connectors

The *circuit* is the path through which current flows from the source back to the source. It includes the conductors, switches, receptacles, and any appliances or lights.

Conductors are wires that carry current. The wires may be enclosed in conduit, a metal sheath, or a plastic sheath, **Figure 31-9**.

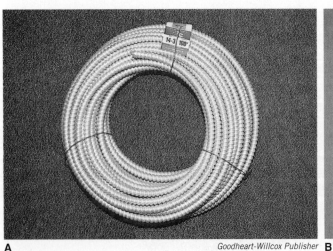

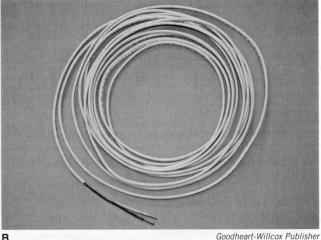

A   *Goodheart-Willcox Publisher* **B**   *Goodheart-Willcox Publisher*

Figure 31-9. Electrical conductors are the copper or aluminum wires that provide a path for current in the building's circuits. Conductors within the various types of cables are available in different sizes and quantities. Codes vary on which types can be used. A—Armored cable, also called BX, has a flexible metal cladding protecting several insulated conductors. B—Nonmetallic (NM) cable is similar to armored cable, but has a plastic protective covering.

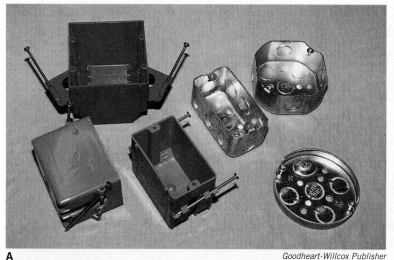

A   *Goodheart-Willcox Publisher*   B   *Stephen P. Kielar*

Figure 31-10. A—Several types of plastic and metal boxes used for either new construction or remodeling work. B—Covers and trim plates protect devices and conductor connections in the boxes.

Various types of boxes and box covers are available, **Figure 31-10**. Their purpose is to enclose devices that control electric current and protect connections between two or more conductors. Covers and trim plates conceal and protect devices, such as switches and receptacles that are placed in boxes.

Circuit breakers and fuses are devices that protect conductors from overload, **Figure 31-11**. Breakers and fuses shut off power to the circuit if there is an excessive current draw. A *short circuit*, which can occur as a result of a connection coming apart or perhaps a rodent chewing through the insulation, causes an increase in the amount of current flowing through the circuit. The circuit breaker or fuse will open the circuit, stopping all current to protect the wires from heating up and causing a fire.

Ground-fault circuit interrupters (GFCIs) are safety devices installed either in an electrical circuit or at an outlet as protection against electrical shock, **Figure 31-12**.

A   *Stephen P. Kielar*

Stephen P. Kielar

Figure 31-11. Circuit breakers protect circuits and equipment from overloads.

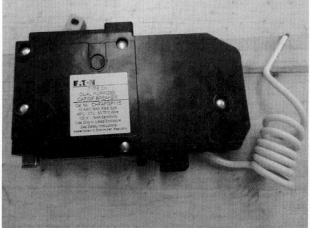

B   *Stephen P. Kielar*

Figure 31-12. Ground-fault circuit interrupters (GFCIs) may be built into circuit breakers or receptacles. A—GFCI protection. B—Combination AFCI and GFCI breaker.

The GFCI is in addition to the circuit breaker. GFCIs are required by electrical code to be in the following areas of the home:

- Bathrooms
- Laundry rooms
- Unfinished basements
- Crawlspaces
- Garages
- Kitchen countertops
- Outdoor receptacles

In addition, all receptacles within 6′ of sinks, tubs, shower stalls, and dishwasher circuits must be protected by a GFCI. A GFCI device compares the current flowing through the hot leg of the circuit with that flowing through the neutral leg. Under normal conditions, the current should be the same in both legs. If anything causes a different path for the current, such as a human body, the GFCI will quickly stop all current flow. It can detect as little as 4 milliamps and can react as quickly as 1/30 of a second. Circuit breakers and fuses protect against fires. GFCIs protect against electric shock.

Arc-fault circuit interrupters (AFCIs) are another type of electrical safety device, **Figure 31-13**. AFCIs help prevent houses from catching fire. An AFCI monitors a circuit and detects erratic forms of current. When electric current jumps through space (or air) from one conductor to another, it causes an arc. Gasoline engines rely on an arc at each spark plug to ignite the fuel in the cylinder. Lightning is another, very powerful type of arc. If uninsulated conductors are close, but not touching, it is possible for the current to jump from one conductor to the other, causing an arc. Arcing does not always mean a significant change in the amount of current flowing, so circuit breakers do not protect against arc faults. Many house fires have been caused by arcing current. The arc-fault circuit interrupter

detects arcing and stops all current before a fire is started. The NEC requires AFCIs on most circuits in a home.

Various types of switches and receptacles are used in residential and light commercial applications, **Figure 31-14**.

Stephen P. Kielar

A

Stephen P. Kielar

B

Figure 31-14. Devices most commonly placed in electrical boxes are switches and receptacles. A—Switches interrupt the path of the current. B—Receptacles allow quick temporary connection of appliances and lights to a circuit.

Stephen P. Kielar

Figure 31-13. Arc-fault circuit interrupters (AFCIs) are required on most of the electrical circuits in a house. This AFCI looks like other circuit breakers from the same manufacturer, except it is much longer. The pigtail of coiled wire at the top attaches the AFCI to the neutral bus bar.

Switches control the current to lights and appliances. *Receptacles* allow convenient connection of appliances and lights to a circuit. Receptacles are sometimes called outlets or convenience outlets.

Conduits or raceways are pipes or enclosed channels through which conductors are run, **Figure 31-15**. *Conduit* is steel or polyvinyl chloride (PVC) rigid tubing. Lengths of conduit can be cut and bent as needed to fit the structure. Conduit lengths are connected with special couplings.

Connectors are fastening devices that ensure tight connections between two or more conductors or lengths of conduit, **Figure 31-16**. Wire nuts form a sound mechanical and electrical connection between conductors. Color coding is used to identify the proper wire nut for appropriate conductor sizes. Couplings are used to connect conduit, while other fittings are used to securely clamp armored or nonmetallic cable to electrical boxes.

31.3 Basic Electrical Wiring Theory

Electricity is generated when a conductor made from a metal such as copper, silver, or aluminum is passed through a magnetic field. This causes electrons to move through the conductor in one direction. The resulting flow of electrons is known as *electric current*. When the conductor (wire) passes through the magnetic field in the opposite direction, the current flows in the opposite direction. This changing of direction is known as *alternating current (AC)*. Alternating current is the form of electric current found in homes.

Goodheart-Willcox Publisher

Figure 31-15. Conduit is used to protect electric wires. Here, several boxes are connected with polyvinyl chloride (PVC) conduit.

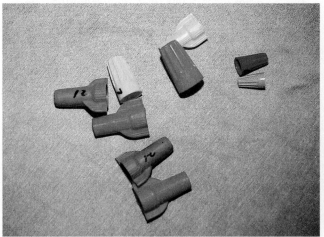

A   *Goodheart-Willcox Publisher*

B   *Goodheart-Willcox Publisher*

Figure 31-16. Fittings connect and hold conductors, cables, and conduit. A—Wire nuts are used to connect two or more conductors. B—Couplings are used to connect lengths of conduit, and connectors are needed to attach conduits and cables to boxes.

Another form of current, used for various purposes, is known as ***direct current (DC)***. In a DC circuit, electrons flow in only one direction. This is the type of current supplied by automotive or flashlight batteries, for example.

Alternating current is generated at an electric power station and sent out through conductors known as transmission lines, **Figure 31-17**. Transformers along the way step up (increase) or step down (reduce) the

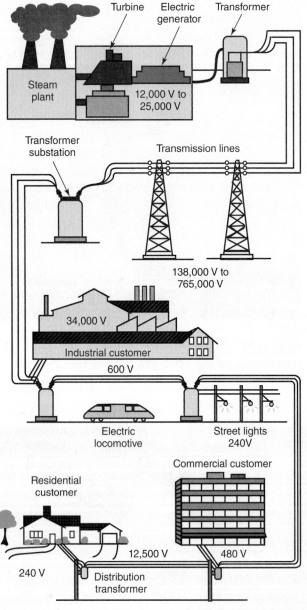

Goodheart-Willcox Publisher

Figure 31-17. This diagram shows a power system delivering electric power from a generating plant to its customers.

voltage as required by power customers. A ***transformer*** is a device that can increase or decrease the voltage applied to it. It does so by way of a magnetic process called ***induction***.

A transformer is made up of two coils. A coil consists of a core of iron with a number of turns of conducting wire wrapped around it. One coil of the transformer has more turns of wire than the other. There is no electrical connection between the two coils. When current enters the primary coil, it sets up a magnetic field in the core. This magnetic field reaches the secondary coil and *induces* a current. The induced current does not have the same voltage because of the difference in number of turns of wire on the two coils.

Step-up transformers used at power stations increase the voltage generated because it is more efficient to send high-voltage current a long distance over conductors. Before the high-voltage current can be delivered to a house, it must pass through a step-down transformer that will reduce the line voltage to 120 or 240 volts. Commercial and industrial customers may use much higher voltages.

Figure 31-18 shows three kinds of transformers for changing voltage. One is called a substation. It serves a whole area, such as a town. The second is a smaller, often pole-mounted distribution transformer that serves one residence or a group of residences. The third type of transformer is used within the house to reduce current to low voltage for devices such as a doorbell or thermostat.

Safety Note

Electrical current can be extremely dangerous if a grounded person accidentally touches bare conductors or other energized parts of a circuit. Always follow these safety rules when you work around electricity:

- Make certain that a circuit is de-energized before working on it; turn off the circuit breaker or remove the fuse.
- If you must work on a live circuit, use only tools with insulated handles.
- Remove all jewelry, watches, rings, and any other metal objects before beginning work.
- Keep work areas dry and clean. Cover wet floors with wood planks.
- Wear rubber boots or rubber-soled shoes.
- Use only wood or fiberglass ladders, never metal.
- Wear clothing that is neither too loose nor too tight.
- Wear safety glasses when using striking tools.

A *Dauphin Island, Goodheart-Willcox Publisher* **B** *Dauphin Island, Goodheart-Willcox Publisher* **C** *Dauphin Island, Goodheart-Willcox Publisher*

Figure 31-18. In an electrical distribution system, voltage is increased and decreased several times before being delivered to customers. A—A substation is a series of transformers that step down voltage before it is delivered to a large number of customers in a neighborhood or town. B—Step-down transformer further reduces voltage. C—Transformers can be very small, such as this 24-volt unit used to power furnace controls and other low-voltage equipment.

From the residential transformer, conductors called a *service* bring electricity to the residence. Service conductors may be strung from poles or buried in the ground, **Figure 31-19**. Two 120-volt conductors and a neutral are supplied to the house, which will provide 240 volts for certain applications, such as an air conditioner or electric hot water heater.

A *Mark Winfrey/Shutterstock.com* **B** *Goodheart-Willcox Publisher*

Figure 31-19. Residential electrical service. A—A trench must be dug to lay down underground service. This trenching machine digs a narrow ditch with a digging chain. The National Electrical Code (NEC) specifies how deeply cables or conduit must be buried. B—Here, an electrician works on an underground service conduit.

31.4 Installing the Service

An electrical *service* consists of the conductors that bring power from a transformer to a building. The service extends only from the transformer, through a metering device, to a distribution or entrance panel located within the dwelling. **Figure 31-20** shows the meter enclosure, where the service first enters the building. The *entrance panel* houses a main breaker and a circuit breaker or fuse for each of a home's circuits, **Figure 31-21**. **Figure 31-22** is an illustration of all the components of the service entrance at the dwelling. Installation of these components is the responsibility of the electrician hired by the homeowner.

31.5 Grounding and Ground Faults

An electrical system is designed so that electrical current safely travels through insulated conductors. When current leaves the conductors, it enters a device that converts electrical energy to provide light, heat, or motion. If a broken device or loose connection causes a current-carrying conductor to contact a path to ground, a ground fault occurs.

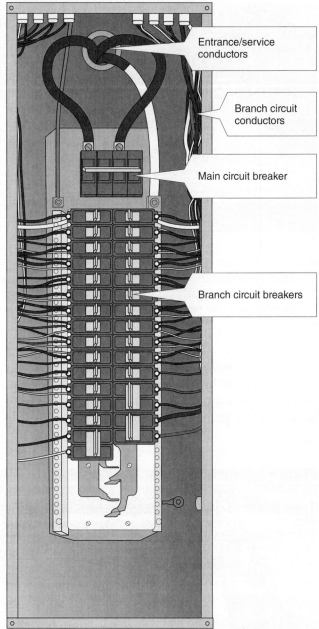

Entrance/service conductors

Branch circuit conductors

Main circuit breaker

Branch circuit breakers

Goodheart-Willcox Publisher

Figure 31-21. The service entrance or distribution panel is located inside the building. Each branch circuit is connected to a circuit breaker. The heavy cables inside the box near the top are the conductors that bring power from the meter. The large black block just below the cables is the main breaker. When connections have been completed, a protective panel cover is placed over the installation.

A   B   *Goodheart-Willcox Publisher*

Figure 31-20. The exterior junction box for electrical service is called a meter socket. A—Meter socket is shown before completion of service. B—When the wiring system of a building is completed, the meter is installed and the system is ready to be used.

When this happens, electrical current follows an unintended path to ground and may energize something that should not be energized. A person coming in contact with the unintended current path may get a dangerous shock. Proper grounding will minimize the possibility of a short circuit causing a shock.

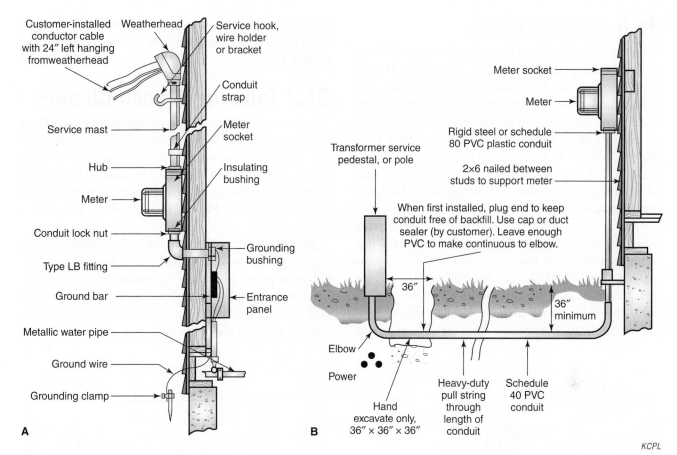

Customer-installed conductor cable with 24″ left hanging from weatherhead

Weatherhead

Service hook, wire holder or bracket

Conduit strap

Service mast

Meter socket

Hub

Insulating bushing

Meter

Conduit lock nut

Type LB fitting

Grounding bushing

Ground bar

Entrance panel

Metallic water pipe

Ground wire

Grounding clamp

A

Transformer service pedestal, or pole

Meter socket

Meter

Rigid steel or schedule 80 PVC plastic conduit

2×6 nailed between studs to support meter

When first installed, plug end to keep conduit free of backfill. Use cap or duct sealer (by customer). Leave enough PVC to make continuous to elbow.

36″

36″ minimum

Elbow

Power

Hand excavate only, 36″ × 36″ × 36″

Heavy-duty pull string through length of conduit

Schedule 40 PVC conduit

B

KCPL

Figure 31-22. Diagrams of two types of three-wire service to a building. A—An above-ground installation. B—An underground installation.

In a properly grounded circuit, every device is connected to ground. The following are ways of providing this connection:

- The use of a ground wire, either bare (uninsulated) or green, that leads from the device to a neutral bar in the service panel
- Using metal boxes and conduit to house the conductors to provide a path for current along the conduit to the neutral bar in the service panel

The neutral bus bar is connected to the grounding bus bar in the service panel. The grounding of the electrical system is completed by connecting the grounding bus bar to a grounding electrode. The grounding electrode may be a metal cold-water pipe or grounding rods driven deep into the soil, **Figure 31-23.**

When a ground fault occurs, the grounding path carries the current back to the panel. The unintended path to ground lowers resistance, which, in turn, causes a surge in power. This causes the circuit breaker to trip or the fuse to blow. At the same time, the faulted current is directed into the earth where no harm results.

Goodheart-Willcox Publisher

Figure 31-23. Electrical circuits must have a grounding system for safety. Copper rods driven 8′ deep into the ground to damp soil provide a satisfactory ground. Another grounding method involves the use of the building's cold water piping.

31.6 Reading Prints

The electrical plan for the house may be a part of the set of drawings or prints prepared by an architect. If there is no electrical print, an electrician will prepare one following all applicable local codes. If no separate electrical plan is required, an electrician may simply mark up the floor plans. Symbols on the electrical print show the types of conductors and devices to install at each location, **Figure 31-24.** All electrical print symbols are standardized so that there can be no misunderstanding, **Figure 31-25.**

After studying the electrical plan, an electrician may draw a wiring layout for each room. This layout shows where in the room electrical devices are located, the size of the conduits (if used), and the quantity, size, and type of conductors used in each run. This layout may also be used in other ways. For example, an electrician may use it as a billing reference or as a way of checking off work that has been completed. **Figure 31-26** shows an example of a cable layout.

31.7 Running Branch Circuits

Think of a circuit as a loop or continuous path from the source of electric power (the panel), through various loads or current-using devices (such as lights or electric motors), and then back to source. The source in this case is the service panel containing the circuit breakers or fuses. **Figure 31-27** shows a sketch of a simple circuit.

Branch circuits are circuits that include all of the loads on one run of conductors. The circuit begins at the distribution panel, where it is connected to a fuse or circuit breaker. An electrician first installs junction, switch, and receptacle boxes. Then, conductors are run (installed) through the building frame into these boxes.

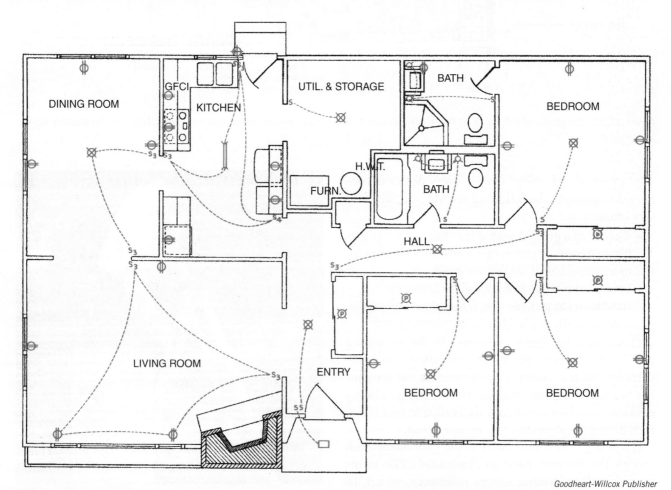

Goodheart-Willcox Publisher

Figure 31-24. Electrical plans such as this one are provided by an architect. The plan serves as a guide to an electrician when installing wiring, devices, and fixtures.

Outlets, general

Ceiling	Wall	
◯	—◯	Outlet
Ⓔ	—Ⓔ	Electrical outlet—used only when circle alone might be confused with other symbols
Ⓕ	—Ⓕ	Fan
Ⓙ	—Ⓙ	Junction box
Ⓛ	—Ⓛ	Lamp holder
ⓁPS	—ⓁPS	Lamp holder, pull switch
Ⓢ		Smoke alarm
Ⓥ	—Ⓥ	Vapor discharge lamp
Ⓒ	—Ⓒ	Clock (indicate voltage)

Outlets, convenience

⊖	Duplex
⊖1.3	Convenience outlet other than duplex
⊖WP	Weatherproof
⊖R	Range
⊖S	Switch and convenience outlet
▷AFCI	Arc-fault protected outlet
⊖GFCI	Ground-fault circuit interrupter
▲	Special purpose outlet (designated in specifications)
◉	Floor outlet

Outlets, switch

S	Single pole switch
S₂	Double pole switch
S₃	Three-way switch
S₄	Four-way switch
S_D	Automatic door switch
Ⓜ	Motion sensor
S_K	Key-operated switch
S_P	Switch and pilot lamp
S_CB	Circuit breaker

Goodheart-Willcox Publisher

Figure 31-25. Electrical symbols are a type of shorthand that tells an electrician which type of device or fixture is to be installed. This is a sample of some symbols for electrical outlets.

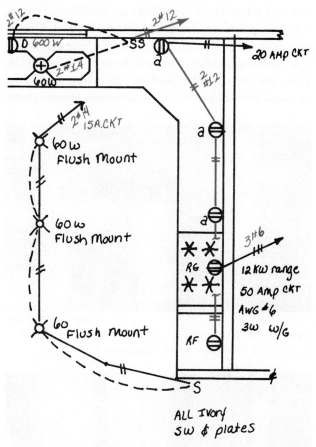

Goodheart-Willcox Publisher

Figure 31-26. A cable layout. Electricians sometimes produce such drawings as a reference while wiring a room.

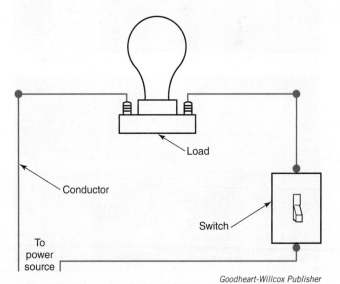

Goodheart-Willcox Publisher

Figure 31-27. A circuit is simply a path from a source of power, through energy-using devices, back to the power source.

In metal-framed structures, holes are already made in the studs and joists. In wood-framed buildings, an electrician drills holes through the wood framing members so that cables or conduits can pass through, **Figure 31-28**. The cables or conduits must be run before the insulation and drywall or plaster are installed, **Figure 31-29**.

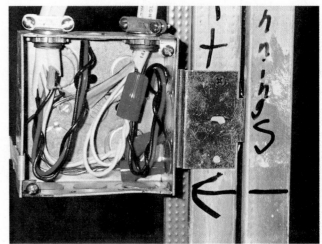

The conductors are run to each box and securely connected to them, **Figure 31-30**.

Electrical cables must also be securely fastened to framing members, as shown in **Figure 31-31**. The NEC requires that fasteners must be installed at intervals not greater than 4 1/2′ and within 12″ of their entrance into an electrical box, provided they are secured at the box with a clamp or cable connector. If the cable is run to a box that does not have any mechanism for securing the cable, it must be stapled within 8″ of the box.

There is always a danger of accidentally piercing the conductors with nails driven into the studs while installing drywall. Metal plates are available to protect the conductors.

Goodheart-Willcox Publisher

Figure 31-28. An electrician often must drill holes through the building frame to run electrical cable or conduit.

A

Goodheart-Willcox Publisher

B

Goodheart-Willcox Publisher

Goodheart-Willcox Publisher

Figure 31-30. Cable or conduit must be securely held to boxes with the proper connectors.

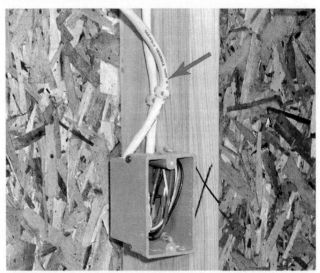

Goodheart-Willcox Publisher

Figure 31-29. Running conductors through wall framing. A—When running nonmetallic electrical cable through steel studs, electrical codes require the use of plastic grommets to protect the cable from abrasion. The grommets are placed in holes prepunched in the studs for wiring runs. B—Running nonmetallic cable through wood studs. Holes should be drilled approximately in the middle of the studs. If the holes are less than 1 1/4″ from the edge of the stud, a steel plate is used to protect the cable from drywall fasteners.

Figure 31-31. Nonmetallic cable must be properly fastened to framing members. Conductors that are secured to the box with clamps or cable connectors must be secured with a cable staple within 12″ of entering a box. For plastic boxes with no clamp or cable connector, the cable must be stapled within 8″ of entering the box (arrow).

31.8 Device Wiring

To complete a circuit, it is sometimes necessary to connect wires to one another. It is also necessary to connect wires to devices such as switches, receptacles, and large appliances. The NEC provides the guidelines on how these connections should be made. Before attempting to connect wires, several inches of the jacket must be stripped away. With the jacket stripped, the insulation is stripped from about 5/8″ of each conductor. To make a connection, twist the bare wires together and then install a wire nut around the twisted wires, **Figure 31-32**.

The terminal connections on switches and receptacles are usually screws. The screws are tightened to grip the conductors. **Figure 31-33** shows the correct method of bending a conductor for attachment to a terminal.

Switches control the current flow through a circuit. The hot, or current-carrying, conductor in the circuit is interrupted by the switch to turn a load on or off, **Figure 31-34**. A single pole switch is simple to wire, but three-way and four-way switches can be more complicated. **Figure 31-35** and **Figure 31-36** illustrate how current flows when three-way switches are operated.

Receptacles allow the transfer of electrical energy from the circuit conductors to lamps, appliances, and a variety of other devices. They have terminals on both sides. The hot wire is attached to the copper-colored terminal. The neutral (white) wire is attached to the lighter, silver-colored terminal, **Figure 31-37**.

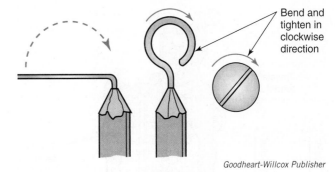

Goodheart-Willcox Publisher

Figure 31-33. The correct method used to prepare a conductor for attachment to a screw-type terminal. Use needle-nosed pliers to form the loop.

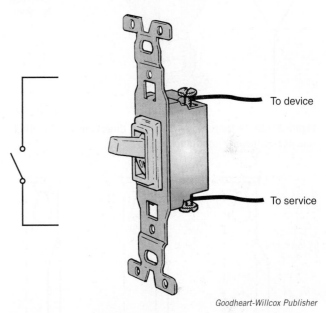

Goodheart-Willcox Publisher

Figure 31-34. A switch interrupts a current-carrying (hot) wire. It is always placed on a black or red wire.

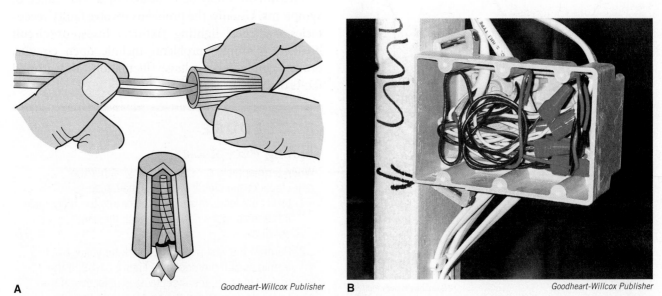

A *Goodheart-Willcox Publisher*   **B** *Goodheart-Willcox Publisher*

Figure 31-32. The proper use of wire nuts to connect bared ends of conductors. A—Twist the conductors together, then insert the bare conductors into the wire nut and twist the nut clockwise until it is tight. B—Connected wires are insulated against accidental shorting by the wire nut.

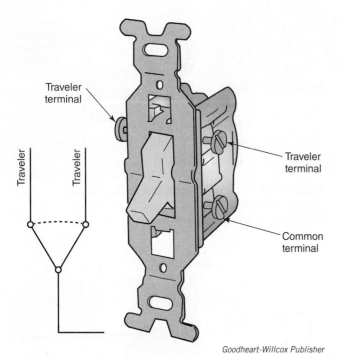

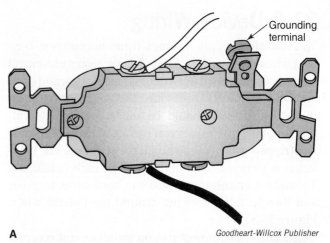

Goodheart-Willcox Publisher

Figure 31-35. A three-way switch can direct current to either one of two traveler conductors.

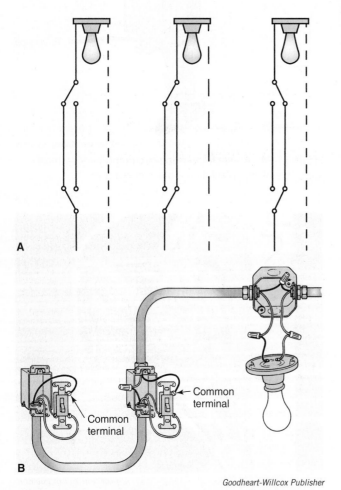

Goodheart-Willcox Publisher

Figure 31-36. Three-way switch installation. A—How a three-way switch arrangement actually works. B—How an electrician might wire it up.

Goodheart-Willcox Publisher

Figure 31-37. Wiring a receptacle. A—Neutral (white) wire is connected to light colored terminals; black or red to darker terminal. B—A receptacle properly wired and installed in a box.

31.9 Electrical Troubleshooting

Electrical issues may be indicated by a wide range of symptoms. Usually, the problems involve faulty receptacles, switches, lighting fixtures, fuses, or circuit breakers. Complex problems include open circuits, broken conductors, voltage fluctuation, and ground faults or current leakage.

PROCEDURE

Testing Receptacles

When a receptacle is not functioning, either the receptacle or the circuit is usually the cause.

1. Insert the leads of a circuit tester into the receptacle. If the tester does not light, check the line conductors.
2. Remove the wall plate and check for voltage at the terminals of the receptacle, **Figure 31-38**. If the tester lights up, the receptacle is defective. If the tester does not light, the problem is in the circuit. Sometimes an open neutral is the problem.

Continued

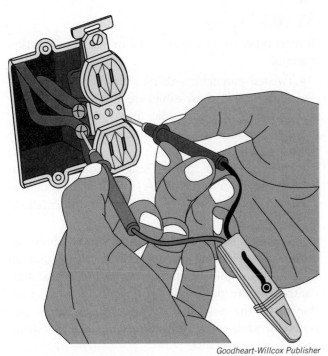

Goodheart-Willcox Publisher

Figure 31-38. To check for voltage, place the probes on the two terminals that have wires attached to them. If the tester lights up, voltage is reaching the device.

PROCEDURE *(continued)*

3. To test for an open neutral, place one test probe in the ground slot and the other in the hot slot of the receptacle, **Figure 31-39**. If the tester lights, indicating voltage, there is an open in the neutral conductor.

4. Check the other receptacles on the circuit between the problem receptacle and the distribution panel, and check the neutral connection at the distribution panel, as well.

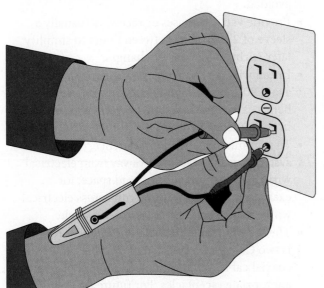

Goodheart-Willcox Publisher

Figure 31-39. Place probes in the hot side slot and the ground. If the tester lights, there is an open in the neutral conductor.

PROCEDURE
Testing Switches

To test switches, remove the cover plate and determine if there is power to the switch.

1. Touch one probe of a circuit tester to the metal box, or the ground wire if the box is plastic, while touching the other probe to the hot side terminal. If the tester lights, the circuit is live.

2. With the switch on, touch the probe to the load terminal of the switch and the other probe to ground. If the tester lights, the switch is good and the fixture or the wiring to the fixture is defective.

PROCEDURE
Testing Fixtures

Nonfunctioning fixtures may be checked by testing whether power reaches the supply conductors at the fixture connection, **Figure 31-40**.

1. If the fixture is controlled by a switch, check to see that the switch is on.

2. Place the probes of a circuit tester on the supply conductors.

3. If the tester lights, voltage is present.

4. Repair or replace the fixture.

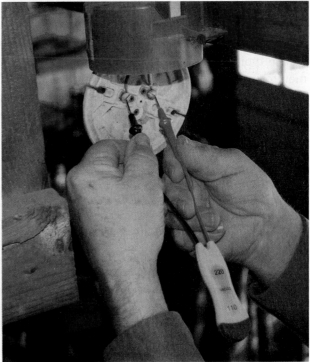

Goodheart-Willcox Publisher

Figure 31-40. Test a fixture by checking if line voltage is reaching the terminals. Place one probe onto each terminal. If the test causes the tester to glow but the fixture is dead, replace or repair the fixture.

31.10 Home Security and Automation Wiring

An increasing number of new homes are being prewired with low-voltage wiring systems to handle anything from a basic home security system to full-scale home automation. The most sophisticated systems include the wiring for the control and distribution of audio and video signals, remote heating/cooling control, telephone and computer network wiring, wiring for Internet access, and security and fire alarm systems. Installing the wiring for modern low-voltage systems is most efficient and least expensive when the walls are still open. Thus, most installations are done in new homes or as part of major remodeling projects where the walls have been stripped down to the studs. Retrofitting of systems into existing structures is both time- and labor-intensive, greatly increasing installation costs.

The "smart house" with advanced communication and automation features is a growing trend among homebuilders. Within a few years, structured wiring may be considered as a basic household system, like plumbing and electrical.

31.10.1 Installation

The installation of home security system wiring is relatively simple. The components of the system include entry/exit keypads used to set or disable the system, door and window switches, motion detectors, security cameras, external lights, audible alarms, the system controller, and associated low-voltage wiring. The system typically includes a telephone connection to a monitoring agency or the local police and fire stations.

Installation of home automation systems, referred to as *structured wiring*, can be considerably more complex. The installation typically involves one or more structured wiring distribution panels and runs of multiple cables from the panels to wall outlets in each room of the house. Distribution panels for structured wiring are similar to standard electrical distribution panels, but contain various electronic modules instead of circuit breakers. In some installations, wiring may be run through conduit. In other installations, bundled cables alone are used. One advantage of using conduit is the ability to later install additional cables.

Pro Tip

To simplify the task of pulling additional cables through a conduit at a later date, pull a length of strong nylon fishing line through each conduit run at the time of original installation. Make each nylon line twice as long as the conduit run and loosely coil the excess at one end. This will allow you to pull additional cable and still have the line in place for later use.

31.10.2 Wiring

Several types of wiring are used in home automation systems:

- Twisted-pair cables—With a number of wire pairs in each cable, these cables are multi-purpose wires capable of carrying audio, high-speed computer data transmissions, telephone, and security system signals.
- Shielded coaxial cable—Designed to handle video signals and Internet access (with a cable modem).
- Fiber-optic cables—Used to transmit information at the speed of light.

Specialized subcontractors usually install both security system wiring and structured wiring, but an increasing number of electricians and electrical subcontractors are performing the work. In most cases, they install the conduit, distribution boxes, cables, and wall outlets following the drawings provided by a structured wiring specialist.

31.10.3 General Considerations

There are some general considerations when installing structured wiring:

- Whenever possible, the structured wiring should be installed after the electrical and plumbing work are done. This avoids potential damage and electrical interference problems.
- To provide the highest quality service, all wiring should be set up as home runs. A *home run* is when the wiring is routed directly from the panel to the outlet. Daisy chains are the connection of several devices in sequence and should be avoided.
- Provide vertical chases or raceways (usually a sleeve of 2″ conduit) between floors to simplify later wiring additions or changes.
- Install wiring runs with extra wire or slack and avoid 90° bends. Tightly stretched cables and tight bends can degrade system performance, especially in computer networks.
- To minimize electrical interference, run all systems wiring at least 12″ away from electrical wiring (in the next joist or stud space, for example). If cables must be run across electrical wires, make sure they cross at a 90° angle.
- The recommended installation for most rooms is two Category 5 (twisted pair) cables and two coaxial cables, all terminated in a wall plate with appropriate receptacles. For future use, two fiber-optic cables are often bundled with the twisted-pair and coaxial cables.

- Place the distribution panel in a central, accessible location such as a closet or mechanical/utility room. Place it 24″ or more from the electrical distribution panel.
- In larger homes, two or more distribution panels with functions divided between them may be used. For example, security and HVAC systems may be in a panel located in the garage or basement. Entertainment, communications, and related functions may be in a panel centrally located in the living area.
- To aid in troubleshooting or system changes, carefully label every cable in the distribution panel with its function and the room where it terminates.

Workplace Skills
Goals and Self-Motivation

Imagine playing a ballgame without points or driving a car without a destination. In some ways, this is like starting your work without setting some goals. Even small goals can bring great meaning and satisfaction to your work. Setting goals can help motivate us to do our best and push ourselves to the next level. Construction business owners know this and set goals for their companies, crews, and employees. It is also important for us to set our own personal and work goals. Setting weekly, daily, and even micro goals for individual tasks can improve your skill level and efficiency, as well as make work far more rewarding.

Construction Careers
Electrician

Electricians install electrical systems in new or existing residential, commercial, and industrial structures. The same electricians are often responsible for installing the low-voltage wiring and devices for alarm and security systems, computer networking, communication systems, and similar applications. Approximately one-quarter of all electricians work in the construction industry, while the remaining three-fourths are employed in maintenance roles. Most large institutions, from hospitals to industrial plants, employ maintenance electricians. An estimated one out of every ten electricians is self-employed.

Electricians must be familiar with all aspects of the National Electrical Code, as well as state and local building codes, and must be able to read construction drawings. Those who work with low-voltage systems need a good working knowledge of electronics. Electricians, especially those employed as maintenance electricians, must be comfortable with the mathematics used in designing and analyzing complex circuits.

Working conditions, especially in construction, are often strenuous—workers stand for long periods and may have to repeatedly climb ladders or scaffolding. Although most jobs are indoors, electricians may also need to work in difficult weather conditions. Safety is a major concern, since the work has the potential for electrical shock and injury from falls. Wearing proper personal protective gear and following good safety practices are very important.

Most electricians learn the trade by completing an apprenticeship of 3–5 years that involves at least 144 hours of classroom instruction and a minimum of 2000 hours of on-the-job training. Apprenticeship programs are operated by a joint effort involving the electrician's union (International Brotherhood of Electrical Workers) and one of several electrical contractors' associations, or by nonunion electrical apprenticeships.

Those who successfully complete the program are qualified to work in either construction or maintenance areas. Smaller numbers of electricians learn their trade through shorter-term training programs or by learning on the job as helpers for experienced electricians.

Most states and local jurisdictions require electricians to be licensed. Licensing is obtained through an examination that typically covers electrical theory and the National Electrical Code. Electricians who work in larger organizations can advance to supervisory positions. Experienced electricians often obtain positions with the government as electrical inspectors. Some choose to open their own electrical contracting businesses.

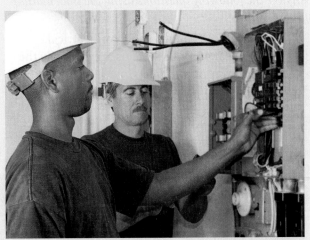

Lisa F. Young/Shutterstock

There is great potential for danger when working with high-voltage components. In order to work safely, electricians must be familiar with complex circuitry and all forms of electronics. They must also keep up to date with the latest local building codes, as well as the National Electrical Code.

Chapter Review

Summary

- While an electrical contractor normally installs the electrical wiring, carpenters need to be aware of the basic components of an electrical installation.
- The standard tools for electrical work include an electrician's hammer, saws, various types of pliers and screwdrivers, drills, measuring and fish tapes, and a wooden or fiberglass ladder. Metal ladders are unsafe around electricity.
- The components of an electrical system include conductors (wires), conduits or raceways, device and junction boxes, covers, switches, receptacles, and wire nuts for joining wire.
- Ground-fault circuit interrupters (GFCIs) are safety devices that protect against electrical shock.
- Arc-fault circuit interrupters (AFCIs) detect and stop erratic current flow to protect against house fires.
- Electricity is transmitted by the power utility over wires to a transformer that steps down the voltage to a level suitable for household use. The wires from the transformer to the building are called the service.
- The service conductors pass through a metering device and terminate in an entrance panel.
- The entrance panel contains circuit breakers or fuses for each of the household electrical branch circuits.
- These circuits are shown on the electrical plan for the house, which is part of the construction drawing set.
- Once an electrical system is installed, it may require troubleshooting, which involves the testing of receptacles, switches, and fixtures.
- Low-voltage systems for home security and home automation are becoming increasingly important in residential building.

Know and Understand

Answer the following questions using the information in this chapter.

1. *True or False?* The installation of electrical, plumbing, and HVAC systems are generally the responsibility of carpenters.

2. The National Electrical Code (NEC) is reviewed and revised every _____ years.
   A. two
   B. three
   C. four
   D. five

3. Which of the following saws is best used to cut metal conduit or cable?
   A. Handsaw.
   B. Jab saw.
   C. Reciprocating saw.
   D. Hacksaw.

4. _____ are good for holding, shaping, and cutting.
   A. Wire strippers
   B. Screwdrivers
   C. Pliers
   D. Wrenches

5. A _____ is a long strip of metal, fiberglass, or nylon wound on a reel for ease of handling.
   A. fish tape
   B. steel tape
   C. circuit tester
   D. drill

6. The purpose of _____ is to enclose devices that control electric current and protect connections between two or more conductors.
   A. circuits
   B. circuit breakers
   C. boxes or box covers
   D. GFCIs

7. Which of the following areas is a GFCI *not* required in a home?
   A. Bathrooms.
   B. Garages.
   C. Crawlspaces.
   D. Patios.

8. _____ control the current to lights and certain appliances.
   A. Outlets
   B. Receptacles
   C. Switches
   D. Conduit

9. Electrical current that goes in the opposite direction of a conductor passing through a magnetic field is known as _____ current.
   A. direct
   B. alternating
   C. induced
   D. arc

10. *True or False?* Transformers are devices that change voltage by means of deduction.

11. Electrical service enters the house through a meter and into a(n) _____.
    A. ductwork
    B. cable
    C. transformer
    D. entrance panel

12. A grounding bus bar is connected to a grounding electrode, which may be a metal cold-water pipe or grounding _____ driven deep into the soil.
    A. rods
    B. cable
    C. boxes
    D. wires

13. *True or False?* An electrical plan is *not* the same as a cable layout.

14. A(n) _____ is used to connect the bared ends of two wires.
    A. wire strip
    B. cap
    C. wire nut
    D. screw

15. *True or False?* Home security and automation systems generally cost less to install before the wall covering has been applied.

16. _____ cable is used in structured wiring systems to carry video signals.
    A. Shielded coaxial
    B. Fiber-optic
    C. Twisted-pair
    D. HDMI

17. To minimize electrical interference, all systems should be wired at least _____″ away from electrical wiring.
    A. 2
    B. 12
    C. 16
    D. 24

Apply and Analyze

1. Why is an electrician's hammer preferred over a carpenter's hammer when attaching boxes?
2. What is included in a building's electrical system?
3. What is the difference between an AFCI and a GFCI?
4. What can result from a ground fault in an electrical system?
5. What is a branch circuit?
6. If the circuit tester does not light when the probes are placed on the terminals of the receptacle, where does the problem lie?
7. In electrical wiring, what is referred to as a *home run*?

Critical Thinking

1. Do you agree with the code requirement for ground-fault circuit interrupters (GFCIs) in unfinished basements and crawl spaces? Why or why not?
2. Study **Figure 31-24**. What are the primary differences in the lighting circuits of the entry, living room, and dining room? Make a list of all the materials needed to install the lighting circuits for each of these three rooms. Include the quantities of each material needed, as well as the type of wire and switch for each room. Some assumptions will need to be made for items such as wire lengths. On your list annotate which room each item will be used in.
3. While building a deck on an existing home, Jared's table saw suddenly stops working mid-cut. He noticed that Kayla was running the circular saw while he was cutting. Both of their saws quit at the same time. What should they do? Is this an indication of faulty wiring? Explain.

Communicating about Carpentry

1. **Art.** Working with a partner, create a model of a building's electrical wiring. Create the model to show the different rooms and appliances that are being powered by the electrical wiring. Create the model so that the parts can be removed. Use your model to demonstrate to the class how electricity flows.
2. **Speaking.** Working in a group, brainstorm ideas for creating classroom tools (posters, flash cards, and/or games, for example) that will help your classmates learn about electrical wiring. Choose the best idea(s), then delegate responsibilities to group members for constructing the tools and presenting the final products to the class.
3. **Listening and Speaking.** Search online for local unions that may have workshops, guest speakers, or presentations on techniques and new materials. Also look online for instruction videos. Your local library may also have various media covering electrical wiring and installation. Watch videos, attend a presentation, and take notes. Listen for terms that you have learned from your textbook. Write a short summary and be prepared to present your findings to the class.

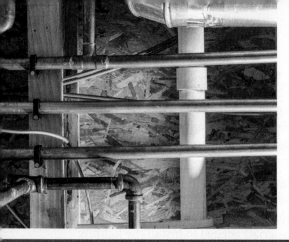

CHAPTER **32**

Plumbing Systems

OBJECTIVES

After studying this chapter, you will be able to:

- Cite codes that govern the installation of plumbing systems.
- Identify the differences between the water supply system and the drainage, waste, and venting (DWV) system.
- List necessary plumbing tools and explain how to use them.
- Describe the different types of materials used in plumbing systems.
- Name and recognize devices and fixtures that are part of the plumbing system.
- Identify materials used to seal various pipes and systems.
- Read plumbing prints.
- Explain the proper design and installation of plumbing systems.
- Describe common sink drain problems and how to repair them.
- Demonstrate a basic understanding of well and pump systems.
- Cite safety measures that plumbers must observe.

TECHNICAL TERMS

bored well	fixture	slip-joint washer
casing	flare fitting	stub-in
compression fitting	gate valve	submersible pump
compression valve	O-ring	sweat soldering
drainage, waste, and venting (DWV) system	pipe compound	Teflon® tape
	plumber's putty	vent stack
drilled well	potable	venting
faucet	rough-in	water supply system

lumbing includes all of the piping and fixtures that supply water for drinking, cooking, bathing, and laundry. Also included are drainage pipes that provide a means of disposing of wastewater, either through a municipal sewer or a local septic system. See **Figure 32-1**.

To install rough plumbing, plumbers must cut holes and notches in the building frame. It is important for a plumber to closely work with a carpenter to be sure that the holes will not weaken joists and studs. The work should be neatly done. Holes should be smooth, cleanly made, and only large enough to receive the pipe. Notches should be

square or rectangular. Where the framing is done with I-joists, pipes can be run through predrilled or perforated openings in the I-joists. Section 602.6 of the International Residential Code (IRC) covers notching studs for plumbing.

Once plumbing is installed, notched framing members should be reinforced with metal strapping. In certain situations, a joist may need to be partially removed. In such cases, a header should be installed to reinforce the floor frame at that point. **Figure 32-2** illustrates how a plumber should handle these situations. Note also how easily plumbing runs are made through truss joists.

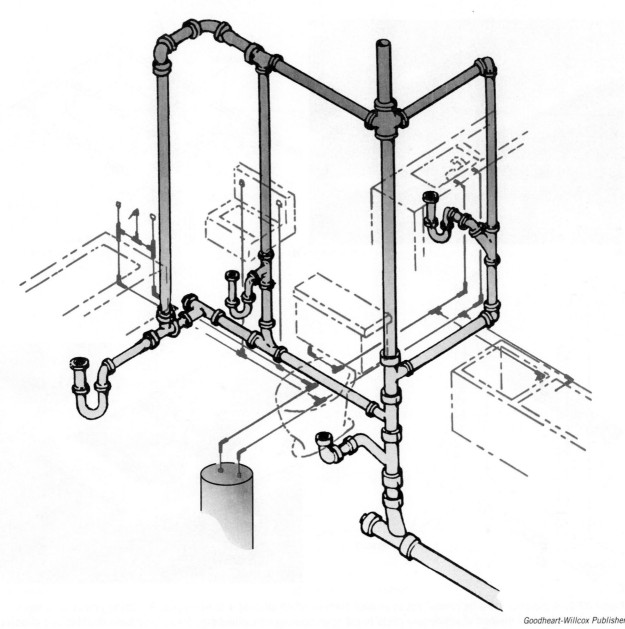

Figure 32-1. A pictorial view of a plumbing system. The large pipes are parts of the drainage, waste, and venting (DWV) system. The upper portions of the DWV comprise the venting. The small pipes are the water supply system.

32.1 Plumbing Regulations

Regulations for installing plumbing systems are covered by several agencies. One is the International Code Council (ICC), which sponsors the International Plumbing Code (IPC). The IPC has been adopted at the state or local level in many states throughout the country. The IRC and the International Building Code, which includes commercial buildings, also each contain several chapters dealing with plumbing. Some states still have plumbing codes that are based on model codes that existed before the IPC was first published. Since codes may vary in their requirements from one community to another, a plumber must always be familiar with the local code.

Plumbing requires a variety of skills. A plumber must use many different kinds of tools, equipment, and materials. Further, plumbers must be skilled in woodworking, metalworking, pipe threading, soldering, and brazing.

Plumbers must make accurate measurements and calculations. They must add and subtract dimensions, figure pipe offsets, and determine the volumes of tanks. As a precaution against costly errors, a good plumber checks each measurement and each calculation twice before marking or cutting.

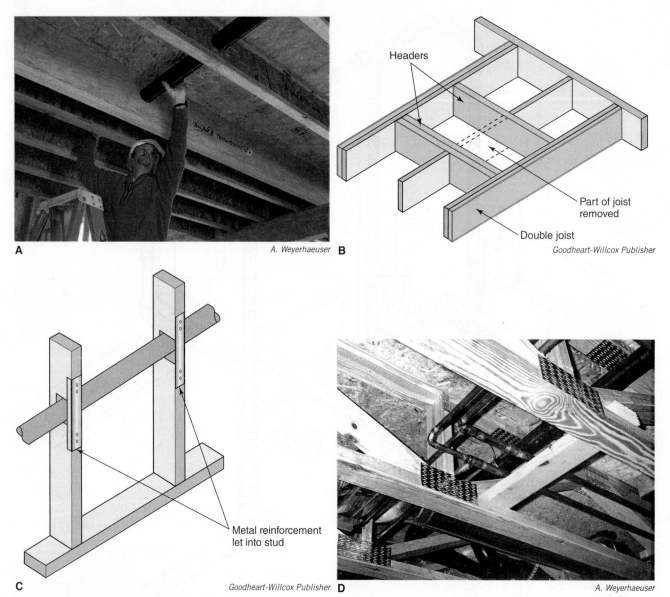

A   *A. Weyerhaeuser*

B   **Headers**   **Part of joist removed**   **Double joist**   *Goodheart-Willcox Publisher*

C   **Metal reinforcement let into stud**   *Goodheart-Willcox Publisher*

D   *A. Weyerhaeuser*

Figure 32-2. A plumber must be careful not to weaken framing when altering it to run pipes. A—Using precut openings in I-joists. B—Approved method of supporting joists to get large openings for plumbing. Joists have been doubled and headers installed. C—Where studs must be notched, metal reinforcing must be installed across the notch. D—When open-truss joists are used, no cutting is necessary. All mechanical systems can be run through the voids in the truss.

32.2 Two Separate Systems

Plumbing includes two subsystems that have important differences:

- Water supply system
- Drainage, waste, and venting (DWV) system

The *water supply system* distributes water under pressure throughout the structure for drinking, cooking, bathing, and laundry. Supply pipes, fittings, and joints must be able to withstand the pressure. This is a two-pipe system. One pipe carries cold water and the other hot water. Supply systems are watertight to prevent leakage.

Drainage piping, commonly referred to as *drainage, waste, and venting (DWV) system*, carries away wastewater and solid waste from bathrooms, kitchens, and laundries. This subsystem is not under pressure; it relies on gravity. DWV systems, like supply systems, are watertight.

Proper venting is extremely important for DWV systems. *Venting* permits air to circulate in the pipes. This prevents backpressure and siphoning of water from traps, **Figure 32-3**. Under certain conditions, it also prevents the introduction of wastewater into the *potable* (fit to drink) water supply. Plumbing codes are very specific about the sizes and locations of vent piping.

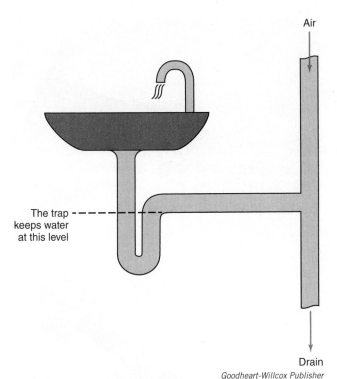

Air

The trap
keeps water
at this level

Drain

Goodheart-Willcox Publisher

Figure 32-3. The vent pipe allows atmospheric air pressure to enter the system, preventing the weight of the water flowing out of the drain from sucking (siphoning) all of the water out of the trap.

32.3 Tools

A plumber's tool kit includes many of the same tools used in carpentry and electrical work. To perform simple plumbing tasks, only a few tools are needed. One of a plumber's most used tools is the wrench, **Figure 32-4**. Some tools designed to perform tasks specific to plumbing are shown in **Figure 32-5**.

Among the more important tools are those used for measuring and producing accurate lines, circles, and other markings. Plumbing measurements must be precisely accurate and the tools must be capable of this accuracy. Measuring tools include rules, tapes, squares, levels, transits, plumb bobs, chalk lines, compasses, and dividers.

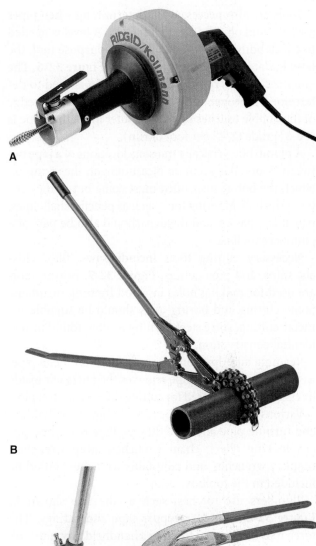

Ridge Tool Company; Reed Manufacturing Co.

Figure 32-5. Certain tools are used only for plumbing.
A—A power auger is used to clear clogs in drains.
B—A chain cutter is used to cut large-diameter soil pipe.
C—This reamer removes burrs from any size of pipe.
D—Pipe wrench pliers grip and turn pipe.

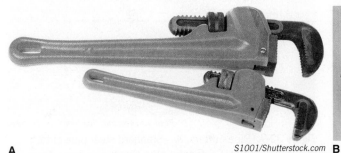

S1001/Shutterstock.com

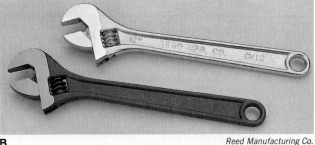

Reed Manufacturing Co.

Figure 32-4. Wrenches are a plumber's most-used tools. A—Pipe wrenches are used to tighten threaded fittings on metal pipe. B—Adjustable wrenches work on fittings with flats.

Tools are also necessary for determining when pipes are level and plumb (exactly vertical). A level is needed to check both conditions. A good all-purpose tool, the level has at least three indicator vials, **Figure 32-6**. The outer set of lines on a level indicator vial is used to determine the proper slope of a drainpipe. When the edge of the bubble touches one of the outer lines, the slope is at a 2° grade (1/4″ per foot of run).

A plumb bob vertically transfers locations of a pipe between floors. For accurate measurement, the string on which the bob is suspended must come out of the center of the bob. Marking tools such as pencils, chalk lines, pencil compasses, and dividers should also be part of a plumber's toolbox.

Necessary cutting tools include saws, files, chisels, snips, and pipe cutters, **Figure 32-7**. Boring tools are used for making holes in wood framing members. Some cutting and boring tools should be suitable for metal-cutting, since steel framing is often found in residential construction.

Reaming and threading tools are used on metal pipe. Reaming the end of a pipe removes the burrs left inside by cutting. Dies are used for cutting threads on steel pipe.

Various types of wrenches are designed for holding and turning pipe and pipe fittings. Pipe wrenches, pliers, locking pliers, chain wrenches, strap wrenches, monkey wrenches, and adjustable wrenches should be included in the toolbox.

Plumbers use torches, such as the one shown in **Figure 32-8**, to solder copper pipe and fittings. The torch may be attached to either a handheld propane tank or, by using a hose, with a much larger tank. There are three fuel choices for plumbers' torches: acetylene, propane, and MAPP. Propane is less expensive than acetylene, but acetylene produces a hotter flame (2500°C when burning in air and with no oxygen added) than propane (1970°C), so less gas is used. MAPP gas burns at just over 2000°C and is the most expensive choice.

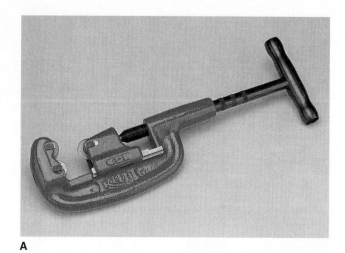

A

B

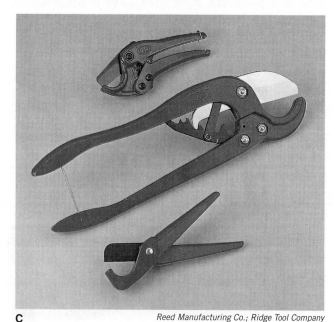

C

Reed Manufacturing Co.; Ridge Tool Company

Figure 32-7. Pipe cutters. A—Standard steel pipe cutter. B—Tubing cutter. C—Plastic pipe shears.

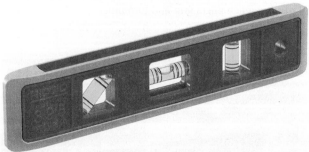

Ridge Tool Company

Figure 32-6. A level is used to find level and plumb. Spirit levels, like this one, are frequently used to ensure that drain pipes have the proper slope to help them drain.

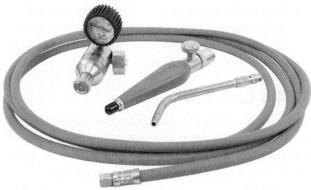

Forney Industries, Inc.

Figure 32-8. A plumber's torch kit may consist of either a torch and a handheld tank or a torch, hose, and regulator attached to a larger tank.

Safety Note
When working with a torch, always be aware of what is behind the workpiece. The heat of a torch flame will quickly ignite many building materials. Make sure the work area is safe and clear of any flammable items. Locate a nearby fire extinguisher before lighting the torch.

32.4 Plumbing Materials

There are several types of plumbing materials:

- Pipe and tubing for supply lines carrying pressurized water to faucets and fixtures
- Large-diameter pipe to drain away wastewater
- Valves and faucets for controlling water supply
- Washers and O-rings for both supply and drainage systems
- Tape, compounds, and putties to help seal connections against leaks

32.4.1 Pipe

Several different types of material are used for pipes and pipe fittings. Depending on codes, supply system pipe may be made of copper, galvanized steel, or plastic. DWV pipe may be made of malleable iron, cast iron, copper, galvanized steel, or plastic.

32.4.2 Copper Tubing

Copper tubing is available as rigid, straight lengths or in flexible coils, **Figure 32-9**. It is sold in a variety of diameters and seven different weights. Only three weights—types K, L, and M—are typically used for water supply lines.

Copper tubing can be cut with a pipe cutter or hacksaw. The flexible tubing is often joined using *compression fittings*, connectors that seal joints by compressing a ring of soft material. The fitting slides over the tubing. Another method is to sweat solder the fittings. Only lead-free solder is permitted, avoiding the hazard of lead getting into the water.

32.4.3 Plastic Pipe

Plastic pipe can be used for both supply and drainage systems. It is available in lengths that can be cut with a handsaw or tubing cutters. It is joined with special fittings or by using a special adhesive.

- Chlorinated polyvinyl chloride (CPVC)—A buff-colored thermoplastic that is rigid, light, easy to handle, and resistant to cracking in freezing conditions. Connections may be threaded or solvent welded. CPVC is suitable for hot water piping and has a rating of 180°F and 100 psi.

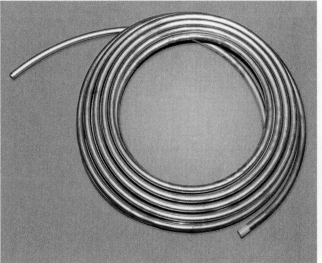

A   *Goodheart-Willcox Publisher*

B   *Goodheart-Willcox Publisher*

Figure 32-9. Typical copper materials. A—Copper tubing. B—Forged copper fittings. These must be sweat soldered to connect to the pipe.

- PEX (cross-linked polyethylene)—This tubing is very popular for water supply piping because it is flexible, lightweight, and connects easily. It can be used for hot and cold water supply lines. Connections are made with crimped connectors, **Figure 32-10**. A centralized manifold is usually installed to distribute the water.

> **Pro Tip**
>
> Polybutylene (PB) tubing was widely used from 1980 to 1995. It was discontinued in 1995 because some chemicals in household water, such as chlorine, oxidized and weakened the PB tubing and fittings. PB is typically, but not always, steel gray in color. Other materials might be steel gray also, so you should not assume that just because tubing is gray in color that it is PB. If PB is encountered in a remodeling job, it should be replaced.

Plastic piping has become popular for drainage systems, as well. It is used in most new construction. There are two major types of plastic piping:

- Acrylonitrile butadiene styrene (ABS)—Resists chemical attack and is inexpensive. It is usually joined with a one-step solvent.

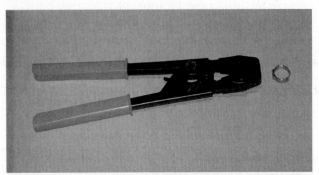

A   *Goodheart-Willcox Publisher*

B   *Goodheart-Willcox Publisher*

Figure 32-10. PEX tubing is frequently used for residential plumbing. A—A special crimping tool is used to tighten the stainless steel clamps that secure the PEX to fittings. B—PEX tubing with fitting.

- Polyvinyl chloride (PVC)—Has lower thermal expansion that makes long runs easier to control. It is joined with a two-step primer/solvent and suitable fittings. See **Figure 32-11**.

32.4.4 Cast Iron

Cast iron was once extensively used in plumbing, and still is used in some special situations. Where it is used, cast iron soil pipe is usually joined with a neoprene sleeve and stainless steel clamps, **Figure 32-12**.

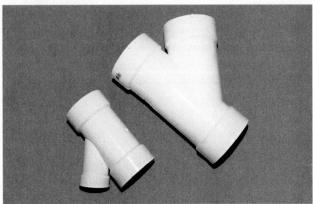

A   *Goodheart-Willcox Publisher*

B   *Goodheart-Willcox Publisher*

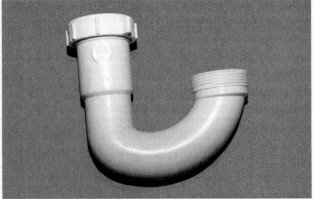

C   *Goodheart-Willcox Publisher*

Figure 32-11. Plastic fittings like these are used in drainage systems. A—Wye. B—Closet flange. C—P-trap.

E.I. duPont de Nemours & Co.

Figure 32-12. The most common means for joining cast iron pipe is with a neoprene sleeve and stainless steel clamps.

32.4.5 Steel Pipe

There are two types of steel pipe used in buildings: galvanized pipe and black pipe. Galvanized pipe is used in water supply plumbing. It is galvanized with a coating of zinc to retard rusting. Galvanized pipe has largely been replaced by copper and plastic for water supply pipes.

Steel pipe is manufactured in standard lengths of 21′ and in diameters from 1/8″ up to 2 1/2″. The ends have a tapered thread so that the joint will seal. Fittings are made of malleable iron to make them more resistant to stress. Joints are sealed by using a putty-like pipe compound or by applying Teflon® tape to threads before assembly. When pipe compound is used on gas pipes, it must be a type that is intended for gas piping. Black pipe is used primarily for carrying gas. The zinc coating on galvanized pipe can flake and choke a gas pipe, making it unsuitable for gas. Black pipe gets its name because it is painted black instead of being galvanized. Black pipe rusts quickly when it is in constant contact with water, so it is not a good choice for water pipes. **Figure 32-13** shows the appearance of galvanized pipe and black pipe.

32.4.6 Valves

Valves are devices that control the flow of water in the water supply system, **Figure 32-14**. They are installed at certain places in the lines so that water can be shut off to drain pipes and appliances. *Gate valves* control water with a flat disk that slides up and down perpendicular to the supply pipe. It is never used to restrict flow; it is either fully open or fully closed. *Compression valves* can stop or restrict water flow. They make use of a tapered plug or a washer that can be fully seated to stop water flow, or opened by different amounts to control the flow.

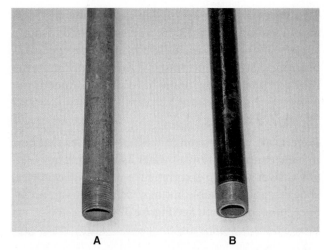

A   **B**

Goodheart-Willcox Publisher

Figure 32-13. Two kinds of steel pipe. A—Galvanized pipe. B—Black pipe.

A   *Goodheart-Willcox Publisher*

B   *Goodheart-Willcox Publisher*

Figure 32-14. Valves interrupt and control water flow. A—A gate valve must always be fully open or fully closed to properly work. B—This compression valve with thread for a hose on the outlet is called a boiler drain.

Two types of compression valves are shown in cutaway view in **Figure 32-15**. Valves are made from a variety of materials including bronze, brass, malleable iron, cast iron, and plastic. Threads may be on the inside or the outside. Some have sweat connections.

32.4.7 Faucets

Faucets are compression or washerless valves that permit the controlled flow of water as needed. They usually deliver water to fixtures such as sinks, lavatories, showers, and bathtubs, but may deliver it to an appliance, hose, or bucket. See **Figure 32-16**.

Safety Note

Anytime plumbing pipes are drained or newly installed, there is a chance that airborne microorganisms will have entered the system. After the system is recharged with water pressure, it should be disinfected. This can be done by running a diluted solution of chlorine laundry bleach through the system. Use caution in handling chlorine bleach, as a few drops can remove the color from clothing. Always wear rubber gloves and safety glasses when handling chlorine and other strong chemical solutions.

32.5 Fixtures

Fixtures are water-using devices, such as sinks, lavatories, bathtubs, bidets, urinals, toilets, or showers. Fixtures receive water from the supply system and have a means for delivering wastewater to the DWV system. Certain water-using appliances also require attachment to plumbing systems. These include clothes washers, dishwashers, and automatic icemakers in refrigerators.

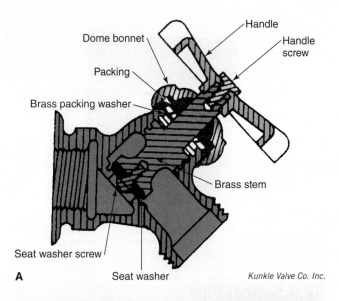

A   *Kunkle Valve Co. Inc.*

B   *American Standard*

Figure 32-16. Faucets control water at a fixture. A—A hose bibb (faucet) is used for outside water supply and is threaded to receive a hose. The flange allows it to fit flush to an outside wall. B—A single-control lavatory faucet.

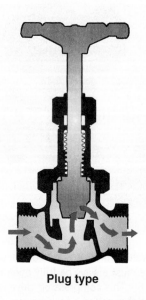

Plug type

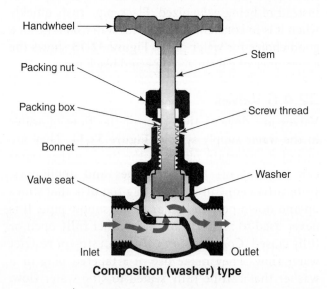

Composition (washer) type

William Powell Co.

Figure 32-15. Two types of compression valves.

32.6 Sealing Plumbing Systems

Some supplies and materials seal the plumbing system against leaks, **Figure 32-17**. *Slip-joint washers* are used to seal P-trap connections beneath sinks. *O-rings* are used on faucets to produce leakproof seals between the faucet body and internal parts. *Teflon® tape* and *pipe compound* are used to seal threaded connections between pipe and other parts. *Plumber's putty* is used to produce a seal on sink rims, strainers, and faucets. A wax ring, **Figure 32-18**, is used to form a seal between the water closet and the closet flange.

A — *Goodheart-Willcox Publisher*
B — *Goodheart-Willcox Publisher*
C — *Goodheart-Willcox Publisher*

Figure 32-17. Sealing plumbing pipes requires different materials and devices. A—Slip-joint washers and a slip-joint nut create a watertight connection between the tailpiece and the P-trap. B—O-rings. C—Teflon® tape, pipe joint compound, and plumber's putty.

A — *Goodheart-Willcox Publisher* B — *Goodheart-Willcox Publisher*

Figure 32-18. The water closet sits on a closet flange, with a wax ring forming a watertight seal. A—Closet flange. B—Wax seal.

32.7 Reading Prints

Certain drawing symbols have been standardized for the plumbing trade. These symbols are a type of shorthand. **Figure 32-19** and **Figure 32-20** show generally-accepted symbols for fixtures, appliances, mechanical equipment, pipes, and fittings.

Since architectural plans may not include pipe drawings, a plumber needs to develop some sketches as a guide to installation. This is especially important if others are working on the same plumbing job. There are three types of drawings or sketches that will be made: riser diagrams, plan view sketches, and isometric sketches. These are shown in **Figure 32-21**.

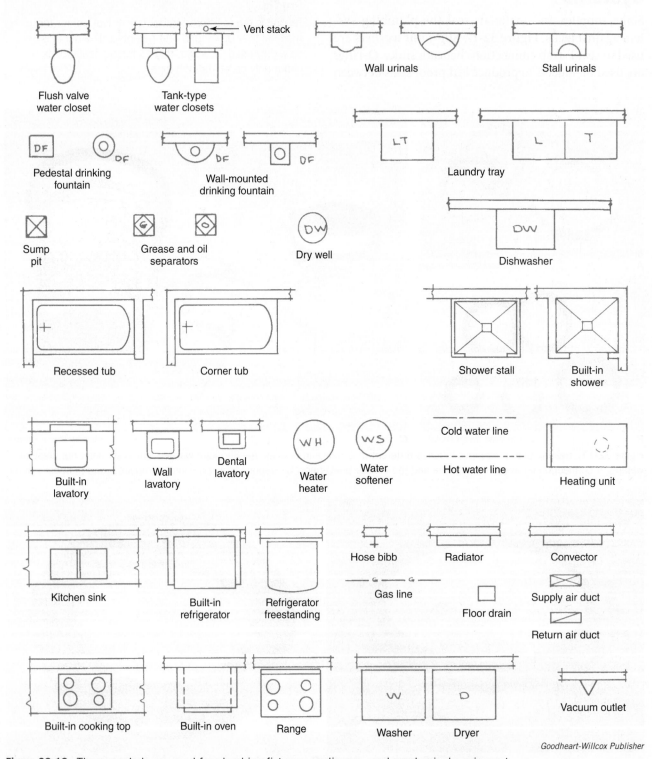

Goodheart-Willcox Publisher

Figure 32-19. These symbols are used for plumbing fixtures, appliances, and mechanical equipment.

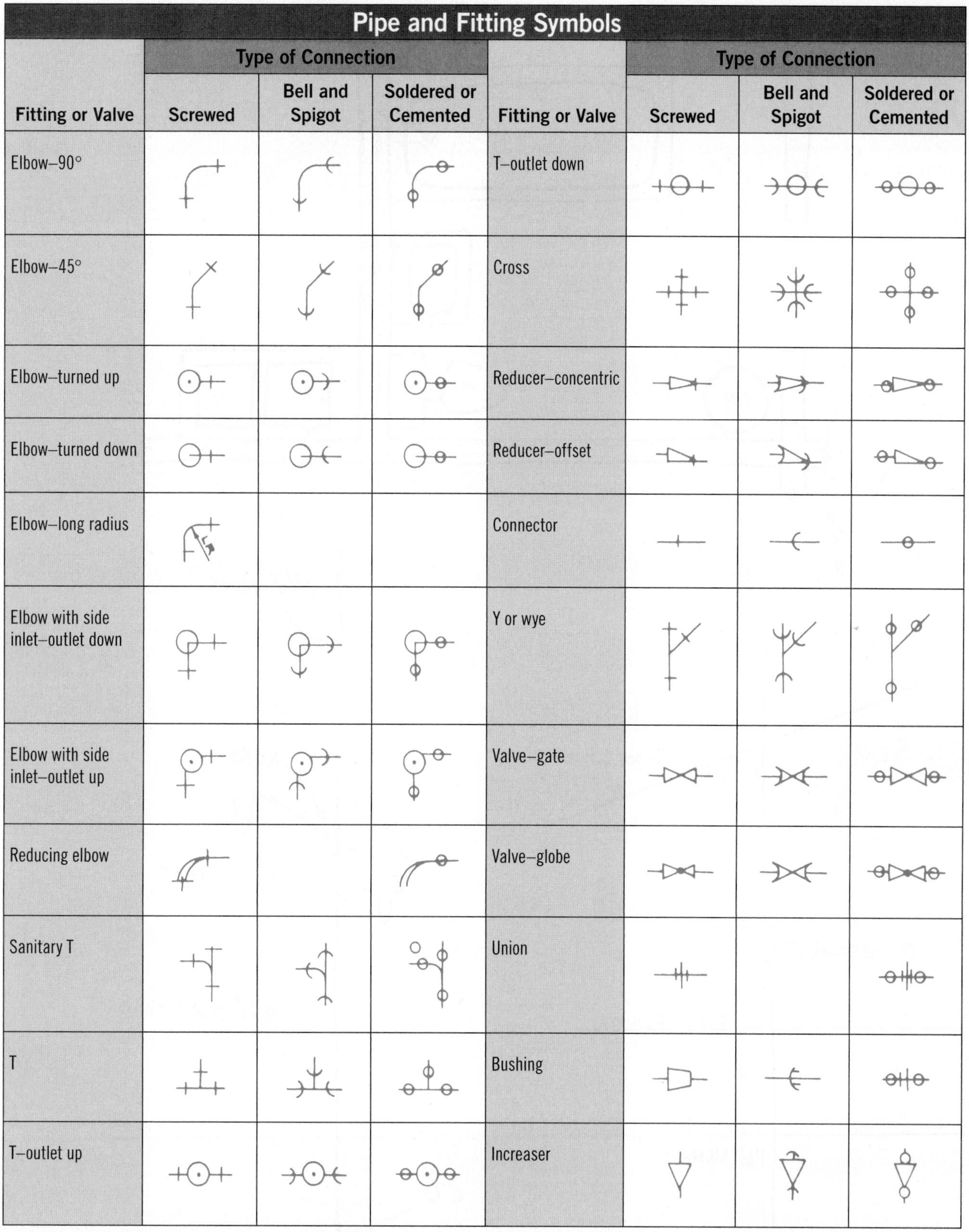

Pipe and Fitting Symbols

Fitting or Valve	Type of Connection			Fitting or Valve	Type of Connection		
	Screwed	Bell and Spigot	Soldered or Cemented		Screwed	Bell and Spigot	Soldered or Cemented
Elbow–90°				T–outlet down			
Elbow–45°				Cross			
Elbow–turned up				Reducer–concentric			
Elbow–turned down				Reducer–offset			
Elbow–long radius				Connector			
Elbow with side inlet–outlet down				Y or wye			
Elbow with side inlet–outlet up				Valve–gate			
Reducing elbow				Valve–globe			
Sanitary T				Union			
T				Bushing			
T–outlet up				Increaser			

Goodheart-Willcox Publisher

Figure 32-20. Pipe and fitting symbols that appear in drawings.

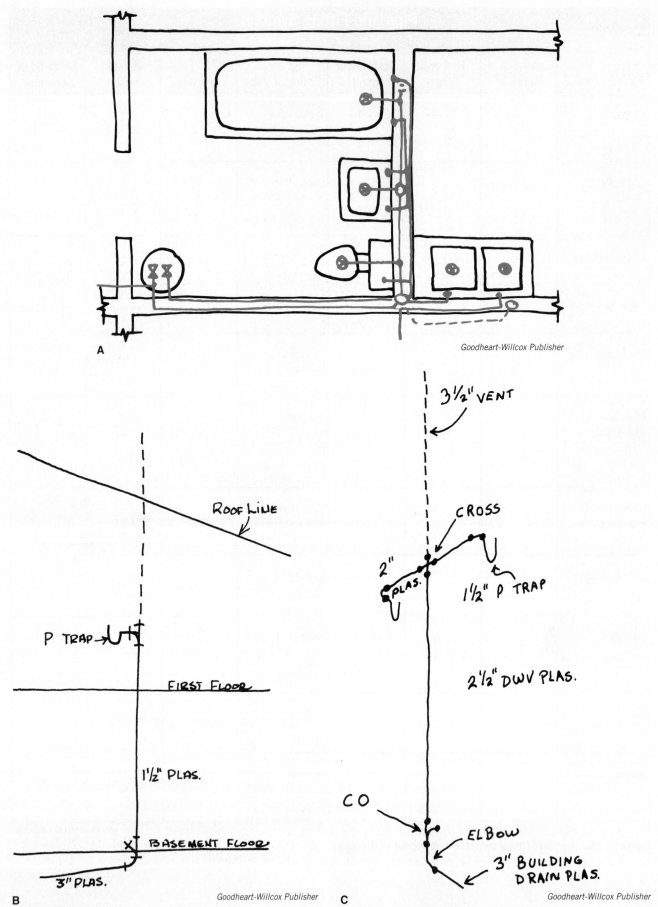

Goodheart-Willcox Publisher

A

3½" VENT

ROOF LINE

CROSS

2"

PLAS.

1½" P TRAP

P TRAP

FIRST FLOOR

2½" DWV PLAS.

1½" PLAS.

CO

ELBOW

BASEMENT FLOOR

3" BUILDING
DRAIN PLAS.

3" PLAS.

B

Goodheart-Willcox Publisher

C

Goodheart-Willcox Publisher

Figure 32-21. Three types of sketches. A—Plan view. B—Riser diagram. C—Isometric sketch.

32.8 Installing Plumbing

Plumbing installation is usually done by a plumbing sub-contractor. It can begin as soon as the building is closed in. Measurements are made to determine exact location of pipes and fixtures. First, measure runs of pipe from face to face. This is the distance from the end of one fitting to the end of the other fitting. Then, add the distance the pipe will extend into the fitting. Use a steel tape or plumber's rule to take careful measurements.

Cutters or saws are used to cut lengths of pipe or tubing. Be certain to ream pipes to remove interior ridges or ragged edges caused by sawing or cutting.

Plumbing is concealed within the building frame. The single exception to this rule is plumbing below first-floor joists. Here, runs can be made below, rather than through joists, unless the basement will be finished as living space. However, use of engineered lumber allows mechanical systems to be run through the open spaces of the truss braces, **Figure 32-22**.

Some bathroom fixtures, such as bathtubs and showers, are too large to fit through doorways of framed-up partitions. They must be placed prior to installation of studs for partitions, **Figure 32-23**.

The first phase of plumbing installation is known as a *rough-in*. Rough-in includes installing all pipes to the point where connections are made with the plumbing fixtures. All runs are placed and tested with pressurized air or water to make sure that no leaks exist. The vent pipes are connected to a large riser, or *vent stack*, that extends through the roof. It serves as an exhaust for any odors from the waste system. It also allows the system to take in air and thus avoid siphoning problems with traps. The pipe ends that extend out from the walls and above the floors are known as *stub-ins*. **Figure 32-24** shows stub-ins where fixtures will be connected.

Goodheart-Willcox Publisher

Figure 32-23. This tub/shower unit was installed before the steel studs for the interior partition were erected.

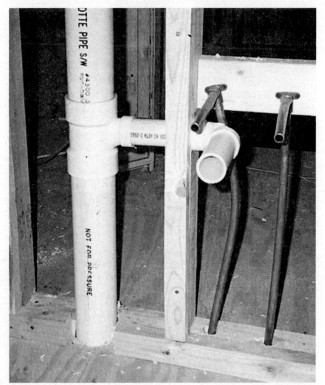

Goodheart-Willcox Publisher

Figure 32-24. Water supply and DWV stub-ins are ready for installation of a bathroom lavatory.

Goodheart-Willcox Publisher

Figure 32-22. The open spaces in truss joists simplify plumbing runs in a basement or crawl space.

32.8.1 Connecting Copper Pipe

Sweat soldering is a joining process for copper pipe, in which the fitting and pipe are heated with a gas flame and solder is drawn into the joint by capillary action. Sweat soldering is relatively easy, but must be carefully done to avoid leaking joints. You will need a gas torch, lead-free solder, paste soldering flux, and a small pad of steel wool or a tube-cleaning brush.

PROCEDURE
Soldering Copper Pipe

1. Ream the cut end of the tubing to remove metal burrs.
2. Clean the end of the copper tubing and the inside of the fitting with steel wool or a cleaning brush. This removes oxidation and reveals bright metal to help achieve a good solder bond.
3. Using a brush, apply paste flux to the cleaned end of the tubing and the inside of the fitting.
4. Slip the fitting over the tubing and heat the joint. It is best to heat the tubing first. Allow the inner cone of the flame to touch the metal. Heat the joint until the flux is smoking.
5. Add solid core solder by touching the end of the solder wire to the joint area at the edge of the fitting, **Figure 32-25**.
6. Feed the solder into the joint as the flame is moved to the center of the fitting. This movement will evenly draw the solder into the fitting, making a leakproof joint. Never try to melt the solder in the flame.
7. Once the solder is drawn into the joint, promptly remove the heat.
8. Wipe excess solder off the joint with a piece of burlap or denim cloth.

32.8.2 Bending and Unrolling Flexible Copper Tubing

Flexible copper tubing can be easily bent to change its direction during an installation. To avoid crimping, which may lead to a leak, slide a tubing spring over the section to be bent. Bend the tube and spring just beyond the desired angle, then back to the correct angle. See **Figure 32-26**. Remove the spring by turning the flared end counterclockwise.

Flexible copper tubing comes in coils. Be careful when uncoiling. A procedure that works well is to lightly place a foot on one end while holding the coil upright. Slowly roll the coil away from the secured end. Move the holding foot along until the proper length of tubing has been unrolled.

Compression or *flare fittings* used with copper tubing require no soldering. The fittings are designed to tightly fit against the tubing and mating parts.

PROCEDURE
Making Compression Joints

1. Slide the flare nut over the tubing, tapered end facing away from the end of the tubing.
2. Select the correct size hole on the flaring tool and slip the tube into it. The hole should be slightly smaller than the outside diameter of the tubing.
3. With the tubing sticking out 1/32″–1/16″ above the bar of the tool, tighten the wing nut so the tubing is tightly held in the bar.
4. Slide the ram over the tube and screw the ram into the end of the tubing. Tighten the ram until a 45° angle flare is made in the tube, **Figure 32-27**.
5. Press the tapered end of the other half of the fitting into the flared end of the tubing.
6. Slide the nut to the tapered fitting and hand tighten.
7. Using two tubing wrenches or open-end wrenches, tighten the assembly until a solid feel is apparent.
8. Test the completed fitting for leaks.

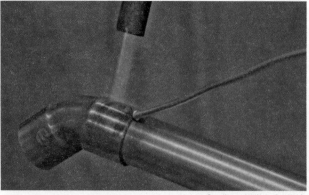

Goodheart-Willcox Publisher

Figure 32-25. Sweat soldering copper tubing. The heated tubing, not the flame, must melt the solder.

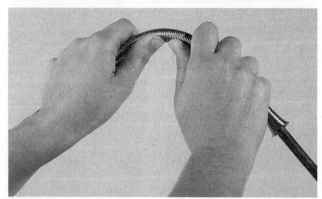

Goodheart-Willcox Publisher

Figure 32-26. Use a tube spring to prevent kinks when bending copper tubing.

32.8.3 Installing Rigid Plastic Supply Pipe

Rigid plastic pipe is lightweight, low in cost, easy to cut, and easy to fit. Two types are available for indoor supply lines. PVC is suited for cold water only. CPVC can be used for both hot and cold water. Either is sold in 10′ or 20′ lengths with sizes 1/2″, 3/4″, and 1″ being most common. The fittings slide over the outside of the pipe and are secured with quick-curing solvent cement. Connections for larger-diameter PVC or ABS plastic DWV piping are made in the same way as the connections for plastic water supply piping.

Pro Tip

To cut plastic pipe with a hacksaw or backsaw, use a miter box to secure a square cut. If a pipe cutter is used, install a special cutter wheel designed for plastic. A plastic tubing cutter is the fastest and most accurate cutting method.

A   *Goodheart-Willcox Publisher*

B   *Goodheart-Willcox Publisher*

Figure 32-27. Flaring copper tubing. A—The tubing end should extend 1/32″–1/16″ above the surface of the flaring tool. B—The finished flare.

PROCEDURE

Installing Push-On Plastic Fittings

1. Remove burrs from the end of the pipe with a knife or sandpaper.
2. Wipe the end with a clean rag to remove any soil. Also clean the inside of the fitting.
3. Remove the glossy finish on the outside of the pipe and the inside of the fitting with sandpaper or a special primer.
4. Slide the fitting onto the pipe until properly seated.
5. Rotate the fitting until it is properly aligned for the run.
6. Draw an alignment mark across the fitting and the pipe.
7. Remove the fitting.
8. Apply a liberal coat of the correct solvent-cement to the outside of the pipe and to the inside of the fitting. Usually, the container has an applicator attached to the cap, **Figure 32-28**. If not, use any soft-bristled brush.
9. Quickly make the connection, giving a quarter turn to evenly spread the cement, **Figure 32-29**. Speed is essential or the connection will be spoiled.
10. Promptly align the marks previously made on the parts. The cement will bond within seconds.

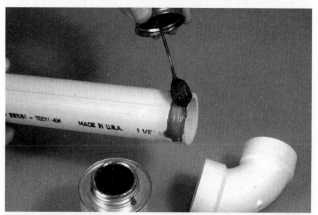

Goodheart-Willcox Publisher

Figure 32-28. Applying solvent to the plastic pipe end.

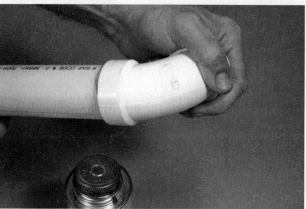

Goodheart-Willcox Publisher

Figure 32-29. Slip the fitting onto the pipe and give it a quarter turn to spread the solvent evenly. Align the fitting with the mark on the pipe.

32.8.4 Installing Galvanized Pipe

Galvanized plumbing pipe is still used, but it is not as popular as copper and plastic. It comes in nominal diameters of 1/4″–2 1/2″ and in 10′ and 21′ lengths. Both ends are threaded. Galvanized connections are one of the most forgiving if you make a mistake in joining. Threaded joints can be taken apart without destroying the parts.

PROCEDURE
Connecting Galvanized Pipe

1. Check lengths to ensure accuracy. A "dry fit" assembly can be made to determine this.
2. Make sure threads are clean. Use a wire brush to remove any bits of metal or other debris.
3. Apply pipe compound or Teflon® tape to exterior threads, **Figure 32-30**.
4. Screw on the fittings finger tight.
5. Final tightening should be made with pipe wrenches, **Figure 32-31**. In long runs, tightening of all joints can be made by starting with the last joint in the run. As each joint is tightened, the force is transferred to the next joint until all are tightened.

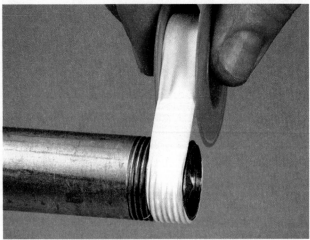

Goodheart-Willcox Publisher

Figure 32-30. Wrap Teflon® tape onto the pipe threads in the same direction that will be used to thread on the fitting.

32.9 Replacing Plumbing Parts

PROCEDURE
Installing a Faucet

To install a faucet, follow the instructions provided by the manufacturer, since there are many types on the market. **Figure 32-32** illustrates these typical installation steps:

1. Clean and dry the mounting surface of the sink or lavatory.
2. Slide the deck gasket over the supply piping and mounting studs until it is seated in the underside of the faucet body.
3. Slip the assembly through the openings in the deck.
4. Install the U-shaped clamp bar (channel up) and secure it with the mounting nuts.
5. Carefully position the faucet and gasket on the deck and tighten mounting nuts.
6. Connect flexible supply lines to the hot and cold supply lines.

32.9.1 Fixing Sink Drain Problems

A sink drain includes several parts. A seat or flange sits inside the bowl and provides a seal. Its lower end has internal threads for connection to the drain body. A drain body threads into the seat and extends to the tailpiece.

Goodheart-Willcox Publisher

Figure 32-31. Pipe wrenches have teeth to grip the fitting or pipe as it is tightened. For final tightening, a second wrench is often used to grip the pipe and prevent rotation.

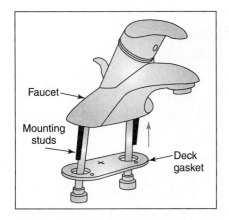 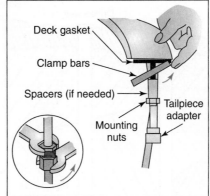 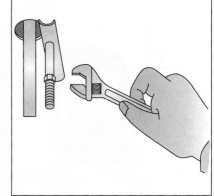

Moen Incorporated

Figure 32-32. Replacing a lavatory faucet.

A rubber gasket slides onto the tailpiece beneath the bowl to provide a seal. A metal washer fits between the seal and a lock nut. A lock nut secures the tailpiece to the bowl and prevents leakage past the flange and the seal. A pivot rod assembly controls the stopper. Slip couplings fasten the parts of the drain together and make the system leak-proof. They are essentially rings with interior threads and are used with washers. A slip coupling and its washer seal the tailpiece to the drain body.

A P-trap connects the tailpiece to the drainpipe and provides a seal to block sewer gases. The trap is designed to remain filled with water so that toxic and disagreeable sewer odors do not enter the house, **Figure 32-33**. As a result of such usage, it is prone to clogging, corrosion, and leaks. To unclog or replace a trap, loosen the couplings that secure it to the tailpiece and drainpipe. Slide the tailpiece down into the P-trap until it clears the sink drain. Rotate the assembly to one side and remove the trap. Clean out the trap or install a new trap. Reverse the procedure to install the trap. Tighten the couplings and test for leaks.

32.10 Wells and Pumps

Many buildings located beyond the lines of municipal water systems have their own wells. Wells require pumps and pressure tanks to maintain a constant water supply.

There are different types of wells, **Figure 32-34**. Dug wells were once common. Wells today, however, are either drilled or bored.

Drilled wells are used where great depths are needed to reach a suitable water table. There are two drilling methods. The percussion method uses a chisel-shaped bit that is raised and lowered by a cable. This creates a pounding action that breaks up the subsurface material. The rotary method requires a tall mast and more complicated drilling equipment, along with a steady flow of water.

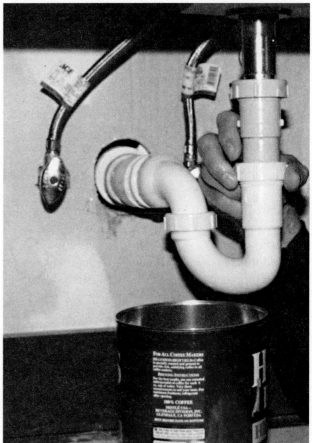

Goodheart-Willcox Publisher

Figure 32-33. P-traps are the most frequent site of clogging and leaking in the DWV system. They are always full of water. Keep a container handy to catch spills when they must be disconnected from the drain.

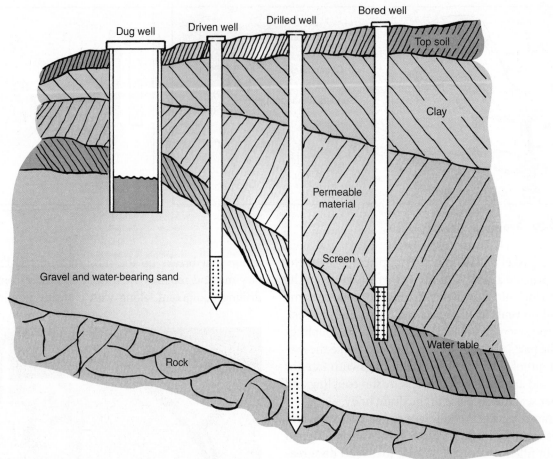

Goodheart-Willcox Publisher

Figure 32-34. Four types of wells. Most wells today are drilled or bored.

The slurry created is pumped to the surface and discarded. This method is used when the water table is far below the surface.

Bored wells are made with an earth auger. The auger cuts and lifts subsurface material in much the same way as a twist drill bit cuts and lifts particles.

The diameter of the well must be large enough to accept a casing. The *casing* is a heavy metal tube that keeps contaminants out of the well water. Galvanized or plastic pipe is installed into the well and a pump is connected to this piping. A *submersible pump* may be installed in the well or an aboveground pump may be placed near the well or in the building.

Also located near the well or the pump is a pressure tank, **Figure 32-35**. Its function is to store a reserve of water under pressure to meet the continuing needs of the building's occupants. The water is under constant pressure so it can be delivered anywhere in the building at sufficient force to operate fixtures and water-using appliances.

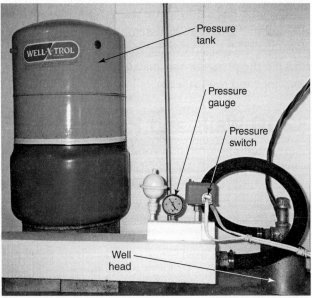

Goodheart-Willcox Publisher

Figure 32-35. A typical setup for a well water system. A submersible pump is located in the well.

Construction Careers
Plumber

Plumbers form the largest portion of a broad trade classification dealing with tradespeople who install, maintain, and repair many different types of piping systems. Other trades within that classification include pipelayers, pipefitters, steamfitters, and sprinkler fitters.

A plumber is generally responsible for planning, installing, and testing the water supply and waste disposal (drain/waste/vent) piping and plumbing fixtures within residential, commercial, and other types of structures. Pipelayers excavate for and install the large underground concrete, clay, cast iron, or plastic pipes for water mains, municipal sewer systems, and oil or gas lines. Pipefitters work primarily in industrial and large commercial structures, installing high- and low-pressure piping used for industrial processes or heating and cooling systems. Steamfitters work primarily in industrial and power generation settings installing and maintaining piping used to move high-pressure gases and liquids. Sprinkler fitters specialize in installing and maintaining automatic fire sprinkler systems.

Approximately 75% of all plumbers work for contractors who do new construction, repair, or maintenance work. About 10% are self-employed, often operating small businesses devoted to plumbing repairs or remodeling work. The remaining 15% of tradespeople in plumbing occupations, primarily pipefitters and steamfitters, work as maintenance personnel in large industrial operations.

Plumbers must be able to read and follow prints, lay out and install water supply and DWV piping, and be skilled in the techniques for cutting and joining pipes made from various materials. They must have the ability to stand for long periods or work in cramped quarters, and have the physical strength to lift and carry piping and associated materials. Although the work is hard and often carried out in uncomfortable conditions, plumbers are among the best-paid members of the building trades.

Like sheet metal workers, plumbers serve an apprenticeship (4–5 years) to prepare for work in the trade. In addition to on-the-job training, apprentices must take at least 144 hours of classroom instruction each year.

Unlike most other building trades, plumbers must be licensed by either state or local authorities. Usually, this involves meeting certain education/experience requirements and passing a written examination. The regulations vary from state to state.

Monkeybusinessimages/iStock/Thinkstock

Plumbers do much more than just crawl under sinks to fix clogged pipes with a few wrench turns. They are familiar with all aspects of a building's water supply and waste disposal. They must be able to read prints to identify problem sources and potential solutions to any water flow issues.

Chapter Review

Summary

- All of the piping and fixtures that supply water for drinking, cooking, bathing, and laundry and the drainage pipes that dispose of wastewater make up the plumbing system.
- To install rough plumbing, plumbers must cut holes and notches in the building frame.
- A plumber must closely work with a carpenter to be sure that the holes will not weaken the joists and studs.
- Plumbing codes govern the installation of plumbing systems.
- The water supply system distributes water under pressure throughout the structure for drinking, cooking, bathing, and laundry.
- The DWV system carries away wastewater and solid waste from bathrooms, kitchens, and laundries. This subsystem is not under pressure; it relies on gravity.
- Plumbing tool kits are generally similar to those used by carpenters and electricians.
- Plumbing supplies include iron, steel, and plastic pipe; copper tubing; valves; faucets; and fixtures such as lavatories or tubs.
- Plumbing materials and supplies that seal the plumbing system against leaks include slip-joint washers, O-rings, Teflon® tape, pipe compound, and plumber's putty.
- Certain drawing symbols, standardized for the plumbing trade, are a type of shorthand that indicate fixtures, appliances, mechanical equipment, pipes, and fittings.
- Specific procedures are used to install and join each type of pipe and tubing.
- Plumbing repairs typically involve fixing leaks, replacing faucets and drain components, and clearing clogged drains.
- Homes located beyond municipal water lines must have a well drilled and a pump installed.
- Safety in plumbing work requires proper protective gear and careful attention to safe work habits.

Know and Understand

Answer the following questions using the information in this chapter.

1. Once plumbing is installed, notched framing members should be reinforced with _____.
   - A. ties
   - B. metal strapping
   - C. screws
   - D. headers

2. *True or False?* Plumbing codes are the same in every community throughout the United States.

3. Plumbing systems are composed of two subsystems: the water supply system and the _____ system.
   - A. ejector
   - B. storm water
   - C. DWV
   - D. irrigation

4. *True or False?* One of a plumber's most used tools is the wrench.

5. A _____ vertically transfers locations of a pipe between floors.
   - A. divider
   - B. plumb bob
   - C. level
   - D. steel tape

6. _____ the end of a pipe removes the burrs left inside by cutting.
   - A. Burring
   - B. Threading
   - C. Reaming
   - D. Soldering

7. Copper tubing is available in rigid, straight lengths or in flexible _____.
   - A. coils
   - B. bundles
   - C. fittings
   - D. lengths

8. What is the maximum temperature at which CPVC can be used?
   - A. 100°F
   - B. 180°F
   - C. 320°F
   - D. 400°F

9. What kind of plastic piping is used to resist chemical attack in drainage systems?
   A. PEX.
   B. PVC.
   C. ABS.
   D. CPVC.

10. Galvanized pipe has largely been replaced by copper and _____ for water supply pipes.
    A. cast iron
    B. plastic
    C. aluminum
    D. steel

11. *True or False?* Gate valves are used to restrict water flow.

12. Sinks, bathtubs, and showers are examples of _____.
    A. appliances
    B. fixtures
    C. faucets
    D. drain devices

13. _____ are used on faucets to produce leakproof seals between the faucet body and internal parts.
    A. Slip-joint washers
    B. Plumber's putty
    C. Teflon® tape
    D. O-rings

14. The first phase of plumbing installation is known as _____.
    A. first-rough
    B. second-rough
    C. rough-in
    D. inspection

15. Compression or _____ fittings used with copper tubing require no soldering.
    A. washerless
    B. flare
    C. galvanized
    D. steel

16. _____ connections are one of the most forgiving if you make a mistake in joining because the threaded joints can be taken apart without destroying the parts.
    A. Galvanized
    B. Copper
    C. Steel
    D. Plastic

17. *True or False?* Drilled wells are made with an earth auger.

18. A metal tube that keeps contaminants out of the well water is called a _____.
    A. submersible pump
    B. casing
    C. pressure pump
    D. DWV pipe

Apply and Analyze

1. Why must the cutting and notching of framing members be carefully done?

2. Why is it important for drainage, waste, and venting (DWV) systems to have proper venting?

3. What are the three weight types of copper tubing used for water supply lines?

4. What is PEX?

5. When sweat soldering copper pipe, how do you ensure a leak-proof joint?

6. How are fittings attached to rigid plastic pipe, such as PVC?

7. Explain the purpose of a P-trap.

Critical Thinking

1. Repairing plumbing can be a messy business. What precautionary measures would you take if you had to repair a water supply system? What about a drainage, waste, and venting (DWV) system?

2. What skills are important for a plumber to have?

3. List three safety and health issues that a plumber might face. What precautionary measures should they take to protect themselves from each?

Communicating about Carpentry

1. **Speaking.** You are a building contractor. You have been hired to inspect a building that was recently renovated to ensure it meets code. The company added a new pool and hot tub, and as a result needed new plumbing. Create your inspection plan in the form of a presentation. Detail what you checked and the specifications you have met. Share the presentation with the class, as though the class were the client who needs approval from the county inspection board.

2. **Reading.** With a partner, create flash cards for the key terms in the chapter. On the front of the card, write the term. On the back of the card, write the phonetic spelling text. (You may use a dictionary for this.) Practice reading aloud the terms, clarifying pronunciations where needed.

3. **Listening and Speaking.** Search online for local unions that may have workshops, guest speakers, or presentations on plumbing techniques and new materials. Also look online for plumbing-related construction videos. Your local library may also have various media covering different techniques. Watch videos, attend a presentation, and take notes. Listen for terms that you have learned from your textbook. Write a short summary and be prepared to present your findings to the class.

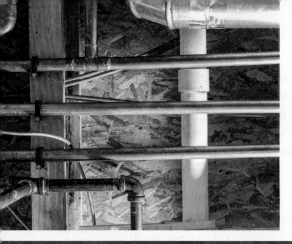

CHAPTER **33**

Heating, Ventilation, and Air Conditioning

OBJECTIVES

After studying this chapter, you will be able to:

- Explain how heat travels through buildings.
- Identify ways to conserve energy in housing.
- Define AFUE and SEER ratings and list the appliances to which they are applied.
- List the characteristics of different central air conditioning systems.
- Describe the functions of HVAC system components.
- Explain the design and operation of HVAC systems.
- Describe automatic controls for heating and cooling systems.

TECHNICAL TERMS

air exchanger
annual fuel utilization efficiency (AFUE)
boiler
central heating
condenser
conduction
convection
cut-in point
cut-out point
differential
direct heating system
duct

electric radiant heat
electric resistance baseboard heating
energy recovery ventilator (ERV)
evaporator
forced-air perimeter system
heat exchanger
heating season performance factor (HSPF)
heat recovery ventilator (HRV)
hydronic perimeter heating system

hydronic radiant heating system
manifold
one-pipe system
plenum
radiation
register
seasonal energy efficiency ratio (SEER)
Second Law of Thermodynamics
thermostat
two-pipe system
zone

Installation of heating, ventilation, and air conditioning (HVAC) is an important building trade. Without it, buildings would not be comfortable places to work or live. They would be either too cold or too hot. Even prehistoric humans with limited technology at hand used open fires to heat their dwellings.

Industry has heavily invested in developing heating and cooling systems. Technology allows for automatic control of systems so that buildings are kept comfortable year-round, regardless of how hot or cold the weather might be.

33.1 Heating and Cooling Principles

To understand how furnaces and air conditioners operate, you must understand how heat moves. This brings up what is called the **Second Law of Thermodynamics**. Simply stated, this law dictates that heat always moves from hot to cold. Consider what happens when you place a metal spoon in a bowl of hot soup. If the spoon is left there for any amount of time, it becomes hotter as the soup becomes colder. Why? Heat in the liquid moved to the cooler spoon.

33.1.1 How Heat Travels

Heating and cooling of buildings depends on the fact that heated matter—whether air, liquid, or solid—passes its heat on to cooler matter. As you learned in Chapter 18, **Thermal and Sound Insulation**, heat travels by three methods: radiation, conduction, and convection. In review:

- *Radiation* moves heat by wave motion, similar to light or radio waves. A source, such as the sun, produces the heat and sends it through the air. The air is not heated, but solid objects that the waves strike absorb the heat.
- *Conduction* is the movement of heat through substances from one end to the other. Heat moving through the metal spoon placed in a bowl of hot soup is a good example of conduction.
- *Convection* is the movement of heat by way of a carrier, such as air or water. This carrier may move naturally or, in the case of furnaces and air conditioners, by the action of fans or blowers.

33.2 Conservation Measures

Fuel costs are constantly rising. Before discussing space conditioning (heating or cooling) mechanisms, it is necessary to consider how to increase their efficiency through conservation measures.

New homes are designed to limit the amount of energy needed to heat and cool them. Savings in heating and cooling come about in several ways. Many of these measures were discussed in Chapter 18, **Thermal and Sound Insulation**.

Furnaces and air conditioners are major users of energy in a home. Assume that the efficiency of a low-efficiency furnace is 80%. For every 100 British thermal units (Btu) of fuel energy consumed, the furnace delivers 80 Btu of heat energy. The other 20 Btu of heat energy from the fuel is lost. Much of this lost heat energy escapes up the chimney. Temperatures in a chimney must usually be around 300°F (150°C) for the chimney to draw (pull combustion gases to the atmosphere). High-efficiency furnaces make use of technology to deliver more heat energy from fuel, **Figure 33-1**. Some systems are rated as high as 98% efficient by the US Department of Energy.

A *heat exchanger* is a section of the furnace where heat from combustion gases is transferred by conduction to the circulating air. Heat exchangers in the high-efficiency units are designed to extract more of the heat of combustion. Spent gases leave a high-efficiency

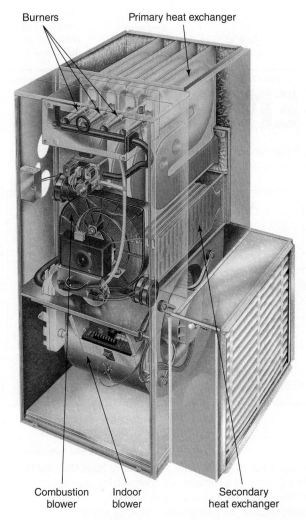

Burners   Primary heat exchanger

Combustion blower   Indoor blower   Secondary heat exchanger

Carrier Corporation, subsidiary of United Technologies Group

Figure 33-1. Components of a high-efficiency gas furnace.

furnace at around 100°F (40°C). The furnace does not use a traditional chimney. Instead, the gases are exhausted to the outside through a small plastic pipe. Also, there is no contact between the inside, conditioned air and the outside air used for combustion.

The organization ASHRAE (originally named the American Society for Heating, Refrigeration, and Air-Conditioning Engineers) develops standards to rate the energy efficiency of heating and cooling appliances. Manufacturers label their heating and cooling appliances with these energy efficiency ratings. The standard used for rating the efficiency of furnaces and boilers is the *annual fuel utilization efficiency (AFUE)* rating. It is a measure of the amount of heat actually delivered to a building compared to the amount of fuel the furnace or boiler uses. The US Department of Energy has determined that all furnaces sold in the United States must have an AFUE rating of at least 80%. **Figure 33-2** is a sample energy efficiency label for a natural gas furnace.

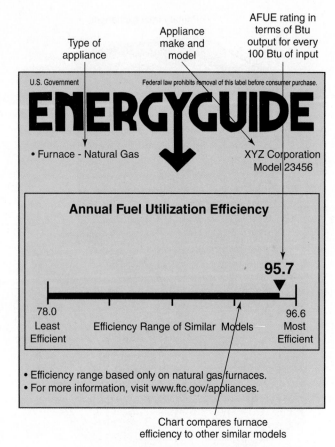

Type of appliance

Appliance make and model

AFUE rating in terms of Btu output for every 100 Btu of input

U.S. Government   Federal law prohibits removal of this label before consumer purchase.

ENERGYGUIDE

• Furnace - Natural Gas

XYZ Corporation
Model 23456

Annual Fuel Utilization Efficiency

95.7

78.0
Least Efficient

Efficiency Range of Similar Models

96.6
Most Efficient

• Efficiency range based only on natural gas furnaces.
• For more information, visit www.ftc.gov/appliances.

Chart compares furnace efficiency to other similar models

United States Federal Trade Commission

Figure 33-2. A sample energy efficiency label showing the AFUE rating of a furnace. All energy-using appliances have a label similar to this one, showing the type of appliance, its manufacturer, and its energy efficiency rating.

Heat pumps use *heating season performance factor (HSPF)* ratings for heating and *seasonal energy efficiency ratio (SEER)* ratings for cooling. HPSF is the ratio of space heating required over a season to the electrical energy used. During the cooling season, the SEER rating is used.

Central air conditioning equipment uses a SEER rating. The higher the SEER, the more efficient the air conditioner. In January 2006, it became mandatory for all cooling units to have a minimum 13 SEER rating. Room air conditioners may use an Energy Efficiency Ratio (EER) rating. The EER is the ratio of the cooling capacity (in Btu per hour) to the power input (in watts). Just like the SEER system, the higher the EER rating, the more efficient the air conditioner. National appliance standards require room air conditioners to have an EER ranging from 8.0 to 9.8 or greater, depending on the type and capacity.

33.3 Heating Systems

Most buildings have central heating. Space heating is the use of heating devices, such as fireplaces, stoves, or portable heaters in individual rooms. In *central heating*, a single, large heating plant produces the heat. The heat is then circulated throughout the structure by perimeter heating or radiant heating. There are several types of circulating systems used with central heating:

- Forced-air perimeter heating
- Hydronic perimeter heating
- Hydronic radiant heating

33.3.1 Forced-Air Perimeter System

In a *forced-air perimeter system*, air is circulated through large ducts that direct warmed air to every room in the building, **Figure 33-3**. The burner delivers high-temperature combustion gases to the furnace heat exchanger. Heated air from the exchanger collects in a large sheet metal chamber called a *plenum*. A blower forces the heated air in the plenum through tubes or passages, called *ducts*, into the living spaces of the building. At the same time, the blower draws cooled air back to the furnace through a series of ducts called the *air return*. An air filter located in the cold air return filters out dust particles picked up from the heated spaces. Refer again to **Figure 33-1**, which shows a cutaway view of a high-efficiency gas furnace.

Furnaces may be oil-fired, gas-fired, or electric. Oil-fired units require a storage tank for the fuel. The tank is usually located either underground outside the building or in the basement near the heating unit. For basement tanks, a filler pipe is usually installed that extends through the wall to the outside for easy access during fuel delivery. Gas-fired units may get their fuel supply directly from a gas main supplied by a gas utility or from a storage tank located on the property. Electric furnaces use the resistance of conductors to produce heat, much like an electric stove or oven.

Registers used with forced air furnaces consist of a grille installed over duct openings in the rooms. They usually have shut-offs for controlling airflow. Some have adjustable vanes that control the direction of the airflow as it exits the duct.

As air is heated, it becomes less dense. That means a cubic foot of warm air is not as heavy as a cubic foot of cooler air. This causes warm air to rise and cooler

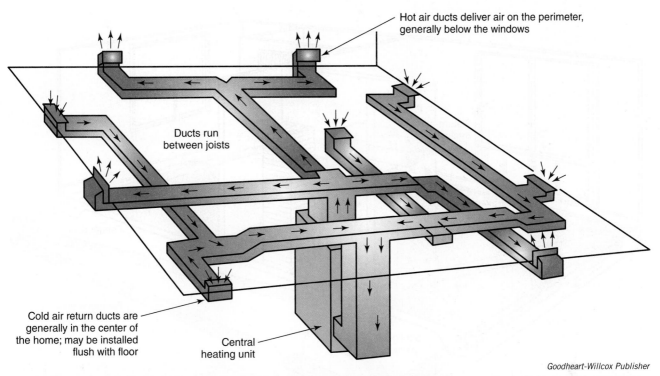

Hot air ducts deliver air on the perimeter, generally below the windows

Ducts run between joists

Cold air return ducts are generally in the center of the home; may be installed flush with floor

Central heating unit

Goodheart-Willcox Publisher

Figure 33-3. A forced-air perimeter heating system. Warm-air ducts deliver heated air to rooms under a slight pressure. The cold air return duct brings air into the furnace for heating each time the circulating blower runs.

air to settle toward the floor. Since warm air rises, heating registers are more efficient if located at floor level, **Figure 33-4.** Since cold air sinks, cold air returns should also be located at floor level. Registers should be placed so that air is not discharged directly on room occupants.

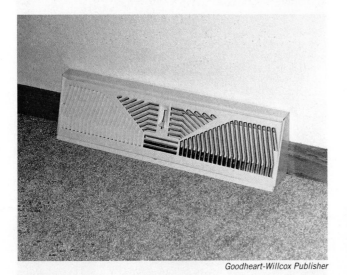

Goodheart-Willcox Publisher

Figure 33-4. Warm air registers usually have vanes to direct flow of heated air and a door that can be positioned to reduce or stop airflow.

Safety Note

A flame that is mostly yellow in natural gas-fired furnaces is often a sign that the fuel is not completely burning, which means high levels of carbon monoxide are being released. Oil furnaces with similar problems can give off an oily odor. Carbon monoxide is a deadly gas created by incomplete combustion. It has no odor. An HVAC specialist should test the furnace for safe operation every year. Also, install carbon monoxide detectors near the furnace and each bedroom. Most states have passed laws requiring carbon monoxide detectors in homes.

Warm-Air Perimeter System

This variation of the forced air system is usually intended for a basement area or for a building with slab-on-grade construction. The ductwork is installed before the slab or the basement floor is placed, **Figure 33-5.** Heated air from the furnace circulates through the duct system embedded in the concrete. This duct system encircles the concrete slab at its outer edges. It is connected to the furnace through a plenum and feeder ducts located beneath the furnace. As in the traditional forced-air system, the warm air is discharged into the rooms through registers.

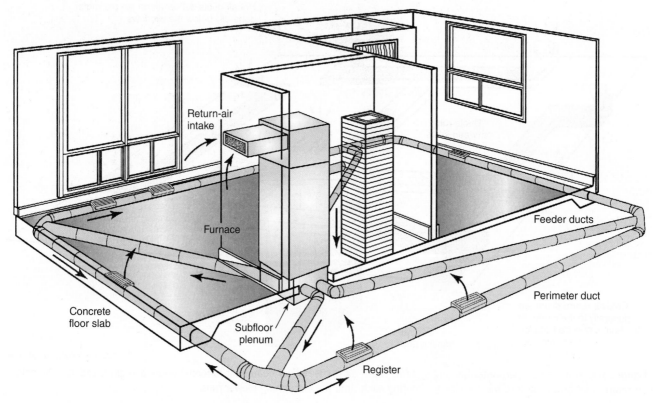

Goodheart-Willcox Publisher

Figure 33-5. A typical warm-air perimeter system in slab-on-grade.

Floor or low sidewall registers are placed along the perimeter of the space. Air returns to the furnace through return-air intakes at high locations on sidewalls or on the sides of the furnace itself.

Ducts may be of sheet metal, vitrified tile, concrete pipe, plastic pipe, or other precast forms. If damp conditions are likely to exist, avoid metal ducts. Metal ducts will corrode and fail over time in damp conditions.

Installing and Maintaining Forced-Air Systems

Installation of a forced-air furnace usually follows the installation of the ductwork. An HVAC technician installs the unit. The installer must be able to make mechanical connections and electrical hookups, **Figure 33-6**. The same person may also install the ductwork. This requires a knowledge of certain metal trades as well as a thorough grounding in electrical principles and wiring practice.

33.3.2 Hydronic Perimeter Heating System

A *hydronic perimeter heating system* uses water as a medium to move the heat from the heating unit, **Figure 33-7**. The heating unit is called a *boiler* for the simple reason that it heats water much like a teakettle sitting on a burner. The heated water is piped to radiators that transfer heat to the rooms.

Goodheart-Willcox Publisher

Figure 33-6. An installer making electrical connections during furnace installation.

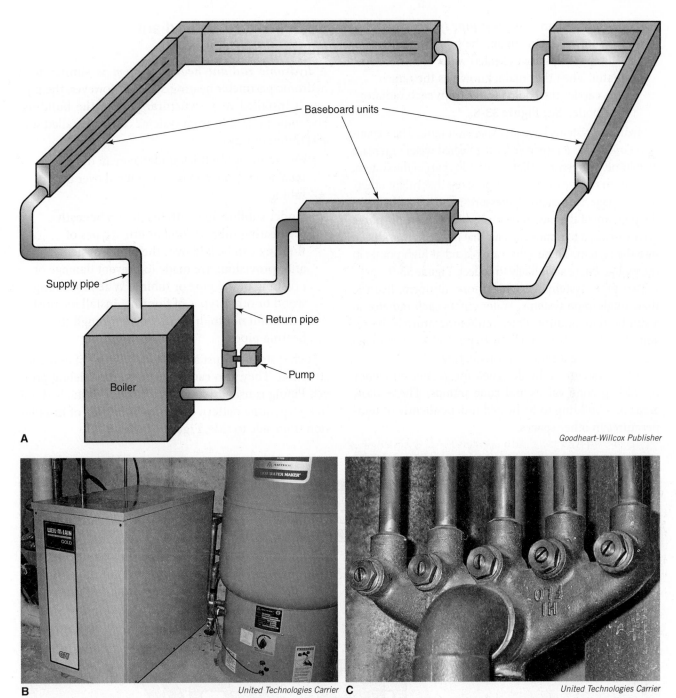

Baseboard units

Supply pipe

Return pipe

Pump

Boiler

A

Goodheart-Willcox Publisher

B   *United Technologies Carrier*

C   *United Technologies Carrier*

Figure 33-7. A—A basic hydronic system. B—This boiler is teamed with a hot water storage unit. The boiler delivers heated water to the tank until it is used by the hydronic heating system. A heat exchanger coil in the tank supplies domestic hot water, eliminating the need for a separate water heater. C—This manifold delivers heated water to five different pipes, creating five different heating zones for the house.

Many hydronic systems include a *manifold*, **Figure 33-7C**. This device balances the delivery of heated water to different *zones* (areas) of the building. The manifold consists of an inlet pipe fitting with several outlets for connecting one pipe with others. Pipes from each outlet are directed to different zones of the building. Each of these zones has its own circulating pump controlled by a thermostat for that zone.

Hydronic systems vary somewhat. There are two basic types of piping layout:

- One-pipe system—In a *one-pipe system*, a single pipe, called a *main*, supplies the heated water to room heating units. The heated water is pumped through the pipe from radiator to radiator. Finally, the water, now cooled, returns to the boiler where it is reheated to repeat the cycle. This type is shown in **Figure 33-7A**.

- Two-pipe system—In a ***two-pipe system***, two mains are used. One main, the *supply main*, simultaneously carries heated water to each room radiator. The other main, known as the *return main*, carries the cooled water from each radiator to the boiler. See **Figure 33-8**.

Hydronic systems have an expansion tank. The expansion tank acts as a reservoir for the heated water's increase in volume. A pressure-relief valve is also part of the system. It opens to release excess water pressure that might otherwise damage the system. A pressure-reducing valve limits the pressure of the incoming freshwater supply. Air purge valves are used to remove any air that has been introduced into the system. These valves are placed at high points in the piping where air is likely to collect, **Figure 33-9**.

Two-pipe systems provide more uniform heating than single-pipe systems. Water enters each radiator at a similar temperature. This even temperature is the result of heated water not having to pass through so many radiators before it returns for reheating.

Either system can be designed for greater efficiency by adding zone valves and zone pumps. These allow areas of a building to be heated independently of temperatures in other spaces.

33.3.3 Hydronic Radiant Heating System

A ***hydronic radiant heating system*** is similar to a hydronic perimeter heating system. However, the piping is installed in a structural part of the building. Hydronic radiant heating systems can be installed several different ways:

- On-grade over a thick underlayment of rigid insulation. Concrete is then poured over the tubing.
- Under subflooring with insulation beneath the heating tubes. Wood or other types of flooring can be laid over the piping system after provisions are made to prevent damage or crushing of the pipe or tubing. When installing wood or other types of flooring, installers must be careful not to drive fasteners through the heating pipes.

Such systems are usually installed as part of new construction. They also can be part of a remodeling project. Piping is usually copper or flexible plastic laid out in a serpentine pattern that covers the area of installation from side to side, **Figure 33-10**.

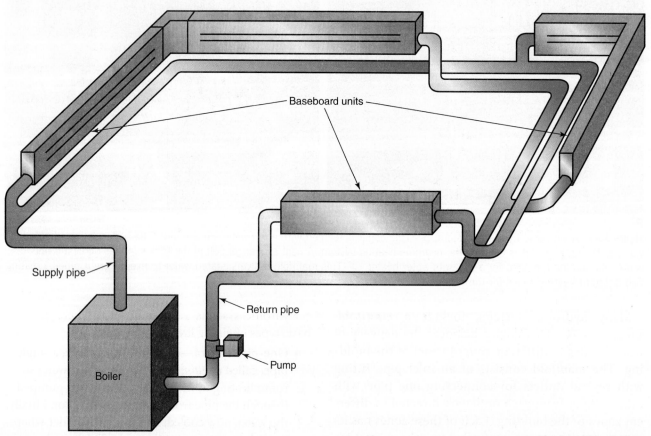

Figure 33-8. A two-pipe hydronic system.

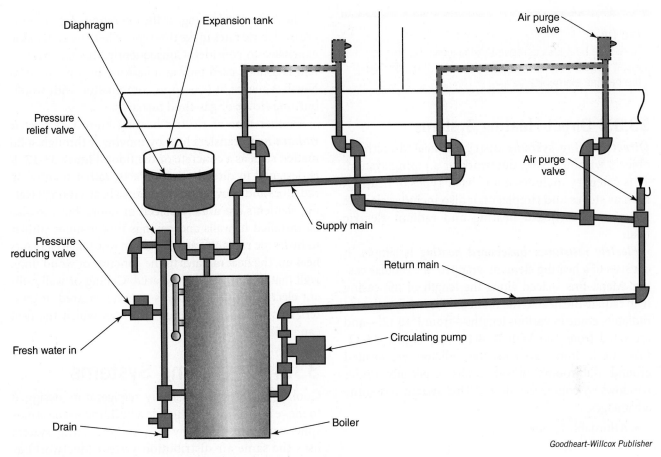

Figure 33-9. The major components of a hydronic heating system.

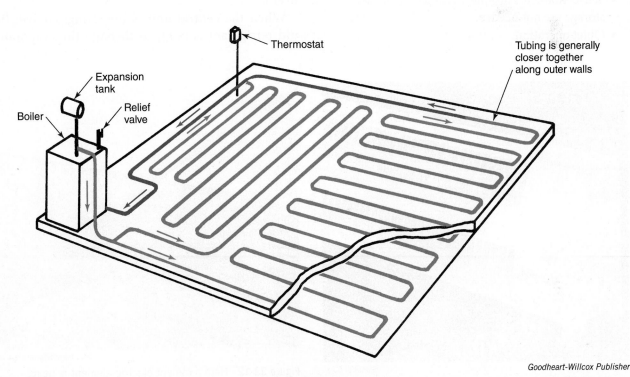

Figure 33-10. This drawing shows a hydronic radiant heating system as it would be installed in a slab floor. Note the location of the thermostat.

33.3.4 Direct Heating Systems

Direct heating systems distribute heat directly to objects without using ductwork or a piping system. This category includes many basic types of heating, such as stoves and fireplaces. It also includes electric resistance baseboard heating and radiant electric heat.

Electric resistance baseboard heating is simple. It consists of a heating element wrapped in a tubular casing. Metal fins spaced along the length of the casing help radiate the heat to the room air. Electric baseboard radiators come in various lengths—from 1′ to 12′—and are rated from 100 W/ft to 400 W/ft, **Figure 33-11**. Like the hydronic systems, the radiators are located around the room's outside walls, especially under windows to counteract drafts. This system has some advantages:

- Affordable to buy.
- Easy to install.
- Easy to zone heat. Some models have thermostatic controls so that rooms can be individually heated.
- Self-contained. Chimneys and piping or fuel storage are not needed.
- Quiet operation.

The main disadvantage is the operating cost. In some areas, electric rates make this type of heating system too expensive to consider. Consequently, this type of system is usually used only as a backup. In some parts of North America, it is used in combination with wood-burning stoves or gas-fired furnaces.

Another type of electric heating known as *electric radiant heat* transfers heat by moving it through solid matter, such as a concrete or tile floor, **Figure 33-12**. It is designed to directly heat objects, rather than using radiators and convection currents to heat a room. Heating elements are usually placed in floors, but can also be installed in walls and ceilings in a manner similar to hydronic radiant systems. As the resistance elements heat up, the heat radiates to the concrete or wood floor, wall material, or ceiling material. Ceiling or wall radiant electric heating elements may be encased in plaster. This provides a medium through which the heat can radiate.

33.4 Air Cooling Systems

Cooling systems are basically refrigerators designed to move heated air from inside a building to the atmosphere outside. In many cases, the cooling system uses the same air distribution system (ductwork) as the heating system, **Figure 33-13**. A system of tubes called an *evaporator* is located in the air conditioning unit where return air from the rooms can pass over it.

When the central unit is operating, cooled, liquid refrigerant is pumped through the evaporator.

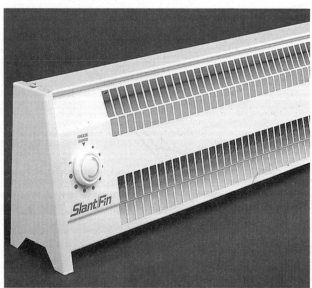

Slant/Fin Corp.

Figure 33-11. An electric baseboard unit is sometimes used as auxiliary heat.

brizmaker/Shutterstock.com

Figure 33-12. Here a radiant electric element is being covered by a self-leveling grout to prepare the floor for its finish material.

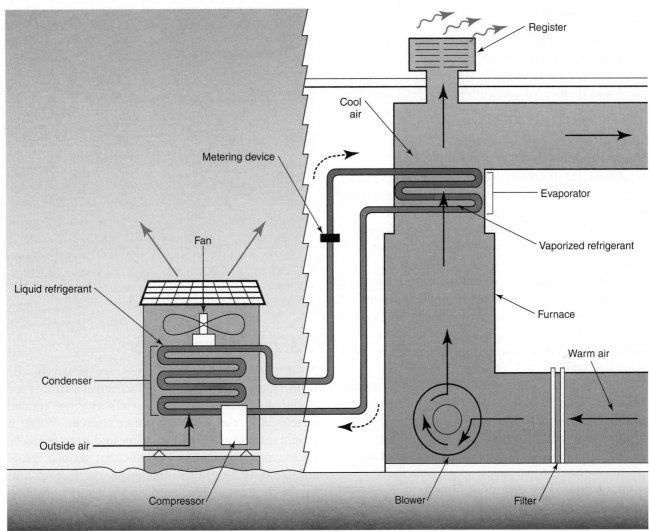

Goodheart-Willcox Publisher

Figure 33-13. A central air-conditioning unit is combined with a forced-air distribution system. A coil located in the furnace receives cold refrigerant. Air forced through the coils is cooled and delivered to rooms by way of the duct system.

Since it is colder than the room air passing over it, the refrigerant absorbs heat from the air and evaporates. (This can be compared to water evaporating as it absorbs heat from a stove and turns to steam.) Then, the hot refrigerant gas is drawn back to the outside unit. Here, it passes through another series of coils and fins called a condenser. The *condenser* acts like the radiator of a car, allowing the heated refrigerant to pass its heat to the atmosphere. In doing so, the refrigerant returns to a liquid state.

From the condenser, the refrigerant is pumped at a high pressure to the evaporator in the furnace. This completes one cycle of the air conditioner. The cycle is continuous as the building's air is passed over the cooling coil. **Figure 33-14** shows an air-conditioning unit located outside a home. It contains the compressor, a condenser, and related valves. The compressor is operated by an electric motor.

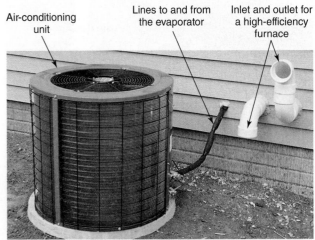

Goodheart-Willcox Publisher

Figure 33-14. A typical central air-conditioner unit located outside of the house. The condenser totally surrounds the compressor and its related components. This arrangement increases the efficiency of the unit.

33.5 Ducts

Ducts can be constructed of various materials, including metal, rigid plastic, or flexible plastic. Ducts should be as unobtrusive as possible. Most ducts are prefabricated and then fitted and cut onsite. Metal duct sections are joined by the installer on-site, **Figure 33-15**.

33.6 Controls

A *thermostat* is an instrument that automatically controls the operation of heating and cooling devices by responding to changes in room temperatures. When the temperature falls below or rises above a set point, a signal is sent to the heating or cooling unit, **Figure 33-16**.

The temperature setting that causes the thermostat to send a signal to the furnace or air conditioner is called the *cut-in point*. The temperature setting that causes the thermostat to stop signaling is called the *cut-out point*. The difference between these two points is called the *differential*.

Programmable thermostats help conserve energy. They can be set to automatically change temperatures at different times of the day. For example, the temperature can be lowered (or raised, if cooling) at night and during daytime hours when occupants are not home. During the hours when people are present, the temperature is increased (or decreased, if cooling). The result can be significant savings in fuel costs.

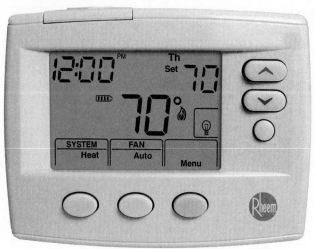

Rheem Manufacturing Company

Figure 33-16. A programmable thermostat helps save energy and fuel costs by adjusting heating or cooling levels for different times of the day.

33.7 Air Exchangers

Most newly constructed homes are relatively airtight. They often need a way to replace indoor air polluted by gases and fumes, such as those created by cooking, cleaning, and other sources. While it would seem that an open window and a fan might be an inexpensive remedy, it is not that simple. While the fresh air might be welcomed, the loss of heat during heating season or of cool air during cooling season is not.

This challenge can be solved with an *air exchanger*, a device that exhausts air from a building and draws in fresh air. There are two types of air exchangers. When used during the heating season, a *heat recovery ventilator (HRV)* uses the heat in the air that is being removed from the building to heat the incoming fresh air. During the cooling season, the HRV transfers incoming heat to the outgoing stale air, maintaining the cool temperatures inside the building, **Figure 33-17**. An *energy recovery ventilator (ERV)* performs the same functions as an HRV, but it also removes humidity from the incoming air to maintain more comfortable humidity levels inside the building, **Figure 33-18**. Testing indicates that indoor air should be exchanged with fresh outdoor air every 2–3 hours. In older, less-tightly built homes, enough fresh air typically leaks in through cracks around doors, windows, and foundations to meet this requirement.

Goodheart-Willcox Publisher

Figure 33-15. Using a metal, strap-type clip called a drive cleat, an installer connects two sections of sheet metal ductwork.

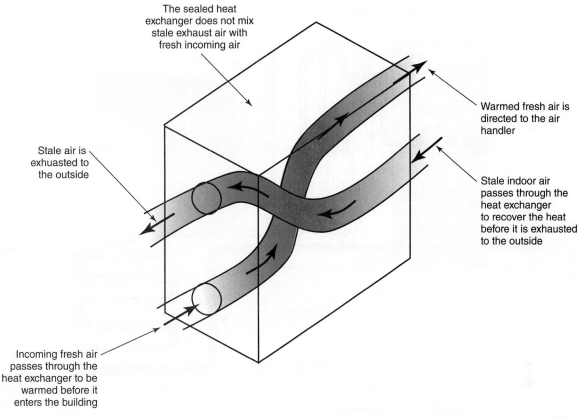

The sealed heat exchanger does not mix stale exhaust air with fresh incoming air

Stale air is exhuasted to the outside

Warmed fresh air is directed to the air handler

Stale indoor air passes through the heat exchanger to recover the heat before it is exhausted to the outside

Incoming fresh air passes through the heat exchanger to be warmed before it enters the building

Goodheart-Willcox Publisher

Figure 33-17. An HRV exchanges the heat in the incoming fresh air with the outgoing stale air to maintain the air temperature inside the building without using additional heating or cooling energy.

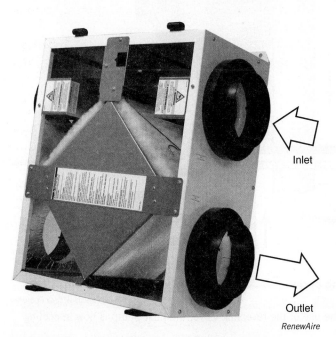

Inlet

Outlet

RenewAire

Figure 33-18. This ERV maintains both the existing air temperature and humidity level inside the building.

33.8 Heat Pumps

A heat pump combines heating and cooling in a single unit. The basic system includes the following:

- A pump or compressor for compressing refrigerant
- An evaporator and condenser
- A reversing valve to switch the functions of the evaporator and condenser
- Pipes or tubing connecting the evaporator and condenser to the compressor

Like an air conditioner, a heat pump uses a refrigerant to collect heat from one place and deliver it to another location. In winter, it collects heat from the outside and delivers it to the inside of a building. In summer, it collects heat from inside of the building and delivers it to the atmosphere.

The secret of heat pump operation is in the reversing valve. By switching the valve, the two chambers are made to switch functions. Thus, in winter, the inside coil becomes a condenser, giving off heat; in the summer, it becomes an evaporator (cooling coil). See **Figure 33-19**.

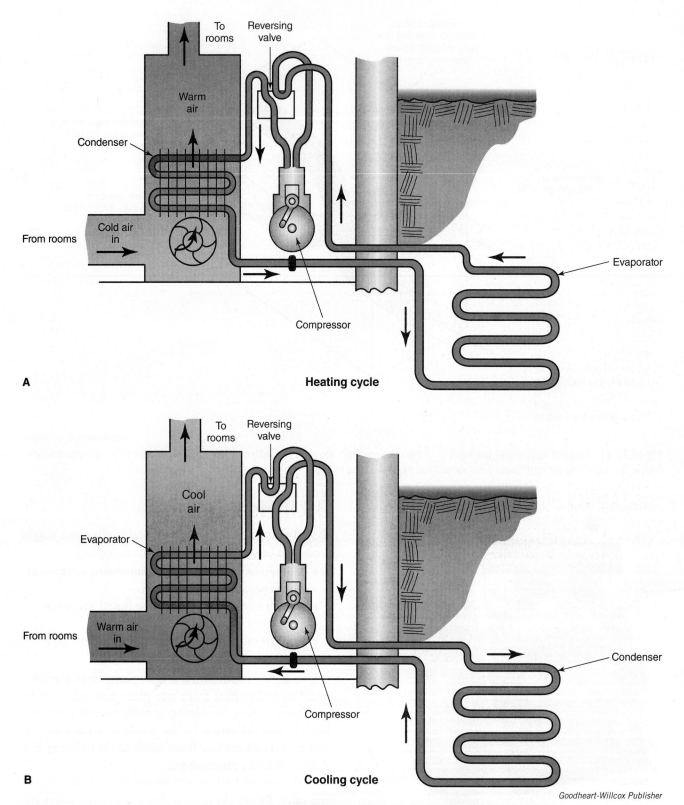

Figure 33-19. A—A heat pump in heating mode. B—Through the use of the reversing valve, the heat pump is now in cooling mode.

Efficiency of the heat pump drops as the outside temperatures dip below 20°F (–5°C). The building must have an auxiliary heating system or the outside coil must be buried in the ground where winter temperatures are higher. As an alternative, the coil can be located in a well. The water temperature in a well may be consistently around 40°F (5°C).

Construction Careers
HVACR Installer

Climate control involves heating, ventilation, and air conditioning (HVAC) of a structure. Without climate control, most homes and commercial buildings would be very uncomfortable or impossible to live and work in. The heating, ventilation, air conditioning, and refrigeration (HVACR) installer is responsible for installing such systems, as well as the refrigeration systems required in many commercial and industrial structures. An HVAC technician also often carries out tasks such as diagnosing and repairing system problems or equipment breakdowns.

In larger contracting companies, installers may specialize in one of the subcategories of this trade: heating (furnace) work, air conditioning, or refrigeration systems. Ventilation is common to both heating and air conditioning work. Installers who are self-employed or who work for smaller contractors and home improvement stores are most likely to work in any of the areas as required. Typically, an installer physically installs the heating, air conditioning, or refrigeration unit; runs any necessary piping or ductwork; and installs and connects electrical wiring as needed. They then test the operation of the system, making any necessary adjustments. On large scale construction projects where union jurisdictions are often involved, the installer may do only portions of the job. Ductwork, piping, and electrical wiring may have to be installed by members of other trades.

Installers may have to work in confined areas, such as crawl spaces or attics, and may have to deal with uncomfortably high or low temperatures. Like most construction trades, HVACR installers must wear appropriate personal protective equipment and carefully follow safety rules. A particular hazard in air conditioning and refrigeration work is the possibility of suffering frostbite or other damage from mishandled refrigerants.

Since HVACR systems are complex and continue to grow more sophisticated, employers seek technicians who have specialized training. High schools, community colleges, and private technical schools offer training ranging from one semester classes to two-year associate degree programs. Formal apprenticeship programs are 3–5 years in length and include both classroom and on-the-job learning. Some technicians begin as helpers and acquire their skills over a period of time working with experienced installers and technicians.

Approximately 50% of HVACR installers and technicians work for heating and cooling contractors. Most of the rest are employed by industrial plants, institutions, and government agencies. About 15% are self-employed.

lisafx/iStock/Thinkstock

HVACR installers can specialize in heating, air conditioning, and/or refrigeration systems. Installers are familiar with all necessary piping and wiring of the HVAC unit. They can work on rooftops where they are exposed to weather conditions, or in cramped quarters such as crawl spaces and attics.

Chapter Review

Summary

- Heating, ventilation, and air conditioning (HVAC) systems keep buildings comfortable for working or living.
- Heat energy moves from warmer to cooler places by means of radiation, conduction, or convection.
- Because energy costs continue to increase, homeowners are concerned about keeping energy costs down and take conservation measures. These include sealing air infiltration points, increasing insulation, and installing the most energy-efficient heating and cooling devices.
- Central heating systems are of three major types: forced air perimeter heating, hydronic perimeter heating, and hydronic radiant heating. Electric resistance heat and radiant electric heating are other types that are sometimes used.
- Forced-air heating and cooling systems distribute conditioned air through a network of ducts serving the building.
- Cooling systems typically have an outdoor condenser unit to disperse heat and an evaporator to absorb heat from the indoor air.
- HVAC systems are controlled by thermostats, which turn furnaces and air conditioners on and off at preset points. Programmable thermostats allow different settings for different times of each day.
- Heat recovery ventilators (HRVs) bring in fresh outdoor air and exhaust polluted indoor air to the outside. Energy recovery ventilators (ERVs) do the same work as HRVs, but also control the humidity of the air being brought into the building.
- Heat pumps are single units that handle both heating and cooling tasks.
- Heating and cooling systems should be checked on a regular basis for efficient and safe operation.

Know and Understand

Answer the following questions using the information in this chapter.

1. _____ is the movement of heat through substances from one end to the other.
   A. Radiation
   B. Conduction
   C. Convection
   D. Thermodynamics

2. *True or False?* A chimney is a section of the furnace where heat from combustion gases is transferred by conduction to the circulating air.

3. The US Department of Energy has determined that all furnaces sold in the United States must have an AFUE rating of at least _____%.
   A. 78
   B. 80
   C. 90
   D. 96

4. In January 2006, it became mandatory for all cooling units to have a minimum _____ SEER rating.
   A. 8
   B. 9.8
   C. 13
   D. 16

5. In _____, a single, large heating plant produces the heat.
   A. space heating
   B. central heating
   C. heat exchange
   D. conduction

6. A(n) _____ heating system is usually installed in a slab-on-grade foundation.
   A. forced-air
   B. hydronic perimeter
   C. hydronic radiant
   D. space

7. Which of the following is *not* a fuel source for a furnace?
   A. Electricity.
   B. Refrigerant.
   C. Gas.
   D. Oil.

8. _____ used with forced air furnaces consist of a grille installed over duct openings in the rooms.
   A. Plenums
   B. Air returns
   C. Registers
   D. Feeder ducts

9. A _____ heats water much like teakettle sitting on a burner in a hydronic perimeter heating system.
   A. manifold
   B. boiler
   C. main
   D. radiator

10. *True or False?* Two-pipe hydronic perimeter systems provide more uniform heating than single-pipe systems.

11. A _____ system does not use ductwork or a piping system to distribute the heat from the source.
    A. hydronic perimeter
    B. hydronic radiant
    C. direct heating
    D. electric resistance baseboard heating

12. *True or False?* In an air-conditioning system, the evaporator is always located where return air can pass over it.

13. A(n) _____ automatically signals the heating system to deliver more heat.
    A. heat pump   C. ERV
    B. HRV   D. thermostat

14. Like an air conditioner, a heat pump uses a(n) _____ to collect heat from one place and deliver it to another location.
    A. exchanger   C. condenser
    B. refrigerant   D. evaporator

15. The secret of heat pump operation is in the _____.
    A. reversing valve   C. heating source
    B. compressor   D. thermostat

Apply and Analyze

1. Explain the three ways that heat travels.
2. What is an AFUE rating and what does it cover?
3. What is the difference between HSPF ratings and SEER ratings?
4. Why is electric resistance baseboard heating not frequently used?
5. How can a programmable thermostat save energy costs?

6. What is the difference between an HRV and an ERV?

Critical Thinking

1. If you were assisting an HVAC technician on the installation of a furnace, what things could you do that would make the job smooth and efficient?
2. Of the various heating and cooling systems you learned about in this chapter, which do you think is the best choice for a new home?
3. What are the implications of owning and living in a home with an HVAC system that performs poorly?

Communicating about Carpentry

1. **Listening, Writing, and Speaking.** Search online for local unions that may have workshops, guest speakers, or presentations on HVAC installation and new materials. Look online for instructional videos. Also, your local library may have various media covering various techniques. Watch videos, attend a presentation, and take notes. Listen for terms that you have learned from your textbook. Write a short summary and be prepared to present your findings to the class.

2. **Speaking and Art.** After reading Chapter 33, you should have a good understanding of HVAC. Create a model or draw a picture with labels describing the different parts and processes of an HVAC system. In your model or picture, explain the results that would occur if there was a breakdown in the process.

Carpentry Math Review

Why Math

All of the building trades and mechanical trades involve using math for numerous tasks. Carpenters use math for such things as calculating the sizes of rafters, girders, and floors. Plumbers use it to determine the necessary sizes of pipes and tanks. Electricians must calculate electrical loads and amperage requirements. Some trades require use of other specialized or more advanced math, but all trades require an understanding of what is presented here.

Calculators

Calculators can be a great time saver, but you should not rely on a calculator to replace your knowledge of basic math. There will be times in the field when a calculator is not available and you will want to do some basic math. However, a calculator can be a handy tool when one is available. Knowing some basic math and using common sense observation can help prevent big errors when using a calculator.

It is important to do operations in the correct order when using a calculator. For example, the formula for the area of a circle requires multiplying 3.1416 by the radius of the circle squared (multiplied by itself). Let's say the radius is 5″. The area is found by multiplying 5×5 (5 squared), then multiplying the result by 3.1416. If you know that 5×5 (written 5^2) is 25, and 3 times 25 is 75, you will know that the answer is slightly more than 75. Now you can check to see if the answer you got on your calculator is close. The correct answer is 78.54 sq in. As stated earlier, it is important to know which order to multiply the numbers. Using this example, if you multiply 3.1416×5 and then multiply that by itself, you'll get 246.7413—not even close! Knowing the correct order to do the operations on paper will help you enter numbers in the correct order on your calculator.

There are several types of calculators one might consider. The most familiar type is a *general calculator*, which allows the user to perform basic math functions.

See **Figure 1**. *Basic calculators* have a memory function that allows the user to store the results of a calculation so that number can be recalled and used in a subsequent calculation. Most also have a % key. When a number is shown in the display, pressing the % key allows the user to then press a number key to calculate that percent of the displayed number. The √ key is for finding the square root of a number. (Squares, roots, and exponents are covered later in this supplement.) The +/− key changes the value of a displayed number from positive to negative or negative to positive. Scientific calculators are more advanced, having the ability to do many trigonometric functions and other advanced operations. Scientific calculators are not necessary for most trades math.

Goodheart-Willcox Publisher

Figure 1

Another type of calculator that can be very useful for trades workers is the *construction calculator*. See **Figure 2**. Construction calculators can be used to convert inches to feet and feet to inches, and calculate rafter lengths and stair stringer lengths. Most allow the user to convert between metric and US Customary measurements. Construction calculators vary from one manufacturer and model to the next, so it is not possible to give a detailed description or instructions here. All calculators come with instructions that explain their functions.

Pro Tip

Always look at your answer and compare it with what might be a reasonable answer. This is the most important tip for calculator use. If the answer does not look reasonable, it is probably not correct.

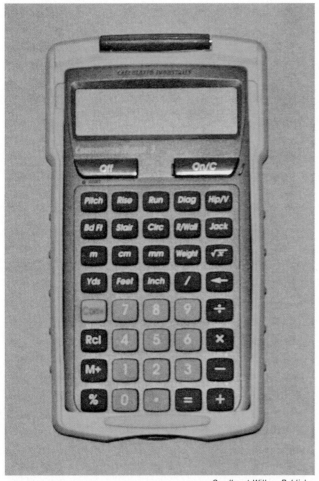

Goodheart-Willcox Publisher

Figure 2

Whole Numbers

Whole numbers are simply numbers without fractions or decimal points, numbers such as 1, 2, 3, 4, etc. Adding, subtracting, multiplying, and dividing whole numbers primarily requires memorizing a few math facts.

Adding and Subtracting Whole Numbers

Example: Adding this column of whole numbers requires memorizing the *sum* of 3 + 5 and the sum of 8 + 2.

$$
\begin{array}{r}
3 \\
5 \\
+\ \ 2 \\
\hline
10
\end{array}
$$

The same type of memorization of math facts is required to subtract whole numbers. We know that the result of subtracting 12 from 37 is 25, because we know that 2 from 7 is 5 and 1 from 3 is 2.

$$
\begin{array}{r}
37 \\
-\ 12 \\
\hline
25
\end{array}
$$

The key to both addition and subtraction is to line up the columns of digits correctly. The whole numbers should be aligned on the right.

In subtraction, if the number being subtracted (the number on the bottom) is larger than the number it is being subtracted from (the number on the top), borrow 10 from the next digit to the left and add it to the one on the right. Write small numerals above the column to help you keep track.

Example:

$$
\begin{array}{r}
{\scriptstyle 2\ 16} \\
\cancel{36} \\
-\ 19 \\
\hline
17
\end{array}
$$

Multiplying Whole Numbers

Multiplication of whole numbers requires memorization of a multiplication table. The only way to get $6 \times 5 = 30$ is to know that multiplication fact or to add $6 + 6 + 6 + 6 + 6$. That becomes way too tedious for bigger multiplication problems. To multiply numbers whose values are ten or more (those with more than one digit), align the digits representing 0 through 9 (the 1s) in the right hand column. Then multiply the top row by the 1s digit in the second row:

Example:

```
    ┌─10s
    │ ┌─1s
    ▼ ▼
    31
 ×  15
   155
```

Next, multiply the top row by the 10s digit in the second row. Because you multiplied by the 10s digit, the *product* (the result of multiplication) is written with its right-most digit in the 10s column:

```
    31
 ×  15
   155
   31
```

If the problem has more digits in the second row, the above steps are repeated for each digit with the products being written in rows beneath one another, with the right-most digit in each row being written in the column for the place it represents: 100s, 1000s, etc.

When all of the multiplication is complete, add the products just as you would for a simple addition problem. The result of this addition is the product (answer) of the multiplication problem.

```
    31
 ×  15
   155
   31
   ───
   465
```

Dividing Whole Numbers

Division of whole numbers is the reverse of multiplication, but the problem must be set up differently. The *dividend* (the number being divided) is written inside the division symbol. The *divisor* (the number the dividend will be divided by) is written to the left of the symbol:

$$\text{divisor} \longrightarrow 7\,\overline{\smash{)}\,28} \longleftarrow \text{dividend}$$

By knowing the multiplication table, we know that $7 \times 4 = 28$. So, if 28 is divided into 7 parts, each part will have 4, or $28 \div 7 = 4$. 4 is the *quotient* (the answer to a division problem) and it is written above the division symbol and above the 1s place of the 28:

$$7\,\overline{\smash{)}\,28}^{\;4} \longleftarrow \text{Quotient}$$

When the dividend is more than 9, the process is divided into steps as follows:

$$4\,\overline{\smash{)}\,320}$$

4 goes into 32 8 times. Write the 8 above the 2 (the right column of the 32). Now multiply 4×8, which is 32. Write the 32 beneath the 32 in the division symbol.

```
     8
4 ) 320
    32
```

Subtract the product of your multiplication (32) from the number above it in the symbol (32). Because 4 goes into 32 exactly 8 times, the numbers are the same, so the result of your subtraction is 0.

```
     8
4 ) 320
    32
    ──
     0
```

Drop the next digit to the right in the dividend (0 in this case) down beside the result of your subtraction. That makes the number at the bottom 00. 4 will not go into 0 (or 00), so the quotient of that step is 0. The quotient (answer) of the division problem is 80.

```
    80
4 ) 320
    32
    ──
    00
```

320 can be divided by 4 80 times. If there are more places under the division symbol, just keep doing the same division, multiplication, subtraction, and drop down for each digit moving to the right.

Example:

```
     102
6 ) 616
    6
    ──
    01
     0
    ───
    016
     12
    ───
      4  ← remainder
```

If the last number produced by the drop-down cannot be divided evenly by the divisor, that number is called the *remainder*. In the example above, the quotient is 102 with a remainder of 4.

Practice

$$\begin{array}{cccccc} 342 & 79 & 68 & 124 & 18 & 213 \\ +\ 16 & +29 & -13 & -\ 35 & \times\ 4 & \times\ 24 \end{array}$$

$$3\,\overline{)36} \qquad 7\,\overline{)214}$$

Fractions

A fraction is a part of something larger. If there are three lag screws in a pound, each lag screw weighs one-third of one pound. One third can be written as:

$$\frac{1}{3} \quad \begin{array}{l} \leftarrow \text{numerator} \\ \leftarrow \text{denominator} \end{array}$$

The number above the fraction bar is called the *numerator*. It indicates how many parts are in the fraction, in this case 1 lag screw. The number below the fraction bar is the *denominator*. The denominator indicates how many parts are in the whole, in this case three lag screws in the whole pound. If there are 50 lag screws in a carton and we take 7 of them out, we have taken this fraction of the lag screws.

$$\frac{7}{50}$$

Equivalent Fractions

If two fractions represent the same value, they are said to be *equivalent fractions*. For example, 1/3 and 2/6 are equivalent fractions because they both represent one-third of the whole. If both the numerator and the denominator of a fraction are multiplied by the same amount, the result is an equivalent fraction.

Example:

$$\frac{1}{3} \times 2 = \frac{2}{6}$$

Adding Fractions

Fractions must have *common denominators* in order to be added. If the denominator of one of the fractions is 8, the other fraction must be written as an equivalent

fraction with a denominator of 8. For example, to add 3/4 and 1/8, write the 3/4 as an equivalent fraction with a denominator of 8.

Example:

$$\frac{3}{4} \times \frac{2}{2} = \frac{6}{8}$$

When both fractions have the same denominator, add the numerators.

$$\frac{6}{8} + \frac{1}{8} = \frac{7}{8}$$

6 eighths plus 1 eighth is a total of 7 eighths.

A common denominator can be found by multiplying all of the denominators in a problem.

Example:

$$\frac{1}{3} + \frac{2}{5} + \frac{9}{14}$$

$$3 \times 5 \times 14 = 210$$

$$\frac{70}{210} + \frac{84}{210} + \frac{135}{210} = \frac{289}{210}$$

In this example, the numerator is larger than the denominator. This is because the value of the fraction is greater than 1. Two hundred eighty-nine is 79 parts larger than the whole of 210 parts. The same number could be written as:

$$1\frac{79}{210}$$

This is called a *mixed number*, because it is made up of a whole number plus a fraction.

Subtracting Fractions

Subtracting fractions is similar to adding them. Both fractions must have a common denominator, then the numerators are subtracted just like whole numbers.

Example:

$$\frac{2}{3} - \frac{1}{4} =$$

Find common denominators and subtract the numerators.

$$\frac{8}{12} - \frac{3}{12} = \frac{5}{12}$$

Multiplying Fractions

To multiply fractions, multiply the numerators. Then multiply the denominators.

Example:

$$\frac{3}{4} \times \frac{2}{5} = \frac{3 \times 2}{4 \times 5} = \frac{6}{20}$$

To make the result easier to work with, always reduce it to its lowest terms. That means to write it as an equivalent fraction with the lowest possible denominator. For example, both 6 and 20 can be divided by 2 to make an equivalent fraction with a smaller denominator.

$$\frac{6 \div 2}{20 \div 2} = \frac{3}{10}$$

In this case, $\frac{3}{10}$ is the lowest terms of $\frac{6}{20}$.

To multiply a fraction by a mixed number, first change the mixed number to a fraction. Then multiply as common fractions. Reduce the product to its lowest terms.

Example:

$$\frac{2}{3} \times 4\frac{1}{2} = \frac{2}{3} \times \frac{9}{2}$$

$$\frac{2}{3} \times \frac{9}{2} = \frac{18}{6}$$

$$\frac{18}{6} = \frac{3}{1} = 3$$

Dividing Fractions

To divide fractions, invert the divisor (swap the numerator and denominator), then multiply as common fractions.

Example:

$$\frac{3}{4} \div \overset{\text{divisor}}{\frac{2}{3}} = \frac{3}{4} \times \frac{3}{2}$$

$$\frac{3 \times 3}{4 \times 2} = \frac{9}{8}$$

In this example, the quotient is a fraction with a larger numerator than its denominator. This indicates that the value is greater than 1. To express this in its simplest form, convert it to a mixed number.

$$\frac{9}{8} = \frac{8+1}{8} = 1\frac{1}{8}$$

Sometimes it is necessary to combine operations in a single problem. Some such problems also involve more than one type of unit, such as inches and feet. The first step in solving such problems is to decide which operation should be done first. It often helps to write the problem in such a way that it states the order of operations. Next, convert everything to the same unit(s). Where the operations will include adding or subtracting fractions, convert them to their least common denominators. Now solve the problem, doing all operations in the planned order. Finally, convert the units to whatever makes sense for the problem. For example, you would not write the area of walls to be painted in square inches.

How long is the shaded space in this drawing?

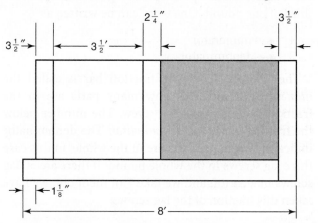

1. Convert the overall length to inches.   96″
2. Subtract $1\frac{1}{8}$″. $95\frac{7}{8}$
3. Convert $3\frac{1}{2}$′ to inches.   42″
4. Add dimensions at top. $3\frac{1}{2}$″ $+ 42$″ $+ 2\frac{1}{4}$′$+ 3\frac{1}{2}$″ $= 51\frac{1}{4}$″
5. Subtract $51\frac{1}{4}$″ from $95\frac{7}{8}$″. $95\frac{7}{8}$″ $- 51\frac{1}{4}$″ $= 44\frac{5}{8}$″

Practice

$$\frac{1}{4} + \frac{5}{8} \qquad \frac{2}{3} + \frac{7}{12}$$

$$\frac{3}{4} - \frac{1}{3} \qquad \frac{13}{16} - \frac{2}{5}$$

$$\frac{1}{2} \times \frac{3}{4} \qquad \frac{2}{3} \times 2\frac{1}{2}$$

$$\frac{1}{3} \div \frac{1}{2} \qquad 1\frac{1}{4} \div \frac{2}{5}$$

Reading a Ruler

Measuring devices such as rulers and tape measures may be marked for measuring inches and fractions of an inch; meters, centimeters, and millimeters; feet, inches and tenths of an inch; or by any other system. The measuring system used to divide the spaces on a measuring device is called the *scale*. The most common linear (in a line) scale in construction uses yards, feet, inches, and fractions of an inch. There are three feet in a yard, 12 inches in a foot, and the inches are most often divided into halves, fourths, eighths, and sixteenths. See **Figure 3**. The longest marks on the scale indicate inches. The inches on a measuring device may be divided into eighths, sixteenths, or even thirty-seconds. The second longest marks on the scale represent halves, the next longest represent fourths, and so on. See **Figure 4**. The first step in reading the scale is to determine what the smallest marks on the scale represent. Count down from the whole inch to the halves, then the quarters, the eighths, sixteenths, and thirty-seconds, if they are used. Then count the number of marks from the last inch mark to the mark you are reading.

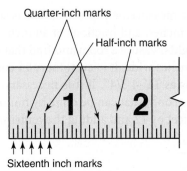

Goodheart-Willcox Publisher

Figure 4

Practice

What measurements are represented by the letters on this figure?

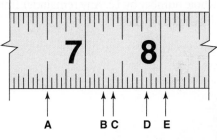

Goodheart-Willcox Publisher

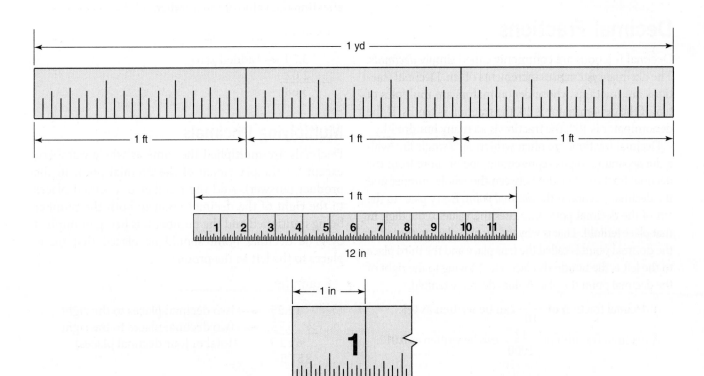

Goodheart-Willcox Publisher

Figure 3

Often, the measurement you will take is several feet, a few more inches, and fractions of an inch. For example, you might measure a room and find that it is 14 feet 4 and 3/4 inches wide. It is customary in construction to write this as 14'-4 3/4". If it is necessary to express this as inches only, that can be done by multiplying the number of feet by 12 and adding the remaining inches.

$$14' - 4\frac{3''}{4}$$

$$14 \times 12 = 168$$

$$168" + 4\frac{3''}{4} = 172\frac{3''}{4}$$

The reverse is also true. If you are given a dimension of 78 1/2 inches, it can be expressed as feet and inches. Divide the whole number by 12. Any remainder from the division will be left as inches.

$$78\frac{1''}{2} = 6' - 6\frac{1''}{2}$$

$$\begin{array}{r} 6 \\ 12\overline{)78} \\ 72 \\ \hline 6 \end{array}$$ – 6 inches left after division

Decimal Fractions

Decimal fractions are commonly called simply *decimals*. The decimal system uses increments of ten. Decimal fractions are fractions whose denominators are multiples of ten. If the denominator is 10, the fraction is tenths. If the denominator is 100, the fraction is so many hundredths.

Decimal fractions are often written on a single line with a dot separating digits representing one or more from the decimal fraction. The dot between the whole number and the decimal fraction is the *decimal point*. Every place to the left of the decimal point increases the value of the digit in that place tenfold. That is why the second place to the left of the decimal point is called the tens place and the third place to the left is the hundreds place, etc. Moving to the right of the decimal point the place values decrease tenfold.

A decimal fraction of $\frac{5}{10}$ can be written as 0.5.

A decimal fraction of $\frac{12}{1000}$ can be written as 0.012.

```
         1 2 3 . 4 5 6
hundreds ──┐ │ │   │ │ └─── thousandths
    tens ──── │   │ └───── hundredths
    ones ─────┘   └─────── tenths
                 └──────── decimal point
```

The value of the number in the previous example is one hundred twenty-three and four hundred fifty-six thousandths.

Adding and Subtracting Decimals

To add decimals, line up the decimal points in a column, add the numbers, and put the decimal point in the result in the decimal point column.

Example:

$$\begin{array}{r} 1.4 \\ 19.2 \\ + \ 31.7 \\ \hline 52.3 \end{array}$$

Subtracting decimals is very similar. Line up the decimal points in the problem and the answer and subtract as usual.

Example:

$$\begin{array}{r} 27.74 \\ - \ 2.23 \\ \hline 25.51 \end{array}$$

If there are more decimal places in the number being subtracted than there are in the number it is being subtracted from, zeroes can be added to the right without affecting the value of the number.

Example:

$$\begin{array}{r} 5.70 \\ - \ 2.02 \\ \hline 3.68 \end{array}$$ ⟵ added zero

Multiplying Decimals

Decimals are multiplied the same as whole numbers, except for the placement of the decimal point in the product (answer). Add the number of decimal places to the right of the decimal point in both the number being multiplied and the number it is being multiplied by. The decimal point should be placed that many places to the left in the product.

Example:

$$\begin{array}{r} 12.25 \\ \times \ 3.75 \\ \hline 6125 \\ 8575 \\ 3675 \\ \hline 45.9375 \end{array}$$

12.25 ⟵ two decimal places to the right
× 3.75 ⟵ two decimal places to the right (total of four decimal places)
45.9375 ⟵ decimal point is four places to the left

Dividing Decimals

Dividing decimals is also much like dividing whole numbers, except for keeping track of the placement of the decimal point. As a reminder, the number being divided is the dividend, the number it is divided by is the divisor, and the answer is the quotient. To start the division problem, move the decimal point in the divisor all the way to the right. Move the decimal point in the dividend the same number of places to the right. Add zeroes to the right of the dividend, if necessary. Divide as you would for whole numbers.

Example:

divisor ⟶ $0.4 \overline{)20}$ ⟵ dividend

$4. \overline{)200.}$ move decimal points

$$4 \overline{)200} \quad \begin{array}{r} 50 \\ \hline 20 \\ \hline 00 \end{array}$$

Practice

2.12	34.09		
17.01	12.125	18.48	134.02
+ 9.05	+ 2.899	− 12.25	− 8.14

5.25	15.34		
× 5	× 6.25	35 ÷ 0.07	2.25 ÷ 0.25

Converting Common Fractions to Decimal Fractions and Rounding Off

To change a decimal fraction to a common fraction, divide the numerator by the denominator.

Example: Change 1/4 to a decimal fraction.

$$4 \overline{)1.00} \quad \begin{array}{r} 0.25 \\ \hline 8 \\ \hline 20 \end{array}$$

$\dfrac{1}{4} = 0.25$

Sometimes the division yields a number with a repeating decimal.

Example: Change 2/3 to a decimal.

$$3 \overline{)2.000} \quad \begin{array}{r} 0.666 \\ \hline 18 \\ \hline 20 \\ 18 \\ \hline 20 \end{array}$$

These numbers should be rounded off to the desired number of places. When rounding off, the last digit should be increased by 1 if the next digit is 5 or more. If the next digit is less than 5, the last digit used stays the same. In the above example, round the answer to two places. The second digit is rounded up to 7, because 0.666 is closer to 0.67 than it is to 0.66.

To convert a mixed number as a decimal, keep the whole number as is and convert the fractional part as above.

Express $12 \dfrac{3}{4}$ as a decimal

$$12 + 4 \overline{)3.00} = 12.75 \quad \begin{array}{r} 0.75 \\ \hline 2\,8 \\ \hline 20 \end{array}$$

Practice

Convert these fractions to decimals and round the answers to three places.

$\dfrac{1}{3}$ $\dfrac{22}{7}$ $\dfrac{10}{15}$

Converting Decimals to Common Fractions

To convert a decimal to a fraction, drop the decimal point and write the given number as the numerator. The denominator will be 10, 100, 100, or 1 with as many 0s as there were places in the decimal number.

Example:

$0.42 = \dfrac{42}{100}$ or $0.125 = \dfrac{125}{1000}$

Metrics in Construction
Reading a Vernier Scale

The vernier scale is named after French mathematician Pierre Vernier. It uses two scales that slide past one another to allow accurate measurements beyond what would be possible with a single scale. The vernier scale principle can be used with inches or the metric system. An inch-based vernier is explained here. The main scale is divided into inches, tenths of an inch, and 25 thousandths of an inch. The vernier scale is divided into slightly smaller increments, so that only one mark on the vernier scale can be aligned with a mark on the main scale. See **Figure 5**.

To read the vernier scale, start with the location of the 0 on the scale. See **Figure 6**. Find the first whole inch mark to the left of the vernier scale 0 mark. In **Figure 6**, that mark is 2″. We will be adding the numbers found in each step, so write down "2″." Next, find the largest 0.1″ mark to the left of the 0 on the

vernier scale. In **Figure 6**, that is the 0.2″ mark. Write that under "2″," with their decimal points aligned for addition later.

```
2.0
0.2
```

Add the decimal point after the top 2 and 0s as necessary to help keep things aligned for later addition. The third step is to find the largest 0.025″ mark on the main scale to the left of the 0 on the vernier scale. In this case, that is the second one after the 0.2″ mark, so it represents 0.025″+ 0.025″ or 0.050″. Write that number in your addition column, adding 0s as necessary.

```
2.000
0.200
0.050
```

The final reading is on the vernier scale. Find a line on the vernier scale that lines up with any mark on the main scale. In **Figure 6**, that is the mark representing 17,

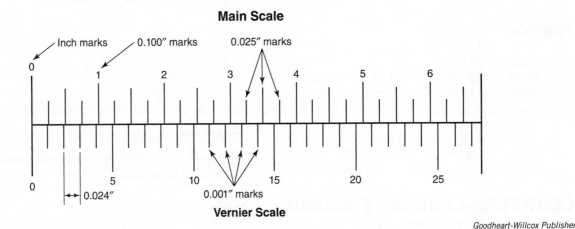

Goodheart-Willcox Publisher

Figure 5

Figure 6

Goodheart-Willcox Publisher

or 0.017″. Write that in your addition column, adding any necessary 0s, and do the addition.

2.000
0.200
0.050
0.017
─────
2.267

The vernier scale in **Figure 6** is reading 2.267″.

Practice

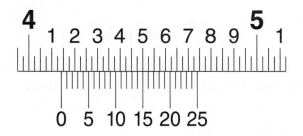

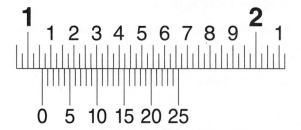

Equations

An *equation* is a mathematical statement that two things have the same or equal value. An equation can be thought of as a mathematical sentence. The words of the sentence are mathematical values called *terms*. An equation is always written with an equals sign (=). For example, 3 + 4 = 7. In that statement 3, 4, and 7 are terms. The statement says that 3 plus 4 has the same value as 7. We work with equations quite often in doing construction math. Many useful formulas are stated as equations, as is evident in the next section. Equations can be used to find the value of one unknown term when the other values in the equation are known. For example, if you know that a truck is loaded with 10 bundles of shingles weighing 80 lb each, an unknown weight of plywood, and the total load is 1000 lb, you can find the weight of the plywood with the following equation:

(80 lb × 10) + weight of plywood = 1000 lb

The (80 lb × 10) represents the total weight of the shingles. It is one of the terms in the equation. It is enclosed in parentheses to indicate that it is a single term that should be computed before the rest of the equation. Whenever a mathematical term such as (80 lb × 10) is enclosed in parentheses, that computation should be done first. Now write the equation with the shingle weight computed:

800 lb + weight of plywood = 1000 lb

When a mathematical operation is done on one side of an equation, the equation remains a true statement if the same thing is done on the other side of the equal sign. If we subtract 800 lb from both sides of our equation, it is still a true equation:

800 lb + weight of plywood − 800 lb = 1000 lb − 800 lb

Weight of plywood = 200 lb

Practice

Find the unknown value in each equation.

cost = ($0.60 − $0.04) × 5

$$X = \frac{3}{4} + 20$$

240 lbs = 2 × weight of crate

$$\frac{1}{4} \div \frac{2}{3} = Y$$

Area Measure

The area of a surface is always measured in units of square inches, square meters, square feet, etc. When a number is squared, that means it is multiplied by itself. For example, 3 squared is 9. Square units are written with a superscript 2, indicating that it is units × units.

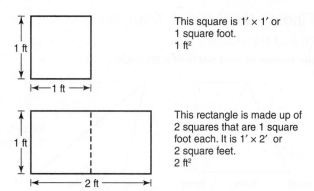

This square is 1′ × 1′ or 1 square foot. 1 ft²

This rectangle is made up of 2 squares that are 1 square foot each. It is 1′ × 2′ or 2 square feet. 2 ft²

Finding the Area of Squares and Rectangles

The area of a square or a rectangle is the number of units it is wide multiplied by the number of units it is long.

Example: Find the area of this rectangle.

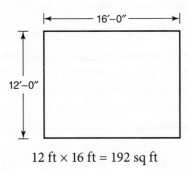

12 ft × 16 ft = 192 sq ft

The width and length must be expressed in the same units.

Example: To find the area of this rectangle, convert all feet to inches, then multiply.

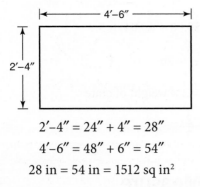

2′–4″ = 24″ + 4″ = 28″

4′–6″ = 48″ + 6″ = 54″

28 in = 54 in = 1512 sq in²

A square foot is 12 inches × 12 inches, or 144 square inches. If an area is given as a large number of square inches, it can be converted to square feet by dividing it by 144.

Example: In the example above, the area of the rectangle is 1512 square inches. If that is divided by 144, we find that it is 10.5 square feet.

Finding the Area of Triangles

To find the area of a triangle, it is necessary to know the names of two parts of a triangle.

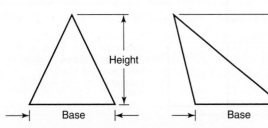

To find the area of any triangle, multiply the height times 1/2 the base.

Example: Find the area of this triangle.

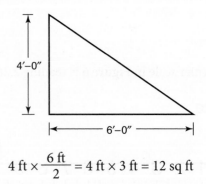

$$4 \text{ ft} \times \frac{6 \text{ ft}}{2} = 4 \text{ ft} \times 3 \text{ ft} = 12 \text{ sq ft}$$

Another way to achieve the same results is to multiply the base times the height, then divide that by 2.

4 ft × 6 ft = 24 sq ft

24 sq ft ÷ 2 = 12 sq ft

Some figures may be made up of squares, rectangles, and triangles of varying sizes. To find the area of such a figure, break it into its various parts and find the area of each part, then add those areas.

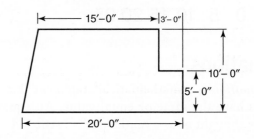

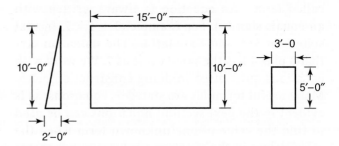

Triangle	Rectangle	Rectangle
10 × 2 = 20	10 × 15 = 150 ft²	3 × 5 = 15 ft²
20 ÷ 2 = 10 ft²		

10 ft² + 150 ft² + 15 ft² = 175 ft²

Practice

Find the areas of the figures.

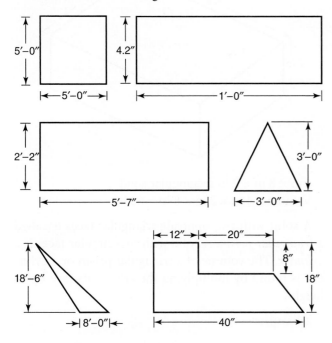

Finding the Circumference and Area of a Circle

The distance from a circle's center point to its outer edge is its *radius*. The total distance across a circle through its center point is its *diameter*.

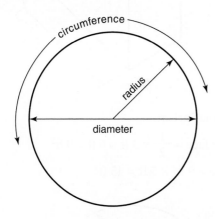

Many calculations involving circles or parts of circles use a constant of approximately 22/7, or 3.1416. The Greek letter π (pronounced "pie") is used to represent this constant. It is a constant because it never changes, regardless of the dimensions of the circle.

The *circumference* of a circle is its perimeter. To find the circumference of a circle, multiply the diameter by π. This is the same as multiplying the radius times 2 and multiplying that product times π.

Example: Find the circumference of a circle with a diameter of 8′.

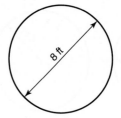

Circumference = π × diameter = 3.1416 × 8 ft
3.1416 × 8 ft = 25.1328 ft
or
Circumference = 2 × 4 ft × π = 8 ft × 3.1416

The area of a circle is found by multiplying π times the radius squared (the radius times the radius).

Example: Find the area of a circle with a radius of 3″.

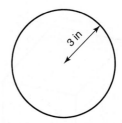

Area = π × radius² (or radius × radius)
Area = 3.1416 × 3 in × 3 in = 3. 1416 × 9 in²
Area = 28.2744 in²

Notice that in both of the examples using π the answer is rounded off to four decimal places. That is because π was rounded to four places. So the answer cannot be accurate to more than that many places.

Many of the shapes encountered in the trades are semicircles or even quarters of a circle. The areas and perimeters of these shapes can be found using the formulas for circles and dividing the result in half for a semicircle or by 4 for a quarter circle.

Example: Find the perimeter of the semi-circular shape.

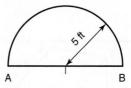

Diameter = 2 × 5 ft= 10 ft
Circumference of circle = π × 10 ft = 31.416 ft
Circumference of circular portion of figure = 15.708 ft
Length of line AB is 10 ft
Add 15.708 ft + 10 ft = 25.708 ft

Practice

Find the perimeter and the area of each of these figures.

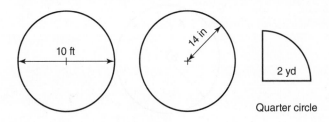

Quarter circle

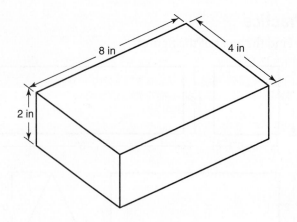

4 in × 8 in × 2 in rectangular solid
Volume = 4 in × 2 in × 8 in = 64 in³

A solid with two opposite triangular faces is called a *triangular prism*. A solid with two circular faces is a *cylinder*. The volume of a triangular prism or a cylinder is found by multiplying the area of its face by its height.

Example:

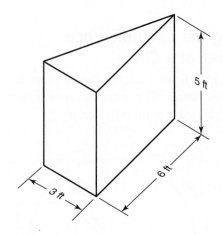

Triangular prism

Area of face = $\frac{1}{2}$ × 3 ft × 6 ft = 9 ft²

Volume = 9 ft² × 5 ft = 45 ft³

Volume Measure

The volume of a solid is always measured in units of cubic inches, cubic meters, cubic feet, etc. When a number is cubed that means it is multiplied by itself, then by itself again. For example, 3 cubed is 27 (3 × 3 × 3). Cubic units are written with a super-script 3, indicating that it is units × units × units.

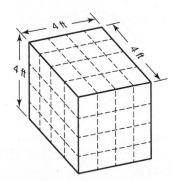

4 ft × 4 ft × 4 ft = 64 cubic ft or 64 ft³

This cube is made up of 64 individual cubes, each measuring 1 foot by 1 foot by 1 foot.

As long as a solid (a three-dimensional shape) has the same size and cross-section shape throughout its depth, its volume can be found by multiplying the area of one surface by the depth from that surface. To find the volume of a *cube* (all edges are the same size) or a rectangular solid (a rectangle with a third dimension) multiply the length, width, and height.

Example:

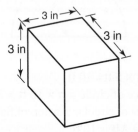

3-inch cube
Volume = 3 in × 3 in × 3 in = 27 in³

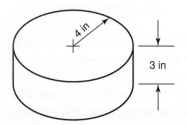

Cylinder
Area of face = π × 4 in × 4 in = 50.2656 in²
Round to 50.27 in²
Volume = 50.27 in² × 3 in = 150.81 in³

Practice

Find the volume of each solid.

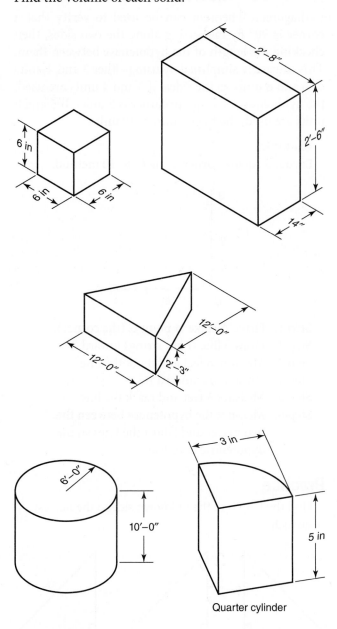

Exponents

When a number is squared or cubed, the little superscript number written to the right is called an *exponent*. For example, in the number 10^2, the exponent is 2, indicating that the number is 10 multiplied by itself. Another way of saying this is "10 to the second power." If the number were to be 10 × 10 × 10, it could be written as 10^3 and it could be called "10 cubed" or "10 to the third power." Exponents of 2 and 3 have names—squared and cubed, respectively—because they are the exponents used

with area and volume measure. Higher exponents are only referred to as powers. For example, 10^5 is read as "10 to the 5th power." It is easier than saying "10 × 10 × 10 × 10 × 10." Both forms of that number equal 100,000.

Practice

- A. What is 12 squared?
- B. What is 8.5 cubed?
- C. What is 4 to the 6th power?
- D. What are two other ways to write 3 × 3 × 3 × 3?

Working with Right Angles

Right angles are common in construction. A *right angle* is a 90°. A triangle having a right angle is called a *right triangle*. Right triangles can be very useful in laying out square (90°) corners and in alternating current electricity. It will be helpful to know a few terms associated with right triangles.

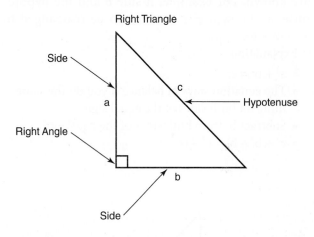

A right triangle can only have one 90° corner. The total of all three corners is always 180°, so if one is 90°, the other two must add up to 90° together.

The *Pythagorean Theorem* is a principle that makes right triangles so handy. Named after the Greek mathematician Pythagoras, the Pythagorean Theorem states that the sum of the squares of the sides of a right triangle is equal to the square of the hypotenuse. To help keep track of the Pythagorean Theorem, it is common to label the two sides a and b and the hypotenuse c. Then the theorem can be stated as an equation—$a^2 + b^2 = c^2$. If the lengths of the two sides of a right triangle are known, the Pythagorean Theorem can be used to find the length of the hypotenuse.

Example:

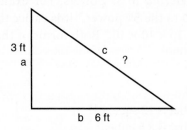

$$a^2 + b^2 = c^2$$
$$3 \times 3 + 6 \times 6 = c^2$$
$$9 + 36 = c^2$$
$$45 = c^2$$

The square root of 45 = the square root of c^2
(This is one time when a calculator will be a great help.)
$$6.7082 = c$$
Rounded to 1 decimal place the hypotenuse is 6.7 ft.

The Pythagorean Theorem can be used to find the length of any side of a right triangle if the other two are known. For example, if side *b* and the hypotenuse are known, $a^2 + b^2 = c^2$ can be rearranged to $a^2 = c^2 - b^2$.

Explanation:

- $a^2 + b^2 = c^2$
- The equation stays in balance if you do the same thing on both sides of the equal sign.
- Subtract b^2 from both sides of the equation.
- $a^2 + b^2 - b^2 = c^2 - b^2$
- $a^2 = c^2 - b^2$

Example:

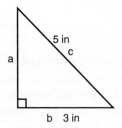

$$a^2 = c^2 - b^2$$
$$a^2 = 25 - 9$$
$$a^2 = 16$$
$$a = 4$$

The same can be done to find side *b* when side *a* and the hypotenuse are known.

Using Pythagorean Theorem to Lay Out Square Corners

Pythagorean Theorem can be used to verify that a corner is 90° by measuring along the two sides, then checking the length of the hypotenuse between them. This is usually simplified by using either 3 and 4 units or 6 and 8 units as the sides. If 3 and 4 units are used, the hypotenuse of a square corner is 5 units. If 6 and 8 units are used, the hypotenuse is 10 units.

Example:
Layout a square corner using 6–8–10 method.

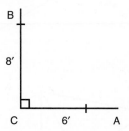

Step 1.   Drive A stake at point C (the corner).
Step 2.   Draw a line (tight string) toward A.
Step 3.   Measure 6 feet and mark the line.
Step 4.   Draw a line toward B.
Step 5.   Measure 8 feet and mark the line.
Step 6.   Measure the hypotenuse between the two marks and adjust the lines so the hypotenuse is 10 feet.

Practice

Find the length of the unknown side to the nearest $\frac{1}{10}$ of an inch.

 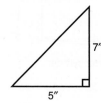 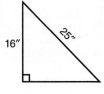

Computing Averages

An average is a typical value of one unit in a group of units. For example, if 4 windows have areas of 12.0 sq ft, 11.2 sq ft, 11.2 sq ft, and 14.5 sq ft; the average size of one of those windows is 12.2 sq ft. The average is computed by adding all of the units and dividing that sum by the number of units in the group.

Example:

12.0 ft²

11.2 ft²

11.2 ft²

14.5 ft²

48.9 ft²

```
       12.22
    4 | 48.90
       4
       08
        8
        09
         8
         10
```

The quotient should be rounded off to the same number of decimal places as is used in the problem.

Average window size is 12.2 sq ft.

Practice

Compute the averages of these groups.

- A. 14, 14.4, 14.5, 15
- B. 80 lb, 83 lb, 88 lb, 79.5 lb, 81.6 lb, 84 lb
- C. 11 cubic yards, 13 cubic yards, 11.5 cubic yards, 12 cubic yards, 12.8 cubic yards

Percent and Percentage

A *percent* is one part in a hundred. One penny is 1% of a dollar. Twenty-five cents is 25% of a dollar. On the other hand, 25 cents is 50% of a half-dollar, because if the half dollar were divided into 100 parts of 1/2 cent each and the quarter were also divided into 1/2-cent increments, the quarter would equal 50 of those 1/2-cent increments.

Think of percent as hundredths. To find a given percentage of an amount, multiply the amount times the desired number of hundredths.

Example:

- Find 12% of $4.40.
- 12% is 0.12 times the whole.
- $4.40 × 0.12 = $0.528, or 53 cents.

Percent is sometimes interchanged with *percentage* in usage, but there is a slight difference. *Percent* should be used when a specific number is used, such as 15% of the labor force. *Percentage* should be used when no specific number is used with the term, such as a large percentage of our homes are green.

Using what was covered in the section on equations and solving for an unknown, it is possible to calculate the whole if you know the percentage.

Example:
What was the total spent on tools if $22.00 was spent on sales tax and the tax rate is 8%?

1. Write an equation with the facts you know: $22.00 = 8% × total.
2. Write percent as hundredths: $22.00 = 0.08 × total.
3. Divide each side of the equation by 0.08: $275 = total.

Practice

A. What is 10 percent of 225?
B. What is 65 percent of $350.50?
C. If merchandise cost $22.25 and the bill comes to $23.14, what percentage was added for sales tax?
D. If a 500-gallon tank is 60 percent full of water, how much water does it contain?

Graphs

Graphs are frequently used to show mathematical data in a more visual way than simply displaying numbers. There are many kinds of graphs. The most common ones are bar graphs (also called bar charts), line graphs, and circle graphs (also called pie graphs). Graphs are sometimes called charts.

Bar Graphs

As their name implies, *bar graphs* use bars to show data. See **Figure 7**. Bar graphs have two dimensions. The horizontal line representing the starting point is called the *x axis*. The vertical line with graduations indicating the height of the bars is called the *y axis*. Every bar graph must have four parts: title, labels, scale, and bars. The *title* tells what is graphed. The *labels* on both the *x* and *y* axes tell what kind of data is being shown. The *scale*, often only on the *y* axis, tells what the units of measure are. The *bars* show the data numbers.

The tops of the bars might not align with the graduations on the *y* axis. For example, the Wednesday bar in **Figure 7** is between the 8-hour and 9-hour graduations. In this case, the value represented by the bar must be extrapolated. To *extrapolate* a value means to estimate it based on where it falls between two known values. We know that the line above the bar represents

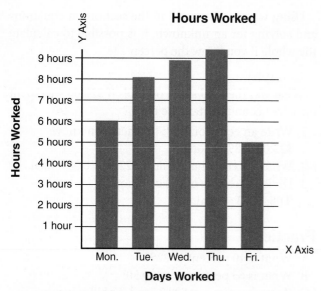

Goodheart-Willcox Publisher

Figure 7

9 hours and the line below the bar top represents 8 hours. The bar is about 3/4 of the way up to the 9-hour line, so the bar represents about 8 3/4 hours.

Some bar graphs have more than one bar for each point on the *x* axis. The bars generally show related data, but with some difference. This is a good way to compare two sets of data.

Example:

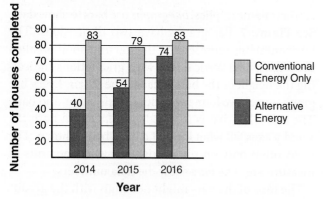

It is easy to see that the builder in the example completed more than twice as many conventionally powered homes as those with alternative energy sources in 2014. By 2016, although the total number of homes increased greatly, the number of homes with alternative energy sources nearly caught up with the number of homes with only conventional energy sources.

Line Graphs

Line graphs are similar to bar graphs, except the data points are connected by a line. The line graph in **Figure 8** shows the rate of output from a pump running at various speeds. The flow increases as the speed of the pump increases, but at somewhere around 600 rpm the rate of increase in the flow begins to drop off. By extrapolation (explained above in the discussion on bar graphs) we can see that at 600 rpm the flow is about 220 gallons per minute. At 700 rpm the flow has only increased to about 235 or 240 gallons per minute. Because this is a line graph, it is easy to see the trend by the slope of the line.

A line graph might use two or more lines (based on separate data points) to show a comparison. The line graph in **Figure 9** compares the fuel efficiency of two different fuels in engines running at different speeds. From this graph it can be seen that if the engine is to be run only at 900 rpm, both fuels have the same run times. If the engine is to be run only at speeds below 900, fuel A is more efficient. If it is to be run only at speeds above 900, fuel B is more efficient.

Circle Graph or Pie Chart

Circle graphs are most commonly called *pie charts* because they resemble a pie divided into slices. A pie chart is useful to show the sizes of the various parts of a whole. See **Figure 10**. A pie chart has three essential parts: title, key, and circle.

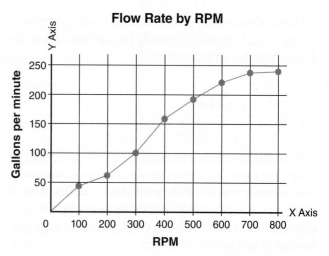

Goodheart-Willcox Publisher

Figure 8

Run Time By Fuel

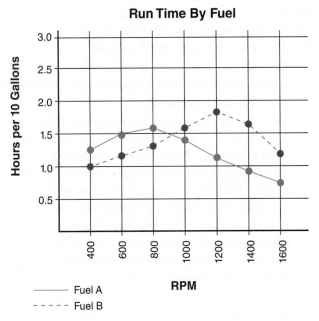

Fuel A
---- Fuel B

Goodheart-Willcox Publisher

Figure 9

The *title* tells what is being graphed or charted. The *key* identifies each of the individual bits of data. The *circle* and its "pie slice" parts makes up the foundation of the chart. If any one of these is missing, the chart is meaningless. For example, without the title, the pie chart in **Figure 10** could be a record of actual spending or a planned budget.

Personal Budget

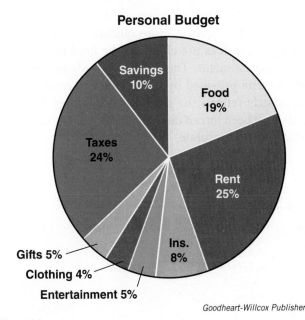

Goodheart-Willcox Publisher

Figure 10

Practice

Housing Starts

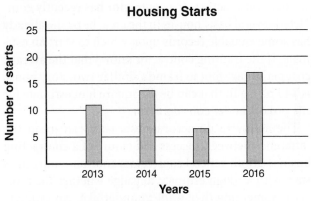

A. What type of graph is shown above?
B. In what year were the smallest number of houses started?
C. How many houses were started in 2014?

Square Feet Installed per Worker per Day

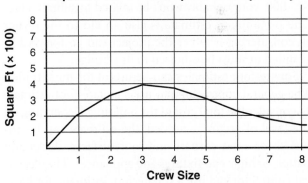

D. What size crews installed the most square feet per worker?
E. With 2 workers, how many square feet can the crew install in 1 day?
F. Is a person working alone more or less productive than each person in a 7-worker crew?
G. What is shown on the x axis of this graph?
H. Sketch a pie chart showing the following data:
  - Building material $4,000
  - Labor $3,000
  - Overhead $2,000
  - Profit $1,000

Estimating

An estimate is not a wild guess, but rather an informed forecast of what it will take to complete a project. Estimates are based on actual costs and man-hours it has taken to do similar jobs. A preliminary estimate may be arrived at by knowing the cost per square foot of similar construction. If a contractor has recently completed several basic houses in the area, he or she already has some valuable records upon which to start an estimate. If the framing contractor knows that the average cost per square foot to frame a similar two-story house is $45 per sq ft, that can be the square foot estimate for the project under consideration.

The ability to do accurate estimates can mean the difference between success and failure of a contracting business. Both general contractors and sub-contractors start with a rough estimate to gauge whether the project is something they want to undertake. An estimate allows them to have an informed conversation with the person who will hire the contractor. Being way off on that first rough estimate might cause you to lose the opportunity to do a more detailed estimate and bid on the job. If the owner expects to pay around $12,000 for HVAC and you are thinking of something closer to $25,000, you probably won't be invited to submit a bid. The preliminary estimate is also an aid to the owner in considering whether the project can go forward as planned or must be adjusted to fit a budget.

Square-foot estimates can be useful for most trades. The general contractor can use a square-foot estimate to consider the total cost of the building. Depending on the level of accuracy needed, the estimate can be adjusted for extra corners and roof surfaces, decks and patios, or high-end fixtures and appliances. The electrical contractor can use a square-foot estimate for basic wiring costs. A square-foot estimate will give the HVAC contractor a basis for estimating how much space must be heated or cooled and approximately how much duct work is involved.

When the time comes to submit a bid for the work, a more detailed estimate will be done. Detailed estimates are used to forecast the actual cost of materials, the cost of labor, the overhead expenses (things such as contractor's office, insurance, and vehicles that are not directly related to the particular project), permits and fees, and profit. The information for a detailed estimate can come from several sources:

- Quantity take-off—This is an item-by-item accounting of the quantity and sizes of all of the materials used for the job.
- Database of manpower and equipment costs—There are companies that publish such databases. Most contractors build their own database as they complete projects. An item on the database might, for example, be the hourly cost of a 110-horsepower bulldozer, an operator, and a laborer; or the cost for two carpenters and a laborer to install one square of shingles on a two-story building with a low-slope roof.
- A thorough review of the plans and specifications by an experienced estimator—The estimator will look for features on the plans that would be exceptions to the information found by the above two methods. For example, special built-in cabinetry might require more time than what is shown on any available database.

Computerized estimating has become the norm. Several companies publish software for various types of estimating.

- General construction for big jobs involving structural steel and iron workers
- Civil construction
- Light-frame construction
- Specialized for individual trades

Using estimating software enables the estimator to complete the estimate more quickly and more accurately, but it does not eliminate the need for a knowledgeable estimator. The estimator needs to review all of the data provided by the software to ensure that it accurately reflects the projects peculiarities. Contractors have been forced out of business because their estimates were inaccurate.

Technical Information

Contents

Standard Abbreviations for Use on Drawings

Above Finished Flooring	AFF	Double Strength Glass	DSG	Lavatory	LAV
Acoustical Tile	AT or ACT	Double-Hung Windows	DHW	Length	LGTH
Aggregate	AGGR	Drain	DR	Level	LVL
Air Conditioning	AIR COND	Drawing	DWG	Light	LT
Air Dried	AD	Dressed and Matched	D&M	Light Switch	LTSW
Alternate	ALT	Each	EA	Linen Closet	L CL
Alternating Current	AC	Edge	EDG	Linoleum	LINO
Aluminum	AL	Edge Grain	EG	Lintel	LNTL
American Institute of Architects	AIA	Electric Panel	EP	Living Room	LR
American Institute of Electrical		Elevation	EL	Low Voltage	LV
Engineers	AIEE	Entrance	ENT	Masonry Opening	MO
American Society for Testing		Excavate	EXC	Mastic	MSTC
and Materials	ASTM	Exhaust Vent	EXHV	Material	MATL
American Standards		Exterior	EXT	Maximum	MAX
Association, Inc.	ASA	Face Brick	FB	Medicine Cabinet	MC
Approximate	APPROX	Federal Housing Authority	FHA	Minimum	MIN
Architectural	ARCH	Finish	FIN	Miscellaneous	MISC
Asbestos	ASB	Finished Floor	FNSHFL or FF	Modular	MOD
Asphalt Roof Shingles	ASPHRS	Fixture	FIX	Molding	MLDG
Basement	BSMT	Flashing	FL	Mortar	MOR
Batter	BAT	Flat Grain	FG	Nominal	NOM
Beam	BM	Flooring	FLG	Nosing	NOS
Better	BTR	Fluorescent	FLUOR	On Center	OC
Beveled	BEV	Flush	FL	Open Web Joint	OWJ
Blocking Board	BLKG	Foot or Feet	FT	Opening	OPNG
Board Foot	BD FT	Footing	FTG	Paint	PNT
Brick	BRK	Foundation	FDN	Pair	PR
British Thermal Unit	BTU	Furring	FUR	Partition	PTN
Building	BLDG	Fuse	FU	Perpendicular	PERP
Bundle	BDL	Gage	GA	Pilaster	P
Cabinet	CAB	Gallon	GAL	Piping	PP
Carpenter	CPNTR	Galvanize	GALV	Plank	PLK
Casing	CSG	Galvanized Iron	GI	Plaster	PLAS
Cement	CEM	Glass	GL	Plate	PL
Cement Floor	CEM FL	Grade	GR	Plate Glass	PL GL
Cement Mortar	CEM MORT	Gypsum	GYP	Plumbing	PLBG
Center Matched	CM	Hardboard	HBD	Power	PWR
Chimney	CHM	Hardwood	HDWD	Precast	PRCST
Circuit	CKT	Header	HDR	Prefabricated	PREFAB
Closet	CL or CLO	Heat Exchanger	HE	Quart	QT
Column	COL	Herringbone	HGBN	Random	RDM
Common	COM	Horsepower	HP	Receptacle	RCPT
Concrete	CONC	Hose Bibb	HB	Recess	REC
Concrete Block	CONC B	Hot Water	HW	Reference	REF
Conduit	CND	Hundred	C	Refrigerator	REF
Construction	CONST	Insulation	INS	Reinforcing	REINF
Counter	CTR	Interior	INT	Revision	REV
Coupling	CPLG	Iron Pipe	IP	Roll Roofing	RR
Cubic Foot	CU FT	Jamb	JMB	Roof	RF
Cubic Yard	CU YD	Joist	J	Rough	RGH
Cutoff Valve	COV	Keyway	KWY	Rough Opening	RO
Diagram	DIAG	Kiln-dried	KD	Saddle	SDL
Diameter	DIA or DIAM	Kitchen	K	Schedule	SCH
Dimension	DIM	Knee Brace	KB	Screen	SCR
Direct Current	DC	Knife Switch	KNSW	Select	SEL
Ditto	DO	Laminate	LAM	Service	SERV
Door	DR	Lath	LTH	Sewer	SEW
Dormer	DRM	Lattice	LTC	Sheathing	SHTHG

Shelving	SHELV	Surfaced Two Sides	S2S	Ventilation	VENT
Shiplap	S/LAP	Switch	SW or S	Volt	V
Siding	SDG	Temperature	TEMP	Voltmeter	VM
Specifications	SPEC	Thermostat	THERMO	Water Closet	WC
Square	SQ	Thick	THK	Water Heater	WH
Square Feet	SQ FT	Thousand	M	Weather Stripping	WS
Stairway	STWY	Timber	TMBR	Weep Hole	WH
Standard	STD	Tongue and Groove	T&G	Weight	WT
Steel	ST or STL	Transformer	XFMR	Welded Wire Fabric	WWF
Stringer	STGR	Truss	TR	Wide Flange	WF or W
Structural	STR	Tubing	TBG	With	W/
Surfaced Four Sides	S4S	Typical	TYP	Without	W/O
Surfaced One Side	S1S	Union	UN	Wood	WD
Surfaced One Side and		Valley	VAL		
Two Edges	S1S2E	Vent Stack	VS		

Slump Testing

While a slump test is normally done in heavy construction work, it is sometimes used in residential construction. A *slump test* measures the consistency, stiffness, and workability of fresh concrete. These characteristics are affected mainly by the amount of water in the mix. Other factors—type of aggregate, air content, admixtures, temperature, and the proportions of all ingredients—also affect slump. Mixing time and standing time also have an effect.

The test is conducted with a slump cone made of metal. It is 12″ high with a diameter of 4″ at the top and 8″ at the bottom. Follow these steps to make the test:

1. With the small diameter up, fill the cone in three layers of equal volume, rodding each layer 25 times.
2. Strike off the top. Then slowly remove the cone with an even motion lasting from 5 to 12 seconds. Do not disturb the mixture or tilt the cone.
3. Turn the cone upside-down alongside the mixture. Immediately measure the slump with the tamping rod and a rule. The test should take no longer than 1 1/2 minutes.

Slump Range for Different Types of Construction		
Type of construction	Maximum slump	Minimum slump
Reinforced footings and foundation walls	3″	1″
Plain footings, caissons, and substructure walls	3″	1″
Beams and reinforced walls	4″	1″
Building columns	4″	1″
Pavement and slabs	3″	1″
Mass concrete	2″	1″

Plywood Siding Joints

Vertical wall joints

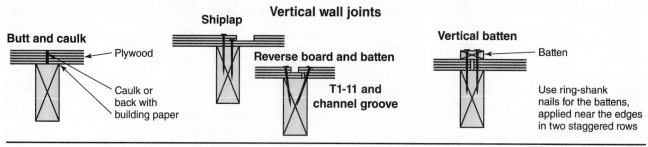

Butt and caulk
- Plywood
- Caulk or back with building paper

Shiplap

Reverse board and batten

T1-11 and channel groove

Vertical batten
- Batten
- Use ring-shank nails for the battens, applied near the edges in two staggered rows

Vertical inside and outside corner joints

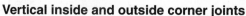

Butt and caulk
- Plywood
- Caulk

Rabbet and caulk
- Plywood
- Rabbet one piece plywood, caulk, and butt

Corner board lap joints
- Plywood
- Corner boards

Horizontal wall joints

Butt and flash
- Plywood
- Flashing (galvanized or aluminum)

Lap plywood
- Plywood
- Lap top plywood over bottom plywood

Shiplap
- Plywood
- Shiplap joint

Horizontal beltline joints

(For multistory buildings, make provisions at horizontal joints for "settling" shrinkage of framing, especially when applying siding direct to studs.)

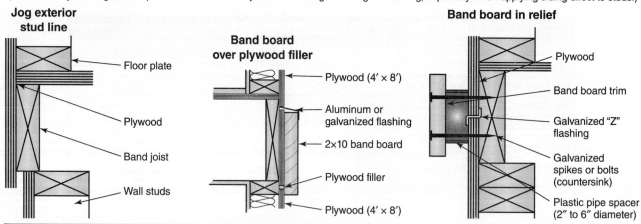

Jog exterior stud line
- Floor plate
- Plywood
- Band joist
- Wall studs

Band board over plywood filler
- Plywood (4′ × 8′)
- Aluminum or galvanized flashing
- 2×10 band board
- Plywood filler
- Plywood (4′ × 8′)

Band board in relief
- Plywood
- Band board trim
- Galvanized "Z" flashing
- Galvanized spikes or bolts (countersink)
- Plastic pipe spacer (2″ to 6″ diameter)

Window details

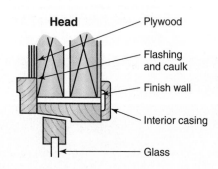
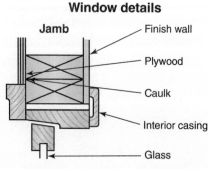
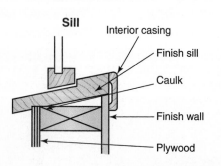

Head
- Plywood
- Flashing and caulk
- Finish wall
- Interior casing
- Glass

Jamb
- Finish wall
- Plywood
- Caulk
- Interior casing
- Glass

Sill
- Interior casing
- Finish sill
- Caulk
- Finish wall
- Plywood

Fire-Resistant Construction

All assemblies shown provide a one-hour fire rating. (APA–The Engineered Wood Association)

One-hour assembly — resilient channel ceiling system

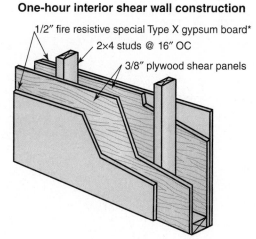

- 5/8″ plywood DFPA underlayment T&G
- Building paper
- 1/2″ Standard grade plywood with exterior glue
- 1/2″ galvanized metal resilient channels at 24″ OC*
- 1/2″ fire resistive Special Type X gypsum board–fasten to channels with self-tapping screws 12″ OC
- Joists 16″ (2×10's min.)

*Channels may be suspended below joists.

One-hour assembly — T-bar grid ceiling system

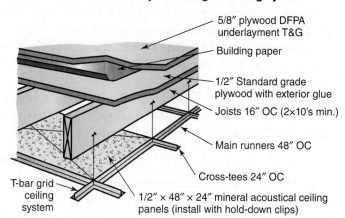

- 5/8″ plywood DFPA underlayment T&G
- Building paper
- 1/2″ Standard grade plywood with exterior glue
- Joists 16″ OC (2×10's min.)
- Main runners 48″ OC
- Cross-tees 24″ OC
- T-bar grid ceiling system
- 1/2″ × 48″ × 24″ mineral acoustical ceiling panels (install with hold-down clips)

One-hour interior shear wall construction

- 1/2″ fire resistive special Type X gypsum board*
- 2×4 studs @ 16″ OC
- 3/8″ plywood shear panels

*Regular 1/2″ gypsum board may be used when mineral wool or glass fiber batts are used in wall cavity.

Insulation batts in wall cavity also used for sound transmission control.

One-hour exterior wall construction

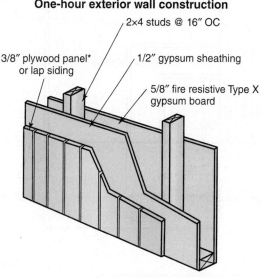

- 2×4 studs @ 16″ OC
- 3/8″ plywood panel* or lap siding
- 1/2″ gypsum sheathing
- 5/8″ fire resistive Type X gypsum board

*Including nominal 3/8″ specialty plywood sidings

Treated stressed-skin panel construction

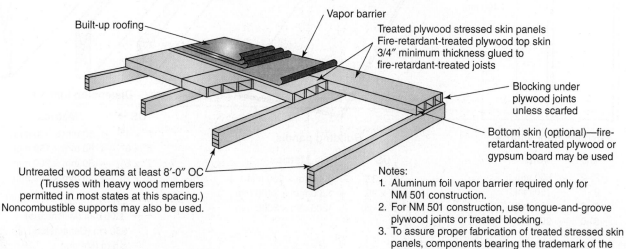

- Built-up roofing
- Vapor barrier
- Treated plywood stressed skin panels Fire-retardant-treated plywood top skin 3/4″ minimum thickness glued to fire-retardant-treated joists
- Blocking under plywood joints unless scarfed
- Bottom skin (optional)—fire-retardant-treated plywood or gypsum board may be used
- Untreated wood beams at least 8′-0″ OC (Trusses with heavy wood members permitted in most states at this spacing.) Noncombustible supports may also be used.

Notes:
1. Aluminum foil vapor barrier required only for NM 501 construction.
2. For NM 501 construction, use tongue-and-groove plywood joints or treated blocking.
3. To assure proper fabrication of treated stressed skin panels, components bearing the trademark of the Plywood Fabricator Service, Inc. are recommended.

Metrics in Construction

Efficient building construction begins with standard sizes for construction parts. Designs geared for mass production are based on standard modules. Layouts (horizontal and vertical) use one specific size. Whole multiples of the size make up all larger measurements.

The US Customary system uses a basic module of 4″. The layout grid is further divided into spaces of 16″, 24″, and 48″. Modular design saves material and time.

In countries using the metric system, the 4″ module is replaced by a 100 mm module. Smaller sizes (submultiples) include 25 mm, 50 mm, and 75 mm. Large modules measure 400 mm, 600 mm, and 1200 mm, as shown in the drawing below.

The 100 mm module is slightly smaller than the 4″ module and the conversion can hardly be detected in small measurements. In a length of 48″, however, the difference is about 3/4″ and a standard 4′ × 8′ plywood panel is about 1 1/2″ longer than the similar metric size (1200 mm × 2400 mm).

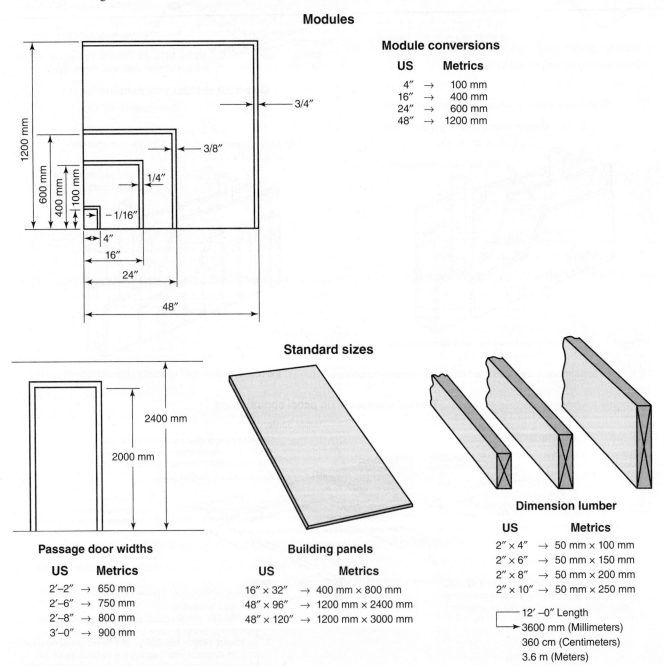

Modules

Module conversions

US		Metrics
4″	→	100 mm
16″	→	400 mm
24″	→	600 mm
48″	→	1200 mm

Passage door widths

US		Metrics
2′–2″	→	650 mm
2′–6″	→	750 mm
2′–8″	→	800 mm
3′–0″	→	900 mm

Standard sizes

Building panels

US		Metrics
16″ × 32″	→	400 mm × 800 mm
48″ × 96″	→	1200 mm × 2400 mm
48″ × 120″	→	1200 mm × 3000 mm

Dimension lumber

US		Metrics
2″ × 4″	→	50 mm × 100 mm
2″ × 6″	→	50 mm × 150 mm
2″ × 8″	→	50 mm × 200 mm
2″ × 10″	→	50 mm × 250 mm

12′ –0″ Length
3600 mm (Millimeters)
360 cm (Centimeters)
3.6 m (Meters)

Metric–Inch Equivalents

Inches			Inches		
Fractions	Decimals	Millimeters	Fractions	Decimals	Millimeters
	.00394	.1	15/32	.46875	11.9063
	.00787	.2		.47244	12.00
	.01181	.3	31/64	.484375	12.3031
1/64	.015625	.3969	1/2	.5000	12.70
	.01575	.4		.51181	13.00
	.01969	.5	33/64	.515625	13.0969
	.02362	.6	17/32	.53125	13.4938
	.02756	.7	35/64	.546875	13.8907
1/32	.03125	.7938		.55118	14.00
	.0315	.8	9/16	.5625	14.2875
	.03543	.9	37/64	.578125	14.6844
	.03937	1.00		.59055	15.00
3/64	.046875	1.1906	19/32	.59375	15.0813
1/16	.0625	1.5875	39/64	.609375	15.4782
5/64	.078125	1.9844	5/8	.625	15.875
	.07874	2.00		.62992	16.00
3/32	.09375	2.3813	41/64	.640625	16.2719
7/64	.109375	2.7781	21/32	.65625	16.6688
	.11811	3.00		.66929	17.00
1/8	.125	3.175	43/64	.671875	17.0657
9/64	.140625	3.5719	11/16	.6875	17.4625
5/32	.15625	3.9688	45/64	.703125	17.8594
	.15748	4.00		.70866	18.00
11/64	.171875	4.3656	23/32	.71875	18.2563
3/16	.1875	4.7625	47/64	.734375	18.6532
	.19685	5.00		.74803	19.00
13/64	.203125	5.1594	3/4	.7500	19.05
7/32	.21875	5.5563	49/64	.765625	19.4469
15/64	.234375	5.9531	25/32	.78125	19.8438
	.23622	6.00		.7874	20.00
1/4	.2500	6.35	51/64	.796875	20.2407
17/64	.265625	6.7469	13/16	.8125	20.6375
	.27559	7.00		.82677	21.00
9/32	.28125	7.1438	53/64	.828125	21.0344
19/64	.296875	7.5406	27/32	.84375	21.4313
5/16	.3125	7.9375	55/64	.859375	21.8282
	.31496	8.00		.86614	22.00
21/64	.328125	8.3344	7/8	.875	22.225
11/32	.34375	8.7313	57/64	.890625	22.6219
	.35433	9.00		.90551	23.00
23/64	.359375	9.1281	29/32	.90625	23.0188
3/8	.375	9.525	59/64	.921875	23.4157
25/64	.390625	9.9219	15/16	.9375	23.8125
	.3937	10.00		.94488	24.00
13/32	.40625	10.3188	61/64	.953125	24.2094
27/64	.421875	10.7156	31/32	.96875	24.6063
	.43307	11.00		.98425	25.00
7/16	.4375	11.1125	63/64	.984375	25.0032
29/64	.453125	11.5094	1	1.0000	25.4000

Conversion Table: US Customary to SI Metric

When You Know:	Multiply By:		To Find:
	Very accurate	Approximate	
Length			
inches	* 25.4		millimeters
inches	* 2.54		centimeters
feet	* 0.3048		meters
feet	* 30.48		centimeters
yards	* 0.9144	0.9	meters
miles	* 1.609344	1.6	kilometers
Weight			
grains	15.43236	15.4	grams
ounces	* 28.349523125	28.0	grams
ounces	* 0.028349523125	0.028	kilograms
pounds	* 0.45359237	0.45	kilograms
short ton	* 0.90718474	0.9	tonnes
Volume			
teaspoons		5.0	milliliters
tablespoons		15.0	milliliters
fluid ounces	29.57353	30.0	milliliters
cups		0.24	liters
pints	* 0.473176473	0.47	liters
quarts	* 0.946352946	0.95	liters
gallons	* 3.785411784	3.8	liters
cubic inches	* 0.016387064	0.02	liters
cubic feet	* 0.028316846592	0.03	cubic meters
cubic yards	* 0.764554857984	0.76	cubic meters
Area			
square inches	* 6.4516	6.5	square centimeters
square feet	* 0.09290304	0.09	square meters
square yards	* 0.83612736	0.8	square meters
square miles		2.6	square kilometers
acres	* 0.40468564224	0.4	hectares
Temperature			
Fahrenheit	*5/9 (after subtracting 32)		Celsius

* = Exact

Conversion Table: SI Metric to US Customary

When You Know:	Multiply By:		To Find:
	Very accurate	Approximate	
Length			
millimeters	0.0393701	0.04	inches
centimeters	0.3937008	0.4	inches
meters	3.280840	3.3	feet
meters	1.093613	1.1	yards
kilometers	0.621371	0.6	miles
Weight			
grains	0.00228571	0.0023	ounces
grams	0.03527396	0.035	ounces
kilograms	2.204623	2.2	pounds
tonnes	1.1023113	1.1	short tons
Volume			
milliliters		0.2	teaspoons
milliliters	0.06667	0.067	tablespoons
milliliters	0.03381402	0.03	fluid ounces
liters	61.02374	61.024	cubic inches
liters	2.113376	2.1	pints
liters	1.056688	1.06	quarts
liters	0.26417205	0.26	gallons
liters	0.03531467	0.035	cubic feet
cubic meters	61023.74	61023.7	cubic inches
cubic meters	35.31467	35.0	cubic feet
cubic meters	1.3079506	1.3	cubic yards
cubic meters	264.17205	264.0	gallons
Area			
square centimeters	0.1550003	0.16	square inches
square centimeters	0.00107639	0.001	square feet
square meters	10.76391	10.8	square feet
square meters	1.195990	1.2	square yards
square kilometers		0.4	square miles
hectares	2.471054	2.5	acres
Temperature			
Celsius	*9/5 (then add 32)		Fahrenheit

* = Exact

Standard Sizes, Counts, and Weights

Nominal size is the definition used by the trade, but it is not always the actual size. Sometimes the actual thickness of hardwood flooring is 1/32″ less than the so-called nominal size.

Actual size is the mill size for thicknesses and face width. *Counted size* determines the board feet in a shipment. Pieces less than 1″ in thickness are considered to be 1″.

Nominal	Actual	Counted	Weights (1000 feet)
Tongue and groove–end-matched			
**3/4×3 1/4″	3/4×3 1/4″	1×4″	2210 lbs.
3/4×2 1/4″	3/4×2 1/4″	1×3″	2020 lbs.
3/4×2″	3/4×2″	1×2 3/4″	1920 lbs.
3/4×1 1/2″	3/4×1 1/2″	1×2 1/4″	1820 lbs.
**3/8×2″	11/32×2″	1×2 1/2″	1000 lbs.
**3/8×1 1/2″	11/32×1 1/2″	1×2″	1000 lbs.
**1/2×2″	15/32×2″	1×2 1/2″	1350 lbs.
**1/2×1 1/2″	15/32×1 1/2″	1×2″	1300 lbs.
Square-edge			
**5/16×2″	5/16×2″	face count	1200 lbs.
**5/16×1 1/2″	5/16×1 1/2″	face count	1200 lbs.

Nail Schedule		
Tongue and groove flooring must be blind-nailed		
3/4×1 1/2, 2 1/4 and 3 1/4″	2″ machine driven fasteners, 7d or 8d screw or cut nail.	10-12″ apart*
3/4×3″** Plank	2″ machine driven fasteners, 7d or 8d screw or cut nail.	8″ apart into and between joists

*If subfloor is 1/2 inch plywood, fasten into each joist, with additional fastening between.
**Plank flooring over 4″ wide must be installed over a subfloor.

Nominal	Actual	Counted	Weights (1000 feet)
Special thickness (T and G–end-matched)			
**33/32×3 1/4″	33/32×3 1/4″	5/4×4″	2400 lbs.
**33/32×2 1/4″	33/32×2 1/4″	5/4×3″	2250 lbs.
**33/32×2″	33/32×2″	5/4×2 3/4″	2250 lbs.
Jointed flooring (i.e., square-edge)			
**3/4×2 1/2″	3/4×2 1/2″	1×3 1/4″	2160 lbs.
**3/4×3 1/4″	3/4×3 1/4″	1×4″	2300 lbs.
**3/4×3 1/2″	3/4×3 1/2″	1×4 1/4″	2400 lbs.
**33/32×2 1/2″	33/32×2 1/2″	5/4×3 1/4″	2500 lbs.
**33/32×3 1/2″	33/32×3 1/2″	5/4×4 1/4″	2600 lbs.

**Special Order Only

Following flooring must be laid on a subfloor		
1/2×1 1/2 & 2″	1 1/2″ machine driven fastener, 5d screw, cut steel or wire casing nail.	10″ apart
3/8×1 1/2 & 2″	1 1/4″ machine driven fastener, or 4d bright wire casing nail.	8″ apart
Square-edge flooring as follows, face-nailed through top face		
5/16×1 1/2 & 2″	1 inch 15 gauge fully barbed flooring brad. 2 nails every 7 inches.	
5/16×1 1/3″	1 inch 15 gauge fully barbed flooring brad. 1 nail every 5 inches on alternate sides of strip.	

National Oak Flooring Manufacturers' Association

Tables showing hardwood flooring grades, sizes, counts, and weights make flooring selection easier. Follow the recommended nailing schedule for best results.

Guide to Hardwood Flooring Grades

Flooring is bundled by averaging the lengths. A bundle may include pieces from 6″ under to 6″ over the nominal length of the bundle. No piece is shorter than 9″. Quantity with length under 4′ is held to stated percentage of total footage.

Unfinished oak flooring (Red and white separated)	Beech, birch, and hard maple	Pecan flooring	Prefinished oak flooring (Red and white separated-graded after finishing)
CLEAR (Plain or quarter sawn)** Best appearance. Best grade, most uniform color, limited small character marks. Bundles 1 1/4′ and up. Average length 3 3/4′.	**FIRST GRADE WHITE HARD MAPLE** (Spec. Order) Same as FIRST GRADE except face all bright sapwood. **FIRST GRADE RED BEECH and BIRCH** (Spec. Order) Same as first grade except face all red heartwood. **FIRST GRADE** Best appearance. Natural color variation.	***FIRST GRADE RED** (Spec. Order) Same as FIRST GRADE except face all heartwood. ***FIRST GRADE WHITE** (Spec. Order) Same as FIRST GRADE except face all bright sapwood. **FIRST GRADE** Excellent appearance. Natural color variation, limited character marks, unlimited sap bdles 2′ and up. 2 and 3′ bdles. up to 25% footage.	***PRIME** (Special Order Only) Excellent appearance. Natural color variation, limited character marks, unlimited sap. Bundles 1 1/4′ and up. Average length 3 1/2′.
ELECT and BETTER (Special Order) A combination of clear and select grades **SELECT** (Plain or quarter sawn)** Excellent appearance. Limited character marks, unlimited sound sap. Bundles 1 1/4′ and up. Average length 3 1/4′.	**SECOND and BETTER GRADE** Excellent appearance. A combination of FIRST and SECOND GRADES Bdles. 2′. and up. 2 and 3′. Bdles up to 40% footage. (NOTE 5% 1 1/4 bdles. allowed in SECOND and BETTER jointed flg. only. **SECOND GRADE** Variegated appearance. Varying sound wood characteristics of species. Bdles. 2′ and up 2 and 3′ bdles. up to 45% footage	***SECOND GRADE RED** (Special Order Only) Same as SECOND GRADE except face all heartwood. **SECOND GRADE** Variegated appearance. Varying sound wood characteristics of species. Bundles 1 1/4′ and up. 1 1/4′ to 3′ bundles as produced up to 40% footage.	**STANDARD and BETTER GRADE** Combination of STANDARD and PRIME. Bundles 1 1/4′ and up. Average length 3 1/2′. **STANDARD GRADE** Variegated appearance. Varying sound wood characteristic of species. A sound floor. Bundles 1 1/4′ and up. Average length 2 3/4′.
NO. 1 COMMON (Red and white may be mixed) Variegated appearance. Light and dark colors; knots, flags, worm holes, and other character marks allowed to provide a variegated appearance, after imperfections are filled and finished. Bundles 1 1/4′ and up. Average length 2 3/4′.	**THIRD and BETTER GRADE** A combination of FIRST, SECOND, and THIRD GRADES. Bundles 1 1/4′ and up. 1 1/4′ and 3′ bundles as produced up to 50% footage. **THIRD GRADE** Rustic appearance All wood characteristics of species. Serviceable, economical floor after filling. Bundles 1 1/4′ and up. 1 1/4′ to 3′ bundles as produced up to 60% footage.	**THIRD GRADE** Rustic appearance All wood characteristics of species. Serviceable, economical floor after filling. Bundles 1 1/4′ and up. 1 1/4′ to 3′ bundles as produced up to 60% footage.	***TAVERN and BETTER GRADE** (Special Order Only) Combination of PRIME, STANDARD, and TAVERN. All wood characteristics of species. Bundles 1 1/4′ and up. Average length 3′. **TAVERN GRADE** Rustic appearance. All wood characteristics of species. Bundles 1 1/4′ and up. Average length 2 1/4′.
NO. 2 COMMON (Red and white may be mixed) Rustic appearance. All wood characteristics of species. A serviceable, economical floor after knot holes, worm holes, checks, and other imperfections are filled and finished. Bundles 1 1/4′ and up. Average length 2 1/4′.			

* 1 1/4′. SHORTS (Red and white may be mixed)
Unique variegated appearance. Lengths 9″–18″. Bundles average nominal 1 1/4′. Production limited.

* NO. 1 COMMON and BETTER SHORTS
A combination grade. CLEAR, SELECT, and NO. 1 COMMON, 9″–18″.

* NO. 2 COMMON SHORTS
Same as NO. 2 COMMON, except length 9″–18″.

** Quarter sawn—Special order only.

* Check with supplier for grade and species available.

NESTED FLOORING: Random-length tongue and groove, end-matched flooring is bundled end to end continuously to form 8 long (nominal) bundles. Regular grade requirements apply.

NESTED FLOORING: If put up in 8′ nested bundles, 9″–18″ pieces will be admitted in 3/4″ × 2 1/4″ as follows in the species of beech, birch, and hard maple: FIRST GRADE, 4 pcs. per bundle; SECOND GRADE, 8 pcs.; THIRD GRADE, as develops. Average lengths: FIRST GRADE, 42″ SECOND GRADE, 33″; THIRD GRADE, 30″.

PREFINISHED BEECH and PECAN FLOORING

* TAVERN and BETTER GRADE (Special order only)

Combination of PRIME, STANDARD, and TAVERN. All wood characteristics of species. Bundles 1 1/4′ and up. Average length 3′.

Common Nail Applications

Joining	Size and type	Placement
Wall framing		
Top plate	8d common	
	16d common	
Header	8d common	
	16d common	
Header to joist	16d common	
Studs	8d common	
	16d common	
Wall sheathing		
Boards	8d common	6″ OC
Plywood (5/16″, 3/8″, 1/2″)	6d common	6″ OC
Plywood (5/8″, 3/4″)	8d common	6″ OC
Fiberboard	1 3/4″ galvanized roofing nail	6″ OC
	8d galvanized common nail	6″ OC
Foamboard	Cap nail, length sufficient for penetration of 1/2″ into framing	12″ OC
Gypsum	1 3/4″ galvanized roofing nail	6″ OC
	8d galvanized common nail	6″ OC
Subflooring	8d common	10″-12″ OC
Underlayment	(1 1/4″ × 14 ga. annular underlayment nail)	6″ OC edges and 12″ OC face
Roof framing		
Rafters, beveled or notched	12d common	
Rafter to joist	16d common	
Joist to rafter and stud	10d common	
Ridge beam	8d & 16d common	
Roof sheathing		
Boards	8d common	
Plywood (5/16″, 3/8″, 1/2″)	6d common	12″ OC and 6″ OC edges
Plywood (5/8″, 3/4″)	8d common	12″ OC and 6″ OC edges
Roofing, asphalt		
New construction shingles and felt	7/8″ through 1 1/2″ galvanized roofing	4 per shingle
Re-roofing application shingles and felt	1 3/4″ or 2″ galvanized roofing	4 per shingle
Roof deck/insulation	Thickness of insulation plus 1″ insulation roof deck nail	
Roofing, wood shingles		
New construction	3d –4d galvanized shingle	2-3 per shingle
Re-roofing application	5d –6d galvanized shingle	2-3 per shingle
Soffit	6d –8d galvanized common	12″ OC max.
Siding*		
Bevel and lap		Consult siding manufacturer's application instructions
Drop and shiplap	Aluminum nails are recommended for optimum performance	
Plywood		
Hardboard	Galvanized hardboard siding nail	Consult siding manufacturer's application instructions
	Galvanized box nail	
Doors, windows, moldings, furring		
Wood strip to masonry	Nail length is determined by thickness of siding and sheathing.	
Wood strip to stud or joist	Nails should penetrate at least 1 1/2″ into solid wood framing.	
Paneling		
Wood	4d–8d casing-finishing	24″ OC
Hardboard	2″ × 16 gage annular	8″ OC
Plywood	3d casing-finishing	8″ OC
Gypsum	1 1/4″ annular drywall	6″ OC
Lathing	4d common blued	4″ OC
Exterior projects		
Decks, patios, etc.	8d–16d hot dipped galvanized common	

*Aluminum nails are recommended for maximum protection from staining
NOTE: Usage may vary somewhat due to regional differences and preferences.

Georgia-Pacific

Drywall Screws

Description	No.	Length	Applications
Bugle Phillips	1E 2E 3E 4E 5R 6R 7R 8R	6×1 6×1 1/8 6×1 1/4 6×1 5/8 6×2 6×2 1/4 6×2 1/2 8×3	For attaching drywall to metal studs from 25 gauge through 20 gauge
Coarse Thread	1C 2C 3C 4C 5C 6C	6×1 6×1 1/8 6×1 1/4 6×1 5/8 6×2 6×2 1/4	For attaching drywall to 25 gauge metal studs, and attaching drywall to wood studs
Pan Framing	19	6×7/16	For attaching stud to track up to 20 gauge
HWH Framing	21 22 35	6×7/16 8×9/16 10×3/4	For attaching stud to track up to 20 gauge where hex head is desired
K-Lath	28	8×9/16	For attaching wire lath, K-lath to 20 gauge studs
Laminating	8	10×1 1/2	Type G laminating screw for attaching gypsum to gypsum, a temporary fastener
Trim Head	9 10	6×1 5/8 6×2 1/4	Trim head screw for attaching wood trim and base to 25 gauge studs

Standard Thickness of Insulating Glass

Unit Construction							
Single glass thickness		Overall unit thickness		Air space		Approximate weight	
Inches	mm	Inches	mm	Inches	mm	lb/ft²	kg/m²
1/8	3	1/2	12	1/4	6	3/27	16
		3/4	19	1/2	12		
3/16	5	5/8	15	1/4	6	4.90	24
		7/8	22	1/2	12		
		1	25	5/8	15		
1/4	6	3/4	19	1/4	6	6.54	32
		1	25	1/2	12		

Libby-Owens-Ford Co., Glass Div.

Welded Wire Fabric Reinforcement

Recommended styles of welded wire fabric reinforcement for concrete		
Type of construction	Recommended style	Remarks
Barbeque foundation slabs	6×6-W2.0'W2.0 to 4×4-W2.9×W2.9	Use heavier style fabric for heavy, massive fireplaces or barbeque pits.
Basement floors	6×6-W1.4'W1.4, 6×6-W2.0×W2.0, or 6×6-W2.9×W2.9	For small areas (15' maximum side dimension), use 6×6-W1.4×W1.4. As a rule of thumb, the larger the area or the poorer the subsoil, the heavier the gauge.
Driveways	6×6-W2.9'W2.9	Continuous reinforcement between 25' to 30' contraction joints.
Residential foundation slabs	6×6-W1.4×W1.4	Use heavier gauge over poorly drained subsoil or when maximum dimension is greater than 15'.
Garage floors	6×6-W2.9×W2.9	Position at midpoint of 5" or 6" thick slab.
Patios and terraces	6×6-W1.4×W1.4	Use 6×6-W2.0×W2.0 if subsoil is poorly drained.
Porch floors A) 6" thick slab up to 6' span B) 6" thick slab up to 8' span	6×6-W2.9×W2.9 4×4-W4.0×W4.0	Position 1" from bottom form to resist tensile stresses.
Sidewalks	6×6-W1.4×W1.4 or 6×6-W2.0×W2.0	Use heavier gauge over poorly drained subsoil. Construct 25' to 30' slabs as for driveways.
Steps (free span)	6×6-W2.9×W2.9	Use heavier style if more than five risers. Position fabric 1" from bottom form.
Steps (on ground)	6×6-W2.0×W2.0	Use 6×6-W2.9×W2.9 for unstable subsoil.

Gypsum Wallboard Application

Thickness	Approximate weight lbs square foot	Size	Location	Application method	Maximum spacing of framing members
1/4"	1.1	4' × 8' to 12'	Over existing walls and ceilings	Horizontal or vertical	
3/8"	1.5	4' × 8' to 14'	Ceilings	Horizontal	16"
3/8"	1.5	4' × 8' to 14'	Sidewalls	Horizontal or vertical	16"
1/2"	2.0	4' × 8' to 14'	Ceilings	Vertical Horizontal	16" 24"
1/2"	2.0	4' × 8' to 14'	Sidewalls	Horizontal or vertical	24"
5/8"	2.5	4' × 8' to 14'	Ceilings	Vertical Horizontal	16" 24"
5/8"	2.5	4' × 8' to 14'	Sidewalls	Horizontal or vertical	24"
1"	4.0	2' × 8' to 12'		For laminated partitions	

Floor Joist Span Data

30 PSF Live Load, 10 PSF Dead Load, Deflection <360										
Species or Group	Grade	2×8			2×10			2×12		
		12" OC	16" OC	24" OC	12" OC	16" OC	24" OC	12" OC	16" OC	24" OC
Douglas Fir and Larch	Sel. Struc.	16-6	15-0	13-1	21-0	19-1	16-8	25-7	23-3	20-3
	No. 1 & Btr.	16-2	14-8	12-10	20-8	18-9	16-1	25-1	22-10	18-8
	No. 1	15-10	14-5	12-4	20-3	18-5	15-0	24-8	21-4	17-5
	No. 2	15-7	14-1	11-6	19-10	17-2	14-1	23-0	19-11	16-3
	No. 3	12-4	10-8	8-8	15-0	13-0	10-7	17-5	15-1	12-4

40 PSF Live Load, 10 PSF Dead Load, Deflection <360										
Species or Group	Grade	2×8			2×10			2×12		
		12" OC	16" OC	24" OC	12" OC	16" OC	24" OC	12" OC	16" OC	24" OC
Douglas Fir and Larch	Sel. Struc.	15-0	13-7	11-11	19-1	17-4	15-2	23-3	21-1	18-5
	No. 1 & Btr.	14-8	13-4	11-8	18-9	17-0	14-5	22-10	20-5	16-8
	No. 1	14-5	13-1	11-0	18-5	16-5	13-5	22-0	19-1	15-7
	No. 2	14-2	12-7	10-3	17-9	15-5	12-7	20-7	17-10	14-7
	No. 3	11-0	9-6	7-9	13-5	11-8	9-6	15-7	13-6	11-0

30 PSF Live Load, 10 PSF Dead Load, Deflection <360										
Species or Group	Grade	2×8			2×10			2×12		
		12" OC	16" OC	24" OC	12" OC	16" OC	24" OC	12" OC	16" OC	24" OC
Southern Pine	Sel. Struc.	16-2	14-8	12-10	20-8	18-9	16-5	25-1	22-10	19-11
	No. 1	15-10	14-5	12-7	20-3	18-5	16-1	24-8	22-5	19-6
	No. 2	15-7	14-2	12-4	19-10	18-0	14-8	24-2	21-1	17-2
	No. 3	13-3	11-6	9-5	15-8	13-7	11-1	18-8	16-2	13-2

40 PSF Live Load, 10 PSF Dead Load, Deflection <360										
Species or Group	Grade	2×8			2×10			2×12		
		12" OC	16" OC	24" OC	12" OC	16" OC	24" OC	12" OC	16" OC	24" OC
Southern Pine	Sel. Struc.	14-8	13-4	11-8	18-9	17-0	14-11	22-10	20-9	18-1
	No. 1	14-5	13-1	11-5	18-5	16-9	14-7	22-5	20-4	17-5
	No. 2	14-2	12-10	11-0	18-0	16-1	13-2	21-9	18-10	15-4
	No. 3	11-11	10-3	8-5	14-0	12-2	9-11	16-8	14-5	11-10

40 PSF Live Load, 10 PSF Dead Load, Deflection <360										
Species or Group	Grade	2×8			2×10			2×12		
		12" OC	16" OC	24" OC	12" OC	16" OC	24" OC	12" OC	16" OC	24" OC
Redwood	Cl. All Heart		7-3	6-0		10-9	8-9		13-6	11-0
	Const. Heart		7-3	6-0		10-9	8-9		13-6	11-0
	Const. Common		7-3	6-0		10-9	8-9		13-6	11-0

Span data is in feet and inches for floor joists of Douglas fir/larch, southern yellow pine, and California redwood. Spans are calculated on the basis of dry sizes with moisture content equal to or less than 19%. Floor joist spans are for a single span.

Ceiling Joist Span Data

Species or Group	Grade	20 PSF Live Load, 10 PSF Dead Load, Deflection <240											
		Drywall ceiling, no future room development, limited attic storage available											
		2×4			2×6			2×8			2×10		
		12" OC	16" OC	24" OC	12" OC	16" OC	24" OC	12" OC	16" OC	24" OC	12" OC	16" OC	24" OC
Douglas Fir and Larch	Sel. Struc.	10-5	9-6	8-3	16-4	14-11	13-0	21-7	19-7	17-1	27-6	25-0	20-11
	No. 1 & Btr.	10-3	9-4	8-1	16-1	14-7	12-0	21-2	18-8	15-3	26-4	22-9	18-7
	No. 1	10-0	9-1	7-8	15-9	13-9	11-2	20-1	17-5	14-2	24-6	21-3	17-4
	No. 2	9-10	8-9	7-2	14-10	12-10	10-6	18-9	16-3	13-3	22-11	19-10	16-3
	No. 3	7-8	6-8	5-5	11-2	9-8	7-11	14-2	12-4	10-0	17-4	15-0	12-3

Species or Group	Grade	20 PSF Live Load, 10 PSF Dead Load, Deflection <240											
		Drywall ceiling, no future room development, limited attic storage available											
		2×4			2×6			2×8			2×10		
		12" OC	16" OC	24" OC	12" OC	16" OC	24" OC	12" OC	16" OC	24" OC	12" OC	16" OC	24" OC
Southern Pine	Sel. Struc.	10-3	9-1	8-1	16-1	14-7	12-9	21-2	19-3	16-10	26-0	24-7	21-6
	No. 1	10-10	9-4	8-0	15-9	14-4	12-6	20-10	18-11	15-11	26-0	23-2	18-11
	No. 2	9-10	8-11	7-8	15-6	13-6	11-0	20-1	17-5	14-2	24-0	20-9	17-0
	No. 3	8-2	7-1	5-9	12-1	10-5	8-6	15-4	13	10-10	18-1	15-8	12-10

Roof Rafter Span Data

Species or Group	Grade	20 PSF Live Load, 10 PSF Dead Load, Deflection <240											
		Roof slope 3:12 or less, light roof covering, no ceiling finish											
		2×6			2×8			2×10			2×12		
		12" OC	16" OC	24" OC	12" OC	16" OC	24" OC	12" OC	16" OC	24" OC	12" OC	16" OC	24" OC
Douglas Fir and Larch	Sel. Struc.	16-4	14-11	13-0	21-7	19-7	17-2	27-6	25-0	21-10	33-6	30-5	26-1
	No. 1 & Btr.	16-1	14-7	12-9	21-2	19-3	16-10	27-1	24-7	20-9	32-11	29-6	24-1
	No. 1	15-9	14-4	12-6	20-10	18-11	15-10	26-6	23-9	19-5	31-10	27-6	22-6
	No. 2	15-6	14-1	11-9	20-5	18-2	14-10	25-8	22-3	18-2	29-9	25-9	21-0
	No. 3	12-6	10-10	8-10	15-10	13-9	11-3	19-5	16-9	13-8	22-6	19-6	15-11

Species or Group	Grade	20 PSF Live Load, 15 PSF Dead Load, Deflection <240											
		Roof slope 3:12 or less, Light roof covering, No ceiling finish											
		2×6			2×8			2×10			2×12		
		12" OC	16" OC	24" OC	12" OC	16" OC	24" OC	12" OC	16" OC	24" OC	12" OC	16" OC	24" OC
Douglas Fir and Larch	Sel. Struc.	16-4	14-11	13-0	21-7	19-7	17-2	27-6	25-0	21-7	33-6	30-5	25-1
	No. 1 & Btr.	16-1	14-7	12-5	21-2	19-3	15-9	27-1	23-7	19-3	31-7	27-4	22-4
	No. 1	15-9	14-3	11-7	20-9	18-0	14-8	25-5	22-0	17-11	29-5	25-6	20-10
	No. 2	15-4	13-3	10-10	19-5	16-10	13-9	23-9	20-7	16-9	27-6	23-10	19-6
	No. 3	11-7	10-1	8-2	14-8	12-9	10-5	17-11	15-7	12-8	20-10	18-0	14-9

Roof Rafter Span Data *(Continued)*

Species or Group	Grade	20 PSF Live Load, 10 PSF Dead Load, Deflection <240											
		Drywall ceiling, light roofing, snow load											
		2×6			2×8			2×10			2×12		
		12″ OC	16″ OC	24″ OC	12″ OC	16″ OC	24″ OC	12″ OC	16″ OC	24″ OC	12″ OC	16″ OC	24″ OC
Southern Pine	Sel. Struc.	16-1	14-7	12-9	21-2	19-3	16-10	26-0	24-7	21-6	26-0	26-0	26-0
	No. 1	15-9	14-4	12-6	20-10	18-11	16-6	26-0	24-1	20-3	26-0	26-0	24-1
	No. 2	15-6	14-1	11-9	20-5	18-6	15-3	25-8	22-3	18-2	26-0	26-0	21-4
	No. 3	12-11	11-2	9-1	16-5	14-3	11-7	19-5	16-10	13-9	23-1	20-0	16-4

Species or Group	Grade	30 PSF Live Load, 15 PSF Dead Load, Deflection <240											
		Drywall ceiling, medium roofing, snow load											
		2×6			2×8			2×10			2×12		
		12″ OC	16″ OC	24″ OC	12″ OC	16″ OC	24″ OC	12″ OC	16″ OC	24″ OC	12″ OC	16″ OC	24″ OC
Southern Pine	Sel. Struc.	14-1	12-9	12-9	18-6	16-10	14-8	23-8	21-6	18-9	26-0	26-0	22-10
	No. 1	13-9	12-6	12-6	18-2	16-6	13-11	23-2	20-3	16-6	26-0	24-1	19-8
	No. 2	13-6	11-9	11-9	17-7	15-3	12-5	21-0	18-2	14-10	24-7	21-4	17-5
	No. 3	10-6	9-1	9-1	13-5	11-7	9-6	15-10	13-9	11-3	18-10	16-4	13-4

Roof Decking Span Data

With a maximum deflection of 1/240th of the span; live load = 20 psf

Thickness in Inches (Nominal)	Lumber Grade	Simple Spans	
		Douglas Fir, Larch, Southern Yellow Pine	Western Red Cedar
		Span	Span
2	Construction	9′-5″	8′-1″
2	Standard	9′-5″	6′-9″
3	Select dex.	15′-3″	13′-0″
3	Compl. dex.	15′-3″	13′-0″
4	Select dex.	20′-3″	17′-3″
4	Compl. dex.	20′-3″	17′-3″

Thickness in Inches (Nominal)	Lumber Grade	Random Lengths	
		Douglas Fir, Larch, Southern Yellow Pine	Western Red Cedar
		Span	Span
2	Construction	10′-3″	8′-10″
2	Standard	10′-3″	6′-9″
3	Select dex.	16′-9″	14′-3″
3	Compl. dex.	16′-9″	13′-6″
4	Select dex.	22′-0″	19′-0″
4	Compl. dex.	22′-0″	18′-0″

Thickness in Inches (Nominal)	Lumber Grade	Comb, Simple, and Two-Span Continuous	
		Douglas Fir, Larch, Southern Yellow Pine	Western Red Cedar
		Span	Span
2	Construction	10′-7″	8′-9″
2	Standard	10′-7″	6′-9″
3	Select dex.	17′-3″	14′-9″
3	Compl. dex.	17′-3″	13′-6″
4	Select dex.	22′-9″	19′-6
4	Compl. dex.	22′-9″	18′-0″

Manufactured 2″ × 4″ Wood Floor Trusses

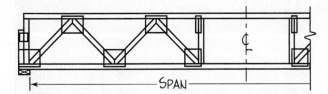

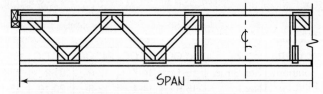

Bottom Chord Bearing Type			
Depth	Clear spans	#Diagonal webs	Camber
12″	7′-2″	4	0.063″
	9′-8″	6	0.063″
	12′-2″	8	0.063″
	14′-8″	10	0.134″
	17′-2″	12	0.237″
	19′-8″	14	0.365″
	21′-4″	16	0.507″
14″	9′-8″	6	0.063″
	12′-2″	8	0.063″
	14′-8″	10	0.095″
	17′-2″	12	0.178″
	19′-8″	14	0.288″
	22′-7″	16	0.449″
	24′-0″	18	0.569″
16″	12′-2″	8	0.065″
	14′-8″	10	0.070″
	17′-2″	12	0.132″
	19′-8″	14	0.228″
	22′-2″	16	0.346″
	25′-1″	18	0.505″
	26′-1″	20	0.596″
18″	14′-8″	10	0.065″
	17′-2″	12	0.120″
	19′-8″	14	0.176″
	22′-2″	16	0.268″
	24′-8″	18	0.367″
	27′-6″	20	0.600″
	27′-10″	22	0.630″
20″	14′-8″	10	0.063″
	17′-2″	12	0.081″
	19′-8″	14	0.140″
	22′-2″	16	0.226″
	24′-8″	18	0.327″
	27′-6″	20	0.451″
	29′-6″	22	0.630″
22″	17′-2″	10	0.066″
	19′-8″	12	0.114″
	22′-2″	14	0.184″
	24′-8″	16	0.266″
	27′-6″	18	0.367″
	30′-0″	20	0.520″
	31′-1″	22	0.630″
24″	17′-2″	12	0.063″
	19′-8″	14	0.095″
	22′-2″	16	0.153″
	24′-8″	18	0.235″
	27′-2″	20	0.325″
	30′-0″	22	0.431″
	32′-6″	24	0.630″

Top Chord Bearing Type			
Depth	Clear spans	#Diagonal webs	Camber
12″	6′-10″	4	0.063″
	9′-4″	6	0.063″
	11′-10″	8	0.063″
	14′-4″	10	0.122″
	16′-10″	12	0.233″
	19′-10″	14	0.376″
	21′-4″	16	0.507″
14″	9′-5″	6	0.063″
	11′-11″	8	0.063″
	14′-5″	10	0.088″
	16′-11″	12	0.167″
	19′-5″	14	0.273″
	21′-4″	16	0.429″
	24′-0″	18	0.569″
16″	12′-0″	8	0.063″
	14′-6″	10	0.067″
	17′-0″	12	0.126″
	19′-6″	14	0.219″
	22′-4″	16	0.337″
	24′-10″	18	0.489″
	26′-1″	20	0.596″
18″	14′-6″	10	0.063″
	17′-0″	12	0.098″
	19′-6″	14	0.170″
	22′-0″	16	0.260″
	24′-10″	18	0.378″
	27′-8″	20	0.617″
	27′-10″	22	0.630″
20″	14′-6″	10	0.063″
	17′-0″	12	0.079″
	19′-6″	14	0.136″
	22′-0″	16	0.221″
	24′-10″	18	0.337″
	27′-4″	20	0.442″
	29′-6″	22	0.630″
22″	17′-1″	12	0.065″
	19′-7″	14	0.112″
	22′-1″	16	0.181″
	24′-10″	18	0.275″
	27′-4″	20	0.381″
	30′-2″	22	0.534″
	31′-1″	24	0.630″
24″	17′-1″	12	0.063″
	19′-7″	14	0.093″
	22′-1″	16	0.150″
	24′-7″	18	0.231″
	27′-5″	20	0.335″
	30′-2″	22	0.443″
	32′-6″	24	0.630″

Wood floor trusses are typically manufactured from #3 southern yellow pine. Pieces are joined together with 18 and 20 gauge galvanized steel plates applied to both faces of the truss at each joint. Where no sheathing is applied directly to top chords, they should be braced at intervals not to exceed 3′-0″. Where no rigid ceiling is applied directly to bottom chords, they should be braced at intervals not to exceed 10′-0″.

Manufactured wood floor trusses are generally spaced 24″ OC and are designed to support various loads. Typical trusses shown here were designed to support 55 psf (live load - 40 psf, dead load - 10 psf, ceiling dead load - 5 psf). A slight bow (camber) is built into each joist so that it will produce a level floor when loaded. Allowable deflection is 1/360 of the span.

Some of the longer trusses require one or more double diagonal webs at both ends. Wood floor trusses are a manufactured product, which must be engineered and produced with a high degree of accuracy to attain the desired performance. See your local manufacturer or lumber company for trusses available in your area.

Courtesy of Trus Joist

Length of Common Rafters

Feet of run	2-in-12 Inclination (set saw at) 9°28' 12.17" per foot of run	2 1/2-in-12 Inclination (set saw at) 11°46' 12.26" per foot of run	3-in-12 Inclination (set saw at) 14°2' 12.37" per foot of run	3 1/2-in-12 Inclination (set saw at) 16°16' 12.5" per foot of run	4-in-12 Inclination (set saw at) 18°26' 12.65" per foot of run	4 1/2-in-12 Inclination (set saw at) 20°33' 12.82" per foot of run
4	4'-0 11/16"	4'-1 1/32"	4'-1 15/32"	4'-2"	4'-2 19/32"	4'-3 9/32"
5	5'-0 27/32"	5'-1 5/16"	5'-1 27/32"	5'-2 1/2"	5'-3 1/4"	5'-4 3/32"
6	6'-1 1/32"	6'-1 9/16"	6'-2 7/32"	6'-3"	6'-3 29/32"	6'-4 15/16"
7	7'-1 3/16"	7'-1 13/16"	7'-2 19/32"	7'-3 1/2"	7'-4 9/16"	7'-5 3/4"
8	8'-1 3/8"	8'-2 3/32"	8'-2 31/32"	8'-4"	8'-5 3/32"	8'-6 9/16"
9	9'-1 17/32"	9'-2 7/32"	9'-3 11/32"	9'-4 1/2"	9'-5 27/32"	9'-7 3/8"
10	10'-1 23/32"	10'-2 19/32"	10'-3 23/32"	10'-5"	10'-6 1/2"	10'-8 7/32"
11	11'-1 7/8"	11'-2 7/8"	11'-4 1/16"	11'-5 1/2"	11'-7 5/32"	11'-9 1/32"
12	12'-2 1/32"	12'-3 1/8"	12'-4 7/16"	12'-6"	12'-7 13/16"	12'-9 27/32"
13	13'-2 7/32"	13'-3 3/8"	13'-4 13/16"	13'-6 1/2"	13'-8 15/32"	13'-10 21/32"
14	14'-2 3/8"	14'-3 21/32"	14'-5 3/16"	14'-7"	14'-9 3/32"	14'-11 1/2"
15	15'-2 9/16"	15'-3 29/32"	15'-5 9/16"	15'-7 1/2"	15'-9 3/4"	16'-0 5/16"
16	16'-2 23/32"	16'-4 5/32"	16'-5 15/16"	16'-8"	16'-10 13/32"	17'-1 1/8"
Inches of run	**These lengths are to be added to those shown above when run involves inches**					
1/4	1/4"	1/4"	1/4"	1/4"	1/4"	9/32"
1/2	1/2"	1/2"	1/2"	17/32"	17/32"	17/32"
1	1"	1"	1 1/32"	1 1/32"	1 1/16"	1 1/16"
2	2 1/32"	2 1/32"	2 1/16"	2 3/32"	2 3/32"	2 1/8"
3	3 1/16"	3 1/16"	3 3/32"	3 1/8"	3 5/32"	3 7/32"
4	4 1/16"	4 3/32"	4 1/8"	4 5/32"	4 7/32"	4 9/32"
5	5 1/16"	5 1/8"	5 5/32"	5 7/32"	5 9/32"	5 11/32"
6	6 3/32"	6 1/8"	6 3/16"	6 1/4"	6 5/16"	6 13/32"
7	7 3/32"	7 5/32"	7 7/32"	7 9/32"	7 3/8"	7 15/32"
8	8 1/8"	8 3/16"	8 1/4"	8 11/32"	8 7/16"	8 17/32"
9	9 1/8"	9 3/16"	9 1/4"	9 3/8"	9 15/32"	9 5/8"
10	10 5/32"	10 7/32"	10 5/16"	10 7/16"	10 17/32"	10 11/16"
11	11 5/32"	11 1/4"	11 11/32"	11 15/32"	11 19/32"	11 3/4"

Feet of run	5-in-12 Inclination (set saw at) 22°37' 13.00" per foot of run	5 1/2-in-12 Inclination (set saw at) 24°37' 13.20" per foot of run	6-in-12 Inclination (set saw at) 26°34' 13.42" per foot of run	6 1/2-in-12 Inclination (set saw at) 28°27' 13.65" per foot of run	7-in-12 Inclination (set saw at) 30°15' 13.89" per foot of run	7 1/2-in-12 Inclination (set saw at) 32°0' 14.15" per foot of run	8-in-12 Inclination (set saw at) 33°41' 14.42" per foot of run
4	4'-4"	4'-4 13/16"	4'-5 11/16"	4'-6 19/32"	4'-7 9/16"	4'-8 19/32"	4'-9 11/16"
5	5'-5"	5'-6"	5'-7 1/8"	5'-8 1/4"	5'-9 15/32"	5'-10 3/4"	6'-0 1/8"
6	6'-6"	6'-7 3/32"	6'-8 1/2"	6'-9 29/32"	6'-11 11/32"	7'-0 29/32"	7'-2 17/32"
7	7'-7"	7'-8 13/32"	7'-9 15/16"	7'-11 9/16"	8'-1 7/32"	8'-3 1/16"	8'-4 15/16"
8	8'-8"	8'-9 19/32"	8'-11 3/8"	9'-1 7/32"	9'-3 1/8"	9'-5 7/32"	9'-7 3/8"
9	9'-9"	9'-10 13/16"	10'-0 25/32"	10'-2 27/32"	10'-5"	10'-7 11/32"	10'-9 25/32"
10	10'-10"	11'-0"	11'-2 7/32"	11'-4 1/2"	11'-6 29/32"	11'-9 1/2"	12'-0 7/32"
11	11'-11"	12'-1 3/32"	12'-3 5/8"	12'-6 5/32"	12'-8 13/16"	12'-11 21/32"	13'-2 5/8"
12	13'-0"	13'-2 13/32"	13'-5 1/32"	13'-7 13/16"	13'-10 11/16"	14'-1 13/16"	14'-5 1/32"
13	14'-1"	14'-3 19/32"	14'-6 15/32"	14'-9 15/32"	15'-0 9/32"	15'-3 31/32"	15'-7 15/32"
14	15'-2"	15'-4 13/16"	15'-7 7/8"	15'-11 1/8"	16'-2 15/32"	16'-6 1/8"	16'-9 7/8"
15	16'-3"	16'-6"	16'-9 5/16"	16'-0 3/4"	17'-4 11/32"	17'-8 1/4"	18'-0 5/16"
16	17'-4"	17'-7 7/32"	17'-10 23/32"	18'-2 13/32"	18'-6 1/4"	18'-10 13/32"	19'-2 23/32"
Inches of run	**These lengths are to be added to those shown above when run involves inches**						
1/4	9/32"	9/32"	9/32"	9/32"	5/16"	5/16"	
1/2	17/32"	9/16"	9/16"	9/16"	1 9/32"	5/8"	
1	1 3/32"	1 3/32"	1 1/8"	1 1/8"	1 3/16"	1 7/32"	
2	2 5/32"	2 7/32"	2 1/4"	2 9/32"	2 11/32"	2 13/32"	
3	3 1/4"	3 5/16"	3 3/8"	3 13/32"	3 17/32"	3 19/32"	
4	4 11/32"	4 13/32"	4 15/32"	4 9/16"	4 23/32"	4 13/16"	
5	5 13/32"	5 1/2"	5 19/32"	5 11/16"	5 29/32"	6"	
6	6 1/2"	6 19/32"	6 23/32"	6 13/16"	7 1/32"	7 7/32"	
7	7 9/16"	7 23/32"	7 13/16"	7 31/32"	8 1/4"	8 13/32"	
8	8 21/32"	8 13/16"	8 15/16"	9 1/8"	9 7/16"	9 5/8"	
9	9 3/4"	9 29/32"	10 5/32"	10 1/4"	10 19/32"	10 13/16"	
10	10 27/32"	11"	11 3/16"	11 3/8"	11 25/32"	1'0"	
11	11 29/32"	12 3/32"	1'-0 5/16"	1'-0 1/2"	1'-0 31/32"	1'-1 7/32"	

Sizes for Performance-Rated I-Joists

Sizes of Performance Rated I-Joists (APA PRI™)				
Joist series	Letter/number designation	Depth (nominal)	Net depth	Width of flang
1×10	C4 (PRI-15)	10″	9 1/2″	1 1/2″
1×10	C6 (PRI-25)	10″	9 1/2″	1 3/4″
1×12	C10 (PRI-15)	12″	11 7/8″	1 1/2″
1×12	C12 (PRI-25)	12″	11 7/8″	1 3/4″
1×14	C14 (PRI-25)	14″	14″	1 3/4″
1×14	C16 (PRI-35)	14″	14″	2 5/16″
1×16	C18 (PRI-25)	16″	16″	1 3/4″
1×16	C20 (PRI-35)	16″	16″	2 5/16″
1×10	S2 (PRI-30)	10″	9 1/2″	2 1/2″
1×10	S4 (PRI-32)	10″	9 1/2″	2 1/2″
1×12	S6 (PRI-30)	12″	11 7/8″	2 1/2″
1×12	S8 (PRI-32)	12″	11 7/8″	2 1/2″
1×12	S10 (PRI-42)	12″	11 7/8″	3 1/2″
1×14	S12 (PRI-32)	14″	14″	2 1/2″
1×14	S14 (PRI-42)	14″	14″	3 1/2″
1×16	S16 (PRI-32)	16″	16″	2 1/2″
1×16	S18 (PRI-42)	16″	16″	3 1/2″

Spans Allowed for Performance-Rated I-Joists

Spans Allowed for APA Performance-Rated I-Joists with Simple-Span OC Spacing Only				
Series and designation	12″ OC	16″ OC	19.2″ OC	24″ OC
1×10-C4 (PRI-15)*	17′-0″	15′-6″	14′-8″	13′-7″
1×10-C6 (PRI-25)	17′-9″	16′-2″	15′-3″	14′-2″
1×12-C10 (PRI-15)*	20′-3″	18′-5″	17′-5″	16′-7″
1×12-C12 (PRI-25)	21′-1″	19′-3″	18′-2″	17′-3″
1×14-C14 (PRI-25)*	24′-0″	21′-10″	21′-0″	19′-8″
1×14-C16 (PRI-35)	25′-11″	23′-7″	22′-2″	21′-0″
1×16-C18 (PRI-25)	26′-7″	24′-8″	23′-4″	20′-2″
1×16-C20 (PRI-35)*	28′-8″	26′-1″	24′-11″	23′-1″
1×10-S2 (PRI-30)	18′-0″	16′-3″	14′-10″	13′-3″
1×10-S4 (PRI-32)*	19′-0″	17′-4″	16′-4″	15′-4″
1×12-S6 (PRI-30)	21′-6″	18′-10″	17′-2″	15′-4″
1×12-S8 (PRI-32)*	22′-8″	20′-8″	19′-6″	18′-3″
1×12-S10 (PRI-42)	24′-11″	22′-8″	21′-4″	19′-11″
1×14-S12 (PRI-32)*	25′-9″	23′-6″	22′-2″	20′-9″
1×14-S14 (PRI-42)	28′-3″	25′-9″	24′-3″	22′-8″
1×16-S16 (PRI-32)*	28′-7″	26′-1″	24′-7″	23′-0″
1×16-S18 (PRI-42)	31′-4″	28′-6″	26′-11″	25′-1″

Notes:
1. Table based on uniform loads (10 psf dead load; 40 psf live load).
2. Clear span allowed here applies only to simple span residential floors.
3. Span limits are for clear distances between support.
4. Minimum end-bearing length shall be 1 3/4″.
5. *These units are the most commonly available.

Clear Spans for Wood I-Beam Rafters

Recommended Clear Spans for Wood I-Beam Rafters, 1 3/4″ Flange (Based on Horizontal Distances and 16″ OC Spacing)					
	9 1/2″ depth			11 7/8″ depth	
No snow load					
Load (psf) live/dead	Slope less than 4/12	Slope 4/12 to 8/12	Slope more than 8/12	Slope less than 4/12	Slope 4/12 to 8/12	Slope more than 8/12
20/10	22′-6″	20′-6″	17′-11″	26′-11″	24′-4″	21′-4″
20/15	21′-4″	19′-4″	16′-9″	25′-6″	22′-11″	20′-0″
20/20	20′-4″	18′-5″	15′-11″	24′-3″	21′-9″	18′-11″
Snow load					
25/10	21′-4″	19′-4″	17′-4″	25′-7″	23′-1″	20′-8″
25/15	20′-4″	18′-4″	16′-3″	24′-4″	21′-11″	19′-6″
30/10	20′-5″	18′-6″	16′-10″	24′-5″	22′-1″	20′-1″
30/15	19′-6″	17′-8″	15′-11″	23′-4″	21′-1″	19′-0″
40/10	18′-6″	17′-0″	15′-11″	22′-2″	20′-4″	19′-1″
40/15	18′-3″	16′-6″	15′-2″	21′-9″	19′-9″	18′-2″
50/10	17′-1″	15′-8″	13′-3″	20′-5″	18′-9″	18′-2″
50/15	17′-1″	15′-7″	14′-7″	20′-5″	18′-8″	17′-5″

Fink Truss Design

Span	Dimensions			Design Stresses					
L	A	B	C	U_1	U_2	L_1	L_2	V_1	D_1
20'-0"	5'-5"	5'-3 5/8"	2'-7 5/8"	1756	1450	1614	1472	430	430
22'-0"	5'-11 1/2"	5'-10 1/16"	2'-10 13/16"	1932	1595	1775	1619	473	473
24'-0"	6'-6"	6'-4 7/16"	3'-2"	2108	1740	1936	1766	516	516
26'-0"	7'-0 1/2"	6'-10 7/8"	3'-5 1/4"	2282	1885	2097	1913	559	559
28'-0"	7'-7"	7'-5 1/4"	3'-8 7/16"	2459	2030	2258	2060	602	602
30'-0"	8'-1 1/2"	7'-11 11/16"	3'-11 5/8"	2634	2175	2420	2207	645	645
32'-0"	8'-8"	8'-6 1/16"	4'-2 13/16"	2810	2320	2581	2354	688	688

Lumber					
	2" × 6"		2" × 4"		Total
Span	No.	Length	No.	Length	F.B.M.
20'-0"	2	12'-0"	2 / 2	12'-0" / 10'-0"	53
22'-0"	2	14'-0"	2 / 2	14'-0" / 10'-0"	60
24'-0"	2	14'-0"	2 / 2	14'-0" / 12'-0"	63
26'-0"	2	16'-0"	2 / 2	16'-0" / 12'-0"	70
	2	16'-0"	2 / 2	16'-0" / 14'-0"	73
30'-0"	2	18'-0"	2 / 2	18'-0" / 14'-0"	79
32'-0"	2	20'-0"	2 / 2	18'-0" / 14'-0"	83

Hardware		
No.	Item	Size
11	Split rings	2 1/2" Diam.
2	Trip-L-Grip	Type A
1	Bolt	1/2" × 7 1/2"
2	Bolts	1/2" × 6"
4	Bolts	1/2" × 4"
14	Washers	1/2"

LUMBER—Lumber shall be of a good grade of sufficient quality to permit the following allowable unit stresses:
c = 900 psi Compression parallel to grain.
f   = 900 psi Extreme fiber in bending.
E = 1,600,000 psi Modulus of elasticity.

CONNECTORS—Timber connectors shall be 2-1/2" diameter split rings and Trip-L-Grip framing anchors.

BOLTS—Bolts shall be 1/2" diameter machine bolts with 2" × 2" × 1/8" plate washers. 2 1/8" diameter cast or malleable iron washers, or ordinary cut washers.

DIMENSIONS—Dimensions shown will provide approximately 1/2" camber at bottom chord panel points. Utilize full uncut length of bottom chord pieces by increasing the spacing of connectors in the splice.

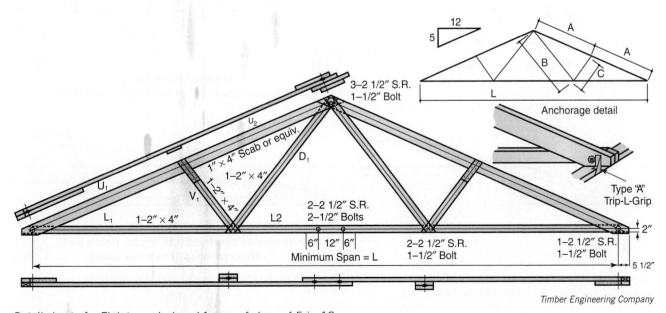

Timber Engineering Company

Detail sheet of a Fink truss designed for a roof slope of 5-in-12.

Roof Trusses

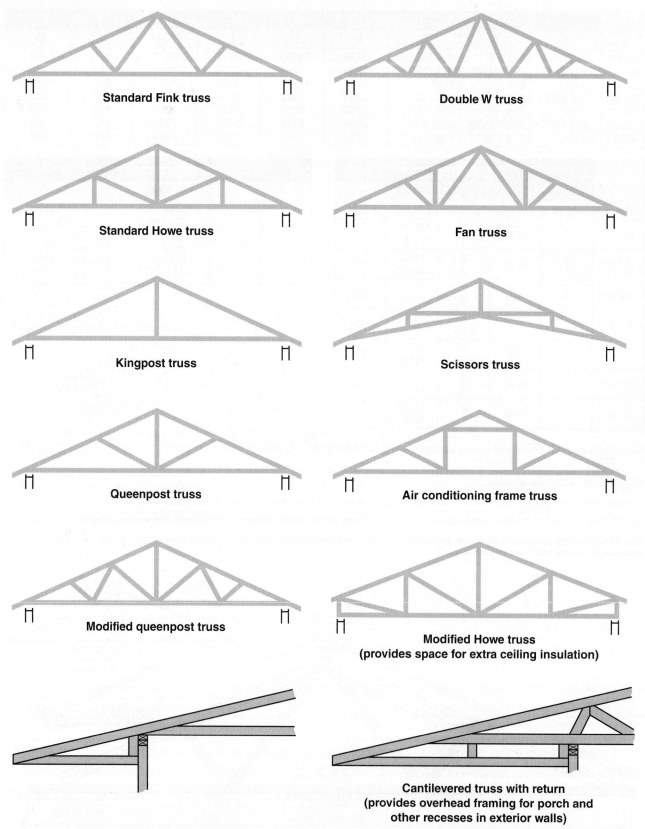

Standard Fink truss

Double W truss

Standard Howe truss

Fan truss

Kingpost truss

Scissors truss

Queenpost truss

Air conditioning frame truss

Modified queenpost truss

**Modified Howe truss
(provides space for extra ceiling insulation)**

**Cantilevered truss with return
(provides overhead framing for porch and
other recesses in exterior walls)**

Types of roof trusses and overhangs. The Fink truss is also called a W truss. Any type of truss should be constructed according to designs developed from engineering data.

Ventilation

Better construction methods and materials used by builders today produce houses that are tighter and more draft-free than those built a half century ago. In such snug shelters, moisture created inside the house often saturates insulation and causes paint to peel off exterior walls.

The use of a vapor barrier on the warm side of walls and ceilings will minimize this condensation, butadequate ventilation is needed to remove themoisture from the house.

FHA requires that attics have a total net free ventilatingarea not less that 1/150 of the square foot area, except that a ratio of 1/300 may be provided if:

(a) a vapor barrier having a transmission rate not exceeding one perm is installed on the warm side of the ceiling, or

(b) at least 50% of the required vent area is provided by ventilators located in the upper portion of the space to be ventilated, with the balance of the required ventilation provided by eave or cornice vents.

There are many types of screened ventilators available, including the handy miniature vents illustrated below, which can be used for problem areas or to supplement larger vents.

Free-Area Ventilation Guide
Square Inches of VentilationRequired for Attic Areas

Width (in feet)	20	22	24	26	28	30	32	34	36	38	40	42
20	192	211	230	250	269	288	307	326	346	365	384	403
22	211	232	253	275	296	317	338	359	380	401	422	444
24	230	253	276	300	323	346	369	392	415	438	461	484
26	250	275	300	324	349	374	399	424	449	474	499	524
28	269	296	323	349	376	403	430	457	484	511	538	564
30	288	317	346	374	403	432	461	490	518	547	576	605
32	307	338	369	399	430	461	492	522	553	584	614	645
34	326	359	392	424	457	490	522	555	588	620	653	685
36	346	380	415	449	484	518	553	588	622	657	691	726
38	365	401	438	474	511	547	584	620	657	693	730	766
40	384	422	461	499	538	576	614	653	691	730	768	806
42	403	444	484	524	564	605	645	685	726	766	806	847
44	422	465	507	549	591	634	676	718	760	803	845	887
46	442	486	530	574	618	662	707	751	795	839	883	927
48	461	507	553	599	645	691	737	783	829	876	922	968
50	480	528	576	624	672	720	768	816	864	912	960	1008

Length (in feet) (vertical axis label)

Using length and width dimensions of each rectangular or square attic space, find one dimension on vertical column, the other dimension on horizontal column. These will intersect at the number of square inches of ventilation required to provide 1/300th.

Installing minature vents

1-Inch
Install in paneling used in basement rooms. Can be painted over to match panel finish.

2-Inch
Used to ventilate stud space. For best results, install in top and bottom of each space.

2 1/2-Inch
Made especially to plug the 2 1/2-in hole cut to blow insulation between studs.

3-Inch
For ventilating rafter space in flat-roof buildings, or other jobs requiring fairly large free area.

4-Inch
For hard-to-reach spots needing large ventilating area. Large enough for venting soffits.

1. Drill or cut a hole the same size as the ventilator.

2. Insert the ventilator. Tension ridges hold ventilator in place...no nails or screws are needed.

3. Tap into place with a hammer, using a wood block to protect the margin. The louvers are recessed so there's no danger of damage during installation.

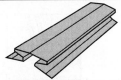
Ridge vent provides 18 sq in of net free area per lineal foot. Installed quickly over a 1 1/2″ gap in the sheathing at the ridge.

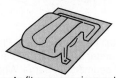
Roof vents fit over openings cut between rafters to pull hot air out of attic.

Triangle vent fits snugly under the roof gable to provide large vent areas at the highest point of the gable end.

Attic vents are installed in the gable end, usually above the level of the probable level of a future ceiling should the attic be finished later.

Soffit vent replaces a portion of the soffit material to provide continuous ventilation along its entire length.

Brick vents are exactly the size of a brick, can be laid in any brick wall. Screened to meet FHA specs.

Cement block vents are designed to be mortared into the same space as an 8″ × 16″ cement block. At least four should be used to vent crawl space.

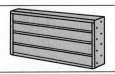

Ventilators and vent applications.

Steps for Installing Insulating Glass

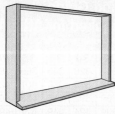

1 – Select a good grade of softwood lumber like ponderosa pine. For a medium-sized frame, use a 1 1/2″ thickness. Make the frame slightly larger than the insulating glass. After assembly, apply a coating of water-repellent preservative.

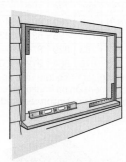

2 – Install frame in opening. Place wedge blocks under sill and at several points around perimeter of frame. When sill is level and frame perfectly square, nail the frame securely to structural members of the building.

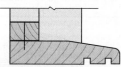

3 – Prepare inside stop members according to recommended sizes. Use miter joints in corners and nail down stops. Also prepare outside stops.

4 – Using recommended grade of glazing sealant, apply thick bed to stops, and top, sides, and bottom of frame. Apply enough material to fill space between frame and glass unit when installation is made.

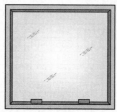

5 – Install neoprene setting blocks on bottom of frame or edge of glass unit as shown. Blocks (use only two) should be moved in from the corners a distance equal to one-quarter of frame width.

6 – Place bottom edge of insulating glass unit on setting blocks as shown. Then carefully press unit into position against stops until proper thickness of glazing sealant is secured around entire perimeter. Small gauge blocks may be helpful.

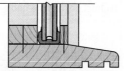

7 – Nail on outside stop. Additional sealant may be required to fill joint. Remove excess sealant from both sides of unit and clean surfaces with an approved solvent.

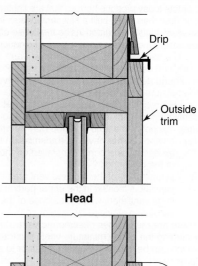

Head

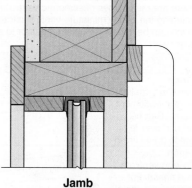

Jamb

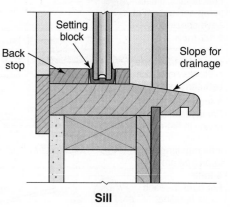

Sill

Details for Constructing a Window Wall

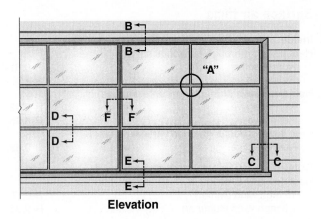

Elevation

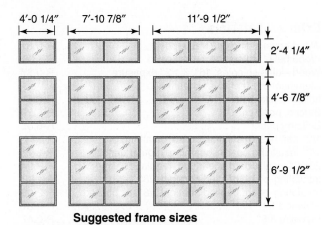

Suggested frame sizes

Glazed metal or wood ventilating units can be installed in any opening as desired without changing the window wall frame. Standard thermopane glass size for window wall construction is 45 1/2″ × 25 1/2″ for non-ventilating units.

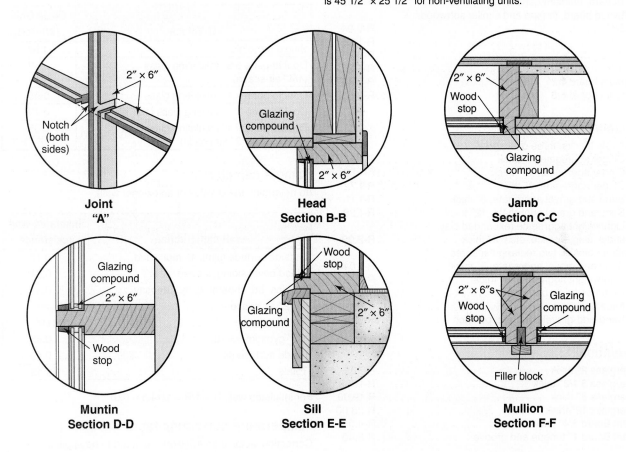

**Joint
"A"**

**Head
Section B-B**

**Jamb
Section C-C**

**Muntin
Section D-D**

**Sill
Section E-E**

**Mullion
Section F-F**

Insulation Values

Exterior materials

Wood bevel siding, 1/2 × 8, lapped R-0.81
Wood bevel siding, 3/4 × 10, lapped R-1.05
Wood siding shingles, 16″, 7 1/2″ exposure R-0.87
Aluminum or Steel, over sheathing,
hollow-backed R-0.61
Stucco, per inch R-0.20
Building paper R-0.06
1/2″ nail-base insulating board sheathing R-1.14
1/2″ insulating board sheathing, regular density R-1.32
25/32″ insul. board sheathing, regular density R-2.04
Insulating-board backed nominal 3/8″ R-1.82
Insulating-board backed nominal 3/8″ foil backed R-2.96
Plywood 1/4″ R-0.31
Plywood 3/8″ R-0.47
Plywood 1/2″ R-0.62
Plywood 5/8″ R-0.78
Hardboard 1/4″ R-0.18
Hardboard, medium density siding 7/16″. R-0.67
Softwood board, fir pine and similar softwoods
    3/4″ R-0.94
    1 1/2″ R-1.89
    2 1/2″ R-3.12
    3 1/2″ R-4.35
Gypsum board 1/2″ R-0.45
Gypsum board 5/8″ R-0.56

Masonry materials

Concrete blocks, three oval cores
    Cinder aggregate, 4″ thick R-1.11
    Cinder aggregate, 12″ thick R-1.89
    Cinder aggregate, 8″ thick R-1.72
    Sand and gravel aggregate, 8″ thick R-1.11
    Sand and gravel aggregate, 12″ thick R-1.28
    Lightweight aggregate (expanded clay,
    shale, slag, pumice, etc.), 8″ thick R-2.00
Concrete blocks, two rectangular cores
    Sand and gravel aggregate, 8″ thick R-1.04
    Lightweight aggregate, 8″ thick R-2.18
    Common brick, per inch R-0.20
    Face brick, per inch R-0.11
    Sand-and-gravel concrete, per inch R-0.08

Insulation

Fiberglass 2″ thick R-7.00
Fiberglass 3 1/2″ thick R-11.00
Fiberglass 6″ thick R-19.00
Fiberglass 12″ thick R-38.00
Foam Board 3/4″ thick R-4.05
Foam Board 1″ tongue and groove R-5.40

Roofing

Asphalt shingles. R-0.44
Wood shingles, plain and plastic film faced. R-0.94

Surface air films

Inside, still air
Heat flow *up* (through horizontal surface)
    Nonreflective. R-0.61
    Reflective R-1.32
Heat flow *down* (through horizontal surface)
    Nonreflective. R-0.92
    Reflective R-4.55
Heat flow *horizontal* (through vertical surface)
    Nonreflective. R-0.68

Outside
Heat flow any direction, surface any position
    15 mph wind (winter). R-0.17
    7.5 mph wind (summer) R-0.25

Glass

U-Values	Glass Only (Winter)
Single-pane glass	1.16
Double-pane 5/8″ insulating glass (1/4″ air space)	0.58
Double-pane × 1″ insulating glass	0.55
Double-pane 1″ insulating glass (1/2″ air space)	0.49
Double-pane × 1″ insulating glass with combination (2″ air space)	0.35

Sample calculation

(to determine the U-value of an exterior wall)

Wall construction	Insulated wall resistance
Outside surface (film), 15 mph wind.	0.17
Wood bevel siding, lapped.	0.81
1/2″ ins. bd. sheathing, reg. density	0.32
3 1/2″ air space	—
R-11 insulation	11.00
1/2″ gypsum board.	0.45
Inside surface (film)	0.45
Totals	14.43

For insulated wall, U = 1/R = 1/14.3 = 0.07

Temperature correction factor

Correction factor is an ASHRAE standard to be applied for
varying outdoor design temperatures. As follows:

If design temperature is:	−20	−10	0	+10	+20
Then correction factor is:	0.778	0.875	1.0	1.167	1.40

Andersen Corp.

Insulation values for common materials. The method used to calculate U- and R-values for a wall can also be used for ceilings and floors. Temperature correction factors are used by designers of heating and cooling systems.

Design Temperatures and Degree Days

Design Temperatures and Degree Days (Heating Season)			
State	City	Outside Design Temperature (°F)	Degree Days (°F-Days)
Alabama	Birmingham	19	2,600
Alaska	Anchorage	−25	10,800
Arizona	Phoenix	31	1,800
Arkansas	Little Rock	19	3,200
California	Los Angeles	41	2,000
California	San Francisco	35	3,000
Colorado	Denver	−2	6,200
Connecticut	Hartford	1	6,200
Florida	Tampa	36	600
Georgia	Atlanta	18	3,000
Idaho	Boise	4	5,800
Illinois	Chicago	−3	6,600
Indiana	Indianapolis	0	5,600
Iowa	Des Moines	−7	6,600
Kansas	Wichita	5	4,600
Kentucky	Louisville	8	4,600
Louisiana	New Orleans	32	1,400
Maryland	Baltimore	12	4,600
Massachusetts	Boston	6	5,600
Michigan	Detroit	4	6,200
Minnesota	Minneapolis	−14	8,400
Mississippi	Jackson	21	2,200
Missouri	St. Louis	4	5,000
Montana	Helena	−17	8,200
Nebraska	Lincoln	−4	5,800
Nevada	Reno	2	6,400
New Hampshire	Concord	−11	7,400
New Mexico	Albuquerque	14	4,400
New York	Buffalo	3	7,000
New York	New York City	12	5,000
North Carolina	Raleigh	16	3,400
North Dakota	Bismarck	−24	8,800
Ohio	Columbus	2	5,600
Oklahoma	Tulsa	12	3,800
Oregon	Portland	21	4,600
Pennsylvania	Philadelphia	11	4,400
Pennsylvania	Pittsburgh	5	6,000
Rhode Island	Providence	6	6,000
South Carolina	Charleston	23	2,000
South Dakota	Sioux Falls	−14	7,800
Tennessee	Chattanooga	15	3,200
Texas	Dallas	19	2,400
Texas	San Antonio	25	1,600
Utah	Salt Lake City	5	6,000
Vermont	Burlington	−12	8,200
Virginia	Richmond	14	3,800
Washington	Seattle	28	5,200
West Virginia	Charleston	9	4,400
Wisconsin	Madison	−9	7,800
Wyoming	Cheyenne	−6	7,400

ASHRAE

This list of U.S. cities, with their outside design temperatures and degree days, is a useful resource for computer-aided energy analysis.

Sound Insulation for Walls and Floors

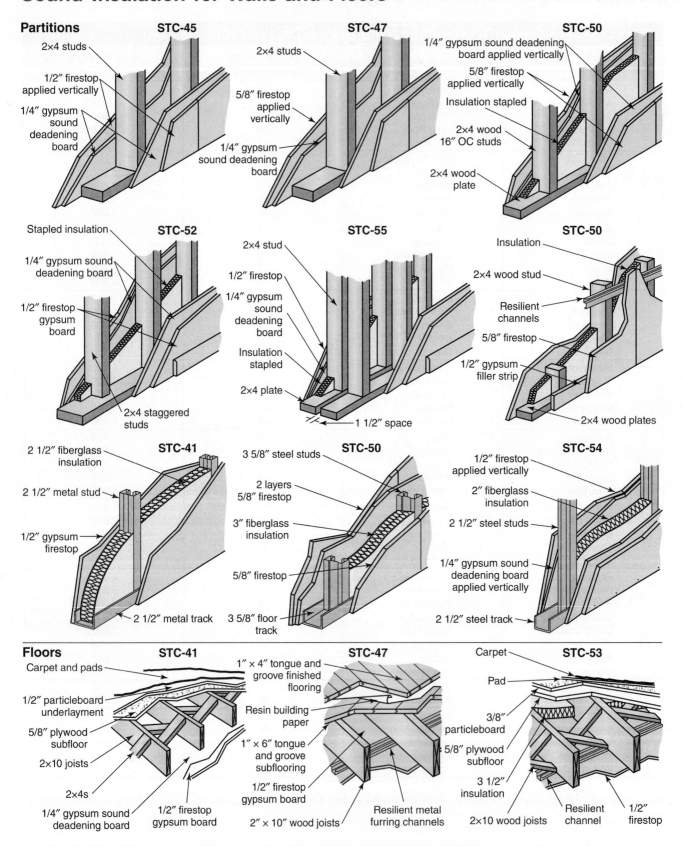

Partitions

STC-45
- 2×4 studs
- 1/2″ firestop applied vertically
- 1/4″ gypsum sound deadening board

STC-47
- 2×4 studs
- 5/8″ firestop applied vertically
- 1/4″ gypsum sound deadening board

STC-50
- 1/4″ gypsum sound deadening board applied vertically
- 5/8″ firestop applied vertically
- Insulation stapled
- 2×4 wood 16″ OC studs
- 2×4 wood plate

STC-52
- Stapled insulation
- 1/4″ gypsum sound deadening board
- 1/2″ firestop gypsum board
- 2×4 staggered studs

STC-55
- 2×4 stud
- 1/2″ firestop
- 1/4″ gypsum sound deadening board
- Insulation stapled
- 2×4 plate
- 1 1/2″ space

STC-50
- Insulation
- 2×4 wood stud
- Resilient channels
- 5/8″ firestop
- 1/2″ gypsum filler strip
- 2×4 wood plates

STC-41
- 2 1/2″ fiberglass insulation
- 2 1/2″ metal stud
- 1/2″ gypsum firestop
- 2 1/2″ metal track

STC-50
- 3 5/8″ steel studs
- 2 layers 5/8″ firestop
- 3″ fiberglass insulation
- 5/8″ firestop
- 3 5/8″ floor track

STC-54
- 1/2″ firestop applied vertically
- 2″ fiberglass insulation
- 2 1/2″ steel studs
- 1/4″ gypsum sound deadening board applied vertically
- 2 1/2″ steel track

Floors

STC-41
- Carpet and pads
- 1/2″ particleboard underlayment
- 5/8″ plywood subfloor
- 2×10 joists
- 2×4s
- 1/4″ gypsum sound deadening board
- 1/2″ firestop gypsum board

STC-47
- 1″ × 4″ tongue and groove finished flooring
- Resin building paper
- 1″ × 6″ tongue and groove subflooring
- 1/2″ firestop gypsum board
- 2″ × 10″ wood joists
- Resilient metal furring channels

STC-53
- Carpet
- Pad
- 3/8″ particleboard
- 5/8″ plywood subfloor
- 3 1/2″ insulation
- 2×10 wood joists
- Resilient channel
- 1/2″ firestop

Georgia-Pacific

How to build wall partition and floor structures with high STC (sound transmission class) ratings. Ratings are based on sound tests conducted according to ASTM-E90.

Branch Pipe Sizes

Fixture	Pipe size	Fixture	Pipe Size
Bathtub	1/2″	Shower	1/2″
Dishwasher	1/2″	Urinal – flush valve	3/4″
Drinking fountain	3/8″	– flush tank	1/2″
Hose bibb	1/2″	Washing machine	1/2″
Kitchen sink	1/2″	Water closet – flush valve	1″
Laundry tray	1/2″	– flush tank	3/8″
Lavatory	1/2″	Water heater	1/2″

Plastic Pipe

Type plastic	Uses	Sizes/grades available	Joining methods	Rigid/flexible
ABS	Drain/waste/vent	1 1/4″–6″ Schedule 40 (Sch. 40)	Solvent cement	Rigid
	Sewer lines	3″–8″ Standard dimension ratio (SDR)		
	Water supply	1/2″–2″ SDR		
PB	Hot and cold water distribution	1/8″–1″ CTS	Insert fittings	Flexible
		3/4″–2″ IPS	Compression fittings	
			Instant connect fittings	
PVC	Water distribution	1/4″–12″ Sch. 40 & 80 and SDR	Solvent welding	Rigid
	Hot and cold water distribution		Sch. 80: threading & flanges	
CPVC	Hot and cold water distribution	1/2″–3/4″	Solvent welding	Rigid
	Process piping	1/2″–8″ Sch. 40 & 80	Threading & flanges	
PE	Water distribution	1/2″–2″ IPS	Compression & flange fittings	Flexible
	Natural gas distribution Oil field piping	1/2″–6″ IPS		
	Water transmission, sewers	3″–48″ SDR	Fusion welding	
SR	Agricultural field drains Storm drains	3″–8″ SDR	Solvent welding	Rigid

Copper Pipe

Type	Color code	Application	Straight lengths	Coils (soft temper only)
K	Green	Underground and Interior service	20 ft in diameters including 8 in Hard and soft temper	60 ft and 100 ft for diameters including 1 in
L	Blue	Aboveground service	20 ft in diameters including 10 in Hard temper only	(same as K)
M	Red	Aboveground water supply drainage, waste, and vent	20 ft in all diameters Hard temper only	Not available
DWV	Yellow	Aboveground drain, waste, and vent piping	20 ft in all diameters 1 1/4 in and greater Hard temper only	Not available

Insulated Wire Coverings

Covering types	Letter designation
Rubber	RH, RHH, RHW
Thermoplastic Compound	TW, THW, TBS, THHN

Conductor Applications

Wire type	Apply where	Temperature maximum	
		°C	°F
RH	Dry/damp	75	167
RHH	Dry/damp	90	194
RHW	Dry/wet	75	167
TW	Dry/wet	60	140
THHN	Dry/damp	90	194
THW	Dry/wet	90	194 Under special conditions
THWN	Dry/wet	75	167
XHHW	Dry/damp	90	194
	Wet	75	167
MI	Dry	90	194 Under special conditions
	Wet	250	482

Conductor Colors

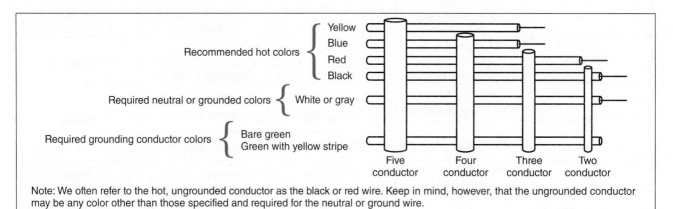

Recommended hot colors { Yellow, Blue, Red, Black

Required neutral or grounded colors { White or gray

Required grounding conductor colors { Bare green, Green with yellow stripe

Five conductor   Four conductor   Three conductor   Two conductor

Note: We often refer to the hot, ungrounded conductor as the black or red wire. Keep in mind, however, that the ungrounded conductor may be any color other than those specified and required for the neutral or ground wire.

Electrical Switches

Type	Purpose
Single-pole, single-throw switch	Controls light or outlet from single location.
Three-way switch	Controls light or outlet from two locations.
Four-way switch	Controls light or outlet from other locations in between pair of three-way switches.
Dimmer switch	Like single-pole, single-throw switch, but also contains rheostat or voltage regulating device, which allows all or only portion of electrical energy to outlet or light fixture.
Pilot-lighted switch	Used to control light or outlet, which is not in sight. It indicates, by use of pilot light, whether power to device is on or off.
Time-delay switch	Used where delayed shut-off is desirable.
Others: Toggle switch, push-button switch, pull chain switch, photoelectric switch, knife switch, tap switch, plate switch, locking switch.	

SkillsUSA—Preparing Students for Leadership in the World of Work

Introduction to SkillsUSA

SkillsUSA is a national organization that brings together students, educators, and industry members dedicated to preparing students for excellence in career and technical occupations. The SkillsUSA Framework supports the acquisition of technical skills grounded in academics, personal skills, and workplace skills. Becoming involved in SkillsUSA is a commitment that can provide many lifelong rewards.

History

In 1965, the Vocational Industrial Clubs of America (VICA) formed to fill demand for a national skills organization that could connect industry professionals, educators, and youth in order to train students for future technical careers. While VICA began with just 200 members from 14 states, it grew quickly and expanded its membership to include college students. By 1969, VICA had more than 82,000 members.

VICA held its first competitive events in 1967, giving student competitors the chance to showcase their technical skill for peers, instructors, and professionals. In 1995, VICA changed the name of its national competition, the US Skill Olympics, to SkillsUSA Championships. In 1998, VICA was renamed VICA-SkillsUSA, and in 2004, the name of the organization was shortened to SkillsUSA.

Membership

SkillsUSA represents over 360,000 members in over 18,000 local chapters across the United States, with industry support from more than 600 corporations, trade associations, and labor unions.

Organization Colors and Relationships

The SkillsUSA colors illustrate the importance of the relationship between the national organization and the individual states and chapters. Red, white, blue, and gold represent the national organization itself. Within this color scheme, red and white represent the individual states and chapters while blue represents their common union. Gold represents the most important part of the organization, the individual member.

Motto, Creed, and Pledge

The SkillsUSA organization lives by its motto, "Preparing for leadership in the world of work," and members follow its creed. In accordance with the creed, SkillsUSA members believe in the dignity of work, the American way of life, fair play, high moral and spiritual standards, and that satisfaction is achieved by good work.

SkillsUSA members also pledge to be productive members of their schools, chapters, and communities:

"Upon my honor, I pledge:

- To prepare myself by diligent study and ardent practice to become a worker whose services will be recognized as honorable by my employer and fellow workers.
- To base my expectations of reward upon the solid foundation of service.
- To honor and respect my vocation in such a way as to bring repute to myself.
- And further, to spare no effort in upholding the ideals of SkillsUSA."

SkillsUSA membership not only helps its members hone their technical skills, it helps them choose their futures. The organization recognizes the importance of diversity in the workforce and gives students from all backgrounds the opportunity to prove their skills and choose their careers. Championship contests are designed to evaluate career readiness, as well as preparedness for applying and interviewing for jobs. In addition to taking assessment tests and career interest inventories, members can try out potential careers in hands-on environments.

SkillsUSA Leadership Opportunities

SkillsUSA promotes good citizenship and expects its members to prepare to become leaders in their fields and their communities. SkillsUSA encourages all its members to contribute to their communities. Chapters can organize community service projects to give back to their communities and submit these projects to championship events.

Mentorship

SkillsUSA members can learn to lead by example through Student2Student mentoring. In these programs, chapters work with middle and elementary schools to help younger students explore future careers. Older students mentor younger students by going on field trips or working on hands-on activities together.

National Officers

Members can learn to lead a community by serving in the House of Delegates and as SkillsUSA National Officers. National officers are elected by vote from the House of Delegates and hold their positions for one year. Potential officers should study the SkillsUSA Leadership Handbook closely and meet the qualifications for candidacy, which include active membership in their chapter and past SkillsUSA leadership experience.

There are strict regulations about where, when, and how candidates may campaign for officership. Campaigning with social media is not allowed, and candidates do the bulk of their campaign work by interacting directly with the delegates. Candidates must give a brief speech, participate in question-and-answer sessions, and attend three Meet the Candidate sessions.

SkillsUSA Competitions

SkillsUSA competitions at the local, state, and national level test for more than a participant's technical skill. The definitive goal of SkillsUSA is preparing members to excel in the workplace. So, participants' preparation, appearance, and behavior are also thoroughly graded as part of a complete work performance. Competitions are held for both technical skills and leadership qualities.

The National SkillsUSA Championship is held annually in Louisville, Kentucky, and more than 6,300 students compete each year in order to earn Skill Point Certificates. Skill Point Certificates are awarded to participants who reach or exceed industry-defined cut scores, regardless of medal standing or contest ranking. The competition holds contests for more than a hundred concentrations in the technical, skilled, and service occupations as well as career and leadership skills. SkillsUSA also sponsors competitions on the local, state, regional, and international levels.

Professional Behavior

Each competition requires careful preparation, both before and during the contest event. Participants should study the regulations for their contest and observe all standards for dress, tools, and other preparations. Contest events have multiple parts. Each event typically has the following four components: a professional development program (PDP) test; a technical skills–related written test; an oral professional assessment or interview; and a submission of a hardcopy résumé. Contestants will then complete industry-defined challenges that are specific to their skill set.

The skills required to net the most points are also those required to do good work on the job. Contestants must know their task, tools, and skill set inside and out in order to perform the task proficiently. Often, projects must be prepared within a set time limit, which requires careful time management. Other events may call for teams that require excellent teamwork and communication to ensure all parts of the project are completed correctly.

Professional Ethics

Participants in SkillsUSA competitions are held to high standards of performance and sportsmanship. The SkillsUSA will not tolerate unethical or disruptive behavior. Interrupting other contestants, tampering with other contestants' work, and other unethical behavior can be grounds for immediate disqualification.

Professional Dress

During the SkillsUSA ceremonies, meetings, and similar functions, students should dress professionally and follow SkillsUSA's guidelines for dress. Both genders are expected to wear black SkillsUSA jackets or red SkillsUSA blazers, windbreakers, or sweaters and dress in a business-formal style. For men, this means a white dress shirt, black dress slacks, black socks, black shoes, and black ties. For women, this means a white blouse or turtleneck with a collar that does not extend over the lapels or neck of her blazer or jacket. Women may wear a black dress skirt or slacks with black shoes. Black sheer (not opaque) or skin-tone seamless hose should be worn with skirts.

SkillsUSA is an organization that asks for a commitment from its members but it provides many rewards. If you are interested in learning more about SkillsUSA, please see their website.

A

admixture. A chemical added to concrete to change the characteristics of the mix. (9)

advanced framing. A framing system that conserves natural resources while optimizing structural performance. (11)

aggregate. A material such as sand, rock, and gravel used to make concrete. (9)

air exchanger. A device that exhausts air from a building and draws fresh air in. (33)

alkaline copper quaternary compound (ACQ). A nontoxic preservative used for treating lumber; requires the use of stainless steel or heavily galvanized fasteners. (30)

all-hard blade. A heat-treated hacksaw blade that is very brittle and easily broken if misused. (4)

alternating current (AC). An electric current used in home wiring, where the flow of electrons regularly reverses direction. (31)

anchor bolt. A bolt embedded in concrete to hold structural members in place. (9)

anchor strap. A strap fastener that is embedded in concrete or masonry walls to hold sills in place. (9)

angle of repose. The greatest slope at which loose material, such as excavated soil, will stand without sliding. (2)

annual fuel utilization efficiency (AFUE). An energy efficiency rating used for furnaces. It is a measure of the amount of heat actually delivered to a building compared to the amount of fuel the furnace or boiler uses. (33)

annular ring. A ring or layer of wood that represents one growth period of a tree. In a cross-section, the rings may indicate the age of the tree. (3)

apprenticeship. A formal method of learning a certain trade, such as carpentry, that involves instruction as well as working and learning on the job in route to certification as a journeyman. (1)

arc-fault circuit interrupter (AFCI). A device that can detect abnormal arcing, which might cause a fire. AFCIs are required in many locations throughout a house. (31)

architectural drawing. A plan that shows the shape, size, and features of a building. (7)

ash dump. A metal frame with pivoted cover provided in the rear hearth for clearing ashes. (25)

asphalt shingle. The most common type of roofing material used today. They are manufactured as strip shingles, interlocking shingles, and large individual shingles. (14)

authority having jurisdiction (AHJ). The government agency that has authority over zoning, building construction, and building safety in a given area. (7)

auto level. An optical device is similar to the builder's level except that it contains an internal compensator that eliminates any minor variant from level. (8)

awning window. A window type with a sash that is hinged at the top, allowing the bottom to swing outward. (16)

B

backfilling. The process of replacing soil around foundations after excavating. (9)

backing sheet. A sheet that is similar to regular laminate without the decorative finish and is usually thinner. Used on unsupported countertop spaces. (23)

backsaw. A saw with a thin blade reinforced with a heavier steel strip along the back edge. The teeth are small (14–16 points), producing fine cuts. It is mostly used for interior finish work. (4)

balloon framing. A type of building construction with upright studs that extend from the foundation sill to the rafter plate. A ribbon supports the second floor joists. This system is not used in new construction. (10)

baluster. The vertical member (spindle) supporting the handrail on open stairs. (21)

balustrade. An assembly with a railing resting on a series of balusters that, in turn, rest on a base, usually the treads. (21)

Note: The number in parentheses following each definition indicates the chapter in which the term can be found.

band joist. A joist set perpendicular to the floor joists and to which the ends of the floor joists are attached. Also called *joist header* or *rim joist*. (10)

banjo. A taping tool used to apply joint compound and tape at the same time. This tool has a reservoir to hold the compound and a reel to hold the tape. (19)

bar clamp. A clamp used for holding larger pieces while they are receiving glue or mechanical fasteners. Also called *pipe clamp*. (4)

baseboard. A finishing board that covers the joint where a wall intersects the floor. (22)

base flashing. The part of chimney flashing that is attached to the roof. (14)

base level line. A level line that is snapped on the wall from the high point around the wall as far as the cabinets will extend. (23)

base shoe. The narrow molding used around the perimeter of a room where the baseboard meets the finish floor. (22)

batten. A horizontal strip placed across the roof in rows to provide a support for clay or cement roofing tile. (14)

batter board. A temporary framework of stakes and ledger boards used to locate corners and building lines when laying out a foundation. (9)

batt insulation. Insulation made of the same fibrous material as blankets, but cut into shorter segments. (18)

bearer. Horizontal member (usually 2×6s) used to connect scaffold uprights. Also called *cross ledger*. (6)

belt sander. A portable tool that consists of a belt of sandpaper that is rotated by a motor. Its size is determined by the width of the belt. (5)

benchmark. A mark on a permanent object fixed to the ground from which grade levels and elevations are taken for construction of a building. Sometimes officially established by government survey. (8)

bevel siding. A wedge-shaped siding used as finish covering on the exterior of a structure. (17)

binder. An ingredient of paint that remains behind after the oils or resins have dried. (24)

bird's mouth. A notch cut on the underside of a rafter to fit the top plate. If the rafter end is flush with the top plate, it is not a full notch. (12)

bisque. The body of a ceramic tile. (20)

blanket insulation. An insulation generally furnished in rolls or strips of convenient length and in various widths suited to standard stud and joist spacing and various wall thicknesses. (18)

blind nailed. A type of nailing concealed by installation of another strip of wood; used in tongue-and-groove flooring. (20)

block flooring. A wood flooring that is cut in square blocks. (20)

block plane. A small surfacing plane that produces a fine, smooth cut, making it suitable for fitting and trimming work. The blade is mounted at a low angle and the bevel of the cutter is turned up. (4)

blower door test. A test of air infiltration done by mounting a large fan in an otherwise sealed doorway. (18)

blue stain. A mark caused by a fungus growth in unseasoned lumber, especially pine. It does not affect the strength of the wood. (3)

board-and-batten siding. A siding application designed around wide, square-edged boards spaced about 1/2″ apart. (17)

board foot. The equivalent of a board 1′ square and 1″ thick. (3)

board-on-board siding. A variation of the board-and-batten siding with boards replacing battens. (17)

boiler. A heating unit for a hydronic heating system. (33)

booking. A method of pasting and folding wallcoverings that allows the paste to soak into the wallcovering. (24)

bored well. A well that is made with an earth auger. (32)

bound water. The moisture contained in the cell walls of wood. (3)

box beam. A beam made of one or more plywood webs connected to lumber flanges. (26)

box cornice. A cornice that encloses the rafters at the eaves. (17)

brake. A piece of equipment used primarily by siding and roofing contractors to bend sheet metal or vinyl. (17)

branch circuit. A single circuit running from a distribution panel. (31)

bridging. A pair of wood or metal pieces fitted from the bottom of one floor joist to the top of adjacent joist, installed between joists to hold joists vertical and help transfer loads from one joist to the next. (10)

British thermal unit (Btu). The amount of heat needed to raise the temperature of 1 lb of water 1°F. (18)

brown coat. A second layer of plaster that is applied on top of the scratch coat to form a base for the finish coat. (19)

buck. A temporary wooden form installed within the forms for poured concrete walls. Its purpose is to provide an opening or void for later installation of features such as windows or doors. (9)

builder's level. An optical device used to determine grade levels and angles for laying out buildings on a site. Also called a *dumpy level* or an *optical level*. (8)

building code. A collection of rules and regulations for construction established by organizations and based on experience and experiment. (7)

building envelope. The separation between the outside environment and the inside environment of an enclosure. (15)

building line. The line marking where the wall of a structure will be located. (8)

building permit. A document issued by building officials certifying that plans meet the requirements of the local building code. The issuing of a permit allows construction to begin. (7)

building science. The application of knowledge of the physical world to building processes and materials. (15)

building site. The place where a building is to be constructed. (8)

build-up strip. Narrow wood piece that is used to attach the countertop to the base cabinets. (23)

built-up roof. A roof covering composed of several layers of roofing felts laminated with coal tar or pitch, mopped over with asphalt, and finished with crushed slag or gravel. Generally used on flat or low-pitched roofs. (14)

built-up stringer. A stringer to which blocking has been added to form a base for adding treads and risers. (21)

butt gauge. A layout tool used to lay out the gain (recess) for hinges. (4)

buttlock. The bottom hook-edge of a panel of vinyl siding or soffit that locks into the previously installed piece. (17)

C

cabinetwork. A term that refers to built-in kitchen and bathroom storage and, in a general way, closet shelving, wardrobe fittings, desks, bookcases, and dressing tables. (23)

CADD-CAM. The combination of computer-aided design and computer-assisted manufacturing. (7)

camber. A slight arch in a beam or other horizontal member so the bottom of the member is straight when under load. (12)

cambium. The layer of wood cells located beneath the bark of a tree. (3)

cantilevered. The state of extending horizontally beyond a supporting surface. (10)

cant strip. A strip of wood, triangular in cross section, used under shingles at gable ends or under the edges of roofing on flat decks. Provides support for the felt layers as they curve from a horizontal to a vertical attitude (14)

cap flashing. Flashing used on chimneys at the roofline to cover base flashing and prevent leaks. Also called *counterflashing*. (14)

capillary action. The action by which moisture passes through fill or building materials. (9)

casein glue. An adhesive of casein and hydrated lime that is suitable for gluing oily woods and for laminating wood with a high moisture content. (3)

casement window. A window in which the sash pivots on its vertical edge, so it may be swung in or out. (16)

casing. A heavy metal tube that keeps contaminants out of the well water. (32)

cat's paw. A prying tool used to pull a nail above the surface of the wood where it can be gripped easily. (4)

caulk. A material used to fill joints, cracks, or holes in construction. (24)

C-clamp. A C-shaped clamping tool that adapts to a wide range of assemblies where parts need to be held together. They are available in sizes from 1″ to 12″. (4)

ceiling frame. The system of support for all components of the ceiling. (11)

ceiling joist. A main ceiling framing member. (11)

cement. A binding material that, when combined with water and aggregate, forms concrete. (9)

cement board. A fireproof, moisture-resistant, fiber-reinforced cement panel used as a base for finishing materials on walls, floors, and countertops. (19)

cementitious. Made of cement. (18)

central heating. A system in which a single large heating plant produces heat, which is then circulated throughout the structure. (33)

chain saw. A gas- or electric-powered saw used to cut heavy timbers, posts, or pilings. Blade lengths vary from about 14″–24″. (5)

chair. A small metal fixture used to hold reinforcing bars away from the ground prior to the pouring of concrete slabs. (9)

chalk line. A tool used to mark long, straight lines. A thin, strong cord that is covered with powdered chalk is held tight and close to the surface. Then it is snapped, driving the chalk onto the surface and forming a distinct mark. (4)

channel rustic siding. Horizontal wood siding that is usually 3/4″ thick and 6″, 8″, or 10″ wide and has shiplap-type joints. (17)

chase. A wood frame jutting from an outside wall that supports a factory-built chimney and fireplace. (25)

chevron paneling. Solid paneling installed diagonally in a chevron pattern. (19)

chimney. A vertical hollow shaft that exhausts smoke and gases from heating units and incinerators. (25)

chop saw. A variation of the power miter saw usually used for cutting metal. (5)

chord. The top or bottom member of a floor or roof truss. Also called *flange*. (10)

chromated copper arsenate (CCA). A wood preservative treatment once widely used for lumber. No longer available for residential use due to toxic emissions. (30)

circuit. A path for electrical power provided by wires (conductors) between a power source and the power-using device (load). (31)

circuit breaker. An overcurrent protection device that opens a circuit, preventing the flow of electricity when an overload or short is detected so that the circuit will not be damaged. (31)

circulator. A factory-built metal unit that increases the heating efficiency of a masonry fireplace. The sides and back are double walled, providing a space where air is heated. Cool air enters this chamber near the floor level. When the air is heated, it rises and returns to the room through registers at a higher level. (25)

Class A fire. A fire that results from burning wood and debris. (2)

Class B fire. A fire that involves highly volatile materials such as gasoline, oil, paints, and oil-soaked rags. (2)

Class C fire. A fire that results from faulty electrical wiring and equipment. (2)

Class D fire. A fire that involves combustible metals, such as magnesium and sodium. (2)

Class K fire. A fire that involves vegetable or cooking oils. (2)

claw hammer. A hammer having a head with one end forked for removing nails. (4)

closed-cut valley. A method of shingling valleys by carrying each course across the valley and onto the adjoining slope. When the adjoining slope is shingled, the overlapping shingles are cut at the valley. (14)

closed panel. A factory-built housing wall panel that is finished on both sides. (27)

closed porch. An attached structure that is framed with studs and sided like the rest of the house. It is more like an extension of the house than a transitional area. (30)

closure block. The last concrete block to be placed in any course. (9)

coatings. A term for all types of finishes, including paints, stains, varnishes, and various synthetic materials. (24)

coefficient of thermal conductivity (k). The amount of heat (Btu) transferred in one hour through 1 sq ft of a given material that is 1″ thick and has a temperature difference between its surfaces of 1°F. (18)

collar tie. A tie beam connecting rafters. It is located considerably above the wall plate. (12)

combination square. A tool that serves a similar purpose to a try square that is also used to lay out miter joints. Its adjustable sliding blade allows it to be conveniently used as a gauging tool. It can be used as a marking gauge to make parallel lines and as a square for 90° and 45° angles. (4)

common difference. The difference in length between any two adjacent studs or rafters when laying out a gable end frame or a hip roof. (12)

common rafter. A rafter connected to both the ridge and the wall plate. (12)

competent person. A person designated by a company to ensure that certain safety standards are being met. (2)

complementary colors. Hues opposite of each other on the color wheel. (24)

composite board. A board consisting of a core of wood between veneered surfaces. (3)

composite decking. A material made with a blend of plastic with wood fibers or sawdust. (30)

compression fitting. A connector that seals joints by compressing a ring of soft material, such as brass. (32)

compression valve. A valve that makes use of a tapered plug or a washer for full seating to stop water flow or partial seating to control the flow. (32)

computer-aided drafting and design (CADD). The process of using computers to create a set of plans and perspective drawings for houses. (7)

computer-aided manufacturing (CAM). A combination of design and manufacturing activities controlled by a computer. (7)

concealed header. A header that is butted against the ends of the joists it is supporting. (29)

concrete masonry unit (CMU). A block made from a mixture of Portland cement and aggregates, such as sand, fine gravel, or crushed stone. They weigh 40–50 lb. Also known as *concrete blocks*. (9)

condensation. The conversion of water vapor into liquid. (15)

condenser. The part of an air conditioning unit that receives hot refrigerant and allows it to release heat to the atmosphere and return to a liquid state. (33)

conductance (C). The amount of heat (Btu) that will flow through a material in one hour per sq ft of surface with 1°F of temperature difference. (18)

conduction. The movement of heat through substances from one end to the other. (18, 33)

conductor. A material, such as a copper wire, that carries an electric current. (31)

conduit. A tube of metal or plastic in which wiring is installed. (31)

conifer. A needle-bearing tree, such as an evergreen. Softwoods come from these trees. (3)

contact cement. A neoprene-rubber-based adhesive that bonds instantly upon contact of parts being fastened. (3)

control layer. A building system that works to mitigate the impact that environmental elements have on a building. (15)

control point. A reference point for determining the elevation of footings, floors, and other parts of a building. Also, the highest elevation on the perimeter of an excavation. (9)

convection. The movement of heat through liquids and gases as a result of heated material rising and being replaced by cooler material. (18, 33)

convection current. The airflow produced when heated air becomes lighter and rises. (18)

coped joint. A joint made in molding by cutting the profile of the molding face on the end of one piece, so it fits snugly against the other. (22)

coping saw. A saw with a thin, flexible blade that is pulled tight by the saw frame. With a blade that has 15 teeth per inch, this saw can make very fine cuts. It is occasionally used for cutting curves. (4)

copper azole (CA). A copper-based, nontoxic wood preservative; requires the use of stainless steel or heavily galvanized fasteners. (30)

corbel. To extend outward from the surface of a masonry wall one or more courses to form a supporting ledge. (25)

core. The innermost layer(s) of plywood between the face veneers of a flush door; may be solid or hollow. (3, 22)

cornice. The exterior trim of a structure at the meeting of the roof and wall that encloses or finishes off the overhang of a roof at the eaves. It usually consists of panels, boards, and moldings. (12, 17)

cove base. A flexible molding glued in place to hide the joint between a tile floor and a wall. (20)

coverage. The amount of weather protection provided by the overlapping of shingles. (14)

crawl space. The space below the first floor of a building directly above the ground. (18)

crawl space foundation. A type of foundation similar to a spread foundation used for a home with a basement, but less deep. The space between the ground and the floor joists is only a few feet. (9)

cripple jack. A rafter that intersects neither the wall plate nor the ridge and is terminated at each end by hip and valley rafters. (12)

cripple stud. A stud used above or below a wall opening. Extends from the header to the top plate or from the sole plate to the rough sill. (11)

cross-band. The inner layer of plywood between the face plies and the core. (3)

cross bridging. A pair of 1×3 or 2×2 lumber set diagonally between the joists to form an X to keep the joists in a vertical position and transfer the load from one joist to the next. (10)

crosscut saw. A saw with pointed teeth designed to cut across the wood grain. (4)

crown molding. Decorative molding applied to the top of a wall to make the transition from the wall to the ceiling. (22)

curing. The process of preventing loss of moisture in fresh concrete to strengthen it and reduce shrinkage, which also reduces cracking. (9)

current. The flow of electrons through a conductor. (5, 31)

cut. The process of removing material to achieve the desired grade. (8)

cut-in point. The temperature setting that causes the thermostat to send a signal to the furnace or air conditioner. (33)

cut-out point. The temperature setting at which a thermostat stops signaling to the furnace or air conditioner. (33)

cut-out stringer. A stair stringer into which the rise and run are cut. (21)

cutting in. The process of painting into the inside corners using a brush. (24)

D

dado. A rectangular horizontal groove cut in stair stringers to support a tread. (30)

damper. A venting device in fireplaces used to control combustion, prevent heat loss, and redirect downdrafts. (25)

dampproofing. The process of preventing soil moisture from penetrating foundations. (9)

darby. A long flat tool used to level the brown coat. (19)

dead bolt. A special door security consisting of a hardened steel bolt and a lock. Lock is operated by a key on the outside and either a key or handle on the inside. Also called a *deadlock* or *throwbolt*. (22)

dead load. The weight of permanent, stationary construction and equipment included in a building. (12)

decay. The disintegration of wood substance due to action of wood-destroying fungi. (3)

decibel (dB). A unit of sound intensity. (18)

deciduous. A broadleaf tree that lose its leaves at the end of the growing season. (3)

defect. An irregularity occurring in or on wood that reduces its strength, durability, or usefulness. (3)

deflection. The state of bending downward at the center of a joist. (10)

degree day. Method of measuring the harshness of climate for insulation and heating purposes. A degree day is the product of one day and the number of degrees the mean temperature is below 65°F. (18)

detail drawing. An enlarged view of a part of the structure that cannot be clearly explained in elevation views or general plan drawings. (7)

dew point. The temperature at which air is sufficiently cooled for water vapor to become liquid. (15)

diagonal paneling. Solid paneling that angles in only one direction. Angles of 22 1/2°, 30°, and 45° are most popular. (19)

differential. The difference between the cut-in point and the cut-out point of a thermostat. (33)

diffusion. The tendency of vapor to move from a higher humidity level to a lower humidity level. (15)

dimension. The lines and text on a drawing that show distance and size. (7)

direct current (DC). An electric current in which the flow of electrons moves in one direction. (31)

direct heating system. A heating system that distributes heat directly to objects without benefit of ducts or a piping system. (33)

discrimination. The act or practice of treating a person or group of people differently because of race, age, religion, or gender. (1)

disodium octaborate tetrahydrate (DOT). A borax-based wood preservative that is nontoxic and does not cause galvanic corrosion to fasteners. (30)

DOE Zero Energy Ready Home Program. A green home building program administered by the US Department of Energy, which achieves energy efficiency that is 40%-50% better than a typical new home. (28)

door casing. Trim applied to each side of the door frame to cover the space between the jambs and the wall surface. (22)

door frame. An assembly of wood parts that form an enclosure and support for a door. (22)

doorstop. A molding nailed to the faces of the door frame jambs to prevent the door from swinging through. (22)

dormer. A framed structure extending from and fastened to a roof slope and having a roof of its own. It provides a wall surface for installing a window. (12)

double-channel lineal. A vinyl siding trim that joins two soffit panels at a transition around a corner or at an end-to-end joint. (17)

double coursing. A process of applying shingle siding with a second layer over the first course. (17)

double-hung window. A window that consists of two sashes that slide up and down past each other in the window frame. (16)

downdraft. A flow of air down a chimney. (25)

draft excluder. A threshold or felt weatherstop. (18)

drainage, waste, and venting (DWV) system. A drainage piping that carries that carries away wastewater and solid waste using gravity, not pressure. (32)

drawn to scale. A reduced drawing that is in exact proportion to the actual size. (7)

drilled well. A well used where great depth is needed to reach a suitable water table. (32)

drip cap. A molding that directs water away from a structure to prevent seepage under the exterior facing material. Applied mainly over window and exterior door frames. (16)

drip edge. A metal edging placed along the rake and eaves before installing shingles. (14)

drop siding. Siding usually 3/4″ thick and 6″ wide; machined into various patterns with tongue-and-groove or shiplap joints. (17)

drywall saw. A saw with large, specially designed teeth for cutting through paper facings, backings, and the gypsum core. (4)

drywall screw shooter. A tool used for drywall installation. Drywall screws are inserted one at a time into the nose. Locators control depth below the face of the drywall. (5)

duct. A tube or passage used to move heated or cooled air from the furnace or air conditioner to living spaces of the building. (33)

dust mask. A woven covering worn over the mouth and nose to filter out dust and larger particles of airborne debris. It is not effective in trapping very small particles, such as paint mists. (2)

E

eaves flashing strip. A roof flashing installed at the eaves where outside temperatures drop to 0°F (–17°C) or colder and there is a possibility of ice forming along the eaves. (14)

eaves trough. A waterway built into the roof surface over the cornice. (14)

edge-grained. A board sawed so that the annular rings form an angle of greater than 45° with the board's surface. (3)

electrical plan. A drawing that shows where electrical equipment is located. Electrical plans include switches, receptacles, light fixtures, smoke alarms, major appliances, and wiring runs. (7)

electric current. The flow of electrons through a conductor. (31)

electric drill. A portable power tool used primarily for drilling holes. Both corded and cordless (battery-powered) models are available. (5)

electric radiant heat. A system that transfers heat by moving it through solid matter, such as a concrete floor. (33)

electric resistance baseboard heating. A system that consists of a heating element wrapped in a tubular casing. Metal fins spaced along its length help radiate the heat to the room air. (33)

electric shock. A discharge of electrical current passing through the body. (2)

elevated deck. A deck that has a supporting ledger physically attached to the building wall. (30)

elevation. A drawing that shows the front, rear, and sides of a building. Also, the height of an object above grade. (7, 8)

emissivity. The ability of a material to absorb heat or emit heat by radiation. (16)

energy recovery ventilator (ERV). An air exchanger that works as an HRV but also removes humidity from the incoming air to maintain more comfortable humidity levels inside the building. (33)

ENERGY STAR®. A program of the US Department of Energy and US Environmental Protection Agency to reduce the energy consumption of devices. (28)

engineered lumber products (ELP). A lumber that has been altered by manufacturing processes. Included are wood I-beams, glue laminated beams, laminated strand lumber, laminated veneer lumber, and parallel strand lumber. (3)

engineered quartz. A countertop material made with chips of quartz and synthetic resin. It makes a durable, attractive countertop that is impervious to stains. (23)

engineered wood flooring. A type of flooring manufactured by laminating several plies and prefinishing the top ply. It is designed to float on the subfloor. (20)

entrance panel. An enclosure where the service ends inside of the house; contains a main breaker and circuit breakers or fuses for all of the home's circuits. (31)

entrepreneur. A person who starts and operates a new business. (1)

epoxy. A polymer (synthetic material) that is often used as an adhesive. (3)

equilibrium moisture content (EMC). The moisture content at which wood neither gains nor loses moisture when surrounded by air at a given relative humidity and temperature. (3)

estimator. A tradesperson who calculates labor, material costs, overhead, and profit before the contractor bids on a job. (1)

euro hinge. A type of cabinet hinge that fits into a pocket in the back of the door and is concealed when the door is closed. (23)

evaporator. The system of coils or tubes in an air conditioning unit that receives the cold refrigerant and allows it to absorb heat and evaporate. (33)

expanded metal lath. A plaster base consisting of a copper alloy steel sheet that is slit and expanded to form openings for keying the plaster. (19)

expanded polystyrene foam (EPS). A type of rigid insulation board that is a petroleum product that uses recycled polystyrene. Small beads formed during manufacturing are blown up like tiny balloons using pentane gas, a blowing agent that is safe for the ozone layer. (15)

exposure. The amount of material exposed to the weather in siding or shingles. (14)

extended rake. A gable overhang. (12)

exterior finish. The exterior materials of a structure. Generally refers to siding, roofing, cornice, and trim members. (17)

exterior insulation and finish systems (EIFS). A multilayered wall system that consists of a layer of insulation board attached to the wall sheathing, a water-resistant base coat with a reinforcing mesh, and a finish coat that resembles stucco or stone. (17)

exterior trim. Parts of the cornice and rake structure that are exposed to view. (17)

extruded polystyrene foam (XPS). A type of rigid insulation board that is impervious to moisture. It has an R-value of R-5 and a perm rating of 1. (15)

F

face. The outer ply of plywood. (3)

face nailing. A fastening method in which the nail is driven perpendicular to and through the surface of a piece. (20)

factory and shop lumber. A grade of lumber intended to be cut up for use in further manufacture. It is graded on the basis of the percentage of the area that will produce a limited number of cuttings of a specified size and quality. (3)

FAS (firsts and seconds). The best grade of hardwood lumber. It specifies that pieces be no less than 6″ wide by 8′ long and yield at least 83 1/3% clear cuttings. (3)

fascia. A wood member nailed to the ends of the rafters and lookouts and used for the outer face of a box cornice. (12, 17)

fascia backer. A 1″ or 2″ wood nailing strip attached to the rafter ends in a box cornice to support the fascia. (17)

faucet. A compression or washerless valve that permit the controlled flow of water as needed. (32)

fiber-cement siding. A strong, durable siding material made from cement reinforced with cellulose fiber. (17)

fiber saturation point. The stage in the drying or wetting of wood at which the cell walls are saturated and the cell cavities are free from water. It is assumed to be 30% moisture content, based on oven-dry weight, and is the point below which shrinkage occurs. (3)

field. The middle area of a sheet of wallboard. (19)

fill. The process of adding material to achieve the desired grade. (8)

filler. A heavy-bodied liquid used to fill open grain in wood before the application of stains or topcoats. (24)

finish coat. A third layer of plaster that is applied when the brown coat has set and is somewhat dry. (19)

finish flooring. Any material used as the final surface of a floor. (20)

finishing sander. A tool used for final sanding, where only a small amount of material needs to be removed. It is also used for cutting down and rubbing finishing coats. (5)

Fink truss. A W-shaped roof truss that can be used for spans up to 50′. Also called *W truss*. (12)

fireclay. Heat-resistant clay material used for flue linings in masonry chimneys. (25)

firestop. A block used in a building wall between studs to prevent the spread of fire and smoke through the air space. (10, 29)

fish tape. A long, flat, flexible steel tape used to pull electrical wires through conduit or walls. (31)

fixed anchor. A metal strap connector used to bind intersecting concrete block walls together. (9)

fixed window. A window that does not open. (16)

fixture. A device that receives water from the water supply system and has a means for discharge into the DWV system. (32)

flare fitting. A fitting that requires no soldering; designed to tightly fit against the tubing and mating parts. (32)

flashing. A sheet metal or other material used in roof and wall construction (especially around chimneys and vents) to prevent rain or other water from entering. (14)

flat bar. A tool ranging from 7″ to 15″ designed for removing larger nails or spikes from lumber of larger dimensions. (4)

flat-grained. A softwood lumber cut so the annular rings form an angle of less than 45° with the board's surface. (3)

flat ICF wall. A solid concrete wall formed by sheets or planks of polystyrene rigid foam. (9)

flat roof. A roof that is either level or pitched only enough to provide for drainage. (12)

flexible-back blade. A hacksaw blade that is hardened only around the teeth. (4)

flexible insulation. Insulation made of loosely felted mats of mineral or vegetable fibers. (18)

floating. The process of smoothing the surface of wet concrete using a large flat tool (float). (9)

floor plan. A line drawing that shows the size and outline of the building and its rooms, including many dimensions to show the location and size of inside partitions, doors, windows, and stairs. (7)

flue. The passage in a chimney through which smoke, gas, and fumes rise. (25)

flue lining. A noncombustible sleeve, usually made of fireclay, that is used to provide a smooth surface inside of a masonry chimney. (25)

flush door. A door that fits into the opening and does not project outward beyond the frame. (16, 22)

foamed-in-place insulation. Insulation materials that can be installed only in open cavities since they expand in volume 100 times beyond the applied thickness. (18)

footprint. The size and shape of a building shown on a plot plan. (7)

forced-air perimeter system. A system where air is circulated through large ducts that direct warmed or cooled air to every room in the building. (33)

form tie. A wood or plastic device that keeps concrete wall forms a uniform distance apart. (9)

foundation plan. A plan showing construction details for a foundation, often combined with basement plans. (7)

framing plan. A drawing showing the size, number, and location of the structural members of a building's frame. (7)

framing square. A square with a short section (called the *tongue*) attached at a right angle to the longer section (called the *blade* or *body*). A number of tables are imprinted on its sides. Also called a *rafter square*. (4)

free water. The moisture contained in the cell cavities of wood. (3)

frequency. The rate at which sound-energized air molecules vibrate; the higher the rate, the more cycles per second (cps). (18)

frieze. A boxed cornice wood trim member attached to the structure. It is used where the soffit and wall meet. (17)

full-adhesive-bonded flooring. A type of sheet vinyl flooring that must be fastened to the substrate with a coat of adhesive covering the entire back surface. (20)

G

gable roof. A roof consisting of a single ridge with slopes in both directions. Made entirely of common rafters. (12)

galvanic corrosion. The eating away of metal fasteners due to a chemical reaction between dissimilar metals. (30)

gambrel roof. A variation of the gable roof, where each slope is broken at midspan. This style is used on two-story construction. (12)

gate valve. A valve that controls water with a flat disk that slides up and down perpendicular to the supply pipe. It is never used to restrict flow; it is either fully open or fully closed. (32)

girder. A principal beam used to support other beams or the ends of joists. (10)

glass block. A window made of two formed pieces of glass fused together to leave an insulating air space between them. (16)

glue-laminated beam (glulam). A beam made by gluing and then applying heavy pressure to a stack of four or more layers of 1 1/2″ thick stock. (3)

grade. The quality of lumber. Also, the height or level of a building site. (8)

grade beam foundation. A foundation type similar to a spread foundation footing, but differs in that it is reinforced with rebar and rests on piling. (9)

grade leveling. The process of finding the difference in the grade between several points or transferring the same level from one point to another. (8)

granite. A hard natural stone that is used for countertops because of its durability and beauty. (23)

gravel stop. A type of drip edge fabricated from galvanized sheet metal. Attached to the roof deck to serve as a trim member and to help keep the mineral surface and asphalt in place. (14)

gravity-feed spray gun. A spray gun with material (paint) cup attached to the top of the gun, so the paint is drawn into the gun by gravity. They are primarily used for light stains, lacquers, and varnishes. (24)

gray water. Any washwater that has been used in the home, except water from toilets. (28)

green certification program. A set of standards for green building practices that are followed by the building team and verified by an independent third party. (28)

green rater. A green expert who is certified by LEED® to score each aspect of a building for LEED certification. Some green raters are employed by providers. (28)

green verifier. A green building expert who is certified by Home Innovation Research Labs to score green homes for the NGBS. (28)

ground-fault circuit interrupter (GFCI). An electrical safety device that can be installed either in an electrical circuit or at an outlet. It is able to detect a difference in the current flowing between the two legs of a circuit and shut off power automatically. To protect against electrical shock, it is used where water is present as an overcurrent protection device that is able to detect a short circuit more rapidly than a fuse or circuit breaker. (2, 31)

ground. 1—The electrical conductor that carries errant current to an earth ground in the event of a malfunction in an appliance, tool, or other load. It may either have green insulation or be bare wire. (2) 2—A strip of wood installed as a guide at the floor line and at openings in a wall used to strike off plaster. (19)

ground-level deck. A deck built close to the ground, not requiring posts for support. (30)

grout. A mixture of cement, sand, and water used to fill gaps between tiles. (20)

gusset. A plywood or steel plate fastened to the outside of a joint to give the joint strength. (12)

gutter. A metal or plastic trough attached to the edge of a roof to collect and conduct water from rain or melting snow. (14)

gypsum lath. A common plaster base consisting of a rigid gypsum filler with a special paper cover. (19)

gypsum wallboard. A wall covering panel consisting of a gypsum core with facing and backing of paper. Also called *drywall* or *sheetrock*. (19)

H

hacksaw. A saw that can be used to cut nails, bolts, other metal fasteners, and metal trim. Most have an adjustable frame, permitting the use of several sizes of blades. (4)

handrail. A pole installed above and parallel to stair steps to act as a support for persons using the stairs. Also called a *stair rail.* (21)

hand screw. A clamp with broad jaws that distributes the pressure over a wide area. Sizes (designated by the width of the jaw opening) range from 4″ to 24″. (4)

hardboard. A board material manufactured of wood fiber formed into a panel with a density of approximately 50–80 lb per cu ft. Also called *fiberboard.* (3, 19)

hard-coat system. A polymer-modified EIFS that is about 1/4″ thick and mechanically attached. (17)

hard hat. Protective headgear that is able to resist breaking when struck with an 80 lb ball dropped from a 5′ height. (2)

header. A horizontal structural member that supports the load over an opening, such as a window or door, or a floor framing member that supports the weight around floor openings. (10, 11)

header block. A concrete block that is dimensioned and shaped to form a shelf in a wall. (9)

head jamb. The part of a door frame that fits between the side jambs, forming the top of the frame. (22)

head joint. The end of the masonry block that butts against the previously laid block. (9)

head lap. The distance in inches from the butt of an overlapping shingle or roll roofing to the top edge of the shingle or roll roofing beneath. (14)

headroom. The clear space between the floor line and ceiling. (21)

hearing protection. Foam ear plugs, earmuffs, or other used to block out loud noise. Should be worn whenever working in the vicinity of loud equipment. (2)

hearth. The part of a fireplace that holds the fuel and contains the fire. (25)

heartwood. The wood extending from the pith or center of the tree to the sapwood. Heartwood cells no longer participate in the life processes of the tree. (3)

heat exchanger. A section of the furnace where heat from combustion gases is transferred by conduction to the circulating air. (33)

heating season performance factor (HSPF). An energy efficiency rating system used for large heating systems that measures the ratio of space heating required over a season to the electrical energy used. (33)

heat recovery ventilator (HRV). An air exchanger used during the heating season that uses the heat in the air that is being removed from the building to heat the incoming fresh air. (33)

heaving. The act of the soil under the footing freezes, forcing the foundation wall upward. (9)

heel. The wide end of a mason's trowel. (17)

herringbone paneling. Angled solid paneling that alternates the direction of the angle at regular intervals. (19)

hinge mortise. A recess cut into the edge of a door to receive the hinge leaf, other hardware, or the end of another structural member. (22, 23)

hip jack. The lower part of a common rafter, but intersects a hip rafter instead of the ridge. (12)

hip jack rafter. A short rafter connecting to the wall plate and a hip rafter. (12)

hip rafter. A rafter that runs from the wall plate to the ridge, always at a 45° angle. (12)

hip roof. A roof that rises from all four sides of a building. (12)

hole. An opening in lumber caused by handling equipment or boring insects and worms. (3)

home run. The wiring routed directly from the panel to the outlet. (31)

honeycombing. The separation of wood fibers inside the tree not always visible in the lumber. (3)

hopper window. A type of window with a sash hinged at the bottom. It swings inward. (16)

horizontal sliding window. A window with two or more sashes, at least one of which moves horizontally within the window frame. (16)

hot conductor. The electrical conductor considered to be supplying power to the load. The insulation on the hot conductor is either black or red. (2)

housed stringer. A stair stringer where the edges of the steps are covered with a board. (21)

housewrap. The plastic sheets used to seal exterior walls against air infiltration. (11)

hue. The class of a color. (24)

hydration. A heat-producing chemical reaction that occurs between cement and water to form concrete. (9)

hydronic perimeter heating system. A heating system that uses water as a medium to move heat from the boiler to the radiators. (33)

hydronic radiant heating system. A heating system that uses water to transfer heat energy from a heat source to tubing located in floors or other structural parts of a building. (33)

hydrophobic. A tendency to repel water. (15)

hypotenuse. The side of a right triangle opposite of the right angle. In roof framing, the length of the rafter is the hypotenuse. (12)

I

ice-and-water barrier. A self-sealing, waterproof covering installed at eaves and in valleys before installing roofing. (14)

I-joist. A beam consisting of flanges (also called *chords*) of structural composite lumber, such as laminated-veneer lumber, or sawn lumber, and a vertical web of oriented strand board, 3/8″ or 7/16″ thick. (3, 10)

impact driver. A compact tool used for driving screws and other fasteners. (5)

impact sounds. The sounds that are carried through a building by the vibrations of the structural materials themselves. (18)

indoor air quality. The amount of pollutants in the air inside a building. (28)

induction. A process by which an electric current is produced when a conductor cuts through a magnetic field. (31)

infiltration. The movement of air into an enclosed space through cracks and other openings. (18)

insulating fiberboard lath. A plaster base that comes in a 3/8″ and 1/2″ thickness with a shiplap edge. It has considerable insulation value and is often used on ceilings or walls adjoining exterior or unheated areas. (19)

insulation. Any material high in resistance to heat, sound, or electrical transmission. (18)

insulation board. The rigid insulation attached to the outside surface of the foundation to insulate floors or foundations of structures without basements. (18)

interior finishing. The installation of cover materials to walls and ceilings. (19)

intermediate colors. A color combination of a primary color and a secondary color already containing that primary color. Intermediate colors are red-violet, blue-violet, blue-green, yellow-green, yellow-orange, and red-orange. (24)

International Energy Conservation Code (IECC). A model code that covers minimum requirements for energy efficiency in new residential and commercial buildings. (28)

International Green Construction Code (IgCC). A model code that covers minimum requirements for improving the environmental and health performance of buildings and their sites. (28)

J

jack post. An adjustable steel post used in place of a stud for shoring. (29)

jalousie window. A window with a series of small horizontal overlapping glass slats that simultaneously rotate when operated. (16)

jamb. The top or side of a door or window frame that is firmly secured to the wall frame. (16)

jamb clip. A steel clip used to hang an interior door, eliminating the need to use shims. (29)

jamb extension. Narrow strips added to a window or doorjamb to increase its width. (16)

jointer. A tool used to dress the edges and ends of boards. They may also be used for planing a face or cutting a rabbet, bevel, chamfer, or taper. The cutter head is cylindrical and usually has three or four knives. As the cylinder rotates at high speed, the knives cut away small chips. (5)

jointing. An operation by which the height of the saw teeth is struck off evenly, usually done before filing. (4)

joist. The horizontal member of the floor frame. (10)

journeyman. A tradesperson who has completed an apprenticeship or, through experience, is qualified to work without supervision or further instruction. (1)

K

kerf. The cut made by a saw blade. (4)

kiln dried. Wood seasoned in a kiln by means of artificial heat, controlled humidity, and air circulation. (3)

king post truss. Roof truss with top and bottom chords and a vertical center post used on shorter spans up to 22′. (12)

king stud. A full-length stud nailed alongside the trimmer at each end of the header. (29)

knot. A lumber defect caused by a branch or limb embedded in the tree and cut through during lumber manufacture. (3)

L

ladder jack. A system of scaffold used with planking between two ladders. Setup requires two sturdy ladders of the same size. (6)

laminated-strand lumber (LSL). A lumber made of 1/32″ × 1″ × 12″ strands of wood bonded with polyurethane adhesive. It is available in two thicknesses: 1 1/4″ and 3 1/2″. Depths vary up to a maximum 16″ and lengths vary up to 35′. Uses include: door and window headers, rim joists in floors, and core stock for flush doors with veneer overlays. (3)

laminated-veneer lumber (LVL). An engineered lumber product that is made up of veneer layers and is designed to carry heavy loads. Produced in a manner much like plywood, except all of the veneer panels have their grain running in the same direction. (3)

laser level. A leveling instrument that emits a level laser beam over 360° without being attended. A receiver attached to a rod and attended by a rod holder makes a sound when the laser beam strikes it. (8)

Leadership in Energy and Environmental Design (LEED®). A program of scoring green building construction and certifying buildings for certain levels of green construction. LEED can be applied to all building classifications. (28)

ledger. A strip attached to vertical framing or structural members to support joists or other horizontal framing. Similar to a ribbon. (10)

ledger board. The horizontal component of batter boards that support the lines set up to locate building lines and corners. (9)

ledger strip. A 1″ or 2″ ribbon horizontally nailed along the wall to support the lookouts. (17)

LEED® for Homes Provider. A green construction expert who is certified by LEED to assist green construction teams with designing and constructing green homes. (28)

level. An instrument designed with a vial filled with a liquid to create a bubble used for laying out vertical and horizontal lines. (4)

leveling rod. A long rod marked off with numbered graduations that is used, along with a builder's level or transit, to sight differences in elevation over long distances. (8)

liability. Legal responsibility. (1)

light tube. A type of skylight, typically about 12″ in diameter, that consists of a clear dome and a tube with a reflective lining. (16)

lignin. A substance in wood that binds cell walls together. (3)

line of sight. A straight line that does not dip, sag, or curve. Any point along a level line of sight is at the same height as any other point. (8)

lintel. A horizontal structural member that supports the load over an opening, such as a door or window. Also called a *header*. (9)

lipped door. A cabinet door rabbeted (partially recessed) along its edges. The resulting lip conceals the joint. (23)

list of materials. A listing of all the materials and assemblies needed to build the structure. Also called a *bill of materials, lumber list,* or *mill list.* (7)

live load. The total of all moving and variable loads that may be placed on a building. (12)

longitudinal beam. A roof beam that runs parallel to the supporting sidewalls and ridge beam. (26)

lookout. A structural member that supports the soffit and runs between the lower end of a rafter and the outside wall. (12, 17)

loose-fill insulation. An insulation used in bulk form and supplied in bags or bales. (18)

lot. A small parcel of land, generally sized for one house or building in a city. (8)

lugged tile. A tile with projections at all edges to maintain proper spacing between individual tiles. (20)

lumber. The name given to natural or engineered products of the sawmill. Logs that have been sawed into usable boards, planks, and timbers. (3)

lumber core. A hardwood plywood that, when compared with veneer core, is easier to cut, has edges that are better for shaping and finishing, and holds nails and screws better. (3)

M

magnetic starter. A safety device on a power tool that automatically turns the switch to the *off* position in case of a power failure. (5)

manifold. A pipe with several outlets that distributes heated water to different areas. (33)

mansard roof. A type of roof that has four sloping sides. However, each of the four sides has a double slope. The lower, outside slope is nearly vertical. The upper slope is slightly pitched. (12)

manual dexterity. The ability to efficiently accomplish tasks requiring use of the hands. (1)

manufactured home. A home built on a chassis and finished and fully equipped at the factory, then pulled to the dealer or directly to the home site. Manufacturers define it as a trailer longer than 28′ and heavier than 4500 lb. (27)

marking gauge. A tool used to lay out parallel lines along the edges of material. It may also be used to transfer a dimension from one place to another and check sizes of material. (4)

masking sounds. The normal sounds within habitable rooms, which tend to hide or mask some of the external sounds entering the room. (18)

mason's line. A reference line (cord) used to keep masonry units aligned as they are laid up. (9)

mastic. A thick adhesive used in construction as a bonding agent for tile and paneling. (3)

measuring tape. A flexible measuring tool in a case with a hand-operated crank for retracting. Typically available in lengths of 50′–300′. (4)

mechanical core. A prefabricated building module that contains one or more of the following utilities: electrical, plumbing, heating, ventilating, and air conditioning. (27)

mechanical system. The installation in a building including plumbing; electrical wiring; and heating, ventilating, and air conditioning. (31)

metal-strap bracing. Used to help exterior walls resist lateral (sideways) loads. Typically made of 2″ wide, 18 or 20 gauge galvanized steel. (11)

metal stud. A stud used with either metal or wood plates. (13)

mineral wool. A material that is spun into fibers from molten rock. (15)

miter saw. A power tool where a motor and blade of this saw are supported on a pivot. A carpenter sets an angle from a scale marked off in degrees, then makes a cut by pulling downward on the handle. (5)

model code. A code developed by a national organization with the intent that it may be adopted in whole or in part by a local community. (7)

modular home. A home consisting of two or more three-dimensional, factory-built units that are assembled on the building site. (27)

module. A three-dimensional assembled housing unit built in a factory and transported to a building site to be combined with other modules. (27)

moisture content (MC). The amount of water contained in wood, expressed as a percentage of the weight of oven-dry wood. (3)

molding. A relatively narrow strip of wood trim, usually with a curved profile throughout its length, used to accent and emphasize the ornamentation of a structure and to conceal surface or angle joints. (22)

mold-resistant wallboard. A type of gypsum wallboard processed to resist the effects of moisture and high humidity. It is not used as a base under ceramic tile and other nonabsorbent finishes used in showers and tub alcoves. (19)

monolithic slab. A term used for concrete construction poured and cast in one unit without joints. (9)

mortar. A combination of cement, sand, lime, and water used to bind masonry blocks. (9, 17)

mortise-and-tenon joint. A type of joint in which a tongue (tenon) on the end of one wood member is inserted into a pocket or slot (mortise) cut into another member. Joints may be glued or fastened with wood pegs or mechanical fasteners. (26)

mosaic tile. A small pieces of tile of different colors laid to form a pattern or picture. (20)

mudsill. Part of the floor frame that rests on the foundation wall to prevent settling. Also called *sill plate*. (6)

mullion. A slender bar forming divisions between units of windows, screens, or similar generally nonstructural frames. (16)

multilevel deck. A deck structure with connected sections built on more than one level. (30)

muntin. A vertical or horizontal sash bar separating different panes of glass in a window. (16)

N

nailer. A piece of lumber, such as 1×6 or 2×4, added as a backing for attaching other members or covering. (11)

nailer board. A wood strip installed on edge at hips and ridges to support trim tiles. (14)

nail gun. A power tool used for driving nails. (5)

nailing strip. Wooden strip cast into a concrete foundation around a window or door opening and used to attach the door or window frame. (9)

nail puller. A specialty tool with a fixed jaw and a movable jaw and lever. A sliding driver on the end of the handle drives the jaws under the nail head. Leverage tightens the jaw's grip on the nail, raising it above the wood's surface. (4)

nail set. Tools designed to drive the heads of casing and finishing nails below the surface of the wood. Tips range in diameter from 1/32″ to 5/32″ in increments of 1/32″. (4)

National Electrical Code (NEC). A collection of rules developed to control and recommend methods of installing electrical systems. (31)

National Green Building Standard (NGBS). A building rating and certification program developed by Home Innovation Research Labs and adopted by the International Code Council. It is also known as *ICC 700*. (28)

neutral conductor. The electrical conductor considered to be carrying current back to the source from the load. The neutral conductor has white insulation. (2)

newel. The main post at the start of a stair and the stiffening post at the landing. (21)

No. 1 common. A low grade of lumber that is expected to yield 66 2/3% clear cuttings. (3)

noise. Unwanted sound, which is a vibration or wave motion that can be heard. (18)

noise-reduction coefficient (NRC). The average percentage of sound absorption at 250, 500, 1000, and 2000 hertz (cycles per second). (18)

nosing. The part of a stair tread that projects beyond the riser. (21)

O

Occupational Safety and Health Administration (OSHA). An agency of the United States Department of Labor formed in 1971 whose charge has been to assure safe and healthful conditions for workers. (2)

oilstone. A stone, lubricated with oil, used for honing edge tools. (4)

one-pipe system. A hydronic heating system in which a single pipe supplies heated water to room heating units. (33)

open cornice. A cornice with no enclosing parts. (17)

open-grain wood. Wood with large pores, such as oak, ash, chestnut, and walnut. (3)

open porch. An attached structure that has no railing—just a deck, roof, and supporting columns. (30)

open time. The amount of time between spreading an adhesive and when the parts must be clamped. (3)

open valley. A method of roofing in which the roof covering is set back from the middle of the flashed valley. (14)

oriented strand board (OSB). A formed panel consisting of layers of compressed strand-like wood particles, arranged at right angles to each other. (3)

O-ring. A ring used on faucets to produce leak-proof seal between the faucet body and internal parts. (32)

oscillating multi-tool. A power tool with a blade that is offset, making it useful for cutting, scraping, sanding, and grinding in difficult to reach places. (5)

OSHA standards. A set of rules that describe the methods that employers must use to protect their employees from hazards. (2)

overlay door. A slab door in cabinetry that covers the cabinet frame. (23)

P

panel door. A style of door that has a frame with separate panels of plywood, hardboard, or steel set into the frame. Also known as a *stile and rail door*. (16, 22)

panelized home. A home built in sections in a factory. Components such as wall sections and roof sections are then shipped to the building site and erected by carpentry crews. (27)

parallel lamination. A laminating process in which all of the veneer panels have their grain running in the same direction used to produce laminated-veneer lumber. (3)

parallel-strand lumber (PSL). An engineered lumber made up of strands of wood up to 8′ long that are bonded together using adhesives, heat, and pressure. (3)

particleboard. A formed panel consisting of wood flakes, shavings, and chips bonded together with a synthetic resin or other added binder. (3)

partition. A wall that subdivides space within any story of a building. (11)

passive house. A home built to a high standard of efficiency, which results in ultra-low energy consumption while maintaining a comfortable interior living environment. (28)

paver tile. A concrete-based masonry unit used for finish flooring and walkways. (20)

pencil compass. A tool used to draw arcs and circles or to step off short distances. (4)

pergola. A structure with an open canopy that provides partial shade from the sun. (30)

perimeter-bonded flooring. A type of sheet vinyl flooring that is fastened down only around the edges and at seams. (20)

permanent wood foundation (PWF). A special building system that saves time because it can be installed in almost any weather. (9)

permeability. The tendency to absorb water. (20)

perm rating. A unit of measure that tells how much water vapor can pass through a material. (15)

phloem. The outside area of the cambium layer of a tree. It develops cells that form the bark. (3)

photosynthesis. The chemical process that plants use to store the sun's energy. Trees use sunlight to convert carbon dioxide and water into leaves and wood. (3)

pictorial sketch. A three-dimensional drawing, much like a photo, that shows how a project will look when built. (7)

pigment. An ingredient of paint that provides color and opacity. (24)

pilaster. A part of a wall that projects not more than one-half its own width beyond the outside or inside face of a wall. Chief purpose is to add strength, but may also be decorative. (9)

pipe compound. A substance used to seal threaded connections between pipe and other parts. (32)

pitch. The ratio of the rise to the span (twice the run). (12)

pitch pocket. A cavity that contains or has contained pitch in solid or liquid form. (3)

plain footing. A footing system that carries light loads and usually does not need reinforcing. (9)

plain-sawed. A hardwood lumber that is cut so that the annular rings form an angle of less than 45° with the surface of the board. (3)

plank. A floor and roof decking material that can be anywhere from 2″ to 4″ thick, depending on the span. Edges are usually tongue-and-groove for a tight, strong surface. (26)

plank flooring. A type of flooring similar to strip flooring, but bored and plugged at the factory to simulate the wooden pegs used to fasten the strips in colonial times. (20)

plaster. A mixture of gypsum and water that can be troweled wet onto interior walls and ceilings. (19)

plastic laminate. Material used as a surface for counters and tops that offers high resistance to wear and is unharmed by boiling water, alcohol, oil, grease, and ordinary household chemicals. (23)

platform. A horizontal section between two flights of stairs. Also called a *landing.* (21)

platform framing. A system of framing a building where the floor joists of each story rest on the top plates of the story below, and the bearing walls and partitions rest on the subfloor of each story. Also called *western framing.* (10)

plenum. A sheet metal chamber in a furnace that collects heated air in preparation for its transfer to rooms. (33)

plies. The layers of asphalt-saturated felts used on built-up roofs. (14)

plinth block. The decorative, carved block set at left and right top corners of window and door trim. (22)

plot plan. A drawing of the view from above a building site. The plan shows distances from a structure to property lines. Sometimes called a *site plan.* (7)

plug. A wooden piece shaped to fit and forced into a defect. (24)

plumb. The quality of being exactly perpendicular or vertical; at a right angle to the horizon or floor. (8)

plumb bob. A device that establishes a vertical line when attached to and suspended from a line. Its weight pulls the line in a true vertical position for layout and checking. (4)

plumber's putty. Putty used on sink rims, strainers, and faucets to produce a seal. (32)

pneumatic tool. A tool that is powered by compressed air. (2)

pocket door. A sliding door suspended on a track that slides into a wall cavity when opened. Also called a *recessed door*. (22)

polyiso. A moisture-resistant insulation material. It uses a blowing agent that slightly depletes the ozone. Its R-value is higher than other rigid foam products. (15)

polyurethane adhesive. A single component, moisture-catalyzed adhesive that will bond with wood of up to 25% moisture content. (3)

polyvinyl resin emulsion adhesive. A wood adhesive intended for interiors. Made from polyvinyl acetates, which are thermoplastic and not suited for temperatures over 165°F (75°C). (3)

post anchor. A bracket that supports the bottom of a post above the floor, protecting the wood from dampness. (10)

post-and-beam construction. A type of construction consisting of posts, beams, and planks that can be spaced farther apart than conventional framing members. (26)

post-and-beam framing. A framing method in which loads are carried by a frame of posts connected with beams, eliminating the need for load-bearing walls. Also called *plank-and-beam framing*. (10)

post-and-beam ICF wall. A poured wall contained by insulated concrete forms. Pour is shaped into vertical or horizontal concrete members more than 12″ apart OC. (9)

potable. Fit to drink. (32)

powder-actuated tool (PAT). A striking tool that uses the force of an exploding cartridge to drive fasteners into concrete and steel. (5)

power block plane. A small power plane that can be used on small surfaces and operated with one hand. (5)

power plane. A power tool that produces finished wood surfaces with speed and accuracy. The motor, which operates at a speed of about 20,000 rpm, drives a spiral cutter. (5)

power stapler. A power tool used for driving staples. (5)

preacher. A small block made from 3/8″ or 1/2″ hardwood and notched to fit over the siding. It is used to mark the siding where it is to be cut off. Also called a *siding gauge*. (17)

precut home. A structure for which lumber is cut, shaped, and labeled to reduce labor and save time on the building site. Manufacturers of this type of house include materials needed to form the outside and inside surfaces. (27)

prehung door unit. A door frame with the door already installed. (22)

premium grade. A special grade of hardwood known as "architectural" or "sequence-matched." It usually requires an order to a plywood mill for a series of matched plywood panels. (3)

pressure differential. The difference in air pressure between the inside and outside of a building envelope. (15)

pressure-feed spray gun. A spray gun that diverts some air pressure into the paint container to force the finishing material into the air-stream. This gun is intended for spraying heavy-bodied paints. (24)

pressure-treated lumber. Wood that has been treated with chemical preservatives to protect it from rot and insects. (2, 3)

primary colors. The three colors (red, blue, and yellow) that can be combined to form all other colors. (24)

print. A copy of the architect's original drawing. Preferred term over blueprint. (7)

property line. The boundary of a site. (8)

pry bar. A prying tool designed for removing larger nails or temporary braces from lumber of larger dimensions. (4)

pull. A piece of hardware that provides a means of grasping and opening a door or drawer. (23)

pull bar. A tool used with a hammer when driving planks together in tight quarters, such as near a wall. (20)

pull saw. A saw with a thin blade and fine teeth that cuts as it is pulled toward the user. (4)

purlin. A horizontal roof member used to support rafters between the plate and ridge board. (12)

putty. A plastic, dough-like material used to fill holes and large depressions in wood surfaces. Also used to fill nail holes before painting. (24)

Q

quarry tile. A tile used in construction as finish flooring and countertops. (20)

quarter-sawed. A hardwood cut so the annular rings form an angle of more than 45° with the surface of the board. (3)

R

racking. In flooring, laying out seven or eight rows of strip flooring prior to nailing them in place. (20)

radial arm saw. A stationary power tool with a motor and blade carried by an overhead arm. The stock being cut is supported on a stationary table. The arm is attached to and supported by a column at the back of the table. By raising or lowering the overhead arm, the depth of cut can be controlled. (5)

radiation. The transmission of heat by wave motion, much like light movement. (18, 33)

radon. An invisible, radioactive gas that is given off by some rocks and soils. It is heavier than air, so it tends to collect near the floor level. (28)

rafter. One of a series of structural members of a roof designed to support roof loads. (12)

rail. A cross or horizontal member of the framework of a sash, door, blind, or other assembly. (22)

rainscreen. A gap between the WRB and the cladding that allows water to drain freely. (15)

rake. The trim members that run parallel to the roof slope and form the finish between the roof and wall at a gable end. (17)

rebar. A steel reinforcing rod placed in concrete. (9)

receiving channel. The area of any vinyl or aluminum siding trim that accepts and conceals the cut end of a siding or soffit panel. (17)

receptacle. A device that allows for current transfer between conductors and appliances. Also called an *outlet* or *convenience outlet*. (31)

reciprocating saw. A portable power tool that has linear (back-and-forth) cutting action. (5)

reflective insulation. A metal foil or foil-surfaced material. The number of reflecting surfaces, not the thickness of the material, determines its insulating value. (18)

refractory mortar. A mortar made with fire clay to resist flame and high temperatures. (25)

register. A grille installed over the end of heating or air conditioning duct to direct the heated air. (33)

reinforced footing. A footing containing steel rebar for added strength. (9)

related colors. Hues that are side by side on the color wheel. (24)

renewable energy. A type of energy from sources that are naturally replenished, such as solar, wind, and water flow. (28)

resistance (R). The reciprocal (inverse) of conductivity or conductance. (18)

respirator. A protective device worn to provide adequate protection from breathing contaminants that are immediately hazardous to your health. (2)

reveal. The amount of setback of a window or door casing from the inside edge of the jamb. (22)

ribbon. A narrow, horizontal board let into a stud or other vertical member of a frame, used as support for joists or other horizontal members. (10, 29)

ridge. The centerline at the junction of the top edges of two roof surfaces. (12)

rigging. Describes the slings, cables, and chains used with a crane to lift heavy objects into place. (6)

rigid insulation board. An insulation manufactured in rigid panels. (15)

ripsaw. A saw with chisel-shaped teeth, which cut best along the grain. (4)

rise. The distance a rafter extends vertically above the wall plate. (12)

riser. The vertical stair member between two consecutive stair treads. (21)

roll roofing. A sheet of asphalt-saturated felt or fiberglass with mineral granules on the top surface. The uncut form of mineral-surfaced shingle material. (14)

roofing bracket. Metal device that attaches securely to a sloped roof and provides a secure, level support for planking. Also called *roof jack*. (6)

roofing tile. A tile manufactured from concrete or molded from hard-burned shale or mixtures of shale and clay. (14)

roof truss. An engineered and prefabricated rafter assembly. The lower chord also serves as a ceiling joist. (12)

rotary hammer drill. A heavy-duty power tool for drilling large holes in concrete and other masonry materials. (5)

rough-in. The first phase of plumbing, which includes installing all pipes to the point where connections are made with the plumbing fixtures. (32)

rough opening. An opening formed by framing members to receive and support a window or door. (11)

router. A power tool used to cut irregular shapes and to form various contours on edges. (5)

run. The horizontal distance of rafters from the outer edge of the wall plate to the center of the ridge. (12)

run of stairs. A series of steps that is a continuous section without breaks formed by landings or other constructions. Also called a *flight of stairs*. (21)

R-value. A unit of measure related to the efficiency of an insulating material. (15)

S

saber saw. A portable power saw useful for a wide range of light work. Also called a *portable jig saw.* (5)

saddle. A small gable roof placed in back of a chimney on a sloping roof to shed water and debris. Also known as a *cricket.* (14)

saddle beam. A beam positioned above the joists it supports. (29)

safety boots. Protective footwear that must be worn on a jobsite where there is risk of injury from falling, from sharp objects, or from electrical hazards. (2)

Safety Data Sheet (SDS). A sheet that describes the hazards, safe handling, and what to do in case of exposure to any product containing a substance that may pose a hazard to human health or safety. (2)

safety factor. A specified number of times (for example, four) in excess of the expected maximum load that a system must be capable of carrying. (2)

safety glasses. A type of eye protection with a safety lens that must withstand the blow of a 1/8″ diameter steel ball dropped from a height of 50″. (2)

safety shoes. An attachment placed on the feet of ladders to enhance stability on uneven or slippery ground. (6)

sanded grout. A type of grout in which cement is combined with sand. (20)

sapwood. The layers of wood next to the bark, usually lighter in color than the heartwood, that are actively involved in the life processes of the tree. More susceptible to decay than heartwood, it is not necessarily weaker or stronger than heartwood of the same species. (3)

sash. The framework that holds the glass in place and forms a tight seal with the frame. (16)

saturated felt. A dry felt soaked with asphalt or coal tar and used under shingles for sheathing paper and as laminations for built-up roofs. (14)

scaffolding. A temporary structure or platform used to support workers and materials during building construction. Also called *staging.* (6)

scarf joint. A mitered-lap joint used where a straight run of molding must be joined. (22)

schedule. A part of the plans, generally presented in tabular form, that provides basic information about related components. (7)

scissors truss. A truss used for buildings having a sloping ceiling. In general, the slope of the bottom chord is 1/2 of the slope of the top chord. (12)

scratch awl. A tool used to scribe lines on the surface of the material. It is also used to mark points and to form starter holes for small screws or nails. (4)

scratch coat. A first layer of plaster that has its surface roughened to provide "tooth" (a roughened surface) for succeeding layers. (19)

screeding. The process of leveling off concrete slabs or plastering on interior walls, using a screed. (9)

screen-grid ICF wall. A poured concrete wall using insulated forms. Resulting wall forms a grid similar to a window screen. (9)

seasonal energy efficiency ratio (SEER). An energy efficiency rating system used for large air-conditioning systems. (33)

seasoning. The process of reducing the moisture content of lumber to the required level specified for its grade and use. (3)

Second Law of Thermodynamics. A scientific law that heat always moves from hot to cold. (33)

secondary color. A color (violet, green, and orange) that is produced when two primary colors are combined. (24)

section drawing. A type of drawing that shows how a part of a structure would look if cut along a given plane. (7)

select. The hardwood grade below first and seconds. (3)

self-adhering tile. A floor tile with factory-applied adhesive. (20)

self-employment. Starting and operating a business of one's own. (1)

self-tapping drywall screw. A screw used to attach wall surface material, such as drywall or paneling, to metal studs. (13)

selvage. The part of the width of roll roofing that is smooth. For example, a 36″ width has a granular surfaced area 17″ wide and a 19″-wide selvage area. (14)

semi-enclosed porch. An attached structure with either an ornamental balustrade or a low wall enclosure. (30)

service. The conductors that bring power from a transformer to a building. (31)

set of plans. Many sheets of drawings bound together. They include various plans (such as the floor plan), drawings of the mechanical systems, elevation drawings, section drawings, and detail drawings. (7)

setback. The minimum distance a building is allowed to be placed to a property line, as established by the local zoning ordinances. (7, 8)

setup line. A chalk line the width of two flooring strips plus 1/2″ away from a partition. (20)

sexual harassment. Any unwelcome sexual advances or requests for sexual favors. (1)

shade. A pure color with black added. Color nearer black in value. (24)

shake. A handsplit shingle. (3)

shear wall. A wall with code-required bracing. (29)

sheathing. A board or prefabricated panel that is attached to the exterior studding or rafters of a structure. (11)

shed roof. A single-slope roof, sometimes called a *lean-to roof.* (12)

shelf. A fixed or adjustable horizontal divider in cabinets designed to provide storage. (23)

shingle butt. The lower, exposed portion of a shingle. (14)

shoring. A method of blocking or strong bracing used to support heavy loads or provide temporary support. (2, 29)

short circuit. An accidental grounding of a current-carrying electrical conductor. Also called a *ground fault.* (31)

side-and-end matched. A strip flooring that is tongue-and-groove on the sides and ends. (20)

side jamb. The part of a door frame that is dadoed to receive the head jamb. (22)

side lap. The overlap distance for side-by-side elements of roofing. (14)

sill. The lowest member of the frame of a structure that rests on the foundation and supports the uprights of the frame. (10)

sill plate. A part of the floor frame that rests on the foundation wall. Also called *mudsill.* (10)

sill sealer. A strip of insulation between the sill plate and foundation. (10)

single coursing. A process of applying shingle siding in single layers. (17)

single-layer construction. An interior finishing process used where economy, fast installation, and fire resistance are important. (19)

siphon-feed spray gun. A spray gun designed for applying light-bodied materials such as shellacs, stains, varnishes, lacquers, and synthetic enamels. (24)

sistering. A process of reinforcing framing members by nailing another member alongside the first member. (29)

six-inch method. A shingling pattern for three-tab shingles where the cutouts break joints on every other course. Cutouts are centered over the tab in the course directly below. (14)

SkillsUSA. An organization that promotes the development of excellence in a variety of occupations, including building trades. (1)

skip sheathing. The board sheathing on a roof spaced to allow wood shingles to dry from the underside. (12)

slab foundation. A poured concrete foundation supported by the soil. (9)

slab-on-grade. Slab foundation that is supported directly on top of the ground. Also called *ground-supported slab.* (9)

sledgehammer. A hammer available in weights from 2 lb to 16 lb that is used to align concrete forms, to drive stakes, and for other heavy work. (4)

sleeper. A wood strip laid over or embedded in a concrete floor to which finish flooring is attached. Also called a *screed.* (20)

sling. A device made of chain, wire rope, or synthetic fibers to hoist a heavy load with a crane. (6)

slip joint washer. A washer used to seal P-trap connections beneath sinks. (32)

slip sheet. A sheet of heavy wrapping paper placed over the base surface when adhering laminates. (23)

slope. The incline of a roof. (12)

smoke chamber. The space extending from the top of the throat to the bottom of the flue in a fire place. (25)

smoke shelf. The horizontal shelf located adjacent to the damper in a fireplace that helps the damper change the direction of the downdraft. (25)

snub cornice. A cornice with no overhang. Rafters end flush with the sidewalls. (17)

soffit. The underside of the members of a building, such as staircases, cornices, beams, and arches. Also the enclosure that fills in the space between the tops of cabinets and the ceiling. Relatively minor in size as compared with ceilings. Also called *drop ceiling* or *furred-down ceiling.* (11, 17)

soft-coat system. A polymer-based EIFS that is typically thin (1/8″), flexible, and attached with adhesives. (17)

sole plate. The horizontal member at the bottom of a wall frame. The sole plate is supported by a wood subfloor, concrete slab, or other closed surface. (11)

solid bridging. A pair of solid pieces of 2″ lumber installed between joists. Often installed above a supporting beam to keep the joist vertical, but also adds rigidity to the floor. (10)

solid-surface. A countertop material made of acrylic or polyester resin with pigment added for color. (23)

sound transmission class (STC). A single number that represents the minimum sound-deadening performance of a wall or floor at all frequencies. (18)

sound transmission loss (STL). The number of decibels sound loses when transmitted through a wall or floor. (18)

span. The distance between structural supports for a horizontal member. (10)

span rating. The maximum distance a structural member can extend without support when the long dimension is at right angles to its supports. (3)

specifications. A written document stipulating the type, quality, and sometimes the quantity of materials and work required for a construction job. (7)

speed square. A triangular measuring tool that can be used to mark any angle for rafter cuts by aligning the desired degree mark on the tool with the edge of the rafter being cut. (4, 12)

splay. The angle of slope of the rear and side walls of a fireplace combustion chamber. (25)

spline. A thin strip of hardwood that fits into mortises or grooves machined into boards that are to be joined. (20)

splits and checks. The separations that run along the grain and across the annular growth rings. (3)

spread footing. A foundation system that transmits the load through the walls, pilasters, columns, or piers to a wide footing. (9)

square. A unit of measure—100 square feet—usually applied to roofing material and to some types of siding. (14)

stairwell. The rough opening in the floor above to provide headroom for stairs. (21)

starter strip. A strip of mineral-surfaced material placed either beneath the first course of shingles to cover the gaps between shingle tabs or at the bottom of a wall to which vinyl siding is attached. (14)

starter track. A channel used at the top and bottom of walls and around openings in an EIFS application. (17)

steel framing member. A framing component manufactured in various widths and gauges and used as a stud, joist, or truss rafter. (13)

steel post. The most popular post for girder and beam support. (10)

stepped footing. A footing that changes grade levels at intervals to accommodate a sloping site. (9)

sticker. A wood strip used to create separation between materials. (3)

stile. The upright or vertical outside pieces of a sash, door, blind, screen, or face frame. (22)

stock plan. A set of plans that is mass-produced to be offered for sale to many clients. (7)

stool. The trim forming the interior sill cap of a window. (22)

storm sash. A sash unit attached to the outside of the window frame. Used to increase the U-factor of the window unit. (16)

story pole. A straightedge, often made from a piece of lumber, marked up to check the height of each course of concrete block. Also used to lay out and transfer measurements for door and window openings, siding and shingle courses, and stairways. (9, 11)

straight run. A stairway that does not change direction. (21)

stringer. A sloping member that supports the risers and treads of stairs. Also called a *carriage*. (21)

strip flooring. A floor consisting of hardwood cut into narrow strips, tongued and grooved, and laid in random pattern of end joints. (20)

strongback. An L-shaped wooden support attached to tops of ceiling joists to strengthen them, maintain spacing, and bring them to the same level. (11)

structural insulated panel (SIP). A panel made of two layers of structural sheathing with insulating foam sandwiched between them. (11)

structurally supported slab. A slab that can be used with other elements such as walls, piers, and footings. (9)

structural sheathing. A type 2 plywood or oriented strand board, normally applied to both sides of the wall frame. Used for permanent bracing of load-bearing walls. (13)

structured wiring. An integrated system of low-voltage wiring and centralized control units used to provide a home with a variety of communication, automation, and security services. (31)

stub-in. Pipe ends that extend out from the walls and above the floors and that are to be connected to fixtures. (32)

stud. One of a series of vertical structural members in walls and partitions. (11)

sub-fascia. A framing member behind the fascia trim, which provides support and backing for fascia trim. (17)

subfloor. A covering of panels nailed to the joists over which a finish floor is laid. (10)

submersible pump. A water pump that is installed in a well. (32)

suspended ceiling. A metal framework designed to support tile or panels. (19)

sweat sheet. A strip of felt or ice-and-water barrier. (14)

sweat soldering. A joining process for copper pipe in which the fitting and pipe are heated with a gas flame and solder is drawn into the joint by capillary action. (32)

switch. A device that controls current to lights and appliances. (31)

symbol. A sign or marking used in plans to represent materials and other items. (7)

systems-built house. A house built of components designed, fabricated, and assembled in a factory. (27)

T

table saw. A stationary power tool with a cutting blade extending out from its flat work surface. (5)

tag line. A rope attached to a load on a crane to control the load and prevent it from swinging and twisting. (6)

tail joist. A short joist spanning from the band joist to the header of a floor opening. (10)

tape measure. A flexible measuring tool in a self-retracting case. The most common lengths are 25′–35′. (4)

T-bevel. A tool with an adjustable blade, which makes it possible to transfer an angle from one place to another. It is useful in laying out cuts for hip and valley rafters. (4)

Teflon® tape. Tape used to seal threaded connections between pipe and other parts. (32)

tension bridging. The use of metal connectors to transfer load from one joist to another. (3)

termination top. The part of the chimney that extends above the roof. (25)

thermoplastic. A material that softens under heat. (3)

thermoset adhesive. An adhesive that is more water- and heat-resistant than the polyvinyl adhesives. In this category are polyurethanes, urea formaldehydes, and resorcinol formaldehydes. (3)

thermostat. An instrument that automatically controls the operation of heating or cooling devices by responding to changes in temperature. (33)

threshold. A beveled or tapered piece attached to each side of a door and is used to close the space between the bottom of a door and the sill or floor underneath. (22)

threshold of pain. The level of loud sounds of over 130 decibels that can cause pain to ears. (18)

throat. The narrowed passage above the hearth and below the damper in a fire place. (25)

tin snips. A tool used to cut asphalt shingles and light sheet metals, such as flashing. (4)

tint. A pure color with white added. Color nearer white in value. (24)

toe. The pointed end of a mason's trowel. (17)

toeboard. A board horizontally fastened to scaffolding slightly above the planking to keep tools and materials from falling on workers below. (6)

top lock. The receiver of a buttlock on the top of a panel of vinyl siding or soffit. Between the nailing hem and the exposed face of the panel. (17)

top plate. The horizontal member at the top of a wall frame. (11)

total heat transmission (U). Measured in Btu per sq ft per hour with 1°F temperature difference for a structure (wall, ceiling, and floor) that may consist of several materials or spaces. (18)

total rise. The vertical distance from one floor to another. (21)

total run. The horizontal distance occupied by the stairs; measured from the foot of the stairs to a point directly beneath where the stairs rest on a floor or landing above. (21)

total station. A leveling instrument similar to a transit in that it measures horizontal and vertical angles, but it also measures distances. It combines optical sensors with electronic circuitry to calculate angles and distances. (8)

tracheid. A long narrow tube or cell that makes up wood. Also called *fibers*. (3)

track. A steel framing member formed into a U-shaped channel that is used at the top and bottom of a wall. The steel studs fit into the track and are secured by screws or welds. (13)

transformer. A device for changing the voltage of an electric current. (31)

transit. An optical leveling instrument commonly used for building layout. (8)

transverse beam. A roof beam that runs perpendicular to the ridge. (26)

tread. The horizontal walking surface of a stair. (21)

trench. An excavation that is deeper than it is wide and is less than 15′ wide at the bottom. (2)

trench box. A steel box, inside of which work can be performed in safety. (2)

trestle jack. A jack used to support low platforms for interior work. (6)

trimmer. The beam or floor joist a header is framed into. Adds strength to the side of the opening. (10)

trimmer stud. The stud a header is framed into. It adds strength to the side of the opening. Also called *jack stud*. (11)

troweling. The process of producing a dense, smooth surface on a concrete floor using a trowel (metal tool with a smooth finish). (9)

truss. A structural unit consisting of such members as beams, bars, and ties arranged to form triangles. Provides rigid support over wide spans with a minimal amount of material. Used to support roofs and joists. (12)

truss plan. A drawing that shows the layout of trusses in a roof frame. (7)

try square. A square with blades 6″–12″ long. Handles are made of wood or metal. These are used to check the squareness of surfaces and edges and to lay out lines perpendicular to an edge. (4)

two-pipe system. A hydronic heating system in which a supply main simultaneously carries heated water to each room unit. A return main carries the cooled water from each radiator to the boiler. (33)

U

U-factor. The rate at which a material conducts heat energy. (9, 16)

undercut. A wood member where the back of the member is slightly shorter than the front surface. (20)

underlayment. A thin cover of asphalt-saturated felt or other material that protects sheathing from moisture until shingles are laid, provides weather protection, and prevents direct contact between shingles and resinous areas in the sheathing. (14)

underpinning. A short wall section between the foundation sill and first floor framing. Also called a *cripple wall*. (10)

unit rise. The height of the stair riser; the vertical distance between two treads. (21)

unit run. The width of a stair tread minus the nosing. (21)

urea-formaldehyde resin glue. A moisture-resistant glue that hardens through chemical action when water is added to the powdered resin. (3)

utility knife. A tool with a sharp, replaceable blade. It is useful for trimming wood and cutting veneer, hardboard, gypsum wallboard, vapor barrier, and house wrap. It is also used to cut batt insulation and for accurate layout. (4)

utility trim. A piece of trim designed to grip the trimmed top edge of a siding panel. (17)

V

valley jack. The upper end of a common rafter, but intersects a valley rafter instead of the plate. (12)

valley rafter. A rafter that forms the intersection of an internal roof angle. (12)

value. The lightness or darkness of a color. (24)

vapor barrier. A watertight material used to prevent the passage of moisture or water vapor. Has a perm rating of less than 1. (15)

vapor drive. The direction in which diffusion is occurring at a given time. (15)

vapor retarder. A material used to resist the passage of moisture or water vapor. Has a perm rating between 0.1 and 1. (15)

veneer core. A hardwood plywood that is less expensive than lumber core, fairly stable, and warp resistant. (3)

veneer wall. A masonry facing, such as a single thickness of brick, applied to a frame building wall. (17)

venting. The part of the drainage piping that permits air to circulate in the pipes. (32)

vent stack. The part of a DWV system that acts as an exhaust for any odors from the waste system, and also allows the system to take in air and thus avoid siphoning problems with traps. (32)

vernier scale. A device on a transit that measures minute portions of an angle. (8)

vinyl siding. Siding manufactured from rigid polyvinyl chloride. (17)

volatile organic compound (VOC). A carbon-containing chemical that is released by many materials, especially paint, that is hazardous to human health. (24)

voltage. The pressure that pushes current through a conductor. (5)

W

waffle-grid ICF wall. An insulated concrete wall system that forms a grid similar to the appearance of a waffle. (9)

waler. A horizontal member used in concrete form construction to stiffen and support the walls of the form. Used to keep the form walls from bending outward under the pressure of poured concrete. (9)

wallpapering. The process of hanging wall-coverings. (24)

wall pocket. An intentional void left in a concrete wall to support a large timber or I-beam. (9)

wall rail. In closed stairs, the support rail that is attached to the wall. (21)

wane. The bark or absence of wood on the edge of lumber. (3)

warp. Any variation from true or plane surface, warp may include any one or combinations of the following: cup, bow, crook, and twist (or wind). (3)

water supply system. A pipe system that distributes water under pressure throughout the structure for drinking, cooking, bathing, and laundry. (32)

waterproofing. The process of preventing both moisture and liquid water from penetrating foundations. (9)

water-resistive barrier (WRB). A membrane or coating behind the exterior cladding that resists bulk water infiltration into the wall system. (15)

WaterSense. A program of the US Environmental Protection Agency to reduce water consumption of plumbing fixtures. (28)

water vapor. A gaseous form of water always present in the air. (15)

web. Material between the chords in trusses or between the flanges in beams. (10)

weld bead. The thickened area of metal forming a joint between two pieces that have been melted together by the welding arc. (13)

wet edge. A paint band narrow enough so that the next band can be completed before the first has dried, eliminating lap marks. (24)

wide flange. A type of steel beam generally used in residential construction. (10)

winder. A wedge-shaped tread installed where stairs turn. (21)

winding stairs. A curving stairway that gradually changes direction; usually circular or elliptical in shape. Also called *geometrical*. (21)

window and door schedule. A list of the doors and windows for the structure. (16)

wind-washing. The movement of unconditioned air through thermal insulation. (15)

wing divider. A tool that is similar to a pencil compass, except with points attached to both legs. The legs can be locked to prevent movement, which would change the measurement being transferred. (4)

wood bleaching. The process of applying bleach to remove some of the natural color from the wood. (24)

wood chisel. A tool used to trim and cut away wood or composition materials to form joints or recesses. (4)

wood lath. A plaster base consisting of thin, narrow strips of soft wood nailed to studs. (19)

wood putty. A mixture of wood and adhesive in powder form. (24)

wood shake. A durable, hand-split roofing material. (14)

wood shingle. An individual wooden piece with a wedge-shaped cross section used for roofing and sidewall applications. Made from cedar or other rot-resistant woods. (17)

working time. The period during which joint compound can be used before it sets. (19)

woven valley. A method of roofing in which shingles are run across valleys. (14)

wrecking bar. A tool used to strip concrete forms, to disassemble scaffolding, and for other rough work involving prying, scraping, and nail pulling. (4)

X

X-bracing. A support system in which steel straps diagonally extend from the top of the wall to the bottom. They are attached to the tracks and studs with screws. (13)

xylem. The inside area of the cambium layer that develops new wood cells. (3)

Z

zero clearance. A quality of a factory-built fireplace that allows the outside of the housing to rest directly on wood floors and close to wood framing members. (25)

zone. An area of a building that can be separately heated. (33)

zoning ordinance. A law put in place by a governing municipality that regulates land usage. (8)

X

Z